Probability and Statistics for Engineers and Scientists

Probability and Statistics for Engineers and Scientists

$2^{Ed.}$

Ronald E. Walpole
Professor of Mathematics and Statistics
Roanoke College

Raymond H. Myers
Professor of Statistics
Virginia Polytechnic Institute
and State University

MACMILLAN PUBLISHING CO., INC.
New York
COLLIER MACMILLAN PUBLISHERS
London

To Janice

Macmillan Publishing Co., Inc.
866 Third Avenue, New York, New York 10022

Collier Macmillan Canada, Ltd.

Library of Congress Cataloging in Publication Data

Library of Congress Cataloging in Publication Data

Walpole, Ronald E.
  Probability and statistics for engineers and
scientists.

  Bibliography: p.
  Includes index.
  1. Engineering—Statistical methods.   I. Myers,
Raymond H., joint author.   II. Title.
TA340.W35   1978   519.2'02'46   77-316
ISBN 0-02-424110-5 (Hardbound)
ISBN 0-02-979870-1 (International Edition)
Printing: 7 8   Year: 0 1 2 3 4

Preface

This second edition, like the first, has been written as an introductory probability and statistics text for students majoring in engineering, mathematics, statistics, or one of the natural sciences. The objectives of the first edition have been maintained. That is, we have endeavored to achieve a balance between theory and applications based upon a prerequisite of a course in differential and integral calculus through partial differentiation and multiple integration.

The major changes in the second edition, which we feel will enhance the book's adaptability to the various scientific areas, are as follows:

1. In response to comments received from reviewers and users of the first edition, certain sections have been reworded and new examples added to clarify some of the explanations that had been troublesome to students.
2. Metric units are now used in all examples and exercises dealing with numerical measurements.
3. The material on nonparametric tests has been greatly expanded and now constitutes a separate chapter.
4. A discussion of tolerance intervals, both parametric and nonparametric, has been added to complement the material previously included on confidence intervals.
5. The concepts of multiple linear regression have been completely rewritten and reorganized into a single chapter. With the increased use of computers, the use of orthogonal polynomials in regression analysis has become somewhat obsolete, and therefore this material has been deleted in this edition.
6. New sections on sequential methods for model selection (stepwise regression) and ridge regression techniques have been included in Chapter 9.
7. Dunnett's test comparing several treatment means with a control is illustrated in the analysis-of-variance chapter.

The text contains sufficient material to allow for flexibility in the length of the course and the selection of topics. For those students who have time for only a one-semester course meeting three hours a week, the authors recommend the study of Chapters 1 through 4 and selected topics from Chapters 5, 6, and 7. Chapter 1 introduces the basic concepts of probability theory using elementary sample spaces. Chapter 2 presents an introduction to discrete and continuous random variables and their probability distributions, joint probability distributions, and mathematical expectations. Chapters 3 and 4 are devoted to a discussion of the particular discrete and continuous probability distributions that the engineer or scientist is most likely to apply in solving the various problems in his field of specialization. Perhaps unusual for a text at this level is the extensive use of transformation theory in our derivations of the sampling distributions in Chapter 5. However, the treatment of estimation procedures and hypothesis testing in Chapters 6 and 7 can only be appreciated and properly understood if one has gained an insight into the mathematical derivation of the test statistics involved.

For those students who wish to continue their training in statistics for an additional semester, the remainder of the text provides an excellent introduction to the study of regression theory, linear models, analysis-of-variance procedures, and the planning and analysis of factorial experiments. As a rule, students who take more than one semester of statistics also enroll in additional courses in mathematics and, perhaps, computer science. We have therefore included the use of matrices in our treatment of multiple and polynomial regression in Chapter 9 and assume the availability of at least a minicomputer. However, since matrix theory is limited primarily to Chapter 9, the professor could either omit this material completely or inject the basic concepts of matrix operations into the lecture sequence without requiring a formal course in matrix theory or linear algebra as a prerequisite.

Throughout the text we have demonstrated each new idea by an example. Only by solving a large number of exercises can the student be expected to develop an understanding of the basic concepts of probability theory and statistics. Therefore, we have included at the end of each chapter numerous exercises, both theoretical and applied, all of which are keyed to answers at the back of the book.

The authors wish to acknowledge their appreciation to all those who assisted in the preparation of this textbook. We are particularly grateful to Diane Milan, Janet Parsons, Lynn Watson, and Kathleen Webster for typing and proofreading the manuscript for the first edition; to Julia Baker and Dale Parris for typing and proofreading the revised second edition and the accompanying *Solutions Supplement*; to Sharon Crews and Susan Crandall for performing many of the computations throughout the text; and to the teachers and reviewers of the first edition for their helpful suggestions in preparing this second edition.

The authors are indebted to the literary executor of the late Sir Ronald A. Fisher, F.R.S., Cambridge, and to Oliver & Boyd Ltd., Edinburgh, for their permission to reprint a table from their book *Statistical Methods for Research Workers*; to Professor E. S. Pearson and the Biometrika trustees for permission to reprint in abridged form Tables 8 and 18 from *Biometrika Tables for*

Statisticians, Vol. I; to Oliver & Boyd Ltd. for permission to reproduce tables from their book *Design and Analysis of Industrial Experiments* by O. L. Davies; to the McGraw-Hill Book Company for permission to reproduce Tables A-25d and A-25e from their book *Introduction to Statistical Analysis* by W. J. Dixon and F. J. Massey, Jr.; to C. Eisenhart, M. W. Hastay, and W. A. Wallis for permission to reproduce two tables from their book *Techniques of Statistical Analysis*. We wish also to express our appreciation for permission to reproduce tables from the *Annals of Mathematical Statistics*, from the *Bulletin of the Educational Research at Indiana University*, from a publication by the American Cyanamid Company, from *Biometrics*, and from *Biometrika*, Vol. 38.

R. E. W.
R. H. M.

Contents

8 Linear Regression and Correlation

9 Multiple Linear Regression

10 Analysis of Variance

11 Factorial Experiments

12 2^k Factorial Experiments

13   Nonparametric Statistics

Probability

1.1 SAMPLE SPACE

The statistician is basically concerned with the presentation and interpretation of *chance outcomes* that occur in a planned study or scientific investigation. For example, he may record the number of accidents that occur monthly at the intersection of Driftwood Lane and Royal Oak Drive, hoping to justify the installation of a traffic light, or he may be interested in the volume of gas released during a chemical experiment when the concentration of an acid is varied. Hence he is usually dealing with counts or numerical measurements. The numerical value so recorded is often called an *observation*.

DEFINITION 1.1 *Recorded information in its original collected form, whether it be counts or measurements, is referred to as* raw data.

The numbers 2, 0, 1, and 2, representing the number of accidents for the first 4 months of the year at the intersection of Driftwood Lane and Royal Oak Drive, constitute a set of raw data. Similarly, the measurements 25, 32, and 19 cubic centimeters of gas in the chemistry experiment are recorded as raw data.

In statistics we use the word *experiment* to describe any process that generates raw data. A very simple example of a statistical experiment might be the tossing of a coin. In this experiment there are only two possible outcomes, heads or tails. Another experiment might be the launching of a missile and observing the velocity at specified times. The opinions of voters concerning a new sales tax can also be considered as observations of an experiment. We are particularly interested in the observations obtained by repeating the experiment several times. In most cases the outcomes will depend on chance and, therefore, cannot be predicted with certainty. If a chemist runs an analysis several times under the same conditions, he will obtain different measurements, indicating an element of chance in the experimental procedure. Even when a coin is tossed repeatedly, we cannot be certain that a given toss will result in a head. However, we do know the entire set of possibilities for each toss.

> DEFINITION 1.2   *The set of all possible outcomes of a statistical experiment is called the* sample space *and is represented by the symbol S.*

Each outcome in a sample space is called an *element* or a *member* of the sample space or simply a *sample point*. If the sample space has a finite number of elements, we may *list* the members separated by commas and enclosed in brackets. Thus the sample space S, of possible outcomes when a coin is tossed, may be written

$$S = \{H, T\},$$

where H and T correspond to "heads" and "tails," respectively. Sample spaces with a large or infinite number of sample points are best described by a *statement* or *rule*. For example, if the possible outcomes of an experiment are the set of cities in the world with a population over 1 million, our sample space is written

$$S = \{x \,|\, x \text{ is a city with a population over 1 million}\},$$

which reads "S is the set of all x such that x is a city with a population over 1 million." The vertical bar is read "such that." Similarly, if S is the set of all points (x, y) on the boundary or the interior of a circle of radius 2 with center at the origin, we write

$$S = \{(x, y) \,|\, x^2 + y^2 \leq 4\}.$$

Whether we describe the sample space by the rule method or by listing the elements, will depend on the specific problem at hand. The rule method has practical advantages, particularly in the many experiments where a listing becomes a very tedious chore.

Example 1.1   Consider the experiment of tossing a die. If we are interested in the number that shows on the top face, the sample space would be

$$S_1 = \{1, 2, 3, 4, 5, 6\}.$$

If we are interested only in whether the number is even or odd, the sample space is simply

$$S_2 = \{\text{even, odd}\}.$$

This example illustrates the fact that more than one sample space can be used to describe the outcomes of an experiment. In this case S_1 provides more information than S_2. If we know which element in S_1 occurs, we can tell which outcome in S_2 occurs; however, a knowledge of what happens in S_2 in no way helps us to know which element in S_1 occurs. In general, it is desirable to use a sample space that gives the most information concerning the outcomes of the experiment.

Example 1.2 Suppose that three items are selected at random from a manufacturing process. Each item is inspected and classified defective or nondefective. The sample space providing the most information would be

$$S_1 = \{NNN, NDN, DNN, NND, DDN, DND, NDD, DDD\}.$$

A second sample space, although it provides less information, might be

$$S_2 = \{0, 1, 2, 3\},$$

where the elements represent no defectives, one defective, two defectives, or three defectives in our random selection of three items.

1.2 EVENTS

In any given experiment we may be interested in the occurrence of certain *events* rather than in the outcome of a specific element in the sample space. For instance, we might be interested in the event A that the outcome when a die is tossed is divisible by 3. This will occur if the outcome is an element of the subset $A = \{3, 6\}$ of the sample space S_1 in Example 1.1. As a further illustration, we might be interested in the event B that the number of defectives is greater than 1 in Example 1.2. This will occur if the outcome is an element of the subset $B = \{DDN, DND, NDD, DDD\}$ of the sample space S_1.

To each event we assign a collection of sample points, which constitute a subset of the sample space. This subset represents all the elements for which the event is true.

DEFINITION 1.3   *An* event *is a subset of a sample space.*

Example 1.3 Given the subset $A = \{t | t < 5\}$ of the sample space $S = \{t | t \geq 0\}$, where t is the life in years of a certain electronic component, A is the event that the component fails before the end of the fifth year.

This example illustrates the fact that corresponding to any subset we can state an event whose elements are the given subset. In practice we usually state the event first and then determine its set.

DEFINITION 1.4 *If an event is a set containing only one element of the sample space, then it is called a* simple event. *A* compound event *is one that can be expressed as the union of simple events.*

Example 1.4 The event of drawing a heart from a deck of cards is the subset $A = \{$heart$\}$ of the sample space $S = \{$heart, spade, club, diamond$\}$. Therefore, A is a simple event. Now the event B of drawing a red card is a compound event, since $B = \{$heart $\cup$ diamond$\} = \{$heart, diamond$\}$.

Note that the union of simple events produces a compound event that is still a subset of the sample space. We should also note that if the 52 cards of the deck were the elements of the sample space rather than the four suits, then the event A of Example 1.4 would be a compound event.

DEFINITION 1.5 *The* null space *or* empty space *is a subset of the sample space that contains no elements. We denote this event by the symbol* $\varnothing$.

If we let A be the event of detecting a microscopic organism by the naked eye in a biological experiment, then $A = \varnothing$. Also, if $B = \{x \mid x$ is a nonprime factor of 7$\}$, then B must be the null set, since the only possible factors of 7 are the prime numbers 1 and 7.

The relationship between events and the corresponding sample space can be illustrated graphically by means of *Venn diagrams*. In a Venn diagram we might let the sample space be a rectangle and represent events by circles drawn

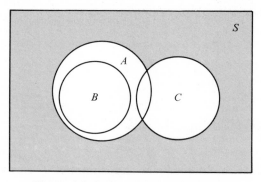

Figure 1.1 Events of the sample space.

inside the rectangle. Thus, in Figure 1.1 we see that events A, B, and C are all subsets of the sample space S. It is also clear that event B is a subset of event A; events B and C have no sample points in common; and events A and C have at least one sample point in common. Figure 1.1 might, therefore, depict a situation in which one selects a card at random from an ordinary deck of 52 playing cards and observes whether the following events occur:

A: the card is red,

B: the card is the jack, queen, or king of diamonds,

C: the card is an ace.

Clearly, the only sample points common to events A and C are the two red aces.

Sometimes it is convenient to shade various areas of the diagram as in Figure 1.2. In this case we take all the students of a certain college to be our sample

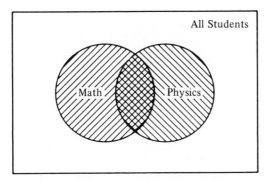

Figure 1.2 Events indicated by shading.

space. The event representing those students taking mathematics has been shaded by drawing straight lines in one direction and the event representing those students studying physics has been shaded in by drawing lines in a different direction. The doubly shaded or crosshatched area represents the event that a student is enrolled in both mathematics and physics, and the unshaded part of the diagram corresponds to the event that a student is studying subjects other than mathematics or physics.

1.3 OPERATIONS WITH EVENTS

We now consider certain operations with events that will result in the formation of new events. These new events will be subsets of the same sample space as the given events.

DEFINITION 1.6   *The intersection of two events A and B, denoted by the symbol $A \cap B$, is the event containing all elements that are common to A and B.*

The elements in the set $A \cap B$ must be those and only those that belong to both A and B. These elements may either be listed or defined by the rule method, that is, $A \cap B = \{x \mid x \in A \text{ and } x \in B\}$, where the symbol $\in$ means "is an element of" or "belongs to." In the Venn diagram in Figure 1.3 the shaded area corresponds to the event $A \cap B$.

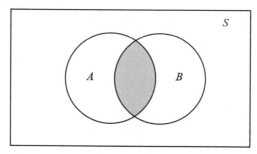

Figure 1.3 Intersection of A and B.

Example 1.5   Let $A = \{1, 2, 3, 4, 5\}$ and $B = \{2, 4, 6, 8\}$; then $A \cap B = \{2, 4\}$.

Example 1.6   Let R be the event that a person selected at random while dining at a local cafeteria is a taxpayer and let S be the event that the person selected is over 65 years of age. Then the event $R \cap S$ is the set of all taxpayers in the cafeteria who are over 65 years of age.

Example 1.7   Let $P = \{a, e, i, o, u\}$ and $Q = \{r, s, t\}$; then $P \cap Q = \varnothing$. That is, P and Q have no elements in common.

In certain statistical experiments it is by no means unusual to define two events A and B that cannot both occur simultaneously. The events A and B are then said to be *mutually exclusive*. Stated more formally, we have the following definition:

DEFINITION 1.7   *Two events A and B are mutually exclusive if $A \cap B = \varnothing$.*

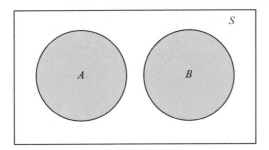

Figure 1.4 Mutually exclusive events.

Two mutually exclusive events A and B are illustrated in the Venn diagram in Figure 1.4. By shading the areas corresponding to the events A and B, we find no overlapping shaded area representing the event $A \cap B$. Hence $A \cap B$ is empty.

Example 1.8 Suppose that a die is tossed. Let A be the event that an even number turns up and B the event that an odd number shows. The events $A = \{2, 4, 6\}$ and $B = \{1, 3, 5\}$ have no points in common, since an even and an odd number cannot occur simultaneously on a single toss of a die. Therefore, $A \cap B = \emptyset$, and consequently the events A and B are mutually exclusive.

Often one is interested in the occurrence of at least one of two events associated with an experiment. Thus in the die-tossing experiment, if $A = \{2, 4, 6\}$ and $B = \{4, 5, 6\}$, we might be interested in either A or B occurring, or both A and B occurring. Such an event, called the *union* of A and B, will occur if the outcome is an element of the subset $\{2, 4, 5, 6\}$.

DEFINITION 1.8   *The union of two events A and B, denoted by the symbol $A \cup B$, is the event containing all the elements that belong to A or to B or to both.*

The elements of $A \cup B$ may be listed or defined by the rule $A \cup B = \{x \,|\, x \in A$ or $x \in B\}$. In the Venn diagram in Figure 1.5 the area representing the elements of the event $A \cup B$ has been shaded.

Example 1.9   Let $A = \{2, 3, 5, 8\}$ and $B = \{3, 6, 8\}$; then $A \cup B = \{2, 3, 5, 6, 8\}$.

Example 1.10   If $M = \{x \,|\, 3 < x < 9\}$ and $N = \{y \,|\, 5 < y < 12\}$,   then $M \cup N = \{z \,|\, 3 < z < 12\}$.

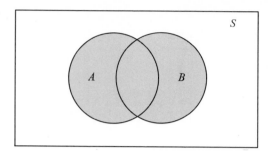

Figure 1.5 Union of A and B.

DEFINITION 1.9   *The complement of an event A with respect to S is the set of all elements of S that are not in A. We denote the complement of A by the symbol A'.* $A \cup A' = S$

The elements of A' may be listed or defined by the rule $A' = \{x \mid x \in S$ and $x \notin A\}$. In the Venn diagram in Figure 1.6, the area representing the elements of the event A' has been shaded.

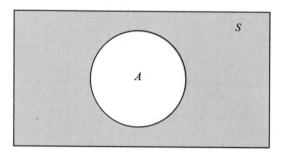

Figure 1.6 Complement of A.

Example 1.11   Let Q be the event that an employee selected at random from a manufacturing firm is a smoker. Then Q' is the event that the employee selected from the firm is a nonsmoker.

Example 1.12   Consider the sample space $S = \{$book, catalyst, cigarette, precipitate, engineer, rivet$\}$. Let $A = \{$catalyst, rivet, book, cigarette$\}$. Then $A' = \{$precipitate, engineer$\}$.

Several results that follow from the above definitions that may easily be verified by means of Venn diagrams are

1. $A \cap \varnothing = \varnothing$.
2. $A \cup \varnothing = A$.
3. $A \cap A' = \varnothing$.
4. $A \cup A' = S$.
5. $S' = \varnothing$.
6. $\varnothing' = S$.
7. $(A')' = A$.

1.4 COUNTING SAMPLE POINTS

One of the problems that the statistician must consider and attempt to evaluate is the element of chance associated with the occurrence of certain events when an experiment is performed. These problems belong in the field of probability, a subject to be introduced in Section 1.5. In many cases we shall be able to solve a probability problem by counting the number of points in the sample space. Knowledge of the actual elements or a listing is not always required. The fundamental principle of counting is stated in the following theorem.

> THEOREM 1.1   *If an operation can be performed in n_1 ways, and if for each of these a second operation can be performed in n_2 ways, then the two operations can be performed together in $n_1 n_2$ ways.*

Example 1.13   How many sample points are in the sample space when a pair of dice are thrown once?

Solution   The first die can land in any one of six ways. For each of these six ways the second die can also land in six ways. Therefore, the pair of dice can land in $(6)(6) = 36$ ways. The student is asked to list these 36 elements in Exercise 4.

Theorem 1.1 may be extended to cover any number of events. The general case is stated in the following theorem.

> THEOREM 1.2   *If an operation can be performed in n_1 ways, and if for each of these a second operation can be performed in n_2 ways, and for each of the first two a third operation can be performed in n_3 ways, and so forth, then the sequence of k operations can be performed in $n_1 n_2 \cdots n_k$ ways.*

Example 1.14   How many lunches are possible consisting of a soup, a sandwich, a dessert, and a drink if one can select from four different soups, three kinds of sandwiches, five desserts, and four drinks?

Solution   The total number of lunches would be $(4)(3)(5)(4) = 240$.

Example 1.15   How many even three-digit numbers can be formed from the digits 1, 2, 5, 6, and 9 if each digit can be used only once?

Solution   Since the number must be even, we have only two choices for the units position. For each of these we have four choices for the hundreds position, and then three choices for the tens position. Therefore, we can form a total of $(2)(4)(3) = 24$ even three-digit numbers.

Frequently, we are interested in a sample space that contains as elements all possible orders or arrangements of a group of objects. For example, we might want to know how many different arrangements are possible for sitting six people around a table, or we might ask how many different orders are possible for drawing 2 lottery tickets from a total of 20. The different arrangements are called *permutations*.

> DEFINITION 1.10   *A permutation is an arrangement of all or part of a set of objects.*

Consider the three letters a, b, and c. The possible permutations are abc, acb, bac, bca, cab, and cba. Thus we see that there are six distinct arrangements. Using Theorem 1.2, we could have arrived at the answer six without actually listing the different orders. There are three positions to be filled from the letters a, b, and c. Therefore, we have three choices for the first position, then two for the second, leaving only one choice for the last position, giving a total of $(3)(2)(1) = 6$ permutations. In general, n distinct objects can be arranged in $n(n - 1)(n - 2) \cdots (3)(2)(1)$ ways. We represent this product by the symbol $n!$, which is read "n factorial." Three objects can be arranged in $3! = (3)(2)(1) = 6$ ways. By definition, $1! = 1$ and $0! = 1$.

> THEOREM 1.3   *The number of permutations of n distinct objects is n!.*

The number of permutations of the four letters a, b, c, and d will be $4! = 24$. Let us now consider the number of permutations that are possible by taking the

four letters two at a time. These would be *ab, ac, ad, ba, ca, da, bc, cb, bd, db, cd,*
and *dc*. Using Theorem 1.2 again, we have two positions to fill with four choices
for the first and then three choices for the second for a total of $(4)(3) = 12$
permutations. In general, n distinct objects taken r at a time can be arranged in
$n(n - 1)(n - 2)\cdots(n - r + 1)$ ways. We represent this product by the symbol
$_nP_r = n!/(n - r)!$.

⟵ REGARDS ORDER

THEOREM 1.4   *The number of permutations of n distinct objects taken r at
a time is*

$$_nP_r = \frac{n!}{(n - r)!}.$$

Example 1.16 Two lottery tickets are drawn from 20 for first and second
prizes. Find the number of sample points in the space *S*.

Solution   The total number of sample points is

$$_{20}P_2 = \frac{20!}{18!} = (20)(19) = 380.$$

Example 1.17 How many ways can a local chapter of the American Chemical
Society schedule three speakers for three different meetings if they are all
available on any of five possible dates?

Solution   The total number of possible schedules is

$$_5P_3 = \frac{5!}{2!} = (5)(4)(3) = 60.$$

Permutations that occur by arranging objects in a circle are called *circular
permutations*. Two circular permutations are not considered different unless
corresponding objects in the two arrangements are preceded or followed by a
different object as we proceed in a clockwise direction. For example, if four
people are playing bridge, we do not have a new permutation if they all move
one position in a clockwise direction. By considering one person in a fixed
position and arranging the other three in 3! ways, we find that there are six
distinct arrangements for the bridge game.

THEOREM 1.5   *The number of permutations of n distinct objects arranged in a circle is* $(n - 1)!$.

So far we have considered permutations of distinct objects. That is, all the objects were completely different or distinguishable. Obviously, if the letters b and c are both equal to x, then the six permutations of the letters a, b, c, become axx, axx, xax, xax, xxa, and xxa, of which only three are distinct. Therefore, with three letters, two being the same, we have $3!/2! = 3$ distinct permutations. With four different letters a, b, c, and d we have 24 distinct permutations. If we let $a = b = x$ and $c = d = y$, we can list only the following: $xxyy, xyxy, yxxy, yyxx, xyyx$, and $yxyx$. Thus we have $4!/2!2! = 6$ distinct permutations.

THEOREM 1.6   *The number of distinct permutations of n things of which* n_1 *are of one kind,* n_2 *of a second kind, ...,* n_k *of a kth kind is*

$$\frac{n!}{n_1!n_2!\cdots n_k!}.$$

Example 1.18   How many different ways can three red, four yellow, and two blue bulbs be arranged in a string of Christmas tree lights with nine sockets?

Solution   The total number of distinct arrangements is

$$\frac{9!}{3!4!2!} = 1260.$$

Often we are concerned with the number of ways of partitioning a set of n objects into r subsets called *cells*. A partition has been achieved if the intersection of every possible pair of the r subsets is the empty set $\varnothing$ and if the union of all subsets gives the original set. The order of the elements within a cell is of no importance. Consider the set $\{a, e, i, o, u\}$. The possible partitions into two cells in which the first cell contains four elements and the second cell one element are $\{(a, e, i, o), (u)\}$, $\{(a, i, o, u), (e)\}$, $\{(e, i, o, u), (a)\}$, $\{(a, e, o, u), (i)\}$, and $\{(a, e, i, u), (o)\}$. We see that there are five such ways to partition a set of five elements into two subsets or cells containing four elements in the first cell and one element in the second.

The number of partitions for this illustration is denoted by the symbol

$$\binom{5}{4, 1} = \frac{5!}{4!1!} = 5,$$

where the top number represents the total number of elements and the bottom numbers represent the number of elements going into each cell. We state this more generally in the following theorem.

> THEOREM 1.7 *The number of ways of partitioning a set of n objects into r cells with n_1 elements in the first cell, n_2 elements in the second, and so forth, is*
>
> $$\binom{n}{n_1, n_2, \ldots, n_r} = \frac{n!}{n_1! n_2! \cdots n_r!},$$
>
> *where $n_1 + n_2 + \cdots + n_r = n$.*

Example 1.19 How many ways can seven scientists be assigned to one triple and two double hotel rooms?

Solution The total number of possible partitions would be

$$\binom{7}{3, 2, 2} = \frac{7!}{3! 2! 2!} = 210.$$

In many problems we are interested in the number of ways of *selecting r* objects from *n* without regard to order. These selections are called *combinations*. A combination is actually a partition with two cells, the one cell containing the *r* objects selected and the other cell containing the $(n - r)$ objects that are left. The number of such combinations, denoted by $\binom{n}{r, n - r}$, is usually shortened to $\binom{n}{r}$, since the number of elements in the second cell must be $n - r$.

‚ WITHOUT REGARD TO ORDER

> THEOREM 1.8 *The number of combinations of n distinct objects taken r at a time is*
>
> $$\binom{n}{r} = \frac{n!}{r!(n - r)!}.$$

Example 1.20 From four chemists and three physicists find the number of committees of three that can be formed consisting of two chemists and one physicist.

Solution   The number of ways of selecting two chemists from four is $\binom{4}{2} = \dfrac{4!}{2!2!} = 6$. The number of ways of selecting one physicist from three is $\binom{3}{1} = \dfrac{3!}{1!2!} = 3$. Using Theorem 1.1, we find the number of committees that can be formed with two chemists and one physicist to be $(6)(3) = 18$.

1.5 PROBABILITY OF AN EVENT

The statistician is basically concerned with drawing conclusions or inferences from experiments involving uncertainties. For these conclusions and inferences to be reasonably accurate, an understanding of probability theory is essential.

What do we mean when we make the statements "John will probably win the tennis match," "I have a fifty-fifty chance of getting an even number when a die is tossed," "I am not likely to win at bingo tonight," or "Most of our graduating class will likely be married within 3 years"? In each case we are expressing an outcome of which we are not certain, but owing to past information or from an understanding of the structure of the experiment, we have some degree of confidence in the validity of the statement.

The mathematical theory of probability for finite sample spaces provides a set of numbers called *weights*, ranging from zero to 1, which provide a means of evaluating the likelihood of occurrence of events resulting from a statistical experiment. To every point in the sample space we assign a weight such that the sum of all the weights is 1. If we have reason to believe that a certain sample point is quite likely to occur when the experiment is conducted, the weight assigned should be close to 1. On the other hand, a weight closer to zero is assigned to a sample point that is unlikely to occur. In many experiments, such as tossing a coin or a die, all the sample points have the same chance of occurring and are assigned equal weights. To points outside the sample space (i.e., for simple events that cannot possibly occur) we assign a weight of zero.

To find the probability of any event A, we sum all the weights assigned to the sample points in A. This sum is called the *measure* of A or the probability of A and is denoted by $P(A)$. Thus the measure of the set $\varnothing$ is zero and the measure of S is 1.

DEFINITION 1.11   *The probability of any event A is the sum of the weights of all sample points in A. Therefore,*

$$0 \le P(A) \le 1, \qquad P(\varnothing) = 0, \qquad and \qquad P(S) = 1.$$

Example 1.21   A coin is tossed twice. What is the probability that at least one head occurs?

Solution   The sample space for this experiment is

$$S = \{HH, HT, TH, TT\}.$$

If the coin is balanced, each of these outcomes would be equally likely to occur. Therefore, we assign a weight of w to each sample point. Then $4w = 1$ or $w = 1/4$. If A represents the event of at least one head occurring, then $P(A) = 3/4$.

Example 1.22   A die is loaded in such a way that an even number is twice as likely to occur as an odd number. If E is the event that a number less than 4 occurs on a single toss of the die, find $P(E)$.

Solution   The sample space is $S = \{1, 2, 3, 4, 5, 6\}$. We assign a weight of w to each odd number and a weight of $2w$ to each even number. Since the sum of the weights must be 1, we have $9w = 1$ or $w = 1/9$. Hence weights of $1/9$ and $2/9$ are assigned to each odd number and even number, respectively. Therefore,

$$P(E) = \tfrac{1}{9} + \tfrac{2}{9} + \tfrac{1}{9} = \tfrac{4}{9}.$$

We can think of weights as being probabilities associated with simple events. If the experiment is of such a nature that we can assume equal weights for the sample points of S, then the probability of any event A is the ratio of the number of elements in A to the number of elements in S.

THEOREM 1.9   *If an experiment can result in any one of N different equally likely outcomes, and if exactly n of these outcomes correspond to event A, then the probability of event A is*

$$P(A) = \frac{n}{N}.$$

Example 1.23   If a card is drawn from an ordinary deck, find the probability that it is a heart.

Solution   The number of possible outcomes is 52, of which 13 are hearts. Therefore, the probability of event A of getting a heart is $P(A) = 13/52 = 1/4$.

If the weights cannot be assumed equal, they must be assigned on the basis of prior knowledge or experimental evidence. For example, if a coin is not balanced, we would estimate the two weights by tossing the coin a large number of times and recording the outcomes. The true weights would be the fractions of heads and tails that occur in the long run. This method of arriving at weights is known as the *relative frequency* definition of probability.

To find a numerical value that represents adequately the probability of winning at tennis, we must depend on our past performance at the game as well as that of our opponent. Similarly, to find the probability that a horse will win a race, we must arrive at a weight based on the previous records of all the horses entered in the race.

1.6 SOME PROBABILITY LAWS

Often it is easier to calculate the probability of some event from known probabilities of other events. This may well be true if the event in question can be represented as the union of two other events or as the complement of some event. Several important laws that frequently simplify the computation of probabilities follow. The first, called the *additive rule*, applies to unions of events.

THEOREM 1.10 *If A and B are any two events, then*

$$P(A \cup B) = P(A) + P(B) - P(A \cap B).$$

Proof Consider the Venn diagram in Figure 1.7. The $P(A \cup B)$ is the sum of the weights of the sample points in $A \cup B$. Now $P(A) + P(B)$ is the sum of all the weights in A plus the sum of all the weights in B. Therefore, we have added the weights in $A \cap B$ twice. Since these weights add up to give $P(A \cap B)$, we

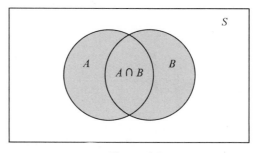

Figure 1.7

must subtract this probability once to obtain the sum of the weights in $A \cup B$, which is $P(A \cup B)$.

COROLLARY 1   *If A and B are mutually exclusive, then*

$$P(A \cup B) = P(A) + P(B).$$

Corollary 1 is an immediate result of Theorem 1.10, since if A and B are mutually exclusive, $A \cap B = \varnothing$ and then $P(A \cap B) = P(\varnothing) = 0$. In general, we write

COROLLARY 2   *If $A_1, A_2, A_3, \ldots, A_n$ are mutually exclusive, then*

$$P(A_1 \cup A_2 \cup \cdots \cup A_n) = P(A_1) + P(A_2) + \cdots + P(A_n).$$

Note that if $A_1, A_2, \ldots, A_n$ is a partition of a sample space S, then

$$\begin{aligned} P(A_1 \cup A_2 \cup \cdots \cup A_n) &= P(A_1) + P(A_2) + \cdots + P(A_n) \\ &= P(S) \\ &= 1. \end{aligned}$$

Example 1.24   The probability that a student passes mathematics is 2/3, and the probability that he passes biology is 4/9. If the probability of passing at least one course is 4/5, what is the probability that he will pass both courses?

Solution   If M is the event "passing mathematics" and B the event "passing biology," then by transposing the terms in Theorem 1.10 we have

$$\begin{aligned} P(M \cap B) &= P(M) + P(B) - P(M \cup B) \\ &= \tfrac{2}{3} + \tfrac{4}{9} - \tfrac{4}{5} \\ &= \tfrac{14}{45}. \end{aligned}$$

Example 1.25   What is the probability of getting a total of 7 or 11 when a pair of dice are tossed?

Solution   Let A be the event that 7 occurs and B the event that 11 comes up. Now a total of 7 occurs for 6 of the 36 sample points and a total of 11 occurs for only 2 of the sample points. Since all sample points are equally likely, we have $P(A) = 1/6$ and $P(B) = 1/18$. The events A and B are mutually exclusive, since a total of 7 and 11 cannot both occur on the same toss. Therefore,

$$P(A \cup B) = P(A) + P(B)$$
$$= \tfrac{1}{6} + \tfrac{1}{18}$$
$$= \tfrac{2}{9}.$$

THEOREM 1.11   *If A and A' are complementary events, then*

$$P(A') = 1 - P(A).$$

Proof Since $A \cup A' = S$ and the sets A and A' are disjoint, then

$$1 = P(S)$$
$$= P(A \cup A')$$
$$= P(A) + P(A').$$

Therefore,

$$P(A') = 1 - P(A).$$

Example 1.26 A coin is tossed six times in succession. What is the probability that at least one head occurs?

Solution Let E be the event that at least one head occurs. The sample space S consists of $2^6 = 64$ sample points since each toss can result in two outcomes. Now, $P(E) = 1 - P(E')$, where E' is the event that no head occurs. This can happen in only one way, when all tosses result in a tail. Therefore, $P(E') = 1/64$ and $P(E) = 1 - (1/64) = 63/64$.

1.7 CONDITIONAL PROBABILITY

The probability of an event B occurring when it is known that some event A has occurred is called a *conditional probability* and is denoted by $P(B|A)$. The symbol $P(B|A)$ is usually read "the probability that B occurs given that A occurs" or simply "the probability of B, given A."

Consider the event B of getting a perfect square when a die is tossed. The die is constructed so that the even numbers are twice as likely to occur as the odd numbers. Based on the sample space $S = \{1, 2, 3, 4, 5, 6\}$, with weights of 1/9

and 2/9 assigned, respectively, to the odd and even numbers, the probability of
B occurring is 1/3. Now suppose that it is known that the toss of the die resulted
in a number greater than 3. We are now dealing with a reduced sample space
$A = \{4, 5, 6\}$, which is a subset of S. To find the probability that B occurs,
relative to the space A, we must first assign new weights to the elements of A
proportional to their original weights such that their sum is 1. Assigning a
weight of w to the odd number in A and a weight of $2w$ to the two even numbers,
we have $5w = 1$ or $w = 1/5$. Relative to the space A, we find that B contains the
single element 4. Denoting this event by the symbol $B|A$, we write $B|A = \{4\}$,
and hence

$$P(B|A) = \tfrac{2}{5}.$$

This example illustrates that events may have different probabilities when
considered relative to different sample spaces.

We can also write

$$P(B|A) = \frac{2}{5} = \frac{2/9}{5/9} = \frac{P(A \cap B)}{P(A)},$$

where $P(A \cap B)$ and $P(A)$ are found from the original sample space S. In other
words, a conditional probability relative to a subspace A of S may be calculated
directly from the sample space S itself.

DEFINITION 1.12   *The conditional probability of B, given A, denoted by*
$P(B|A)$*, is defined by*

$$P(B|A) = \frac{P(A \cap B)}{P(A)} \qquad if \quad P(A) > 0.$$

As a further illustration, suppose that our sample space S represents the adults
in a small town who have completed the requirements for a college degree.
We shall categorize them according to sex and employment status:

	Employed	Unemployed
Male	460	40
Female	140	260

One of these individuals is to be selected at random for a tour throughout the country to publicize the advantages of establishing new industries in the town. We shall be concerned with the following events:

M: a man is chosen,

E: the chosen one is employed.

Using the reduced sample space E, we find that

$$P(M\,|\,E) = \tfrac{460}{600} = \tfrac{23}{30}.$$

EMPLOYED MEN

$\dfrac{P(M \cap E)}{P(E)}$

TOTAL EMPLOYED

Let $n(A)$ denote the number of elements in any set A. Using this notation, we can write

$$P(M\,|\,E) = \frac{n(E \cap M)}{n(E)} = \frac{n(E \cap M)/n(S)}{n(E)/n(S)} = \frac{P(E \cap M)}{P(E)},$$

where $P(E \cap M)$ and $P(E)$ are found from the original sample space S. To verify this result, note that

$$P(E) = \tfrac{600}{900} = \tfrac{2}{3},$$

and

$$P(E \cap M) = \tfrac{460}{900} = \tfrac{23}{45}.$$

Hence

$$P(M\,|\,E) = \frac{23/45}{2/3} = \frac{23}{30},$$

as before.

Multiplying the formula in Definition 1.12 by $P(A)$, we obtain the following important *multiplication theorem*.

THEOREM 1.12   *If in an experiment the events A and B can both occur, then*

$$P(A \cap B) = P(A)P(B\,|\,A). = P(B)P(A|B)$$

Thus the probability that both A and B occur is equal to the probability that A occurs multiplied by the probability that B occurs, given that A occurs.

To illustrate the use of Theorem 1.12, suppose that we have a fuse box containing 20 fuses, of which 5 are defective. If 2 fuses are selected at random and removed from the box in succession without replacing the first, what is the probability that both fuses are defective? To answer this question, we shall let A be the event that the first fuse is defective and B the event that the second fuse is defective and then interpret $A \cap B$ as the event that A occurs and then B occurs after A has occurred. The probability of first removing a defective fuse is 1/4 and then the probability of removing a second defective fuse from the remaining 4 is 4/19. Hence $P(A \cap B) = (1/4)(4/19) = 1/19$.

Generalizing Theorem 1.12, we write

THEOREM 1.13   *If in an experiment the events* $A_1, A_2, A_3, \ldots,$ *can occur, then*

$$P(A_1 \cap A_2 \cap A_3 \cap \cdots) = P(A_1)P(A_2|A_1)P(A_3|A_1 \cap A_2) \cdots.$$

If in the illustration above the first fuse is replaced and the fuses thoroughly rearranged before the second is removed, then the probability of a defective fuse on the second selection is still 1/4. That is, $P(B|A) = P(B)$. When this is true, the events A and B are said to be *independent*.

DEFINITION 1.13   *The events A and B are* independent *if, and only if,*

$$P(A \cap B) = P(A)P(B).$$

Example 1.27   A pair of dice are thrown twice. What is the probability of getting totals of 7 and 11?

Solution   Let A_1, A_2, B_1, and B_2 be the respective independent events that a 7 occurs on the first throw, a 7 occurs on the second throw, an 11 occurs on the first throw, and an 11 occurs on the second throw. We are interested in the probability of the union of the mutually exclusive events $A_1 \cap B_2$ and $B_1 \cap A_2$. Therefore,

$$
\begin{aligned}
P[(A_1 \cap B_2) \cup (B_1 \cap A_2)] &= P(A_1 \cap B_2) + P(B_1 \cap A_2) \\
&= P(A_1)P(B_2) + P(B_1)P(A_2) \\
&= (\tfrac{1}{6})(\tfrac{1}{18}) + (\tfrac{1}{18})(\tfrac{1}{6}) \\
&= \tfrac{1}{54}.
\end{aligned}
$$

1.8 BAYES' RULE

Let us return to the illustration on Section 1.7, where an individual is being selected at random from the adults of a small town to tour the country and publicize the advantages of establishing new industries in the town. At that time we had no difficulty in establishing the fact that $P(E) = 2/3$, where E is the event that the one chosen is employed. Suppose that we are given the additional information that 36 of those employed and 12 of those unemployed are members of the Rotary Club. What is the probability that the individual selected is employed if it is known that the person belongs to the Rotary Club?

Let A be the event that the person selected is a member of the Rotary Club. The conditional probability that we seek is then given by

$$P(E|A) = \frac{P(E \cap A)}{P(A)}.$$

Referring to Figure 1.8, we can write A as the union of the two mutually exclusive events $E \cap A$ and $E' \cap A$. Hence

$$A = (E \cap A) \cup (E' \cap A),$$

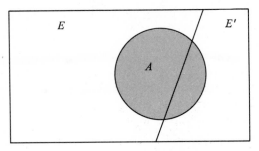

Figure 1.8 Venn diagram showing the events A, E, and E'.

and, by Corollary 1 of Theorem 1.10,

$$P(A) = P(E \cap A) + P(E' \cap A).$$

We can now write

$$P(E|A) = \frac{P(E \cap A)}{P(E \cap A) + P(E' \cap A)}.$$

The data of Section 1.7, together with the additional information about set A, enable one to compute

$$P(E \cap A) = \frac{36}{900} = \frac{1}{25}$$

$$P(E' \cap A) = \frac{12}{900} = \frac{1}{75}$$

$$P(E|A) = \frac{1/25}{1/25 + 1/75} = \frac{3}{4}.$$

A generalization of the foregoing procedure leads to the following theorem, called *Bayes' rule*.

THEOREM 1.14 (BAYES' RULE)   *Let* $\{B_1, B_2, \ldots, B_n\}$ *be a set of events forming a partition of the sample space S, where* $P(B_i) \neq 0$, *for i* $= 1, 2, \ldots, n$. *Let A be any event of S such that* $P(A) \neq 0$. *Then, for k* $= 1, 2, \ldots, n$,

$$P(B_k|A) = \frac{P(B_k \cap A)}{\sum\limits_{i=1}^{n} P(B_i \cap A)} = \frac{P(B_k)P(A|B_k)}{\sum\limits_{i=1}^{n} P(B_i)P(A|B_i)}.$$

Proof   Consider the Venn diagram in Figure 1.9. The event A is seen to be the union of the mutually exclusive events $B_1 \cap A, B_2 \cap A, \ldots, B_n \cap A$. That is,

$$A = (B_1 \cap A) \cup (B_2 \cap A) \cup \cdots \cup (B_n \cap A).$$

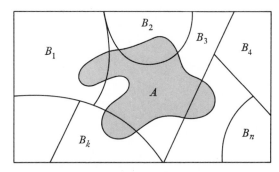

Figure 1.9 Partitioning of the sample space S.

Using Corollary 2 of Theorem 1.10, we have

$$P(A) = P(B_1 \cap A) + P(B_2 \cap A) + \cdots + P(B_n \cap A)$$

$$= \sum_{i=1}^{n} P(B_i \cap A).$$

By the definition of conditional probability,

$$P(B_k | A) = \frac{P(B_k \cap A)}{P(A)}$$

$$= \frac{P(B_k \cap A)}{\sum\limits_{i=1}^{n} P(B_i \cap A)}.$$

Applying Theorem 1.12 to both numerator and denominator, we obtain the alternative form,

$$P(B_k | A) = \frac{P(B_k)P(A | B_k)}{\sum\limits_{i=1}^{n} P(B_i)P(A | B_i)},$$

which completes the proof.

Example 1.28   Three members of a private country club have been nominated for the office of president. The probability that Mr. Adams will be elected is 0.3, the probability that Mr. Brown will be elected is 0.5, and the probability that Mr. Cooper will be elected is 0.2. Should Mr. Adams be elected, the probability for an increase in membership fees is 0.8. Should Mr. Brown or Mr. Cooper be elected, the corresponding probabilities for an increase in fees are 0.1 and 0.4. If someone is considering joining the club but delays his decision for several weeks only to find out that the fees have been increased, what is the probability that Mr. Cooper was elected president of the club?

Solution   We consider the following events:

A: the person elected increased fees,

B_1:   Mr. Adams is elected,

B_2:   Mr. Brown is elected,

B_3:   Mr. Cooper is elected.

Using Bayes' rule, we write

$$P(B_3|A) = \frac{P(B_3 \cap A)}{P(B_1 \cap A) + P(B_2 \cap A) + P(B_3 \cap A)}.$$

Now

$$P(B_1 \cap A) = P(B_1)P(A|B_1) = (0.3)(0.8) = 0.24$$
$$P(B_2 \cap A) = P(B_2)P(A|B_2) = (0.5)(0.1) = 0.05$$
$$P(B_3 \cap A) = P(B_3)P(A|B_3) = (0.2)(0.4) = 0.08.$$

Hence

$$P(B_3|A) = \frac{0.08}{0.24 + 0.05 + 0.08} = \frac{8}{37}.$$

In view of the fact that fees have increased, this result suggests that Mr. Cooper is probably not the president of the club.

EXERCISES   WHEN WORDED WITH OR => BOTH EVENTS CAN HAPPEN

1. List the elements of each of the following sample spaces:
   (a) The set of integers between 1 and 50 divisible by 7.
   (b) The set $S = \{x|x^2 + x - 6 = 0\}$.
   (c) The set of outcomes when a die and a coin are tossed simultaneously.
   (d) The set $S = \{x|x \text{ is a continent}\}$.
   (e) The set $S = \{x|2x - 4 = 0 \text{ and } x > 5\}$.

2. Use the rule method to describe the sample space S consisting of all points in the first quadrant inside a circle of radius 3 with center at the origin.

3. Which of the following events are equal?
   (a) $A = \{1, 3\}$.
   (b) $B = \{x|x \text{ is a number on a die}\}$.
   (c) $C = \{x|x^2 - 4x + 3 = 0\}$.
   (d) $D = \{x|x \text{ is the number of heads when six coins are tossed}\}$.

4. An experiment involves tossing a pair of dice, one green and one red, and recording the numbers that come up.
   (a) List the elements of the sample space S.
   (b) List the elements of S corresponding to event A that the sum is less than 5.
   (c) List the elements of S corresponding to event B that a 6 occurs on either die.
   (d) List the elements of S corresponding to event C that a 2 comes up on the green die.
   (e) Construct a Venn diagram to show the relationship among the events A, B, C, and S.

5. An experiment consists of flipping a coin and then flipping it a second time if a head occurs. If a tail occurs on the first flip, then a die is tossed once.
   (a) List the elements of the sample space S.
   (b) List the elements of S corresponding to event A that a number less than 4 occurred on the die.
   (c) List the elements of S corresponding to event B that two tails occurred.

6. An experiment consists of asking three women at random if they wash their dishes with brand X detergent.
   (a) List the elements of a sample space S using the letter Y for "yes" and N for "no."
   (b) List the elements of S corresponding to event E that at least two of the women use brand X.
   (c) Define an event that has as its elements the points $\{YYY, NYY, YYN, NYN\}$.

7. Four students are selected at random from a chemistry class and classified as male or female. List the elements of the sample space S_1 using the letter M for "male" and F for "female." Define a second sample space S_2 where the elements represent the number of females selected.

8. The résumés of two male applicants for a college teaching position in psychology are placed in the same file as the résumés of two female applicants. Two positions become available and the first, at the rank of assistant professor, is filled by selecting one of the four applicants at random. The second position, at the rank of instructor, is then filled by selecting at random one of the remaining three applicants.
   (a) List the elements of the sample space S.
   (b) List the elements of S corresponding to event A that the position of assistant professor is filled by a male applicant.
   (c) List the elements of S corresponding to event B that exactly one of the two positions is filled by a male applicant.
   (d) List the elements of S corresponding to event C that neither position was filled by a male applicant.
   (e) Construct a Venn diagram to show the relationship among the events A, B, C, and S.

9. List the elements of S corresponding to the event $A \cap C$ in Exercise 4.

10. List the elements of S in Exercise 8 corresponding to the events $A \cap B$ and $A \cup C$.

11. If $S = \{0, 1, 2, 3, 4, 5, 6, 7, 8, 9\}$ and $A = \{0, 2, 4, 6, 8\}, B = \{1, 3, 5, 7, 9\}, C = \{2, 3, 4, 5\}$, and $D = \{1, 6, 7\}$, list the elements of the sets corresponding to the following events:
   (a) $A \cup C$.   (b) $A \cap B$.   (c) C'.
   (d) $(C' \cap D) \cup B$.   (e) $(S \cap C)'$.   (f) $A \cap C \cap D'$.

12. Consider the sample space

$$S = \{\text{copper, sodium, nitrogen, potassium, uranium, oxygen, zinc}\}$$

and the events

$$A = \{\text{copper, sodium, zinc}\},$$
$$B = \{\text{sodium, nitrogen, potassium}\},$$
$$C = \{\text{oxygen}\}.$$

List the elements of the sets corresponding to the following events:
   (a) A'.   (b) $A \cup C$.   (c) $(A \cap B') \cup C'$.
   (d) $(B' \cap C')$.   (e) $A \cap B \cap C$.   (f) $(A' \cup B') \cap (A' \cap C)$.

13. If $P = \{x | 1 < x < 9\}$ and $Q = \{y | y < 5\}$, find $P \cup Q$ and $P \cap Q$.

14. Let A, B, and C be events relative to the sample space S. Using Venn diagrams, shade the areas representing the following events:
    (a) $(A \cap B)'$.　　(b) $(A \cup B)'$.　　(c) $(A \cap C) \cup B$.

15. (a) How many ways can five people be lined up to get on a bus?
    (b) If a certain two persons refuse to follow each other, how many ways are possible?

16. A college freshman must take a science course, a social studies course, and a mathematics course. If he may select any of three sciences, any of four social studies, and any of two mathematics courses, how many ways can he arrange his program?

17. In how many different ways can an eight-question true–false examination be answered?

18. How many distinct permutations can be made from the letters of the word "statistics"?

19. How many ways can the five starting positions on a basketball team be filled with nine men who can play any of the postions?

20. (a) How many three-digit numbers can be formed from the digits 0, 1, 2, 3, 4, and 5 if each digit can be used only once?
    (b) How many of these are odd numbers?
    (c) How many are greater than 330?

21. A contractor wishes to build five houses, each different in design. In how many ways can he place these homes on a street if three lots are on one side of the street and two lots are on the opposite side?

22. In how many ways can four boys and three girls sit in a row if the boys and girls must alternate?

23. In how many ways can six trees be planted in a circle?

24. In how many ways can two oaks, three pines, and two maples be arranged in a straight line if one does not distinguish between trees of the same kind?

25. A college plays eight football games during a season. In how many ways can the team end the season with four wins, three losses, and one tie?

26. Nine people are going on a skiing trip in three cars that will hold 2, 4, and 5 passengers, respectively. How many ways is it possible to transport the nine people to the ski lodge?

27. From a group of five men and three women, how many committees of three people are possible
    (a) With no restrictions?
    (b) With two men and one woman?
    (c) With one man and two women if a certain woman must be on the committee?

28. How many bridge hands are possible containing five spades, three diamonds, three clubs, and two hearts?

29. From three red, four green, and five yellow apples, how many selections consisting of six apples are possible if two of each color are to be selected?

30. A shipment of 10 television sets contains 3 defective sets. In how many ways can a hotel purchase 4 of these sets and receive at least 2 of the defective sets?

31. Three men are seeking public office. Candidates A and B are given about the same chance of winning but candidate C is given twice the chance of either A or B. What is the probability that C wins? What is the probability that A does not win?

32. Find the probability of event A in Exercise 4.

33. A box contains 500 envelopes, of which 50 contain $100 in cash, 100 contain $25, and 350 contain $10. An envelope may be purchased for $25. What is the sample space for the different amounts of money? Assign weights to the sample points and then find the probability that the first envelope purchased contains less than $100.

34. A pair of dice are tossed. What is the probability of getting a total of 5? At most a total of 4?

35. In a poker hand consisting of five cards, what is the probability of holding
    (a) Two aces and two kings?
    (b) Five spades?

36. If three books are picked at random from a shelf containing four novels, three books of poems, and a dictionary, what is the probability that
    (a) The dictionary is selected?
    (b) Two novels and one book of poems are selected?

37. Two cards are drawn in succession from a deck without replacement. Use Theorem 1.8 to find the probability that both cards are greater than 2 and less than 9.

38. If A and B are mutually exclusive events and $P(A) = 0.4$ and $P(B) = 0.5$, find
    (a) $P(A \cup B)$.   (b) $P(A')$.   (c) $P(A' \cap B)$.

39. A town has two fire engines operating independently. The probability that a specific fire engine is available when needed is 0.99.
    (a) What is the probability that neither is available when needed?
    (b) What is the probability that a fire engine is available when needed?

40. A pair of dice are thrown. If it is known that one die shows a 4, what is the probability that
    (a) The other die shows a 5?
    (b) The total of both dice is greater than 7?

41. Repeat Exercise 37 by using Theorem 1.12.

42. If the probability that Tom will be alive in 20 years is 0.6 and the probability that Jim will be alive in 20 years is 0.9, what is the probability that neither will be alive in 20 years?

43. In a high school graduating class of 100 students, 42 studied mathematics, 68 studied psychology, 54 studied history, 22 studied both mathematics and history, 25 studied both mathematics and psychology, 7 studied history and neither mathematics nor psychology, 10 studied all three subjects, and 8 did not take any of the three subjects. If a student is selected at random, find
    (a) The probability that he takes history and psychology but not mathematics.
    (b) The probability that if he is enrolled in history, he takes all three subjects.
    (c) The probability that he takes mathematics only.

44. The probability that a married man watches a certain television show is 0.4 and the probability that a married woman watches the show is 0.5. The probability that a man watches the show, given that his wife does, is 0.7. Find
    (a) The probability that a married couple watch the show.
    (b) The probability that a wife watches the show given that her husband does.
    (c) The probability that at least one person of a married couple will watch the show.

45. A basketball player sinks 50% of his shots. What is the probability that he makes exactly three of his next four shots?

46. A coin is biased so that a head is twice as likely to occur as a tail. If the coin is tossed three times, what is the probability of getting exactly two tails?

47. One bag contains four white balls and three black balls, and a second bag contains three white balls and five black balls. One ball is drawn from the first bag and placed unseen in the second bag. What is the probability that a ball now drawn from the second bag is black?

48. From a box containing five black balls and three green balls, three balls are drawn in succession, each ball being replaced in the box before the next draw is made. What is the probability that all three balls are the same color? What is the probability that each color is represented?

49. A real estate man has eight master keys to open several new homes. Only one master key will open any given house. If 40% of these homes are usually left unlocked, what is the probability that the real estate man can get into a specific home if he selects three master keys at random before leaving the office?

50. Suppose that colored balls are distributed in three indistinguishable boxes as follows:

	Box 1	Box 2	Box 3
Red	2	4	3
White	3	1	4
Blue	5	3	3

A box is selected at random from which a ball is selected at random and it is observed to be red. What is the probability that box 3 was selected?

51. A commuter owns two cars, one a compact and one a standard model. About three fourths of the time he uses the compact to travel to work and about one fourth of the time the larger car is used. When he uses the compact car he usually gets home by 5:30 P.M. about 75% of the time; if he uses the standard-sized car he gets home by 5:30 P.M. about 60% of the time (but he enjoys the air conditioner in the larger car). If he gets home at 5:35 P.M., what is the probability that he used the compact car?

52. A truth serum given to a suspect is known to be 90% reliable when the person is guilty and 99% reliable when the person is innocent. In other words, 10% of the guilty are judged innocent by the serum and 1% of the innocent are judged guilty. If the suspect was selected from a group of suspects of which only 5% have ever committed a crime, and the serum indicates that he is guilty, what is the probability that he is innocent?

Random Variables

2

2.1 CONCEPT OF A RANDOM VARIABLE

The term *statistical experiment* has been used to describe any process by which several chance measurements are obtained. Often we are not interested in the details associated with each sample point but only in some numerical description of the outcome. For example, the sample space giving a detailed description of each possible outcome when one tosses a coin three times may be written

$$S = \{HHH, HHT, HTH, THH, HTT, THT, TTH, TTT\}.$$

If one is concerned only with the number of heads that fall, then a numerical value of 0, 1, 2, or 3 will be assigned to each sample point.

The numbers 0, 1, 2, and 3 are random *observations* determined by the outcome of the experiment. They may be looked upon as the values assumed by some *random variable X*, which in this case represents the number of heads when a coin is tossed three times.

> DEFINITION 2.1 *A function whose value is a real number determined by each element in the sample space is called a* random variable.

We shall use a capital letter, say X, to denote a random variable and its corresponding small letter, x in this case, for one of its values. In the coin-tossing illustration above, we notice that the random variable X assumes the value 2 for all elements in the subset

$$E = \{HHT, HTH, THH\}$$

of the sample space S. That is, each possible value of X represents an event that is a subset of the sample space for the given experiment.

Example 2.1   Two balls are drawn in succession without replacement from an urn containing four red balls and three black balls. The possible outcomes and the values y of the random variable Y, where Y is the number of red balls, are

Simple event	y
RR	2
RB	1
BR	1
BB	0

Example 2.2   A stockroom clerk returns three safety helmets at random to three steel mill employees who had previously checked them. If Smith, Jones, and Brown, in that order, receive one of the three hats, list the sample points for the possible orders of returning the helmets and find the values m of the random variable M that represents the number of correct matches.

Solution   If S, J, and B stand for Smith's, Jones', and Brown's helmets, respectively, then the possible arrangements in which the helmets may be returned and the number of correct matches are

Simple event	m
SJB	3
SBJ	1
JSB	1
JBS	0
BSJ	0
BJS	1

In each of the two preceding examples the sample space contains a finite number of elements. On the other hand, when a die is thrown until a 5 occurs, we obtain a sample space with an unending sequence of elements,

$$S = \{F, NF, NNF, NNNF, \ldots\},$$

where F and N represent, respectively, the occurrence and nonoccurrence of a 5. But even in this experiment, the number of elements can be equated to the number of whole numbers and in this sense can be counted.

> DEFINITION 2.2   *If a sample space contains a finite number of possibilities or an unending sequence with as many elements as there are whole numbers, it is called a discrete sample space, and a random variable defined over this space is called a discrete random variable.*

The outcomes of some statistical experiments may be neither finite nor countable. Such is the case, for example, when one conducts an investigation measuring the distances that a certain make of automobile will travel over a prescribed test course on 5 liters of gasoline. Assuming distance to be a variable measured to any degree of accuracy, then clearly we have an infinite number of possible distances in the sample space that cannot be equated to the number of whole numbers. Also, if one were to record the length of time for a chemical reaction to take place, once again the possible time intervals making up our sample space are infinite in number and uncountable. We see now that all sample spaces need not be discrete.

any measured quantity

> DEFINITION 2.3   *If a sample space contains an infinite number of possibilities equal to the number of points on a line segment, it is called a continuous sample space and a random variable defined over this space is called a continuous random variable.*

In most practical problems continuous random variables represent *measured* data, such as all possible heights, weights, temperatures, distances, or life periods, whereas discrete random variables represent *count* data, such as the number of defectives in a sample of k items or the number of highway fatalities per year in a given state. Note that the random variables Y and M of Examples 2.1 and 2.2 both represent count data, Y the number of red balls and M the number of correct hat matches.

2.2 DISCRETE PROBABILITY DISTRIBUTIONS

A discrete random variable assumes each of its values with a certain probability. In the case of tossing a coin three times, the variable X, representing the number of heads, assumes the value 2 with probability 3/8, since three of the eight equally likely sample points result in two heads and one tail. If one assumes equal weights for the simple events in Example 2.2, the probability that no employee gets back his right helmet, that is, the probability that M assumes the value zero, is 1/3. The possible values m of M and their probabilities are given by

m	0	1	3
$P(M = m)$	$\frac{1}{3}$	$\frac{1}{2}$	$\frac{1}{6}$

Note that the values of m exhaust all possible cases and hence the probabilities add to 1.

Frequently, it is convenient to represent all the probabilities of a random variable X by a formula. Such a formula would necessarily be a function of the numerical values x that we shall denote by $f(x)$, $g(x)$, $r(x)$, and so forth. Hence we write $f(x) = P(X = x)$; that is $f(3) = P(X = 3)$.

DEFINITION 2.4   *The function $f(x)$ is a* probability function *or a* probability distribution *of the discrete random variable X if, for each possible outcome x,*

1. $f(x) \geq 0$.

2. $\sum_x f(x) = 1$.

3. $P(X = x) = f(x)$.

Example 2.3   Find the probability distribution of the sum of the numbers when a pair of dice are tossed.

Solution   Let X be a random variable whose values x are the possible totals. Then x can be any integer from 2 to 12. Two dice can fall in $(6)(6) = 36$ ways, each with probability 1/36. The $P(X = 3) = 2/36$, since a total of 3 can occur in only two ways. Consideration of the other cases leads to the following probability distribution:

x	2	3	4	5	6	7	8	9	10	11	12
$f(x)$	$\frac{1}{36}$	$\frac{2}{36}$	$\frac{3}{36}$	$\frac{4}{36}$	$\frac{5}{36}$	$\frac{6}{36}$	$\frac{5}{36}$	$\frac{4}{36}$	$\frac{3}{36}$	$\frac{2}{36}$	$\frac{1}{36}$

Example 2.4   Find a formula for the probability distribution of the number of heads when a coin is tossed four times.

Solution   Since there are $2^4 = 16$ points in the sample space representing equally likely outcomes, the denominator for all probabilities, and therefore for our function, will be 16. To obtain the number of ways of getting, say three heads, we need to consider the number of ways of partitioning four outcomes into two cells with three heads assigned to one cell and a tail assigned to the other. This can be done in $\binom{4}{3} = 4$ ways. In general, x heads and $4 - x$ tails can occur in $\binom{4}{x}$ ways, where x can be 0, 1, 2, 3, or 4. Thus the probability distribution $f(x) = P(X = x)$ is

$$f(x) = \frac{\binom{4}{x}}{16}, \qquad x = 0, 1, 2, 3, 4.$$

DEFINITION 2.5 *The cumulative distribution $F(x)$ of a discrete random variable X with probability distribution $f(x)$ is given by*

$$F(x) = P(X \leq x) = \sum_{t \leq x} f(t).$$

For the random variable M, the number of correct matches, we have

$$F(2.4) = P(M \leq 2.4) = f(0) + f(1) = (\tfrac{1}{3}) + (\tfrac{1}{2}) = \tfrac{5}{6}.$$

The cumulative distribution of M is given by

$$F(m) = \begin{cases} 0 & \text{for } m < 0 \\ \tfrac{1}{3} & \text{for } 0 \leq m < 1 \\ \tfrac{5}{6} & \text{for } 1 \leq m < 3 \\ 1 & \text{for } m \geq 3. \end{cases}$$

One should pay particular notice to the fact that the cumulative distribution is defined not only for the values assumed by the given random variable but for all real numbers.

Example 2.5 Find the cumulative distribution of the random variable X in Example 2.4. Using $F(x)$, verify that $f(2) = 3/8$.

Solution   Direct calculations of the probability distribution of Example 2.4 give $f(0) = 1/16$, $f(1) = 1/4$, $f(2) = 3/8$, $f(3) = 1/4$, and $f(4) = 1/16$. Therefore,

$$F(0) = f(0) = \tfrac{1}{16}$$

$$F(1) = f(0) + f(1) = \tfrac{5}{16}$$

$$F(2) = f(0) + f(1) + f(2) = \tfrac{11}{16}$$

$$F(3) = f(0) + f(1) + f(2) + f(3) = \tfrac{15}{16}$$

$$F(4) = f(0) + f(1) + f(2) + f(3) + f(4) = 1.$$

Hence

$$F(x) = \begin{cases} 0 & \text{for } x < 0 \\ \tfrac{1}{16} & \text{for } 0 \le x < 1 \\ \tfrac{5}{16} & \text{for } 1 \le x < 2 \\ \tfrac{11}{16} & \text{for } 2 \le x < 3 \\ \tfrac{15}{16} & \text{for } 3 \le x < 4 \\ 1 & \text{for } x \ge 4. \end{cases}$$

Now

$$f(2) = F(2) - F(1) = \tfrac{11}{16} - \tfrac{5}{16} = \tfrac{3}{8}.$$

It is often helpful to look at a probability distribution in graphic form. One might plot the points $(x, f(x))$ of Example 2.4 to obtain Figure 2.1. By joining the points to the x axis either with a dashed or solid line, we obtain what is commonly called a *bar chart*. Figure 2.1 makes it very easy to see what values of X are most likely to occur, and it also indicates a perfectly symmetric situation in this case.

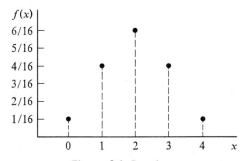

Figure 2.1 Bar chart.

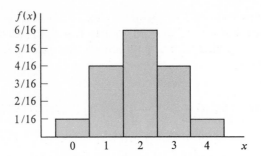

Figure 2.2 Probability histogram.

Instead of plotting the points $(x, f(x))$, we more frequently construct rectangles, as in Figure 2.2. Here the rectangles are constructed so that their bases of equal width are centered at each value x and their heights are equal to the corresponding probabilities given by $f(x)$. The bases are constructed so as to leave no space between the rectangles. Figure 2.2 is called a *probability histogram*.

Since each base in Figure 2.2 has unit width, the $P(X = x)$ is equal to the area of the rectangle centered at x. Even if the bases were not of unit width, we could adjust the heights of the rectangles to give areas that would still equal the probabilities of X assuming any of its values x. This concept of using areas to represent probabilities is necessary for our consideration of the probability distribution of a continuous random variable.

The graph of the cumulative distribution of Example 2.5, which appears as a step function in Figure 2.3, is obtained by plotting the points $(x, F(x))$.

Certain probability distributions are applicable to more than one physical situation. The probability distribution of Example 2.4, for example, also applies to the random variable Y, where Y is the number of red cards that occur when

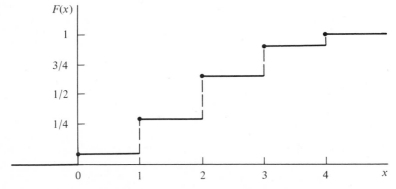

Figure 2.3 Discrete cumulative distribution.

four cards are drawn at random from a deck in succession with each card replaced and the deck shuffled before the next drawing. Special discrete distributions that can be applied to many different experimental situations will be considered in Chapter 3.

2.3 CONTINUOUS PROBABILITY DISTRIBUTIONS

A continuous random variable has a probability of zero of assuming exactly any of its values. Consequently, its probability distribution cannot be given in tabular form. At first this may seem startling, but it becomes more plausible when we consider a particular example. Let us discuss a random variable whose values are the heights of all people over 21 years of age. Between any two values, say 163.5 and 164.5 centimeters, or even 163.99 and 164.01 centimeters, there are an infinite number of heights, one of which is 164 centimeters. The probability of selecting a person at random who is exactly 164 centimeters tall and not one of the infinitely large set of heights so close to 164 centimeters that you cannot humanly measure the difference is remote, and thus we assign a probability of zero to the event. It follows that

$$P(a < X \leq b) = P(a < X < b) + P(X = b)$$
$$= P(a < X < b).$$

That is, it does not matter whether we include an end point of the interval or not.

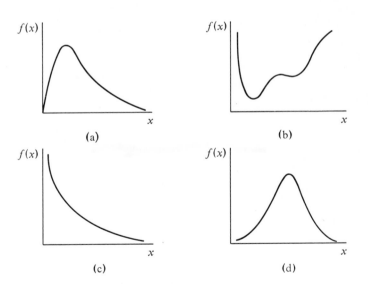

Figure 2.4 Typical density functions.

While the probability distribution of a continuous random variable cannot be presented in tabular form, it does have a formula. As before, we shall designate the probability distribution of X by the functional notation $f(x)$. In dealing with continuous variables, $f(x)$ is usually called the *density function*. Since X is defined over a continuous sample space, it is possible for $f(x)$ to have a finite number of discontinuities. However, most density functions that have practical applications in the analysis of statistical data are continuous and their graphs may, for example, take one of the forms shown in Figure 2.4.

A probability density function is constructed so that the area under its curve bounded by the x axis is equal to 1 when computed over the range of X for which $f(x)$ is defined. If $f(x)$ is represented in Figure 2.5, then the probability

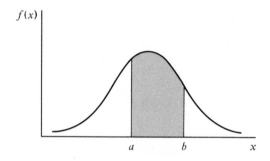

Figure 2.5 $P(a < X < b)$.

that X assumes a value between a and b is equal to the shaded area under the density function between the ordinates at $x = a$ and $x = b$, and from integral calculus is given by

$$P(a < X < b) = \int_a^b f(x)\,dx.$$

Areas have been computed and put in tabular form for those density functions that are used most frequently in conducting experiments. Because areas represent probabilities and probabilities are positive numerical values, the density function must be entirely above the x axis. If the range of X for which $f(x)$ is defined is a finite interval, it is always possible to extend the interval to include the entire set of real numbers by defining $f(x)$ to be zero at all points in the extended portions of the interval.

DEFINITION 2.6   *The function $f(x)$ is a* probability density function *for the continuous random variable X, defined over the set of real numbers R, if*

   1. $f(x) \geq 0$ *for all $x \in R$.*

   2. $\int_{-\infty}^{\infty} f(x)\, dx = 1.$

   3. $P(a < X < b) = \int_{a}^{b} f(x)\, dx.$

Example 2.6   Let the random variable X have the probability density function

$$f(x) = \frac{x^2}{3}, \qquad -1 < x < 2$$

$$= 0, \qquad \text{elsewhere.}$$

1. Verify condition 2 of Definition 2.6.
2. Find $P(0 < X \leq 1)$.

Solution

1. $\displaystyle \int_{-\infty}^{\infty} f(x)\, dx = \int_{-1}^{2} \frac{x^2}{3}\, dx = \frac{x^3}{9} \Big|_{-1}^{2} = \frac{8}{9} + \frac{1}{9} = 1.$

2. $\displaystyle P(0 < X \leq 1) = \int_{0}^{1} \frac{x^2}{3}\, dx = \frac{x^3}{9} \Big|_{0}^{1} = \frac{1}{9}.$

DEFINITION 2.7.   *The* cumulative distribution $F(x)$ *of a continuous random variable X with density function $f(x)$ is given by*

$$F(x) = P(X \leq x) = \int_{-\infty}^{x} f(t)\, dt.$$

As an immediate consequence of Definition 2.7 one can write the two results

$$P(a < X < b) = F(b) - F(a)$$

and

$$f(x) = \frac{dF(x)}{dx}$$

if the derivative exists.

Example 2.7   For the density function of Example 2.6 find $F(x)$ and use it to evaluate $P(0 < X \le 1)$.

Solution

$$F(x) = \int_{-\infty}^{x} f(t)\, dt = \int_{-1}^{x} \frac{t^2}{3}\, dt = \frac{t^3}{9} \Big|_{-1}^{x} = \frac{x^3 + 1}{9}.$$

Therefore,

$$P(0 < X \le 1) = F(1) - F(0) = \tfrac{2}{9} - \tfrac{1}{9} = \tfrac{1}{9},$$

which agrees with the result obtained by using the density function in Example 2.6.

2.4 EMPIRICAL DISTRIBUTIONS

Usually, in an experiment involving a continuous random variable the density function $f(x)$ is unknown and its form is assumed. For the choice of $f(x)$ to be reasonably valid, good judgment based on all available information is needed in its selection. Statistical data, generated in large masses, can be very useful in studying the behavior of the distribution if presented in the form of a *relative frequency distribution*. Such an arrangement is obtained by grouping the data into classes and determining the proportion of measurements in each of the classes.

To illustrate the construction of a relative frequency distribution, consider the data of Table 2.1, which represent the lives of 40 similar car batteries recorded to the nearest tenth of a year. The batteries were guaranteed to last 3 years.

We must first decide on the number of classes into which the data are to be grouped. This is done arbitrarily although we are guided by the amount of data available. Usually, we choose between 5 and 20 class intervals. The smaller the number of data available, the smaller is our choice for the number of classes. For the data of Table 2.1 let us choose 7 class intervals. The class width must be large enough so that 7 class intervals accommodate all the data. To determine

Table 2.1 Car Battery Lives

2.2	4.1	3.5	4.5	3.2	3.7	3.0	2.6
3.4	1.6	3.1	3.3	3.8	3.1	4.7	3.7
2.5	4.3	3.4	3.6	2.9	3.3	3.9	3.1
3.3	3.1	3.7	4.4	3.2	4.1	1.9	3.4
4.7	3.8	3.2	2.6	3.9	3.0	4.2	3.5

the approximate class width, we divide the difference between the largest and smallest measurements by the number of intervals. Therefore, in our example the class width can be no less than $(4.7 - 1.6)/7 = 0.443$. In practice, it is desirable to choose equal class widths having the same number of significant places as the given data. Denoting this width by c, we choose $c = 0.5$. If we begin the lowest interval at 1.5, the second class would begin at 2.0, and so forth. The relative frequency distribution for the data of Table 2.1, showing the midpoints of each class interval, is given in Table 2.2.

The information provided by a relative frequency distribution in tabular form is easier to grasp if presented graphically. Using the midpoints of each interval and the corresponding relative frequencies, we construct a *relative frequency histogram* (Figure 2.6) in exactly the same manner that we constructed the probability histogram of Section 2.2.

In Section 2.2 we suggested that the heights of the rectangles be adjusted so that the areas would represent probabilities. Once this is done, the vertical axis may be omitted. If we wish to estimate the probability distribution $f(x)$ of a continuous random variable X by a smooth curve as in Figure 2.7, it is important that the rectangles of the relative frequency histogram be adjusted so that the total area is equal to 1.

Table 2.2 Relative Frequency Distribution of Battery Lives

Class interval	Class midpoint	Frequency f	Relative frequency
1.5–1.9	1.7	2	0.050
2.0–2.4	2.2	1	0.025
2.5–2.9	2.7	4	0.100
3.0–3.4	3.2	15	0.375
3.5–3.9	3.7	10	0.250
4.0–4.4	4.2	5	0.125
4.5–4.9	4.7	3	0.075

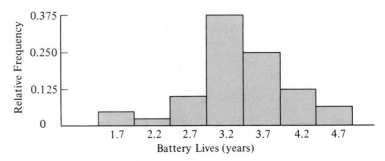

Figure 2.6 Relative frequency histogram.

The probability that a battery lasts between 3.45 and 4.45 years when selected at random from the infinite line of production of such batteries is given by the shaded area under the curve. Our estimated probability based on the recorded lives of the 40 batteries would be the sum of the areas contained in the rectangles between 3.45 and 4.45.

Although we have drawn an estimate of the shape of $f(x)$ in Figure 2.7, we still have no knowledge of its formula or equation and therefore cannot find the area that has been shaded. To help understand the method of estimating the formula for $f(x)$, let us recall some elementary analytic geometry. Parabolas, hyperbolas, circles, ellipses, and so forth, all have well-known forms of equations, and in each case we would recognize their graphs. Thinking in reverse, if we had only their graphs, but recognized their form, then it is not difficult to estimate the unknown constants or parameters and arrive at the exact equation. For example, if the curve appeared to have the form of a parabola, then we know it has an equation of the form $f(x) = ax^2 + bx + c$, where a, b, and c are parameters that can be determined by various estimation procedures.

Many continuous distributions can be represented graphically by the characteristic bell-shaped curve of Figure 2.7. The equation of the probability

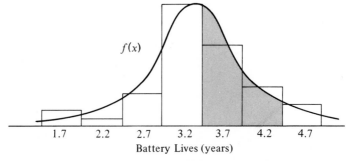

Figure 2.7 Estimating the probability density function.

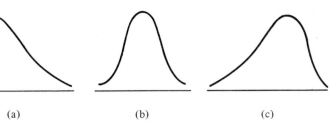

(a)   (b)   (c)

Figure 2.8 Skewness of data.

density function $f(x)$ in this case is as well known as that of a parabola or circle and depends only on the determination of two parameters. Once these parameters are estimated from the data, we can write the estimated equation, and then, using appropriate tables, find any probabilities we choose.

A distribution is said to be symmetric if it can be folded along an axis so that the two sides coincide. A distribution that lacks symmetry with respect to a vertical axis is said to be *skewed*. The distribution illustrated in Figure 2.8(a) is said to be skewed to the right, since it has a long right tail and a much shorter left tail. In Figure 2.8(b) we see that the distribution is symmetric, while in Figure 2.8(c) it is skewed to the left.

Histograms can have almost any shape or form. Some of the possible density functions that might arise were illustrated in Figure 2.4. In Chapter 4 we shall consider most of the important density functions that are used in engineering and scientific investigations.

The cumulative distribution of X, where X represents the life of the car battery, can be estimated geometrically using the data of Table 2.2. To construct such a graph, we first arrange our data as in Table 2.3, a *relative cumulative*

Table 2.3 Relative Cumulative Frequency Distribution of Battery Lives

Class boundaries	Relative cumulative frequency
Less than 1.45	0.000
Less than 1.95	0.050
Less than 2.45	0.075
Less than 2.95	0.175
Less than 3.45	0.550
Less than 3.95	0.800
Less than 4.45	0.925
Less than 4.95	1.000

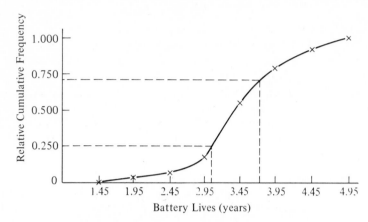

Figure 2.9 Continuous cumulative distribution.

frequency distribution, and then plot the relative cumulative frequency less than any upper class boundary against the upper class boundary as in Figure 2.9. We estimate $F(x)$ by drawing a smooth curve through the points.

Percentile, decile, and quartile points may be read quickly from the cumulative distribution. In Figure 2.9 the dashed lines indicate that the twenty-fifth percentile or first quartile and the seventh decile are approximately 3.05 and 3.70 years, respectively. This means that 25% or one fourth of all the batteries of this type are expected to last less than 3.05 years, while 70% of such batteries can be expected to last less than 3.70 years.

2.5 JOINT PROBABILITY DISTRIBUTIONS

Our study of random variables and their probability distributions in the preceding sections was restricted to one-dimensional sample spaces in that we recorded outcomes of an experiment assumed by a single random variable. There will be many situations, however, where we may find it desirable to record the simultaneous outcomes of several random variables. For example, we might measure the amount of precipitate P and volume V of gas released from a controlled chemical experiment giving rise to a two-dimensional sample space consisting of the outcomes (p, v), or one might be interested in the hardness H and tensile strength T of cold-drawn copper resulting in the outcomes (h, t). In a study to determine the likelihood of success in college, based on high school data, one might use a three-dimensional sample space and record for each individual his aptitude test score, high school rank in class, and grade-point average at the end of the freshman year in college.

If X and Y are two random variables, the probability distribution for their simultaneous occurrence can be represented in functional notation by $f(x, y)$. It is customary to refer to $f(x, y)$ as the *joint probability distribution* of X and Y. Hence in the discrete case where a listing is possible, $f(x, y) = P(X = x, Y = y)$. That is, $f(x, y)$ gives the probability that the outcomes x and y occur at the same time. For example, if a television set is to be serviced and X represents the age of the set and Y represents the number of defective tubes in the set, then $f(5, 3)$ is the probability that the television set is 5 years old and needs three new tubes.

DEFINITION 2.8   *The function $f(x, y)$ is a joint probability function of the discrete random variables X and Y if*

1. $f(x, y) \geq 0$ *for all (x, y).*

2. $\sum_x \sum_y f(x, y) = 1.$

3. $P[(X, Y) \in A] = \sum\sum_A f(x, y)$

for any region A in the xy plane.

Example 2.8   Two refills for a ballpoint pen are selected at random from a box that contains three blue refills, two red refills, and three green refills. If X is the number of blue refills and Y is the number of red refills selected, find (1) the joint probability function $f(x, y)$, and (2) $P[(X, Y) \in A]$, where A is the region $\{(x, y)|x + y \leq 1\}$.

Solution   1. The possible pairs of values (x, y) are $(0, 0), (0, 1), (1, 0), (1, 1), (0, 2),$ and $(2, 0)$. Now, $f(0, 1)$, for example, represents the probability that a red and a green refill are selected. The total number of equally likely ways of selecting

$$n = 8$$
$$r = 2$$

$$\binom{n}{r} = \frac{n!}{r!(n-r)!} = \frac{8!}{2!(8-2)!} = 28$$

Table 2.4 Joint Probability Distribution for Example 2.8

y \ x	0	1	2
0	$\frac{3}{28}$	$\frac{9}{28}$	$\frac{3}{28}$
1	$\frac{3}{14}$	$\frac{3}{14}$	
2	$\frac{1}{28}$		

any two refills from the eight is $\binom{8}{2} = 28$. The number of ways of selecting one

red from two red refills and one green from three green refills is $\binom{2}{1}\binom{3}{1} = 6$.

Hence $f(0, 1) = 6/28 = 3/14$. Similar calculations yield the probabilities for the other cases, which are presented in Table 2.4. In Chapter 3 it will become clear that the joint probability distribution of Table 2.4 can be represented by the formula

$$f(x, y) = \frac{\binom{3}{x}\binom{2}{y}\binom{3}{2 - x - y}}{\binom{8}{2}}, \qquad \begin{array}{l} x = 0, 1, 2 \\ y = 0, 1, 2 \\ 0 \leq x + y \leq 2. \end{array}$$

2. $P[(X, Y) \in A] = P(X + Y \leq 1)$
$= f(0, 0) + f(0, 1) + f(1, 0)$
$= \frac{3}{28} + \frac{3}{14} + \frac{9}{28}$
$= \frac{9}{14}.$

DEFINITION 2.9 *The function $f(x, y)$ is a* joint density function *of the* continuous *random variables X and Y if*

1. $f(x, y) \geq 0$ *for all (x, y).*

2. $\displaystyle\int_{-\infty}^{\infty} \int_{-\infty}^{\infty} f(x, y)\, dx\, dy = 1.$

3. $P[(X, Y) \in A] = \displaystyle\iint_{A} f(x, y)\, dx\, dy$

for any region A in the xy plane.

Example 2.9 Consider the joint density function

$$f(x, y) = \frac{x(1 + 3y^2)}{4}, \qquad 0 < x < 2, 0 < y < 1$$

$$= 0, \qquad\qquad \text{elsewhere.}$$

1. Verify condition 2 of Definition 2.9.
2. Find $P[(X, Y) \in A]$, where A is the region $\{(x, y)|0 < x < 1, 1/4 < y < 1/2\}$.

Solution

1. $\displaystyle\int_{-\alpha}^{\infty}\int_{-\alpha}^{\infty} f(x,y)\,dx\,dy = \int_0^1\int_0^2 \frac{x(1+3y^2)}{4}\,dx\,dy$

$\displaystyle\qquad = \int_0^1 \frac{x^2}{8} + \frac{3x^2y^2}{8}\Big|_{x=0}^{x=2}\,dy$

$\displaystyle\qquad = \int_0^1\left(\frac{1}{2} + \frac{3y^2}{2}\right)dy = \frac{y}{2} + \frac{y^3}{2}\Big|_0^1$

$\displaystyle\qquad = \frac{1}{2} + \frac{1}{2} = 1.$

2. $\displaystyle P[(X,Y)\in A] = P\left(0 < X < 1, \frac{1}{4} < Y < \frac{1}{2}\right)$

$\displaystyle\qquad = \int_{1/4}^{1/2}\int_0^1 \frac{x(1+3y^2)}{4}\,dx\,dy$

$\displaystyle\qquad = \int_{1/4}^{1/2} \frac{x^2}{8} + \frac{3x^2y^2}{8}\Big|_{x=0}^{x=1}\,dy$

$\displaystyle\qquad = \int_{1/4}^{1/2}\left(\frac{1}{8} + \frac{3y^2}{8}\right)dy = \frac{y}{8} + \frac{y^3}{8}\Big|_{1/4}^{1/2}$

$\displaystyle\qquad = \left(\frac{1}{16} + \frac{1}{64}\right) - \left(\frac{1}{32} + \frac{1}{512}\right) = \frac{23}{512}.$

Given the probability distribution $f(x,y)$ of the random variables X and Y, the probability distributions of X alone and Y alone are given by

$$g(x) = \sum_y f(x,y)$$

$$h(y) = \sum_x f(x,y)$$

DISCRETE

for the discrete case and by

$$g(x) = \int_{-\infty}^{\infty} f(x,y)\,dy$$

$$h(y) = \int_{-\infty}^{\infty} f(x,y)\,dx$$

CONTINUOUS

for the continuous case. We define $g(x)$ and $h(y)$ to be the *marginal distributions* of X and Y, respectively. The fact that these marginal distributions are indeed the probability distributions of the individual variables can easily be verified by showing that the conditions of Definition 2.4 or Definition 2.6 are satisfied. For example, in the continuous case,

$$\int_{-\infty}^{\infty} g(x)\, dx = \int_{-\infty}^{\infty} \int_{-\infty}^{\infty} f(x, y)\, dy\, dx = 1$$

and

$$P(a < X < b) = P(a < X < b, -\infty < Y < \infty)$$

$$= \int_{a}^{b} \int_{-\infty}^{\infty} f(x, y)\, dy\, dx$$

$$= \int_{a}^{b} g(x)\, dx.$$

In Section 2.1 we stated that the value x of the random variable X represents an event that is a subset of the sample space. Using the definition of conditional probability as given in Chapter 1,

$$P(B|A) = \frac{P(A \cap B)}{P(A)}, \qquad P(A) > 0,$$

where A and B are now the events defined by $X = x$ and $Y = y$, respectively,

$$P(Y = y | X = x) = \frac{P(X = x, Y = y)}{P(X = x)}$$

$$= \frac{f(x, y)}{g(x)}, \qquad g(x) > 0,$$

when X and Y are discrete random variables.

It is not difficult to show that the function $f(x, y)/g(x)$, which is strictly a function of y with x fixed, satisfies all the conditions of a probability distribution. Writing this probability distribution as $f(y|x)$, we have

$$f(y|x) = \frac{f(x, y)}{g(x)}, \qquad g(x) > 0,$$

which is called the *conditional distribution* of the *discrete* random variable Y, given that $X = x$. Similarly, we define $f(x|y)$ to be the conditional distribution of the *discrete* random variable X, given that $Y = y$, and write

$$f(x|y) = \frac{f(x, y)}{h(y)}, \qquad h(y) > 0.$$

The *conditional probability density function* of the *continuous* random variable X, given that $Y = y$, is, by definition,

$$f(x|y) = \frac{f(x, y)}{h(y)}, \qquad h(y) > 0,$$

while the conditional probability density function of the *continuous* random variable Y, given that $X = x$, is defined to be

$$f(y|x) = \frac{f(x, y)}{g(x)}, \qquad g(x) > 0.$$

If one wished to find the probability that the continuous random variable X falls between a and b when it is known that $Y = y$, we evaluate

$$P(a < X < b | Y = y) = \int_a^b f(x|y)\, dx.$$

Example 2.10   Referring to Example 2.8, find $f(x|1)$ and

$$P(X = 0 | Y = 1).$$

Solution   First we find that

$$h(1) = \sum_{x=0}^{2} f(x, 1) = \tfrac{3}{14} + \tfrac{3}{14} + 0 = \tfrac{3}{7}.$$

Now

$$f(x|1) = \frac{f(x, 1)}{h(1)} = \frac{7}{3} f(x, 1), \qquad x = 0, 1, 2.$$

Therefore,

$$f(0|1) = \tfrac{7}{3} f(0, 1) = (\tfrac{7}{3})(\tfrac{3}{14}) = \tfrac{1}{2}$$
$$f(1|1) = \tfrac{7}{3} f(1, 1) = (\tfrac{7}{3})(\tfrac{3}{14}) = \tfrac{1}{2}$$
$$f(2|1) = \tfrac{7}{3} f(2, 1) = (\tfrac{7}{3})(0) = 0$$

and the conditional distribution of X, given that $Y = 1$, is

x	0	1	2
$f(x\|1)$	$\frac{1}{2}$	$\frac{1}{2}$	0

Finally,

$$P(X = 0 | Y = 1) = f(0|1) = \tfrac{1}{2}.$$

Example 2.11   The joint density function of the random variables X and Y is given by

$$f(x, y) = 8xy, \qquad 0 < x < 1, 0 < y < x$$
$$= 0, \qquad \text{elsewhere.}$$

Find $g(x)$, $h(y)$, $f(y|x)$, and $P(Y < 1/8 | X = 1/2)$.

Solution   By definition,

$$g(x) = \int_{-\infty}^{\infty} f(x, y)\, dy = \int_{0}^{x} 8xy\, dy$$

$$= 4xy^2 \Big|_{y=0}^{y=x} = 4x^3, \qquad 0 < x < 1,$$

and

$$h(y) = \int_{-\infty}^{\infty} f(x, y)\, dx = \int_{y}^{1} 8xy\, dx$$

$$= 4x^2y \Big|_{x=y}^{x=1} = 4y(1 - y^2), \qquad 0 < y < 1.$$

Now,

$$f(y|x) = \frac{f(x, y)}{g(x)} = \frac{8xy}{4x^3} = \frac{2y}{x^2}, \qquad 0 < y < x,$$

and

$$P(Y < \tfrac{1}{8} | X = \tfrac{1}{2}) = \int_{0}^{1/8} 8y\, dy = \tfrac{1}{16}.$$

Example 2.12   Find $g(x)$, $h(y)$, $f(x|y)$, and

$$P(\tfrac{1}{4} < X < \tfrac{1}{2} | Y = \tfrac{1}{3})$$

for the density function of Example 2.9.

Solution   By definition,

$$g(x) = \int_{-\infty}^{\infty} f(x, y)\, dy = \int_{0}^{1} \frac{x(1 + 3y^2)}{4}\, dy$$

$$= \frac{xy}{4} + \frac{xy^3}{4}\Big|_{y=0}^{y=1} = \frac{x}{2}, \qquad 0 < x < 2,$$

and

$$h(y) = \int_{-\infty}^{\infty} f(x, y)\, dx = \int_{0}^{2} \frac{x(1 + 3y^2)}{4}\, dx$$

$$= \frac{x^2}{8} + \frac{3x^2 y^2}{8}\Big|_{x=0}^{x=2} = \frac{1 + 3y^2}{2}, \qquad 0 < y < 1.$$

Therefore,

$$f(x|y) = \frac{f(x, y)}{h(y)} = \frac{x(1 + 3y^2)/4}{(1 + 3y^2)/2} = \frac{x}{2}, \qquad 0 < x < 2,$$

and

$$P\left(\frac{1}{4} < X < \frac{1}{2}\,\middle|\, Y = \frac{1}{3}\right) = \int_{1/4}^{1/2} \frac{x}{2}\, dx = \frac{3}{64}.$$

If $f(x|y)$ does not depend on y, as was the case in Example 2.12, then $f(x|y) = g(x)$ and $f(x, y) = g(x)h(y)$. The proof follows by substituting

$$f(x, y) = f(x|y)h(y)$$

into the marginal distribution of X. That is,

$$g(x) = \int_{-\infty}^{\infty} f(x, y)\, dy = \int_{-\infty}^{\infty} f(x|y)h(y)\, dy.$$

If $f(x|y)$ does not depend on y, we may write

$$g(x) = f(x|y) \int_{-\infty}^{\infty} h(y)\, dy.$$

Now

$$\int_{-\infty}^{\infty} h(y)\, dy = 1$$

since $h(y)$ is the probability density function of Y. Therefore,

$$g(x) = f(x|y)$$

and then

$$f(x, y) = g(x)h(y),$$

which leads to the following definition.

DEFINITION 2.10   *Let X and Y be two random variables, discrete or continuous, with joint probability distribution $f(x, y)$ and marginal distributions $g(x)$ and $h(y)$, respectively. The random variables X and Y are said to be statistically independent if and only if*

$$f(x, y) = g(x)h(y)$$

for all (x, y).

   The continuous random variables of Example 2.12 are statistically independent, since the product of the two marginal distributions gives the joint density function. This is obviously not the case, however, for the continuous variables of Example 2.11. Checking for statistical independence of discrete random variables requires a more thorough investigation, since it is possible to have the product of the marginal distributions equal to the joint probability distribution for some but not all combinations of (x, y). If you can find any point (x, y) for which $f(x, y)$ is defined such that $f(x, y) \neq g(x)h(y)$, the discrete

$$g(x) = \int f(x,y)\, dy$$

$$h(y) = \int f(x,y)\, dx$$

variables X and Y are not statistically independent. Consider, for example, the three probabilities $f(0, 1)$, $g(0)$, and $h(1)$ in Example 2.8. From Table 2.4 we find

$$f(0, 1) = \tfrac{3}{14}$$

$$g(0) = \sum_{y=0}^{2} f(0, y) = \tfrac{3}{28} + \tfrac{3}{14} + \tfrac{1}{28} = \tfrac{5}{14}$$

$$h(1) = \sum_{x=0}^{2} f(x, 1) = \tfrac{3}{14} + \tfrac{3}{14} + 0 = \tfrac{3}{7}.$$

Clearly,

$$f(0, 1) \neq g(0)h(1),$$

and therefore X and Y are not statistically independent.

All the preceding definitions concerning two random variables can be generalized to the case of n random variables. Let $f(x_1, x_2, \ldots, x_n)$ be the joint probability function of the random variables $X_1, X_2, \ldots, X_n$. The marginal distribution of X_1, for example, is given by

$$g(x_1) = \sum_{x_2} \cdots \sum_{x_n} f(x_1, x_2, \ldots, x_n)$$

for the discrete case and by

$$g(x_1) = \int_{-\infty}^{\infty} \cdots \int_{-\infty}^{\infty} f(x_1, x_2, \ldots, x_n) \, dx_2 \, dx_3 \cdots dx_n$$

for the continuous case. We can now obtain *joint marginal distributions* such as $\phi(x_1, x_2)$, where

$$\phi(x_1, x_2) = \sum_{x_3} \cdots \sum_{x_n} f(x_1, x_2, \ldots, x_n) \qquad \text{(discrete case)}$$

$$= \int_{-\infty}^{\infty} \cdots \int_{-\infty}^{\infty} f(x_1, x_2, \ldots, x_n) \, dx_3 \, dx_4 \cdots dx_n \qquad \text{(continuous case)}.$$

One could consider numerous conditional distributions. For example, the *joint conditional distribution* of X_1, X_2, and X_3 given that $X_4 = x_4$, $X_5 = x_5, \ldots, X_n = x_n$ is written

$$f(x_1, x_2, x_3 \mid x_4, x_5, \ldots, x_n) = \frac{f(x_1, x_2, \ldots, x_n)}{g(x_4, x_5, \ldots, x_n)},$$

where $g(x_4, x_5, \ldots, x_n)$ is the joint marginal distribution of the random variables $X_4, X_5, \ldots, X_n$.

A generalization of Definition 2.10 leads to the following definition for the mutually statistical independence of the variables $X_1, X_2, \ldots, X_n$.

DEFINITION 2.11   *Let $X_1, X_2, \ldots, X_n$ be n random variables, discrete or continuous, with joint probability distribution $f(x_1, x_2, \ldots, x_n)$ and marginal distributions $f_1(x_1), f_2(x_2), \ldots, f_n(x_n)$, respectively. The random variables $X_1, X_2, \ldots, X_n$ are said to be mutually statistically independent if and only if*

$$f(x_1, x_2, \ldots, x_n) = f_1(x_1)f_2(x_2) \cdots f_n(x_n).$$

Example 2.13   Let X_1, X_2, and X_3 be three mutually statistically independent random variables and let each have probability density function

$$f(x) = e^{-x}, \qquad x > 0$$
$$= 0, \qquad \text{elsewhere.}$$

Find $P(X_1 < 2, 1 < X_2 < 3, X_3 > 2)$.

Solution   The joint probability density function of X_1, X_2, and X_3 is

$$f(x_1, x_2, x_3) = f(x_1)f(x_2)f(x_3)$$
$$= e^{-x_1}e^{-x_2}e^{-x_3}$$
$$= e^{-x_1-x_2-x_3}, \qquad x_1 > 0, x_2 > 0, x_3 > 0.$$

Hence

$$P(X_1 < 2, 1 < X_2 < 3, X_3 > 2) = \int_2^\infty \int_1^3 \int_0^2 e^{-x_1-x_2-x_3} \, dx_1 \, dx_2 \, dx_3$$
$$= (1 - e^{-2})(e^{-1} - e^{-3})e^{-2}$$
$$= 0.0376.$$

2.6 MATHEMATICAL EXPECTATION

If two coins are tossed 16 times and X is the number of heads that occur per toss, then the values of X can be 0, 1, and 2. Suppose that the experiment yields no heads, one head, and two heads a total of 4, 7, and 5 times, respectively. The average number of heads per toss of the two coins is then

$$\frac{(0)(4) + (1)(7) + (2)(5)}{16} = (0)\left(\frac{4}{16}\right) + (1)\left(\frac{7}{16}\right) + (2)\left(\frac{5}{16}\right)$$

$$= 1.06.$$

This is an average value and is not necessarily a possible outcome for the experiment. For instance, a salesman's average monthly income is not likely to be equal to any of his monthly paychecks.

The numbers 4/16, 7/16, and 5/16 are the fractions of the total tosses resulting in zero, one, and two heads, respectively. These fractions are also the relative frequencies for the different outcomes.

Let us now consider the problem of calculating the average number of heads per toss that we might expect in the long run. We denote this expected value or mathematical expectation by $E(X)$. From the relative frequency definition of probability we can, in the long run, expect no heads about one fourth of the time, one head about one half of the time, and two heads about one fourth of the time. Therefore,

$$E(X) = (0)(\tfrac{1}{4}) + (1)(\tfrac{1}{2}) + (2)(\tfrac{1}{4}) = 1.$$

This means that a person who throws two coins over and over again will, on the average, get one head per toss.

This illustration suggests that the mean or expected value of any random variable may be obtained by multiplying each value of the random variable by its corresponding probability and summing the results. This is true, of course, only if the variable is discrete. In the case of continuous random variables, the definition of mathematical expection is essentially the same with summations being replaced by integrals.

DEFINITION 2.12 *Let X be a random variable with probability distribution* $f(x)$. *The expected value of X or the mathematical expectation of X is*

$$E(X) = \sum_{x} xf(x) \qquad \text{if X is discrete}$$

$$= \int_{-\infty}^{\infty} xf(x)\,dx \qquad \text{if X is continuous.}$$

Example 2.14 Find the expected number of chemists on a committee of size 3 selected at random from four chemists and three biologists.

Solution   Let X represent the number of chemists on the committee. The probability distribution of X is given by

$$f(x) = \frac{\binom{4}{x}\binom{3}{3-x}}{\binom{7}{3}}, \qquad x = 0, 1, 2, 3.$$

A few simple calculations yield $f(0) = 1/35$, $f(1) = 12/35$, $f(2) = 18/35$, and $f(3) = 4/35$. Therefore,

$$E(X) = (0)(\tfrac{1}{35}) + (1)(\tfrac{12}{35}) + (2)(\tfrac{18}{35}) + (3)(\tfrac{4}{35})$$
$$= \tfrac{12}{7} = 1.7.$$

Thus, if a committee of size 3 is selected at random over and over again from four chemists and three biologists, it would contain, on the average, 1.7 chemists.

Example 2.15   In a gambling game a man is paid $5 if he gets all heads or all tails when three coins are tossed and he pays out $3 if either one or two heads show. What is his expected gain?

Solution   The random variable of interest is Y, the amount he can win. The possible values of Y are 5 and -3, with probabilities 1/4 and 3/4, respectively. Therefore,

$$E(Y) = (5)(\tfrac{1}{4}) + (-3)(\tfrac{3}{4}) = -1.$$

In this game the gambler will, on the average, lose $1 per toss.

A game is considered "fair" if the gambler will, on the average, come out even. Therefore, an expected gain of zero defines a fair game.

Example 2.16   Let X be the random variable that denotes the life in hours of a certain type of tube. The probability density function is given by

$$f(x) = \frac{20,000}{x^3}, \qquad x > 100$$

$$= 0, \qquad \text{elsewhere.}$$

Find the expected life of this type of tube.

Solution   Using Definition 2.12, we have

$$E(X) = \int_{100}^{\infty} x \frac{20,000}{x^3} \, dx$$

$$= \int_{100}^{\infty} \frac{20,000}{x^2} \, dx \quad = \quad -\frac{20,000}{X} \Big|_{100}^{\infty}$$

$$= 200.$$

Therefore, we can expect this type of tube to last, on the average, 200 hours.

Now let us consider a function $g(X)$ of the random variable X. That is, each value of $g(X)$ is determined by knowing the values of X. For instance, $g(X)$ might be X^2 or $3X - 1$, so that whenever X assumes the value 2, $g(X)$ assumes the value $g(2)$. In particular, if X is a discrete random variable with probability distribution $f(x)$, $x = -1, 0, 1, 2$, and $g(X) = X^2$, then

$$P[g(X) = 0] = P(X = 0) = f(0)$$

$$P[g(X) = 1] = P(X = -1) + P(X = 1) = f(-1) + f(1)$$

$$P[g(X) = 4] = P(X = 2) = f(2).$$

By Definition 2.12,

$$E[g(X)] = \sum_{g(x)} g(x) P[g(X) = g(x)]$$

$$= 0P[g(X) = 0] + 1P[g(X) = 1] + 4P[g(X) = 4]$$

$$= 0f(0) + 1[f(-1) + f(1)] + 4f(2)$$

$$= \sum_{x} g(x) f(x).$$

This result is generalized in Theorem 2.1 for both discrete and continuous random variables.

THEOREM 2.1   *Let X be a random variable with probability distribution $f(x)$. The expected value of the function $g(X)$ is*

$$E[g(X)] = \sum_{x} g(x) f(x) \qquad \textit{if X is discrete}$$

$$= \int_{-\infty}^{\infty} g(x) f(x) \, dx \qquad \textit{if X is continuous.}$$

Example 2.17   Let X be a random variable with probability distribution as follows:

x	0	1	2	3
$f(x)$	$\frac{1}{3}$	$\frac{1}{2}$	0	$\frac{1}{6}$

Find the expected value of $Y = (X - 1)^2$.

Solution   By Theorem 2.1, we write

$$E[(X - 1)^2] = \sum_{x=0}^{3} (x - 1)^2 f(x)$$
$$= (-1)^2 f(0) + (0)^2 f(1) + (1)^2 f(2) + (2)^2 f(3)$$
$$= (1)(\tfrac{1}{3}) + (0)(\tfrac{1}{2}) + (1)(0) + (4)(\tfrac{1}{6})$$
$$= 1.$$

Example 2.18   Let X be a random variable with density function

$$f(x) = \frac{x^2}{3}, \qquad -1 < x < 2$$
$$= 0, \qquad \text{elsewhere.}$$

Find the expected value of $g(X) = 2X - 1$.

Solution   By Theorem 2.1, we have

$$E(2X - 1) = \int_{-1}^{2} \frac{(2x - 1)x^2}{3} \, dx$$
$$= \tfrac{1}{3} \int_{-1}^{2} (2x^3 - x^2) \, dx \quad = \frac{1}{3}\left[\frac{2x^4}{4} - \frac{x^3}{3}\right]\Bigg|_{-1}^{2}$$
$$= \tfrac{3}{2}.$$

We shall now extend our concept of mathematical expectation to the case of two random variables X and Y with joint probability distribution $f(x, y)$.

DEFINITION 2.13   *Let X and Y be random variables with joint probability distribution f(x, y). The* expected value *of the function g(X, Y) is*

$$E[g(X, Y)] = \sum_x \sum_y g(x, y) f(x, y) \qquad \text{if X and Y are discrete.}$$

$$= \int_{-\infty}^{\infty} \int_{-\infty}^{\infty} g(x, y) f(x, y) \, dx \, dy \qquad \text{if X and Y are continuous.}$$

Generalization of Definition 2.13 for the calculation of mathematical expectations of functions of several random variables is straightforward.

Example 2.19   Let X and Y be random variables with joint probability distribution given by Table 2.4. Find the expected value of $g(X, Y) = XY$.

Solution   By Definition 2.13, we write

$$E(XY) = \sum_{x=0}^{2} \sum_{y=0}^{2} xy f(x, y)$$

$$= (0)(0)f(0, 0) + (0)(1)f(0, 1) + (0)(2)f(0, 2)$$
$$+ (1)(0)f(1, 0) + (1)(1)f(1, 1)$$
$$+ (2)(0)f(2, 0)$$
$$= f(1, 1) = \tfrac{3}{14}.$$

Example 2.20   Find $E(Y/X)$ for the density function

$$f(x, y) = \frac{x(1 + 3y^2)}{4}, \qquad 0 < x < 2, 0 < y < 1$$

$$= 0, \qquad\qquad\quad \text{elsewhere.}$$

Solution   We have

$$E\left(\frac{Y}{X}\right) = \int_0^1 \int_0^2 \frac{y(1 + 3y^2)}{4} \, dx \, dy$$

$$= \int_0^1 \frac{y + 3y^3}{2} \, dy$$

$$= \frac{5}{8}.$$

Note that if $g(x, y) = X$ in Definition 2.13, we have

$$E(X) = \sum_x \sum_y xf(x, y) = \sum_x xg(x) \qquad \text{(discrete case)}$$

$$= \int_{-\infty}^{\infty} \int_{-\infty}^{\infty} xf(x, y) \, dx \, dy = \int_{-\infty}^{\infty} xg(x) \, dx \qquad \text{(continuous case)},$$

where $g(x)$ is the marginal distribution of X. Therefore, in calculating $E(X)$ over a two-dimensional space, one may use either the joint probability distribution of X and Y or the marginal distribution of X.

Similarly, we define

$$E(Y) = \sum_x \sum_y yf(x, y) = \sum_y yh(y) \qquad \text{(discrete case)}$$

$$= \int_{-\infty}^{\infty} \int_{-\infty}^{\infty} yf(x, y) \, dx \, dy = \int_{-\infty}^{\infty} yh(y) \, dy \qquad \text{(continuous case)},$$

where $h(y)$ is the marginal distribution of the random variable Y.

2.7 LAWS OF EXPECTATION

We shall now develop some useful laws that will simplify the calculations of mathematical expectations. These laws or theorems will permit us to calculate expectations in terms of other expectations that are either known or easily computed. All the results are valid for both discrete and continuous random variables. Proofs will be given only for the continuous case.

THEOREM 2.2 *If a and b are constant, then*

$$E(aX + b) = aE(X) + b.$$

Proof By the definition of an expected value,

$$E(aX + b) = \int_{-\infty}^{\infty} (ax + b)f(x) \, dx$$

$$= a \int_{-\infty}^{\infty} xf(x) \, dx + b \int_{-\infty}^{\infty} f(x) \, dx.$$

The first integral on the right is $E(X)$ and the second integral equals 1. Therefore, we have

$$E(aX + b) = aE(X) + b.$$

COROLLARY 1   *Setting $a = 0$, we see that $E(b) = b$.*

COROLLARY 2   *Setting $b = 0$, we see that $E(aX) = aE(X)$.*

THEOREM 2.3   *The expected value of the sum or difference of two or more functions of a random variable X is the sum or difference of the expected values of the functions. That is,*

$$E[g(X) \pm h(X)] = E[g(X)] \pm E(h(X)].$$

Proof By definition,

$$E[g(X) \pm h(X)] = \int_{-\infty}^{\infty} [g(x) \pm h(x)] f(x)\, dx$$

$$= \int_{-\infty}^{\infty} g(x) f(x)\, dx \pm \int_{-\infty}^{\infty} h(x) f(x)\, dx$$

$$= E[g(X)] \pm E[h(X)].$$

Example 2.21   In Example 2.17 we could write

$$E[(X - 1)^2] = E(X^2 - 2X + 1) = E(X^2) - 2E(X) + E(1).$$

From Corollary 1, $E(1) = 1$, and by direct computation

$$E(X) = (0)(\tfrac{1}{3}) + (1)(\tfrac{1}{2}) + (2)(0) + (3)(\tfrac{1}{6}) = 1$$
$$E(X^2) = (0)(\tfrac{1}{3}) + (1)(\tfrac{1}{2}) + (4)(0) + (9)(\tfrac{1}{6}) = 2.$$

Hence

$$E[(X - 1)^2] = 2 - (2)(1) + 1 = 1,$$

as before.

Example 2.22   In Example 2.18 we may prefer to write

$$E(2X - 1) = 2E(X) - 1.$$

Now

$$E(X) = \int_{-1}^{2} x\left(\frac{x^2}{3}\right) dx = \int_{-1}^{2} \frac{x^3}{3} dx = \frac{5}{4}.$$

Therefore,

$$E(2X - 1) = (2)(\tfrac{5}{4}) - 1 = \tfrac{3}{2},$$

as before.

Suppose that we have two random variables X and Y with joint probability distribution $f(x, y)$. Two additional laws that will be very useful in succeeding chapters involve the expected values of the sum, difference, and product of these two random variables. First, however, let us prove a theorem on the expected value of the sum or difference of functions of the given variables. This, of course, is merely an extension of Theorem 2.3.

THEOREM 2.4   *The expected value of the sum or difference of two or more functions of the random variables X and Y is the sum or difference of the expected values of the functions. That is,*

$$E[g(X, Y) \pm h(X, Y)] = E[g(X, Y)] \pm E[h(X, Y)].$$

Proof   By Definition 2.13,

$$E[g(X, Y) \pm h(X, Y)] = \int_{-\infty}^{\infty} \int_{-\infty}^{\infty} [g(x, y) \pm h(x, y)] f(x, y) \, dx \, dy$$

$$= \int_{-\infty}^{\infty} \int_{-\infty}^{\infty} g(x, y) f(x, y) \, dx \, dy$$

$$\pm \int_{-\infty}^{\infty} \int_{-\infty}^{\infty} h(x, y) f(x, y) \, dx \, dy$$

$$= E[g(X, Y)] \pm E[h(X, Y)].$$

COROLLARY *Setting $g(X, Y) = X$ and $h(X, Y) = Y$, we see that*

$$E(X \pm Y) = E(X) \pm E(Y).$$

If X represents the daily production of some item from machine A and Y the daily production of the same kind of item from machine B, then $X + Y$ represents the total number of items produced daily from both machines. The Corollary of Theorem 2.4 states that the average daily production for both machines is equal to the sum of the average daily production of each machine.

THEOREM 2.5 *Let X and Y be two independent random variables. Then*

$$E(XY) = E(X)E(Y).$$

Proof By Definition 2.13,

$$E(XY) = \int_{-\infty}^{\infty} \int_{-\infty}^{\infty} xy f(x, y) \, dx \, dy.$$

Since X and Y are independent, we may write

$$f(x, y) = g(x)h(y),$$

where $g(x)$ and $h(y)$ are the marginal distributions of X and Y, respectively. Hence

$$E(XY) = \int_{-\infty}^{\infty} \int_{-\infty}^{\infty} xy g(x)h(y) \, dx \, dy$$

$$= \int_{-\infty}^{\infty} xg(x) \, dx \int_{-\infty}^{\infty} yh(y) \, dy$$

$$= E(X)E(Y).$$

Example 2.23 Let X and Y be independent random variables with joint probability distribution

$$f(x, y) = \frac{x(1 + 3y^2)}{4}, \qquad 0 < x < 2, 0 < y < 1$$

$$= 0, \qquad\qquad\qquad \text{elsewhere.}$$

Verify Theorem 2.5.

Solution Now

$$E(XY) = \int_0^1 \int_0^2 \frac{x^2 y(1 + 3y^2)}{4}\, dx\, dy$$

$$= \int_0^1 \frac{x^3 y(1 + 3y^2)}{12} \Big|_{x=0}^{x=2} dy$$

$$= \int_0^1 \frac{2y(1 + 3y^2)}{3}\, dy$$

$$= \frac{5}{6}$$

$$E(X) = \int_0^1 \int_0^2 \frac{x^2(1 + 3y^2)}{4}\, dx\, dy$$

$$= \int_0^1 \frac{x^3(1 + 3y^2)}{12} \Big|_{x=0}^{x=2} dy$$

$$= \int_0^1 \frac{2(1 + 3y^2)}{3}\, dy$$

$$= \frac{4}{3}$$

$$E(Y) = \int_0^1 \int_0^2 \frac{xy(1 + 3y^2)}{4}\, dx\, dy$$

$$= \int_0^1 \frac{x^2 y(1 + 3y^2)}{8} \Big|_{x=0}^{x=2} dy$$

$$= \int_0^1 \frac{y(1 + 3y^2)}{2}\, dy$$

$$= \frac{5}{8}.$$

Hence

$$E(X)E(Y) = (\tfrac{4}{3})(\tfrac{5}{8}) = \tfrac{5}{6} = E(XY).$$

2.8 SPECIAL MATHEMATICAL EXPECTATIONS ✸

If $g(X) = X^k$, Theorem 2.1 yields an expected value called the kth _moment about the origin_ of the random variable X, which we denote by μ'_k. Therefore,

$$\mu'_k = E(X^k) = \sum_x x^k f(x) \qquad \text{if } X \text{ is discrete}$$

$$= \int_{-\infty}^{\infty} x^k f(x)\, dx \qquad \text{if } X \text{ is continuous.}$$

Note that when $k = 0$ we have $\mu'_0 = E(X^0) = E(1) = 1$, since

$$E(1) = \sum_x f(x) = 1 \qquad \text{if } X \text{ is discrete}$$

$$= \int_{-\infty}^{\infty} f(x)\, dx = 1 \qquad \text{if } X \text{ is continuous.}$$

When $k = 1$ we have $\mu'_1 = E(X)$, which is just the expected value of the random variable X itself. Because the first moment about the origin of a random variable X is somewhat special in that it represents the _mean_ of the random variable, we shall write it as μ_X or simply μ. Thus

$$\mu = \mu'_1 = E(X). \qquad\qquad ✰$$

If $g(X) = (X - \mu)^k$, Theorem 2.1 yields an expected value called the kth _moment about the mean_ of the random variable X, which we denote by μ_k. Therefore,

$$\mu_k = E[(X - \mu)^k] = \sum_x (x - \mu)^k f(x) \qquad \text{if } X \text{ is discrete}$$

$$= \int_{-\infty}^{\infty} (x - \mu)^k f(x)\, dx \qquad \text{if } X \text{ is continuous.}$$

The second moment about the mean, μ_2, is of special importance because it tells us something about the variability of the measurements about the mean.

$$E(x) = \int x\, f(x)\,dx, \text{ or } \sum x f(x)$$
$$\text{MEAN} = \mu = E(x)$$

We shall henceforth call μ_2 the *variance* of the random variable X and denote it by σ_X^2, or simply σ^2. Thus

$$\sigma^2 = \mu_2 = E[(X - \mu)^2].$$

The positive square root of the variance is a measure called the *standard deviation*.

An alternative and preferred formula for σ^2 is given in the following theorem.

THEOREM 2.6   *The variance of a random variable X is given by*

$$\sigma^2 = E(X^2) - \mu^2.$$

Proof

$$\begin{aligned}
\sigma^2 &= E[(X - \mu)^2] \\
&= E(X^2 - 2\mu X + \mu^2) \\
&= E(X^2) - 2\mu E(X) + E(\mu^2) \\
&= E(X^2) - \mu^2,
\end{aligned}$$

since $\mu = E(X)$ by definition and $E(\mu^2) = \mu^2$ by Theorem 2.2, Corollary 1.

Example 2.24   Calculate the variance of X, where X is the number of chemists on a committee of size 3 selected at random from four chemists and three biologists.

Solution   In Example 2.14 we showed that $\mu = 12/7$. Now

$$E(x) = 0\left(\tfrac{1}{35}\right) + (1)\left(\tfrac{12}{35}\right) + (2)\left(\tfrac{18}{35}\right) + (3)\left(\tfrac{4}{35}\right)$$

FROM 2.14

$$E(X^2) = (0)(\tfrac{1}{35}) + (1)(\tfrac{12}{35}) + (4)(\tfrac{18}{35}) + (9)(\tfrac{4}{35})$$
$$= \tfrac{24}{7}.$$

Therefore,

$$\sigma^2 = \tfrac{24}{7} - (\tfrac{12}{7})^2 = \tfrac{24}{49}.$$

Example 2.25   Find the mean and variance of the random variable X, where X has the density function

$$\begin{aligned}
f(x) &= 2(x - 1), & 1 < x < 2 \\
&= 0, & \text{elsewhere.}
\end{aligned}$$

$$E(X^2) = \int x^2 f(x)\, dx$$

Solution

$$\mu = E(X) = 2 \int_1^2 x(x-1)\, dx = \tfrac{5}{3}$$

and

$$E(X^2) = 2 \int_1^2 x^2(x-1)\, dx = \tfrac{17}{6}.$$

Therefore,

$$\sigma^2 = \tfrac{17}{6} - (\tfrac{5}{3})^2 = \tfrac{1}{18}.$$

If $g(X, Y) = (X - \mu_X)(Y - \mu_Y)$, where $\mu_X = E(X)$ and $\mu_Y = E(Y)$, Definition 2.13 yields an expected value called the covariance of X and Y, which we denote by σ_{XY} or $\mathrm{cov}(X, Y)$. Therefore,

$$\sigma_{XY} = E[(X - \mu_X)(Y - \mu_Y)]$$
$$= \sum_x \sum_y (x - \mu_X)(y - \mu_Y) f(x, y) \qquad \text{if } X \text{ and } Y \text{ are discrete}$$
$$= \int_{-\infty}^{\infty} \int_{-\infty}^{\infty} (x - \mu_X)(y - \mu_Y) f(x, y)\, dx\, dy \qquad \text{if } X \text{ and } Y \text{ are continuous.}$$

The covariance will be positive when high values of X are associated with high values of Y and low values of X are associated with low values of Y. If low values of X are associated with high values of Y, and vice versa, then the covariance will be negative. When X and Y are statistically independent, it can be shown that the covariance is zero (see Theorem 2.11, Corollary 1). The converse, however, is not generally true. Two variables may have zero covariance and still not be statistically independent.

The alternative and preferred formula for σ_{XY} is given in the following theorem.

THEOREM 2.7 *The covariance of two random variables X and Y with means μ_X and μ_Y, respectively, is given by*

$$\sigma_{XY} = E(XY) - \mu_X \mu_Y.$$

Proof

$$\sigma_{XY} = E[(X - \mu_X)(Y - \mu_Y)]$$
$$= E(XY - \mu_X Y - \mu_Y X + \mu_X \mu_Y)$$
$$= E(XY) - \mu_X E(Y) - \mu_Y E(X) + E(\mu_X \mu_Y)$$
$$= E(XY) - \mu_X \mu_Y,$$

since $\mu_X = E(X)$ and $\mu_Y = E(Y)$ by definition and $E(\mu_X \mu_Y) = \mu_X \mu_Y$ by Theorem 2.2, Corollary 1.

Example 2.26   Referring to the joint probability distribution of Example 2.8 and to the computations of Example 2.19, we see that $E(XY) = \frac{3}{14}$. Now

$$\mu_X = E(X) = \sum_{x=0}^{2} \sum_{y=0}^{2} xf(x, y) = \sum_{x=0}^{2} xg(x)$$

$$= (0)(\tfrac{10}{28}) + (1)(\tfrac{15}{28}) + (2)(\tfrac{3}{28})$$

$$= \tfrac{3}{4}$$

and

$$\mu_Y = E(Y) = \sum_{x=0}^{2} \sum_{y=0}^{2} yf(x, y) = \sum_{y=0}^{2} yh(y)$$

$$= (0)(\tfrac{15}{28}) + (1)(\tfrac{3}{7}) + (2)(\tfrac{1}{28})$$

$$= \tfrac{1}{2}.$$

Therefore,

$$\sigma_{XY} = E(XY) - \mu_X \mu_Y$$
$$= \tfrac{3}{14} - (\tfrac{3}{4})(\tfrac{1}{2})$$
$$= -\tfrac{9}{56}.$$

Example 2.27   Let the random variables X and Y have the joint probability density function

$$f(x, y) = 2, \qquad 0 < x < y, 0 < y < 1$$
$$= 0, \qquad \text{elsewhere.}$$

Find σ_{XY}.

Solution   We compute

$$\mu_X = E(X) = \int_0^1 \int_0^y 2x \, dx \, dy = \tfrac{1}{3}$$

$$\mu_Y = E(Y) = \int_0^1 \int_0^y 2y \, dx \, dy = \tfrac{2}{3}$$

and

$$E(XY) = \int_0^1 \int_0^y 2xy \, dx \, dy = \tfrac{1}{4}.$$

Then

$$\sigma_{XY} = E(XY) - \mu_X \mu_Y$$
$$= \tfrac{1}{4} - (\tfrac{1}{3})(\tfrac{2}{3})$$
$$= \tfrac{1}{36}.$$

2.9 PROPERTIES OF THE VARIANCE

We shall now prove four theorems that are useful in calculating variances or standard deviations. If we let $g(X)$ be a function of the random variable X, then the mean and variance of $g(X)$ will be denoted by $\mu_{g(X)}$ and $\sigma^2_{g(X)}$, respectively.

THEOREM 2.8   *Let X be a random variable with probability distribution $f(x)$. The variance of the function $g(X)$ is*

$$\sigma^2_{g(X)} = E[\{g(X) - \mu_{g(X)}\}^2].$$

Proof   Since $g(X)$ is a random variable, the result follows from the definition of the variance.

THEOREM 2.9   *If X is a random variable and b is a constant, then*

$$\sigma^2_{X+b} = \sigma^2_X = \sigma^2.$$

Proof

$$\sigma^2_{X+b} = E[\{(X + b) - \mu_{X+b}\}^2].$$

Now

$$\mu_{X+b} = E(X + b) = E(X) + b = \mu + b$$

by Theorem 2.2. Therefore,

$$\begin{aligned}
\sigma^2_{X+b} &= E[(X + b - \mu - b)^2] \\
&= E[(X - \mu)^2] \\
&= \sigma^2.
\end{aligned}$$

This theorem states that the variance is unchanged if a constant is added to or subtracted from a random variable. The addition or subtraction of a constant simply shifts the values of X to the right or to the left but does not change their variability.

THEOREM 2.10   *If X is a random variable and a is any constant, then*

$$\sigma^2_{aX} = a^2\sigma^2_X = a^2\sigma^2.$$

Proof

$$\sigma^2_{aX} = E[\{aX - \mu_{aX}\}^2].$$

Now

$$\mu_{aX} = E(aX) = aE(X) = a\mu$$

by Theorem 2.2, Corollary 2. Therefore,

$$\begin{aligned}
\sigma^2_{aX} &= E[(aX - a\mu)^2] \\
&= a^2E[(X - \mu)^2] \\
&= a^2\sigma^2.
\end{aligned}$$

Therefore, if a random variable is multiplied or divided by a constant, the variance is multiplied or divided by the square of the constant.

THEOREM 2.11   *If X and Y are random variables with joint probability distribution $f(x, y)$, then*

$$\sigma^2_{aX+bY} = a^2\sigma^2_X + b^2\sigma^2_Y + 2ab\sigma_{XY}.$$

Proof

$$\sigma^2_{aX+bY} = E[(aX + bY) - \mu_{aX+bY}]^2.$$

Now

$$\mu_{aX+bY} = E(aX + bY) = aE(X) + bE(Y) = a\mu_X + b\mu_Y,$$

by using Theorem 2.3 followed by Theorem 2.2, Corollary 2. Therefore,

$$
\begin{aligned}
\sigma^2_{aX+bY} &= E\{[(aX + bY) - (a\mu_X + b\mu_Y)]^2\} \\
&= E\{[a(X - \mu_X) + b(Y - \mu_Y)]^2\} \\
&= a^2 E[(X - \mu_X)^2] + b^2 E[(Y - \mu_Y)^2] + 2ab E[(X - \mu_X)(Y - \mu_Y)] \\
&= a^2\sigma^2_X + b^2\sigma^2_Y + 2ab\sigma_{XY}.
\end{aligned}
$$

COROLLARY 1   *If X and Y are independent random variables, then*

$$\sigma^2_{aX+bY} = a^2\sigma^2_X + b^2\sigma^2_Y.$$

The result given in Corollary 1 is obtained from Theorem 2.11 by proving the covariance of the independent variables X and Y to be zero. Hence, from Theorem 2.7,

$$
\begin{aligned}
\sigma_{XY} &= E(XY) - \mu_X\mu_Y \\
&= 0,
\end{aligned}
$$

since $E(XY) = E(X)E(Y)$ for independent variables.

COROLLARY 2   *If X and Y are independent random variables, then*

$$\sigma^2_{aX-bY} = a^2\sigma^2_X + b^2\sigma^2_Y.$$

Corollary 2 follows by replacing b by $-b$ in Corollary 1.

Example 2.28   If X and Y are random variables with variances $\sigma_X^2 = 2$, $\sigma_Y^2 = 4$, and covariance $\sigma_{XY} = -2$, find the variance of the random variable $Z = 3X - 4Y + 8$.

Solution

$$
\begin{aligned}
\sigma_Z^2 &= \sigma_{3X-4Y+8}^2 \\
&= \sigma_{3X-4Y}^2 && \text{by Theorem 2.9} \\
&= 9\sigma_X^2 + 16\sigma_Y^2 - 24\sigma_{XY} && \text{by Theorem 2.11} \\
&= (9)(2) + (16)(4) - (24)(-2) \\
&= 130.
\end{aligned}
$$

Example 2.29   If X and Y are independent random variables with variances $\sigma_X^2 = 1$ and $\sigma_Y^2 = 2$, find the variance of the random variable $Z = 3X - 2Y + 5$.

Solution

$$
\begin{aligned}
\sigma_Z^2 &= \sigma_{3X-2Y+5}^2 \\
&= \sigma_{3X-2Y}^2 && \text{by Theorem 2.9} \\
&= 9\sigma_X^2 + 4\sigma_Y^2 && \text{by Theorem 2.11, Corollary 2} \\
&= (9)(1) + (4)(2) \\
&= 17.
\end{aligned}
$$

2.10 CHEBYSHEV'S THEOREM

In Section 2.8 we stated that the variance of a random variable tells us something about the variability of the observations about the mean. If a random variable has a small variance or standard deviation, we would expect most of the values to be grouped around the mean. Therefore, the probability that a random variable assumes a value within a certain interval about the mean is greater than for a similar random variable with a larger standard deviation. If we think of probability in terms of area, we would expect a continuous distribution with a small standard deviation to have most of its area close to μ, as in Figure 2.10(a).

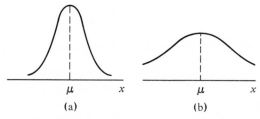

Figure 2.10 Variability of continuous observations about the mean.

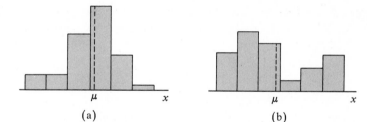

Figure 2.11 Variability of discrete observations about the mean.

However, a large value of σ indicates a greater variability, and therefore we would expect the area to be more spread out, as in Figure 2.10(b).

We can argue the same way for a discrete distribution. The area in the probability histogram in Figure 2.11(b) is spread out much more than that of Figure 2.11(a), indicating a more variable distribution of measurements or outcomes.

The Russian mathematician Chebyshev discovered that the fraction of the area between any two values symmetric about the mean is related to the standard deviation. Since the area under a probability distribution curve or in a probability histogram add to 1, the area between any two numbers is the probability of the random variable assuming a value between these numbers.

The following theorem, due to Chebyshev, gives a conservative estimate of the probability of a random variable falling within k standard deviations of its mean. We shall give the proof only for the continuous case, leaving the discrete case as an exercise.

CHEBYSHEV'S THEOREM   *The probability that any random variable X falls within k standard deviations of the mean is* at least $(1 - 1/k^2)$. *That is,*

$$P(\mu - k\sigma < X < \mu + k\sigma) \geq 1 - \frac{1}{k^2}.$$

Proof   By our previous definition of the variance of X we can write

$$\sigma^2 = E[(X - \mu)^2]$$

$$= \int_{-\infty}^{\infty} (x - \mu)^2 f(x)\, dx$$

$$= \int_{-\infty}^{\mu - k\sigma} (x - \mu)^2 f(x)\, dx + \int_{\mu - k\sigma}^{\mu + k\sigma} (x - \mu)^2 f(x)\, dx + \int_{\mu + k\sigma}^{\infty} (x - \mu)^2 f(x)\, dx$$

$$\geq \int_{-\infty}^{\mu - k\sigma} (x - \mu)^2 f(x)\, dx + \int_{\mu + k\sigma}^{\infty} (x - \mu)^2 f(x)\, dx,$$

since the second of the three integrals is nonnegative. Now, since $|x - \mu| \geq k\sigma$ wherever $x \geq \mu + k\sigma$ or $x \leq \mu - k\sigma$, we have $(x - \mu)^2 \geq k^2\sigma^2$ in both remaining integrals. It follows that

$$\sigma^2 \geq \int_{-\infty}^{\mu - k\sigma} k^2\sigma^2 f(x) \, dx + \int_{\mu + k\sigma}^{\infty} k^2\sigma^2 f(x) \, dx$$

and that

$$\int_{-\infty}^{\mu - k\sigma} f(x) \, dx + \int_{\mu + k\sigma}^{\infty} f(x) \, dx \leq \frac{1}{k^2}.$$

Hence

$$P(\mu - k\sigma < X < \mu + k\sigma) = \int_{\mu - k\sigma}^{\mu + k\sigma} f(x) \, dx \geq 1 - \frac{1}{k^2}$$

and the theorem is established.

For $k = 2$ the theorem states that the random variable X has a probability of at least $1 - (1/2)^2 = 3/4$ of falling within two standard deviations of the mean. That is, three fourths or more of the observations of any distribution lie in the interval $\mu \pm 2\sigma$. Similarly, the theorem says that at least eight ninths of the observations of any distribution fall in the interval $\mu \pm 3\sigma$.

Example 2.30   A random variable X has a mean $\mu = 8$, a variance $\sigma^2 = 9$, and an unknown probability distribution. Find (1) $P(-4 < X < 20)$, and (2) $P(|X - 8| \geq 6)$.

Solution

1. $P(-4 < X < 20 = P[8 - (4)(3) < X < 8 + (4)(3)]$
   $\geq \frac{15}{16}. = 1 - \frac{1}{k^2} = 1 - \frac{1}{4^2}$
2. $P(|X - 8| \geq 6) = 1 - P(|X - 8| < 6)$
   $= 1 - P(-6 < X - 8 < 6)$
   $= 1 - P[8 - (2)(3) < X < 8 + (2)(3)]$
   $\leq \frac{1}{4}.$

Chebyshev's theorem holds for any distribution of observations and, for this reason, the results are usually weak. The value given by the theorem is a lower bound only. That is, we know that the probability of a random variable falling within two standard deviations of the mean can be *no less* than 3/4, but we never know how much more it might actually be. Only when the probability distribution is known can we determine exact probabilities.

EXERCISES

1. Classify the following random variables as discrete or continuous.
   X: the number of automobile accidents per year in Virginia.
   Y: the length of time to play 18 holes of golf.
   M: the amount of milk produced yearly by a particular cow.
   N: the number of eggs laid each month by one hen.
   P: the number of building permits issued each month in a certain city.
   Q: the weight of grain produced per acre.

2. From a box containing four black balls and two green balls, three balls are drawn in succession, each ball being replaced in the box before the next draw is made. Find the probability distribution for the number of green balls.

3. Find the probability distribution for the number of jazz records when four records are selected at random from a collection consisting of five jazz records, two classical records, and three polka records. Express your results by means of a formula.

4. Find a formula for the probability distribution of the random variable X representing the outcome when a single die is rolled once.

5. A shipment of six television sets contains two defective sets. A hotel makes a random purchase of three of the sets. If X is the number of defective sets purchased by the hotel, find the probability distribution of X. Express the results graphically as a probability histogram.

6. A coin is biased so that a head is twice as likely to occur as a tail. If the coin is tossed three times, find the probability distribution for the number of heads. Construct a probability histogram for the distribution.

7. Find the cumulative distribution of the random variable X in Exercise 5. Using $F(x)$, find
   (a) $P(X = 1)$.   (b) $P(0 < X \leq 2)$.

8. Construct a graph of the cumulative distribution of Exercise 7.

9. Find the cumulative distribution of the random variable X representing the number of heads in Exercise 6. Using $F(x)$, find
   (a) $P(1 \leq X < 3)$.   (b) $P(X > 2)$.

10. Construct a graph of the cumulative distribution of Exercise 9.

11. A continuous random variable X that can assume values between $x = 1$ and $x = 3$ has a density function given by $f(x) = 1/2$.
    (a) Show that the area under the curve is equal to 1.
    (b) Find $P(2 < X < 2.5)$.
    (c) Find $P(X \leq 1.6)$.

12. A continuous random variable X that can assume values between $x = 2$ and $x = 5$ has a density function given by $f(x) = 2(1 + x)/27$. Find
    (a) $P(X < 4)$.   (b) $P(3 < X < 4)$.

13. For the density function of Exercise 11, find $F(x)$ and use it to evaluate $P(2 < X < 2.5)$.

14. For the density function of Exercise 12, find $F(x)$ and use it to evaluate $P(3 \le X < 4)$.

15. Consider the density function   *(cdf.)*

$$f(x) = k\sqrt{x}, \qquad 0 < x < 1$$
$$= 0, \qquad \text{elsewhere.}$$

(a) Evaluate k.

(b) Find $F(x)$ and use it to evaluate $P(0.3 < X < 0.6)$.

16. The following scores represent the final examination grade for an elementary statistics course:

23	60	79	32	57	74	52	70	82	36
80	77	81	95	41	65	92	85	55	76
52	10	64	75	78	25	80	98	81	67
41	71	83	54	64	72	88	62	74	43
60	78	89	76	84	48	84	90	15	79
34	67	17	82	69	74	63	80	85	61

Using 10 intervals with the lowest starting at 9,

(a) Set up a relative frequency distribution.

(b) Construct a relative frequency histogram.

(c) Construct a smoothed relative cumulative frequency distribution.

(d) Estimate the first quartile and the seventh decile.

17. The following data represent the length of life in years, measured to the nearest tenth, of a random sample of 30 similar fuel pumps:

2.0	3.0	0.3	3.3	1.3	0.4
0.2	6.0	5.5	6.5	0.2	2.3
1.5	4.0	5.9	1.8	4.7	0.7
4.5	0.3	1.5	0.5	2.5	5.0
1.0	6.0	5.6	6.0	1.2	0.2

Using six intervals with the lowest starting at 0.1,

(a) Set up a relative frequency distribution.

(b) Construct a relative frequency histogram.

(c) Construct a smoothed relative cumulative frequency distribution.

(d) Estimate the value below which two thirds of the values fall.

18. From a sack of fruit containing three oranges, two apples, and three bananas a random sample of four pieces of fruit is selected. If X is the number of oranges and Y is the number of apples in the sample, find

(a) The joint probability distribution of X and Y.

(b) $P[(X, Y) \in A]$, where A is the region $\{(x, y)|x + y \le 2\}$.

19. Two random variables have the joint density given by

$$f(x, y) = 4xy, \qquad 0 < x < 1, 0 < y < 1$$
$$= 0, \qquad \text{elsewhere.}$$

(a) Find the probability that $0 \leq X \leq 3/4$ and $1/8 \leq Y \leq 1/2$.

(b) Find the probability that $Y > X$.

20. Two random variables have the joint density given by

$$f(x, y) = k(x^2 + y^2), \qquad 0 < x < 2, 1 < y < 4$$
$$= 0, \qquad\qquad \text{elsewhere.}$$

(a) Find k.

(b) Find the probability that $1 < X < 2$ and $2 < Y \leq 3$.

(c) Find the probability that $1 \leq X \leq 2$.

(d) Find the probability that $X + Y > 4$.

21. Let X and Y have the joint density function

$$f(x, y) = \frac{1}{y}, \qquad 0 < x < y, 0 < y < 1$$
$$= 0, \qquad \text{elsewhere.}$$

Find $P(X + Y > 1/2)$.

22. Referring to Exercise 18, find

(a) $f(y|2)$. \hspace{4cm} (b) $P(Y = 0 | X = 2)$.

23. Suppose that X and Y have the following joint probability distribution:

y \ x	1	2	3
1	0	$\frac{1}{6}$	$\frac{1}{12}$
2	$\frac{1}{5}$	$\frac{1}{9}$	0
3	$\frac{2}{15}$	$\frac{1}{4}$	$\frac{1}{18}$

Evaluate the marginal and conditional probability distributions.

24. Suppose that X and Y have the following joint probability function:

y \ x	2	4
1	0.10	0.15
3	0.20	0.30
5	0.10	0.15

Find the marginal probability distributions and determine whether X and Y are independent.

25. Determine whether the two random variables of Exercise 19 are dependent or independent.
26. Determine whether the two random variables of Exercise 20 are dependent or independent.
27. Given the joint density function

$$f(x, y) = \frac{6 - x - y}{8}, \qquad 0 < x < 2, 2 < y < 4$$

$$= 0, \qquad\qquad \text{elsewhere,}$$

find $P(1 < Y < 3 | X = 2)$.

28. If X and Y have the joint density function

$$f(x, y) = e^{-(x+y)}, \qquad x > 0, y > 0$$

$$= 0, \qquad\qquad \text{elsewhere,}$$

find $P(0 < X < 1 | Y = 2)$.

29. The joint probability density function of the random variables X and Y is given by

$$f(x, y) = 2, \qquad 0 < x < y < 1$$

$$= 0, \qquad \text{elsewhere.}$$

(a) Determine if X and Y are independent.
(b) Find $P(1/4 < X < 1/2 | Y = 3/4)$.

30. The joint density function of the random variables X and Y is given by

$$f(x, y) = 6x, \qquad 0 < x < 1, 0 < y < 1 - x$$

$$= 0, \qquad \text{elsewhere.}$$

(a) Show that X and Y are not independent.
(b) Find $P(X > 0.3 | Y = 0.5)$.

31. If X, Y, and Z have the joint probability density function

$$f(x, y, z) = kxy^2z, \qquad 0 < x < 1, 0 < y < 1, 0 < z < 2$$

$$= 0, \qquad\qquad \text{elsewhere,}$$

(a) Find k.
(b) Find $P(X < 1/4, Y > 1/2, 1 < Z < 2)$.

32. The joint probability density function of the random variables X, Y, and Z is given by

$$f(x, y, z) = \frac{4xyz^2}{9}, \qquad 0 < x < 1, 0 < y < 1, 0 < z < 3$$

$$= 0, \qquad\qquad \text{elsewhere.}$$

Find
(a) The joint marginal density function of Y and Z.
(b) The marginal density of Y.
(c) $P(1/4 < X < 1/2, Y > 1/3, 1 < Z < 2)$.
(d) $P(0 < X < 1/2 | Y = 1/4, Z = 2)$.

33. Find the expected value of the random variable X in Exercise 5.

34. The probability distribution of the discrete random variable X is

$$f(x) = \binom{3}{x}\left(\frac{1}{4}\right)^x\left(\frac{3}{4}\right)^{3-x}, \qquad x = 0, 1, 2, 3.$$

Find $E(X)$.

35. Find the expected value of the random variable X having the density function

$$\begin{aligned} f(x) &= 2(1 - x), && 0 < x < 1 \\ &= 0, && \text{elsewhere.} \end{aligned}$$

36. The density function of coded measurements of pitch diameter of threads of a fitting is given by

$$\begin{aligned} f(x) &= \frac{4}{\pi(1 + x^2)}, && 0 < x < 1 \\ \\ &= 0, && \text{elsewhere.} \end{aligned}$$

Find $E(X)$.

37. A continuous random variable X has the density function

$$\begin{aligned} f(x) &= e^{-x}, && x > 0 \\ &= 0, && \text{elsewhere.} \end{aligned}$$

Find the expected value of $g(X) = e^{2X/3}$.

38. By investing in a particular stock a man can make a profit in 1 year of $3000 with probability 0.3 or take a loss of $1000 with probability 0.7. What is his mathematical expectation?

39. In a gambling game a man is paid $2 if he draws a jack or queen and $5 if he draws a king or ace from an ordinary deck of 52 playing cards. If he draws any other card, he loses. How much should he pay to play if the game is fair?

40. A race car driver wishes to insure his car for the racing season for $10,000. The insurance company estimates a total loss may occur with probability 0.002, a 50% loss with probability 0.01, and a 25% loss with probability 0.1. Ignoring all other partial losses, what premium should the insurance company charge each season to make a profit of $100?

41. Let X represent the outcome when a balanced die is tossed. Find $E(Y)$, where $Y = 2X^2 - 5$.

42. Find the expected value of $g(X) = X^2$, where X has the density function of Exercise 35.

43. Let X and Y have the joint probability distribution of Exercise 24. Find the expected value of $g(X, Y) = XY^2$.

44. Let X and Y be random variables with joint density function

$$f(x, y) = 4xy, \quad 0 < x < 1, 0 < y < 1$$
$$= 0, \quad \text{elsewhere.}$$

Find the expected value of $Z = \sqrt{X^2 + Y^2}$.

45. Suppose that X and Y are independent random variables with probability densities $g(x) = 8/x^3$, $x > 2$, and $h(y) = 2y, 0 < y < 1$. Find the expected value of $Z = XY$.

46. Let X be a random variable with the following probability distribution:

x	-3	6	9
$P(X = x)$	$\frac{1}{6}$	$\frac{1}{2}$	$\frac{1}{3}$

Find $E(X)$ and $E(X^2)$ and then, using the laws of expectation, evaluate
(a) $E[(2X + 1)^2]$.
(b) $E[\{X - E(X)\}^2]$.

47. Let X represent the number that occurs when a red die is tossed and Y the number that occurs when a green die is tossed. Find
(a) $E(X + Y)$.   (b) $E(X - Y)$.   (c) $E(XY)$.

48. Referring to the random variables whose joint distribution is given in Exercise 24, find
(a) $E(2X - 3Y)$.
(b) $E(XY)$.

49. If a random variable X is such that $E[(X - 1)^2] = 10, E[(X - 2)^2] = 6$, find μ and σ^2.

50. Find the variance of the random variable X of Exercise 5.

51. Let X be a random variable with the following probability distribution:

x	-2	3	5
$P(X = x)$	0.3	0.2	0.5

Find the mean and standard deviation of X.

52. From a group of five men and three women a committee of size 3 is selected at random. If X represents the number of women on the committee, find the mean and variance of X.

53. Find the variance of the random variable X of Exercise 35.

54. Compute the $P(\mu - 2\sigma < X < \mu + 2\sigma)$, where X has the density function

$$f(x) = 6x(1 - x), \quad 0 < x < 1$$
$$= 0, \quad \text{elsewhere.}$$

55. Find the covariance of the random variables X and Y of Exercise 18.

56. Find the covariance of the random variables X and Y of Exercise 23.

57. Find the covariance of the random variables X and Y of Exercise 20.

58. Find the covariance of the random variables X and Y having the joint probability density function

$$f(x, y) = x + y, \qquad 0 < x < 1, 0 < y < 1$$
$$= 0, \qquad \text{elsewhere.}$$

59. Show that cov $(aX, bY) = ab$ cov (X, Y).

60. Let X represent the number that occurs when a green die is tossed and Y the number that occurs when a red die is tossed. Find the variance of the random variable
    (a) $2X - Y$.   (b) $X + 3Y - 5$.

61. If X and Y are independent random variables with variances $\sigma_X^2 = 5$ and $\sigma_Y^2 = 3$, find the variance of the random variable $Z = -2X + 4Y - 3$.

62. Repeat Exercise 61 if X and Y are not independent and $\sigma_{XY} = 1$.

63. A random variable X has a mean $\mu = 12$, a variance $\sigma^2 = 9$, and an unknown probability distribution. Using Chebyshev's theorem, find
    (a) $P(6 < X < 18)$.   (b) $P(3 < X < 21)$.

64. A random variable X has a mean $\mu = 10$ and a variance $\sigma^2 = 4$. Using Chebyshev's theorem, find
    (a) $P(|X - 10| \geq 3)$.
    (b) $P(|X - 10| < 3)$.
    (c) $P(5 < X < 15)$.
    (d) The value of c such that $P(|X - 10| \geq c) \leq 0.04$.

65. Prove Chebyshev's theorem when X is a discrete random variable.

$\longrightarrow \# 52)\qquad f(x) = \dfrac{\binom{3}{x}\binom{5}{3-x}}{\binom{8}{3}}$

3

Some Discrete Probability Distributions

3.1 INTRODUCTION

In Chapter 2 a discrete probability distribution was represented graphically, in tabular form, and, if convenient, by means of a formula. No matter what method of presentation is used, the behavior of a random variable is described. Many random variables associated with statistical experiments have similar properties and can be described by essentially the same probability distribution. For example, all random variables representing the number of successes in n independent trials of an experiment, where the probability of a success is constant for all n trials, have the same general type of behavior and therefore can be represented by a single formula. Thus, if in firing a rifle at a target a direct hit is considered a success and the probability of a hit remains constant for successive firings, the formula for the distribution of hits in five firings of the rifle has the same structure as the formula for the distribution of 4's in seven tosses of a die, where the occurrence of a 4 is considered a success.

Frequent reference will be made throughout the text to the moments of a probability distribution. By this we shall mean the moments of any random variable having that particular probability distribution. Therefore, the mean or variance of a given probability distribution is defined to be the mean or variance of any random variable having that distribution.

Care should be exercised in choosing the probability distribution that correctly describes the observations being generated by the experiment. In this chapter we shall investigate several important discrete probability distributions that describe most random variables encountered in practice.

3.2 UNIFORM DISTRIBUTION

The simplest of all discrete probability distributions is one where the random variable assumes all its values with equal probability. Such a probability distribution is called the *uniform distribution*.

UNIFORM DISTRIBUTION   *If the random variable X assumes the values* $x_1, x_2, \ldots, x_k$, *with equal probability, then the discrete uniform distribution is given by*

$$f(x; k) = \frac{1}{k}, \qquad x = x_1, x_2, \ldots, x_k.$$

We have used the notation $f(x; k)$ instead of $f(x)$ to indicate that the uniform distribution depends on the *parameter k*.

Example 3.1   When a die is tossed, each element of the sample space $S = \{1, 2, 3, 4, 5, 6\}$ occurs with probability 1/6. Therefore, we have a uniform distribution, with $f(x; 6) = 1/6, x = 1, 2, 3, 4, 5, 6$.

Example 3.2   Suppose that an employee is selected at random from a staff of 10 to supervise a certain project. Each employee has the same probability, 1/10, of being selected. Assuming that the employees have been numbered in same way from 1 to 10, the distribution is uniform with $f(x; 10) = 1/10$, $x = 1, 2, \ldots, 10$.

The graphic representation of the uniform distribution by means of a histogram always turns out to be a set of rectangles with equal heights. The histogram for Example 3.1 is shown in Figure 3.1.

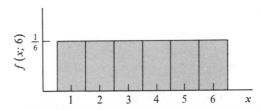

Figure 3.1 Histogram for the tossing of a die.

THEOREM 3.1   *The mean and variance of the discrete uniform distribution* $f(x; k)$ *are*

$$\mu = \frac{\sum\limits_{i=1}^{k} x_i}{k} \qquad and \qquad \sigma^2 = \frac{\sum\limits_{i=1}^{k} (x_i - \mu)^2}{k}.$$

Proof   By definition

$$\mu = E(X) = \sum_{i=1}^{k} x_i \, f(x_i; k) = \sum_{i=1}^{k} \frac{x_i}{k}$$

$$= \frac{\sum_{i=1}^{k} x_i}{k}.$$

Also, by definition,

$$\sigma^2 = E[(X - \mu)^2] = \sum_{i=1}^{k} (x_i - \mu)^2 f(x_i; k)$$

$$= \sum_{i=1}^{k} \frac{(x_i - \mu)^2}{k} = \frac{\sum_{i=1}^{k} (x_i - \mu)^2}{k}.$$

Example 3.3   Referring to Example 3.1, we find that

$$\mu = \frac{1 + 2 + 3 + 4 + 5 + 6}{6} = 3.5$$

and

$$\sigma^2 = \frac{(1 - 3.5)^2 + (2 - 3.5)^2 + \cdots + (6 - 3.5)^2}{6} = \frac{35}{12}.$$

3.3 BINOMIAL AND MULTINOMIAL DISTRIBUTIONS

An experiment often consists of repeated trials, each with two possible outcomes, which may be labeled *success* or *failure*. This is true in testing items as they come off an assembly line, where each test or trial may indicate a defective or a non-defective item. We may choose to define either outcome as a success. It is also true if five cards are drawn in succession from an ordinary deck and each trial is labeled a success or failure depending on whether the card is red or black. If each card is replaced and the deck shuffled before the next drawing, then the two experiments described have similar properties in that the repeated trials are independent and the probability of a success remains constant, from trial to trial. Experiments of this type are known as *binomial experiments*. Observe in the card-drawing example that the probabilities of a success for repeated trials change if the cards are not replaced. That is, the probability of selecting a red card on the first draw is 1/2, but on the second draw it is a conditional probability having a value of 26/51 or 25/51, depending on the color that occurred on the first draw. This, then, would no longer be considered a binomial experiment.

A binomial experiment is one that possesses the following properties:

1. The experiment consists of n repeated trials.
2. Each trial results in an outcome that may be classified as a success or a failure.
3. The probability of success, denoted by p, remains constant from trial to trial.
4. The repeated trials are independent.

Consider the binomial experiment where three items are selected at random from a manufacturing process, inspected, and classified defective or non-defective. A defective item is designated a success. The number of successes is a random variable X assuming integral values from zero through 3. The eight possible outcomes and the corresponding values of X are

Outcome	x
NNN	0
NDN	1
NND	1
DNN	1
NDD	2
DND	2
DDN	2
DDD	3

Since the items are selected independently from a process that we shall assume produces 25 % defectives, the

$$P(NDN) = P(N)P(D)P(N) = (\tfrac{3}{4})(\tfrac{1}{4})(\tfrac{3}{4}) = \tfrac{9}{64}.$$

Similar calculations yield the probabilities for the other possible outcomes. The probability distribution of X is therefore given by

x	0	1	2	3
$f(x)$	$\tfrac{27}{64}$	$\tfrac{27}{64}$	$\tfrac{9}{64}$	$\tfrac{1}{64}$

DEFINITION 3.1   *The number X of successes in n trials of a binomial experiment is called a* binomial random variable.

The probability distribution of the binomial variable X is called the *binomial distribution* and will be denoted by $b(x; n, p)$, since its values depend on the number of trials and the probability of a success on a given trial. Thus for the probability distribution of X, the number of defectives,

$$P(X = 2) = f(2) = b(2; 3, \tfrac{1}{4}) = \tfrac{9}{64}.$$

Let us now generalize the above illustration to yield a formula for $b(x; n, p)$. That is, we wish to find a formula that gives the probability of x successes in n trials for a binomial experiment. First, consider the probability of x successes and $n - x$ failures in a specified order. Since the trials are independent, we can multiply all the probabilities corresponding to the different outcomes. Each success occurs with probability p and each failure with probability $q = 1 - p$. Therefore, the probability for the specified order is $p^x q^{n-x}$. We must now determine the total number of sample points in the experiment that have x successes and $n - x$ failures. This number is equal to the number of partitions of n outcomes into two groups with x in one group and $n - x$ in the other and is given by $\binom{n}{x}$. Because these partitions are mutually exclusive, we add the probabilities of all the different partitions to obtain the general formula, or simply multiply $p^x q^{n-x}$ by $\binom{n}{x}$.

BINOMIAL DISTRIBUTION   *If a binomial trial can result in a success with probability p and a failure with probability $q = 1 - p$, then the probability distribution of the binomial random variable X, the number of successes in n independent trials, is*

$$b(x; n, p) = \binom{n}{x} p^x q^{n-x}, \qquad x = 0, 1, 2, \ldots, n.$$

Note that when $n = 3$ and $p = 1/4$, the probability distribution of X, the number of defectives, may be written as

$$b\left(x; 3, \frac{1}{4}\right) = \binom{3}{x}\left(\frac{1}{4}\right)^x\left(\frac{3}{4}\right)^{3-x}, \qquad x = 0, 1, 2, 3,$$

rather than in the tabular form above.

Example 3.4   The probability that a certain kind of component will survive a given shock test is 3/4. Find the probability that exactly two of the next four components tested survive.

Solution   Assuming the tests are independent and $p = 3/4$ for each of the four tests,

$$\frac{4!}{2!(4-2)!}$$

$$b\left(2; 4, \frac{3}{4}\right) = \binom{4}{2}\left(\frac{3}{4}\right)^2\left(\frac{1}{4}\right)^2$$

$$= \frac{4!}{2!2!} \cdot \frac{3^2}{4^4}$$

$$= \frac{27}{128}.$$

The binomial distribution derives its name from the fact that the $n + 1$ terms in the binomial expansion of $(q + p)^n$ correspond to the values of $b(x; n, p)$ for $x = 0, 1, 2, \ldots, n$. That is,

$$(q + p)^n = \binom{n}{0}q^n + \binom{n}{1}pq^{n-1} + \binom{n}{2}p^2q^{n-2} + \cdots + \binom{n}{n}p^n$$

$$= b(0; n, p) + b(1; n, p) + b(2; n, p) + \cdots + b(n; n, p).$$

Since $p + q = 1$, we see that $\sum_{x=0}^{n} b(x; n, p) = 1$, a condition that must hold for any probability distribution.

Frequently, we are interested in problems where it is necessary to find $P(X < r)$ or $P(a \leq X \leq b)$. Fortunately, binomial sums $B(r; n, p) = \sum_{x=0}^{r} b(x; n, p)$ are available and are given in Table II (see the Statistical Tables at the end of the book) for samples of size $n = 5, 10, 15,$ and 20, and selected values of p from 0.1 to 0.90. We illustrate the use of Table II with the following example.

Example 3.5   The probability that a patient recovers from a rare blood disease is 0.4. If 15 people are known to have contracted this disease, what is the probability that (1) at least 10 survive, (2) from 3 to 8 survive, (3) exactly 5 survive?

Solution

1. Let X be the number of people that survive. Then

$$P(X \geq 10) = 1 - P(X < 10)$$

$$= 1 - \sum_{x=0}^{9} b(x; 15, 0.4)$$

$$= 1 - 0.9662$$

$$= 0.0338.$$

2.

$$P(3 \leq X \leq 8) = \sum_{x=3}^{8} b(x; 15, 0.4)$$

$$= \sum_{x=0}^{8} b(x; 15, 0.4) - \sum_{x=0}^{2} b(x; 15, 0.4)$$

$$= 0.9050 - 0.0271$$

$$= 0.8779.$$

3.

$$P(X = 5) = b(5; 15, 0.4)$$

$$= \sum_{x=0}^{5} b(x; 15, 0.4) - \sum_{x=0}^{4} b(x; 15, 0.4)$$

$$= 0.4032 - 0.2173$$

$$= 0.1859.$$

THEOREM 3.2   *The mean and variance of the binomial distribution $b(x; n, p)$ are*

$$\mu = np \qquad and \qquad \sigma^2 = npq.$$

Proof   Let the outcome on the jth trial be represented by the random variable I_j, which assumes the values zero and 1 with probabilities q and p, respectively. This is called a *Bernoulli variable* or perhaps more appropriately, an *indicator variable*, since $I_j = 0$ indicates a failure and $I_j = 1$ indicates a success.

Therefore, in a binomial experiment the number of successes can be written as the sum of the n independent indicator variables. Hence

$$X = I_1 + I_2 + \cdots + I_n.$$

The mean of any I_j is $E(I_j) = 0 \cdot q + 1 \cdot p = p$. Therefore, using the corollary of Theorem 2.4, the mean of the binomial distribution is

$$\mu = E(X) = E(I_1) + E(I_2) + \cdots + E(I_n)$$
$$= \underbrace{p + p + \cdots + p}_{n \text{ terms}}$$
$$= np.$$

The variance of any I_j is given by

$$\sigma_{I_j}^2 = E[(I_j - p)^2] = E(I_j^2) - p^2 = (0)^2 q + (1)^2 p - p^2 = p(1-p) = pq.$$

Therefore, by Theorem 2.11, Corollary 1, the variance of the binomial distribution is

$$\sigma_X^2 = \sigma_{I_1}^2 + \sigma_{I_2}^2 + \cdots + \sigma_{I_n}^2$$
$$= \underbrace{pq + pq + \cdots + pq}_{n \text{ terms}}$$
$$= npq.$$

Example 3.6 Using Chebyshev's theorem, find and interpret the interval $\mu \pm 2\sigma$ for Example 3.5.

Solution Since Example 3.5 was a binomial experiment with $n = 15$ and $p = 0.4$, by Theorem 3.2 we have

$$\mu = (15)(0.4) = 6 \quad \text{and} \quad \sigma^2 = (15)(0.4)(0.6) = 3.6.$$

Taking the square root of 3.6, we find that $\sigma = 1.897$. Hence the required interval is $6 \pm (2)(1.897)$, or from 2.206 to 9.794. Chebyshev's theorem states that the recovery rate of 15 patients subjected to the given disease has a probability of at least $3/4$ of falling between 2.206 and 9.794.

The binomial experiment becomes a *multinomial experiment* if we let each trial have more than two possible outcomes. Hence the classification of a manufactured product as being light, heavy, or acceptable and the recording of accidents at a certain intersection according to the day of the week constitute multinomial experiments. The drawing of a card from a deck *with replacement* is also a multinomial experiment if the four suits are the outcomes of interest.

In general, if a given trial can result in any one of k possible outcomes $E_1, E_2, \ldots, E_k$ with probabilities $p_1, p_2, \ldots, p_k$, then the *multinomial distribution* will give the probability that E_1 occurs x_1 times, E_2 occurs x_2 times, $\ldots$, E_k

occurs x_k times in n independent trials, where $x_1 + x_2 + \cdots + x_k = n$. We shall denote this joint probability distribution by $f(x_1, x_2, \ldots, x_k; p_1, p_2, \ldots, p_k, n)$. Clearly, $p_1 + p_2 + \cdots + p_k = 1$, since the result of each trial must be one of the k possible outcomes.

To derive the general formula, we proceed as in the binomial case. Since the trials are independent, any specified order yielding x_1 outcomes for E_1, x_2 for $E_2, \ldots, x_k$ for E_k will occur with probability $p_1^{x_1} p_2^{x_2} \cdots p_k^{x_k}$. The total number of orders yielding similar outcomes for the n trials is equal to the number of partitions of n items into k groups with x_1 in the first group, x_2 in the second group, $\ldots, x_k$ in the kth group. This can be done in

$$\binom{n}{x_1, x_2, \ldots, x_k} = \frac{n!}{x_1! x_2! \cdots x_k!}.$$

ways. Since all the partitions are mutually exclusive and occur with equal probability, we obtain the multinomial distribution by multiplying the probability for a specified order by the total number of partitions.

MULTINOMIAL DISTRIBUTION   *If a given trial can result in the k outcomes $E_1, E_2, \ldots, E_k$ with probabilities $p_1, p_2, \ldots, p_k$, then the probability distribution of the random variables $X_1, X_2, \ldots, X_k$, representing the number of occurrences for $E_1, E_2, \ldots, E_k$ in n independent trials is*

$$f(x_1, x_2, \ldots, x_k; p_1, p_2, \ldots, p_k, n) = \binom{n}{x_1, x_2, \ldots, x_k} p_1^{x_1} p_2^{x_2} \cdots p_k^{x_k}$$

with

$$\sum_{i=1}^{k} x_i = n \quad and \quad \sum_{i=1}^{k} p_i = 1.$$

The multinomial distribution derives its name from the fact that the terms of the multinomial expansion of $(p_1 + p_2 + \cdots + p_k)^n$ correspond to all the possible values of $f(x_1, x_2, \ldots, x_k; p_1, p_2, \ldots, p_k, n)$.

Example 3.7   If a pair of dice is tossed six times, what is the probability of obtaining a total of 7 or 11 twice, a matching pair once, and any other combination three times?

Solution   We list the following possible events,

E_1:   a total of 7 or 11 occurs,

E_2:   a matching pair occurs,

E_3:   neither a pair nor a total of 7 or 11 occurs.

The corresponding probabilities for a given trial are $p_1 = 2/9$, $p_2 = 1/6$, and $p_3 = 11/18$. These values remain constant for all six trials. Using the multinomial distribution with $x_1 = 2, x_2 = 1$, and $x_3 = 3$, the required probability is

$$f\left(2, 1, 3; \frac{2}{9}, \frac{1}{6}, \frac{11}{18}, 6\right) = \binom{6}{2, 1, 3}\left(\frac{2}{9}\right)^2\left(\frac{1}{6}\right)^1\left(\frac{11}{18}\right)^3$$

$$= \frac{6!}{2!1!3!} \cdot \frac{2^2}{9^2} \cdot \frac{1}{6} \cdot \frac{11^3}{18^3}$$

$$= 0.1127.$$

3.4 HYPERGEOMETRIC DISTRIBUTION READ ONLY OR OMIT

In Section 3.3 we saw that the binomial distribution did not apply if we wished to find the probability of observing 3 red cards in five draws from an ordinary deck of 52 playing cards unless each card is replaced and the deck reshuffled before the next drawing is made. To solve the problem of sampling without replacement, let us restate the problem. If 5 cards are drawn at random, we are interested in the probability of selecting 3 red cards from the 26 available and 2 black cards from the 26 black cards available in the deck. There are $\binom{26}{3}$ ways of selecting 3 red cards and for each of these ways we can choose 2 black cards in $\binom{26}{2}$ ways. Therefore, the total number of ways to select 3 red and 2 black cards in five draws is the product $\binom{26}{3}\binom{26}{2}$. The total number of ways to select any 5 cards from the 52 that are available is $\binom{52}{5}$. Hence the probability of selecting 5 cards without replacement of which 3 are red and 2 are black is given by

$$\frac{\binom{26}{3}\binom{26}{2}}{\binom{52}{5}} = \frac{(26!/3!23!)(26!/2!24!)}{(52!/5!47!)} = 0.3251.$$

In general we are interested in the probability of selecting x successes from the k items labeled success and $n - x$ failures from the $N - k$ items labeled failures when a random sample of size n is selected from N items. This is known as a *hypergeometric experiment*.

A hypergeometric experiment is one that possesses the following two properties:

1. A random sample of size n is selected from N items.
2. k of the N items may be classified as successes and $N - k$ are classified as failures.

DEFINITION 3.2   *The number X of successes in a hypergeometric experiment is called a* hypergeometric random variable.

The probability distribution of the hypergeometric variable X is called the *hypergeometric distribution* and will be denoted by $h(x; N, n, k)$, since its values depend on the number of successes k in the set N from which we select n items.

Example 3.8   A committee of size 5 is to be selected at random from three chemists and five physicists. Find the probability distribution for the number of chemists on the committee.

Solution   Let the random variable X be the number of chemists on the committee. The two properties of a hypergeometric experiment are satisfied. Hence

$$P(X = 0) = h(0; 8, 5, 3) = \frac{\binom{3}{0}\binom{5}{5}}{\binom{8}{5}} = \frac{1}{56}$$

$$P(X = 1) = h(1; 8, 5, 3) = \frac{\binom{3}{1}\binom{5}{4}}{\binom{8}{5}} = \frac{15}{56}$$

$$P(X = 2) = h(2; 8, 5, 3) = \frac{\binom{3}{2}\binom{5}{3}}{\binom{8}{5}} = \frac{30}{56}$$

$$P(X = 3) = h(3; 8, 5, 3) = \frac{\binom{3}{3}\binom{5}{2}}{\binom{8}{5}} = \frac{10}{56}.$$

In tabular form the hypergeometric distribution of X is as follows:

x	0	1	2	3
$h(x; 8, 5, 3)$	$\frac{1}{56}$	$\frac{15}{56}$	$\frac{30}{56}$	$\frac{10}{56}$

It is not difficult to see that the probability distribution can be given by the formula

$$h(x; 8, 5, 3) = \frac{\binom{3}{x}\binom{5}{5-x}}{\binom{8}{5}}, \qquad x = 0, 1, 2, 3.$$

Let us now generalize Example 3.8 to find a formula for $h(x; N, n, k)$. The total number of samples of size n chosen from N items is $\binom{N}{n}$. These samples are assumed to be equally likely. There are $\binom{k}{x}$ ways of selecting x successes from the k that are available and for each of these ways we can choose the $n - x$ failures in $\binom{N-k}{n-x}$ ways. Thus the total number of favorable samples among the $\binom{N}{n}$ possible samples is given by $\binom{k}{x}\binom{N-k}{n-x}$. Hence we have the following definition.

> **HYPERGEOMETRIC DISTRIBUTION** *The probability distribution of the hypergeometric random variable X, the number of successes in a random sample of size n selected from N items of which k are labeled* success *and* $N - k$ *labeled* failure, *is*
>
> $$h(x; N, n, k) = \frac{\binom{k}{x}\binom{N-k}{n-x}}{\binom{N}{n}}, \qquad x = 0, 1, 2, \ldots, n.$$

Example 3.9 Lots of 40 components each are called acceptable if they contain no more than three defectives. The procedure for sampling the lot is to select 5 components at random and to reject the lot if a defective is found. What is the probability that exactly one defective will be found in the sample if there are three defectives in the entire lot?

Solution   Using the hypergeometric distribution with $n = 5$, $N = 40$, $k = 3$, and $x = 1$, we find the probability of obtaining one defective to be

$$h(1; 40, 5, 3) = \frac{\binom{3}{1}\binom{37}{4}}{\binom{40}{5}} = 0.3011.$$

THEOREM 3.3   *The mean and variance of the hypergeometric distribution* $h(x; N, n, k)$ *are*

$$\mu = \frac{nk}{N}$$

and

$$\sigma^2 = \frac{N - n}{N - 1} \cdot n \cdot \frac{k}{N}\left(1 - \frac{k}{N}\right).$$

Proof   To find the mean of the hypergeometric distribution, we write

$$E(X) = \sum_{x=0}^{n} x \frac{\binom{k}{x}\binom{N-k}{n-x}}{\binom{N}{n}}$$

$$= k \sum_{x=1}^{n} \frac{(k-1)!}{(x-1)!(k-x)!} \cdot \frac{\binom{N-k}{n-x}}{\binom{N}{n}}$$

$$= k \sum_{x=1}^{n} \frac{\binom{k-1}{x-1}\binom{N-k}{n-x}}{\binom{N}{n}}.$$

Letting $y = x - 1$, we find that this becomes

$$E(X) = k \sum_{y=0}^{n-1} \frac{\binom{k-1}{y}\binom{N-k}{n-1-y}}{\binom{N}{n}}.$$

Writing

$$\binom{N-k}{n-1-y} = \binom{(N-1)-(k-1)}{n-1-y}$$

and

$$\binom{N}{n} = \frac{N!}{n!(N-n)!} = \frac{N}{n}\binom{N-1}{n-1},$$

we obtain

$$E(X) = \frac{nk}{N}\sum_{y=0}^{n-1}\frac{\binom{k-1}{y}\binom{(N-1)-(k-1)}{n-1-y}}{\binom{N-1}{n-1}},$$

$$= \frac{nk}{N},$$

since the summation represents the total of all probabilities in a hypergeometric experiment when $n-1$ items are selected at random from $N-1$, of which $k-1$ are labeled success.

To find the variance of the hypergeometric distribution, we proceed along the same steps as above to obtain

$$E[X(X-1)] = \frac{k(k-1)n(n-1)}{N(N-1)}.$$

Now, by Theorem 2.6,

$$\sigma^2 = E(X^2) - \mu^2$$

$$= E[X(X-1)] + \mu - \mu^2$$

$$= \frac{k(k-1)n(n-1)}{N(N-1)} + \frac{nk}{N} - \frac{n^2k^2}{N^2}$$

$$= \frac{nk(N-k)(N-n)}{N^2(N-1)}$$

$$= \frac{N-n}{N-1} \cdot n \cdot \frac{k}{N}\left(1 - \frac{k}{N}\right).$$

Example 3.10 Using Chebyshev's theorem, find and interpret the interval $\mu \pm 2\sigma$ for Example 3.9.

Solution Since Example 3.9 was a hypergeometric experiment with $N = 40$, $n = 5$, and $k = 3$, then by Theorem 3.3 we have

$$\mu = \frac{(5)(3)}{40} = \frac{3}{8} = 0.375$$

and

$$\sigma^2 = \left(\frac{40 - 5}{39}\right)(5)\left(\frac{3}{40}\right)\left(1 - \frac{3}{40}\right)$$

$$= 0.3113.$$

Taking the square root of 0.3113, we find that $\sigma = 0.558$. Hence the required interval is $0.375 \pm (2)(0.558)$, or from -0.741 to 1.491. Chebyshev's theorem states that the number of defectives obtained when 5 components are selected at random from a lot of 40 components of which three are defective has a probability of at least 3/4 of falling between -0.741 and 1.491. That is, at least three fourths of the time, the 5 components include less than two defectives.

If n is small relative to N, the probability for each drawing will change only slightly. Hence we essentially have a binomial experiment and can approximate the hypergeometric distribution by using the binomial distribution with $p = k/N$. The mean and variance can also be approximated by the formulas

$$\mu = np = \frac{nk}{N}$$

$$\sigma^2 = npq = n \cdot \frac{k}{N}\left(1 - \frac{k}{N}\right).$$

Comparing these formulas with those of Theorem 3.3, we see that the mean is the same while the variance differs by a correction factor of $(N - n)/(N - 1)$. This is negligible when n is small relative to N.

Example 3.11 A manufacturer of automobile tires reports that among a shipment of 5000 sent to a local distributor, 1000 are slightly blemished. If one purchases 10 of these tires at random from the distributor, what is the probability that exactly 3 will be blemished?

Solution Since $N = 5000$ is large relative to the sample size $n = 10$, we shall approximate the desired probability by using the binomial distribution. The probability of obtaining a blemished tire is 0.2. Therefore, the probability of obtaining exactly 3 blemished tires is

$$\left[\sum_{x=0}^{2} \binom{10}{x} .2^{x} (1-.2)^{n-x} \right. $$

$$h(3; 5000, 10, 1000) \simeq b(3; 10, 0.2)$$

pg 86

$$= \sum_{x=0}^{3} b(x; 10, 0.2) - \sum_{x=0}^{2} b(x; 10, 0.2)$$

$$= 0.8791 - 0.6778$$

$$= 0.2013.$$

The hypergeometric distribution can be extended to treat the case where the N items can be partitioned into k cells $A_1, A_2, \ldots, A_k$ with a_1 elements in the first cell, a_2 elements in the second cell, $\ldots, a_k$ elements in the kth cell. We are now interested in the probability that a random sample of size n yields x_1 elements from A_1, x_2 elements from $A_2, \ldots,$ and x_k elements from A_k. Let us represent this probability by

$$f(x_1, x_2, \ldots, x_k; a_1, a_2, \ldots, a_k, N, n).$$

To obtain a general formula, we note that the total number of samples that can be chosen of size n from N items is still $\binom{N}{n}$. There are $\binom{a_1}{x_1}$ ways of selecting x_1 items from the items in A_1, and for each of these we can choose x_2 items from the items in A_2 in $\binom{a_2}{x_2}$ ways. Therefore, we can select x_1 items from A_1 and x_2 items from A_2 in $\binom{a_1}{x_1}\binom{a_2}{x_2}$ ways. Continuing in this way, we can select all n items consisting of x_1 from A_1, x_2 from $A_2, \ldots,$ and x_k from A_k in $\binom{a_1}{x_1}\binom{a_2}{x_2}\cdots\binom{a_k}{x_k}$ ways. The required probability distribution is now defined as follows.

EXTENSION OF THE HYPERGEOMETRIC DISTRIBUTION   *If N items can be partitioned into the k cells $A_1, A_2, \ldots, A_k$ with $a_1, a_2, \ldots, a_k$ elements, respectively, then the probability distribution of the random variables $X_1, X_2, \ldots, X_k$, representing the number of elements selected from $A_1, A_2, \ldots, A_k$ in a random sample of size n is*

$$f(x_1, x_2, \ldots, x_k; a_1, a_2, \ldots, a_k, N, n) = \frac{\binom{a_1}{x_1}\binom{a_2}{x_2} \cdots \binom{a_k}{x_k}}{\binom{N}{n}}$$

with

$$\sum_{i=1}^{k} x_i = n \quad and \quad \sum_{i=1}^{k} a_i = N.$$

Example 3.12   A group of 10 individuals is being used in a biological case study. The group contains 3 people with blood type O, 4 with blood type A, and 3 with blood type B. What is the probability that a random sample of 5 will contain 1 person with blood type O, 2 people with blood type A, and 2 people with blood type B?

Solution   Using the extension of the hypergeometric distribution with $x_1 = 1$, $x_2 = 2$, $x_3 = 2$, $a_1 = 3$, $a_2 = 4$, $a_3 = 3$, $N = 10$, and $n = 5$, the desired probability is

$$f(1, 2, 2; 3, 4, 3, 10, 5) = \frac{\binom{3}{1}\binom{4}{2}\binom{3}{2}}{\binom{10}{5}} = \frac{3}{14}.$$

3.5 POISSON DISTRIBUTION

Experiments yielding numerical values of a random variable X, the number of successes occurring during a given time interval or in a specified region, are often called *Poisson experiments*. The given time interval may be of any length, such as a minute, a day, a week, a month, or even a year. Hence a Poisson experiment might generate observations for the random variable X representing the number of telephone calls per hour received by an office, the number of days school is closed due to snow during the winter, or the number of postponed games due to rain during a baseball season. The specified region could be a line segment, an area, a volume, or perhaps a piece of material. In this case X might represent the number of field mice per acre, the number of bacteria in a given culture, or the number of typing errors per page.

A Poisson experiment is one that possesses the following properties:

1. The number of successes occurring in one time interval or specified region are independent of those occurring in any other disjoint time interval or region of space.
2. The probability of a single success occurring during a very short time interval or in a small region is proportional to the length of the time interval or the size of the region and does not depend on the number of successes occurring outside this time interval or region.
3. The probability of more than one success occurring in such a short time interval or falling in such a small region is negligible.

DEFINITION 3.3   *The number X of successes in a Poisson experiment is called a* Poisson random variable.

The probability distribution of the Poisson variable X is called the *Poisson distribution* and will be denoted by $p(x; \mu)$, since its values depend only on μ, the average number of successes occurring in the given time interval or specified region. The derivation of the formula for $p(x; \mu)$ based on the properties that have been listed for a Poisson experiment is beyond the scope of this text. We list the result in the following definition.

POISSON DISTRIBUTION   *The probability distribution of the Poisson random variable X, representing the number of successes occurring in a given time interval or specified region, is*

$$p(x; \mu) = \frac{e^{-\mu}\mu^x}{x!}, \qquad x = 0, 1, 2, \ldots,$$

where μ is the average number of successes occurring in the given time interval or specified region and $e = 2.71828\ldots$

Table III (see Statistical Tables) contains Poisson probability sums $P(r; \mu) = \sum_{x=0}^{r} p(x; \mu)$ for a few selected values of μ ranging from 0.1 to 18. We illustrate the use of this table with the following two examples.

Example 3.13   The average number of radioactive particles passing through a counter during 1 millisecond in a laboratory experiment is four. What is the probability that six particles enter the counter in a given millisecond?

Solution   Using the Poisson distribution with $x = 6$ and $\mu = 4$, we find from Table III that

$$p(6; 4) = \frac{e^{-4}4^6}{6!} = \sum_{x=0}^{6} p(x; 4) - \sum_{x=0}^{5} p(x; 4) = 0.8893 - 0.7851$$

$$= 0.1042.$$

Example 3.14   The average number of oil tankers arriving each day at a certain port city is known to be 10. The facilities at the port can handle at most 15 tankers per day. What is the probability that on a given day tankers will have to be sent away?

Solution   Let X be the number of tankers arriving each day. Then, using Table III we have

$$P(X > 15) = 1 - P(X \le 15)$$

$$= 1 - \sum_{x=0}^{15} p(x; 10)$$

$$= 1 - 0.9513$$

$$= 0.0487.$$

THEOREM 3.4   *The mean and variance of the Poisson distribution $p(x; \mu)$ both have the value μ.*

Proof   To verify that the mean is indeed μ, we write

$$E(X) = \sum_{x=0}^{\infty} x \cdot \frac{e^{-\mu}\mu^x}{x!} = \sum_{x=1}^{\infty} x \cdot \frac{e^{-\mu}\mu^x}{x!}$$

$$= \mu \sum_{x=1}^{\infty} \frac{e^{-\mu}\mu^{x-1}}{(x-1)!}.$$

Now, let $y = x - 1$ to give

$$E(X) = \mu \sum_{y=0}^{\infty} \frac{e^{-\mu}\mu^y}{y!} = \mu,$$

since

$$\sum_{y=0}^{\infty} \frac{e^{-\mu}\mu^y}{y!} = \sum_{y=0}^{\infty} p(y; \mu) = 1.$$

The variance of the Poisson distribution is obtained by first finding

$E[x(x-1)] = E(x^2) - E(x)$ VARIANCE $= E(x^2) - \mu^2$

↑ mean

$$E[X(X - 1)] = \sum_{x=0}^{\infty} x(x - 1) \cdot \frac{e^{-\mu}\mu^x}{x!} = \sum_{x=2}^{\infty} x(x - 1) \frac{e^{-\mu}\mu^x}{x!}$$

$$= \mu^2 \sum_{x=2}^{\infty} \frac{e^{-\mu}\mu^{x-2}}{(x - 2)!}.$$

Setting $y = x - 2$, we have

$$E[X(X - 1)] = \mu^2 \sum_{y=0}^{\infty} \frac{e^{-\mu}\mu^y}{y!} = \mu^2. \; = E(x^2) - E(x)$$

Hence

$$\sigma^2 = E[X(X - 1)] + \mu - \mu^2$$
$$= \mu^2 + \mu - \mu^2$$
$$= \mu.$$

In Example 3.13 where $\mu = 4$, we also have $\sigma^2 = 4$ and hence $\sigma = 2$. Using Chebyshev's theorem, we can state that our random variable has a probability of at least 3/4 of falling in the interval $\mu \pm 2\sigma = 4 \pm (2)(2)$, or from zero to 8. Therefore, we conclude that at least three fourths of the time the number of radioactive particles entering the counter will be anywhere from zero to 8 during a given millisecond.

We shall now derive the Poisson distribution as a limiting form of the binomial distribution when $n \to \infty$, $p \to 0$, and np remains constant. Hence, if n is large and p is close to zero, the Poisson distribution can be used, with $\mu = np$, to approximate binomial probabilities. If p is close to 1, we can still use the Poisson distribution to approximate binomial probabilities by interchanging what we have defined to be a success and a failure, thereby changing p to a value close to zero.

THEOREM 3.5   *Let X be a binomial random variable with probability distribution $b(x; n, p)$. When $n \to \infty$, $p \to 0$, and $\mu = np$ remains constant,*

$$b(x; n, p) \to p(x; \mu).$$

Proof   The binomial distribution can be written

$$b(x; n, p) = \binom{n}{x} p^x q^{n-x}$$

$$= \frac{n!}{x!(n-x)!} p^x (1 - p)^{n-x}$$

$$= \frac{n(n-1)\cdots(n-x+1)}{x!} p^x (1 - p)^{n-x}.$$

Substituting $p = \mu/n$, we have

$$b(x; n, p) = \frac{n(n-1)\cdots(n-x+1)}{x!} \left(\frac{\mu}{n}\right)^x \left(1 - \frac{\mu}{n}\right)^{n-x}$$

$$= 1\left(1 - \frac{1}{n}\right)\cdots\left(1 - \frac{x-1}{n}\right)\frac{\mu^x}{x!}\left(1 - \frac{\mu}{n}\right)^n \left(1 - \frac{\mu}{n}\right)^{-x}.$$

As $n \to \infty$ while x and μ remain constant,

$$\lim_{n \to \infty} 1\left(1 - \frac{1}{n}\right)\cdots\left(1 - \frac{x-1}{n}\right) = 1,$$

$$\lim_{n \to \infty}\left(1 - \frac{\mu}{n}\right)^{-x} = 1,$$

and from the definition of the number e,

$$\lim_{n \to \infty}\left(1 - \frac{\mu}{n}\right)^n = \lim_{n \to \infty}\left\{\left[1 + \frac{1}{(-n)/\mu}\right]^{-n/\mu}\right\}^{-\mu} = e^{-\mu}.$$

Hence, under the given limiting conditions

$$b(x; n, p) \to \frac{e^{-\mu}\mu^x}{x!}, \qquad x = 0, 1, 2, \ldots.$$

Example 3.15   In a certain manufacturing process in which glass items are being produced, defects or bubbles occur, occasionally rendering the piece undesirable for marketing. It is known that on the average 1 in every 1000 of these items produced has one or more bubbles. What is the probability that a random sample of 8000 will yield fewer than 7 items possessing bubbles?

Solution   This is essentially a binomial experiment with $n = 8000$ and $p = 0.001$. Since p is very close to zero and n is quite large, we shall approximate with the Poisson distribution using $\mu = (8000)(0.001) = 8$. Hence, if X represents the number of bubbles, we have

$$P(X < 7) = \sum_{x=0}^{6} b(x; 8000, 0.001)$$

$$\simeq \sum_{x=0}^{6} p(x; 8)$$

$$= 0.3134.$$

READ ONLY

IMPORTANT

3.6 NEGATIVE BINOMIAL AND GEOMETRIC DISTRIBUTIONS

Let us consider an experiment in which the properties are the same as those listed for a binomial experiment, with the exception that the trials will be repeated until a *fixed* number of successes occur. Therefore, instead of finding the probability of x successes in n trials, where n is fixed, we are now interested in the probability that the kth success occurs on the xth trial. Experiments of this kind are called *negative binomial experiments*.

As an illustration, consider the use of a drug that is known to be effective in 60% of the cases in which it is used. The use of the drug will be considered a success if it is effective in bringing some degree of relief to the patient. We are interested in finding the probability that the fifth patient to experience relief is the seventh patient to receive the drug during a given week. Designating a success by S and a failure by F, a possible order of achieving the desired result is $SFSSSFS$, which occurs with probability $(0.6)(0.4)(0.6)(0.6)(0.6)(0.4)(0.6) = (0.6)^5(0.4)^2$. We could list all possible orders by rearranging the F's and S's except for the last outcome, which must be the fifth success. The total number of possible orders is equal to the number of partitions of the first six trials into two groups with two failures assigned to the one group and the four successes assigned to the other group. This can be done in $\binom{6}{4} = 15$ mutually exclusive ways. Hence, if X represents the outcome on which the fifth success occurs, then

$$P(X = 7) = \binom{6}{4}(0.6)^5(0.4)^2 = 0.1866.$$

DEFINITION 3.4   *The number X of trials to produce k successes in a negative binomial experiment is called a* negative binomial variable.

The probability distribution of the negative binomial variable X is called the *negative binomial distribution* and will be denoted by $b^*(x; k, p)$, since its values depend on the number of successes desired and the probability of a success on a given trial. To obtain the general formula for $b^*(x; k, p)$, consider the probability of a success on the xth trial preceded by $k - 1$ successes and $x - k$ failures in some specified order. Since the trials are independent, we can multiply all the probabilities corresponding to each desired outcome. Each success occurs with probability p and each failure with probability $q = 1 - p$. Therefore, the probability for the specified order, ending in a success, is $p^{k-1}q^{x-k}p = p^kq^{x-k}$. The total number of sample points in the experiment ending in a success, after the occurrence of $k - 1$ successes and $x - k$ failures in any order, is equal to the number of partitions of $x - 1$ trials into two groups with $k - 1$ successes corresponding to one group and $x - k$ failures corresponding to the other group. This number is given by the term $\binom{x-1}{k-1}$, each mutually exclusive and occurring with equal probability p^kq^{x-k}. We obtain the general formula by multiplying p^kq^{x-k} by $\binom{x-1}{k-1}$.

NEGATIVE BINOMIAL DISTRIBUTION   *If repeated independent trials can result in a success with probability p and a failure with probability $q = 1 - p$, then the probability distribution of the random variable X, the number of the trial on which the kth success occurs, is given by*

$$b^*(x; k, p) = \binom{x-1}{k-1}p^kq^{x-k}, \qquad x = k, k+1, k+2, \ldots.$$

Example 3.16   Find the probability that a person tossing three coins will get either all heads or all tails for the second time on the fifth toss.

Solution Using the negative binomial distribution with $x = 5$, $k = 2$, and $p = 1/4$, we have

$$b^*\left(5; 2, \frac{1}{4}\right) = \binom{4}{1}\left(\frac{1}{4}\right)^2\left(\frac{3}{4}\right)^3$$

$$= \frac{4!}{1!\,3!} \cdot \frac{3^3}{4^5}$$

$$= \frac{27}{256}.$$

The negative binomial distribution derives its name from the fact that each term in the expansion of $p^k(1 - q)^{-k}$ corresponds to the values of $b^*(x; k, p)$ for $x = k, k + 1, k + 2, \dots$.

If we consider the special case of the negative binomial distribution where $k = 1$, we have a probability distribution for the number of trials required for a single success. An example would be the tossing of a coin until a head occurs. We might be interested in the probability that the first head occurs on the fourth toss. The negative binomial distribution reduces to the form $b^*(x; 1, p) = pq^{x-1}$, $x = 1, 2, 3, \dots$. Since the successive terms constitute a geometric progression, it is customary to refer to this special case as the *geometric distribution* and denote it by $g(x; p)$.

GEOMETRIC DISTRIBUTION   *If repeated independent trials can result in a success with probability p and a failure with probability $q = 1 - p$, then the probability distribution of the random variable X, the number of the trial on which the first success occurs, is given by*

$$g(x; p) = pq^{x-1}, \qquad x = 1, 2, 3, \dots.$$

Example 3.17 In a certain manufacturing process it is known that, on the average, 1 in every 100 items is defective. What is the probability that 5 items are inspected before a defective item is found?

Solution Using the geometric distribution with $x = 5$ and $p = 0.01$, we have

$$g(5; 0.01) = (0.01)(0.99)^4$$
$$= 0.0096.$$

DO ALL ODDS

EXERCISES

1. Find a formula for the distribution of the random variable X representing the number on a tag drawn at random from a box containing 10 tags numbered 1 to 10. What is the probability that the number drawn is less than 4?

2. A roulette wheel is divided into 25 sectors of equal area numbered from 1 to 25. Find a formula for the probability distribution of X, the number that occurs when the wheel is spun.

1-9

3. Find the mean and variance of the random variable X of Exercise 1.

4. A baseball player's batting average is 0.250. What is the probability that he gets exactly one hit in his next four times at bat?

5. If we define the random variable X to be equal to the number of heads that occur when a balanced coin is flipped once, find the probability distribution of X. What two well-known distributions describe the values of X?

6. In testing a certain kind of truck tire over a rugged terrain, it is found that 25% of the trucks fail to complete the test run without a blowout. What is the probability that from 5 to 10 of the next 15 trucks tested have flat tires?

6-9

7. The probability that a patient recovers from a delicate heart operation is 0.9. What is the probability that exactly five of the next seven patients having this operation survive?

8. A traffic control engineer reports that 75% of the vehicles passing through a check point are from within the state. What is the probability that at least three of the next five vehicles are from out of the state?

9. A survey of the residents in a United States city showed that 20% preferred a white telephone over any other color available. What is the probability that more than half of the next 20 telephones installed in this city will be white?

10. OMIT It is known that 75% of mice inoculated with a serum are protected from a certain disease. If three mice are inoculated, what is the probability that at most two of the mice contract the disease?

11. If X represents the number of out-of-state vehicles in Exercise 8 when five vehicles are checked, find the probability distribution of X. Using Chebyshev's theorem, find and interpret $\mu \pm 2\sigma$.

12. Suppose that airplane engines operate independently in flight and fail with probability $q = 1/5$. Assuming that a plane makes a safe flight if at least half of its engines run, determine whether a four-engine plane or a two-engine plane has the highest probability for a successful flight.

13. Repeat Exercise 12 for $q = 1/2$ and $q = 1/3$.

14. In Exercise 6 how many of the 15 trucks tested would you expect to have flat tires? Using Chebyshev's theorem, find and interpret the interval $\mu \pm 2\sigma$.

15. A card is drawn from a well-shuffled deck of 52 playing cards, the result recorded, and the card replaced. If the experiment is repeated five times, what is the probability of obtaining two spades and one heart?

16. OMIT The surface of a circular dart board has a small center circle called the bull's-eye and 20 pie-shaped regions numbered from 1 to 20. Each of the pie-shaped regions is

further divided into three parts such that a person throwing a dart that lands on a specified number scores the value of the number, double the number, or triple the number, depending on which of the three parts the dart falls. If a person hits the bull's-eye with probability 0.01, hits a double with probability 0.10, hits a triple with probability 0.05, and misses the dart board with probability 0.02, what is the probability that seven throws will result in no bull's-eyes, no triples, a double twice, and a complete miss once?

17. According to the theory of genetics, a certain cross of guinea pigs will result in red, black, and white offspring in the ratio 8:4:4. Find the probability that among eight offspring five will be red, two black, and one white.

18. To avoid detection at customs, a traveler has placed six narcotic tablets in a bottle containing nine vitamin pills that are similar in appearance. If the customs official selects three of the tablets at random for analysis, what is the probability that the traveler will be arrested for illegal possession of narcotics?

19. A company is interested in evaluating its current inspection procedure on shipments of 50 identical items. The procedure is to take a sample of 5 and pass the shipment if no more than 2 are found to be defective. What proportion of 20% defective shipments will be accepted?

20. A manufacturing company uses an acceptance scheme on production items before they are shipped. The plan is a two-stage one. Boxes of 25 are readied for shipment and a sample of 3 are tested for defectives. If any defectives are found, the entire box is sent back for 100% screening. If no defectives are found, the box is shipped.
    (a) What is the probability that a box containing three defectives will be shipped?
    (b) What is the probability that a box containing only one defective will be sent back for screening?

21. Suppose that the manufacturing company in Exercise 20 decided to change its acceptance scheme. Under the new scheme an inspector takes one at random, inspects it, and then replaces it in the box; a second inspector does likewise. Finally, a third inspector goes through the same procedure. The box is not shipped if any of the three find a defective. Answer parts (a) and (b) of Exercise 20 under this new plan.

22. A homeowner plants six bulbs selected at random from a box containing five tulip bulbs and four daffodil bulbs. What is the probability that he planted two daffodil bulbs and four tulip bulbs?

23. Find the probability of obtaining 2 ones, 1 two, 1 three, 2 fours, 3 fives, and 1 six in 10 rolls of a balanced die?

24. A random committee of size 3 is selected from four doctors and two nurses. Write a formula for the probability distribution of the random variable X representing the number of doctors on the committee. Find $P(2 \le X \le 3)$.

25. From a lot of 10 missiles, 4 are selected at random and fired. If the lot contains 3 defective missiles that will not fire, what is the probability that
    (a) All 4 will fire?
    (b) At most 2 will not fire?

26. In Exercise 25 how many defective missiles might we expect to be included among the 4 that are selected? Use Chebyshev's theorem to describe the variability of the number

of defective missiles included when 4 are selected from several lots each of size 10 containing 3 defective missiles.

27. If a person is dealt 13 cards from an ordinary deck of 52 playing cards several times, how many hearts per hand can he expect? Between what two values would you expect the number of hearts to fall at least 75% of the time?

28. It is estimated that 4000 of the 10,000 voting residents of a town are against a new sales tax. If 15 eligible voters are selected at random and asked their opinion, what is the probability that at most 7 favor the new tax?

29. An annexation suit is being considered against a county subdivision of 1200 residences by a neighboring city. If the occupants of half the residences object to being annexed, what is the probability that in a random sample of 10 at least 3 favor the annexation suit?

30. Find the probability of being dealt a bridge hand of 13 cards containing 5 spades, 2 hearts, 3 diamonds, and 3 clubs.

31. A foreign student club lists as its members two Canadians, three Japanese, five Italians, and two Germans. Find the probability that all nationalities are represented if a committee of size 4 is selected at random.

32. An urn contains three green balls, two blue balls, and four red balls. In a random sample of five balls, find the probability that both blue balls and at least one red ball are selected.

33. On the average a certain intersection results in three traffic accidents per week. What is the probability that exactly five accidents will occur at this intersection in any given week?

34. On the average a secretary makes two typing errors per page. What is the probability that she makes
    (a) Four or more errors on the next page she types?
    (b) No errors?

35. A certain area of the eastern United States is, on the average, hit by six hurricanes per year. Find the probability that in a given year
    (a) Fewer than four hurricanes will hit this area.
    (b) Anywhere from six to eight hurricanes will hit the area.

36. In an inventory study it was determined that on the average demands for a particular item at a warehouse were made five times per day. What is the probability that on a given day this item is requested
    (a) More than five times?
    (b) Not at all?

37. The probability that a person dies from a certain respiratory infection is 0.002. Find the probability that fewer than 5 of the next 2000 so infected will die.

38. Suppose than on the average 1 person in 1000 makes a numerical error in preparing their income tax returns. If 10,000 forms are selected at random and examined, find the probability that 6, 7, or 8 of the forms will be in error.

39. Using Chebyshev's theorem, find and interpret the interval $\mu \pm 2\sigma$ for Exercise 37.

40. The probability that a person will install a black telephone in a residence is estimated to be 0.3. Find the probability that the tenth phone installed in a new subdivision is the fifth black phone.

41. A scientist inoculates several mice, one at a time, with a disease germ until he finds two that have contracted the disease. If the probability of contracting the disease is 1/6, what is the probability that eight mice are required?

42. Three people toss a coin and the odd man pays for the coffee. If the coins all turn up the same, they are tossed again. Find the probability that fewer than four tosses are needed.

43. Find the probability that a person flipping a coin gets the third head on the seventh flip.

44. The probability that a student pilot passes the written test for his private pilot's license is 0.7. Find the probability that a person passes the test
(a) On the third try.
(b) Before the fourth try.

4

Some Continuous Probability Distributions

4.1 NORMAL DISTRIBUTION

The most important continuous probability distribution in the entire field of statistics is the *normal distribution*. Its graph, called the *normal curve*, is the bell-shaped curve of Figure 4.1 that describes the distribution of so many sets of data which occur in nature, industry, and research. In 1733 DeMoivre developed the mathematical equation of the normal curve. This provided a basis for much of the theory of inductive statistics. The normal distribution is often referred to as the *Gaussian distribution* in honor of Gauss (1777–1855), who also derived its equation from a study of errors in repeated measurements of the same quantity.

A random variable X having the bell-shaped distribution of Figure 4.1 is called a *normal random variable*. The mathematical equation for the probability distribution of the continuous normal variable depends on the two parameters μ and σ, its mean and standard deviation. Hence we shall denote the density function of X by $n(x; \mu, \sigma)$.

NORMAL DISTRIBUTION   *The density function of the normal random variable X, with mean μ and variance σ^2, is*

$$n(x; \mu, \sigma) = \frac{1}{\sqrt{2\pi}\sigma} e^{-(1/2)[(x-\mu)/\sigma]^2}, \qquad -\infty < x < \infty,$$

where $\pi = 3.14159\ldots$ and $e = 2.71828\ldots$.

Once μ and σ are specified, the normal curve is completely determined. For example, if $\mu = 50$ and $\sigma = 5$, then the ordinates $n(x; 50, 5)$ can easily be computed for various values of x and the curve drawn. In Figure 4.2 we have sketched two normal curves having the same standard deviation but different

110

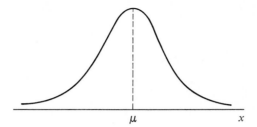

Figure 4.1 The normal curve.

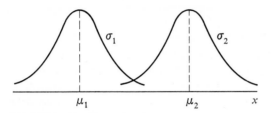

Figure 4.2 Normal curves with $\mu_1 < \mu_2$ and $\sigma_1 = \sigma_2$.

means. The two curves are identical in form but are centered at different positions along the horizontal axis.

In Figure 4.3 we have sketched two normal curves with the same mean but different standard deviations. This time we see that the two curves are centered at exactly the same position on the horizontal axis, but the curve with the larger standard deviation is lower and spreads out farther. Remember that the area under a probability curve must be equal to 1 and therefore the more variable the set of observations the lower and wider the corresponding curve will be.

Figure 4.4 shows the results of sketching two normal curves having different means and different standard deviations. Clearly, they are centered at different positions on the horizontal axis and their shapes reflect the two different values of σ.

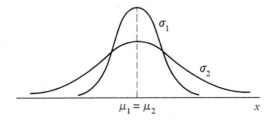

Figure 4.3 Normal curves with $\mu_1 = \mu_2$ and $\sigma_1 < \sigma_2$.

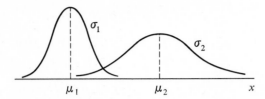

Figure 4.4 Normal curves with $\mu_1 < \mu_2$ and $\sigma_1 < \sigma_2$.

From an inspection of the graphs and by examination of the first and second derivatives of $n(x; \mu, \sigma)$, we list the following properties of the normal curve:

1. The *mode*, which is the point on the horizontal axis where the curve is a maximum, occurs at $x = \mu$.
2. The curve is symmetric about a vertical axis through the mean μ.
3. The curve has its points of inflection at $x = \mu \pm \sigma$, is concave downward if $\mu - \sigma < X < \mu + \sigma$, and is concave upward otherwise.
4. The normal curve approaches the horizontal axis asymptotically as we proceed in either direction away from the mean.
5. The total area under the curve and above the horizontal axis is equal to 1.

We shall now show that the parameters μ and σ^2 are indeed the mean and the variance of the normal distribution. To evaluate the mean, we write

$$E(X) = \frac{1}{\sqrt{2\pi}\,\sigma} \int_{-\infty}^{\infty} x e^{-(1/2)[(x-\mu)/\sigma]^2}\, dx.$$

Setting $z = (x - \mu)/\sigma$ and $dx = \sigma\, dz$, we obtain

$$E(X) = \frac{1}{\sqrt{2\pi}} \int_{-\infty}^{\infty} (\mu + \sigma z) e^{-z^2/2}\, dz$$

$$= \mu \frac{1}{\sqrt{2\pi}} \int_{-\infty}^{\infty} e^{-z^2/2}\, dz + \frac{\sigma}{\sqrt{2\pi}} \int_{-\infty}^{\infty} z e^{-z^2/2}\, dz.$$

The first integral is μ times the area under a normal curve with mean zero and variance 1, and hence equal to μ. By straightforward integration or from the fact that the integrand is an odd function, the second integral is equal to zero. Hence

$$E(X) = \mu.$$

The variance of the normal distribution is given by

$$E[(X - \mu)^2] = \frac{1}{\sqrt{2\pi}\,\sigma} \int_{-\infty}^{\infty} (x - \mu)^2 e^{-(1/2)[(x - \mu)/\sigma]^2}\, dx.$$

Again setting $z = (x - \mu)/\sigma$ and $dx = \sigma\, dz$, we obtain

$$E[(X - \mu)^2] = \frac{\sigma^2}{\sqrt{2\pi}} \int_{-\infty}^{\infty} z^2 e^{-z^2/2}\, dz.$$

Integrating by parts with $u = z$ and $dv = ze^{-z^2/2}$ so that $du = dz$ and $v = -e^{-z^2/2}$, we find that

$$E[(X - \mu)^2] = \frac{\sigma^2}{\sqrt{2\pi}} \left(-ze^{-z^2/2} \Big|_{-\infty}^{\infty} + \int_{-\infty}^{\infty} e^{-z^2/2}\, dz \right)$$

$$= \sigma^2(0 + 1)$$

$$= \sigma^2.$$

4.2 AREAS UNDER THE NORMAL CURVE

The curve of any continuous probability distribution or density function is constructed so that the area under the curve bounded by the two ordinates $x = x_1$ and $x = x_2$ equals the probability that the random variable X assumes a value between $x = x_1$ and $x = x_2$. Thus for the normal curve in Figure 4.5,

$$P(x_1 < X < x_2) = \int_{x_1}^{x_2} n(x; \mu, \sigma)\, dx$$

$$= \frac{1}{\sqrt{2\pi}\,\sigma} \int_{x_1}^{x_2} e^{-(1/2)[(x - \mu)/\sigma]^2}\, dx$$

is represented by the area of the shaded region.

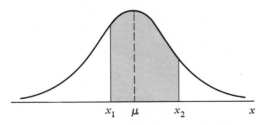

Figure 4.5 $P(x_1 < X < x_2) =$ area of the shaded region.

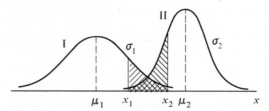

Figure 4.6 $P(x_1 < X < x_2)$ for different normal curves.

In Figures 4.2, 4.3, and 4.4 we saw how the normal curve is dependent on the mean and the standard deviation of the distribution under investigation. The area under the curve between any two ordinates must then also depend on the values μ and σ. This is evident in Figure 4.6, where we have shaded regions corresponding to $P(x_1 < X < x_2)$ for two curves with different means and variances. The $P(x_1 < X < x_2)$, where X is the random variable describing distribution I, is indicated by the shaded region where the lines slope up to the right. If X is the random variable describing distribution II, then $P(x_1 < X < x_2)$ is given by the shaded region where the lines slope down to the right. Obviously, the two shaded regions are different in size; therefore, the probability associated with each distribution will be different.

The difficulty encountered in solving integrals of normal density functions necessitates the tabulation of normal curve areas for quick reference. However, it would be a hopeless task to attempt to set up separate tables for every conceivable value of μ and σ. Fortunately, we are able to transform all the observations of any normal random variable X to a new set of observations of a normal random variable Z with mean zero and variance 1. This can be done by means of the transformation

$$Z = \frac{X - \mu}{\sigma}.$$

Whenever X assumes a value x, the corresponding value of Z is given by $z = (x - \mu)/\sigma$. Therefore, if X falls between the values $x = x_1$ and $x = x_2$, the random variable Z will fall between the corresponding values $z_1 = (x_1 - \mu)/\sigma$ and $z_2 = (x_2 - \mu)/\sigma$. Consequently, we may write

$$P(x_1 < X < x_2) = \frac{1}{\sqrt{2\pi}\,\sigma} \int_{x_1}^{x_2} e^{-(1/2)[(x-\mu)/\sigma]^2} \, dx$$

$$= \frac{1}{\sqrt{2\pi}} \int_{z_1}^{z_2} e^{-z^2/2} \, dz$$

$$= \int_{z_1}^{z_2} n(z; 0, 1) \, dz = P(z_1 < Z < z_2),$$

where Z is seen to be a normal random variable with mean zero and variance 1.

DEFINITION 4.1   *The distribution of a normal random variable with mean zero and variance* 1 *is called a* standard normal distribution.

The original and transformed distributions are illustrated in Figure 4.7. Since all the values of X falling between x_1 and x_2 have corresponding z values between z_1 and z_2, the area under the X curve between the ordinates $x = x_1$ and $x = x_2$ in Figure 4.7 equals the area under the Z curve between the transformed ordinates $z = z_1$ and $z = z_2$.

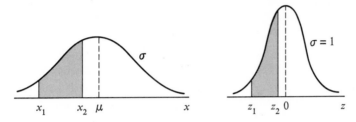

Figure 4.7 The original and transformed normal distributions.

We have now reduced the required number of tables of normal curve areas to one, that of the standard normal distribution. Table IV (see Statistical Tables) gives the area under the standard normal curve corresponding to $P(Z < z)$ for values of z from -3.4 to 3.4. To illustrate the use of this table, let us find the probability that Z is less than 1.74. First, we locate a value of z equal to 1.7 in the left column, then move across the row to the column under 0.04, where we read 0.9591. Therefore, $P(Z < 1.74) = 0.9591$.

Example 4.1   Given a normal distribution with $\mu = 50$ and $\sigma = 10$, find the probability that X assumes a value between 45 and 62.

Solution   The z values corresponding to $x_1 = 45$ and $x_2 = 62$ are

$$z_1 = \frac{45 - 50}{10} = -0.5$$

$$z_2 = \frac{62 - 50}{10} = 1.2.$$

Therefore,

$$P(45 < X < 62) = P(-0.5 < Z < 1.2).$$

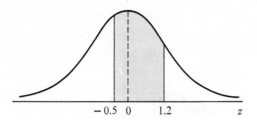

Figure 4.8 Area for Example 4.1.

The $P(-0.5 < Z < 1.2)$ is given by the area of the shaded region in Figure 4.8. This area may be found by subtracting the area to the left of the ordinate $z = -0.5$ from the entire area to the left of $z = 1.2$. Using Table IV, we have

$$
\begin{aligned}
P(45 < X < 62) &= P(-0.5 < Z < 1.2) \\
&= P(Z < 1.2) - P(Z < -0.5) \\
&= 0.8849 - 0.3085 \\
&= 0.5764.
\end{aligned}
$$

According to Chebyshev's theorem, the probability that a random variable assumes a value within 2 standard deviations of the mean is at least $3/4$. If the random variable has a normal distribution, the z values corresponding to $x_1 = \mu - 2\sigma$ and $x_2 = \mu + 2\sigma$ are easily computed to be

$$
z_1 = \frac{(\mu - 2\sigma) - \mu}{\sigma} = -2
$$

and

$$
z_2 = \frac{(\mu + 2\sigma) - \mu}{\sigma} = 2.
$$

Hence

$$
\begin{aligned}
P(\mu - 2\sigma < X < \mu + 2\sigma) &= P(-2 < Z < 2) \\
&= P(Z < 2) - P(Z < -2) \\
&= 0.9772 - 0.0228 \\
&= 0.9544,
\end{aligned}
$$

which is a much stronger statement than that given by Chebyshev's theorem.

Some of the many problems where the normal distribution is applicable are treated in the following examples. The use of the normal curve to approximate binomial probabilities will be considered in Section 4.3.

Example 4.2   A certain type of storage battery lasts on the average 3.0 years, with a standard deviation of 0.5 year. Assuming that the battery lives are normally distributed, find the probability that a given battery will last less than 2.3 years.

Solution   First construct a diagram such as Figure 4.9, showing the given distribution of battery lives and the desired area. To find the $P(X < 2.3)$, we need to evaluate the area under the normal curve to the left of 2.3. This is

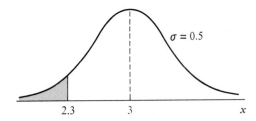

$\sigma = 0.5$

2.3   3 x

Figure 4.9 Area for Example 4.2.

accomplished by finding the area to the left of the corresponding z value. Hence we find that

$$z = \frac{2.3 - 3}{0.5} = -1.4,$$

and then using Table IV we have

$$P(X < 2.3) = P(Z < -1.4)$$
$$= 0.0808.$$

Example 4.3   An electrical firm manufactures light bulbs that have a length of life that is normally distributed with mean equal to 800 hours and a standard deviation of 40 hours. Find the probability that a bulb burns between 778 and 834 hours.

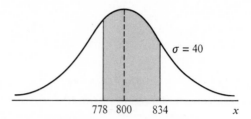

Figure 4.10 Area for Example 4.3.

Solution The distribution of light bulbs is illustrated in Figure 4.10. The z values corresponding to $x_1 = 778$ and $x_2 = 834$ are

$$z_1 = \frac{778 - 800}{40} = -0.55$$

$$z_2 = \frac{834 - 800}{40} = 0.85.$$

Hence

$$\begin{aligned}
P(778 < X < 834) &= P(-0.55 < Z < 0.85) \\
&= P(Z < 0.85) - P(Z < -0.55) \\
&= 0.8023 - 0.2912 \\
&= 0.5111.
\end{aligned}$$

Example 4.4 Gauges are used to reject all components in which a certain dimension is not within the specification $1.50 \pm d$. It is known that this measurement is normally distributed with mean 1.50 and standard deviation 0.2. Determine the value d such that the specifications "cover" 95% of the measurements.

Solution Examples 4.2 and 4.3 were solved by going first from a value of x to a z value and then computing the desired area. In this problem we reverse the process and begin with a known area or probability, find the z value, and then determine x from the formula $x = \sigma z + \mu$. We know from Table IV that

$$P(-1.96 < Z < 1.96) = 0.95.$$

Thus

$$1.50 + d = (0.2)(1.96) + 1.50,$$

from which we obtain

$$d = (0.2)(1.96) = 0.392.$$

An illustration of the specifications is given in Figure 4.11.

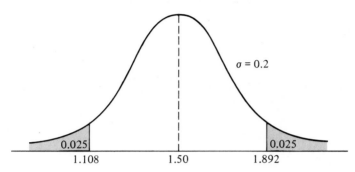

Figure 4.11 Specifications for Example 4.4.

Example 4.5 A certain machine makes electrical resistors having a mean resistance of 40 ohms and a standard deviation of 2 ohms. Assuming that the resistance follows a normal distribution and can be measured to any degree of accuracy, what percentage of resistors will have a resistance that exceeds 43 ohms?

Solution A percentage is found by multiplying the relative frequency by 100%. Since the relative frequency for an interval is equal to the probability of falling in the interval, we must find the area to the right of $x = 43$ in Figure 4.12. This can be done by transforming $x = 43$ to the corresponding z value,

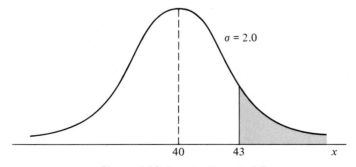

Figure 4.12 Area for Example 4.5.

obtaining the area to the left of z from Table IV, and then subtracting this area from 1. We find

$$z = \frac{43 - 40}{2} = 1.5.$$

Hence

$$P(X > 43) = P(Z > 1.5)$$
$$= 1 - P(Z < 1.5)$$
$$= 1 - 0.9332$$
$$= 0.0668.$$

Therefore, 6.68 % of the resistors will have a resistance exceeding 43 ohms.

Example 4.6   Find the percentage of resistances exceeding 43 ohms in Example 4.5 if the resistance is measured to the nearest ohm.

Solution   This problem differs from Example 4.5 in that we now assign a measurement of 43 ohms to all resistors whose resistances are greater than 42.5 and less than 43.5. We are actually approximating a discrete distribution by means of a continuous normal distribution. The required area is the region shaded to the right of 43.5 in Figure 4.13. We now find that

$$z = \frac{43.5 - 40}{2} = 1.75.$$

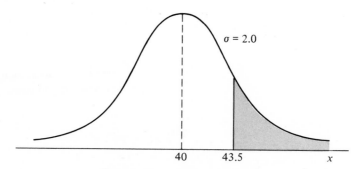

Figure 4.13 Area for Example 4.6.

Hence

$$P(X > 43.5) = P(Z > 1.75)$$
$$= 1 - P(Z < 1.75)$$
$$= 1 - 0.9599$$
$$= 0.0401.$$

Therefore, 4.01% of the resistances exceed 43 ohms when measured to the nearest ohm. The difference 6.68% - 4.01% = 2.67% between this answer and that of Example 4.5 represents all those resistors having a resistance greater than 43 and less than 43.5 that are now being recorded as 43 ohms.

Example 4.7 The quality grade-point averages of 300 college freshmen follow approximately a normal distribution with a mean of 2.1 and a standard deviation of 0.6. How many of these freshmen would you expect to have a score between 2.5 and 3.5 inclusive if the grade-point averages are computed to the nearest tenth?

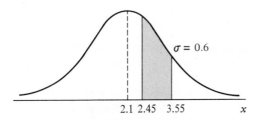

$\sigma = 0.6$

2.1 2.45   3.55 x

Figure 4.14 Area for Example 4.7.

Solution   Since the scores are recorded to the nearest tenth, we require the area between $x_1 = 2.45$ and $x_2 = 3.55$, as indicated in Figure 4.14. The corresponding z values are

$$z_1 = \frac{2.45 - 2.1}{0.6} = 0.58$$

$$z_2 = \frac{3.55 - 2.1}{0.6} = 2.42.$$

Therefore,

$$P(2.45 < X < 3.55) = P(0.58 < Z < 2.42)$$
$$= P(Z < 2.42) - P(Z < 0.58)$$
$$= 0.9922 - 0.7190$$
$$= 0.2732.$$

Hence 27.32%, or approximately 82 of the 300 freshmen, should have a score between 2.5 and 3.5 inclusive.

4.3 NORMAL APPROXIMATION TO THE BINOMIAL

Probabilities associated with binomial experiments are readily obtainable from the formula $b(x; n, p)$ of the binomial distribution or from Table II (see Statistical Tables) when n is small. If n is not listed in any available table, we must compute the binomial probabilities by approximation procedures. In Section 3.5 we illustrated how the Poisson distribution can be used to approximate binomial probabilities when n is large and p is close to zero or 1. Both the binomial and Poisson distributions are discrete. The first application of a continuous probability distribution to approximate probabilities over a discrete sample space was demonstrated in Examples 4.6 and 4.7, where the normal curve was used. We shall now state a theorem that allows us to use areas under the normal curve to approximate binomial probabilities when n is sufficiently large.

THEOREM 4.1   *If X is a binomial random variable with mean $\mu = np$ and variance $\sigma^2 = npq$, then the limiting form of the distribution of*

$$Z = \frac{X - np}{\sqrt{npq}},$$

as $n \to \infty$, is the standardized normal distribution $n(z; 0, 1)$.

It turns out that the proper normal distribution provides a very accurate approximation to the binomial distribution when n is large and p is close to $1/2$. In fact, even when n is small and p is not extremely close to zero or 1, the approximation is fairly good.

To investigate the normal approximation to the binomial distribution, we first draw the histogram for $b(x; 15, 0.4)$ and then superimpose the particular normal curve having the same mean and variance as the binomial variable X. Hence we draw a normal curve with

$$\mu = np = (15)(0.4) = 6$$

and

$$\sigma^2 = npq = (15)(0.4)(0.6) = 3.6.$$

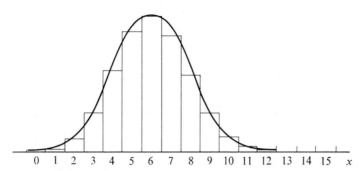

Figure 4.15 Normal curve approximation of $b(x; 15, 0.4)$.

The histogram of $b(x; 15, 0.4)$ and the corresponding superimposed normal curve, which is completely determined by its mean and variance, are illustrated in Figure 4.15.

The exact probability of the binomial random variable X assuming a given value x is equal to the area of the rectangle whose base is centered at x. For example, the exact probability that X assumes the value 4 is equal to the area of the rectangle with base centered at $x = 4$. Using the formula for the binomial distribution, we find this area to be

$$b(4; 15, 0.4) = 0.1268.$$

This probability is approximately equal to the area of the shaded region under the normal curve between the two ordinates $x_1 = 3.5$ and $x_2 = 4.5$ in Figure 4.16. Converting to z values, we have

$$z_1 = \frac{3.5 - 6}{1.9} = -1.316$$

$$z_2 = \frac{4.5 - 6}{1.9} = -0.789.$$

If X is a binomial random variable and Z a standard normal variable,

$$
\begin{aligned}
P(X = 4) &= b(4; 15, 0.4) \\
&\simeq P(-1.316 < Z < -0.789) \\
&= P(Z < -0.789) - P(Z < -1.316) \\
&= 0.2151 - 0.0941 \\
&= 0.1210.
\end{aligned}
$$

This agrees very closely with the exact value of 0.1268.

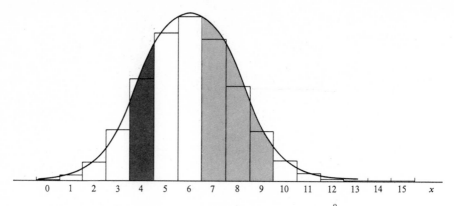

Figure 4.16 Normal approximation for $b(4; 15, 0.4)$ and $\sum_{x=7}^{9} b(x; 15, 0.4)$.

The normal approximation is most useful in calculating binomial sums for large values of n, which, without tables of binomial sums, is an impossible task. Referring to Figure 4.16, we might be interested in the probability that X assumes a value from 7 to 9 inclusive. The exact probability is given by

$$P(7 \leq X \leq 9) = \sum_{x=7}^{9} b(x; 15, 0.4)$$

$$= \sum_{x=0}^{9} b(x; 15, 0.4) - \sum_{x=0}^{6} b(x; 15, 0.4)$$

$$= 0.9662 - 0.6098$$

$$= 0.3564,$$

which is equal to the sum of the areas of the rectangles with bases centered at $x = 7$, 8, and 9. For the normal approximation we find the area of the shaded region under the curve between the ordinates $x_1 = 6.5$ and $x_2 = 9.5$ in Figure 4.16. The corresponding z values are

$$z_1 = \frac{6.5 - 6}{1.9} = 0.263$$

$$z_2 = \frac{9.5 - 6}{1.9} = 1.842.$$

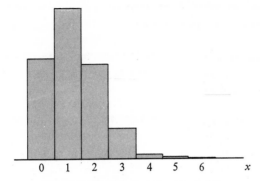

Figure 4.17 Histogram for $b(x; 6, 0.2)$.

Now

$$P(7 \leq X \leq 9) \simeq P(0.263 < Z < 1.842)$$
$$= P(Z < 1.842) - P(Z < 0.263)$$
$$= 0.9673 - 0.6037$$
$$= 0.3636.$$

Once again, the normal curve approximation provides a value that agrees very closely to the exact value of 0.3564. The degree of accuracy, which depends on how well the curve fits the histogram, will increase as n increases. This is particularly true when p is not very close to $1/2$ and the histogram is no longer symmetric. Figures 4.17 and 4.18 show the histograms for $b(x; 6, 0.2)$ and $b(x; 15, 0.2)$, respectively. It is evident that a normal curve would fit the histogram when $n = 15$ considerably better than when $n = 6$.

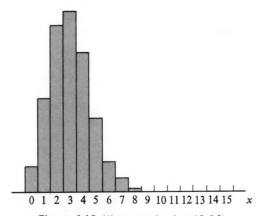

Figure 4.18 Histogram for $b(x; 15, 0.2)$.

In summary, we use the normal approximation to evaluate binomial probabilities whenever p is not close to zero or 1. The approximation is excellent when n is large and fairly good for small values of n if p is reasonably close to 1/2. One possible guide to determine when the normal approximation may be used is provided by calculating np and nq. If both np and nq are greater than 5, the approximation will be good.

Example 4.8   A process yields 10% defective items. If 100 items are randomly selected from the process, what is the probability that the number of defectives exceeds 13?

Solution   The number of defectives has the binomial distribution with parameters $n = 100$ and $p = 0.1$. Since the sample size is large, we should obtain fairly accurate results using the normal curve approximation with

$$\mu = np = (100)(0.1) = 10$$

$$\sigma = \sqrt{npq} = \sqrt{(100)(0.1)(0.9)} = 3.0.$$

To obtain the desired probability, we have to find the area to the right of $x = 13.5$. The z value corresponding to 13.5 is

$$z = \frac{13.5 - 10}{3} = 1.167$$

and the probability that the number of defectives exceeds 13 is given by the area of the shaded region in Figure 4.19. Hence, if X represents the number of defectives, then

$$P(X > 13) = \sum_{x=14}^{100} b(x; 100, 0.1)$$

$$\simeq P(Z > 1.167)$$

$$= 1 - P(Z < 1.167)$$

$$= 1 - 0.8784$$

$$= 0.1216.$$

Example 4.9   A multiple-choice quiz has 200 questions each with four possible answers of which only one is the correct answer. What is the probability that sheer guesswork yields from 25 to 30 correct answers for 80 of the 200 problems about which the student has no knowledge?

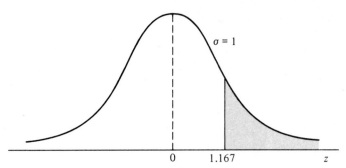

Figure 4.19 Area for Example 4.8.

Solution The probability of a correct answer for each of the 80 questions is $p = 1/4$. If X represents the number of correct answers due to guesswork, then

$$P(25 \le X \le 30) = \sum_{x=25}^{30} b(x; 80, \tfrac{1}{4}).$$

Using the normal curve approximation with

$$\mu = np = (80)(\tfrac{1}{4}) = 20$$

$$\sigma = \sqrt{npq} = \sqrt{(80)(\tfrac{1}{4})(\tfrac{3}{4})} = 3.87,$$

we need the area between $x_1 = 24.5$ and $x_2 = 30.5$. The corresponding z values are

$$z_1 = \frac{24.5 - 20}{3.87} = 1.163$$

$$z_2 = \frac{30.5 - 20}{3.87} = 2.713.$$

The probability of correctly guessing from 25 to 30 questions is given by the area of the shaded region in Figure 4.20. From Table IV we find that

$$P(25 \le X \le 30) = \sum_{x=25}^{30} b(x; 80, \tfrac{1}{4})$$

$$\simeq P(1.163 < Z < 2.713)$$

$$= P(Z < 2.713) - P(Z < 1.163)$$

$$= 0.9966 - 0.8776$$

$$= 0.1190.$$

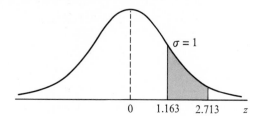

Figure 4.20 Area for Example 4.9.

4.4 GAMMA, EXPONENTIAL, AND CHI-SQUARE DISTRIBUTIONS

Although the normal distribution can be used to solve many problems in engineering and science, there are still numerous situations that require a different type of density function. Three such density functions, the *gamma*, *exponential*, and *chi-square distributions*, will be discussed in this section. Additional densities will be presented in Section 4.5 and in Sections 5.7 and 5.8.

The gamma distribution derives its name from the well-known *gamma function*, studied in many areas of mathematics. Before we proceed to the gamma distribution, let us review this function and some of its important properties.

DEFINITION 4.2 *The gamma function is defined by*

$$\Gamma(\alpha) = \int_0^\infty x^{\alpha-1} e^{-x} \, dx$$

for $\alpha > 0$.

Integrating by parts with $u = x^{\alpha-1}$ and $dv = e^{-x} \, dx$, we obtain

$$\Gamma(\alpha) = -e^{-x} x^{\alpha-1} \Big|_0^\infty + \int_0^\infty e^{-x}(\alpha-1) x^{\alpha-2} \, dx$$

$$= (\alpha-1) \int_0^\infty e^{-x} x^{\alpha-2} \, dx,$$

which yields the recursion formula

$$\Gamma(\alpha) = (\alpha-1)\Gamma(\alpha-1).$$

Repeated application of the recursion formula gives

$$\Gamma(\alpha) = (\alpha - 1)(\alpha - 2)\Gamma(\alpha - 2)$$
$$= (\alpha - 1)(\alpha - 2)(\alpha - 3)\Gamma(\alpha - 3),$$

and so forth. Note that when $\alpha = n$, where n is a positive integer,

$$\Gamma(n) = (n - 1)(n - 2)\cdots\Gamma(1).$$

However, by Definition 4.2,

$$\Gamma(1) = \int_0^\infty e^{-x}\,dx = 1$$

and hence

$$\boxed{\Gamma(n) = (n - 1)!.}$$

One important property of the gamma function, left for the student to verify (see Exercise 25), is that $\Gamma(1/2) = \sqrt{\pi}$.

We shall now include the gamma function in our definition of the gamma distribution.

GAMMA DISTRIBUTION   *The continuous random variable X has a* gamma *distribution, with parameters α and β, if its density function is given by*

$$f(x) = \frac{1}{\beta^\alpha \Gamma(\alpha)} x^{\alpha - 1} e^{-x/\beta}, \qquad x > 0$$

$$= 0, \qquad\qquad\qquad elsewhere,$$

where $\alpha > 0$ and $\beta > 0$.

Graphs of several gamma distributions are shown in Figure 4.21 for certain specified values of the parameters α and β. The special gamma distribution for which $\alpha = 1$ is called the *exponential distribution.*

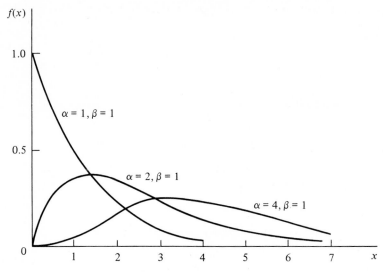

Figure 4.21 Gamma distributions.

$\alpha = 1$

EXPONENTIAL DISTRIBUTION   *The continuous random variable X has an* exponential distribution, *with parameter β, if its density function is given by*

$$f(x) = \frac{1}{\beta} e^{-x/\beta}, \qquad x > 0$$

$$= 0, \qquad\qquad elsewhere,$$

where β > 0.

The exponential distribution has many applications in the field of statistics, particularly in the areas of *reliability theory* and *waiting times* or *queueing problems*. One interesting application of the exponential distribution to reliability theory is given in the following example.

Example 4.10   Suppose that a system contains a certain type of component whose time in years to failure is given by the random variable *T*, distributed exponentially with parameter $\beta = 5$. If five of these components are installed in different systems, what is the probability that at least two are still functioning at the end of 8 years?

Solution The probability that a given component is still functioning after 8 years is given by

$$P(T > 8) = \tfrac{1}{5} \int_8^\infty e^{-t/5}\, dt$$

$$= e^{-8/5}$$

$$\simeq 0.2.$$

Let X represent the number of components functioning after 8 years. Then, using the binomial distribution,

$$P(X \geq 2) = \sum_{x=2}^{5} b(x; 5, 0.2)$$

$$= 1 - \sum_{x=0}^{1} b(x; 5, 0.2)$$

$$= 1 - 0.7373$$

$$= 0.2627.$$

A second special case of the gamma distribution is obtained by letting $\alpha = v/2$ and $\beta = 2$, where v is a positive integer. The probability density so obtained is called the *chi-square distribution* with v *degrees of freedom*.

CHI-SQUARE DISTRIBUTION   *The continuous random variable X has a chi-square distribution, with v degrees of freedom, if its density function is given by*

$$f(x) = \frac{1}{2^{v/2}\Gamma(v/2)} x^{v/2-1} e^{-x/2}, \qquad x > 0$$

$$= 0, \qquad\qquad\qquad\qquad elsewhere,$$

where v is a positive integer.

The chi-square distribution is one of our main tools in the area of hypothesis testing, a subject to be studied in later chapters. We shall now derive the mean and variance of these three density functions.

THEOREM 4.2 *The mean and variance of the gamma distribution are*

$$\mu = \alpha\beta \qquad and \qquad \sigma^2 = \alpha\beta^2.$$

Proof The rth moment about the origin for the gamma distribution is

$$\mu_r' = E(X^r) = \frac{1}{\beta^\alpha \Gamma(\alpha)} \int_0^\infty x^{r+\alpha-1} e^{-x/\beta}\, dx.$$

Now, let $y = x/\beta$, to give

$$\mu_r' = \frac{\beta^r}{\Gamma(\alpha)} \int_0^\infty y^{r+\alpha-1} e^{-y}\, dy$$

$$= \frac{\beta^r \Gamma(\alpha + r)}{\Gamma(\alpha)}.$$

Then

$$\mu = \mu_1' = \frac{\beta \Gamma(\alpha + 1)}{\Gamma(\alpha)} = \alpha\beta$$

and

$$\sigma^2 = \mu_2' - \mu^2 = \frac{\beta^2 \Gamma(\alpha + 2)}{\Gamma(\alpha)} - \alpha^2\beta^2$$

$$= \alpha\beta^2.$$

COROLLARY 1 *The mean and variance of the exponential distribution are*

$$\mu = \beta \qquad and \qquad \sigma^2 = \beta^2.$$

COROLLARY 2 *The mean and variance of the chi-square distribution are*

$$\mu = v \qquad and \qquad \sigma^2 = 2v.$$

4.5 WEIBULL DISTRIBUTION ◌̇MIT

Modern technology has enabled us to design many complicated systems whose operation, or perhaps safety, depend on the reliability of the various components making up the systems. For example, a fuse may burn out, a steel column may buckle, or a heat-sensing device may fail. Identical components subjected to identical environmental conditions will fail at different and unpredictable times. The *time to failure* or *life length* of a component, measured from some specified time until it fails, is represented by the continuous random variable T with probability density function $f(t)$. One distribution that has been used extensively in recent years to deal with such problems as reliability and life testing is the *Weibull distribution*.

WEIBULL DISTRIBUTION *The continuous random variable T has a* Weibull *distribution, with parameters α and β, if its density function is given by*

$$f(t) = \alpha\beta t^{\beta-1}e^{-\alpha t^{\beta}}, \qquad t > 0$$
$$= 0, \qquad\qquad elsewhere,$$

where $\alpha > 0$ and $\beta > 0$.

The graphs of the Weibull distribution for various values of the parameters α and β are illustrated in Figure 4.22. We see that the curves change in shape

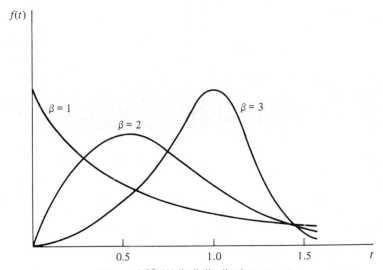

Figure 4.22 Weibull distributions ($\alpha = 1$).

considerably for different values of the parameters, particularly the parameter β. If we let $\beta = 1$, the Weibull distribution reduces to the exponential distribution. For values of $\beta > 1$, the curves become somewhat bell-shaped and resemble the normal curves, but display some skewness.

The mean and variance of the Weibull distribution are stated in the following theorem. The reader is asked to provide the proof in Exercise 28.

THEOREM 4.3   *The mean and variance of the Weibull distribution are*

$$\mu = \alpha^{-1/\beta}\Gamma\left(1 + \frac{1}{\beta}\right)$$

$$\sigma^2 = \alpha^{-2/\beta}\left\{\Gamma\left(1 + \frac{2}{\beta}\right) - \left[\Gamma\left(1 + \frac{1}{\beta}\right)\right]^2\right\}.$$

To apply the Weibull distribution to reliability theory, let us first define the reliability of a component or product as the *probability that it will function properly for at least a specified time under specified experimental conditions.* Therefore, if $R(t)$ is defined to be the reliability of the given component at time t, we may write

$$R(t) = P(T > t)$$

$$= \int_t^\infty f(t)\,dt$$

$$= 1 - F(t),$$

where $F(t)$ is the cumulative distribution of T. The conditional probability that a component will fail in the interval from $T = t$ to $T = t + \Delta t$, given that it survived to time t, is given by

$$\frac{F(t + \Delta t) - F(t)}{R(t)}.$$

Dividing this ratio by Δt and taking the limit as $\Delta t \to 0$, we get the *failure rate*, denoted by $Z(t)$. Hence

$$Z(t) = \lim_{\Delta t \to 0} \frac{F(t + \Delta t) - F(t)}{\Delta t} \frac{1}{R(t)}$$

$$= \frac{F'(t)}{R(t)} = \frac{f(t)}{R(t)} = \frac{f(t)}{1 - F(t)},$$

which expresses the failure rate in terms of the distribution of the time to failure.

From the fact that $R(t) = 1 - F(t)$ and then $R'(t) = -F'(t)$, we can write the differential equation

$$Z(t) = \frac{-R'(t)}{R(t)} = \frac{-d[\ln R(t)]}{dt}$$

and then solving,

$$\ln R(t) = - \int Z(t)\, dt$$

or

$$R(t) = e^{-\int Z(t)\, dt} + c,$$

where c satisfies the initial assumption that $R(0) = 1$ or $F(0) = 1 - R(0) = 0$. Thus we see that a knowledge of either the density function $f(t)$ or the failure rate $Z(t)$ uniquely determines the other.

Example 4.11   Show that the failure-rate function is given by

$$Z(t) = \alpha \beta t^{\beta - 1}, \qquad t > 0,$$

if and only if the time to failure distribution is the Weibull distribution

$$f(t) = \alpha \beta t^{\beta - 1} e^{-\alpha t^{\beta}}, \qquad t > 0.$$

Solution   Assume that $Z(t) = \alpha \beta t^{\beta - 1}, t > 0$. Then we can write

$$f(t) = Z(t)R(t),$$

where

$$R(t) = e^{-\int Z(t)\, dt} = e^{-\int \alpha \beta t^{\beta - 1}\, dt} = e^{-\alpha t^{\beta}} + c.$$

From the condition that $R(0) = 1$, we find that $c = 0$. Hence

$$R(t) = e^{-\alpha t^{\beta}}$$

and

$$f(t) = \alpha \beta t^{\beta - 1} e^{-\alpha t^{\beta}}, \qquad t > 0.$$

Now, if we assume

$$f(t) = \alpha\beta t^{\beta-1} e^{-\alpha t^{\beta}}, \qquad t > 0,$$

then $Z(t)$ is determined by writing

$$Z(t) = \frac{f(t)}{R(t)},$$

where

$$R(t) = 1 - F(t) = 1 - \int_0^t \alpha\beta x^{\beta-1} e^{-\alpha x^{\beta}} \, dx$$

$$= 1 + \int_0^t d e^{-\alpha x^{\beta}}$$

$$= e^{-\alpha t^{\beta}}.$$

Then

$$Z(t) = \frac{\alpha\beta t^{\beta-1} e^{-\alpha t^{\beta}}}{e^{-\alpha t^{\beta}}} = \alpha\beta t^{\beta-1}, \qquad t > 0.$$

In Example 4.11 the failure rate is seen to decrease with time if $\beta < 1$, increases with time if $\beta > 1$, and is constant if $\beta = 1$. In view of the fact that the Weibull distribution with $\beta = 1$ reduces to the exponential distribution, the assumption of constant failure rate is often referred to as the *exponential assumption*.

EXERCISES

1. Given a normal distribution with $\mu = 40$ and $\sigma = 6$, find
   (a) The area below 32.
   (b) The area above 27.
   (c) The area between 42 and 51.
   (d) The point that has 45% of the area below it.
   (e) The point that has 13% of the area above it.
2. Given a normal distribution with $\mu = 200$ and $\sigma^2 = 100$, find
   (a) The area below 214.
   (b) The area above 179.

(c) The area between 188 and 206.

(d) The point that has 80% of the area below it.

(e) The two points containing the middle 75% of the area.

3. Given the normally distributed variable X with mean 18 and standard deviation 2.5, find

(a) $P(X < 15)$.

(b) The value of k such that $P(X < k) = 0.2578$.

(c) $P(17 < X < 21)$.

(d) The value of k such that $P(X > k) = 0.1539$.

4. A soft drink machine is regulated so that it discharges an average of 207 milliliters per cup. If the amount of drink is normally distributed with standard deviation equal to 15 milliliters,

(a) What fraction of the cups will contain more than 231 milliliters?

(b) What is the probability that a cup contains between 198 and 216 milliliters?

(c) How many cups will likely overflow if 237-milliliter cups are used for the next 1000 drinks?

(d) Below what value do we get the smallest 25% of the drinks?

5. The finished inside diameter of a piston ring is normally distributed with a mean of 10 centimeters and a standard deviation of 0.03 centimeter.

(a) What proportion of rings will have inside diameters exceeding 10.075 centimeters?

(b) What is the probability that a piston ring will have an inside diameter between 9.97 and 10.03 centimeters?

(c) Below what value of inside diameter will 15% of the piston rings fall?

6. If a set of grades on a statistics examination are approximately normally distributed with a mean of 74 and a standard deviation of 7.9, find

(a) The lowest passing grade if the lowest 10% of the students are given F's.

(b) The highest B if the top 5% of the students are given A's.

7. The heights of 1000 students are normally distributed with a mean of 174.5 centimeters and a standard deviation of 6.9 centimeters. Assuming that the heights are recorded to the nearest half-centimeter, how many of these students would you expect to have heights

(a) Less than 160.0 centimeters?

(b) Between 171.5 and 182.0 centimeters inclusive?

(c) Equal to 175.0 centimeters?

(d) Greater than or equal to 188 centimeters?

interpret to mean inclusive

8. In a mathematics examination the average grade was 82 and the standard deviation was 5. All students with grades from 88 to 94 received a grade of B. If the grades are approximately normally distributed and eight students received a B grade, how many students took the examination?

9. A company pays its employees an average wage of $5.25 per hour with a standard deviation of 60 cents. If the wages are approximately normally distributed,

(a) What percent of the workers receive wages between $4.75 and $5.69 per hour inclusive?

(b) The highest 5% of the hourly wages are greater than what amount?

10. The tensile strength of a certain metal component is normally distributed with a mean of 10,000 kilograms per square centimeter and a standard deviation of 100 kilograms

per square centimeter. Measurements are recorded to the nearest 50 kilograms per square centimeter.

(a) What proportion of these components exceed 10,150 kilograms per square centimeter in tensile strength?

(b) If specifications require that all components have tensile strength between 9800 and 10,200 kilograms per square centimeter inclusive, what proportion of pieces would we expect to scrap?

11. If a set of observations are normally distributed, what percent of these differ from the mean by
   (a) More than 1.3σ?
   (b) Less than 0.52σ?

12. The IQ's of 600 applicants to a certain college are approximately normally distributed with a mean of 115 and a standard deviation of 12. If the college requires an IQ of at least 95, how many of these students will be rejected on this basis regardless of their other qualifications?

13. The average rainfall, recorded to the nearest hundredth of a centimeter, in Roanoke, Virginia, for the month of March is 9.22 centimeters. Assuming a normal distribution with a standard deviation of 2.83 centimeters, find the probability that next March Roanoke receives
   (a) Less than 1.84 centimeters of rain.
   (b) More than 5 centimeters but not over 7 centimeters.
   (c) More than 13.8 centimeters.

14. The average life of a certain type of small motor is 10 years with a standard deviation of 2 years. The manufacturer replaces free all motors that fail while under guarantee. If he is willing to replace only 3% of the motors that fail, how long a guarantee should he offer? Assume that the lives of the motors follow a normal distribution.

15. Find the error in approximating $\sum_{x=1}^{4} b(x; 20, 0.1)$ by the normal curve approximation.

16. A coin is tossed 400 times. Use the normal curve approximation to find the probability of obtaining
   (a) Between 185 and 210 heads inclusive.
   (b) Exactly 205 heads.
   (c) Less than 176 or more than 227 heads.

17. A pair of dice is rolled 180 times. What is the probability that a total of 7 occurs
   (a) At least 25 times?
   (b) Between 33 and 41 times inclusive?
   (c) Exactly 30 times?

18. The probability that a patient recovers from a delicate heart operation is 0.9. What is the probability that between 84 and 95 inclusive of the next 100 patients having this operation survive?

19. A drug manufacturer claims that a certain drug cures a blood disease on the average 80% of the time. To check the claim, government testers used the drug on a sample of 100 individuals and decided to accept the claim if 75 or more are cured.
   (a) What is the probability that the claim will be rejected when the cure probability is, in fact, 0.8?

(b) What is the probability that the claim will be accepted by the government when the cure probability is as low as 0.7?

20. A survey of the residents in a United States city showed that 20% preferred a white telephone over any other color available. What is the probability that between 170 and 185 inclusive of the next 1000 telephones installed in this city will be white?

21. One sixth of the male freshmen entering a large state school are out-of-state students. If the students are assigned at random to the dormitories, 180 to a building, what is the probability that in a given dormitory at least one fifth of the students are from out of state?

22. A certain pharmaceutical company knows that, on the average, 5% of a certain type of pill has an ingredient that is below the minimum strength and thus unacceptable. What is the probability that fewer than 10 in a sample of 200 pills will be unacceptable?

23. If a random variable X has the gamma distribution with $\alpha = 2$ and $\beta = 1$, find $P(1.8 < X < 2.4)$.

24. In a certain city, the daily consumption of water (in millions of liters) follows approximately a gamma distribution with $\alpha = 2$ and $\beta = 3$. If the daily capacity of this city is 9 million liters of water, what is the probability that on any given day the water supply is inadequate?

25. Use the gamma function, with $x = y^2/2$, to show that $\Gamma(1/2) = \sqrt{\pi}$.

26. The length of time for one individual to be served at a cafeteria is a random variable having an exponential distribution with a mean of 4 minutes. What is the probability that a person is served in less than 3 minutes on at least 4 of the next 6 days?

27. The life in years of a certain type of electrical switch has an exponential distribution with a failure rate of $\beta = 2$. If 100 of these switches are installed in different systems, what is the probability that at most 30 fail during the first year?

28. Derive the mean and variance of the Weibull distribution.

29. The lives of a certain automobile seal have the Weibull distribution with failure rate $Z(t) = 1/\sqrt{t}$. Find the probability that such a seal is still in use after 4 years.

30. The continuous random variable X has the *beta distribution* with parameters α and β, if its density function is given by

$$f(x) = \frac{\Gamma(\alpha + \beta)}{\Gamma(\alpha)\Gamma(\beta)} x^{\alpha - 1}(1 - x)^{\beta - 1}, \qquad 0 < x < 1$$

$$= 0, \qquad\qquad\qquad\qquad\qquad \text{elsewhere,}$$

where $\alpha > 0$ and $\beta > 0$. If the proportion of a brand of television sets requiring service during the first year of operation is a random variable having a beta distribution with $\alpha = 3$ and $\beta = 2$, what is the probability that at least 80% of the new models sold this year of this brand will require service during their first year of operation?

Functions of Random Variables

5.1 TRANSFORMATIONS OF VARIABLES

Frequently in statistics, one encounters the need to derive the probability distribution of a function of one or more random variables. For example, suppose that X is a discrete random variable with probability distribution $f(x)$ and suppose further that $Y = u(X)$ defines a one-to-one transformation between the values of X and Y. We wish to find the probability distribution of Y. It is important to note that the one-to-one transformation implies that each value x is related to one, and only one, value $y = u(x)$ and that each value y is related to one, and only one, value $x = w(y)$, where $w(y)$ is obtained by solving $y = u(x)$ for x in terms of y.

From our discussion of discrete probability distributions in Chapter 2 it is clear that the random variable Y assumes the value y when X assumes the value $w(y)$. Consequently, the probability distribution of Y is given by

$$g(y) = P(Y = y) = P[X = w(y)] = f[w(y)].$$

THEOREM 5.1 *Suppose that X is a discrete random variable with probability distribution $f(x)$. Let $Y = u(X)$ define a one-to-one transformation between the values of X and Y so that the equation $y = u(x)$ can be uniquely solved for x in terms of y, say $x = w(y)$. Then the probability distribution of Y is*

$$g(y) = f[w(y)].$$

Example 5.1 Let X be a geometric random variable with probability distribution $f(x) = \frac{3}{4}(\frac{1}{4})^{x-1}$, $x = 1, 2, 3, \ldots$. Find the probability distribution of the random variable $Y = X^2$.

Solution Since the values of X are all positive, the transformation defines a one-to-one correspondence between the x and y values, $y = x^2$ and $x = \sqrt{y}$. Hence

$$g(y) = f(\sqrt{y}) = \tfrac{3}{4}(\tfrac{1}{4})^{\sqrt{y}-1}, \qquad y = 1, 4, 9, \ldots$$
$$ = 0, \qquad\qquad\qquad\quad \text{elsewhere.}$$

Consider a problem where X_1 and X_2 are two discrete random variables with joint probability distribution $f(x_1, x_2)$ and we wish to find the joint probability distribution $g(y_1, y_2)$ of the two new random variables $Y_1 = u_1(X_1, X_2)$ and $Y_2 = u_2(X_1, X_2)$, which define a one-to-one transformation between the set of points (x_1, x_2) and (y_1, y_2). Solving the equations $y_1 = u_1(x_1, x_2)$ and $y_2 = u_2(x_1, x_2)$ simultaneously, we obtain the unique inverse solutions $x_1 = w_1(y_1, y_2)$ and $x_2 = w_2(y_1, y_2)$. Hence the random variables Y_1 and Y_2 assume the values y_1 and y_2, respectively, when X_1 assumes the value $w_1(y_1, y_2)$ and X_2 assumes the value $w_2(y_1, y_2)$. The joint probability distribution of Y_1 and Y_2 is then

$$g(y_1, y_2) = P(Y_1 = y_1, Y_2 = y_2)$$
$$= P[X_1 = w_1(y_1, y_2), X_2 = w_2(y_1, y_2)]$$
$$= f[w_1(y_1, y_2), w_2(y_1, y_2)].$$

THEOREM 5.2 *Suppose that X_1 and X_2 are discrete random variables with joint probability distribution $f(x_1, x_2)$. Let $Y_1 = u_1(X_1, X_2)$ and $Y_2 = u_2(X_1, X_2)$ define a one-to-one transformation between the points (x_1, x_2) and (y_1, y_2) so that the equations $y_1 = u_1(x_1, x_2)$ and $y_2 = u_2(x_1, x_2)$ may be uniquely solved for x_1 and x_2 in terms of y_1 and y_2, say $x_1 = w_1(y_1, y_2)$ and $x_2 = w_2(y_1, y_2)$. Then the joint probability distribution of Y_1 and Y_2 is*

$$g(y_1, y_2) = f[w_1(y_1, y_2), w_2(y_1, y_2)].$$

Theorem 5.2 is extremely useful in finding the distribution of some random variable $Y_1 = u_1(X_1, X_2)$, where X_1 and X_2 are discrete random variables with joint probability distribution $f(x_1, x_2)$. We simply define a second function, say $Y_2 = u_2(X_1, X_2)$, maintaining a one-to-one correspondence between the points (x_1, x_2) and (y_1, y_2), and obtain the joint probability distribution $g(y_1, y_2)$. The distribution of Y_1 is just the marginal distribution of $g(y_1, y_2)$, found by summing over the y_2 values. Denoting the distribution of Y_1 by $h(y_1)$, we can then write

$$h(y_1) = \sum_{y_2} g(y_1, y_2).$$

Example 5.2   Let X_1 and X_2 be two independent random variables having Poisson distributions with parameters μ_1 and μ_2, respectively. Find the distribution of the random variable $Y_1 = X_1 + X_2$.

Solution   Since X_1 and X_2 are independent, we can write

$$f(x_1, x_2) = f(x_1)f(x_2)$$
$$= \frac{e^{-\mu_1}\mu_1^{x_1}}{x_1!} \frac{e^{-\mu_2}\mu_2^{x_2}}{x_2!}$$
$$= \frac{e^{-(\mu_1 + \mu_2)}\mu_1^{x_1}\mu_2^{x_2}}{x_1!x_2!},$$

where $x_1 = 0, 1, 2, \ldots$ and $x_2 = 0, 1, 2, \ldots$. Let us now define a second random variable, say $Y_2 = X_2$. The inverse functions are given by $x_1 = y_1 - y_2$ and $x_2 = y_2$. Using Theorem 5.2, we find the joint probability distribution of Y_1 and Y_2 to be

$$g(y_1, y_2) = \frac{e^{-(\mu_1 + \mu_2)}\mu_1^{y_1 - y_2}\mu_2^{y_2}}{(y_1 - y_2)!y_2!},$$

where $y_1 = 0, 1, 2, \ldots$ and $y_2 = 0, 1, 2, \ldots, y_1$. Note that since $x_1 > 0$, the transformation $x_1 = y_1 - x_2$ implies that x_2 and hence y_2 must always be less than or equal to y_1. Consequently, the marginal probability distribution of Y_1 is

$$h(y_1) = \sum_{y_2=0}^{y_1} g(y_1, y_2)$$
$$= e^{-(\mu_1 + \mu_2)} \sum_{y_2=0}^{y_1} \frac{\mu_1^{y_1 - y_2}\mu_2^{y_2}}{(y_1 - y_2)!y_2!}$$
$$= \frac{e^{-(\mu_1 + \mu_2)}}{y_1!} \sum_{y_2=0}^{y_1} \frac{y_1!}{y_2!(y_1 - y_2)!} \mu_1^{y_1 - y_2}\mu_2^{y_2}$$
$$= \frac{e^{-(\mu_1 + \mu_2)}}{y_1!} \sum_{y_2=0}^{y_1} \binom{y_1}{y_2} \mu_1^{y_1 - y_2}\mu_2^{y_2}.$$

Recognizing this sum as the binomial expansion of $(\mu_1 + \mu_2)^{y_1}$, we obtain

$$h(y_1) = \frac{e^{-(\mu_1 + \mu_2)}(\mu_1 + \mu_2)^{y_1}}{y_1!}, \qquad y_1 = 0, 1, 2, \ldots,$$

from which we conclude that the sum of the two independent random variables having Poisson distributions, with parameters μ_1 and μ_2, has a Poisson distribution with parameter $\mu_1 + \mu_2$.

To find the probability distribution of the random variable $Y = u(X)$ when X is a continuous random variable and the transformation is one to one, we shall need Theorem 5.3.

THEOREM 5.3 *Suppose that X is a* continuous *random variable with probability distribution $f(x)$. Let $Y = u(X)$ define a one-to-one correspondence between the values of X and Y so that the equation $y = u(x)$ can be uniquely solved for x in terms of y, say $x = w(y)$. Then the probability distribution of Y is*

$$g(y) = f[w(y)]|J|,$$

where $J = w'(y)$ and is called the Jacobian *of the transformation.*

Proof 1. Suppose that $y = u(x)$ is an increasing function as in Figure 5.1. Then we see that whenever Y falls between a and b, the random variable X

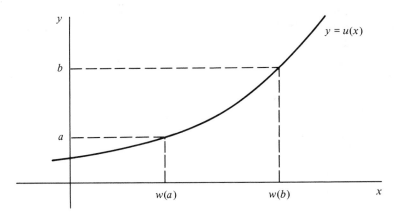

Figure 5.1 Increasing function.

must fall between $w(a)$ and $w(b)$. Hence

$$P(a < Y < b) = P[w(a) < X < w(b)]$$

$$= \int_{w(a)}^{w(b)} f(x)\, dx.$$

Changing the variable of integration from x to y by the relation $x = w(y)$, we obtain $dx = w'(y)\, dy$, and hence

$$P(a < Y < b) = \int_a^b f[w(y)]w'(y)\, dy.$$

Since the integral gives the desired probability for every $a < b$ within the permissible set of y values, then the probability distribution of Y is

$$g(y) = f[w(y)]w'(y) = f[w(y)]J.$$

If we recognize $J = w'(y)$ as the reciprocal of the slope of the tangent line to the curve of the increasing function $y = u(x)$, it is then obvious that $J = |J|$. Hence

$$g(y) = f[w(y)]|J|.$$

2. Suppose that $y = u(x)$ is a decreasing function as in Figure 5.2. Then we write

$$P(a < Y < b) = P[w(b) < X < w(a)]$$

$$= \int_{w(b)}^{w(a)} f(x)\, dx.$$

Again changing the variable of integration to y, we obtain

$$P(a < Y < b) = \int_b^a f[w(y)]w'(y)\, dy$$

$$= -\int_a^b f[w(y)]w'(y)\, dy,$$

from which we conclude that

$$g(y) = -f[w(y)]w'(y) = -f[w(y)]J.$$

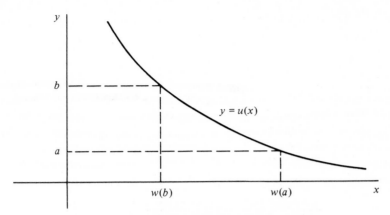

Figure 5.2 Decreasing function.

In this case the slope of the curve is negative and $J = -|J|$. Hence

$$g(y) = f[w(y)]|J|,$$

as before.

Example 5.3 Let X be a continuous random variable with probability distribution

$$f(x) = \frac{x}{12}, \qquad 1 < x < 5$$

$$= 0, \qquad \text{elsewhere.}$$

Find the probability distribution of the random variable $Y = 2X - 3$.

Solution The inverse solution of $y = 2x - 3$ yields $x = (y + 3)/2$, from which we obtain $J = dx/dy = 1/2$. Therefore, using Theorem 5.3, we find the density function of Y to be

$$g(y) = \frac{(y + 3)/2}{12}\left(\frac{1}{2}\right) = \frac{y + 3}{48}, \qquad -1 < y < 7$$

$$= 0, \qquad\qquad\qquad \text{elsewhere.}$$

To find the joint probability distribution of the random variables $Y_1 = u_1(X_1, X_2)$ and $Y_2 = u_2(X_1, X_2)$ when X_1 and X_2 are continuous and the transformation is one to one, we need an additional theorem, analogous to Theorem 5.2, which we state without proof.

THEOREM 5.4   *Suppose that* X_1 *and* X_2 *are continuous random variables with joint probability distribution* $f(x_1, x_2)$. *Let* $Y_1 = u_1(X_1, X_2)$ *and* $Y_2 = u_2(X_1, X_2)$ *define a one-to-one transformation between the points* (x_1, x_2) *and* (y_1, y_2) *so that the equations* $y_1 = u_1(x_1, x_2)$ *and* $y_2 = u_2(x_1, x_2)$ *may be uniquely solved for* x_1 *and* x_2 *in terms of* y_1 *and* y_2, *say* $x_1 = w_1(y_1, y_2)$ *and* $x_2 = w_2(y_1, y_2)$. *Then the joint probability distribution of* Y_1 *and* Y_2 *is*

$$g(y_1, y_2) = f[w_1(y_1, y_2), w_2(y_1, y_2)]|J|,$$

where the Jacobian is the 2×2 *determinant*

$$J = \begin{vmatrix} \partial x_1/\partial y_1 & \partial x_1/\partial y_2 \\ \partial x_2/\partial y_1 & \partial x_2/\partial y_2 \end{vmatrix}$$

and $\partial x_1/\partial y_1$ *is simply the derivative of* $x_1 = w_1(y_1, y_2)$ *with respect to* y_1 *with* y_2 *held constant, referred to in calculus as the partial derivative of* x_1 *with respect to* y_1. *The other partial derivatives are defined in a similar manner.*

Example 5.4   Let X_1 and X_2 be two continuous random variables with joint probability distribution

$$f(x_1, x_2) = 4x_1x_2, \qquad 0 < x_1 < 1, 0 < x_2 < 1$$
$$= 0, \qquad \text{elsewhere.}$$

Find the joint probability distribution of $Y_1 = X_1^2$ and $Y_2 = X_1 X_2$.

Solution   The inverse solutions of $y_1 = x_1^2$ and $y_2 = x_1 x_2$ are $x_1 = \sqrt{y_1}$ and $x_2 = y_2/\sqrt{y_1}$, from which we obtain

$$J = \begin{vmatrix} \frac{1}{2}\sqrt{y_1} & 0 \\ -y_2/2y_1^{3/2} & 1/\sqrt{y_1} \end{vmatrix} = \frac{1}{2y_1}.$$

The transformation is one to one, mapping the points $\{(x_1, x_2)|0 < x_1 < 1, 0 < x_2 < 1\}$ into the set $\{(y_1, y_2)|y_2^2 < y_1 < 1, 0 < y_2 < 1\}$. From Theorem

5.4 the joint probability distribution of Y_1 and Y_2 is

$$g(y_1, y_2) = 4(\sqrt{y_1})\frac{y_2}{\sqrt{y_1}}\frac{1}{2y_1}$$

$$= \frac{2y_2}{y_1}, \qquad y_2^2 < y_1 < 1, 0 < y_2 < 1$$

$$= 0, \qquad \text{elsewhere.}$$

Problems frequently arise when we wish to find the probability distribution of the random variable $Y = u(X)$ when X is a continuous random variable and the transformation is not one to one. That is, to each value x there corresponds exactly one value y, but to each y value there corresponds more than one x value. For example, suppose that $f(x)$ is positive over the interval $-1 < x < 2$ and zero elsewhere. Consider the transformation $y = x^2$. In this case $x = \pm\sqrt{y}$ for $0 < y < 1$ and $x = \sqrt{y}$ for $1 < y < 4$. For the interval $1 < y < 4$, the probability distribution of Y is found as before, using Theorem 5.3. That is,

$$g(y) = f[w(y)]|J| = \frac{f(\sqrt{y})}{2\sqrt{y}}, \qquad 1 < y < 4.$$

However, when $0 < y < 1$, we may partition the interval $-1 < x < 1$ to obtain the two inverse functions

$$x = -\sqrt{y}, \qquad -1 < x < 0$$

$$= \sqrt{y}, \qquad 0 < x < 1.$$

Then to every y value there corresponds a single x value for each partition. From Figure 5.3 we see that

$$P(a < Y < b) = P(-\sqrt{b} < X < -\sqrt{a}) + P(\sqrt{a} < X < \sqrt{b})$$

$$= \int_{-\sqrt{b}}^{-\sqrt{a}} f(x)\, dx + \int_{\sqrt{a}}^{\sqrt{b}} f(x)\, dx.$$

Changing the variable of integration from x to y, we obtain

$$P(a < Y < b) = \int_b^a f(-\sqrt{y})J_1\, dy + \int_a^b f(\sqrt{y})J_2\, dy$$

$$= -\int_a^b f(-\sqrt{y})J_1\, dy + \int_a^b f(\sqrt{y})J_2\, dy,$$

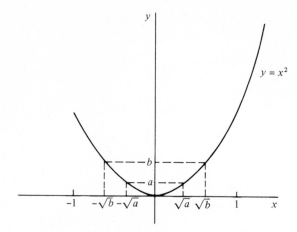

Figure 5.3 Decreasing and increasing function.

where

$$J_1 = \frac{d(-\sqrt{y})}{dy} = \frac{-1}{2\sqrt{y}} = -|J_1|$$

and

$$J_2 = \frac{d(\sqrt{y})}{dy} = \frac{1}{2\sqrt{y}} = |J_2|.$$

Hence we can write

$$P(a < Y < b) = \int_a^b [f(-\sqrt{y})|J_1| + f(\sqrt{y})|J_2|]\, dy,$$

and then

$$g(y) = f(-\sqrt{y})|J_1| + f(\sqrt{y})|J_2|$$
$$= \frac{[f(-\sqrt{y}) + f(\sqrt{y})]}{2\sqrt{y}}, \qquad 0 < y < 1.$$

The probability distribution of Y for $0 < y < 4$ may now be written

$$g(y) = \frac{[f(-\sqrt{y}) + f(\sqrt{y})]}{2\sqrt{y}}, \qquad 0 < y < 1$$

$$= \frac{f(\sqrt{y})}{2\sqrt{y}}, \qquad 1 < y < 4$$

$$= 0, \qquad \text{elsewhere.}$$

This procedure for finding $g(y)$ when $0 < y < 1$ is generalized in Theorem 5.5 for k inverse functions. For transformations not one to one of functions of several variables, the reader is referred to *Introduction to Mathematical Statistics* by Hogg and Craig (see the Bibliography).

THEOREM 5.5   *Suppose that X is a continuous random variable with probability distribution $f(x)$. Let $Y = u(X)$ define a transformation between the values of X and Y that is not one to one. If the interval over which X is defined can be partitioned into k mutually disjoint sets such that each of the inverse functions $x_1 = w_1(y)$, $x_2 = w_2(y)$, ..., $x_k = w_k(y)$ of $y = u(x)$ defines a one-to-one correspondence, then the probability distribution of Y is*

$$g(y) = \sum_{i=1}^{k} f[w_i(y)] |J_i|,$$

where $J_i = w_i'(y)$, $i = 1, 2, \ldots, k$.

Example 5.5   Show that $Y = (X - \mu)^2/\sigma^2$ has a chi-square distribution with 1 degree of freedom when X has a normal distribution with mean μ and variance σ^2.

Solution   Let $Z = (X - \mu)/\sigma$, where the random variable Z has the standard normal distribution

$$f(z) = \frac{1}{\sqrt{2\pi}} e^{-z^2/2}, \qquad -\infty < z < \infty.$$

We shall now find the distribution of the random variable $Y = Z^2$. The inverse solutions of $y = z^2$ are $z = \pm\sqrt{y}$. Designating $z_1 = -\sqrt{y}$ and $z_2 = \sqrt{y}$, $J_1 = -1/2\sqrt{y}$ and $J_2 = 1/2\sqrt{y}$. Hence, by Theorem 5.5, we have

$$g(y) = \frac{1}{\sqrt{2\pi}} e^{-y/2} \left| \frac{-1}{2\sqrt{y}} \right| + \frac{1}{\sqrt{2\pi}} e^{-y/2} \left| \frac{1}{2\sqrt{y}} \right|$$

$$= \frac{1}{2^{1/2}\sqrt{\pi}} y^{1/2-1} e^{-y/2}, \qquad y > 0.$$

Since $g(y)$ is a density function, it follows that

$$1 = \frac{1}{2^{1/2}\sqrt{\pi}} \int_0^\infty y^{1/2-1} e^{-y/2} \, dy$$

$$= \frac{\Gamma(1/2)}{\sqrt{\pi}} \int_0^\infty \frac{1}{2^{1/2}\Gamma(1/2)} y^{1/2-1} e^{-y/2} \, dy$$

$$= \frac{\Gamma(1/2)}{\sqrt{\pi}},$$

the integral being the area under a gamma probability curve with parameters $\alpha = 1/2$ and $\beta = 2$. Therefore, $\sqrt{\pi} = \Gamma(1/2)$ and the probability distribution of Y is given by

$$g(y) = \frac{1}{2^{1/2}\Gamma(1/2)} y^{1/2-1} e^{-y/2}, \qquad y > 0,$$

which is seen to be a chi-square distribution with 1 degree of freedom.

5.2 MOMENT-GENERATING FUNCTIONS

Although the method of transforming variables provides an effective way of finding the distribution of a function of several random variables, there is an alternative and often preferred procedure when the function in question is the sum of independent random variables. We shall refer to this method as the *moment-generating function* technique.

DEFINITION 5.1   *The moment-generating function of the random variable X is given by $E(e^{tX})$ and is denoted by $M_X(t)$. Hence*

$$M_X(t) = E(e^{tX})$$

$$= \sum_{i=1}^{n} e^{tx_i} f(x_i) \qquad \text{if X is discrete}$$

$$= \int_{-\infty}^{\infty} e^{tx} f(x)\, dx \qquad \text{if X is continuous.}$$

Moment-generating functions will exist only if the sum or integral of Definition 5.1 converges. If a moment-generating function of a random variable X does exist, it can be used to generate all the moments of that variable. The method is described in Theorem 5.6.

THEOREM 5.6   *Let X be a random variable with moment-generating function $M_X(t)$. Then*

$$\left. \frac{d^r M_X(t)}{dt^r} \right|_{t=0} = \mu_r'.$$

Proof   Assuming that we can differentiate inside summation and integral signs, we obtain

$$\frac{d^r M_X(t)}{dt^r} = \sum_{i=1}^{n} x_i^r e^{tx_i} f(x_i) \qquad \text{(discrete case)}$$

$$= \int_{-\infty}^{\infty} x^r e^{tx} f(x)\, dx \qquad \text{(continuous case)}.$$

Setting $t = 0$, we see that both cases reduce to $E(X^r) = \mu_r'$.

Example 5.6   Find the moment-generating function of the binomial random variable X and then use it to verify that $\mu = np$ and $\sigma^2 = npq$.

Solution   From Definition 5.1 we have

$$M_X(t) = \sum_{x=0}^{n} e^{tx} \binom{n}{x} p^x q^{n-x}$$

$$= \sum_{x=0}^{n} \binom{n}{x} (pe^t)^x q^{n-x}.$$

Recognizing this last sum as the binomial expansion of $(pe^t + q)^n$, we obtain

$$M_X(t) = (pe^t + q)^n.$$

Now

$$\frac{dM_X(t)}{dt} = n(pe^t + q)^{n-1} pe^t$$

and

$$\frac{d^2 M_X(t)}{dt^2} = np[e^t(n-1)(pe^t + q)^{n-2} pe^t + (pe^t + q)^{n-1} e^t].$$

Setting $t = 0$, we get

$$\mu_1' = np \quad\quad \text{and} \quad\quad \mu_2' = np[(n-1)p + 1].$$

Therefore,

$$\mu = \mu_1' = np$$

and

$$\sigma^2 = \mu_2' - \mu^2 = np(1 - p) = npq,$$

which agrees with the results obtained in Chapter 3.

Example 5.7   Show that the moment-generating function of the random variable X having a normal probability distribution with mean μ and variance σ^2 is given by $M_X(t) = e^{\mu t + \sigma^2 t^2/2}$.

Solution   From Definition 5.1 the moment-generating function of the normal random variable X is

$$M_X(t) = \int_{-\infty}^{\infty} e^{tx} \frac{1}{\sqrt{2\pi}\sigma} e^{-(1/2)[(x-\mu)/\sigma]^2} \, dx$$

$$= \int_{-\infty}^{\infty} \frac{1}{\sqrt{2\pi}\sigma} e^{-[x^2 - 2(\mu + t\sigma^2)x + \mu^2]/2\sigma^2} \, dx.$$

Completing the square in the exponent, we can write

$$x^2 - 2(\mu + t\sigma^2)x + \mu^2 = [x - (\mu + t\sigma^2)]^2 - 2\mu t\sigma^2 - t^2\sigma^4$$

and then

$$M_X(t) = \int_{-\infty}^{\infty} \frac{1}{\sqrt{2\pi}\sigma} e^{-\{[x-(\mu+t\sigma^2)]^2 - 2\mu t\sigma^2 - t^2\sigma^4\}/2\sigma^2} \, dx$$

$$= e^{\mu t + \sigma^2 t^2/2} \int_{-\infty}^{\infty} \frac{1}{\sqrt{2\pi}\sigma} e^{-(1/2)\{[x-(\mu+t\sigma^2)]/\sigma\}^2} \, dx.$$

Let $w = [x - (\mu + t\sigma^2)]/\sigma$; then $dx = \sigma \, dw$ and

$$M_X(t) = e^{\mu t + \sigma^2 t^2/2} \int_{-\infty}^{\infty} \frac{1}{\sqrt{2\sigma}} e^{-w^2/2} \, dw$$

$$= e^{\mu t + \sigma^2 t^2/2},$$

since the last integral represents the area under a standard normal density curve and hence equals 1.

Example 5.8   Show that the moment-generating function of the random variable X having a chi-square distribution with v degrees of freedom is $M_X(t) = (1 - 2t)^{-v/2}$.

Solution   The chi-square distribution was obtained as a special case of the gamma distribution by setting $\alpha = v/2$ and $\beta = 2$. Substituting for $f(x)$ in Definition 5.1, we obtain

$$M_X(t) = \int_0^{\infty} e^{tx} \frac{1}{2^{v/2}\Gamma(v/2)} x^{v/2 - 1} e^{-x/2} \, dx$$

$$= \frac{1}{2^{v/2}\Gamma(v/2)} \int_0^{\infty} x^{v/2 - 1} e^{-x(1 - 2t)/2} \, dx.$$

Writing $y = x(1 - 2t)/2$ and $dx = [2/(1 - 2t)]\, dy$, we get

$$M_X(t) = \frac{1}{2^{v/2}\Gamma(v/2)} \int_0^\infty \left(\frac{2y}{1 - 2t}\right)^{v/2 - 1} e^{-y} \frac{2}{1 - 2t}\, dy$$

$$= \frac{1}{\Gamma(v/2)(1 - 2t)^{v/2}} \int_0^\infty y^{v/2 - 1} e^{-y}\, dy$$

$$= (1 - 2t)^{-v/2},$$

since the last integral equals $\Gamma(v/2)$.

The properties of moment-generating functions discussed in the following four theorems will be of particular importance in determining the distributions of linear combinations of independent random variables. In keeping with the mathematical scope of this text, we state Theorem 5.7 without proof.

THEOREM 5.7 (UNIQUENESS THEOREM) *Let X and Y be two random variables with moment-generating functions $M_X(t)$ and $M_Y(t)$, respectively. If $M_X(t) = M_Y(t)$ for all values of t, then X and Y have the same probability distribution.*

THEOREM 5.8 $M_{X+a}(t) = e^{at}M_X(t)$.

Proof

$$M_{X+a}(t) = E[e^{t(X + a)}]$$
$$= e^{at}E(e^{tX}) = e^{at}M_X(t).$$

THEOREM 5.9 $M_{aX}(t) = M_X(at)$.

Proof

$$M_{aX}(t) = E[e^{t(aX)}] = E[e^{(at)X}]$$
$$= M_X(at).$$

THEOREM 5.10   *If X_1 and X_2 are independent random variables with moment-generating functions $M_{X_1}(t)$ and $M_{X_2}(t)$, respectively, and $Y = X_1 + X_2$, then*

$$M_Y(t) = M_{X_1 + X_2}(t) = M_{X_1}(t)M_{X_2}(t).$$

Proof

$$M_Y(t) = E(e^{tY}) = E[e^{t(X_1 + X_2)}]$$

$$= \int_{-\infty}^{\infty} \int_{-\infty}^{\infty} e^{t(x_1 + x_2)} f(x_1, x_2)\, dx_1\, dx_2.$$

Since the variables are independent, we have $f(x_1, x_2) = g(x_1)h(x_2)$ and then

$$M_Y(t) = \int_{-\infty}^{\infty} e^{tx_1} g(x_1)\, dx_1 \int_{-\infty}^{\infty} e^{tx_2} h(x_2)\, dx_2$$

$$= M_{X_1}(t)M_{X_2}(t).$$

The proof for the discrete case is obtained in a similar manner by replacing integrals with summations.

One might use Theorems 5.7 and 5.10 along with the result of Exercise 14 as an alternative method to Example 5.2 in finding the distribution of the sum of two independent Poisson random variables. For example, if X_1 and X_2 are independent Poisson variables with moment-generating functions given by

$$M_{X_1}(t) = e^{\mu_1(e^t - 1)} \qquad \text{and} \qquad M_{X_2}(t) = e^{\mu_2(e^t - 1)},$$

respectively, then according to Theorem 5.10, the moment-generating function of the random variable $Y_1 = X_1 + X_2$ is

$$M_{Y_1}(t) = M_{X_1}(t)M_{X_2}(t)$$
$$= e^{\mu_1(e^t - 1)}e^{\mu_2(e^t - 1)}$$
$$= e^{(\mu_1 + \mu_2)(e^t - 1)},$$

which we immediately identify as the moment-generating function of a random variable having a Poisson distribution with the parameter $\mu_1 + \mu_2$. Hence, according to Theorem 5.7, we again conclude that the sum of two independent random variables having Poisson distributions, with parameters μ_1 and μ_2, has a Poisson distribution with parameter $\mu_1 + \mu_2$.

In applied statistics one frequently needs to know the probability distribution of a linear combination of independent normal random variables. Let us obtain the distribution of the random variable $Y = a_1 X_1 + a_2 X_2$ when X_1 is a normal variable with mean μ_1 and variance σ_1^2 and X_2 is also a normal variable but independent of X_1, with mean μ_2 and variance σ_2^2. First, by Theorem 5.10, we find

$$M_Y(t) = M_{a_1 X_1}(t) M_{a_2 X_2}(t),$$

and then, using Theorem 5.9,

$$M_Y(t) = M_{X_1}(a_1 t) M_{X_2}(a_2 t).$$

Substituting $a_1 t$ for t in the moment-generating function of the normal distribution derived in Example 5.7 and then $a_2 t$ for t, we have

$$M_Y(t) = e^{a_1 \mu_1 t + a_1^2 \sigma_1^2 t^2/2} e^{a_2 \mu_2 t + a_2^2 \sigma_2^2 t^2/2}$$
$$= e^{(a_1 \mu_1 + a_2 \mu_2)t + (a_1^2 \sigma_1^2 + a_2^2 \sigma_2^2)t^2/2},$$

which we recognize as the moment-generating function of a distribution that is normal with mean $a_1 \mu_1 + a_2 \mu_2$ and variance $a_1^2 \sigma_1^2 + a_2^2 \sigma_2^2$.

Generalizing to the case of n independent normal variables, we state the following result.

THEOREM 5.11   *If $X_1, X_2, \ldots, X_n$ are independent random variables having normal distributions with means $\mu_1, \mu_2, \ldots, \mu_n$ and variances $\sigma_1^2, \sigma_2^2, \ldots, \sigma_n^2$, respectively, then the random variable*

$$Y = a_1 X_1 + a_2 X_2 + \cdots + a_n X_n$$

has a normal distribution with mean

$$\mu_Y = a_1 \mu_1 + a_2 \mu_2 + \cdots + a_n \mu_n$$

and variance

$$\sigma_Y^2 = a_1^2 \sigma_1^2 + a_2^2 \sigma_2^2 + \cdots + a_n^2 \sigma_n^2.$$

It is now evident that the Poisson distribution and the normal distribution possess a reproductive property in that the sums of independent random variables having either of these distributions is a random variable that also has the same type of distribution. This reproductive property is also possessed by the chi-square distribution.

THEOREM 5.12   *If* $X_1, X_2, \ldots, X_n$ *are mutually independent random variables that have, respectively, chi-square distributions with* $v_1, v_2, \ldots, v_n$ *degrees of freedom, then the random variable*

$$Y = X_1 + X_2 + \cdots + X_n$$

has a chi-square distribution with $v = v_1 + v_2 + \cdots + v_n$ *degrees of freedom.*

Proof   By Theorem 5.10,

$$M_Y(t) = M_{X_1}(t)M_{X_2}(t) \cdots M_{X_n}(t).$$

From Example 5.8,

$$M_{X_i}(t) = (1 - 2t)^{-v_i/2}, \qquad i = 1, 2, \ldots, n.$$

Therefore,

$$M_Y(t) = (1 - 2t)^{-v_1/2}(1 - 2t)^{-v_2/2} \cdots (1 - 2t)^{-v_n/2}$$
$$= (1 - 2t)^{-(v_1 + v_2 + \cdots + v_n)/2},$$

which we recognize as the moment-generating function of a chi-square distribution with $v = v_1 + v_2 + \cdots + v_n$ degrees of freedom.

COROLLARY   *If* $X_1, X_2, \ldots, X_n$ *are independent random variables having identical normal distributions with mean* μ *and variance* σ^2, *then the random variable*

$$Y = \sum_{i=1}^{n} \left(\frac{X_i - \mu}{\sigma} \right)^2$$

has a chi-square distribution with $v = n$ *degrees of freedom.*

This corollary is an immediate consequence of Example 5.5, which states that each of the n independent random variables $[(X_i - \mu)/\sigma]^2, i = 1, 2, \ldots, n,$ has a chi-square distribution with 1 degree of freedom.

A 5.3 RANDOM SAMPLING

The outcome of a statistical experiment may be recorded either as a numerical value or as a descriptive representation. When a pair of dice are tossed and the total is the outcome of interest, we record a numerical value. However, if the students in a certain school are given blood tests and the type of blood is of interest, then a descriptive representation might be the most useful. A person's blood can be classified in eight ways. It must be AB, A, B, or O, with a plus or minus sign, depending on the presence or absence of the Rh antigen.

The statistician works primarily with numerical observations. For the experiment involving the blood types he will probably let numbers 1 to 8 represent each blood type and then record the appropriate number for each student. In the classification of blood types we can have only as many observations as there are students in the school. The experiment, therefore, results in a finite number of observations. In the die-tossing experiment we are interested in recording the total that occurs. Therefore, if we toss the dice indefinitely, we obtain an infinite set of values, each representing the result of a single toss of a pair of dice.

The totality of observations with which we are concerned, whether finite or infinite, constitute what we call a *population*. In past years the word *population* referred to observations obtained from statistical studies involving people. Today the statistician uses the term to refer to observations relevant to anything of interest, whether it be groups of people, animals, or objects.

DEFINITION 5.2   *A* population *consists of the totality of the observations with which we are concerned.*

The number of observations in the population is defined to be the *size* of the population. If there are 600 students in the school that are classified according to blood type, we say we have a population of size 600. The die-tossing experiment generates a population whose size is infinite. The numbers on the cards in a deck, the heights of residents in a certain city, and the lengths of fish in a particular lake are examples of populations with finite size. In each case the total number of observations is a finite number. The observations obtained by measuring the atmospheric pressure every day from the past on into the future or all measurements on the depth of a lake from any conceivable position are examples of populations whose sizes are infinite. Some finite populations are so large that in theory we assume them to be infinite. This is true if you consider the population of lives of a certain type of storage battery being manufactured for mass distribution throughout the country.

Each observation in a population is a value of a random variable X having some probability distribution $f(x)$. If one is inspecting items coming off an assembly line for defects, then each observation in the population might be a value zero or 1 of the binomial random variable X with probability distribution

$$b(x; 1, p) = p^x q^{1-x}, \qquad x = 0, 1,$$

where zero indicates a nondefective item and 1 indicates a defective item. Of course, it is assumed that p, the probability of any item being defective, remains constant from trial to trial. In the blood-type experiment the random variable X represents the type of blood by assuming a value from 1 to 8. Each student is given one of the values of the discrete random variable. The lives of the storage batteries are values assumed by a continuous random variable having perhaps a normal distribution. When we speak hereafter about a "binomial population," a "normal population," or, in general, the "population $f(x)$," we shall mean a population whose observations are values of a random variable having a binomial distribution, a normal distribution, or the probability distribution $f(x)$. Hence the mean and variance of a random variable or probability distribution are also referred to as the mean and variance of the corresponding population.

The statistician is interested in arriving at conclusions concerning unknown population parameters. In a normal population, for example, the parameters μ and σ^2 may be unknown and are to be estimated from the information provided by a sample selected from the population. This takes us into the theory of sampling. If our inferences are to be accurate, we must understand the relation of a sample to its population. Certainly, the sample should be representative of the population. It should be a *random sample* in the sense that the observations are made independently and at random.

In selecting a random sample of size n from a population $f(x)$, let us define the random variable X_i, $i = 1, 2, \ldots, n$, to represent the ith measurement or sample value that we observe. The random variables $X_1, X_2, \ldots, X_n$ will then constitute a random sample from the population $f(x)$ with numerical values $x_1, x_2, \ldots, x_n$ if the measurements are obtained by repeating the experiment n independent times under essentially the same conditions. Because of the identical conditions under which the elements of the sample are selected, it is reasonable to assume that the n random variables $X_1, X_2, \ldots, X_n$ are independent and that each has the same probability distribution $f(x)$. That is, the probability distributions of $X_1, X_2, \ldots, X_n$ are, respectively, $f(x_1), f(x_2)$, $\ldots, f(x_n)$ and their joint probability distribution is

$$f(x_1, x_2, \ldots, x_n) = f(x_1)f(x_2) \cdots f(x_n).$$

The concept of a random sample is defined formally in the following definition.

DEFINITION 5.3   *Let $X_1, X_2, \ldots, X_n$ be n independent random variables each having the same probability distribution $f(x)$. We then define $X_1, X_2, \ldots, X_n$ to be a random sample of size n from the population $f(x)$ and write its joint probability distribution as*

$$f(x_1, x_2, \ldots, x_n) = f(x_1)f(x_2) \cdots f(x_n).$$

If one makes a random selection of $n = 8$ storage batteries from a manufacturing process, which has maintained the same specifications, and records the length of life for each battery with the first measurement x_1 being a value of X_1, the second measurement x_2 a value of X_2, and so forth, then $x_1, x_2, \ldots, x_8$ are the values of the random sample $X_1, X_2, \ldots, X_8$. If we assume the population of battery lives to be normal, the possible values of any X_i, $i = 1, 2, \ldots, 8$, will be precisely the same as those in the original population, and hence X_i has the same identical normal distribution as X.

5.4 SAMPLING THEORY

Our main purpose in selecting random samples is to elicit information about the unknown population parameters. Suppose that we wish to arrive at a conclusion concerning the proportion of people in the United States who prefer a certain brand of coffee. It would be impossible to question every American and compute the parameter representing the true proportion. Instead, a large random sample is selected and the proportion of this sample favoring the brand of coffee in question is calculated. This value is now used to make some inference concerning the true proportion.

A value computed from a sample is called a *statistic*. Since many random samples are possible from the same population, we would expect the statistic to vary somewhat from sample to sample. Hence a statistic is a *random variable*.

DEFINITION 5.4   *A statistic is a random variable that depends only on the observed random sample.*

A statistic is usually represented by ordinary Latin letters. The sample proportion in the preceding illustration is a statistic that is commonly represented by $\hat{P}$. The value of the random variable $\hat{P}$ for the given sample is denoted by $\hat{p}$. To use $\hat{p}$ to estimate, with some degree of accuracy, the true proportion p of people in the United States who prefer the given brand of coffee, we must first know more about the probability distribution of the statistic $\hat{P}$.

In Chapter 2 we introduced the two parameters μ and σ^2, which measure the center and the variability of a probability distribution. These are constant population parameters and are in no way affected or influenced by the observations of a random sample. We shall, however, define some important statistics that describe corresponding measures of a random sample. The most commonly used statistics for measuring the center of a set of data, arranged in order of magnitude, are the *mean*, *median*, and *mode*. The most important of these and the one we shall consider first is the mean.

DEFINITION 5.5   *If* $X_1, X_2, \ldots, X_n$ *represent a random sample of size n, then the* sample mean *is defined by the statistic*

$$\bar{X} = \frac{\sum\limits_{i=1}^{n} X_i}{n}.$$

Note that the statistic $\bar{X}$ assumes the value $\bar{x} = \sum\limits_{i=1}^{n} x_i/n$ when X_1 assumes the value x_1, X_2 assumes the value x_2, and so forth.

Example 5.9   Find the mean of the random sample whose observations are 20, 27, and 25.

Solution   The observed value $\bar{x}$ of the statistic $\bar{X}$ is

$$\bar{x} = \frac{20 + 27 + 25}{3} = 24.$$

The second most useful statistic for measuring the center of a set of data is the median. We shall designate the median by the symbol $\tilde{X}$.

DEFINITION 5.6 *If* $X_1, X_2, \ldots, X_n$ *represent a random sample of size n, arranged in increasing order of magnitude, then the* sample median *is defined by the statistic*

$$\tilde{X} = X_{(n+1)/2} \qquad \qquad \text{if n is odd}$$

$$= \frac{X_{n/2} + X_{(n/2)+1}}{2} \qquad \text{if n is even.}$$

Example 5.10 Find the median for the random sample whose observations are 8, 3, 9, 5, 6, 8, and 5.

Solution Arranging the observations in order of magnitude, 3, 5, 5, 6, 8, 8, 9, gives $\tilde{x} = 6$.

Example 5.11 Find the median for the random sample whose observations are 10, 8, 4, and 7.

Solution Arranging the observations in order of magnitude, 4, 7, 8, 10, the median is the arithmetic mean of the two middle values. Therefore, $\tilde{x} = (7 + 8)/2 = 7.5$.

The third and final statistic for measuring the center of a random sample that we shall discuss is the mode, designated by the statistic M.

DEFINITION 5.7 *If* $X_1, X_2, \ldots, X_n$, *not necessarily all different, represent a random sample of size n, then the* mode M *is that value of the sample that occurs most often or with the greatest frequency. The mode may not exist, and when it does it is not necessarily unique.*

Example 5.12 The mode of the random sample whose observations are 2, 4, 4, 5, 6, 6, 6, 7, 7, and 8 is $m = 6$.

Example 5.13 The observations 3, 4, 4, 4, 4, 6, 7, 7, 8, 8, 8, 8, and 9 have two modes, 4 and 8, since both 4 and 8 occur with the greatest frequency. The distribution of the sample is said to be *bimodal*.

When the mode could be either of two adjacent numbers arranged in order of magnitude, we take the arithmetic mean of the two numbers as the mode.

Therefore, the modes of the observations 3, 5, 5, 5, 6, 6, 6, 7, 9, 9, and 9 are $(5 + 6)/2 = 5.5$ and 9.

In summary, let us consider the relative merits of the mean, median, and mode. The mean is the most commonly used measure of central tendency in statistics. It is easy to calculate and it employs all available information. The distributions of sample means are well known, and consequently the methods used in statistical inference are based on the sample mean. The only real disadvantage to the mean is that it may be affected adversely by extreme values. That is, if most contributions to a charity are less than $5, then a very large contribution, say $10,000, would produce an average donation that is considerably higher than the majority of gifts.

The median has the advantage of being easy to compute. It is not influenced by extreme values and would give a truer average in the case of the charitable contributions. In dealing with samples selected from populations, the sample means will not vary as much from sample to sample as will the medians. Therefore, if we are attempting to estimate the center of a population based on a sample value, the mean is more stable than the median. Hence a sample mean is likely to be closer to the population mean than the sample median would be to the population median.

The mode is the least used measure of the three. For small sets of data its value is almost useless, if in fact it exists at all. Only in the case of a large mass of data does it have a significant meaning. Its only advantage is that it requires no calculation.

The three statistics defined above do not by themselves give an adequate description of the distribution of our data. We need to know how the observations spread out from the average. It is quite possible to have two sets of observations with the same mean or median that differ considerably in the variability or their measurements about the average.

Consider the following measurements, in centiliters, for two samples of orange juice bottled by two different companies, A and B:

Sample A	75	80	74	83	86
Sample B	86	80	69	71	94

Both samples have the same mean, 80. It is quite obvious that company A bottles orange juice with a more uniform content than company B. We say the variability or the dispersion of the observations from the average is less for sample A than for sample B. Therefore, in buying orange juice we would feel more confident that the bottle we select will be closer to the advertised average if we buy from company A.

The most important statistics for measuring the variability of a random sample are the *range* and the *variance*. The simplest of these to compute is the range.

DEFINITION 5.8   *The range of a random sample* $X_1, X_2, \ldots, X_n$, *arranged in increasing order of magnitude, is defined by the statistic* $X_n - X_1$.

Example 5.14   The range of the set of observations 10, 12, 12, 18, 19, 22, and 24 is $24 - 10 = 14$.

In the case of the companies bottling orange juice, the range for company A is 12 compared to a range of 25 for company B, indicating a greater spread in the values for company B.

The range is a poor measure of variability, particularly if the size of the sample is large. It considers only the extreme values and tells us nothing about the distribution of values in between. Consider, for example, the following two sets of data both with a range of 12:

$$3, \quad 4, \quad 5, \quad 6, \quad 8, \quad 9, \quad 10, \quad 12, \quad 15$$
$$3, \quad 8, \quad 8, \quad 9, \quad 9, \quad 9, \quad 10, \quad 10, \quad 15.$$

In the first set the mean and median are both 8, but the numbers vary over the entire interval from 3 to 15. In the second set the mean and median are both 9, but most of the values are closer to the average. Although the range fails to measure this variability between the upper and lower observations, it does have some useful applications. In industry the range for measurements on items coming off an assembly line might be specified in advance. As long as all measurements fall within the specified range the process is said to be in control.

To overcome the disadvantage of the range, we shall consider a measure of variability, the *sample variance*, that considers the position of each observation relative to the sample mean.

DEFINITION 5.9   *If* $X_1, X_2, \ldots, X_n$ *represent a random sample of size n, then the* sample variance *is defined by the statistic*

$$S^2 = \frac{\sum\limits_{i=1}^{n} (X_i - \bar{X})^2}{n - 1}.$$

The computed value of S^2 for a given sample is denoted by s^2. Note that S^2 is essentially defined to be the average of the squares of the deviations of the observations from their mean. The reason for using $n - 1$ as a divisor rather than the more obvious choice n will become apparent in Chapter 6.

THEOREM 5.13   *If S^2 is the variance of a random sample of size n, we may write*

$$S^2 = \frac{n \sum\limits_{i=1}^{n} X_i^2 - \left(\sum\limits_{i=1}^{n} X_i \right)^2}{n(n-1)}.$$

Proof   By definition,

$$S^2 = \frac{\sum\limits_{i=1}^{n} (X_i - \bar{X})^2}{n-1}$$

$$= \frac{\sum\limits_{i=1}^{n} (X_i^2 - 2\bar{X}X_i + \bar{X}^2)}{n-1}$$

$$= \frac{\sum\limits_{i=1}^{n} X_i^2 - 2\bar{X} \sum\limits_{i=1}^{n} X_i + n\bar{X}^2}{n-1}.$$

Replacing $\bar{X}$ by $\sum\limits_{i=1}^{n} X_i/n$ and multiplying numerator and denominator by n, we obtain the more useful computational formula

$$S^2 = \frac{n \sum\limits_{i=1}^{n} X_i^2 - \left(\sum\limits_{i=1}^{n} X_i \right)^2}{n(n-1)}.$$

The sample standard deviation, denoted by S, is defined to be the positive square root of the sample variance.

Example 5.15   Find the variance of the sample whose observations are 3, 4, 5, 6, 6, and 7.

Solution   We find that $\sum\limits_{i=1}^{6} x_i^2 = 171$, $\sum\limits_{i=1}^{6} x_i = 31$, $n = 6$. Hence

$$s^2 = \frac{(6)(171) - (31)^2}{(6)(5)} = \frac{13}{6}.$$

The field of inductive statistics is basically concerned with generalizations and predictions. Generalizations from a statistic to a parameter can be made with confidence only if we understand the fluctuating behavior of our statistic when computed for different random samples from the same population. The distribution of the statistic in question will depend on the size of the population, the size of the samples, and the method of choosing the random samples. If the size of the population is large or infinite, the statistic has the same distribution whether we sample with or without replacement. On the other hand, sampling with replacement from a small finite population gives a slightly different distribution for the statistic than if we sample without replacement. Sampling with replacement from a finite population is equivalent to sampling from an infinite population, since there is no limit on the possible size of the sample selected.

DEFINITION 5.10 *The probability distribution of a statistic is called a* sampling distribution.

DEFINITION 5.11 *The standard deviation of the sampling distribution of a statistic is called the* standard error *of the statistic.*

The probability distribution of $\bar{X}$ is called the *sampling distribution of the mean*, and the standard error of the mean is the standard deviation of the sampling distribution of $\bar{X}$. Every sample of size n selected from a specified population provides a value s of the statistic S, the sample standard deviation. The standard error of the sample standard deviation is then the standard deviation of the statistic S.

In the remainder of this chapter we shall study several important sampling distributions of frequently used statistics. The applications of these sampling distributions to problems of statistical inference will be considered in Chapters 6 and 7.

5.5 SAMPLING DISTRIBUTIONS OF MEANS

The first important sampling distribution to be considered is that of the mean $\bar{X}$. Suppose that a random sample of n observations is taken from a normal population with mean μ and variance σ^2. Each observation X_i, $i = 1, 2, \ldots, n$, of the random sample will then have the same normal distribution as the population being sampled. Hence, by the reproductive property of the normal distribution established in Theorem 5.11, we conclude that

$$\bar{X} = \frac{X_1 + X_2 + \cdots + X_n}{n}$$

has a normal distribution with mean

$$\mu_{\bar{X}} = \frac{\mu + \mu + \cdots + \mu}{n} = \mu$$

and variance

$$\sigma_{\bar{X}}^2 = \frac{\sigma^2 + \sigma^2 + \cdots + \sigma^2}{n^2} = \frac{\sigma^2}{n}.$$

If we are sampling from a population with unknown distribution, either finite or infinite, the sampling distribution of $\bar{X}$ will still be approximately normal with mean μ and variance σ^2/n provided the sample size is large. This amazing result is an immediate consequence of the following theorem, called the *central limit theorem*. The proof is outlined in Exercise 30.

THEOREM 5.14   *If $\bar{X}$ is the mean of a random sample of size n taken from a population with mean μ and finite variance σ^2, then the limiting form of the distribution of*

$$Z = \frac{\bar{X} - \mu}{\sigma/\sqrt{n}},$$

as $n \to \infty$, is the standardized normal distribution $n(z; 0, 1)$.

The normal approximation for $\bar{X}$ will generally be good if $n \geq 30$ regardless of the shape of the population. If $n < 30$, the approximation is good only if the population is not too different from a normal population. If the population is known to be normal, the sampling distribution of $\bar{X}$ will follow a normal distribution exactly, no matter how small the size of the samples.

Example 5.16   An electrical firm manufactures light bulbs that have a length of life that is approximately normally distributed, with mean equal to 800 hours and a standard deviation of 40 hours. Find the probability that a random sample of 16 bulbs will have an average life of less than 775 hours.

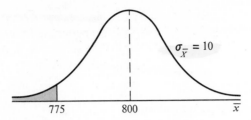

Figure 5.4 Area for Example 5.16.

Solution The sampling distribution of $\overline{X}$ will be approximately normal, with $\mu_{\overline{x}} = 800$ and $\sigma_{\overline{x}} = 40/\sqrt{16} = 10$. The desired probability is given by the area of the shaded region in Figure 5.4. Corresponding to $\bar{x} = 775$, we find that

$$z = \frac{775 - 800}{10} = -2.5,$$

and therefore

$$P(\overline{X} < 775) = P(Z < -2.5)$$
$$= 0.006.$$

Example 5.17 Given the discrete uniform population

$$f(x) = \tfrac{1}{4}, \qquad x = 0, 1, 2, 3$$
$$= 0, \qquad \text{elsewhere,}$$

find the probability that a random sample of size 36, selected with replacement, will yield a sample mean greater than 1.4 but less than 1.8 if the mean is measured to the nearest tenth.

Solution Calculating the mean and variance of the uniform distribution by means of the formulas in Theorem 3.1, we find that

$$\mu = \frac{0 + 1 + 2 + 3}{4} = \frac{3}{2}$$

and

$$\sigma^2 = \frac{(0 - 3/2)^2 + (1 - 3/2)^2 + (2 - 3/2)^2 + (3 - 3/2)^2}{4}$$
$$= \frac{5}{4}.$$

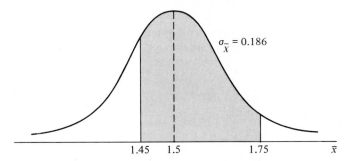

Figure 5.5 Area for Example 5.17.

The sampling distribution of $\bar{X}$ may be approximated by the normal distribution with mean $\mu_{\bar{x}} = 3/2$ and variance $\sigma_{\bar{x}}^2 = \sigma^2/n = 5/144$. Taking the square root, we find the standard deviation to be $\sigma_{\bar{x}} = 0.186$. The probability that $\bar{X}$ is greater than 1.4 but less than 1.8 is given by the area of the shaded region in Figure 5.5. The z values corresponding to $\bar{x}_1 = 1.45$ and $\bar{x}_2 = 1.75$ are

$$z_1 = \frac{1.45 - 1.5}{0.186} = -0.269 \quad \text{from pg } 116$$

$$z_2 = \frac{1.75 - 1.5}{0.186} = 1.344.$$

Therefore,

$$
\begin{aligned}
P(1.4 < \bar{X} < 1.8) &\simeq P(-0.269 < Z < 1.344) \\
&= P(Z < 1.344) - P(Z < -0.269) \\
&= 0.9105 - 0.3932 \\
&= 0.5173.
\end{aligned}
$$

Suppose that we now have two populations, the first with mean μ_1 and variance σ_1^2, and the second with mean μ_2 and variance σ_2^2. Let the statistic $\bar{X}_1$ represent the mean of a random sample of size n_1 selected from the first population, and the statistic $\bar{X}_2$ represent the mean of a random sample selected from the second population, independent of the sample from the first population. What can we say about the sampling distribution of the difference $\bar{X}_1 - \bar{X}_2$ for repeated samples of size n_1 and n_2? According to Theorem 5.14, the variables $\bar{X}_1$ and $\bar{X}_2$ are both approximately normally distributed with means μ_1 and μ_2 and variances σ_1^2/n_1 and σ_2^2/n_2, respectively. This approximation improves as n_1 and n_2 increase. By choosing independent samples from the two populations the variables $\bar{X}_1$ and $\bar{X}_2$ will be independent, and then using Theorem 5.11, with $a_1 = 1$ and $a_2 = -1$, we conclude that $\bar{X}_1 - \bar{X}_2$ is approximately normally distributed with mean

$$\mu_{\bar{X}_1 - \bar{X}_2} = \mu_{\bar{X}_1} - \mu_{\bar{X}_2} = \mu_1 - \mu_2$$

and variance

$$\sigma^2_{\bar{X}_1 - \bar{X}_2} = \sigma^2_{\bar{X}_1} + \sigma^2_{\bar{X}_2} = \frac{\sigma_1^2}{n_1} + \frac{\sigma_2^2}{n_2}.$$

THEOREM 5.15 *If independent samples of size n_1 and n_2 are drawn at random from two populations, discrete or continuous, with means μ_1 and μ_2 and variances σ_1^2 and σ_2^2, respectively, then the sampling distribution of the differences of means, $\bar{X}_1 - \bar{X}_2$, is approximately normally distributed with mean and variance given by*

$$\mu_{\bar{X}_1 - \bar{X}_2} = \mu_1 - \mu_2$$

$$\sigma^2_{\bar{X}_1 - \bar{X}_2} = \frac{\sigma_1^2}{n_1} + \frac{\sigma_2^2}{n_2}.$$

Hence

$$Z = \frac{(\bar{X}_1 - \bar{X}_2) - (\mu_1 - \mu_2)}{\sqrt{(\sigma_1^2/n_1) + (\sigma_2^2/n_2)}}$$

is approximately a standard normal variable.

If both n_1 and n_2 are greater than or equal to 30, the normal approximation for the distribution of $\bar{X}_1 - \bar{X}_2$ is very good.

Example 5.18 The television picture tubes of manufacturer A have a mean lifetime of 6.5 years and a standard deviation of 0.9 year, while those of manufacturer B have a mean lifetime of 6.0 years and a standard deviation of 0.8 year. What is the probability that a random sample of 36 tubes from manufacturer A will have a mean lifetime that is at least 1 year more than the mean lifetime of a sample of 49 tubes from manufacturer B?

Solution We are given the following information:

	Population 1	Population 2
	$\mu_1 = 6.5$	$\mu_2 = 6.0$
	$\sigma_1 = 0.9$	$\sigma_2 = 0.8$
	$n_1 = 36$	$n_2 = 49$

Using Theorem 5.15, the sampling distribution of $\bar{X}_1 - \bar{X}_2$ will have a mean and standard deviation given by

$$\mu_{\bar{X}_1 - \bar{X}_2} = 6.5 - 6.0 = 0.5$$

$$\sigma_{\bar{X}_1 - \bar{X}_2} = \sqrt{\frac{0.81}{36} + \frac{0.64}{49}} = 0.189.$$

The probability that the mean of 36 tubes from manufacturer A will be at least 1 year longer than the mean of 49 tubes from manufacturer B is given by

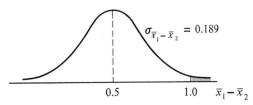

$$\sigma_{\bar{X}_1 - \bar{X}_2} = 0.189$$

0.5   1.0 $\bar{x}_1 - \bar{x}_2$

Figure 5.6 Area for Example 5.18.

the area of the shaded region in Figure 5.6. Corresponding to the value $\bar{x}_1 - \bar{x}_2 = 1.0$, we find that

$$z = \frac{1.0 - 0.5}{0.189} = 2.646,$$

and hence

$$P(\bar{X}_1 - \bar{X}_2 \geq 1.0) = P(Z > 2.646)$$
$$= 1 - P(Z < 2.646)$$
$$= 1 - 0.9959$$
$$= 0.0041.$$

5.6 SAMPLING DISTRIBUTION OF $(n - 1)S^2/\sigma^2$

If a random sample of size n is drawn from a normal population with mean μ and variance σ^2, and the sample variance s^2 is computed, we obtain a value of the statistic S^2. The sampling distribution of S^2 has little practical application in statistics. Instead, we shall consider the distribution of the random variable $(n - 1)S^2/\sigma^2$.

By the addition and subtraction of the sample mean $\bar{X}$, it is easy to see that

$$\sum_{i=1}^{n} (X_i - \mu)^2 = \sum_{i=1}^{n} [(\bar{X}_i - \bar{X}) + (\bar{X} - \mu)]^2$$

$$= \sum_{i=1}^{n} (X_i - \bar{X})^2 + \sum_{i=1}^{n} (\bar{X} - \mu)^2 + 2(\bar{X} - \mu) \sum_{i=1}^{n} (X_i - \bar{X})$$

$$= \sum_{i=1}^{n} (X_i - \bar{X})^2 + n(\bar{X} - \mu)^2.$$

Dividing each term of the equality by σ^2 and substituting $(n - 1)S^2$ for $\sum_{i=1}^{n} (X_i - \bar{X})^2$, we obtain

$$\frac{\sum_{i=1}^{n} (X_i - \mu)^2}{\sigma^2} = \frac{(n - 1)S^2}{\sigma^2} + \frac{(\bar{X} - \mu)^2}{\sigma^2/n}.$$

Now, according to the corollary of Theorem 5.12 we know that $\sum_{i=1}^{n} (X_i - \mu)^2/\sigma^2$ is a chi-square random variable with n degrees of freedom. The second term on the right of the equality is the square of a standard normal variable, since $\bar{X}$ is a normal random variable with mean $\mu_{\bar{X}} = \mu$ and variance $\sigma_{\bar{X}}^2 = \sigma^2/n$. Therefore, we may conclude from Example 5.5 that $(\bar{X} - \mu)^2/(\sigma^2/n)$ is a chi-square random variable with 1 degree of freedom. Using advanced techniques beyond the scope of this text, one can also show that the two chi-square variables $\sum_{i=1}^{n} (X_i - \mu)^2/\sigma^2$ and $(\bar{X} - \mu)^2/(\sigma^2/n)$ are independent. Owing to the reproductive property of independent chi-square random variables, established in Theorem 5.12, it would seem reasonable to assume that $(n - 1)S^2/\sigma^2$ is also a chi-square random variable with $v = n - 1$ degrees of freedom. We state this result, without formal proof, in the following theorem.

THEOREM 5.16   *If S^2 is the variance of a random sample of size n taken from a normal population having the variance σ^2, then the random variable*

$$X^2 = \frac{(n-1)S^2}{\sigma^2}$$

has a chi-square distribution with $v = n - 1$ degrees of freedom.

The values of the random variable X^2 are calculated from each sample by the formula

$$\chi^2 = \frac{(n-1)s^2}{\sigma^2}.$$

The probability that a random sample produces a χ^2 value greater than some specified value is equal to the area under the curve to the right of this value. It is customary to let χ_α^2 represent the χ^2 value above which we find an area of α. This is illustrated by the shaded region in Figure 5.7.

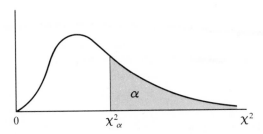

Figure 5.7 Tabulated values of the chi-square distribution.

Table VI (see Statistical Tables) gives values of χ_α^2 for various values of α and v. The areas α are the column headings, the degrees of freedom v are given in the left column, and the table entries are the χ^2 values. Hence the χ^2 value with 7 degrees of freedom, leaving an area of 0.05 to the right, is $\chi_{0.05}^2 = 14.067$. Because of lack of symmetry, we must also use the tables to find $\chi_{0.95}^2 = 2.167$ for $v = 7$.

Exactly 95% of a chi-square distribution with $n - 1$ degrees of freedom lies between $\chi_{0.975}^2$ and $\chi_{0.025}^2$. A χ^2 value falling to the right of $\chi_{0.025}^2$ is not likely to occur unless our assumed value of σ^2 is too small. Similarly, a χ^2 value falling to the left of $\chi^2_{0.975}$ is unlikely unless our assumed value of σ^2 is too large. In other words, it is possible to have a χ^2 value to the left of $\chi^2_{0.975}$ or to the right of $\chi^2_{0.025}$ when σ^2 is correct, but if this should occur, it is more probable that the assumed value of σ^2 is in error.

Example 5.19 A manufacturer of car batteries guarantees that his batteries will last, on the average, 3 years with a standard deviation of 1 year. If five of these batteries have lifetimes of 1.9, 2.4, 3.0, 3.5, and 4.2 years, is the manufacturer still convinced that his batteries have a standard deviation of 1 year?

Solution We first find the sample variance:

$$s^2 = \frac{(5)(48.26) - (15)^2}{(5)(4)} = 0.815. \qquad formula \\ pg \ 165$$

Then

$$\chi^2 = \frac{(4)(0.815)}{1} = 3.26$$

is a value from a chi-square distribution with 4 degrees of freedom. Since 95% of the χ^2 values with 4 degrees of freedom fall between 0.484 and 11.143, the computed value with $\sigma^2 = 1$ is reasonable, and therefore the manufacturer has no reason to suspect that the standard deviation is other than 1 year.

5.7 t DISTRIBUTION

Most of the time we are not fortunate enough to know the variance of the population from which we select our random samples. For samples of size $n \geq 30$, a good estimate of σ^2 is provided by calculating a value for S^2. What then happens to our statistic $(\bar{X} - \mu)/(\sigma/\sqrt{n})$ of Theorem 5.14 if we replace σ^2 by S^2? As long as S^2 provides a good estimate of σ^2 and does not vary from sample to sample, which is usually the case for $n \geq 30$, the distribution of the statistic $(\bar{X} - \mu)/(S/\sqrt{n})$ is still approximately distributed as a standard normal variable Z.

If the sample size is small ($n < 30$), the values of S^2 fluctuate considerably from sample to sample (see Exercise 36) and the distribution of the random variable $(\overline{X} - \mu)/(S/\sqrt{n})$ is no longer a standard normal distribution. We are now dealing with the distribution of a statistic that we shall call *T*, where

$$T = \frac{\overline{X} - \mu}{S/\sqrt{n}}.$$

In deriving the sampling distribution of *T*, we shall assume our random sample was selected from a normal population. We can then write

$$T = \frac{(\overline{X} - \mu)/(\sigma/\sqrt{n})}{\sqrt{S^2/\sigma^2}} = \frac{Z}{\sqrt{V/(n-1)}},$$

where

$$Z = \frac{\overline{X} - \mu}{\sigma/\sqrt{n}}$$

has the standard normal distribution, and

$$V = \frac{(n-1)S^2}{\sigma^2}$$

has a chi-square distribution with $v = n - 1$ degrees of freedom. In sampling from normal populations, one can show that $\overline{X}$ and S^2 are independent, and consequently so are *Z* and *V*. We are now in a position to derive the distribution of *T*.

THEOREM 5.17 *Let Z be a standard normal random variable and V a chi-square random variable with v degrees of freedom. If Z and V are independent, then the distribution of the random variable T, where*

$$T = \frac{Z}{\sqrt{V/v}},$$

is given by

$$h(t) = \frac{\Gamma[(v+1)/2]}{\Gamma(v/2)\sqrt{\pi v}}\left(1 + \frac{t^2}{v}\right)^{-(v+1)/2}, \qquad -\infty < t < \infty.$$

This is known as the t *distribution with v degrees of freedom.*

Proof   Since Z and V are independent random variables, their joint probability distribution is given by the product of the distribution of Z and V. That is,

$$f(z, v) = \frac{1}{\sqrt{2\pi}} e^{-z^2/2} \frac{1}{2^{v/2}\Gamma(v/2)} v^{v/2 - 1} e^{-v/2}, \qquad -\infty < z < \infty, 0 < v < \infty$$

$$= 0, \qquad\qquad\qquad\qquad\qquad\qquad\qquad \text{elsewhere.}$$

Let us define a second random variable $U = V$. The inverse solutions of $t = z/\sqrt{v/v}$ and $u = v$ are $z = t\sqrt{u}/\sqrt{v}$ and $v = u$, from which we obtain

$$J = \begin{vmatrix} \sqrt{u}/\sqrt{v} & t/2\sqrt{uv} \\ 0 & 1 \end{vmatrix} = \frac{\sqrt{u}}{\sqrt{v}}.$$

The transformation is one to one, mapping the points $\{(z, v)| -\infty < z < \infty, 0 < v < \infty\}$ into the set $\{(t, u)| -\infty < t < \infty, 0 < u < \infty\}$. Using Theorem 5.4, we find the joint probability distribution of T and U to be

$$g(t, u) = \frac{1}{\sqrt{2\pi}\, 2^{v/2}\Gamma(v/2)} u^{v/2 - 1} e^{-\{(u/2)[1 + (t^2/v)]\}} \frac{\sqrt{u}}{\sqrt{v}}, \qquad -\infty < t < \infty,$$

$$0 < u < \infty$$

$$= 0, \qquad\qquad\qquad\qquad\qquad\qquad\qquad \text{elsewhere.}$$

Integrating out u, we find that the distribution of T is given by

$$h(t) = \int_0^\infty g(t, u)\, du$$

$$= \int_0^\infty \frac{1}{\sqrt{2\pi v}\, 2^{v/2}\Gamma(v/2)} u^{[(v + 1)/2] - 1} e^{-\{(u/2)[1 + (t^2/v)]\}}\, du.$$

Let us substitute $z = u(1 + t^2/v)/2$ and $du = dz/(1 + t^2/v)$ to give

$$h(t) = \frac{1}{\sqrt{2\pi v}\, 2^{v/2}\Gamma(v/2)} \int_0^\infty \left(\frac{2z}{1 + t^2/v}\right)^{[(v + 1)/2] - 1} e^{-z} \left(\frac{2}{1 + t^2/v}\right) dz$$

$$= \frac{1}{\Gamma(v/2)\sqrt{\pi v}} \left(1 + \frac{t^2}{v}\right)^{-[(v + 1)/2]} \int_0^\infty z^{[(v + 1)/2] - 1} e^{-z}\, dz$$

$$= \frac{\Gamma[(v + 1)/2]}{\Gamma(v/2)\sqrt{\pi v}} \left(1 + \frac{t^2}{v}\right)^{-(v + 1)/2}, \qquad -\infty < t < \infty$$

$$= 0, \qquad\qquad\qquad\qquad\qquad\qquad\qquad \text{elsewhere.}$$

The probability distribution of *T* was first published in 1908 in a paper by W. S. Gosset. At the time, Gosset was employed by an Irish brewery that disallowed publication of research by members of its staff. To circumvent this restriction he published his work secretly under the name "Student." Consequently, the distribution of *T* is usually called the *Student t distribution*, or simply the *t distribution*. In deriving the equation of this distribution, Gosset assumed the samples were selected from a normal population. Although this would seem to be a very restrictive assumption, it can be shown that nonnormal populations possessing bell-shaped distributions will still provide values of *T* that approximate the *t* distribution very closely.

The distribution of *T* is similar to the distribution of *Z* in that they both are symmetric about a mean of zero. Both distributions are bell-shaped, but the *t* distribution is more variable, owing to the fact that the *T* values depend on the fluctuations of two quantities, $\overline{X}$ and S^2, whereas the *Z* values depend only on the changes of $\overline{X}$ from sample to sample. The distribution of *T* differs from that of *Z* in that the variance of *T* depends on the sample size *n* and is always greater than 1. Only when the sample size $n \to \infty$ will the two distributions become the same. In Figure 5.8 we show the relationship between a standard normal distribution ($v = \infty$) and *t* distributions with 2 and 5 degrees of freedom.

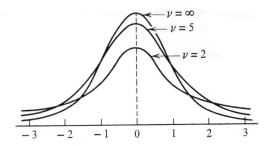

Figure 5.8 *t* distribution curves for $v = 2, 5,$ and ∞.

The probability that a random sample produces a value $t = (\overline{x} - \mu)/(s/\sqrt{n})$ falling between any two specified values is equal to the area under the curve of the *t* distribution between the two ordinates corresponding to the specified values. It would be a tedious task to attempt to set up separate tables giving the areas between every conceivable pair of ordinates for all values of $n \leq 30$. Table V (see Statistical Tables) gives only those *t* values above which we find a specified area α, where α is 0.1, 0.05, 0.025, 0.01, or 0.005. This table is set up differently than the table of normal curve areas in that the areas are now the column headings and the entries are the *t* values. The left column gives the

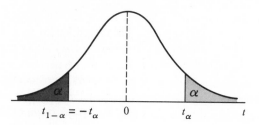

Figure 5.9 Symmetry property of the t distribution.

degrees of freedom. It is customary to let t_α represent the t value above which we find an area equal to α. Hence the t value with 10 degrees of freedom leaving an area of 0.025 to the right is $t = 2.228$. Since the t distribution is symmetric about a mean of zero, we have $t_{1-\alpha} = -t_\alpha$; that is, the t value leaving an area of $1 - \alpha$ to the right and therefore an area of α to the left is equal to the negative t value that leaves an area of α in the right tail of the distribution (see Figure 5.9). For a t distribution with 10 degrees of freedom, we have $t_{0.975} = -t_{0.025} = -2.228$. This means that the t value of a random sample of size 11, selected from a normal population, will fall between -2.228 and 2.228 with probability equal to 0.95.

Exactly 95% of a t distribution with $n - 1$ degrees of freedom lies between $-t_{0.025}$ and $t_{0.025}$. Therefore, a t value falling below $-t_{0.025}$ or above $t_{0.025}$ would tend to make one believe that either a rare event has taken place or our assumption about μ is in error. The importance of μ will determine the length of the interval for an acceptable t value. In other words, if you do not mind having the true mean slightly different from what you claim it to be, you might choose a wide interval from $-t_{0.01}$ to $t_{0.01}$ in which the t value should fall. A t value falling at either end of the interval, but within the interval, would lead us to believe that our assumed value for μ is correct, although it is very probable that some other value close to μ is the true value. If μ is to be known with a high degree of accuracy, a short interval such as $-t_{0.05}$ to $t_{0.05}$ should be used. In this case a t value falling outside the interval would lead you to believe that the assumed value of μ is in error when it is entirely possible that it is correct. The problems connected with the establishment of proper intervals in testing hypotheses concerning the parameter μ will be treated in Chapter 7.

Example 5.20   A manufacturer of light bulbs claims that his bulbs will burn on the average 500 hours. To maintain this average, he tests 25 bulbs each month. If the computed t value falls between $-t_{0.05}$ and $t_{0.05}$, he is satisfied with his claim. What conclusion should he draw from a sample that has a mean $\bar{x} = 518$ hours and a standard deviation $s = 40$ hours? Assume the distribution of burning times to be approximately normal.

Solution   From Table V we find that $t_{0.05} = 1.711$ for 24 degrees of freedom. Therefore, the manufacturer is satisfied with his claim if a sample of 25 bulbs yields a t value between -1.711 and 1.711. If $\mu = 500$, then

$$t = \frac{518 - 500}{40/\sqrt{25}} = 2.25,$$

a value well above 1.711. The probability of obtaining a t value, with $v = 24$, equal to or greater than 2.25 is approximately 0.02. If $\mu > 500$, the value of t computed from the sample would be more reasonable. Hence the manufacturer is likely to conclude that his bulbs are a better product than he thought.

5.8 *F* DISTRIBUTION

One of the most important distributions in applied statistics is the *F distribution*. The statistic *F* is defined to be the ratio of two independent chi-square random variables, each divided by their degrees of freedom. Hence we can write

$$F = \frac{U/v_1}{V/v_2},$$

where U and V are independent random variables having chi-square distributions with v_1 and v_2 degrees of freedom, respectively. We shall now derive the sampling distribution of F.

THEOREM 5.18   *Let U and V be two independent random variables having chi-square distributions with v_1 and v_2 degrees of freedom, respectively. Then the distribution of the random variable*

$$F = \frac{U/v_1}{V/v_2}$$

is given by

$$h(f) = \frac{\Gamma[(v_1 + v_2)/2](v_1/v_2)^{v_1/2}}{\Gamma(v_1/2)\Gamma(v_2/2)} \frac{f^{v_1/2 - 1}}{(1 + v_1 f/v_2)^{(v_1 + v_2)/2}}, \quad 0 < f < \infty$$

$$= 0, \qquad\qquad\qquad\qquad\qquad\qquad\qquad\qquad elsewhere.$$

This is known as the F *distribution with v_1 and v_2 degrees of freedom.*

Proof The joint probability distribution of the independent random variables U and V is given by

$$\phi(u, v) = r(u)s(v),$$

where $r(u)$ and $s(v)$ represent the distributions of U and V, respectively. Hence

$$\phi(u, v) = \frac{1}{2^{v_1/2}\Gamma(v_1/2)} u^{v_1/2 - 1} e^{-u/2} \frac{1}{2^{v_2/2}\Gamma(v_2/2)} v^{v_2/2 - 1} e^{-v/2}$$

$$= \frac{1}{2^{(v_1 + v_2)/2}\Gamma(v_1/2)\Gamma(v_2/2)} u^{v_1/2 - 1} v^{v_2/2 - 1} e^{-(u + v)/2}, \quad \begin{matrix} 0 < u < \infty, \\ 0 < v < \infty, \end{matrix}$$

$$= 0, \qquad\qquad\qquad\qquad\qquad\qquad\qquad\qquad\qquad \text{elsewhere.}$$

Let us define a second random variable $W = V$. The inverse solutions of $f = (u/v_1)/(v/v_2)$ and $w = v$ are $u = (v_1/v_2)fw$ and $v = w$, from which we obtain

$$J = \begin{vmatrix} (v_1/v_2)w & (v_1/v_2)f \\ 0 & 1 \end{vmatrix} = \frac{v_1}{v_2} w.$$

The transformation is one to one, mapping the points $\{(u, v) | 0 < u < \infty, 0 < v < \infty\}$ into the set $\{(f, w) | 0 < f < w, 0 < w < \infty\}$. Using Theorem 5.4, we find that the joint probability distribution of F and W is

$$g(f, w) = \frac{1}{2^{(v_1 + v_2)/2}\Gamma(v_1/2)\Gamma(v_2/2)} \left(\frac{v_1 fw}{v_2}\right)^{v_1/2 - 1} w^{v_2/2 - 1} e^{-(w/2)[(v_1 f/v_2) + 1]} \frac{v_1 w}{v_2},$$

$$0 < f < \infty, 0 < w < \infty$$

$$= 0, \qquad \text{elsewhere.}$$

The distribution of F is then given by the marginal distribution

$$h(f) = \int_0^\infty g(f, w)\, dw$$

$$= \frac{(v_1/v_2)^{v_1/2} f^{v_1/2 - 1}}{2^{(v_1 + v_2)/2}\Gamma(v_1/2)\Gamma(v_2/2)} \int_0^\infty w^{[(v_1 + v_2)/2] - 1} e^{-(w/2)[(v_1 f/v_2) + 1]}\, dw.$$

Substituting $z = (w/2)[(v_1 f/v_2) + 1]$ and $dw = [2/(v_1 f/v_2 + 1)]\,dz$, we obtain

$$h(f) = \frac{(v_1/v_2)^{v_1/2} f^{v_1/2 - 1}}{2^{(v_1 + v_2)/2}\Gamma(v_1/2)\Gamma(v_2/2)} \int_0^\infty \left(\frac{2z}{v_1 f/v_2 + 1}\right)^{(v_1 + v_2)/2 - 1} e^{-z} \frac{2}{v_1 f/v_2 + 1}\,dz$$

$$= \frac{(v_1/v_2)^{v_1/2} f^{v_1/2 - 1}}{\Gamma(v_1/2)\Gamma(v_2/2)(1 + v_1 f/v_2)^{(v_1 + v_2)/2}} \int_0^\infty z^{(v_1 + v_2)/2 - 1} e^{-z}\,dz$$

$$= \frac{\Gamma[(v_1 + v_2)/2](v_1/v_2)^{v_1/2}}{\Gamma(v_1/2)\Gamma(v_2/2)} \frac{f^{v_1/2 - 1}}{(1 + v_1 f/v_2)^{(v_1 + v_2)/2}}, \qquad 0 < f < \infty$$

$$= 0, \qquad\qquad\qquad\qquad\qquad\qquad\qquad\qquad \text{elsewhere.}$$

The number of degrees of freedom associated with the chi-square random variable appearing in the numerator of F is always stated first, followed by the number of degrees of freedom associated with the chi-square random variable appearing in the denominator. Thus the curve of the F distribution depends not only on the two parameters v_1 and v_2 but also on the order in which we state them. Once these two values are given, we can identify the curve. Typical F curves are shown in Figure 5.10.

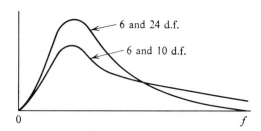

Figure 5.10 Typical F distributions.

Let us define f_α to be the particular value f of the random variable F above which we find an area equal to α. This is illustrated by the shaded region in Figure 5.11. Table VII (see Statistical Tables) gives values of f_α only for $\alpha = 0.05$ and $\alpha = 0.01$ for various combinations of the degrees of freedom v_1 and v_2. Hence the f value with 6 and 10 degrees of freedom, leaving an area of 0.05 to the right, is $f_{0.05} = 3.22$. By means of the following theorem, Table VII can be used to find values of $f_{0.95}$ and $f_{0.99}$. The proof is left for the reader.

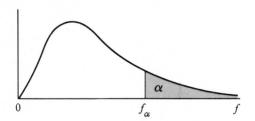

Figure 5.11 Tabulated values of the F distribution.

THEOREM 5.19   *Writing $f_\alpha(v_1, v_2)$ for f_α with v_1 and v_2 degrees of freedom, we obtain*

$$f_{1-\alpha}(v_1, v_2) = \frac{1}{f_\alpha(v_2, v_1)}.$$

Thus the f value with 6 and 10 degrees of freedom, leaving an area of 0.95 to the right, is

$$f_{0.95}(6, 10) = \frac{1}{f_{0.05}(10, 6)} = \frac{1}{4.06} = 0.246.$$

Suppose that random samples of size n_1 and n_2 are selected from two normal populations with variances σ_1^2 and σ_2^2, respectively. From Theorem 5.16, we know that

$$X_1^2 = \frac{(n_1 - 1)S_1^2}{\sigma_1^2}$$

and

$$X_2^2 = \frac{(n_2 - 1)S_2^2}{\sigma_2^2}$$

are random variables having chi-square distributions with $v_1 = n_1 - 1$ and $v_2 = n_2 - 1$ degrees of freedom. Furthermore, since the samples are selected at random, we are dealing with independent random variables and then using Theorem 5.18, with $X_1^2 = U$ and $X_2^2 = V$, we obtain the following result.

> THEOREM 5.20 *If S_1^2 and S_2^2 are the variances of independent random samples of size n_1 and n_2 taken from normal populations with variances σ_1^2 and σ_2^2, respectively, then*
>
> $$F = \frac{S_1^2/\sigma_1^2}{S_2^2/\sigma_2^2} = \frac{\sigma_2^2 S_1^2}{\sigma_1^2 S_2^2}$$
>
> *has an F distribution with $v_1 = n_1 - 1$ and $v_2 = n_2 - 1$ degrees of freedom.*

In Chapters 6 and 7 we shall use Theorem 5.20 to make inferences concerning the variances of two normal populations. The F distribution is applied primarily, however, in the analysis-of-variance procedures of Chapters 10, 11, and 12, where we wish to test the equality of several means simultaneously.

EXERCISES   1-7

1. Let X be a random variable with probability distribution

$$f(x) = \tfrac{1}{3}, \qquad x = 1, 2, 3$$
$$= 0, \qquad \text{elsewhere.}$$

Find the probability distribution of the random variable $Y = 2X - 1$.

2. Let X be a binomial random variable with probability distribution

$$f(x) = \binom{3}{x}\left(\frac{2}{5}\right)^x\left(\frac{3}{5}\right)^{3-x}, \qquad x = 0, 1, 2, 3$$
$$= 0, \qquad \text{elsewhere.}$$

Find the probability distribution of the random variable $Y = X^2$.

3. Let X_1 and X_2 be discrete random variables with the multinomial distribution

$$f(x_1, x_2) = \binom{2}{x_1, x_2, 2 - x_1 - x_2}\left(\frac{1}{4}\right)^{x_1}\left(\frac{1}{3}\right)^{x_2}\left(\frac{5}{12}\right)^{2 - x_1 - x_2}$$

for $x_1 = 0, 1, 2$; $x_2 = 0, 1, 2$; $x_1 + x_2 \le 2$; and zero elsewhere. Find the joint probability distribution of $Y_1 = X_1 + X_2$ and $Y_2 = X_1 - X_2$.

4. Let X_1 and X_2 be discrete random variables with joint probability distribution

$$f(x_1, x_2) = \frac{x_1 x_2}{18}, \qquad x_1 = 1, 2, x_2 = 1, 2, 3$$
$$= 0, \qquad \text{elsewhere.}$$

Find the probability distribution of the random variable $Y = X_1 X_2$.

5. Let X have the probability distribution

$$f(x) = 1, \qquad 0 < x < 1$$
$$= 0, \qquad \text{elsewhere.}$$

Show that the random variable $Y = -2 \ln X$ has a chi-square distribution with 2 degrees of freedom.

6. Given the random variable X with probability distribution

$$f(x) = 2x, \qquad 0 < x < 1$$
$$= 0, \qquad \text{elsewhere,}$$

find the probability distribution of Y, where $Y = 8X^3$.

7. The speed of a molecule in a uniform gas at equilibrium is a random variable V whose probability distribution is given by

$$f(v) = kv^2 e^{-bv^2}, \qquad v > 0$$
$$= 0, \qquad \text{elsewhere,}$$

where k is an appropriate constant and b depends on the absolute temperature and mass of the molecule. Find the probability distribution of the kinetic energy of the molecule W, where $W = mV^2/2$.

8. Let X_1 and X_2 be independent random variables each having the probability distribution

$$f(x) = e^{-x}, \qquad x > 0$$
$$= 0, \qquad \text{elsewhere.}$$

Show that the random variables Y_1 and Y_2 are independent when $Y_1 = X_1 + X_2$ and $Y_2 = X_1/(X_1 + X_2)$.

9. A current of I amperes flowing through a resistance of R ohms varies according to the probability distribution

$$f(i) = 6i(1 - i), \qquad 0 < i < 1$$
$$= 0, \qquad \text{elsewhere.}$$

If the resistance varies independently of the current according to the probability distribution

$$g(r) = 2r, \qquad 0 < r < 1$$
$$= 0, \qquad \text{elsewhere,}$$

find the probability distribution for the power $W = I^2 R$ watts.

10. Let X be a random variable with probability distribution

NOT
Bi VARIATE

$$f(x) = \frac{1 + x}{2}, \quad -1 < x < 1$$

$$= 0, \quad \text{elsewhere.}$$

Find the probability distribution of the random variable $Y = X^2$.

11. Let X have the probability distribution

AND
OUT

$$f(x) = \frac{2(x + 1)}{9}, \quad -1 < x < 2$$

$$= 0, \quad \text{elsewhere.}$$

Find the probability distribution of the random variable $Y = \dot{X}^2$.

12. Given the discrete uniform distribution

$$f(x) = \frac{1}{n}, \quad x = 1, 2, 3, \ldots, n$$

$$= 0, \quad \text{elsewhere,}$$

show that the moment-generating function of X is

$$M_X(t) = \frac{e^t(1 - e^{nt})}{n(1 - e^t)}.$$

13. Show that the moment-generating function of the geometric random variable is

$$M_X(t) = \frac{pe^t}{1 - qe^t}$$

and then use $M_X(t)$ to find the mean and variance of the geometric distribution.

14. Show that the moment-generating function of the random variable X having a Poisson distribution with parameter μ is $M_X(t) = e^{\mu(e^t - 1)}$. Using $M_X(t)$, find the mean and variance of the Poisson distribution.

15. The moment-generating function of a certain Poisson random variable X is given by

$$M_X(t) = e^{4(e^t - 1)}.$$

Find $P(\mu - 2\sigma < X < \mu + 2\sigma)$.

16. Using the moment-generating function of Example 5.8, show that the mean and variance of the chi-square distribution with v degrees of freedom are, respectively, v and $2v$.

17. In a random sample of 18 students at Roanoke College the following numbers of days absent were recorded for the previous semester: 1, 3, 4, 0, 4, 2, 3, 1, 2, 3, 0, 4, 1, 1, 1, 5, 1, and 0. Find the mean, median, and mode.

18. The numbers of trout caught by eight fishermen on the first day of the trout season are 7, 4, 6, 7, 4, 4, 8, and 7. If these eight values represent the catch of a random sample of fishermen at Smith Mountain Lake, define a suitable population. If the values represent the catch of a random sample of fishermen at various lakes and streams in Montgomery County, define a suitable population. Find the mean, median, and mode for the data.

19. Find the mean, median, and mode for the sample 18, 10, 11, 98, 22, 15, 11, 25, and 17. Which value appears to be the best measure of central tendency? Give reasons for your preference.

20. Calculate the range and standard deviation for the data of Exercise 17.

21. The grade-point averages of 15 college seniors selected at random from the graduating class are as follows:

2.3	3.4	2.9
2.6	2.1	2.4
3.1	2.7	2.6
1.9	2.0	3.6
2.1	1.8	2.1.

Calculate the standard deviation.

22. Show that the sample variance is unchanged if a constant is added to or subtracted from each value in the sample.

23. If each observation in a sample is multiplied by k, show that the sample variance becomes k^2 times its original value.

24. (a) Calculate the variance of the sample 3, 5, 8, 7, 5, and 7.
(b) Without calculating, state the variance of the sample 6, 10, 16, 14, 10, and 14.
(c) Without calculating, state the variance of the sample 25, 27, 30, 29, 27, and 29.

25. If all possible samples of size 16 are drawn from a normal population with mean equal to 50 and standard deviation equal to 5, what is the probability that a sample mean $\bar{X}$ will fall in the interval from $\mu_{\bar{X}} - 1.9\sigma_{\bar{X}}$ to $\mu_{\bar{X}} - 0.4\sigma_{\bar{X}}$? Assume that the sample means can be measured to any degree of accuracy.

26. If the size of a sample is 36 and the standard error of the mean is 2, what must the size of the sample become if the standard error is to be reduced to 1.2?

27. A soft-drink machine is regulated so that the amount of drink dispensed is approximately normally distributed with a mean of 207 milliliters per cup and a standard deviation equal to 15 milliliters. Periodically, the machine is checked by taking a sample of nine drinks and computing the average content. If the mean, $\bar{X}$, of the nine drinks falls within the interval $\mu_{\bar{X}} \pm 2\sigma_{\bar{X}}$, the machine is thought to be operating satisfactorily; otherwise, adjustments must be made. What action should one take if a sample of nine drinks has a mean content of 219 milliliters?

28. The heights of 1000 students are approximately normally distributed with a mean of 174.5 centimeters and a standard deviation of 6.9 centimeters. If 200 random samples of size 25 are drawn from this population and the means recorded to the nearest tenth of a centimeter, determine
(a) The expected mean and standard deviation of the sampling distribution of the mean.

(b) The number of sample means that fall between 172.5 and 175.8 centimeters inclusive.

(c) The number of sample means falling below 172.0 centimeters.

29. By expanding e^{tx} in a Maclaurin series and integrating term by term, show that

$$M_X(t) = \int_{-\infty}^{\infty} e^{tx} f(x)\, dx$$

$$= 1 + \mu t + \mu_2' \frac{t^2}{2!} + \cdots + \mu_r' \frac{t^r}{r!} + \cdots .$$

30. *Central Limit Theorem*

(a) Using Theorems 5.8, 5.9, and 5.10, show that

$$M_{(\bar{X}-\mu)/(\sigma/\sqrt{n})}(t) = e^{-\mu\sqrt{n}t/\sigma} \left[M_X\left(\frac{t}{\sigma\sqrt{n}}\right) \right]^n,$$

where $\bar{X}$ is the mean of a random sample of size n from a population $f(x)$ with mean μ and variance σ^2, and hence

$$\ln M_{(\bar{X}-\mu)/(\sigma/\sqrt{n})}(t) = \frac{-\mu\sqrt{nt}}{\sigma} + n \ln M_X\left(\frac{t}{\sigma\sqrt{n}}\right).$$

(b) Use the result of Exercise 29 to expand $M_X(t/\sigma\sqrt{n})$ as an infinite series in powers of t. We can then write $M_X(t/\sigma\sqrt{n}) = 1 + v$, where v is an infinite series.

(c) Assuming n sufficiently large, expand $\ln(1 + v)$ in a Maclaurin series and then show that

$$\lim_{n\to\infty} \ln M_{(\bar{X}-\mu)/(\sigma/\sqrt{n})}(t) = \frac{t^2}{2}$$

and hence

$$\lim_{n\to\infty} M_{(\bar{X}-\mu)/(\sigma/\sqrt{n})}(t) = e^{t^2/2}.$$

31. A random sample of size 25 is taken from a normal population having a mean of 80 and a standard deviation of 5. A second random sample of size 36 is taken from a different normal population having a mean of 75 and a standard deviation of 3. Find the probability that the sample mean computed from the 25 measurements will exceed the sample mean computed from the 36 measurements by at least 3.4 but less than 5.9. Assume the means to be measured to the nearest tenth.

32. The mean score for freshmen on an aptitude test, at a certain college, is 540, with a standard deviation of 50. What is the probability that two groups of students selected

at random, consisting of 32 and 50 students, respectively, will differ in their mean scores by

(a) More than 20 points?

(b) An amount between 5 and 10 points?

Assume the means to be measured to any degree of accuracy.

33. For a chi-square distribution find
    (a) $\chi^2_{0.01}$ with $v = 18$.
    (b) $\chi^2_{0.975}$ with $v = 29$.
    (c) χ^2_a such that $P(X^2 < \chi^2_a) = 0.99$ with $v = 4$.

34. Find the probability that a random sample of 25 observations, from a normal population with variance $\sigma^2 = 6$, will have a variance s^2
    (a) Greater than 9.1.
    (b) Between 3.462 and 10.745.
    Assume the sample variances to be continuous measurements.

35. A placement test has been given for the past 5 years to college freshmen with a mean $\mu = 74$ and a variance $\sigma^2 = 8$. Would a school consider these values valid today if 20 students obtained a mean $\bar{x} = 72$ and a variance $s^2 = 16$ on this test?

36. Show that the variance of S^2 decreases as n becomes large. [*Hint*: First find the variance of $(n - 1)S^2/\sigma^2$.]

37. (a) Find $t_{0.025}$ when $v = 17$.
    (b) Find $t_{0.99}$ when $v = 10$.
    (c) Find t_a such that $P(-t_a < T < t_a) = 0.90$ when $v = 23$.

38. A normal population with unknown variance has a mean of 20. Is one likely to obtain a random sample of size 9 from this population with a mean of 24 and a standard deviation of 4.1? If not, what conclusion would you draw?

39. A cigarette manufacturer claims that his cigarettes have an average tar content of 18.3 milligrams. If a random sample of eight cigarettes of this type have tar contents of 20, 17, 21, 19, 22, 21, 20, and 16 milligrams, would you agree with the manufacturer's claim?

40. For an F distribution find
    (a) $f_{0.05}$ with $v_1 = 7$ and $v_2 = 15$.
    (b) $f_{0.05}$ with $v_1 = 15$ and $v_2 = 7$.
    (c) $f_{0.01}$ with $v_1 = 24$ and $v_2 = 19$.
    (d) $f_{0.95}$ with $v_1 = 19$ and $v_2 = 24$.
    (e) $f_{0.99}$ with $v_1 = 28$ and $v_2 = 12$.

41. If S_1^2 and S_2^2 represent the variances of independent random samples of size $n_1 = 25$ and $n_2 = 31$, taken from normal populations with variances $\sigma_1^2 = 10$ and $\sigma_2^2 = 15$, respectively, find the $P(S_1^2/S_2^2 > 1.26)$.

42. If S_1^2 and S_2^2 represent the variances of independent random samples of size $n_1 = 8$ and $n_2 = 12$, taken from normal populations with equal variances, find the $P(S_1^2/S_2^2 < 4.89)$. May want to take reciprocal & reverse the deg. of freedom

6

Estimation Theory

6.1 INTRODUCTION

The theory of *statistical inference* may be defined to be those methods by which one makes inferences or generalizations about a population. The trend of today is to distinguish between the *classical method* of estimating a population parameter, whereby inferences are based strictly on information obtained from a random sample selected from the population, and the *Bayesian method*, which utilizes prior subjective knowledge about the probability distribution of the unknown parameters in conjunction with the information provided by the sample data. Throughout most of this chapter we shall obtain classical estimates of unknown population parameters such as the mean, proportion, and the standard deviation by computing statistics from random samples and applying the theory of sampling distributions from Chapter 5. For completeness, the Bayesian approach to statistical decision theory is presented in Sections 6.10 and 6.11.

Statistical inference may be divided into two major areas: *estimation* and *tests of hypotheses*. We shall treat these two areas separately, dealing with the theory of estimation in this chapter and the theory of hypothesis testing in Chapter 7. To distinguish clearly between the two areas, consider the following examples. A candidate for public office may wish to estimate the true proportion of voters favoring him by obtaining the opinions from a random sample of 100 eligible voters. The fraction of voters in the sample favoring the candidate could be used as an estimate of the true proportion of the population of voters. A knowledge of the sampling distribution of a proportion enables one to establish the degree of accuracy of our estimate. This problem falls in the area of estimation.

Now consider the case in which a housewife is interested in finding out whether brand A floor wax is more scuff-resistant than brand B floor wax. She might hypothesize that brand A is better than brand B and, after proper testing, accept or reject this hypothesis. In this example we do not attempt to estimate a parameter, but instead we try to arrive at a correct decision about a prestated hypothesis. Once again we are dependent on sampling theory to provide us with some measure of accuracy for our decision.

6.2 CLASSICAL METHODS OF ESTIMATION

An estimate of a population parameter may be given as a *point estimate* or as an *interval estimate*. A point estimate of some population parameter θ is a single value $\hat{\theta}$ of a statistic $\hat{\Theta}$. For example, the value $\bar{x}$ of the statistic $\bar{X}$, computed from a sample of size n, is a point estimate of the population parameter μ.

The statistic that one uses to obtain a point estimate is called an *estimator* or a *decision function*. Hence the decision function S, which is a function of the random sample, is an estimator of σ and the estimate s is the "action" taken. Different samples will generally lead to different actions or estimates.

DEFINITION 6.1 *The set of all possible actions that can be taken in an estimation problem is called the* action space *or* decision space.

An estimator is not expected to estimate the population parameter without error. We do not expect $\bar{X}$ to estimate μ exactly, but we certainly hope that it is not too far off. For a particular sample it is possible to obtain a closer estimate of μ by using the sample median $\tilde{X}$ as an estimator. Consider, for instance, a sample consisting of the values 2, 5, and 11 from a population whose mean is 4 but supposedly unknown. We would estimate μ to be $\bar{x} = 6$, using the sample mean as our estimate, or $\tilde{x} = 5$, using the sample median as our estimate. In this case the estimator $\tilde{X}$ produces an estimate closer to the true parameter than that of the estimator $\bar{X}$. On the other hand, if our random sample contains the values 2, 6, and 7, then $\bar{x} = 5$ and $\tilde{x} = 6$, so that $\bar{X}$ is now the better estimator. Not knowing the true value of μ, we must decide in advance whether to use $\bar{X}$ or $\tilde{X}$ as our estimator.

What are the desirable properties of a "good" decision function that would influence us to choose one estimator rather than another? Let $\hat{\Theta}$ be an estimator whose value $\hat{\theta}$ is a point estimate of some unknown population parameter θ. Certainly, we would like the sampling distribution of $\hat{\Theta}$ to have a mean equal to the parameter estimated. An estimator possessing this property is said to be *unbiased*.

DEFINITION 6.2 *A statistic $\hat{\Theta}$ is said to be an* unbiased *estimator of the parameter θ if $\mu_{\hat{\Theta}} = E(\hat{\Theta}) = \theta$.*

Example 6.1 Show that S^2 is an unbiased estimator of the parameter σ^2.

Solution   Let us write

$$\sum_{i=1}^{n}(X_i - \bar{X})^2 = \sum_{i=1}^{n}[(X_i - \mu) - (\bar{X} - \mu)]^2$$

$$= \sum_{i=1}^{n}(X_i - \mu)^2 - 2(\bar{X} - \mu)\sum_{i=1}^{n}(X_i - \mu) + n(\bar{X} - \mu)^2$$

$$= \sum_{i=1}^{n}(X_i - \mu)^2 - n(\bar{X} - \mu)^2.$$

Now

$$E(S^2) = E\left[\frac{\sum_{i=1}^{n}(X_i - \bar{X})^2}{n - 1}\right]$$

$$= \frac{1}{n - 1}\left[\sum_{i=1}^{n}E(X_i - \mu)^2 - nE(\bar{X} - \mu)^2\right]$$

$$= \frac{1}{n - 1}\left(\sum_{i=1}^{n}\sigma_{X_i}^2 - n\sigma_{\bar{X}}^2\right).$$

However,

$$\sigma_{X_i}^2 = \sigma^2 \qquad \text{for } i = 1, 2, \ldots, n$$

and

$$\sigma_{X_i}^2 \; (\text{not } \bar{X}_i)$$

$$\sigma_{\bar{X}}^2 = \frac{\sigma^2}{n}.$$

Therefore,

$$E(S^2) = \frac{1}{n - 1}\left(n\sigma^2 - n\frac{\sigma^2}{n}\right) = \sigma^2.$$

Although S^2 is an unbiased estimator of σ^2, S, on the other hand, is a biased estimator of σ with the bias becoming insignificant for large samples.

If $\hat{\Theta}_1$ and $\hat{\Theta}_2$ are two unbiased estimators of the same population parameter θ, we would choose the estimator whose sampling distribution has the smallest variance. Hence, if $\sigma_{\hat{\Theta}_1}^2 < \sigma_{\hat{\Theta}_2}^2$, we say that $\hat{\Theta}_1$ is a *more efficient* estimator than $\hat{\Theta}_2$.

DEFINITION 6.3   *If we consider all possible unbiased estimators of some parameter θ, the one with the smallest variance is called the* most efficient estimator *of θ.*

In Figure 6.1 we illustrate the sampling distributions of three different estimators $\hat{\Theta}_1$, $\hat{\Theta}_2$, and $\hat{\Theta}_3$, all estimating θ. It is clear that only $\hat{\Theta}_1$ and $\hat{\Theta}_2$ are unbiased since their distributions are centered at θ. The estimator $\hat{\Theta}_1$ has a smaller variance than $\hat{\Theta}_2$ and is therefore more efficient. Hence our choice for an estimator of θ, among the three considered, would be $\hat{\Theta}_1$.

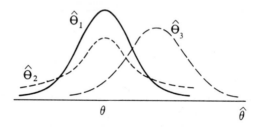

Figure 6.1 Sampling distributions of different estimators of θ.

An interval estimate of a population parameter θ is an interval of the form $\hat{\theta}_1 < \theta < \hat{\theta}_2$, where $\hat{\theta}_1$ and $\hat{\theta}_2$ depend on the value of the statistic $\hat{\Theta}$ for a particular sample and also on the sampling distribution of $\hat{\Theta}$. For the parameter μ we shall see that $\hat{\theta}_1 = x - k$ and $\hat{\theta}_2 - x + k$, where k is determined from the sampling distribution of X. Now, rather than claim x to be exactly equal to μ, we would feel more confident in stating that

$$x - k < \mu < x + k$$

Thus a random sample of SAT verbal scores for students of the entering freshman class might produce an interval from 530 to 550 within which we expect to find the true average of all SAT verbal scores for the freshman class. The values of the end points, 530 and 550, will depend on the computed sample mean $\bar{x}$ and the sampling distribution of $\overline{X}$. As the sample size increases, we

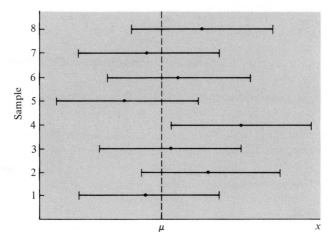

Figure 6.2 Interval estimates of μ for different samples.

know that $\sigma_{\bar{X}}^2 = \sigma^2/n$ decreases, and consequently our estimate is likely to be closer to the parameter μ, resulting in a shorter interval. Thus the interval estimate indicates, by its length, the accuracy of the point estimate.

Different samples yield different $\bar{X}$ values, and therefore produce different interval estimates of the parameter μ as shown in Figure 6.2. The circular dots at the center of each interval indicate the position of the point estimate $\bar{X}$ for each random sample. Most of the intervals are seen to contain μ, but not in every case. Note that all these intervals are of the same width, since their widths depend only on the choice of k once $\bar{X}$ is determined. The larger the value we choose for k, the wider we make all the intervals and the more confident we can be that the particular sample selected will produce an interval that contains the unknown parameter.

From the sampling distribution of $\hat{\Theta}$ we shall be able to determine $\hat{\theta}_1$ and $\hat{\theta}_2$ such that $P(\hat{\Theta}_1 < \theta < \hat{\Theta}_2)$ is equal to any value we care to specify. If, for instance, we find $\hat{\theta}_1$ and $\hat{\theta}_2$ such that $P(\hat{\Theta}_1 < \theta < \hat{\Theta}_2) = 0.95$, then we have a probability equal to 0.95 of selecting a random sample that will produce an interval that contains θ. This interval, computed from the selected random sample, is called a 95% *confidence interval*. In other words, we are 95% confident that our computed interval does, in fact, contain the true population parameter. Corresponding to a probability of 0.99, we would obtain a larger value of k and hence a wider interval. The wider the confidence interval is, the more confident we can be that the given interval contains the unknown parameter. Of course, it is better to be 95% confident that the average life of a certain television transistor is between 6 and 7 years than to be 99% confident that it is between 3 and 10 years. Ideally, we prefer a short interval with a high degree of confidence. Sometimes, restrictions on the size of our sample prevent us from achieving short intervals without sacrificing some of our degree of confidence.

Generally speaking, the distribution of $\hat{\Theta}$ enables us to compute k so that

$$P(\hat{\Theta} - k < \theta < \hat{\Theta} + k) = 1 - \alpha, \qquad 0 < \alpha < 1.$$

The interval computed from a particular sample is then called a $(1 - \alpha)100\%$ *confidence interval*. Thus, when $\alpha = 0.05$, we have a 95% confidence interval, and when $\alpha = 0.01$, we obtain a wider 99% confidence interval. The fraction $1 - \alpha$ is called the *confidence coefficient*, and the end points, $\hat{\theta} - k$ and $\hat{\theta} + k$, are called the *confidence limits* or *fiducial limits*.

6.3 ESTIMATING THE MEAN

A point estimator of the population mean μ is given by the statistic $\overline{X}$. The sampling distribution of $\overline{X}$ is centered at μ and in most applications the variance is smaller than that of any other estimator. Thus the sample mean $\bar{x}$ will be used as a point estimate for the population mean μ. Recall that $\sigma_{\overline{X}}^2 = \sigma^2/n$, so that a large sample will yield a value of $\overline{X}$ that comes from a sampling distribution with a small variance. Hence $\overline{X}$ is likely to estimate μ very closely when n is large.

Let us now consider the interval estimate of μ. If our sample is selected from a normal population or, failing this, if n is sufficiently large, we can establish a confidence interval for μ by considering the sampling distribution of $\overline{X}$. According to the central limit theorem, we can expect the sampling distribution of $\overline{X}$ to be approximately normally distributed with mean $\mu_{\overline{X}} = \mu$ and standard deviation $\sigma_{\overline{X}} = \sigma/\sqrt{n}$. Writing $z_{\alpha/2}$ for the z value above which we find an area of $\alpha/2$, we can see from Figure 6.3 that

$$P(-z_{\alpha/2} < Z < z_{\alpha/2}) = 1 - \alpha,$$

where

$$Z = \frac{\overline{X} - \mu}{\sigma/\sqrt{n}}.$$

Hence

$$P\left(-z_{\alpha/2} < \frac{\overline{X} - \mu}{\sigma/\sqrt{n}} < z_{\alpha/2}\right) = 1 - \alpha.$$

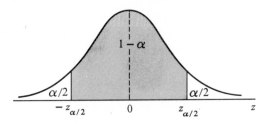

Figure 6.3 $P(-z_{\alpha/2} < Z < z_{\alpha/2}) = 1 - \alpha$.

Multiplying each term in the inequality by $\sigma/\sqrt{n}$, and then subtracting $\overline{X}$ from each term and multiplying by -1 (reversing the sense of the inequalities), we obtain

$$P\left(\overline{X} - z_{\alpha/2}\frac{\sigma}{\sqrt{n}} < \mu < \overline{X} + z_{\alpha/2}\frac{\sigma}{\sqrt{n}}\right) = 1 - \alpha.$$

A random sample of size n is selected from a population whose variance σ^2 is known and the mean $\bar{x}$ is computed to give the $(1 - \alpha)100\%$ confidence interval:

$$\bar{x} - z_{\alpha/2}\frac{\sigma}{\sqrt{n}} < \mu < \bar{x} + z_{\alpha/2}\frac{\sigma}{\sqrt{n}}.$$

CONFIDENCE INTERVAL FOR μ; σ KNOWN   *A $(1 - \alpha)100\%$ confidence interval for μ is*

$$\bar{x} - z_{\alpha/2}\frac{\sigma}{\sqrt{n}} < \mu < \bar{x} + z_{\alpha/2}\frac{\sigma}{\sqrt{n}},$$

where $\bar{x}$ is the mean of a sample of size n from a population with known variance σ^2 and $z_{\alpha/2}$ is the value of the standard normal distribution leaving an area of $\alpha/2$ to the right.

For small samples selected from nonnormal populations, we cannot expect our degree of confidence to be accurate. However, for samples of size $n \geq 30$, regardless of the shape of most populations, sampling theory guarantees good results.

To compute a $(1 - \alpha)100\%$ confidence interval for μ we have assumed that σ is known. Since this is generally not the case, we shall replace σ by the sample standard deviation s, provided that $n \geq 30$.

Example 6.2   The mean and standard deviation for the quality point averages of a random sample of 36 college seniors are calculated to be 2.6 and 0.3, respectively. Find the 95% and 99% confidence intervals for the mean of the entire senior class.

Solution   The point estimate of μ is $\bar{x} = 2.6$. Since the sample size is large, the standard deviation σ can be approximated by $s = 0.3$. The z value, leaving an area of 0.025 to the right and therefore an area of 0.975 to the left, is $z_{0.025} = 1.96$ (Table IV). Hence the 95% confidence interval is

$$\underset{\overset{|}{\bar{x}}}{2.6} - \underset{\overset{|}{z_{\alpha/2}}}{(1.96)}\overset{\left(\frac{s}{\sqrt{n}}\right)}{\left(\frac{0.3}{\sqrt{36}}\right)} < \mu < 2.6 + (1.96)\left(\frac{0.3}{\sqrt{36}}\right),$$

which reduces to

$$2.50 < \mu < 2.70.$$

To find a 99% confidence interval, we find the z value leaving an area of 0.005 to the right and 0.995 to the left. Therefore, using Table IV again, $z_{0.005} = 2.575$, and the 99% confidence interval is

$$2.6 - (2.575)\left(\frac{0.3}{\sqrt{36}}\right) < \mu < 2.6 + (2.575)\left(\frac{0.3}{\sqrt{36}}\right),$$

or simply

$$2.47 < \mu < 2.73.$$

We now see that a longer interval is required to estimate μ with a higher degree of accuracy.

The $(1 - \alpha)100\%$ confidence interval provides an estimate of the accuracy of our point estimate. If μ is actually the center value of the interval, then $\bar{x}$ estimates μ without error. Most of the time, however, $\bar{x}$ will not be exactly equal to μ and the point estimate is in error. The size of this error will be the difference between μ and $\bar{x}$, and we can be $(1 - \alpha)100\%$ confident that this difference will be less than $z_{\alpha/2}\sigma/\sqrt{n}$. We can readily see this if we draw a diagram of the confidence interval as in Figure 6.4.

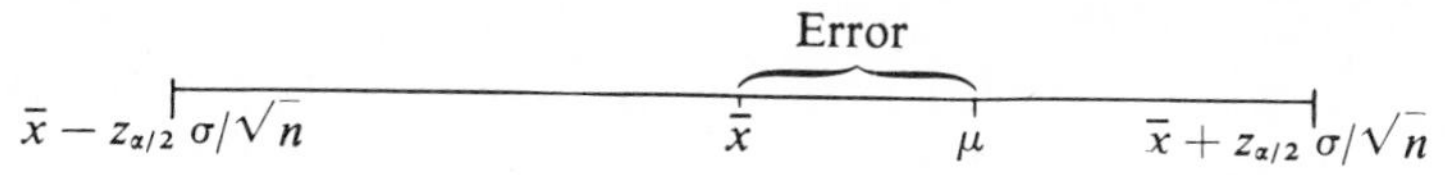

Error

$$\bar{x} - z_{\alpha/2}\, \sigma/\sqrt{n} \qquad\qquad\qquad \bar{x} \qquad \mu \qquad\qquad \bar{x} + z_{\alpha/2}\, \sigma/\sqrt{n}$$

Figure 6.4 Error in estimating μ by $\bar{x}$.

THEOREM 6.1 *If $\bar{x}$ is used as an estimate of μ, we can be $(1 - \alpha)100\%$ confident that the error will be less than $z_{\alpha/2}\,\sigma/\sqrt{n}$.*

In Example 6.2 we are 95% confident that the sample mean $\bar{x} = 2.6$ differs from the true mean μ by an amount less than 0.1 and 99% confident that the difference is less than 0.13.

Frequently, we wish to know how large a sample is necessary to ensure that the error in estimating μ will be less than a specified amount e. By Theorem 6.1 this means we must choose n such that $z_{\alpha/2}\,\sigma/\sqrt{n} = e$.

THEOREM 6.2 *If $\bar{x}$ is used as an estimate of μ, we can be $(1 - \alpha)100\%$ confident that the error will be less than a specified amount e when the sample size is*

$$n = \left(\frac{z_{\alpha/2}\,\sigma}{e}\right)^2.$$

e = error

Strictly speaking, the formula in Theorem 6.2 is applicable only if we know the variance of the population from which we are to select our sample. Lacking this information, we could take a preliminary sample of size $n \geq 30$ to provide an estimate of σ; then, using Theorem 6.2, we could determine approximately how many observations are needed to provide the desired degree of accuracy.

Example 6.3 How large a sample is required in Example 6.2 if we want to be 95% confident that our estimate of μ is off by less than 0.05?

Solution The sample standard deviation $s = 0.3$ obtained from the preliminary sample of size 36 will be used for σ. Then, by Theorem 6.2,

$$n = \left[\frac{(1.96)(0.3)}{0.05}\right]^2 = 138.3.$$

Therefore, we can be 95% confident that a random sample of size 139 will provide an estimate $\bar{x}$ differing from μ by an amount less than 0.05.

Frequently, we are attempting to estimate the mean of a population when the variance is unknown and it is impossible to obtain a sample of size $n \geq 30$. Cost can often be a factor that limits our sample size. As long as our population is approximately bell-shaped, confidence intervals can be computed when σ^2 is unknown and the sample size is small by using the sampling distribution of T, where

$$T = \frac{\bar{X} - \mu}{S/\sqrt{n}}.$$

The procedure is the same as for large samples except that we use the t distribution in place of the standard normal.

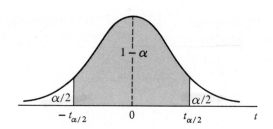

Figure 6.5 $P(-t_{\alpha/2} < T < t_{\alpha/2}) = 1 - \alpha$.

Referring to Figure 6.5, we can assert that

$$P(-t_{\alpha/2} < T < t_{\alpha/2}) = 1 - \alpha,$$

where $t_{\alpha/2}$ is the t value with $n - 1$ degrees of freedom, above which we find an area of $\alpha/2$. Because of symmetry, an equal area of $\alpha/2$ will fall to the left of $-t_{\alpha/2}$. Substituting for T, we write

$$P\left(-t_{\alpha/2} < \frac{\bar{X} - \mu}{S/\sqrt{n}} < t_{\alpha/2}\right) = 1 - \alpha.$$

Multiplying each term in the inequality by $S/\sqrt{n}$, and then subtracting $\bar{X}$ from each term and multiplying by -1, we obtain

$$P\left(\bar{X} - \frac{t_{\alpha/2}\,S}{\sqrt{n}} < \mu < \bar{X} + \frac{t_{\alpha/2}\,S}{\sqrt{n}}\right) = 1 - \alpha.$$

For our particular random sample of size n, the mean $\bar{x}$ and standard deviation s are computed and the $(1 - \alpha)100\%$ confidence interval is given by

$$\bar{x} - \frac{t_{\alpha/2}\,s}{\sqrt{n}} < \mu < \bar{x} + \frac{t_{\alpha/2}\,s}{\sqrt{n}}.$$

CONFIDENCE INTERVAL FOR μ; σ UNKNOWN AND $n < 30$ *A* $(1 - \alpha)100\%$
confidence interval for μ is

$$\bar{x} - \frac{t_{\alpha/2}\,s}{\sqrt{n}} < \mu < \bar{x} + \frac{t_{\alpha/2}\,s}{\sqrt{n}},$$

where $\bar{x}$ and s are the mean and standard deviation, respectively, of a sample of size $n < 30$ from an approximate normal population, and $t_{\alpha/2}$ is the value of the t distribution, with $v = n - 1$ degrees of freedom, leaving an area of $\alpha/2$ to the right.

Example 6.4   The contents of seven similar containers of sulfuric acid are 9.8, 10.2, 10.4, 9.8, 10.0, 10.2, and 9.6 liters. Find a 95% confidence interval for the mean of all such containers, assuming an approximate normal distribution.

Solution   The sample mean and standard deviation for the given data are

$$\bar{x} = 10.0 \quad \text{and} \quad s = 0.283.$$

Using Table V, we find $t_{0.025} = 2.447$ for $v = 6$ degrees of freedom. Hence the 95% confidence interval for μ is

$$10.0 - (2.447)\left(\frac{0.283}{\sqrt{7}}\right) < \mu < 10.0 + (2.447)\left(\frac{0.283}{\sqrt{7}}\right),$$

which reduces to

$$9.74 < \mu < 10.26.$$

6.4 ESTIMATING THE DIFFERENCE BETWEEN TWO MEANS

If we have two populations with means μ_1 and μ_2 and variances σ_1^2 and σ_2^2, respectively, a point estimator of the difference between μ_1 and μ_2 is given by the statistic $\overline{X}_1 - \overline{X}_2$. Therefore, to obtain a point estimate of $\mu_1 - \mu_2$, we shall select two independent random samples, one from each population, of size n_1 and n_2, and compute the difference, $\bar{x}_1 - \bar{x}_2$, of the sample means.

If our independent samples are selected from normal populations, or failing this, if n_1 and n_2 are both greater than 30, we can establish a confidence interval for $\mu_1 - \mu_2$ by considering the sampling distribution of $\overline{X}_1 - \overline{X}_2$.

According to Theorem 5.15, we can expect the sampling distribution of $\overline{X}_1 - \overline{X}_2$ to be approximately normally distributed with mean $\mu_{\overline{X}_1 - \overline{X}_2} = \mu_1 - \mu_2$ and standard deviation $\sigma_{\overline{X}_1 - \overline{X}_2} = \sqrt{(\sigma_1^2/n_1) + (\sigma_2^2/n_2)}$. Therefore, we can assert with a probability of $1 - \alpha$ that the standard normal variable

$$Z = \frac{(\overline{X}_1 - \overline{X}_2) - (\mu_1 - \mu_2)}{\sqrt{(\sigma_1^2/n_1) + (\sigma_2^2/n_2)}}$$

will fall between $-z_{\alpha/2}$ and $z_{\alpha/2}$. Referring once again to Figure 6.3, we write

$$P(-z_{\alpha/2} < Z < z_{\alpha/2}) = 1 - \alpha.$$

Substituting for Z, we state equivalently that

$$P\left[-z_{\alpha/2} < \frac{(\overline{X}_1 - \overline{X}_2) - (\mu_1 - \mu_2)}{\sqrt{(\sigma_1^2/n_1) + (\sigma_2^2/n_2)}} < z_{\alpha/2}\right] = 1 - \alpha.$$

Multiplying each term of the inequality by $\sqrt{(\sigma_1^2/n_1) + (\sigma_2^2/n_2)}$, and then subtracting $\overline{X}_1 - \overline{X}_2$ from each term and multiplying by -1, we have

$$P\left[(\overline{X}_1 - \overline{X}_2) - z_{\alpha/2}\sqrt{\frac{\sigma_1^2}{n_1} + \frac{\sigma_2^2}{n_2}} < \mu_1 - \mu_2 < (\overline{X}_1 - \overline{X}_2)\right.$$

$$\left. + z_{\alpha/2}\sqrt{\frac{\sigma_1^2}{n_1} + \frac{\sigma_2^2}{n_2}}\right] = 1 - \alpha.$$

For any two independent random samples of size n_1 and n_2 selected from two populations whose variances σ_1^2 and σ_2^2 are known, the difference of the sample means, $\bar{x}_1 - \bar{x}_2$, is computed and the $(1 - \alpha)100\%$ confidence interval is given by

$$(\bar{x}_1 - \bar{x}_2) - z_{\alpha/2}\sqrt{\frac{\sigma_1^2}{n_1} + \frac{\sigma_2^2}{n_2}} < \mu_1 - \mu_2 < (\bar{x}_1 - \bar{x}_2) + z_{\alpha/2}\sqrt{\frac{\sigma_1^2}{n_1} + \frac{\sigma_2^2}{n_2}}.$$

CONFIDENCE INTERVAL FOR $\mu_1 - \mu_2$; σ_1^2 AND σ_2^2 KNOWN A $(1 - \alpha)100\%$
confidence interval for $\mu_1 - \mu_2$ is

$$(\bar{x}_1 - \bar{x}_2) - z_{\alpha/2}\sqrt{\frac{\sigma_1^2}{n_1} + \frac{\sigma_2^2}{n_2}} < \mu_1 - \mu_2 < (\bar{x}_1 - \bar{x}_2) + z_{\alpha/2}\sqrt{\frac{\sigma_1^2}{n_1} + \frac{\sigma_2^2}{n_2}},$$

where $\bar{x}_1$ and $\bar{x}_2$ are means of independent random samples of size n_1 and n_2 from populations with known variances σ_1^2 and σ_2^2, respectively, and $z_{\alpha/2}$ is the value of the standard normal curve leaving an area of $\alpha/2$ to the right.

The degree of confidence is exact when samples are selected from normal populations. For nonnormal populations we obtain an approximate confidence interval that is very good when both n_1 and n_2 exceed 30. As before, if σ_1^2 and σ_2^2 are unknown and our samples are sufficiently large, we may replace σ_1^2 by s_1^2 and σ_2^2 by s_2^2 without appreciably affecting the confidence interval.

Example 6.5   A standardized chemistry test was given to 50 girls and 75 boys. The girls made an average grade of 76 with a standard deviation of 6, while the boys made an average grade of 82 with a standard deviation of 8. Find a 96% confidence interval for the difference $\mu_1 - \mu_2$, where μ_1 is the mean score of all boys and μ_2 is the mean score of all girls who might take this test.

Solution   The point estimate of $\mu_1 - \mu_2$ is $\bar{x}_1 - \bar{x}_2 = 82 - 76 = 6$. Since n_1 and n_2 are both large, we can substitute $s_1 = 8$ for σ_1 and $s_2 = 6$ for σ_2. Using $\alpha = 0.04$, we find $z_{0.02} = 2.054$ from Table IV. Hence substituting in the formula

$$(\bar{x}_1 - \bar{x}_2) - z_{\alpha/2}\sqrt{\frac{\sigma_1^2}{n_1} + \frac{\sigma_2^2}{n_2}} < \mu_1 - \mu_2 < (\bar{x}_1 - \bar{x}_2) + z_{\alpha/2}\sqrt{\frac{\sigma_1^2}{n_1} + \frac{\sigma_2^2}{n_2}}$$

yields the 96% confidence interval

$$6 - 2.054\sqrt{\tfrac{64}{75} + \tfrac{36}{50}} < \mu_1 - \mu_2 < 6 + 2.054\sqrt{\tfrac{64}{75} + \tfrac{36}{50}},$$

or

$$3.42 < \mu_1 - \mu_2 < 8.58.$$

This procedure for estimating the difference between two means is applicable if σ_1^2 and σ_2^2 are known or can be estimated from large samples. If the sample sizes are small, we must again resort to the t distribution to provide confidence intervals that are valid when the populations are approximately normally distributed.

Let us now assume σ_1^2 and σ_2^2 are unknown and that n_1 and n_2 are small (<30). If $\sigma_1^2 = \sigma_2^2 = \sigma^2$, we obtain a standard normal variable in the form

$$Z = \frac{(\bar{X}_1 - \bar{X}_2) - (\mu_1 - \mu_2)}{\sqrt{\sigma^2[(1/n_1) + (1/n_2)]}}.$$

According to Theorem 5.16 the random variables $(n_1 - 1)S_1^2/\sigma^2$ and $(n_2 - 1)S_2^2/\sigma^2$ have chi-square distributions with $n_1 - 1$ and $n_2 - 1$ degrees of freedom, respectively. Furthermore, they are independent chi-square variables since the random samples were selected independently. Consequently, their sum

$$V = \frac{(n_1 - 1)S_1^2}{\sigma^2} + \frac{(n_2 - 1)S_2^2}{\sigma^2} = \frac{(n_1 - 1)S_1^2 + (n_2 - 1)S_2^2}{\sigma^2}$$

has a chi-square distribution with $v = n_1 + n_2 - 2$ degrees of freedom.

Replacing Z and V in Theorem 5.17 by the preceding expressions, we obtain the statistic

$$T = \frac{(\bar{X}_1 - \bar{X}_2) - (\mu_1 - \mu_2)}{\sqrt{\sigma^2[(1/n_1) + (1/n_2)]}} \bigg/ \sqrt{\frac{(n_1 - 1)S_1^2 + (n_2 - 1)S_2^2}{\sigma^2(n_1 + n_2 - 2)}},$$

which has the t distribution with $v = n_1 + n_2 - 2$ degrees of freedom.

A point estimate of the unknown common variance σ^2 can be obtained by pooling the sample variances. Denoting the pooled estimator by S_p^2, we write

$$S_p^2 = \frac{(n_1 - 1)S_1^2 + (n_2 - 1)S_2^2}{n_1 + n_2 - 2}.$$

Substituting S_p^2 in the T statistic, we obtain the less cumbersome form

$$T = \frac{(\bar{X}_1 - \bar{X}_2) - (\mu_1 - \mu_2)}{S_p\sqrt{(1/n_1) + (1/n_2)}}.$$

Using the statistic T, we have

$$P(-t_{\alpha/2} < T < t_{\alpha/2}) = 1 - \alpha,$$

where $t_{\alpha/2}$ is the t value with $n_1 + n_2 - 2$ degrees of freedom, above which we find an area of $\alpha/2$. Substituting for T in the inequality, we write

$$P\left[-t_{\alpha 2} < \frac{(\bar{X}_1 - \bar{X}_2) - (\mu_1 - \mu_2)}{S_p\sqrt{(1/n_1) + (1/n_2)}} < t_{\alpha 2}\right] = 1 - \alpha.$$

Multiplying each term of the inequality by $S_p\sqrt{(1/n_1) + (1/n_2)}$, and then subtracting $\bar{X}_1 - \bar{X}_2$ from each term and multiplying by -1, we obtain

$$P\left[(\bar{X}_1 - \bar{X}_2) - t_{\alpha/2} S_p\sqrt{\frac{1}{n_1} + \frac{1}{n_2}} < \mu_1 - \mu_2 < (\bar{X}_1 - \bar{X}_2)\right.$$

$$\left. + t_{\alpha/2} S_p\sqrt{\frac{1}{n_1} + \frac{1}{n_2}}\right] = 1 - \alpha.$$

For any two independent random samples of size n_1 and n_2 selected from two normal populations, the difference of the sample means, $\bar{x}_1 - \bar{x}_2$, and the pooled standard deviation, s_p, are computed and the $(1 - \alpha)100\%$ confidence interval is given by

$$(\bar{x}_1 - \bar{x}_2) - t_{\alpha/2} s_p\sqrt{\frac{1}{n_1} + \frac{1}{n_2}} < \mu_1 - \mu_2 < (\bar{x}_1 - \bar{x}_2) + t_{\alpha/2} s_p\sqrt{\frac{1}{n_1} + \frac{1}{n_2}}.$$

SMALL-SAMPLE CONFIDENCE INTERVAL FOR $\mu_1 - \mu_2$; $\sigma_1^2 = \sigma_2^2$ **BUT UNKNOWN**
A $(1 - \alpha)100\%$ *confidence interval for* $\mu_1 - \mu_2$ *is*

$$(\bar{x}_1 - \bar{x}_2) - t_{\alpha/2} s_p\sqrt{\frac{1}{n_1} + \frac{1}{n_2}} < \mu_1 - \mu_2 < (\bar{x}_1 - \bar{x}_2) + t_{\alpha/2} s_p\sqrt{\frac{1}{n_1} + \frac{1}{n_2}},$$

where $\bar{x}_1$ *and* $\bar{x}_2$ *are the means of small independent samples of size* n_1 *and* n_2, *respectively, from approximate normal distributions,* s_p *is the pooled standard deviation, and* $t_{\alpha/2}$ *is the value of the t distribution with* $v = n_1 + n_2 - 2$ *degrees of freedom, leaving an area of* $\alpha/2$ *to the right.*

Example 6.6   In a batch chemical process, two catalysts are being compared for their effect on the output of the process reaction. A sample of 12 batches are prepared using catalyst 1 and a sample of 10 batches were obtained using catalyst 2. The 12 batches for which catalyst 1 was used gave an average yield of 85 with a sample standard deviation of 4, while the average for the second sample gave an average of 81 and a sample standard deviation of 5. Find a 90% confidence interval for the difference between the population means, assuming the populations are approximately normally distributed with equal variances.

Solution   Let μ_1 and μ_2 represent the population means of all yields using catalyst 1 and catalyst 2, respectively. We wish to find a 90% confidence interval for $\mu_1 - \mu_2$. Our point estimate of $\mu_1 - \mu_2$ is $\bar{x}_1 - \bar{x}_2 = 85 - 81 = 4$. The pooled estimate, s_p^2, of the common variance, σ^2, is

$$s_p^2 = \frac{(n_1 - 1)s_1^2 + (n_2 - 1)s_2^2}{n_1 + n_2 - 2}$$

$$= \frac{(11)(16) + (9)(25)}{12 + 10 - 2} = 20.05.$$

Taking the square root, $s_p = 4.478$. Using $\alpha = 0.1$, we find in Table V that $t_{0.05} = 1.725$ for $v = n_1 + n_2 - 2 = 20$ degrees of freedom. Therefore, substituting in the formula

$$(\bar{x}_1 - \bar{x}_2) - t_{\alpha/2} s_p \sqrt{\frac{1}{n_1} + \frac{1}{n_2}} < \mu_1 - \mu_2 < (\bar{x}_1 - \bar{x}_2) + t_{\alpha/2} s_p \sqrt{\frac{1}{n_1} + \frac{1}{n_2}}$$

we obtain the 90% confidence interval

$$4 - (1.725)(4.478)\sqrt{\tfrac{1}{12} + \tfrac{1}{10}} < \mu_1 - \mu_2 < 4 + (1.725)(4.478)\sqrt{\tfrac{1}{12} + \tfrac{1}{10}},$$

which simplifies to

$$0.69 < \mu_1 - \mu_2 < 7.31.$$

Hence we are 90% confident that the interval from 0.69 to 7.31 contains the true difference of the yields for the two catalysts. The fact that both confidence limits are positive indicates that catalyst 1 is superior to catalyst 2.

The procedure for constructing confidence intervals for $\mu_1 - \mu_2$ from small samples assumes the populations to be normal and the population variances to be equal. Slight departures from either of these assumptions do not seriously alter the degree of confidence for our interval. A procedure will be presented in Chapter 7 for testing the equality of two unknown population variances based on the information provided by the sample variances. If the population variances are considerably different, we still obtain good results when the populations are normal, provided that $n_1 = n_2$. Therefore, in a planned experiment, one should make every effort to equalize the size of the samples.

Let us now consider the problem of finding an interval estimate of $\mu_1 - \mu_2$ for small samples when the unknown population variances are not likely to be equal, and it is impossible to select samples of equal size. The statistic most often used in this case is

$$T' = \frac{(\bar{X}_1 - \bar{X}_2) - (\mu_1 - \mu_2)}{\sqrt{(S_1^2/n_1) + (S_2^2/n_2)}},$$

which has approximately a t distribution with v degrees of freedom, where

$$v = \frac{(s_1^2/n_1 + s_2^2/n_2)^2}{[(s_1^2/n_1)^2/(n_1 - 1)] + [(s_2^2/n_2)^2/(n_2 - 1)]}.$$

Since v is seldom an integer, we round it off to the nearest whole number.

Using the statistic T', we write

$$P(-t_{\alpha/2} < T' < t_{\alpha/2}) \simeq 1 - \alpha,$$

where $t_{\alpha/2}$ is the value of the t distribution with v degrees of freedom, above which we find an area of $\alpha/2$. Substituting for T' in the inequality, and following the exact steps as before, we state the final result.

SMALL-SAMPLE CONFIDENCE INTERVAL FOR $\mu_1 - \mu_2$; $\sigma_1^2 \neq \sigma_2^2$ AND UN-KNOWN   *An approximate* $(1 - \alpha)100\%$ *confidence interval for* $\mu_1 - \mu_2$ *is*

$$(\bar{x}_1 - \bar{x}_2) - t_{\alpha/2}\sqrt{\frac{s_1^2}{n_1} + \frac{s_2^2}{n_2}} < \mu_1 - \mu_2 < (\bar{x}_1 - \bar{x}_2) + t_{\alpha/2}\sqrt{\frac{s_1^2}{n_1} + \frac{s_2^2}{n_2}},$$

where $\bar{x}_1$ and s_1^2, and $\bar{x}_2$ and s_2^2, are the means and variances of small independent samples of size n_1 and n_2, respectively, from approximate normal distributions, and $t_{\alpha/2}$ is the value of the t distribution with

$$v = \frac{(s_1^2/n_1 + s_2^2/n_2)^2}{[(s_1^2/n_1)^2/(n_1 - 1)] + [(s_2^2/n_2)^2/(n_2 - 1)]}$$

degrees of freedom, leaving an area of $\alpha/2$ to the right.

Example 6.7 Records for the past 15 years have shown the average rainfall in a certain region of the country for the month of May to be 4.93 centimeters, with a standard deviation of 1.14 centimeters. A second region of the country has had an average rainfall in May of 2.64 centimeters of rain, with a standard deviation of 0.66 centimeter during the past 10 years. Find a 95% confidence interval for the difference of the true average rainfalls in these two regions, assuming that the observations came from normal populations with different variances.

Solution For the first region we have $\bar{x}_1 = 4.93$, $s_1 = 1.14$, and $n_1 = 15$. For the second region $\bar{x}_2 = 2.64$, $s_2 = 0.66$, and $n_2 = 10$. We wish to find a 95% confidence interval for $\mu_1 - \mu_2$. Since the population variances are assumed to be unequal and our sample sizes are not the same, we can only find an approximate 95% confidence interval based on the t distribution with v degrees of freedom, where

$$v = \frac{(s_1^2/n_1 + s_2^2/n_2)^2}{[(s_1^2/n_1)^2/(n_1 - 1)] + [(s_2^2/n_2)^2/(n_2 - 1)]}$$

$$= \frac{(1.14^2/15 + 0.66^2/10)^2}{[(1.14^2/15)^2/14] + [(0.66^2/10)^2/9]}$$

$$= 22.7 \simeq 23.$$

Our point estimate of $\mu_1 - \mu_2$ is $\bar{x}_1 - \bar{x}_2 = 4.93 - 2.64 = 2.29$. Using $\alpha = 0.05$, we find in Table V that $t_{0.025} = 2.069$ for $v = 23$ degrees of freedom. Therefore, substituting in the formula

$$(\bar{x}_1 - \bar{x}_2) - t_{\alpha/2}\sqrt{\frac{s_1^2}{n_1} + \frac{s_2^2}{n_2}} < \mu_1 - \mu_2 < (\bar{x}_1 - \bar{x}_2) + t_{\alpha/2}\sqrt{\frac{s_1^2}{n_1} + \frac{s_2^2}{n_2}}$$

we obtain the approximate 95% confidence interval

$$2.29 - 2.069\sqrt{\frac{1.14^2}{15} + \frac{0.66^2}{10}} < \mu_1 - \mu_2 < 2.29 + 2.069\sqrt{\frac{1.14^2}{15} + \frac{0.66^2}{10}},$$

which simplifies to

$$2.02 < \mu_1 - \mu_2 < 2.56.$$

Hence we are 95% confident that the interval from 2.02 to 2.56 contains the true difference of the average rainfall for the two regions.

We conclude this section by considering estimation procedures for the difference of two means when the samples are not independent and the variances of the two populations are not necessarily equal. This will be true if the observations in the two samples occur in pairs so that the two observations are related. For instance, if we run a test on a new diet using 15 individuals, the weights before and after completion of the test form our two samples. Observations in the two samples made on the same individual are related and hence form a pair. To determine if the diet is effective, we must consider the differences d_i of paired observations. These differences are the values of a random sample $D_1, D_2, \ldots, D_n$ from a population that we shall assume to be normal with mean μ_D and unknown variance σ_D^2. We estimate σ_D^2 by s_d^2, the variance of the differences constituting the sample. Therefore, s_d^2 is a value of the statistic S_d^2 that fluctuates from sample to sample. The point estimator of $\mu_1 - \mu_2 = \mu_D$ is given by $\bar{D}$.

A $(1 - \alpha)100\%$ confidence interval for μ_D can be established by writing

$$P(-t_{\alpha/2} < T < t_{\alpha/2}) = 1 - \alpha,$$

where

$$T = \frac{\bar{D} - \mu_D}{S_d/\sqrt{n}}$$

and $t_{\alpha/2}$, as before, is a value of the t distribution with $n - 1$ degrees of freedom.

It is now a routine procedure to replace T, by its definition, in the above inequality and carry out the mathematical steps that lead to the $(1 - \alpha)100\%$ confidence interval:

$$\bar{d} - t_{\alpha/2} \frac{s_d}{\sqrt{n}} < \mu_D < \bar{d} + t_{\alpha/2} \frac{s_d}{\sqrt{n}}.$$

CONFIDENCE INTERVAL FOR $\mu_1 - \mu_2 = \mu_D$ FOR PAIRED OBSERVATIONS
A $(1 - \alpha)100\%$ *confidence interval for* μ_D *is*

$$\bar{d} - t_{\alpha/2} \frac{s_d}{\sqrt{n}} < \mu_D < \bar{d} + t_{\alpha/2} \frac{s_d}{\sqrt{n}},$$

where $\bar{d}$ *and* s_d *are the mean and standard deviation of the differences of n pairs of measurements and* $t_{\alpha/2}$ *is the value of the t distribution with* $v = n - 1$ *degrees of freedom, leaving an area of* $\alpha/2$ *to the right.*

Example 6.8   Twenty college freshmen were divided into 10 pairs, each member of the pair having approximately the same IQ. One of each pair was selected at random and assigned to a mathematics section using programmed materials only. The other member of each pair was assigned to a section in which the professor lectured. At the end of the semester each group was given the same examination and the following results were recorded:

Pair	Programmed materials	Lectures	d
1	76	81	-5
2	60	52	8
3	85	87	-2
4	58	70	-12
5	91	86	5
6	75	77	-2
7	82	90	-8
8	64	63	1
9	79	85	-6
10	88	83	5

Find a 98% confidence interval for the true difference in the average grades of the two learning procedures.

Solution   We wish to find a 98% confidence interval for $\mu_1 - \mu_2$, where μ_1 and μ_2 represent the average grades of all students by the programmed and lecture method of presentation, respectively. Since the observations are paired, $\mu_1 - \mu_2 = \mu_D$. The point estimate of μ_D is given by $\bar{d} = -1.6$. The variance s_d^2 of the sample differences is

$$s_d^2 = \frac{n \sum d_i^2 - (\sum d_i)^2}{n(n-1)}$$

$$= \frac{(10)(392) - (-16)^2}{(10)(9)} = 40.7.$$

Taking the square root, we obtain $s_d = 6.38$. Using $\alpha = 0.02$, we find in Table V that $t_{0.01} = 2.821$ for $v = n - 1 = 9$ degrees of freedom. Therefore, substituting in the formula

$$\bar{d} - t_{\alpha/2}\frac{s_d}{\sqrt{n}} < \mu_D < \bar{d} + t_{\alpha/2}\frac{s_d}{\sqrt{n}},$$

we obtain the 98 % confidence interval,

$$-1.6 - (2.821)\frac{6.38}{\sqrt{10}} < \mu_D < -1.6 + (2.821)\frac{6.38}{\sqrt{10}}$$

or simply

$$-7.29 < \mu_D < 4.09.$$

Hence we are 98 % confident that the interval from -7.29 to 4.09 contains the true difference of the average grades for the two methods of instruction. Since this interval allows for the possibility of μ_D being equal to zero, we are unable to state that one method of instruction is better than the other, even though this particular sample of differences shows the lecture procedure to be superior.

6.5 ESTIMATING A PROPORTION

A point estimator of the proportion p in a binomial experiment is given by the statistic $\hat{P} = X/n$. Therefore, the sample proportion $\hat{p} = x/n$ will be used as the point estimate for the parameter p.

   If the unknown proportion p is not expected to be too close to zero or 1, we can establish a confidence interval for p by considering the sampling distribution of $\hat{P}$, which, of course, is the same as that of the random variable X except for a change in scale. Hence, by Theorem 4.1, the distribution of $\hat{P}$ is approximately normally distributed with mean

$$\mu_{\hat{P}} = E(\hat{P}) = E\left(\frac{X}{n}\right) = \frac{np}{n} = p$$

and variance

$$\sigma_{\hat{P}}^2 = \sigma_{X/n}^2 = \frac{\sigma_X^2}{n^2} = \frac{npq}{n^2} = \frac{pq}{n}.$$

Therefore, we can assert that

$$P(-z_{\alpha/2} < Z < z_{\alpha/2}) = 1 - \alpha,$$

where

$$Z = \frac{\hat{P} - p}{\sqrt{pq/n}}.$$

and $z_{\alpha/2}$ is the value of the standard normal curve above which we find an area of $\alpha/2$. Substituting for Z, we write

$$P\left(-z_{\alpha/2} < \frac{\hat{P} - p}{\sqrt{pq/n}} < z_{\alpha/2}\right) = 1 - \alpha.$$

Multiplying each term of the inequality by $\sqrt{pq/n}$, and then subtracting $\hat{P}$ and multiplying by -1, we obtain

$$P\left(\hat{P} - z_{\alpha/2}\sqrt{\frac{pq}{n}} < p < \hat{P} + z_{\alpha/2}\sqrt{\frac{pq}{n}}\right) = 1 - \alpha.$$

It is difficult to manipulate the inequalities so as to obtain a random interval whose end points are independent of p, the unknown parameter. When n is large, very little error is introduced by substituting the point estimate $\hat{p} = x/n$ for the p under the radical sign. Then we can write

$$P\left(\hat{P} - z_{\alpha/2}\sqrt{\frac{\hat{p}\hat{q}}{n}} < p < \hat{P} + z_{\alpha/2}\sqrt{\frac{\hat{p}\hat{q}}{n}}\right) \simeq 1 - \alpha.$$

For our particular random sample of size n, the sample proportion $\hat{p} = x/n$ is computed, and the approximate $(1 - \alpha)100\%$ confidence interval for p is given by

$$\hat{p} - z_{\alpha/2}\sqrt{\frac{\hat{p}\hat{q}}{n}} < p < \hat{p} + z_{\alpha/2}\sqrt{\frac{\hat{p}\hat{q}}{n}}.$$

CONFIDENCE INTERVAL FOR p; $n \geq 30$ *A $(1 - \alpha)100\%$ confidence interval for the binomial parameter p is approximately*

$$\hat{p} - z_{\alpha/2}\sqrt{\frac{\hat{p}\hat{q}}{n}} < p < \hat{p} + z_{\alpha/2}\sqrt{\frac{\hat{p}\hat{q}}{n}},$$

where $\hat{p}$ is the proportion of successes in a random sample of size n, $\hat{q} = 1 - \hat{p}$, and $z_{\alpha/2}$ is the value of the standard normal curve leaving an area of $\alpha/2$ to the right.

The method for finding a confidence interval for the binomial parameter p is also applicable when the binomial distribution is being used to approximate the hypergeometric distribution, that is, when n is small relative to N as illustrated in Example 6.9.

Example 6.9 In a random sample of $n = 500$ families owning television sets in the city of Hamilton, Canada, it was found that $x = 340$ owned color sets. Find a 95% confidence interval for the actual proportion of families in this city with color sets.

Solution The point estimate of p is $\hat{p} = 340/500 = 0.68$. Using Table IV, we find $z_{0.025} = 1.96$. Therefore, substituting in the formula

$$\hat{p} - z_{\alpha/2}\sqrt{\frac{\hat{p}\hat{q}}{n}} < p < \hat{p} + z_{\alpha/2}\sqrt{\frac{\hat{p}\hat{q}}{n}},$$

we obtain the 95% confidence interval,

$$0.68 - 1.96\sqrt{\frac{(0.68)(0.32)}{500}} < p < 0.68 + 1.96\sqrt{\frac{(0.68)(0.32)}{500}},$$

which simplifies to

$$0.64 < p < 0.72.$$

If p is the center value of a $(1 - \alpha)100\%$ confidence interval, then $\hat{p}$ estimates p without error. Most of the time, however, $\hat{p}$ will not be exactly equal to p and the point estimate is in error. The size of this error will be the difference between p and $\hat{p}$, and we can be $(1 - \alpha)100\%$ confident that this difference will be less than $z_{\alpha/2}\sqrt{\hat{p}\hat{q}/n}$. We can readily see this if we draw a diagram of the confidence interval as in Figure 6.6.

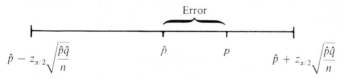

Figure 6.6 Error in estimating p by $\hat{p}$.

THEOREM 6.3 *If $\hat{p}$ is used as an estimate of p, we can be $(1 - \alpha)100\%$ confident that the error will be less than $z_{\alpha/2}\sqrt{\hat{p}\hat{q}/n}$.*

In Example 6.9 we are 95% confident that the sample proportion $\hat{p} = 0.32$ differs from the true proportion p by an amount less than 0.04.

Let us now determine how large a sample is necessary to ensure that the error in estimating p will be less than a specified amount e. By Theorem 6.3, this means we must choose n such that $z_{\alpha/2}\sqrt{\hat{p}\hat{q}/n} = e$.

THEOREM 6.4 *If $\hat{p}$ is used as an estimate of p, we can be $(1 - \alpha)100\%$ confident that the error will be less than a specified amount e when the sample size is*

$$n = \frac{z_{\alpha/2}^2 \hat{p}\hat{q}}{e^2}. \qquad e = \text{ERROR}$$

Theorem 6.4 is somewhat misleading in that we must use $\hat{p}$ to determine the sample size n, but $\hat{p}$ is computed from the sample. If a crude estimate of p can be made without taking a sample, we could use this value for $\hat{p}$ and then determine n. Lacking such an estimate, we could take a preliminary sample of size $n \geq 30$ to provide an estimate of p. Then, using Theorem 6.4 we could determine approximately how many observations are needed to provide the desired degree of accuracy.

An upper bound for n can be established for any degree of confidence by noting that $\hat{p}\hat{q} = \hat{p}(1 - \hat{p})$, which must be at most equal to $1/4$, since $\hat{p}$ must lie between 0 and 1. This fact may be verified by completing the square. Hence,

$$\hat{p}(1 - \hat{p}) = -(\hat{p}^2 - \hat{p}) = \tfrac{1}{4} - (\hat{p}^2 - \hat{p} + \tfrac{1}{4})$$
$$= \tfrac{1}{4} - (\hat{p} - \tfrac{1}{2})^2,$$

which is always less than $1/4$ except when $\hat{p} = 1/2$ and then $\hat{p}\hat{q} = 1/4$.

THEOREM 6.5 *If $\hat{p}$ is used as an estimate of p, we can be at least $(1 - \alpha)100\%$ confident that the error will be less than a specified amount e when the sample size is*

$$n = \frac{z_{\alpha/2}^2}{4e^2}.$$

Example 6.10   How large a sample is required in Example 6.9 if we want to be (1) 95% confident that our estimate of p is off by less than 0.02, and (2) at least 95% confident?

Solution   1. Let us treat the 500 families as a preliminary sample providing an estimate $\hat{p} = 0.32$. Then, by Theorem 6.4,

$$n = \frac{(1.96)^2(0.32)(0.68)}{(0.02)^2} = 2090.$$

Therefore, if we base our estimate of p on a random sample of size 2090, we can be 95% confident that our sample proportion will not differ from the true proportion by more than 0.02.

2. According to Theorem 6.5, we can be at least 95% confident that our sample proportion will not differ from the true proportion by more than 0.02 if we choose a sample of size

$$n = \frac{(1.96)^2}{4(0.02)^2} = 2401.$$

Comparing the two parts of Example 6.10, we see that information concerning p, provided by a preliminary sample or perhaps from past experience, enables us to choose a smaller sample.

6.6 ESTIMATING THE DIFFERENCE BETWEEN TWO PROPORTIONS

Consider independent samples of size n_1 and n_2 selected at random from two binomial populations with means $n_1 p_1$ and $n_2 p_2$ and variances $n_1 p_1 q_1$ and $n_2 p_2 q_2$, respectively. We denote the proportion of successes in each sample by $\hat{p}_1$ and $\hat{p}_2$. A point estimator of the difference between the two proportions $p_1 - p_2$ is given by the statistic $\hat{P}_1 - \hat{P}_2$.

A confidence interval for $p_1 - p_2$ can be established by considering the sampling distribution of $\hat{P}_1 - \hat{P}_2$. From Section 6.5 we know that $\hat{P}_1$ and $\hat{P}_2$ are each approximately normally distributed, with means p_1 and p_2 and variances $p_1 q_1 / n_1$ and $p_2 q_2 / n_2$, respectively. By choosing independent samples from the two populations, the variables $\hat{P}_1$ and $\hat{P}_2$ will be independent, and then by the reproductive property of the normal distribution established in Theorem 5.11, we conclude that $\hat{P}_1 - \hat{P}_2$ is approximately normally distributed with mean

$$\mu_{\hat{P}_1 - \hat{P}_2} = p_1 - p_2$$

and variance

$$\sigma^2_{\hat{P}_1 - \hat{P}_2} = \frac{p_1 q_1}{n_1} + \frac{p_2 q_2}{n_2}.$$

Therefore, we can assert that

$$P(-z_{\alpha/2} < Z < z_{\alpha/2}) = 1 - \alpha,$$

where

$$Z = \frac{(\hat{P}_1 - \hat{P}_2) - (p_1 - p_2)}{\sqrt{(p_1 q_1/n_1) + (p_2 q_2/n_2)}}.$$

and $z_{\alpha/2}$ is a value of the standard normal curve above which we find an area of $\alpha/2$. Substituting for Z, we write

$$P\left[-z_{\alpha/2} < \frac{(\hat{P}_1 - \hat{P}_2) - (p_1 - p_2)}{\sqrt{(p_1 q_1/n_1) + (p_2 q_2/n_2)}} < z_{\alpha/2}\right] = 1 - \alpha.$$

Multiplying each term of the inequality by $\sqrt{(p_1 q_1/n_1) + (p_2 q_2/n_2)}$, and then subtracting $\hat{P}_1 - \hat{P}_2$ and multiplying by -1, we obtain

$$P\left[(\hat{P}_1 - \hat{P}_2) - z_{\alpha/2}\sqrt{\frac{p_1 q_1}{n_1} + \frac{p_2 q_2}{n_2}} < p_1 - p_2 < (\hat{P}_1 - \hat{P}_2)\right.$$

$$\left. + z_{\alpha/2}\sqrt{\frac{p_1 q_1}{n_1} + \frac{p_2 q_2}{n_2}}\right] = 1 - \alpha.$$

If n_1 and n_2 are both large, we replace p_1 and p_2 under the radical sign by their estimates $\hat{p}_1 = x_1/n_1$ and $\hat{p}_2 = x_2/n_2$. Then

$$P\left[(\hat{P}_1 - \hat{P}_2) - z_{\alpha/2}\sqrt{\frac{\hat{p}_1 \hat{q}_1}{n_1} + \frac{\hat{p}_2 \hat{q}_2}{n_2}} < p_1 - p_2 < (\hat{P}_1 - \hat{P}_2)\right.$$

$$\left. + z_{\alpha/2}\sqrt{\frac{\hat{p}_1 \hat{q}_1}{n_1} + \frac{\hat{p}_2 \hat{q}_2}{n_2}}\right] \simeq 1 - \alpha.$$

For any two independent random samples of size n_1 and n_2, selected from two binomial populations, the difference of the sample proportions, $\hat{p}_1 - \hat{p}_2$, is computed and the $(1 - \alpha)100\%$ confidence interval is given by

$$(\hat{p}_1 - \hat{p}_2) - z_{\alpha/2}\sqrt{\frac{\hat{p}_1\hat{q}_1}{n_1} + \frac{\hat{p}_2\hat{q}_2}{n_2}} < p_1 - p_2 < (\hat{p}_1 - \hat{p}_2) + z_{\alpha/2}\sqrt{\frac{\hat{p}_1\hat{q}_1}{n_1} + \frac{\hat{p}_2\hat{q}_2}{n_2}}.$$

CONFIDENCE INTERVAL FOR $p_1 - p_2$; n_1 AND $n_2 \geq 30$ $A(1 - \alpha)100\%$ *confidence interval for the difference of two binomial parameters, $p_1 - p_2$, is approximately*

$$(\hat{p}_1 - \hat{p}_2) - z_{\alpha/2}\sqrt{\frac{\hat{p}_1\hat{q}_1}{n_1} + \frac{\hat{p}_2\hat{q}_2}{n_2}} < p_1 - p_2 < (\hat{p}_1 - \hat{p}_2)$$

$$+ z_{\alpha/2}\sqrt{\frac{\hat{p}_1\hat{q}_1}{n_1} + \frac{\hat{p}_2\hat{q}_2}{n_2}},$$

where $\hat{p}_1$ and $\hat{p}_2$ are the proportion of successes in random samples of size n_1 and n_2, respectively, $\hat{q}_1 = 1 - \hat{p}_1$ and $\hat{q}_2 = 1 - \hat{p}_2$, and $z_{\alpha/2}$ is the value of the standard normal curve leaving an area of $\alpha/2$ to the right.

Example 6.11   A certain change in a manufacturing procedure for component parts is being considered. Samples are taken using both the existing and the new procedure in order to determine if the new procedure results in an improvement. If 75 of 1500 items from the existing procedure were found to be defective and 80 of 2000 items from the new procedure were found to be defective, find a 90% confidence interval for the true difference in the fraction of defectives between the existing and the new process.

Solution   Let p_1 and p_2 be the true proportions of defectives for the existing and new procedures, respectively. Hence $\hat{p}_1 = 75/1500 = 0.05$ and $\hat{p}_2 = 80/2000 = 0.04$, and the point estimate of $p_1 - p_2$ is $\hat{p}_1 - \hat{p}_2 = 0.05 - 0.04 = 0.01$. Using Table IV, we find $z_{0.05} = 1.645$. Therefore, substituting into this formula we obtain the 90% confidence interval

$$0.01 - 1.645\sqrt{\frac{(0.05)(0.95)}{1500} + \frac{(0.04)(0.96)}{2000}} < p_1 - p_2$$

$$< 0.01 + 1.645\sqrt{\frac{(0.05)(0.95)}{1500} + \frac{(0.04)(0.96)}{2000}},$$

which simplifies to

$$-0.0017 < p_1 - p_2 < 0.0217.$$

Since the interval contains the value zero, there is no reason to believe that the new procedure produced a significant decrease in the proportion of defectives over the existing method.

6.7 ESTIMATING THE VARIANCE

An unbiased point estimate of the population variance σ^2 is provided by the sample variance s^2. Hence the statistic S^2 is called an *estimator* of σ^2.

An interval estimate of σ^2 can be established by using the statistic

$$X^2 = \frac{(n-1)S^2}{\sigma^2}.$$

According to Theorem 5.16, the statistic X^2 has a chi-square distribution with $n-1$ degrees of freedom when samples are chosen from a normal population.

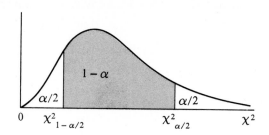

Figure 6.7 $P(\chi^2_{1-\alpha/2} < X^2 < \chi^2_{\alpha/2}) = 1 - \alpha.$

We may write (see Figure 6.7)

$$P(\chi^2_{1-\alpha/2} < X^2 < \chi^2_{\alpha/2}) = 1 - \alpha,$$

where $\chi^2_{1-\alpha/2}$ and $\chi^2_{\alpha/2}$ are values of the chi-square distribution with $n-1$ degrees of freedom, leaving areas of $1 - \alpha/2$ and $\alpha/2$, respectively, to the right. Substituting for X^2, we write

$$P\left[\chi^2_{1-\alpha/2} < \frac{(n-1)S^2}{\sigma^2} < \chi^2_{\alpha/2}\right] = 1 - \alpha.$$

Dividing each term in the inequality by $(n - 1)S^2$, and then inverting each term (thereby changing the sense of the inequalities), we obtain

$$P\left[\frac{(n - 1)S^2}{\chi^2_{\alpha/2}} < \sigma^2 < \frac{(n - 1)S^2}{\chi^2_{1-\alpha/2}}\right] = 1 - \alpha.$$

For our particular sample of size n, the sample variance s^2 is computed, and the $(1 - \alpha)100\%$ confidence interval is given by

$$\frac{(n - 1)s^2}{\chi^2_{\alpha/2}} < \sigma^2 < \frac{(n - 1)s^2}{\chi^2_{1-\alpha/2}}.$$

CONFIDENCE INTERVAL FOR σ^2 *A $(1 - \alpha)100\%$ confidence interval for the variance σ^2 of a normal population is*

$$\frac{(n - 1)s^2}{\chi^2_{\alpha/2}} < \sigma^2 < \frac{(n - 1)s^2}{\chi^2_{1-\alpha/2}},$$

where s^2 is the variance of a random sample of size n, and $\chi^2_{\alpha/2}$ and $\chi^2_{1-\alpha/2}$ are the values of a chi-square distribution with $\nu = n - 1$ degrees of freedom, leaving areas of $\alpha/2$ and $1 - \alpha/2$, respectively, to the right.

Example 6.12   The following are the weights, in decigrams, of 10 packages of grass seed distributed by a certain company: 46.4, 46.1, 45.8, 47.0, 46.1, 45.9, 45.8, 46.9, 45.2, and 46.0. Find a 95% confidence interval for the variance of all such packages of grass seed distributed by this company.

Solution   First we find

$$s^2 = \frac{n \sum_{i=1}^{n} x_i^2 - \left(\sum_{i=1}^{n} x_i\right)^2}{n(n - 1)} = \frac{(10)(21{,}273.12) - (461.2)^2}{(10)(9)}$$

$$= 0.286.$$

To obtain a 95% confidence interval, we choose $\alpha = 0.05$. Then, using Table VI with $\nu = 9$ degrees of freedom we find $\chi^2_{0.025} = 19.023$ and $\chi^2_{0.975} = 2.700$. Substituting in the formula

$$\frac{(n - 1)s^2}{\chi^2_{\alpha/2}} < \sigma^2 < \frac{(n - 1)s^2}{\chi^2_{1-\alpha/2}},$$

we obtain the 95% confidence interval

$$\frac{(9)(0.286)}{19.023} < \sigma^2 < \frac{(9)(0.286)}{2.700},$$

or simply

$$0.135 < \sigma^2 < 0.953.$$

6.8 ESTIMATING THE RATIO OF TWO VARIANCES

A point estimate of the ratio of two population variances σ_1^2/σ_2^2 is given by the ratio s_1^2/s_2^2 of the sample variances. Hence the statistic S_1^2/S_2^2 is called an estimator of σ_1^2/σ_2^2.

If σ_1^2 and σ_2^2 are the variances of normal populations, we can establish an interval estimate of σ_1^2/σ_2^2 by using the statistic

$$F = \frac{\sigma_2^2 S_1^2}{\sigma_1^2 S_2^2}.$$

According to Theorem 5.20 the random variable F has an F distribution with $v_1 = n_1 - 1$ and $v_2 = n_2 - 1$ degrees of freedom. Therefore, we may write (see Figure 6.8)

$$P[f_{1-\alpha/2}(v_1, v_2) < F < f_{\alpha/2}(v_1, v_2)] = 1 - \alpha,$$

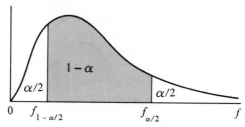

Figure 6.8 $P[f_{1-\alpha/2}(v_1, v_2) < F < f_{\alpha/2}(v_1, v_2)] = 1 - \alpha.$

where $f_{1-\alpha/2}(v_1, v_2)$ and $f_{\alpha/2}(v_1, v_2)$ are the values of the F distribution with v_1 and v_2 degrees of freedom, leaving areas of $1 - \alpha/2$ and $\alpha/2$, respectively, to the right. Substituting for F, we write

$$P\left[f_{1-\alpha/2}(v_1, v_2) < \frac{\sigma_2^2 S_1^2}{\sigma_1^2 S_2^2} < f_{\alpha/2}(v_1, v_2)\right] = 1 - \alpha.$$

Multiplying each term in the inequality by S_2^2/S_1^2, and then inverting each term (again changing the sense of the inequalities), we obtain

$$P\left[\frac{S_1^2}{S_2^2}\frac{1}{f_{\alpha/2}(v_1, v_2)} < \frac{\sigma_1^2}{\sigma_2^2} < \frac{S_1^2}{S_2^2}\frac{1}{f_{1-\alpha/2}(v_1, v_2)}\right] = 1 - \alpha.$$

The results of Theorem 5.19 enable us to replace $f_{1-\alpha/2}(v_1, v_2)$ by $1/f_{\alpha/2}(v_2, v_1)$. Therefore,

$$P\left[\frac{S_1^2}{S_2^2}\frac{1}{f_{\alpha/2}(v_1, v_2)} < \frac{\sigma_1^2}{\sigma_2^2} < \frac{S_1^2}{S_2^2}f_{\alpha/2}(v_2, v_1)\right] = 1 - \alpha.$$

For any two independent random samples of size n_1 and n_2 selected from two normal populations, the ratio of the sample variances, s_1^2/s_2^2, is computed and the $(1 - \alpha)100\%$ confidence interval for σ_1^2/σ_2^2 is

$$\frac{s_1^2}{s_2^2}\frac{1}{f_{\alpha/2}(v_1, v_2)} < \frac{\sigma_1^2}{\sigma_2^2} < \frac{s_1^2}{s_2^2}f_{\alpha/2}(v_2, v_1).$$

CONFIDENCE INTERVAL FOR σ_1^2/σ_2^2 *A $(1 - \alpha)100\%$ confidence interval for the ratio σ_1^2/σ_2^2 is*

$$\frac{s_1^2}{s_2^2}\frac{1}{f_{\alpha/2}(v_1, v_2)} < \frac{\sigma_1^2}{\sigma_2^2} < \frac{s_1^2}{s_2^2}f_{\alpha/2}(v_2, v_1),$$

where s_1^2 and s_2^2 are the variances of independent samples of size n_1 and n_2, respectively, from normal populations, $f_{\alpha/2}(v_1, v_2)$ is a value of the F distribution with $v_1 = n_1 - 1$ and $v_2 = n_2 - 1$ degrees of freedom, leaving an area of $\alpha/2$ to the right, and $f_{\alpha/2}(v_2, v_1)$ is a similar f value with $v_2 = n_2 - 1$ and $v_1 = n_1 - 1$ degrees of freedom.

Example 6.13   A standardized placement test in mathematics was given to 25 boys and 16 girls. The boys made an average grade of 82 with a standard deviation of 8, while the girls made an average grade of 78 with a standard deviation of 7. Find a 98% confidence interval for σ_1^2/σ_2^2 and σ_1/σ_2, where σ_1^2 and σ_2^2 are the variances of the populations of grades for all boys and girls, respectively, who at some time have taken or will take this test.

Solution We have $n_1 = 25$, $n_2 = 16$, $s_1 = 8$, and $s_2 = 7$. For a 98 % confidence interval, $\alpha = 0.02$. Using Table VII, $f_{0.01}(24,15) = 3.29$, and $f_{0.01}(15, 24) = 2.89$. Substituting in the formula

$$\frac{s_1^2}{s_2^2} \frac{1}{f_{\alpha/2}(v_1, v_2)} < \frac{\sigma_1^2}{\sigma_2^2} < \frac{s_1^2}{s_2^2} f_{\alpha/2}(v_2, v_1),$$

we obtain the 98 % confidence interval

$$\frac{64}{49}\left(\frac{1}{3.29}\right) < \frac{\sigma_1^2}{\sigma_2^2} < \frac{64}{49}(2.89),$$

which simplifies to

$$0.397 < \frac{\sigma_1^2}{\sigma_2^2} < 3.775.$$

Taking square roots of the confidence limits, we find that a 98 % confidence interval for σ_1/σ_2 is

$$0.630 < \frac{\sigma_1}{\sigma_2} < 1.943.$$

6.9 TOLERANCE LIMITS

In Section 6.2 we defined a confidence interval for a parameter θ to be an interval of the form $\hat{\theta} - k < \theta < \hat{\theta} + k$, where the point estimate $\hat{\theta}$ is computed from the particular sample and the value of k depends on the sampling distribution of the estimator $\hat{\Theta}$ of the parameter θ. That is, we computed confidence limits so that a specified fraction of the intervals computed from all possible samples of the same size contain the population parameter θ. For example, when all possible samples of the same size n are selected from a normal distribution, 95 % of the intervals determined by the confidence limits $\bar{x} \pm 1.96 \, \sigma/\sqrt{n}$ will contain the parameter μ. This fact was demonstrated in Figure 6.2. Hence we are 95 % confident that the interval $\bar{x} \pm 1.96 \, \sigma/\sqrt{n}$, computed from a particular sample, will contain the parameter μ.

Industrial engineers often find it more important to establish a confidence interval not containing the single value μ, but rather a confidence interval that will contain a fixed proportion or percentage of the measurements corresponding to some characteristic of a manufactured product. Such intervals are called *tolerance intervals* and their end points are called *tolerance limits*. One would expect a 95% tolerance interval that contains 0.90 or 90% of the measurements to be wider than the corresponding 95% confidence interval containing only the single parameter μ.

If we can assume from past experience that the measurements of our product are normally distributed with known mean μ and known standard deviation σ, the tolerance limits containing a specified proportion $1 - \alpha$ of the measurements are of the form $\mu \pm z_{\alpha/2}\sigma$, where $z_{\alpha/2}$ is determined from the table of normal curve areas. For a tolerance interval containing 0.90 of the measurements, $z_{0.05} = 1.645$. Unfortunately, in most industrial processes the values of μ and σ are seldom known. Nevertheless, tolerance limits can be defined by substituting the sample estimates $\bar{x}$ and s for μ and σ, respectively, and forming the interval $\bar{x} \pm ks$, where k is obtained from Table VIII (see Statistical Tables).

TOLERANCE LIMITS *For a normal distribution of measurements with unknown mean μ and unknown standard deviation σ, tolerance limits are given by $\bar{x} \pm ks$, where k is determined so that one can assert with $100\gamma\%$ confidence that the given limits contain* at least *the proportion $1 - \alpha$ of the measurements.*

Table VIII gives values of k for $1 - \alpha = 0.90$, 0.95, 0.99; $\gamma = 0.95$, 0.99; and for selected values of n from 2 to 1000.

Example 6.14 A machine is producing metal pieces that are cylindrical in shape. A sample of these pieces is taken and the diameters are found to be 1.01, 0.97, 1.03, 1.04, 0.99, 0.98, 0.99, 1.01, and 1.03 centimeters. Find the 99% tolerance limits that will contain 95% of the metal pieces produced by this machine, assuming an approximate normal distribution.

Solution The sample mean and standard deviation for the given data are

$$\bar{x} = 1.0056 \quad \text{and} \quad s = 0.0245.$$

From Table VIII for $n = 9$, $\gamma = 0.99$, and $1 - \alpha = 0.95$, we find $k = 4.550$. Hence the 99% tolerance limits are given by

$$1.0056 \pm (4.550)(0.0245),$$

or 0.894 and 1.117. That is, we are 99% confident that the tolerance interval from 0.894 to 1.117 will contain 95% of the metal pieces produced by this machine. It is interesting to note that the corresponding 99% confidence interval for μ

(see Exercise 9) has a lower limit of 0.978 and an upper limit of 1.033, verifying our earlier statement that a tolerance interval must necessarily be longer than a confidence interval with the same degree of confidence.

6.10 BAYESIAN METHODS OF ESTIMATION

The classical methods of estimation that we have studied so far are based solely on information provided by the random sample. These methods essentially interpret probabilities as relative frequencies. For example, in arriving at a 95% confidence interval for μ, we interpret the statement $P(-1.96 < Z < 1.96) = 0.95$ to mean that 95% of the time in repeated experiments Z will fall between -1.96 and 1.96. Probabilities of this type that can be interpreted in the frequency sense will be referred to as *objective probabilities*. The Bayesian approach to statistical methods of estimation combines sample information with other available prior information that may appear to be pertinent. The probabilities associated with this prior information are called *subjective probabilities*, in that they measure a person's *degree of belief* in a proposition. The person uses his own experience and knowledge as the basis for arriving at a subjective probability.

   Consider the problem of finding a point estimate of the parameter θ for the population $f(x; \theta)$. The classical approach would be to take a random sample of size n and substitute the information provided by the sample into the appropriate estimator or decision function. Thus in the case of a binomial population $b(x; n, p)$ our estimate of the proportion of successes would be $\hat{p} = x/n$. Suppose, however, that additional information is given about θ, namely, that it is known to vary according to the probability distribution $f(\theta)$. That is, we are now assuming θ to be a value of a random variable Θ with probability distribution $f(\theta)$ and we wish to estimate the particular value θ for the population from which we selected our random sample. We define $f(\theta)$ to be the *prior distribution* for the unknown variable parameter Θ. Note that $f(\theta)$ expresses our degree of belief of the location of Θ prior to sampling. Bayesian techniques use the prior distribution $f(\theta)$ along with the joint distribution of the sample, $f(x_1, x_2, \ldots, x_n; \theta)$ to compute the *posterior distribution* $f(\theta | x_1, x_2, \ldots, x_n)$. The posterior distribution consists of information from both the subjective prior distribution and the objective sampling distribution and expresses our degree of belief of the location of Θ after we have observed the sample.

   Let us write $f(x_1, x_2, \ldots, x_n | \theta)$ instead of $f(x_1, x_2, \ldots, x_n; \theta)$ for the joint probability distribution of our sample whenever we wish to indicate that the parameter is also a random variable. The joint distribution of the sample $X_1, X_2, \ldots, X_n$ and the parameter Θ is then

$$f(x_1, x_2, \ldots, x_n, \theta) = f(x_1, x_2, \ldots, x_n | \theta) f(\theta),$$

from which we readily obtain the marginal distribution

$$g(x_1, x_2, \ldots, x_n) = \sum_\theta f(x_1, x_2, \ldots, x_n, \theta) \qquad \text{(discrete case)}$$

$$= \int_{-\infty}^{\infty} f(x_1, x_2, \ldots, x_n, \theta) \, d\theta \qquad \text{(continuous case)}.$$

Hence the posterior distribution may be written

$$f(\theta \,|\, x_1, x_2, \ldots, x_n) = \frac{f(x_1, x_2, \ldots, x_n, \theta)}{g(x_1, x_2, \ldots, x_n)}.$$

DEFINITION 6.4　*The mean of the posterior distribution* $f(\theta \,|\, x_1, x_2, \ldots, x_n)$, *denoted by* θ^*, *is called the* Bayes estimate *for* θ.

Example 6.15　Using a random sample of size 2, estimate the proportion p of defectives produced by a machine when we assume our prior distribution to be

p	0.1	0.2
$f(p)$	0.6	0.4

Solution　Let X be the number of defectives in our sample. Then the probability distribution for our sample is

$$f(x \,|\, p) = b(x; n, p) = \binom{2}{x} p^x q^{2-x}, \qquad x = 0, 1, 2.$$

From the fact that $f(x, p) = f(x \,|\, p) f(p)$, we can set up the following table:

	x		
$f(x, p)$	0	1	2
p			
0.1	0.486	0.108	0.006
0.2	0.256	0.128	0.016

The marginal distribution for X is then

x	0	1	2
$g(x)$	0.742	0.236	0.022

We obtain the posterior distribution from the formula $f(p|x) = f(x, p)/g(x)$. Hence we have

p	0.1	0.2
$f(p|x = 0)$	0.655	0.345

p	0.1	0.2
$f(p|x = 1)$	0.458	0.542

p	0.1	0.2
$f(p|x = 2)$	0.273	0.727

from which we get

$$
\begin{aligned}
p^* &= (0.1)(0.655) + (0.2)(0.345) = 0.1345, &&\text{if } x = 0 \\
&= (0.1)(0.458) + (0.2)(0.542) = 0.1542, &&\text{if } x = 1 \\
&= (0.1)(0.273) + (0.2)(0.727) = 0.1727, &&\text{if } x = 2.
\end{aligned}
$$

Example 6.16 Repeat Example 6.15 using the uniform prior distribution $f(p) = 1, 0 < p < 1$.

Solution As before, we find that

$$
f(x|p) = \binom{2}{x} p^x q^{2-x}, \qquad x = 0, 1, 2.
$$

Now

$$
\begin{aligned}
f(x, p) &= f(x|p)f(p) \\
&= \binom{2}{x} p^x q^{2-x} \\
&= (1 - p)^2, &&\text{for } x = 0, 0 < p < 1 \\
&= 2p(1 - p), &&\text{for } x = 1, 0 < p < 1 \\
&= p^2, &&\text{for } x = 2, 0 < p < 1
\end{aligned}
$$

and the marginal distribution for X is obtained by evaluating the integral

$$g(x) = \int_0^1 (1 - p)^2 \, dp = \tfrac{1}{3}, \qquad \text{for } x = 0$$

$$= \int_0^1 2p(1 - p) \, dp = \tfrac{1}{3}, \qquad \text{for } x = 1$$

$$= \int_0^1 p^2 \, dp = \tfrac{1}{3}, \qquad \text{for } x = 2.$$

The posterior distribution is then

$$f(p|x) = \frac{f(x, p)}{g(x)}$$

$$= 3\binom{2}{x} p^x q^{2 - x}$$

$$= 3(1 - p)^2, \qquad x = 0, 0 < p < 1$$
$$= 6p(1 - p), \qquad x = 1, 0 < p < 1$$
$$= 3p^2, \qquad x = 2, 0 < p < 1$$

from which we evaluate the point estimate of our parameter to be

$$p^* = 3 \int_0^1 p(1 - p)^2 \, dp = \tfrac{1}{4}, \qquad \text{if } x = 0$$

$$= 6 \int_0^1 p^2(1 - p) \, dp = \tfrac{1}{2}, \qquad \text{if } x = 1$$

$$= 3 \int_0^1 p^2 \, dp = 1, \qquad \text{if } x = 2.$$

Comparing these estimates with the values obtained by classical procedures, we see that p^* and $\hat{p}$ are equivalent if $x = 1$ or $x = 2$ but that $\hat{p} = 0$ for $x = 0$.

A $(1 - \alpha)100\%$ Bayesian interval for the parameter θ can be constructed by finding an interval centered at the posterior mean that contains $(1 - \alpha)100\%$ of the posterior probability.

DEFINITION 6.5   *The interval $a < \theta < b$ will be called a $(1 - \alpha)100\%$ Bayes interval for θ if*

$$\int_{\theta^*}^{b} f(\theta \,|\, x_1, x_2, \ldots, x_n) \, d\theta = \int_{a}^{\theta^*} f(\theta \,|\, x_1, x_2, \ldots, x_n) \, d\theta = \frac{1 - \alpha}{2}.$$

Example 6.17   An electrical firm manufactures light bulbs that have a length of life that is approximately normally distributed with a standard deviation of 100 hours. Prior experience leads us to assume μ to be a value of a normal random variable M with a mean equal to 800 hours and a standard deviation of 10 hours. If a random sample of 25 bulbs have an average life of 780 hours, find a 95% Bayes interval for μ.

Solution   Multiplying the density of our sample

$$f(x_1, x_2, \ldots, x_{25} \,|\, \mu) = \frac{1}{(2\pi)^{25/2} \cdot 100^{25}} \exp\left[-\left(\frac{1}{2}\right) \sum_{i=1}^{25} \left(\frac{x_i - \mu}{100}\right)^2\right],$$

$$-\infty < x_i < \infty, \quad i = 1, 2, \ldots, 25$$

by our prior

$$f(\mu) = \frac{1}{\sqrt{2\pi} \cdot 10} e^{-(1/2)[(\mu - 800)/10]^2}, \quad -\infty < \mu < \infty,$$

we obtain the joint density of the random sample and M. That is,

$$f(x_1, x_2, \ldots, x_{25}, \mu) = \frac{1}{(2\pi)^{13} \cdot 10^{51}}$$

$$\times \exp\left[-\left(\frac{1}{2}\right)\left\{\sum_{i=1}^{25} \left(\frac{x_i - \mu}{100}\right)^2 + \left(\frac{\mu - 800}{10}\right)^2\right\}\right].$$

In Section 5.6 we established the identity

$$\sum_{i=1}^{n} (x_i - \mu)^2 = \sum_{i=1}^{n} (x_i - \bar{x})^2 + n(\bar{x} - \mu)^2,$$

which enables us to write

$$f(x_1, x_2, \ldots, x_{25}, \mu) = \frac{1}{(2\pi)^{13} \cdot 10^{51}} \exp\left[-\left(\frac{1}{2}\right) \sum_{i=1}^{25} \left(\frac{x_i - 780}{100}\right)^2 \right]$$

$$\times \, e^{-(1/2)\{25[(780-\mu)/100]^2 + [(\mu-800)/10]^2\}}.$$

Completing the square in the second exponent, we have

$$25\left(\frac{780 - \mu}{100}\right)^2 + \left(\frac{\mu - 800}{10}\right)^2 = \frac{\mu^2 - 1592\mu + 635{,}280}{80}$$

$$= \frac{(\mu - 796)^2 + 1664}{80}.$$

The joint density of the sample and M can now be written

$$f(x_1, x_2, \ldots, x_{25}, \mu) = K e^{-(1/2)[(\mu - 796)/\sqrt{80}]^2},$$

where K is a function of the sample values. The marginal distribution of the sample is then

$$g(x_1, x_2, \ldots, x_{25}) = K\sqrt{2\pi}\,\sqrt{80} \int_{-\infty}^{\infty} \frac{1}{\sqrt{2\pi}\,\sqrt{80}}\, e^{-(1/2)[(\mu - 796)/\sqrt{80}]^2}\, d\mu$$

$$= K\sqrt{2\pi}\,\sqrt{80},$$

and the posterior distribution is

$$f(\mu \mid x_1, x_2, \ldots, x_{25}) = \frac{f(x_1, x_2, \ldots, x_{25}, \mu)}{g(x_1, x_2, \ldots, x_{25})}$$

$$= \frac{1}{\sqrt{2\pi}\,\sqrt{80}}\, e^{-(1/2)[(\mu - 796)/\sqrt{80}]^2}, \qquad -\infty < \mu < \infty,$$

which is normal with mean $\mu^* = 796$ and standard deviation $\sigma^* = \sqrt{80}$. The 95% Bayes interval for μ is then given by

$$\mu^* - 1.96\sigma^* < \mu < \mu^* + 1.96\sigma^*.$$

That is,

$$796 - 1.96\sqrt{80} < \mu < 796 + 1.96\sqrt{80}$$

or

$$778.5 < \mu < 813.5.$$

Ignoring the prior information about μ and comparing this result with that given by the classical 95% confidence interval

$$780 - (1.96)\tfrac{100}{5} < \mu < 780 + (1.96)\tfrac{100}{5}$$

or

$$740.8 < \mu < 819.2,$$

we notice that the Bayes interval is shorter than the classical confidence interval.

6.11 DECISION THEORY

In our discussion of the classical approach to point estimation, we adopted the criterion that selects the decision function that is most efficient. That is, we choose from all possible unbiased estimators the one with the smallest variance as our "best" estimator. In *decision theory* we also take into account the rewards for making correct decisions and the penalties for making incorrect decisions. This leads to a new criterion that chooses the decision function $\hat{\Theta}$ that penalizes us the least when the action taken is incorrect. It is convenient now to introduce a *loss function* whose values depend on the true value of the parameter θ and the action $\hat{\theta}$. This is usually written in functional notation as $L(\hat{\Theta}; \theta)$. In many decision-making problems it is desirable to use a loss function of the form

$$L(\hat{\Theta}; \theta) = |\hat{\Theta} - \theta|$$

or perhaps

$$L(\hat{\Theta}; \theta) = (\hat{\Theta} - \theta)^2$$

in arriving at a choice between two or more decision functions.

Since θ is unknown, it must be assumed that it can equal any of several possible values. The set of all possible values that θ can assume is called the *parameter space*. For each possible value of θ in the parameter space, the loss function will vary from sample to sample. We define the *risk function* for the decision function $\hat{\Theta}$ to be the expected value of the loss function when the value of the parameter is θ and denote this function by $R(\hat{\Theta}; \theta)$. Hence we have

$$R(\hat{\Theta}; \theta) = E[L(\hat{\Theta}; \theta)].$$

One method of arriving at a choice between $\hat{\Theta}_1$ and $\hat{\Theta}_2$ as an estimator for θ would be to apply the *minimax criterion*. Essentially, we determine the maximum value of $R(\hat{\Theta}_1; \theta)$ and the maximum value of $R(\hat{\Theta}_2; \theta)$ in the parameter space and then choose the decision function that provided the minimum of these two maximum risks.

Example 6.18   According to the minimax criterion, is $\bar{X}$ or $\tilde{X}$ a better estimator of the mean μ of a normal population with known variance σ^2, based on a random sample of size n when the loss function is of the form $L(\hat{\Theta}; \theta) = (\hat{\Theta} - \theta)^2$?

Solution   The loss function corresponding to $\bar{X}$ is given by

$$L(\bar{X}; \mu) = (\bar{X} - \mu)^2.$$

Hence the risk function is

$$R(\bar{X}; \mu) = E[(\bar{X} - \mu)^2] = \frac{\sigma^2}{n}$$

for every μ in the parameter space. Similarly, one can show that the risk function corresponding to $\tilde{X}$ is given by

$$R(\tilde{X}; \mu) = E[(\tilde{X} - \mu)^2] \simeq \frac{\pi\sigma^2}{2n}$$

for every μ in the parameter space. In view of the fact that $\sigma^2/n < \pi\sigma^2/2n$, the minimax criterion selects $\bar{X}$ rather than $\tilde{X}$ as the better estimator for μ.

In some practical situations we may have additional information concerning the unknown parameter θ. For example, suppose that we wish to estimate the binomial parameter p, the proportion of defectives produced by a machine during a certain day when we know that p varies from day to day. If we can write down the prior distribution $f(p)$, then it is possible to determine the expected value of the risk function for each decision function. The expected risk corresponding to the estimator $\hat{P}$, often referred to as the *Bayes risk*, is written $B(\hat{P}) = E[R(\hat{P}; P)]$, where we are now treating the true proportion of defectives as a random variable. In general, when the unknown parameter is treated as a random variable with a prior distribution given by $f(\theta)$, the Bayes risk in estimating θ by means of the estimator $\hat{\Theta}$ is given by

$$B(\hat{\Theta}) = E[R(\hat{\Theta}; \theta)] = \sum_i R(\hat{\Theta}; \theta_i) f(\theta_i) \qquad \text{(discrete case)}$$

$$= \int_{-\infty}^{\infty} R(\hat{\Theta}; \theta) f(\theta) \, d\theta \qquad \text{(continuous case)}.$$

The decision function $\hat{\Theta}$ that minimizes $B(\hat{\Theta})$ is the *optimal estimator* of θ. We shall make no attempt in this book to derive an optimal estimator, but instead we shall employ the Bayes risk to establish a criterion for choosing between two estimators.

> BAYES' CRITERION *Let* $\hat{\Theta}_1$ *and* $\hat{\Theta}_2$ *be two estimators of the unknown parameter* θ, *which may be looked upon as a value of the random variable* Θ *with probability distribution* $f(\theta)$. *If* $B(\hat{\Theta}_1) < B(\hat{\Theta}_2)$, *then* $\hat{\Theta}_1$ *is selected as the better estimator for* θ.

The foregoing discussion on decision theory might better be understood if one considers the following two examples.

Example 6.19   Suppose that a friend has three similar coins except for the fact that the first one has two heads, the second one has two tails, and the third one is honest. We wish to estimate which coin our friend is flipping on the basis of two flips of the coin. Let θ be the number of heads on the coin. Consider two decision functions $\hat{\Theta}_1$ and $\hat{\Theta}_2$, where $\hat{\Theta}_1$ is the estimator that assigns to θ the number of heads that occur when the coin is flipped twice and $\hat{\Theta}_2$ is the estimator that assigns the value of 1 to θ no matter what the experiment yields. If the loss function is of the form $L(\hat{\Theta}; \theta) = (\hat{\Theta} - \theta)^2$, which estimator is better according to the minimax procedure?

Solution   For the estimator $\hat{\Theta}_1$, the loss function assumes the values $L(\hat{\theta}_1; \theta) = (\hat{\theta}_1 - \theta)^2$, where $\hat{\theta}_1$ may be 0, 1, or 2, depending on the true value of θ. Clearly, if $\theta = 0$ or 2, both flips will yield all tails or all heads and our decision will be a correct one. Hence $L(0; 0) = 0$ and $L(2; 2) = 0$, from which one may easily conclude that $R(\hat{\Theta}_1; 0) = 0$ and $R(\hat{\Theta}_1; 2) = 0$. However, when $\theta = 1$ we could obtain 0, 1, or 2 heads in the two flips with probabilities 1/4, 1/2, and 1/4, respectively. In this case we have $L(0; 1) = 1$, $L(1; 1) = 0$, and $L(2; 1) = 1$, from which we find that

$$R(\hat{\Theta}_1; 1) = 1 \times \tfrac{1}{4} + 0 \times \tfrac{1}{2} + 1 \times \tfrac{1}{4} = \tfrac{1}{2}.$$

For the estimator $\hat{\Theta}_2$, the loss function assumes values given by $L(\hat{\theta}_2; \theta) = (\hat{\theta}_2 - \theta)^2 = (1 - \theta)^2$. Hence $L(1; 0) = 1$, $L(1; 1) = 0$, and $L(1; 2) = 1$, and the corresponding risks are $R(\hat{\Theta}_2; 0) = 1$, $R(\hat{\Theta}_2; 1) = 0$, and $R(\hat{\Theta}_2; 2) = 1$. Since the maximum risk is 1/2 for the estimator $\hat{\Theta}_1$ compared to a maximum risk of 1 for $\hat{\Theta}_2$, the minimax criterion selects $\hat{\Theta}_1$ as the better of the two estimators.

Example 6.20   Referring to Example 6.19, let us suppose that our friend flips the honest coin 80% of the time and the other two coins each about 10% of the time. Does the Bayes criterion select $\hat{\Theta}_1$ or $\hat{\Theta}_2$ as the better estimator?

Solution   The parameter Θ may now be treated as a random variable with the following probability distribution:

θ	0	1	2
$f(\theta)$	0.1	0.8	0.1

For the estimator $\hat{\Theta}_1$, the Bayes risk is

$$B(\hat{\Theta}_1) = R(\hat{\Theta}_1; 0)f(0) + R(\hat{\Theta}_1; 1)f(1) + R(\hat{\Theta}_1; 2)f(2)$$
$$= (0)(0.1) + \tfrac{1}{2}(0.8) + (0)(0.1) = 0.4.$$

Similarly, for the estimator $\hat{\Theta}_2$, we have

$$B(\hat{\Theta}_2) = R(\hat{\Theta}_2; 0)f(0) + R(\hat{\Theta}_2; 1)f(1) + R(\hat{\Theta}_2; 2)f(2)$$
$$= (1)(0.1) + (0)(0.8) + (1)(0.1) = 0.2.$$

Since $B(\hat{\Theta}_2) < B(\hat{\Theta}_1)$, the Bayes criterion selects $\hat{\Theta}_2$ as the better estimator for the parameter θ.

odds 23-31

EXERCISES   Prob   1-3, 5, 7, 9   1-13 all   35-41

1. Let $S'^2 = \sum_{i=1}^{n} (X_i - \bar{X})^2/n$. Show that $E(S'^2) = [(n-1)/n]\sigma^2$, and hence S'^2 is a biased estimator for σ^2.

2. An electrical firm manufactures light bulbs that have a length of life that is approximately normally distributed with a standard deviation of 40 hours. If a sample of 30 bulbs has an average life of 780 hours, find a 96% confidence interval for the population mean of all bulbs produced by this firm.

3. A soft-drink machine is regulated so that the amount of drink dispensed is approximately normally distributed with a standard deviation equal to 1.5 deciliters. Find a 95% confidence interval for the mean of all drinks dispensed by this machine if a random sample of 36 drinks had an average content of 22.5 deciliters.

4. The heights of a random sample of 50 college students showed a mean of 174.5 centimeters and a standard deviation of 6.9 centimeters.
   (a) Construct a 98% confidence interval for the mean height of all college students.
   (b) What can we assert with 98% confidence about the possible size of our error if we estimate the mean height of all college students to be 174.5 centimeters?

5. A random sample of 100 automobile owners shows that an automobile is driven on the average 23,500 kilometers per year, in the state of Virginia, with a standard deviation of 3900 kilometers.
   (a) Construct a 99 % confidence interval for the average number of miles an automobile is driven annually in Virginia.
   (b) What can we assert with 99 %, confidence about the possible size of our error if we estimate the average number of miles driven by car owners in Virginia as 23,500 kilometers per year?

6. How large a sample is needed in Exercise 2 if we wish to be 96% confident that our sample mean will be within 10 hours of the true mean?

7. How large a sample is needed in Exercise 3 if we wish to be 95% confident that our sample mean will be within 0.9 deciliter of the true mean?

8. An efficiency expert wishes to determine the average time that it takes to drill three holes in a certain metal clamp. How large a sample will he need to be 95% confident that his sample mean will be within 15 seconds of the true mean? Assume that it is known from previous studies that $\sigma = 40$ seconds.

9. A machine is producing metal pieces that are cylindrical in shape. A sample of pieces is taken and the diameters are 1.01, 0.97, 1.03, 1.04, 0.99, 0.98, 0.99, 1.01, and 1.03 centimeters. Find a 99% confidence interval for the mean diameter of pieces from this machine, assuming an approximate normal distribution.

10. A random sample of size 20 from a normal distribution has a mean $\bar{x} = 32.8$ and a standard deviation $s = 4.51$. Construct a 95% confidence interval for μ.

11. A random sample of eight cigarettes of a certain brand has an average tar content of 18.6 milligrams and a standard deviation of 2.4 milligrams. Construct a 99% confidence interval for the true average tar content of this particular brand of cigarettes, assuming an approximate normal distribution.

12. A random sample of 12 shearing pins are taken in a study of the Rockwell hardness of the head on the pin. Measurements on the Rockwell hardness were made for each of the 12, yielding an average value of 48.50 with a sample standard deviation of 1.5. Assuming the measurements to be normally distributed, construct a 90% confidence interval for the mean Rockwell hardness.

13. A random sample of size $n_1 = 25$ taken from a normal population with a standard deviation $\sigma_1 = 5$ has a mean $\bar{x}_1 = 80$. A second random sample of size $n_2 = 36$, taken from a different normal population with a standard deviation $\sigma_2 = 3$, has a mean $\bar{x}_2 = 75$. Find a 94% confidence interval for $\mu_1 - \mu_2$.

14. Two kinds of thread are being compared for strength. Fifty pieces of each type of thread are tested under similar conditions. Brand A had an average tensile strength of 78.3 kilograms with a standard deviation of 5.6 kilograms, while brand B had an average tensile strength of 87.2 kilograms with a standard deviation of 6.3 kilograms. Construct a 95% confidence interval for the difference of the population means.

15. A study was made to determine if a certain metal treatment has any effect on the amount of metal removed in a pickling operation. A random sample of 100 pieces was immersed in a bath for 24 hours without the treatment, yielding an average of 12.2 millimeters of metal removed and a sample standard deviation of 1.1 millimeters. A second sample of 200 pieces was exposed to the treatment followed by the 24-hour

immersion in the bath, resulting in an average removal of 9.1 millimeters of metal with a sample standard deviation of 0.9 millimeter. Compute a 98% confidence interval estimate for the difference between the population means. Does the treatment appear to reduce the mean amount of metal removed?

16. Given two random samples of size $n_1 = 9$ and $n_2 = 16$, from two independent normal populations, with $\bar{x}_1 = 64$, $\bar{x}_2 = 59$, $s_1 = 6$, and $s_2 = 5$, find a 95% confidence interval for $\mu_1 - \mu_2$, assuming that $\sigma_1 = \sigma_2$.

17. Students may choose between a 3-semester-hour course in physics without labs and a 4-semester-hour course with labs. The final written examination is the same for each section. If 12 students in the section with labs made an average examination grade of 84 with a standard deviation of 4, and 18 students in the section without labs made an average grade of 77 with a standard deviation of 6, find a 99% confidence interval for the difference between the average grades for the two courses. Assume the populations to be approximately normally distributed with equal variances.

18. A taxi company is trying to decide whether to purchase brand A or brand B tires for its fleet of taxis. To estimate the difference in the two brands, an experiment is conducted using 12 of each brand. The tires are run until they wear out. The results are brand A: $\bar{x}_1 = 36,300$ kilometers, $s_1 = 5000$ kilometers; brand B: $\bar{x}_2 = 38,100$ kilometers, $s_2 = 6100$ kilometers. Compute a 95% confidence interval for $\mu_1 - \mu_2$, assuming the populations to be approximately normally distributed.

19. The following data represent the running times of films produced by two motion-picture companies.

			Time (minutes)				
Company I	103	94	110	87	98		
Company II	97	82	123	92	175	88	118

Compute a 90% confidence interval for the difference between the average running times of films produced by the two companies. Assume that the running times are approximately normally distributed.

20. The government awarded grants to the agricultural departments of nine universities to test the yield capabilities of two new varieties of wheat. Each variety was planted on plots of equal area at each university and the yields, in kilograms per plot, recorded as follows:

	University								
	1	2	3	4	5	6	7	8	9
Variety 1	38	23	35	41	44	29	37	31	38
Variety 2	45	25	31	38	50	33	36	40	43

Find a 95% confidence interval for the mean difference between the yields of the two varieties, assuming the distributions of yields to be approximately normal. Explain why pairing is necessary in this problem.

21. Referring to Exercise 18, find a 99% confidence interval for $\mu_1 - \mu_2$ if a tire from each company is assigned at random to the rear wheels of eight taxis and the following results recorded:

Brand A	Brand B
34,400	36,700
45,500	46,800
36,700	37,700
32,000	31,100
48,400	47,800
32,800	36,400
38,100	38,900
30,100	31,500

22. It is claimed that a new diet will reduce a person's weight by 4.5 kilograms on the average in a period of 2 weeks. The weights of seven women who followed this diet were recorded before and after a 2-week period.

	Woman						
	1	2	3	4	5	6	7
Weight before	58.5	60.3	61.7	69.0	64.0	62.6	56.7
Weight after	60.0	54.9	58.1	62.1	58.5	59.9	54.4

Test a manufacturer's claim by computing a 95% confidence interval for the mean difference in the weight. Assume the distributions of weights to be approximately normal.

23. (a) A random sample of 200 voters is selected and 114 are found to support an annexation suit. Find the 96% confidence interval for the fraction of the voting population favoring the suit.
    (b) What can we assert with 96% confidence about the possible size of our error if we estimate the fraction of voters favoring the annexation suit to be 0.57?

24. (a) A random sample of 500 cigarette smokers is selected and 86 are found to have a preference for brand X. Find the 90% confidence interval for the fraction of the population of cigarette smokers who prefer brand X.
    (b) What can we assert with 90% confidence about the possible size of our error if we estimate the fraction of cigarette smokers who prefer brand X to be 0.172?

25. Compute a 98% confidence interval for the proportion of defective items in a process when it is found that a sample of size 100 yields 8 defectives.

26. A certain new rocket-launching system is being considered for deployment of small short-range launches. The existing system has $p = 0.8$ as the probability of a successful launch. A sample of 40 experimental launches is made with the new system and 34 are successful.
    (a) Give a point estimate of the probability of a successful launch using the new system.
    (b) Construct a 95% confidence interval for this probability.
    (c) Does the evidence strongly indicate that the new system is better? Explain.

27. How large a sample is needed in Exercise 23 if we wish to be 96% confident that our sample proportion will be within 0.02 of the true fraction of the voting population?

28. How large a sample is needed in Exercise 25 if we wish to be 98% confident that our sample proportion will be within 0.05 of the true proportion defective?

29. A study is to be made to estimate the percentage of citizens in a town who favor having their water fluoridated. How large a sample is needed if one wishes to be at least 95% confident that our estimate is within 1% of the true percentage?

30. A study is to be made to estimate the proportion of housewives who own an automatic dryer. How large a sample is needed if one wishes to be at least 99% confident that our estimate differs from the true proportion by an amount less than 0.01?

31. A certain geneticist is interested in the proportion of males and females in the population that have a certain minor blood disorder. In a random sample of 1000 males, 250 are found to be afflicted, whereas 275 of 1000 females tested appear to have the disorder. Compute a 95% confidence interval for the difference between the proportion of males and females that have the blood disorder.

32. A cigarette-manufacturing firm claims that its brand A line of cigarettes outsells its brand B line by 8%. If it is found that 42 of 200 smokers prefer brand A and 18 of 150 smokers prefer brand B, compute a 94% confidence interval for the difference between the proportions of sales of the two brands and decide if the 8% difference is a valid claim.

33. A clinical trial is conducted to determine if a certain type of inoculation has an effect on the incidence of a certain disease. A sample of 1000 rats was kept in a controlled environment for a period of 1 year and 500 of the rats were given the inoculation. Of the group not given the drug, there were 120 incidences of the disease, while 98 of the inoculated group contracted it. If we call p_1 the probability of incidence of the disease in uninoculated rats and p_2 the probability of incidence after receiving the drug, compute a 90% confidence interval for $p_1 - p_2$.

34. A study is made to determine if a cold climate results in more students being absent from school during a semester than a warmer climate. Two groups of students are selected at random, one group from Vermont and the other group from Georgia. Of the 300 students from Vermont, 64 were absent at least 1 day during the semester, and of the 400 students from Georgia, 51 were absent 1 or more days. Find a 95% confidence interval for the difference between the fractions of the students who are absent in the two states.

35. Construct a 99% confidence interval for σ^2 in Exercise 9.

36. Construct a 95% confidence interval for σ in Exercise 10.

37. Construct a 99% confidence interval for σ in Exercise 11.

38. Construct a 90% confidence interval for σ^2 in Exercise 12.

39. Construct a 98% confidence interval for σ_1/σ_2 in Exercise 16. Were we justified in assuming $\sigma_1 = \sigma_2$?

40. Construct a 90% confidence interval for σ_1^2/σ_2^2 in Exercise 18.

41. Construct a 90% confidence interval for σ_1^2/σ_2^2 in Exercise 19.

42. A random sample of 25 cigarettes of a certain brand has an average nicotine content of 1.3 milligrams and a standard deviation of 0.17 milligram. Find the 95% tolerance limits that will contain 90% of the nicotine contents for this brand of cigarettes, assuming the measurements to be normally distributed.

43. The following measurements were recorded for the drying time, in hours, of a certain brand of latex paint:

3.4	2.5	4.8	2.9	3.6
2.8	3.3	5.6	3.7	2.8
4.4	4.0	5.2	3.0	4.8

Assuming that the measurements represent a random sample from a normal population, find the 99% tolerance limits that will contain 95% of the drying times.

44. Referring to Exercise 5, construct a 99% tolerance interval containing 99% of the miles driven by automobiles annually in Virginia. Assume the distribution of measurements to be approximately normal.

45. Referring to Exercise 12, construct a 95% interval containing 90% of the measurements.

46. Using a random sample of size 2, estimate the proportion p of defectives produced by a machine when we assume the prior distribution to be

p	0.05	0.10	0.15
$f(p)$	0.3	0.5	0.2

47. Repeat Exercise 46 using the uniform prior distribution $f(p) = 10, 0.05 < p < 0.15$.

48. The time of burn for the first stage of a rocket is a normal random variable with a standard deviation of 0.8 minute. Assume a normal prior distribution for μ with a mean of 8 minutes and a standard deviation of 0.2 minute. If 10 of these rockets are fired and the first stage has an average burning time of 9 minutes, find a 95% Bayes interval for μ.

49. Suppose that in Example 6.17 the electrical firm does not have enough prior information regarding the population mean length of life to be able to assume a normal distribution for μ. The firm believes, however, that μ is surely between 770 and 830 hours and it is felt that a more realistic Bayesian approach would be to assume the prior distribution

$$f(\mu) = \tfrac{1}{60}, \qquad 770 < \mu < 830.$$

If a random sample of 25 bulbs gives an average life of 780 hours, find the posterior distribution $f(\mu \mid x_1, x_2, \ldots, x_{25})$.

50. Suppose that the time to failure T of a certain hinge is an exponential random variable with probability density

$$f(t) = \theta e^{-\theta t}, \qquad t > 0.$$

From prior experience we are led to believe that θ is a value of an exponential random variable with probability density

$$f(\theta) = 2e^{-2\theta}, \qquad \theta > 0.$$

If we have a sample of n observations on T, show that the posterior distribution of Θ is a gamma distribution with parameters $\alpha = n + 1$ and $\beta = 1 / \left(\sum_{i=1}^{n} t_i + 2 \right)$.

51. We wish to estimate the binomial parameter p by the decision function $\hat{P}$, the proportion of successes in a binomial experiment consisting of n trials. Find $R(\hat{P}; p)$ when the loss function is of the form $L(\hat{P}; p) = (\hat{P} - p)^2$.

52. Suppose that an urn contains three balls, of which θ are red and the remainder black, where θ can vary from zero to 3. We wish to estimate θ by selecting two balls in succession without replacement. Let $\hat{\Theta}_1$ be the decision function that assigns to θ the value zero if neither ball is red, the value 1 if the first ball only is red, the value 2 if the second ball only is red, and the value 3 if both balls are red. Using a loss function of the form $L(\hat{\Theta}_1; \theta) = |\hat{\Theta}_1 - \theta|$, find $R(\hat{\Theta}_1; \theta)$.

53. In Exercise 52 consider the estimator $\hat{\Theta}_2 = X(X + 1)/2$, where X is the number of red balls in our sample. Find $R(\hat{\Theta}_2; \theta)$.

54. Use the minimax criterion to determine whether the estimator $\hat{\Theta}_1$ of Exercise 52 or the estimator $\hat{\Theta}_2$ of Exercise 53 is the better estimator.

55. Use the Bayes criterion to determine whether the estimator $\hat{\Theta}_1$ of Exercise 52 or the estimator $\hat{\Theta}_2$ of Exercise 53 is the better estimator, given the following additional information:

θ	0	1	2	3
$f(\theta)$	0.1	0.5	0.1	0.3

7

Tests of Hypotheses

7.1 STATISTICAL HYPOTHESES

The testing of statistical hypotheses is perhaps the most important area of decision theory. First, let us define precisely what we mean by a statistical hypothesis.

> DEFINITION 7.1 *A* statistical hypothesis *is an assumption or statement, which may or may not be true, concerning one or more populations.*

The truth or falsity of a statistical hypothesis is never known with certainty unless we examine the entire population. This, of course, would be impractical in most situations. Instead, we take a random sample from the population of interest and use the information contained in this sample to decide whether the hypothesis is likely to be true or false. Evidence from the sample that is inconsistent with the stated hypothesis leads to a rejection of the hypothesis, whereas evidence supporting the hypothesis leads to its acceptance. We should make it clear at this point that the acceptance of a statistical hypothesis is a result of insufficient evidence to reject it and does not necessarily imply that it is true. For example, in tossing a coin 100 times we might test the hypothesis that the coin is balanced. In terms of population parameters, we are testing the hypothesis that the proportion of heads is $p = 0.5$ if the coin were tossed indefinitely. An outcome of 48 heads would not be surprising if the coin is balanced. Such a result would surely support the hypothesis $p = 0.5$. One might argue that such an occurrence is also consistent with the hypothesis that $p = 0.45$. Thus, in accepting the hypothesis, the only thing we can be reasonably certain about is that the true proportion of heads is not a great deal different from one half. If the 100 trials had resulted in only 35 heads, we would then have evidence to support the rejection of our hypothesis. In view of the fact that the probability of obtaining 35 or fewer heads in 100 tosses of a balanced coin is approximately

238

0.002, either a very rare event has occurred or we are right in concluding that $p \neq 0.5$.

Although we shall use the terms *accept* and *reject* frequently throughout this chapter, it is important to understand that the rejection of a hypothesis is to conclude that it is false, while the acceptance of a hypothesis merely implies that we have no evidence to believe otherwise. Because of this terminology, the statistician or experimenter should always state as his hypothesis that which he hopes to reject. If he is interested in a new cold vaccine, he should assume that it is no better than the vaccine now on the market and then set out to reject this contention. Similarly, to prove that one teaching technique is superior to another, we test the hypothesis that there is no difference in the two techniques.

Hypotheses that we formulate with the hope of rejecting are called *null hypotheses* and are denoted by H_0. The rejection of H_0 leads to the acceptance of an *alternative hypothesis* denoted by H_1. Hence, if H_0 is the null hypothesis $p = 0.5$ for a binomial population, the alternative hypothesis H_1 might be $p = 0.75$, $p > 0.5$, $p < 0.5$, or $p \neq 0.5$.

7.2 TYPE I AND TYPE II ERRORS

To illustrate the concepts used in testing a statistical hypothesis about a population, consider the following example. A certain type of cold vaccine is known to be only 25% effective after a period of 2 years. To determine if a new and somewhat more expensive vaccine is superior in providing protection against the same virus for a longer period of time, 20 people are chosen at random and inoculated. If 9 or more of those receiving the new vaccine surpass the 2-year period without contracting the virus, the new vaccine will be considered superior to the one presently in use. The choice of the number 9 is somewhat arbitrary but appears reasonable in that it represents a modest gain over the 5 people that could be expected to receive protection if the 20 people had been inoculated with the vaccine already in use. We are essentially testing the null hypothesis that the new vaccine is equally effective after a period of 2 years as the one now commonly used against the alternative hypothesis that the new vaccine is in fact superior. This is equivalent to testing the hypothesis that the binomial parameter for the probability of a success on a given trial is $p = 1/4$ against the alternative that $p > 1/4$. This is usually written as follows:

$$H_0: \quad p = 1/4,$$

$$H_1: \quad p > 1/4.$$

The statistic on which we base our decision is X, the number of people in our sample who receive protection from the new vaccine for a period of at least 2 years. The possible values, from zero to 20, are divided into two groups: those numbers less than 9 and those greater than or equal to 9. All possible scores above 8.5 constitute the *critical region* and all possible scores below 8.5 determine the *acceptance region*. The number 8.5 separating these two regions is called the *critical value*. If the statistic X falls in the critical region, we reject H_0 in favor of the alternative hypothesis H_1. If X falls in the acceptance region, we accept H_0.

The decision procedure just described could lead to either of two wrong conclusions. For instance, the new vaccine may be no better than the one now in use and, for this particular randomly selected group of individuals, 9 or more people may surpass the 2-year period without contracting the virus. We would be committing an error by rejecting H_0 in favor of H_1 when, in fact, H_0 is true. Such an error is called a *type I error*.

> **DEFINITION 7.2** *A type I error has been committed if we reject the null hypothesis when it is true.*

A second kind of error is committed if fewer than 9 of the group surpass the 2-year period successfully and we conclude that the new vaccine is no better when it actually is. In this case we would accept H_0 when it is false. This is called a *type II error*.

> **DEFINITION 7.3** *A type II error has been committed if we accept the null hypothesis when it is false.*

The probability of committing a type I error is called *the level of significance* of the test and is denoted by α. In our example, a type I error will occur when 9 or more individuals surpass the 2-year period without contracting the virus using a new vaccine that is actually equivalent to the one in use. Hence, if X is the number of individuals who remain healthy for at least 2 years,

$$\alpha = P(\text{type I error})$$
$$= P(X \geq 9 | p = \tfrac{1}{4})$$
$$= \sum_{x=9}^{20} b(x; 20, \tfrac{1}{4})$$
$$= 1 - \sum_{x=0}^{8} b(x; 20, \tfrac{1}{4})$$
$$= 1 - 0.9591$$
$$= 0.0409.$$

We say that the null hypothesis, $p = 1/4$, is being tested at the $\alpha = 0.0409$ level of significance. Sometimes the level of significance is called the *size* of the critical region. A critical region of size 0.0409 is very small and therefore it is unlikely that a type I error will be committed. Consequently, it would be most unusual for 9 or more individuals to remain immune to a virus for a 2-year period using a new vaccine that is essentially equivalent to the one now on the market.

The probability of committing a type II error, denoted by β, is impossible to compute unless we have a specific alternative hypothesis. If we test the null hypothesis that $p = 1/4$ against the alternative hypothesis that $p = 1/2$, then we are able to compute the probability of accepting H_0 when it is false. We simply find the probability of obtaining fewer than 9 in the group that surpass the 2-year period when $p = 1/2$. In this case

$$\beta = P(\text{type II error})$$
$$= P(X < 9 | p = \tfrac{1}{2})$$
$$= \sum_{x=0}^{8} b(x; 20, \tfrac{1}{2})$$
$$= 0.2517.$$

This is a rather high probability indicating a poor test procedure. It is quite likely that we shall reject the new vaccine when, in fact, it is superior to that now in use. Ideally, we like to use a test procedure for which both the type I and type II errors are small.

It is possible that the director of the testing program is willing to make a type II error if the more expensive vaccine is not significantly superior. The only time he wishes to guard against the type II error is when the true value of p is at least 0.7. If $p = 0.7$, this test procedure gives

$$\beta = P(\text{type II error})$$
$$= P(X < 9 | p = 0.7)$$
$$= \sum_{x=0}^{8} b(x; 20, 0.7)$$
$$= 0.0051.$$

With such a small probability of committing a type II error, it is extremely unlikely that the new vaccine would be rejected when it is 70% effective after a period of 2 years. As the alternative hypothesis approaches unity, the value of β diminishes to zero.

Let us assume that the director of the testing program is unwilling to commit a type II error when the alternative hypothesis $p = 1/2$ is true even though we have found the probability of such an error to be $\beta = 0.2517$. A reduction in β is always possible by increasing the size of the critical region. For example, consider what happens to the values of α and β when we change our critical value to 7.5 so that all scores of 8 or more fall in the critical region and those below 8 fall in the acceptance region. Now, in testing $p = 1/4$ against the alternative hypothesis that $p = 1/2$, we find that

$$\alpha = \sum_{x=8}^{20} b(x; 20, \tfrac{1}{4})$$

$$= 1 - \sum_{x=0}^{7} b(x; 20, \tfrac{1}{4})$$

$$= 1 - 0.8982$$

$$= 0.1018$$

and

$$\beta = \sum_{x=0}^{7} b(x; 20, \tfrac{1}{2})$$

$$= 0.1316.$$

By adopting a new decision procedure we have reduced the probability of committing a type II error at the expense of increasing the probability of committing a type I error. For a fixed sample size, a decrease in the probability of one error will usually result in an increase in the probability of the other error. Fortunately, the probability of committing both types of error can be reduced by increasing the sample size. Consider the same problem using a random sample of 100 individuals. If 37 or more of the group surpass the 2-year period, we reject the null hypothesis that $p = 1/4$ and accept the alternative hypothesis $p > 1/4$. The critical value is now 36.5. All possible scores above 36.5 constitute the critical region and all possible scores below 36.5 fall in the acceptance region.

To determine the probability of committing a type I error, we shall use the normal curve approximation with

$$\mu = np = (100)(\tfrac{1}{4}) = 25$$

and

$$\sigma = \sqrt{npq} = \sqrt{(100)(\tfrac{1}{4})(\tfrac{3}{4})} = 4.33.$$

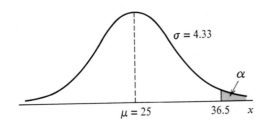

Figure 7.1 Probability of a type I error.

Referring to Figure 7.1,

$$\alpha = P(\text{type I error})$$
$$= P(X > 36.5 | H_0 \text{ is true}).$$

The z value corresponding to $x = 36.5$ is

$$z = \frac{36.5 - 25}{4.33} = 2.656.$$

Therefore,

$$\alpha = P(Z > 2.656)$$
$$= 1 - P(Z < 2.656)$$
$$= 1 - 0.9961$$
$$= 0.0039.$$

If H_0 is false and the true value of H_1 is $p = 1/2$, we can determine the probability of a type II error using the normal curve approximation with

$$\mu = np = (100)(\tfrac{1}{2}) = 50$$

and

$$\sigma = \sqrt{npq} = \sqrt{(100)(\tfrac{1}{2})(\tfrac{1}{2})} = 5.$$

The probability of falling in the acceptance region when H_1 is true is given by the area of the shaded region in Figure 7.2. Hence

$$\beta = P(\text{type II error})$$
$$= P(X < 36.5 | H_1 \text{ is true}).$$

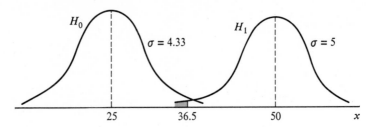

Figure 7.2 Probability of a type II error.

The z value corresponding to $x = 36.5$ is

$$z = \frac{36.5 - 50}{5} = -2.7.$$

Therefore,

$$\beta = P(Z < -2.7)$$
$$= 0.0035.$$

Obviously, the type I and type II errors will rarely occur if the experiment consists of 100 individuals.

   The concepts discussed above can easily be seen graphically when the population is continuous. Consider the null hypothesis that the average weight of male students in a certain college is 68 kilograms against the alternative hypothesis that it is unequal to 68. That is, we wish to test

$$H_0: \quad \mu = 68,$$
$$H_1: \quad \mu \neq 68.$$

The alternative hypothesis allows for the possibility that $\mu < 68$ or $\mu > 68$.

   Assume the standard deviation of the population of weights to be $\sigma = 3.6$. Our decision statistic, based on a sample of size $n = 36$, will be $\bar{X}$, the most efficient estimator of μ. From Chapter 5 we know that the sampling distribution of $\bar{X}$ is approximately normally distributed with standard deviation $\sigma_{\bar{X}} = \sigma/\sqrt{n} = 3.6/6 = 0.6$.

   A sample mean that falls close to the hypothesized value of 68 would be considered evidence in favor of H_0. On the other hand, a sample mean that is considerably less than or more than 68 would be evidence inconsistent with H_0 and therefore favoring H_1. A critical region, indicated by the shaded area in Figure 7.3, is arbitrarily chosen to be $\bar{X} < 67$ and $\bar{X} > 69$. The acceptance

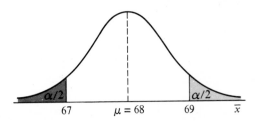

Figure 7.3 Critical region for testing $\mu = 68$ versus $\mu \neq 68$.

region will therefore be $67 < \bar{X} < 69$. Hence, if our sample mean $\bar{x}$ falls inside the critical region, H_0 is rejected; otherwise, we accept H_0.

The probability of committing a type I error, or the level of significance of our test, is equal to the sum of the areas that have been shaded in each tail of the distribution in Figure 7.3. Therefore,

$$\alpha = P(\bar{X} < 67 \,|\, H_0 \text{ is true}) + P(\bar{X} > 69 \,|\, H_0 \text{ is true}).$$

The z values corresponding to $\bar{x}_1 = 67$ and $\bar{x}_2 = 69$ when H_0 is true are

$$z_1 = \frac{67 - 68}{0.6} = -1.67$$

$$z_2 = \frac{69 - 68}{0.6} = 1.67.$$

Therefore,

$$\begin{aligned}\alpha &= P(Z < -1.67) + P(Z > 1.67) \\ &= 2P(Z < -1.67) \\ &= 0.0950.\end{aligned}$$

Thus 9.5% of all samples of size 36 would lead us to reject $\mu = 68$ kilograms when it is true. To reduce α, we have a choice of increasing the sample size or widening the acceptance region. Suppose that we increase the sample size to $n = 64$. Then $\sigma_{\bar{x}} = 3.6/8 = 0.45$. Now

$$z_1 = \frac{67 - 68}{0.45} = -2.22$$

$$z_2 = \frac{69 - 68}{0.45} = 2.22.$$

Hence

$$\alpha = P(Z < -2.22) + P(Z > 2.22)$$
$$= 2P(Z < -2.22)$$
$$= 0.0264.$$

The reduction in α is not sufficient by itself to guarantee a good testing procedure. We must evaluate β for various alternative hypotheses that we feel should be accepted if true. Therefore, if it is important to reject H_0 when the true mean is some value $\mu \geq 70$ or $\mu \leq 66$, then the probability of committing a type II error should be computed and examined for the alternatives $\mu = 66$ and $\mu = 70$. Because of symmetry, it is only necessary to consider the probability of accepting the null hypothesis that $\mu = 68$ when the alternative $\mu = 70$ is true. A type II error will result when the sample mean $\bar{x}$ falls between 67 and 69 when H_1 is true. Therefore, referring to Figure 7.4,

$$\beta = P(67 < \bar{X} < 69 | H_1 \text{ is true}).$$

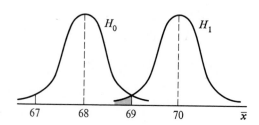

Figure 7.4 Type II error for testing $\mu = 68$ versus $\mu = 70$.

The z values corresponding to $\bar{x}_1 = 67$ and $\bar{x}_2 = 69$ when H_1 is true are

$$z_1 = \frac{67 - 70}{0.45} = -6.67$$

$$z_2 = \frac{69 - 70}{0.45} = -2.22.$$

Therefore,

$$\beta = P(-6.67 < Z < -2.22)$$
$$= P(Z < -2.22) - P(Z < -6.67)$$
$$= 0.0132 - 0.0000$$
$$= 0.0132.$$

If the true value of μ is the alternative $\mu = 66$, the value of β will again be 0.0132. For all possible values of $\mu < 66$ or $\mu > 70$, the value of β will be even smaller when $n = 64$, and consequently there would be little chance of accepting H_0 when it is false.

The probability of committing a type II error increases rapidly when the true value of μ approaches, but is not equal to, the hypothesized value. Of course, this is usually the situation where we do not mind making a type II error. For example, if the alternative hypothesis $\mu = 68.5$ is true, we do not mind committing a type II error by concluding that the true answer is $\mu = 68$. The probability of making such an error will be high when $n = 64$. Referring to Figure 7.5, we have

$$\beta = P(67 < \bar{X} < 69 \,|\, H_1 \text{ is true}).$$

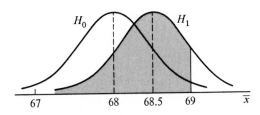

Figure 7.5 Type II error for testing $\mu = 68$ versus $\mu = 68.5$.

The z values corresponding to $\bar{x}_1 = 67$ and $\bar{x}_2 = 69$ when $\mu = 68.5$ are

$$z_1 = \frac{67 - 68.5}{0.45} = -3.33$$

$$z_2 = \frac{69 - 68.5}{0.45} = 1.11.$$

Therefore,

$$\begin{aligned}
\beta &= P(-3.33 < Z < 1.11) \\
&= P(Z < 1.11) - P(Z < -3.33) \\
&= 0.8665 - 0.0004 \\
&= 0.8661.
\end{aligned}$$

The preceding examples illustrate the following important properties:

1. The type I error and type II error are related. A decrease in the probability of one generally results in an increase in the probability of the other.
2. The size of the critical region, and therefore the probability of committing a type I error, can always be reduced by adjusting the critical value(s).
3. An increase in the sample size n will reduce α and β simultaneously.
4. If the null hypothesis is false, β is a maximum when the true value of a parameter is close to the hypothesized value. The greater the distance between the true value and the hypothesized value, the smaller β will be.

7.3 ONE-TAILED AND TWO-TAILED TESTS

A test of any statistical hypothesis where the alternative is *one-sided* such as

$$H_0: \quad \theta = \theta_0,$$
$$H_1: \quad \theta > \theta_0,$$

or perhaps

$$H_0: \quad \theta = \theta_0,$$
$$H_1: \quad \theta < \theta_0,$$

is called a *one-tailed test*. The critical region for the alternative hypothesis $\theta > \theta_0$ lies entirely in the right tail of the distribution, while the critical region for the alternative hypothesis $\theta < \theta_0$ lies entirely in the left tail. A one-sided test was used by the director of the testing program to test the hypothesis $p = 1/4$ against the one-sided alternative $p > 1/4$ for the binomial distribution.

A test of any statistical hypothesis where the alternative is *two-sided* such as

$$H_0: \quad \theta = \theta_0,$$
$$H_1: \quad \theta \neq \theta_0,$$

is called a *two-tailed test*. The alternative hypothesis states that either $\theta < \theta_0$ or $\theta > \theta_0$. Values in both tails of the distribution constitute the critical region. A two-tailed test was used to test the null hypothesis that $\mu = 68$ kilograms against the two-tailed alternative $\mu \neq 68$ for the continuous population of male students' weights.

Whether one sets up a one-sided or a two-sided alternative hypothesis will depend on the conclusion to be drawn if H_0 is rejected. The location of the critical region can be determined only after H_1 has been stated. For example, in testing a new drug, one sets up the hypothesis that it is no better than similar drugs now on the market and tests this against the alternative hypothesis that the new drug is superior. Such an alternative hypothesis will result in a one-tailed test with the critical region in the right tail. However, if we wish to determine whether two teaching procedures are equally effective, the alternative hypothesis should allow for either procedure to be superior. Hence the test is two-tailed with the critical region divided so as to fall in the extreme left and right tails of the distribution.

In testing hypotheses about discrete populations, the critical region is chosen arbitrarily and its size determined. If the size α is too large, it can be reduced by making an adjustment in the critical value. It may be necessary to increase the sample size to offset the increase that automatically occurs in β. In testing hypotheses about continuous populations, it is customary to choose the value of α to be 0.05 or 0.01 and then find the critical value(s). For example, in a two-tailed test at the 0.05 level of significance the critical values for a statistic having a standard normal distribution will be $-z_{0.025} = -1.96$ and $z_{0.025} = 1.96$. In terms of z values, the critical region of size 0.05 will be $Z < -1.96$ and $Z > 1.96$ and the acceptance region is $-1.96 < Z < 1.96$. A test is said to be *significant* if the null hypothesis is rejected at the 0.05 level of significance, and is considered *highly significant* if the null hypothesis is rejected at the 0.01 level of significance.

In the remaining sections of this chapter we shall consider several special tests of hypotheses that are frequently used by statisticians.

7.4 TESTS CONCERNING MEANS AND VARIANCES

Consider the problem of testing the hypothesis that the mean μ of a population, with known variance σ^2, equals a specified value μ_0 against the two-sided alternative that the mean is not equal to μ_0; that is, we shall test

$$H_0: \quad \mu = \mu_0,$$
$$H_1: \quad \mu \neq \mu_0.$$

An appropriate statistic on which we base our decision criterion is the random variable $\overline{X}$. From Chapter 5 we already know that the sampling distribution of $\overline{X}$ is approximately normally distributed with mean $\mu_{\overline{X}} = \mu$ and variance $\sigma_{\overline{X}}^2 = \sigma^2/n$, where μ and σ^2 are the mean and variance of the population from which we select random samples of size n. If we use a significance level of α, it is possible to find two critical values $\bar{x}_1$ and $\bar{x}_2$ such that the interval $\bar{x}_1 < \overline{X} < \bar{x}_2$ defines the acceptance region and the two tails of the distribution, $\overline{X} < \bar{x}_1$ and $\overline{X} > \bar{x}_2$, constitute the critical region.

The critical region can be given in terms of z values by means of the transformation

$$z = \frac{\bar{x} - \mu_0}{\sigma/\sqrt{n}}.$$

Hence, for an α level of significance, the critical values of the random variable Z, corresponding to $\bar{x}_1$ and $\bar{x}_2$, are shown in Figure 7.6 to be

$$-z_{\alpha/2} = \frac{\bar{x}_1 - \mu_0}{\sigma/\sqrt{n}}$$

$$z_{\alpha/2} = \frac{\bar{x}_2 - \mu_0}{\sigma/\sqrt{n}}.$$

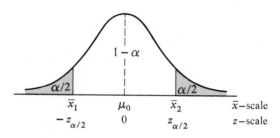

Figure 7.6 Critical region for alternative hypothesis $\mu \neq \mu_0$.

From the population we select a random sample of size n and compute the sample mean $\bar{x}$. If $\bar{x}$ falls in the acceptance region, $\bar{x}_1 < \bar{X} < \bar{x}_2$, then

$$z = \frac{\bar{x} - \mu_0}{\sigma/\sqrt{n}}$$

will fall in the region $-z_{\alpha/2} < Z < z_{\alpha/2}$ and we conclude that $\mu = \mu_0$; otherwise, we reject H_0 and accept the alternative hypothesis that $\mu \neq \mu_0$. The critical region is usually stated in terms of Z rather than $\bar{X}$.

The test procedure just described is equivalent to finding a $(1 - \alpha)100\%$ confidence interval for μ and accepting H_0 if μ_0 lies in the interval. If μ_0 lies outside the interval, we reject H_0 in favor of the alternative hypothesis H_1. Consequently, when one makes inferences about the mean μ from a population with known variance σ^2, whether it be by the construction of a confidence interval or through the testing of a statistical hypothesis, the same statistic $Z = (\bar{X} - \mu)/(\sigma/\sqrt{n})$ is employed.

In general, if one uses a statistic to construct a confidence interval for a parameter θ, whether it be the statistic Z, T, X^2, or F, we can use that same statistic to test the hypothesis that the parameter equals some specified value θ_0 against an appropriate alternative. Of course, all the underlying assumptions made in Chapter 6 relative to the use of a given statistic will apply to the tests described here. This essentially means that all our samples are selected from approximately normal populations. However, a Z statistic may be used to test hypotheses about means from nonnormal populations when $n \geq 30$.

In Table 7.1 we list the statistic used to test a specified hypothesis H_0 and give the appropriate critical regions for one- and two-sided alternative hypotheses H_1. The steps for testing a hypothesis concerning a population parameter θ against some alternative hypothesis may be summarized as follows:

1. H_0: $\theta = \theta_0$.
2. H_1: Alternatives are $\theta < \theta_0$, $\theta > \theta_0$, or $\theta \neq \theta_0$.
3. Choose a level of significance equal to α.
4. Select the appropriate test statistic and establish the critical region.
5. Compute the value of the statistic from a random sample of size n.
6. Conclusion: Reject H_0 if the statistic has a value in the critical region; otherwise accept H_0.

Example 7.1   A manufacturer of sports equipment has developed a new synthetic fishing line that he claims has a mean breaking strength of 8 kilograms with a standard deviation of 0.5 kilogram. Test the hypothesis that $\mu = 8$ kilograms against the alternative that $\mu \neq 8$ kilograms if a random sample of 50 lines is tested and found to have a mean breaking strength of 7.8 kilograms. Use a 0.01 level of significance.

Solution
1. H_0: $\mu = 8$ kilograms.
2. H_1: $\mu \neq 8$ kilograms.
3. $\alpha = 0.01$.
4. Critical region: $Z < -2.575$ and $Z > 2.575$, where

$$Z = \frac{\bar{X} - \mu_0}{\sigma/\sqrt{n}}.$$

Table 7.1 Tests Concerning Means and Variances

H_0	Test statistic	H_1	Critical region
$\mu = \mu_0$	$Z = \dfrac{\bar{X} - \mu_0}{\sigma/\sqrt{n}}$; σ known	$\mu < \mu_0$ $\mu > \mu_0$ $\mu \neq \mu_0$	$Z < -z_\alpha$ $Z > z_\alpha$ $Z < -z_{\alpha/2}$ and $Z > z_{\alpha/2}$
$\mu = \mu_0$	$T = \dfrac{\bar{X} - \mu_0}{S/\sqrt{n}}$; $v = n - 1$, σ unknown	$\mu < \mu_0$ $\mu > \mu_0$ $\mu \neq \mu_0$	$T < -t_\alpha$ $T > t_\alpha$ $T < -t_{\alpha/2}$ and $T > t_{\alpha/2}$
$\mu_1 - \mu_2 = d_0$	$Z = \dfrac{(\bar{X}_1 - \bar{X}_2) - d_0}{\sqrt{(\sigma_1^2/n_1) + (\sigma_2^2/n_2)}}$; σ_1 and σ_2 known	$\mu_1 - \mu_2 < d_0$ $\mu_1 - \mu_2 > d_0$ $\mu_1 - \mu_2 \neq d_0$	$Z < -z_\alpha$ $Z > z_\alpha$ $Z < -z_{\alpha/2}$ and $Z > z_{\alpha/2}$
$\mu_1 - \mu_2 = d_0$	$T = \dfrac{(\bar{X}_1 - \bar{X}_2) - d_0}{S_p\sqrt{(1/n_1) + (1/n_2)}}$; $v = n_1 + n_2 - 2$, $\sigma_1 = \sigma_2$ but unknown, $S_p^2 = \dfrac{(n_1 - 1)S_1^2 + (n_2 - 1)S_2^2}{n_1 + n_2 - 2}$	$\mu_1 - \mu_2 < d_0$ $\mu_1 - \mu_2 > d_0$ $\mu_1 - \mu_2 \neq d_0$	$T < -t_\alpha$ $T > t_\alpha$ $T < -t_{\alpha/2}$ and $T > t_{\alpha/2}$
$\mu_1 - \mu_2 = d_0$	$T' = \dfrac{(\bar{X}_1 - \bar{X}_2) - d_0}{\sqrt{(S_1^2/n_1) + (S_2^2/n_2)}}$; $v = \dfrac{(s_1^2/n_1 + s_2^2/n_2)^2}{\dfrac{(s_1^2/n_1)^2}{n_1 - 1} + \dfrac{(s_2^2/n_2)^2}{n_2 - 1}}$; $\sigma_1 \neq \sigma_2$ and unknown	$\mu_1 - \mu_2 < d_0$ $\mu_1 - \mu_2 > d_0$ $\mu_1 - \mu_2 \neq d_0$	$T' < -t_\alpha$ $T' > t_\alpha$ $T' < -t_{\alpha/2}$ and $T' > t_{\alpha/2}$
$\mu_D = d_0$	$T = \dfrac{\bar{D} - d_0}{S_d/\sqrt{n}}$; $v = n - 1$, paired observations	$\mu_D < d_0$ $\mu_D > d_0$ $\mu_D \neq d_0$	$T < -t_\alpha$ $T > t_\alpha$ $T < -t_{\alpha/2}$ and $T > t_{\alpha/2}$
$\sigma^2 = \sigma_0^2$	$X^2 = \dfrac{(n - 1)S^2}{\sigma_0^2}$; $v = n - 1$	$\sigma^2 < \sigma_0^2$ $\sigma^2 > \sigma_0^2$ $\sigma^2 \neq \sigma_0^2$	$X^2 < \chi_{1-\alpha}^2$ $X^2 > \chi_\alpha^2$ $X^2 < \chi_{1-\alpha/2}^2$ and $X^2 > \chi_{\alpha/2}^2$
$\sigma_1^2 = \sigma_2^2$	$F = S_1^2/S_2^2$; $v_1 = n_1 - 1$ and $v_2 = n_2 - 1$	$\sigma_1^2 < \sigma_2^2$ $\sigma_1^2 > \sigma_2^2$ $\sigma_1^2 \neq \sigma_2^2$	$F < f_{1-\alpha}(v_1, v_2)$ $F > f_\alpha(v_1, v_2)$ $F < f_{1-\alpha/2}(v_1, v_2)$ and $F > f_{\alpha/2}(v_1, v_2)$

5. Computations:

$$\bar{x} = 7.8 \text{ kilograms,} \qquad n = 50$$

$$z = \frac{7.8 - 8}{0.5/\sqrt{50}} = -2.828.$$

6. Conclusion: Reject H_0 and conclude that the average breaking strength is not equal to 8 but is, in fact, less than 8 kilograms.

Example 7.2 The average length of time for students to register for fall classes at a certain college has been 50 minutes with a standard deviation of 10 minutes. A new registration procedure using modern computing machines is being tried. If a random sample of 12 students had an average registration time of 42 minutes with a standard deviation of 11.9 minutes under the new system, test the hypothesis that the population mean is now less than 50, using a level of significance of (1) 0.05, and (2) 0.01. Assume the population of times to be normal.

Solution
1. H_0: $\mu = 50$ minutes.
2. H_1: $\mu < 50$ minutes.
3. (1) $\alpha = 0.05$, (2) $\alpha = 0.01$.
4. Critical region: (1) $T < -1.796$, (2) $T < -2.718$, where $T = (\bar{X} - \mu_0)/(S/\sqrt{n})$ with $v = 11$ degrees of freedom.
5. Computations: $\bar{x} = 42$ minutes, $s = 11.9$ minutes, and $n = 12$. Hence

$$t = \frac{42 - 50}{11.9/\sqrt{12}} = -2.33.$$

6. Conclusion: Reject H_0 at the 0.05 level of significance but not at the 0.01 level. This essentially means that the true mean is likely to be less than 50 minutes but does not differ sufficiently to warrant the high cost that would be required to operate a computer.

Example 7.3 An experiment was performed to compare the abrasive wear of two different laminated materials. Twelve pieces of material 1 were tested, by exposing each piece to a machine measuring wear. Ten pieces of material 2 were similarly tested. In each case, the depth of wear was observed. The samples of material 1 gave an average (coded) wear of 85 units with a standard deviation of 4, while the samples of material 2 gave an average of 81 and a standard deviation of 5. Test the hypothesis that the two types of material exhibit the same mean abrasive wear at the 0.10 level of significance. Assume the populations to be approximately normal with equal variances.

Solution Let μ_1 and μ_2 represent the population means of material 1 and material 2, respectively. Using the six-step procedure, we have

1. H_0: $\mu_1 = \mu_2$ or $\mu_1 - \mu_2 = 0$.
2. H_1: $\mu_1 \neq \mu_2$ or $\mu_1 - \mu_2 \neq 0$.
3. $\alpha = 0.10$.
4. Critical region: $T < -1.725$ and $T > 1.725$, where

$$T = \frac{(\bar{X}_1 - \bar{X}_2) - d_0}{S_p\sqrt{(1/n_1) + (1/n_2)}}$$

with $v = 20$ degrees of freedom.
5. Computations: $\bar{x}_1 = 85$, $s_1 = 4$, $n_1 = 12$, and $\bar{x}_2 = 81$, $s_2 = 5$, $n_2 = 10$. Hence

$$S_p = \sqrt{\frac{(11)(16) + (9)(25)}{12 + 10 - 2}} = 4.478$$

$$t = \frac{(85 - 81) - 0}{4.478\sqrt{(1/12) + (1/10)}} = 2.07.$$

6. Conclusion: Reject H_0 and conclude that the two materials do not exhibit the same abrasive wear.

Example 7.4 Five samples of a ferrous-type substance are to be used to determine if there is a difference between a laboratory chemical analysis and an X-ray fluorescence analysis of the iron content. Each sample was split into two subsamples and the two types of analysis were applied. Following are the coded data showing the iron content analysis:

	Sample				
Analysis	1	2	3	4	5
X-ray	2.0	2.0	2.3	2.1	2.4
Chemical	2.2	1.9	2.5	2.3	2.4

Assuming the populations normal, test at the 0.05 level of significance whether the two methods of analysis give, on the average, the same result.

Solution　Let μ_1 and μ_2 be the average iron content determined by the chemical and X-ray analyses, respectively. We proceed as follows:

1. H_0:　$\mu_1 = \mu_2$ or $\mu_D = 0$.
2. H_1:　$\mu_1 \neq \mu_2$ or $\mu_D \neq 0$.
3. $\alpha = 0.05$.
4. Critical region: $T < -2.776$ and $T > 2.776$, where $T = (\bar{D} - d_0)/(S_d/\sqrt{n})$ with $v = 4$ degrees of freedom.
5. Computations:

X-ray	Chemical	d_i	d_i^2
2.0	2.2	−0.2	0.04
2.0	1.9	0.1	0.01
2.3	2.5	−0.2	0.04
2.1	2.3	−0.2	0.04
2.4	2.4	0.0	0.00
		−0.5	0.13

We find $\bar{d} = -0.5/5 = -0.1$ and $s_d^2 = [(5)(0.13) - (-0.5)^2]/(5)(4) = 0.02$. Taking the square root, we have $s_d = 0.14142$. Hence

$$t = \frac{-0.1 - 0}{0.14142/\sqrt{5}} = -1.6.$$

6. Conclusion: Accept H_0 and conclude that the two methods of analysis are not significantly different.

Example 7.5　A manufacturer of car batteries claims that the life of his batteries is approximately normally distributed with a standard deviation equal to 0.9 year. If a random sample of 10 of these batteries has a standard deviation of 1.2 years, do you think that $\sigma > 0.9$ year? Use a 0.05 level of significance.

Solution
1. H_0:　$\sigma^2 = 0.81$.
2. H_1:　$\sigma^2 > 0.81$.
3. $\alpha = 0.05$.
4. Critical region: $X^2 > 16.919$, where $X^2 = (n - 1)S^2/\sigma_0^2$ with $v = 9$ degrees of freedom.

5. Computations: $s^2 = 1.44$, $n = 10$, and

$$\chi^2 = \frac{(9)(1.44)}{0.81} = 16.0.$$

6. Conclusion: Accept H_0 and conclude that there is no reason to doubt that the standard deviation is 0.9 year.

7.5 CHOICE OF SAMPLE SIZE FOR TESTING MEANS

The significance level for testing a statistical hypothesis is normally controlled by the experimenter, while β, or the *power* of the test defined by $1 - \beta$, is controlled by using the proper sample size. In this section we shall discuss the choice of sample size for tests involving one and two population means. For situations in which the normal distribution is used and the population variance or variances are known, it is a simple matter to determine the sample size necessary to attain the desired power.

Suppose that we wish to test the hypothesis

$$H_0: \quad \mu = \mu_0,$$
$$H_1: \quad \mu > \mu_0,$$

with a significance level α when the variance σ^2 is known. For a specific alternative, say $\mu = \mu_0 + \delta$, the power of our test is given by

$$1 - \beta = P\left[\frac{\overline{X} - \mu_0}{\sigma/\sqrt{n}} > z_\alpha | H_1 \text{ is true}\right]$$

$$= P\left[\frac{\overline{X} - (\mu_0 + \delta)}{\sigma/\sqrt{n}} > z_\alpha - \frac{\delta}{\sigma/\sqrt{n}} | \mu = \mu_0 + \delta\right].$$

Under the alternative hypothesis $\mu = \mu_0 + \delta$, the statistic

$$\frac{\overline{X} - (\mu_0 + \delta)}{\sigma/\sqrt{n}}$$

is a standard normal variable. Therefore,

$$1 - \beta = P\left(Z > z_\alpha - \frac{\delta\sqrt{n}}{\sigma}\right),$$

from which we conclude that

$$-z_\beta = z_\alpha - \frac{\delta\sqrt{n}}{\sigma}$$

and hence

$$n = \frac{(z_\alpha + z_\beta)^2\sigma^2}{\delta^2},$$

a result that is also true when the alternative hypothesis is $\mu < \mu_0$.

In the case of a two-tailed test we obtain the power $1 - \beta$ for a specified alternative when

$$n \simeq \frac{(z_{\alpha/2} + z_\beta)^2\sigma^2}{\delta^2}.$$

Example 7.6   Suppose that we wish to test the hypothesis

$$H_0: \quad \mu = 68,$$
$$H_1: \quad \mu > 68,$$

with a significance level $\alpha = 0.05$ when $\sigma = 5$. Find the sample size required if the power of our test is to be 0.95 when the true mean is 69.

Solution   Since $\alpha = \beta = 0.05$, we have $z_\alpha = z_\beta = 1.645$. For the alternative $\mu = 69$, we take $\delta = 1$ and then

$$n = \frac{(1.645 + 1.645)^2(25)}{1} = 270.6.$$

Therefore, it requires 271 observations if the test is to reject the null hypothesis 95% of the time when, in fact, μ is as large as 69.

A similar procedure can be used to determine the sample size $n = n_1 = n_2$ required for a specific power of the test in which two population means are being compared. For example, suppose that we wish to test the hypothesis

$$H_0: \quad \mu_1 = \mu_2,$$
$$H_1: \quad \mu_1 \neq \mu_2,$$

when σ_1 and σ_2 are known. For a specific alternative, say $\mu_1 - \mu_2 = \delta$, the power of our test is given by

$$1 - \beta = P\left[\frac{|\bar{X}_1 - \bar{X}_2|}{\sqrt{(\sigma_1^2 + \sigma_2^2)/n}} > z_{\alpha/2}|\mu_1 - \mu_2 = \delta\right].$$

Therefore,

$$\beta = P\left[-z_{\alpha/2} < \frac{\bar{X}_1 - \bar{X}_2}{\sqrt{(\sigma_1^2 + \sigma_2^2)/n}} < z_{\alpha/2}|\mu_1 - \mu_2 = \delta\right]$$

$$= P\left[-z_{\alpha/2} - \frac{\delta}{\sqrt{(\sigma_1^2 + \sigma_2^2)/n}} < \frac{\bar{X}_1 - \bar{X}_2 - \delta}{\sqrt{(\sigma_1^2 + \sigma_2^2)/n}} < z_{\alpha/2}\right.$$

$$\left. - \frac{\delta}{\sqrt{(\sigma_1^2 + \sigma_2^2)/n}}|\mu_1 - \mu_2 = \delta\right].$$

Under the alternative hypothesis $\mu_1 - \mu_2 = \delta$, the statistic

$$\frac{\bar{X}_1 - \bar{X}_2 - \delta}{\sqrt{(\sigma_1^2 + \sigma_2^2)/n}}$$

is a standard normal variable. Therefore,

$$\beta = P\left[-z_{\alpha/2} - \frac{\delta}{\sqrt{(\sigma_1^2 + \sigma_2^2)/n}} < Z < z_{\alpha/2} - \frac{\delta}{\sqrt{(\sigma_1^2 + \sigma_2^2)/n}}\right],$$

from which we conclude that

$$-z_\beta \simeq z_{\alpha/2} - \frac{\delta}{\sqrt{(\sigma_1^2 + \sigma_2^2)/n}}$$

and hence

$$n \simeq \frac{(z_{\alpha/2} + z_\beta)^2(\sigma_1^2 + \sigma_2^2)}{\delta^2}.$$

For the one-tailed test, the expression for the required sample size when $n = n_1 = n_2$ is given by

$$n = \frac{(z_\alpha + z_\beta)^2(\sigma_1^2 + \sigma_2^2)}{\delta^2}.$$

When the population variance (or variances in the two-sample situation) is unknown, the choice of sample size is not straightforward. In testing the hypothesis $\mu = \mu_0$ when the true value is $\mu = \mu_0 + \delta$, the statistic

$$\frac{\overline{X} - (\mu_0 + \delta)}{S/\sqrt{n}}$$

does not follow the t distribution, as one might expect, but instead follows the *noncentral t distribution*. However, tables or charts based on the noncentral t distribution do exist for determining the appropriate sample size if some estimate of σ is available or if δ is a multiple of σ. Table IX (see Statistical Tables) gives the sample sizes needed to control the values of α and β for various values of

$$\Delta = \frac{|\delta|}{\sigma} = \frac{|\mu - \mu_0|}{\sigma}$$

for both one-tailed and two-tailed tests. In the case of the two-sample t test in which the variances are unknown but assumed equal, we obtain the sample sizes $n = n_1 = n_2$ needed to control the values of α and β for various values of

$$\Delta = \frac{|\delta|}{\sigma} = \frac{|\mu_1 - \mu_2|}{\sigma}$$

from Table X.

Example 7.7   In comparing the performance of two catalysts on the effect of a reaction yield, a two-sample t test is to be conducted with $\alpha = 0.05$. The variances in the yields are considered to be the same for the two catalysts. How large a sample for each catalyst is needed to test the hypothesis

$$H_0: \quad \mu_1 = \mu_2,$$
$$H_1: \quad \mu_1 \neq \mu_2,$$

if it is essential to detect a difference of 0.8σ between the catalysts with probability 0.9?

Solution   From Table X, with $\alpha = 0.05$ for a two-tailed test, $\beta = 0.1$, and

$$\Delta = \frac{|0.8\sigma|}{\sigma} = 0.8,$$

we find the required sample size to be $n = 34$.

7.6 TESTS CONCERNING PROPORTIONS

Tests of hypotheses concerning proportions are required in many areas. The politician is certainly interested in knowing what fraction of the voters will favor him in the next election. All manufacturing firms are concerned about the proportion of defectives when a shipment is made. The gambler depends on a knowledge of the proportion of outcomes that he considers favorable.

We shall consider the problem of testing the hypothesis that the proportion of successes in a binomial experiment equals some specified value. That is, we are testing the null hypothesis H_0 that $p = p_0$, where p is the parameter of the binomial distribution. The alternative hypothesis may be one of the usual one-sided or two-sided alternatives: $p < p_0$, $p > p_0$, or $p \neq p_0$.

The appropriate statistic on which we base our decision criterion is the binomial random variable X, although we could just as well use the statistic $\hat{P} = X/n$. Values of X that are far from the mean $\mu = np_0$ will lead to the rejection of the null hypothesis. To test the hypothesis

$$H_0: \quad p = p_0,$$
$$H_1: \quad p < p_0,$$

we use the binomial distribution with $p = p_0$ and $q = 1 - p_0$ to determine $P(X \leq x \mid H_0$ is true). The value x is the number of successes in our sample of size n. If $P(X \leq x \mid H_0$ is true) $< \alpha$, our test is significant at the α level and we reject H_0 in favor of H_1. Similarly, to test the hypothesis

$$H_0: \quad p = p_0,$$
$$H_1: \quad p > p_0,$$

we find $P(X \geq x \mid H_0$ is true) and reject H_0 in favor of H_1 if this probability is less than α. Finally, to test the hypothesis

$$H_0: \quad p = p_0,$$
$$H_1: \quad p \neq p_0,$$

at the α level of significance, we reject H_0 when $x < np_0$ and $P(X \leq x \mid H_0$ is true) $< \alpha/2$ or when $x > np_0$ and $P(X \geq x \mid H_0$ is true) $< \alpha/2$.

The steps for testing a hypothesis about a proportion against various alternatives are now summarized:

1. H_0: $p = p_0$.
2. H_1:   Alternatives are $p < p_0$, $p > p_0$, or $p \neq p_0$.
3. Choose a level of significance equal to α.
4. Critical region:
   (a) All x values such that $P(X \leq x | H_0$ is true$) < \alpha$ for the alternative $p < p_0$,
   (b) All x values such that $P(X \geq x | H_0$ is true$) < \alpha$ for the alternative $p > p_0$.
   (c) All x values such that $P(X \leq x | H_0$ is true$) < \alpha/2$ when $x < np_0$, and all x values such that $P(X \geq x | H_0$ is true$) < \alpha/2$ when $x > np_0$, for the alternative $p \neq p_0$.
5. Computations: Find x and compute the appropriate probability.
6. Conclusion: Reject H_0 if x falls in the critical region; otherwise, accept H_0.

Example 7.8   A pheasant hunter claims that he hits 80% of the birds he shoots at. Would you agree with this claim if on a given day be brings down 9 of the 15 pheasants he shoots at? Use a 0.05 level of significance.

Solution   We follow the six-step procedure:

1. H_0: $p = 0.8$.
2. H_1: $p \neq 0.8$.
3. $\alpha = 0.05$.
4. Critical region: All x values such that $P(X \leq x | H_0$ is true$) < 0.025$.
5. Computations: We have $x = 9$ and $n = 15$. Therefore, using Table II,

$$P(X \leq 9 | p = 0.8) = \sum_{x=0}^{9} b(x; 15, 0.8)$$

$$= 0.0611 > 0.025.$$

6. Conclusion: Accept H_0 and conclude that there is no reason to doubt the hunter's claim.

In Section 4.3 we saw that binomial probabilities were obtainable from the actual binomial formula or from Table II when n is small. For large n, approximation procedures are required. When the hypothesized value p_0 is very close to zero or 1, the Poisson distribution, with parameter $\mu = np_0$, may be used. The normal curve approximation is usually preferred for large n and is very accurate as long as p_0 is not extremely close to zero or 1. Using the normal approximation, we base our decision criterion on the standard normal variable

$$Z = \frac{\hat{P} - p_0}{\sqrt{p_0 q_0/n}} = \frac{X - np_0}{\sqrt{np_0 q_0}}.$$

Hence, for a two-tailed test at the α level of significance, the critical region is $Z < -z_{\alpha/2}$ and $Z > z_{\alpha/2}$. For the one-sided alternative $p < p_0$, the critical region is $Z < -z_\alpha$, and for the alternative $p > p_0$, the critical region is $Z > z_\alpha$.

To test a hypothesis about a proportion using the normal curve approximation, we proceed as follows:

1. H_0: $p = p_0$.
2. H_1:   Alternatives are $p < p_0$, $p > p_0$, or $p \neq p_0$.
3. Choose a level of significance equal to α.
4. Critical region:
   (a) $Z < -z_\alpha$ for the alternative $p < p_0$.
   (b) $Z > z_\alpha$ for the alternative $p > p_0$.
   (c) $Z < -z_{\alpha/2}$ and $Z > z_{\alpha/2}$ for the alternative $p \neq p_0$.
5. Computations: Find x from a sample of size n, and then compute

$$z = \frac{x - np_0}{\sqrt{np_0 q_0}}.$$

6. Conclusion: Reject H_0 if z falls in the critical region; otherwise, accept H_0.

Example 7.9   A manufacturing company has submitted a claim that 90% of items produced by a certain process are nondefective. An improvement in the process is being considered that they feel will lower the proportion of defectives below the current 10%. In an experiment 100 items are produced with the new process and 5 are defective. Is this evidence sufficient to conclude that the method has been improved? Use a 0.05 level of significance.

Solution   As usual, we follow the six-step procedure:

1. H_0: $p = 0.9$.
2. H_1: $p > 0.9$.
3. $\alpha = 0.05$.
4. Critical region: $Z > 1.645$.

5. Computations: $x = 95$, $n = 100$, $np_0 = (100)(0.95) = 95$, and

$$z = \frac{95 - 90}{\sqrt{(100)(0.90)(0.10)}} = 1.67.$$

6. Conclusion: Reject H_0 and conclude that the improvement has reduced the proportion of defectives.

7.7 TESTING THE DIFFERENCE BETWEEN TWO PROPORTIONS

Situations often arise where we wish to test the hypothesis that two proportions are equal. For example, we might try to prove that the proportion of doctors who are pediatricians in one state is equal to the proportion of pediatricians in another state. A person may decide to give up smoking only if he is convinced that the proportion of smokers with lung cancer exceeds the proportion of nonsmokers with lung cancer.

In general, we wish to test the null hypothesis

$$H_0: \quad p_1 = p_2 = p$$

against some suitable alternative. The parameters p_1 and p_2 are the two population proportions of the attribute under investigation. The statistic on which we base our decision criterion is the random variable $\hat{P}_1 - \hat{P}_2$. Independent samples of size n_1 and n_2 are selected at random from two binomial populations and the proportion of successes $\hat{P}_1$ and $\hat{P}_2$ for the two samples are computed. From Section 6.6 we know that the statistic

$$Z = \frac{\hat{P}_1 - \hat{P}_2}{\sqrt{(p_1 q_1/n_1) + (p_2 q_2/n_2)}} = \frac{\hat{P}_1 - \hat{P}_2}{\sqrt{pq[(1/n_1) + (1/n_2)]}}$$

has a standard normal distribution when H_0 is true and n_1 and n_2 are large. To compute a value of Z, we must estimate the parameter p appearing in the radical. Pooling the data from both samples, we write

$$\hat{p} = \frac{x_1 + x_2}{n_1 + n_2},$$

where x_1 and x_2 are the number of successes in each of the two samples. Substituting $\hat{p}$ for p the statistic Z assumes the form

$$Z = \frac{\hat{P}_1 - \hat{P}_2}{\sqrt{\hat{p}\hat{q}[(1/n_1) + (1/n_2)]}},$$

where $\hat{q} = 1 - \hat{p}$. The critical regions for the appropriate alternative hypotheses are set up before using the critical points of the standard normal curve.

To test the hypothesis that two proportions are equal when the samples are large, we proceed by the following six steps:

1. H_0: $p_1 = p_2$.
2. H_1: Alternatives are $p_1 < p_2$, $p_1 > p_2$, or $p_1 \neq p_2$.
3. Choose a level of significance equal to α.
4. Critical region:
    (a) $Z < -z_\alpha$ for the alternative $p_1 < p_2$.
    (b) $Z > z_\alpha$ for the alternative $p_1 > p_2$.
    (c) $Z < -z_{\alpha/2}$ and $Z > z_{\alpha/2}$ for the alternative $p_1 \neq p_2$.
5. Computations: Compute $\hat{p}_1 = x_1/n_1$, $\hat{p}_2 = x_2/n_2$, and $\hat{p} = (x_1 + x_2)/(n_1 + n_2)$ and then find

$$z = \frac{\hat{p}_1 - \hat{p}_2}{\sqrt{\hat{p}\hat{q}[(1/n_1) + (1/n_2)]}}.$$

6. Conclusion: Reject H_0 if z falls in the critical region; otherwise, accept H_0.

Example 7.10 A vote is to be taken among the residents of a town and the surrounding county to determine whether a proposed chemical plant should be constructed. The construction site is within the town limits and for this reason many voters in the county feel that the proposal will pass because of the large proportion of town voters who favor the construction. To determine if there is a significant difference in the proportion of town voters and county voters favoring the proposal, a poll is taken. If 120 of 200 town voters favor the proposal and 240 of 500 county residents favor it, would you agree that the proportion of town voters favoring the proposal is higher than the proportion of county voters? Use a 0.025 level of significance.

Solution Let p_1 and p_2 be the true proportion of voters in the town and county, respectively, favoring the proposal. We now follow the six-step procedure:

1. H_0: $p_1 = p_2$.
2. H_1: $p_1 > p_2$.

3. $\alpha = 0.025$.
4. Critical region: $Z > 1.96$.
5. Computations:

$$\hat{p}_1 = \frac{x_1}{n_1} = \frac{120}{200} = 0.60$$

$$\hat{p}_2 = \frac{x_2}{n_2} = \frac{240}{500} = 0.48$$

$$\hat{p} = \frac{x_1 + x_2}{n_1 + n_2} = \frac{120 + 240}{200 + 500} = 0.51.$$

Therefore,

$$z = \frac{0.60 - 0.48}{\sqrt{(0.51)(0.49)[(1/200) + (1/500)]}} = 2.9.$$

6. Conclusion: Reject H_0 and agree that the proportion of town voters favoring the proposal is higher than the proportion of county voters.

7.8 GOODNESS-OF-FIT TEST

Throughout this chapter we have been concerned with the testing of statistical hypotheses about single population parameters such as μ, σ^2, and p. Now we shall consider a test to determine if a population has a specified theoretical distribution. The test is based on how good a fit we have between the frequency of occurrence of observations in an observed sample and the expected frequencies obtained from the hypothesized distribution.

To illustrate, consider the tossing of a die. We hypothesize that the die is honest, which is equivalent to testing the hypothesis that the distribution of outcomes is uniform. Suppose that the die is tossed 120 times and each outcome is recorded. Theoretically, if the die is balanced, we would expect each face to occur 20 times. The results are given in Table 7.2. By comparing the observed

Table 7.2 Observed and Expected Frequencies of 120 Tosses of a Die

		Faces				
	1	2	3	4	5	6
Observed	20	22	17	18	19	24
Expected	20	20	20	20	20	20

frequencies with the corresponding expected frequencies, we must decide whether these discrepancies are likely to occur as a result of sampling fluctuations and the die is balanced or the die is not honest and the distribution of outcomes is not uniform. It is common practice to refer to each possible outcome of an experiment as a cell. Hence in our illustration we have six cells. The appropriate statistic on which we base our decision criterion for an experiment involving k cells is defined by the following theorem.

THEOREM 7.1   *A goodness-of-fit test between observed and expected frequencies is based on the quantity*

$$\chi^2 = \sum_{i=1}^{k} \frac{(o_i - e_i)^2}{e_i},$$

where χ^2 is a value of the random variable X^2 whose sampling distribution is approximated very closely by the chi-square distribution. The symbols o_i and e_i represent the observed and expected frequencies, respectively, for the ith cell.

If the observed frequencies are close to the corresponding expected frequencies, the χ^2 value will be small, indicating a good fit. If the observed frequencies differ considerably from the expected frequencies, the χ^2 value will be large and the fit is poor. A good fit leads to the acceptance of H_0, whereas a poor fit leads to its rejection. The critical region will, therefore, fall in the right tail of the chi-square distribution. For a level of significance equal to α, we find the critical value χ_α^2 from Table VI, and then $X^2 > \chi_a^2$ constitutes the critical region. The decision criterion described here should not be used unless each of the expected frequencies is at least equal to 5.

The number of degrees of freedom associated with the chi-square distribution used here depends on two factors: the number of cells in the experiment, and the number of quantities obtained from the observed data that are necessary in the calculation of the expected frequencies. We arrive at this number by the following theorem:

THEOREM 7.2   *The number of degrees of freedom in a chi-square goodness-of-fit test is equal to the number of cells minus the number of quantities obtained from the observed data that are used in the calculations of the expected frequencies.*

The only quantity provided by the observed data, in computing expected frequencies for the outcome when a die is tossed, is the total frequency. Hence, according to our definition, the computed χ^2 value has $6 - 1 = 5$ degrees of freedom.

From Table 7.2 we find the χ^2 value to be

$$\chi^2 = \frac{(20 - 20)^2}{20} + \frac{(22 - 20)^2}{20} + \frac{(17 - 20)^2}{20}$$

$$+ \frac{(18 - 20)^2}{20} + \frac{(19 - 20)^2}{20} + \frac{(24 - 20)^2}{20}$$

$$= 1.7.$$

Using Table VI, we find $\chi^2_{0.05} = 11.070$ for $v = 5$ degrees of freedom. Since 1.7 is less than the critical value, we fail to reject H_0 and conclude that the distribution is uniform. In other words, the die is balanced.

As a second illustration let us test the hypothesis that the frequency distribution of battery lives given in Table 2.2 may be approximated by the normal distribution. The expected frequencies for each class (cell), listed in Table 7.3, are obtained from a normal curve having the same mean and standard deviation as our sample. From the data of Table 2.1 we find that the sample of 40 batteries has a mean $\bar{x} = 3.4125$ and a standard deviation $s = 0.703$. These values will be used for μ and σ in computing z values corresponding to the class boundaries. The z value corresponding to the boundaries of the fourth class, for example, are

$$z_1 = \frac{2.95 - 3.4125}{0.703} = -0.658$$

$$z_2 = \frac{3.45 - 3.4125}{0.703} = 0.053.$$

From Table IV we find the area between $z_1 = -0.658$ and $z_2 = 0.053$ to be

$$\text{area} = P(-0.658 < Z < 0.053)$$

$$= P(Z < 0.053) - P(Z < -0.658)$$

$$= 0.5211 - 0.2552$$

$$= 0.2659.$$

Table 7.3 Observed and Expected Frequencies of Battery Lives Assuming Normality

Class boundaries	o_i	e_i
1.45–1.95	2 ⎫	0.8 ⎫
1.95–2.45	1 ⎬ 7	2.7 ⎬ 10.3
2.45–2.95	4 ⎭	6.8 ⎭
2.95–3.45	15	10.6
3.45–3.95	10	10.3
3.95–4.45	5 ⎫ 8	6.1 ⎫ 8.4
4.45–4.95	3 ⎭	2.3 ⎭

Hence the expected frequency for the fourth class is

$$e_4 = (0.2659)(40) = 10.6.$$

The expected frequency for the first class interval is obtained by using the total area under the normal curve to the left of the boundary 1.95. For the last class interval, we use the total area to the right of the boundary 4.45. All other expected frequencies are determined by the method described for the fourth class. Note that we have combined adjacent classes in Table 7.3 where the expected frequencies are less than 5. Consequently, the total number of intervals is reduced from 7 to 4. The χ^2 value is then given by

$$\chi^2 = \frac{(7 - 10.1)^2}{10.3} + \frac{(15 - 10.6)^2}{10.6} + \frac{(10 - 10.3)^2}{10.3} + \frac{(8 - 8.3)^2}{8.4}$$

$$= 2.911.$$

The number of degrees of freedom for this test will be $4 - 3 = 1$, since three quantities, the total frequency, mean, and standard deviation of the observed data, were required to find the expected frequencies. Since the computed χ^2 value is less than $\chi^2_{0.05} = 3.841$ for 1 degree of freedom, we have no reason to reject the null hypothesis and conclude that the normal distribution provides a good fit for the distribution of battery lives.

7.9 TEST FOR INDEPENDENCE

The chi-square test procedure discussed in Section 7.8 can also be used to test the hypothesis of independence of two variables. Suppose that we wish to

Table 7.4 2 × 3 Contingency Table

	Protestant	Catholic	Jewish	Total
East coast	182	215	203	600
West coast	154	136	110	400
Total	336	351	313	1000

study the relationship between religious affiliation and geographical region. Two groups of people are chosen at random, one from the East Coast and one from the West Coast of the United States, and each person is classified as Protestant, Catholic, or Jewish. The observed frequencies are presented in Table 7.4, which is known as a *contingency table*.

A contingency table with r rows and c columns is referred to as an $r \times c$ table. The symbol $r \times c$ is read "r by c." The row and column totals in Table 7.4 are called *marginal frequencies*. To test the null hypothesis H_0 of independence between a person's religious faith and the region where he lives, we must first find the expected frequencies for each cell of Table 7.4 under the assumption that H_0 is true.

Let us define the following events:

P: an individual selected from our sample is Protestant.

C: an individual selected from our sample is Catholic.

J: an individual selected from our sample is Jewish.

E: an individual selected from our sample lives on the East Coast.

W: an individual selected from our sample lives on the West Coast.

Using the marginal frequencies, we can list the following probabilities:

$$P(P) = \frac{336}{1000}, \quad P(C) = \frac{351}{1000}, \quad P(J) = \frac{313}{1000}, \quad P(E) = \frac{600}{1000},$$

$$P(W) = \frac{400}{1000}.$$

Now, if H_0 is true and the two variables are independent, we should have

$$P(P \cap E) = P(P)P(E) = \left(\frac{336}{1000}\right)\left(\frac{600}{1000}\right)$$

$$P(P \cap W) = P(P)P(W) = \left(\frac{336}{1000}\right)\left(\frac{400}{1000}\right)$$

$$P(C \cap E) = P(C)P(E) = \left(\frac{351}{1000}\right)\left(\frac{600}{1000}\right)$$

$$P(C \cap W) = P(C)P(W) = \left(\frac{351}{1000}\right)\left(\frac{400}{1000}\right)$$

$$P(J \cap E) = P(J)P(E) = \left(\frac{313}{1000}\right)\left(\frac{600}{1000}\right)$$

$$P(J \cap W) = P(J)P(W) = \left(\frac{313}{1000}\right)\left(\frac{400}{1000}\right).$$

The expected frequencies are obtained by multiplying each cell probability by the total number of observations. Thus the expected number of Protestants living on the East Coast in our sample will be

$$\left(\frac{336}{1000}\right)\left(\frac{600}{1000}\right)(1000) = \frac{(336)(600)}{1000} = 202$$

when H_0 is true. The general formula for obtaining the expected frequency of any cell is given by

$$e = \frac{RC}{T},$$

where R and C are the corresponding row and column totals and T is the grand total of all the observed frequencies. The expected frequencies for each cell are recorded in parentheses beside the actual observed value in Table 7.5. Note that the expected frequencies in any row or column add up to the appropriate marginal total. In our example we need to compute only the two expected frequencies in the top row of Table 7.5 and then find the others by subtraction. By using three marginal totals and the grand total to arrive at the expected frequencies, we have lost 4 degrees of freedom, leaving a total of 2. A simple formula providing the correct number of degrees of freedom is given by $v = (r - 1)(c - 1)$. Hence, for our example, $v = (2 - 1)(3 - 1) = 2$ degrees of freedom.

Table 7.5 Observed and Expected Frequencies

	Protestant	*Catholic*	*Jewish*	*Total*
East coast	182 (202)	215 (211)	203 (187)	600
West coast	154 (134)	136 (140)	110 (126)	400
Total	336	351	313	1000

To test the null hypothesis of independence, we use the following decision criterion:

TEST FOR INDEPENDENCE *Calculate*

$$\chi^2 = \sum_i \frac{(o_i - e_i)^2}{e_i},$$

where the summation extends over all cells in the $r \times c$ contingency table. If $\chi^2 > \chi_\alpha^2$, reject the null hypothesis of independence at the α level of significance; otherwise, accept the null hypothesis. The number of degrees of freedom is

$$v = (r - 1)(c - 1).$$

Applying this criterion to our example, we find that

$$\chi^2 = \frac{(182 - 202)^2}{202} + \frac{(215 - 211)^2}{211} + \frac{(203 - 187)^2}{187}$$

$$+ \frac{(154 - 134)^2}{134} + \frac{(136 - 140)^2}{140} + \frac{(110 - 126)^2}{126}$$

$$= 8.556.$$

From Table VI we find that $\chi_{0.05}^2 = 5.991$ for $v = (2 - 1)(3 - 1) = 2$ degrees of freedom. The null hypothesis is rejected at the 0.05 level of significance and we conclude that religious faith and the region where one lives are not independent.

The chi-square statistic for testing independence is also applicable when testing the hypothesis that k binomial populations have the same parameter p.

This is, therefore, an extension of the test presented in Section 7.7 for the difference between two proportions to the differences among k proportions. Hence we are interested in testing the hypothesis

$$H_0: \quad p_1 = p_2 = \cdots = p_k = p$$

against the alternative hypothesis that the population proportions are not all equal, which is equivalent to testing that the number of successes or failures is independent of the sample chosen. To perform this test, we first select independent random samples of size $n_1, n_2, \ldots, n_k$ from the k populations and arrange the data as in a $2 \times k$ contingency table (Table 7.6). The expected cell frequencies are calculated as before and substituted together with the observed frequencies into the preceding chi-square formula for independence with $v = k - 1$ degrees of freedom. By selecting an appropriate critical region, one can now reach a conclusion concerning H_0.

It is important to remember that the statistic on which we base our decision has a distribution that is only approximated by the chi-square distribution. The computed χ^2 values depend on the cell frequencies and consequently are discrete. The continuous chi-square distribution seems to approximate the discrete sampling distribution of X^2 very well, provided the number of degrees of freedom is greater than 1. In a 2×2 contingency table, where we have only 1 degree of freedom, a correction called *Yates' correction for continuity* is applied. The corrected formula then becomes

$$\chi^2 \text{ (corrected)} = \sum_i \frac{(|o_i - e_i| - 0.5)^2}{e_i}.$$

If the expected cell frequencies are large, the corrected and uncorrected results are almost the same. When the expected frequencies are between 5 and 10, Yates' correction should be applied. For expected frequencies less than 5, the Fisher–Irwin exact test should be used. A discussion of this test may be

Table 7.6 k Independent Binomial Samples

	Sample			
	1	*2*	$\cdots$	*k*
Success	x_1	x_2	$\cdots$	x_k
Failures	$n_1 - x_1$	$n_2 - x_2$	$\cdots$	$n_k - x_k$

found in *Basic Concepts of Probability and Statistics* by Hodges and Lehmann (see the Bibliography). The Fisher–Irwin test may be avoided, however, by choosing a larger sample.

EXERCISES

1. The proportion of adults living in a small town who are college graduates is estimated to be $p = 0.3$. To test this hypothesis a random sample of 15 adults is selected. If the number of college graduates in our sample is anywhere from 2 to 7, we shall accept the null hypothesis that $p = 0.3$; otherwise, we shall conclude that $p \neq 0.3$. Evaluate α assuming $p = 0.3$. Evaluate β for the alternatives $p = 0.2$ and $p = 0.4$. Is this a good test procedure?

2. The proportion of families buying milk from company A in a certain city is believed to be $p = 0.6$. If a random sample of 10 families shows that 3 or less buy milk from company A, we shall reject the hypothesis that $p = 0.6$ in favor of the alternative, $p < 0.6$. Find the probability of committing a type I error if the true proportion is $p = 0.6$. Evaluate the probability of committing a type II error for the alternatives $p = 0.3$, $p = 0.4$, and $p = 0.5$.

3. In a large experiment to determine the success of a new drug, 400 patients with a certain disease are to be given the drug. If more than 300 but less than 340 patients are cured, we shall conclude that the drug is 80% effective. Find the probability of committing a type I error. What is the probability of committing a type II error if the new drug is only 70% effective?

4. A new cure has been developed for a certain type of cement that results in a compressive strength of 5000 kilograms per square centimeter and a standard deviation of 120. To test the hypothesis that $\mu = 5000$ against the alternative that $\mu < 5000$, a random sample of 50 pieces of cement are tested. The critical region is defined to be $\bar{X} < 4970$. Find the probability of committing a type I error. Evaluate β for the alternatives $\mu = 4970$ and $\mu = 4960$.

5. An electrical firm manufactures light bulbs that have a length of life that is approximately normally distributed with a mean of 800 hours and a standard deviation of 40 hours. Test the hypothesis that $\mu = 800$ hours against the alternative $\mu \neq 800$ hours if a random sample of 30 bulbs has an average life of 788 hours. Use a 0.04 level of significance.

6. A random sample of 36 drinks from a soft-drink machine has an average content of 21.9 deciliters, with a standard deviation of 1.42 deciliters. Test the hypothesis that $\mu = 22.2$ deciliters against the alternative hypothesis, $\mu < 22.2$, at the 0.05 level of significance.

7. The average height of males in the freshman class of a certain college has been 174.5 centimeters, with a standard deviation of 6.9 centimeters. Is there reason to believe that there has been a change in the average height if a random sample of 50 males in the present freshman class has an average height of 177.2 centimeters? Use a 0.02 level of significance.

8. It is claimed that an automobile is driven on the average less than 20,000 kilometers per year. To test this claim, a random sample of 100 automobile owners are asked to keep a record of the kilometers they travel. Would you agree with this claim if the random sample showed an average of 23,500 kilometers and a standard deviation of 3900 kilometers? Use a 0.01 level of significance.

9. Test the hypothesis that the average content of containers of a particular lubricant is 10 liters if the contents of a random sample of 10 containers are 10.2, 9.7, 10.1, 10.3, 10.1, 9.8, 9.9, 10.4, 10.3, and 9.8 liters. Use a 0.01 level of significance and assume that the distribution of contents is normal.

10. A random sample of size 20 from a normal distribution has a mean $\bar{x} = 32.8$ and a standard deviation $s = 4.51$. Does this suggest, at the 0.05 level of significance, that the population mean is greater than 30?

11. A random sample of eight cigarettes of a certain brand has an average tar content of 18.6 milligrams and a standard deviation of 2.4 milligrams. Is this in line with the manufacturer's claim that the average tar content does not exceed 17.5 milligrams? Use a 0.01 level of significance and assume the distribution of tar contents to be normal.

12. A male student will spend, on the average, $8.00 for a Saturday evening fraternity party. Test the hypothesis at the 0.1 level of significance that $\mu = \$8.00$ against the alternative $\mu \neq \$8.00$ if a random sample of 12 male students attending a homecoming party showed an average expenditure of $8.90 with a standard deviation of $1.75. Assume that the expenses are approximately normally distributed.

13. A random sample of size $n_1 = 25$, taken from a normal population with a standard deviation $\sigma_1 = 5.2$, has a mean $\bar{x}_1 = 81$. A second random sample of size $n_2 = 36$, taken from a different normal population with a standard deviation $\sigma_2 = 3.4$, has a mean $\bar{x}_2 = 76$. Test the hypothesis, at the 0.06 level of significance, that $\mu_1 = \mu_2$ against the alternative $\mu_1 \neq \mu_2$.

14. A manufacturer claims that the average tensile strength of thread A exceeds the average tensile strength of thread B by at least 12 kilograms. To test his claim, 50 pieces of each type of thread are tested under similar conditions. Type A thread had an average tensile strength of 86.7 kilograms with a standard deviation of 6.28 kilograms, while type B thread had an average tensile strength of 77.8 kilograms with a standard deviation of 5.61 kilograms. Test the manufacturer's claim using a 0.05 level of significance.

15. A study was made to estimate the difference in salaries of college professors in the private and state colleges of Virginia. A random sample of 100 professors in private colleges showed an average 9-month salary of $15,000 with a standard deviation of $1300. A random sample of 200 professors in state colleges showed an average salary of $15,900 with a standard deviation of $1400. Test the hypothesis that the average salary for professors teaching in state colleges does not exceed the average salary for professors teaching in private colleges by more than $500. Use a 0.02 level of significance.

16. Given two random samples of size $n_1 = 11$ and $n_2 = 14$, from two independent normal populations, with $\bar{x}_1 = 75$, $\bar{x}_2 = 60$, $s_1 = 6.1$, and $s_2 = 5.3$, test the hypothesis at the 0.05 level of significance that $\mu_1 = \mu_2$ against the alternative that $\mu_1 \neq \mu_2$. Assume that the population variances are equal.

17. A study is made to see if increasing the substrate concentration has an appreciable effect on the velocity of a chemical reaction. With the substrate concentration of 1.5 moles per liter, the reaction was run 15 times with an average velocity of 7.5 micromoles per 30 minutes and a standard deviation of 1.5. With a substrate concentration of 2.0 moles per liter, 12 runs were made yielding an average velocity of 8.8 micromoles per 30 minutes and a sample standard deviation of 1.2. Would you say that the increase in substrate concentration increases the mean velocity by as much as 0.5 micromole per 30 minutes? Use a 0.01 level of significance and assume the populations to be approximately normally distributed with equal variances.

18. A large automobile manufacturing company is trying to decide whether to purchase brand A or brand B tires for its new models. To help arrive at a decision an experiment is conducted using 12 of each brand. The tires are run until they wear out. The results are

$$\text{brand A:} \ \bar{x}_1 = 37{,}900 \ \text{kilometers,} \ s_1 = 5100 \ \text{kilometers}$$

$$\text{brand B:} \ \bar{x}_2 = 39{,}800 \ \text{kilometers,} \ s_2 = 5900 \ \text{kilometers}$$

Test the hypothesis at the 0.05 level of significance that there is no difference in the two brands of tires. Assume the populations to be approximately normally distributed.

19. The following data represent the running times of films produced by two motion-picture companies:

	Time (minutes)						
Company 1	102	86	98	109	92		
Company 2	81	165	97	134	92	87	114

Test the hypothesis that the average running time of films produced by company 2 exceeds the average running time of films produced by company 1 by 10 minutes against the one-sided alternative that the difference is more than 10 minutes. Use a 0.1 level of significance and assume the distributions of times to be approximately normal.

20. In Exercise 20, Chapter 6, use the t distribution to test the hypothesis, at the 0.05 level of significance, that the average yields of the two varieties of wheat are equal against the alternative hypothesis that they are unequal.

21. In Exercise 21, Chapter 6, use the t distribution to test the hypothesis, at the 0.01 level of significance, that $\mu_1 = \mu_2$ against the alternative hypothesis that $\mu_1 < \mu_2$.

22. In Exercise 22, Chapter 6, use the t distribution to test the hypothesis, at the 0.05 level of significance, that the diet reduces a person's weight by 4.5 kilograms on the average against the alternative hypothesis that the mean difference in weight is less than 4.5 kilograms.

23. Test the hypothesis that $\sigma^2 = 0.03$ against the alternative hypothesis that $\sigma^2 \neq 0.03$ in Exercise 9. Use a 0.01 level of significance.

24. Test the hypothesis that $\sigma = 6$ against the alternative that $\sigma < 6$ in Exercise 10. Use a 0.05 level of significance.

25. Test the hypothesis that $\sigma^2 = 2.3$ against the alternative that $\sigma^2 \neq 2.3$ in Exercise 11. Use a 0.05 level of significance.

26. Test the hypothesis that $\sigma = 1.40$ against the alternative that $\sigma > 1.40$ in Exercise 12. Use a 0.01 level of significance.

27. Test the hypothesis that $\sigma_1^2 = \sigma_2^2$ against the alternative that $\sigma_1^2 > \sigma_2^2$ in Exercise 16. Use a 0.01 level of significance.

28. Test the hypothesis that $\sigma_1 = \sigma_2$ against the alternative that $\sigma_1 < \sigma_2$ in Exercise 18. Use a 0.05 level of significance.

29. Test the hypothesis that $\sigma_1^2 = \sigma_2^2$ against the alternative that $\sigma_1^2 \neq \sigma_2^2$ in Exercise 19. Use a 0.10 level of significance.

30. How large a sample is required in Exercise 6 if the power of our test is to be 0.90 when the true mean is 21.3? Assume that $\sigma = 1.42$.

31. How large a sample is required in Exercise 7 if the power of our test is to be 0.95 when the true average height differs from 174.5 by 3.1 centimeters?

32. How large should the samples be in Exercise 14 if the power of our test is to be 0.95 when the true difference between thread types A and B is 8 kilograms?

33. How large a sample is required in Exercise 11 if the power of our test is to be 0.8 when the true tar content exceeds the hypothesized value by 1.2σ?

34. On testing

$$H_0: \quad \mu = 14,$$
$$H_1: \quad \mu \neq 14,$$

an $\alpha = 0.05$ level t test is being considered. What sample size is necessary in order that the probability is 0.1 of falsely accepting H_0 when the true population mean differs from 14 by 0.5? From a preliminary sample we estimate σ to be 1.25.

35. In testing the hypothesis

$$H_0: \quad \sigma^2 = 2,$$
$$H_1: \quad \sigma^2 > 2,$$

at the 0.05 level of significance with a sample of size 15, find the power of the test when the true variance is actually $\sigma^2 = 3.5$.

36. It is believed that at least 60% of the residents in a certain area favor an annexation suit by a neighboring city. What conclusion would you draw if only 110 in a sample of 200 voters favor the suit? Use a 0.04 level of significance.

37. A manufacturer of cigarettes claims that 20% of the cigarette smokers prefer brand X. To test this claim, a random sample of 20 cigarette smokers is selected and the smokers are asked what brand they prefer. If 6 of the 20 smokers prefer brand X, what conclusion do we draw? Use a 0.01 level of significance.

38. A new radar device is being considered for a certain defense missile system. The system is checked by experimenting with actual aircraft in which a *kill* or a *no kill* is

simulated. If in 300 trials, 250 kills occur, accept or reject, at the 0.04 level of significance, the claim that the probability of a kill with the new system does not exceed the 0.8 probability of the existing device.

39. At a certain college it is estimated that fewer than 25% of the students have cars on campus. Does this seem to be a valid estimate if in a random sample of 90 college students, 28 are found to have cars? Use a 0.05 level of significance.

40. In a study to estimate the proportion of housewives who own an automatic dryer, it is found that 63 of 100 urban residents have a dryer and 59 of 125 suburban residents own a dryer. Is there a significant difference between the proportion of urban and suburban housewives who own an automatic dryer? Use a 0.04 level of significance.

41. A cigarette manufacturing firm distributes two brands of cigarettes. If it is found that 56 of 200 smokers prefer brand A and that 29 of 150 smokers prefer brand B, can we conclude at the 0.06 level of significance that brand A outsells brand B?

42. A random sample of 100 men and 100 women at a southern college is asked if they have an automobile on campus. If 31 of the men and 24 of the women have cars, can we conclude that more men than women have cars on campus? Use a 0.01 level of significance.

43. A study is made to determine if a cold climate contributes more to absenteeism from school during a semester than a warmer climate. Two groups of students are selected at random, one group from Maine and the other from Alabama. Of the 300 students from Maine, 72 were absent at least 1 day during the semester, and of the 400 students from Alabama, 70 were absent 1 or more days. Can we conclude that a colder climate results in a greater number of students being absent from school at least 1 day during the semester? Use a 0.05 level of significance.

44. A die is tossed 180 times with the following results:

x	1	2	3	4	5	6
f	28	36	36	30	27	23

Is this a balanced die? Use a 0.01 level of significance.

45. In 100 tosses of a coin, 63 heads and 37 tails are observed. Is this a balanced coin? Use a 0.05 level of significance.

46. Three marbles are selected from an urn containing five red marbles and three green marbles. After recording the number X of red marbles, the marbles are replaced in the urn and the experiment repeated 112 times. The results obtained are as follows:

x	0	1	2	3
f	1	31	55	25

Test the hypothesis at the 0.05 level of significance that the recorded data may be fitted by the hypergeometric distribution $h(x: 8, 3, 5)$, $x = 0, 1, 2, 3$.

47. Three cards are drawn from an ordinary deck of playing cards, with replacement, and the number Y of spades is recorded. After repeating the experiment 64 times, the following outcomes were recorded:

y	0	1	2	3
f	21	31	12	0

Test the hypothesis at the 0.01 level of significance that the recorded data may be fitted by the binomial distribution $b(y; 3, 1/4)$, $y = 0, 1, 2, 3$.

48. The grades in a statistics course for a particular semester were as follows:

Grade	A	B	C	D	F
f	14	18	32	20	16

Test the hypothesis, at the 0.05 level of significance, that the distribution of grades is uniform.

49. A coin is thrown until a head occurs and the number X of tosses recorded. After repeating the experiment 256 times, we obtained the following results:

x	1	2	3	4	5	6	7	8
f	136	60	34	12	9	1	3	1

Test the hypothesis at the 0.05 level of significance that the observed distribution of X may be fitted by the geometric distribution $g(x; 1/2)$, $x = 1, 2, 3. \ldots$

50. In Exercise 16, Chapter 2, test the goodness of fit between the observed frequencies and the expected normal frequencies, using a 0.05 level of significance.

51. In an experiment to study the dependence of hypertension on smoking habits, the following data were taken on 180 individuals:

	Nonsmokers	Moderate smokers	Heavy smokers
Hypertension	21	36	30
No hypertension	48	26	19

Test the hypothesis that the presence or absence of hypertension is independent of smoking habits. Use a 0.05 level of significance.

52. A random sample of 200 married men, all retired, were classified according to education and number of children:

	Number of children		
Education	0–1	2–3	Over 3
Elementary	14	37	32
Secondary	19	42	17
College	12	17	10

Test the hypothesis, at the 0.05 level of significance, that the size of a family is independent of the level of education attained by the father.

53. In a shop study, a set of data was collected to determine whether or not the proportion of defectives produced by workers was the same for the day, evening, or night shift worked. The following data were collected on the items produced:

	Shift		
	Day	Evening	Midnight
Defective	45	55	70
Nondefective	905	890	870

What is your conclusion? Use an $\alpha = 0.025$ level of significance.

Linear Regression and Correlation

8.1 LINEAR REGRESSION

Often in practice one is called upon to solve problems involving sets of variables when it is known that there exists some inherent relationship among the variables. For example, in an industrial situation it may be known that the tar content in the outlet stream in a chemical process is related to the inlet temperature. It may be of interest to develop a method of prediction, that is, a procedure for estimating the tar content for various levels of the inlet temperature from experimental information. The statistical aspect of the problem then becomes one of arriving at the best estimate of the relationship between the variables.

For this example and most applications there is a clear distinction between the variables as far as their role in the experimental process is concerned. Quite often there is a single *dependent variable* or response Y, which is uncontrolled in the experiment. This response depends on one or more *independent variables*, say, $x_1, x_2, \ldots, x_k$, which are measured with negligible error and indeed are often controlled in the experiment. Thus the independent variables $x_1, x_2, \ldots, x_k$ are *not* random variables but are fixed quantities preselected by the investigator and have no distributional properties. In the example cited earlier, inlet temperature is the independent variable and tar content is the response Y. The relationship, fitted to a set of experimental data, is characterized by a prediction equation called a *regression equation*. In the case of a single Y and a single x, the situation becomes a regression of Y on x. For k independent variables, we speak in terms of a regression of Y on $x_1, x_2, \ldots, x_k$. A chemical engineer may, in fact, be concerned with the amount of hydrogen lost from samples of a particular metal when the material is placed in storage. In this case there may be two inputs, storage time x_1 in hours and storage temperature x_2 in degrees centigrade. The response would then be hydrogen loss Y in parts per million.

In this chapter we shall deal with the topic of *simple linear regression*, treating only the case of a single independent variable. For the case of more than one independent variable, the reader is referred to Chapter 9. Let us denote a random sample of size n by the set $\{(x_i, y_i); i = 1, 2, \ldots, n\}$. If additional samples were taken using exactly the same values of x, we would expect the

280

Table 8.1 Experimental Data

i	1	2	3	4	5	6	7	8	9
x_i	1.5	1.8	2.4	3.0	3.5	3.9	4.4	4.8	5.0
y_i	4.8	5.7	7.0	8.3	10.9	12.4	13.1	13.6	15.3

y values to vary. Hence the value y_i in the ordered pair (x_i, y_i) is a value of some random variable Y_i. For convenience we define $Y|x$ to be the random variable Y corresponding to a fixed value x and denote its probability distribution by $f(y|x)$. Clearly, then, if $x = x_i$, the symbol $Y|x_i$ represents the random variable Y_i.

The term *linear regression* implies that the mean of $Y|x$ is linearly related to x in the usual slope-intercept form:

$$\mu_{Y|x} = \alpha + \beta x,$$

where α and β are parameters to be estimated from the sample data. Denoting their estimates by a and b, respectively, the estimated response $\hat{y}$ is obtained from the sample regression line

$$\hat{y} = a + bx.$$

Consider the experimental data of Table 8.1, which has been plotted in Figure 8.1 to give a *scatter diagram*. The assumption of linearity appears to be reasonable.

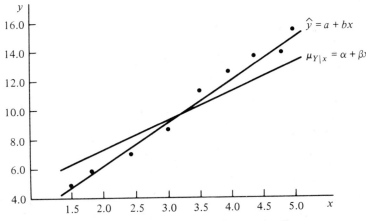

Figure 8.1 Scatter diagram with regression lines.

The sample regression line and a hypothetical true regression line have been drawn on the scatter diagram of Figure 8.1. The agreement between the sample line and the unknown hypothetical line should be good if we have a large amount of data available.

In the following sections we shall develop procedures for finding estimators of the *regression coefficients*, α and β, in order that the regression equation can be used for predicting or estimating the mean response for specific values of the independent variable x.

8.2 SIMPLE LINEAR REGRESSION

For the case where there is a single x and a single Y, the data take the form of pairs of observations $\{(x_i, y_i); i = 1, 2, \ldots, n\}$. If the values for x are controlled, that is, the experiment is *designed*, then the experimental process is to fix or choose the x_i values in advance and observe the corresponding y_i values.

Assuming that all the means, $\mu_{Y|x}$, fall on a straight line, the random variable $Y_i = Y|x_i$ may then be written

$$Y_i = \mu_{Y|x_i} + E_i = \alpha + \beta x_i + E_i,$$

where the random variable E_i must necessarily have a mean of zero. Each observation (x_i, y_i) in our sample satisfies the relation

$$y_i = \alpha + \beta x_i + \varepsilon_i,$$

where ε_i is the value assumed by E_i when Y_i takes on the value y_i.

Similarly, using the estimated regression line

$$\hat{y} = a + bx,$$

each pair of observations satisfies the relation

$$y_i = a + bx_i + e_i,$$

where e_i is called the *residual*. The difference between e_i and ε_i is clearly shown in Figure 8.2.

We shall find a and b, the estimates of α and β, so that the sum of the squares of the residuals is a minimum. The residual sum of squares is often called the sum of squares of the errors about the regression line and denoted by SSE.

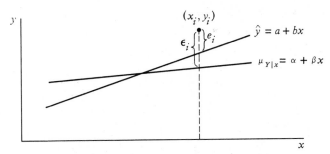

Figure 8.2 Comparing ε_i with the residual e_i.

This minimization procedure for estimating parameters is called the *method of least squares*. Hence we shall find a and b so as to minimize

$$SSE = \sum_{i=1}^{n} e_i^2 = \sum_{i=1}^{n} (y_i - a - bx_i)^2.$$

Differentiating SSE with respect to a and b, we have

$$\frac{\partial(SSE)}{\partial a} = -2 \sum_{i=1}^{n} (y_i - a - bx_i)$$

$$\frac{\partial(SSE)}{\partial b} = -2 \sum_{i=1}^{n} (y_i - a - bx_i)x_i.$$

Setting the partial derivatives equal to zero and rearranging the terms, we obtain the equations (called the *normal equations*)

$$na + b \sum_{i=1}^{n} x_i = \sum_{i=1}^{n} y_i$$

$$a \sum_{i=1}^{n} x_i + b \sum_{i=1}^{n} x_i^2 = \sum_{i=1}^{n} x_i y_i,$$

which yield

$$b = \frac{n \sum_{i=1}^{n} x_i y_i - \left(\sum_{i=1}^{n} x_i \right) \left(\sum_{i=1}^{n} y_i \right)}{n \sum_{i=1}^{n} x_i^2 - \left(\sum_{i=1}^{n} x_i \right)^2}.$$

From the first of these two normal equations we can write

$$a = \bar{y} - b\bar{x}.$$

ESTIMATION OF THE PARAMETERS α AND β   *The regression line*

$$\mu_{Y|x} = \alpha + \beta x$$

is estimated from the sample $\{(x_i, y_i); i = 1, 2, \ldots, n\}$ by the line

$$\hat{y} = a + bx,$$

where

$$b = \frac{n\sum_{i=1}^{n} x_i y_i - \left(\sum_{i=1}^{n} x_i\right)\left(\sum_{i=1}^{n} y_i\right)}{n\sum_{i=1}^{n} x_i^2 - \left(\sum_{i=1}^{n} x_i\right)^2},$$

$$a = \bar{y} - b\bar{x}.$$

The calculations of a and b using the data of Table 8.1 are illustrated by the following example.

Example 8.1   Estimate the regression line for the data of Table 8.1.

Solution   Using a desk calculator, we find that

$$\sum_{i=1}^{9} x_i = 30.3, \qquad \sum_{i=1}^{9} y_i = 91.1, \qquad \sum_{i=1}^{9} x_i y_i = 345.09,$$

$$\sum_{i=1}^{9} x_i^2 = 115.11, \qquad \bar{x} = 3.3667, \qquad \bar{y} = 10.1222.$$

Therefore,

$$b = \frac{(9)(345.09) - (30.3)(91.1)}{(9)(115.11) - (30.3)^2} = 2.9303,$$

$$a = 10.1222 - (2.9303)(3.3667) = 0.2568.$$

Thus the estimated regression line is given by

$$\hat{y} = 0.2568 + 2.9303x.$$

By substituting any two of the given values of x into this equation, say $x_1 = 1.5$ and $x_2 = 5.0$, we obtain the ordinates $\hat{y}_1 = 4.7$ and $\hat{y}_2 = 14.9$. The sample regression line in Figure 8.1 was drawn by connecting these two points with a straight line.

8.3 PROPERTIES OF THE LEAST SQUARES ESTIMATORS

In addition to the assumptions that the error term in the model

$$Y_i = \alpha + \beta x_i + E_i$$

is a random variable with mean zero, suppose that we make the further assumption that each E_i is normally distributed with the same variance σ^2 and that $E_1, E_2, \ldots, E_n$ are independent from run to run in the experiment. With this normality assumption on the E_i's, we have a procedure for finding the means and variances for the estimators of α and β.

It is important to remember that our values of a and b are only estimates of the true parameters α and β based on a given sample of n observations. The different estimates of α and β that could be computed by drawing several samples of size n may be thought of as values assumed by the random variables A and B.

Since the values of x remain fixed, the values of A and B depend on the variations in the values of y, or, more precisely, on the values of the random variables $Y_1, Y_2, \ldots, Y_n$. The distributional assumptions on the E_i's implies that the Y_i's, $i = 1, 2, \ldots, n$, are likewise independently distributed, each with probability distribution $n(y_i; \alpha + \beta x_i, \sigma)$, and since the estimator

$$B = \frac{n\sum\limits_{i=1}^{n} x_i Y_i - \left(\sum\limits_{i=1}^{n} x_i\right)\left(\sum\limits_{i=1}^{n} Y_i\right)}{n\sum\limits_{i=1}^{n} x_i^2 - \left(\sum\limits_{i=1}^{n} x_i\right)^2}$$

$$= \frac{\sum\limits_{i=1}^{n} (x_i - \bar{x})(Y_i - \bar{Y})}{\sum\limits_{i=1}^{n} (x_i - \bar{x})^2}$$

$$= \frac{\sum\limits_{i=1}^{n} (x_i - \bar{x}) Y_i}{\sum\limits_{i=1}^{n} (x_i - \bar{x})^2}$$

is a linear function of the random variables $Y_1, Y_2, \ldots, Y_n$ with coefficients

$$a_i = \frac{x_i - \bar{x}}{\displaystyle\sum_{i=1}^{n} (x_i - \bar{x})^2}, \qquad i = 1, 2, \ldots, n,$$

we may conclude from Theorem 5.11 that it is normally distributed with mean

$$\mu_B = E(B) = \frac{\displaystyle\sum_{i=1}^{n} (x_i - \bar{x}) E(Y_i)}{\displaystyle\sum_{i=1}^{n} (x_i - \bar{x})^2}$$

$$= \frac{\displaystyle\sum_{i=1}^{n} (x_i - \bar{x})(\alpha + \beta x_i)}{\displaystyle\sum_{i=1}^{n} (x_i - \bar{x})^2} = \beta$$

and variance

$$\sigma_B^2 = \frac{\displaystyle\sum_{i=1}^{n} (x_i - \bar{x})^2 \sigma_{Y_i}^2}{\left[\displaystyle\sum_{i=1}^{n} (x_i - \bar{x})^2\right]^2} = \frac{\sigma^2}{\displaystyle\sum_{i=1}^{n} (x_i - \bar{x})^2}.$$

It can be shown (Exercise 7) that the random variable A is also normally distributed with mean

$$\mu_A = \alpha$$

and variance

$$\sigma_A^2 = \frac{\displaystyle\sum_{i=1}^{n} x_i^2}{n \displaystyle\sum_{i=1}^{n} (x_i - \bar{x})^2} \sigma^2.$$

To be able to draw inferences on α and β, it becomes necessary to arrive at an estimate of the parameter σ^2 appearing in the two preceding variance formulas for A and B. In deriving such an estimate it is advantageous, from a theoretical point of view, to introduce the notation

$$S_{xx} = \sum_{i=1}^{n}(x_i - \bar{x})^2 = \sum_{i=1}^{n}x_i^2 - \frac{\left(\sum_{i=1}^{n}x_i\right)^2}{n}$$

$$S_{yy} = \sum_{i=1}^{n}(y_i - \bar{y})^2 = \sum_{i=1}^{n}y_i^2 - \frac{\left(\sum_{i=1}^{n}y_i\right)^2}{n}$$

$$S_{xy} = \sum_{i=1}^{n}(x_i - \bar{x})(y_i - \bar{y}) = \sum_{i=1}^{n}x_iy_i - \frac{\left(\sum_{i=1}^{n}x_i\right)\left(\sum_{i=1}^{n}y_i\right)}{n}.$$

Now we may write the error sum of squares as follows:

$$SSE = \sum_{i=1}^{n}(y_i - a - bx_i)^2$$

$$= \sum_{i=1}^{n}[(y_i - \bar{y}) - b(x_i - \bar{x})]^2$$

$$= \sum_{i=1}^{n}(y_i - \bar{y})^2 - 2b\sum_{i=1}^{n}(x_i - \bar{x})(y_i - \bar{y}) + b^2\sum_{i=1}^{n}(x_i - \bar{x})^2$$

$$= S_{yy} - 2bS_{xy} + b^2S_{xx}$$

$$= S_{yy} - bS_{xy},$$

the final step following from the fact that $b = S_{xy}/S_{xx}$. We may now prove the following theorem.

THEOREM 8.1   *An unbiased estimate of σ^2 is given by*

$$s^2 = \frac{SSE}{n-2} = \frac{S_{yy} - bS_{xy}}{n-2}.$$

Proof Interpreting the error sum of squares as a random variable whose values will vary if the experiment were repeated several times, we may write

$$SSE = S_{YY} - BS_{xY}$$

$$= S_{YY} - B^2 S_{xx} \qquad \text{(since } S_{xY} = BS_{xx})$$

$$= \sum_{i=1}^{n} (Y_i - \bar{Y})^2 - B^2 \sum_{i=1}^{n} (x_i - \bar{x})^2$$

$$= \sum_{i=1}^{n} Y_i^2 - n\bar{Y}^2 - \left(\sum_{i=1}^{n} x_i^2 - n\bar{x}^2 \right) B^2.$$

Now, taking expected values, we have

$$E(SSE) = \sum_{i=1}^{n} E(Y_i^2) - nE(\bar{Y}^2) - \left(\sum_{i=1}^{n} x_i^2 - n\bar{x}^2 \right) E(B^2).$$

By Theorem 2.6 we may substitute the quantities

$$E(Y_i^2) = \sigma_{Y_i}^2 + \mu_{Y_i}^2,$$

$$E(\bar{Y}^2) = \sigma_{\bar{Y}}^2 + \mu_{\bar{Y}}^2,$$

$$E(B^2) = \sigma_B^2 + \mu_B^2,$$

into the preceding equation to give

$$E(SSE) = \sum_{i=1}^{n} (\sigma_{Y_i}^2 + \mu_{Y_i}^2) - n(\sigma_{\bar{Y}}^2 + \mu_{\bar{Y}}^2) - \left(\sum_{i=1}^{n} x_i^2 - n\bar{x}^2 \right) (\sigma_B^2 + \mu_B^2)$$

$$= n\sigma^2 + \sum_{i=1}^{n} (\alpha + \beta x_i)^2 - n \left[\frac{\sigma^2}{n} + (\alpha + \beta\bar{x})^2 \right]$$

$$- \left(\sum_{i=1}^{n} x_i^2 - n\bar{x}^2 \right) \left(\frac{\sigma^2}{S_{xx}} + \beta^2 \right).$$

Setting $S_{xx} - \sum_{i=1}^{n} x_i^2 - n\bar{x}^2$ and simplifying, we obtain $E(SSE) = (n - 2)\sigma^2$. Therefore,

$$E(S^2) = \frac{E(SSE)}{n - 2} = \sigma^2$$

and s^2 is an unbiased estimate of σ^2.

8.4 CONFIDENCE LIMITS AND TESTS OF SIGNIFICANCE

Aside from merely estimating the linear relationship between x and Y for purposes of prediction, the experimenter may also be interested in drawing certain inferences about the slope, intercept, or the general quality of the estimated regression line. Too often, in fact, regression results are reported by the scientist without any reference to how well b estimates β or how well the regression line will eventually predict response.

Since B is a normal random variable and $(n - 2)S^2/\sigma^2$ is a chi-square variable with $n - 2$ degrees of freedom, Theorem 5.17 assures us that the statistic

$$T = \frac{(B - \beta)/(\sigma/\sqrt{S_{xx}})}{S/\sigma}$$

$$= \frac{B - \beta}{S/\sqrt{S_{xx}}}$$

has a t distribution with $n - 2$ degrees of freedom. The statistic T can be used to construct a $(1 - \alpha)100\%$ confidence interval for the coefficient β.

CONFIDENCE INTERVAL FOR β *A $(1 - \alpha)100\%$ confidence interval for the parameter β in the regression line $\mu_{Y|x} = \alpha + \beta x$ is*

$$b - \frac{t_{\alpha/2} s}{\sqrt{S_{xx}}} < \beta < b + \frac{t_{\alpha/2} s}{\sqrt{S_{xx}}},$$

where $t_{\alpha/2}$ is a value of the t distribution with $n - 2$ degrees of freedom.

Example 8.2   Find a 95% confidence interval for β in the regression line $\mu_{Y|x} = \alpha + \beta x$ based on the data in Table 8.1.

Solution   In Example 8.1 we found that

$$\sum_{i=1}^{9} x_i = 30.3, \qquad \sum_{i=1}^{9} x_i^2 = 115.11, \qquad \sum_{i=1}^{9} y_i = 91.1, \qquad \sum_{i=1}^{9} x_i y_i = 345.09.$$

Referring to the data in Table 8.1, we now find $\sum_{i=1}^{9} y_i^2 = 1036.65$. Therefore,

$$S_{xx} = 115.11 - \frac{(30.3)^2}{9} = 13.10$$

$$S_{yy} = 1036.65 - \frac{(91.1)^2}{9} = 114.52$$

$$S_{xy} = 345.09 - \frac{(30.3)(91.1)}{9} = 38.39.$$

Recall that $b = 2.9303$. Hence

$$s^2 = \frac{S_{yy} - bS_{xy}}{n - 2}$$

$$= \frac{114.52 - (2.9303)(38.39)}{7} = 0.2894.$$

Therefore, taking square roots we obtain $\sqrt{S_{xx}} = 3.6194$ and $s = 0.5380$. Using Table V, we find $t_{0.025} = 2.365$ for 7 degrees of freedom. Therefore, a 95% confidence interval for β is given by

$$2.9305 - \frac{(2.365)(0.5380)}{(3.6194)} < \beta < 2.9305 + \frac{(2.365)(0.5380)}{(3.6194)},$$

which simplifies to

$$2.579 < \beta < 3.282.$$

To test the null hypothesis H_0 that $\beta = \beta_0$ against a suitable alternative, we again use the t distribution with $n - 2$ degrees of freedom to establish a critical region and then base our decision on the value of

$$t = \frac{b - \beta_0}{s/\sqrt{S_{xx}}}.$$

The method is illustrated in the following example.

Example 8.3 Using the estimated value $b = 2.9303$ of Example 8.1, test the hypothesis that $\beta = 2.5$ at the 0.01 level of significance against the alternative that $\beta > 2.5$.

Solution

1. H_0: $\beta = 2.5$.
2. H_1: $\beta > 2.5$.
3. Choose a 0.01 level of significance.
4. Critical region: $T > 2.998$.
5. Computations:

$$t = \frac{2.9303 - 2.5}{0.5380/3.6194} = 2.8948.$$

6. Conclusion: Accept H_0 and conclude that β does not differ significantly from 2.5.

Confidence intervals and hypothesis testing on the coefficient α may be established from the fact that A is also normally distributed. It is not difficult to show that

$$T = \frac{A - \alpha}{S\sqrt{\sum\limits_{i=1}^{n} x_i^2/nS_{xx}}}$$

has a t distribution with $n - 2$ degrees of freedom from which we may construct a $(1 - \alpha)100\%$ confidence interval for α.

CONFIDENCE INTERVAL FOR α *A $(1 - \alpha)100\%$ confidence interval for the parameter α in the regression line $\mu_{Y|x} = \alpha + \beta x$ is*

$$a - \frac{t_{\alpha/2}\, S\sqrt{\sum\limits_{i=1}^{n} x_i^2}}{\sqrt{nS_{xx}}} < \alpha < a + \frac{t_{\alpha/2}\, S\sqrt{\sum\limits_{i=1}^{n} x_i^2}}{\sqrt{nS_{xx}}},$$

where $t_{\alpha/2}$ is a value of the t distribution with $n - 2$ degrees of freedom.

To test the null hypothesis H_0 that $\alpha = \alpha_0$ against a suitable alternative, we can use the t distribution with $n - 2$ degrees of freedom to establish a critical region and then base our decision on the value of

$$t = \frac{a - \alpha_0}{S\sqrt{\sum\limits_{i=1}^{n} x_i^2/nS_{xx}}}.$$

Quite often it is important for the experimenter to attach a confidence interval on the *mean response* at some fixed level of x not necessarily one of the prechosen values. Suppose that we are interested in the mean of Y for $x = x_0$; we are then estimating $\mu_{Y|x_0} = \alpha + \beta x_0$ with the point estimator $\hat{Y}_0 = A + Bx_0$.

Since $E(\hat{Y}_0) = E(A + Bx_0) = \alpha + \beta x_0$, the estimator $\hat{Y}_0$ is unbiased with variance

$$\sigma_{\hat{Y}_0}^2 = \sigma_{A+Bx_0}^2 = \sigma_{\bar{Y}+B(x_0-\bar{x})}^2$$

$$= \sigma^2 \left[\frac{1}{n} + \frac{(x_0 - \bar{x})^2}{S_{xx}}\right],$$

the latter following from the fact that cov $(\bar{Y}, B) = 0$ (see Exercise 6). Thus the $(1 - \alpha)100\%$ confidence interval on the mean response $\mu_{Y|x_0}$ can now be constructed from the statistic

$$T = \frac{\hat{Y}_0 - \mu_{Y|x_0}}{S\sqrt{(1/n) + [(x_0 - \bar{x})^2/S_{xx}]}},$$

which has a t distribution with $n - 2$ degrees of freedom.

CONFIDENCE INTERVAL FOR $\mu_{Y|x_0}$ A $(1 - \alpha)100\%$ *confidence interval for the mean response $\mu_{Y|x_0}$ is given by*

$$\hat{y}_0 - t_{\alpha/2} s \sqrt{\frac{1}{n} + \frac{(x_0 - \bar{x})^2}{S_{xx}}} < \mu_{Y|x_0} < \hat{y}_0 + t_{\alpha/2} s \sqrt{\frac{1}{n} + \frac{(x_0 - \bar{x})^2}{S_{xx}}},$$

where $t_{\alpha/2}$ is a value of the t distribution with $n - 2$ degrees of freedom.

Example 8.4   Using the data of Table 8.1, construct 95% confidence limits for the mean response $\mu_{Y|x}$.

Solution   From the regression equation we find for $x_0 = 2$, say,

$$\hat{y}_0 = 0.2568 + (2.9303)(2) = 6.1174.$$

Previously, we had $\bar{x} = 3.3667$, $S_{xx} = 13.10$, $s = 0.5380$, and $t_{0.025} = 2.365$ for 7 degrees of freedom. Therefore, a 95% confidence interval for $\mu_{Y|2}$ is given by

$$6.1174 - (2.365)(0.5380)\sqrt{\frac{1}{9} + \frac{(2 - 3.3667)^2}{13.10}} < \mu_{Y|2} < 6.1174$$

$$+ (2.365)(0.5380)\sqrt{\frac{1}{9} + \frac{(2 - 3.3667)^2}{13.10}}$$

or simply

$$5.4765 < \mu_{Y|2} < 6.7583.$$

Repeating the previous calculations for each of several different values of x_0, one can obtain the corresponding confidence limits on each $\mu_{Y|x_0}$. Figure 8.3 displays the data points, the estimated regression line, and the upper and lower confidence limits on the mean of $Y|x$.

Another type of interval that is often misinterpreted and confused with that given for $\mu_{Y|x}$ is the prediction interval on a future observed response. Actually, in many instances the prediction interval is more relevant to the scientist or engineer than the confidence interval on the mean. In the tar content–inlet temperature example, there would certainly be interest not only in estimating the mean tar content at a specific temperature but also in constructing a confidence interval for predicting the actual amount of tar content at the given temperature for some future measurement.

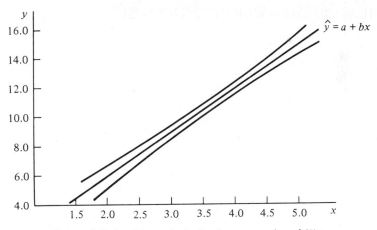

Figure 8.3 Confidence limits for the mean value of $Y|x$.

Suppose that we begin with the distributional properties of the differences between the ordinates $\hat{y}_0$, obtained from the computed regression line in repeated sampling, and the corresponding true ordinate y_0 at $x = x_0$. We may think of the difference $\hat{y}_0 - y_0$ as a value of the random variable $\hat{Y}_0 - Y_0$, whose sampling distribution can be shown to be normal with mean

$$\mu_{\hat{Y}_0 - Y_0} = E(\hat{Y}_0 - Y_0)$$
$$= E[A + Bx_0 - (\alpha + \beta x_0 + E_0)]$$
$$= 0$$

and variance

$$\sigma^2_{\hat{Y}_0 - Y_0} = \sigma^2_{A + Bx_0 - E_0}$$
$$= \sigma^2_{\bar{Y} + B(x_0 - \bar{x}) - E_0}$$
$$= \sigma^2 \left[1 + \frac{1}{n} + \frac{(x_0 - \bar{x})^2}{S_{xx}} \right].$$

Thus the $(1 - \alpha)100\%$ confidence interval for a single predicted value y_0 can be constructed from the statistic

$$T = \frac{\hat{Y}_0 - Y_0}{S\sqrt{1 + (1/n) + [(x_0 - \bar{x})^2/S_{xx}]}},$$

which has a t distribution with $n - 2$ degrees of freedom.

CONFIDENCE INTERVAL FOR y_0 $A(1 - \alpha)100\%$ *confidence interval for a single response* y_0 *is given by*

$$\hat{y}_0 - t_{\alpha/2} s \sqrt{1 + \frac{1}{n} + \frac{(x_0 - \bar{x})^2}{S_{xx}}} < y_0 < \hat{y}_0 + t_{\alpha/2} s \sqrt{1 + \frac{1}{n} + \frac{(x_0 - \bar{x})^2}{S_{xx}}},$$

where $t_{\alpha/2}$ *is a value of the t distribution with* $n - 2$ *degrees of freedom.*

Example 8.5   Using the data of Table 8.1, construct a 95% confidence interval for y_0 when $x_0 = 2$.

Solution   We have $n = 9$, $x_0 = 2$, $\bar{x} = 3.3667$, $\hat{y}_0 = 6.1171$, $S_{xx} = 13.10$, $s = 0.5380$, and $t_{0.025} = 2.365$ for 7 degrees of freedom. Therefore, a 95% confidence interval for y_0 is given by

$$6.1174 - (2.365)(0.5380)\sqrt{1 + \frac{1}{9} + \frac{(2 - 3.3667)^2}{13.10}} < y_0 < 6.1174$$

$$+ (2.365)(0.5380)\sqrt{1 + \frac{1}{9} + \frac{(2 - 3.3667)^2}{13.10}},$$

which simplifies to

$$4.6927 < y_0 < 7.5421.$$

8.5 CHOICE OF A REGRESSION MODEL

Much of what has been presented to this point on regression involving a single independent variable depends on the assumption that the model chosen is correct, the presumption that $\mu_{Y|x}$ is related to x linearly in the parameters. Certainly, one would not expect the prediction of the response to be good if there are several independent variables, not considered in the model, that are affecting the response and are varying in the system. In addition, the prediction would certainly be inadequate if the true structure relating $\mu_{Y|x}$ to x is extremely nonlinear in the range of the variables considered.

Often the simple linear regression model is used even though it is known that the model is something other than linear or that the true structure is unknown. This approach is often sound, particularly when the range of x is narrow. Thus the model used becomes an approximating function that one hopes is an adequate representation of the true picture in the region of interest. One should note, however, the effect of an inadequate model on the results presented thus far. For example, if the true model, unknown to the experimenter, is linear in more than one x, say,

$$\mu_{Y|x_1, x_2} = \alpha + \beta x_1 + \gamma x_2,$$

then the ordinary least squares estimate $b = S_{xy}/S_{xx}$, calculated by only considering x_1 in the experiment, is, under general circumstances, a biased estimate of the coefficient β, the bias being a function of the additional coefficient γ (see Exercise 16). Also, the estimate s^2 for σ^2 is biased due to the additional variable.

8.6 ANALYSIS-OF-VARIANCE APPROACH

Often the problem of analyzing the quality of the estimated regression line is handled through an *analysis-of-variance* approach. This is merely a procedure whereby the total variation in the dependent variable is subdivided into meaningful components that are then observed and treated in a systematic fashion. The analysis of variance, discussed extensively in Chapter 10, is a powerful tool that is used in many applications.

Suppose that we have n experimental data points in the usual form (x_i, y_i) and that the regression line is estimated. In our estimation of σ^2 in Section 8.3 we established the identity

$$S_{yy} = bS_{xy} + SSE.$$

So we have achieved a partitioning of the *total corrected sum of squares of y* into two components that should reflect particular meaning to the experimenter. We shall indicate this partitioning symbolically as

$$SST = SSR + SSE.$$

The first component on the right is called the *regression sum of squares* and it reflects the amount of variation in the y values explained by the model, in this case the postulated straight line. The second component is just the familiar error sum of squares, which reflects variation about the regression line.

Since SSR/σ^2 and SSE/σ^2 are values of independent chi-square variables with 1 and $n - 2$ degrees of freedom, respectively, it follows that SST/σ^2 is also a value of a chi-square variable with $n - 1$ degrees of freedom. To test the hypothesis

$$H_0: \quad \beta = 0,$$
$$H_1: \quad \beta \neq 0,$$

where the null hypothesis essentially says that variation in Y is not explained by the straight line but rather by chance or random fluctuations, we compute

$$f = \frac{SSR/1}{SSE/(n - 2)} = \frac{SSR}{s^2}$$

and reject H_0 at the α level of significance when $f > f_\alpha(1, n - 2)$.

Table 8.2 Analysis of Variance for Testing $\beta = 0$

Source of variation	Sum of squares	Degrees of freedom	Mean square	Computed f
Regression	SSR	1	SSR	SSR/s^2
Error	SSE	$n - 2$	$s^2 = \dfrac{SSE}{n-2}$	
Total	SST	$n - 1$		

In practice, one first computes

$$SST = S_{yy}$$
$$SSR = bS_{xy}$$

and then, making use of the previous sum of squares identity, obtains

$$SSE = SST - SSR.$$

The computations are usually summarized by means of an *analysis-of-variance table* as indicated in Table 8.2.

When the null hypothesis is rejected, that is, when the computed F statistic exceeds the critical value $f_\alpha(1, n - 2)$, we conclude that there is a significant amount of variation in the response accounted for by the postulated model, the straight-line function. If the F statistic is in the acceptance region, we conclude that the data did not reflect sufficient evidence to support the model postulated.

In Section 8.4 a procedure was given whereby the statistic

$$T = \frac{B - \beta_0}{S/\sqrt{S_{xx}}}$$

was used to test the hypothesis

$$H_0: \quad \beta = \beta_0,$$
$$H_1: \quad \beta \neq \beta_0,$$

where T follows the t distribution with $n - 2$ degrees of freedom. The hypothesis is rejected if $|T| > t_{\alpha/2}$ for an α level of significance. It is interesting to note that in the special case in which we are testing

$$H_0: \quad \beta = 0,$$

$$H_1: \quad \beta \neq 0,$$

the value of our T statistic becomes

$$t = \frac{b}{s/\sqrt{S_{xx}}}$$

and the hypothesis is identical to that being tested in Table 8.2. Namely, the null hypothesis states that the variation in the response is due merely to chance. The analysis of variance uses the F distribution rather than the t distribution, but *the two procedures are identical*. This we can see by writing

$$t^2 = \frac{b^2 S_{xx}}{s^2} = \frac{b S_{xy}}{s^2} = \frac{SSR}{s^2},$$

which is identical to the f value used in the analysis of variance. The basic relationship between the t distribution with v degrees of freedom and the F distribution with 1 and v degrees of freedom is given by

$$t_{\alpha/2}^2 = f_\alpha(1, v).$$

8.7 REPEATED MEASUREMENTS ON THE RESPONSE

Quite often it is advantageous for the experimenter to obtain repeated observations for each value of x. While it is not necessary to have these repetitions in order to estimate α and β, nevertheless it does enable the experimenter to obtain quantitative information concerning the appropriateness of the model. In fact, if repeated observations are at his disposal, the experimenter can make a significance test to aid in determining whether or not the model is adequate.

Let us select a random sample of n observations using k distinct values of x, say $x_1, x_2, \ldots, x_k$, such that the sample contains n_1 observed values of the random variable Y_1 corresponding to x_1, n_2 observed values of Y_2 corresponding to $x_2, \ldots, n_k$ observed values of Y_k corresponding to x_k. Of necessity, $n = \sum_{i=1}^{k} n_i$.

We define

$$y_{ij} = \text{the } j\text{th value of the random variable } Y_i,$$

$$y_{i\cdot} = T_{i\cdot} = \sum_{j=1}^{n_i} y_{ij},$$

$$\bar{y}_{i\cdot} = \frac{T_{i\cdot}}{n_i}.$$

Hence, if $n_4 = 3$ measurements of Y are made corresponding to $x = x_4$, we would indicate these observations by $y_{41}, y_{42},$ and y_{43}. Then

$$T_{4\cdot} = y_{41} + y_{42} + y_{43}.$$

The error sum of squares consists of two parts: the amount due to the variation between the values of Y within given values of x and a component that is normally called the *lack of fit* contribution. The first component reflects mere random variation or *pure experimental error*, while the second component is a measure of the systematic variation brought about by higher-order terms. In our case these are terms in x other than the linear or *first-order contribution*. Note that in choosing a linear model we are essentially assuming that this second component does not exist and hence our error sum of squares is completely due to random errors. If this should be the case, then $s^2 = SSE/(n-2)$ is an unbiased estimate of σ^2. However, if the model does not adequately fit the data, then the error sum of squares is inflated and produces a biased estimate of σ^2. Whether or not the model fits the data, an unbiased estimate of σ^2 can always be obtained when we have repeated observations simply by computing

$$s_i^2 = \frac{\sum_{j=1}^{n_i} (y_{ij} - \bar{y}_{i\cdot})^2}{n_i - 1}; \qquad i = 1, 2, \ldots, k.$$

for each of the k distinct values of x and then pooling these variances to give

$$s^2 = \frac{\sum_{i=1}^{k} (n_i - 1)s_i^2}{n - k} = \frac{\sum_{i=1}^{k} \sum_{j=1}^{n_i} (y_{ij} - \bar{y}_{i\cdot})^2}{n - k}.$$

The numerator of s^2 is a measure of the pure experimental error. A computational procedure for separating the error sum of squares into the two components representing pure error and lack of fit is as follows:

1. Compute the pure error sum of squares

$$\sum_{i=1}^{k} \sum_{j=1}^{n_i} (y_{ij} - \bar{y}_{i.})^2 = \sum_{i=1}^{k} \sum_{j=1}^{n_i} y_{ij}^2 - \sum_{i=1}^{k} \frac{T_i^2}{n_i}.$$

This sum of squares has $n - k$ degrees of freedom associated with it and the resulting mean square is our unbiased estimate s^2 of σ^2.

2. Subtract the pure error sum of squares from the error sum of squares, SSE, thereby obtaining the sum of squares due to lack of fit. The degrees of freedom for lack of fit are also obtained by simply subtracting $(n - 2) - (n - k) = k - 2$.

The computations required for testing hypotheses in a regression problem with repeated measurements on the response may be summarized as shown in Table 8.3.

The concept of lack of fit is extremely important in applications of regression analysis. In fact, the need to construct or design an experiment that will account for lack of fit becomes more critical as the problem and the underlying mechanism involved become more complicated. Surely, one cannot always be certain that his postulated structure, in this case the linear regression model, is correct or even an adequate representation. The following example shows how the

Table 8.3 Analysis of Variance for Repeated Measurements on the Response

Source of variation	Sum of squares	Degrees of freedom	Mean square	Computed f
Regression	SSR	1	SSR	$\dfrac{SSR}{s^2}$
Error	SSE	$n - 2$		
Lack of fit	$SSE - SSE(\text{pure})$	$k - 2$	$\dfrac{SSE - SSE(\text{pure})}{k - 2}$	$\dfrac{SSE - SSE(\text{pure})}{s^2(k - 2)}$
Pure error	$SSE(\text{pure})$	$n - k$	$s^2 = \dfrac{SSE(\text{pure})}{n - k}$	
Total	SST	$n - 1$		

error sum of squares is partitioned into the two components representing pure error and lack of fit. The adequacy of the model is tested at the α level of significance by comparing the lack-of-fit mean square divided by s^2 with $f_\alpha(k - 2, n - k)$.

Example 8.6 Observations on the yield of a chemical reaction taken at various temperatures were recorded as follows:

y (%)	x (°C)	y (%)	x (°C)
77.4	150	88.9	250
76.7	150	89.2	250
78.2	150	89.7	250
84.1	200	94.8	300
84.5	200	94.7	300
83.7	200	95.9	300

Estimate the linear model $\mu_{Y|x} = \alpha + \beta x$ and test for lack of fit.

Solution We have $n_1 = n_2 = n_3 = n_4 = 3$. Therefore,

$$S_{yy} = \sum_{i=1}^{4} \sum_{j=1}^{3} y_{ij}^2 - \frac{\left(\sum_{i=1}^{4} \sum_{j=1}^{3} y_{ij} \right)^2}{12}$$

$$= 90{,}265.5200 - 89{,}752.4033$$

$$= 513.1167$$

$$S_{xx} = \sum_{i=1}^{4} n_i x_i^2 - \frac{\left(\sum_{i=1}^{4} n_i x_i \right)^2}{12}$$

$$= 645{,}000 - 607{,}500$$

$$= 37{,}500$$

$$S_{xy} = \sum_{i=1}^{4} \sum_{j=1}^{3} x_i y_{ij} - \frac{\left(\sum_{i=1}^{4} n_i x_i \right)\left(\sum_{i=1}^{4} \sum_{j=1}^{3} y_{ij} \right)}{12}$$

$$= 237{,}875 - 233{,}505$$

$$= 4370,$$

$$\bar{y} = 86.4833 \quad \text{and} \quad \bar{x} = 225.$$

The regression coefficients are then given by

$$b = \frac{4370}{37,500} = 0.1165$$

and

$$a = 86.4833 - (0.1165)(225) = 60.2708.$$

Hence our estimated regression line is

$$\hat{y} = 60.2708 + 0.1165x.$$

To test for lack of fit, we proceed in the usual manner:

1. H_0:   the regression is linear in x.
2. H_1:   the regression is nonlinear in x.
3. Choose of 0.05 level of significance.
4. Critical region: $F > 4.46$ with 2 and 8 degrees of freedom.
5. Computations: We have

$$SST = S_{yy} = 513.1167$$

$$SSR = bS_{xy} = (0.1165)(4370) = 509.1050$$

$$SSE = S_{yy} - bS_{xy} = 4.0117.$$

To compute the pure error sum of squares, we first write

$$x_1 = 150 \qquad T_1. = 232.3$$
$$x_2 = 200 \qquad T_2. = 252.3$$
$$x_3 = 250 \qquad T_3. = 267.8$$
$$x_4 = 300 \qquad T_4. = 285.4.$$

Therefore,

$$SSE\text{(pure)} = 90,265.52 - \frac{232.3^2 + 252.3^2 + 267.8^2 + 285.4^2}{3}$$

$$= 2.66.$$

These results and the remaining computations are exhibited in Table 8.4.

Table 8.4 Analysis of Variance on Yield-Temperature Data

Source of variation	Sum of squares	Degrees of freedom	Mean square	Computed f
Regression	509.1050	1	509.1050	1531.60
Error	4.0117	10		
Lack of fit	{1.3517	{2	0.6758	2.03
Pure error	{2.6600	{8	0.3324	
Total	513.1167	11		

6. Conclusion: The partitioning of the total variation in this manner reveals a significant variation accounted for by the linear model and an insignificant amount of variation due to lack of fit. Thus the experimental data do not seem to suggest the need to consider terms higher than first order in the model and the null hypothesis is accepted.

8.8 CORRELATION

Up to this point we have assumed that the independent variable x is controlled and therefore is not a random variable. In fact, in this context, x is often called a *mathematical variable*, which, in the sampling process, is measured with negligible error. In many applications of regression techniques it is more realistic to assume that both X and Y are random variables and the measurements $\{(x_i, y_i); i = 1, 2, \ldots, n\}$ are observations from a joint density function $f(x, y)$. For example, in an archeological study, it might be assumed that two measurements on a particular kind of bone in the adult body are both random variables and follow a bivariate distribution.

It is often assumed that the conditional distribution $f(y|x)$ of Y, for fixed values of X, is normal with mean $\mu_{Y|x} = \alpha + \beta x$ and variance $\sigma^2_{Y|x} = \sigma^2$ and that X is likewise normally distributed with mean μ_X and variance σ^2_X. The joint density of X and Y is then given by

$$f(x, y) = n(y|x; \alpha + \beta x, \sigma)n(x; \mu_X, \sigma_X)$$

$$= \frac{1}{2\pi\sigma_X\sigma} \exp\left\{-\left(\frac{1}{2}\right)\left[\left(\frac{y - (\alpha + \beta x)}{\sigma}\right)^2 + \left(\frac{x - \mu_X}{\sigma_X}\right)^2\right]\right\},$$

for $-\infty < x < \infty$ and $-\infty < y < \infty$.

Writing the random variable Y in the form

$$Y = \alpha + \beta X + E,$$

where X is now a random variable independent of the random error E, we have

$$\mu_Y = \alpha + \beta \mu_X$$

$$\sigma_Y^2 = \sigma^2 + \beta^2 \sigma_X^2.$$

Substituting into the above expression for $f(x, y)$, we obtain the *bivariate normal distribution*

$$f(x, y) = \frac{1}{2\pi\sigma_X\sigma_Y\sqrt{1 - \rho^2}}$$

$$\times \exp\left\{-\frac{1}{2(1 - \rho^2)}\left[\left(\frac{x - \mu_X}{\sigma_X}\right)^2 - 2\rho\left(\frac{x - \mu_X}{\sigma_X}\right)\left(\frac{y - \mu_Y}{\sigma_Y}\right) + \left(\frac{y - \mu_Y}{\sigma_Y}\right)^2\right]\right\},$$

for $-\infty < x < \infty$ and $-\infty < y < \infty$, where

$$\rho^2 = 1 - \frac{\sigma^2}{\sigma_Y^2} = \beta^2 \frac{\sigma_X^2}{\sigma_Y^2}.$$

The constant ρ is called the *correlation coefficient* and plays a major role in many bivariate data analysis problems. It is important for the reader to understand the physical interpretation of this correlation coefficient and the distinction between correlation and regression. The term *regression* still has meaning here. In fact, the straight line given by $\mu_{Y|x} = \alpha + \beta x$ is still called the regression line as before, and the estimates of α and β are identical to those given in Section 8.2. The value of ρ is zero when $\beta = 0$, which results when there essentially is no linear regression; that is, the regression line is horizontal and any knowledge of X is useless in predicting Y. Since $\sigma_Y^2 \geq \sigma^2$, we must have $-1 \leq \rho \leq 1$. Values of $\rho = \pm 1$ only occur when $\sigma^2 = 0$, in which case we have a perfect linear relationship between the two variables. Thus a value of ρ equal to $+1$ implies a perfect linear relationship with a positive slope, while a value of ρ equal to -1 results from a perfect linear relationship with a negative slope. It might be said then that sample estimates of ρ close to unity in magnitude imply good correlation or *linear association* between X and Y, while values near zero indicate little or no correlation.

To obtain a sample estimate of ρ, recall from Section 8.3 that the error sum of squares is given by

$$SSE = S_{yy} - bS_{xy}.$$

Dividing both sides of this equation by S_{yy} and replacing b by S_{xy}/S_{xx}, we obtain the relation

$$b^2 \frac{S_{xx}}{S_{yy}} = 1 - \frac{SSE}{S_{yy}}.$$

The value of $b^2 S_{xx}/S_{yy}$ is zero when $b = 0$, which will occur when the sample points show no linear relationship. Since $S_{yy} \geq SSE$, we conclude that $b^2 S_{xx}/S_{yy}$ must be between zero and 1. Consequently, $b\sqrt{S_{xx}/S_{yy}}$ must range from -1 to $+1$, negative values corresponding to lines with negative slopes and positive values to lines with positive slopes. A value of -1 or $+1$ will occur when $SSE = 0$, but this is the case where all sample points lie in a straight line. Hence a perfect linear relationship exists between X and Y when $b\sqrt{S_{xx}/S_{yy}} = \pm 1$. Clearly, the quantity $b\sqrt{S_{xx}/S_{yy}}$, which we shall henceforth designate as r, can be used as an estimate of the correlation coefficient ρ.

CORRELATION COEFFICIENT   *The measure ρ of linear relationship between two variables X and Y is estimated by the* sample correlation coefficient r, *where*

$$r = b\sqrt{\frac{S_{xx}}{S_{yy}}} = \frac{S_{xy}}{\sqrt{S_{xx} S_{yy}}}.$$

For values of r between -1 and $+1$ we must be careful in our interpretation. For example, values of r equal to 0.3 and 0.6 only mean that we have two positive correlations, one somewhat stronger than the other. It is wrong to conclude that $r = 0.6$ indicates a linear relationship twice as good as that indicated by the value $r = 0.3$. On the other hand, if we consider r^2, then $100 \times r^2 \%$ of the variation in the values of Y may be accounted for by the linear relationship with the variable X. Thus a correlation of 0.6 means that 36% of the variation of the random variable Y is accounted for by differences in the variable X.

Example 8.7 In a study of the correlation between the amount of rainfall and the quantity of air pollution removed, the following data were collected:

x, daily rainfall (0.01 centimeter)	y, particulate removed (micrograms per cubic meter)
4.3	126
4.5	121
5.9	116
5.6	118
6.1	114
5.2	118
3.8	132
2.1	141
7.5	108

Compute and interpret the sample correlation coefficient.

Solution From the data we find

$$S_{xx} = 19.2600, \qquad S_{yy} = 804.2222, \qquad S_{xy} = -121.8000.$$

Therefore,

$$r = \frac{-121.8000}{\sqrt{(19.2600)(804.2222)}} = -0.9786.$$

A correlation coefficient of -0.9786 indicates a very good linear relationship between X and Y. Since $r^2 = 0.9581$, we can say that approximately 96% of the variation in the values of Y is accounted for by a linear relationship with X.

A test of the hypothesis

$$H_0: \quad \rho = \rho_0,$$

$$H_1: \quad \rho \neq \rho_0,$$

is easily conducted from the sample information. For observations from the bivariate normal distribution, the quantity

$$\frac{1}{2} \ln \left(\frac{1 + r}{1 - r} \right)$$

is a value of a random variable that follows approximately the normal distribution with mean $(1/2) \ln [(1 + \rho)/(1 - \rho)]$ and variance $1/(n - 3)$. Thus the test procedure is to compute

$$z = \frac{\sqrt{n - 3}}{2} \left[\ln \left(\frac{1 + r}{1 - r} \right) - \ln \left(\frac{1 + \rho_0}{1 - \rho_0} \right) \right]$$

$$= \frac{\sqrt{n - 3}}{2} \ln \left[\frac{(1 + r)(1 - \rho_0)}{(1 - r)(1 + \rho_0)} \right]$$

and compare with the critical points of the standard normal distribution.

Example 8.8   For the data of Example 8.7, test the null hypothesis that there is no linear association between the variables. Use a 0.05 level of significance.

Solution

1. H_0: $\rho = 0$.
2. H_1: $\rho \neq 0$.
3. $\alpha = 0.05$.
4. Critical region: $Z < -1.96$ and $Z > 1.96$.
5. Computations:

$$z = \frac{\sqrt{6}}{2} \ln \left(\frac{0.0214}{1.9786} \right) = -5.55.$$

6. Conclusion: Reject the hypothesis of no linear relationship.

It should be pointed out that in correlation studies, as in linear regression problems, the results obtained are only as good as the model that is assumed. In the correlation techniques studied here, a bivariate normal density is assumed for the variables X and Y, with the mean value of Y at each x value being linearly related to x. To observe the suitability of the linearity assumption, a preliminary

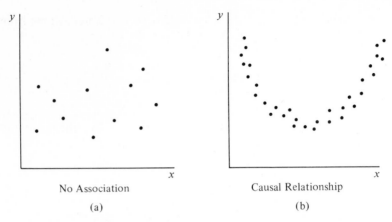

No Association   Causal Relationship

(a)   (b)

Figure 8.4 Scatter diagrams showing zero correlation.

plotting of the experimental data is often helpful. A value of the sample cor-relation coefficient close to zero will result from data that display a strictly random effect as in Figure 8.4(a), thus implying little or no causal relationship. It is important to remember that the correlation coefficient between two variables is a measure of their linear relationship, and that a value of $r = 0$ implies a lack of linearity and not a lack of association. Hence, if a strong quadratic relationship exists between X and Y as indicated in Figure 8.4(b), we shall still obtain a zero correlation indicating a nonlinear relationship.

EXERCISES

1. Consider the following experimental data:

x	y	x	y
10.0	18.7	13.5	22.4
10.5	21.5	14.0	23.3
11.0	18.5	14.5	19.6
11.5	19.6	15.0	23.8
12.0	18.2	15.5	21.7
12.5	20.8	16.0	23.2
13.0	21.6		

(a) Estimate the linear regression line.
(b) Graph the line on a scatter diagram.
(c) Find a point estimate of $\mu_{Y|12}$.

2. A study was made on the amount of converted sugar in a certain process at various temperatures. The data were coded and recorded as follows:

x, temperature	y, converted sugar
1.0	8.1
1.1	7.8
1.2	8.5
1.3	9.8
1.4	9.5
1.5	8.9
1.6	8.6
1.7	10.2
1.8	9.3
1.9	9.2
2.0	10.5

(a) Estimate the linear regression line.
(b) Estimate the amount of converted sugar produced when the coded temperature is 1.75.

3. In a certain type of test specimen, the normal stress on a specimen is known to be functionally related to the shear resistance. The following is a set of coded experimental data on the two variables:

x, normal stress	y, shear resistance
26.8	26.5
25.4	27.3
28.9	24.2
23.6	27.1
27.7	23.6
23.9	25.9
24.7	26.3
28.1	22.5
26.9	21.7
27.4	21.4
22.6	25.8
25.6	24.9

(a) Estimate the regression line $\mu_{Y|x} = \alpha + \beta x$.
(b) Estimate the shear resistance for a normal stress of 24.5 kilograms per square centimeter.

4. The following data are the amounts, y, of an unconverted substance from six similar chemical reactions after x minutes:

x (minutes)	y (milligrams)
1	23.5
2	16.9
2	17.5
3	14.0
5	9.8
5	8.9

(a) Fit a curve of the form $\mu_{Y|x} = \gamma \delta^x$ by means of the nonlinear sample regression equation $\hat{y} = cd^x$. [*Hint*: Write

$$\log \hat{y} = \log c + (\log d)x$$
$$= a + bx,$$

where $a = \log c$ and $b = \log d$, and then estimate a and b by the formulas of Section 8.2 using the sample points $(x_i, \log y_i)$.]

(b) Estimate the amount of unconverted substance in this type of reaction after 4 minutes have elapsed.

5. The pressure P of a gas corresponding to various volumes V was recorded as follows:

V (cm^3)	50	60	70	90	100
P (kg/cm^2)	64.7	51.3	40.5	25.9	7.8

The ideal gas law is given by the equation $PV^\gamma = C$, where γ and C are constants.

(a) Following the suggested procedure of Exercise 4, find the least squares estimates of γ and C from the given data.

(b) Estimate P when $V = 80$ cubic centimeters.

6. For a simple linear regression model

$$Y_i = \alpha + \beta x_i + E_i, \qquad i = 1, 2, \ldots, n,$$

where the E_i's are independent and normally distributed with zero means and equal variances σ^2, show that $\bar{Y}$ and

$$B = \frac{\sum_{i=1}^{n}(x_i - \bar{x})Y_i}{\sum_{i=1}^{n}(x_i - \bar{x})^2}$$

have zero covariance.

7. Show that A, the least squares estimator of the intercept α in the equation $\mu_{Y|x} = \alpha + \beta x$, is normally distributed with mean α and variance

$$\sigma_A^2 = \frac{\sum\limits_{i=1}^{n} x_i^2}{n \sum\limits_{i=1}^{n} (x_i - \bar{x})^2} \sigma^2.$$

8. Test the hypothesis that $\beta = 0$ in Exercise 1, using a 0.05 level of significance, against the alternative that $\beta > 0$.

9. Construct 95% confidence intervals for α and β in Exercise 2.

10. Construct 99% confidence intervals for α and β in Exercise 3.

11. Construct 95% confidence intervals for the mean response $\mu_{Y|x}$ and for the single predicted value y_0, corresponding to $x = 24.5$ in Exercise 3.

12. Graph the regression line and the 95% confidence bands for the mean response $\mu_{Y|x}$ for the data of Exercise 2.

13. Construct a 95% confidence interval for the amount of converted sugar corresponding to $x = 1.6$ in Exercise 2.

14. (a) Find the least squares estimate for the parameter β in the linear equation $\mu_{Y|x} = \beta x$.
    (b) Estimate the regression line passing through the origin for the following data:

x	0.5	1.5	3.2	4.2	5.1	6.5
y	1.3	3.4	6.7	8.0	10.0	13.2

15. Suppose it is not known in Exercise 14 whether or not the true regression should pass through the origin. Estimate the general linear model $\mu_{Y|x} = \alpha + \beta x$ and test the hypothesis that $\alpha = 0$ at the 0.10 level of significance against the alternative that $\alpha \neq 0$.

16. Suppose an experimenter postulates a model of the type

$$Y_i = \alpha + \beta x_{1i} + E_i, \qquad i = 1, 2, \ldots, n,$$

when in fact, an additional variable, say x_2, also contributes linearly to the response. The true model is then given by

$$Y_i = \alpha + \beta x_{1i} + \gamma x_{2i} + E_i, \qquad i = 1, 2, \ldots, n.$$

Compute the expected value of the estimator

$$B = \frac{\sum\limits_{i=1}^{n} (x_{1i} - \bar{x}_1) Y_i}{\sum\limits_{i=1}^{n} (x_{1i} - \bar{x}_1)^2}.$$

17. Use an analysis-of-variance approach to test the hypothesis that $\beta = 0$ against the alternative hypothesis $\beta \neq 0$ in Exercise 2 at the 0.05 level of significance.

18. The amounts of a chemical compound y, which were dissolved in 100 grams of water at various temperatures, x, were recorded as follows:

x (°C)		y (grams)	
0	8	6	8
15	12	10	14
30	25	21	24
45	31	33	28
60	44	39	42
75	48	51	44

(a) Find the equation of the regression line.
(b) Estimate the amount of chemical that will dissolve in 100 grams of water at 50°C.
(c) Test the hypothesis that $\alpha = 6$, using a 0.01 level of significance, against the alternative that $\alpha \neq 6$.
(d) Test whether the linear model is adequate.

19. The amounts of solids removed from a particular material when exposed to drying periods of different lengths are as follows:

x (hours)	y (grams)	
4.4	13.1	14.2
4.5	9.0	11.5
4.8	10.4	11.5
5.5	13.8	14.8
5.7	12.7	15.1
5.9	9.9	12.7
6.3	13.8	16.5
6.9	16.4	15.7
7.5	17.6	16.9
7.8	18.3	17.2

(a) Estimate the linear regression line.
(b) Test whether the linear model is adequate.
(c) Construct a 90% confidence interval for the slope β.

20. Compute and interpret the correlation coefficient for the following data:

x	4	5	9	14	18	22	24
y	16	22	11	16	7	3	17

21. Compute and interpret the correlation coefficient for the following grades of six students selected at random:

Mathematics grade	70	92	80	74	65	83
English grade	74	84	63	87	78	90

22. Compute the correlation coefficient for the random variables of Exercise 1 and test the hypothesis that $\rho = 0$ against the alternative that $\rho \neq 0$. Use a 0.05 level of significance.

9

Multiple Linear Regression

9.1 INTRODUCTION

In most research problems where regression analysis is applied, more than one independent variable is needed in the regression model. The complexity of most scientific mechanisms is such that in order to be able to predict an important response, a *multiple regression model* is needed. When this model is linear in the coefficients, it is called a *multiple linear regression model*. For the case of k independent variables $x_1, x_2, \ldots, x_k$, the mean of $Y \mid x_1, x_2, \ldots, x_k$ is given by the multiple linear regression model

$$\mu_{Y \mid x_1, x_2, \ldots, x_k} = \beta_0 + \beta_1 x_1 + \cdots + \beta_k x_k$$

and the estimated response is obtained from the sample regression equation

$$\hat{y} = b_0 + b_1 x_1 + \cdots + b_k x_k,$$

where each regression coefficient β_i is estimated by b_i from the sample data using the method of least squares. As in the case of a single independent variable, the multiple linear regression model can often be an adequate representation of a more complicated structure within certain ranges of the independent variables.

Similar least squares techniques can also be applied in estimating the coefficients when the linear model involves, say, powers and products of the independent variables. For example, when $k = 1$, the experimenter may feel that the means $\mu_{Y \mid x}$ do not fall on a straight line but are more appropriately described by the *polynomial regression model*

$$\mu_{Y \mid x} = \beta_0 + \beta_1 x + \beta_2 x^2 + \cdots + \beta_r x^r,$$

and the estimated response is obtained from the polynomial regression equation

$$\hat{y} = b_0 + b_1 x + b_2 x^2 + \cdots + b_r x^r.$$

314

Confusion arises occasionally when we speak of a polynomial model as a linear model. However, statisticians normally refer to a linear model as one in which the parameters occur linearly, regardless of how the independent variables enter the model. An example of a nonlinear model is the *exponential relationship* given by

$$\mu_{Y|x} = \alpha\beta^x,$$

which is estimated by the regression equation

$$\hat{y} = ab^x.$$

There are many phenomena in science and engineering that are inherently nonlinear in nature and, when the true structure is known, an attempt should certainly be made to fit the actual model. The literature on estimation by least squares of nonlinear models is voluminous and, except for Exercises 4 and 5 of Chapter 8, we do not attempt to cover the subject in this text. A student who wants a good account of some of the aspects of this subject should consult *Applied Regression Analysis* by Draper and Smith (see the Bibliography).

9.2 ESTIMATING THE COEFFICIENTS

In this section we shall first obtain the least squares estimates of the parameters β_0, β_1, and β_2 in a multiple linear regression model for the case of two independent variables x_1 and x_2. Next we shall fit a quadratic model in terms of the single variable x as a special case of the multiple linear regression model by setting $x_1 = x$ and $x_2 = x^2$. The general linear model involving several independent variables is discussed in Section 9.3.

Suppose that we wish to fit the plane

$$\mu_{Y|x_1, x_2} = \beta_0 + \beta_1 x_1 + \beta_2 x_2$$

to the data points $\{(x_{1i}, x_{2i}, y_i); i = 1, 2, \ldots, n$ and $n > 2\}$, where y_i is the observed response corresponding to the values x_{1i} and x_{2i} of the two independent variables x_1 and x_2. Each observation (x_{1i}, x_{2i}, y_i) satisfies the equation

$$\begin{aligned} y_i &= \beta_0 + \beta_1 x_{1i} + \beta_2 x_{2i} + \varepsilon_i \\ &= b_0 + b_1 x_{1i} + b_2 x_{2i} + e_i, \end{aligned}$$

where ε_i and e_i are the random error and residual, respectively, associated with the response y_i. In using the concept of least squares to arrive at estimates b_0, b_1, and b_2, we minimize the expression

$$SSE = \sum_{i=1}^{n} e_i^2 = \sum_{i=1}^{n} (y_i - b_0 - b_1 x_{1i} - b_2 x_{2i})^2.$$

Differentiating SSE in turn with respect to b_0, b_1, and b_2, and equating to zero, we generate the set of three normal equations

$$nb_0 + b_1 \sum_{i=1}^{n} x_{1i} + b_2 \sum_{i=1}^{n} x_{2i} = \sum_{i=1}^{n} y_i$$

$$b_0 \sum_{i=1}^{n} x_{1i} + b_1 \sum_{i=1}^{n} x_{1i}^2 + b_2 \sum_{i=1}^{n} x_{1i} x_{2i} = \sum_{i=1}^{n} x_{1i} y_i$$

$$b_0 \sum_{i=1}^{n} x_{2i} + b_1 \sum_{i=1}^{n} x_{1i} x_{2i} + b_2 \sum_{i=1}^{n} x_{2i}^2 = \sum_{i=1}^{n} x_{2i} y_i.$$

These equations can be solved for b_1 and b_2 by any appropriate method for handling three linear equations in three unknowns, such as Cramer's rule, and then b_0 is obtained from the first of the three normal equations by observing that

$$b_0 = \bar{y} - b_1 \bar{x}_1 - b_2 \bar{x}_2.$$

Example 9.1 An experiment was conducted to determine if the weight of an animal can be predicted after a given period of time on the basis of the initial weight of the animal and the amount of feed eaten. The following data, measured in kilograms, were recorded:

Final weight, y	Initial weight, x_1	Feed eaten, x_2
95	42	272
77	33	226
80	33	259
100	45	292
97	39	311
70	36	183
50	32	173
80	41	236
92	40	230
84	38	235

Fit a multiple linear regression equation of the form $\mu_{Y|x_1,x_2} = \beta_0 + \beta_1 x_1 + \beta_2 x_2$ and then predict the final weight of an animal having an initial weight of 35 kg that is fed 250 kg of feed.

Solution From the given data we find

$$\sum_{i=1}^{10} x_{1i} = 379 \qquad \sum_{i=1}^{10} x_{2i} = 2417 \qquad \sum_{i=1}^{10} y_i = 825$$

$$\sum_{i=1}^{10} x_{1i}^2 = 14{,}533 \qquad \sum_{i=1}^{10} x_{2i}^2 = 601{,}365 \qquad \sum_{i=1}^{10} x_{1i} x_{2i} = 92{,}628$$

$$\sum_{i=1}^{10} x_{1i} y_i = 31{,}726 \qquad \sum_{i=1}^{10} x_{2i} y_i = 204{,}569 \qquad n = 10.$$

Inserting these values into the simultaneous equations above, we obtain

$$10 b_0 + \quad 379 b_1 + \quad 2417 b_2 = \quad 825$$

$$379 b_0 + 14{,}533 b_1 + \quad 92{,}628 b_2 = 31{,}726$$

$$2417 b_0 + 92{,}628 b_1 + 601{,}365 b_2 = 204{,}569.$$

The solutions of this set of equations yields the unique estimates

$$b_0 = -22.9915, \qquad b_1 = 1.3957, \qquad b_2 = 0.2176.$$

Therefore, our regression equation is

$$\hat{y} = -22.9915 + 1.3957 x_1 + 0.2176 x_2.$$

For an animal with an initial weight of $x_1 = 35$ kg that is fed $x_2 = 250$ kg of feed over the given time period, we would predict a final weight of

$$\hat{y} = -22.9915 + (1.3957)(35) + (0.2176)(250)$$
$$= 103 \text{ kg}.$$

Now suppose that we wish to fit the quadratic equation

$$\mu_{Y|x} = \beta_0 + \beta_1 x + \beta_2 x^2$$

to the n pairs of observations $\{(x_i, y_i); i = 1, 2, \ldots, n$ and $n > 2\}$. Setting $x_1 = x$ and $x_2 = x^2$ in the set of normal equations for the multiple linear regression model, we obtain the equations

$$nb_0 + b_1 \sum_{i=1}^{n} x_i + b_2 \sum_{i=1}^{n} x_i^2 = \sum_{i=1}^{n} y_i$$

$$b_0 \sum_{i=1}^{n} x_i + b_1 \sum_{i=1}^{n} x_i^2 + b_2 \sum_{i=1}^{n} x_i^3 = \sum_{i=1}^{n} x_i y_i$$

$$b_0 \sum_{i=1}^{n} x_i^2 + b_1 \sum_{i=1}^{n} x_i^3 + b_2 \sum_{i=1}^{n} x_i^4 = \sum_{i=1}^{n} x_i^2 y_i,$$

which are solved as before for b_0, b_1 and b_2.

Example 9.2   Given the data

x	0	1	2	3	4	5	6	7	8	9
y	9.1	7.3	3.2	4.6	4.8	2.9	5.7	7.1	8.8	10.2

fit a regression curve of the form $\mu_{Y|x} = \beta_0 + \beta_1 x + \beta_2 x^2$ and then estimate $\mu_{Y|2}$.

Solution   From the data given, we find

$$\sum_{i=1}^{10} x_i = 45 \qquad \sum_{i=1}^{10} x_i^2 = 297 \qquad \sum_{i=1}^{10} x_i^3 = 2025 \qquad \sum_{i=1}^{10} x_i^4 = 15{,}332$$

$$\sum_{i=1}^{10} y_i = 53.7 \qquad \sum_{i=1}^{10} x_i y_i = 307.3 \qquad \sum_{i=1}^{10} x_i^2 y_i = 2153.3 \qquad n = 10.$$

Solving the normal equations

$$10b_0 + \quad 45b_1 + \quad\quad 297b_2 = \quad 53.7$$

$$45b_0 + \quad 297b_1 + \quad 2025b_2 = \quad 307.3$$

$$297b_0 + 2025b_1 + 15{,}332b_2 = 2153.3,$$

we obtain

$$b_0 = 8.697, \qquad b_1 = -2.341, \qquad b_2 = 0.288.$$

Therefore,

$$\hat{y} = 8.697 - 2.341x + 0.288x^2.$$

When $x = 2$ our estimate of $\mu_{Y|2}$ is

$$\hat{y} = 8.697 - (2.341)(2) + (0.288)(2^2)$$
$$= 5.2.$$

9.3 ESTIMATING THE COEFFICIENTS USING MATRICES

In fitting a multiple linear regression equation, particularly when the number of variables exceeds two, a knowledge of matrix theory can facilitate the mathematical manipulations considerably. Suppose that the experimenter has data of the type

$$\{(x_{1i}, x_{2i}, \ldots, x_{ki}, y_i); \qquad i = 1, 2, \ldots, n \text{ and } n > k\},$$

where y_i is the observed response to the values $x_{1i}, x_{2i}, \ldots, x_{ki}$ of the k independent variables $x_1, x_2, \ldots, x_k$. Each observation $(x_{1i}, x_{2i}, \ldots, x_{ki}, y_i)$ satisfies the equation

$$y_i = \beta_0 + \beta_1 x_{1i} + \beta_2 x_{2i} + \cdots + \beta_k x_{ki} + \varepsilon_i$$
$$= b_0 + b_1 x_{1i} + b_2 x_{2i} + \cdots + b_k x_{ki} + e_i,$$

where ε_i and e_i are the random error and residual, respectively, associated with the response y_i. We can use the concept of least squares again to arrive at the estimates $b_0, b_1, b_2, \ldots, b_k$. Consider the expression

$$SSE = \sum_{i=1}^{n} e_i^2$$

$$= \sum_{i=1}^{n} (y_i - b_0 - b_1 x_{1i} - b_2 x_{2i} - \cdots - b_k x_{ki})^2,$$

which is to be minimized. Differentiating SSE in turn with respect to $b_0, b_1, \ldots, b_k$ and equating to zero, we generate the set of normal equations

$$nb_0 + b_1 \sum_{i=1}^{n} x_{1i} + b_2 \sum_{i=1}^{n} x_{2i} + \cdots + b_k \sum_{i=1}^{n} x_{ki} = \sum_{i=1}^{n} y_i$$

$$b_0 \sum_{i=1}^{n} x_{1i} + b_1 \sum_{i=1}^{n} x_{1i}^2 + b_2 \sum_{i=1}^{n} x_{1i}x_{2i} + \cdots + b_k \sum_{i=1}^{n} x_{1i}x_{ki} = \sum_{i=1}^{n} x_{1i}y_i$$

$$\vdots \qquad \vdots \qquad \vdots \qquad \vdots \qquad \vdots$$

$$b_0 \sum_{i=1}^{n} x_{ki} + b_1 \sum_{i=1}^{n} x_{ki}x_{1i} + b_2 \sum_{i=1}^{n} x_{ki}x_{2i} + \cdots + b_k \sum_{i=1}^{n} x_{ki}^2 = \sum_{i=1}^{n} x_{ki}y_i.$$

We now define a matrix $\mathbf{A}$ such that

$$\mathbf{A} = \mathbf{X'X},$$

where

$$\mathbf{X} = \begin{bmatrix} 1 & x_{11} & x_{21} & \cdots & x_{k1} \\ 1 & x_{12} & x_{22} & \cdots & x_{k2} \\ \vdots & \vdots & \vdots & & \vdots \\ 1 & x_{1n} & x_{2n} & \cdots & x_{kn} \end{bmatrix}.$$

Apart from the initial element, the ith row of $\mathbf{X}$ represents those x values that give rise to the response y_i. Writing

$$\mathbf{A} = \begin{bmatrix} n & \sum_{i=1}^{n} x_{1i} & \sum_{i=1}^{n} x_{2i} & \cdots & \sum_{i=1}^{n} x_{ki} \\ \sum_{i=1}^{n} x_{1i} & \sum_{i=1}^{n} x_{1i}^2 & \sum_{i=1}^{n} x_{1i}x_{2i} & \cdots & \sum_{i=1}^{n} x_{1i}x_{ki} \\ \vdots & \vdots & \vdots & & \vdots \\ \sum_{i=1}^{n} x_{ki} & \sum_{i=1}^{n} x_{ki}x_{1i} & \sum_{i=1}^{n} x_{ki}x_{2i} & \cdots & \sum_{i=1}^{n} x_{ki}^2 \end{bmatrix},$$

$$\mathbf{b} = \begin{bmatrix} b_0 \\ b_1 \\ \vdots \\ b_k \end{bmatrix}, \qquad \mathbf{g} = \begin{bmatrix} g_0 = \sum_{i=1}^{n} y_i \\ g_1 = \sum_{i=1}^{n} x_{1i}y_i \\ \vdots \\ g_k = \sum_{i=1}^{n} x_{ki}y_i \end{bmatrix},$$

the normal equations can be put in the matrix form

$$\mathbf{Ab} = \mathbf{g}.$$

If the matrix $\mathbf{A}$ is nonsingular, we can write the solution for the regression coefficients as

$$\mathbf{b} = \mathbf{A}^{-1}\mathbf{g}.$$

Thus one can obtain the prediction equation or regression equation by solving a set of $k + 1$ equations in a like number of unknowns. This involves the inversion of the $k + 1$ by $k + 1$ matrix $\mathbf{A}$. Techniques for inverting this matrix are explained in most textbooks on elementary determinants and matrices. There are many high-speed computer programs available for multiple regression problems, programs that not only give estimates of the regression coefficients, but also obtain other information relevant in making inferences concerning the regression equation.

Example 9.3   The percent survival of a certain type of animal semen after storage was measured at various combinations of concentrations of three materials used to increase chance of survival. The data are as follows:

y (% survival)	x_1 (weight %)	x_2 (weight %)	x_3 (weight %)
25.5	1.74	5.30	10.80
31.2	6.32	5.42	9.40
25.9	6.22	8.41	7.20
38.4	10.52	4.63	8.50
18.4	1.19	11.60	9.40
26.7	1.22	5.85	9.90
26.4	4.10	6.62	8.00
25.9	6.32	8.72	9.10
32.0	4.08	4.42	8.70
25.2	4.15	7.60	9.20
39.7	10.15	4.83	9.40
35.7	1.72	3.12	7.60
26.5	1.70	5.30	8.20

Estimate the multiple linear regression model for the given data.

Solution   From the experimental data we list the following sums of squares and products:

$$\sum_{i=1}^{13} y_i = 377.5 \qquad \sum_{i=1}^{13} y_i^2 = 11{,}400.15 \qquad \sum_{i=1}^{13} x_{1i} = 59.43$$

$$\sum_{i=1}^{13} x_{2i} = 81.82 \qquad \sum_{i=1}^{13} x_{3i} = 115.40 \qquad \sum_{i=1}^{13} x_{1i}^2 = 394.7255$$

$$\sum_{i=1}^{13} x_{2i}^2 = 576.7264 \qquad \sum_{i=1}^{13} x_{3i}^2 = 1035.9600 \qquad \sum_{i=1}^{13} x_{1i} y_i = 1877.567$$

$$\sum_{i=1}^{13} x_{2i} y_i = 2246.661 \qquad \sum_{i=1}^{13} x_{3i} y_i = 3337.780 \qquad \sum_{i=1}^{13} x_{1i} x_{2i} = 360.6621$$

$$\sum_{i=1}^{13} x_{1i} x_{3i} = 522.0780 \qquad \sum_{i=1}^{13} x_{2i} x_{3i} = 728.3100 \qquad n = 13.$$

The least squares estimating equations, $\mathbf{Ab} = \mathbf{g}$, are given by

$$\begin{bmatrix} 13 & 59.43 & 81.82 & 115.40 \\ 59.43 & 394.7255 & 360.6621 & 522.0780 \\ 81.82 & 360.6621 & 576.7264 & 728.3100 \\ 115.40 & 522.0780 & 728.3100 & 1035.9600 \end{bmatrix} \begin{bmatrix} b_0 \\ b_1 \\ b_2 \\ b_3 \end{bmatrix} = \begin{bmatrix} 377.5 \\ 1877.567 \\ 2246.661 \\ 3337.780 \end{bmatrix}.$$

From a computer readout we obtain the elements of the inverse matrix

$$\mathbf{A}^{-1} = \begin{bmatrix} 8.0648 & -0.0826 & -0.0942 & -0.7905 \\ -0.0826 & 0.0085 & 0.0017 & 0.0037 \\ -0.0942 & 0.0017 & 0.0166 & -0.0021 \\ -0.7905 & 0.0037 & -0.0021 & 0.0886 \end{bmatrix},$$

and then using the relation $\mathbf{b} = \mathbf{A}^{-1}\mathbf{g}$ the computer printout gives

$$b_0 = 39.1574, \qquad b_1 = 1.0161, \qquad b_2 = -1.8616, \qquad b_3 = -0.3433.$$

Hence our estimated regression equation is

$$\hat{y} = 39.1574 + 1.0161x_1 - 1.8616x_2 - 0.3433x_3.$$

For the case of a single independent variable, the *degree* of the best fitting polynomial can often be determined by plotting a scatter diagram of the data obtained from an experiment that yields n pairs of observations of the form $\{(x_i, y_i); i = 1, 2, \ldots, n\}$. Suppose a model is postulated so that each observation satisfies the equation

$$y_i = \beta_0 + \beta_1 x_i + \beta_2 x_i^2 + \cdots + \beta_r x_i^r + \varepsilon_i$$
$$= b_0 + b_1 x_i + b_2 x_i^2 + \cdots + b_r x_i^r + e_i,$$

where r is the degree of the polynomial, and ε_i and e_i are again the random error and residual associated with the response y_i. Here, the number of pairs, n, must be at least as large as $r + 1$, the number of parameters to be estimated. As in Section 9.2 we notice that the polynomial model can be considered a special case of the more general multiple linear regression model, where we set $x_1 = x$, $x_2 = x^2, \ldots, x_r = x^r$. Therefore, the set of normal equations assumes the form

$$
\begin{bmatrix}
n & \sum_{i=1}^{n} x_i & \sum_{i=1}^{n} x_i^2 & \cdots & \sum_{i=1}^{n} x_i^r \\
\sum_{i=1}^{n} x_i & \sum_{i=1}^{n} x_i^2 & \sum_{i=1}^{n} x_i^3 & \cdots & \sum_{i=1}^{n} x_i^{r+1} \\
\sum_{i=1}^{n} x_i^2 & \sum_{i=1}^{n} x_i^3 & \sum_{i=1}^{n} x_i^4 & \cdots & \sum_{i=1}^{n} x_i^{r+2} \\
\vdots & \vdots & \vdots & & \vdots \\
\sum_{i=1}^{n} x_i^r & \sum_{i=1}^{n} x_i^{r+1} & \sum_{i=1}^{n} x_i^{r+2} & \cdots & \sum_{i=1}^{n} x_i^{2r}
\end{bmatrix}
\begin{bmatrix}
b_0 \\ b_1 \\ b_2 \\ \vdots \\ b_r
\end{bmatrix}
=
\begin{bmatrix}
\sum_{i=1}^{n} y_i \\ \sum_{i=1}^{n} x_i y_i \\ \vdots \\ \sum_{i=1}^{n} x_i^r y_i
\end{bmatrix}.
$$

Solving these $r + 1$ equations, we obtain the estimates $b_0, b_1, \ldots, b_r$ and thereby generate the polynomial regression prediction equation

$$\hat{y} = b_0 + b_1 x + b_2 x^2 + \cdots + b_r x^r.$$

The procedure for fitting a polynomial regression model can be generalized to the case of more than one independent variable. In fact, the student of regression analysis should at this stage have the facility for fitting any linear model in, say, k independent variables. Suppose, for example, that we have a response Y with $k = 2$ independent variables and a quadratic model is postulated of the type

$$y_i = \beta_0 + \beta_1 x_{1i} + \beta_2 x_{2i} + \beta_{11} x_{1i}^2 + \beta_{22} x_{2i}^2 + \beta_{12} x_{1i} x_{2i} + \varepsilon_i,$$

where y_i, $i = 1, 2, \ldots, n$, is the response to the combination (x_{1i}, x_{2i}) of the independent variables in the experiment. In this situation n must be at least 6, since there are six parameters to estimate by the least squares procedure. In addition, since the model contains quadratic terms in both variables, at least three levels of each variable must be used. The reader should easily verify that the least squares normal equations $(\mathbf{X'X})\mathbf{b} = \mathbf{g}$ are given by

$$
\begin{bmatrix}
n & \sum_{i=1}^{n} x_{1i} & \sum_{i=1}^{n} x_{2i} & \sum_{i=1}^{n} x_{1i}^2 & \sum_{i=1}^{n} x_{2i}^2 & \sum_{i=1}^{n} x_{1i}x_{2i} \\
\sum_{i=1}^{n} x_{1i} & \sum_{i=1}^{n} x_{1i}^2 & \sum_{i=1}^{n} x_{1i}x_{2i} & \sum_{i=1}^{n} x_{1i}^3 & \sum_{i=1}^{n} x_{1i}x_{2i}^2 & \sum_{i=1}^{n} x_{1i}^2 x_{2i} \\
\sum_{i=1}^{n} x_{2i} & \sum_{i=1}^{n} x_{1i}x_{2i} & \sum_{i=1}^{n} x_{2i}^2 & \sum_{i=1}^{n} x_{1i}^2 x_{2i} & \sum_{i=1}^{n} x_{2i}^3 & \sum_{i=1}^{n} x_{1i}x_{2i}^2 \\
\sum_{i=1}^{n} x_{1i}^2 & \sum_{i=1}^{n} x_{1i}^3 & \sum_{i=1}^{n} x_{1i}^2 x_{2i} & \sum_{i=1}^{n} x_{1i}^4 & \sum_{i=1}^{n} x_{1i}^2 x_{2i}^2 & \sum_{i=1}^{n} x_{1i}^3 x_{2i} \\
\sum_{i=1}^{n} x_{2i}^2 & \sum_{i=1}^{n} x_{1i}x_{2i}^2 & \sum_{i=1}^{n} x_{2i}^3 & \sum_{i=1}^{n} x_{1i}^2 x_{2i}^2 & \sum_{i=1}^{n} x_{2i}^4 & \sum_{i=1}^{n} x_{1i}x_{2i}^3 \\
\sum_{i=1}^{n} x_{1i}x_{2i} & \sum_{i=1}^{n} x_{1i}^2 x_{2i} & \sum_{i=1}^{n} x_{1i}x_{2i}^2 & \sum_{i=1}^{n} x_{1i}^3 x_{2i} & \sum_{i=1}^{n} x_{1i}x_{2i}^3 & \sum_{i=1}^{n} x_{1i}^2 x_{2i}^2
\end{bmatrix}
\times
\begin{bmatrix}
b_0 \\
b_1 \\
b_2 \\
b_{11} \\
b_{22} \\
b_{12}
\end{bmatrix}
=
\begin{bmatrix}
\sum_{i=1}^{n} y_i \\
\sum_{i=1}^{n} x_{1i}y_i \\
\sum_{i=1}^{n} x_{2i}y_i \\
\sum_{i=1}^{n} x_{1i}^2 y_i \\
\sum_{i=1}^{n} x_{2i}^2 y_i \\
\sum_{i=1}^{n} x_{1i}x_{2i}y_i
\end{bmatrix}.
$$

Example 9.4 The following data represent the percent of impurities that occurs at various temperatures and sterilizing times in a reaction associated with the manufacturing of a certain beverage.

Sterilizing time, x_2 (minutes)	Temperature, x_1 (°C)		
	75	100	125
15	14.05	10.55	7.55
	14.93	9.48	6.59
20	16.56	13.63	9.23
	15.85	11.75	8.78
25	22.41	18.55	15.93
	21.66	17.98	16.44

Estimate the regression coefficients in the model

$$\mu_{Y|x_1, x_2} = \beta_0 + \beta_1 x_1 + \beta_2 x_2 + \beta_{11} x_1^2 + \beta_{22} x_2^2 + \beta_{12} x_1 x_2.$$

Solution If we define $x_3 = x_1^2$, $x_4 = x_2^2$, and $x_5 = x_1 x_2$, then the computer program that was used in Example 9.3 can once again be used to invert the 6×6 **A** matrix. From the equation $\mathbf{b} = \mathbf{A}^{-1}\mathbf{g}$ we obtain

$$b_0 = 56.4668 \qquad b_{11} = 0.00081$$
$$b_1 = -0.36235 \qquad b_{22} = 0.08171$$
$$b_2 = -2.75299 \qquad b_{12} = 0.00314,$$

and our estimated regression equation is

$$\hat{y} = 56.4668 - 0.36235 x_1 - 2.75299 x_2 + 0.00081 x_1^2$$
$$+ 0.08171 x_2^2 + 0.00314 x_1 x_2.$$

Many of the principles and procedures associated with the estimation of polynomial regression functions fall into the category of *response surface methodology*, a collection of techniques that have been used quite successfully by scientists and engineers in many fields. Such problems as selecting a proper experimental design, particularly in cases where a large number of variables are in the model, and choosing "optimum" operating conditions on $x_1, x_2, \ldots, x_k$

are often approached through the use of these methods. For an extensive exposure the reader is referred to *Response Surface Methodology* by Myers (see the Bibliography).

9.4 PROPERTIES OF THE LEAST SQUARES ESTIMATORS

The means and variances of the estimators $B_0, B_1, \ldots, B_k$ are easily obtained under certain assumptions on the random errors $E_1, E_2, \ldots, E_k$ that are identical to those made in the case of simple linear regression. Suppose that we assume these errors to be independent, each with zero mean and variance σ^2. It can then be shown that $B_0, B_1, \ldots, B_k$ are, respectively, unbiased estimators of the regression coefficients $\beta_0, \beta_1, \ldots, \beta_k$. In addition, the variances of the B's are obtained through the elements of the inverse of the $\mathbf{A}$ matrix. One will note that the off-diagonal elements of $\mathbf{A} = \mathbf{X'X}$ represent sums of products of elements in the columns of $\mathbf{X}$, while the diagonal elements of $\mathbf{A}$ represent sums of squares of elements in the columns of $\mathbf{X}$. The inverse matrix, $\mathbf{A}^{-1}$, apart from the multiple σ^2, represents the *variance–covariance matrix* of the estimated regression coefficients. That is, the elements of the matrix $\mathbf{A}^{-1}\sigma^2$ display the variances of $B_0, B_1, \ldots, B_k$ on the main diagonal and covariances on the off diagonal. For example, in a $k = 2$ multiple linear regression problem, we might write

$$\mathbf{A}^{-1} = \begin{bmatrix} c_{00} & c_{01} & c_{02} \\ c_{10} & c_{11} & c_{12} \\ c_{20} & c_{21} & c_{22} \end{bmatrix}$$

with the elements below the main diagonal determined through the symmetry of the matrix. Then we can write

$$\sigma_{B_i}^2 = c_{ii}\sigma^2, \qquad\qquad i = 0, 1, 2$$

$$\sigma_{B_i B_j} = \mathrm{cov}\,(B_i, B_j) = c_{ij}\sigma^2, \qquad i \neq j.$$

Of course, the estimates of the variances and hence the standard errors of these estimators are obtained by replacing σ^2 with the appropriate estimate obtained through experimental data. An unbiased estimate of σ^2 is once again defined in terms of the error sum of squares, which is computed using the formula established in Theorem 9.1.

THEOREM 9.1   *For the linear regression equation*

$$y_i = b_0 + b_1 x_{1i} + \cdots + b_k x_{ki} + e_i,$$

$i = 1, 2, \ldots, n,$ *the error sum of squares may be written*

$$SSE = SST - SSR,$$

where

$$SST = S_{yy}$$

and

$$SSR = \sum_{j=0}^{k} b_j g_j - \frac{\left(\sum_{i=1}^{n} y_i\right)^2}{n}.$$

Proof   Let us first consider the case of $k = 2$ independent variables and then generalize to the general linear model containing k independent variables. Following the basic procedure outlined in Section 8.3, we may write

$$SSE = \sum_{i=1}^{n} (y_i - b_0 - b_1 x_{1i} - b_2 x_{2i})^2$$

$$= \sum_{i=1}^{n} [y_i - (b_0 + b_1 x_{1i} + b_2 x_{2i})]^2$$

$$= \sum_{i=1}^{n} [y_i^2 - y_i(b_0 + b_1 x_{1i} + b_2 x_{2i}) + (b_0 + b_1 x_{1i} + b_2 x_{2i})^2$$

$$- y_i(b_0 + b_1 x_{1i} + b_2 x_{2i})]$$

$$= \sum_{i=1}^{n} y_i^2 - b_0 \sum_{i=1}^{n} y_i - b_1 \sum_{i=1}^{n} x_{1i} y_i - b_2 \sum_{i=1}^{n} x_{2i} y_i$$

$$+ b_0 \sum_{i=1}^{n} (b_0 + b_1 x_{1i} + b_2 x_{2i} - y_i)$$

$$+ b_1 \sum_{i=1}^{n} (b_0 x_{1i} + b_1 x_{1i}^2 + b_2 x_{1i} x_{2i} - x_{1i} y_i)$$

$$+ b_2 \sum_{i=1}^{n} (b_0 x_{2i} + b_1 x_{1i} x_{2i} + b_2 x_{2i}^2 - x_{2i} y_i).$$

Using the normal equations of Section 9.2, we see that the last three sums are each equal to zero. Therefore,

$$SSE = \sum_{i=1}^{n} y_i^2 - b_0 \sum_{i=1}^{n} y_i - b_1 \sum_{i=1}^{n} x_{1i} y_i - b_2 \sum_{i=1}^{n} x_{2i} y_i$$

$$= S_{yy} - \left[b_0 g_0 + b_1 g_1 + b_2 g_2 - \frac{\left(\sum_{i=1}^{n} y_i \right)^2}{n} \right].$$

Generalizing to k independent variables, we have

$$SSE = S_{yy} - \left[\sum_{j=0}^{k} b_j g_j - \frac{\left(\sum_{i=1}^{n} y_i \right)^2}{n} \right]$$

$$= SST - SSR.$$

There are k degrees of freedom associated with SSR in this general case. As always, SST has $n - 1$ degrees of freedom and therefore, after subtraction, SSE has $n - k + 1$ degrees of freedom. Thus our estimate of σ^2 is given by

$$s^2 = \frac{SSE}{n - k - 1}.$$

Example 9.5   For the data of Example 9.3, estimate σ^2.

Solution   The error sum of squares with $n - k - 1 = 9$ degrees of freedom can be found by writing

$$SSE = SST - SSR,$$

where

$$SST = S_{yy} = 11,400.15 - \frac{(377.5)^2}{13} = 438.13$$

and

$$SSR = \sum_{j=0}^{3} b_j g_j - \frac{\left(\sum\limits_{i=1}^{13} y_i\right)^2}{13}$$

$$= (39.1574)(377.5) + (1.0161)(1877.567)$$

$$+ (-1.8616)(2246.6610) + (-0.3433)(3337.780) - \frac{(377.5)^2}{13}$$

$$= 399.45,$$

with 3 degrees of freedom. Our estimate of σ^2 is then given by

$$s^2 = \frac{438.13 - 399.45}{9} = 4.298.$$

The error and regression sums of squares take on the same meaning that they did in the special case of a single independent variable. While these two portions give an intuitive indication of how adequate the model is, no form of hypothesis testing or confidence interval estimation can be accomplished without making some assumption about the distribution of the errors E_i. In Section 9.5 we outline certain methods for finding confidence intervals and testing hypotheses that can be helpful in the multiple linear regression analysis.

9.5 INFERENCES IN MULTIPLE LINEAR REGRESSION

One of the most useful inferences that can be made regarding the ability of the regression equation to predict the response $\hat{y}_0$ corresponding to the values $x_{10}, x_{20}, \ldots, x_{k0}$ is the confidence interval on the mean response $\mu_{Y|x_{10}, x_{20}, \ldots, x_{k0}}$. In vector notation we are then interested in constructing a confidence interval on the mean response for the set of conditions given by $\mathbf{x}_0' = [1, x_{10}, x_{20}, \ldots, x_{k0}]$. We augment the conditions on the x's by the number 1 in order to facilitate using matrix notation. As in the $k = 1$ case, if we make the additional assumption that the errors are independent and normally distributed, then the B_j's are normal, with mean, variances, and covariances as indicated in Section 9.4, and hence

$$\hat{Y}_0 = B_0 + \sum_{j=1}^{k} B_j x_{j0}$$

is likewise normally distributed and is, in fact, an unbiased estimator for the *mean response* on which we are attempting to attach confidence intervals. The variance of $\hat{Y}_0$ written in matrix notation simply as a function of σ^2, A^{-1}, and the condition vector $\mathbf{x}_0'$, is

$$\sigma_{\hat{Y}_0}^2 = \sigma^2 \mathbf{x}_0' A^{-1} \mathbf{x}_0.$$

If this expression is expanded for a given case, say $k = 2$, it is easily seen that it appropriately accounts for the variances and covariances of the B_j's. After replacing σ^2 by s^2 as given in Section 9.4, the $100(1 - \alpha)\%$ confidence interval on $\mu_{Y|x_{10}, x_{20}, \ldots, x_{k0}}$ can be constructed from the statistic

$$T = \frac{\hat{Y}_0 - \mu_{Y|x_{10}, x_{20}, \ldots, x_{k0}}}{S\sqrt{\mathbf{x}_0' A^{-1} \mathbf{x}_0}},$$

which has a t distribution with $n - k - 1$ degrees of freedom.

CONFIDENCE INTERVAL FOR $\mu_{Y|x_{10}, x_{20}, \ldots, x_{k0}}$ $A (1 - \alpha)100\%$ *confidence interval for the mean response* $\mu_{Y|x_{10}, x_{20}, \ldots, x_{k0}}$ *is given by*

$$\hat{y}_0 - t_{\alpha/2} s\sqrt{\mathbf{x}_0' A^{-1} \mathbf{x}_0} < \mu_{Y|x_{10}, x_{20}, \ldots, x_{k0}} < \hat{y}_0 + t_{\alpha/2} s\sqrt{\mathbf{x}_0' A^{-1} \mathbf{x}_0},$$

where $t_{\alpha/2}$ *is a value of the t distribution with* $n - k - 1$ *degrees of freedom.*

Example 9.6   Using the data of Example 9.3, construct a 95% confidence interval for the mean response when $x_1 = 3$, $x_2 = 8$, and $x_3 = 9$.

Solution   From the regression equation of Example 9.3 the estimated response when $x_1 = 3$, $x_2 = 8$, and $x_3 = 9$ is

$$\hat{y}_0 = 39.1574 + (1.0161)(3) - (1.8616)(8) - (0.3433)(9) = 24.2232.$$

Next we find

$$\mathbf{x}_0' A^{-1} \mathbf{x}_0 = [1, 3, 8, 9] \begin{bmatrix} 8.064 & -0.0826 & -0.0942 & -0.7905 \\ -0.0826 & 0.0085 & 0.0017 & 0.0037 \\ -0.0942 & 0.0017 & 0.0166 & -0.0021 \\ -0.7905 & 0.0037 & -0.0021 & 0.0886 \end{bmatrix} \begin{bmatrix} 1 \\ 3 \\ 8 \\ 9 \end{bmatrix}$$

$$= 0.1267.$$

Previously, we found $s^2 = 4.298$ or $s = 2.073$, and using Table V we see that $t_{0.025} = 2.262$ for 9 degrees of freedom. Therefore, a 95% confidence interval for $\mu_{Y|3,8,9}$ is given by

$$24.2232 - (2.262)(2.073)\sqrt{0.1267} < \mu_{Y|3,8,9} < 24.2232$$
$$+ (2.262)(2.073)\sqrt{0.1267}$$

or simply

$$22.5541 < \mu_{Y|3,8,9} < 25.8923.$$

A confidence interval for a single predicted response $\hat{y}_0$ is once again established by considering the differences $\hat{y}_0 - y_0$ of the random variable $\hat{Y}_0 - Y_0$. The sampling distribution can be shown to be normal with mean

$$\mu_{\hat{Y}_0 - Y_0} = 0$$

and variance

$$\sigma^2_{\hat{Y}_0 - Y_0} = \sigma^2[1 + \mathbf{x}'_0 \mathbf{A}^{-1} \mathbf{x}_0].$$

Thus the $(1 - \alpha)100\%$ confidence interval for a single predicted value y_0 can be constructed from the statistic

$$T = \frac{\hat{Y}_0 - Y_0}{S\sqrt{1 + \mathbf{x}'_0 \mathbf{A}^{-1} \mathbf{x}_0}},$$

which has a t distribution with $n - k - 1$ degrees of freedom.

CONFIDENCE INTERVAL FOR y_0 $A (1 - \alpha)100\%$ *confidence interval for a single response y_0 is given by*

$$\hat{y}_0 - t_{\alpha/2} s\sqrt{1 + \mathbf{x}'_0 \mathbf{A}^{-1} \mathbf{x}_0} < y_0 < \hat{y}_0 + t_{\alpha/2} s\sqrt{1 + \mathbf{x}'_0 \mathbf{A}^{-1} \mathbf{x}_0},$$

where $t_{\alpha/2}$ is a value of the t distribution with $n - k - 1$ degrees of freedom.

Example 9.7   Using the data of Example 9.3, construct a 95% confidence interval for the predicted response when $x_1 = 3$, $x_2 = 8$, and $x_3 = 9$.

Solution   Referring to the results of Example 9.6, we find that the 95 % confidence interval for the response y_0 when $x_1 = 3$, $x_2 = 8$, and $x_3 = 9$ is

$$24.2232 - (2.262)(2.073)\sqrt{1.1267} < y_0 < 24.2232 + (2.262)(2.073)\sqrt{1.1267},$$

which reduces to

$$19.2459 < y_0 < 29.2005.$$

A knowledge of the distributions of the individual coefficient estimators enables the experimenter to construct confidence intervals for the coefficients and to test hypotheses about them. One recalls that the B_j's ($j = 0, 1, 2, \ldots, k$) are normally distributed with mean β_j and variance $c_{jj}\sigma^2$. Thus we can use the statistic

$$T = \frac{B_j - \beta_j}{S\sqrt{c_{jj}}}$$

with $n - k - 1$ degrees of freedom to test hypotheses and construct confidence intervals on β_j. For example, if we wish to test

$$H_0: \quad \beta_j = \beta_{j0},$$
$$H_1: \quad \beta_j \neq \beta_{j0},$$

we compute the statistic

$$t = \frac{b_j - \beta_{j0}}{s\sqrt{c_{jj}}}$$

and accept H_0 if

$$-t_{\alpha/2} < t < t_{\alpha/2},$$

where $t_{\alpha/2}$ has $n - k - 1$ degrees of freedom.

Example 9.8   For the model of Example 9.3, test the hypothesis that $\beta_2 = -2.5$ at the 0.05 level of significance against the alternative that $\beta_2 > -2.5$.

Solution
1. $H_0: \quad \beta_2 = -2.5.$
2. $H_1: \quad \beta_2 > -2.5.$

3. Choose a 0.05 level of significance.
4. Critical region: $T > 1.833$.
5. Computations:

$$t = \frac{b_2 - \beta_{20}}{s\sqrt{c_{22}}} = \frac{-1.8616 + 2.5}{2.073\sqrt{0.0166}} = 2.391.$$

6. Conclusion: Reject H_0 and conclude that $\beta_2 > -2.5$.

9.6 ADEQUACY OF THE MODEL

In many regression situations, individual coefficients are of importance to the experimenter. For example, in an economics application, $\beta_1, \beta_2, \ldots$, might have some particular significance, and thus confidence intervals and tests of hypotheses on these parameters are of interest to the economist. However, consider an industrial chemical situation in which the postulated model assumes that reaction yield is dependent linearly on reaction temperature and concentration of a certain catalyst. It is probably known that this is not the true model but an adequate representation, so the interest is likely to be not in the individual parameters but rather in the ability of the entire function to predict the true response in the range of the variables considered. Therefore, in this situation, one would put more emphasis on $\sigma_{\hat{Y}}^2$, confidence intervals on the mean response, and so forth, and likely deemphasize inferences on individual parameters.

The experimenter using regression analysis is also interested in deletion of variables when the situation dictates that, in addition to arriving at a workable prediction equation, he must find the "best regression" involving only variables that are useful predictors. There are a number of computer programs available for the practitioner that sequentially arrive at the so-called best regression equation depending on certain criteria. We shall discuss this further in Section 9.8.

One criterion that is commonly used to illustrate the adequacy of a fitted regression model is the *coefficient of multiple determination*:

$$R^2 = \frac{SSR}{SST} = \frac{\displaystyle\sum_{j=0}^{k} b_j g_j - \left(\displaystyle\sum_{i=1}^{n} y_i\right)^2 \bigg/ n}{S_{yy}}.$$

This quantity merely indicates what proportion of the total variation in the response Y is explained by the fitted model. Often an experimenter will report $R^2 \times 100\%$ and interpret the result as percentage variation explained by the postulated model. The square root of R^2 is called the *multiple correlation coefficient* between Y and the set $x_1, x_2, \ldots, x_k$. In Example 9.3 the value of R^2 indicating the proportion of variation explained by the three independent variables x_1, x_2, and x_3 is found to be

$$R^2 = \frac{SSR}{SST} = \frac{399.45}{438.13} = 0.9117,$$

which means that 91.17% of the variation in Y has been explained by the linear regression model.

The regression sum of squares can be used to give some indication concerning whether or not the model is an adequate explanation of the true situation. One can test the hypothesis H_0 that the regression is significant by merely forming the ratio

$$f = \frac{SSR/k}{SSE/(n - k - 1)} = \frac{SSR/k}{s^2}$$

and rejecting H_0 at the α level of significance when $f > f_\alpha(k, n - k - 1)$. For the data of Example 9.3 we obtain

$$f = \frac{399.45/3}{4.298} = 30.98.$$

This value of F exceeds the tabulated critical point of the F distribution for 3 and 9 degrees of freedom at the $\alpha = 0.01$ level. The result here should not be misinterpreted. Although it does indicate that the regression explained by the model is significant, this does not rule out the possibility that

1. The linear regression model in this set of x's is not the only model that can be used to explain the data; indeed, there might be other models with transformations on the x's that might give a larger value of the F statistic.
2. The model might have been more effective with the inclusion of other variables in addition to x_1, x_2, and x_3 or perhaps with the deletion of one or more of the variables in the model.

The addition of any single variable to a regression system will increase the regression sum of squares and thus reduce the error sum of squares. Consequently, we must decide whether the increase in regression is sufficient to warrant using it in the model. As one might expect, the use of unimportant variables can reduce the effectiveness of the prediction equation by increasing the variance of the estimated response. We shall pursue this point further by considering the importance of x_3 in Example 9.3. Initially, we can test

$$H_0: \quad \beta_3 = 0,$$
$$H_1: \quad \beta_3 \neq 0,$$

by using the t distribution with 9 degrees of freedom. We have

$$t = \frac{b_3 - 0}{s\sqrt{c_{33}}} = \frac{-0.3433}{\sqrt{4.298c_{33}}} = \frac{-0.3433}{2.073\sqrt{0.0886}} = -0.556,$$

which indicates that β_3 does not differ significantly from zero, and hence one may very well feel justified in removing x_3 from the model. Suppose that we consider the regression of Y on the set (x_1, x_2), the least squares normal equations now reducing to

$$\begin{bmatrix} 13 & 59.43 & 81.82 \\ 59.43 & 394.7255 & 360.6621 \\ 81.82 & 360.6621 & 576.7264 \end{bmatrix} \begin{bmatrix} b_0 \\ b_1 \\ b_2 \end{bmatrix} = \begin{bmatrix} 377.75 \\ 1877.5670 \\ 2246.6610 \end{bmatrix}.$$

The estimated regression coefficients for this reduced model are given by

$$b_0 = 36.094, \qquad b_1 = 1.031, \qquad b_2 = -1.870,$$

and the resulting regression sum of squares with 2 degrees of freedom is as follows:

$$R(\beta_1, \beta_2) = \sum_{j=0}^{2} b_j g_j - \frac{\left(\sum_{i=1}^{13} y_i\right)^2}{13}$$

$$= 398.12.$$

Here we use the notation $R(\beta_1, \beta_2)$ to indicate the regression sum of squares of the restricted model and it is not to be confused with SSR, the regression sum of squares of the original model with 3 degrees of freedom. The new error sum of squares is then given by

$$SST - R(\beta_1, \beta_2) = 438.13 - 398.12$$
$$= 40.01,$$

and the resulting error mean square with 10 degrees of freedom becomes

$$s^2 = \frac{40.01}{10} = 4.001.$$

The amount of variation in the response, the percent survival, which is attributed to x_3, the weight percent of the third additive, in the presence of the variables x_1 and x_2, is given by

$$R(\beta_3|\beta_1, \beta_2) = SSR - R(\beta_1, \beta_2)$$
$$= 399.45 - 398.12$$
$$= 1.33,$$

which represents a small proportion of the entire regression variation. This amount of added regression is statistically insignificant as indicated by our previous test on β_3. An equivalent test involves the formation of the ratio

$$f = \frac{R(\beta_3|\beta_1, \beta_2)}{s^2}$$

$$= \frac{1.33}{4.298} = 0.309,$$

which is a value of the F distribution with 1 and 9 degrees of freedom. Recall that the basic relationship between the t distribution with v degrees of freedom and the F distribution with 1 and v degrees of freedom is given by

$$t_{\alpha/2}^2 = f_\alpha(1, v)$$

and we note that the f value of 0.309 is indeed the square of the t value of 0.556.

We can provide additional support for deleting x_3 from the model by considering $\sigma_{\hat{Y}}^2$ under both the full and reduced regression equation. We first note that the estimate of σ^2 was reduced from 4.298 to 4.001 by deleting x_3 from the model. In the case of the full model the variance of the estimated response $\hat{Y}_0$ when $x_1 = 4.00$, $x_2 = 5.5$, and $x_3 = 8.90$, is

$$\sigma_{\hat{Y}_0}^2 = s^2 \mathbf{x}_0' \mathbf{A}^{-1} \mathbf{x}_0$$

$$= 4.298[1, 4.00, 5.5, 8.90]$$

$$\times \begin{bmatrix} 8.064 & -0.0826 & -0.0942 & -0.7905 \\ -0.0826 & 0.0085 & 0.0017 & 0.0037 \\ -0.0942 & 0.0017 & 0.0166 & -0.0021 \\ -0.7905 & 0.0037 & -0.0021 & 0.0886 \end{bmatrix} \begin{bmatrix} 1 \\ 4.00 \\ 5.5 \\ 8.90 \end{bmatrix}$$

$$= 0.3936.$$

For the reduced model our point of interest becomes $x_0' = [1, 4.00, 5.5]$, our A matrix has been reduced to a 3×3 matrix with inverse given by

$$
A^{-1} = \begin{bmatrix} 1.0114 & -0.0494 & -0.1126 \\ -0.0494 & 0.0083 & 0.0018 \\ -0.1126 & 0.0018 & 0.0166 \end{bmatrix},
$$

and the variance of the estimated response is now

$$
\sigma_{\hat{Y}_0}^2 = 4.00[1, 4.00, 5.5] \begin{bmatrix} 1.0114 & -0.0494 & -0.1126 \\ -0.0494 & 0.0083 & 0.0018 \\ -0.1126 & 0.0018 & 0.0166 \end{bmatrix} \begin{bmatrix} 1 \\ 4.00 \\ 5.5 \end{bmatrix}
$$

$$
= 0.3668,
$$

which represents a reduction over that found for the complete model.

To generalize the above concepts, one can assess the work of an independent variable x_i in the general multiple linear regression model

$$
\mu_{Y|x_1, x_2, \ldots, x_k} = \beta_0 + \beta_1 x_1 + \cdots + \beta_k x_k
$$

by observing the amount of regression attributed to x_i over that attributed to the other variables, that is, the regression on x_i *adjusted* for the other variables. This is computed by subtracting the regression sum of squares for a model with x_i removed, from SSR. For example, we say that x_1 is assessed by calculating

$$
R(\beta_1 | \beta_2, \beta_3, \ldots, \beta_k) = SSR - R(\beta_2, \beta_3, \ldots, \beta_k),
$$

where $R(\beta_2, \beta_3, \ldots, \beta_k)$ is the regression sum of squares with $\beta_1 x_1$ removed from the model. To test the hypothesis

$$
H_0: \quad \beta_1 = 0,
$$
$$
H_1: \quad \beta_1 \neq 0,
$$

compute

$$
f = \frac{R(\beta_1 | \beta_2, \beta_3, \ldots, \beta_k)}{s^2}
$$

and compare with $f_\alpha(1, n - k - 1)$.

In a similar manner we can test for the significance of a *set* of the variables. For example, to investigate simultaneously the importance of including x_1 and x_2 in the model, we test the hypothesis

$$H_0: \quad \beta_1 = \beta_2 = 0,$$

$$H_1: \quad \beta_1 \text{ and } \beta_2 \text{ are not both zero,}$$

by computing

$$f = \frac{[R(\beta_1, \beta_2 | \beta_3, \beta_4, \ldots, \beta_k)]/2}{s^2}$$

$$= \frac{[SSR - R(\beta_3, \beta_4, \ldots, \beta_k)]/2}{s^2}$$

and comparing with $f_\alpha(2, n - k - 1)$. The number of degrees of freedom associated with the numerator, in this case 2, equals the number of variables in the set.

9.7 SPECIAL CASE OF ORTHOGONALITY

Prior to our original development of the general linear regression problem, the assumption was made that the independent variables were measured without error and are often controlled by the experimenter. Quite often they occur as a result of an elaborately *designed experiment*. In fact, one can increase the effectiveness of the resulting prediction equation with the use of a suitable experimental plan.

Suppose that we once again consider the **X** matrix as defined in Section 9.3. We can rewrite it to read

$$\mathbf{X} = [\mathbf{1}, \mathbf{x}_1, \mathbf{x}_2, \ldots, \mathbf{x}_k],$$

where **1** represents a column of ones and $\mathbf{x}_j$ is a column vector representing the levels of x_j. If

$$\mathbf{x}_p' \mathbf{x}_q = 0, \quad p \neq q,$$

the variables x_p and x_q are said to be *orthogonal* to each other. There are certain obvious advantages to having a completely orthogonal situation whereby $\mathbf{x}_p' \mathbf{x}_q = 0$ for all possible p and q, $p \neq q$, and, in addition,

$$\sum_{i=1}^{n} x_{ji} = 0, \quad j = 1, 2, \ldots, k.$$

The resulting $\mathbf{X}'\mathbf{X}$ is a diagonal matrix and the normal equations in Section 9.3 reduce to

$$nb_0 = \sum_{i=1}^{n} y_i$$

$$b_1 \sum_{i=1}^{n} x_{1i}^2 = \sum_{i=1}^{n} x_{1i} y_i$$

$$\vdots \qquad \vdots$$

$$b_k \sum_{i=1}^{n} x_{ki}^2 = \sum_{i=1}^{n} x_{ki} y_i.$$

The most important advantage is that one is easily able to partition *SSR* into *single-degree-of-freedom components*, each of which corresponds to the amount of variation in *Y* accounted for by a given controlled variable. Solving the normal equations for $b_0, b_1, \ldots, b_k$, we can write

$$SSR = b_0 g_0 + b_1 g_1 + \cdots + b_k g_k - \frac{\left(\sum\limits_{i=1}^{n} y_i\right)^2}{n}$$

$$= b_1 g_1 + b_2 g_2 + \cdots + b_k g_k$$

$$= \frac{\left(\sum\limits_{i=1}^{n} x_{1i} y_i\right)^2}{\sum\limits_{i=1}^{n} x_{1i}^2} + \frac{\left(\sum\limits_{i=1}^{n} x_{2i} y_i\right)^2}{\sum\limits_{i=1}^{n} x_{2i}^2} + \cdots + \frac{\left(\sum\limits_{i=1}^{n} x_{ki} y_i\right)^2}{\sum\limits_{i=1}^{n} x_{ki}^2}$$

$$= R(\beta_1) + R(\beta_2) + \cdots + R(\beta_k).$$

The quantity $R(\beta_i)$ is the amount of regression sum of squares associated with a model involving a single independent variable x_i.

To test simultaneously for the significance of a set of *m* variables in an orthogonal situation, the regression sum of squares becomes

$$R(\beta_1, \beta_2, \ldots, \beta_m | \beta_{m+1}, \beta_{m+2}, \ldots, \beta_k) = R(\beta_1) + R(\beta_2) + \cdots + R(\beta_m)$$

and simplifies to

$$R(\beta_1 | \beta_2, \beta_3, \ldots, \beta_k) = R(\beta_1)$$

when evaluating a single independent variable. Therefore, the contribution of a given variable or set of variables is essentially found by *ignoring* the other

Table 9.1 Analysis of Variance for Orthogonal Variables

Source of variation	Sum of squares	Degrees of freedom	Mean square	Computed f
β_1	$R(\beta_1) = \dfrac{\left(\sum\limits_{i=1}^{n} x_{1i} y_i\right)^2}{\sum\limits_{i=1}^{n} x_{1i}^2}$	1	$R(\beta_1)$	$\dfrac{R(\beta_1)}{s^2}$
β_2	$R(\beta_2) = \dfrac{\left(\sum\limits_{i=1}^{n} x_{2i} y_i\right)^2}{\sum\limits_{i=1}^{n} x_{2i}^2}$	1	$R(\beta_2)$	$\dfrac{R(\beta_2)}{s^2}$
$\vdots$	$\vdots$	$\vdots$	$\vdots$	$\vdots$
β_k	$R(\beta_k) = \dfrac{\left(\sum\limits_{i=1}^{n} x_{ki} y_i\right)^2}{\sum\limits_{i=1}^{n} x_{ki}^2}$	1	$R(\beta_k)$	$\dfrac{R(\beta_k)}{s^2}$
Error	SSE	$n-k-1$	$s^2 = \dfrac{SSE}{n-k-1}$	
Total	$SST = S_{yy}$	$n-1$		

variables in the model. Independent evaluations of the worth of the individual variables are accomplished using analysis-of-variance techniques as given in Table 9.1. The total variation in the response is partitioned into single-degree-of-freedom components plus the error term with $n - k - 1$ degrees of freedom. Each computed f value is used to test one of the hypotheses

$$\left.\begin{array}{l} H_0: \quad \beta_i = 0 \\ H_1: \quad \beta_i \neq 0 \end{array}\right\} \quad i = 1, 2, \ldots, k$$

by comparing with the critical point $f_\alpha(1, n - k - 1)$.

Example 9.9 Suppose that a scientist takes experimental data on the radius of a propellant grain Y as a function of powder temperature x_1, extrusion rate x_2, and die temperature x_3. Fit a linear regression model for predicting grain radius and determine the effectiveness of each variable in the model. The data are given as follows:

Grain radius	Powder temperature	Extrusion rate	Die temperature
82	150 (−1)	12 (−1)	220 (−1)
93	190 (1)	12 (−1)	220 (−1)
114	150 (−1)	24 (1)	220 (−1)
124	150 (−1)	12 (−1)	250 (1)
111	190 (1)	24 (1)	220 (−1)
129	190 (1)	12 (−1)	250 (1)
157	150 (−1)	24 (1)	250 (1)
164	190 (1)	24 (1)	250 (1)

Solution   Note that each variable is controlled at two levels and the experiment represents each of the eight possible combinations. The data on the independent variables are coded for convenience by means of the following formulas:

$$x_1 = \frac{\text{powder temperature} - 170}{20}$$

$$x_2 = \frac{\text{extrusion rate} - 18}{6}$$

$$x_3 = \frac{\text{die temperature} - 235}{15}.$$

The resulting levels of x_1, x_2, and x_3 take on the values -1 and $+1$ as indicated in the table of data. This particular *experimental design* affords the orthogonality that we are illustrating here. A more thorough treatment of this type of experimental layout will be given in Chapter 12. The **X** matrix is given by

$$\mathbf{X} = \begin{bmatrix} 1 & -1 & -1 & -1 \\ 1 & 1 & -1 & -1 \\ 1 & -1 & 1 & -1 \\ 1 & -1 & -1 & 1 \\ 1 & 1 & 1 & -1 \\ 1 & 1 & -1 & 1 \\ 1 & -1 & 1 & 1 \\ 1 & 1 & 1 & 1 \end{bmatrix}$$

and it is easy to verify the orthogonality conditions.

One can now compute the coefficients

$$b_0 = \frac{\sum\limits_{i=1}^{8} y_i}{8} = 121.75$$

$$b_1 = \frac{\sum\limits_{i=1}^{8} x_{1i}y_i}{\sum\limits_{i=1}^{8} x_{1i}^2} = \frac{20}{8} = 2.5$$

$$b_2 = \frac{\sum\limits_{i=1}^{8} x_{2i}y_i}{\sum\limits_{i=1}^{8} x_{2i}^2} = \frac{118}{8} = 14.75$$

$$b_3 = \frac{\sum\limits_{i=1}^{8} x_{3i}y_i}{\sum\limits_{i=1}^{8} x_{3i}^2} = \frac{174}{8} = 21.75,$$

so in terms of the *coded* variables, the prediction equation is given by

$$y = 121.75 + 2.5x_1 + 14.75x_2 + 21.75x_3.$$

Table 9.2 Analysis of Variance on Grain Radius Data

Source of variation	Sum of squares	Degrees of freedom	Mean square	Computed f
β_1	$\dfrac{(20)^2}{8} = 50$	1	50	2.16
β_2	$\dfrac{(118)^2}{8} = 1740.50$	1	1740.50	75.26
β_3	$\dfrac{(174)^2}{8} = 3784.50$	1	3784.50	163.65
Error	92.5	4	23.1250	
Total	5667.50	7		

The analysis-of-variance table showing independent contributions to SSR for each variable is given in Table 9.2. The results, when compared to the $f_{0.05}(1, 4)$ critical point of 7.71, indicate that x_1 does not contribute significantly at the 0.05 level, while variables x_2 and x_3 are significant. In this example the estimate for σ^2 is 23.1250. As in the single-independent-variable case, it should be pointed out that this estimate does not solely contain experimental error variation unless the postulated model is correct. Otherwise, the estimate is "contaminated" by lack of fit in addition to pure error, and the lack of fit can be separated only if one obtains multiple experimental observations at the various (x_1, x_2, x_3) combinations.

9.8 SEQUENTIAL METHODS FOR MODEL SELECTION

At times the significance tests outlined in Section 9.6 are quite adequate in determining which variables should be used in the final regression model. These tests are certainly effective if the experiment can be planned and the variables are orthogonal to each other. Even if the variables are not orthogonal, the individual t tests can be of some use in many problems where the number of variables under investigation is small. However, there are many problems in which it is necessary to use more elaborate techniques for screening variables, particularly when the experiment exhibits a substantial deviation from orthogonality. Useful measures of the amount of *multicollinearity* (linear dependency) among the independent variables are provided by the sample correlation coefficients $r_{x_i x_j}$. Since we are concerned only with linear dependency among independent variables, no confusion will result if we drop the x's from our notation and simply write $r_{x_i x_j} = r_{ij}$, where

$$r_{ij} = \frac{S_{ij}}{\sqrt{S_{ii} S_{jj}}}.$$

It should be noted that the r_{ij}'s do not give true estimates of population correlation coefficients in the strict sense, since the x's are actually not random variables in the context discussed here. Thus the term "correlation," albeit standard, is perhaps a misnomer.

When one or more of these sample correlation coefficients deviate substantially from zero, it can be quite difficult to find the most effective subset of variables for inclusion in our prediction equation. In fact, for some problems the multicollinearity will be so extreme that a suitable predictor cannot be found unless all possible subsets of the variables are investigated. Of course, the latter approach is prohibitive for a large number of variables. An informative discussion of model selection in regression by Thompson and Cady is cited in the Bibliography.

The user of multiple linear regression attempts to accomplish one of three objectives:

1. Obtain estimates of individual coefficients in a complete model.
2. Screen variables to determine which have a significant effect on the response.
3. Arrive at the most effective prediction equation.

In (1) it is known a priori that all variables are to be included in the model. In (2) prediction is secondary, while in (3) individual regression coefficients are not as important as the quality of the estimated response $\hat{y}$. In each of the situations above, multicollinearity in the experiment can have a profound effect on the success of the regression.

In this section some standard sequential procedures for selecting variables are discussed. They are based on the notion that a single variable or a collection of variables should not appear in the estimating equation unless they result in a significant increase in the regression sum of squares or, equivalently, a significant increase in R^2, the coefficient of multiple determination.

Consider the data in Table 9.3 in which measurements were taken on 9 infants. The purpose of the experiment was to arrive at a suitable estimating equation relating the length of an infant to all or a subset of the independent variables. The sample correlation coefficients, indicating the linear dependency among the independent variables, are displayed in the symmetric matrix

$$
\begin{array}{cccc}
x_1 & x_2 & x_3 & x_4
\end{array}
$$
$$
\begin{bmatrix}
1.0000 & 0.9523 & 0.5340 & 0.3900 \\
0.9523 & 1.0000 & 0.2626 & 0.1549 \\
0.5340 & 0.2626 & 1.0000 & 0.7847 \\
0.3900 & 0.1549 & 0.7847 & 1.0000
\end{bmatrix}.
$$

Note that there appears to be an appreciable amount of multicollinearity. Using the least squares technique outlined in Section 9.3, the estimated regression equation using the complete model was fitted and is given by

$$
\hat{y} = 7.1475 + 0.1000x_1 + 0.7264x_2 + 3.0758x_3 - 0.0300x_4.
$$

The value of s^2 with 4 degrees of freedom is 0.7414, and the value for the coefficient of determination for this model is found to be 0.9907. Regression sum of squares measuring the variation attributed to each individual variable in the presence of the others, and the corresponding t values, are given in Table 9.4.

Table 9.3 Data Relating to Infant Length

Infant length, y (cm)	Age, x_1 (days)	Length at birth, x_2 (cm)	Weight at birth, x_3 (kg)	Chest size at birth, x_4 (cm)
57.5	78	48.2	2.75	29.5
52.8	69	45.5	2.15	26.3
61.3	77	46.3	4.41	32.2
67.0	88	49.0	5.52	36.5
53.5	67	43.0	3.21	27.2
62.7	80	48.0	4.32	27.7
56.2	74	48.0	2.31	28.3
68.5	94	53.0	4.30	30.3
69.2	102	58.0	3.71	28.7

A two-tailed critical region with 4 degrees of freedom at the 0.05 level of significance is given by $|T| > 2.776$. Of the four computed t values, only variable x_3 appears to be significant. However, it should be recalled that while the t statistic described in Section 9.6 measures the worth of a variable adjusted for all other variables, it does not detect the potential importance of a variable in combination with a subset of the variables. For example, consider the model with only the variables x_2 and x_3 in the equation. The data analysis gives the regression function

$$\hat{y} = 2.1833 + 0.9576x_2 + 3.3253x_3,$$

with $R^2 = 0.9905$, certainly not a substantial reduction from $R^2 = 0.9907$ for the complete model. However, unless the performance characteristics of this particular combination had been observed, one would not be aware of its predictive potential. This, of course, lends support for a methodology that observes all possible regressions, or a systematic sequential procedure designed to test several subsets.

Table 9.4 t Values for the Regression Data of Table 9.3

Variable x_1	Variable x_2	Variable x_3	Variable x_4
$R(\beta_1 \mid \beta_2\beta_3\beta_4)$	$R(\beta_2 \mid \beta_1\beta_3\beta_4)$	$R(\beta_3 \mid \beta_1\beta_2\beta_4)$	$R(\beta_4 \mid \beta_1\beta_2\beta_3)$
$= 0.0644$	$= 0.6334$	$= 6.2523$	$= 0.0241$
$t = 0.2947$	$t = 0.9243$	$t = 2.9040$	$t = -0.1805$

One standard procedure for searching for the "optimum subset" of variables in the absence of orthogonality is a technique called *stepwise regression*. It is based on the procedure of sequentially introducing the variables into the model one at a time. The description of the stepwise routine will be better understood by the reader if the methods of *forward selection* and *backward elimination* are described first.

Forward selection is based on the notion that variables should be inserted one at a time until a satisfactory regression equation is found. The procedure is as follows:

1. Choose the variable that gives the largest regression sum of squares when performing a simple linear regression with y, or equivalently, that which gives the largest value of R^2. We shall call this initial variable x_1.
2. Choose the variable that when inserted in the model gives the largest increase in R^2, in the presence of x_1, over the R^2 found in step 1. This, of course, is the variable x_j, for which

$$R(\beta_j|\beta_1) = R(\beta_1, \beta_j) - R(\beta_1)$$

is largest. Let us call this variable x_2. The regression model with x_1 and x_2 is then fitted and R^2 observed.
3. Choose the variable x_j that gives the largest value of

$$R(\beta_j|\beta_1, \beta_2) = R(\beta_1, \beta_2, \beta_j) - R(\beta_1\beta_2),$$

again resulting in the largest increase of R^2 over that given in step 2. Calling this variable x_3, we now have a regression model involving x_1, x_2, and x_3.

This process is continued until the most recent variable inserted fails to induce a significant increase in the explained regression. Such an increase can be determined at each step by using the appropriate F test or t test. For example, in step 2 the value

$$f = \frac{R(\beta_2|\beta_1)}{s^2}$$

can be determined to test the appropriateness of x_2 in the model. Here the value of s^2 is the error mean square for the model containing the variables x_1 and x_2. Likewise, in step 3 the ratio

$$f = \frac{R(\beta_3|\beta_1, \beta_2)}{s^2}$$

tests the appropriateness of x_3 in the model. Now, however, the value for s^2 is the error mean square for the model that contains the three variables x_1, x_2, and x_3. If $f < f_\alpha(1, n - 3)$ at step 2, for a prechosen significant level, x_2 is not included and the process is terminated, resulting in a simple linear equation relating y and x_1. However, if $f > f_\alpha(1, n - 3)$, we proceed to step 3. Again, if $f < f_\alpha(1, n - 4)$ at step 3, x_3 is not included and the process is terminated with the appropriate regression equation containing the variables x_1 and x_2.

Backward elimination involves the same concepts as *forward selection* except that one begins with all the variables in the model. Suppose, for example, that there are five variables under consideration. The steps are as follows:

1. Fit a regression equation with all five variables included in the model. Choose the variable that gives the smallest value of the regression sum of squares *adjusted for the others.* Suppose this variable is x_2. Remove x_2 from the model if

$$f = \frac{R(\beta_2 \mid \beta_1, \beta_3, \beta_4, \beta_5)}{s^2}$$

   is insignificant.
2. Fit a regression equation using the remaining variables x_1, x_3, x_4, and x_5 and repeat step 1. Suppose that variable x_5 is chosen this time. Once again if

$$f = \frac{R(\beta_5 \mid \beta_1, \beta_3, \beta_4)}{s^2}$$

   is insignificant, the variable x_5 is removed from the model. At each step the s^2 used in the F test is the error mean square for the regression model at that stage.

This process is repeated until at some step the variable with the smallest adjusted regression sum of squares does not result in a significant f value for some predetermined significance level.

Stepwise regression is accomplished with a slight but important modification of the forward selection procedure. The modification involves further testing at each stage to ensure the continued effectiveness of variables that had been inserted into the model at an earlier stage. This represents an improvement over forward selection, since it is quite possible that a variable entering the regression equation at an early stage might have been rendered unimportant or redundant because of relationships that exist between it and other variables entering at later stages. Therefore, at a stage in which a new variable has been entered into the regression equation through a significant increase in R^2 as determined by the F test, all the variables already in the model are subjected to

F tests (or, equivalently to t tests) in light of this new variable, and are deleted if they do not display a significant f value. The procedure is continued until a stage is reached in which no additional variables can be inserted or deleted. We illustrate the stepwise procedure in the following example.

Example 9.10   Using the techniques of stepwise regression, fit a linear regression model for predicting the length of infants for the data of Table 9.3.

Solution   1. In considering each variable separately, four individual simple linear regression equations are fitted. The following pertinent regression sum of squares are computed:

$$R(\beta_1) = 288.1468 \qquad R(\beta_2) = 215.3013$$

$$R(\beta_3) = 186.1065 \qquad R(\beta_4) = 100.8594.$$

Variable x_1 very clearly gives the largest regression sum of squares. The error mean square for the equation involving only x_1 is $s^2 = 4.7276$, and since

$$f = \frac{R(\beta_1)}{s^2} = \frac{288.1468}{4.7276} = 60.9500,$$

which exceeds $f_{0.05}(1, 7) = 5.59$, the variable x_1 is entered into the model.
  2. Three regression equations are fitted at this stage, all containing x_1. The important results for the combinations (x_1, x_2), (x_1, x_3), and (x_1, x_4) are:

$$R(\beta_2 | \beta_1) = 23.8703, \qquad R(\beta_3 | \beta_1) = 29.3086, \qquad R(\beta_4 | \beta_1) = 13.8178.$$

Variable x_3 displays the largest regression sum of squares in the presence of x_1. The regression involving x_1 and x_3 gives a new value of $s^2 = 0.6307$, and since

$$f = \frac{R(\beta_3 | \beta_1)}{s^2} = \frac{29.3086}{0.6307} = 46.47,$$

which exceeds $f_{0.05}(1, 6) = 5.99$, the variable x_3 is included along with x_1 in the model. Now we must subject x_1 in the presence of x_3 to a significance test. We find that $R(\beta_1 | \beta_3) = 131.349$, and hence

$$f = \frac{R(\beta_1 | \beta_3)}{s^2} = \frac{131.349}{0.6307} = 208.26,$$

which is highly significant. Therefore, x_1 is retained along with x_3.

3. With x_1 and x_3 already in the model, we now require $R(\beta_2|\beta_1, \beta_3)$ and $R(\beta_4|\beta_1, \beta_3)$ in order to determine which, if any, of the remaining two variables is entered at this stage. From the regression analysis using x_2 along with x_1 and x_3 we find $R(\beta_2|\beta_1, \beta_3) = 0.7948$, and when x_4 is used along with x_1 and x_3, we obtain $R(\beta_4|\beta_1, \beta_3) = 0.1855$. The value of s^2 is 0.5979 for the (x_1, x_2, x_3) combination and 0.7198 for the (x_1, x_2, x_4) combination. Since neither f value is significant at the $\alpha = 0.05$ level, the final regression model includes only the variables x_1 and x_3. The estimating equation is found to be

$$\hat{y} = 20.8753 + 0.3976x_1 + 2.1180x_3$$

and the coefficient of determination for this model is $R^2 = 0.9882$.

While (x_1, x_3) is the combination chosen by stepwise regression, it is not necessarily the combination of two variables that gives the largest value of R^2. In fact, we have already observed that the combination (x_2, x_3) gives an $R^2 = 0.9905$. Of course, the stepwise procedure never actually observed this combination. A rational argument could be made that there is actually a negligible difference in performance between these two estimating equations, at least in terms of *per cent variation explained*. It is interesting to observe, however, that the *backward elimination* procedure gives the combination (x_2, x_3) in the final equation (see Exercise 28).

The main function of each of the procedures outlined in this section is to expose the variables to a systematic methodology designed to ensure the eventual inclusion of the best combinations of the variables. Obviously, there is no assurance that this will happen in all problems, and, of course, it is possible that the multicollinearity is so extensive that one has no alternative but to resort to estimation procedures other than least squares. These estimation procedures are discussed in the next section.

9.9 RIDGE REGRESSION

Sequential methods for arriving at a suitable regression equation cannot be used if one requires that all variables in the experiment contribute to the predicted response $\hat{y}$. In most cases we simply proceed according to the methods outlined in Section 9.3. However, if there is an excessive amount of multi-collinearity among the independent variables, the **A** matrix approaches a near-singular condition, resulting in extremely large values along the diagonal of $\mathbf{A}^{-1}$. In other words, the least squares procedure yields unbiased estimators for the regression coefficients but estimators with variances that can be quite large. Since the correlation among the independent variables is often a natural phenomenon, one cannot always alleviate the difficulty brought about by the multicollinearity simply by changing the experimental plan. One solution to the problem is to abandon the usual least squares procedure and resort to biased estimation techniques. In using a biased estimation procedure, we are essentially willing to allow for a certain amount of bias in our estimates in order to reduce the variances of our estimators.

The biased estimates obtained here for the regression coefficients $\beta_0, \beta_1, \ldots, \beta_k$ in the model

$$y = \beta_0 + \beta_1 x_1 + \beta_2 x_2 + \cdots + \beta_k x_k + \varepsilon,$$

denoted by $b_0^*, b_1^*, \ldots, b_k^*$ and called *ridge regression estimates*, are generated through an intuitively appealing constrained least squares approach in which we minimize the error sum of squares subject to the constraints that $b_j^* = \rho_j$, $j = 0, 1, 2, \ldots, k$, where the ρ_j's are finite positive constants. That is, the estimates will be least squares estimates but we are essentially restricting the magnitudes of the coefficients from becoming too large. Excessively large coefficients are, of course, a direct result of multicollinearity.

Suppose that we write the multiple linear regression model as

$$\mathbf{y} = \mathbf{X}\boldsymbol{\beta} + \boldsymbol{\varepsilon},$$

where $\mathbf{X}$ is defined as in Section 9.3 and $\boldsymbol{\beta}$ is given by

$$\boldsymbol{\beta} = \begin{bmatrix} \beta_0 \\ \beta_1 \\ \vdots \\ \beta_k \end{bmatrix}.$$

Recall from the theory of matrix algebra that since $\mathbf{A}$ is a real k-square symmetric matrix with *eigenvalues* or *characteristic roots* $\lambda_1, \lambda_2, \ldots, \lambda_k$, then there exists a real orthogonal matrix $\mathbf{P}$ such that

$$\mathbf{P}'\mathbf{A}\mathbf{P} = \mathbf{P}\mathbf{A}\mathbf{P}' = \text{diag}(\lambda_1, \lambda_2, \ldots, \lambda_k).$$

The rows of $\mathbf{P}$ are the *normalized eigenvectors* of the $\mathbf{A}$ matrix. Methods for obtaining the matrix $\mathbf{P}$ are demonstrated in most matrix algebra textbooks; however, for our purposes $\mathbf{P}$ is more easily obtained by using one of the computer programs that are readily available for high-speed computers.

Since $\mathbf{P}$ is orthogonal, the multiple linear regression model can be written in *canonical form* as

$$\mathbf{y} = \mathbf{X}^*\boldsymbol{\alpha} + \boldsymbol{\varepsilon},$$

where

$$X^* = XP'$$

and

$$\alpha = \begin{bmatrix} \alpha_0 \\ \alpha_1 \\ \vdots \\ \alpha_k \end{bmatrix} = P\beta.$$

The estimates $\hat{\alpha}_0^*, \hat{\alpha}_1^*, \ldots, \hat{\alpha}_k^*$ of $\alpha_1, \alpha_2, \ldots, \alpha_k$, respectively, are related to the ridge regression estimates by the equation

$$\hat{\alpha}^* = Pb^*.$$

Once the $\hat{\alpha}_j^*$'s are determined, the ridge regression estimates are found by writing

$$b^* = P'\hat{\alpha}^*.$$

Our problem therefore reduces to finding estimates of the $\hat{\alpha}_j^*$'s.

As a result of the constraints applied to the b_j^*'s the relationship $\hat{\alpha}^* = Pb^*$ necessarily restricts the magnitudes of the $\hat{\alpha}_j^*$'s from becoming too large. To minimize the error sum of squares for the canonical model

$$y = X^*\hat{\alpha}^* + e,$$

subject to the $k + 1$ constraints on the b_j^*'s and hence on the $\hat{\alpha}_j^*$'s, requires the use of $k + 1$ *Lagrange multipliers*, which we shall designate as $d_0, d_1, d_2, \ldots, d_k$. The procedure is somewhat involved and for this reason we omit the details. However, once the derivatives with respect to the unknown parameters are equated to zero, we obtain the system of equations

$$(A^* + D)\hat{\alpha}^* = g^*,$$

in which

$$A^* = X^{*\prime}X^*$$

and

$$g^* = X^{*\prime}y.$$

The solution of this system of equations leads to the estimates of the ridge regression coefficients as stated in Definition 9.1.

DEFINITION 9.1 *The generalized ridge regression estimates are given by*

$$\mathbf{b}^* = \mathbf{P}'\hat{\boldsymbol{\alpha}}^*,$$

where

$$\hat{\boldsymbol{\alpha}}^* = (\mathbf{A}^* + \mathbf{D})^{-1}\mathbf{g}^*$$

and

$$\mathbf{D} = \text{diag}\,(d_0, d_1, \ldots, d_k)$$

with $d_j > 0$ for $j = 0, 1, 2, \ldots, k$.

By restricting the magnitudes of our coefficients in the minimization procedure, we have, in effect, added constants to the diagonal elements of $\mathbf{A}^*$ and thereby introduced the bias into our estimates. However, the addition of these constants to the diagonal elements of $\mathbf{A}^*$ causes the matrix to behave as if the variables are orthogonal to each other. Consequently, by requiring the d_j's to be positive in Definition 9.1, the elements on the diagonal of $(\mathbf{A}^* + \mathbf{D})^{-1}$ will be smaller, indicating greater stability in the estimates of the coefficients.

Definition 9.1 outlines the method for finding generalized ridge regression estimates but gives no insight into the specific numerical values of the d_j's that one should add to the main diagonal elements of the $\mathbf{A}^*$ matrix. It would seem reasonable that the optimal values assigned to the d_j's should yield $\hat{\alpha}_j^*$'s, and hence b_j^*'s, that minimize the quantity

$$\sum_{j=0}^{k} E(B_j^* - \beta_j)^2.$$

This is accomplished when the d_j's are given by

$$d_j = \frac{\sigma^2}{\alpha_j^2}, \qquad j = 0, 1, 2, \ldots, k.$$

Unfortunately, σ^2 and the α_j's are unknown and estimates must be used. In practice one estimates σ^2 by s^2 using ordinary least squares procedures. The estimates of the α_j's are of course given in Definition 9.1 once the d_j's are known.

To circumvent this seemingly hopeless situation, one must resort to the following iterative procedure.

Generalized Ridge Regression

1. Using ordinary least squares procedures on the canonical model, estimate the α_j's by computing

$$\hat{\alpha} = A^{*-1}g^{*}$$

and estimate σ^2 by s^2.

2. Use the value of s^2 and the $\hat{\alpha}_j$'s from step 1 to compute

$$\hat{d}_j = \frac{s^2}{\hat{\alpha}_j^2}, \quad j = 0, 1, 2, \ldots, k.$$

3. Use the $\hat{d}_j$'s to solve the expression

$$\alpha^{*} = (A^{*} + D)^{-1}g^{*}$$

and thus obtain initial estimates of the $\hat{\alpha}_j^{*}$'s. Next compute

$$\hat{\alpha}^{*\prime}\hat{\alpha}^{*} = \sum_{j=0}^{k} \hat{\alpha}_j^{*2}.$$

4. Repeat steps 2 and 3 using the $\hat{\alpha}_j^{*}$'s from step 3 and again compute $\hat{\alpha}^{*\prime}\hat{\alpha}^{*}$.
5. Continue this iterative procedure and terminate only when stability is achieved in $\hat{\alpha}^{*\prime}\hat{\alpha}^{*}$.
6. The generalized ridge regression coefficients are now computed from the formula $b^{*} = P'\hat{\alpha}^{*}$.

Example 9.11   Compute the generalized regression estimates for the parameters of the multiple linear regression model

$$\mu_{Y|x_1, x_2, x_3, x_4} = \beta_0 + \beta_1 x_1 + \beta_2 x_2 + \beta_3 x_3 + \beta_4 x_4$$

for the data of Table 9.3.

Solution   Suppose that we code the data by letting

$$x_j' = \frac{x_j - \bar{x}_j}{\sqrt{n-1}\, s_j}, \quad j = 1, 2, 3, 4,$$

where

$$s_j^2 = \frac{\sum_{i=1}^{n} (x_{ji} - \bar{x}_j)^2}{n - 1},$$

and find the generalized ridge regression estimates for the equation

$$\mu_{Y|x_1, x_2, x_3, x_4} = \beta_0' + \beta_1' x_1' + \beta_2' x_2' + \beta_3' x_3' + \beta_4' x_4'.$$

If this form we find that the first column of X (column of ones) is orthogonal to the other columns. Consequently, the ridge regression estimate b_0^* of β_0' is simply $\bar{y} = 60.9667$, the average of the y values, while b_1^*, b_2^*, b_3^*, and b_4^* can be found from the iterative procedure described above, but with the first column of X deleted. Now the resulting 4×4 A matrix is just the correlation matrix given in Example 9.10.

To find the x' values of the X matrix and the corresponding correlation matrix A for the data of Table 9.3, we must first compute

$$\bar{x}_1 = 81.0000 \qquad s_1 = 11.5866$$
$$\bar{x}_2 = 48.7778 \qquad s_2 = 4.3979$$
$$\bar{x}_3 = 3.1311 \qquad s_3 = 1.1178$$
$$\bar{x}_4 = 29.6333 \qquad s_4 = 3.1149.$$

Then the P matrix, whose rows are the normalized eigenvectors of the correlation matrix, is given by

$$P = \begin{bmatrix} 0.5729 & 0.4776 & 0.7987 & 0.4416 \\ -0.3586 & -0.5778 & 0.4592 & 0.5716 \\ 0.0381 & -0.1453 & 0.7071 & -0.6910 \\ -0.7361 & 0.6457 & 0.2011 & 0.0294 \end{bmatrix}.$$

Next one can determine $X^* = X'P$ and $A^* = X^{*'}X^*$, and then proceed to step 1 of the iterative procedure. Solving $\hat{\alpha} = A^{*-1}g^*$ gives

$$\hat{\alpha}_1 = 10.9281$$
$$\hat{\alpha}_2 = -2.0833$$
$$\hat{\alpha}_3 = 5.8716$$
$$\hat{\alpha}_4 = 5.3672.$$

In Section 9.8 we found the error mean square for the complete model to be $s^2 = 0.7414$. Therefore, from step 2,

$$\hat{d}_1 = 0.0062$$

$$\hat{d}_2 = 0.1708$$

$$\hat{d}_3 = 0.0215$$

$$\hat{d}_4 = 0.0257.$$

Steps 3, 4, and 5 are now carried out and the results displayed in Table 9.5.

As a result of the convergence in $\hat{a}^{*\prime}\hat{a}^*$, the process was terminated after the third iteration. From step 6 the generalized ridge regression coefficients are found to be

$$\hat{b}_1^* = 7.0687$$

$$\hat{b}_2^* = 5.4613$$

$$\hat{b}_3^* = 8.3069$$

$$\hat{b}_4^* = 0.2284,$$

and the estimated regression equation, in terms of the coded variables, is given by

$$\hat{y} = 60.9667 + 7.0687x_1' + 5.4613x_2' + 8.3069x_3' + 0.2284x_4',$$

or, converting back to the original variables, assumes the form

$$\hat{y} = 13.0851 + 0.2157x_1 + 0.4390x_2 + 2.6274x_3 + 0.0259x_4.$$

The generalized ridge regression described here is one of several formal procedures for biased estimation, the purpose of which is to reduce the variance of the estimators of the regression coefficients, even though the resulting estimators are biased. Frequently, one obtains ridge regression estimates by minimizing the error sum of squares for the model

$$\mathbf{y} = \mathbf{X}\mathbf{b}^* + \mathbf{e}$$

subject to the single constraint that $\sum_{j=0}^{k} b_j^{*2} = \rho$, where ρ is again a finite positive constant. The method of Lagrange multipliers requires the differentiation of

$$L = \sum_{i=1}^{n}(y_i - b_0^* - b_1^*x_{1i} - b_2^*x_{2i} - \cdots - b_k^*x_{ki})^2 + d\left(\sum_{j=0}^{k} b_j^{*2} - \rho\right)$$

Table 9.5 Iterations in Generalized Ridge Regression

Iteration	$\hat{d}_1$	$\hat{d}_2$	$\hat{d}_3$	$\hat{d}_4$	$\hat{\alpha}_1^*$	$\hat{\alpha}_2^*$	$\hat{\alpha}_3^*$	$\hat{\alpha}_4^*$	$\hat{\alpha}^{*\prime}\hat{\alpha}^*$
1	0.0062	0.1708	0.0215	0.0257	10.9017	−1.8291	5.3221	0.6011	150.8787
2	0.0062	0.2216	0.0262	2.0521	10.9015	−1.7650	5.2161	0.0085	149.1657
3	0.0062	0.2380	0.0272	10.320	10.9015	−1.7453	5.1923	0.0000	148.8488

with respect to $b_0, b_1, \ldots, b_k$. When these derivatives are equated to zero, we obtain the system of equations

$$(\mathbf{A} + d\mathbf{I})\mathbf{b}^* = \mathbf{g},$$

which can be solved for the ridge regression estimates of the coefficients in terms of the Lagrange multiplier d.

DEFINITION 9.2   Ridge regression estimates *are given by*

$$\mathbf{b}^* = (\mathbf{A} + d\mathbf{I})^{-1}\mathbf{g},$$

where $d > 0$.

In Definition 9.2 the ridge regression estimates are computed for various increasing values of d, beginning with $d = 0$, until we determine a value of d for which all the regression coefficients appear to have stabilized. Several calculations may be required before the estimates of the coefficients reach stability. By plotting the values of the coefficients against the successive values of d, we obtain a curve referred to as the *ridge trace*.

The purpose of the ridge trace is to indicate, for a given set of data, a set of estimates that are reasonable. Sometimes, unfortunately, it is difficult to determine an appropriate value of d for which the estimates of all the coefficients have stabilized. The method for obtaining the ridge regression estimates as stated in Definition 9.2 is illustrated in Example 9.12.

Table 9.6 Ridge Regression Estimates

d	b_1^*	b_2^*	b_3^*	b_4^*
0.000	3.2803	9.0361	9.7245	−0.2647
0.004	5.4449	7.1270	9.0451	−0.2790
0.012	6.3413	6.3189	8.6327	−0.1789
0.020	6.5993	6.0695	8.4053	−0.0634
0.100	6.7426	5.6694	7.1994	0.8067
0.200	6.5000	5.4970	6.3675	1.3844
0.300	6.2533	5.3273	5.8221	1.7013
0.400	6.0225	5.1583	5.4220	1.8870

Example 9.12 Use Definition 9.2 to estimate the ridge regression coefficients of the parameters β_1', β_2', β_3', and β_4' in Example 9.11 for various values of d and then plot the ridge trace.

Solution The computations were carried out on a computer and are summarized in Table 9.6.

The ridge trace is illustrated in Figure 9.1. It was decided that all the coefficients had sufficiently stabilized at $d = 0.1$. Therefore, the ridge regression estimates of the coefficients of the coded variables are given by

$$b_1^* = 6.7426$$
$$b_2^* = 5.6694$$
$$b_3^* = 7.1994$$
$$b_4^* = 0.8067.$$

As one would expect, the estimates are not the same as those computed by the generalization ridge regression procedure. Observe from the ridge trace how relatively unstable all the estimates seem to be in the vicinity of $d = 0$, where the b_j^*'s are the unbiased least squares estimates.

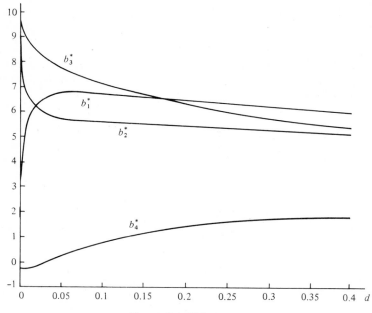

Figure 9.1 Ridge trace.

EXERCISES

1. Given the data

y	2	5	7	8	5
x_1	8	8	6	5	3
x_2	0	1	1	3	4

   estimate the multiple linear regression equation $\mu_{Y|x_1,x_2} = \beta_0 + \beta_1 x_1 + \beta_2 x_2$.

2. The following data represent a set of 10 experimental runs in which two independent variables x_1 and x_2 are controlled and values of a response, y, are observed:

y	x_1	x_2
61.5	2400	54.5
61.2	2450	56.4
32.0	2500	43.2
52.5	2700	65.2
31.5	2750	45.5
22.5	2800	47.5
53.0	2900	65.0
56.8	3000	66.5
34.8	3100	57.3
52.7	3200	68.0

   Estimate the multiple linear regression equation $\mu_{Y|x_1,x_2} = \beta_0 + \beta_1 x_1 + \beta_2 x_2$.

3. A set of experimental runs was made to determine a way of predicting coking time y at various levels of oven width x_1 and flue temperature x_2. The coded data were recorded as follows:

y	x_1	x_2
6.40	1.32	1.15
15.05	2.69	3.40
18.75	3.56	4.10
30.25	4.41	8.75
44.85	5.35	14.82
48.94	6.20	15.15
51.55	7.12	15.32
61.50	8.87	18.18
100.44	9.80	35.19
111.42	10.65	40.40

   Estimate the multiple regression equation $\mu_{Y|x_1,x_2} = \beta_0 + \beta_1 x_1 + \beta_2 x_2$.

4. Given the data

x	0	1	2	3	4	5	6	7
y	4.6	4.2	6.5	8.7	9.0	7.3	5.5	3.2

(a) Fit a regression curve of the form $\mu_{Y|x} = \beta_0 + \beta_1 x + \beta_2 x^2$.
(b) Estimate Y when $x = 5$.

5. An experiment was conducted in order to determine if cerebral blood flow in human beings can be predicted from arterial oxygen tension (millimeters of mercury). Fifteen patients were used in the study and the following data were observed:

Blood flow, y	Arterial oxygen tension, x
84.33	603.40
87.80	582.50
82.20	556.20
78.21	594.60
78.44	558.90
80.01	575.20
83.53	580.10
79.46	451.20
75.22	404.00
76.58	484.00
77.90	452.40
78.80	448.40
80.67	334.80
86.60	320.30
78.20	350.30

Estimate the quadratic regression equation

$$\mu_{Y|x} = \beta_0 + \beta_1 x + \beta_2 x^2.$$

6. The following is a set of coded experimental data on the compressive strength of a particular alloy at various values of the concentration of some additive:

Concentration, x	Compressive strength, y		
10.0	25.2	27.3	28.7
15.0	29.8	31.1	27.8
20.0	31.2	32.6	29.7
25.0	31.7	30.1	32.3
30.0	29.4	30.8	32.8

(a) Estimate the quadratic regression equation $\mu_{Y|x} = \beta_0 + \beta_1 x + \beta_2 x^2$.

(b) Test for lack of fit of the model.

7. Given the data

x	0	1	2	3	4	5	6
y	1	4	5	3	2	3	4

(a) Fit the cubic model $\mu_{Y|x} = \beta_0 + \beta_1 x + \beta_2 x^2 + \beta_3 x^3$.

(b) Predict Y when $x = 2$.

8. The following data resulted from 15 experimental runs made on four independent variables and a single response y.

y	x_1	x_2	x_3	x_4
14.8	7.8	4.3	11.5	6.3
12.1	6.9	3.9	14.3	7.4
19.0	9.3	8.4	9.4	5.9
14.5	6.8	10.3	15.2	8.7
16.6	11.7	6.4	8.8	9.1
17.2	8.5	5.7	9.8	5.6
17.5	12.6	6.8	11.2	6.8
14.1	7.5	4.2	10.9	7.4
13.8	8.4	7.3	14.7	8.2
14.7	11.3	8.8	15.1	9.2
17.7	10.7	3.6	8.7	4.7
17.0	7.3	4.9	8.6	5.5
17.6	8.4	7.3	9.3	6.6
16.3	6.7	9.7	10.8	8.7
18.2	9.6	8.4	11.9	5.4

Estimate the multiple linear regression model relating y to x_1, x_2, x_3, and x_4.

9. Consider the following two-way table of experimental data:

x_1	x_2 0.5	1.0	1.5
1.0	18.0	20.9	19.8
	18.5	21.3	18.9
2.0	17.5	18.8	18.2
	16.8	18.5	17.9

Estimate the regression coefficients in the model

$$\mu_{Y|x_1,x_2} = \beta_0 + \beta_1 x_1 + \beta_2 x_2 + \beta_{22} x_2^2 + \beta_{12} x_1 x_2.$$

10. The personnel department of a certain industrial firm used 12 subjects in a study to determine the relationship between job performance rating (y) and scores of four tests. The data are as follows:

y	x_1	x_2	x_3	x_4
11.2	56.5	71.0	38.5	43.0
14.5	59.5	72.5	38.2	44.8
17.2	69.2	76.0	42.5	49.0
17.8	74.5	79.5	43.4	56.3
19.3	81.2	84.0	47.5	60.2
24.5	88.0	86.2	47.4	62.0
21.2	78.2	80.5	44.5	58.1
16.9	69.0	72.0	41.8	48.1
14.8	58.1	68.0	42.1	46.0
20.0	80.5	85.0	48.1	60.3
13.2	58.3	71.0	37.5	47.1
22.5	84.0	87.2	51.0	65.2

Estimate the regression coefficients in the model

$$\mu_{Y|x_1,x_2,x_3,x_4} = \beta_0 + \beta_1 x_1 + \beta_2 x_2 + \beta_3 x_3 + \beta_4 x_4.$$

11. For the data of Exercise 2, estimate σ^2.

12. For the data of Exercise 8, estimate σ^2.

13. Obtain estimates of the variances and the covariance of the estimators B_1 and B_2 of Exercise 2.

14. Referring to Exercise 8, find estimates of
    (a) $\sigma_{B_3}^2$.   (b) $\mathrm{cov}\,(B_1, B_4)$.

15. Using the data of Exercise 2 and the estimate of σ from Exercise 11, construct 95% confidence intervals for the predicted response and the mean response when $x_1 = 2500$ and $x_2 = 48.0$.

16. For the model of Exercise 5, test the hypothesis that $\beta_2 = 0$ at the 0.05 level of significance against the alternative that $\beta_2 \neq 0$.

17. For Exercise 6, using the quadratic model, construct a 90% confidence interval for the mean compressive strength when the concentration is $x = 19.5$.

18. Using the data of Exercise 8 and the estimate of σ^2 from Exercise 12, construct 95% confidence intervals for the predicted response and the mean response when $x_1 = 8.2$, $x_2 = 6.0$, $x_3 = 10.3$, and $x_4 = 5.8$.

19. For the model of Exercise 2, use the T statistic to test the hypothesis that $\beta_1 = 0$ at the 0.05 level of significance against the alternative that $\beta_1 \neq 0$.

20. For the model of Exercise 3, test the hypothesis that $\beta_1 = 2$ at the 0.05 level of significance against the alternative that $\beta_1 \neq 2$.

21. Compute and interpret the coefficient of multiple determination for the variables of Exercise 3.

22. Test whether the regression explained by the model of Exercise 3 is significant at the 0.01 level of significance.

23. Test whether the regression explained by the model of Exercise 8 is significant at the 0.05 level of significance.

24. For the model of Exercise 8, test the hypothesis

$$H_0: \quad \beta_1 = \beta_2 = 0$$

$$H_1: \quad \beta_1 \text{ and } \beta_2 \text{ are not both zero.}$$

25. Repeat Exercise 19 using an F statistic.

26. A small experiment is conducted to fit a multiple regression equation relating the yield y to temperature x_1, reaction time x_2, and concentration of one of the reactants x_3. Two levels of each variable were chosen and measurements corresponding to the coded independent variables were recorded as follows:

y	x_1	x_2	x_3
7.6	-1	-1	-1
8.4	1	-1	-1
9.2	-1	1	-1
10.3	-1	-1	1
9.8	1	1	-1
11.1	1	-1	1
10.2	-1	1	1
12.6	1	1	1

(a) Using the coded variables, estimate the multiple linear regression equation

$$\mu_{Y|x_1, x_2, x_3} = \beta_0 + \beta_1 x_1 + \beta_2 x_2 + \beta_3 x_3.$$

(b) Partition SSR, the regression sum of squares, into three single-degree-of-freedom components attributable to x_1, x_2, and x_3, respectively. Show an analysis of-variance table, indicating significance tests on each variable.

27. For the data of Exercise 10,
(a) Use the techniques of *forward selection* with a 0.05 level of significance to fit a linear regression equation.
(b) Use the techniques of *backward elimination* with $\alpha = 0.05$ to find a suitable prediction equation.
(c) Use the techniques of *stepwise regression* to fit a linear regression model. Let $\alpha = 0.05$.

28. Use the techniques of *backward elimination* with $\alpha = 0.05$ to find a suitable prediction equation for the data of Table 9.3.

29. Compute the generalized ridge regression estimates for the parameters of the multiple linear regression equation

$$\mu_{Y|x_1, x_2, x_3, x_4} = \beta_0 + \beta_1 x_1 + \beta_2 x_2 + \beta_3 x_3 + \beta_4 x_4$$

for the data of Exercise 10.

30. Use Definition 9.2 to compute ridge regression estimates for the model in Exercise 29. Plot the ridge trace.

10
Analysis of Variance

10.1 INTRODUCTION

In the estimation and hypotheses testing material covered in Chapters 6 and 7, we were restricted in each case to considering no more than two parameters. Such was the case, for example, in testing for the equality of two population means using independent samples from normal populations with common but unknown variance, where it was necessary to obtain a pooled estimate of σ^2. It would seem obvious and desirable that the reader be able to extend the techniques developed so far to cover tests of hypotheses in which there are, say, k population means.

Analysis of variance is certainly not a new technique to the reader if he has followed the material on regression theory. We used the analysis-of-variance approach to partition the total sum of squares into a portion due to regression and a portion due to error. Further, we were able, in some cases, to conveniently partition SSR into meaningful components for the purpose of testing relevant hypotheses on the parameters in the model. The term *analysis of variance* describes a technique whereby the total variation is being analyzed or divided into meaningful components.

Regression problems in which the model contains *quantitative variables* (like those we have discussed to this point) are not the only type in which the analysis of variance plays an important role. In this chapter we shall present and study other types of models in which this technique is used. The degree of difficulty of the analysis depends on the complexity of the problem.

Suppose in an industrial experiment that an engineer is interested in how the mean absorption of moisture in concrete varies among five different concrete aggregates. The samples are exposed to moisture for 48 hours. It is decided that 6 samples are to be tested for each aggregate, requiring a total of 30 samples to be tested. The data are recorded in Table 10.1.

The model for this situation may be considered as follows. There are six observations taken from each of five populations with means $\mu_1, \mu_2, \ldots, \mu_5$, respectively. We might wish to test

$$H_0: \quad \mu_1 = \mu_2 = \cdots = \mu_5,$$

$$H_1: \quad \text{at least two of the means are not equal.}$$

Table 10.1 Absorption of Moisture in Concrete Aggregates

	Aggregate (weight %)					
	1	*2*	*3*	*4*	*5*	
	551	595	639	417	563	
	457	580	615	449	631	
	450	508	511	517	522	
	731	583	573	438	613	
	499	633	648	415	656	
	632	517	677	555	679	
Total	3,320	3,416	3,663	2,791	3,664	16,854
Mean	553.33	569.33	610.50	465.17	610.67	561.80

In addition, we might be interested in making individual comparisons among these five population means.

In the analysis-of-variance procedure, it is assumed that whatever variation exists between the aggregate averages is attributed to (1) variation in absorption among observations within aggregate types and (2) variation due to aggregate types, that is, due to differences in the chemical composition of the aggregates. The *within-aggregate variation* is, of course, brought about by various causes. Perhaps humidity and temperature conditions were not kept entirely constant throughout the experiment. It is possible that there was a certain amount of heterogeneity in the batches of raw materials that were used. At any rate, we shall call the within-sample variation *chance* or *random variation*, and part of the goal of the analysis of variance is to determine if the differences between the five sample means are what one would expect due to random variation alone or if indeed there is also a contribution from the systematic variation attributed to the aggregate types. The procedure essentially, then, separates the total variability into the following important two components:

1. Variability between aggregates, measuring systematic and random variation.
2. Variability within aggregates, measuring only random variation.

There remains then the task of determining if component 1 is significantly larger than component 2.

Many pointed questions appear at this stage concerning the preceding problem. For example, how many samples must be tested in each aggregate?

This is a question that continually haunts the practitioner. In addition, what if the within-sample variation is so large that it is difficult for a statistical procedure to detect the systematic differences? Can we systematically control extraneous sources of variation and thus remove them from the portion we call random variation? We shall attempt to answer these and other questions in the following sections.

10.2 ONE-WAY CLASSIFICATION

Random samples of size n are selected from each of k populations. The k different populations are often classified according to different treatments or groups. Today the term *treatment* is used very generally to refer to the various classifications, whether they be aggregates, analysts, different fertilizers, or different regions of the country. It will be assumed that the k populations are independent and normally distributed with means $\mu_1, \mu_2, \ldots, \mu_k$ and common variance σ^2. We wish to derive appropriate methods for testing the hypothesis

$$H_0: \quad \mu_1 = \mu_2 = \cdots = \mu_k,$$
$$H_1: \quad \text{at least two of the means are not equal.}$$

Let y_{ij} denote the jth observation from the ith treatment and arrange the data as in Table 10.2. Here, $T_i.$ is the total of all observations in the sample from the ith treatment, $\bar{y}_i.$ is the mean of all observations in the sample from the ith treatment, $T..$ is the total of all nk observations, and $\bar{y}..$ is the mean of all nk

Table 10.2 k Random Samples

	Treatment					
	1	*2*	$\cdots$	i	$\cdots$	k
	y_{11}	y_{21}	$\cdots$	y_{i1}	$\cdots$	y_{k1}
	y_{12}	y_{22}	$\cdots$	y_{i2}	$\cdots$	y_{k2}
	$\vdots$	$\vdots$		$\vdots$		$\vdots$
	y_{1n}	y_{2n}	$\cdots$	y_{in}	$\cdots$	y_{kn}
Total	$T_1.$	$T_2.$	$\cdots$	$T_i.$	$\cdots$	$T_k.$ $T..$
Mean	$\bar{y}_1.$	$\bar{y}_2.$	$\cdots$	$\bar{y}_i.$	$\cdots$	$\bar{y}_k.$ $\bar{y}..$

observations. Each observation may be written in the form

$$y_{ij} = \mu_i + \varepsilon_{ij},$$

where ε_{ij} measures the deviation of the jth observation of the ith sample from the corresponding treatment mean. The ε_{ij} term represents random error and plays the same role as the error terms in the regression models. An alternative and preferred form of this equation is obtained by substituting $\mu_i = \mu + \alpha_i$, where μ is defined to be the mean of all the μ_i's; that is,

$$\mu = \frac{\sum_{i=1}^{k} \mu_i}{k}.$$

Hence we may write

$$y_{ij} = \mu + \alpha_i + \varepsilon_{ij},$$

subject to the constraint that $\sum_{i=1}^{k} \alpha_i = 0$. It is customary to refer to α_i as the *effect* of the ith treatment.

The null hypothesis that the k population means are equal against the alternative that at least two of the means are unequal may now be replaced by the equivalent hypothesis,

$$H_0: \quad \alpha_1 = \alpha_2 = \cdots = \alpha_k = 0,$$

$$H_1: \quad \text{at least one of the } \alpha_i\text{'s is not equal to zero.}$$

Our test will be based on a comparison of two independent estimates of the common population variance σ^2. These estimates will be obtained by splitting the total variability of our data into two components.

THEOREM 10.1 *Sum of Squares Identity*

$$\sum_{i=1}^{k} \sum_{j=1}^{n} (y_{ij} - \bar{y}_{..})^2 = n \sum_{i=1}^{k} (\bar{y}_{i.} - \bar{y}_{..})^2 + \sum_{i=1}^{k} \sum_{j=1}^{n} (y_{ij} - \bar{y}_{i.})^2.$$

Proof

$$\sum_{i=1}^{k} \sum_{j=1}^{n} (y_{ij} - \bar{y}_{..})^2 = \sum_{i=1}^{k} \sum_{j=1}^{n} [(\bar{y}_{i.} - \bar{y}_{..}) + (y_{ij} - \bar{y}_{i.})]^2$$

$$= \sum_{i=1}^{k} \sum_{j=1}^{n} [(\bar{y}_{i.} - \bar{y}_{..})^2 + 2(\bar{y}_{i.} - \bar{y}_{..})(y_{ij} - \bar{y}_{i.})$$
$$+ (y_{ij} - \bar{y}_{i.})^2]$$

$$= \sum_{i=1}^{k} \sum_{j=1}^{n} (\bar{y}_{i.} - \bar{y}_{..})^2$$

$$+ 2 \sum_{i=1}^{k} \sum_{j=1}^{n} (\bar{y}_{i.} - \bar{y}_{..})(y_{ij} - \bar{y}_{i.})$$

$$+ \sum_{i=1}^{k} \sum_{j=1}^{n} (y_{ij} - \bar{y}_{i.})^2.$$

The middle term is zero, since

$$\sum_{j=1}^{n} (y_{ij} - \bar{y}_{i.}) = \sum_{j=1}^{n} y_{ij} - n\bar{y}_{i.} = \sum_{j=1}^{n} y_{ij} - n\frac{\sum_{j=1}^{n} y_{ij}}{n} = 0.$$

The first sum does not have j as a subscript and therefore may be written as

$$\sum_{i=1}^{k} \sum_{j=1}^{n} (\bar{y}_{i.} - \bar{y}_{..})^2 = n \sum_{i=1}^{k} (\bar{y}_{i.} - \bar{y}_{..})^2.$$

Hence

$$\sum_{i=1}^{k} \sum_{j=1}^{n} (y_{ij} - \bar{y}_{..})^2 = n \sum_{i=1}^{k} (\bar{y}_{i.} - \bar{y}_{..})^2 + \sum_{i=1}^{k} \sum_{j=1}^{n} (y_{ij} - \bar{y}_{i.})^2.$$

It will be convenient in what follows to identify the terms of the sum of squares identity by the following notation:

$$SST = \sum_{i=1}^{k} \sum_{j=1}^{n} (y_{ij} - \bar{y}_{..})^2 = \text{total sum of squares}$$

$$SSA = n \sum_{i=1}^{k} (\bar{y}_{i.} - \bar{y}_{..})^2 = \text{treatment sum of squares}$$

$$SSE = \sum_{i=1}^{k} \sum_{j=1}^{n} (y_{ij} - \bar{y}_{i.})^2 = \text{error sum of squares.}$$

The sum of squares identity can then be represented symbolically by

$$SST = SSA + SSE.$$

As implied earlier, we need to compare the appropriate measure of the between-treatment variation with the within-treatment variation in order to detect significant differences in the observations due to the treatment effects. Suppose that we look at the expected value of the treatment sum of squares.

THEOREM 10.2

$$E(SSA) = (k - 1)\sigma^2 + n\sum_{i=1}^{k}\alpha_i^2.$$

Proof Looking upon SSA as a random variable whose values would undoubtedly vary if the experiment were repeated several times, we now write

$$SSA = n\sum_{i=1}^{k}(\bar{Y}_{i.} - \bar{Y}_{..})^2.$$

From the model

$$Y_{ij} = \mu + \alpha_i + E_{ij}$$

we obtain

$$\bar{Y}_{i.} = \mu + \alpha_i + \bar{E}_{i.}.$$

$$\bar{Y}_{..} = \mu + \bar{E}_{..},$$

since $\sum_{i=1}^{k}\alpha_i = 0$. Hence

$$SSA = n\sum_{i=1}^{k}(\alpha_i + \bar{E}_{i.} - \bar{E}_{..})^2$$

and

$$E(SSA) = n\sum_{i=1}^{k}\alpha_i^2 + n\sum_{i=1}^{k}E(\bar{E}_{i.}^2) - nkE(\bar{E}_{..}^2) + 2n\sum_{i=1}^{k}\alpha_i E(\bar{E}_{i.}).$$

Recalling that the E_{ij}'s are independent variables with mean zero and variance σ^2, we find that

$$E(\bar{E}_{i.}^2) = \frac{\sigma^2}{n}, \qquad E(\bar{E}_{..}^2) = \frac{\sigma^2}{nk}, \qquad E(\bar{E}_{i.}) = 0.$$

Therefore,

$$E(SSA) = n \sum_{i=1}^{k} \alpha_i^2 + k\sigma^2 - \sigma^2$$

$$= (k-1)\sigma^2 + n \sum_{i=1}^{k} \alpha_i^2.$$

One estimate of σ^2, based on $k-1$ degrees of freedom, is given by the treatment mean square

$$s_1^2 = \frac{SSA}{k-1}.$$

If H_0 is true and each α_i in Theorem 10.2 is set equal to zero, we see that

$$E\left(\frac{SSA}{k-1}\right) = \sigma^2$$

and s_1^2 is an unbiased estimate of σ^2. However, if H_1 is true, we have

$$E\left(\frac{SSA}{k-1}\right) = \sigma^2 + \frac{n \sum_{i=1}^{k} \alpha_i^2}{k-1}$$

and s_1^2 estimates σ^2 plus an additional term, which measures variation due to the systematic effects.

A second and independent estimate of σ^2, based on $k(n-1)$ degrees of freedom, is the familiar formula

$$s^2 = \frac{SSE}{k(n-1)}.$$

The estimate s^2 is unbiased regardless of the truth or falsity of the null hypothesis (see Exercise 1). It is important to note that the sum of squares identity has not only partitioned the total variability of the data, but also the total number of degrees of freedom. That is,

$$nk - 1 = k - 1 + k(n - 1).$$

When H_0 is true, the ratio

$$f = \frac{s_1^2}{s^2}$$

is a value of the random variable F having the F distribution with $k - 1$ and $k(n - 1)$ degrees of freedom. Since s_1^2 overestimates σ^2 when H_0 is false, we have a one-tailed test with the critical region entirely in the right tail of the distribution. The null hypothesis H_0 is rejected at the α level of significance when

$$f > f_\alpha[k - 1, k(n - 1)].$$

In practice one usually computes SST and SSA first and then, making use of the sum of squares identity, obtains

$$SSE = SST - SSA.$$

The previously defined formulas for SST and SSA are not in the best computational form. Equivalent and preferred formulas are given by

$$SST = \sum_{i=1}^{k} \sum_{j=1}^{n} y_{ij}^2 - \frac{T_{..}^2}{nk}$$

and

$$SSA = \frac{\sum_{i=1}^{k} T_{i.}^2}{n} - \frac{T_{..}^2}{nk}.$$

The computations in an analysis-of-variance problem are usually summarized in tabular form as shown in Table 10.3.

Example 10.1 Test the hypothesis $\mu_1 = \mu_2 = \cdots = \mu_5$ at the 0.05 level of significance for the data of Table 10.1 on absorption of moisture by various types of cement aggregates.

Table 10.3 Analysis of Variance for the One-Way Classification

Source of variation	Sum of squares	Degrees of freedom	Mean square	Computed f
Treatments	SSA	$k - 1$	$s_1^2 = \dfrac{SSA}{k - 1}$	$\dfrac{s_1^2}{s^2}$
Error	SSE	$k(n - 1)$	$s^2 = \dfrac{SSE}{k(n - 1)}$	
Total	SST	$nk - 1$		

Solution

1. H_0: $\mu_1 = \mu_2 = \cdots = \mu_5$.
2. H_1: at least two of the means are not equal.
3. $\alpha = 0.05$.
4. Critical region: $F > 2.76$ with $v_1 = 4$ and $v_2 = 25$ degrees of freedom.
5. Computations:

$$SST = 551^2 + 457^2 + \cdots + 679^2 - \frac{16,854^2}{30}$$

$$= 9,677,954 - 9,468,577$$

$$= 209,377$$

$$SSA = \frac{3320^2 + 3416^2 + \cdots + 3664^2}{6} - 9,468,577$$

$$= 85,356$$

$$SSE = 209,377 - 85,356 = 124,021.$$

These results and the remining computations are exhibited in Table 10.4.

Table 10.4 Analysis of Variance for the Data of Table 10.1

Source of variation	Sum of squares	Degrees of freedom	Mean square	Computed f
Aggregates	85,356	4	21,339	4.30
Error	124,021	25	4,961	
Total	209,377	29		

6. Conclusion: Reject H_0 and conclude that the aggregates do not have the same mean absorption.

In experimental work one often loses some of the desired observations. For example, an experiment might be conducted to determine if college students obtain different grades on the average for classes meeting at different times of the day. Because of dropouts during the semester it is entirely possible to conclude the experiment with unequal numbers of students in the various sections. The previous analysis for equal sample size will still be valid by slightly modifying the sum of squares formulas. We now assume the k random samples to be of size $n_1, n_2, \ldots, n_k$, respectively, with $N = \sum\limits_{i=1}^{k} n_i$. The computational formulas for SST and SSA are given by

$$SST = \sum_{i=1}^{k} \sum_{j=1}^{n_i} y_{ij}^2 - \frac{T_{..}^2}{N}$$

and

$$SSA = \sum_{i=1}^{k} \frac{T_{i.}^2}{n_i} - \frac{T_{..}^2}{N}.$$

As before, we find SSE by subtraction. The degrees of freedom are partitioned in the same way: $N - 1$ for SST, $k - 1$ for SSA, and $N - 1 - (k - 1) = N - k$ for SSE.

In concluding our discussion on the analysis of variance for the one-way classification, we state the advantages of choosing equal sample sizes over the choice of unequal sample sizes. The first advantage is that the f ratio is insensitive to slight departures from the assumption of equal variances for the k populations when the samples are of equal size. Second, the choice of equal sample size minimizes the probability of committing a type II error. Finally, the computation of SSA is simplified if the sample sizes are equal.

10.3 TESTS FOR THE EQUALITY OF SEVERAL VARIANCES

We have already stated in Section 10.2 that the f ratio obtained from the analysis-of-variance procedure is insensitive to slight departures from the assumption of equal variances for the k populations when the samples are of equal size. This is not the case, however, if the sample sizes are unequal or if one variance is much larger than the others. Consequently, one may wish to test the hypothesis

$$H_0: \quad \sigma_1^2 = \sigma_2^2 = \cdots = \sigma_k^2$$

against the alternative

$$H_1: \quad \text{the variances are not all equal.}$$

The test most often used, called *Bartlett's test*, is based on a statistic whose sampling distribution is approximated very closely by the chi-square distribution when the k random samples are drawn from independent normal populations.

First, we compute the k sample variances $s_1^2, s_2^2, \ldots, s_k^2$ from samples of size $n_1, n_2, \ldots, n_k$, with $\sum_{i=1}^{k} n_i = N$. Second, combine the sample variances to give the pooled estimate

$$s_p^2 = \frac{\sum_{i=1}^{k} (n_i - 1) s_i^2}{N - k},$$

which is equivalent to the formula $SSE/(N - k)$ of Section 10.2. Now

$$b = 2.3026 \, \frac{q}{h},$$

where

$$q = (N - k) \log s_p^2 - \sum_{i=1}^{k} (n_i - 1) \log s_i^2$$

and

$$h = 1 + \frac{1}{3(k - 1)} \left(\sum_{i=1}^{k} \frac{1}{n_i - 1} - \frac{1}{N - k} \right),$$

is a value of the random variable B having approximately the chi-square distribution with $k - 1$ degrees of freedom.

The quantity q is large when the sample variances differ greatly and is equal to zero when all the sample variances are equal. Hence we reject H_0 at the α level

Table 10.5 Hypothetical
Samples

	Sample		
	A	B	C
	4	5	8
	7	1	6
	6	3	8
	6	5	9
		3	5
		4	
Total	23	21	36	40

of significance only when $b > \chi_\alpha^2$. The hazard in applying Bartlett's test is that it is very sensitive to the normality assumption. In fact, a value of $B > \chi_\alpha^2$ may result from a deviation from normality as well as from heterogeneity of variances.

Example 10.2 Use Bartlett's test to test the hypothesis of equal variances for the three populations from which the hypothetical data of Table 10.5 were selected.

Solution
1. H_0: $\sigma_1^2 = \sigma_2^2 = \sigma_3^2$.
2. H_1: The variances are not all equal.
3. $\alpha = 0.05$.
4. Critical region: $B > 5.991$.
5. Computations: Referring to Table 10.5, we have $n_1 = 4$, $n_2 = 6$, $n_3 = 5$, $N = 15$, and $k = 3$. First compute

$$s_1^2 = 1.583, \qquad s_2^2 = 2.300, \qquad s_3^2 = 2.700,$$

and then

$$s_p^2 = \frac{(3)(1.583) + (5)(2.300) + (4)(2.700)}{12} = 2.254.$$

Now

$$q = 12 \log 2.254 - (3 \log 1.583 + 5 \log 2.300 + 4 \log 2.700)$$
$$= (12)(0.3530) - [(3)(0.1995) + (5)(0.3617) + (4)(0.4314)]$$
$$= 0.1034$$

$$h = 1 + \tfrac{1}{6}(\tfrac{1}{3} + \tfrac{1}{5} + \tfrac{1}{4} - \tfrac{1}{12}) = 1.1167.$$

Hence

$$b = \frac{(2.3026)(0.1034)}{1.1167} = 0.213.$$

6. Conclusion: Accept the hypothesis and conclude that the variances of the three populations are equal.

Although Bartlett's test is most often used in testing for homogeneity of variances, there are other methods available. A method due to Cochran provides a computationally simple procedure, but it is restricted to situations in which the sample sizes are equal. Cochran's test is particularly useful in detecting if one variance is much larger than the others. The statistic that is used is given by

$$G = \frac{\text{largest } S_i^2}{\sum\limits_{i=1}^{k} S_i^2}$$

and the hypothesis of equality of variances is rejected if $g > g_\alpha$, where the value of g_α is obtained from Table XI (see Statistical Tables).

To illustrate Cochran's test, let us refer again to the data of Table 10.1 on the absorption of moisture in concrete aggregates. Were we justified in assuming equal variances when we performed the analysis of variance in Example 10.1? We find that

$$s_1^2 = 12{,}134, \qquad s_2^2 = 2303, \qquad s_3^2 = 3594, \qquad s_4^2 = 3319, \qquad s_5^2 = 3455.$$

Therefore,

$$g = \frac{12{,}134}{24{,}805} = 0.4892,$$

which does not exceed the tabled value $g_{0.05} = 0.5065$. Hence we conclude that the assumption of equal variances is reasonable.

10.4 SINGLE-DEGREE-OF-FREEDOM COMPARISONS

The analysis of variance in a one-way classification or the one-factor experiment, as it is often called, merely indicates whether or not the hypothesis of equal treatment means can be rejected. Usually, an experimenter would prefer his analysis to probe deeper than this. For instance, in Example 10.1, by rejecting the null hypothesis we concluded that the means are not all equal, but we still do not know where the differences exist among the aggregates. The engineer might have the feeling from the outset that aggregates 1 and 2 should have similar absorption properties due to similar composition and that the same is true for aggregates 3 and 5 but that the two groups possibly differ a great deal. It would seem, then, appropriate to test the hypothesis

$$H_0: \quad \mu_1 + \mu_2 - \mu_3 - \mu_5 = 0,$$

$$H_1: \quad \mu_1 + \mu_2 - \mu_3 - \mu_5 \neq 0.$$

We notice that the hypothesis is a linear function of the population means in which the coefficients sum to zero.

> DEFINITION 10.1   *Any linear function of the form*
>
> $$\omega = \sum_{i=1}^{k} c_i \mu_i, \qquad where \ \sum_{i=1}^{k} c_i = 0,$$
>
> *is called a* comparison *or* contrast *in the treatment means.*

The experimenter can often make multiple comparisons by testing the significance of contrasts in the treatment means, that is, by testing a hypothesis of the type

$$H_0: \quad \sum_{i=1}^{k} c_i \mu_i = 0,$$

$$H_1: \quad \sum_{i=1}^{k} c_i \mu_i \neq 0,$$

where $\sum_{i=1}^{k} c_i = 0$. The test is conducted by first computing a similar contrast in the sample means,

$$w = \sum_{i=1}^{k} c_i \bar{y}_{i\cdot\cdot}.$$

Since $\bar{Y}_1., \bar{Y}_2., \ldots, \bar{Y}_k.$ are independent random variables having a normal distribution with means $\mu_1, \mu_2, \ldots, \mu_k$ and variances $\sigma^2/n_1, \sigma^2/n_2, \ldots, \sigma^2/n_k$, respectively, Theorem 5.11 assures us that w is a value of the normal random variable W with mean

$$\mu_W = \sum_{i=1}^{k} c_i \mu_i$$

and variance

$$\sigma_W^2 = \sigma^2 \sum_{i=1}^{k} \frac{c_i^2}{n_i}.$$

Therefore, when H_0 is true, $\mu_W = 0$ and, by Example 5.5, the statistic

$$\frac{\left(\sum_{i=1}^{k} c_i \bar{Y}_i.\right)^2}{\sigma^2 \sum_{i=1}^{k} (c_i^2/n_i)}$$

is distributed as a chi-square random variable with 1 degree of freedom. Our hypothesis is tested at the α level of significance by computing

$$f = \frac{\left(\sum_{i=1}^{k} c_i \bar{y}_i.\right)^2}{s^2 \sum_{i=1}^{k} (c_i^2/n_i)} = \frac{\left[\sum_{i=1}^{k} (c_i T_i./n_i)\right]^2}{s^2 \sum_{i=1}^{k} (c_i^2/n_i)} = \frac{SSw}{s^2},$$

where f is a value of the random variable F having the F distribution with 1 and $N - k$ degrees of freedom and

$$SSw = \frac{\left[\sum_{i=1}^{k} (c_i T_i./n_i)\right]^2}{\sum_{i=1}^{k} (c_i^2/n_i)}.$$

When the sample sizes are all equal to n,

$$SSw = \frac{\left(\sum_{i=1}^{k} c_i T_i.\right)^2}{n \sum_{i=1}^{k} c_i^2}.$$

The quantity SSw, called the *contrast sum of squares*, indicates the portion of SSA that is explained by the contrast in question.

DEFINITION 10.2   *The two contrasts*

$$\omega_1 = \sum_{i=1}^{k} b_i \mu_i \quad and \quad \omega_2 = \sum_{i=1}^{k} c_i \mu_i$$

are said to be orthogonal *if* $\sum_{i=1}^{k} b_i c_i/n_i = 0$ *or when the n_i's are all equal to n if*

$$\sum_{i=1}^{k} b_i c_i = 0.$$

If ω_1 and ω_2 are orthogonal, then the quantities SSw_1 and SSw_2 are components of SSA each with a single degree of freedom. The treatment sum of squares with $k - 1$ degrees of freedom can be partitioned into at most $k - 1$ independent single-degree-of-freedom contrast sum of squares satisfying the identity

$$SSA = SSw_1 + SSw_2 + \cdots + SSw_{k-1}$$

if the contrasts are orthogonal to each other.

Example 10.3   Referring to Example 10.1, find the contrast sum of squares corresponding to the orthogonal contrasts

$$\omega_1 = \mu_1 + \mu_2 - \mu_3 - \mu_5$$
$$\omega_2 = \mu_1 + \mu_2 + \mu_3 + \mu_5 - 4\mu_4$$

and carry out appropriate tests of significance.

Solution   It is obvious that the two contrasts are orthogonal since $(1)(1) + (1)(1) + (-1)(1) + (0)(-4) + (-1)(1) = 0$. The second contrast indicates a comparison between aggregates 1, 2, 3, and 5, and aggregate 4. One can write down two additional contrasts orthogonal to the first two such as

$$\omega_3 = \mu_1 - \mu_2 \quad \text{(aggregate 1 versus aggregate 2)}$$
$$\omega_4 = \mu_3 - \mu_5 \quad \text{(aggregate 3 versus aggregate 5)}.$$

From the data of Table 10.1, we have

$$SSw_1 = \frac{(3320 + 3416 - 3663 - 3664)^2}{6[(1)^2 + (1)^2 + (-1)^2 + (-1)^2]} = 14{,}553$$

$$SSw_2 = \frac{[3320 + 3416 + 3663 + 3664 - 4(2791)]^2}{6[(1)^2 + (1)^2 + (1)^2 + (1)^2 + (-4)^2]} = 70{,}035.$$

A more extensive analysis-of-variance table is then given by Table 10.6. We note that the two contrast sum of squares account for nearly all the aggregate sum of squares. While there is a significant difference between aggregates in their absorption properties, the contrast w_1 is not significant when compared to the critical value $f_{0.05}(1, 25) = 4.24$. However, the f value of 14.12 for w_2 is significant and the hypothesis

$$H_0: \quad \mu_1 + \mu_2 + \mu_3 + \mu_5 = 4\mu_4$$

is rejected.

Orthogonal contrasts are used when the experimenter is interested in partitioning the treatment variation into independent components. There are several choices available in selecting the orthogonal contrasts except for the last one. Normally, the experimenter would have certain contrasts that are of interest to him. Such was the case in our example, where chemical composition suggested that aggregates (1, 2) and (3, 5) constitute distinct groups with different absorption properties, a postulation that was not supported by the significance test. However, the second comparison supports the conclusion that aggregate 4 seems to "stand out" from the rest. In this case the complete partitioning of

Table 10.6 Analysis of Variance Using Orthogonal Contrasts

Source of variation	Sum of squares	Degrees of freedom	Mean square	Computed f
Aggregates	85,356	4	21,339	4.30
(1, 2) vs. (3, 5)	$\begin{cases}14{,}553\\70{,}035\end{cases}$	$\begin{cases}1\\1\end{cases}$	$\begin{cases}14{,}553\\70{,}035\end{cases}$	2.93
(1, 2, 3, 5) vs. 4				14.12
Error	124,021	25	4,961	
Total	209,377	29		

SSA was not necessary, since two of the four possible independent comparisons accounted for a majority of the variation in treatments.

10.5 MULTIPLE-RANGE TEST

We have discussed procedures whereby multiple comparisons among the treatment means can be made after the analysis of variance has indicated that the means do differ significantly. These comparisons are made by testing the significance of particular linear combinations of the treatment means, these linear combinations representing meaningful contrasts. There are other test procedures in the form of *multiple-range tests* that are often applied for somewhat similar purposes. These tests divide the k population means into subgroups such that any two means in a subgroup do not differ significantly. Essentially, then, one is testing all possible hypotheses of the type $\mu_i - \mu_j = 0$ at a single controlled significance level α. The test that we shall study in this section is called *Duncan's multiple-range test*.

Let us assume that the k random samples are all of equal size n. The range of any subset of p sample means must exceed a certain value before we consider any of the p population means to be different. This value is called the *least significant range* for the p means and is denoted by R_p, where

$$R_p = r_p \sqrt{\frac{s^2}{n}}.$$

The sample variance s^2, which is an estimate of the common variance σ^2, is obtained from the error mean square in the analysis of variance. The values of the quantity r_p, called the *least significant studentized range*, depend on the desired level of significance and the number of degrees of freedom of the error mean square. These values may be obtained from Table XII for $p = 2, 3, \ldots, 10$ means.

To illustrate the multiple-range test procedure, let us consider a hypothetical example in which six treatments are compared in a one-way classification with five observations per treatment. The error mean square, obtained from the analysis-of-variance table, is $s^2 = 2.45$ with 24 degrees of freedom. First, we arrange the sample means in increasing order of magnitude:

$\bar{y}_2.$	$\bar{y}_5.$	$\bar{y}_1.$	$\bar{y}_3.$	$\bar{y}_6.$	$\bar{y}_4.$
14.50	16.75	19.84	21.12	22.90	23.20

Let $\alpha = 0.05$. Then the values of r_p are obtained from Table XII, with $v = 24$ degrees of freedom, for $p = 2, 3, 4, 5,$ and 6. Finally, we obtain R_p by multiplying each r_p by $\sqrt{s^2/n} = \sqrt{2.45/5} = 0.7$. The results of these computations are summarized as follows:

p	2	3	4	5	6
r_p	2.919	3.066	3.160	3.226	3.276
R_p	2.043	2.146	2.212	2.258	2.293

Comparing these least significant ranges with the differences in ordered means, we arrive at the following conclusions:

1. Since $\bar{y}_4. - \bar{y}_2. = 8.70 > R_6 = 2.293$, we conclude that $\bar{y}_4.$ and $\bar{y}_2.$ are significantly different.
2. Comparing $\bar{y}_4. - \bar{y}_5.$ and $\bar{y}_6. - \bar{y}_2.$ with R_5, we conclude that $\bar{y}_4.$ is significantly greater than $\bar{y}_5.$ and $\bar{y}_6.$ is significantly greater than $\bar{y}_2..$
3. Comparing $\bar{y}_4. - \bar{y}_1., \bar{y}_6 - \bar{y}_5.,$ and $\bar{y}_3. - \bar{y}_2.$ with R_4, we conclude that each difference is significant.
4. Comparing $\bar{y}_4. - \bar{y}_3., \bar{y}_6. - \bar{y}_1., \bar{y}_3. - \bar{y}_5.,$ and $\bar{y}_1. - \bar{y}_2.$ with R_3, we find all differences significant except for $\bar{y}_4. - \bar{y}_3..$ Therefore, $\bar{y}_4., \bar{y}_3.,$ and $\bar{y}_6.$ constitute a subset of homogeneous means.
5. Comparing $\bar{y}_3. - \bar{y}_1., \bar{y}_1. - \bar{y}_5.,$ and $\bar{y}_5. - \bar{y}_2.,$ with R_2, we conclude that only $\bar{y}_3.$ and $\bar{y}_1.$ are not significantly different.

It is customary to summarize the above conclusions by drawing a line under any subset of adjacent means that are not significantly different. Thus we have

$\bar{y}_2.$	$\bar{y}_5.$	$\bar{y}_1.$	$\bar{y}_3.$	$\bar{y}_6.$	$\bar{y}_4.$
14.50	16.75	19.84	21.12	22.90	23.20

One can immediately observe from this manner of presentation that $\mu_1 = \mu_3$, $\mu_3 = \mu_4, \mu_3 = \mu_6,$ and $\mu_4 = \mu_6$, while all other pairs of population means are considered significantly different.

10.6 COMPARING TREATMENTS WITH A CONTROL

In many scientific and engineering problems one is not interested in drawing inferences regarding all possible comparisons among the treatment means of the type $\mu_i - \mu_j$. Rather, the experiment often dictates the need for comparing simultaneously each *treatment* with a *control*. A test procedure for determining significant differences between each treatment mean and the control, at a single joint significance level α, has been developed by C. W. Dunnett. To illustrate Dunnett's procedure, let us consider the experimental data of Table 10.7 for the

Table 10.7 Yield of Reaction

Control	Catalyst 1	Catalyst 2	Catalyst 3
50.7	54.1	52.7	51.2
51.5	53.8	53.9	50.8
49.2	53.1	57.0	49.7
53.1	52.5	54.1	48.0
52.7	54.0	52.5	47.2
$\bar{y}_{0\cdot} = 51.44$	$\bar{y}_{1\cdot} = 53.50$	$\bar{y}_{2\cdot} = 54.04$	$\bar{y}_{3\cdot} = 49.38$

one-way classification in which the effect of three catalysts on the yield of a reaction is being studied. A fourth treatment, no catalyst, was used as a control. In general we wish to test the k hypotheses

$$\left.\begin{array}{l} H_0: \quad \mu_0 = \mu_i \\ H_1: \quad \mu_0 \neq \mu_i \end{array}\right\} i = 1, 2, \ldots, k,$$

where μ_0 represents the mean yield for the population of measurements in which the control is used. The usual analysis-of-variance assumptions, as outlined in Section 10.2 are expected to remain valid. To test the null hypotheses specified by H_0 against two-sided alternatives for an experimental situation in which there are k treatments, excluding the control, and n observations per treatment, we first calculate the values

$$d_i = \frac{\bar{y}_{i\cdot} - \bar{y}_{0\cdot}}{\sqrt{2s^2/n}}, \qquad i = 1, 2, \ldots, k.$$

The sample variance s^2 is obtained, as before, from the error mean square in the analysis of variance. Now, the critical region for rejecting H_0, at the α level of significance, is established by the inequality

$$|d_i| > d_{\alpha/2}(k, v),$$

where v is the number of degrees of freedom for the error mean square. The values of the quantity $d_{\alpha/2}(k, v)$ for a two-tailed test are given in Table XIII (see Statistical Tables) for $\alpha = 0.05$ and $\alpha = 0.01$ for various values of k and v.

Example 10.4   For the data of Table 10.7, test hypotheses comparing each catalyst with the control, using two-sided alternatives. Choose $\alpha = 0.05$ as the joint significance level.

Solution The error sum of squares with 16 degrees of freedom is obtained from the analysis-of-variance table using all $k + 1$ treatments or by direct computation from the formula

$$SSE = \sum_{i=1}^{k+1} \sum_{j=1}^{n} y_{ij}^2 - \frac{\sum\limits_{i=1}^{k+1} T_{i\cdot}^2}{n}$$

$$= 54{,}371.960 - 54{,}335.148$$

$$= 36.812.$$

Then the error mean square is given by

$$s^2 = \frac{36.812}{16} = 2.30075$$

and

$$\sqrt{\frac{2s^2}{n}} = \sqrt{\frac{(2)(2.30075)}{5}} = 0.9593.$$

Hence

$$d_1 = \frac{53.50 - 51.44}{0.9593} = 2.147,$$

$$d_2 = \frac{54.04 - 51.44}{0.9593} = 2.710,$$

$$d_3 = \frac{49.38 - 51.44}{0.9593} = -2.147.$$

From Table XIII the critical value for $\alpha = 0.05$ is found to be $d_{0.025}(3, 16) = 2.59$. Since $|d_1| < 2.59$, $|d_2| > 2.59$, and $|d_3| < 2.59$, we conclude that only the mean yield for catalyst 2 is significantly different from the mean yield of the reaction using the control.

Many practical applications dictate the need for a one-tailed test in comparing treatments with a control. Certainly, when a pharmacologist is concerned with the comparison of various dosages of a drug on the effect of reducing cholesterol level and his control is zero dosage, it is of interest to determine if each dosage produces a significantly larger reduction than that of the control. Table XIV gives the critical values $d_\alpha(k, \nu)$ for one-sided alternatives.

10.7 COMPARING A SET OF TREATMENTS IN BLOCKS

It often becomes necessary in analysis-of-variance problems to design the experiment in such a way that the experimental error variation due to extraneous sources can be systematically controlled. In the preceding development of the one-way analysis-of-variance problem, it was assumed that conditions remained relatively homogeneous for the *experimental units* used in the various test runs. For example, in a chemical experiment designed to determine if there is a difference in mean reaction yield among four catalysts, samples of materials to be tested are drawn from the same batches of raw materials, while other conditions, such as temperature and concentration of reactants, are held constant. In this case time of the experimental runs might represent the experimental units, and if the experimenter feels that there could possibly be a slight time effect, he would *randomize* the assignment of the catalysts to the runs to counteract the possible trend. This type of experimental strategy, whereby the treatments (catalysts) are assigned randomly to the experimental units, is called a *completely randomized design*. As a second example of such a design, consider an experiment to compare four methods of measuring a particular physical property of a fluid substance. Suppose the sampling process is destructive; that is, once a sample of the substance has been measured by one method, it cannot be measured again by one of the other methods. If it is decided that five measurements are to be taken for each method, then 20 samples of the material are selected from a large batch *at random* and are used in the experiment to compare the four measuring devices. The experimental units are the randomly selected samples. Any variation from sample to sample will appear in the error variation, as measured by s^2 in the analysis.

If the variation due to heterogeneity in experimental units is so large that the sensitivity of detecting treatment differences is reduced due to an inflated value of s^2, a better plan might be to "block off" variation due to these units and thus reduce the extraneous variation to that accounted for by smaller or more homogeneous blocks. The simplest design calling for this strategy is a *randomized block design*. For example, suppose in the previous catalyst illustration it is known a priori that there definitely is a significant day-to-day effect on the yield and that we can measure yield for four catalysts on a given day. Rather than assign the four catalysts to the 20 test runs completely at random, we choose, say, 5 days and run each of the four catalysts on each day, randomly assigning the catalysts to the runs within days. In this way the day-to-day variation is removed in the analysis and consequently the experimental error, which still includes any time trend *within days*, more accurately represents chance variation. Each day is referred to as a *block*.

The classical example, using a randomized block design, is an agricultural experiment in which different fertilizers are being compared for their ability to increase the yield of a particular crop. Rather than assign fertilizers at random to many plots over a large area of variable soil composition, one should assign the fertilizers to smaller blocks comprised of homogeneous plots. The variation between these blocks, which is most likely significant compared to the uniformity of the plots within a block, is then removed from the experimental error in the analysis of variance.

It should be obvious to the reader that the most straightforward of the randomized block designs is one in which we randomly assign each treatment once to every block. Such an experimental layout is called a *randomized complete block design*, each block constituting a single *replication* of the treatments.

10.8 RANDOMIZED COMPLETE BLOCK DESIGNS

A typical layout for the randomized complete block design using three treatments in four blocks is as follows:

Block 1	Block 2	Block 3	Block 4
t_2	t_1	t_1	t_2
t_1	t_3	t_2	t_1
t_3	t_2	t_3	t_3

The t's denote the assignment to blocks of each of the three treatments. Of course, the true allocation to units within blocks is random.

Let us generalize and consider the case of k treatments assigned to b blocks. The observation y_{ij} denotes the response of the ith treatment applied to the jth block. It will be assumed that the y_{ij}, $i = 1, 2, \ldots, k$ and $j = 1, 2, \ldots, b$, are values of independent random variables having normal distributions with mean μ_{ij} and common variance σ^2. Let us define

$$\bar{y}_{i.} = i\text{th treatment mean}$$

$$\bar{y}_{..} = \text{overall mean}$$

$$\bar{y}_{.j} = j\text{th block mean}$$

$$T_{i.} = i\text{th treatment total}$$

$$T_{.j} = j\text{th block total}$$

$$T_{..} = \text{overall total.}$$

The average of the population means for the ith treatment, $\mu_{i\cdot}$, is defined by

$$\mu_{i\cdot} = \frac{\sum\limits_{j=1}^{b} \mu_{ij}}{b}.$$

Similarly, the average of the population means for the jth block, $\mu_{\cdot j}$, is defined by

$$\mu_{\cdot j} = \frac{\sum\limits_{i=1}^{k} \mu_{ij}}{k},$$

and the average of the bk population means, μ, is defined by

$$\mu = \frac{\sum\limits_{i=1}^{k} \sum\limits_{j=1}^{b} \mu_{ij}}{bk}.$$

To determine if part of the variation in our observations is due to differences among the treatments, we consider the test

$$H_0' : \quad \mu_{1\cdot} = \mu_{2\cdot} = \cdots = \mu_{k\cdot} = \mu,$$

$$H_1' : \quad \text{the } \mu_{i\cdot}\text{'s are not all equal.}$$

Similarly, to determine if part of the variation is due to differences among the blocks, we consider the test

$$H_0'' : \quad \mu_{\cdot 1} = \mu_{\cdot 2} = \cdots = \mu_{\cdot b} = \mu,$$

$$H_1'' : \quad \text{the } \mu_{\cdot j}\text{'s are not all equal.}$$

Each observation may be written in the form

$$y_{ij} = \mu_{ij} + \varepsilon_{ij},$$

where ε_{ij} measures the deviation of the observed value y_{ij} from the population mean μ_{ij}. The preferred form of this equation is obtained by substituting

$$\mu_{ij} = \mu + \alpha_i + \beta_j,$$

where α_i is, as before, the effect of the ith treatment and β_j is the effect of the jth block. It is assumed that the treatment and block effects are additive. Hence we may write

$$y_{ij} = \mu + \alpha_i + \beta_j + \varepsilon_{ij}.$$

Notice that the model resembles that of the one-way classification, the essential difference being the introduction of the block effect β_j. The basic concept is much like that of the one-way classification except that we must account in the analysis for the additional effect due to blocks since we are now systematically controlling variation in *two directions*. If we now impose the restrictions that

$$\sum_{i=1}^{k} \alpha_i = 0 \quad \text{and} \quad \sum_{j=1}^{b} \beta_j = 0,$$

then

$$\mu_{i\cdot} = \frac{\sum_{j=1}^{b} (\mu + \alpha_i + \beta_j)}{b} = \mu + \alpha_i$$

and

$$\mu_{\cdot j} = \frac{\sum_{i=1}^{k} (\mu + \alpha_i + \beta_j)}{k} = \mu + \beta_j.$$

The null hypothesis that the k treatment means $\mu_{i\cdot}$ are equal, and therefore equal to μ, is now equivalent to testing the hypothesis

$$H_0': \quad \alpha_1 = \alpha_2 = \cdots = \alpha_k = 0,$$

$H_1':$ at least one of the α_i's is not equal to zero.

Similarly, the null hypothesis that the b block means $\mu_{\cdot j}$ are equal is equivalent to testing the hypothesis

$$H_0'': \quad \beta_1 = \beta_2 = \cdots = \beta_b = 0,$$

$H_1'':$ at least one of the β_j's is not equal to zero.

Each of these tests will be based on a comparison of independent estimates of the common population variance σ^2. These estimates will be obtained by splitting the total sum of squares of our data into three components by means of the following identity.

THEOREM 10.3   *Sum of Squares Identity*

$$\sum_{i=1}^{k} \sum_{j=1}^{b} (y_{ij} - \bar{y}_{..})^2 = b \sum_{i=1}^{k} (\bar{y}_{i.} - \bar{y}_{..})^2 + k \sum_{j=1}^{b} (\bar{y}_{.j} - \bar{y}_{..})^2$$

$$+ \sum_{i=1}^{k} \sum_{j=1}^{b} (y_{ij} - \bar{y}_{i.} - \bar{y}_{.j} + \bar{y}_{..})^2.$$

Proof

$$\sum_{i=1}^{k} \sum_{j=1}^{b} (y_{ij} - \bar{y}_{..})^2 = \sum_{i=1}^{k} \sum_{j=1}^{b} [(\bar{y}_{i.} - \bar{y}_{..}) + (\bar{y}_{.j} - \bar{y}_{..}) + (y_{ij} - \bar{y}_{i.} - \bar{y}_{.j} + \bar{y}_{..})]^2$$

$$= \sum_{i=1}^{k} \sum_{j=1}^{b} (\bar{y}_{i.} - \bar{y}_{..})^2 + \sum_{i=1}^{k} \sum_{j=1}^{b} (\bar{y}_{.j} - \bar{y}_{..})^2$$

$$+ \sum_{i=1}^{k} \sum_{j=1}^{b} (y_{ij} - \bar{y}_{i.} - \bar{y}_{.j} + \bar{y}_{..})^2$$

$$+ 2 \sum_{i=1}^{k} \sum_{j=1}^{b} (\bar{y}_{i.} - \bar{y}_{..})(\bar{y}_{.j} - \bar{y}_{..})$$

$$+ 2 \sum_{i=1}^{k} \sum_{j=1}^{b} (\bar{y}_{i.} - \bar{y}_{..})(y_{ij} - \bar{y}_{i.} - \bar{y}_{.j} + \bar{y}_{..})$$

$$+ 2 \sum_{i=1}^{k} \sum_{j=1}^{b} (\bar{y}_{.j} - \bar{y}_{..})(y_{ij} - \bar{y}_{i.} - \bar{y}_{.j} + \bar{y}_{..}).$$

The cross-product terms are all equal to zero. Hence

$$\sum_{i=1}^{k} \sum_{j=1}^{b} (y_{ij} - \bar{y}_{..})^2 = b \sum_{i=1}^{k} (\bar{y}_{i.} - \bar{y}_{..})^2 + k \sum_{j=1}^{b} (\bar{y}_{.j} - \bar{y}_{..})^2$$

$$+ \sum_{i=1}^{k} \sum_{j=1}^{b} (y_{ij} - \bar{y}_{i.} - \bar{y}_{.j} + \bar{y}_{..})^2.$$

The sum of squares identity may be represented symbolically by

$$SST = SSA + SSB + SSE,$$

where

$$SST = \sum_{i=1}^{k} \sum_{j=1}^{b} (y_{ij} - \bar{y}_{..})^2 = \text{total sum of squares}$$

$$SSA = b \sum_{i=1}^{k} (\bar{y}_{i.} - \bar{y}_{..})^2 = \text{treatment sum of squares}$$

$$SSB = k \sum_{j=1}^{b} (\bar{y}_{.j} - \bar{y}_{..})^2 = \text{block sum of squares}$$

$$SSE = \sum_{i=1}^{k} \sum_{j=1}^{b} (y_{ij} - \bar{y}_{i.} - \bar{y}_{.j} + \bar{y}_{..})^2 = \text{error sum of squares.}$$

Following the procedure outlined in Theorem 10.2, where we interpret the sum of squares as functions of the independent random variables $Y_{11}, Y_{12}, \ldots, Y_{kb}$, we can show that the expected values of the treatment, block, and error sum of squares are given by

$$E(SSA) = (k - 1)\sigma^2 + b \sum_{i=1}^{k} \alpha_i^2$$

$$E(SSB) = (b - 1)\sigma^2 + k \sum_{j=1}^{b} \beta_j^2$$

$$E(SSE) = (b - 1)(k - 1)\sigma^2.$$

One estimate of σ^2, based on $k - 1$ degrees of freedom, is given by

$$s_1^2 = \frac{SSA}{k - 1}.$$

If the treatment effects $\alpha_1 = \alpha_2 = \cdots = \alpha_k = 0$, s_1^2 is an unbiased estimate of σ^2. However, if the treatment effects are not all zero, we have

$$E\left(\frac{SSA}{k - 1}\right) = \sigma^2 + \frac{b \sum_{i=1}^{k} \alpha_i^2}{k - 1}$$

and s_1^2 overestimates σ^2. A second estimate of σ^2, based on $b - 1$ degrees of freedom, is given by

$$s_2^2 = \frac{SSB}{b - 1}.$$

The estimate s_2^2 is an unbiased estimate of σ^2 when the block effects $\beta_1 = \beta_2 = \cdots = \beta_b = 0$. If the block effects are not all zero, then

$$E\left(\frac{SSB}{b-1}\right) = \sigma^2 + \frac{k \sum\limits_{j=1}^{b} \beta_j^2}{b-1}$$

and s_2^2 will overestimate σ^2. A third estimate of σ^2, based on $(k-1)(b-1)$ degrees of freedom and independent of s_1^2 and s_2^2, is given by

$$s^2 = \frac{SSE}{(k-1)(b-1)},$$

which is unbiased regardless of the truth or falsity of either null hypothesis.

To test the null hypothesis that the treatment effects are all equal to zero, we compute the ratio

$$f_1 = \frac{s_1^2}{s^2},$$

which is a value of the random variable F_1 having the F distribution with $k-1$ and $(k-1)(b-1)$ degrees of freedom when the null hypothesis is true. The null hypothesis is rejected at the α level of significance when $f_1 > f_\alpha[k-1, (k-1)(b-1)]$.

Similarly, to test the null hypothesis that the block effects are all equal to zero, we compute the ratio

$$f_2 = \frac{s_2^2}{s^2},$$

which is a value of the random variable F_2 having the F distribution with $b-1$ and $(k-1)(b-1)$ degrees of freedom when the null hypothesis is true. In this case the null hypothesis is rejected at the α level of significance when $f_2 > f_\alpha[b-1, (k-1)(b-1)]$.

In practice we first compute SST, SSA, and SSB, and then obtain SSE by subtraction by the formula

$$SSE = SST - SSA - SSB.$$

The degrees of freedom associated with SSE are also usually obtained by subtraction. It is not difficult to verify the identity

$$(k-1)(b-1) = (kb-1) - (k-1) - (b-1).$$

Table 10.8 Analysis of Variance for the Randomized Complete Block Design

Source of variation	Sum of squares	Degrees of freedom	Mean square	Computed f
Treatments	SSA	$k - 1$	$s_1^2 = \dfrac{SSA}{k-1}$	$f_1 = \dfrac{s_1^2}{s^2}$
Blocks	SSB	$b - 1$	$s_2^2 = \dfrac{SSB}{b-1}$	$f_2 = \dfrac{s_2^2}{s^2}$
Error	SSE	$(k-1)(b-1)$	$s^2 = \dfrac{SSE}{(b-1)(k-1)}$	
Total	SST	$bk - 1$		

Preferred computational formulas for the sums of squares are given as follows:

$$SST = \sum_{i=1}^{k} \sum_{j=1}^{b} y_{ij}^2 - \frac{T_{..}^2}{bk}$$

$$SSA = \frac{\sum_{i=1}^{k} T_{i.}^2}{b} - \frac{T_{..}^2}{bk}$$

$$SSB = \frac{\sum_{j=1}^{b} T_{.j}^2}{k} - \frac{T_{..}^2}{bk}.$$

The computations in an analysis-of-variance problem for a randomized complete block design may be summarized as shown in Table 10.8.

Example 10.5　Four different machines are to be considered in the assembling of a particular product. It is decided that six different operators are to be used in an experiment to compare the machines. The operation of the machines requires a certain amount of physical dexterity and it is known that there is a difference among the operators in the speed with which they operate the machines. The basic measurements on which the machines are to be compared is *time in seconds* to completion. The data are given in Table 10.9. (a) Test the hypothesis H_0, at the 0.05 level of significance, that the machines perform at the same mean rate of speed. (b) Test the hypothesis H_0'' that the operators perform at the same mean rate of speed.

Table 10.9 Time in Seconds to Assemble Product

| Machine | Operator | | | | | | Total |
	1	*2*	*3*	*4*	*5*	*6*	
1	42.5	39.3	39.6	39.9	42.9	43.6	247.8
2	39.8	40.1	40.5	42.3	42.5	43.1	248.3
3	40.2	40.5	41.3	43.4	44.9	45.1	255.4
4	41.3	42.2	43.5	44.2	45.9	42.3	259.4
Total	163.8	162.1	164.9	169.8	176.2	174.1	1010.9

Solution

1. (a) H_0': $\alpha_1 = \alpha_2 = \alpha_3 = \alpha_4 = 0$ (machine effects are zero).
   (b) H_0'': $\beta_1 = \beta_2 = \cdots = \beta_6 = 0$ (operator effects are zero).
2. (a) H_1': At least one of the α_i's is not equal to zero.
   (b) H_1'': At least one of the β_j's is not equal to zero.
3. $\alpha = 0.05$.
4. Critical regions: (a) $F_1 > 3.29$. (b) $F_2 > 2.90$.
5. Computations:

$$SST = 42.5^2 + 39.8^2 + \cdots + 42.3^2 - \frac{1010.9^2}{24} = 81.86$$

$$SSA = \frac{247.8^2 + 248.3^2 + 255.4^2 + 259.4^2}{6} - \frac{1010.9^2}{24} = 15.93$$

$$SSB = \frac{163.8^2 + 162.1^2 + \cdots + 174.1^2}{4} - \frac{1010.9^2}{24} = 42.09$$

$$SSE = 81.86 - 15.93 - 42.09 = 23.84.$$

These and the remaining computations are exhibited in Table 10.10.

Table 10.10 Analysis of Variance for the Data of Table 10.9

Source of variation	Sum of squares	Degrees of freedom	Mean square	Computed f
Machines	15.93	3	5.31	3.34
Operators	42.09	5	8.42	5.30
Error	23.84	15	1.59	
Total	81.86	23		

6. Conclusion:
   (a) Reject H_0' and conclude that the machines do not perform at the same mean rate of speed.
   (b) Reject H_0'', as expected, and conclude that the operators do not perform at the same mean rate of speed.

10.9 ADDITIONAL REMARKS CONCERNING RANDOMIZED COMPLETE BLOCK DESIGNS

In Chapter 7 we presented a procedure for comparing means when the observations were *paired*. The procedure involved "subtracting out" the effect due to the homogeneous pair and thus working with differences. This is a special case of a randomized complete block design with $k = 2$ treatments. The n homogeneous units to which the treatments were assigned take on the role of blocks.

If there is heterogeneity in the experimental units, the experimenter should not be mislead into believing that it is always advantageous to reduce the experimental error through the use of small homogeneous blocks. Indeed, there may be instances where it would not be desirable to block. The purpose in reducing the error variance is to increase the *sensitivity* of the test for detecting differences in the treatment means. This is reflected in the power of the test procedure. (The power of the analysis-of-variance test procedure is discussed more extensively in Section 10.12.) The power for detecting certain differences among the treatment means increases with a decrease in the error variance. However, the power is also affected by the degrees of freedom with which this variance is estimated, and blocking reduces the degrees of freedom that are available from $k(b - 1)$ for the one-way classification to $(k - 1)(b - 1)$. So one could lose power by blocking if there is not a significant reduction in the error variance.

Another important assumption that is implicit in writing the model for a randomized complete block design is that the treatment and block effects were additive. This is equivalent to stating that $\mu_{ij} - \mu_{ij'} = \mu_{i'j} - \mu_{i'j'}$ or $\mu_{ij} - \mu_{i'j} = \mu_{ij'} - \mu_{i'j'}$ for every value of i, i', j, and j'. That is, the difference between the population means for blocks j and j' is the same for every treatment and the difference between the population means for treatments i and i' is the same for every block. Referring to Example 10.5, we see that if operator 3 is 0.5 second faster on the average than operator 2 when machine 1 is used, then operator 3 will still be 0.5 second faster on the average than operator 2 when machine 2, 3, or 4 is used. Similarly, if operator 1 is 1.2 seconds faster on the average using machine 2 than on machine 4, then operator 2, 3, ..., 6 will also be 1.2 seconds faster on the average using machine 2 than on machine 4.

In many experiments the assumption of additivity does not hold and the analysis of Section 10.8 leads to erroneous conclusions. Suppose, for instance, that operator 3 is 0.5 second faster on the average than operator 2 when machine 1 is used but is 0.2 second slower on the average than operator 2 when machine 2 is used. The operators and machines are now said to *interact*.

An inspection of Table 10.9 suggests the presence of interaction. This apparent interaction may be real or it may be due to experimental error. The analysis of Example 10.5 was based on the assumption that the apparent interaction was due entirely to experimental error. If the total variability of our data was in part due to an interaction effect, this source of variation remained a part of the error sum of squares, causing the error mean square to overestimate σ^2, and thereby increased the probability of committing a type II error. We have, in effect, assumed an incorrect model. If we let $(\alpha\beta)_{ij}$ denote the interaction effect of the ith treatment and the jth block, we can write a more appropriate model in the form

$$y_{ij} = \mu + \alpha_i + \beta_j + (\alpha\beta)_{ij} + \varepsilon_{ij},$$

on which we impose the additional restrictions $\sum_{i=1}^{k} (\alpha\beta)_{ij} = \sum_{j=1}^{b} (\alpha\beta)_{ij} = 0$. One can now very easily verify that

$$E\left[\frac{SSE}{(b-1)(k-1)}\right] = \sigma^2 + \frac{\sum_{i=1}^{k} \sum_{j=1}^{b} (\alpha\beta)_{ij}^2}{(b-1)(k-1)}.$$

Thus the error mean square is seen to be a biased estimate of σ^2 when existing interaction has been ignored. It would seem necessary at this point to arrive at a procedure for the detection of interaction for cases where there is suspicion that it exists. Such a procedure requires the availability of an unbiased and independent estimate of σ^2. Unfortunately, the randomized block design does not lend itself to such a test unless the experimental setup is altered. This is discussed extensively in Chapter 11.

10.10 RANDOM EFFECTS MODELS

Throughout this chapter we have dealt with analysis-of-variance procedures in which the primary goal was to study the effect on some response of certain fixed or predetermined treatments. Experiments in which the treatments or treatment levels are preselected by the experimenter as opposed to being chosen randomly are called *fixed effects experiments* or *model I experiments*. For the fixed effects model, inferences were made only on those particular treatments used in the experiment.

It is often important that the experimenter be able to draw inferences about a population of treatments by means of an experiment in which the treatments used are chosen randomly from the population. For example, a biologist may be interested in whether or not there is a significant variance in some physiological characteristic due to animal type. The animal types actually used in the experiment are then chosen randomly and represent the treatment effects. A chemist may be interested in studying the effect of analytical laboratories on the chemical analysis of a substance. He is not concerned with particular laboratories but rather with a large population of laboratories. He might then select a group of laboratories at random and allocate samples to each for analysis. The statistical inference would then involve (1) testing whether or not the laboratories contribute a nonzero variance to the analytical results, and (2) estimating the variance due to laboratories and the variance within laboratories.

The one-way random effects model, often referred to as *model II*, is written like the fixed effects model but with the terms taking on different meanings. The response

$$y_{ij} = \mu + \alpha_i + \varepsilon_{ij}$$

is now a value of the random variable

$$Y_{ij} = \mu + A_i + E_{ij}$$

with $i = 1, 2, \ldots, k$ and $j = 1, 2, \ldots, n$, where the A_i's are normally and independently distributed with mean zero and variance σ_α^2 and are independent of the E_{ij}'s. As for the fixed effects model, the E_{ij}'s are also normally and independently distributed with mean zero and variance σ^2. Note that for a model II experiment the random variable $\sum_{i=1}^{k} A_i$ assumes the value $\sum_{i=1}^{k} \alpha_i \neq 0$.

THEOREM 10.4   *For the random effects one-way analysis-of-variance model*

$$E(SSA) = (k - 1)\sigma^2 + n(k - 1)\sigma_\alpha^2$$

and

$$E(SSE) = k(n - 1)\sigma^2.$$

Proof From the model

$$Y_{ij} = \mu + A_i + E_{ij}$$

we obtain

$$\bar{Y}_{i\cdot} = \mu + A_i + \bar{E}_{i\cdot},$$
$$\bar{Y}_{\cdot\cdot} = \mu + \bar{A}_{\cdot} + \bar{E}_{\cdot\cdot}$$

Hence

$$SSA = n\sum_{i=1}^{k}(\bar{Y}_{i\cdot} - \bar{Y}_{\cdot\cdot})^2$$

$$= n\sum_{i=1}^{k}[(A_i - \bar{A}_{\cdot}) + (\bar{E}_{i\cdot} - \bar{E}_{\cdot\cdot})]^2$$

and

$$E(SSA) = n\sum_{i=1}^{k}E(A_i^2) - nkE(\bar{A}_{\cdot}^2) + n\sum_{i=1}^{k}E(\bar{E}_{i\cdot}^2) - nkE(\bar{E}_{\cdot\cdot}^2)$$

$$= nk\sigma_\alpha^2 - n\sigma_\alpha^2 + k\sigma^2 - \sigma^2$$

$$= (k-1)\sigma^2 + n(k-1)\sigma_\alpha^2.$$

Following the same steps as above, we also find that

$$SSE = \sum_{i=1}^{k}\sum_{j=1}^{n}(Y_{ij} - \bar{Y}_{i\cdot})^2$$

$$= \sum_{i=1}^{k}\sum_{j=1}^{n}(E_{ij} - \bar{E}_{i\cdot})^2$$

and therefore

$$E(SSE) = \sum_{i=1}^{k}\sum_{j=1}^{n}E(E_{ij}^2) - n\sum_{i=1}^{k}E(\bar{E}_{i\cdot}^2)$$

$$= nk\sigma^2 - k\sigma^2$$

$$= k(n-1)\sigma^2.$$

Table 10.11 Expected Mean Squares for the One-Way Classification

Source of variation	Degrees of freedom	Mean squares	Expected mean squares	
			Model I	Model II
Treatments	$k - 1$	s_1^2	$\sigma^2 + \dfrac{n \sum_{i=1}^{k} \alpha_i^2}{k - 1}$	$\sigma^2 + n\sigma_\alpha^2$
Error	$k(n - 1)$	s^2	σ^2	σ^2
Total	$nk - 1$			

Table 10.11 shows the expected mean squares for both model I and model II. The computations for model II are carried out in exactly the same way as for model I. That is, the sum-of-squares, degrees-of-freedom, and mean-square columns in an analysis-of-variance table are the same for both models.

In the random effects model we are interested in testing the hypothesis

$$H_0: \quad \sigma_\alpha^2 = 0,$$
$$H_1: \quad \sigma_\alpha^2 \neq 0.$$

It is obvious from Table 10.11 that s_1^2 and s^2 are both estimates of σ^2 when H_0 is true and that the ratio

$$f = \frac{s_1^2}{s^2}$$

is a value of the random variable F having the F distribution with $k - 1$ and $k(n - 1)$ degrees of freedom. The null hypothesis is rejected at the α level of significance when

$$f > f_\alpha[k - 1, k(n - 1)].$$

Table 10.11 can also be used to estimate the *variance components* σ^2 and σ_α^2. Since s_1^2 estimates $\sigma^2 + n\sigma_\alpha^2$ and s^2 estimates σ^2,

$$\hat{\sigma}^2 = s^2$$
$$\hat{\sigma}_\alpha^2 = \frac{s_1^2 - s_2^2}{n}.$$

Example 10.6 The following data are coded observations on the yield of a chemical process using five batches of raw material selected randomly:

	\multicolumn{5}{c}{Batch}				
	1	2	3	4	5
	9.7	10.4	15.9	8.6	9.7
	5.6	9.6	14.4	11.1	12.8
	8.4	7.3	8.3	10.7	8.7
	7.9	6.8	12.8	7.6	13.4
	8.2	8.8	7.9	6.4	8.3
	7.7	9.2	11.6	5.9	11.7
	8.1	7.6	9.8	8.1	10.7
Total	55.6	59.7	80.7	58.4	75.3	329.7

Show that the batch variance component is significantly greater than zero and obtain its estimate.

Solution The total, batch, and error sum of squares are given by

$$SST = 9.7^2 + 5.6^2 + \cdots + 10.7^2 - \frac{329.7^2}{35}$$

$$= 194.64$$

$$SSA = \frac{55.6^2 + 59.7^2 + \cdots + 75.3^2}{7} - \frac{329.7^2}{35}$$

$$= 72.60$$

$$SSE = 194.64 - 72.60 = 122.04.$$

These results, with the remaining computations, are given in Table 10.12. The f ratio is significant at the $\alpha = 0.05$ level, indicating that the hypothesis of a zero batch component is rejected. An estimate of the batch variance component is given by

$$\hat{\sigma}_{\alpha}^2 = \frac{18.15 - 4.07}{7} = 2.01.$$

Table 10.12 Analysis of Variance for Example 10.6

Source of variation	Sum of squares	Degrees of freedom	Mean square	Computed f
Batches	72.60	4	18.15	4.46
Error	122.04	30	4.07	
Total	194.64	34		

In a randomized complete block experiment in which the blocks represent days it is conceivable that the experimenter would like his results to apply not only to the actual days used in the analysis but to every day in the year. He would then select the days on which he runs his experiment as well as the treatments at random and use the random effects model

$$Y_{ij} = \mu + A_i + B_j + E_{ij},$$

$i = 1, 2, \ldots, k$ and $j = 1, 2, \ldots, b$, with the A_i, B_j, and E_{ij} being independent random variables with means zero and variances σ_α^2, σ_β^2, and σ^2, respectively. The expected mean squares for model II are obtained using the same procedure as for the one-way classification and are presented along with those for model I in Table 10.13.

Table 10.13 Expected Mean Squares for the Randomized Complete Block Design

Source of variation	Degrees of freedom	Mean square	Expected mean squares Model I	Expected mean squares Model II
Treatments	$k - 1$	s_1^2	$\sigma^2 + \dfrac{b \sum\limits_{i=1}^{k} \alpha_i^2}{k - 1}$	$\sigma^2 + b\sigma_\alpha^2$
Blocks	$b - 1$	s_2^2	$\sigma^2 + \dfrac{k \sum\limits_{j=1}^{b} \beta_j^2}{b - 1}$	$\sigma^2 + k\sigma_\beta^2$
Error	$(b - 1)(k - 1)$	s^2	σ^2	σ^2
Total	$bk - 1$			

Again the computations for the individual sum of squares and degrees of freedom are identical to those of the fixed effects model. The hypothesis

$$H_0: \quad \sigma_\alpha^2 = 0,$$
$$H_1: \quad \sigma_\alpha^2 \neq 0,$$

is carried out by computing

$$f = \frac{s_1^2}{s^2}$$

and rejecting H_0 when $f > f_\alpha[k - 1, (b - 1)(k - 1)]$. Similarly, we can test

$$H_0: \quad \sigma_\beta^2 = 0,$$
$$H_1: \quad \sigma_\beta^2 \neq 0,$$

by comparing

$$f = \frac{s_2^2}{s^2}$$

with the critical point $f_\alpha[b - 1, (b - 1)(k - 1)]$.

The unbiased estimates of the variance components are given by

$$\hat{\sigma}^2 = s^2$$

$$\hat{\sigma}_\alpha^2 = \frac{s_1^2 - s^2}{b}$$

$$\hat{\sigma}_\beta^2 = \frac{s_2^2 - s^2}{k}.$$

10.11 REGRESSION APPROACH TO ANALYSIS OF VARIANCE

So far, we have treated the regression models and the analysis-of-variance models as two separate and unrelated topics. Although this has become the accepted approach in dealing with these procedures on an elementary level, one can treat an analysis-of-variance model as a special case of a multiple linear regression model. In this section we shall show the relationship between the two models and indicate how the analysis of variance techniques can be developed through a regression approach.

Suppose that we consider two models, the multiple linear regression model

$$y_i = \beta_0 + \beta_1 x_{1i} + \beta_2 x_{2i} + \cdots + \beta_k x_{ki} + \varepsilon_i$$

and the one-way classification analysis-of-variance model

$$y_{ij} = \mu + \alpha_i + \varepsilon_{ij}.$$

Traditionally, the two are presented as methods for handling different practical problems, the regression model being a means of arriving at a procedure for predicting some response as a function of one or more quantitative independent variables, and the analysis-of-variance model for arriving at significance tests on multiple population means. However, any mathematical model that is linear in the parameters, such as the analysis-of-variance model, can be considered a special case of the multiple linear regression model. We can use conventional matrix notation to describe how each observation is expressed as a function of the parameters for the two models. For the regression model

$$\mathbf{y} = \mathbf{X}\boldsymbol{\beta} + \boldsymbol{\varepsilon},$$

or, more explicitly,

$$
\begin{bmatrix} y_1 \\ y_2 \\ \vdots \\ y_n \end{bmatrix} =
\begin{bmatrix}
1 & x_{11} & x_{21} & \cdots & x_{k1} \\
1 & x_{12} & x_{22} & \cdots & x_{k2} \\
\vdots & \vdots & \vdots & & \vdots \\
1 & x_{1n} & x_{2n} & \cdots & x_{kn}
\end{bmatrix}
\begin{bmatrix} \beta_0 \\ \beta_1 \\ \vdots \\ \beta_k \end{bmatrix} +
\begin{bmatrix} \varepsilon_1 \\ \varepsilon_2 \\ \vdots \\ \varepsilon_n \end{bmatrix},
$$

where the **y** vector on the left of the equality sign is the array of responses in the experiment. The **X** matrix has already been described and used in Section 9.3. The $\boldsymbol{\beta}$ vector is the vector of parameters appearing in the model, and the $\boldsymbol{\varepsilon}$ vector completes the model by the addition of the random error. The reader will recall that the least squares estimates $b_0, b_1, \ldots, b_k$ of the parameters $\beta_0, \beta_1, \ldots, \beta_k$ are obtained by solving the equation

$$\mathbf{Ab} = \mathbf{g},$$

where $\mathbf{A} = \mathbf{X'X}$ is a nonsingular matrix and $\mathbf{g} = \mathbf{X'y}$ is a vector whose elements are sums of products of elements in the columns of **X** and the elements in the vector **y**. Thus the estimates are given by

$$\mathbf{b} = \mathbf{A}^{-1}\mathbf{g}.$$

Consider now the analysis-of-variance model in matrix form:

$$
\begin{bmatrix}
y_{11} \\
y_{12} \\
\vdots \\
y_{1n} \\
\hline
y_{21} \\
y_{22} \\
\vdots \\
y_{2n} \\
\hline
\vdots \\
\hline
y_{k1} \\
y_{k2} \\
\vdots \\
y_{kn}
\end{bmatrix}
=
\begin{bmatrix}
1 & 1 & 0 & \cdots & 0 \\
1 & 1 & 0 & \cdots & 0 \\
\vdots & \vdots & \vdots & & \vdots \\
1 & 1 & 0 & \cdots & 0 \\
\hline
1 & 0 & 1 & \cdots & 0 \\
1 & 0 & 1 & \cdots & 0 \\
\vdots & \vdots & \vdots & & \vdots \\
1 & 0 & 1 & \cdots & 0 \\
\hline
\vdots & \vdots & \vdots & & \vdots \\
\hline
1 & 0 & 0 & \cdots & 1 \\
1 & 0 & 0 & \cdots & 1 \\
\vdots & \vdots & \vdots & & \vdots \\
1 & 0 & 0 & \cdots & 1
\end{bmatrix}
\begin{bmatrix}
\mu \\
\alpha_1 \\
\alpha_2 \\
\vdots \\
\alpha_k
\end{bmatrix}
+
\begin{bmatrix}
\varepsilon_{11} \\
\varepsilon_{12} \\
\vdots \\
\varepsilon_{1n} \\
\hline
\varepsilon_{21} \\
\varepsilon_{22} \\
\vdots \\
\varepsilon_{2n} \\
\hline
\vdots \\
\hline
\varepsilon_{k1} \\
\varepsilon_{k2} \\
\vdots \\
\varepsilon_{kn}
\end{bmatrix}.
$$

Again each observation is expressed as a function of the parameters. Here the very important **X** matrix, the matrix of experimental conditions, consists of ones and zeros. Similar formulations can be written for the randomized complete block model.

Let us apply the least squares approach to the one-way analysis-of-variance model. The normal equations are given by

$$
\begin{bmatrix}
nk & n & n & \cdots & n \\
n & n & 0 & \cdots & 0 \\
n & 0 & n & \cdots & 0 \\
\vdots & \vdots & \vdots & & \vdots \\
n & 0 & 0 & \cdots & n
\end{bmatrix}
\begin{bmatrix}
\hat{\mu} \\
\hat{\alpha}_1 \\
\hat{\alpha}_2 \\
\vdots \\
\hat{\alpha}_k
\end{bmatrix}
=
\begin{bmatrix}
T_{..} \\
T_{1.} \\
T_{2.} \\
\vdots \\
T_{k.}
\end{bmatrix}.
$$

At this stage it is simple to illustrate why a distinction is made in the presentation of the two models. The last k columns of the **A** matrix for the analysis-of-variance model add to the first column and thus the matrix is *singular*, implying that there is no unique solution to the estimating equations. This initially seems like a serious drawback as far as the model is concerned. In fact, we say that the parameters in the model are not *estimable*. The reader will recall that significance tests were performed on the population means, $\mu_1 = \mu + \alpha_1, \mu_2 = \mu + \alpha_2, \ldots, \mu_k = \mu + \alpha_k$, and, in formulating the test procedure, the linear constraint $\sum_{i=1}^{k} \alpha_i = 0$ was applied. Thus the α_i's take on the role of deviations (plus or minus) of the treatment or population means from the overall mean μ. Testing equality of population means then becomes equivalent to testing that the α_i's ($i = 1, 2, \ldots, k$) are all zero.

With the constraint that the α_i's sum to zero, the estimating equations can be solved to yield

$$\hat{\mu} = \frac{T_{..}}{nk} = \bar{y}_{..}$$

$$\hat{\alpha}_i = \frac{T_{i.}}{n} - \frac{T_{..}}{nk} = \bar{y}_{i.} - \bar{y}_{..}, \qquad i = 1, 2, \ldots, k.$$

While these estimates are not unique, since they are dependent on the constraint that was applied to the α_i's, they do give us a basis for using the general regression procedure outlined in Section 9.6 to determine if the deletion of the α_i's from the model significantly increases the error sum of squares, thereby providing us, in the regression context, with a test of hypothesis of no significant treatment effects.

If we approach the hypothesis testing problem for the one-way analysis-of-variance model following the multiple regression procedures in Chapter 9, we might begin by computing the regression sum of squares for the parameters $\alpha_1, \alpha_2, \ldots, \alpha_k$. These parameters take on the same role as the coefficients $\beta_1, \beta_2, \ldots, \beta_k$ in the multiple linear regression model. We would then compute the regression sum of squares

$$R(\alpha, \alpha_2, \ldots, \alpha_k) = SSR$$

$$= b_0 g_0 + b_1 g_1 + \cdots + b_k g_k - \frac{\left(\sum_{i=1}^{k} \sum_{j=1}^{n} y_{ij} \right)^2}{nk}$$

$$= b_1 g_1 + \cdots + b_k g_k$$

$$= \hat{\alpha}_1 g_1 + \cdots + \hat{\alpha}_k g_k.$$

The right side of the estimating equations give $g_1 = T_1.., g_2 = T_2.., \ldots, g_k = T_k..$ Hence

$$R(\alpha_1, \alpha_2, \ldots, \alpha_k) = \sum_{i=1}^{k} \left(\frac{T_{i.}}{n} - \frac{T_{..}}{nk} \right) T_{i.}$$

$$= \sum_{i=1}^{k} \frac{T_{i.}^2}{n} - \frac{T_{..}^2}{nk}$$

$$= SSA.$$

with $k - 1$ degrees of freedom rather than k. One degree of freedom is lost on account of the single linear restraint imposed on the treatment effects. The error sum of squares with $(nk - 1) - (k - 1) = k(n - 1)$ degrees of freedom is given by

$$SSE = SST - R(\alpha_1, \alpha_2, \ldots, \alpha_k)$$
$$= SST - SSA,$$

which is identical to the expression developed earlier in this chapter.

The hypothesis that the regression on the α_i's is insignificant; that is, $\alpha_i = 0$ for all i's is tested by forming the ratio

$$f = \frac{R(\alpha_1, \alpha_2, \ldots, \alpha_k)/(k - 1)}{SSE/k(n - 1)} = \frac{SSA/(k - 1)}{s^2}.$$

A value of $f > f_\alpha[(k - 1), k(n - 1)]$ implies that regression is significantly increased and consequently the error sum of squares is significantly decreased by including the treatment effects in the model.

The regression approach to analysis-of-variance-type models can be extended to the randomized complete block design and to the factorial models of Chapter 11.

10.12 POWER OF ANALYSIS-OF-VARIANCE TESTS

As we indicated earlier, the research worker is often plagued by the problem of not knowing how large a sample to choose. In conducting a one-way fixed effects analysis of variance with n observations per treatment, the main objective is to test the hypothesis of equality of treatment means,

$$H_0: \quad \alpha_1 = \alpha_2 = \cdots = \alpha_k = 0,$$

$$H_1: \quad \text{at least one of the } \alpha_i\text{'s are not equal to zero.}$$

Quite often, however, the experimental error variance, σ^2, is so large that the test procedure will be insensitive to actual differences among the k treatment means. In Section 10.2 the expected values of the mean squares for the one-way model were given by

$$E(S_1^2) = E\left(\frac{SSA}{k - 1}\right) = \sigma^2 + \frac{n \sum_{i=1}^{k} \alpha_i^2}{k - 1}$$

$$E(S^2) = E\left(\frac{SSE}{k(n - 1)}\right) = \sigma^2.$$

Thus for a given deviation from the null hypothesis H_0, as measured by $n \sum_{i=1}^{k} \alpha_i^2 / (k - 1)$, large values of σ^2 decrease the chance of obtaining a value $f = s_1^2/s^2$ that is in the critical region for the test. The sensitivity of the test describes the ability of the procedure to detect differences in the population means and is measured by the power of the test (see Section 7.5), which is merely $1 - \beta$, where β is the probability of accepting a false hypothesis. We can interpret the power for our analysis-of-variance tests, then, as the probability that the F statistic is in the critical region when, in fact, the null hypothesis is false and the treatment means do differ. For the one-way analysis-of-variance test, the power, $1 - \beta$, is given by

$$1 - \beta = P\left[\frac{S_1^2}{S^2} > f_\alpha(v_1, v_2) \Big| H_1 \text{ is true} \right]$$

$$= P\left[\frac{S_1^2}{S^2} > f_\alpha(v_1, v_2) \Bigg| \frac{n \sum_{i=1}^{k} \alpha_i^2}{k - 1} \right].$$

The term $f_\alpha(v_1, v_2)$ is, of course, the upper tail critical point of the F distribution with v_1 and v_2 degrees of freedom. For given values of $\sum_{i=1}^{k} \alpha_i^2/(k - 1)$ and σ^2, the power can be increased by using a larger sample size n. The problem becomes one of designing the experiment with a value of n so that the power requirements are met. For example, we might require that for specific values of $\sum_{i=1}^{k} \alpha_i^2 \neq 0$ and σ^2, the hypothesis be rejected with probability 0.9. When the power of the test is low, it severely limits the scope of the inferences that can be drawn from the experimental data.

Fixed Effects Case

In the analysis of variance the power depends on the distribution of the F ratio under the alternative hypothesis that the treatment means differ. Therefore, in the case of the one-way fixed effects model, we require the distribution of S_1^2/S^2 when, in fact, $\sum_{i=1}^{k} \alpha_i^2 \neq 0$. Of course, when the hypothesis is true, $\alpha_i = 0$ for $i = 1, 2, \ldots, k$, and the statistic follows the F distribution with $k - 1$ and $N - k$ degrees of freedom. If $\sum_{i=1}^{k} \alpha_i^2 \neq 0$, the ratio follows a *noncentral F distribution*.

The basic random variable of the noncentral F is denoted by F'. Let $f'_\alpha(v_1, v_2, \lambda)$ be a value of F' with parameters v_1, v_2, and λ. The parameters v_1 and v_2 of the distribution are the degrees of freedom associated with S_1^2 and S^2, respectively, and λ is called the *noncentrality parameter*. When $\lambda = 0$, the noncentral F simply reduces to the ordinary F distribution with v_1 and v_2 degrees of freedom.

For the fixed effects, one-way analysis of variance with sample sizes $n_1, n_2, \ldots, n_k$, we define

$$\lambda = \frac{\sum_{i=1}^{k} n_i \alpha_i^2}{2\sigma^2}.$$

If we have tables of the noncentral F at our disposal, the power for detecting a particular alternative is obtained by evaluating the following probability:

$$1 - \beta = P\left[\frac{S_1^2}{S} > f_\alpha(k-1, N-k)\,\middle|\,\lambda = \frac{\sum_{i=1}^{k} n_i \alpha_i^2}{2\sigma^2}\right]$$

$$= P[F' > f_\alpha(k-1, N-k)].$$

Although the noncentral F is normally defined in terms of λ, it is more convenient, for purposes of tabulation, to work with

$$\phi^2 = \frac{2\lambda}{v_1 + 1}.$$

Table XV (see Statistical Tables) gives graphs of the power of the analysis of variance as a function of ϕ for various values of v_1, v_2, and the significance level α. These *power charts* can be used not only for the fixed effects models discussed in this chapter but also for the multifactor models of Chapter 11. It remains now to give a procedure whereby the noncentrality parameter λ, and thus ϕ, can be found for these fixed effects cases.

The noncentrality parameter λ can be written in terms of the *expected values of the numerator mean square* of the F ratio in the analysis of variance. We have

$$\lambda = \frac{v_1[E(S_i^2)]}{2\sigma^2} - \frac{v_1}{2}$$

and thus

$$\phi^2 = \frac{[E(S_i^2) - \sigma^2]}{\sigma^2}\frac{v_1}{v_1 + 1}.$$

Table 10.14 Noncentrality Parameter λ and ϕ^2 for Fixed Effects Models

	One-way classification	Randomized complete block
λ	$\dfrac{\sum\limits_{i=1}^{k} n_i \alpha_i^2}{2\sigma^2}$	$\dfrac{b \sum\limits_{i=1}^{k} \alpha_i^2}{2\sigma^2}$
ϕ^2	$\dfrac{\sum\limits_{i=1}^{k} n_i \alpha_i^2}{k\sigma^2}$	$\dfrac{b \sum\limits_{i=1}^{k} \alpha_i^2}{k\sigma^2}$

Expressions for λ and ϕ^2 for the one-way classification and the randomized complete block design are given in Table 10.14.

Note from Table XV that for given values of v_1 and v_2, the power of the test increases with increasing values of ϕ. The value of λ depends, of course, on σ^2, and in a practical problem one may often need to substitute the error mean square as an estimate in determining ϕ^2.

Example 10.7 In a randomized block experiment four treatments are to be compared in six blocks, resulting in 15 degrees of freedom for error. Are six blocks sufficient if the power of our test for detecting differences among the treatment means, at the 0.05 level of significance, is to be at least 0.8 when the true means are $\mu_1. = 5.0$, $\mu_2. = 7.0$, $\mu_3. = 4.0$, and $\mu_4. = 4.0$? An estimate of σ^2 to be used in the computation of the power is given by $\hat{\sigma}^2 = 2.0$.

Solution Recall that the treatment means are given by $\mu_i. = \mu + \alpha_i$. If we invoke the restriction that $\sum\limits_{i=1}^{4} \alpha_i = 0$, we have

$$\mu = \frac{\sum\limits_{i=1}^{4} \mu_i.}{4} = 5.0,$$

and then $\alpha_1 = 0$, $\alpha_2 = 2.0$, $\alpha_3 = -1.0$, and $\alpha_4 = -1.0$. Therefore,

$$\phi^2 = \frac{b \sum\limits_{i=1}^{k} \alpha_i^2}{k\sigma^2} = \frac{(6)(6)}{(4)(2)} = 4.5,$$

from which we obtain $\phi = 2.121$. Using Table XV, the power is found to be approximately 0.89 and thus the power requirements are met. This means that if the value of $\sum_{i=1}^{4} \alpha_i^2 = 6$ and $\sigma^2 = 2.0$, the use of six blocks will result in rejecting the hypothesis of equal treatment means with probability 0.89.

Random Effects Case

In the fixed effects case, the computation of power requires the use of the non-central F distribution. Such is not the case in the random effects model. In fact, the power is computed very simply by the use of the standard F tables. Consider, for example, the one-way random effects model, n observations per treatment, with the hypothesis

$$H_0: \quad \sigma_\alpha^2 = 0,$$
$$H_1: \quad \sigma_\alpha^2 \neq 0.$$

When H_1 is true, the ratio

$$f = \frac{SSA[(k-1)(\sigma^2 + n\sigma_\alpha^2)]}{SSE/k(n-1)\sigma^2} = \frac{s_1^2}{s^2(1 + n\sigma_\alpha^2/\sigma^2)}$$

is a value of the random variable F having the F distribution with $k-1$ and $k(n-1)$ degrees of freedom. The problem becomes one, then, of determining the probability of rejecting H_0 under the condition that the true treatment variance component $\sigma_\alpha^2 \neq 0$. We have then

$$1 - \beta = P\left\{ \frac{S_1^2}{S^2} > f_\alpha[(k-1), k(n-1)] \,\middle|\, H_1 \text{ is true} \right\}$$

$$= P\left\{ \frac{S_1^2}{S^2(1 + n\sigma_\alpha^2/\sigma^2)} > \frac{f_\alpha[(k-1), k(n-1)]}{1 + n\sigma_\alpha^2/\sigma^2} \right\}$$

$$= P\left\{ F > \frac{f_\alpha[(k-1), k(n-1)]}{1 + n\sigma_\alpha^2/\sigma^2} \right\}.$$

Note that as n increases, the value $f_\alpha[(k-1), k(n-1)]/(1 + n\sigma_\alpha^2/\sigma^2)$ approaches zero, resulting in an increase in the power of the test. An illustration of the power for this kind of situation is given in Figure 10.1. The crosshatched area is the significance level α, while the entire shaded area is the power of the test.

Example 10.8   Suppose in a one-way classification it is of interest to test for the significance of the variance component σ_α^2. Four treatments are to be used

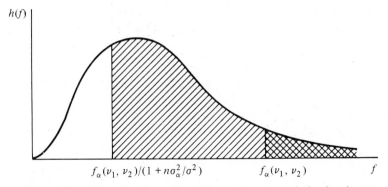

Figure 10.1 Power for the random effects one-way analysis of variance.

in the experiment with five observations per treatment. What will be the probability of rejecting the hypothesis $\sigma_\alpha^2 = 0$, when in fact the treatment variance component is $(3/4)\sigma^2$?

Solution Using an $\alpha = 0.05$ significance level, we have

$$1 - \beta = P\left\{F > \frac{f_{0.05}(3, 16)}{1 + (5)(3)/4}\right\}$$

$$= P\left[F > \frac{f_{0.05}(3, 16)}{4.75}\right]$$

$$= P\left(F > \frac{3.24}{4.75}\right)$$

$$= P(F > 0.864).$$

Using Theorem 5.19 and then Table A-7c of *Introduction to Statistical Analysis* by Dixon and Massey, we see that

$$1 - \beta \simeq 0.46.$$

Therefore, only about 46% of the time will the test procedure detect a variance component that is $(3/4)\sigma^2$.

EXERCISES

1. Show that the error mean square

$$s^2 = \frac{SSE}{k(n-1)}$$

for the analysis of variance in a one-way classification is an unbiased estimate of σ^2.

2. Show that the computing formula for SSA, in the analysis of variance of the one-way classification, is equivalent to the corresponding term in the identity of Theorem 10.1.

3. Six machines are being considered for use in manufacturing rubber seals. The machines are being compared with respect to tensile strength of the product. A random sample of four seals from each machine is used to determine whether or not the mean tensile strength varies from machine to machine. The following are the tensile strength measurements in kilograms per square centimeter $\times 10^{-1}$:

		Machine			
1	*2*	*3*	*4*	*5*	*6*
17.5	16.4	20.3	14.6	17.5	18.3
16.9	19.2	15.7	16.7	19.2	16.2
15.8	17.7	17.8	20.8	16.5	17.5
18.6	15.4	18.9	18.9	20.5	20.1

Perform the analysis of variance at the 0.05 level of significance and indicate whether or not the treatment means differ significantly.

4. Three sections of the same elementary mathematics course are taught by three teachers. The final grades were recorded as follows:

	Teacher	
A	*B*	*C*
73	88	68
89	78	79
82	48	56
43	91	91
80	51	71
73	85	71
66	74	87
60	77	41
45	31	59
93	78	68
36	62	53
77	76	79
	96	15
	80	
	56	

Is there a significant difference in the average grades given by the three teachers? Use a 0.05 level of significance.

5. Test for homogeneity of variances in Exercise 4. Use a 0.05 level of significance.

6. Four laboratories are being used to perform chemical analyses. Samples of the same material are sent to the laboratories for analysis as part of the study to determine whether or not they give, on the average, the same results. The analytical results for the four laboratories are as follows:

	Laboratory		
A	B	C	D
58.7	62.7	55.9	60.7
61.4	64.5	56.1	60.3
60.9	63.1	57.3	60.9
59.1	59.2	55.2	61.4
58.2	60.3	58.1	62.3

(a) Use Bartlett's test to show that the within-laboratory variances are not significantly different at the $\alpha = 0.05$ level of significance.

(b) Perform the analysis of variance and give conclusions concerning the laboratories.

(c) Extend the analysis of variance to make significance tests on the following contrasts: (1) B versus A, C, D; (2) C versus A and D; and (3) A versus D.

7. The number of bacteria in six containers of milk were recorded by each of four observers. The bacteria counts are as follows:

	Observer		
A	B	C	D
230	184	205	196
241	72	156	210
336	214	308	284
128	348	118	312
253	68	247	125
124	330	104	99

Use Cochran's test at the 0.05 level of significance to test for homogeneity of variances.

8. An investigation was conducted to determine the source of reduction in yield of a certain chemical product. It was known that the loss in yield occurred in the mother liquor, that is, the material removed at the filtration stage. It was felt that different blends of the original material may result in different yield reductions at the mother

liquor stage. The following are results of the per cent reduction for three batches at each of four preselected blends:

	Blend		
1	*2*	*3*	*4*
25.6	25.2	20.8	31.6
24.3	28.6	26.7	29.8
27.9	24.7	22.2	34.3

(a) Perform the analysis of variance at the $\alpha = 0.05$ level of significance.
(b) Use Duncan's multiple-range test to determine which blends differ.

9. In the following biological experiment four concentrations of a certain chemical are used to enhance the growth in centimeters of a certain type of plant over time. Five plants are used at each concentration and the growth in each plant is measured. The following growth data are taken. A control (no chemical) is also applied.

	Concentration			
Control	*1*	*2*	*3*	*4*
5.9	8.2	7.7	6.9	6.8
6.1	8.7	8.4	5.8	7.3
6.9	9.4	8.6	7.2	6.3
5.7	9.2	8.1	6.8	6.9
6.1	8.6	8.0	7.4	7.1

Use Dunnett's one-sided test at the $\alpha = 0.05$ level of significance to simultaneously compare the concentrations with the control.

10. Four kinds of fetilizer $f_1, f_2, f_3,$ and f_4 are used to study the yield of beans. The soil is divided into three blocks each containing four homogenous plots. The yields in kilograms per plot and the corresponding treatments are as follows:

Block 1	Block 2	Block 3
$f_1 = 42.7$	$f_3 = 50.9$	$f_4 = 51.1$
$f_3 = 48.5$	$f_1 = 50.0$	$f_2 = 46.3$
$f_4 = 32.8$	$f_2 = 38.0$	$f_1 = 51.9$
$f_2 = 39.3$	$f_4 = 40.2$	$f_3 = 53.5$

(a) Conduct an analysis of variance using the randomized complete block model.

(b) Use single-degree-of-freedom contrasts to make the following comparisons among the fertilizers: (1) (f_1, f_3) versus (f_2, f_4); (2) f_1 versus f_3.

11. Show that the computing formula for SSB, in the analysis of variance of the randomized complete block design, is equivalent to the corresponding term in the identity of Theorem 10.3.

12. For the randomized block design with k treatments and b blocks, show that

$$E(SSB) = (b - 1)\sigma^2 + k \sum_{j=1}^{b} \beta_j^2.$$

13. An experiment is conducted in which four treatments are to be compared in five blocks. The following data are generated:

			Block		
Treatment	1	2	3	4	5
1	12.8	10.6	11.7	10.7	11.0
2	11.7	14.2	11.8	9.9	13.8
3	11.5	14.7	13.6	10.7	15.9
4	12.6	16.5	15.4	9.6	17.1

Perform the analysis of variance, separating out the treatment, block, and error sums of squares. Use a 0.05 level of significance to test the hypothesis that there is no difference between the treatment means.

14. Three catalysts are used in a chemical process with a control (no catalyst) being included. The following are yield data from the process:

		Catalyst	
Control	1	2	3
74.5	77.5	81.5	78.1
76.1	82.0	82.3	80.2
75.9	80.6	81.4	81.5
78.1	84.9	79.5	83.0
76.2	81.0	83.0	82.1

Use Dunnett's test at the $\alpha = 0.01$ level of significance to determine if a significantly higher yield is obtained with the catalysts than with no catalyst.

15. The following data show the effect of four operators, chosen randomly, on the output of a particular machine:

	Operator		
1	2	3	4
175.4	168.5	170.1	175.2
171.7	162.7	173.4	175.7
173.0	165.0	175.7	180.1
170.5	164.1	170.7	183.7

Perform a model II analysis of variance and compute an estimate of the operator variance component and the experimental error variance component.

16. Assuming a random effects model for Exercise 13, compute estimates of the treatment and block variance components.

17. (a) Using a regression approach for the randomized complete block design, obtain the normal equations $\mathbf{Ab} = \mathbf{g}$ in matrix form.

(b) Show that $R(\beta_1, \beta_2, \ldots, \beta_b | \alpha_1, \alpha_2, \ldots, \alpha_k) = SSB$.

18. In Exercise 13, if one uses an $\alpha = 0.05$ level test, how many blocks are needed in order that we accept the hypothesis of equality of treatment means with probability 0.1 when, in fact,

$$\frac{\sum\limits_{i=1}^{4} \alpha_i^2}{\sigma^2} = 2.0?$$

19. In Exercise 15, if we are interested in testing for the significance of the operator variance component, do we have large enough samples to ensure with a probability as large as 0.95 a significant variance component if the true σ_α^2 is $1.5\sigma^2$? If not, how many runs are necessary for each operator? Use a 0.05 level of significance.

11

Factorial Experiments

11.1 TWO-FACTOR EXPERIMENTS

Consider a situation in which it is of interest to study the effect of *two factors* A and B on some response. For example, in a chemical experiment we would like to simultaneously vary the reaction pressure and reaction time and study the effect of each on the yield. In a biological experiment, it is of interest to study the effect of drying time and temperature on the amount of solids (percent by weight) left in samples of yeast. The term *factor* is used in a general sense to denote any feature of the experiment such as temperature, time, or pressure that may be varied from trial to trial. We define the *levels* of a factor to be the actual values used in the experiment.

In each of these cases it is important not only to determine if the two factors have an influence on the response but also if there is a significant interaction between the two factors. As far as terminology is concerned, the experiment described here is a two-way classification or a two-factor experiment with a completely randomized design. That is, no restrictions such as blocking are made on the experimental units. In the case of the yeast example, the various treatment combinations of temperature and drying time would be assigned randomly to the samples of yeast.

11.2 INTERACTION IN TWO-FACTOR EXPERIMENTS

In the randomized block model discussed previously it was assumed that one observation on each treatment is taken in each block. If the model assumption is correct, that is, if blocks and treatments are the only real effects and interaction does not exist, the expected value of the error mean square is the experimental error variance σ^2. Suppose, however, that there is interaction occurring between treatments and blocks as indicated by the model

$$y_{ij} = \mu + \alpha_i + \beta_j + (\alpha\beta)_{ij} + \varepsilon_{ij}$$

417

of Section 10.9. The expected value of the error mean square was then given as

$$E\left[\frac{SSE}{(b-1)(k-1)}\right] = \sigma^2 + \frac{\sum_{i=1}^{k}\sum_{j=1}^{b}(\alpha\beta)_{ij}^2}{(b-1)(k-1)}.$$

The treatment and block effects do not appear in the expected error mean square but the interaction effects do. Thus if there is interaction in the model, the error mean square reflects variation due to experimental error *plus* an interaction contribution and, for this experimental plan, there is no way of separating them.

From an experimenter's point of view it should seem necessary to arrive at a significance test on the existence of interaction by separating true error variation from that due to interaction. The effects of factors A and B, often called the *main effects*, take on a different meaning in the presence of interaction. In the previous biological example the effect that drying time has on the amount of solids left in the yeast might very well depend on the temperature that the samples are exposed to. In general, there could very well be experimental situations in which factor A has a positive effect on the response at one level of factor B, while at a different level of factor B the effect of A is negative. We use the term *positive effect* here to indicate that the yield or response increases as the levels of a given factor increase according to some defined order. In the same sense a *negative effect* corresponds to a decrease in yield for increasing levels of the factor. Consider, for example, the following hypothetical data taken on two factors each at three levels:

		B		
A	b_1	b_2	b_3	*Total*
a_1	4.4	8.8	5.2	18.4
a_2	7.5	8.5	2.4	18.4
a_3	9.7	7.9	0.8	18.4
Total	21.6	25.2	8.4	55.2

Clearly, the effect of A is positive at b_1 and negative at b_3. These differences in the levels of A at different levels of B are of interest to the experimenter but an ordinary significance test on factor A would yield a value of zero for SSA, since the totals for each level of A are all of the same magnitude. We say, then, that the presence of interaction is *masking* the effect of factor A. Therefore, if we consider the average influence of A, over all three levels of B, there is no effect. However, this is most likely not what is pertinent to the experimenter.

We suggest that the experimenter first attempt to detect interaction with a test of significance. Then if interaction is not significant, he would proceed to make tests on the effects of the main factors. If the data indicate the presence of interaction, this might well dictate the need to observe the influence of each factor at fixed levels of the other.

Interaction and experimental error are separated in the two-factor experiment only if multiple observations are taken at the various treatment combinations. To ease the computations involved, there should be the same number, n, of observations at each combination. These should be true replications, not just repeated measurements. For example, in the yeast illustration, if we take $n = 2$ observations at each combination of temperature and drying time, there should be two separate samples and not merely repeated measurements on the same sample. This is important because now the measure of experimental error comes from variation between readings *within* treatment combinations and thus indicates true or pure experimental error.

11.3 TWO-FACTOR ANALYSIS OF VARIANCE

To present general formulas for the analysis of variance of a two-factor experiment using repeated observations, we shall consider the case of n replications of the treatment combinations determined by a levels of factor A and b levels of factor B. The observations may be classified by means of a rectangular array in which the rows represent the levels of factor A and the columns represent the levels of factor B. Each treatment combination defines a cell in our array. Thus we have ab cells, each cell containing n observations. Denoting the kth observation taken at the ith level of factor A and the jth level of factor B by y_{ijk}, the abn observations are shown in Table 11.1.

The observations in the (ij)th cell constitute a random sample of size n from a population that is assumed to be normally distributed with mean μ_{ij} and variance σ^2. All ab populations are assumed to have the same variance σ^2. Let us define the following useful symbols, some of which are used in Table 11.1:

$T_{ij.}$ = sum of the observations in the (ij)th cell

$T_{i..}$ = sum of the observations for the ith level of factor A

$T_{.j.}$ = sum of the observations for the jth level of factor B

$T_{...}$ = sum of all abn observations

$\bar{y}_{ij.}$ = mean of the observations in the (ij)th cell

$\bar{y}_{i..}$ = mean of the observations for the ith level of factor A

$\bar{y}_{.j.}$ = mean of the observations for the jth level of factor B

$\bar{y}_{...}$ = mean of all abn observations.

Table 11.1 Two-Factor Experiment with n Replications

	B					
A	1	2	$\cdots$	b	Total	Mean
1	y_{111} y_{112} $\vdots$ y_{11n}	y_{121} y_{122} $\vdots$ y_{12n}	$\cdots$ $\cdots$ $\cdots$	y_{1b1} y_{1b2} $\vdots$ y_{1bn}	$T_{1..}$	$\bar{y}_{1..}$
2	y_{211} y_{212} $\vdots$ y_{21n}	y_{221} y_{222} $\vdots$ y_{22n}	$\cdots$ $\cdots$ $\cdots$	y_{2b1} y_{2b2} $\vdots$ y_{2bn}	$T_{2..}$	$\bar{y}_{2..}$
$\vdots$	$\vdots$	$\vdots$		$\vdots$	$\vdots$	$\vdots$
a	y_{a11} y_{a12} $\vdots$ y_{a1n}	y_{a21} y_{a22} $\vdots$ y_{a2n}	$\cdots$ $\cdots$ $\cdots$	y_{ab1} y_{ab2} $\vdots$ y_{abn}	$T_{a..}$	$\bar{y}_{a..}$
Total	$T_{.1.}$	$T_{.2.}$	$\cdots$	$T_{.b.}$	$T_{...}$	
Mean	$\bar{y}_{.1.}$	$\bar{y}_{.2.}$	$\cdots$	$\bar{y}_{.b.}$		$\bar{y}_{...}$

Each observation in Table 11.1 may be written in the form

$$y_{ijk} = \mu_{ij} + \varepsilon_{ijk},$$

where ε_{ijk} measures the deviations of the observed y_{ijk} values in the (ij)th cell from the population mean μ_{ij}. If we let $(\alpha\beta)_{ij}$ denote the interaction effect of the ith level of factor A and the jth level of factor B, α_i the effect of the ith level of factor A, β_j the effect of the jth level of factor B, and μ the overall mean, we can write

$$\mu_{ij} = \mu + \alpha_i + \beta_j + (\alpha\beta)_{ij},$$

and then

$$y_{ijk} = \mu + \alpha_i + \beta_j + (\alpha\beta)_{ij} + \varepsilon_{ijk},$$

on which we impose the restrictions

$$\sum_{i=1}^{a} \alpha_i = 0, \qquad \sum_{j=1}^{b} \beta_j = 0, \qquad \sum_{i=1}^{a} (\alpha\beta)_{ij} = 0, \qquad \sum_{j=1}^{b} (\alpha\beta)_{ij} = 0.$$

The three hypotheses to be tested are as follows:

1. H'_0: $\alpha_1 = \alpha_2 = \cdots = \alpha_a = 0$,
   H'_1:   at least one of the α_i's is not equal to zero.
2. H''_0: $\beta_1 = \beta_2 = \cdots = \beta_b = 0$,
   H''_1:   at least one of the β_j's is not equal to zero.
3. H'''_0: $(\alpha\beta)_{11} = (\alpha\beta)_{12} = \cdots = (\alpha\beta)_{ab} = 0$,
   H'''_1:   at least one of the $(\alpha\beta)_{ij}$'s is not equal to zero.

Each of these tests will be based on a comparison of independent estimates of σ^2 provided by the splitting of the total sum of squares of our data into four components by means of the following identity.

THEOREM 11.1   *Sum of Squares Identity*

$$\sum_{i=1}^{a} \sum_{j=1}^{b} \sum_{k=1}^{n} (y_{ijk} - \bar{y}...)^2 = bn \sum_{i=1}^{a} (\bar{y}_{i..} - \bar{y}...)^2$$

$$+ an \sum_{j=1}^{b} (\bar{y}_{.j.} - \bar{y}...)^2$$

$$+ n \sum_{i=1}^{a} \sum_{j=1}^{b} (\bar{y}_{ij.} - \bar{y}_{i..} - \bar{y}_{.j.} + \bar{y}...)^2$$

$$+ \sum_{i=1}^{a} \sum_{j=1}^{b} \sum_{k=1}^{n} (y_{ijk} - \bar{y}_{ij.})^2.$$

Proof

$$\sum_{i=1}^{a} \sum_{j=1}^{b} \sum_{k=1}^{n} (y_{ijk} - \bar{y}...)^2 = \sum_{i=1}^{a} \sum_{j=1}^{b} \sum_{k=1}^{n} [(\bar{y}_{i..} - \bar{y}...) + (\bar{y}_{.j.} - \bar{y}...)$$

$$+ (\bar{y}_{ij.} - \bar{y}_{i..} - \bar{y}_{.j.} + \bar{y}...) + (y_{ijk} - \bar{y}_{ij.})]^2$$

$$= \sum_{i=1}^{a} \sum_{j=1}^{b} \sum_{k=1}^{n} (\bar{y}_{i..} - \bar{y}...)^2$$

$$+ \sum_{i=1}^{a} \sum_{j=1}^{b} \sum_{k=1}^{n} (\bar{y}_{.j.} - \bar{y}...)^2$$

$$+ \sum_{i=1}^{a} \sum_{j=1}^{b} \sum_{k=1}^{n} (\bar{y}_{ij.} - \bar{y}_{i..} - \bar{y}_{.j.} + \bar{y}...)^2$$

$$+ \sum_{i=1}^{a} \sum_{j=1}^{b} \sum_{k=1}^{n} (y_{ijk} - \bar{y}_{ij.})^2$$

$$+ \text{6 cross-product terms.}$$

The cross-product terms are all equal to zero. Hence

$$\sum_{i=1}^{a}\sum_{j=1}^{b}\sum_{k=1}^{n}(y_{ijk} - \bar{y}...)^2 = bn\sum_{i=1}^{a}(\bar{y}_{i..} - \bar{y}...)^2 + an\sum_{j=1}^{b}(\bar{y}_{.j.} - \bar{y}...)^2$$

$$+ n\sum_{i=1}^{a}\sum_{j=1}^{b}(\bar{y}_{ij.} - \bar{y}_{i..} - \bar{y}_{.j.} + \bar{y}...)^2$$

$$+ \sum_{i=1}^{a}\sum_{j=1}^{b}\sum_{k=1}^{n}(y_{ijk} - \bar{y}_{ij.})^2.$$

Symbolically, we write the sum of squares identity as

$$SST = SSA + SSB + SS(AB) + SSE,$$

where SSA and SSB are called the sum of squares for the main effects A and B, respectively, $SS(AB)$ is called the interaction sum of squares for A and B, and SSE is the error sum of squares. The degrees of freedom are partitioned according to the identity

$$abn - 1 = (a - 1) + (b - 1) + (a - 1)(b - 1) + ab(n - 1).$$

Dividing each of the sum of squares on the right side of the sum of squares identity by their corresponding number of degrees of freedom, we obtain the four independent estimates

$$s_1^2 = \frac{SSA}{a - 1}, \quad s_2^2 = \frac{SSB}{b - 1}, \quad s_3^2 = \frac{SS(AB)}{(a - 1)(b - 1)}, \quad s^2 = \frac{SSE}{ab(n - 1)}$$

of σ^2. If we interpret the sum of squares as functions of the independent random variables $Y_{111}, Y_{112}, \ldots, Y_{abn}$, it is not difficult to verify that

$$E(S_1^2) = E\left[\frac{SSA}{a - 1}\right] = \sigma^2 + \frac{nb\sum_{i=1}^{a}\alpha_i^2}{a - 1}$$

$$E(S_2^2) = E\left[\frac{SSB}{b - 1}\right] = \sigma^2 + \frac{na\sum_{j=1}^{b}\beta_j^2}{b - 1}$$

$$E(S_3^2) = E\left[\frac{SS(AB)}{(a - 1)(b - 1)}\right] = \sigma^2 + \frac{n\sum_{i=1}^{a}\sum_{j=1}^{b}(\alpha\beta)_{ij}^2}{(a - 1)(b - 1)}$$

$$E(S^2) = E\left[\frac{SSE}{ab(n - 1)}\right] = \sigma^2,$$

from which we immediately conclude that all four estimates of σ^2 are unbiased when H_0', H_0'', and H_0''' are true.

To test the hypothesis H_0', that the effects of factors A are all equal to zero, we compute the ratio

$$ f_1 = \frac{s_1^2}{s^2}, $$

which is a value of the random variable F_1 having the F distribution with $a - 1$ and $ab(n - 1)$ degrees of freedom when H_0' is true. The null hypothesis is rejected at the α level of significance when $f_1 > f_\alpha[a - 1, ab(n - 1)]$. Similarly, to test the hypothesis H_0'', that the effects of factor B are all equal to zero, we compute the ratio

$$ f_2 = \frac{s_2^2}{s^2}, $$

which is a value of the random variable F_2 having the F distribution with $b - 1$ and $ab(n - 1)$ degrees of freedom when H_0'' is true. This hypothesis is rejected at the α level of significance when $f_2 > f_\alpha[b - 1, ab(n - 1)]$. Finally, to test the hypothesis H_0''', that the interaction effects are all equal to zero, we compute the ratio

$$ f_3 = \frac{s_3^2}{s^2}, $$

Table 11.2 Analysis of Variance for the Two-Factor Experiment with n Replications

Source of variation	Sum of squares	Degrees of freedom	Mean square	Computed f
Main effect				
A	SSA	$a - 1$	$s_1^2 = \dfrac{SSA}{a - 1}$	$f_1 = \dfrac{s_1^2}{s^2}$
B	SSB	$b - 1$	$s_2^2 = \dfrac{SSB}{b - 1}$	$f_2 = \dfrac{s_2^2}{s^2}$
Two-factor interaction				
AB	SS(AB)	$(a - 1)(b - 1)$	$s_3^2 = \dfrac{SS(AB)}{(a - 1)(b - 1)}$	$f_3 = \dfrac{s_3^2}{s^2}$
Error	SSE	$ab(n - 1)$	$s^2 = \dfrac{SSE}{ab(n - 1)}$	
Total	SST	$abn - 1$		

which is a value of the random variable F_3 having the F distribution with $(a - 1)(b - 1)$ and $ab(n - 1)$ degrees of freedom when H_0''' is true. We conclude that interaction is present when $f_3 > f_\alpha[(a - 1)(b - 1), ab(n - 1)]$.

As indicated in Section 11.2, it is advisable to conduct the test for interaction before attempting to draw inferences on the main effects. If interaction is not significant, the experimenter can proceed to test the main effects. However, a significant interaction could very well imply that the data should be analyzed in a somewhat different manner—perhaps observing the effect of factor A at fixed levels of factor B, and so forth.

The computations in an analysis-of-variance problem, for a two-factor experiment with n replications, are usually summarized as in Table 11.2.

The sums of squares are usually obtained by constructing the following table of totals:

	B				
A	1	2	$\cdots$	b	Total
1	$T_{11\cdot}$	$T_{12\cdot}$	$\cdots$	$T_{1b\cdot}$	$T_{1\cdot\cdot}$
2	$T_{21\cdot}$	$T_{22\cdot}$	$\cdots$	$T_{2b\cdot}$	$T_{2\cdot\cdot}$
$\vdots$	$\vdots$	$\vdots$		$\vdots$	$\vdots$
a	$T_{a1\cdot}$	$T_{a2\cdot}$	$\cdots$	$T_{ab\cdot}$	$T_{a\cdot\cdot}$
Total	$T_{\cdot1\cdot}$	$T_{\cdot2\cdot}$	$\cdots$	$T_{\cdot b\cdot}$	$T_{\cdot\cdot\cdot}$

and using the following computational formulas:

$$SST = \sum_{i=1}^{a} \sum_{j=1}^{b} \sum_{k=1}^{n} y_{ijk}^2 - \frac{T_{\cdots}^2}{abn}$$

$$SSA = \frac{\sum_{i=1}^{a} T_{i\cdots}^2}{bn} - \frac{T_{\cdots}^2}{abn}$$

$$SSB = \frac{\sum_{j=1}^{b} T_{\cdot j\cdot}^2}{an} - \frac{T_{\cdots}^2}{abn}$$

$$SS(AB) = \frac{\sum_{i=1}^{a} \sum_{j=1}^{b} T_{ij\cdot}^2}{n} - \frac{\sum_{i=1}^{a} T_{i\cdots}^2}{bn} - \frac{\sum_{j=1}^{b} T_{\cdot j\cdot}^2}{an} + \frac{T_{\cdots}^2}{abn}.$$

As before, SSE is obtained by subtraction using the formula

$$SSE = SST - SSA - SSB - SS(AB).$$

Table 11.3 Propellant Burning Rates

Missile system	Propellant type			
	b_1	b_2	b_3	b_4
a_1	34.0	30.1	29.8	29.0
	32.7	32.8	26.7	28.9
a_2	32.0	30.2	28.7	27.6
	33.2	29.8	28.1	27.8
a_3	28.4	27.3	29.7	28.8
	29.3	28.9	27.3	29.1

Example 11.1   In an experiment conducted to determine which of three different missile systems is preferable, the propellant burning rate for 24 static firings was measured. Four different propellant types were used. The experiment yielded duplicate observations of burning rates at each combination of the treatments. The data, after coding, are given in Table 11.3.

Use a 0.05 level of significance to test the following hypotheses: (a) H_0': There is no difference in the mean propellant burning rates when different missile systems are used. (b) H_0'': There is no difference in the mean propellant burning rates of the four propellant types. (c) H_0''': There is no interaction between the different missile systems and the different propellant types.

Solution
1. (a) H_0': $\alpha_1 = \alpha_2 = \alpha_3 = 0$.
   (b) H_0'': $\beta_1 = \beta_2 = \beta_3 = \beta_4 = 0$.
   (c) H_0''': $(\alpha\beta)_{11} = (\alpha\beta)_{12} = \cdots = (\alpha\beta)_{34} = 0$.
2. (a) H_1':   at least one of the α_i's is not equal to zero.
   (b) H_1'': at least one of the β_j's is not equal to zero.
   (c) H_1''': at least one of the $(\alpha\beta)_{ij}$'s is not equal to zero.
3. $\alpha = 0.05$.
4. Critical regions: (a) $F_1 > 3.89$, (b) $F_2 > 3.49$, (c) $F_3 > 3.00$.
5. Computations: From Table 11.3 we first construct the following table of totals:

	b_1	b_2	b_3	b_4	Total
a_1	66.7	62.9	56.5	57.9	244.0
a_2	65.2	60.0	56.8	55.4	237.4
a_3	57.7	56.2	57.0	57.9	228.8
Total	189.6	179.1	170.3	171.2	710.2

Now

$$SST = 34.0^2 + 32.7^2 + \cdots + 29.1^2 - \frac{710.2^2}{24}$$

$$= 21,107.68 - 21,016.00 = 91.68$$

$$SSA = \frac{244.0^2 + 237.4^2 + 228.8^2}{8} - \frac{710.2^2}{24}$$

$$= 21,030.52 - 21,016.00 = 14.52$$

$$SSB = \frac{189.6^2 + 179.1^2 + 170.3^2 + 171.2^2}{6} - \frac{710.2^2}{24}$$

$$= 21,056.08 - 21,016.00 = 40.08$$

$$SS(AB) = \frac{66.7^2 + 65.2^2 + \cdots + 57.9^2}{2} - 21,030.52$$

$$- 21,056.08 + 21,016.00$$

$$= 22.17$$

$$SSE = 91.68 - 14.52 - 40.08 - 22.17 = 14.91.$$

These results, with the remaining computations, are given in Table 11.4.
6. Conclusions:
   (a) Accept H_0''' and conclude that there is no interaction between the different missile systems and the different propellant types.
   (b) Reject H_0' and conclude that different missile systems result in different mean propellant burning rates.
   (c) Reject H_0'' and conclude that the mean propellant burning rates are not the same for the four propellant types.

Table 11.4 Analysis of Variance for the Data of Table 11.3

Source of variation	Sum of squares	Degrees of freedom	Mean square	Computed f
Missile system	14.52	2	7.26	5.85
Propellant type	40.08	3	13.36	10.77
Interaction	22.17	6	3.70	2.98
Error	14.91	12	1.24	
Total	91.68	23		

Example 11.2 Referring to Example 11.1, choose two orthogonal contrasts to partition the sum of squares for the missile systems into single-degree-of-freedom components to be used in comparing systems 1 and 2 with 3 and system 1 versus system 2.

Solution The contrast for comparing systems 1 and 2 with 3 is given by

$$\omega_1 = \mu_{1.} + \mu_{2.} - 2\mu_{3..}$$

A second contrast, orthogonal to ω_1, for comparing system 1 with system 2, is given by $\omega_2 = \mu_{1.} - \mu_{2..}$. The single-degree-of-freedom sum of squares are as follows:

$$SSw_1 = \frac{[244.0 + 237.4 - (2)(228.8)]^2}{(8)[(1)^2 + (1)^2 + (-2)^2]} = 11.80$$

$$SSw_2 = \frac{(244.0 - 237.4)^2}{(8)[(1)^2 + (-1)^2]} = 2.72.$$

Notice that $SSw_1 + SSw_2 = SSA$, as expected. The computed f values corresponding to w_1 and w_2 are, respectively,

$$f_1 = \frac{11.80}{1.24} = 9.5$$

and

$$f_2 = \frac{2.72}{1.24} = 2.2.$$

Compared to the critical value $f_{0.05}(1, 12) = 4.75$, we find f_1 to be significant. Thus the first contrast indicates that the hypothesis

$$H_0: \quad \frac{\mu_{1.} + \mu_{2.}}{2} = \mu_{3.}.$$

is rejected. Since $f_2 < 4.75$, the mean burning rates of the first and second systems are not significantly different.

If the hypothesis of no interaction in Example 11.1 is true, as indicated by the f ratio in the analysis-of-variance table, we are able to make the *general* comparisons of Example 11.2 regarding our missile systems rather than separate comparisons for each propellant. Similarly, we might make general comparisons among the propellants rather than separate comparisons for each missile system. For example, we could compare propellants 1 and 2 with 3 and 4 and also propellant 1 versus propellant 2. The resulting f ratios, each with 1 and 12 degrees of freedom, turn out to be 24.86 and 7.41, respectively, and both are significant at the 0.05 level of significance. The indication is then that propellant 1 gives the highest mean burning rate. A prudent experimenter might be somewhat cautious in making overall conclusions in a problem such as this one where the f ratio for interaction is barely below the 0.05 critical value. This is far from overwhelming evidence that interaction between the factors does not exist. In fact, a quick inspection of the cell totals points out possible evidence of interaction. For example, the overall evidence, 189.6 versus 171.2, certainly indicates that propellant 1 is superior, in terms of a higher burning rate, to propellant 4. However, if we restrict ourselves to system 3, where we have a total of 57.7 for propellant 1 as opposed to 57.9 for propellant 4, there appears to be little or no difference between propellants 1 and 4. In fact, there appears to be a stabilization of burning rates for the different propellants if we operate with system 3. There is certainly overall evidence, 244.0 versus 228.8, that indicates that system 1 gives a higher burning rate than system 3, but if we restrict ourselves to propellant 4, this conclusion does not appear to hold.

11.4 THREE-FACTOR EXPERIMENTS

In this section we consider an experiment with three factors A, B, and C at a, b, and c levels, respectively, in a completely randomized experimental design. Assume again that we have n observations for each of the abc treatment combinations. We shall proceed to outline significance tests for the three main effects and interactions involved. It is hoped that the reader can then use the description given here to generalize the analysis to $k > 3$ factors.

The model for the three-factor experiment is given by

$$y_{ijkl} = \mu + \alpha_i + \beta_j + \gamma_k + (\alpha\beta)_{ij} + (\alpha\gamma)_{ik} + (\beta\gamma)_{jk} + (\alpha\beta\gamma)_{ijk} + \varepsilon_{ijkl},$$

$i = 1, 2, \ldots, a$; $j = 1, 2, \ldots, b$; $k = 1, 2, \ldots, c$; and $l = 1, 2, \ldots, n$—where α_i, β_j, and γ_k are the main effects; $(\alpha\beta)_{ij}$, $(\alpha\gamma)_{ik}$, and $(\beta\gamma)_{jl}$ are the two-factor interaction effects that have the same interpretation as in the two-factor experiment. The term $(\alpha\beta\gamma)_{ijk}$ is called the *three-factor interaction effect*, a term that represents a nonadditivity of the $(\alpha\beta)_{ij}$ over the different levels of the factor C. As before, the sum of all main effects is zero and the sum over any subscript of the two- and three-factor interaction effects is zero. In many experimental situations these higher-order interactions are insignificant and their mean squares reflect only random variation, but we shall outline the analysis in its most general detail.

Again, in order that valid significance tests can be made, we must assume that the errors are values of independent and normally distributed random variables, each with zero mean and common variance σ^2.

The general philosophy concerning the analysis is the same as that discussed for the one- and two-factor experiments. The sum of squares is partitioned into eight terms, each representing a source of variation from which we obtain independent estimates of σ^2 when all the main effects and interaction effects are zero. If the effects of any given factor or interaction are not all zero, then the mean square will estimate the error variance plus a component due to the systematic effect in question.

Let us proceed directly to the computational procedure for obtaining the sums of squares in the three-factor analysis of variance. We require the following notation:

$T_{....}$ = sum of all $abcn$ observations

$T_{i...}$ = sum of the observations for the ith level of factor A

$T_{.j..}$ = sum of the observations for the jth level of factor B

$T_{..k.}$ = sum of the observations for the kth level of factor C

$T_{ij..}$ = sum of the observations for the ith level of A and the jth level of B

$T_{i\cdot k\cdot}$ = sum of the observations for the ith level of A and the kth level of C

$T_{\cdot jk\cdot}$ = sum of the observations for the jth level of B and kth level of C

$T_{ijk\cdot}$ = sum of the observations for the (ijk)th treatment combination.

In practice it is advantageous to construct the following two-way tables of totals and subtotals:

		B				
A	1	2	$\cdots$	b	Total	
1	$T_{11k\cdot}$	$T_{12k\cdot}$	$\cdots$	$T_{1bk\cdot}$	$T_{1\cdot k\cdot}$	
2	$T_{21k\cdot}$	$T_{22k\cdot}$	$\cdots$	$T_{2bk\cdot}$	$T_{2\cdot k\cdot}$	
$\vdots$	$\vdots$	$\vdots$		$\vdots$	$\vdots$	$k = 1, 2, \ldots, c$
a	$T_{a1k\cdot}$	$T_{a2k\cdot}$	$\cdots$	$T_{abk\cdot}$	$T_{a\cdot k\cdot}$	
Total	$T_{\cdot 1k\cdot}$	$T_{\cdot 2k\cdot}$	$\cdots$	$T_{\cdot bk\cdot}$	$T_{\cdot\cdot k\cdot}$	

A	B				Total
	1	2	$\cdots$	b	
1	$T_{11\cdot\cdot}$	$T_{12\cdot\cdot}$	$\cdots$	$T_{1b\cdot\cdot}$	$T_{1\cdots}$
2	$T_{21\cdot\cdot}$	$T_{22\cdot\cdot}$	$\cdots$	$T_{2b\cdot\cdot}$	$T_{2\cdots}$
$\vdots$	$\vdots$	$\vdots$		$\vdots$	$\vdots$
a	$T_{a1\cdot\cdot}$	$T_{a2\cdot\cdot}$	$\cdots$	$T_{ab\cdot\cdot}$	$T_{a\cdots}$
Total	$T_{\cdot1\cdot\cdot}$	$T_{\cdot2\cdot\cdot}$	$\cdots$	$T_{\cdot b\cdot\cdot}$	$T_{\cdots\cdot}$

A	C				Total
	1	2	$\cdots$	c	
1	$T_{1\cdot1\cdot}$	$T_{1\cdot2\cdot}$	$\cdots$	$T_{1\cdot c\cdot}$	$T_{1\cdots}$
2	$T_{2\cdot1\cdot}$	$T_{2\cdot2\cdot}$	$\cdots$	$T_{2\cdot c\cdot}$	$T_{2\cdots}$
$\vdots$	$\vdots$	$\vdots$		$\vdots$	$\vdots$
a	$T_{a\cdot1\cdot}$	$T_{a\cdot2\cdot}$	$\cdots$	$T_{a\cdot c\cdot}$	$T_{a\cdots}$
Total	$T_{\cdot\cdot1\cdot}$	$T_{\cdot\cdot2\cdot}$	$\cdots$	$T_{\cdot\cdot c\cdot}$	$T_{\cdots\cdot}$

B	C				Total
	1	2	$\cdots$	c	
1	$T_{\cdot11\cdot}$	$T_{\cdot12\cdot}$	$\cdots$	$T_{\cdot1c\cdot}$	$T_{\cdot1\cdot\cdot}$
2	$T_{\cdot21\cdot}$	$T_{\cdot22\cdot}$	$\cdots$	$T_{\cdot2c\cdot}$	$T_{\cdot2\cdot\cdot}$
$\vdots$	$\vdots$	$\vdots$		$\vdots$	$\vdots$
b	$T_{\cdot b1\cdot}$	$T_{\cdot b2\cdot}$	$\cdots$	$T_{\cdot bc\cdot}$	$T_{\cdot b\cdot\cdot}$
Total	$T_{\cdot\cdot1\cdot}$	$T_{\cdot\cdot2\cdot}$	$\cdots$	$T_{\cdot\cdot c\cdot}$	$T_{\cdots\cdot}$

The sums of squares are computed by substituting the appropriate totals into the following computational formulas:

$$SST = \sum_{i=1}^{a} \sum_{j=1}^{b} \sum_{k=1}^{c} \sum_{l=1}^{n} y_{ijkl}^2 - \frac{T_{....}^2}{abcn}$$

$$SSA = \frac{\sum_{i=1}^{a} T_{i...}^2}{bcn} - \frac{T_{....}^2}{abcn}$$

$$SSB = \frac{\sum_{j=1}^{b} T_{\cdot j..}^2}{acn} - \frac{T_{....}^2}{abcn}$$

$$SSC = \frac{\sum_{k=1}^{c} T_{..k\cdot}^2}{abn} - \frac{T_{....}^2}{abcn}$$

$$SS(AB) = \frac{\sum_{i=1}^{a} \sum_{j=1}^{b} T_{ij..}^2}{cn} - \frac{\sum_{i=1}^{a} T_{i...}^2}{bcn} - \frac{\sum_{j=1}^{b} T_{\cdot j..}^2}{acn} + \frac{T_{....}^2}{abcn}$$

$$SS(AC) = \frac{\sum_{i=1}^{a} \sum_{k=1}^{c} T_{i\cdot k\cdot}^2}{bn} - \frac{\sum_{i=1}^{a} T_{i...}^2}{bcn} - \frac{\sum_{k=1}^{c} T_{..k\cdot}^2}{abn} + \frac{T_{....}^2}{abcn}$$

$$SS(BC) = \frac{\sum_{j=1}^{b} \sum_{k=1}^{c} T_{\cdot jk\cdot}^2}{an} - \frac{\sum_{j=1}^{b} T_{\cdot j..}^2}{acn} - \frac{\sum_{k=1}^{c} T_{..k\cdot}^2}{abn} + \frac{T_{....}^2}{abcn}$$

$$SS(ABC) = \frac{\sum_{i=1}^{a} \sum_{j=1}^{b} \sum_{k=1}^{c} T_{ijk\cdot}^2}{n} - \frac{\sum_{i=1}^{a} \sum_{j=1}^{b} T_{ij..}^2}{cn} - \frac{\sum_{i=1}^{a} \sum_{k=1}^{c} T_{i\cdot k\cdot}^2}{bn}$$
$$- \frac{\sum_{j=1}^{b} \sum_{k=1}^{c} T_{\cdot jk\cdot}^2}{an} + \frac{\sum_{i=1}^{a} T_{i...}^2}{bcn} + \frac{\sum_{j=1}^{b} T_{\cdot j..}^2}{acn} + \frac{\sum_{k=1}^{c} T_{..k\cdot}^2}{abn} - \frac{T_{....}^2}{abcn},$$

and SSE, as usual, is obtained by subtraction. The computations in an analysis-of-variance problem, for a three-factor experiment with n replications, are summarized in Table 11.5.

Table 11.5 Analysis of Variance for the Three-Factor Experiment with n Replications

Source of variation	Sum of squares	Degrees of freedom	Mean square	Computed f
Main effect				
A	SSA	$a-1$	s_1^2	$f_1 = \dfrac{s_1^2}{s^2}$
B	SSB	$b-1$	s_2^2	$f_2 = \dfrac{s_2^2}{s^2}$
C	SSC	$c-1$	s_3^2	$f_3 = \dfrac{s_3^2}{s^2}$
Two-factor interaction				
AB	$SS(AB)$	$(a-1)(b-1)$	s_4^2	$f_4 = \dfrac{s_4^2}{s^2}$
AC	$SS(AC)$	$(a-1)(c-1)$	s_5^2	$f_5 = \dfrac{s_5^2}{s^2}$
BC	$SS(BC)$	$(b-1)(c-1)$	s_6^2	$f_6 = \dfrac{s_6^2}{s^2}$
Three-factor interaction				
ABC	$SS(ABC)$	$(a-1)(b-1)(c-1)$	s_7^2	$f_7 = \dfrac{s_7^2}{s^2}$
Error	SSE	$abc(n-1)$	s^2	
Total	SST	$abcn-1$		

For the three-factor experiment with a single replicate we may use the analysis of Table 11.5 by setting $n = 1$ and using the ABC interaction for our error sum of squares. In this case we are assuming that the ABC interaction is zero and that $SS(ABC)$ represents variation due only to experimental error and thereby provides an estimate of the error variance.

Example 11.3 In the production of a particular material three variables are of interest: A the operator effect (three operators), B the catalyst used in the experiment (three catalysts), and C the washing time of the product following the cooling process (15 minutes and 20 minutes). Three runs were made at each combination of factors. It was felt that all interactions among the factors should be studied. The coded yields are as follows:

				Washing time, C			
		15 minutes			20 minutes		
		B			B		
A	1	2	3	1	2	3
1	10.7	10.3	11.2	10.9	10.5	12.2
	10.8	10.2	11.6	12.1	11.1	11.7
	11.3	10.5	12.0	11.5	10.3	11.0
2	11.4	10.2	10.7	9.8	12.6	10.8
	11.8	10.9	10.5	11.3	7.5	10.2
	11.5	10.5	10.2	10.9	9.9	11.5
3	13.6	12.0	11.1	10.7	10.2	11.9
	14.1	11.6	11.0	11.7	11.5	11.6
	14.5	11.5	11.5	12.7	10.9	12.2

Perform an analysis of variance to test for significant effects.

Solution   First we construct the following two-way tables:

C (15 minutes)	B			Total
	1	2	3	
A				
1	32.8	31.0	34.8	98.6
2	34.7	31.6	31.4	97.7
3	42.2	35.1	33.6	110.9
Total	109.7	97.7	99.8	307.2

C	B			Total
(20 minutes)	1	2	3	
A				
1	34.5	31.9	34.9	101.3
2	32.0	30.0	32.5	94.5
3	35.1	32.6	35.7	103.4
Total	101.6	94.5	103.1	299.2

A	B			Total
	1	2	3	
1	67.3	62.9	69.7	199.9
2	66.7	61.6	63.9	192.2
3	77.3	67.7	69.3	214.3
Total	211.3	192.2	202.9	606.4

A	C		Total
	1	2	
1	98.6	101.3	199.9
2	97.7	94.5	192.2
3	110.9	103.4	214.3
Total	307.2	299.2	606.4

B	C		Total
	1	2	
1	109.7	101.6	211.3
2	97.7	94.5	192.2
3	99.8	103.1	202.9
Total	307.2	299.2	606.4

Now

$$SST = 10.7^2 + 10.8^2 + \cdots + 12.2^2 - \frac{606.4^2}{54}$$

$$= 6872.84 - 6809.65 = 63.19$$

$$SSA = \frac{199.9^2 + 192.2^2 + 214.3^2}{18} - \frac{606.4^2}{54}$$

$$= 6823.63 - 6809.65 = 13.98$$

$$SSB = \frac{211.3^2 + 192.2^2 + 202.9^2}{18} - \frac{606.4^2}{54}$$

$$= 6819.83 - 6809.65 = 10.18$$

$$SSC = \frac{307.2^2 + 299.2^2}{27} - \frac{606.4^2}{54}$$

$$= 6810.83 - 6809.65 = 1.18$$

$$SS(AB) = \frac{67.3^2 + 66.7^2 + \cdots + 69.3^2}{6} - 6823.63 - 6819.83 + 6809.65$$

$$= 4.78$$

$$SS(AC) = \frac{98.6^2 + 97.7^2 + \cdots + 103.4^2}{9} - 6823.63 - 6810.83 + 6809.65$$

$$= 2.92$$

$$SS(BC) = \frac{109.7^2 + 97.7^2 + \cdots + 103.1^2}{9} - 6819.83 - 6810.83 + 6809.65$$

$$= 3.64$$

$$SS(ABC) = \frac{32.8^2 + 34.7^2 + \cdots + 35.7^2}{3} - 6838.59 - 6827.73 - 6824.65$$

$$+ 6823.63 + 6819.83 + 6810.83 - 6809.65$$

$$= 4.89$$

$$SSE = 63.19 - 13.98 - 10.18 - 1.18 - 4.78 - 2.92 - 3.64 - 4.89$$

$$= 21.62.$$

These results, with the remaining computations, are given in Table 11.6. None of the interactions show a significant effect at the $\alpha = 0.05$ level. The operator and catalyst effects are significant, while the washing time has no significant effect on the yield for the range used in the experiment.

Table 11.6 Analysis of Variance for Example 11.3

Source of variation	Sum of squares	Degrees of freedom	Mean square	Computed f
Main effects				
A	13.98	2	6.99	11.65
B	10.18	2	5.09	8.48
C	1.18	1	1.18	1.97
Two-factor interaction				
AB	4.78	4	1.20	2.00
AC	2.92	2	1.46	2.43
BC	3.64	2	1.82	3.03
Three-factor interaction				
ABC	4.89	4	1.22	2.03
Error	21.62	36	0.60	
Total	63.19	53		

11.5 DISCUSSION OF SPECIFIC MULTIFACTOR MODELS

We have described the three-factor model and its analysis in the most general form by including all possible interactions in the model. Of course, there are many situations in which it is known a priori that the model should not contain certain interactions. We can then take advantage of this knowledge by combining or pooling the sums of squares corresponding to negligible interactions with the error sum of squares to form a new estimator for σ^2 with a larger number of degrees of freedom. For example, in a metallurgy experiment designed to study the effect on film thickness of three important processing variables, suppose it is known that factor A, acid concentration, does not interact with factors B and C. The sums of squares SSA, SSB, SSC, and $SS(BC)$ are computed using the methods described in Section 11.4. The mean squares for the remaining effects will now all independently estimate the error variance σ^2. Therefore, we form our new error mean square by pooling $SS(AB)$, $SS(AC)$, $SS(ABC)$, and SSE, along with the corresponding degrees of freedom. The resulting denominator for the significance tests is then the error mean square given by

$$s^2 = \frac{SS(AB) + SS(AC) + SS(ABC) + SSE}{(a-1)(b-1) + (a-1)(c-1) + (a-1)(b-1)(c-1) + abc(n-1)}.$$

Computationally, of course, one obtains the pooled sum of squares and the pooled degrees of freedom by subtraction once SST and the sums of squares for

Table 11.7 Analysis of Variance with Factor A Noninteracting

Source of variation	Sum of squares	Degrees of freedom	Mean square	Computed f
Main effect				
A	SSA	$a - 1$	s_1^2	$f_1 = \dfrac{s_1^2}{s^2}$
B	SSB	$b - 1$	s_2^2	$f_2 = \dfrac{s_2^2}{s^2}$
C	SSC	$c - 1$	s_3^2	$f_3 = \dfrac{s_3^2}{s^2}$
Two-factor interaction				
BC	$SS(BC)$	$(b - 1)(c - 1)$	s_4^2	$f_4 = \dfrac{s_4^2}{s^2}$
Error	SSE	Subtraction	s^2	
Total	SST	$abcn - 1$		

the existing effects are computed. The analysis-of-variance table would then take the form of Table 11.7.

In our analysis of the two-factor experiment in Section 11.3 a completely randomized design was used. By interpreting the levels of factor A in Table 11.7 as different blocks, we then have the analysis-of-variance procedure for a two-factor experiment in a randomized block design. For example, if we interpret

Table 11.8 Analysis of Variance for a Two-Factor Experiment in a Randomized Block Design

Source of variation	Sum of squares	Degrees of freedom	Mean square	Computed f
Blocks	13.98	2	6.99	9.45
Main effect				
B	10.18	2	5.09	6.88
C	1.18	1	1.18	1.59
Two-factor interaction				
BC	3.64	2	1.82	2.46
Error	34.21	46	0.74	
Total	63.19	53		

the operators in Example 11.3 as blocks and assume no interaction between blocks and the other two factors, the analysis of variance takes the form of Table 11.8 rather than that given in Table 11.6. The reader can easily verify that the error mean square is also given by

$$s^2 = \frac{4.78 + 2.92 + 4.89 + 21.62}{4 + 2 + 4 + 36} = 0.74,$$

which demonstrates the pooling of the sums of squares for the nonexisting interaction effects.

11.6 MODEL II FACTORIAL EXPERIMENTS

In a two-factor experiment with random effects we have the model

$$Y_{ijk} = \mu + A_i + B_j + (AB)_{ij} + E_{ijk},$$

$i = 1, 2, \ldots, a; j = 1, 2, \ldots, b;$ and $k = 1, 2, \ldots, n,$ where the A_i, B_j, $(AB)_{ij}$, and E_{ijk} are independent random variables with zero means and variances σ_α^2, σ_β^2, $\sigma_{\alpha\beta}^2$, and σ^2, respectively. The sum of squares for the model II experiments are computed in exactly the same way as for the model I experiments. We are now interested in testing hypotheses of the form

$$H_0': \sigma_\alpha^2 = 0, \qquad H_0'': \sigma_\beta^2 = 0, \qquad H_0''': \sigma_{\alpha\beta}^2 = 0,$$

$$H_1': \sigma_\alpha^2 \neq 0, \qquad H_1'': \sigma_\beta^2 \neq 0, \qquad H_1''': \sigma_{\alpha\beta}^2 \neq 0,$$

where the denominator in the f ratio is not necessarily the error mean square. The appropriate denominator can be determined by examining the expected values of the various mean squares. These are shown in Table 11.9.

Table 11.9 Expected Mean Squares for a Model II Two-Factor Experiment

Source of variation	Degrees of freedom	Mean square	Expected mean square
A	$a - 1$	s_1^2	$\sigma^2 + n\sigma_{\alpha\beta}^2 + bn\sigma_\alpha^2$
B	$b - 1$	s_2^2	$\sigma^2 + n\sigma_{\alpha\beta}^2 + an\sigma_\beta^2$
AB	$(a - 1)(b - 1)$	s_3^2	$\sigma^2 + n\sigma_{\alpha\beta}^2$
Error	$ab(n - 1)$	s^2	σ^2
Total	$abn - 1$		

From Table 11.9 we see that H_0' and H_0'' are tested by using s_3^2 in the denominator of the f ratio, while H_0''' is tested using s^2 in the denominator. The unbiased estimates of the variance components are given by

$$\hat{\sigma}^2 = s^2$$

$$\hat{\sigma}_{\alpha\beta}^2 = \frac{s_3^2 - s^2}{n}$$

$$\hat{\sigma}_{\alpha}^2 = \frac{s_1^2 - s_3^2}{bn}$$

$$\hat{\sigma}_{\beta}^2 = \frac{s_2^2 - s_3^2}{an}.$$

The expected mean squares for the three-factor experiment with random effects in a completely randomized design are shown in Table 11.10. It is evident from the expected mean squares of Table 11.10 that one can form appropriate f ratios for testing all two-factor and three-factor interaction variance components. However, to test a hypothesis of the form

$$H_0: \quad \sigma_{\alpha}^2 = 0,$$

$$H_1: \quad \sigma_{\alpha}^2 \neq 0,$$

there appears to be no appropriate f ratio unless we have found one or more of the two-factor interaction variance components not significant. Suppose, for example, that we have compared s_5^2 with s_7^2 and found $\sigma_{\alpha\gamma}^2$ to be negligible.

Table 11.10 Expected Mean Squares for a Model II Three-Factor Experiment

Source of variation	Degrees of freedom	Mean square	Expected mean square
A	$a-1$	s_1^2	$\sigma^2 + n\sigma_{\alpha\beta\gamma}^2 + cn\sigma_{\alpha\beta}^2 + bn\sigma_{\alpha\gamma}^2 + bcn\sigma_{\alpha}^2$
B	$b-1$	s_2^2	$\sigma^2 + n\sigma_{\alpha\beta\gamma}^2 + cn\sigma_{\alpha\beta}^2 + an\sigma_{\beta\gamma}^2 + acn\sigma_{\beta}^2$
C	$c-1$	s_3^2	$\sigma^2 + n\sigma_{\alpha\beta\gamma}^2 + bn\sigma_{\alpha\gamma}^2 + an\sigma_{\beta\gamma}^2 + abn\sigma_{\gamma}^2$
AB	$(a-1)(b-1)$	s_4^2	$\sigma^2 + n\sigma_{\alpha\beta\gamma}^2 + cn\sigma_{\alpha\beta}^2$
AC	$(a-1)(c-1)$	s_5^2	$\sigma^2 + n\sigma_{\alpha\beta\gamma}^2 + bn\sigma_{\alpha\gamma}^2$
BC	$(b-1)(c-1)$	s_6^2	$\sigma^2 + n\sigma_{\alpha\beta\gamma}^2 + an\sigma_{\beta\gamma}^2$
ABC	$(a-1)(b-1)(c-1)$	s_7^2	$\sigma^2 + n\sigma_{\alpha\beta\gamma}^2$
Error	$abc(n-1)$	s^2	σ^2
Total	$abcn-1$		

We could then argue that the term $\sigma_{\alpha\gamma}^2$ should be dropped from all the expected mean squares of Table 11.10; then the ratio s_1^2/s_4^2 provides a test for the significance of the variance component σ_{α}^2. Therefore, if we are to test hypotheses concerning the variance component of the main effects it is necessary first to investigate the significance of the two-factor interaction components. An approximate test derived by Satterthwaite may be used when certain two-factor interaction variance components are found to be significant and hence must remain a part of the expected mean square.

Example 11.4 In a study to determine which are the important sources of variation in an industrial process, three measurements are taken on yield for three operators chosen randomly and four batches of raw materials chosen randomly. It was decided that a significance test should be made at the 0.05 level of significance to determine if the variance components due to batches, operators, and interaction are significant. In addition, estimates of variance components are to be computed. The data are as follows, with the response being percent by weight:

| | | Batches | | |
Operators	1	2	3	4
	66.9	68.3	69.0	69.3
1	68.1	67.4	69.8	70.9
	67.2	67.7	67.5	71.4
	66.3	68.1	69.7	69.4
2	65.4	66.9	68.8	69.6
	65.8	67.6	69.2	70.0
	65.6	66.0	67.1	67.9
3	66.3	66.9	66.2	68.4
	65.2	67.3	67.4	68.7

Solution The sums of squares are found in the usual way with the results given by

$$SSA \text{ (operators)} = 21.45$$

$$SSB \text{ (batches)} = 52.21$$

$$SS(AB) \text{ (interaction)} = 5.04$$

$$SSE \text{ (error)} = 8.27.$$

Table 11.11 Analysis of Variance for Example 11.4

Source of variation	Sum of squares	Degrees of freedom	Mean square	Computed f
Operators	21.45	2	10.45	12.44
Batches	52.21	3	17.40	20.71
Interaction	5.04	6	0.84	2.47
Error	8.27	24	0.34	
Total	86.97	35		

All other computations are carried out and exhibited in Table 11.11. Since $f_{0.05}(2,24) = 3.40, f_{0.05}(3,24) = 3.01$, and $f_{0.05}(6,24) = 2.51$ we find the operator and batch variance components to be significant, while the interaction variance is not significant at the $\alpha = 0.05$ level. Estimates of the main effect variance components are given by

$$\hat{\sigma}_\alpha^2 = \frac{10.45 - 0.84}{12} = 0.82$$

$$\hat{\sigma}_\beta^2 = \frac{17.40 - 0.84}{9} = 1.84.$$

11.7 CHOICE OF SAMPLE SIZE

Our study of factorial experiments throughout this chapter has been restricted to the use of a completely randomized design with the exception of Section 11.5, where we demonstrated the analysis of a two-factor experiment in a randomized block design. The completely randomized design is very easy to lay out and the analysis is simple to perform; however, it should be used only when the number of treatment combinations is small and the experimental material is homogeneous. Although the randomized block design is ideal for dividing a large group of heterogeneous units into subgroups of homogeneous units, it is generally difficult to obtain uniform blocks with enough units to which a large number of treatment combinations may be assigned. This disadvantage may be overcome by choosing a design from the catalog of *incomplete block designs*. These designs allow one to investigate differences among t treatments arranged in b blocks containing k experimental units, where $k < t$.

Once the experimenter has selected a completely randomized design, he must decide if the number of replications is sufficient to yield tests in the analysis

Table 11.12 Noncentrality Parameter λ and ϕ^2 for Two-Factor and Three-Factor Models

	Two-factor experiments		Three-factor experiments		
	A	B	A	B	C
λ	$\dfrac{bn \sum_{i=1}^{a} \alpha_i^2}{2\sigma^2}$	$\dfrac{an \sum_{j=1}^{b} \beta_j^2}{2\sigma^2}$	$\dfrac{bcn \sum_{i=1}^{a} \alpha_i^2}{2\sigma^2}$	$\dfrac{acn \sum_{j=1}^{b} \beta_j^2}{2\sigma^2}$	$\dfrac{abn \sum_{k=1}^{c} \gamma_k^2}{2\sigma^2}$
ϕ^2	$\dfrac{bn \sum_{i=1}^{a} \alpha_i^2}{a\sigma^2}$	$\dfrac{an \sum_{j=1}^{b} \beta_j^2}{b\sigma^2}$	$\dfrac{bcn \sum_{i=1}^{a} \alpha_i^2}{a\sigma^2}$	$\dfrac{acn \sum_{j=1}^{b} \beta_j^2}{b\sigma^2}$	$\dfrac{abn \sum_{k=1}^{c} \gamma_k^2}{c\sigma^2}$

of variance with high power. If not, he must add additional replications, which in turn may force him into a randomized complete block design. Had he started with a randomized block design, it would still be necessary to determine if the number of blocks is sufficient to yield powerful tests. Basically, then, we are back to the question of sample size.

The power of a fixed effects test or the probability of rejecting H_0 when the alternative H_1 is true, for a given sample size, is found from Table XV by computing the noncentrality parameter λ and the function ϕ discussed in Section 10.12. Expressions for λ and ϕ^2 for the two-factor and three-factor fixed effects experiments are given in Table 11.12.

The results of Section 10.12 for the random effects model can be extended very easily to the two- and three-factor models. Once again the general procedure is based on the values of the expected mean squares. For example, if we are testing $\sigma_\alpha^2 = 0$ in a two-factor experiment by computing the ratio s_1^2/s_3^2 (see Table 11.9), then

$$f = \frac{s_1^2/(\sigma^2 + n\sigma_{\alpha\beta}^2 + bn\sigma_\alpha^2)}{s_3^2/(\sigma^2 + n\sigma_{\alpha\beta}^2)}$$

is a value of the random variable F having the F distribution with $a - 1$ and $(a - 1)(b - 1)$ degrees of freedom, and the power of the test is given by

$$1 - \beta = P\left\{ \frac{S_1^2}{S_3^2} > f_\alpha[(a - 1), (a - 1)(b - 1)] \,\middle|\, \sigma_\alpha^2 \neq 0 \right\}$$

$$= P\left\{ F > \frac{f_\alpha[(a - 1), (a - 1)(b - 1)](\sigma^2 + n\sigma_{\alpha\beta}^2)}{\sigma^2 + n\sigma_{\alpha\beta}^2 + bn\sigma_\alpha^2} \right\}.$$

EXERCISES

1. An experiment was conducted to study the effect of temperature and type of oven on the life of a particular component being tested. Four types of ovens and three temperature levels were used in the experiment. Twenty-four pieces were assigned randomly, two to each combination of treatments, and the following results recorded:

Temperature (degrees)	O_1	O_2	O_3	O_4
	\multicolumn{4}{c}{Oven}			

Temperature (degrees)	O_1	O_2	O_3	O_4
500	227	214	225	260
	221	259	236	229
550	187	181	232	246
	208	179	198	273
600	174	198	178	206
	202	194	213	219

(a) Test the hypothesis of no interaction between type of oven and temperature at the $\alpha = 0.05$ level of significance.

(b) If the significance test in part (a) is accepted, test the hypotheses that the ovens and temperatures have no effect on the life of the component.

2. A study was made to determine if humidity conditions have an effect on the force required to pull apart pieces of glued plastic. Three types of plastic were tested using four different levels of humidity. The results, in kilograms, are given as follows:

Plastic type	30%	50%	70%	90%
	\multicolumn{4}{c}{Humidity}			

Plastic type	30%	50%	70%	90%
A	39.0	33.1	33.8	33.0
	42.8	37.8	30.7	32.9
B	36.9	27.2	29.7	28.5
	41.0	26.8	29.1	27.9
C	27.4	29.2	26.7	30.9
	30.3	29.9	32.0	31.5

(a) Assuming a model I experiment, perform an analysis of variance and test the hypothesis of no interaction between humidity and plastic type at the 0.05 level of significance.

(b) Perform an analysis of variance using only plastics A and B, and once again test for the presence of interaction.

(c) Use a single-degree-of-freedom comparison to evaluate the force required at 30% humidity versus (50%, 70%, 90%).

(d) Repeat part (c) using only plastic C.

In all tests use s^2 from the overall analysis of variance of part (a) as the denominator in the f ratio.

3. The following data are taken in a study involving three factors A, B, and C, all fixed effects:

| | C_1 | | | C_2 | | | C_3 | | |
	B_1	B_2	B_3	B_1	B_2	B_3	B_1	B_2	B_3
A_1	15.0	14.8	15.9	16.8	14.2	13.2	15.8	15.5	19.2
	18.5	13.6	14.8	15.4	12.9	11.6	14.3	13.7	13.5
	22.1	12.2	13.6	14.3	13.0	10.1	13.0	12.6	11.1
A_2	11.3	17.2	16.1	18.9	15.4	12.4	12.7	17.3	7.8
	14.6	15.5	14.7	17.3	17.0	13.6	14.2	15.8	11.5
	18.2	14.2	13.4	16.1	18.6	15.2	15.9	14.6	12.2

(a) Perform tests of significance on all interactions at the $\alpha = 0.05$ level.

(b) Perform tests of significance on the main effects at the $\alpha = 0.05$ level.

(c) Give an explanation of how a significant interaction has masked the effect of factor C.

4. Consider an experimental situation involving factors A, B, and C, where we assume a three-way fixed effects model of the form

$$y_{ijkl} = \mu + \alpha_i + \beta_j + \gamma_k + (\beta\gamma)_{jk} + \varepsilon_{ijkl}.$$

All other interactions are considered to be nonexistent or negligible. The data were recorded as follows:

| | B_1 | | | B_2 | | |
	C_1	C_2	C_3	C_1	C_2	C_3
A_1	4.0	3.4	3.9	4.4	3.1	3.1
	4.9	4.1	4.3	3.4	3.5	3.7
A_2	3.6	2.8	3.1	2.7	2.9	3.7
	3.9	3.2	3.5	3.0	3.2	4.2
A_3	4.8	3.3	3.6	3.6	2.9	2.9
	3.7	3.8	4.2	3.8	3.3	3.5
A_4	3.6	3.2	3.2	2.2	2.9	3.6
	3.9	2.8	3.4	3.5	3.2	4.3

(a) Perform a test of significance on the BC interaction at the $\alpha = 0.05$ level.

(b) Perform tests of significance on the main effects A, B, and C using a pooled error mean square at the $\alpha = 0.05$ level.

5. To estimate the various components of variability in a filtration process, the per cent of material lost in the mother liquor was measured for 12 experimental conditions, three runs on each condition. Three filters and four operators were selected at random to use in the experiment resulting in the following measurements:

	Operator			
Filter	1	2	3	4
1	16.2	15.9	15.6	14.9
	16.8	15.1	15.9	15.2
	17.1	14.5	16.1	14.9
2	16.6	16.0	16.1	15.4
	16.9	16.3	16.0	14.6
	16.8	16.5	17.2	15.9
3	16.7	16.5	16.4	16.1
	16.9	16.9	17.4	15.4
	17.1	16.8	16.9	15.6

(a) Test the hypothesis of no interaction variance component between filters and operators at the $\alpha = 0.05$ level of significance.

(b) Test the hypotheses that the operators and the filters have no effect on the variability of the filtration process.

(c) Estimate the components of variance due to filters, operators, and experimental error.

6. Consider the following analysis of variance for a model II experiment:

Source of variation	Degrees of freedom	Mean square
A	3	140
B	1	480
C	2	325
AB	3	15
AC	6	24
BC	2	18
ABC	6	2
Error	24	5
Total	47	

Test for significant variance components among all main effects and interaction effects at the 0.01 level of significance (a) using a pooled estimate of error when appropriate and (b) not pooling sums of squares of insignificant effects.

7. Are two observations for each treatment combination in Exercise 4 sufficient if the power of our test for detecting differences among the levels of factor C at the 0.05 level of significance is to be at least 0.8 when $\gamma_1 = -0.2$, $\gamma_2 = 0.4$, and $\gamma_3 = -0.2$? Use the same pooled estimate of σ^2 that was used in the analysis of variance.

8. Using the estimates of the variance components in Exercise 5, evaluate the power when we test the variance component due to filters to be zero.

2ᵏ Factorial Experiments

12

12.1 INTRODUCTION

In almost any experimental study in which statistical procedures are applied to a collection of scientific data, the methods involve performing certain operations or computations on the sample information, followed by the drawing of inferences about the population or populations studied. Often there are characteristics of the experiment that are subject to the control of the experimenter, quantities such as sample size, number of levels of the factors, treatment combinations to be used, and so forth. These *experimental parameters* can often have a great effect on the precision with which hypotheses are tested or estimation is accomplished.

We have already been exposed to certain experimental design concepts. The sampling plan for the simple t test on the mean of a normal population and also the analysis of variance involve randomly allocating pre-chosen treatments to experimental units. The randomized block design, where treatments are assigned to units within relatively homogeneous blocks, involves restricted randomization.

In this chapter we give special attention to experimental designs in which the experimental plan calls for the study of the effect on a response of k factors, each at two levels. These are commonly known as 2^k factorial experiments. We often denote the levels as "high" and "low," even though this notation may be arbitrary in the case of qualitative variables. The complete factorial design requires that each level of every factor occur with each level of every other factor, giving a total of 2^k treatment combinations. We shall denote the higher levels of the factors $A, B, C, \ldots$ by the letters $a, b, c, \ldots$ and the lower levels of each factor by the notation (1). In the presence of other letters we omit the symbol (1). For example, the treatment combination in a 2^4 experiment that contains the high levels of factors B and C and the low levels of factors A and D is written simply as bc. The treatment combination that consists of the low level of all factors in the experiment is denoted by the symbol (1). In the case of a 2^3 experiment, the eight possible treatment combinations are (1), a, b, c, ab, ac, bc, and abc.

447

The factorial experiment allows the effect of each and every factor to be estimated and tested independently through the usual analysis of variance. In addition, the interaction effects are easily assessed. The disadvantage, of course, with the factorial experiment is the excessive amount of experimentation that is required. For example, if it is desired to study the effect of eight variables, $2^8 = 256$ treatment combinations are required. In many instances we can obtain considerable information by using only a fraction of the experimental runs. This type of design is called a *fractional factorial design* and will be considered in Sections 12.6 and 12.7.

12.2 ANALYSIS OF VARIANCE

Consider initially a 2^2 factorial plan in which there are n experimental observations per treatment combination. Extending our previous notation, we now interpret the symbols (1), a, b, and ab to be the total yields for each of the four treatment combinations. Table 12.1 gives a two-way table of these total yields.

Let us define the following contrasts among the treatment totals:

$$A \text{ contrast} = ab + a - b - (1)$$
$$B \text{ contrast} = ab - a + b - (1)$$
$$AB \text{ contrast} = ab - a - b + (1).$$

Clearly, there will be exactly one single-degree-of-freedom contrast for the means of each factor A and B, which we shall write as

$$w_A = \frac{ab + a - b - (1)}{2n} = \frac{A \text{ contrast}}{2n}$$

$$w_B = \frac{ab + b - a - (1)}{2n} = \frac{B \text{ contrast}}{2n}.$$

The contrast w_A is seen to be the difference between the mean response at the low and high levels of factor A. In fact, we call w_A the *main effect* of A. Similarly, w_B is the main effect of factor B. Apparent interaction in the data is observed by inspecting the difference between $ab - b$ and $a - (1)$ or between $ab - a$ and $b - (1)$ in Table 12.1. Hence a third contrast in the treatment totals, orthogonal to these main effect contrasts, is the *interaction effect*, given by

$$w_{AB} = \frac{ab - a - b + (1)}{2n} = \frac{AB \text{ contrast}}{2n}.$$

Table 12.1 2^2 Factorial Experiment

		B		Mean
		(1)	b	$\dfrac{b + (1)}{2n}$
A		a	ab	$\dfrac{ab + a}{2n}$
Mean		$\dfrac{a + (1)}{2n}$	$\dfrac{ab + b}{2n}$	

We take advantage of the fact that in the 2^2 factorial, or for that matter in the general 2^k factorial experiment, each main effect and interaction effect has associated with it a single degree of freedom. Therefore, we can write $2^k - 1$ orthogonal single-degree-of-freedom contrasts in the treatment combinations, each representing variation due to some main or interaction effect. Thus, under the usual independence and normality assumptions in the experimental model, we can make tests to determine if the contrast reflects systematic variation or merely chance or random variation. The sums of squares for each contrast is found by following the procedures given in Section 10.4. Writing $T_{1..} = b + (1)$, $T_{2..} = ab + a$, $c_1 = -1$, and $c_2 = 1$, where $T_{1..}$ and $T_{2..}$ are the totals of $2n$ observations, we have

$$SSA = SSw_A = \frac{\left(\sum\limits_{i=1}^{2} c_i T_{i..}\right)^2}{2n \sum\limits_{i=1}^{2} c_i^2}$$

$$= \frac{[ab + a - b - (1)]^2}{2^2 n} = \frac{(A \text{ contrast})^2}{2^2 n},$$

with 1 degree of freedom. Similarly, we find that

$$SSB = \frac{[ab + b - (1) - a]^2}{2^2 n} = \frac{(B \text{ contrast})^2}{2^2 n}$$

$$SS(AB) = \frac{[ab + (1) - a - b]^2}{2^2 n} = \frac{(AB \text{ contrast})^2}{2^2 n},$$

Table 12.2 Signs for Contrasts in a 2^2 Factorial Experiment

Treatment combination	Factorial effect		
	A	B	AB
(1)	$-$	$-$	$+$
a	$+$	$-$	$-$
b	$-$	$+$	$-$
ab	$+$	$+$	$+$

each with 1 degree of freedom, while the error sum of squares, with $2^2(n-1)$ degrees of freedom, is obtained by subtraction from the formula

$$SSE = SST - SSA - SSB - SS(AB).$$

In computing the sums of squares for the main effects A and B and the interaction effect AB, it is convenient to present the total yields of the treatment combinations along with the appropriate algebraic signs for each contrast as in Table 12.2. The main effects are obtained as simple comparisons between the low and high levels. Therefore, we assign a positive sign to the treatment combination that is at the high level of a given factor and a negative sign to the treatment combination at the lower level. The positive and negative signs for the interaction effect are obtained by multiplying the corresponding signs of the contrasts of the interacting factors.

Let us now consider an experiment using three factors A, B, and C with levels (1), a; (1), b; and (1), c, respectively. This is a 2^3 factorial experiment giving the eight treatment combinations (1), a, b, c, ab, ac, bc, and abc. The treatment combinations and the appropriate algebraic signs for each contrast used in computing the sums of squares for the main effects and interaction effects are presented in Table 12.3.

An inspection of Table 12.3 reveals that for the 2^3 experiment any two contrasts among the seven are orthogonal and therefore the seven effects are assessed independently. The sum of squares for, say, the ABC interaction with 1 degree of freedom is given by

$$SS(ABC) = \frac{[abc + a + b + c - (1) - ab - ac - bc]^2}{2^3 n}.$$

For a 2^k factorial experiment the single-degree-of-freedom sums of squares for the main effects and interaction effects are obtained by squaring the appro-

Table 12.3 Signs for Contrasts in a 2^3 Factorial Experiment

Treatment combination	Factorial effect						
	A	*B*	*C*	*AB*	*AC*	*BC*	*ABC*
(1)	−	−	−	+	+	+	−
a	+	−	−	−	−	+	+
b	−	+	−	−	+	−	+
c	−	−	+	+	−	−	+
ab	+	+	−	+	−	−	−
ac	+	−	+	−	+	−	−
bc	−	+	+	−	−	+	−
abc	+	+	+	+	+	+	+

priate contrasts in the treatment totals and dividing by $2^k n$, where n is the number of replications of the treatment combinations.

The orthogonality property has the same importance here as it did in the material on comparisons discussed in Chapter 11. Orthogonality of contrasts implies that the estimated effects and thus the sums of squares are independent. This independence is easily illustrated in a 2^3 factorial experiment if the yields, with factor A at its high level, are increased by an amount x in Table 12.3. Only the A contrast leads to a larger sum of squares since the x effect cancels out in the formation of the six remaining contrasts as a result of the two positive and two negative signs associated with treatment combinations in which A is at the high level.

12.3 YATES' TECHNIQUE FOR COMPUTING CONTRASTS

It is laborious to write out the table of positive and negative signs for large experiments. A systematic tabular technique for deriving the factorial effects has been developed by Yates. The treatment combinations and the observations must be written down in *standard form*. For one factor the standard form is (1), *a*. For two factors we add *b* and *ab*, derived by multiplying the first two treatment combinations by the additional letter *b*. For three factors we add *c, ac, bc,* and *abc*, derived by multiplying the first four treatment combinations by the additional letter *c*, and so on. In the case of three factors the standard order is then

$$(1), \quad a, \quad b, \quad ab, \quad c, \quad ac, \quad bc, \quad abc.$$

Table 12.4 Yates' Technique for a 2^3 Factorial Experiment

Treatment combination	(1)	(2)	(3)	Identifica
(1)	(1) + a	(1) + a + b + ab	(1) + a + b + ab + c + ac + bc + abc	Total
a	b + ab	c + ac + bc + abc	a − (1) + ab − b + ac − c + abc − bc	A cont
b	c + ac	a − (1) + ab − b	b + ab − (1) − a + bc + abc − c − ac	B cont
ab	bc + abc	ac − c + abc − bc	ab − b − a + (1) + abc − bc − ac + c	AB cont
c	a − (1)	b + ab − (1) − a	c + ac + bc + abc − (1) − a − b − ab	C cont
ac	ab − b	bc + abc − c − ac	ac − c + abc − bc − a + (1) − ab + b	AC cont
bc	ac − c	ab − b − a + (1)	bc + abc − c − ac − b − ab + (1) + a	BC cont
abc	abc − bc	abc − bc − ac + c	abc − bc − ac + c − ab + b + a − (1)	ABC cont

Yates' method is carried out in the following steps:

1. Place the treatment combinations and the corresponding total yields in a column in standard order.
2. Obtain the top half of a column marked (1) by adding the first two yields, then the next two, and so on. The bottom half is obtained by subtracting the first from the second of each of these same pairs.

Table 12.5 Yates' Technique for a 2^4 Example

Treatment combination	Treatment total	(1)	(2)	(3)	(4)	Sum of squares
(1)	58.9	118.8	222.0	420.6	865.0	
a	59.9	103.2	198.6	444.4	− 19.2	11.52
b	50.1	98.9	228.3	1.0	− 19.6	12.00
ab	53.1	99.7	216.1	− 20.2	15.8	7.80
c	48.2	116.0	4.0	− 14.8	− 35.6	39.61
ac	50.7	112.3	− 3.0	− 4.8	9.8	3.00
bc	52.6	108.6	− 18.5	− 6.0	19.0	11.28
abc	47.1	107.5	− 1.7	21.8	− 8.8	2.42
d	65.2	1.0	− 15.6	− 23.4	23.8	17.70
ad	50.8	3.0	0.8	− 12.2	− 21.2	14.05
bd	58.2	2.5	− 3.7	− 7.0	10.0	3.13
abd	54.1	− 5.5	− 1.1	16.8	27.8	24.15
cd	57.6	− 14.4	2.0	16.4	11.2	3.92
acd	51.0	− 4.1	− 8.0	2.6	23.8	17.70
bcd	51.3	− 6.6	10.3	− 10.0	− 13.8	5.95
abcd	56.2	4.9	11.5	1.2	11.2	3.92

3. Repeat the operation using the results in column (1) to obtain column (2). This operation is continued until we have k columns for a 2^k experiment.
4. The first value of the kth column will be the grand total of the yields in the experiment. Each remaining number will be a contrast in the treatment totals. Finally, the sum of squares for the main effects and interaction effects are obtained by squaring the entries in column (k) and dividing by $2^k n$, where n is the number of replications and 2^k is the sum of the squares of the coefficients of the individual contrasts.

As an illustration we outline the procedure in Table 12.4 for the 2^3 factorial experiment.

Example 12.1   In a metallurgy experiment it is desired to test the effect of four factors and their interactions on the concentration (percent by weight) of a particular phosphorus compound in casting material. The variables are (A), per cent phosphorus in the refinement; (B), per cent remelted material; (C), fluxing time; and (D), holding time. The four factors are varied in a 2^4 factorial experiment with two castings taken and the content measured at each treatment combination. Using Yates' technique, perform the analysis of variance of the following data:

Treatment combination	Weight % of phosphorus compound		
	Replication 1	Replication 2	Total
(1)	30.3	28.6	58.9
a	28.5	31.4	59.9
b	24.5	25.6	50.1
ab	25.9	27.2	53.1
c	24.8	23.4	48.2
ac	26.9	23.8	50.7
bc	24.8	27.8	52.6
abc	22.2	24.9	47.1
d	31.7	33.5	65.2
ad	24.6	26.2	50.8
bd	27.6	30.6	58.2
abd	26.3	27.8	54.1
cd	29.9	27.7	57.6
acd	26.8	24.2	51.0
bcd	26.4	24.9	51.3
abcd	26.9	29.3	56.2
Total	428.1	436.9	865.0

Solution Table 12.5 gives the 16 treatment totals and outlines Yates' technique for computing the individual sums of squares. The total sum of squares is given by

$$SST = 30.3^2 + 28.5^2 + \cdots + 29.3^2 - \frac{865^2}{32} = 217.51.$$

We can now set up the analysis of variance as in Table 12.6. Note that the interactions *BC*, *AD*, *ABD*, and *ACD* are significant when compared with $f_{0.05}(1, 16) = 4.49$. The tests on the main effects (which in the presence of interactions may be regarded as the effects *averaged* over the levels of the other factors) indicate significance in each case.

Very often the experimenter knows in advance that certain interactions in a 2^k factorial experiment are negligible and should not be included in the model.

Table 12.6 Analysis of Variance for the Data of Table 12.5

Source of variation	Sum of squares	Degrees of freedom	Mean square	Computed f
Main effect				
A	11.52	1	11.52	4.68
B	12.00	1	12.00	4.90
C	39.61	1	39.61	16.10
D	17.70	1	17.70	7.20
Two-factor interaction				
AB	7.80	1	7.80	3.17
AC	3.00	1	3.00	1.22
AD	14.05	1	14.05	5.71
BC	11.28	1	11.28	4.59
BD	3.13	1	3.13	1.27
CD	3.92	1	3.92	1.59
Three-factor interaction				
ABC	2.42	1	2.42	0.98
ABD	24.15	1	24.15	9.82
ACD	17.70	1	17.70	7.20
BCD	5.95	1	5.95	2.42
Four-factor interaction				
ABCD	3.92	1	3.92	1.59
Error	39.36	16	2.46	
Total	217.51	31		

For example, in a 2^4 factorial experiment he may postulate a model that contains only two-factor interaction effects and then pool the sums of squares and corresponding degrees of freedom of the remaining higher-ordered interactions with the pure error. In fact this is often done in lieu of taking several replications.

12.4 FACTORIAL EXPERIMENTS IN INCOMPLETE BLOCKS

The 2^k factorial experiment lends itself to partitioning into *incomplete blocks*. For a k-factor experiment, it is often useful to use a design in 2^p blocks $(p < k)$ when the entire 2^k treatment combinations cannot be applied under homogeneous conditions. The disadvantage with this experimental setup is that certain effects are completely sacrificed as a result of the blocking, the amount of sacrifice depending on the number of blocks required. For example, suppose that the eight treatment combinations in a 2^3 factorial experiment must be run in two blocks of size 4. One possible arrangement is given by

Block 1	*Block 2*
(1)	a
ab	b
ac	c
bc	abc

If one assumes the usual model with the additive effect, this effect cancels out in the formation of the contrasts on all effects except ABC. To illustrate, let x denote the contribution to the yield due to the difference between blocks. Writing the yields as

Block 1	*Block 2*
(1)	$a + x$
ab	$b + x$
ac	$c + x$
bc	$abc + x$

we see that the ABC contrast and also the contrast comparing the two blocks are both given by

$$ABC \text{ contrast} = (abc + x) + (c + x) + (b + x) + (a + x) - (1) - ab$$
$$- ac - bc$$
$$= abc + a + b + c - (1) - ab - ac - bc + 4x.$$

Therefore, we are measuring the ABC effect plus the block effect and there is no way of assessing the ABC interaction effect independent of blocks. We say then that the ABC interaction is *completely confounded with blocks*. By necessity, information on ABC has been sacrificed. On the other hand, the block effect cancels out in the formation of all other contrasts. For example, the A contrast is given by

$$A \text{ contrast} = (abc + x) + (a + x) + ab + ac - (b + x) - (c + x) - bc - (1)$$
$$= abc + a + ab + ac - b - c - bc - (1),$$

as in the case of a completely randomized design. We say that the effects A, B, C, AB, AC, and BC are orthogonal to blocks. Generally, for a 2^k factorial experiment in 2^p blocks, the number of effects confounded with blocks is 2^{p-1}, which is equivalent to the degrees of freedom for blocks.

When two blocks are to be used with a 2^k factorial, one effect, usually a high-order interaction, is chosen as the *defining contrast*. This effect is to be confounded with blocks. The additional $2^k - 2$ effects are orthogonal with the defining contrast and thus with blocks.

Suppose that we represent the defining contrast as $A^{\gamma_1}B^{\gamma_2}C^{\gamma_3}\ldots$, where γ_i is either zero or 1. This generates the expression

$$L = \gamma_1 + \gamma_2 + \cdots + \gamma_k,$$

which in turn is evaluated for each of the 2^k treatment combinations by setting γ_i equal to zero or 1 according as to whether the treatment combination contains the ith factor at its high or low level. The L values are then reduced (modulo 2) to either zero or 1 and thereby determine to which block the treatment combinations are assigned. In other words, the treatment combinations are divided into two blocks according to whether the L values leave a remainder of zero or 1 when divided by 2.

Example 12.2   Determine the values of L (modulo 2) for a 2^3 factorial experiment when the defining contrast is ABC.

Solution   With ABC the defining contrast, we have

$$L = \gamma_1 + \gamma_2 + \gamma_3,$$

which is applied to each treatment combination as follows:

$$
\begin{aligned}
(1): \quad & L = 0 + 0 + 0 = 0 = 0 && \text{(modulo 2)} \\
a: \quad & L = 1 + 0 + 0 = 1 = 1 && \text{(modulo 2)} \\
b: \quad & L = 0 + 1 + 0 = 1 = 1 && \text{(modulo 2)} \\
ab: \quad & L = 1 + 1 + 0 = 2 = 0 && \text{(modulo 2)} \\
c: \quad & L = 0 + 0 + 1 = 1 = 1 && \text{(modulo 2)} \\
ac: \quad & L = 1 + 0 + 1 = 2 = 0 && \text{(modulo 2)} \\
bc: \quad & L = 0 + 1 + 1 = 2 = 0 && \text{(modulo 2)} \\
abc: \quad & L = 1 + 1 + 1 = 3 = 1 && \text{(modulo 2)}.
\end{aligned}
$$

The blocking arrangement, in which ABC is confounded, is given as before by

Block 1	Block 2
(1)	a
ab	b
ac	c
bc	abc

The A, B, C, AB, AC, and BC effects and sums of squares are computed in the usual way, ignoring blocks.

The block containing the treatment combination (1) in Example 12.2 is called the *principal block*. This block forms an algebraic group with respect to multiplication when the exponents are reduced to the modulo 2 base. For example, the property of closure holds since $(ab)(bc) = ab^2c = ac$, $(ab)(ab) = a^2b^2 = (1)$, and so forth.

If the experimenter is required to allocate the treatment combinations to four blocks, two defining contrasts are chosen by the experimenter. A third effect, known as their *generalized interaction*, is automatically confounded with blocks, these three effects corresponding to the three degrees of freedom for blocks. The procedure for constructing the design is best explained through an example. Suppose it is decided that for a 2^4 factorial AB and CD are the defining contrasts. The third effect confounded, their generalized interaction, is formed by multiplying together the initial two modulo 2. Thus the effect

$$(AB)(CD) = ABCD$$

is also confounded with blocks. We construct the design by calculating the expressions

$$L_1 = \gamma_1 + \gamma_2 \quad (AB)$$
$$L_2 = \gamma_3 + \gamma_4 \quad (CD)$$

modulo 2 for each of the 16 treatment combinations to generate the following blocking scheme:

Block 1	*Block 2*	*Block 3*	*Block 4*
(1)	a	c	ac
ab	b	abc	bc
cd	acd	d	ad
abcd	bcd	abd	bd
$L_1 = 0$	$L_1 = 1$	$L_1 = 0$	$L_1 = 1$
$L_2 = 0$	$L_2 = 0$	$L_2 = 1$	$L_2 = 1$

A shortcut procedure can be used to construct the remaining blocks after the principal block has been generated. We begin by placing any treatment combination not in the principal block in a second block and build the block by multiplying (modulo 2) by the treatment combinations in the principal block. In the preceding example the second, third, and fourth blocks are generated as follows:

Block 2	*Block 3*	*Block 4*
$a(1) = a$	$c(1) = c$	$ac(1) = ac$
$a(ab) = b$	$c(ab) = abc$	$ac(ab) = bc$
$a(cd) = acd$	$c(cd) = d$	$ac(cd) = ad$
$a(abcd) = bcd$	$c(abcd) = abd$	$ac(abcd) = bd$

The analysis for the case of four blocks is quite simple. All effects that are orthogonal to blocks (those other than the defining contrasts) are computed in the usual fashion. In fact, Yates' technique can be used on the entire experiment, but the sums of squares for the three confounded effects are then added together to form the sum of squares due to blocks.

The general scheme for the 2^k factorial experiment in 2^p blocks is not difficult. We select p defining contrasts such that none is the generalized interaction of any two in the group. Since there are $2^p - 1$ degrees of freedom for blocks, we have $2^p - 1 - p$ additional effects confounded with blocks. For example, in a 2^6 factorial experiment in eight blocks, we might choose ACF, $BCDE$, and $ABDF$ as the defining contrasts. Then

$$(ACF)(BCDE) = ABDEF$$

$$(ACF)(ABDF) = BCD$$

$$(BCDE)(ABDF) = ACEF$$

$$(ACF)(BCDE)(ABDF) = E$$

are the additional four effects confounded with blocks. This is not a desirable blocking scheme, since one of the confounded effects is the main effect E. The design is constructed by evaluating

$$L_1 = \gamma_1 + \gamma_3 + \gamma_6$$

$$L_2 = \gamma_2 + \gamma_3 + \gamma_4 + \gamma_5$$

$$L_3 = \gamma_1 + \gamma_2 + \gamma_4 + \gamma_6$$

and assigning treatment combinations to blocks according to the following scheme:

Block 1: $L_1 = 0$, $L_2 = 0$, $L_3 = 0$

Block 2: $L_1 = 0$, $L_2 = 0$, $L_3 = 1$

Block 3: $L_1 = 0$, $L_2 = 1$, $L_3 = 0$

Block 4: $L_1 = 0$, $L_2 = 1$, $L_3 = 1$

Block 5: $L_1 = 1$, $L_2 = 0$, $L_3 = 0$

Block 6: $L_1 = 1$, $L_2 = 0$, $L_3 = 1$

Block 7: $L_1 = 1$, $L_2 = 1$, $L_3 = 0$

Block 8: $L_1 = 1$, $L_2 = 1$, $L_3 = 1$.

The shortcut procedure that was illustrated for the case of four blocks also applies here. Hence we can construct the remaining seven blocks from the principal block.

Example 12.3   It is of interest to study the effect of five factors on some response with the assumption that interactions involving three, four, and five of the factors are negligible. We shall divide the 32 treatment combinations into four blocks using the defining contrasts *BCDE* and *ABCD*. Thus $(BCDE)(ABCD) = AE$ is also confounded with blocks. The experimental design and the observations are given in Table 12.7.

   The allocation of treatment combinations to experimental units within blocks is, of course, random. By pooling the unconfounded three, four, and five factor interactions to form the error term, perform the analysis of variance for the data of Table 12.7.

Solution   The sums of squares for each of the 31 contrasts are computed by Yates' method and the block sum of squares is found to be

$$SS(\text{blocks}) = SS(ABCD) + SS(BCDE) + SS(AE)$$
$$= 7.538.$$

The analysis of variance is given in Table 12.8. None of the two-factor interactions are significant at the $\alpha = 0.05$ level when compared to $f_{0.05}(1, 14) = 4.60$.

Table 12.7 Data for a 2^5 Experiment in Four Blocks

Block 1	*Block 2*	*Block 3*	*Block 4*
(1) = 30.6	a = 32.4	b = 32.6	e = 30.7
bc = 31.5	abc = 32.4	c = 31.9	bce = 31.7
bd = 32.4	abd = 32.1	d = 33.3	bde = 32.2
cd = 31.5	acd = 35.3	bcd = 33.0	cde = 31.8
abe = 32.8	be = 31.5	ae = 32.0	ab = 32.0
ace = 32.1	ce = 32.7	abce = 33.1	ac = 33.1
ade = 32.4	de = 33.4	abde = 32.9	ad = 32.2
abcde = 31.8	bcde = 32.9	acde = 35.0	abcd = 32.3

Table 12.8 Analysis of Variance for the Data of Table 12.7

Source of variation	Sum of squares	Degrees of freedom	Mean square	Computed f
Main effect				
A	3.251	1	3.251	6.32
B	0.320	1	0.320	0.62
C	1.361	1	1.361	2.64
D	4.061	1	4.061	7.89
E	0.005	1	0.005	0.01
Two-factor interaction				
AB	1.531	1	1.531	2.97
AC	1.125	1	1.125	2.18
AD	0.320	1	0.320	0.62
BC	1.201	1	1.201	2.33
BD	1.711	1	1.711	3.32
BE	0.020	1	0.020	0.04
CD	0.045	1	0.045	0.09
CE	0.001	1	0.001	0.002
DE	0.001	1	0.001	0.002
Blocks (ABCD, BCDE, AE)	7.538	3	2.513	4.88
Error	7.208	14	0.515	

The main effects A and D are significant and both give positive effects on the response as we go from the low to the high level. Notice that the block effects are also significant when compared to $f_{0.05}(3, 14) = 3.34$. However, there is no way to determine whether the significant block effects are due to actual differences in the blocks or perhaps due to a significant interaction that has been confounded with blocks.

12.5 PARTIAL CONFOUNDING

It is possible to confound any effect with blocks by the methods described in Section 12.4. Suppose we consider a 2^3 factorial experiment in two blocks with three complete replications. If ABC is confounded with blocks in all three replicates, we can proceed as before and determine single-degree-of-freedom sums of squares for all main effects and two-factor interaction effects. The sum of squares for blocks has 5 degrees of freedom, leaving $23 - 5 - 6 = 12$ degrees of freedom for error.

Now let us confound ABC in one replicate, AC in the second, and BC in the third. The plan for this type of experiment would be as follows:

Block			Block			Block	
1	*2*		*1*	*2*		*1*	*2*
abc	ab		abc	ab		abc	ab
a	ac		ac	bc		bc	ac
b	bc		b	a		a	b
c	(1)		(1)	c		(1)	c

Replicate 1   Replicate 2   Replicate 3
ABC confounded AC confounded BC confounded

The effects ABC, AC, and BC are said to be *partially* confounded with blocks. These three effects can be estimated from two of the three replicates. The ratio 2/3 serves as a measure of the extent of the confounding. Yates calls this ratio the *relative information* on the confounded effects. This ratio gives the amount of information available on the partially confounded effect relative to that available on an unconfounded effect.

Table 12.9 Analysis of Variance with Partial Confounding

Source of variation	Degrees of freedom
Blocks	5
A	1
B	1
C	1
AB	1
AC	1'
BC	1'
ABC	1'
Error	11
Total	23

The analysis-of-variance layout is given in Table 12.9. The sums of squares for blocks and for the unconfounded effects A, B, C, and AB are found in the usual way. The sums of squares for AC, BC, and ABC are computed from the two replicates in which the particular effect is not confounded. One must be careful to divide by 16 instead of 24 when obtaining the sums of squares for the partially confounded effects, since we are only using 16 observations. In Table 12.9 the primes are inserted with the degrees of freedom as a reminder that these effects are partially confounded and require special calculations.

12.6 FRACTIONAL FACTORIAL EXPERIMENTS

The 2^k factorial experiment can become quite demanding, in terms of the number of experimental units required, when k is large. One of the real advantages with this experimental plan is that it allows a degree of freedom for each interaction. However, in many experimental situations, it is known that certain interactions are negligible, and thus it would be a waste of experimental effort to use the complete factorial experiment. In fact, the experimenter may have an economic constraint that disallows taking observations at all of the 2^k treatment combinations. When k is large, we can often make use of a *fractional factorial experiment* in which perhaps one half, one fourth, or even one eighth of the total factorial plan is actually carried out.

The construction of the half-replicate design is identical to the allocation of the 2^k factorial experiment into two blocks. We begin by selecting a defining contrast that is to be completely sacrificed. We then construct the two blocks accordingly and choose either of them as the experimental plan.

Consider a 2^4 factorial experiment in which we wish to use a half-replicate. The defining contrast $ABCD$ is chosen and thus an appropriate experimental plan would be to select the principal block consisting of the following treatment combinations:

$$\{(1), ab, ac, ad, bc, bd, cd, abcd\}.$$

With this plan, we have contrasts on all effects except $ABCD$. Clearly,

$$A \text{ contrast} = ab + ac + ad + abcd - (1) - bc - bd - cd$$

$$AB \text{ contrast} = abcd + ab + (1) + cd - ac - ad - bc - bd,$$

with similar expressions for the contrasts of the remaining main effects and interaction effects. However, with no more than 8 of the 16 observations in our fractional design, only 7 of the 14 unconfounded contrasts are orthogonal. Consider, for example, the CD contrast given by

$$CD \text{ contrast} = abcd + cd + (1) + ab - ac - ad - bc - bd.$$

Observe that this is also the single-degree-of-freedom contrast for AB. The word *aliases* is given to two factorial effects that have the same contrast. Therefore, AB and CD are aliases. In the 2^k factorial experiments, the alias of any factorial effect is its generalized interaction with the defining contrast. For example, if $ABCD$ is the defining contrast, then the alias of A is $A(ABCD) = BCD$. It can be seen then that the complete alias structure in a half-replicate of a 2^4 factorial experiment, using $ABCD$ as the defining contrast, is (the symbol $\equiv$ implies *aliased with*)

$$A \equiv BCD$$

$$B \equiv ACD$$

$$C \equiv ABD$$

$$D \equiv ABC$$

$$AB \equiv CD$$

$$AC \equiv BD$$

$$AD \equiv BC.$$

Without supplementary statistical evidence, there is no way of explaining which of two aliased effects are actually providing the influence on the response. In a sense they *share a degree of freedom*. Herein lies the disadvantage in fractional factorial experiments. They have their greatest use when k is quite large and there is some a priori knowledge concerning the interactions. In the example presented, the main effects can be estimated if the three factor interactions are known to be negligible. For testing purposes the only possible procedure, in the absence of either an outside measure of experimental error or a replication of the experiment, would be to pool the sums of squares associated with the two-factor interactions. This, of course, is desirable only if these interactions represent negligible effects.

The construction of the 1/4 fraction or quarter-replicate is identical to the procedure whereby one assigns 2^k treatment combinations to four blocks. This involves the sacrificing of two defining contrasts along with their generalized interaction. Any of the four resulting blocks serves as an appropriate set of experimental runs. Each effect has three aliases, which are given by the generalized interaction with the three defining contrasts. Suppose in a 1/4 fraction of a 2^6 factorial experiment, we use $ACEF$ and $BDEF$ as the defining contrast, resulting in

$$(ACEF)(BDEF) = ABCD$$

also being sacrificed. Using $L_1 = 0$, $L_2 = 0$ (modulo 2), where

$$L_1 = \gamma_1 + \gamma_3 + \gamma_5 + \gamma_6$$
$$L_2 = \gamma_2 + \gamma_4 + \gamma_5 + \gamma_6,$$

we have an appropriate set of experimental runs given by

$\{(1), abcd, ef, abcdef, cde, cdf, abe, abf, acef, bdef, ac, bd, adf, ade, bcf, bce\}$

and the alias structure for the main effects is written

$$A \equiv CEF \equiv ABDEF \equiv BCD$$
$$B \equiv ABCEF \equiv DEF \equiv ACD$$
$$C \equiv AEF \equiv BCDEF \equiv ABD$$
$$D \equiv ACDEF \equiv BEF \equiv ABC$$
$$E \equiv ACF \equiv BDF \equiv ABCDE$$
$$F \equiv ACE \equiv BDE \equiv ABCDF,$$

each with a single degree of freedom. For the two-factor interactions,

$$AB \equiv BCEF \equiv ADEF \equiv CD$$
$$AC \equiv EF \equiv ABCDEF \equiv BD$$
$$AD \equiv CDEF \equiv ABEF \equiv BC$$
$$AE = CF \equiv ABDF \equiv BCDE$$
$$AF \equiv CE \equiv ABDE \equiv BCDF$$
$$BE \equiv ABCF \equiv DF \equiv ACDE$$
$$BF \equiv ABCE \equiv DE \equiv ACDF.$$

Here, of course, there is some aliasing among the two-factor interactions. The remaining two degrees of freedom are accounted for by the following groups:

$$ADF \equiv CDE \equiv ABE \equiv BCF$$
$$ABF \equiv BCE \equiv ADE \equiv CDF.$$

It becomes evident that one should always be aware of what the alias structure is for a fractional factorial experiment before he finally recommends the experimental plan. Proper choice of defining contrasts is important, since it dictates the alias structure. For example, if one would like to study main effects and all two-factor interactions in an experiment involving eight factors and it is known that interactions involving three or more factors are negligible, a very practical design would be one in which the defining contrasts are $ACEGH$ and $BDEFGH$, resulting in a third,

$$(ACEGH)(BDEFGH) = ABCDF.$$

All main effects and two-factor interactions are not aliased with one another and are therefore estimable. The analysis of variance would contain the following:

Main effects	8 single degrees of freedom
Two-factor interactions	28 single degrees of freedom
Error	27 pooled degrees of freedom
Total	63 degrees of freedom

For the 1/8 and higher fractional factorials, the method of constructing the design generalizes. Of course, the aliasing can become quite extensive. For example, with a 1/8 fraction, each effect has seven aliases. The design is constructed by selecting three defining contrasts as if eight blocks were being constructed. Four additional effects are sacrificed and any one of the eight blocks can be properly used as the design.

12.7 ANALYSIS OF FRACTIONAL FACTORIAL EXPERIMENTS

The difficulty in making formal significance tests using data from fractional factorial experiments lies in the determination of the proper error term. Unless there are data available from prior experiments, the error must come from a pooling of contrasts representing effects that are presumed to be negligible.

Sums of squares for individual effects are found using essentially the same procedures given for the complete factorial. One can form a contrast in the treatment combinations by constructing the usual table of positive and negative signs. For example, for a half-replicate of a 2^3 factorial experiment, with ABC the defining contrast, one possible set of treatment combinations and the appropriate algebraic signs for each contrast used in computing the sums of squares for the various effects are presented in Table 12.10.

Note that in Table 12.10 the A and BC contrasts are identical, illustrating the aliasing. Also, $B \equiv AC$ and $C \equiv AB$. In this situation we have three orthogonal contrasts representing the 3 degrees of freedom available. If two observations are obtained for each of the four treatment combinations, we would then have an estimate of the error variance with 4 degrees of freedom. Assuming the

Table 12.10 Signs for Contrasts in a Half-Replicate of a
            2^3 Factorial Experiment

Treatment combination	Factorial effect						
	A	B	C	AB	AC	BC	ABC
a	$+$	$-$	$-$	$-$	$-$	$+$	$+$
b	$-$	$+$	$-$	$-$	$+$	$-$	$+$
c	$-$	$-$	$+$	$+$	$-$	$-$	$+$
abc	$+$	$+$	$+$	$+$	$+$	$+$	$+$

interaction effects to be negligible, we could test all the main effects for significance.

The sum of squares for any main effect, say A, is given by

$$SSA = \frac{(a - b - c + abc)^2}{2^2 n}.$$

In general, the single-degree-of-freedom sum of squares for any effect in a 2^{-p} fraction of a 2^k factorial experiment ($k > p$), is obtained by squaring contrasts in the treatment totals selected and dividing by $2^{k-p}n$, where n is the number of replications of these treatment combinations.

Example 12.4   Suppose that we wish to use a half-replicate to study the effects of five factors, each at two levels, on some response and it is known that whatever the effect of each factor, it will be constant for each level of the other factors. Let the defining contrast be $ABCDE$, causing main effects to be aliased with four-factor interactions. The pooling of contrasts involving interactions provides $15 - 5 = 10$ degrees of freedom for error. Perform an analysis of variance on the following data, testing all main effects for significance at the 0.05 level:

Treatment	Response	Treatment	Response
a	11.3	bcd	14.1
b	15.6	abe	14.2
c	12.7	ace	11.7
d	10.4	ade	9.4
e	9.2	bce	16.2
abc	11.0	bde	13.9
abd	8.9	cde	14.7
acd	9.6	$abcde$	13.2

Solution The sums of squares for the main effects are

$$SSA = \frac{(11.3 - 15.6 - \cdots - 14.7 + 13.2)^2}{2^{5-1}} = \frac{(-17.5)^2}{16} = 19.14$$

$$SSB = \frac{(-11.3 + 15.6 - \cdots - 14.7 + 13.2)^2}{2^{5-1}} = \frac{(18.1)^2}{16} = 20.48$$

$$SSC = \frac{(-11.3 - 15.6 + \cdots + 14.7 + 13.2)^2}{2^{5-1}} = \frac{(10.3)^2}{16} = 6.63$$

$$SSD = \frac{(-11.3 - 15.6 - \cdots + 14.7 + 13.2)^2}{2^{5-1}} = \frac{(-7.7)^2}{16} = 3.71$$

$$SS(E) = \frac{(-11.3 - 15.6 - \cdots + 14.7 + 13.2)^2}{2^{5-1}} = \frac{(8.9)^2}{16} = 4.95,$$

where the factor E is enclosed in parentheses to avoid confusion with the error sum of squares. The total sum of squares is

$$SST = 11.3^2 + 15.6^2 + \cdots + 13.2^2 - \frac{196.1^2}{16} = 85.74.$$

All other calculations and tests of significance are summarized in Table 12.11. The tests indicate that factor A has a significant negative effect on the response, while factor B has a significant positive effect. Factors C, D, and E are not significant at the 0.05 level.

Table 12.11 Analysis of Variance for the Data of a Half-Replicate of a 2^5 Factorial Experiment

Source of variation	Sum of squares	Degrees of freedom	Mean square	Computed f
Main effect				
A	19.14	1	19.14	6.21
B	20.48	1	20.48	6.65
C	6.63	1	6.63	2.15
D	3.71	1	3.71	1.20
E	4.95	1	4.95	1.61
Error	30.83	10	3.08	
Total	85.74	15		

EXERCISES

1. The following data were obtained from a 2^3 factorial experiment replicated three times:

Treatment combination	Replicate 1	Replicate 2	Replicate 3
(1)	12	19	10
a	15	20	16
b	24	16	17
ab	23	17	27
c	17	25	21
ac	16	19	19
bc	24	23	29
abc	28	25	20

Evaluate the sums of squares for all factorial effects by the contrast method.

2. The effects of four factors on some response are to be studied. Each factor is varied at two levels in a 2^4 factorial arrangement and the following data recorded:

Treatment combination	Response
(1)	23.8
a	19.6
b	29.9
ab	25.7
c	26.5
ac	22.6
bc	32.6
abc	28.6
d	21.6
ad	17.5
bd	27.5
abd	23.7
cd	24.6
acd	20.9
bcd	31.1
abcd	26.7

Assuming all three- and four-factor interactions to be negligible, analyze the given data by Yates' technique.

3. A preliminary experiment is conducted to study the effects of four factors and their interactions on the output of a certain machining operation. Two runs are made at each of the treatment combinations in order to supply a measure of pure experimental error. Two levels of each factor are used, resulting in the following data:

Treatment combination	Replicate 1	Replicate 2
(1)	7.9	9.6
a	9.1	10.2
b	8.6	5.8
c	10.4	12.0
d	7.1	8.3
ab	11.1	12.3
ac	16.4	15.5
ad	7.1	8.7
bc	12.6	15.2
bd	4.7	5.8
cd	7.4	10.9
abc	21.9	21.9
abd	9.8	7.8
acd	13.8	11.2
bcd	10.2	11.1
abcd	12.8	14.3

Use Yates' method to make tests on all main effects and interactions.

4. In a 2^3 factorial experiment with three replications, show the block arrangement and indicate by means of an analysis-of-variance table the effects to be tested and their degrees of freedom, when the AB interaction is confounded with blocks.

5. The following coded data represent the strength of a certain type of bread-wrapper stock produced under 16 different conditions, the latter representing two levels of

Operator 1	Operator 2
(1) = 18.8	a = 14.7
ab = 16.5	b = 15.1
ac = 17.8	c = 14.7
bc = 17.3	abc = 19.0
d = 13.5	ad = 16.9
abd = 17.6	bd = 17.5
acd = 18.5	cd = 18.2
bcd = 17.6	abcd = 20.1

each of four process variables. An operator effect was introduced into the model since it was necessary to obtain half the experimental runs under operator 1 and half under operator 2. It was felt that operators do have an effect on the quality of the product.

(a) Assuming all interactions are negligible, make significance tests for the factors A, B, C, and D.

(b) What interaction is confounded with operators?

6. Divide the treatment combinations of a 2^4 factorial experiment into four blocks by confounding ABC and ABD. What additional effect is also confounded with blocks?

7. An experiment was conducted to determine the breaking strength of a certain alloy containing five metals, A, B, C, D, and E. Two different percentages of each metal were used in forming the $2^5 = 32$ different alloys. Since only eight alloys could be tested on a given day, the experiment was conducted over a period of 4 days in which the $ABDE$ and the AE effects were confounded with days. The experimental data were recorded as follows:

Treatment combination	Breaking strength	Treatment combination	Breaking strength
(1)	21.4	e	29.5
a	32.5	ae	31.3
b	28.1	be	33.0
ab	25.7	abe	23.7
c	34.2	ce	26.1
ac	34.0	ace	25.9
bc	23.5	bce	35.2
abc	24.7	abce	30.4
d	32.6	de	28.5
ad	29.0	ade	36.2
bd	30.1	bde	24.7
abd	27.3	abde	29.0
cd	22.0	cde	31.3
acd	35.8	acde	34.7
bcd	26.8	bcde	26.8
abcd	36.4	abcde	23.7

(a) Set up the blocking scheme for the 4 days.

(b) What additional effect is confounded with days?

(c) Use Yates' technique to obtain the sums of squares for all main effects.

8. By confounding ABC in two replicates and AB in the third, show the block arrangement and the analysis-of-variance table for a 2^3 factorial experiment with three replicates. What is the relative information on the confounded effects?

9. The following experiment was run to study main effects and all interactions. Four factors are used at two levels each. The experiment is replicated and two blocks are

necessary in each replication. The data are as follows:

Replicate I		Replicate II	
Block 1	Block 2	Block 3	Block 4
(1) = 17.1	a = 15.5	(1) = 18.7	a = 17.0
d = 16.8	b = 14.8	ab = 18.6	b = 17.1
ab = 16.4	c = 16.2	ac = 18.5	c = 17.2
ac = 17.2	ad = 17.2	ad = 18.7	d = 17.6
bc = 16.8	bd = 18.3	bc = 18.9	abc = 17.5
abd = 18.1	cd = 17.3	bd = 17.0	abd = 18.3
acd = 19.1	abc = 17.7	cd = 18.7	acd = 18.4
bcd = 18.4	abcd = 19.2	abcd = 19.8	bcd = 18.3

(a) What effect is confounded with blocks in the first replication of the experiment? In the second replication?

(b) Conduct an appropriate analysis of variance showing tests on all main effects and interaction effects.

10. Construct a design involving 12 runs in which two factors are varied at two levels each. You are further restricted in that blocks of size 2 must be used and you must be able to make significance tests on both main effects and the interaction effect.

11. List the aliases for the various effects in a 2^5 factorial experiment when the defining contrast is $ACDE$.

12. Construct a 1/4 fraction of a 2^6 factorial design using $ABCD$ and $BDEF$ as the defining contrasts. Show what effects are aliased with the six main effects.

13. Show the blocking scheme for a 2^7 factorial experiment in eight blocks of size 16 each, using $ABCD$, $CDEFG$, and BDF as defining contrasts. Indicate what interactions are completely sacrificed in the experiment.

14. Seven factors are varied at two levels in an experiment involving only 16 trials. A 1/8 fraction of a 2^7 factorial experiment is used with the defining contrasts being ACD, BEF, and CEG. The data are as follows:

Treatment combination	Response	Treatment combination	Response
(1)	13.6	acg	31.1
ad	28.7	cdg	32.0
abce	33.1	beg	32.8
cdef	33.6	adefg	35.3
acef	33.7	efg	32.4
bcde	34.2	abdeg	35.3
abdf	32.5	bcdfg	35.6
bf	27.8	abcfg	35.1

Perform an analysis of variance on all seven main effects, assuming that interactions are negligible.

13
Nonparametric Statistics

13.1 INTRODUCTION

Most of the hypothesis-testing procedures discussed so far in this book are based on the assumption that the random samples are selected from normal populations. Fortunately, most of these tests are still reasonably reliable for slight departures from normality, particularly when the sample size is large. In this chapter we shall consider a number of test procedures that assume no knowledge whatsoever about the distribution of the underlying population. A test performed without this information is called a *nonparametric* or *distribution-free test*.

Nonparametric tests have gained a certain appeal in recent years for several reasons. First, the computations involved are usually very quick and easy to carry out. Second, the data need not be quantitative measurements but could be in the form of qualitative responses such as "defective" versus "nondefective," "yes" versus "no," and so forth, or frequently are values of an ordinal scale to which we assign ranks. On an ordinal scale the subjects are ranked according to a specified order, and a nonparametric test analyzes the various ranks. For example, two judges might rank five brands of premium beer by assigning a rank of 1 to the brand believed to have the best overall quality, a rank of 2 to the second best, and so forth. A nonparametric test could then be used to determine whether there is any agreement between the two judges. A third and perhaps the most important advantage in using nonparametric tests is that they are encumbered with less restrictive assumptions than their parametric counterparts. Such tests usually only assume the underlying distributions to be continuous and symmetrical.

We should also point out that there are a number of disadvantages associated with nonparametric tests. Primarily, they do not utilize all the information provided by the sample. As a result of this wastefulness, a nonparametric test will be less efficient than the corresponding parametric procedure when both methods are applicable. Consequently, a nonparametric test will require a larger sample size than will the corresponding parametric test in order to achieve the same probability of committing a type II error.

473

In summary, if a parametric and a nonparametric test are both applicable to the same set of data, one should always use the more efficient parametric technique. However, recognizing the fact that the assumptions of normality often cannot be justified, and also the fact that we do not always have quantitative measurements, it is fortunate that statisticians have provided us with a number of useful nonparametric procedures.

13.2 WILCOXON TWO-SAMPLE TEST

The procedures discussed in Section 7.4 for testing hypotheses about the difference between two means are valid only if the populations are approximately normal or if the samples are large. In 1945 Frank Wilcoxon proposed a very simple nonparametric procedure for the comparison of two continuous populations when only small independent samples are available and the populations from which they are selected are nonnormal. This procedure is now referred to as the *Wilcoxon two-sample test* or the *Wilcoxon rank-sum test*.

We shall test the null hypothesis H_0 that $\mu_1 = \mu_2$ against some suitable alternative. First we select a random sample from each of the populations. Let n_1 be the number of observations in the smaller sample, and n_2 the number of observations in the larger sample. When the samples are of equal size, n_1 and n_2 may be randomly assigned. Arrange the $n_1 + n_2$ observations of the combined samples in ascending order and substitute a rank of $1, 2, \ldots, n_1 + n_2$ for each observation. In the case of ties (identical observations), we replace the observations by the mean of the ranks that the observations would have if they were distinguishable. For example, if the seventh and eighth observations are identical, we would assign a rank of 7.5 to each of the two observations.

The sum of the ranks corresponding to the n_1 observations in the smaller sample is denoted by w_1. Similarly, the value w_2 represents the sum of the n_2 ranks corresponding to the larger sample. The total $w_1 + w_2$ depends only on the number of observations in the two samples and is in no way affected by the results of the experiment. Hence, if $n_1 = 3$ and $n_2 = 4$, then $w_1 + w_2 = 1 + 2 + \cdots + 7 = 28$, regardless of the numerical values of the observations. In general,

$$w_1 + w_2 = \frac{(n_1 + n_2)(n_1 + n_2 + 1)}{2},$$

the arithmetic sum of the integers $1, 2, \ldots, n_1 + n_2$. Once we have determined w_1 it may be easier to find w_2 by the formula

$$w_2 = \frac{(n_1 + n_2)(n_1 + n_2 + 1)}{2} - w_1.$$

In choosing repeated samples of size n_1 and n_2, we would expect w_1, and therefore w_2, to vary. Thus we may think of w_1 and w_2 as values of the random variables W_1 and W_2, respectively. The null hypothesis $\mu_1 = \mu_2$ will be rejected in favor of the alternative $\mu_1 < \mu_2$ only if w_1 is small and w_2 is large. Likewise, the alternative $\mu_1 > \mu_2$ can be accepted only if w_1 is large and w_2 is small. For a two-tailed test, we may reject H_0 in favor of H_1 if w_1 is small and w_2 is large or if w_1 is large and w_2 is small. As a result of symmetry in the distributions of W_1 and W_2, upper tail probabilities may be obtained from the lower tail probabilities. Hence no matter what the alternative hypothesis may be, we reject the null hypothesis when the smaller of w_1 and w_2 is sufficiently small. Suppose that for a given experiment that $w_1 < w_2$. Knowing the distribution of W_1, we can determine $P(W_1 \leq w_1 | H_0 \text{ is true})$. If this probability is less than or equal to 0.05, our test is significant and we would reject H_0 in favor of the appropriate one-sided alternative. When the probability does not exceed 0.01, the test is highly significant. In the case of a two-tailed test symmetry permits us to base our decision on the value of $2P(W_1 \leq w_1 | H_0 \text{ is true})$. Therefore, when $2P(W_1 \leq w_1 | H_0 \text{ is true}) < 0.05$, the test is significant and we conclude that $\mu_1 \neq \mu_2$.

The distribution of W_1, when H_0 is true, is based on the fact that all the observations in the smaller sample could be assigned ranks at random as long as their sum is less than or equal to w_1. The total number of ways of assigning $n_1 + n_2$ ranks to n_1 observations so that the sum of the ranks does not exceed w_1 is denoted by $n(W_1 \leq w_1)$. There are $\binom{n_1 + n_2}{n_1}$ equally likely ways to assign the $n_1 + n_2$ ranks to n_1 observations giving all possible values of W_1. Hence

$$P(W_1 \leq w_1 | H_0 \text{ is true}) = \frac{n(W_1 \leq w_1)}{\binom{n_1 + n_2}{n_1}}, \qquad \text{for } n_1 \leq n_2.$$

It is possible to find $n(W_1 \leq w_1)$ for any given test by listing all the cases and counting them. Thus when $n_1 = 3$ and $n_2 = 5$, the number of cases where the sum of the ranks in the smaller sample is less than or equal to 8 may be listed as follows:

$$1 + 2 + 3 = 6$$

$$1 + 2 + 4 = 7$$

$$1 + 3 + 4 = 8$$

$$1 + 2 + 5 = 8.$$

Therefore, there are 4 favorable cases out of a possible $\binom{8}{3} = 56$ equally likely cases. Hence

$$P(W_1 \leq 8 | H_0 \text{ is true}) = \tfrac{4}{56} = 0.0714.$$

For the case where $w_2 < w_1$, we could proceed as above to determine $P(W_2 \leq w_2 | H_0 \text{ is true})$. However, in either case it is usually easier to find the desired probability by using Table XVI when n_2 does not exceed eight. Table XVI is based upon the statistic U, the minimum of U_1 and U_2, where

$$U_1 = W_1 - \frac{n_1(n_1 + 1)}{2}$$

and

$$U_2 = W_2 - \frac{n_2(n_2 + 1)}{2}.$$

If $P(U \leq u | H_0 \text{ is true}) \leq \alpha$, our test is significant and we reject H_0 in favor of the appropriate one-sided alternative. For a two-tailed test, our test is significant when $2P(U \leq u | H_0 \text{ is true}) \leq \alpha$, in which case we accept the alternative hypothesis that $\mu_1 \neq \mu_2$.

In the preceding illustration, where we had $n_1 = 3$, $n_2 = 5$, and $w_1 = 8$, we find $w_2 = [(8)(9)/2] - 8 = 28$, and then

$$u_1 = 8 - [(3)(4)/2] = 2$$
$$u_2 = 28 - [(5)(6)/2] = 13.$$

Using Table XVI, with $u = 2$, we have

$$P(U \leq 2 | H_0 \text{ is true}) = 0.071,$$

which agrees with the previous answer. If, for the same illustration, $w_1 = 7$ so that $u = 1$, we find

$$P(U \leq 1 | H_0 \text{ is true}) = 0.036,$$

which is significant for a one-tailed test at the 0.05 level but not at the 0.01 level. For a two-tailed test, the probability that the sample means differ by an amount as great as or greater than that observed is

$$2P(U \leq 1 | H_0 \text{ is true}) = (2)(0.036) = 0.072,$$

from which we conclude that H_0 is true.

When n_2 is between 9 and 20, Table XVII may be used. If the observed value of U is less than or equal to the tabled value, the null hypothesis may be rejected at the level of significance indicated by the table. Table XVII gives critical values of U for levels of significance equal to 0.001, 0.01, 0.025, and 0.05 for a one-tailed test. In the case of a two-tailed test the critical values of U correspond to the 0.002, 0.02, 0.05, and 0.1 levels of significance. When n_1 and n_2 increase in size, the sampling distribution of U_1 approaches the normal distribution with mean

$$\mu_{U_1} = \frac{n_1 n_2}{2}$$

and variance

$$\sigma_{U_1}^2 = \frac{n_1 n_2 (n_1 + n_2 + 1)}{12}.$$

Consequently, when n_2 is greater than 20, one could use the statistic $Z = (U_1 - \mu_{U_1})/\sigma_{U_1}$ for our test, with the critical region falling in either or both tails of the standard normal distribution, depending on the form of H_1.

To test the null hypothesis that the means of two nonnormal populations are equal when only small independent samples are available, we proceed by the following steps.

1. H_0: $\mu_1 = \mu_2$.
2. H_1: Alternatives are $\mu_1 < \mu_2$, $\mu_1 > \mu_2$, or $\mu_1 \neq \mu_2$.
3. Choose a level of significance equal to α.
4. Critical region:
   (a) All u values for which $P(U \leq u \mid H_0$ is true$) < \alpha$ when $n_2 \leq 8$ and the test is one-tailed.
   (b) All u values for which $2P(U \leq u \mid H_0$ is true$) < \alpha$ when $n_2 \leq 8$ and the test is two-tailed.
   (c) All u values less than or equal to the appropriate critical value in Table XVII when $9 \leq n_2 \leq 20$.
5. Compute w_1, w_2, u_1, and u_2 from independent samples of size n_1 and n_2, where $n_1 \leq n_2$. Using the smaller of u_1 and u_2 for u, determine whether u falls in the acceptance or critical region.
6. Conclusion: Reject H_0 if u falls in the critical region; otherwise, accept H_0.

Example 13.1   To find out whether a new serum will arrest leukemia, nine mice, which have all reached an advanced stage of the disease, are selected.

five mice receive the treatment and four do not. The survival times, in years, from the time the experiment commenced are:

Treatment	2.1	5.3	1.4	4.6	0.9
No treatment	1.9	0.5	2.8	3.1	

At the 0.05 level of significance, can the serum be said to be effective?

Solution   We follow the six-step procedure above with $n_1 = 4$ and $n_2 = 5$.

1. $H_0: \mu_1 = \mu_2$.
2. $H_1: \mu_1 < \mu_2$.
3. $\alpha = 0.05$.
4. Critical region: All u values for which $P(U \le u | H_0$ is true$) < 0.05$.
5. Computations: The observations are arranged in ascending order and ranks from 1 to 9 assigned.

Original data	0.5	0.9	1.4	1.9	2.1	2.8	3.1	4.6	5.3
Ranks	1	2	3	4	5	6	7	8	9

The treatment observations are underscored for identification purposes. Now

$$w_1 = 1 + 4 + 6 + 7 = 18$$

and

$$w_2 = [(9)(10)/2] - 18 = 27.$$

Therefore,

$$u_1 = 18 - [(4)(5)/2] = 8$$

$$u_2 = 27 - [(5)(6)/2] = 12,$$

so that $u = 8$. Since $P(U \le 8 | H_0$ is true$) = 0.365 > 0.05$, the value $u = 8$ falls in the acceptance region.

6. Conclusion: Accept H_0 and conclude that the serum does not prolong life by arresting leukemia.

Example 13.2 The nicotine content of two brands of cigarettes, measured in milligrams, was found to be as follows:

Brand A	2.1	4.0	6.3	5.4	4.8	3.7	6.1	3.3		
Brand B	4.1	0.6	3.1	2.5	4.0	6.2	1.6	2.2	1.9	5.4

Test the hypothesis, at the 0.05 level of significance, that the average nicotine contents of the two brands are equal against the alternative that they are unequal.

Solution We proceed by the six-step rule with $n_1 = 8$ and $n_2 = 10$.

1. H_0: $\mu_1 = \mu_2$.
2. H_1: $\mu_1 \neq \mu_2$.
3. $\alpha = 0.05$.
4. Critical region: $U \leq 17$ (from Table XVII).
5. Computations: The observations are arranged in ascending order and ranks from 1 to 18 assigned.

Original data	Ranks
0.6	1
1.6	2
1.9	3
2.1	4
2.2	5
2.5	6
3.1	7
3.3	8
3.7	9
4.0	10.5
4.0	10.5
4.1	12
4.8	13
5.4	14.5
5.4	14.5
6.1	16
6.2	17
6.3	18

The ranks of the observations belonging to the smaller sample are underscored. Now

$$w_1 = 4 + 8 + 9 + 10.5 + 13 + 14.5 + 18 = 93$$

and

$$w_2 = \left[\frac{(18)(19)}{2}\right] - 93 = 78.$$

Therefore,

$$u_1 = 93 - \left[\frac{(8)(9)}{2}\right] = 57$$

$$u_2 = 78 - \left[\frac{(10)(11)}{2}\right] = 23,$$

so that $u = 33$.

6. Conclusion: Accept H_0 and conclude that there is no difference in the average nicotine contents of the two brands of cigarettes.

The use of the Wilcoxon two-sample test is not restricted to nonnormal populations. It can be used in place of the t test when the populations are normal, although the probability of committing a type II error will be larger. The Wilcoxon two-sample test is always superior to the t test for nonnormal populations.

13.3 SIGN TEST

Assume that n pairs of observations are selected from two nonnormal populations defined over a *continuous* sample space. For large n the distribution of the mean of the differences of the matched pairs of observations in repeated sampling is approximately normal, and tests of hypotheses concerning the two population means may be carried out using the statistic

$$T = \frac{\bar{D} - d_0}{S_d/\sqrt{n}},$$

as given in Table 7.1. However, if $n < 30$ and the population of differences is decidedly nonnormal, we must resort to a nonparametric test. Perhaps the easiest and quickest to perform is a test called the *sign test*. In testing the null hypothesis H_0 that $\mu_1 = \mu_2$ or $\mu_D = 0$, each difference d_i of the paired observations is assigned a *plus* or *minus* sign, depending on whether d_i is positive or negative. If the null hypothesis is true and the populations are symmetric, the sum of the plus signs should be approximately equal to the sum of the minus signs. When one sign appears more frequently than it should, based on chance alone, we reject the hypothesis that the population means are equal.

To provide an appropriate test statistic for the sign test, let r_- and r_+ represent the number of minus and plus signs, respectively, in our random sample of paired observations. Our test statistic is then defined to be the random variable R which assumes the value r in a particular experiment, where

$$r = \text{smaller of } r_- \text{ and } r_+.$$

The sign test is applicable only in situations where ties or zero differences between the paired observations cannot occur. Although a zero difference is theoretically impossible since the populations are continuous, nevertheless in practice zero differences often do occur from a lack of precision in recording our data. When ties or zeros are observed, they must be excluded from the analysis and the number of paired observations is correspondingly reduced. Hence, for a particular experiment, we have $n = r_- + r_+$.

If the null hypothesis that $\mu_1 = \mu_2$ is true, the probability that a difference for a matched pair results in either a plus sign or a minus sign is equal to 1/2. Consequently, the test statistic R has a binomial probability distribution with parameter $p = 1/2$ when H_0 is true from which levels of significance for both one-sided and two-sided alternatives can be computed. For example, in testing

$$H_0: \quad \mu_1 - \mu_2 = 0,$$
$$H_1: \quad \mu_1 - \mu_2 < 0,$$

we will reject H_0 in favor of H_1 only if the proportion of plus signs is sufficiently less than 1/2, that is, when $r = r_+$ is small. Hence a critical region is established by forming the inequality $R < r^*$, where r^* is an appropriate positive integer less than $n/2$ so as to give a reasonable value for the significance level,

$$\alpha = P(R < r^* | H_0 \text{ is true}).$$

Because of the discreteness of the binomial distribution, we cannot expect to find values of r^* that establish customary critical regions of size exactly equal to 0.01 or 0.05. However, this is usually a problem only for small sample sizes. For example, when $n = 15$ and $r^* = 5$, we find from Table II (see Statistical Tables) that

$$\alpha = P(R < 5) = \sum_{x=0}^{4} b(x; 15, 0.5) = 0.0592,$$

so that the critical region $R < 5$ results in a statistical test at the 0.0592 level of significance. However, if $n = 6$ and $r^* = 1$, we find that

$$\alpha = P(R < 1) = b(0; 6, 0.5) = 0.0156$$

whereas with $r^* = 2$ we get

$$\alpha = P(R < 2) = b(0; 6, 0.5) + b(1; 15, 0.5)$$
$$= 0.1094,$$

so that neither value of r^* produces a significance level of size α close to 0.05.

To test the hypothesis

$$H_0: \quad \mu_1 - \mu_2 = 0,$$
$$H_1: \quad \mu_1 - \mu_2 > 0,$$

we reject H_0 in favor of H_1 only if the proportion of *minus* signs is sufficiently less than 1/2, that is, when $r = r_-$ is small. Hence the critical region of size α is once again established by forming the interval $R < r^*$, where α is computed as before by the binomial probability distribution. Finally, to test the hypothesis

$$H_0: \quad \mu_1 - \mu_2 = 0,$$
$$H_1: \quad \mu_1 - \mu_2 \neq 0,$$

we reject H_0 in favor of H_1 when the proportion of *plus* signs (or *minus* signs) is significantly less than or greater than 1/2. This, of course, is equivalent to either r_- or r_+ being sufficiently small or sufficiently large. However, symmetry allows us to again use a one-sided critical region $R < r^*$ for this two-tailed test with the significance level now given by

$$\alpha = 2P(R < r^* | H_0 \text{ is true}).$$

Since the values of α are computed from a binomial distribution with parameter $p = 1/2$, one could use the normal curve to approximate α as discussed in Section 4.3, provided that $n > 10$. Hence for $n = 15$ and $r^* = 5$,

$$\alpha = P(R < 5) \simeq P(X < 4.5),$$

where X is a normal random variable with mean

$$\mu = np = (15)(0.5) = 7.5$$

and standard deviation

$$\sigma = \sqrt{npq} = \sqrt{(15)(0.5)(0.5)} = 1.936.$$

Therefore,

$$z = \frac{4.5 - 7.5}{1.936} = -1.55$$

and

$$\alpha = P(R < 5) \simeq P(Z < -1.55)$$
$$= 0.0606,$$

which agrees very closely with the exact value of 0.0592 derived above.

Significance levels corresponding to various critical regions can be obtained quickly from Table XVIII (see Statistical Tables), where $P(R < r^* | H_0$ is true) is given for meaningful values of r^* and values of $n = 5, 6, \ldots, 25$. Since H_0 is always rejected in favor of every H_1 when $R < r^*$, Table XVIII gives significance levels only for one-sided alternatives. The table entries must be doubled to provide the correct significance levels for two-sided alternatives. For example, if $n = 21$, a one-tailed test will reject H_0 when $R < 7$ at the 0.039 level of significance, whereas a two-tailed test will reject H_0 in favor of H_1 when $R < 6$ at the $(2)(0.013) = 0.026$ level of significance.

Example 13.3 A taxi company is trying to decide whether the use of radial tires instead of regular belted tires improves fuel economy. Twelve cars were equipped with radial tires and driven over a prescribed test course. Without changing drivers, the same cars were then equipped with the regular belted tires and driven once again over the test course. The gasoline consumption, in kilometers per liter, was recorded as follows:

Car	Kilometers per liter	
	Radial tires	Belted tires
1	4.2	4.1
2	4.7	4.9
3	6.6	6.2
4	7.0	6.9
5	6.7	6.8
6	4.5	4.4
7	5.7	5.7
8	6.0	5.8
9	7.4	6.9
10	4.9	4.7
11	6.1	6.0
12	5.2	4.9

Can we conclude that cars equipped with radial tires give better fuel economy than those equipped with regular belted tires?

Solution Let μ_1 and μ_2 represent the mean kilometers per liter for cars equipped with radial and belted tires, respectively. Examination of the data indicates nine plus signs, two minus signs, and one zero. Therefore, with $n = 11$, after throwing out the zero, we proceed as follows:

1. H_0: $\mu_1 - \mu_2 = 0$.
2. H_1: $\mu_1 - \mu_2 > 0$.
3. $\alpha = 0.033$ (from Table XVIII).
4. Critical region: $R < 3$.
5. Computations: $r_+ = 9, r_- = 2$, so that $r = 2$, the smaller of r_+ and r_-.
6. Conclusion: Reject H_0 and conclude that, on the average, radial tires do improve fuel economy.

The sign test for paired observations may also be used to test the null hypothesis that $\mu_1 - \mu_2 = d_0$. We simply assign plus or minus signs after each d_i has been adjusted by subtracting d_0 and then proceed as before. One can also use the sign test to test the hypothesis $\mu = \mu_0$ for a random sample from a single population. In this case we assign plus and minus signs to the differences $(x_i - \mu_0)$ and apply the same procedure as above.

Not only is the sign test one of our simplest nonparametric procedures to apply, it has the additional advantage of being applicable to dichotomous data that cannot be recorded on a numerical scale but can be represented by positive and negative responses. For example, the sign test is applicable in experiments where a qualitative response such as "hit" or "miss" is recorded, and in sensory-type experiments where a plus or minus sign is recorded depending on whether the taste tester correctly or incorrectly identifies the desired ingredient.

13.4 WILCOXON TEST FOR PAIRED OBSERVATIONS

The sign test shows, by the assigned plus or minus sign, which member of a pair of observations is the larger, but it does not indicate the magnitude of the difference. A test utilizing both direction and magnitude was proposed in 1945 by Wilcoxon and is now commonly referred to as the *Wilcoxon test for paired observations*. Wilcoxon's test is more sensitive than the sign test in detecting a difference in the population means and therefore will be considered in detail.

To test the hypothesis that $\mu_1 = \mu_2$ by the Wilcoxon test, first discard all differences equal to zero and then rank the remaining d_i's without regard to sign. A rank of 1 is assigned to the smallest d_i in absolute value, a rank of 2 to the next smallest, and so on. When the absolute value of two or more differences is the same, assign to each the average of the ranks that would have been assigned if the differences were distinguishable. If there is no difference between the two population means, the total of the ranks corresponding to the positive differences should be almost equal to the total of the ranks corresponding to the negative differences. Let us represent these totals by w_+ and w_-, respectively. We shall designate the smaller of the w_+ and w_- by w and find the probability of obtaining, by chance alone, a value less than or equal to w when H_0 is true.

In selecting repeated samples of paired observations, we would expect w to vary. Thus we may think of w as a value of some random variable W. Once the distribution of W is known, we can determine $P(W \leq w|H_0$ is true). For a level of significance equal to α, we reject H_0 when

$$P(W \leq w|H_0 \text{ is true}) < \alpha$$

and accept the appropriate one-sided alternative. In the case of a two-tailed test, we reject H_0 at the α level of significance in favor of the alternative two-sided hypothesis $\mu_1 \neq \mu_2$ when

$$2P(W \leq w|H_0 \text{ is true}) < \alpha.$$

If we assume that there is no difference in the population means, each d_i is just as likely to be positive as it is to be negative. Thus there are two equally likely ways for a given rank to receive a sign. For n differences, there are 2^n equally likely ways for the n ranks to receive signs. Let $n(W \leq w)$ be the number of the 2^n ways of assigning signs to the n ranks such that the value of W does not exceed w. Then

$$P(W \leq w|H_0 \text{ is true}) = \frac{n(W \leq w)}{2^n}.$$

Consider, for example, the case of $n = 6$ matched pairs that yield a value $w = 5$. What is the probability that $W \leq 5$ when the two population means are equal? The sets of ranks whose total does not exceed 5 may be listed as follows:

Value of W	Sets of ranks totaling W
0	$\varnothing$
1	$\{1\}$
2	$\{2\}$
3	$\{3\}, \{1, 2\}$
4	$\{4\}, \{1, 3\}$
5	$\{5\}, \{1, 4\}, \{2, 3\}$

Therefore, $n(W \le 5) = 10$ out of a possible $2^6 = 64$ equally likely cases. Hence

$$P(W \le 5 | H_0 \text{ is true}) = \tfrac{10}{64} = 0.1563,$$

a result that is quite likely to occur when $\mu_1 = \mu_2$.

When $5 \le n \le 30$, Table XIX (see Statistical Tables) gives approximate critical values of W for levels of significance equal to 0.01, 0.025, and 0.05 for a one-tailed test and equal to 0.02, 0.05, and 0.10 for a two-tailed test. In the preceding example for which $n = 6$, Table XIX shows that a value of $W \le 2$ is required for a one-tailed test to be significant at the 0.05 level. When $n > 30$ the sampling distribution of W_+ approaches the normal distribution with mean

$$\mu_{W_+} = \frac{n(n + 1)}{4}$$

and variance

$$\sigma_{W_+}^2 = \frac{n(n + 1)(2n + 1)}{24}.$$

In this case the statistic $Z = (W_+ - \mu_{W_+})/\sigma_{W_+}$ can be used to determine the critical region for our test.

The Wilcoxon test for paired observations may also be used to test the null hypothesis that $\mu_1 - \mu_2 = \mu_D = d_0$. We simply apply the same procedure as before after each d_i is adjusted by subtracting d_0. Therefore, to test a hypothesis about the difference between the means of two populations whose distributions are unknown, where the observations occur in pairs and the sample size is small, we proceed by the following six steps:

1. H_0: $\mu_1 - \mu_2 = \mu_D = d_0$.
2. H_1: Alternatives are $\mu_1 - \mu_2 < d_0, \mu_1 - \mu_2 > d_0$, or $\mu_1 - \mu_2 \ne d_0$.
3. Choose a level of significance equal to α.
4. Critical region:
   (a) All w values for which $P(W \le w | H_0 \text{ is true}) < \alpha$ when $n < 5$ and the test is one-tailed.
   (b) All w values for which $2P(W \le w | H_0 \text{ is true}) < \alpha$ when $n < 5$ and the test is two-tailed.
   (c) All w values less than or equal to the appropriate critical value in Table XIX when $5 \le n \le 30$.
5. Rank the n differences, $d_i - d_0$, without regard to sign, and then compute w.
6. Conclusion: Reject H_0 if w falls in the critical region; otherwise, accept H_0.

Example 13.4 It is claimed that a college senior can increase his score in the major field area of the graduate record examination by at least 50 points if he is provided sample problems in advance. To test this claim, 20 college seniors were divided into 10 pairs such that each matched pair had almost the same overall quality point average for their first 3 years in college. Sample problems and answers were provided at random to one member of each pair 1 week prior to the examination. The following examination scores were recorded:

	Pair									
	1	*2*	*3*	*4*	*5*	*6*	*7*	*8*	*9*	*10*
With sample problems	531	621	663	579	451	660	591	719	543	575
Without sample problems	509	540	688	502	424	683	568	748	530	524

Test the null hypothesis at the 0.05 level of significance that sample problems increase the scores by 50 points against the alternative hypothesis that the increase is less than 50 points.

Solution Let μ_1 and μ_2 represent the mean score of all students taking the test in question with and without sample problems, respectively. We follow the six-step procedure already outlined:

1. H_0: $\mu_1 - \mu_2 = 50$.
2. H_1: $\mu_1 - \mu_2 < 50$.
3. $\alpha = 0.05$.
4. Critical region: Since $n = 10$, Table XIX shows the critical region to be $W \leq 11$.
5. Computations:

	Pair									
	1	*2*	*3*	*4*	*5*	*6*	*7*	*8*	*9*	*10*
d_i	22	81	-25	77	27	-23	23	-29	13	51
$d_i - d_0$	-28	31	-75	27	-23	-73	-27	-79	-37	1
Ranks	5	6	9	3.5	2	8	3.5	10	7	1

Now $w_+ = 10.5$ and $w_- = 44.5$, so that $w = 10.5$, the smaller of w_+ and w_-.

6. Conclusion: Reject H_0 and conclude that the sample problems do not, on the average, increase one's graduate record score by as much as 50 points.

13.5 RUNS TEST

In applying the many statistical concepts that were discussed throughout this text, it was always assumed that our sample data had been collected by some randomization procedure. The *runs test*, based on the order in which the sample observations are obtained, is a useful technique for testing the null hypothesis H_0 that the observations have indeed been drawn at random.

To illustrate the runs test, let us suppose that 12 people have been polled to find out if they use a certain product. One would seriously question the assumed randomness of the sample if all 12 people were of the same sex. We shall designate a male and female by the symbols M and F, respectively, and record the outcomes according to their sex in the order in which they occur. A typical sequence for the experiment might be

$$\underbrace{M\ \ M}\ \ \underbrace{F\ \ F\ \ F}\ \ \underbrace{M}\ \ \underbrace{F\ \ F}\ \ \underbrace{M\ \ M\ \ M\ \ M},$$

where we have grouped subsequences of similar symbols. Such groupings are called *runs*.

> DEFINITION 13.1 *A run is a subsequence of one or more identical symbols representing a common property of the data.*

Regardless of whether our sample measurements represent qualitative or quantitative data, the runs test divides the data into two mutually exclusive categories: male or female; defective or nondefective; heads or tails; above or below the median; and so forth. Consequently, a sequence will always be limited to two distinct symbols. Let n_1 be the number of symbols associated with the category that occurs the least and n_2 be the number of symbols that belong to the other category. Then the sample size $n = n_1 + n_2$.

For the $n = 12$ symbols in our poll we have five runs with the first containing two M's, the second containing three F's, and so on. If the number of runs is larger or smaller than what we would expect by chance, the hypothesis that the sample was drawn at random should be rejected. Certainly, a sample resulting in only two runs,

$$M\ \ M\ \ M\ \ M\ \ M\ \ M\ \ M\ \ F\ \ F\ \ F\ \ F\ \ F,$$

or the reverse, is most unlikely to occur from a random selection process. Such a result indicates that the first seven people interviewed were all males followed by five females. Likewise, if the sample resulted in the maximum number of 12 runs, as in the alternating sequence

$$M \quad F \quad M \quad F \quad M \quad F \quad M \quad F \quad M \quad F \quad M \quad F,$$

we would again be suspicious of the order in which the individuals were selected for the poll.

The runs test for randomness is based on the random variable V, the total number of runs that occur in the complete sequence of our experiment. A rejection region for a two-tailed test would take the form $V \leq a$ and $V \geq b$. The sampling distribution of V is known, thus enabling us to establish significance levels for our test. Values of $P(V \leq a | H_0$ is true) are given in Table XX (see Statistical Tables) for values of n_1 and n_2 less than or equal to 10. Rejection regions for both one-tailed and two-tailed tests can be obtained using these tabled values. In the poll taken above we exhibit a total of five F's and seven M's. Hence with $n_1 = 5$, $n_2 = 7$, and $v = 5$, we find from Table XX that $P(V \leq 5 | H_0$ is true) = 0.197 > 0.025. That is, the value $v = 5$ is reasonable when H_0 is true, and therefore we have insufficient evidence to reject the hypothesis of randomness in our sample. Critical values for a two-tailed test for a significance level of exactly 0.05 or 0.01 cannot be found from Table XX. However, for $n_1 = 5$ and $n_2 = 7$ we note that $P(V \leq 3 | H_0$ is true) = 0.015 and $P(V \geq 11 | H_0$ is true) = $1 - P(V \leq 10 | H_0$ is true) = $1 - 0.992 = 0.008$. Therefore, we would reject the hypothesis of randomness when $V \leq 3$ or $V \geq 11$ at the $\alpha = 0.015 + 0.008 = 0.023$ level of significance. As we have already seen, the value $v = 5$ for our example falls in the acceptance region.

The runs test can also be used to detect departures in randomness of a sequence of quantitative measurements over time, caused by trends or periodicities. Replacing each measurement in the order in which they are collected by a *plus* symbol if it falls above the median, by a *minus* symbol if it falls below the median, and omitting all measurements that are exactly equal to the median, we generate a sequence of plus and minus symbols that are tested for randomness as illustrated in the following example.

Example 13.5   A machine is adjusted to dispense acrylic paint thinner into a container. Would you say that the amount of paint thinner being dispensed by this machine varies randomly if the contents of the next 15 containers are measured and found to be 3.6, 3.9, 4.1, 3.6, 3.8, 3.7, 3.4, 4.0, 3.8, 4.1, 3.9, 4.0, 3.8, 4.2, and 4.1 liters?

Solution   For the given sample we find $\tilde{x} = 3.9$. Replacing each measurement by the symbol " + " if it falls above 3.9, by the symbol " − " if it falls below 3.9, and omitting the two measurements that equal 3.9, we obtain the sequence

$$- \quad + \quad - \quad - \quad - \quad - \quad + \quad + \quad + \quad + \quad - \quad + \quad +$$

for which $n_1 = 6$, $n_2 = 7$, and $v = 6$. Consulting Table XX, we find that $P(V \leq 6 | H_0$ is true$) = 0.298 > 0.025$. Therefore, we accept the hypothesis that the sequence of measurements vary randomly.

The runs test, although less powerful, can also be used as an alternative to the Wilcoxon two-sample test to test the claim that two random samples come from populations having the same distributions and therefore equal means. If the populations are symmetric, rejection of the claim of equal distributions is equivalent to accepting the alternative hypothesis that the means are not equal. In performing the test, we first combine the observations from both samples and arrange them in ascending order. Now assign the letter A to each observation taken from one of the populations and the letter B to each observation from the second population, thereby generating a sequence consisting of the symbols A and B.

For the survival times of the leukemia patients of Example 13.1, we have

$$
\begin{array}{ccccccccc}
0.5 & 0.9 & 1.4 & 1.9 & 2.1 & 2.8 & 3.1 & 4.6 & 5.3 \\
B & A & A & B & A & B & B & A & A,
\end{array}
$$

resulting in $v = 6$ runs. If the two symmetric populations have equal means the observations from the two samples will be intermingled resulting in many runs. However, if the population means are significantly different we would expect most of the observations for one of the two samples to be smaller than those for the other sample. In the extreme case where the populations do not overlap, we would obtain a sequence of the form

$$
A \ A \ A \ A \ A \ B \ B \ B \ B \quad \text{or} \quad B \ B \ B \ B \ A \ A \ A \ A \ A,
$$

and in either case there are only two runs. Consequently, the hypothesis of equal population means will be rejected when V is small and falls in the critical region $V \leq a$, implying a one-tailed test.

Returning to the data of Example 13.1, for which $n_1 = 4$, $n_2 = 5$, and $v = 6$, we find from Table XX that $P(V \leq 6 | H_0$ is true$) = 0.786$ and therefore accept the null hypothesis of equal means. In order to reject H_0 even at the 0.071 level of significance, a value of $V \leq 3$ would be required. Hence we conclude that the new serum does not prolong life by arresting leukemia.

When n_1 and n_2 increase in size, the sampling distribution of V approaches the normal distribution with mean

$$
\mu_V = \frac{2n_1 n_2}{n_1 + n_2} + 1
$$

and variance

$$\sigma_V^2 = \frac{2n_1 n_2 (2n_1 n_2 - n_1 - n_2)}{(n_1 + n_2)^2 (n_1 + n_2 - 1)}.$$

Consequently, when n_1 and n_2 are both greater than 10, one could use the statistic

$$Z = \frac{V - \mu_V}{\sigma_V}$$

to establish the critical region for the runs test.

13.6 TOLERANCE LIMITS

Tolerance limits for a normal distribution of measurements were discussed in Chapter 6. In this section we shall consider a method for constructing tolerance intervals that are independent of the shape of the underlying distribution. As one might suspect, for a reasonable degree of confidence they will be substantially longer than those constructed assuming normality, and the sample size required is generally very large. Nonparametric tolerance limits are stated in terms of the smallest and largest observations in our sample.

TWO-SIDED TOLERANCE LIMITS   *For any distribution of measurements, two-sided tolerance limits are given by the smallest and largest observations in a sample of size n, where n is determined so that one can assert with $100\gamma\%$ confidence that* at least *the proportion $1 - \alpha$ of the distribution is included between the sample extremes.*

Table XXI (see Statistical Tables) gives required sample sizes for selected values of γ and $1 - \alpha$. For example, when $\gamma = 0.99$ and $1 - \alpha = 0.95$, we must choose a random sample of size $n = 130$ in order to be 99% confident that at least 95% of the distribution of measurements is included between the sample extremes.

Instead of determining the sample size n such that a specified proportion of measurements are contained between the sample extremes, it is desirable in many industrial processes to determine the sample size such that a fixed proportion of the population falls below the largest (or above the smallest) observation in the sample. Such limits are called one-sided tolerance limits.

ONE-SIDED TOLERANCE LIMITS *For any distribution of measurements, a one-sided tolerance limit is given by the smallest (largest) observation in a sample of size n, where n is determined so that one can assert with* $100\gamma\%$ *confidence that* at least *the proportion* $1 - \alpha$ *of the distribution will exceed the smallest (be less than the largest) observation in the sample.*

Table XXII (see Statistical Tables) gives required sample sizes corresponding to selected values of γ and $1 - \alpha$. Hence, when $\gamma = 0.95$ and $1 - \alpha = 0.70$, we must choose a sample of size $n = 9$ in order to be 95% confident that 70% of our distribution of measurements will exceed the smallest observation in the sample.

13.7 RANK CORRELATION COEFFICIENT

In Chapter 8 we used the sample correlation coefficient r to measure the linear relationship between two continuous variables X and Y. If ranks $1, 2, \ldots, n$ are assigned to the x observations in order of magnitude and similarly to the y observations, and if these ranks are then substituted for the actual numerical values into the formula for r, we obtain the nonparametric counterpart of the conventional correlation coefficient. A correlation coefficient calculated in this manner is known as the *Spearman rank correlation coefficient* and denoted by r_s. When there are no ties among either set of measurements, the formula for r_s reduces to a much simpler expression, which we now state.

SPEARMAN RANK CORRELATION COEFFICIENT *A nonparametric measure of association between two variables X and Y is given by the rank correlation coefficient*

$$r_s = 1 - \frac{6 \sum_{i=1}^{n} d_i^2}{n(n^2 - 1)},$$

where d_i is the difference between the ranks assigned to x_i and y_i, and n is the number of pairs of data.

In practice the preceding formula is also used when there are ties either among the x or y observations. The ranks for tied observations are assigned as in the Wilcoxon test for paired observations by averaging the ranks that would have been assigned if the observations were distinguishable.

The value of r_s will usually be close to the value obtained by finding r based on numerical measurements and is interpreted in much the same way. As before, the values of r_s will range from -1 to $+1$. A value of $+1$ or -1 indicates perfect association between X and Y, the plus sign occurring for identical rankings and the minus sign occurring for reverse rankings. When r_s is close to zero, we would conclude that the variables are uncorrelated.

Example 13.6 The following figures, released by the Federal Trade Commission, show the milligrams of tar and nicotine found in 10 brands of cigarettes.

Cigarette brand	Tar content	Nicotine content
Viceroy	14	0.9
Marlboro	17	1.1
Chesterfield	28	1.6
Kool	17	1.3
Kent	16	1.0
Raleigh	13	0.8
Old Gold	24	1.5
Philip Morris	25	1.4
Oasis	18	1.2
Players	31	2.0

Calculate the rank correlation coefficient to measure the degree of relationship between tar and nicotine content in cigarettes.

Solution Let X and Y represent the tar and nicotine contents, respectively. First we assign ranks to each set of measurements with the rank of 1 assigned to the lowest number in each set, the rank of 2 to the second lowest number in each set, and so forth, until the rank of 10 is assigned to the largest number. Table 13.1 shows the individual rankings of the measurements and the differences in ranks for the 10 pairs of observations. Substituting into the formula for r_s, we find that

$$r_s = 1 - \frac{(6)(5.5)}{(10)(100 - 1)} = 0.9667,$$

indicating a high positive correlation between the amount of tar and nicotine found in cigarettes.

Some advantages in using r_s rather than r do exist. For instance, we no longer assume the underlying relationship between X and Y to be linear and therefore, when the data possess a distinct curvilinear relationship, the rank correlation

Table 13.1 Rankings for Tar and Nicotine Contents

Cigarette brand	x_i	y_i	d_i
Viceroy	2	2	0
Marlboro	4.5	4	0.5
Chesterfield	9	9	0
Kool	4.5	6	−1.5
Kent	3	3	0
Raleigh	1	1	0
Old Gold	7	8	−1
Philip Morris	8	7	1
Oasis	6	5	1
Players	10	10	0

coefficient will likely be more reliable than the conventional measure. A second advantage in using the rank correlation coefficient is the fact that no assumptions of normality are made concerning the distributions of X and Y. Perhaps the greatest advantage occurs when one is unable to make meaningful numerical measurements but nevertheless can establish rankings. Such is the case, for example, when different judges rank a group of individuals according to some attribute. The rank correlation coefficient can be used in this situation as a measure of the consistency of the two judges.

To test the significance of the rank correlation coefficient, one needs to consider the distribution of the r_s values under the assumption that X and Y are independent. Critical values for $\alpha = 0.05$, 0.025, 0.01, and 0.005 have been calculated and are given in Table XXIII (see Statistical Tables). This table is set up similar to the table of critical values for the t distribution except for the left column, which now gives the number of pairs of observations rather than the degrees of freedom. Since the distribution of the r_s values is symmetric about $r_s = 0$, the r_s value that leaves an area of α to the left is equal to the negative of the r_s value that leaves an area of α to the right. For a two-sided alternative hypothesis, the critical region of size α falls equally in the two tails of the distribution. For a test in which the alternative hypothesis is negative, the critical region is entirely in the left tail of the distribution, and when the alternative is positive, the critical region is placed entirely in the right tail. In Example 13.6 the critical value for testing the null hypothesis H_0 that the rank correlation coefficient is zero against the alternative hypothesis H_1 that it is greater than zero, with $\alpha = 0.01$ and $n = 10$, is 0.745. That is, we reject H_0 if $r_s > 0.745$, and since our calculated value was $r_s = 0.9667$, we conclude at the 0.01 level of significance that a high positive correlation does exist between the amount of tar and nicotine found in cigarettes.

When X and Y are independent, it can be shown that the distribution of the r_s values approaches a normal distribution with a mean of zero and a standard deviation of $1/\sqrt{n-1}$ as n increases. Consequently, when n exceeds the values given in Table XXIII, one could test for a significant correlation by computing

$$z = \frac{r_s - 0}{1/\sqrt{n-1}} = r_s\sqrt{n-1}$$

and comparing with critical values obtained from the standard normal curve.

The test for independence of two *continuous variables* presented in this section is a simplified alternative to the more cumbersome chi-square procedure using contingency tables outlined in Section 7.9. Unfortunately, all too often the values of the two random variables must fall into certain established categories and therefore cannot be measured on a continuous scale, thereby necessitating the more complex calculations associated with contingency tables.

13.8 KRUSKAL-WALLIS TEST

The Kruskal–Wallis test, also called the Kruskal–Wallis H test, is a generalization of the Wilcoxon two-sample test to the case of $k > 2$ samples. It is used to test the null hypothesis H_0 that k independent samples are from identical populations. Introduced in 1952 by W. H. Kruskal and W. A. Wallis, the test is an alternative nonparametric procedure to the F test for testing the equality of means in the one-factor analysis of variance when the experimenter wishes to avoid the assumption that the samples were selected from normal populations.

Let n_i $(i = 1, 2, \ldots, k)$ be the number of observations in the ith sample. First we combine all k samples and arrange the $n = n_1 + n_2 + \cdots + n_k$ observations in ascending order, substituting the appropriate rank from $1, 2, \ldots, n$ for each observation. In the case of ties (identical observations), we follow the usual procedure of replacing the observations by the means of the ranks that the observations would have if they were distinguishable. The sum of the ranks corresponding to the n_i observations in the ith sample is denoted by the random variable R_i. Now let us consider the statistic

$$H = \frac{12}{n(n+1)} \sum_{i=1}^{k} \frac{R_i^2}{n_i} - 3(n+1),$$

which is approximated very well by a chi-square distribution with $k - 1$ degrees of freedom when H_0 is true and if each sample consists of at least five observations. Note that the statistic H assumes the value h, where

$$h = \frac{12}{n(n+1)} \sum_{i=1}^{k} \frac{r_i^2}{n_i} - 3(n+1),$$

when R_1 assumes the value r_1, R_2 assumes the value r_2, and so forth. The fact that h is large when the independent samples come from populations that are not identical allows us to establish the following decision criterion for testing H_0:

KRUSKAL–WALLIS TEST *To test the null hypothesis H_0 that k independent samples are from identical populations, compute*

$$h = \frac{12}{n(n + 1)} \sum_{i=1}^{k} \frac{r_i^2}{n_i} - 3(n + 1).$$

If h falls in the critical region $H > \chi_\alpha^2$ with $v = k - 1$ degrees of freedom, reject H_0 at the α level of significance; otherwise, accept H_0.

Example 13.7 In an experiment to determine which of three different missile systems is preferable, the propellant burning rate was measured. The data, after coding, are given in Table 13.2.

Table 13.2 Propellant Burning Rates

	Missile system	
1	*2*	*3*
24.0	23.2	18.4
16.7	19.8	19.1
22.8	18.1	17.3
19.8	17.6	17.3
18.9	20.2	19.7
	17.8	18.9
		18.8
		19.3

Use the Kruskal–Wallis test and a significance level of $\alpha = 0.05$ to test the hypothesis that the propellant burning rates are the same for the three missile systems.

Solution In Table 13.3 we convert the 19 observations to ranks and sum the ranks for each missile system.

Table 13.3 Ranks for Propellant Burning Rates

	Missile system	
1	*2*	*3*
19	18	7
1	14.5	11
17	6	2.5
14.5	4	2.5
9.5	16	13
$r_1 = 61.0$	5	9.5
	$r_2 = 63.5$	8
		12
		$r_3 = 65.5$

Now, substituting $n_1 = 5$, $n_2 = 6$, $n_3 = 8$, and $r_1 = 61.0$, $r_2 = 63.5$, $r_3 = 65.5$, our test statistic H assumes the value

$$h = \frac{12}{(19)(20)} \left(\frac{61.0^2}{5} + \frac{63.5^2}{6} + \frac{65.5^2}{8} \right) - (3)(20)$$

$$= 1.6586.$$

Using a 0.05 level of significance with $k - 1 = 2$ degrees of freedom, we find $\chi^2_{0.05} = 5.991$. Since $h = 1.6586$ does not fall in the critical region $H > 5.991$, we have insufficient evidence to reject the hypothesis that the propellant burning rates are the same for the three missile systems.

EXERCISES

1. A cigarette manufacturer claims that the tar content of brand B cigarettes is lower than that of brand A. To test this claim, the following determinations of tar content, in milligrams, were recorded:

Brand A	12	9	13	11	14
Brand B	8	10	7		

Use the Wilcoxon two-sample test with $\alpha = 0.05$ to test whether the claim is valid.

2. The following data represent the number of hours that two different types of scientific pocket calculators operate before a recharge is required:

Calculator A	5.5	5.6	6.3	4.6	5.3	5.0	6.2	5.8	5.1
Calculator B	3.8	4.8	4.3	4.2	4.0	4.9	4.5	5.2	4.5

Use the Wilcoxon two-sample test with $\alpha = 0.01$ to determine if calculator A operates longer than calculator B on a full battery charge.

3. The following data represent the weights of personal luggage carried on a large aircraft by the members of two baseball clubs:

Club A	34	39	41	28	33	
Club B	36	40	35	31	39	36

Use the Wilcoxon two-sample test with $\alpha = 0.05$ to test the hypothesis that the two clubs carry the same amount of luggage on the average against the alternative hypothesis that the average weight of luggage for club B is greater than that of club A.

4. A fishing line is being manufactured by two processes. To determine if there is a difference in the mean breaking strength of the lines, 10 pieces by each process are selected and tested for breaking strength. The results are as follows:

Process 1	10.4	9.8	11.5	10.0	9.9	9.6	10.9	11.8	9.3	10.7
Process 2	8.7	11.2	9.8	10.1	10.8	9.5	11.0	9.8	10.5	9.9

Use the Wilcoxon two-sample test with $\alpha = 0.1$ to determine if there is a difference between the mean breaking strengths of the lines manufactured by the two processes.

5. From a mathematics class of 12 equally capable students using programmed materials, 5 are selected at random and given additional instruction by the teacher. The results on the final examination were as follows:

	Grade						
Additional instruction	87	69	78	91	80		
No additional instruction	75	88	64	82	93	79	67

Use the Wilcoxon two-sample test with $\alpha = 0.05$ to determine if the additional instruction affects the average grade.

6. A paint supplier claims that a new additive will reduce the drying time of his acrylic paint. To test this claim, 10 panels of wood are painted, one half of each panel with

paint containing the regular additive and the other half with paint containing the new additive. The drying times, in hours, were recorded as follows:

	Drying time (hours)	
Panel	New additive	Regular additive
1	6.4	6.6
2	5.8	5.8
3	7.4	7.8
4	5.5	5.7
5	6.3	6.0
6	7.8	8.4
7	8.6	8.8
8	8.2	8.4
9	7.0	7.3
10	4.9	5.8

Use the sign test, with α not exceeding 0.05, to test the hypothesis that the new additive is no better than the regular additive in reducing the drying time of this kind of paint.

7. Two types of instruments for measuring the amount of sulfur monoxide in the atmosphere are being compared in an air-pollution experiment. The following readings were recorded daily for a period of 2 weeks:

	Sulfur monoxide	
Day	Instrument A	Instrument B
1	0.96	0.87
2	0.82	0.74
3	0.75	0.63
4	0.61	0.55
5	0.89	0.76
6	0.64	0.70
7	0.81	0.69
8	0.68	0.57
9	0.65	0.53
10	0.84	0.88
11	0.59	0.51
12	0.94	0.79
13	0.91	0.84
14	0.77	0.63

Perform a sign test to determine whether the different instruments lead to different results. Use a significance level not exceeding 0.01.

8. In Exercise 22, Chapter 6, use the sign test to test the hypothesis, at a level of significance close to 0.05, that the diet reduces a person's weight by 4.5 kilograms on the average, against the alternative hypothesis that the mean difference in weight is less than 4.5 kilograms. Compare your conclusion with that of Exercise 22.

9. A food inspector examined 15 jars of a certain brand of jam to determine the percent of foreign impurities. The following data were recorded: 2.4, 2.3, 1.7, 1.7, 2.3, 1.2, 1.1, 3.6, 3.1, 1.0, 4.2, 1.6, 2.5, 2.4, 2.3. Test the hypothesis that the average percent of impurities in this brand of jam is 2.5%. Use a level of significance close to 0.01.

10. An international electronics firm is considering an expense-free vacation trip for its senior executives and their families. In order to determine a preference between a week in Hawaii or a week in Spain, a random sample of 18 executives were asked their preference. Use the sign test with α close to 0.05 to test the hypothesis that the two locations are equally preferred by the executives if 5 of the 18 stated that they preferred Spain.

11. The weights of four people before they stopped smoking and 5 weeks after they stopped smoking, in kilograms, are as follows:

	Individual			
	1	2	3	4
Before	66	80	69	52
After	71	82	68	56

Use the Wilcoxon test for paired observations to test the hypothesis, at the 0.05 level of significance, that giving up smoking has no effect on a person's weight against the alternative that one's weight increases if he quits smoking.

12. Analyze the data of Exercise 6 using the Wilcoxon test for paired observations. How does your conclusion compare with that of Exercise 6?

13. In Exercise 20, Chapter 6, use the Wilcoxon test for paired observations to test the hypothesis, at the 0.05 level of significance, that the average yields of the two varieties of wheat are equal against the alternative hypothesis that they are unequal. Compare your conclusion with that of Exercise 20.

14. In Exercise 21, Chapter 6, use the Wilcoxon test for paired observations to test the hypothesis, at the 0.01 level of significance, that $\mu_1 = \mu_2$ against the alternative hypothesis that $\mu_1 < \mu_2$. Compare your conclusions with that of Exercise 21.

15. A random sample of 15 adults living in a small town are selected to estimate the proportion of voters favoring a certain candidate for mayor. Each individual was also asked if he or she was a college graduate. Letting Y and N designate the responses of "yes" and "no" to the education question, the following sequence was obtained:

$$N \ N \ N \ N \ Y \ Y \ N \ Y \ Y \ N \ Y \ N \ N \ N \ N.$$

Use the runs test to determine if the sequence supports the contention that the sample was selected at random.

16. A silver-plating process is being used to coat a certain type of serving tray. When the process is in control, the thickness of the silver on the trays will vary randomly following a normal distribution with a mean of 0.02 millimeter and a standard deviation of 0.005 millimeter. Suppose that the next 12 trays examined show the following thicknesses of silver: 0.019, 0.021, 0.020, 0.019, 0.020, 0.018, 0.023, 0.021, 0.024, 0.022, 0.023, 0.022. Use the runs test to determine if the fluctuations in thickness from one tray to another is random.

17. Use the runs test to test the hypothesis of Exercise 2.

18. In an industrial production line, items are inspected periodically for defectives. The following is a sequence of defective items, D, and nondefective items, N, produced by this production line:

    D D N N N D N N D D N N N N N D D D N N D N N N N D N D.

    Use the large-sample theory for the runs test, with a significance level of 0.05, to determine whether the defectives are occurring at random or not.

19. Assuming the measurements of Exercise 17, Chapter 2, were recorded in successive rows from left to right as they were collected, use the runs test, with $\alpha = 0.05$, to test the hypothesis that the data represent a random sample.

20. How large a sample is required to be 95% confident that at least 85% of the distribution of measurements is included between the sample extremes?

21. What is the probability that the range of a random sample of size 24 includes at least 90% of the population?

22. How large a sample is required to be 99% confident that at least 80% of the population will be less than the largest observation in the sample?

23. What is the probability that at least 95% of a population will exceed the smallest value in a random sample of size $n = 135$?

24. The following table gives the recorded grades for 10 students on a midterm test and the final examination in a calculus course:

Student	Midterm test	Final examination
L.S.A.	84	73
W.P.B.	98	63
R.W.K.	91	87
J.R.L.	72	66
J.K.L.	86	78
D.L.P.	93	78
B.L.P.	80	91
D.W.M.	0	0
M.N.M.	92	88
R.H.S.	87	77

(a) Calculate the rank correlation coefficient.

(b) Test the hypothesis that the rank correlation coefficient is zero against the alternative that it is greater than zero. Use $\alpha = 0.025$.

25. Calculate the rank correlation coefficient for the data of Exercise 1, Chapter 8, and test the hypothesis at the 0.05 level of significance that the variables are independent against the alternative that they are not independent. Compare your results with those obtained in Exercise 22, Chapter 8.

26. Calculate the rank correlation coefficient for the data of Exercise 5, Chapter 9.

27. A consumer panel tested nine makes of microwave ovens for overall quality. The ranks assigned by the panel and the suggested retail prices were as follows:

Manufacturer	Panel rating	Suggested price
A	6	$480
B	9	395
C	2	575
D	8	550
E	5	510
F	1	545
G	7	400
H	4	465
I	3	420

Is there a significant relationship between the quality and the price of a microwave oven?

28. Two judges at a college homecoming parade ranked eight floats in the following order:

	Float							
	1	2	3	4	5	6	7	8
Judge A	5	8	4	3	6	2	7	1
Judge B	7	5	4	2	8	1	6	3

(a) Calculate the rank correlation coefficient.

(b) Test the hypothesis that the rank correlation coefficient is zero against the alternative hypothesis that it is greater than zero. Use $\alpha = 0.05$.

29. Test the hypothesis that X and Y are independent against the alternative that they are dependent if for a sample of size $n = 50$ pairs of observations we find that $r_s = -0.29$. Use $\alpha = 0.05$.

30. The following data represent the operating times in hours for three types of scientific pocket calculators before a recharge is required:

	Calculator	
A	B	C
4.9	5.5	6.4
6.1	5.4	6.8
4.3	6.2	5.6
4.6	5.8	6.5
5.3	5.5	6.3
	5.2	6.6
	4.8	

Use the Kruskal–Wallis test, at the 0.01 level of significance, to test the hypothesis that the operating times for all three calculators are equal.

31. Random samples of four brands of cigarettes were tested for tar content. The following figures show the milligrams of tar found in the 16 cigarettes tested:

Brand A	Brand B	Brand C	Brand D
14	16	16	17
10	18	15	20
11	14	14	19
13	15	12	21

Use the Kruskal–Wallis test, at the 0.05 level of significance, to test whether there is a significant difference in tar content among the four brands of cigarettes.

32. In Exercise 4, Chapter 10, use the Kruskal–Wallis test, at the 0.05 level of significance, to determine if the grade distribution given by the three teachers differ significantly.

33. In Exercise 6, Chapter 10, use the Kruskal–Wallis test, at the 0.05 level of significance, to determine if the chemical analyses performed by the four laboratories give, on the average, the same results.

Bibliography

BENNETT, C. A., AND N. L. FRANKLIN. *Statistical Analysis in Chemistry and the Chemical Industry*. New York: John Wiley & Sons, Inc., 1954.

BOWKER, A. H., AND G. J. LIEBERMAN. *Engineering Statistics*, 2nd ed. Englewood Cliffs, N.J.: Prentice-Hall, Inc., 1972.

BROWNLEE, K. A. *Statistical Theory and Methodology in Science and Engineering*, 2nd ed. New York: John Wiley & Sons, Inc., 1965.

CHEW, V. *Experimental Designs in Industry*. New York: John Wiley & Sons, Inc., 1958.

COCHRAN, W. G. "Some Consequences When the Assumptions for the Analysis of Variance Are Not Satisfied." *Biometrics*, Vol. 3, 1947.

COCHRAN, W. G. "Some Methods for Strengthening the Common Chi-Square Tests." *Biometrics*, Vol. 10, 1954.

COCHRAN, W. G., AND G. M. COX. *Experimental Designs*, 2nd ed. New York: John Wiley & Sons, Inc., 1957.

DAVIES, O. L. *The Design and Analysis of Industrial Experiments*. New York: Hafner Publishing Co., 1956.

DIXON, W. J., AND F. J. MASSEY, JR. *Introduction to Statistical Analysis*, 3rd ed. New York: McGraw-Hill Book Company, 1969.

DRAPER, N., AND H. SMITH. *Applied Regression Analysis*. New York: John Wiley & Sons, Inc., 1966.

FREUND, J. E. *Mathematical Statistics*, 2nd ed. Englewood Cliffs, N.J.: Prentice-Hall, Inc., 1971.

GUENTHER, W. C. *Analysis of Variance*. Englewood Cliffs, N.J.: Prentice-Hall, Inc., 1964.

GUENTHER, W. C. *Concepts of Statistical Inference*. New York: McGraw-Hill Book Company, 1965.

GUTTMAN, I., AND S. S. WILKS. *Introductory Engineering Statistics*. New York: John Wiley & Sons, Inc., 1965.

HICKS, C. R. *Fundamental Concepts in the Design of Experiments*. New York: Holt, Rinehart and Winston, Inc., 1964.

HODGES, J. L., JR., AND E. L. LEHMANN. *Basic Concepts of Probability and Statistics*, 2nd ed. San Francisco: Holden-Day, Inc., 1970.

HOERL, A. E., AND R. W. KENNARD. "Ridge Regression: Applications to Nonorthogonal Problems." *Technometrics*, Vol. 12, no. 1, 1970.

HOGG, R. V., AND A. T. CRAIG. *Introduction to Mathematical Statistics*, 3rd ed. New York: Macmillan Publishing Co., Inc., 1970.

JOHNSON, N. L., AND F. C. LEONE. *Statistics and Experimental Design: In Engineering and the Physical Sciences*, Vols. I and II. New York: John Wiley & Sons, Inc., 1964.

KEMPTHORNE, O. *The Design and Analysis of Experiments*. New York: John Wiley & Sons, Inc., 1952.

LARSON, H. J. *Introduction to Probability Theory and Statistical Inference*. New York: John Wiley & Sons, Inc., 1969.

LI, C. C. *Introduction to Experimental Statistics*. New York: McGraw-Hill Book Company, 1964.

LI, J. C. R. *Introduction to Statistical Inference*, 2nd ed. Ann Arbor, Mich.: J. W. Edwards, Publisher, Inc., 1961.

MENDENHALL, W. *An Introduction to Linear Models and the Design and Analysis of Experiments*. Belmont, Calif.: Wadsworth Publishing Company, Inc., 1967.

MEYER, P. L. *Introductory Probability and Statistical Applications*. Reading, Mass.: Addison-Wesley Publishing Company, Inc., 1965.

MILLER, I., AND J. E. FREUND. *Probability and Statistics for Engineers*, 2nd ed. Englewood Cliffs, N.J.: Prentice-Hall, Inc., 1977.

MYERS, R. H. *Response Surface Methodology*. Boston: Allyn and Bacon, Inc., 1971.

NOETHER, G. E. *Introduction to Statistics: A Nonparametric Approach*, 2nd ed. Boston: Houghton Mifflin Company, 1976.

OSTLE, B. *Statistics in Research*, 2nd ed. Ames, Iowa: The Iowa State University Press, 1964.

PEARSON, E. S., AND H. O. HARTLEY. *Biometrika Tables for Statisticians*, Vol. I, 3rd ed. Cambridge: Cambridge University Press, 1966.

SATTERTHWAITE, F. E. "An Approximate Distribution of Estimates of Variance Components." *Biometrics*, Vol. 2, 1946.

SCHEFFÉ, H. *The Analysis of Variance*. New York: John Wiley & Sons, Inc., 1959.

SNEDECOR, G., AND W. G. COCHRAN. *Statistical Methods*, 6th ed. Ames, Iowa: The Iowa State University Press, 1967.

STEEL, R. G. D., AND J. H. TORRIE. *Principles and Procedures of Statistics*. New York: McGraw-Hill Book Company, 1960.

THOMPSON, W. O., AND F. B. CADY. *Proceedings of the University of Kentucky Conference on Regression with a Large Number of Predictor Variables*, Lexington, 1973.

WALPOLE, R. E. *Introduction to Statistics*, 2nd ed. New York: Macmillan Publishing Co., Inc., 1974.

WINER, B. J. *Statistical Principles in Experimental Design*, 2nd ed. New York: McGraw-Hill Book Company, 1971.

Statistical Tables

Table I Squares and Square Roots

n	n^2	$\sqrt{n}$	$\sqrt{10n}$	n	n^2	$\sqrt{n}$	$\sqrt{10n}$
1.0	1.00	1.000	3.162	5.5	30.25	2.345	7.416
1.1	1.21	1.049	3.317	5.6	31.36	2.366	7.483
1.2	1.44	1.095	3.464	5.7	32.49	2.387	7.550
1.3	1.69	1.140	3.606	5.8	33.64	2.408	7.616
1.4	1.96	1.183	3.742	5.9	34.81	2.429	7.681
1.5	2.25	1.225	3.873	6.0	36.00	2.449	7.746
1.6	2.56	1.265	4.000	6.1	37.21	2.470	7.810
1.7	2.89	1.304	4.123	6.2	38.44	2.490	7.874
1.8	3.24	1.342	4.243	6.3	39.69	2.510	7.937
1.9	3.61	1.378	4.359	6.4	40.96	2.530	8.000
2.0	4.00	1.414	4.472	6.5	42.25	2.550	8.062
2.1	4.41	1.449	4.583	6.6	43.56	2.569	8.124
2.2	4.84	1.483	4.690	6.7	44.89	2.588	8.185
2.3	5.29	1.517	4.796	6.8	46.24	2.608	8.246
2.4	5.76	1.549	4.899	6.9	47.61	2.627	8.307
2.5	6.25	1.581	5.000	7.0	49.00	2.646	8.367
2.6	6.76	1.612	5.099	7.1	50.41	2.665	8.426
2.7	7.29	1.643	5.196	7.2	51.84	2.683	8.485
2.8	7.84	1.673	5.292	7.3	53.29	2.702	8.544
2.9	8.41	1.703	5.385	7.4	54.76	2.720	8.602
3.0	9.00	1.732	5.477	7.5	56.25	2.739	8.660
3.1	9.61	1.761	5.568	7.6	57.76	2.757	8.718
3.2	10.24	1.789	5.657	7.7	59.29	2.775	8.775
3.3	10.89	1.817	5.745	7.8	60.84	2.793	8.832
3.4	11.56	1.844	5.831	7.9	62.41	2.811	8.888
3.5	12.25	1.871	5.916	8.0	64.00	2.828	8.944
3.6	12.96	1.897	6.000	8.1	65.61	2.846	9.000
3.7	13.69	1.924	6.083	8.2	67.24	2.864	9.055
3.8	14.44	1.949	6.164	8.3	68.89	2.881	9.110
3.9	15.21	1.975	6.245	8.4	70.56	2.898	9.165
4.0	16.00	2.000	6.325	8.5	72.25	2.915	9.220
4.1	16.81	2.025	6.403	8.6	73.96	2.933	9.274
4.2	17.64	2.049	6.481	8.7	75.69	2.950	9.327
4.3	18.49	2.074	6.557	8.8	77.44	2.966	9.381
4.4	19.36	2.098	6.633	8.9	79.21	2.983	9.434
4.5	20.25	2.121	6.708	9.0	81.00	3.000	9.487
4.6	21.16	2.145	6.782	9.1	82.81	3.017	9.539
4.7	22.09	2.168	6.856	9.2	84.64	3.033	9.592
4.8	23.04	2.191	6.928	9.3	86.49	3.050	9.644
4.9	24.01	2.214	7.000	9.4	88.36	3.066	9.695
5.0	25.00	2.236	7.071	9.5	90.25	3.082	9.747
5.1	26.01	2.258	7.141	9.6	92.16	3.098	9.798
5.2	27.04	2.280	7.211	9.7	94.09	3.114	9.849
5.3	28.09	2.302	7.280	9.8	96.04	3.130	9.899
5.4	29.16	2.324	7.348	9.9	98.01	3.146	9.950

Table II Binomial Probability Sums $\sum_{x=0}^{r} b(x; n, p)$

						p					
n	r	0.10	0.20	0.25	0.30	0.40	0.50	0.60	0.70	0.80	0.90
5	0	0.5905	0.3277	0.2373	0.1681	0.0778	0.0312	0.0102	0.0024	0.0003	0.0000
	1	0.9185	0.7373	0.6328	0.5282	0.3370	0.1875	0.0870	0.0308	0.0067	0.0005
	2	0.9914	0.9421	0.8965	0.8369	0.6826	0.5000	0.3174	0.1631	0.0579	0.0086
	3	0.9995	0.9933	0.9844	0.9692	0.9130	0.8125	0.6630	0.4718	0.2627	0.0815
	4	1.0000	0.9997	0.9990	0.9976	0.9898	0.9688	0.9222	0.8319	0.6723	0.4095
	5	1.0000	1.0000	1.0000	1.0000	1.0000	1.0000	1.0000	1.0000	1.0000	1.0000
10	0	0.3487	0.1074	0.0563	0.0282	0.0060	0.0010	0.0001	0.0000	0.0000	0.0000
	1	0.7361	0.3758	0.2440	0.1493	0.0464	0.0107	0.0017	0.0001	0.0000	0.0000
	2	0.9298	0.6778	0.5256	0.3828	0.1673	0.0547	0.0123	0.0016	0.0001	0.0000
	3	0.9872	0.8791	0.7759	0.6496	0.3823	0.1719	0.0548	0.0106	0.0009	0.0000
	4	0.9984	0.9672	0.9219	0.8497	0.6331	0.3770	0.1662	0.0474	0.0064	0.0002
	5	0.9999	0.9936	0.9803	0.9527	0.8338	0.6230	0.3669	0.1503	0.0328	0.0016
	6	1.0000	0.9991	0.9965	0.9894	0.9452	0.8281	0.6177	0.3504	0.1209	0.0128
	7	1.0000	0.9999	0.9996	0.9984	0.9877	0.9453	0.8327	0.6172	0.3222	0.0702
	8	1.0000	1.0000	1.0000	0.9999	0.9983	0.9893	0.9536	0.8507	0.6242	0.2639
	9	1.0000	1.0000	1.0000	1.0000	0.9999	0.9990	0.9940	0.9718	0.8926	0.6513
	10	1.0000	1.0000	1.0000	1.0000	1.0000	1.0000	1.0000	1.0000	1.0000	1.0000
15	0	0.2059	0.0352	0.0134	0.0047	0.0005	0.0000	0.0000	0.0000	0.0000	0.0000
	1	0.5490	0.1671	0.0802	0.0353	0.0052	0.0005	0.0000	0.0000	0.0000	0.0000
	2	0.8159	0.3980	0.2361	0.1268	0.0271	0.0037	0.0003	0.0000	0.0000	0.0000
	3	0.9444	0.6482	0.4613	0.2969	0.0905	0.0176	0.0019	0.0001	0.0000	0.0000
	4	0.9873	0.8358	0.6865	0.5155	0.2173	0.0592	0.0094	0.0007	0.0000	0.0000
	5	0.9978	0.9389	0.8516	0.7216	0.4032	0.1509	0.0338	0.0037	0.0001	0.0000
	6	0.9997	0.9819	0.9434	0.8689	0.6098	0.3036	0.0951	0.0152	0.0008	0.0000
	7	1.0000	0.9958	0.9827	0.9500	0.7869	0.5000	0.2131	0.0500	0.0042	0.0000
	8	1.0000	0.9992	0.9958	0.9848	0.9050	0.6964	0.3902	0.1311	0.0181	0.0003
	9	1.0000	0.9999	0.9992	0.9963	0.9662	0.8491	0.5968	0.2784	0.0611	0.0023
	10	1.0000	1.0000	0.9999	0.9993	0.9907	0.9408	0.7827	0.4845	0.1642	0.0127
	11	1.0000	1.0000	1.0000	0.9999	0.9981	0.9824	0.9095	0.7031	0.3518	0.0556
	12	1.0000	1.0000	1.0000	1.0000	0.9997	0.9963	0.9729	0.8732	0.6020	0.1841
	13	1.0000	1.0000	1.0000	1.0000	1.0000	0.9995	0.9948	0.9647	0.8329	0.4510
	14	1.0000	1.0000	1.0000	1.0000	1.0000	1.0000	0.9995	0.9953	0.9648	0.7941
	15	1.0000	1.0000	1.0000	1.0000	1.0000	1.0000	1.0000	1.0000	1.0000	1.0000
20	0	0.1216	0.0115	0.0032	0.0008	0.0000	0.0000	0.0000	0.0000	0.0000	0.0000
	1	0.3917	0.0692	0.0243	0.0076	0.0005	0.0000	0.0000	0.0000	0.0000	0.0000
	2	0.6769	0.2061	0.0913	0.0355	0.0036	0.0002	0.0000	0.0000	0.0000	0.0000
	3	0.8670	0.4114	0.2252	0.1071	0.0160	0.0013	0.0001	0.0000	0.0000	0.0000
	4	0.9568	0.6296	0.4148	0.2375	0.0510	0.0059	0.0003	0.0000	0.0000	0.0000
	5	0.9887	0.8042	0.6172	0.4164	0.1256	0.0207	0.0016	0.0000	0.0000	0.0000
	6	0.9976	0.9133	0.7858	0.6080	0.2500	0.0577	0.0065	0.0003	0.0000	0.0000
	7	0.9996	0.9679	0.8982	0.7723	0.4159	0.1316	0.0210	0.0013	0.0000	0.0000
	8	0.9999	0.9900	0.9591	0.8867	0.5956	0.2517	0.0565	0.0051	0.0001	0.0000
	9	1.0000	0.9974	0.9861	0.9520	0.7553	0.4119	0.1275	0.0171	0.0006	0.0000
	10	1.0000	0.9994	0.9961	0.9829	0.8725	0.5881	0.2447	0.0480	0.0026	0.0000
	11	1.0000	0.9999	0.9991	0.9949	0.9435	0.7483	0.4044	0.1133	0.0100	0.0001
	12	1.0000	1.0000	0.9998	0.9987	0.9790	0.8684	0.5841	0.2277	0.0321	0.0004
	13	1.0000	1.0000	1.0000	0.9997	0.9935	0.9423	0.7500	0.3920	0.0867	0.0024
	14	1.0000	1.0000	1.0000	1.0000	0.9984	0.9793	0.8744	0.5836	0.1958	0.0113
	15	1.0000	1.0000	1.0000	1.0000	0.9997	0.9941	0.9490	0.7625	0.3704	0.0432
	16	1.0000	1.0000	1.0000	1.0000	1.0000	0.9987	0.9840	0.8929	0.5886	0.1330
	17	1.0000	1.0000	1.0000	1.0000	1.0000	0.9998	0.9964	0.9645	0.7939	0.3231
	18	1.0000	1.0000	1.0000	1.0000	1.0000	1.0000	0.9995	0.9924	0.9308	0.6083
	19	1.0000	1.0000	1.0000	1.0000	1.0000	1.0000	1.0000	0.9992	0.9885	0.8784
	20	1.0000	1.0000	1.0000	1.0000	1.0000	1.0000	1.0000	1.0000	1.0000	1.0000

Table III Poisson Probability Sums $\sum\limits_{x=0}^{r} p(x; \mu)$

r	μ								
	0.1	0.2	0.3	0.4	0.5	0.6	0.7	0.8	0.9
0	0.9048	0.8187	0.7408	0.6730	0.6065	0.5488	0.4966	0.4493	0.4066
1	0.9953	0.9825	0.9631	0.9384	0.9098	0.8781	0.8442	0.8088	0.7725
2	0.9998	0.9989	0.9964	0.9921	0.9856	0.9769	0.9659	0.9526	0.9371
3	1.0000	0.9999	0.9997	0.9992	0.9982	0.9966	0.9942	0.9909	0.9865
4		1.0000	1.0000	0.9999	0.9998	0.9996	0.9992	0.9986	0.9977
5				1.0000	1.0000	1.0000	0.9999	0.9998	0.9997
6							1.0000	1.0000	1.0000

r	μ								
	1.0	1.5	2.0	2.5	3.0	3.5	4.0	4.5	5.0
0	0.3679	0.2231	0.1353	0.0821	0.0498	0.0302	0.0183	0.0111	0.0067
1	0.7358	0.5578	0.4060	0.2873	0.1991	0.1359	0.0916	0.0611	0.0404
2	0.9197	0.8088	0.6767	0.5438	0.4232	0.3208	0.2381	0.1736	0.1247
3	0.9810	0.9344	0.8571	0.7576	0.6472	0.5366	0.4335	0.3423	0.2650
4	0.9963	0.9814	0.9473	0.8912	0.8153	0.7254	0.6288	0.5321	0.4405
5	0.9994	0.9955	0.9834	0.9580	0.9161	0.8576	0.7851	0.7029	0.6160
6	0.9999	0.9991	0.9955	0.9858	0.9665	0.9347	0.8893	0.8311	0.7622
7	1.0000	0.9998	0.9989	0.9958	0.9881	0.9733	0.9489	0.9134	0.8666
8		1.0000	0.9998	0.9989	0.9962	0.9901	0.9786	0.9597	0.9319
9			1.0000	0.9997	0.9989	0.9967	0.9919	0.9829	0.9682
10				0.9999	0.9997	0.9990	0.9972	0.9933	0.9863
11				1.0000	0.9999	0.9997	0.9991	0.9976	0.9945
12					1.0000	0.9999	0.9997	0.9992	0.9980
13						1.0000	0.9999	0.9997	0.9993
14							1.0000	0.9999	0.9998
15								1.0000	0.9999
16									1.0000

Table III Poisson Probability Sums $\sum\limits_{x=0}^{r} p(x;\mu)$ (*continued*)

					μ				
r	5.5	6.0	6.5	7.0	7.5	8.0	8.5	9.0	9.5
0	0.0041	0.0025	0.0015	0.0009	0.0006	0.0003	0.0002	0.0001	0.0001
1	0.0266	0.0174	0.0113	0.0073	0.0047	0.0030	0.0019	0.0012	0.0008
2	0.0884	0.0620	0.0430	0.0296	0.0203	0.0138	0.0093	0.0062	0.0042
3	0.2017	0.1512	0.1118	0.0818	0.0591	0.0424	0.0301	0.0212	0.0149
4	0.3575	0.2851	0.2237	0.1730	0.1321	0.0996	0.0744	0.0550	0.0403
5	0.5289	0.4457	0.3690	0.3007	0.2414	0.1912	0.1496	0.1157	0.0885
6	0.6860	0.6063	0.5265	0.4497	0.3782	0.3134	0.2562	0.2068	0.1649
7	0.8095	0.7440	0.6728	0.5987	0.5246	0.4530	0.3856	0.3239	0.2687
8	0.8944	0.8472	0.7916	0.7291	0.6620	0.5925	0.5231	0.4557	0.3918
9	0.9462	0.9161	0.8774	0.8305	0.7764	0.7166	0.6530	0.5874	0.5218
10	0.9747	0.9574	0.9332	0.9015	0.8622	0.8159	0.7634	0.7060	0.6453
11	0.9890	0.9799	0.9661	0.9466	0.9208	0.8881	0.8487	0.8030	0.7520
12	0.9955	0.9912	0.9840	0.9730	0.9573	0.9362	0.9091	0.8758	0.8364
13	0.9983	0.9964	0.9929	0.9872	0.9784	0.9658	0.9486	0.9261	0.8981
14	0.9994	0.9986	0.9970	0.9943	0.9897	0.9827	0.9726	0.9585	0.9400
15	0.9998	0.9995	0.9988	0.9976	0.9954	0.9918	0.9862	0.9780	0.9665
16	0.9999	0.9998	0.9996	0.9990	0.9980	0.9963	0.9934	0.9889	0.9823
17	1.0000	0.9999	0.9998	0.9996	0.9992	0.9984	0.9970	0.9947	0.9911
18		1.0000	0.9999	0.9999	0.9997	0.9994	0.9987	0.9976	0.9957
19			1.0000	1.0000	0.9999	0.9997	0.9995	0.9989	0.9980
20					1.0000	0.9999	0.9998	0.9996	0.9991
21						1.0000	0.9999	0.9998	0.9996
22							1.0000	0.9999	0.9999
23								1.0000	0.9999
24									1.0000

					μ				
r	10.0	11.0	12.0	13.0	14.0	15.0	16.0	17.0	18.0
0	0.0000	0.0000	0.0000						
1	0.0005	0.0002	0.0001	0.0000	0.0000				
2	0.0028	0.0012	0.0005	0.0002	0.0001	0.0000	0.0000		
3	0.0103	0.0049	0.0023	0.0010	0.0005	0.0002	0.0001	0.0000	0.0000
4	0.0293	0.0151	0.0076	0.0037	0.0018	0.0009	0.0004	0.0002	0.0001
5	0.0671	0.0375	0.0203	0.0107	0.0055	0.0028	0.0014	0.0007	0.0003
6	0.1301	0.0786	0.0458	0.0259	0.0142	0.0076	0.0040	0.0021	0.0010
7	0.2202	0.1432	0.0895	0.0540	0.0316	0.0180	0.0100	0.0054	0.0029
8	0.3328	0.2320	0.1550	0.0998	0.0621	0.0374	0.0220	0.0126	0.0071
9	0.4579	0.3405	0.2424	0.1658	0.1094	0.0699	0.0433	0.0261	0.0154
10	0.5830	0.4599	0.3472	0.2517	0.1757	0.1185	0.0774	0.0491	0.0304
11	0.6968	0.5793	0.4616	0.3532	0.2600	0.1848	0.1270	0.0847	0.0549
12	0.7916	0.6887	0.5760	0.4631	0.3585	0.2676	0.1931	0.1350	0.0917
13	0.8645	0.7813	0.6815	0.5730	0.4644	0.3632	0.2745	0.2009	0.1426
14	0.9165	0.8540	0.7720	0.6751	0.5704	0.4657	0.3675	0.2808	0.2081
15	0.9513	0.9074	0.8444	0.7636	0.6694	0.5681	0.4667	0.3715	0.2867
16	0.9730	0.9441	0.8987	0.8355	0.7559	0.6641	0.5660	0.4677	0.3750
17	0.9857	0.9678	0.9370	0.8905	0.8272	0.7489	0.6593	0.5640	0.4686
18	0.9928	0.9823	0.9626	0.9302	0.8826	0.8195	0.7423	0.6550	0.5622
19	0.9965	0.9907	0.9787	0.9573	0.9235	0.8752	0.8122	0.7363	0.6509
20	0.9984	0.9953	0.9884	0.9750	0.9521	0.9170	0.8682	0.8055	0.7307
21	0.9993	0.9977	0.9939	0.9859	0.9712	0.9469	0.9108	0.8615	0.7991
22	0.9997	0.9990	0.9970	0.9924	0.9833	0.9673	0.9418	0.9047	0.8551
23	0.9999	0.9995	0.9985	0.9960	0.9907	0.9805	0.9633	0.9367	0.8989
24	1.0000	0.9998	0.9993	0.9980	0.9950	0.9888	0.9777	0.9594	0.9317
25		0.9999	0.9997	0.9990	0.9974	0.9938	0.9869	0.9748	0.9554
26		1.0000	0.9999	0.9995	0.9987	0.9967	0.9925	0.9848	0.9718
27			0.9999	0.9998	0.9994	0.9983	0.9959	0.9912	0.9827
28			1.0000	0.9999	0.9997	0.9991	0.9978	0.9950	0.9897
29				1.0000	0.9999	0.9996	0.9989	0.9973	0.9941
30					0.9999	0.9998	0.9994	0.9986	0.9967
31					1.0000	0.9999	0.9997	0.9993	0.9982
32						1.0000	0.9999	0.9996	0.9990
33							0.9999	0.9998	0.9995
34							1.0000	0.9999	0.9998
35								1.0000	0.9999
36									0.9999
37									1.0000

Table IV
Areas Under the Normal Curve

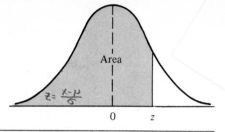

$z = \frac{x-\mu}{\sigma}$

z	0.00	0.01	0.02	0.03	0.04	0.05	0.06	0.07	0.08	0.09
−3.4	0.0003	0.0003	0.0003	0.0003	0.0003	0.0003	0.0003	0.0003	0.0003	0.0002
−3.3	0.0005	0.0005	0.0005	0.0004	0.0004	0.0004	0.0004	0.0004	0.0004	0.0003
−3.2	0.0007	0.0007	0.0006	0.0006	0.0006	0.0006	0.0006	0.0005	0.0005	0.0005
−3.1	0.0010	0.0009	0.0009	0.0009	0.0008	0.0008	0.0008	0.0008	0.0007	0.0007
−3.0	0.0013	0.0013	0.0013	0.0012	0.0012	0.0011	0.0011	0.0011	0.0010	0.0010
−2.9	0.0019	0.0018	0.0017	0.0017	0.0016	0.0016	0.0015	0.0015	0.0014	0.0014
−2.8	0.0026	0.0025	0.0024	0.0023	0.0023	0.0022	0.0021	0.0021	0.0020	0.0019
−2.7	0.0035	0.0034	0.0033	0.0032	0.0031	0.0030	0.0029	0.0028	0.0027	0.0026
−2.6	0.0047	0.0045	0.0044	0.0043	0.0041	0.0040	0.0039	0.0038	0.0037	0.0036
−2.5	0.0062	0.0060	0.0059	0.0057	0.0055	0.0054	0.0052	0.0051	0.0049	0.0048
−2.4	0.0082	0.0080	0.0078	0.0075	0.0073	0.0071	0.0069	0.0068	0.0066	0.0064
−2.3	0.0107	0.0104	0.0102	0.0099	0.0096	0.0094	0.0091	0.0089	0.0087	0.0084
−2.2	0.0139	0.0136	0.0132	0.0129	0.0125	0.0122	0.0119	0.0116	0.0113	0.0110
−2.1	0.0179	0.0174	0.0170	0.0166	0.0162	0.0158	0.0154	0.0150	0.0146	0.0143
−2.0	0.0228	0.0222	0.0217	0.0212	0.0207	0.0202	0.0197	0.0192	0.0188	0.0183
−1.9	0.0287	0.0281	0.0274	0.0268	0.0262	0.0256	0.0250	0.0244	0.0239	0.0233
−1.8	0.0359	0.0352	0.0344	0.0336	0.0329	0.0322	0.0314	0.0307	0.0301	0.0294
−1.7	0.0446	0.0436	0.0427	0.0418	0.0409	0.0401	0.0392	0.0384	0.0375	0.0367
−1.6	0.0548	0.0537	0.0526	0.0516	0.0505	0.0495	0.0485	0.0475	0.0465	0.0455
−1.5	0.0668	0.0655	0.0643	0.0630	0.0618	0.0606	0.0594	0.0582	0.0571	0.0559
−1.4	0.0808	0.0793	0.0778	0.0764	0.0749	0.0735	0.0722	0.0708	0.0694	0.0681
−1.3	0.0968	0.0951	0.0934	0.0918	0.0901	0.0885	0.0869	0.0853	0.0838	0.0823
−1.2	0.1151	0.1131	0.1112	0.1093	0.1075	0.1056	0.1038	0.1020	0.1003	0.0985
−1.1	0.1357	0.1335	0.1314	0.1292	0.1271	0.1251	0.1230	0.1210	0.1190	0.1170
−1.0	0.1587	0.1562	0.1539	0.1515	0.1492	0.1469	0.1446	0.1423	0.1401	0.1379
−0.9	0.1841	0.1814	0.1788	0.1762	0.1736	0.1711	0.1685	0.1660	0.1635	0.1611
−0.8	0.2119	0.2090	0.2061	0.2033	0.2005	0.1977	0.1949	0.1922	0.1894	0.1867
−0.7	0.2420	0.2389	0.2358	0.2327	0.2296	0.2266	0.2236	0.2206	0.2177	0.2148
−0.6	0.2743	0.2709	0.2676	0.2643	0.2611	0.2578	0.2546	0.2514	0.2483	0.2451
−0.5	0.3085	0.3050	0.3015	0.2981	0.2946	0.2912	0.2877	0.2843	0.2810	0.2776
−0.4	0.3446	0.3409	0.3372	0.3336	0.3300	0.3264	0.3228	0.3192	0.3156	0.3121
−0.3	0.3821	0.3783	0.3745	0.3707	0.3669	0.3632	0.3594	0.3557	0.3520	0.3483
−0.2	0.4207	0.4168	0.4129	0.4090	0.4052	0.4013	0.3974	0.3936	0.3897	0.3859
−0.1	0.4602	0.4562	0.4522	0.4483	0.4443	0.4404	0.4364	0.4325	0.4286	0.4247
−0.0	0.5000	0.4960	0.4920	0.4880	0.4840	0.4801	0.4761	0.4721	0.4681	0.4641
0.0	0.5000	0.5040	0.5080	0.5120	0.5160	0.5199	0.5239	0.5279	0.5319	0.5359
0.1	0.5398	0.5438	0.5478	0.5517	0.5557	0.5596	0.5636	0.5675	0.5714	0.5753
0.2	0.5793	0.5832	0.5871	0.5910	0.5948	0.5987	0.6026	0.6064	0.6103	0.6141
0.3	0.6179	0.6217	0.6255	0.6293	0.6331	0.6368	0.6406	0.6443	0.6480	0.6517
0.4	0.6554	0.6591	0.6628	0.6664	0.6700	0.6736	0.6772	0.6808	0.6844	0.6879
0.5	0.6915	0.6950	0.6985	0.7019	0.7054	0.7088	0.7123	0.7157	0.7190	0.7224
0.6	0.7257	0.7291	0.7324	0.7357	0.7389	0.7422	0.7454	0.7486	0.7517	0.7549
0.7	0.7580	0.7611	0.7642	0.7673	0.7704	0.7734	0.7764	0.7794	0.7823	0.7852
0.8	0.7881	0.7910	0.7939	0.7967	0.7995	0.8023	0.8051	0.8078	0.8106	0.8133
0.9	0.8159	0.8186	0.8212	0.8238	0.8264	0.8289	0.8315	0.8340	0.8365	0.8389
1.0	0.8413	0.8438	0.8461	0.8485	0.8508	0.8531	0.8554	0.8577	0.8599	0.8621
1.1	0.8643	0.8665	0.8686	0.8708	0.8729	0.8749	0.8770	0.8790	0.8810	0.8830
1.2	0.8849	0.8869	0.8888	0.8907	0.8925	0.8944	0.8962	0.8980	0.8997	0.9015
1.3	0.9032	0.9049	0.9066	0.9082	0.9099	0.9115	0.9131	0.9147	0.9162	0.9177
1.4	0.9192	0.9207	0.9222	0.9236	0.9251	0.9265	0.9278	0.9292	0.9306	0.9319
1.5	0.9332	0.9345	0.9357	0.9370	0.9382	0.9394	0.9406	0.9418	0.9429	0.9441
1.6	0.9452	0.9463	0.9474	0.9484	0.9495	0.9505	0.9515	0.9525	0.9535	0.9545
1.7	0.9554	0.9564	0.9573	0.9582	0.9591	0.9599	0.9608	0.9616	0.9625	0.9633
1.8	0.9641	0.9649	0.9656	0.9664	0.9671	0.9678	0.9686	0.9693	0.9699	0.9706
1.9	0.9713	0.9719	0.9726	0.9732	0.9738	0.9744	0.9750	0.9756	0.9761	0.9767
2.0	0.9772	0.9778	0.9783	0.9788	0.9793	0.9798	0.9803	0.9808	0.9812	0.9817
2.1	0.9821	0.9826	0.9830	0.9834	0.9838	0.9842	0.9846	0.9850	0.9854	0.9857
2.2	0.9861	0.9864	0.9868	0.9871	0.9875	0.9878	0.9881	0.9884	0.9887	0.9890
2.3	0.9893	0.9896	0.9898	0.9901	0.9904	0.9906	0.9909	0.9911	0.9913	0.9916
2.4	0.9918	0.9920	0.9922	0.9925	0.9927	0.9929	0.9931	0.9932	0.9934	0.9936
2.5	0.9938	0.9940	0.9941	0.9943	0.9945	0.9946	0.9948	0.9949	0.9951	0.9952
2.6	0.9953	0.9955	0.9956	0.9957	0.9959	0.9960	0.9961	0.9962	0.9963	0.9964
2.7	0.9965	0.9966	0.9967	0.9968	0.9969	0.9970	0.9971	0.9972	0.9973	0.9974
2.8	0.9974	0.9975	0.9976	0.9977	0.9977	0.9978	0.9979	0.9979	0.9980	0.9981
2.9	0.9981	0.9982	0.9982	0.9983	0.9984	0.9984	0.9985	0.9985	0.9986	0.9986
3.0	0.9987	0.9987	0.9987	0.9988	0.9988	0.9989	0.9989	0.9989	0.9990	0.9990
3.1	0.9990	0.9991	0.9991	0.9991	0.9992	0.9992	0.9992	0.9992	0.9993	0.9993
3.2	0.9993	0.9993	0.9994	0.9994	0.9994	0.9994	0.9994	0.9995	0.9995	0.9995
3.3	0.9995	0.9995	0.9995	0.9996	0.9996	0.9996	0.9996	0.9996	0.9996	0.9997
3.4	0.9997	0.9997	0.9997	0.9997	0.9997	0.9997	0.9997	0.9997	0.9997	0.9998

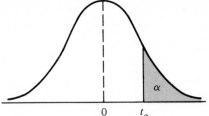

$\sqrt{} = n-1$

ν	α				
	0.10	0.05	0.025	0.01	0.005
1	3.078	6.314	12.706	31.821	63.657
2	1.886	2.920	4.303	6.965	9.925
3	1.638	2.353	3.182	4.541	5.841
4	1.533	2.132	2.776	3.747	4.604
5	1.476	2.015	2.571	3.365	4.032
6	1.440	1.943	2.447	3.143	3.707
7	1.415	1.895	2.365	2.998	3.499
8	1.397	1.860	2.306	2.896	3.355
9	1.383	1.833	2.262	2.821	3.250
10	1.372	1.812	2.228	2.764	3.169
11	1.363	1.796	2.201	2.718	3.106
12	1.356	1.782	2.179	2.681	3.055
13	1.350	1.771	2.160	2.650	3.012
14	1.345	1.761	2.145	2.624	2.977
15	1.341	1.753	2.131	2.602	2.947
16	1.337	1.746	2.120	2.583	2.921
17	1.333	1.740	2.110	2.567	2.898
18	1.330	1.734	2.101	2.552	2.878
19	1.328	1.729	2.093	2.539	2.861
20	1.325	1.725	2.086	2.528	2.845
21	1.323	1.721	2.080	2.518	2.831
22	1.321	1.717	2.074	2.508	2.819
23	1.319	1.714	2.069	2.500	2.807
24	1.318	1.711	2.064	2.492	2.797
25	1.316	1.708	2.060	2.485	2.787
26	1.315	1.706	2.056	2.479	2.779
27	1.314	1.703	2.052	2.473	2.771
28	1.313	1.701	2.048	2.467	2.763
29	1.311	1.699	2.045	2.462	2.756
inf.	1.282	1.645	1.960	2.326	2.576

† From Table IV of R. A. Fisher, *Statistical Methods for Research Workers*, published by Oliver & Boyd, Edinburgh, by permission of the author and publishers.

Table VI† Critical Values
of the Chi-Square Distribution

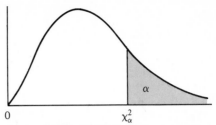

ν	α							
	0.995	0.99	0.975	0.95	0.05	0.025	0.01	0.005
1	0.0^4393	0.0^3157	0.0^3982	0.0^2393	3.841	5.024	6.635	7.879
2	0.0100	0.0201	0.0506	0.103	5.991	7.378	9.210	10.597
3	0.0717	0.115	0.216	0.352	7.815	9.348	11.345	12.838
4	0.207	0.297	0.484	0.711	9.488	11.143	13.277	14.860
5	0.412	0.554	0.831	1.145	11.070	12.832	15.086	16.750
6	0.676	0.872	1.237	1.635	12.592	14.449	16.812	18.548
7	0.989	1.239	1.690	2.167	14.067	16.013	18.475	20.278
8	1.344	1.646	2.180	2.733	15.507	17.535	20.090	21.955
9	1.735	2.088	2.700	3.325	16.919	19.023	21.666	23.589
10	2.156	2.558	3.247	3.940	18.307	20.483	23.209	25.188
11	2.603	3.053	3.816	4.575	19.675	21.920	24.725	26.757
12	3.074	3.571	4.404	5.226	21.026	23.337	26.217	28.300
13	3.565	4.107	5.009	5.892	22.362	24.736	27.688	29.819
14	4.075	4.660	5.629	6.571	23.685	26.119	29.141	31.319
15	4.601	5.229	6.262	7.261	24.996	27.488	30.578	32.801
16	5.142	5.812	6.908	7.962	26.296	28.845	32.000	34.267
17	5.697	6.408	7.564	8.672	27.587	30.191	33.409	35.718
18	6.265	7.015	8.231	9.390	28.869	31.526	34.805	37.156
19	6.844	7.633	8.907	10.117	30.144	32.852	36.191	38.582
20	7.434	8.260	9.591	10.851	31.410	34.170	37.566	39.997
21	8.034	8.897	10.283	11.591	32.671	35.479	38.932	41.401
22	8.643	9.542	10.982	12.338	33.924	36.781	40.289	42.796
23	9.260	10.196	11.689	13.091	35.172	38.076	41.638	44.181
24	9.886	10.856	12.401	13.848	36.415	39.364	42.980	45.558
25	10.520	11.524	13.120	14.611	37.652	40.646	44.314	46.928
26	11.160	12.198	13.844	15.379	38.885	41.923	45.642	48.290
27	11.808	12.879	14.573	16.151	40.113	43.194	46.963	49.645
28	12.461	13.565	15.308	16.928	41.337	44.461	48.278	50.993
29	13.121	14.256	16.047	17.708	42.557	45.722	49.588	52.336
30	13.787	14.953	16.791	18.493	43.773	46.979	50.892	53.672

ν= DEGREES OF FREEDOM

Table VII† Critical
Values of the F Distribution

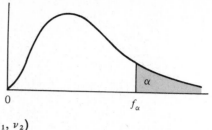

PROBABILITY OF
THE SHADED AREA $f_{0.05}(\nu_1, \nu_2)$

ν_2	ν_1								
	1	2	3	4	5	6	7	8	9
1	161.4	199.5	215.7	224.6	230.2	234.0	236.8	238.9	240.5
2	18.51	19.00	19.16	19.25	19.30	19.33	19.35	19.37	19.38
3	10.13	9.55	9.28	9.12	9.01	8.94	8.89	8.85	8.81
4	7.71	6.94	6.59	6.39	6.26	6.16	6.09	6.04	6.00
5	6.61	5.79	5.41	5.19	5.05	4.95	4.88	4.82	4.77
6	5.99	5.14	4.76	4.53	4.39	4.28	4.21	4.15	4.10
7	5.59	4.74	4.35	4.12	3.97	3.87	3.79	3.73	3.68
8	5.32	4.46	4.07	3.84	3.69	3.58	3.50	3.44	3.39
9	5.12	4.26	3.86	3.63	3.48	3.37	3.29	3.23	3.18
10	4.96	4.10	3.71	3.48	3.33	3.22	3.14	3.07	3.02
11	4.84	3.98	3.59	3.36	3.20	3.09	3.01	2.95	2.90
12	4.75	3.89	3.49	3.26	3.11	3.00	2.91	2.85	2.80
13	4.67	3.81	3.41	3.18	3.03	2.92	2.83	2.77	2.71
14	4.60	3.74	3.34	3.11	2.96	2.85	2.76	2.70	2.65
15	4.54	3.68	3.29	3.06	2.90	2.79	2.71	2.64	2.59
16	4.49	3.63	3.24	3.01	2.85	2.74	2.66	2.59	2.54
17	4.45	3.59	3.20	2.96	2.81	2.70	2.61	2.55	2.49
18	4.41	3.55	3.16	2.93	2.77	2.66	2.58	2.51	2.46
19	4.38	3.52	3.13	2.90	2.74	2.63	2.54	2.48	2.42
20	4.35	3.49	3.10	2.87	2.71	2.60	2.51	2.45	2.39
21	4.32	3.47	3.07	2.84	2.68	2.57	2.49	2.42	2.37
22	4.30	3.44	3.05	2.82	2.66	2.55	2.46	2.40	2.34
23	4.28	3.42	3.03	2.80	2.64	2.53	2.44	2.37	2.32
24	4.26	3.40	3.01	2.78	2.62	2.51	2.42	2.36	2.30
25	4.24	3.39	2.99	2.76	2.60	2.49	2.40	2.34	2.28
26	4.23	3.37	2.98	2.74	2.59	2.47	2.39	2.32	2.27
27	4.21	3.35	2.96	2.73	2.57	2.46	2.37	2.31	2.25
28	4.20	3.34	2.95	2.71	2.56	2.45	2.36	2.29	2.24
29	4.18	3.33	2.93	2.70	2.55	2.43	2.35	2.28	2.22
30	4.17	3.32	2.92	2.69	2.53	2.42	2.33	2.27	2.21
40	4.08	3.23	2.84	2.61	2.45	2.34	2.25	2.18	2.12
60	4.00	3.15	2.76	2.53	2.37	2.25	2.17	2.10	2.04
120	3.92	3.07	2.68	2.45	2.29	2.17	2.09	2.02	1.96
∞	3.84	3.00	2.60	2.37	2.21	2.10	2.01	1.94	1.88

† Reproduced from Table 18 of *Biometrika Tables for Statisticians*, Vol. I, by permission of E. S. Pearson and the Biometrika Trustees.

Table VII Critical Values of the F Distribution (*continued*)

$$f_{0.05}(\nu_1, \nu_2)$$

ν_2	ν_1									
	10	12	15	20	24	30	40	60	120	∞
1	241.9	243.9	245.9	248.0	249.1	250.1	251.1	252.2	253.3	254.3
2	19.40	19.41	19.43	19.45	19.45	19.46	19.47	19.48	19.49	19.50
3	8.79	8.74	8.70	8.66	8.64	8.62	8.59	8.57	8.55	8.53
4	5.96	5.91	5.86	5.80	5.77	5.75	5.72	5.69	5.66	5.63
5	4.74	4.68	4.62	4.56	4.53	4.50	4.46	4.43	4.40	4.36
6	4.06	4.00	3.94	3.87	3.84	3.81	3.77	3.74	3.70	3.67
7	3.64	3.57	3.51	3.44	3.41	3.38	3.34	3.30	3.27	3.23
8	3.35	3.28	3.22	3.15	3.12	3.08	3.04	3.01	2.97	2.93
9	3.14	3.07	3.01	2.94	2.90	2.86	2.83	2.79	2.75	2.71
10	2.98	2.91	2.85	2.77	2.74	2.70	2.66	2.62	2.58	2.54
11	2.85	2.79	2.72	2.65	2.61	2.57	2.53	2.49	2.45	2.40
12	2.75	2.69	2.62	2.54	2.51	2.47	2.43	2.38	2.34	2.30
13	2.67	2.60	2.53	2.46	2.42	2.38	2.34	2.30	2.25	2.21
14	2.60	2.53	2.46	2.39	2.35	2.31	2.27	2.22	2.18	2.13
15	2.54	2.48	2.40	2.33	2.29	2.25	2.20	2.16	2.11	2.07
16	2.49	2.42	2.35	2.28	2.24	2.19	2.15	2.11	2.06	2.01
17	2.45	2.38	2.31	2.23	2.19	2.15	2.10	2.06	2.01	1.96
18	2.41	2.34	2.27	2.19	2.15	2.11	2.06	2.02	1.97	1.92
19	2.38	2.31	2.23	2.16	2.11	2.07	2.03	1.98	1.93	1.88
20	2.35	2.28	2.20	2.12	2.08	2.04	1.99	1.95	1.90	1.84
21	2.32	2.25	2.18	2.10	2.05	2.01	1.96	1.92	1.87	1.81
22	2.30	2.23	2.15	2.07	2.03	1.98	1.94	1.89	1.84	1.78
23	2.27	2.20	2.13	2.05	2.01	1.96	1.91	1.86	1.81	1.76
24	2.25	2.18	2.11	2.03	1.98	1.94	1.89	1.84	1.79	1.73
25	2.24	2.16	2.09	2.01	1.96	1.92	1.87	1.82	1.77	1.71
26	2.22	2.15	2.07	1.99	1.95	1.90	1.85	1.80	1.75	1.69
27	2.20	2.13	2.06	1.97	1.93	1.88	1.84	1.79	1.73	1.67
28	2.19	2.12	2.04	1.96	1.91	1.87	1.82	1.77	1.71	1.65
29	2.18	2.10	2.03	1.94	1.90	1.85	1.81	1.75	1.70	1.64
30	2.16	2.09	2.01	1.93	1.89	1.84	1.79	1.74	1.68	1.62
40	2.08	2.00	1.92	1.84	1.79	1.74	1.69	1.64	1.58	1.51
60	1.99	1.92	1.84	1.75	1.70	1.65	1.59	1.53	1.47	1.39
120	1.91	1.83	1.75	1.66	1.61	1.55	1.50	1.43	1.35	1.25
∞	1.83	1.75	1.67	1.57	1.52	1.46	1.39	1.32	1.22	1.00

Table VII Critical Values of the F Distribution (*continued*)

$$f_{0.01}(\nu_1, \nu_2)$$

ν_2	ν_1								
	1	2	3	4	5	6	7	8	9
1	4052	4999.5	5403	5625	5764	5859	5928	5981	6022
2	98.50	99.00	99.17	99.25	99.30	99.33	99.36	99.37	99.39
3	34.12	30.82	29.46	28.71	28.24	27.91	27.67	27.49	27.35
4	21.20	18.00	16.69	15.98	15.52	15.21	14.98	14.80	14.66
5	16.26	13.27	12.06	11.39	10.97	10.67	10.46	10.29	10.16
6	13.75	10.92	9.78	9.15	8.75	8.47	8.26	8.10	7.98
7	12.25	9.55	8.45	7.85	7.46	7.19	6.99	6.84	6.72
8	11.26	8.65	7.59	7.01	6.63	6.37	6.18	6.03	5.91
9	10.56	8.02	6.99	6.42	6.06	5.80	5.61	5.47	5.35
10	10.04	7.56	6.55	5.99	5.64	5.39	5.20	5.06	4.94
11	9.65	7.21	6.22	5.67	5.32	5.07	4.89	4.74	4.63
12	9.33	6.93	5.95	5.41	5.06	4.82	4.64	4.50	4.39
13	9.07	6.70	5.74	5.21	4.86	4.62	4.44	4.30	4.19
14	8.86	6.51	5.56	5.04	4.69	4.46	4.28	4.14	4.03
15	8.68	6.36	5.42	4.89	4.56	4.32	4.14	4.00	3.89
16	8.53	6.23	5.29	4.77	4.44	4.20	4.03	3.89	3.78
17	8.40	6.11	5.18	4.67	4.34	4.10	3.93	3.79	3.68
18	8.29	6.01	5.09	4.58	4.25	4.01	3.84	3.71	3.60
19	8.18	5.93	5.01	4.50	4.17	3.94	3.77	3.63	3.52
20	8.10	5.85	4.94	4.43	4.10	3.87	3.70	3.56	3.46
21	8.02	5.78	4.87	4.37	4.04	3.81	3.64	3.51	3.40
22	7.95	5.72	4.82	4.31	3.99	3.76	3.59	3.45	3.35
23	7.88	5.66	4.76	4.26	3.94	3.71	3.54	3.41	3.30
24	7.82	5.61	4.72	4.22	3.90	3.67	3.50	3.36	3.26
25	7.77	5.57	4.68	4.18	3.85	3.63	3.46	3.32	3.22
26	7.72	5.53	4.64	4.14	3.82	3.59	3.42	3.29	3.18
27	7.68	5.49	4.60	4.11	3.78	3.56	3.39	3.26	3.15
28	7.64	5.45	4.57	4.07	3.75	3.53	3.36	3.23	3.12
29	7.60	5.42	4.54	4.04	3.73	3.50	3.33	3.20	3.09
30	7.56	5.39	4.51	4.02	3.70	3.47	3.30	3.17	3.07
40	7.31	5.18	4.31	3.83	3.51	3.29	3.12	2.99	2.89
60	7.08	4.98	4.13	3.65	3.34	3.12	2.95	2.82	2.72
120	6.85	4.79	3.95	3.48	3.17	2.96	2.79	2.66	2.56
∞	6.63	4.61	3.78	3.32	3.02	2.80	2.64	2.51	2.41

Table VII Critical Values of the F Distribution (*continued*)

$$f_{0.01}(\nu_1, \nu_2)$$

ν_2	ν_1									
	10	12	15	20	24	30	40	60	120	∞
1	6056	6106	6157	6209	6235	6261	6287	6313	6339	6366
2	99.40	99.42	99.43	99.45	99.46	99.47	99.47	99.48	99.49	99.50
3	27.23	27.05	26.87	26.69	26.60	26.50	26.41	26.32	26.22	26.13
4	14.55	14.37	14.20	14.02	13.93	13.84	13.75	13.65	13.56	13.46
5	10.05	9.89	9.72	9.55	9.47	9.38	9.29	9.20	9.11	9.02
6	7.87	7.72	7.56	7.40	7.31	7.23	7.14	7.06	6.97	6.88
7	6.62	6.47	6.31	6.16	6.07	5.99	5.91	5.82	5.74	5.65
8	5.81	5.67	5.52	5.36	5.28	5.20	5.12	5.03	4.95	4.86
9	5.26	5.11	4.96	4.81	4.73	4.65	4.57	4.48	4.40	4.31
10	4.85	4.71	4.56	4.41	4.33	4.25	4.17	4.08	4.00	3.91
11	4.54	4.40	4.25	4.10	4.02	3.94	3.86	3.78	3.69	3.60
12	4.30	4.16	4.01	3.86	3.78	3.70	3.62	3.54	3.45	3.36
13	4.10	3.96	3.82	3.66	3.59	3.51	3.43	3.34	3.25	3.17
14	3.94	3.80	3.66	3.51	3.43	3.35	3.27	3.18	3.09	3.00
15	3.80	3.67	3.52	3.37	3.29	3.21	3.13	3.05	2.96	2.87
16	3.69	3.55	3.41	3.26	3.18	3.10	3.02	2.93	2.84	2.75
17	3.59	3.46	3.31	3.16	3.08	3.00	2.92	2.83	2.75	2.65
18	3.51	3.37	3.23	3.08	3.00	2.92	2.84	2.75	2.66	2.57
19	3.43	3.30	3.15	3.00	2.92	2.84	2.76	2.67	2.58	2.49
20	3.37	3.23	3.09	2.94	2.86	2.78	2.69	2.61	2.52	2.42
21	3.31	3.17	3.03	2.88	2.80	2.72	2.64	2.55	2.46	2.36
22	3.26	3.12	2.98	2.83	2.75	2.67	2.58	2.50	2.40	2.31
23	3.21	3.07	2.93	2.78	2.70	2.62	2.54	2.45	2.35	2.26
24	3.17	3.03	2.89	2.74	2.66	2.58	2.49	2.40	2.31	2.21
25	3.13	2.99	2.85	2.70	2.62	2.54	2.45	2.36	2.27	2.17
26	3.09	2.96	2.81	2.66	2.58	2.50	2.42	2.33	2.23	2.13
27	3.06	2.93	2.78	2.63	2.55	2.47	2.38	2.29	2.20	2.10
28	3.03	2.90	2.75	2.60	2.52	2.44	2.35	2.26	2.17	2.06
29	3.00	2.87	2.73	2.57	2.49	2.41	2.33	2.23	2.14	2.03
30	2.98	2.84	2.70	2.55	2.47	2.39	2.30	2.21	2.11	2.01
40	2.80	2.66	2.52	2.37	2.29	2.20	2.11	2.02	1.92	1.80
60	2.63	2.50	2.35	2.20	2.12	2.03	1.94	1.84	1.73	1.60
120	2.47	2.34	2.19	2.03	1.95	1.86	1.76	1.66	1.53	1.38
∞	2.32	2.18	2.04	1.88	1.79	1.70	1.59	1.47	1.32	1.00

Table VIII† Tolerance Factors for Normal Distributions

$v = 0.95$

n	0.90	0.95	0.99
		$1 - \alpha$	
2	32.019	37.674	48.430
3	8.380	9.916	12.861
4	5.369	6.370	8.299
5	4.275	5.079	6.634
6	3.712	4.414	5.775
7	3.369	4.007	5.248
8	3.136	3.732	4.891
9	2.967	3.532	4.631
10	2.839	3.379	4.433
11	2.737	3.259	4.277
12	2.655	3.162	4.150
13	2.587	3.081	4.044
14	2.529	3.012	3.955
15	2.480	2.954	3.878
16	2.437	2.903	3.812
17	2.400	2.858	3.754
18	2.366	2.819	3.702
19	2.337	2.784	3.656
20	2.310	2.752	3.615
25	2.208	2.631	3.457
30	2.140	2.549	3.350
35	2.090	2.490	3.272
40	2.052	2.445	3.213

$v = 0.99$

n	0.90	0.95	0.99
		$1 - \alpha$	
2	160.193	188.491	242.300
3	18.930	22.401	29.055
4	9.398	11.150	14.527
5	6.612	7.855	10.260
6	5.337	6.345	8.301
7	4.613	5.488	7.187
8	4.147	4.936	6.468
9	3.822	4.550	5.966
10	3.582	4.265	5.594
11	3.397	4.045	5.308
12	3.250	3.870	5.079
13	3.130	3.727	4.893
14	3.029	3.608	4.737
15	2.945	3.507	4.605
16	2.872	3.421	4.492
17	2.808	3.345	4.393
18	2.753	3.279	4.307
19	2.703	3.221	4.230
20	2.659	3.168	4.161
25	2.494	2.972	3.904
30	2.385	2.841	3.733
35	2.306	2.748	3.611
40	2.247	2.677	3.518

† Adapted from C. Eisenhart, M. W. Hastay, and W. A. Wallis, *Techniques of Statistical Analysis*, Chapter 2, McGraw-Hill Book Company, New York, 1947. Used with permission of McGraw-Hill Book Company.

Table VIII Tolerance Factors for Normal Distributions (*continued*)

$v = 0.95$

n	1 − α		
	0.90	0.95	0.99
45	2.021	2.408	3.165
50	1.996	2.379	3.126
55	1.976	2.354	3.094
60	1.958	2.333	3.066
65	1.943	2.315	3.042
70	1.929	2.299	3.021
75	1.917	2.285	3.002
80	1.907	2.272	2.986
85	1.897	2.261	2.971
90	1.889	2.251	2.958
95	1.881	2.241	2.945
100	1.874	2.233	2.934
150	1.825	2.175	2.859
200	1.798	2.143	2.816
250	1.780	2.121	2.788
300	1.767	2.106	2.767
400	1.749	2.084	2.739
500	1.737	2.070	2.721
600	1.729	2.060	2.707
700	1.722	2.052	2.697
800	1.717	2.046	2.688
900	1.712	2.040	2.682
1000	1.709	2.036	2.676
∞	1.645	1.960	2.576

$v = 0.99$

n	1 − α		
	0.90	0.95	0.99
45	2.200	2.621	3.444
50	2.162	2.576	3.385
55	2.130	2.538	3.335
60	2.103	2.506	3.293
65	2.080	2.478	3.257
70	2.060	2.454	3.225
75	2.042	2.433	3.197
80	2.026	2.414	3.173
85	2.012	2.397	3.150
90	1.999	2.382	3.130
95	1.987	2.368	3.112
100	1.977	2.355	3.096
150	1.905	2.270	2.983
200	1.865	2.222	2.921
250	1.839	2.191	2.880
300	1.820	2.169	2.850
400	1.794	2.138	2.809
500	1.777	2.117	2.783
600	1.764	2.102	2.763
700	1.755	2.091	2.748
800	1.747	2.082	2.736
900	1.741	2.075	2.726
1000	1.736	2.068	2.718
∞	1.645	1.960	2.576

Table IX† Sample Size for the *t* Test of the Mean

The left‑hand value column is labelled **Value of** $\Delta = \dfrac{\mu - \mu_0}{\sigma}$.

Δ	α=0.005 / 0.01					α=0.01 / 0.02					α=0.025 / 0.05					α=0.05 / 0.1					Δ
β =	0.01	0.05	0.1	0.2	0.5	0.01	0.05	0.1	0.2	0.5	0.01	0.05	0.1	0.2	0.5	0.01	0.05	0.1	0.2	0.5	
0.05																					0.05
0.10																					0.10
0.15																				122	0.15
0.20										139					99					70	0.20
0.25					110					90				128	64			139	101	45	0.25
0.30				134	78				115	63			119	90	45		122	97	71	32	0.30
0.35			125	99	58			109	85	47		109	88	67	34		90	72	52	24	0.35
0.40		115	97	77	45		101	85	66	37	117	84	68	51	26	101	70	55	40	19	0.40
0.45		92	77	62	37	110	81	68	53	30	93	67	54	41	21	80	55	44	33	15	0.45
0.50	100	75	63	51	30	90	66	55	43	25	76	54	44	34	18	65	45	36	27	13	0.50
0.55	83	63	53	42	26	75	55	46	36	21	63	45	37	28	15	54	38	30	22	11	0.55
0.60	71	53	45	36	22	63	47	39	31	18	53	38	32	24	13	46	32	26	19	9	0.60
0.65	61	46	39	31	20	55	41	34	27	16	46	33	27	21	12	39	28	22	17	8	0.65
0.70	53	40	34	28	17	47	35	30	24	14	40	29	24	19	10	34	24	19	15	8	0.70
0.75	47	36	30	25	16	42	31	27	21	13	35	26	21	16	9	30	21	17	13	7	0.75
0.80	41	32	27	22	14	37	28	24	19	12	31	22	19	15	9	27	19	15	12	6	0.80
0.85	37	29	24	20	13	33	25	21	17	11	28	21	17	13	8	24	17	14	11	6	0.85
0.90	34	26	22	18	12	29	23	19	16	10	25	19	16	12	7	21	15	13	10	5	0.90
0.95	31	24	20	17	11	27	21	18	14	9	23	17	14	11	7	19	14	11	9	5	0.95
1.00	28	22	19	16	10	25	19	16	13	9	21	16	13	10	6	18	13	11	8	5	1.00
1.1	24	19	16	14	9	21	16	14	12	8	18	13	11	9	6	15	11	9	7		1.1
1.2	21	16	14	12	8	18	14	12	10	7	15	12	10	8	5	13	10	8	6		1.2
1.3	18	15	13	11	8	16	13	11	9	6	14	10	9	7		11	8	7	6		1.3
1.4	16	13	12	10	7	14	11	10	9	6	12	9	8	7		10	8	7	5		1.4
1.5	15	12	11	9	7	13	10	9	8	6	11	8	7	6		9	7	6			1.5
1.6	13	11	10	8	6	12	10	9	7	5	10	8	7	6		8	6	6			1.6
1.7	12	10	9	8	6	11	9	8	7		9	7	6	5		8	6	5			1.7
1.8	12	10	9	8	6	10	8	7	7		8	7	6			7	6				1.8
1.9	11	9	8	7	6	10	8	7	6		8	6	6			7	5				1.9
2.0	10	8	8	7	5	9	7	7	6		7	6	5			6					2.0
2.1	10	8	7	7		8	7	6	6		7	6				6					2.1
2.2	9	8	7	6		8	7	6	5		7	6				6					2.2
2.3	9	7	7	6		8	6	6			6	5				5					2.3
2.4	8	7	7	6		7	6	6			6										2.4
2.5	8	7	6	6		7	6	6			6										2.5
3.0	7	6	6	5		6	5	5			5										3.0
3.5	6	5	5			5															3.5
4.0	6																				4.0

The column headings above read, under **Level of *t*-test**: Single‑sided test / Double‑sided test —
α = 0.005 (α = 0.01), α = 0.01 (α = 0.02), α = 0.025 (α = 0.05), α = 0.05 (α = 0.1).

† Reproduced with permission from O. L. Davies, ed., *Design and Analysis of Industrial Experiments*, Oliver & Boyd, Edinburgh, 1956.

Table X† Sample Size for the *t* Test of the Difference Between Two Means

Level of *t*-test

Single-sided test: α = 0.005 | α = 0.01 | α = 0.025 | α = 0.05
Double-sided test: α = 0.01 | α = 0.02 | α = 0.05 | α = 0.1

Value of $\Delta = \dfrac{\mu_1 - \mu_2}{\sigma}$

	α=0.005 / α=0.01					α=0.01 / α=0.02					α=0.025 / α=0.05					α=0.05 / α=0.1					
$\beta =$	0.01	0.05	0.1	0.2	0.5	0.01	0.05	0.1	0.2	0.5	0.01	0.05	0.1	0.2	0.5	0.01	0.05	0.1	0.2	0.5	
0.05																					0.05
0.10																					0.10
0.15																					0.15
0.20																			137		0.20
0.25															124				88		0.25
0.30										123					87				61		0.30
0.35					110					90					64				102	45	0.35
0.40					85					70				100	50			108	78	35	0.40
0.45				118	68				101	55			105	79	39		108	86	62	28	0.45
0.50				96	55			106	82	45		106	86	64	32		88	70	51	23	0.50
0.55			101	79	46		106	88	68	38		87	71	53	27	112	73	58	42	19	0.55
0.60		101	85	67	39		90	74	58	32	104	74	60	45	23	89	61	49	36	16	0.60
0.65		87	73	57	34	104	77	64	49	27	88	63	51	39	20	76	52	42	30	14	0.65
0.70	100	75	63	50	29	90	66	55	43	24	76	55	44	34	17	66	45	36	26	12	0.70
0.75	88	66	55	44	26	79	58	48	38	21	67	48	39	29	15	57	40	32	23	11	0.75
0.80	77	58	49	39	23	70	51	43	33	19	59	42	34	26	14	50	35	28	21	10	0.80
0.85	69	51	43	35	21	62	46	38	30	17	52	37	31	23	12	45	31	25	18	9	0.85
0.90	62	46	39	31	19	55	41	34	27	15	47	34	27	21	11	40	28	22	16	8	0.90
0.95	55	42	35	28	17	50	37	31	24	14	42	30	25	19	10	36	25	20	15	7	0.95
1.00	50	38	32	26	15	45	33	28	22	13	38	27	23	17	9	33	23	18	14	7	1.00
1.1	42	32	27	22	13	38	28	23	19	11	32	23	19	14	8	27	19	15	12	6	1.1
1.2	36	27	23	18	11	32	24	20	16	9	27	20	16	12	7	23	16	13	10	5	1.2
1.3	31	23	20	16	10	28	21	17	14	8	23	17	14	11	6	20	14	11	9	5	1.3
1.4	27	20	17	14	9	24	18	15	12	8	20	15	12	10	6	17	12	10	8	4	1.4
1.5	24	18	15	13	8	21	16	14	11	7	18	13	11	9	5	15	11	9	7	4	1.5
1.6	21	16	14	11	7	19	14	12	10	6	16	12	10	8	5	14	10	8	6	4	1.6
1.7	19	15	13	10	7	17	13	11	9	6	14	11	9	7	4	12	9	7	6	3	1.7
1.8	17	13	11	10	6	15	12	10	8	5	13	10	8	6	4	11	8	7	5		1.8
1.9	16	12	11	9	6	14	11	9	8	5	12	9	7	6	4	10	7	6	5		1.9
2.0	14	11	10	8	6	13	10	9	7	5	11	8	7	6	4	9	7	6	4		2.0
2.1	13	10	9	8	5	12	9	8	7	5	10	8	6	5	3	8	6	5	4		2.1
2.2	12	10	8	7	5	11	9	7	6	4	9	7	6	5		8	6	5	4		2.2
2.3	11	9	8	7	5	10	8	7	6	4	9	7	6	5		7	5	5	4		2.3
2.4	11	9	8	6	5	10	8	7	6	4	8	6	5	4		7	5	4	4		2.4
2.5	10	8	7	6	4	9	7	6	5	4	8	6	5	4		6	5	4	3		2.5
3.0	8	6	6	5	4	7	6	5	4	3	6	5	4	4		5	4	3			3.0
3.5	6	5	5	4	3	6	5	4	4		5	4	4	3		4	3				3.5
4.0	6	5	4	4		5	4	4	3		4	4	3			4					4.0

† Reproduced with permission from O. L. Davies, ed., *Design and Analysis of Industrial Experiments*, Oliver & Boyd, Edinburgh, 1956.

Table XI† Critical Values for Cochran's Test

$\alpha = 0.01$

k \ n	2	3	4	5	6	7	8	9	10	11	17	37	145	∞
2	0.9999	0.9950	0.9794	0.9586	0.9373	0.9172	0.8988	0.8823	0.8674	0.8539	0.7949	0.7067	0.6062	0.5000
3	0.9933	0.9423	0.8831	0.8335	0.7933	0.7606	0.7335	0.7107	0.6912	0.6743	0.6059	0.5153	0.4230	0.3333
4	0.9676	0.8643	0.7814	0.7212	0.6761	0.6410	0.6129	0.5897	0.5702	0.5536	0.4884	0.4057	0.3251	0.2500
5	0.9279	0.7885	0.6957	0.6329	0.5875	0.5531	0.5259	0.5037	0.4854	0.4697	0.4094	0.3351	0.2644	0.2000
6	0.8828	0.7218	0.6258	0.5635	0.5195	0.4866	0.4608	0.4401	0.4229	0.4084	0.3529	0.2858	0.2229	0.1667
7	0.8376	0.6644	0.5685	0.5080	0.4659	0.4347	0.4105	0.3911	0.3751	0.3616	0.3105	0.2494	0.1929	0.1429
8	0.7945	0.6152	0.5209	0.4627	0.4226	0.3932	0.3704	0.3522	0.3373	0.3248	0.2779	0.2214	0.1700	0.1250
9	0.7544	0.5727	0.4810	0.4251	0.3870	0.3592	0.3378	0.3207	0.3067	0.2950	0.2514	0.1992	0.1521	0.1111
10	0.7175	0.5358	0.4469	0.3934	0.3572	0.3308	0.3106	0.2945	0.2813	0.2704	0.2297	0.1811	0.1376	0.1000
12	0.6528	0.4751	0.3919	0.3428	0.3099	0.2861	0.2680	0.2535	0.2419	0.2320	0.1961	0.1535	0.1157	0.0833
15	0.5747	0.4069	0.3317	0.2882	0.2593	0.2386	0.2228	0.2104	0.2002	0.1918	0.1612	0.1251	0.0934	0.0667
20	0.4799	0.3297	0.2654	0.2288	0.2048	0.1877	0.1748	0.1646	0.1567	0.1501	0.1248	0.0960	0.0709	0.0500
24	0.4247	0.2871	0.2295	0.1970	0.1759	0.1608	0.1495	0.1406	0.1338	0.1283	0.1060	0.0810	0.0595	0.0417
30	0.3632	0.2412	0.1913	0.1635	0.1454	0.1327	0.1232	0.1157	0.1100	0.1054	0.0867	0.0658	0.0480	0.0333
40	0.2940	0.1915	0.1508	0.1281	0.1135	0.1033	0.0957	0.0898	0.0853	0.0816	0.0668	0.0503	0.0363	0.0250
60	0.2151	0.1371	0.1069	0.0902	0.0796	0.0722	0.0668	0.0625	0.0594	0.0567	0.0461	0.0344	0.0245	0.0167
120	0.1225	0.0759	0.0585	0.0489	0.0429	0.0387	0.0357	0.0334	0.0316	0.0302	0.0242	0.0178	0.0125	0.0083
∞	0	0	0	0	0	0	0	0	0	0	0	0	0	0

† Reproduced from C. Eisenhart, M. W. Hastay, and W. A. Wallis, *Techniques of Statistical Analysis*, Chapter 15, McGraw-Hill Book Company, New York, 1947. Used with permission of McGraw-Hill Book Company.

Table XI Critical Values for Cochran's Test (continued)

$$\alpha = 0.05$$

k \ n	2	3	4	5	6	7	8	9	10	11	17	37	145	∞
2	0.9985	0.9750	0.9392	0.9057	0.8772	0.8534	0.8332	0.8159	0.8010	0.7880	0.7341	0.6602	0.5813	0.5000
3	0.9669	0.8709	0.7977	0.7457	0.7071	0.6771	0.6530	0.6333	0.6167	0.6025	0.5466	0.4748	0.4031	0.3333
4	0.9065	0.7679	0.6841	0.6287	0.5895	0.5598	0.5365	0.5175	0.5017	0.4884	0.4366	0.3720	0.3093	0.2500
5	0.8412	0.6838	0.5981	0.5441	0.5065	0.4783	0.4564	0.4387	0.4241	0.4118	0.3645	0.3066	0.2513	0.2000
6	0.7808	0.6161	0.5321	0.4803	0.4447	0.4184	0.3980	0.3817	0.3682	0.3568	0.3135	0.2612	0.2119	0.1667
7	0.7271	0.5612	0.4800	0.4307	0.3974	0.3726	0.3535	0.3384	0.3259	0.3154	0.2756	0.2278	0.1833	0.1429
8	0.6798	0.5157	0.4377	0.3910	0.3595	0.3362	0.3185	0.3043	0.2926	0.2829	0.2462	0.2022	0.1616	0.1250
9	0.6385	0.4775	0.4027	0.3584	0.3286	0.3067	0.2901	0.2768	0.2659	0.2568	0.2226	0.1820	0.1446	0.1111
10	6.6020	0.4450	0.3733	0.3311	0.3029	0.2823	0.2666	0.2541	0.2439	0.2353	0.2032	0.1655	0.1308	0.1000
12	0.5410	0.3924	0.3264	0.2880	0.2624	0.2439	0.2299	0.2187	0.2098	0.2020	0.1737	0.1403	0.1100	0.0833
15	0.4709	0.3346	0.2758	0.2419	0.2195	0.2034	0.1911	0.1815	0.1736	0.1671	0.1429	0.1144	0.0889	0.0667
20	0.3894	0.2705	0.2205	0.1921	0.1735	0.1602	0.1501	0.1422	0.1357	0.1303	0.1108	0.0879	0.0675	0.0500
24	0.3434	0.2354	0.1907	0.1656	0.1493	0.1374	0.1286	0.1216	0.1160	0.1113	0.0942	0.0743	0.0567	0.0417
30	0.2929	0.1980	0.1593	0.1377	0.1237	0.1137	0.1061	0.1002	0.0958	0.0921	0.0771	0.0604	0.0457	0.0333
40	0.2370	0.1576	0.1259	0.1082	0.0968	0.0887	0.0827	0.0780	0.0745	0.0713	0.0595	0.0462	0.0347	0.0250
60	0.1737	0.1131	0.0895	0.0765	0.0682	0.0623	0.0583	0.0552	0.0520	0.0497	0.0411	0.0316	0.0234	0.0167
120	0.0998	0.0632	0.0495	0.0419	0.0371	0.0337	0.0312	0.0292	0.0279	0.0266	0.0218	0.0165	0.0120	0.0083
∞	0	0	0	0	0	0	0	0	0	0	0	0	0	0

Table XII† Least Significant Studentized Ranges r_p

$$\alpha = 0.05$$

ν	\multicolumn{9}{c}{p}								
	2	3	4	5	6	7	8	9	10
1	17.97	17.97	17.97	17.97	17.97	17.97	17.97	17.97	17.97
2	6.085	6.085	6.085	6.085	6.085	6.085	6.085	6.085	6.085
3	4.501	4.516	4.516	4.516	4.516	4.516	4.516	4.516	4.516
4	3.927	4.013	4.033	4.033	4.033	4.033	4.033	4.033	4.033
5	3.635	3.749	3.797	3.814	3.814	3.814	3.814	3.814	3.814
6	3.461	3.587	3.649	3.680	3.694	3.697	3.697	3.697	3.697
7	3.344	3.477	3.548	3.588	3.611	3.622	3.626	3.626	3.626
8	3.261	3.399	3.475	3.521	3.549	3.566	3.575	3.579	3.579
9	3.199	3.339	3.420	3.470	3.502	3.523	3.536	3.544	3.547
10	3.151	3.293	3.376	3.430	3.465	3.489	3.505	3.516	3.522
11	3.113	3.256	3.342	3.397	3.435	3.462	3.480	3.493	3.501
12	3.082	3.225	3.313	3.370	3.410	3.439	3.459	3.474	3.484
13	3.055	3.200	3.289	3.348	3.389	3.419	3.442	3.458	3.470
14	3.033	3.178	3.268	3.329	3.372	3.403	3.426	3.444	3.457
15	3.014	3.160	3.250	3.312	3.356	3.389	3.413	3.432	3.446
16	2.998	3.144	3.235	3.298	3.343	3.376	3.402	3.422	3.437
17	2.984	3.130	3.222	3.285	3.331	3.366	3.392	3.412	3.429
18	2.971	3.118	3.210	3.274	3.321	3.356	3.383	3.405	3.421
19	2.960	3.107	3.199	3.264	3.311	3.347	3.375	3.397	3.415
20	2.950	3.097	3.190	3.255	3.303	3.339	3.368	3.391	3.409
24	2.919	3.066	3.160	3.226	3.276	3.315	3.345	3.370	3.390
30	2.888	3.035	3.131	3.199	3.250	3.290	3.322	3.349	3.371
40	2.858	3.006	3.102	3.171	3.224	3.266	3.300	3.328	3.352
60	2.829	2.976	3.073	3.143	3.198	3.241	3.277	3.307	3.333
120	2.800	2.947	3.045	3.116	3.172	3.217	3.254	3.287	3.314
∞	2.772	2.918	3.017	3.089	3.146	3.193	3.232	3.265	3.294

† Abridged from H. Leon Harter, "Critical Values for Duncan's New Multiple Range Test," *Biometrics*, Vol. 16, no. 4, 1960, by permission of the author and the editor.

Table XII Least Significant Studentized Ranges r_p (continued)

$$\alpha = 0.01$$

ν	\multicolumn{9}{c}{p}								
	2	3	4	5	6	7	8	9	10
1	90.03	90.03	90.03	90.03	90.03	90.03	90.03	90.03	90.03
2	14.04	14.04	14.04	14.04	14.04	14.04	14.04	14.04	14.04
3	8.261	8.321	8.321	8.321	8.321	8.321	8.321	8.321	8.321
4	6.512	6.677	6.740	6.756	6.756	6.756	6.756	6.756	6.756
5	5.702	5.893	5.989	6.040	6.065	6.074	6.074	6.074	6.074
6	5.243	5.439	5.549	5.614	5.655	5.680	5.694	5.701	5.703
7	4.949	5.145	5.260	5.334	5.383	5.416	5.439	5.454	5.464
8	4.746	4.939	5.057	5.135	5.189	5.227	5.256	5.276	5.291
9	4.596	4.787	4.906	4.986	5.043	5.086	5.118	5.142	5.160
10	4.482	4.671	4.790	4.871	4.931	4.975	5.010	5.037	5.058
11	4.392	4.579	4.697	4.780	4.841	4.887	4.924	4.952	4.975
12	4.320	4.504	4.622	4.706	4.767	4.815	4.852	4.883	4.907
13	4.260	4.442	4.560	4.644	4.706	4.755	4.793	4.824	4.850
14	4.210	4.391	4.508	4.591	4.654	4.704	4.743	4.775	4.802
15	4.168	4.347	4.463	4.547	4.610	4.660	4.700	4.733	4.760
16	4.131	4.309	4.425	4.509	4.572	4.622	4.663	4.696	4.724
17	4.099	4.275	4.391	4.475	4.539	4.589	4.630	4.664	4.693
18	4.071	4.246	4.362	4.445	4.509	4.560	4.601	4.635	4.664
19	4.046	4.220	4.335	4.419	4.483	4.534	4.575	4.610	4.639
20	4.024	4.197	4.312	4.395	4.459	4.510	4.552	4.587	4.617
24	3.956	4.126	4.239	4.322	4.386	4.437	4.480	4.516	4.546
30	3.889	4.056	4.168	4.250	4.314	4.366	4.409	4.445	4.477
40	3.825	3.988	4.098	4.180	4.244	4.296	4.339	4.376	4.408
60	3.762	3.922	4.031	4.111	4.174	4.226	4.270	4.307	4.340
120	3.702	3.858	3.965	4.044	4.107	4.158	4.202	4.239	4.272
∞	3.643	3.796	3.900	3.978	4.040	4.091	4.135	4.172	4.205

Table XIII† Values of $d_{\alpha/2}(k, v)$ for Two-Sided Comparisons Between k Treatments and a Control

$$\alpha = 0.05$$

v	\multicolumn{9}{c}{k = number of treatment means (excluding control)}
	1	2	3	4	5	6	7	8	9
5	2.57	3.03	3.29	3.48	3.62	3.73	3.82	3.90	3.97
6	2.45	2.86	3.10	3.26	3.39	3.49	3.57	3.64	3.71
7	2.36	2.75	2.97	3.12	3.24	3.33	3.41	3.47	3.53
8	2.31	2.67	2.88	3.02	3.13	3.22	3.29	3.35	3.41
9	2.26	2.61	2.81	2.95	3.05	3.14	3.20	3.26	3.32
10	2.23	2.57	2.76	2.89	2.99	3.07	3.14	3.19	3.24
11	2.20	2.53	2.72	2.84	2.94	3.02	3.08	3.14	3.19
12	2.18	2.50	2.68	2.81	2.90	2.98	3.04	3.09	3.14
13	2.16	2.48	2.65	2.78	2.87	2.94	3.00	3.06	3.10
14	2.14	2.46	2.63	2.75	2.84	2.91	2.97	3.02	3.07
15	2.13	2.44	2.61	2.73	2.82	2.89	2.95	3.00	3.04
16	2.12	2.42	2.59	2.71	2.80	2.87	2.92	2.97	3.02
17	2.11	2.41	2.58	2.69	2.78	2.85	2.90	2.95	3.00
18	2.10	2.40	2.56	2.68	2.76	2.83	2.89	2.94	2.98
19	2.09	2.39	2.55	2.66	2.75	2.81	2.87	2.92	2.96
20	2.09	2.38	2.54	2.65	2.73	2.80	2.86	2.90	2.95
24	2.06	2.35	2.51	2.61	2.70	2.76	2.81	2.86	2.90
30	2.04	2.32	2.47	2.58	2.66	2.72	2.77	2.82	2.86
40	2.02	2.29	2.44	2.54	2.62	2.68	2.73	2.77	2.81
60	2.00	2.27	2.41	2.51	2.58	2.64	2.69	2.73	2.77
120	1.98	2.24	2.38	2.47	2.55	2.60	2.65	2.69	2.73
∞	1.96	2.21	2.35	2.44	2.51	2.57	2.61	2.65	2.69

† Reproduced from Charles W. Dunnett, "New Tables for Multiple Comparison with a Control," *Biometrics*, Vol. 20, no. 3, 1964, by permission of the author and the editor.

Table XIII Values of $d_{\alpha/2}(k, v)$ for Two-Sided Comparisons Between k Treatments and a Control (*continued*)

$$\alpha = 0.01$$

v	k = number of treatment means (excluding control)								
	1	2	3	4	5	6	7	8	9
5	4.03	4.63	4.98	5.22	5.41	5.56	5.69	5.80	5.89
6	3.71	4.21	4.51	4.71	4.87	5.00	5.10	5.20	5.28
7	3.50	3.95	4.21	4.39	4.53	4.64	4.74	4.82	4.89
8	3.36	3.77	4.00	4.17	4.29	4.40	4.48	4.56	4.62
9	3.25	3.63	3.85	4.01	4.12	4.22	4.30	4.37	4.43
10	3.17	3.53	3.74	3.88	3.99	4.08	4.16	4.22	4.28
11	3.11	3.45	3.65	3.79	3.89	3.98	4.05	4.11	4.16
12	3.05	3.39	3.58	3.71	3.81	3.89	3.96	4.02	4.07
13	3.01	3.33	3.52	3.65	3.74	3.82	3.89	3.94	3.99
14	2.98	3.29	3.47	3.59	3.69	3.76	3.83	3.88	3.93
15	2.95	3.25	3.43	3.55	3.64	3.71	3.78	3.83	3.88
16	2.92	3.22	3.39	3.51	3.60	3.67	3.73	3.78	3.83
17	2.90	3.19	3.36	3.47	3.56	3.63	3.69	3.74	3.79
18	2.88	3.17	3.33	3.44	3.53	3.60	3.66	3.71	3.75
19	2.86	3.15	3.31	3.42	3.50	3.57	3.63	3.68	3.72
20	2.85	3.13	3.29	3.40	3.48	3.55	3.60	3.65	3.69
24	2.80	3.07	3.22	3.32	3.40	3.47	3.52	3.57	3.61
30	2.75	3.01	3.15	3.25	3.33	3.39	3.44	3.49	3.52
40	2.70	2.95	3.09	3.19	3.26	3.32	3.37	3.41	3.44
60	2.66	2.90	3.03	3.12	3.19	3.25	3.29	3.33	3.37
120	2.62	2.85	2.97	3.06	3.12	3.18	3.22	3.26	3.29
∞	2.58	2.79	2.92	3.00	3.06	3.11	3.15	3.19	3.22

Table XIV† Values of $d_\alpha(k, v)$ for One-Sided Comparisons Between k Treatments and a Control

$$\alpha = 0.05$$

v	\multicolumn{9}{c}{k = number of treatment means (excluding control)}
	1	2	3	4	5	6	7	8	9
5	2.02	2.44	2.68	2.85	2.98	3.08	3.16	3.24	3.30
6	1.94	2.34	2.56	2.71	2.83	2.92	3.00	3.07	3.12
7	1.89	2.27	2.48	2.62	2.73	2.82	2.89	2.95	3.01
8	1.86	2.22	2.42	2.55	2.66	2.74	2.81	2.87	2.92
9	1.83	2.18	2.37	2.50	2.60	2.68	2.75	2.81	2.86
10	1.81	2.15	2.34	2.47	2.56	2.64	2.70	2.76	2.81
11	1.80	2.13	2.31	2.44	2.53	2.60	2.67	2.72	2.77
12	1.78	2.11	2.29	2.41	2.50	2.58	2.64	2.69	2.74
13	1.77	2.09	2.27	2.39	2.48	2.55	2.61	2.66	2.71
14	1.76	2.08	2.25	2.37	2.46	2.53	2.59	2.64	2.69
15	1.75	2.07	2.24	2.36	2.44	2.51	2.57	2.62	2.67
16	1.75	2.06	2.23	2.34	2.43	2.50	2.56	2.61	2.65
17	1.74	2.05	2.22	2.33	2.42	2.49	2.54	2.59	2.64
18	1.73	2.04	2.21	2.32	2.41	2.48	2.53	2.58	2.62
19	1.73	2.03	2.20	2.31	2.40	2.47	2.52	2.57	2.61
20	1.72	2.03	2.19	2.30	2.39	2.46	2.51	2.56	2.60
24	1.71	2.01	2.17	2.28	2.36	2.43	2.48	2.53	2.57
30	1.70	1.99	2.15	2.25	2.33	2.40	2.45	2.50	2.54
40	1.68	1.97	2.13	2.23	2.31	2.37	2.42	2.47	2.51
60	1.67	1.95	2.10	2.21	2.28	2.35	2.39	2.44	2.48
120	1.66	1.93	2.08	2.18	2.26	2.32	2.37	2.41	2.45
∞	1.64	1.92	2.06	2.16	2.23	2.29	2.34	2.38	2.42

† Reproduced from Charles W. Dunnett, "A Multiple Comparison Procedure for Comparing Several Treatments with a Control," *J. Amer. Stat. Assoc.*, Vol. 50, 1096–1121, 1955, by permission of the author and the editor.

Table XIV Values of $d_t(k, v)$ for One-Sided Comparisons Between k Treatments and a Control (*continued*)

$$\alpha = 0.01$$

v	k = number of treatment means (excluding control)								
	1	2	3	4	5	6	7	8	9
5	3.37	3.90	4.21	4.43	4.60	4.73	4.85	4.94	5.03
6	3.14	3.61	3.88	4.07	4.21	4.33	4.43	4.51	4.59
7	3.00	3.42	3.66	3.83	3.96	4.07	4.15	4.23	4.30
8	2.90	3.29	3.51	3.67	3.79	3.88	3.96	4.03	4.09
9	2.82	3.19	3.40	3.55	3.66	3.75	3.82	3.89	3.94
10	2.76	3.11	3.31	3.45	3.56	3.64	3.71	3.78	3.83
11	2.72	3.06	3.25	3.38	3.48	3.56	3.63	3.69	3.74
12	2.68	3.01	3.19	3.32	3.42	3.50	3.56	3.62	3.67
13	2.65	2.97	3.15	3.27	3.37	3.44	3.51	3.56	3.61
14	2.62	2.94	3.11	3.23	3.32	3.40	3.46	3.51	3.56
15	2.60	2.91	3.08	3.20	3.29	3.36	3.42	3.47	3.52
16	2.58	2.88	3.05	3.17	3.26	3.33	3.39	3.44	3.48
17	2.57	2.86	3.03	3.14	3.23	3.30	3.36	3.41	3.45
18	2.55	2.84	3.01	3.12	3.21	3.27	3.33	3.38	3.42
19	2.54	2.83	2.99	3.10	3.18	3.25	3.31	3.36	3.40
20	2.53	2.81	2.97	3.08	3.17	3.23	3.29	3.34	3.38
24	2.49	2.77	2.92	3.03	3.11	3.17	3.22	3.27	3.31
30	2.46	2.72	2.87	2.97	3.05	3.11	3.16	3.21	3.24
40	2.42	2.68	2.82	2.92	2.99	3.05	3.10	3.14	3.18
60	2.39	2.64	2.78	2.87	2.94	3.00	3.04	3.08	3.12
120	2.36	2.60	2.73	2.82	2.89	2.94	2.99	3.03	3.06
∞	2.33	2.56	2.68	2.77	2.84	2.89	2.93	2.97	3.00

Table XV† Power of the Analysis-of-Variance Test

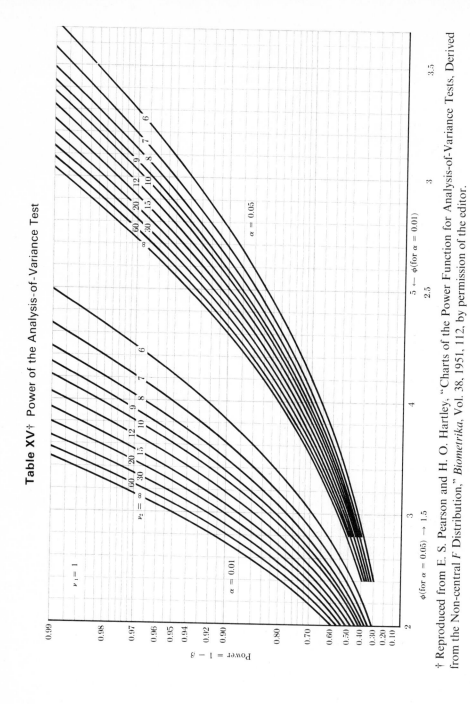

† Reproduced from E. S. Pearson and H. O. Hartley, "Charts of the Power Function for Analysis-of-Variance Tests, Derived from the Non-central *F* Distribution," *Biometrika*, Vol. 38, 1951, 112, by permission of the editor.

Table XV Power of the Analysis-of-Variance Test (*continued*)

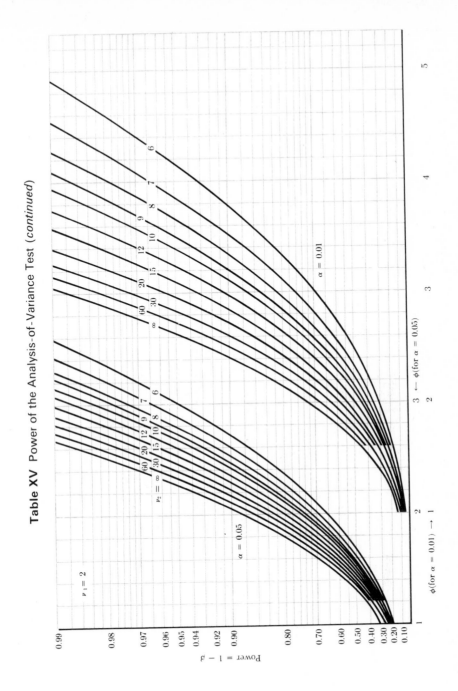

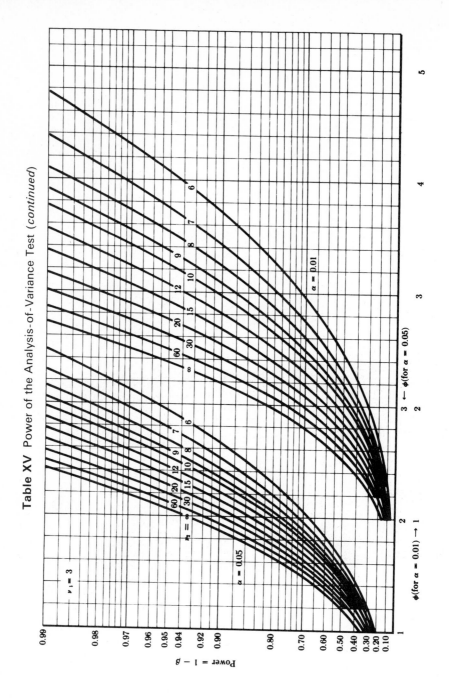

Table XV Power of the Analysis-of-Variance Test (*continued*)

Table XV Power of the Analysis-of-Variance Test (*continued*)

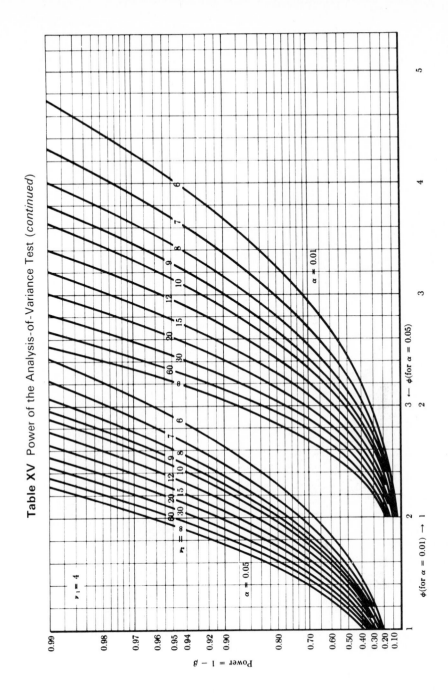

535

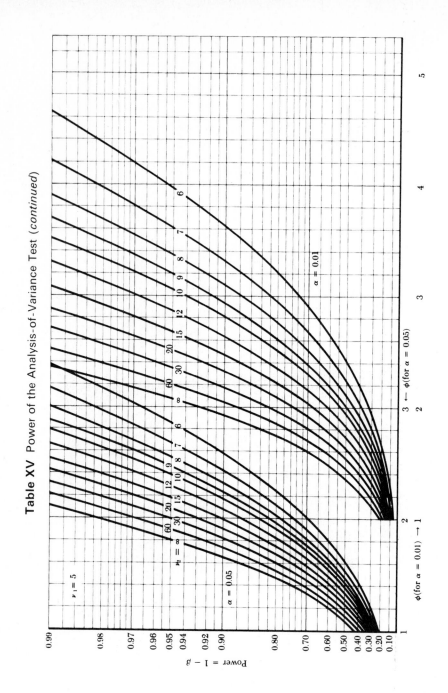

Table XV Power of the Analysis-of-Variance Test (*continued*)

Table XV Power of the Analysis-of-Variance Test (*continued*)

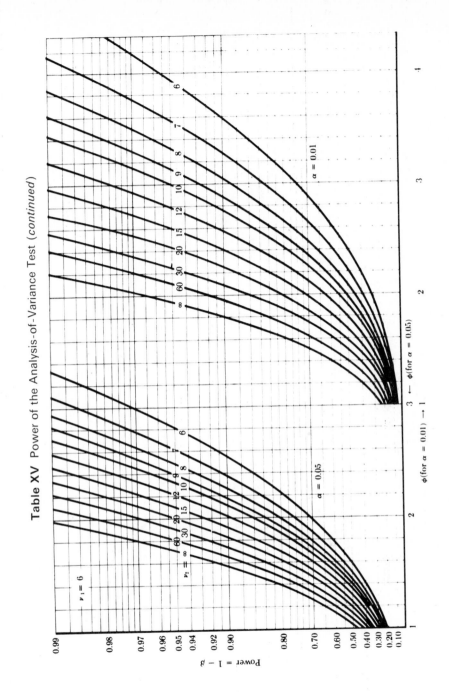

537

Table XV Power of the Analysis-of-Variance Test (*continued*)

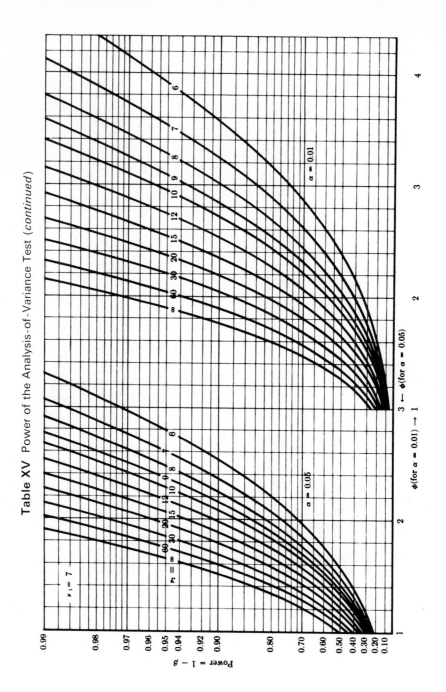

538

Table XV Power of the Analysis-of-Variance Test (*continued*)

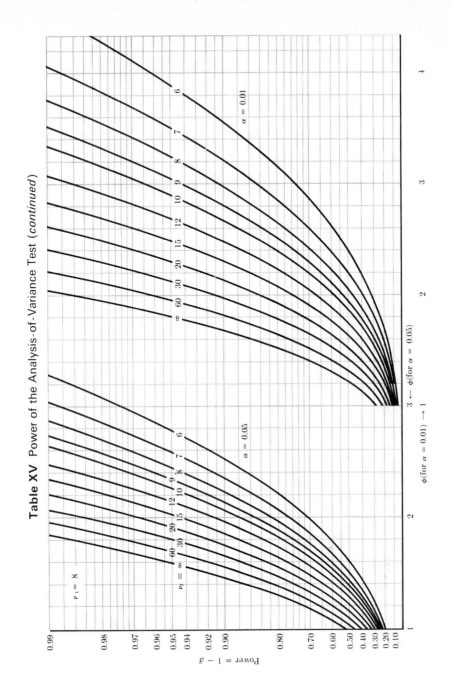

Table XVI† $P(U \leq u | H_0$ is true) in the Wilcoxon Two-Sample Test

$n_2 = 3$

	n_1		
u	1	2	3
0	0.250	0.100	0.050
1	0.500	0.200	0.100
2	0.750	0.400	0.200
3		0.600	0.350
4			0.500
5			0.650

$n_2 = 4$

	n_1			
u	1	2	3	4
0	0.200	0.067	0.028	0.014
1	0.400	0.133	0.057	0.029
2	0.600	0.267	0.114	0.057
3		0.400	0.200	0.100
4		0.600	0.314	0.171
5			0.429	0.243
6			0.571	0.343
7				0.443
8				0.557

$n_2 = 5$

	n_1				
u	1	2	3	4	5
0	0.167	0.047	0.018	0.008	0.004
1	0.333	0.095	0.036	0.016	0.008
2	0.500	0.190	0.071	0.032	0.016
3	0.667	0.286	0.125	0.056	0.028
4		0.429	0.196	0.095	0.048
5		0.571	0.286	0.143	0.075
6			0.393	0.206	0.111
7			0.500	0.278	0.155
8			0.607	0.365	0.210
9				0.452	0.274
10				0.548	0.345
11					0.421
12					0.500
13					0.579

† Reproduced from H. B. Mann and D. R. Whitney, "On a Test of Whether One of Two Random Variables Is Stochastically Larger Than the Other," *Ann. Math. Stat.*, Vol. 18, 1947, 52–54, by permission of the authors and the publisher.

Table XVI $P(U \leq u | H_0$ is true) in the Wilcoxon Two-Sample Test
(*continued*)

$$n_2 = 6$$

	n_1					
u	1	2	3	4	5	6
0	0.143	0.036	0.012	0.005	0.002	0.001
1	0.286	0.071	0.024	0.010	0.004	0.002
2	0.428	0.143	0.048	0.019	0.009	0.004
3	0.571	0.214	0.083	0.033	0.015	0.008
4		0.321	0.131	0.057	0.026	0.013
5		0.429	0.190	0.086	0.041	0.021
6		0.571	0.274	0.129	0.063	0.032
7			0.357	0.176	0.089	0.047
8			0.452	0.238	0.123	0.066
9			0.548	0.305	0.165	0.090
10				0.381	0.214	0.120
11				0.457	0.268	0.155
12				0.545	0.331	0.197
13					0.396	0.242
14					0.465	0.294
15					0.535	0.350
16						0.409
17						0.469
18						0.531

Table XVI $P(U \leq u | H_0$ is true) in the Wilcoxon Two-Sample Test
(*continued*)

$$n_2 = 7$$

u	\multicolumn{7}{c}{n_1}						
	1	2	3	4	5	6	7
0	0.125	0.028	0.008	0.003	0.001	0.001	0.000
1	0.250	0.056	0.017	0.006	0.003	0.001	0.001
2	0.375	0.111	0.033	0.012	0.005	0.002	0.001
3	0.500	0.167	0.058	0.021	0.009	0.004	0.002
4	0.625	0.250	0.092	0.036	0.015	0.007	0.003
5		0.333	0.133	0.055	0.024	0.011	0.006
6		0.444	0.192	0.082	0.037	0.017	0.009
7		0.556	0.258	0.115	0.053	0.026	0.013
8			0.333	0.158	0.074	0.037	0.019
9			0.417	0.206	0.101	0.051	0.027
10			0.500	0.264	0.134	0.069	0.036
11			0.583	0.324	0.172	0.090	0.049
12				0.394	0.216	0.117	0.064
13				0.464	0.265	0.147	0.082
14				0.538	0.319	0.183	0.104
15					0.378	0.223	0.130
16					0.438	0.267	0.159
17					0.500	0.314	0.191
18					0.562	0.365	0.228
19						0.418	0.267
20						0.473	0.310
21						0.527	0.355
22							0.402
23							0.451
24							0.500
25							0.549

Table XVI $P(U \le u \mid H_0$ is true$)$ in the Wilcoxon Two-Sample Test

(*continued*)

$$n_2 = 8$$

	n_1							
u	1	2	3	4	5	6	7	8
0	0.111	0.022	0.006	0.002	0.001	0.000	0.000	0.000
1	0.222	0.044	0.012	0.004	0.002	0.001	0.000	0.000
2	0.333	0.089	0.024	0.008	0.003	0.001	0.001	0.000
3	0.444	0.133	0.042	0.014	0.005	0.002	0.001	0.001
4	0.556	0.200	0.067	0.024	0.009	0.004	0.002	0.001
5		0.267	0.097	0.036	0.015	0.006	0.003	0.001
6		0.356	0.139	0.055	0.023	0.010	0.005	0.002
7		0.444	0.188	0.077	0.033	0.015	0.007	0.003
8		0.556	0.248	0.107	0.047	0.021	0.010	0.005
9			0.315	0.141	0.064	0.030	0.014	0.007
10			0.387	0.184	0.085	0.041	0.020	0.010
11			0.461	0.230	0.111	0.054	0.027	0.014
12			0.539	0.285	0.142	0.071	0.036	0.019
13				0.341	0.177	0.091	0.047	0.025
14				0.404	0.217	0.114	0.060	0.032
15				0.467	0.262	0.141	0.076	0.041
16				0.533	0.311	0.172	0.095	0.052
17					0.362	0.207	0.116	0.065
18					0.416	0.245	0.140	0.080
19					0.472	0.286	0.168	0.097
20					0.528	0.331	0.198	0.117
21						0.377	0.232	0.139
22						0.426	0.268	0.164
23						0.475	0.306	0.191
24						0.525	0.347	0.221
25							0.389	0.253
26							0.433	0.287
27							0.478	0.323
28							0.522	0.360
29								0.399
30								0.439
31								0.480
32								0.520

Table XVII† Critical Values of U in the Wilcoxon Two-Sample Test

One-Tailed Test at $\alpha = 0.001$ or Two-Tailed Test at $\alpha = 0.002$

n_1	n_2											
	9	10	11	12	13	14	15	16	17	18	19	20
1												
2												
3									0	0	0	0
4		0	0	0	1	1	1	2	2	3	3	3
5	1	1	2	2	3	3	4	5	5	6	7	7
6	2	3	4	4	5	6	7	8	9	10	11	12
7	3	5	6	7	8	9	10	11	13	14	15	16
8	5	6	8	9	11	12	14	15	17	18	20	21
9	7	8	10	12	14	15	17	19	21	23	25	26
10	8	10	12	14	17	19	21	23	25	27	29	32
11	10	12	15	17	20	22	24	27	29	32	34	37
12	12	14	17	20	23	25	28	31	34	37	40	42
13	14	17	20	23	26	29	32	35	38	42	45	48
14	15	19	22	25	29	32	36	39	43	46	50	54
15	17	21	24	28	32	36	40	43	47	51	55	59
16	19	23	27	31	35	39	43	48	52	56	60	65
17	21	25	29	34	38	43	47	52	57	61	66	70
18	23	27	32	37	42	46	51	56	61	66	71	76
19	25	29	34	40	45	50	55	60	66	71	77	82
20	26	32	37	42	48	54	59	65	70	76	82	88

One-Tailed Test at $\alpha = 0.01$ or Two-Tailed Test at $\alpha = 0.02$

n_1	n_2											
	9	10	11	12	13	14	15	16	17	18	19	20
1												
2					0	0	0	0	0	0	1	1
3	1	1	1	2	2	2	3	3	4	4	4	5
4	3	3	4	5	5	6	7	7	8	9	9	10
5	5	6	7	8	9	10	11	12	13	14	15	16
6	7	8	9	11	12	13	15	16	18	19	20	22
7	9	11	12	14	16	17	19	21	23	24	26	28
8	11	13	15	17	20	22	24	26	28	30	32	34
9	14	16	18	21	23	26	28	31	33	36	38	40
10	16	19	22	24	27	30	33	36	38	41	44	47
11	18	22	25	28	31	34	37	41	44	47	50	53
12	21	24	28	31	35	38	42	46	49	53	56	60
13	23	27	31	35	39	43	47	51	55	59	63	67
14	26	30	34	38	43	47	51	56	60	65	69	73
15	28	33	37	42	47	51	56	61	66	70	75	80
16	31	36	41	46	51	56	61	66	71	76	82	87
17	33	38	44	49	55	60	66	71	77	82	88	93
18	36	41	47	53	59	65	70	76	82	88	94	100
19	38	44	50	56	63	69	75	82	88	94	101	107
20	40	47	53	60	67	73	80	87	93	100	107	114

† Adapted and abridged from Tables 1, 3, 5, and 7 of D. Auble, "Extended Tables for the Mann–Whitney Statistic," *Bull. Inst. Educ. Res. Indiana Univ.*, Vol. 1, no. 2, 1953, by permission of the director.

Table XVII Critical Values of U in the Wilcoxon Two-Sample Test (*continued*)

One-Tailed Test at $\alpha = 0.025$ or Two-Tailed Test at $\alpha = 0.05$

n_1	n_2 9	10	11	12	13	14	15	16	17	18	19	20
1												
2	0	0	0	1	1	1	1	1	2	2	2	2
3	2	3	3	4	4	5	5	6	6	7	7	8
4	4	5	6	7	8	9	10	11	11	12	13	13
5	7	8	9	11	12	13	14	15	17	18	19	20
6	10	11	13	14	16	17	19	21	22	24	25	27
7	12	14	16	18	20	22	24	26	28	30	32	34
8	15	17	19	22	24	26	29	31	34	36	38	41
9	17	20	23	26	28	31	34	37	39	42	45	48
10	20	23	26	29	33	36	39	42	45	48	52	55
11	23	26	30	33	37	40	44	47	51	55	58	62
12	26	29	33	37	41	45	49	53	57	61	65	69
13	28	33	37	41	45	50	54	59	63	67	72	76
14	31	36	40	45	50	55	59	64	67	74	78	83
15	34	39	44	49	54	59	64	70	75	80	85	90
16	37	42	47	53	59	64	70	75	81	86	92	98
17	39	45	51	57	63	67	75	81	87	93	99	105
18	42	48	55	61	67	74	80	86	93	99	106	112
19	45	52	58	65	72	78	85	92	99	106	113	119
20	48	55	62	69	76	83	90	98	105	112	119	127

One-Tailed Test at $\alpha = 0.05$ or Two-Tailed Test at $\alpha = 0.10$

n_1	n_2 9	10	11	12	13	14	15	16	17	18	19	20
1											0	0
2	1	1	1	2	2	2	3	3	3	4	4	4
3	3	4	5	5	6	7	7	8	9	9	10	11
4	6	7	8	9	10	11	12	14	15	16	17	18
5	9	11	12	13	15	16	18	19	20	22	23	25
6	12	14	16	17	19	21	23	25	26	28	30	32
7	15	17	19	21	24	26	28	30	33	35	37	39
8	18	20	23	26	28	31	33	36	39	41	44	47
9	21	24	27	30	33	36	39	42	45	48	51	54
10	24	27	31	34	37	41	44	48	51	55	58	62
11	27	31	34	38	42	46	50	54	57	61	65	69
12	30	34	38	42	47	51	55	60	64	68	72	77
13	33	37	42	47	51	56	61	65	70	75	80	84
14	36	41	46	51	56	61	66	71	77	82	87	92
15	39	44	50	55	61	66	72	77	83	88	94	100
16	42	48	54	60	65	71	77	83	89	95	101	107
17	45	51	57	64	70	77	83	89	96	102	109	115
18	48	55	61	68	75	82	88	95	102	109	116	123
19	51	58	65	72	80	87	94	101	109	116	123	130
20	54	62	69	77	84	92	100	107	115	123	130	138

Table XVIII $P(R < r^* | H_0$ is true) in the Sign Test

n	\multicolumn{9}{c}{r^*}
	1	2	3	4	5	6	7	8	9
5	0.031								
6	0.016	0.109							
7	0.008	0.063							
8	0.004	0.035	0.145						
9	0.002	0.020	0.090						
10	0.001	0.011	0.055						
11		0.006	0.033	0.113					
12		0.003	0.019	0.073					
13		0.002	0.011	0.046	0.133				
14		0.001	0.006	0.029	0.090				
15			0.004	0.018	0.059	0.151			
16			0.002	0.011	0.038	0.105			
17			0.001	0.006	0.025	0.072			
18				0.004	0.015	0.048	0.119		
19				0.002	0.010	0.032	0.084		
20				0.001	0.006	0.021	0.058		
21					0.004	0.013	0.039	0.095	
22					0.002	0.008	0.026	0.067	
23					0.001	0.005	0.017	0.047	0.105
24						0.003	0.011	0.032	0.076
25						0.002	0.007	0.022	0.054

Table XIX† Critical Values of W in the Wilcoxon Test for Paired Observations

n	One-sided $\alpha = 0.01$ Two-sided $\alpha = 0.02$	One-sided $\alpha = 0.025$ Two-sided $\alpha = 0.05$	One-sided $\alpha = 0.05$ Two-sided $\alpha = 0.10$
5			1
6		1	2
7	0	2	4
8	2	4	6
9	3	6	8
10	5	8	11
11	7	11	14
12	10	14	17
13	13	17	21
14	16	21	26
15	20	25	30
16	24	30	36
17	28	35	41
18	33	40	47
19	38	46	54
20	43	52	60
21	49	59	68
22	56	66	75
23	62	73	83
24	69	81	92
25	77	90	101
26	85	98	110
27	93	107	120
28	102	117	130
29	111	127	141
30	120	137	152

† Reproduced from F. Wilcoxon and R. A. Wilcox, *Some Rapid Approximate Statistical Procedures*, American Cyanamid Company, Pearl River, N.Y., 1964, by permission of the American Cyanamid Company.

Table XX† $P(V \le a | H_0$ is true) in the Runs Test

(n_1, n_2)	\multicolumn{9}{c}{a}

(n_1, n_2)	2	3	4	5	6	7	8	9	10
(2, 3)	0.200	0.500	0.900	1.000					
(2, 4)	0.133	0.400	0.800	1.000					
(2, 5)	0.095	0.333	0.714	1.000					
(2, 6)	0.071	0.286	0.643	1.000					
(2, 7)	0.056	0.250	0.583	1.000					
(2, 8)	0.044	0.222	0.533	1.000					
(2, 9)	0.036	0.200	0.491	1.000					
(2, 10)	0.030	0.182	0.455	1.000					
(3, 3)	0.100	0.300	0.700	0.900	1.000				
(3, 4)	0.057	0.200	0.543	0.800	0.971	1.000			
(3, 5)	0.036	0.143	0.429	0.714	0.929	1.000			
(3, 6)	0.024	0.107	0.345	0.643	0.881	1.000			
(3, 7)	0.017	0.083	0.283	0.583	0.833	1.000			
(3, 8)	0.012	0.067	0.236	0.533	0.788	1.000			
(3, 9)	0.009	0.055	0.200	0.491	0.745	1.000			
(3,10)	0.007	0.045	0.171	0.455	0.706	1.000			
(4, 4)	0.029	0.114	0.371	0.629	0.886	0.971	1.000		
(4, 5)	0.016	0.071	0.262	0.500	0.786	0.929	0.992	1.000	
(4, 6)	0.010	0.048	0.190	0.405	0.690	0.881	0.976	1.000	
(4, 7)	0.006	0.033	0.142	0.333	0.606	0.833	0.954	1.000	
(4, 8)	0.004	0.024	0.109	0.279	0.533	0.788	0.929	1.000	
(4, 9)	0.003	0.018	0.085	0.236	0.471	0.745	0.902	1.000	
(4, 10)	0.002	0.014	0.068	0.203	0.419	0.706	0.874	1.000	
(5, 5)	0.008	0.040	0.167	0.357	0.643	0.833	0.960	0.992	1.000
(5, 6)	0.004	0.024	0.110	0.262	0.522	0.738	0.911	0.976	0.998
(5, 7)	0.003	0.015	0.076	0.197	0.424	0.652	0.854	0.955	0.992
(5, 8)	0.002	0.010	0.054	0.152	0.347	0.576	0.793	0.929	0.984
(5, 9)	0.001	0.007	0.039	0.119	0.287	0.510	0.734	0.902	0.972
(5, 10)	0.001	0.005	0.029	0.095	0.239	0.455	0.678	0.874	0.958
(6, 6)	0.002	0.013	0.067	0.175	0.392	0.608	0.825	0.933	0.987
(6, 7)	0.001	0.008	0.043	0.121	0.296	0.500	0.733	0.879	0.966
(6, 8)	0.001	0.005	0.028	0.086	0.226	0.413	0.646	0.821	0.937
(6, 9)	0.000	0.003	0.019	0.063	0.175	0.343	0.566	0.762	0.902
(6, 10)	0.000	0.002	0.013	0.047	0.137	0.288	0.497	0.706	0.864
(7, 7)	0.001	0.004	0.025	0.078	0.209	0.383	0.617	0.791	0.922
(7, 8)	0.000	0.002	0.015	0.051	0.149	0.296	0.514	0.704	0.867
(7, 9)	0.000	0.001	0.010	0.035	0.108	0.231	0.427	0.622	0.806
(7, 10)	0.000	0.001	0.006	0.024	0.080	0.182	0.355	0.549	0.743
(8, 8)	0.000	0.001	0.009	0.032	0.100	0.214	0.405	0.595	0.786
(8, 9)	0.000	0.001	0.005	0.020	0.069	0.157	0.319	0.500	0.702
(8, 10)	0.000	0.000	0.003	0.013	0.048	0.117	0.251	0.419	0.621
(9, 9)	0.000	0.000	0.003	0.012	0.044	0.109	0.238	0.399	0.601
(9, 10)	0.000	0.000	0.002	0.008	0.029	0.077	0.179	0.319	0.510
(10, 10)	0.000	0.000	0.001	0.004	0.019	0.051	0.128	0.242	0.414

† Reproduced from C. Eisenhart and F. Swed, "Tables for Testing Randomness of Grouping in a Sequence of Alternatives," *Ann. Math. Stat.*, Vol. 14, 1943, by permission of the editor.

Table XX $P(V \le a \mid H_0$ is true) in the Runs Test (*continued*)

(n_1, n_2)	11	12	13	14	15	16	17	18	19	20
(2, 3)										
(2, 4)										
(2, 5)										
(2, 6)										
(2, 7)										
(2, 8)										
(2, 9)										
(2, 10)										
(3, 3)										
(3, 4)										
(3, 5)										
(3, 6)										
(3, 7)										
(3, 8)										
(3, 9)										
(3, 10)										
(4, 4)										
(4, 5)										
(4, 6)										
(4, 7)										
(4, 8)										
(4, 9)										
(4, 10)										
(5, 5)										
(5, 6)	1.000									
(5, 7)	1.000									
(5, 8)	1.000									
(5, 9)	1.000									
(5, 10)	1.000									
(6, 6)	0.998	1.000								
(6, 7)	0.992	0.999	1.000							
(6, 8)	0.984	0.998	1.000							
(6, 9)	0.972	0.994	1.000							
(6, 10)	0.958	0.990	1.000							
(7, 7)	0.975	0.996	0.999	1.000						
(7, 8)	0.949	0.988	0.998	1.000	1.000					
(7, 9)	0.916	0.975	0.994	0.999	1.000					
(7, 10)	0.879	0.957	0.990	0.998	1.000					
(8, 8)	0.900	0.968	0.991	0.999	1.000	1.000				
(8, 9)	0.843	0.939	0.980	0.996	0.999	1.000	1.000			
(8, 10)	0.782	0.903	0.964	0.990	0.998	1.000	1.000			
(9, 9)	0.762	0.891	0.956	0.988	0.997	1.000	1.000	1.000		
(9, 10)	0.681	0.834	0.923	0.974	0.992	0.999	1.000	1.000	1.000	
(10, 10)	0.586	0.758	0.872	0.949	0.981	0.996	0.999	1.000	1.000	1.000

Table XXI† Sample Size for Two-Sided Nonparametric Tolerance Limits

	γ					
$1 - \alpha$	0.50	0.70	0.90	0.95	0.99	0.995
0.995	336	488	777	947	1,325	1,483
0.99	168	244	388	473	662	740
0.95	34	49	77	93	130	146
0.90	17	24	38	46	64	72
0.85	11	16	25	30	42	47
0.80	9	12	18	22	31	34
0.75	7	10	15	18	24	27
0.70	6	8	12	14	20	22
0.60	4	6	9	10	14	16
0.50	3	5	7	8	11	12

† Reproduced from Tables A-25d of Wilfrid J. Dixon and Frank J. Massey, Jr., *Introduction to Statistical Analysis*, 3rd ed., McGraw-Hill Book Company, New York, 1969. Used with permission of McGraw-Hill Book Company.

Table XXII† Sample Size for One-Sided Nonparametric Tolerance Limits

	γ				
$1 - \alpha$	0.50	0.70	0.95	0.99	0.995
0.995	139	241	598	919	1,379
0.99	69	120	299	459	688
0.95	14	24	59	90	135
0.90	7	12	29	44	66
0.85	5	8	19	29	43
0.80	4	6	14	21	31
0.75	3	5	11	17	25
0.70	2	4	9	13	20
0.60	2	3	6	10	14
0.50	1	2	5	7	10

† Reproduced from Tables A-25e of Wilfrid J. Dixon and Frank J. Massey, Jr., *Introduction to Statistical Analysis*, 3rd ed., McGraw-Hill Book Company, New York, 1969. Used with permission of McGraw-Hill Book Company.

Table XXIII† Critical Values of Spearman's Rank Correlation
Coefficient

n	$\alpha = 0.05$	$\alpha = 0.025$	$\alpha = 0.01$	$\alpha = 0.005$
5	0.900	—	—	—
6	0.829	0.886	0.943	—
7	0.714	0.786	0.893	—
8	0.643	0.738	0.833	0.881
9	0.600	0.683	0.783	0.833
10	0.564	0.648	0.745	0.794
11	0.523	0.623	0.736	0.818
12	0.497	0.591	0.703	0.780
13	0.475	0.566	0.673	0.745
14	0.457	0.545	0.646	0.716
15	0.441	0.525	0.623	0.689
16	0.425	0.507	0.601	0.666
17	0.412	0.490	0.582	0.645
18	0.399	0.476	0.564	0.625
19	0.388	0.462	0.549	0.608
20	0.377	0.450	0.534	0.591
21	0.368	0.438	0.521	0.576
22	0.359	0.428	0.508	0.562
23	0.351	0.418	0.496	0.549
24	0.343	0.409	0.485	0.537
25	0.336	0.400	0.475	0.526
26	0.329	0.392	0.465	0.515
27	0.323	0.385	0.456	0.505
28	0.317	0.377	0.448	0.496
29	0.311	0.370	0.440	0.487
30	0.305	0.364	0.432	0.478

† Reproduced from E. G. Olds, "Distribution of Sums of Squares of
Rank Differences for Small Samples," *Ann. Math. Stat.*, Vol. 9, 1938, by
permission of the editor.

Answers to Exercises

CHAPTER 1

1. (a) {7, 14, 21, 28, 35, 42, 49}.
   (b) {−3, 2}.
   (c) {H1, H2, H3, H4, H5, H6, T1, T2, T3, T4, T5, T6}.
   (d) {North America, South America, Europe, Asia, Africa, Australia, Antarctica}.
   (e) ∅.

2. $\{(x, y) \mid x^2 + y^2 < 9; x > 0, y > 0\}$.

3. $A = C$.

4. (a)

Green	|	|	Red	|	|	|
	1	*2*	*3*	*4*	*5*	*6*
1	(1, 1)	(1, 2)	(1, 3)	(1, 4)	(1, 5)	(1, 6)
2	(2, 1)	(2, 2)	(2, 3)	(2, 4)	(2, 5)	(2, 6)
3	(3, 1)	(3, 2)	(3, 3)	(3, 4)	(3, 5)	(3, 6)
4	(4, 1)	(4, 2)	(4, 3)	(4, 4)	(4, 5)	(4, 6)
5	(5, 1)	(5, 2)	(5, 3)	(5, 4)	(5, 5)	(5, 6)
6	(6, 1)	(6, 2)	(6, 3)	(6, 4)	(6, 5)	(6, 6)

   (b) $A = \{(1, 1), (1, 2), (1, 3), (2, 1), (3, 1), (2, 2)\}$.
   (c) $B = \{(1, 6), (2, 6), (3, 6), (4, 6), (5, 6), (6, 6),$
      $(6, 1), (6, 2), (6, 3), (6, 4), (6, 5)\}$.
   (d) $C = \{(2, 1), (2, 2), (2, 3), (2, 4), (2, 5), (2, 6)\}$.
   (e)

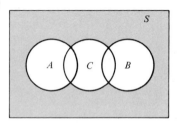

5. (a) $S = \{HH, HT, T1, T2, T3, T4, T5, T6\}$.
   (b) $A = \{T1, T2, T3\}$.
   (c) $B = \varnothing$.

6. (a) $S = \{YYY, YYN, YNY, NYY, YNN, NYN, NNY, NNN\}$.
   (b) $E = \{YYY, YYN, YNY, NYY\}$.
   (c) One possible event: "The second woman interviewed uses brand X."

7. $S_1 = \{MMMM, MMMF, MMFM, MFMM, FMMM, MMFF, MFMF, MFFM, FMFM, FFMM, FMMF, MFFF, FMFF, FFMF, FFFM, FFFF\}; S_2 = \{0, 1, 2, 3, 4\}$.

8. (a) $S = \{M_1M_2, M_1F_1, M_1F_2, M_2M_1, M_2F_1, M_2F_2, F_1M_1, F_1M_2, F_1F_2, F_2M_1, F_2M_2, F_2F_1\}$.
   (b) $A = \{M_1M_2, M_1F_1, M_1F_2, M_2M_1, M_2F_1, M_2F_2\}$.
   (c) $B = \{M_1F_1, M_1F_2, M_2F_1, M_2F_2, F_1M_1, F_1M_2, F_2M_1, F_2M_2\}$.
   (d) $C = \{F_1F_2, F_2F_1\}$.
   (e)

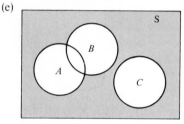

9. $A \cap C = \{(2, 1), (2, 2)\}$.

10. $A \cap B = \{M_1F_1, M_1F_2, M_2F_1, M_2F_2\}$.
    $A \cup C = \{M_1M_2, M_1F_1, M_1F_2, M_2M_1, M_2F_1, M_2F_2, F_1F_2, F_2F_1\}$.

11. (a) $\{0, 2, 3, 4, 5, 6, 8\}$. (b) $\varnothing$. (c) $\{0, 1, 6, 7, 8, 9\}$. (d) $\{1, 3, 5, 6, 7, 9\}$.
    (e) $\{0, 1, 6, 7, 8, 9\}$. (f) $\{2, 4\}$.

12. (a) {nitrogen, potassium, uranium, oxygen}.
    (b) {copper, sodium, zinc, oxygen}.
    (c) {copper, sodium, nitrogen, potassium, uranium, zinc}.
    (d) {copper, uranium, zinc}. (e) $\varnothing$. (f) {oxygen}.

13. $P \cup Q = \{z \mid z < 9\}; P \cap Q = \{z \mid 1 < z < 5\}$.

14. (a)

$(A \cap B)'$

(b)

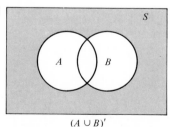

$(A \cup B)'$

(c)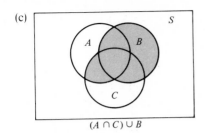

$(A \cap C) \cup B$

15. (a) 120. (b) 72.

16. 24.

17. 256.

18. 50,400.

19. 15,120.

20. (a) 100. (b) 48. (c) 48.

21. 120.

22. 144.

23. 120.

24. 210.

25. 280.

26. 3654.   ? 336

27. (a) 56. (b) 30. (c) 10.

28. 8,211,173,256.

29. 180.

30. 70.

31. $P(C) = 1/2$; $P(A') = 3/4$.

32. 1/6.

33. $S = \{\$10, \$25, \$100\}$; 9/10.

34. 1/9; 1/6.

35. (a) 33/54,145. (b) 33/66,640.

36. (a) 3/8. (b) 9/28.

37. 46/221.

38. (a) 0.9. (b) 0.6. (c) 0.5.

39. (a) 0.0001. (b) 0.9999.

40. (a) 2/11. (b) 5/11.

41. 46/221.

42. 0.04.

43. (a) 1/4. (b) 5/27. (c) 1/20.

44. (a) 0.35. (b) 0.875. (c) 0.55.

45. 1/4.

46. 2/9.

47. 38/63.

48. 19/64; 45/64.

49. 5/8.

50. 3/10.

51. 15/23.

52. 0.174.

CHAPTER 2

1. Discrete; continuous; continuous; discrete; discrete; continuous.

2.

x	0	1	2	3
$P(X = x)$	8/27	4/9	2/9	1/27

3. $f(x) = \dfrac{\dbinom{5}{x}\dbinom{5}{4-x}}{\dbinom{10}{4}}$, $x = 0, 1, 2, 3, 4.$

4. $f(x) = 1/6, x = 1, 2, \ldots, 6.$

5.

x	0	1	2
$P(X = x)$	1/5	3/5	1/5

6.

x	0	1	2	3
$P(X = x)$	1/27	2/9	4/9	8/27

7.
$$F(x) = \begin{cases} 0 & \text{for } x < 0 \\ 1/5 & \text{for } 0 \le x < 1 \\ 4/5 & \text{for } 1 \le x < 2 \\ 1 & \text{for } x \ge 2. \end{cases}$$

(a) 3/5. (b) 4/5.

9.
$$F(x) = \begin{cases} 0 & \text{for } x < 0 \\ 1/27 & \text{for } 0 \le x < 1 \\ 7/27 & \text{for } 1 \le x < 2 \\ 19/27 & \text{for } 2 \le x < 3 \\ 1 & \text{for } x \ge 3. \end{cases}$$

(a) 2/3.   (b) 8/27.

11. (b) 1/4.   (c) 0.3.

12. (a) 16/27.   (b) 1/3.

13. $F(x) = (x - 1)/2$; 1/4.

14. $F(x) = (x + 4)(x - 2)/27$; 1/3.

15. (a) 3/2.   (b) $F(x) = x^{3/2}$; 0.3004.

16. (d) 55; 79.

17. (d) 4.5.

18. (a) $f(1, 0) = 3/70, f(2, 0) = 9/70, f(3, 0) = 3/70$
    $f(0, 1) = 2/70, f(1, 1) = 18/70, f(2, 1) = 18/70, f(3, 1) = 2/70$
    $f(0, 2) = 3/70, f(1, 2) = 9/70, f(2, 2) = 3/70.$
(b) 1/2.

19. (a) 135/1024.   (b) 1/2.

20. (a) 1/50.   (b) 13/75.   (c) 14/25.   (d) 8/15.

21. $1 - \dfrac{n^2}{2} = 0.6534.$

22. (a)

y	0	1	2
$f(y\,\|\,2)$	3/10	3/5	1/10

(b) 3/10.

23.

x	1	2	3
$g(x)$	1/3	19/36	5/36

y	1	2	3
$h(y)$	1/4	14/45	79/180

x	1	2	3
$f(x\,\|\,1)$	0	2/3	1/3
$f(x\,\|\,2)$	9/14	5/14	0
$f(x\,\|\,3)$	24/79	45/79	10/79

y	1	2	3
$f(y\,\|\,1)$	0	3/5	2/5
$f(y\,\|\,2)$	6/19	4/19	9/19
$f(y\,\|\,3)$	3/5	0	2/5

24.

x	2	4
$g(x)$	0.40	0.60

y	1	3	5
$h(y)$	0.25	0.50	0.25

X and Y are independent.

25. Independent.

26. Dependent.

27. 3/4.

28. 0.6321.

29. (a) Dependent.   (b) 1/3.

30. (b) 0.64.

31. (a) 3.   (b) 21/512.

32. (a) $g(y, z) = 2yz^2/9, 0 < y < 1, 0 < z < 3.$
(b) $h(y) = 2y, 0 < y < 1.$
(c) 7/162.
(d) 1/4.

33. 1.

34. 3/4.

35. 1/3.

36. $(\ln 4)/\pi$.

37. 3.

38. $200.

39. $1.08.

40. $420.

41. 76/3.

42. 1/6.

43. 35.2

44. 0.9752.

45. 8/3.

46. (a) 209.   (b) 65/4.

47. (a) 7.   (b) 0.   (c) 12.25.

48. (a) -2.60.   (b) 9.60.

49. $\mu = 7/2; \sigma^2 = 15/4.$

50. 2/5.

51. $\mu = 2.5; \sigma = 3.041.$

52. $\mu = 1.125; \sigma^2 = 0.502.$

53. 1/18.

54. 0.9839.

55. $-3/14$.

56. -0.1244.

57. -0.036.

58. $-1/144$.

60. (a) $175/12$.   (b) $175/6$.

61. 68.

62. 52.

63. (a) At least $3/4$.   (b) At least $8/9$.

64. (a) At most $4/9$.   (b) At least $5/9$.
       (c) At least $21/25$.   (d) 10.

CHAPTER 3

1. $f(x) = 1/10$, $x = 1, 2, \ldots, 10$; $P(X < 4) = 3/10$.

2. $f(x) = 1/25$, $x = 1, 2, \ldots, 25$.

3. $\mu = 5.5$; $\sigma^2 = 8.25$.

4. 0.4219.

5. Uniform and binomial.

6. 0.3134.

7. 0.1240.

8. 0.1035.

9. 0.0006.

10. $63/64$.

11. $f(x) = \binom{5}{x}\left(\frac{1}{4}\right)^x\left(\frac{3}{4}\right)^{5-x}$, $x = 0, 1, 2, \ldots, 5$; $\mu \pm 2\sigma = 1.25 \pm 1.936$.

12. Four-engine plane.

13. Two-engine plane when $q = 1/2$; either plane when $q = 1/3$.

14. 3.75; $\mu \pm 2\sigma = 3.75 \pm 3.35$.

15. $15/128$.

16. 0.0095.

17. $21/256$.

18. $53/65$.

19. 0.9517.

20. (a) 77/115.   (b) 3/25.

21. (a) 0.6815.   (b) 0.1153.

22. 5/14.

23. 0.0025.

24. $f(x) = \dfrac{\dbinom{4}{x}\dbinom{2}{3-x}}{\dbinom{6}{3}}, \; x = 1, 2, 3;$

$P(2 \leq X \leq 3) = 4/5.$

25. (a) 1/6.   (b) 203/210.

26. 1.2.

27. $3.25; 0.52 - 5.98.$

28. 0.2131.

29. 0.9453.

30. 0.0129.

31. 4/33.

32. 17/63.

33. 0.1008.

34. (a) 0.1429.   (b) 0.1353.

35. (a) 0.1512.   (b) 0.4015.

36. (a) 0.3840.   (b) 0.0067.

37. 0.6288.

38. 0.2657.

39. $0 - 8.$

40. 0.0515.

41. 0.0651.

42. 63/64.

43. 0.1172.

44. (a) 0.0630.   (b) 0.9730.

CHAPTER 4

1. (a) 0.0913.   (b) 0.9849.   (c) 0.3362.   (d) 39.244.   (e) 46.756.

2. (a) 0.9192.   (b) 0.9821.   (c) 0.6106.   (d) 208.42.   (e) 188.5; 211.5.

3. (a) 0.1151.   (b) 16.375.   (c) 0.5403.   (d) 20.55.

4. (a) 0.0548.   (b) 0.4514.   (c) 23.   (d) 196.875 milliliters.

5. (a) 0.0062.   (b) 0.6826.   (c) 9.969 centimeters.

6. (a) 64.   (b) 86.

7. (a) 16.   (b) 551.   (c) 29.   (d) 27.

8. 62.

9. (a) 57.11%.   (b) $6.23.

10. (a) 0.0401.   (b) 0.0244.

11. (a) 19.36%.   (b) 39.70%.

12. 26.

13. (a) 0.0045.   (b) 0.1486.   (c) 0.0526.

14. 6.238 years.

15. 0.0018.

16. (a) 0.7925.   (b) 0.0352.   (c) 0.0101.

17. (a) 0.8643.   (b) 0.2978.   (c) 0.0796.

18. 0.9515.

19. (a) 0.0846.   (b) 0.1630.

20. 0.1179.

21. 0.1357.

22. 0.4356.

23. $2.8e^{-1.8} - 3.4e^{-2.4} = 0.1545$.

24. $4e^{-3} = 0.1992$.

26. $\sum_{x=4}^{6} \binom{6}{x}(1 - e^{-3/4})^x(e^{-3/4})^{6-x} = 0.3968$.

27. 0.0350.

29. $e^{-4} = 0.0183$.

30. 0.18.

CHAPTER 5

1. $g(y) = 1/3, y = 1, 3, 5.$

2. $g(y) = \left(\dfrac{3}{\sqrt{y}}\right)\left(\dfrac{2}{5}\right)^{\sqrt{y}}\left(\dfrac{3}{5}\right)^{3-\sqrt{y}}, y = 0, 1, 4, 9.$

3. $g(y_1, y_2) = \left(\dfrac{y_1 + y_2}{2}, \dfrac{y_1 - y_2}{2}, 2 - y_1\right)\left(\dfrac{1}{4}\right)^{(y_1 + y_2)/2}\left(\dfrac{1}{3}\right)^{(y_1 - y_2)/2}\left(\dfrac{5}{12}\right)^{2-y_1};$

   $y_1 = 0, 1, 2; y_2 = -2, -1, 0, 1, 2; y_2 \le y_1; y_1 + y_2 = 0, 2, 4.$

4.

y	1	2	3	4	6
$h(y)$	1/18	2/9	1/6	2/9	1/3

6. $g(y) = 1/6y^{1/3}, 0 < y < 8.$

7. Gamma distribution with $\alpha = 3/2$ and $\beta = m/2b.$

9. $h(w) = 6 + 6w - 12w^{1/2}, 0 < w < 1.$

10. $g(y) = 1/2\sqrt{y}, 0 < y < 1.$

11. $g(y) = 2/9\sqrt{y}, 0 < y < 1,$
    $= (\sqrt{y} + 1)/9\sqrt{y}, 1 < y < 4.$

13. $\mu = 1/p, \sigma^2 = q/p^2.$

14. Both equal $\mu.$

15. 0.9306.

17. $\bar{x} = 2; \tilde{x} = 1.5; m = 1.$

18. $\bar{x} = 5.875; \tilde{x} = 6.5; m = 4$ and $7.$

19. $\bar{x} = 25.2; \tilde{x} = 17; m = 11.$

20. Range $= 5; s = 1.57.$

21. $s = 0.55.$

24. (a) 3.367. (b) 13.468. (c) 3.367.

25. 0.3159.

26. 100.

27. Adjust the machine.

28. (a) $\mu_{\bar{x}} = 174.5; \sigma_{\bar{x}} = 1.38.$ (b) 153. (c) 6.

31. 0.7064.

32. (a) 0.0772. (b) 0.2814.

33. (a) 34.805. (b) 16.047. (c) 13.277.

34. (a) 0.05.   (b) 0.94.

35. Not valid.

37. (a) 2.110.   (b) -2.764.   (c) 1.714.

38. No; $\mu > 20$.

39. Yes.

40. (a) 2.71.   (b) 3.51.   (c) 2.92.   (d) 0.47.   (e) 0.34.

41. 0.05.

42. 0.99.

CHAPTER 6

2. $765 < \mu < 795$.

3. $22.01 < \mu < 22.99$.

4. (a) $172.23 < \mu < 176.77$.
(b) $e < 2.27$.

5. (a) $22{,}496 < \mu < 24{,}504$.
(b) $e < 1004$.

6. 68.

7. 11.

8. 28.

9. $0.978 < \mu < 1.033$.

10. $30.69 < \mu < 34.91$.

11. $15.63 < \mu < 21.57$.

12. $47.722 < \mu < 49.278$.

13. $2.9 < \mu_1 - \mu_2 < 7.1$.

14. $6.56 < \mu_1 - \mu_2 < 11.24$.

15. $2.80 < \mu_1 - \mu_2 < 3.40$; yes.

16. $0.3 < \mu_1 - \mu_2 < 9.7$.

17. $1.5 < \mu_1 - \mu_2 < 12.5$.

18. $-6522 < \mu_1 - \mu_2 < 2922$.

19. $-11.9 < \mu_{II} - \mu_I < 36.5$.

20. $-0.7 < \mu_D < 6.3$.

21. $-2912 < \mu_D < 687$.

22. $0.990 < \mu_D < 6.124$.

23. (a) $0.498 < p < 0.642$.
(b) $e < 0.072$.

24. (a) $0.1442 < p < 0.1998$.
(b) $e < 0.0278$.

25. $0.017 < p < 0.143$.

26. (a) 0.85. (b) $0.739 < p < 0.961$.
(c) No.

27. 2586.

28. 160.

29. 9604.

30. 16,577.

31. $-0.0136 < p_F - p_M < 0.0636$.

32. $0.016 < p_A - p_B < 0.164$.

33. $0.0011 < p_1 - p_2 < 0.0869$.

34. $0.0284 < p_1 - p_2 < 0.1416$.

35. $0.00022 < \sigma^2 < 0.00357$.

36. $3.430 < \sigma < 6.587$.

37. $1.410 < \sigma < 6.385$.

38. $1.258 < \sigma^2 < 5.410$.

39. $0.600 < \sigma_1/\sigma_2 < 2.819$; yes.

40. $0.238 < \sigma_1^2/\sigma_2^2 < 1.895$.

41. $0.016 < \sigma_1^2/\sigma_2^2 < 0.454$.

42. 0.925 to 1.675.

43. 0.382 to 7.192.

44. 11,426 to 35,574.

45. 44.52 to 52.48.

46. $p^* = 0.0923$ for $x = 0$
      $= 0.1066$ for $x = 1$
      $= 0.1183$ for $x = 2$.

47. $p^* = 0.0982$ for $x = 0$
      $= 0.1075$ for $x = 1$
      $= 0.1157$ for $x = 2$.

48. $8.077 < \mu < 8.692$.

49. $f(\mu|x_1, x_2, \dots, x_{25}) = \dfrac{1}{\sqrt{2\pi}\,13.706} e^{-1/2[(\mu-780)/20]2}$, $770 < \mu < 830$.

51. $R(\hat{P}; p) = pq/n$.

52.
$$R(\Theta; \theta) = \begin{cases} 0 & \text{for } \theta = 0 \\ 2/3 & \text{for } \theta = 1 \\ 2/3 & \text{for } \theta = 2 \\ 0 & \text{for } \theta = 3. \end{cases}$$

53.
$$R(\Theta_2; \theta) = \begin{cases} 0 & \text{for } \theta = 0 \\ 1/3 & \text{for } \theta = 1 \\ 1 & \text{for } \theta = 2 \\ 0 & \text{for } \theta = 3. \end{cases}$$

54. $\hat{\Theta}_1$.

55. $\hat{\Theta}_2$.

CHAPTER 7

1. $\alpha = 0.0853$; $\beta = 0.8287$; $\beta = 0.7817$; not a good test procedure.

2. $\alpha = 0.0548$; $\beta = 0.3504$; $\beta = 0.6177$; $\beta = 0.8281$.

3. $\alpha = 0.0146$; $\beta = 0.0126$.

4. $\alpha = 0.0386$; $\beta = 0.2779$; $\beta = 0.5$.

5. $z = -1.643$; accept H_0.

6. $z = -1.268$; accept H_0.

7. $z = 2.716$; $\mu > 174.5$.

8. $z = 8.974$; $\mu > 20{,}000$.

9. $t = 0.771$; accept H_0.

10. $t = 2.776$; $\mu > 30$.

11. $t = 1.296$; valid claim.

12. $t = 1.781$; accept H_0.

13. $z = 4.222$; $\mu_1 > \mu_2$.

14. $z = -2.603$; $\mu_A - \mu_B < 12$.

15. $z = 2.448$; $\mu_1 - \mu_2 > \$500$.

16. $t = 6.575$; $\mu_1 > \mu_2$.

17. $t = 1.501$; yes.

18. $t = -0.706$; accept H_0.

19. $t' = 0.215$; accept H_0.

20. $t = 1.821$; accept H_0.

21. $t = -2.164$; accept H_0.

22. $t = -0.899$; accept H_0.

23. $\chi^2 = 18.120$; accept H_0.

24. $\chi^2 = 10.735$; accept H_0.

25. $\chi^2 = 17.530$; reject H_0.

26. $\chi^2 = 17.188$; accept H_0.

27. $f = 1.325$; accept H_0.

28. $f = 0.747$; accept H_0.

29. $f = 0.086$; reject H_0.

30. 21.

31. 78.

32. 12.

33. 10.

34. 68.

35. $-\beta = 0.955$.

36. $z = -1.443$; $p = 0.6$.

37. Valid claim.

38. $z = 1.443$; accept H_0.

39. $z = 1.339$; valid estimate.

40. $z = 2.395$; yes.

41. $z = 1.878$; yes.

42. $z = 1.109$; no.

43. $z = 2.090$; yes.

44. $\chi^2 = 4.467$; yes.

45. $\chi^2 = 6.76$; no.

46. $\chi^2 = 1.667$; accept H_0.

47. $\chi^2 = 2.326$; accept H_0.

48. $\chi^2 = 10.000$; reject H_0.

49. $\chi^2 = 2.571$; accept H_0.

50. $\chi^2 = 9.335$; not normal.

51. $\chi^2 = 13.629$; not independent.

52. $\chi^2 = 9.048$; independent.

53. $\chi^2 = 6.484$; accept H_0.

CHAPTER 8

1. (a) $\hat{y} = 12.2784 + 0.6703x$.
   (c) 20.32.

2. (a) $\hat{y} = 6.4136 + 1.8091x$.
   (b) 9.5795.

3. (a) $\hat{y} = 42.5830 - 0.6859x$.
   (b) 25.7784.

4. (a) $\hat{y} = (27.542)(0.804)^x$.
   (b) 11.5 milligrams.

5. (a) $\hat{\gamma} = 2.660$; $\hat{C} = 2.63 \times 10^6$.
   (b) 22.9 pounds per square inch.

8. $t = 3.07$; reject H_0.

9. $4.323 < \alpha < 8.505$; $0.445 < \beta < 3.173$.

10. $21.959 < \alpha < 63.207$; $-1.478 < \beta < 0.106$.

11. $24.444 < \mu_{Y|24.5} < 27.113$; $21.888 < y_0 < 29.669$.

13. $7.808 < y_0 < 10.808$.

14. (a) $b = \sum_{i=1}^{n} x_i y_i \Big/ \sum_{i=1}^{n} x_i^2$. (b) $\hat{y} = 2.003x$.

15. $\hat{y} = 0.349 + 1.929x$; $t = 1.395$; accept H_0.

16. $E(B) = \beta + \gamma \sum_{i=1}^{n} (x_{1i} - \bar{x}_1)x_{2i} \Big/ \sum_{i=1}^{n} (x_{1i} - \bar{x}_1)^2$.

17. $f = 9.00$; reject H_0.

18. (a) $\hat{y} = 5.8253 + 0.567x$. (b) 34.205 grams. (c) $t = -0.163$; accept H_0.
   (d) $f = 1.58$; regression is linear.

19. (a) $\hat{y} = 3.1266 + 1.8429x$. (b) $f = 2.60$; accept H_0.
   (c) $1.2722 < \beta < 2.4136$.

20. $r = -0.526$.

21. $r = 0.240$.

22. $r = 0.679$; $z = 2.62$; reject H_0.

CHAPTER 9

1. $\hat{y} = 4.4882 - 0.0394x_1 + 0.6378x_2$.

2. $\hat{y} = 55.2266 - 0.0378x_1 + 1.6816x_2$.

3. $\hat{y} = 0.5801 + 2.712x_1 + 2.05x_2$.

4. (a) $\hat{y} = 3.4167 + 2.7262x - 0.3905x^2$.
(b) 7.2852.

5. $\hat{y} = 141.6112 - 0.2819x + 0.0003x^2$.

6. (a) $\hat{y} = 19.03353 + 1.00854x - 0.02038x^2$.
(b) $f = 0.023$; model is adequate.

7. (a) $\hat{y} = 1.0714 + 4.6032x - 1.8452x^2 + 0.1945x^3$.
(b) 4.453.

8. $\hat{y} = 19.98518 + 0.30363x_1 + 0.59635x_2 - 0.49706x_3 - 0.70377x_4$.

9. $\hat{y} = 14.30100 - 1.41687x_1 + 14.69890x_2 - 6.69953x_2^2 - 0.19995x_1x_2$.

10. $\hat{y} = 3.3205 + 0.4210x_1 - 0.2958x_2 + 0.0164x_3 + 0.1247x_4$.

11. 34.3699.

12. 0.00106.

13. $\hat{\sigma}_{B_1}^2 = 0.000071$; $\hat{\sigma}_{B_2}^2 = 0.063523$; $\hat{\sigma}_{B_1 B_2} = -0.001134$.

14. (a) $\hat{\sigma}_{B_2}^2 = 0.00002$. (b) $\hat{\sigma}_{B_1 B_4} = -0.000003$.

15. $26.2353 < y_0 < 57.1515$; $34.8580 < \mu_{Y|2500, 48.0} < 48.5288$.

16. $f = 8.16$; reject H_0.

17. $29.930 < \mu_{Y|19.5} < 31.970$.

18. $16.7749 < y_0 < 16.9291$; $16.8260 < \mu_{Y|8.2, 6.0, 10.3, 5.8} < 16.8780$.

19. $t = -4.48$; reject H_0.

20. $t = 3.55$; reject H_0.

21. $R^2 = 0.9997$.

22. $f = 12,865$; regression is significant.

23. $f = 12,589$; regression is significant.

24. $f = 11,040$; reject H_0.

25. $f = 20.07$; reject H_0.

26. (a) $\hat{y} = 9.9 + 0.575x_1 + 0.550x_2 + 1.150x_3$.
 (b) $f = 7.69$ for β_1; not significant.
  $f = 7.04$ for β_2; not significant.
  $f = 30.77$ for β_3; significant at the 0.01 level.

27. (a) $\hat{y} = -6.3359 + 0.3374x_1$.
 (b) Same as (a).
 (c) Same as (a).

28. $\hat{y} = 2.1833 + 0.9576x_2 + 3.3253x_3$.

29. $\hat{y} = -4.2468 + 0.3561x_1 - 0.1415x_2 + 0.1966x_3 - 0.0185x_4$.

CHAPTER 10

3. $f = 0.307$; no significant difference.

4. $f = 0.464$; no significant difference.

5. $b = 0.189$; variances are equal.

6. (a) $b \doteq 3.80$; variances are equal.
 (b) $f = 13.33$; reject H_0.
 (c) (1) $f = 14.27$; reject H_0. (2) $f = 23.23$; reject H_0.
  (3) $f = 2.48$; accept H_0.

7. $g = 0.4237$; variances are equal.

8. (a) $f = 7.10$; reject H_0. (b) Blend 4 differs significantly from all others.

9. $d_1 = 9.088$; significant.
 $d_2 = 6.8498$; significant.
 $d_3 = 2.3059$; significant.
 $d_4 = 2.5093$; significant.

10. (a) $f(\text{blocks}) = 8.30$; significant at the 0.05 level.
  $f(\text{fertilizer}) = 6.11$; significant at the 0.05 level.
 (b) (1) $f = 17.37$; significant at the 0.01 level.
  (2) $f = 0.96$; not significant.

13. $f(\text{blocks}) = 4.86$; significant. $f(\text{treatments}) = 3.33$; no significant difference.

14. $d_1 = 4.2439$; significant.
 $d_2 = 4.5301$; significant.
 $d_3 = 4.0586$; significant.

15. $\hat{\sigma}_\alpha^2 = 28.91$; $s^2 = 8.32$.

16. $\hat{\sigma}_\alpha^2 = 1.08$; $\hat{\sigma}_\beta^2 = 2.25$.

18. 9.

19. No; 16.

CHAPTER 11

1. (a) $f = 1.63$; accept H_0.
   (b) f(temperatures) $= 8.13$; reject H_0.
   f(ovens) $= 5.18$; reject H_0.

2. (a) $f = 5.29$; reject H_0.   (b) $f = 1.97$; accept H_0.
   (c) $f = 33.94$; reject H_0.   (d) $f = 0.498$; accept H_0.

3. (a) AB: $f = 3.83$; reject H_0.   (b) A: $f = 0.54$; accept H_0.
   AC: $f = 3.79$; reject H_0. B: $f = 6.85$; reject H_0.
   BC: $f = 1.31$; accept H_0. C: $f = 2.15$; accept H_0.
   ABC: $f = 1.63$; accept H_0.

4. (a) $f = 2.66$; accept H_0.
   (b) A: $f = 3.37$; reject H_0.
   B: $f = 5.67$; reject H_0.
   C: $f = 4.85$; reject H_0.
   The pooled error includes BC.

5. (a) $f = 1.49$; accept H_0.
   (b) f(operators) $= 12.45$; reject H_0.
   f(filters) $= 8.39$; reject H_0.
   (c) $\hat{\sigma}_\alpha^2 = 0.17$ (filters);
   $\hat{\sigma}_\beta^2 = 0.3514$ (operators);
   $s^2 = 0.1867$.

6. (a) $\hat{\sigma}_\beta^2, \hat{\sigma}_\gamma^2, \hat{\sigma}_{\alpha\gamma}^2$ are significant.   (b) $\hat{\sigma}_\gamma^2, \hat{\sigma}_{\alpha\gamma}^2$ are significant.

7. Yes.

8. 0.59.

CHAPTER 12

1. $SSA = 2.6667$, $SSB = 170.6667$, $SSC = 104.1667$, $SS(AB) = 1.5000$,
   $SS(AC) = 42.6667$, $SS(BC) = 0.0000$, $SS(ABC) = 1.5000$.

2. Significant effects:
   A: $f = 1940.64$
   B: $f = 4411.62$
   C: $f = 1098.38$
   D: $f = 458.50$
   CD: $f = 5.38$.

3. Significant effects:
   A: $f = 57.85$ D: $f = 44.72$ AD: $f = 4.85$ CD: $f = 6.52$.
   B: $f = 7.52$ AB: $f = 6.94$ BC: $f = 10.96$
   C: $f = 127.87$ AC: $f = 7.08$ BD: $f = 4.85$

Insignificant effects:

ABC:	$f = 1.26$	BCD:	$f = 1.20$
ABD:	$f = 1.14$	$ABCD$:	$f = 0.87.$
ACD:	$f = 1.20$		

4. A, B, C, AC, BC, and ABC each with 1 degree of freedom can be tested using an error mean square with 12 degrees of freedom.

5. (a) A: $f = 1.55$; not significant.
B: $f = 1.27$; not significant.
C: $f = 3.49$; not significant.
D: $f = 0.79$; not significant.
(b) ABC

6.

Block 1	Block 2	Block 3	Block 4
(1)	c	d	a
ab	abc	ac	b
acd	ad	bc	cd
bcd	bd	abd	$abcd$

CD is also confounded.

7. (a)

Block 1	Block 2	Block 3	Block 4
(1)	a	b	ab
c	ac	bc	abc
ae	e	abe	be
bd	abd	d	ad
ace	ce	$abce$	bce
bcd	$abcd$	cd	acd
$abde$	bde	ade	de
$abcde$	$bcde$	$acde$	cde

(b) BD.
(c) $SSA = 21.9453$; $SSC = 2.4753$; $SS(E) = 1.0878$;
$SSB = 40.2753$; $SSD = 7.7028$.

8.

Block			Block			Block	
1	2		1	2		1	2
abc	ab		abc	ab		(1)	a
a	ac		a	ac		c	b
b	bc		b	bc		ab	ac
c	(1)		c	(1)		abc	bc

Replicate 1	Replicate 2	Replicate 3
ABC confounded	ABC confounded	AB confounded

Analysis of Variance

Source of variation	Degrees of freedom	
Blocks	5	
A	1	
B	1	Relative information on
C	1	$ABC = 1/3$.
AB	1′	Relative information on
AC	1	$AB = 2/3$.
BC	1	
ABC	1′	
Error	11	
Total	23	

9. (a) ABC; $ABCD$.

(b)

A:	$f = 3.35$	AC: $f = 0.54$	ABC:	$f = 1.94$
B:	$f = 0.84$	AD: $f = 1.20$	ABD:	$f = 0.54$
C:	$f = 7.53$	BC: $f = 0.84$	ACD:	$f = 0.03$
D:	$f = 13.38$	BD: $f = 0.54$	BCD:	$f = 0.30$
AB:	$f = 0.84$	CD: $f = 0.30$	$ABCD$:	$f = 0.89$

10.

Block	
1	2
(1)	a
ab	b

Replicate 1

Block	
1	2
a	(1)
ab	b

Replicate 2

Block	
1	2
(1)	ab
a	b

Replicate 3

11.
$A \equiv CDE$ $AB \equiv BCDE$ $BD \equiv ABCE$
$B \equiv ABCDE$ $AC \equiv DE$ $BE \equiv ABCD$
$C \equiv ADE$ $AD \equiv CE$ $ABC \equiv BDE$
$D \equiv ACE$ $AE \equiv CD$ $ABD \equiv BCE$
$E \equiv ACD$ $BC \equiv ABDE$ $ABE \equiv BCD$.

12.
$A \equiv BCD \equiv ABDEF \equiv CEF$
$B \equiv ACD \equiv DEF \equiv ABCEF$
$C \equiv ABD \equiv BCDEF \equiv AEF$
$D \equiv ABC \equiv BEF \equiv ACDEF$
$E \equiv ABCDE \equiv BDF \equiv ACF$
$F \equiv ABCDF \equiv BDE \equiv ACE$.

13. Interactions sacrificed: $ABCD, CDEFG, BDF, ABEFG, ACF, BCEG, ADEG$.

14. A: $f = 0.48$ E: $f = 5.39$
      B: $f = 1.35$ F: $f = 1.09$
      C: $f = 3.03$ G: $f = 4.36$.
      D: $f = 1.94$

CHAPTER 13

1. $u = 1$; claim is valid.

2. $u = 5$; A operates longer.

3. $u = 12.5$; $\mu_A = \mu_B$.

4. $u = 43.5$; $\mu_1 = \mu_2$.

5. $u = 15$; $\mu_1 = \mu_2'$.

6. $r = 1$; reject H_0.

7. $r = 2$; $\mu_A = \mu_B$.

8. $r = 3$; accept H_0.

9. $r = 3$; accept H_0.

10. $r = 5$; accept H_0.

11. $P(W \le 1 \,|\, H_0$ is true$) = 0.125$; accept H_0.

12. $w = 5.5$; reject H_0.

13. $w = 8.5$; accept H_0.

14. $w = 4$; accept H_0.

15. $v = 7$; random sample.

16. $v = 2$; reject randomness at $\alpha = 0.016$.

17. $v = 6$; $\mu_A = \mu_B$.

18. $z = -0.549$; random sample.

19. $z = 1.115$; random sample.

20. 30.

21. 0.70.

22. 21.

23. 0.995.

24. (a) 0.2394.   (b) Accept H_0.

25. $r_s = 0.7047$; reject H_0.

26. $r_s = 0.2571$.

27. $r_S = -0.4667$; no significant relationship.

28. (a) 0.7143.   (b) Reject H_0.

29. $z = -2.03$; reject H_0.

30. $h = 10.4737$; reject H_0.

31. $h = 11.2721$; reject H_0.

32. $h = 1.0743$; no significant difference.

33. $h = 12.8257$; reject H_0.

Index

Dark Princess
A Romance

Dark Princess
A Romance

W. E. B. DU BOIS

New Introduction by
Herbert Aptheker

KRAUS-THOMSON ORGANIZATION LIMITED, MILLWOOD, N.Y.

Introduction copyright, 1974 by Herbert Aptheker
    all rights reserved
First Printing 1974
Second Printing 1976

Library of Congress Cataloging in Publication Data

Du Bois, William Edward Burghardt, 1868-1963.
    Dark princess, a romance.

    Reprint of the 1928 ed. published by Harcourt,
Brace, New York.
    Includes bibliographical references.
    I. Title.
PZ3.D8525Dar9   [PS3507.U147]   813'.5'2   74-7248
ISBN 0-527-25295-6

Printed in the United States of America

INTRODUCTION

In the mid-1920s, Dr. Du Bois stood without rival as the outstanding Black person in the United States; indeed, in the world perhaps only Gandhi was a more famous figure among the so-called colored peoples.

Du Bois was at that time the decisive personality in the widely-known and increasingly effective National Association for the Advancement of Colored People (NAACP), and his name was synonymous with that of *The Crisis*, the magazine he had founded in 1910 and was then editing and which reached each month about 60,000 subscribers or some 100,000 readers. He also had a world-wide reputation as the founder of the Pan-African Movement, whose first, second and third Congresses had met in 1919, 1921 and 1923 in Paris, Brussels, London, and Lisbon.

By 1924, eight of his books had been published. Several of them had had a wide sale; this was especially true of *The Souls of Black Folk* (1903), *The Negro* (1915), and *Darkwater* (1920), while in 1924 under the auspices of the Knights of Columbus, *The Gift of Black Folk* had appeared. In these years his essays appeared frequently in significant magazines of national circulation, such as *The New Republic, The Nation, The Forum, The American Mercury, Foreign Affairs,* and *The North American Review.*

It is in this period also that the impact of what was known as the Harlem Renaissance was being felt and Du Bois—with Alain Locke—was the parent of that movement. Du Bois was both mentor and the inspiration of that Renaissance, with the example of his life and work and the influence of his spoken and printed words. Again, his *Crisis*, with its poetry, drama, essay contests and its review section, offered opportunity and criticism to the younger Black creative artists surging forward, while in his own efforts towards a Black theatre he further challenged and stimulated such artists.

The influence was of course mutual; Du Bois could not help being stimulated and uplifted by the appearance of such figures as Paul

Robeson, Langston Hughes, Countee Cullen (briefly married to his daughter), Jean Toomer, Jessie Fauset, Claude McKay, Roland Hayes, Rudolph Fisher, Walter White, James Weldon Johnson and Josephine Baker.

The 1920s saw Du Bois not only visiting the European cities already mentioned but also, for the first time, the Soviet Union, which made a profound impression upon him. Also making a decisive impact was the visit in 1923-24 to Africa when Du Bois took the opportunity afforded him by his appointment as U. S. Minister Plenipotentiary and Envoy Extraordinary to the inauguration of Liberia's President to visit much of the western portion of that enormous continent.

It is in the 1920s also that the existence of Black people as *people* began to dawn with new force and scope upon widening segments of the white population. This helps explain why Du Bois was welcomed as a contributor to the magazines mentioned earlier. Indeed, it was not too rare at this time for magazines to devote entire issues to what was called "the Negro Problem." This was true, for example, of *The Survey Graphic, The World Tomorrow* and the *Annals* of the American Academy of Political and Social Science. White scholars, writers, artists, in increasing numbers, in addition to those who had been part of such efforts as the NAACP, reflected a serious interest in the Afro-American; it is the 1920s that witnessed this in connection with Eugene O'Neill, Carl Van Vechten, Michael Gold, Ridgely Torrence, Dorothy Canfield, Waldo Frank, Henry Mencken, V.F. Calverton and Franz Boas, Melville J. Herskovits and Frank Tannenbaum.

At the same time, the 1920s was a period of frequent lynchings and pogroms, of a powerful KKK and the resurgence of racist practices and myths as reflected in the immigration laws of the time and books by people like Madison Grant, then a Trustee of the American Museum of Natural History and Chairman of the New York Zoological Society. Grant's *The Passing of the Great Race*, first issued in 1916, had a very wide sale; similarly, Lothrop Stoddard's vastly influential *The Rising Tide of Color Against White Supremacy* was published by Scribner's in 1920, with an introduction by Grant.

Du Bois was at the center of all these developments[1] and his whole purpose in life was to defeat the forces of racism and obscurantism, of repression and oppression. Logically, then, he conceived of the novel form as being elastic enough and, if successful, attractive enough to

---

[1] Du Bois debated Stoddard in Chicago in March, 1929 before an audience of five thousand people. The debate was widely reported in the press and was issued in pamphlet form that year.

make some contribution to that purpose. He had tried to do that after the crushing of Populism, the institutionalizing of peonage and the legalization of disfranchisement and segregation with his *Quest of the Silver Fleece* (1911). Perhaps now—with the slaughter of World War I, the murders of the Red Summer of 1919, the propaganda of the Grants and Stoddards, and, on the other hand, the mounting resistance to racism apparent in the 1920s, such a novel with such a purpose would find a publisher and would make an impact.

But with fifteen years having passed since *Fleece* was published, the world and Du Bois had changed significantly. Du Bois—with such experiences as the First All-Races Congress of 1911, the war years, the Pan-African movement, the NAACP effort and the travels abroad—thought increasingly in global terms and had shed much of the benevolent, elitist, philosophically idealist approach that had marked his earlier period. By 1911 he thought of himself as a socialist and while his ideas as to what socialism meant changed through the years, his commitment to socialism and his conviction that capitalism was essentially anti-rational and anti-human never thereafter left him.

* * *

Certain experiences specific to Du Bois personally are consequential in the creation of *Dark Princess*. Thus, in the novel he describes the ecstasy he felt upon first seeing the Princess: "Many, many times in after years he tried to catch and rebuild that first wildly beautiful phantasy which the girl's face stirred in him. He knew well that no human being could be quite as beautiful as she looked to him then" (p. 8). This recalls Du Bois' description of himself as a student when ". . . at Fisk at the first dinner I saw opposite me a girl of whom I have often said, no human being could possibly be as beautiful as she seemed to my young eyes that far-off September night of 1885." These words are repeated in Du Bois' final book, with this additional sentence: "She was the great aunt of Lena Horne and fair as Lena Horne is, Lena Calhoun was far more beautiful."[2]

Mary White Ovington, in a review of the *Princess* book, suggests another source for its heroine. Miss Ovington participated in the First Universal Races Congress in 1911; present and prominent in its proceedings was Dr. Du Bois. Mary White Ovington wrote:[3]

---

[2] Du Bois, *Dusk of Dawn* (1940), p. 24; *Autobiography* (1968), p. 108.

[3] *Chicago Bee*, August 4, 1928.

I think I saw the dark Indian Princess in 1911 as she came down the steps of the ballroom at the last meeting of the First Universal Races Congress in London. I thought her the loveliest person there, except perhaps the darker daughter of the Haitian President, Légitime. And by the Princess' side was one of the most distinguished men of the Conference, Burghardt Du Bois. They were talking earnestly, of course of the race problem. Did this Indian Princess remain in the American Negro's memory to become the Titania of his Midsummer Night's Dream?

Certainly that 1911 conference made a deep impression upon Du Bois and at it he met peoples from every corner of the earth. At an earlier all-Black conference also held in London in 1900, Du Bois had had the opportunity of meeting delegates from the West Indies and Africa; he later recorded some discussion with another participant at this meeting which is germane to a theme in *Dark Princess*. This occurred in Du Bois' review of a book published in 1929; its chief author was a Black man named Harry Dean.

Du Bois, commenting that the book tells of the adventures "of a Negro sea captain in Africa and on the Seven Seas in his attempts to found an Ethiopian Empire," added:[4]

> I saw Harry Dean first in 1900. It was in London at the time of the World Fair in Paris, when a Pan-African Conference had been called—the first of the Pan-African meetings. Dean was bitter. He wanted to lead a black army across the straits of Gibraltar. I saw his point of view, but I did not think the scheme was practical.

Just prior to this 1900 experience, a novel by an Afro-American was published which had as its theme an elaborate and nation-wide secret effort by Black people to establish a Black-ruled state in Texas. In some respects the book reminds one of *Dark Princess* and one of its main characters, Bernard Belgrave, is a brown-skinned man, a Harvard graduate, one who had attracted national attention because of a commencement address at that university—all of which strongly suggested Du Bois.[5]

[4] The book reviewed was Harry Dean, with Sterling North, *The Pedro Gorino* (Boston, Houghton Mifflin, 1929); this appeared in *The Crisis*, November, 1929; 36: 376. Of Dean's Ethiopian Empire idea, Du Bois added: "Perhaps his dream goes in some respects beyond the facts, but it is all worth reading." In the book itself, Dean wrote: "I would instigate a movement to rehabilitate Africa and found such an Ethiopian Empire as the world has never seen. . . I dreamed of downfall for the imperialists. . .", (cited work, pp. 67, 68).

[5] Sutton Griggs, *Imperium in Imperio*, originally published by the author in 1899, reprinted, with a preface by Hugh M. Gloster by Arno Press, New York, 1969. It is of some interest that Griggs had studied at the Theological Seminary of Virginia Union Seminary in Richmond. The president of that seminary, Lewis Garnett Jordan, was quite militant and Du Bois knew him;

Advocates and prophets of a liberated Africa—"Africa for Africans"—were all well-known to Du Bois. This included Alexander Crummell, to whom Du Bois devoted a chapter in his *Souls of Black Folk* (1903), Henry M. Turner, Edward W. Blyden, Joseph Booth, Casely Hayford. Around 1912, Du Bois had discussed with Booth, during one of the latter's periodic visits to the United States, "the possibility of a native revolt in the South and Central African regions." And, of course, Du Bois followed with the closest interest all news of organization, strikes and uprisings in Africa and in other parts of the colonial world.[6] In 1912, Du Bois attended a conference on Africa held at Tuskegee; his own Niagara Movement, founded in 1905, had had a Pan-African Department. With the end of World War I came Du Bois' historic work in connection with the Pan-African Congresses, beginning in 1919; as the leader of this effort, Du Bois was in constant communication with African and African-derived leaders and peoples.

By 1906, in his Niagara Movement address, Du Bois was writing: "The Slav is raising in his might, the yellow millions are tasting liberty, the black Africans are writhing toward the light, and everywhere the laborer, with ballot in his hand, is voting open the gates of Opportunity and Peace." By the time World War I had ended, Du Bois was writing, in letters of fire—and the italics are his:

> What, then, is this dark world thinking? It is thinking that as wild and awful as this shameful war was, *it is nothing to compare with that fight for freedom which black and brown and yellow men must and will make unless their oppression and humiliation and insult at the hands of the* White World *cease.* The Dark World *is going to submit just as long as it must and not one moment longer.*

---

also at the end of the 19th century one of the students at that school was John Chilembwe, destined to lead an uprising against colonial rule in Africa in 1915. On this, see the present writer's introduction to Du Bois' *The Negro* (part of the Kraus-Thomson series of Du Bois' works). Robert Bone, in *The Negro Novel in America* (New Haven, Yale University Press, 1958), p. 100, calls attention to the Griggs novel in terms of *Dark Princess*. Earlier in the 1890s, there had been fairly strong efforts to found an all-Black state in Oklahoma—see my *Afro-American History: The Modern Era* (N.Y., Citadel, 1970), pp. 115-117.

[6] For the Booth conversation, see George Shepperson and Thomas Price, *Independent African: John Chilembwe* (Edinburgh, the University Press, 1959), p. 184. Du Bois in *The Horizon* and the early years of *The Crisis* paid very close attention to events in Africa, Asia, and Latin America. See my *Annotated Bibliography of the Published Writings of W.E.B. Du Bois* (Kraus-Thomson, 1973) for specific references. Note that in 1910 John Buchan's novelized account of an African uprising, *Prester John*, appeared and became a best-seller. See also, J. Ayodele Langley, *Pan-Africanism and Nationalism in West Africa, 1900-1945* (London, Oxford University Press, 1973).

By 1924 Du Bois was writing of a "tremendous and sometimes almost fanatic increase of race pride"; was warning that "white people do not sense this" and suggesting that this sense of outrage and this gap of ignorance would induce not only hatred but, possibly, open war. "Of course," Du Bois continued, "it is impossible for twelve million men to fight a hundred million—but can they not hate the harder for their very impotence? Whether they migrate, die or live, can they not add the red flame of their bitter hatred to all the mounting bill of deviltry which the dark world holds against the white? No—there's no hurry; it will not happen in our day. No. But it will happen."[7]

In words and sentences such as these, one can see the theme of *Dark Princess* form itself in Du Bois' mind; the last phrases will evoke in the mind of the reader the figure of Perigua, the terrorist in the novel.

Frequently in *The Crisis* Du Bois presented quotations from the world press, joined together by his own phrase or sentence, conveying some sense of a particularly important development. One such collection, which he entitled, "The Hegemony of Race," was published in that magazine's March, 1926 issue (31: 248-249). Preceding the selections, Du Bois wrote: "It is gradually being borne in upon our whiter brethren that this matter of ruling the colored races is not going to be the same easy parade in the future that it has been in the past." Then appears a lengthy extract from a recent widely reprinted essay by David Lloyd George, the former British Prime Minister, containing these sentences:

> In Asia the brown and yellow races are seeking to throw off the dominion or domination of the pale-faced foreigner. In India there is undoubted unrest. In Africa a formidable war is even now in process between the invading Europeans and the native. . .

The popular magazine, *Liberty*, owned by the Chicago *Tribune*, is quoted: "China and India are rising. North Africa is in arms. . .Soviet Russia is the enemy of every white government." Confirmatory reports from the Irish and South African press also are quoted; and this from one Lord Willingdon, formerly a British official in India: "There can be no question that, for the first time in human history, this cry of the dark against the pretensions of the white is an organized and practical determination. Once it was a mere protest, now it is a crusade."

The next issue of *The Crisis*, April, 1926, contains a long and

---

[7]For sources and additional relevant material see Chapter 11 in my *Afro-American History: The Modern Era* (1971).

quite favorable account of the October, 1925 meeting in Chicago of the American Negro Labor Congress, a Marxist and Communist effort to develop a militant anti-imperialist and trade union movement in the United States. This was by Abram L. Harris; earlier Du Bois himself, in an editorial entitled "The Black Man and Labor," had welcomed that Congress and insisted that Black people in particular should pay sympathetic attention to "the astounding effort of Soviet Russia to reorganize the industrial world" and should investigate all effort at basic change "whether it springs from Russia, China or the South Seas" (*The Crisis*, December, 1925; 31: 60).

These passages convey the entire scope and purpose of *Dark Princess* and they appeared just at the time that Du Bois had turned to the actual creation of that novel.

A few more suggestions as to aspects of the novel may be offered. The centrality of Indian figures in the unfolding of the world-wide conspiracy may well have been suggested to Du Bois not only by the realities of developments in India after the First World War but also the wide publicity given, especially in 1917 and 1918, to the efforts of Indians in the United States to further the cause of independence from Great Britain. This resulted in arrests, deportations, prolonged court cases and even pitched gun battles.[8]

The central role of Black Pullman porters in the novel and the descriptions of their work and attitudes reflect the efforts to unionize those workers then going forward under the leadership of A. Philip Randolph and with the repeatedly published strong report of Dr. Du Bois. Occasionally, in the past, Du Bois had used Pullman coaches and porters as scenes for and actors in short stories.[9]

A passage in a piece by Du Bois commenting on Henry Lincoln Johnson, a leading Black Republican politician of Georgia, who died in 1925, reflects certain concepts that Du Bois held concerning such people, given the realities of society in the United States, and also suggests something of the character of the Black political boss of Chicago Du Bois depicts in *Dark Princess.* Du Bois wrote[10] that

[8]The *New York Times* reported all this as "German plots": see its issues, in particular, of early March, 1917; April 8-9, 1917; much of May, 1917, July, 1917, October 19, 1917; and various issues from February 19, 1918 to March 28, 1919. A recent study is Kalyan Kumar Banerjee, *Indian Freedom Movement: Revolutionaries in America* (Calcutta, Jijnasa Pub., 1969).

[9]Occasionally Du Bois wrote at some length about Pullman porters in ways that recall passages in *Princess*. For example, see *The Crisis*, January, 1926; 31: 113, and April, 1926; 31: 271. For an example of a short story using Pullman locale and personnel and again suggesting passages to appear in his *Princess*, see "The Case," in *The Horizon*, July 1907; Vol. II, no. 1.

[10]*The Crisis*, November, 1925; 31: 11.

. . .looking on politics not as a matter of ideal but of hard, bitter, conscience-less fighting mainly between unprincipled and selfish men. . .the man who enters [politics in the United States] must strip himself—strip his body and soul, his mind and his conscience and go in to kill and be killed. . . at times he [Johnson] gained much for us, at other times he yielded or lost. . . fighting in the jungles of Georgia.

Du Bois' conception of the role of art also is significant in appreciating his efforts in fiction-writing. With characteristic clarity and forcefulness he expressed this in a paper entitled "Criteria of Negro Art" which he delivered at the 1926 Annual Meeting of the NAACP held in Chicago. The artist, Du Bois said, must serve truth; in serving that he serves Beauty and serving both he also serves Justice. He continued:[11]

Thus all Art is propaganda and ever must be, despite the wailing of the purists. I stand in utter shamelessness and say that whatever art I have for writing has been used always for propaganda for gaining the right of black folk to love and enjoy. I do not care a damn for any art that is not used for propaganda.

* * *

Certain particulars in the novel merit elucidation. With the growing urbanization of Black people, it was clear—and Du Bois led in pointing this out[12] —that political offices, including those on a national level, would soon be within their grasp. But that, in a novel written in 1927, Du Bois should have placed the successful Black candidate for Congress in Chicago—which happened in November, 1928 when Oscar DePriest (1871-1951) was elected to the House of Representatives from there—reflected Du Bois' unparalleled knowledge of Black life mixed, it must be confessed, with some luck. But in addition, Du Bois' novel pictured this as coming about because of the sudden death of the white incumbent; in actuality, Martin B. Madden (1855-1928), who had represented the district without a break from 1905, had just won the 1928 primary and was certain to again win the election, died suddenly in Congress on April 27, 1928, so that a Black man replaced him as the Republican candidate and thus was elected. This surely reflected sheer good fortune on Du Bois' part!

[11] *The Crisis*, October, 1926; 32: 296.

[12] In *The Crisis*, February, 1918; 15: 164, Du Bois suggested that with hard work it might soon be possible to elect Black Congressmen from New York, Illinois, Pennsylvania, Ohio and New Jersey.

It is worth adding that a weekly paper, dated April 28, 1928—that is, one day after Congressman Madden's death—headlined the fact that "Madden Is Rep. Doolittle in Du Bois' New Book" and there is no doubt that it was true.[13]

Another prophetic confirmation of a central incident in the novel occurred within a few months following its publication. This concerned the opening of the book where Matthew Towns learns from the Dean of the University of Manhattan that he may not pursue his medical studies since, because of racism, the obstetrics department will not have him. On September 25, 1928, Du Bois wrote to Harcourt, Brace—the novel's publisher—as follows:[14]

> The Sloane Hospital [in New York City] has just made almost exactly the kind of discrimination against a colored nurse which I tell of in DARK PRINCESS, as exercised against a young colored Doctor, and I may say that I had in mind this very hospital.

* * *

Additional elements in the novel are clarified by various letters and manuscripts in the Du Bois Papers. Thus, a Mrs. William C. Kenyon of Chicago wrote Du Bois on May 3, 1934 that she found *Dark Princess* "an exceedingly interesting book." She wondered if his scenes dealing with subway excavation work were based on such labor that went on at Wacker Drive in Chicago and asked further for "anything from you that directly contacts Chicago" so far as the novel was concerned. Du Bois' response, dated May 17, 1934, discloses that "the excavation which I described. . .took place in front of my residence in St. Nicholas Avenue, New York." He continued: "This description I transferred to Chicago. The only thing that connects the novel with Chicago is my general knowledge of the colored group there, and the description of the court house. That was true to life."

---

[13] *The* (Baltimore) *Afro-American,* stated date, p. 5. For later studies of details of politics in Chicago and the ghetto shortly after World War I, see Allan H. Spear, *Black Chicago* (Univ. of Chicago Press, 1967), pp. 189-192; and Arvarh E. Strickland, *History of the Chicago Urban League* (Urbana, Univ. of Illinois Press, 1966), pp. 56-103. See also the account of DePriest's life by K.C. Folkes in Walter Christmas, ed., *Negroes in Public Affairs and Government* (Yonkers, N.Y., Educational Heritage, 1966), I, pp. 98-101.

[14] Letter in custody of editor; now in Library of University of Massachusetts, Amherst. The woman involved was Gladys L. Catchings, supervising head of the obstetrical department, Freedmen's Hospital, Washington; she had been accepted in July, 1928 for post-graduate study in obstetrics at Sloane Maternity Hospital, affiliated with Columbia University. Protest by the NAACP was in vain. There are rather full accounts in the New York *Herald Tribune*, September 25 and 26, 1928, on page 3 in both instances.

While working on the novel, probably in 1927, Du Bois sent a detailed inquiry to some friend in Chicago concerning not only the court house—and asking for a photograph of it—but also about various other neighborhoods and churches in Chicago. Also, on October 19, 1927, Du Bois wrote to Carl D. Thompson, secretary of the League for Public Ownership, in Chicago. He explained that at the suggestion of "my friend, Robert Morss Lovett," he was requesting information concerning "the public service situation in Chicago and especially the activities of Samuel Insull." Responding to a request from Mr. Thompson for clarification as to just what Du Bois wanted, the latter replied on November 2, 1927, explaining that a part of a novel he was doing "deals with the Negro in Chicago politics." In this connection he had "woven. . .something about the way in which the men back of the movement to monopolize public service utilities have influenced Negro voters." Further, "I have gone on to state politics to show how members of the legislature have been gradually coralled into the movement for the monopoly of super-power. And in the question of electing a Negro congressman, I have indicated the same sinister influence in United States politics." Du Bois concluded by remarking that he wanted his novel to accord with "actual facts."

On November 8, 1927, Mr. Thompson replied from Chicago; he stated that Du Bois' letter of November 2 provided him with "a much clearer idea of just what you want in connection with your forthcoming book." He went on:

> I am of the opinion that there are no published articles or printed material of any kind that will give you an inside view of the political situation here in Chicago. It is altogether different in fact and reality from what it appears upon the surface and to the outside world.

Mr. Thompson expressed pleasure that Du Bois was working on this, stated he wanted to discuss it further with "dependable sources" and that he would write to Du Bois again "a little later."

In a letter of November 17, thanking Thompson for the above, Du Bois went on to write:

> I am sorry to give you all this trouble, but my writing of fiction, as well as other forms of literature, is for propaganda and reform. I make no bone of saying that art that isn't propaganda doesn't interest me.
>
> Particularly do I want to make clear in my novel the economic foundation of race problems. My novel deals with a young Negro who becomes a

part of the colored political machine in Chicago. Being a man of education and ability he becomes eventually a member of the legislature and is even mentioned for Congress.

It is here that I want to bring in the role which organized public service is playing in politics. First; I show that a political organization which formerly depended chiefly upon graft from gambling, bootlegging and crime, began to realize that organized business was willing to pay for privilege. When my hero gets into legislature he is induced to vote for street railway consolidation in Chicago because he thinks that the terms are about as good as he can get. Then before he knows it, his vote is pledged for certain super-power projects; a general consolidation of state and inter-state organizations to furnish electric power and monopolize various kinds of public service. Here he balks. He is not squeamish or moral about ordinary graft, but this project does not seem to him nearly as fair as a roulette wheel on the South side. Nevertheless, he yields and then with his probable nomination to Congress he faces the prospect of national super-power projects, and finally for personal as well as moral reasons, refuses to pledge himself and loses the nomination.

How close to reality this was, Du Bois wanted to know, "so that I could work in enough realism to make my message clear."

This letter brought a brief response, dated November 22, 1927. Mr. Thompson wrote that he now had "a very much clearer idea of just what you wanted." He continued: "I shall see some of my friends who know more about these matters than I within the next day or so and will then try to write you further about it." He concluded: "I think you have a splendid scheme for your novel."

So far as the Du Bois Papers show, however, that concluded Mr. Thompson's correspondence with the Doctor. It is possible, of course, that there was further writing from Mr. Thompson but Du Bois was generally so careful to retain his letters that this is not probable.[15]

The Thompson correspondence is typical of Du Bois' efforts in the writing of his books. With *Dark Princess* additional examples occur. Thus, on October 21, 1927, Du Bois addressed a letter to "The Warden"

[15]Carl Dean Thompson (1870-1949), a graduate of the Chicago Theological School, served briefly as a minister of a Congressional Church. He was a member of the Wisconsin House (1907-1909) and city clerk of Milwaukee (1910-1911). He was a founding secretary in 1914 of the Public Ownership League of America and edited its monthly, *Public Ownership*. Among the original leaders of the League were Amos Pinchot, Lynn J. Frazier, William Lemke, John Brophy, and Charles Edward Russell. Robert Morss Lovett (1870-1950) was a professor of English, an author, an editor of the *New Republic* and a classmate at Harvard of Du Bois. Viking published his autobiography—*All Our Years*—in 1948. Samuel Insull's activities in public utilities and politics were the subject of sensational investigations and court proceedings widely publicized in the press beginning in 1926. Insull gave heavily to charities in the Chicago ghetto, and was concerned about its rising political strength and the role of DePriest. See Forrest McDonald, *Insull* (University of Chicago Press, 1962), especially pp. 219, 254-275.

of the Leavenworth Federal Penitentiary asking if it was plausible that a prisoner confined there after being sentenced in 1924 would be given the designation Number 3277. The warden—one T. B. White, at the time—seems not to have answered.[16] In another instance, in an inter-office memo at the NAACP, Du Bois asked (October 18, 1927) Herbert J. Seligmann for some information on paintings, especially modern art—which, he added, he had seen in Moscow. And of his dear friend in Boston, musician Maude Cuney Hare, he asked for an opinion—on October 11, 1927—concerning references he was making in his novel to music; in this Du Bois refers to hearing "colored audiences in country districts" who "take the regular English tune but divide it up with pauses and musical quavers until it makes an extraordinary impression."

To his Indian friend and active fighter for independence, Lajpat Rai (1865-1928), Du Bois enclosed passages from his descriptions of scenes in the Indian continent and asked for criticisms and corrections, which he received. In this correspondence, Du Bois makes it clear that the Bwodpur of his novel is meant to be Nepal and Bhupal is Bhutan. Lajpat Rai returned the manuscript pages with some marginal corrections in a letter dated Calcutta, December 21, 1927, concerning Indian attitude towards widowhood and other aspects of Indian belief and culture, all of which were accepted by Du Bois for his final manuscript.

* * *

One reference in the novel (p. 20) is somewhat obscure. Mention is made "of the work on cancer by your Peyton Rous in Carrel's laboratory." Dr. Rous (1879-1970) was a Nobel Prize winner (1966) in medicine; his work on cancer was associated with investigating tumor-producing viruses. Dr. Alexis Carrel (1873-1944), co-developer of the artificial heart, was a Nobel Prize winner in medicine in 1912; Rous and Carrel both worked at the Rockefeller Institute for many years.

Occasional sentences evoke direct memories of Du Bois in terms of his own life. Thus, on page 77, Towns learns that the wife of the lynching's victim was not permitted to look upon his corpse, since, as Perigua says, "He didn't have no face." Du Bois comments: "Something died in Matthew that day." Very much the same thing occurred to Du Bois when he was on his way to the editor of a leading daily in Atlanta, Georgia to protest a recent lynching. Du Bois turned back on

---

[16]Perhaps he did; if so the letter does not survive. But Du Bois gives Towns number 1,277 and sends him to jail in Joliet, Illinois and not in Leavenworth—see p. 105.

learning that the victim's knuckles were on display in the window of a butcher shop in the very street upon which he was walking.[17]

At another point (p. 136) Matthew is told, "The most hopeless of deaths is the death of Faith." Du Bois bore this idea—the principled rejection of cynicism—firmly in his mind throughout his life. In his last message, left in safekeeping to his wife to be opened at the time of his death, occur the lines: "One thing alone I charge you. As you live, believe in life. . .The only possible death is to lose belief in this truth. . . ."

The awakening that the realities of the Great War brought to the Princess in this novel (p. 233) and her rejection therefore of the mythology about its character also reflect Du Bois' own attitudes and experiences in connection with that war—all of which made a profound impression upon him.

As for the rest, the book is before the reader. But let the reader in particular note Du Bois' poetic delight in colors—something very meaningful to him, as in his love of flowers; his enormous pride in his own people and his emphasis upon what is today called "soul" as an especially distinguishing feature among them; his love for—almost veneration of—Black women, as for example, Towns' mother; his sense of nationality when it came to the Afro-American people; and, in particular, his defense of democracy and his impassioned rejection of elitism in all forms, leading him in this book to express his belief in both the possibility of and the need for socialism.

* * *

The sense of optimism, the affirmation of the usefulness and rationality of struggle, the insistence upon the possibility of bringing about a decent global society—all of which are themes in *Dark Princess*—are certainly not the themes of the dominant white novelists or playwrights of the 1920's in the United States. On the contrary, they are the voices of the "Lost Generation." Sinclair Lewis' *Babbitt* (1922) and *Elmer Gantry* (1927) were depictions of despair and corruption, while the hero in *Arrowsmith* (1925) must hide himself in a laboratory to escape social decay and to save his own soul. Alienation, loneliness and social corruption are basic to Hemingway's *The Sun Also Rises* (1926) and *Farewell to Arms* (1929). The same characteristics appear

[17]The incident is in his *Autobiography* (1968 edition), p. 222. Dr. Du Bois once used these very words—"something died within me that day"—in describing this episode to the present writer.

in the 1920s novels of, for example, F. Scott Fitzgerald and William Faulkner and the plays of Eugene O'Neill.

These are absent in Du Bois (as in the poetry of Langston Hughes, Countee Cullen, Claude McKay in the same period) and I think this is because of Du Bois' commitment to the cause of his people's liberation and therefore to the cause of human liberation, and because Du Bois was an active participant individually and as part of a collective in this cause. This is a basic source of Du Bois' eternal contemporaneity; moods of despair appear and reappear and the best of the literature reflecting this may endure, but the need for human emancipation certainly persists; works contributing to or celebrating efforts towards this end have a way of remaining relevant in content, however dated may be aspects of form.

* * *

On December 15, 1927, Du Bois sent along—as he put it, in a letter to Harcourt, Brace—"the final manuscript of 'Dark Princess'." In this letter, he remarked that "some time ago" he had sent "a general statement about the novel for your use" and he wondered if this were suitable. Actually, in Du Bois' papers are two such statements; phrases from them do appear in some published newspaper notices of the book and so one must believe that Harcourt, Brace did send out releases based upon these manuscripts prepared by Du Bois. They are part of the novel's history and convey something of Du Bois' intent.

The first manuscript reads as follows:

> "Dark Princess" is a story of the great movement of the darker races for self-expression and self-determination. It stretches from Banares to Chicago and from Berlin to Atlanta, but centers in two persons, a Princess of India seeking light and freedom for her people and an American Negro seeking first to escape his problem and then to understand it. The main movement of the story takes place in America where the hero is first a medical student and then a Pullman Porter. Through entanglement in radical propaganda he is sent to jail but pardoned by the efforts of a leading Negro politician of Chicago. He becomes a ward politician in that city, then a member of the legislature and finally, through the manipulation of a shrewd woman whom he has married, has the nomination to Congress in his grasp. But his political success has meant his moral degradation, and when the Indian Princess with her great dream of world unity for the darker races dramatically reappears in his life, he surrenders all to become a common laborer. Groping thus in the depth for light and understanding, these two fighters for the common good are finally called to the supreme sacrifice of parting but in the end are brought back together in the person of a child who is crowned in Virginia, "Messenger and Messiah of all the Darker Worlds!"

The second manuscript is slightly longer and more revealing.

DARK PRINCESS is a romance with a message. Its first aim is to tell a story: the story of a colored medical student whom race prejudice forces out of his course of study. He tries to run away from America and then learns of a union of darker peoples against white imperialism. It is not a union of hatred or of offense, but of defense and self-development of the best in all races. But the problem is how such a movement can be set going.

Matthew is sent back to America by the Dark Princess to find out how Negro Americans can aid. Disappointed and out of touch with his foreign friends, he turns from his main object and seeks murder and revenge. This lands him in jail.

He is freed by the political influence of colored Chicago and he tries to avoid all ambition and to become simply a ward politician. He is forced from this by his own conscience, by political preferment and finally, by the reappearance of the Indian Princess whom he met and loved in Europe. He gives up politics, and together they search for a way of rebuilding the world across the color line. They are compelled to part as Matthew becomes a common laborer and she awaits the call back to her throne. They are finally united by the birth of a child who is heralded as "Messenger and Messiah to all the darker worlds!"

This is the story. But the second and deeper aim of the book is to outline the reaction of the difficulties and realities of race prejudices upon many sorts of people—ambitious black American youth, educated Asiatics, selfish colored politicians, ambitious self-seekers of all races. The book is thus a first-hand frank and sincere study of human characters both sides of the color line.

* * *

Once the manuscript was in hand, production proceeded with speed. By mid-February, 1928, Du Bois was informed that Harcourt, Brace had received corrected proofs from him. His changes—as was his wont—were extensive and so revised galleys were sent to him. He got these back to the publishers early in March and on April 14, 1928 the publishers wrote to Du Bois that they were sending him two copies of the book, whose publication date was April 19 and whose price was $2.

Among the Du Bois Papers is a single typed sheet signed by him and headed ABOUT THIS WRITING. It is not dated but surely was written soon after *Dark Princess* appeared, I should guess sometime in the summer of 1928. It reads:

Letters that come to me each day tell me that this book of mine is doing some small part of the life task I have set before me in making men realize that the persons portrayed here are human beings. They are not super-human demi-gods living in upper air and touching earth but softly here and again. They are not fools and devils even when they do foolish and devil-ish things. All of them, puppets though they be, called up for an hour's dreaming, typify the forces of flesh and blood striving in all this land for spiritual freedom and unfolding. They are not, I hasten to add, universally typical. They are but single types chosen out from thousands. Other types I will choose and set before you in some new and modest COMEDIE HUMAINE if you continue to feel interested to listen. And listen you must, now or then, to me or to others, for the telling of this marvelous human tale which the African Slave Trade arranged in moving picture for America. What after all is greater than the art of the story teller? Other arts I have tried—figures and essay and speech—but found none so subtly satisfying, so responsive to the touch as this, if haply I prove my soul strong and cunning to the task. For, then, the courage you give me in this effort receive my hand, good reader.[18]

As we shall see, the book was widely commented upon but its sale was not considerable. The first report, dated October 25, 1928 covered sales as of the end of June, that is to say, for two and a half months after its publication. This showed a total sale of 2,267 copies, which meant that Du Bois was over fifty dollars short of having earned in royalties his advance of five hundred dollars. Statements for the rest of 1928 and for 1929 seem to be lost, but those covering the six months ending June 30, 1930 show a sale of 40 copies; the six months ending December 31, 1931 a sale of 39 copies. Statements covering the six months ending June 30, 1932, December 31, 1932 and June 30, 1933 show a total—for those eighteen months—of 49 copies sold.

Not surprisingly, therefore, Harcourt, Brace wrote Du Bois on March 21, 1933 that "the sales now seem to indicate that the demand for the book is not sufficient to render its publication profitable." This letter also disclosed that from the original printing, 490 copies were on hand. Exactly what the original printing was is not known; but it prob-ably came to about five thousand copies so that something like 4,500 copies were sold in the five years of its first printing.

Having seen the first report, Du Bois wrote Alfred Harcourt, October 25, 1928, that "I am disappointed as I presume you are." He went on to seek Mr. Harcourt's "frank advice" as to whether or not there existed "a market for my fiction." An especially informative paragraph followed:

[18]What may have been the occasion or the purpose of this statement, I do not know. It is pos-sible that it was prepared for a German-language edition; in the Du Bois letters there is evidence that Dr. Anna Nussbaum of Vienna had proposed such an undertaking. Du Bois wrote of this to Alfred Harcourt on August 10, 1928, but it did not eventuate.

I have planned a series of novels on the Negro race; a sort of short black Comedie Humaine. "Dark Princess" was the first and I am working on a second, "Bethesda, A.M.E.," which is the 100 years' history of a Negro church. I have others more or less definitely in mind. The question is: is there a public for this work sufficiently large to pay for its publication? My own Negro public is growing, but it is still too small to be depended on for more than a small part of the total support.[19]

Du Bois closed by declaring he would consider it "a great favor" if Harcourt would think this over and discuss it with others and then, "if you have the time, I could come in and talk with you." Perhaps this discussion took place. In any case, Du Bois was not to publish another novel until, in the ninth decade of his life, he produced the trilogy, *The Black Flame.*

* * *

Dark Princess was widely reviewed in the United States. The notices in the white press were mixed; those of particular interest are summarized herewith.

A very young Tess Slesinger, having just obtained her bachelor's degree at Columbia University, gave the novel a one-paragraph notice in the *New York Post*, May 12, 1928. "Mr. Du Bois," she wrote, "in this novel of a negro's struggle in a white world, has unfortunately adopted white ideals, which he confuses with dress shirts and a parlor knowledge of the arts." Having disposed of this "confusion," the reviewer went on to correct another confusion suffered by Du Bois; he mistakenly thought that "natives of India" were colored. Still, "Occasionally the very real problems rouse our indignant sympathy." However, and this is the conclusion: "But the truth is obscured by sentimental melodrama, in which Mr. Du Bois himself seems to find a consoling ration of sour grapes."[20] That a twenty-two year old white youngster would write this

[19]In terms of interest in *Dark Princess*, on July 23, 1928, Harcourt wrote to Du Bois that the Micheaux Film Corporation had offered "to buy the Negro photoplay rights" to the novel for $350. Du Bois on August 6 replied favorably, wanted some control over the content of the product and suggested that it would be wise to ask for payment in advance. Further information on this has not been found but in the account of Oscar Micheaux in *Who's Who in Colored America, 1930-32* (3rd edition, Brooklyn, N.Y., 1933), p. 302 "Dark Princess" is included among the films he had produced. Micheaux (1884-1951), born in Illinois, had been a Pullman porter, farmer, rancher, author and movie producer. His "Homesteader" seems to have been the first full-length all-Black motion picture.

[20]The review is initialed "T.S." but other reviews signed Tess Slesinger appear in this paper during the same period. Tess Slesinger (later Mrs. Henry Solow) was born in 1905 and received a bachelor's degree in 1927. Simon & Schuster published a book of her stories and a novel by her; she is best known as a screenwriter, having collaborated in this work on "A Tree Grows in Brooklyn," "The Good Earth," "Remember the Day," and other successes.

sort of thing in the United States almost fifty years ago is perhaps not too startling, but that the editors of the *Post* would print it is somewhat surprising.

Of an altogether different order was the rather long review given the book by Robert O. Ballou, in the *Chicago News*, May 9, 1928. Aspects of the work he found puzzling, and its greatest weakness was "a too great romanticism." But there was much more to the book, Mr. Ballou thought: its penetrating picture of Chicago's politics—this in a Chicago newspaper—and its many pages containing "an immense and haunting beauty," as for example those (pp. 264-266) describing the building of the subway. Above all, concluded Ballou, the novel was new in its scope and sweep and the way in which the life and work of a Black man was placed within the context of life in the United States as a whole.[21]

An exceedingly thoughtful—and favorable—review of the book appeared in the *Rocky Mountain News* of Denver, May 20, 1928. The reviewer—one George V. Burns, not otherwise identified—thought the novel "a remarkable presentation of the desires of the Negro race, of all colored races"; it dealt, however, he wrote, "with the basic elements of life in such a way as to make it universal." A summary and long quotations follow and then this conclusion:

> It is a searching exposition of that age-old cry of man for an answer to life, not alone of the dominating white man, but of the man who not only must solve the forces of life, but also must contend with the restrictions imposed on him by fellow human beings.

A personal note was struck in a review published in a small-town mid-Western newspaper, the Bloomington, Indiana *Pantagraph*, May 13, 1928. The writer was the Rev. Rupert Holloway:[22]

> It happened a year or two ago that the author of this book came to Bloomington to deliver a lecture. The reviewer found difficulty in obtaining hotel accommodation for him, because Dr. Du Bois happens to be a Negro.

The point of the book, said this reviewer, was to bring to its reader some awareness of "the movement among the dark races for self-expression and self-determination." After a summary of the novel's plot, this reviewer remarked on the "rare wizardry" of Du Bois' writings

---

[21] Mr. Ballou (1892) was editor of the *Ben Franklin Monthly* in Chicago (1922-1927), author of several books—especially in religion—and a publisher.

[22] The Bobbs-Merrill Company in Indianapolis published *The Unknown Garden* by the Rev. Mr. Holloway in 1941.

and that—with his earlier *Darkwater*—it seemed to "express the aspiration of the author's race."

Another churchman, Hubert C. Herring, writing in *The Congregationalist* (July 12, 1928), thought the book—which he mistitled *Black Princess*—"comes close to being a great novel." Certainly, he wrote, Du Bois "tears his pages from his own bitter soul." Mr. Herring noted especially the novel's account of radicalism among Black people and thought that its attractiveness lay in the social evils of the nation; he added: "the fact that more Negroes refuse to listen to their honeyed words reflects great credit upon their innate good sense." The book, concluded the reviewer, was a good one for every churchman, indeed, "for everyone who hopes that the Church may take the leadership in making America a land undimmed by human tears."[23]

Several of the reviewers saw the book, even more explicitly than did Mr. Herring, as offering a warning to the dominant white world. This is the main point, for example, in the book's review in the *Annals* of the American Academy of Political and Social Science, November, 1928 (140: 347-348), by Jane Reitell.[24] As befits the periodical for which she was writing, the reviewer stressed that "to the student of sociology, the facts in the book are more interesting than the fancies"; she suggested, too, that the novel "would have gained in effectiveness if the author had restrained his sense of the dramatic a trifle." She thought that Du Bois' including Chinese, Japanese, Hindu and Black people "under one color badge is stretching a point, and it is a point to which the races themselves might take serious objection." Still, the reviewer returned to "the impressiveness of this scholarly book." For herself, she found that "after reading it one has the feeling that the much written of white man's burden is sentimental twaddle, but that the black man's burden of suffering is a cross which sometimes must be borne on the shoulders of humanity."

Similarly, Charles C. Vance, reviewing the book in *The West Virginian*, published in Fairmont (May 10, 1928), after summarizing its plot, tells the reader: "His book is a strong one and, whatever your beliefs on the race question are, it is one that will at least waken your mind to the problems that must be solved within the next few years."[25]

---

[23] Hubert Clinton Herring, born in Iowa in 1889, was a Congregational minister in Wisconsin Kansas, 1913-1924 and director of social action activities of the National Congregational Church, 1924-1929. Later his interest turned to Latin America and his books on this area published in the 1940s and 1950s are well known.

[24] Jane Reitell was co-author (with Orton Lowe) of *Pennsylvania, A Story of our Domain*, published in Boston in 1927 by R.G. Badger; further information has not been obtained.

[25] The review is signed C.C.V., but in the Du Bois Papers, this is identified as above; no information on Mr. Vance has been found.

In some ways, one of the most penetrating contemporary reviews, and most glowing, too, came from one signing himself or herself C.G.; this appeared in the *Honolulu Star Bulletin* of June 16, 1928. This review emphasizes Du Bois' "poetic intensity" in its discussion of the novel's form; as to its content it sees that the main point is the dramatization of the world-wide drive on the part of colonial peoples and those racially oppressed for "self-expression and self-determination." Furthermore, the reviewer sees in the book "Du Bois' old white-hot indignation against racial oppression and also his fierce pride in the beauty and deep tragic joy and rich loveliness of the dark peoples."

*   *   *

Interest naturally attaches to the response from some of the more potent periodicals of the period.

The reviewer in *The New Republic* (August 22, 1928; 56: 27) on the whole reacted positively to *Dark Princess*; he made a point that bears repeating: "It is worth noting that this author is one of the few who can write about minority and 'queer' races as if they were men rather than types. He is conscious of the individual as the individual is of himself."[26]

The reviewer in the *New York Sun* (May 19, 1928)—then an important newspaper—found the book to be an expression "of negro aspiration with a chip of provocation well in evidence on its shoulder." As to the reflection of discontent among the colored peoples of the world and the projection of efforts to overcome subjugation, the reviewer was certain "that many, many negroes will be quietly going home to Harlem blissfully unaware" of all this travail and possible future glory. All this is discerned, says the reviewer "when viewed by a naked and unprejudiced white eye"—such as he possesses![27]

In the book section of the Saturday edition of the Boston *Evening Transcript* (July 14, 1928), a more insightful estimate is given. We are offered, says this reviewer, "a too melodramatic story" but its "great value lies in the revelation it gives to white readers of many grades of negro life and many problems which are totally unknown to them." Moreover, "What impresses one about 'Dark Princess' is the author's intention to link the American negro with all the other peoples of the

---

[26]The review is initialed M.P.L. This is almost certainly M.P. Levy who occasionally reviewed for this magazine and was the author of novels, such as *Wedding* and *Matrix*.

[27]This was Frederick H. Martens (1874-1932), privately educated in New York City and editor and author in the field of music.

world who are not distinctly white. For this reason the book demands a serious reading."[28]

The book review section of the Sunday *New York Times* (May 13, 1928), in an unsigned review of five paragraphs, uses four to summarize the plot—with touches of what presumably was meant to be irony—and then concludes:

> The book is well written, but there is enough material in it for several novels, and the plot is flamboyant and unconvincing. The theme of miscegenation is so decidedly controversial that to those who believe in the preservation of racial purity the book will fail to lessen their prejudice. There is, however, real meat in the "Dark Princess" and such proof of the author's power that it seems a pity he is not using his talent to show the natural ability of the colored man or his nobility of character, as in "Porgy," rather than to dwell, oversensitively, on social injustices which are inevitable in any period of racial transition and development—of white or black.

* * *

Notice of *Dark Princess* in white Border State and southern newspapers was sparse and intensely hostile. The reviewer—A. B. Bernd—in the Macon (Georgia) *Telegraph*, June 3, 1928, took a strange tack to open his remarks. The literature "of the new Negro" tends to be completely "given over to propaganda" but this novel by Du Bois seeks "not the furthering of a cause but the telling of a story." "Possibly," says this perceptive person, Du Bois' "original intent was to gild the philosophic pill of dusky solidarity"; but, if that was so, "he has wielded the gilt-brush so lavishly and so haphazardly that what is left of the kernel is robbed of all efficacy."

The reviewer—Rachel Dyas—of the book for the Asheville (North Carolina) *Times*, July 29, 1928, had no doubt that propaganda was its nature: "The book is intended as a terrific indictment of the white world for its disregard of the black. But so blurred are the outlines, so trivial the real gist, that the story blubbers down to drivelings of incoherent hate." The book, indeed, "illustrates the truth of the worst possible arraignment ever uttered against the negro as a race." Possibly Du Bois may be "considered as a producer of literature" but it was of the order just described. In general: "There are many highly acclaimed negro writers, nowadays. Black is a fashionable color in belles lettres. There are also flea circuses. Both are interesting but scarcely art."

---

[28]The review is signed "D.L.M." This almost certainly is Dorothea Lawrence Mann, whose full name is occasionally found in this paper as a reviewer in this period. She wrote for such magazines as *Bookman* and *Harpers.* In 1927 Doubleday, Doran published her *Ellen Glasgow.*

Not quite as venomous was the review in the St. Louis *Globe-Democrat* of June 9, 1928. The plot is more or less summarized and then appears this conclusion: "Throughout the book there is a petulance and an arrogance that his characters try to justify but cannot, since they are but spokesmen for Mr. Du Bois behind them."[29]

* * *

Writers in the Black press responded enthusiastically to *Dark Princess*. George S. Schuyler, in the "Chats About Books" column that he then regularly conducted, thought that "In this novel Dr. Du Bois is at his best, which is a whole lot better than the best of nine-tenths of the white and black writers of today." Alice Dunbar-Nelson, who published a column called "As In A Looking Glass," devoted most of one to the novel, which she found "complete and eminently soul-satisfying." Du Bois was especially successful in his depiction of Sara, this reviewer thought. She knew Du Bois and added this comment: "Dr. Du Bois is always somewhat bitter, even though the advancing years are tempering his acidity with some of the alkali of pity for the non-understanding, and he can more gracefully act the showman's part than he could some years ago." The main thing, she added, however, was the story and in *Dark Princess* one had a gripping and dynamic tale that, once begun, could not be put down. All in all, Alice Dunbar-Nelson found the novel "a soul-satisfying piece of creative writing."[30]

An unsigned reviewer in the magazine published at Du Bois' first college—*The Greater Fisk Herald*—recommended the novel as "refreshingly different in that it wanders from the 'inevitable Harlem'." Melodramatic treatment hurt the book, this reviewer thought, but he ended as he had begun—"It lets Harlem rest for a season and this alone should make the 'Dark Princess' a best-seller"—alas, an erroneous prophecy.

Aubrey Bowser, reviewing the book at length in the New York

---

[29]The reviewer was Eda Heijermans Houwink; she was the author of a pamphlet, "The Place of Case Work in the Public Assistance Program," published in Chicago in 1941 by the American Public Welfare Association. At that time she was field work supervisor, School of Social Service Administration, University of Chicago; she was a graduate of Washington University in St. Louis.

[30]Schuyler's review appeared in the Little Rock, *Arkansas Survey*, November 11, 1928; Alice Dunbar-Nelson's in the *Washington Eagle*, May 11, 1928. Alice Dunbar-Nelson (1875-1935), for a time the wife of Paul Laurence Dunbar, was an author, editor and social worker of distinction and headed the English department of a high school in Delaware for eighteen years. She was active in anti-war and anti-jim crow struggles. George S. Schuyler (born 1895), author, editor and leading columnist for many Black newspapers, has lately been identified with far-Right organizations in the United States.

Amsterdam News, May 9, 1928, began with a perceptive paragraph recalling that it was Du Bois who had successfully challenged the Tuskegee Machine and that it was Du Bois whose work and life were fountains from which gushed the then-current Harlem Renaissance. Mr. Bowser continued that he had thought Du Bois' earlier novel, *The Quest of the Silver Fleece* (1911) was "good, even beautiful," but "it does not compare with *Dark Princess.*" The latter, he felt, "is the first Negro novel of any considerable scope" and "after all the garbage that has been offered in the name of Negro literature it is refreshing" to be able to turn to it. A long summary of the book follows, with the injunction to the reader that "you cannot afford to miss it."[31]

Allison Davis reviewed the book at length in Du Bois' own *Crisis* (October, 1928; 35: 339-340). For him, "Courage is the real theme of *Dark Princess*, courage emphasized by symbols which are more representative than any yet chosen from Negro life." But it is a mature courage filled with the experience of sorrow, he continues:

> Courage and the discipline of suffering must be equally effective in moulding the new Negro. There is a great suffering behind us. We should be unified by this common tragedy. It is this last note which makes *Dark Princess* a sane and balanced work, purged of any rash and strident 'aggressiveness.' In fact, Dr. Du Bois has revealed the futility of mere 'protest' and 'aggressiveness' by the satiric but tragic portrait of Perigua, who ended by blowing himself out of the 'problem' with dynamite, while the Kluxers he meant to kill were unhurt.

The exposure of the meanness and hollowness of the Black bourgeoisie is a feature of the book, Davis remarks; he especially commends the portrait of Sara. He notes that his own Black college students have told him that they do not read Black authors and he is certain that the richer Black people also do not; "Consequently Negro books are written for white readers, just as Negro schools are run for white patrons, and Negro spirituals are sung in New York to please white audiences." In the face of this, Davis concludes: "*Dark Princess* will not appeal to the same public which enjoyed *Nigger Heaven* and *Home to Harlem*. All those

---

[31] Aubrey Bowser graduated from Harvard in 1907; for a time, beginning in 1919, he edited a weekly magazine, *The Rainbow*, in New York City. He was the son-in-law of T. Thomas Fortune; see Emma L. Thornbrough's biography of Fortune (University of Chicago Press, 1972), p. 350. Dewey R. Jones, reviewing the book in the *Chicago Defender*, August 1, 1928, was much less impressed with it, though he ended: "It is a book worth reading and you'll enjoy it."

who have a high faith in the destiny and nature of the Negro, therefore, ought to read it."[32]

In *The Southern Workman*, published by Hampton Institute in Virginia, the review is positive though somewhat hesitant in terms of the militancy manifested by Du Bois. The reviewer manages not to convey a central theme of the novel—worldwide unity of colored peoples in opposition to white racist domination. He puts it, rather, in terms of a depiction of the stymied effort of "the Negro in his uphill climb to success by way of the leading professions"—which was more descriptive of the vision of Hampton than of *Dark Princess*! Still, the novel is recommended, for those "who read this book intelligently and sympathetically must think a little more seriously and constructively" about the "way out."[33]

Perhaps the most thoughtful and thought-provoking review came from the pen of Alain Locke, who, with Du Bois and James Weldon Johnson, formed the leading triumvirate among Black intelligentsia and who, perhaps in some ways even more than Dr. Du Bois, was the inspirer of the younger members of the Harlem Renaissance movement. Locke's review[34] is fairly long; perhaps the heart of it is in this passage:

> We must thank the talented author of this book for breaking ground for this skyscraper problem novel of the Negro intellectual and the world radical. Chiefly because of the still latent possibilities of the theme it is to be regretted that, as a novel, 'Dark Princess' is not wholly successful. As a document, however, it should be widely read.

Locke concluded that, though "half spoiled by oversophistication," the novel "will be interesting and revealing reading to the white reader who has yet few ways of looking into the many closed chambers of Negro life or of seeing into the dilemmas of the intellectual Negro

---

[32] Allison Davis, a graduate of Amherst and then teaching at Morgan State College in Baltimore, contributed a significant essay, "Our Negro 'Intellectuals'," to *The Crisis*, August, 1928; here he expatiates on the views only broached in the above review.

[33] The review is in the issue of August, 1928, pp. 334-335; it is initialled W.A.A., meaning William Anthony Aery, a graduate of Columbia (1904) and at this time director of the School of Education at Hampton Institute.

[34] Locke (1886-1954) the first Black Rhodes Scholar, taught for many years at Howard University. He wrote widely on themes of music, art and aesthetics and edited the very influential book, *The New Negro* (1925)—to which Du Bois contributed. His review appeared in New York *Herald Tribune Books*, May 20, 1928. Locke sent a manuscript copy of the review to Du Bois with a covering letter dated May 13, 1928, closing: "I hope it also reflects in a small measure my great personal respect for your work as a writer and your services for cultural self-expression among us."

mind and heart." Further, for Black readers, Locke wrote, the novel was "an intriguing problem study of the cross currents and paradoxes of the thinking Negro who today faces the problems of two worlds and two loyalties."[35]

* * *

At various times in his marvelously full life, Dr. Du Bois spoke with special fondness of certain of his books—among them *The Souls of Black Folk*, his life of *John Brown* and—quite often—*Dark Princess: A Romance*, which appeared soon after his sixtieth birthday and which conveyed both his "Dream of the Spirit" and "the Pain of the Bone."

February, 1974   Herbert Aptheker

---

[35] Of some interest is the review of the novel by Leon Whipple, in *Opportunity*, the organ of the National Urban League, in August, 1928 (6: 244). Whipple, a white man, was born in 1882 in St. Louis and taught English for some time at the University of Missouri and University of Virginia. He then entered journalism and served with leading Virginia newspapers; thereafter he was a professor of journalism at New York University for many years. He was very active in civil liberties efforts and his *Story of Civil Liberty in the United States* (1927) remains important. He died in New York City in 1964. Whipple thought Du Bois "has undertaken too heavy a load for even his real talents." He then summarized the novel's action and offered some opinions as to its alleged "theses."

DARK PRINCESS

DARK PRINCESS

A Romance

By

W. E. BURGHARDT DU BOIS
Author of "Darkwater"

New York
HARCOURT, BRACE AND COMPANY

PRINTED IN THE U. S. A. BY
QUINN & BODEN COMPANY, INC.
RAHWAY, N. J.

To

Her High Loveliness

TITANIA XXVII

BY HER OWN GRACE

QUEEN OF FAERIE,

Commander of the Bath; Grand Medallion of
Merit; Litterarum Humanarum Doctor; Fidei
Extensor; etc., etc.

OF WHOSE FAITH AND FOND AFFECTION

THIS ROMANCE

WAS SURELY BORN

CONTENTS

PART I
THE EXILE

PART I

THE EXILE

August, 1923

Summer is come with bursting flower and promise of perfect fruit. Rain is rolling down Nile and Niger. Summer sings on the sea where giant ships carry busy worlds, while mermaids swarm the shores. Earth is pregnant. Life is big with pain and evil and hope. Summer in blue New York; summer in gray Berlin; summer in the red heart of the world!

I

MATTHEW TOWNS was in a cold white fury. He stood on the deck of the *Orizaba* looking down on the flying sea. In the night America had disappeared and now there was nothing but waters heaving in the bright morning. There were many passengers walking, talking, laughing; but none of them spoke to Matthew. They spoke about him, noting his tall, lean form and dark brown face, the stiff, curled mass of his sinewy hair, and the midnight of his angry eyes.

They spoke about him, and he was acutely conscious of every word. Each word heard and unheard pierced him and quivered in the quick. Yet he leaned stiff and grim, gazing into the sea, his back toward all. He saw the curled grace of the billows, the changing blues and greens; and he saw, there at the edge of the world, certain shining shapes that leapt and played.

Then they changed—always they changed; and there arose the great cool height of the room at the University of Manhattan. Again he stood before the walnut rail that separated student and Dean. Again he felt the bewilderment, the surge of hot surprise.

"I cannot register at all for obstetrics?"

3

"No," said the Dean quietly, his face growing red. "I'm sorry, but the committee—"

"But—but—why, I'm Towns. I've already finished two years—I've ranked my class. I took honors—why—I— This is my Junior year—I must—"

He was sputtering with amazement.

"I'm sorry."

"Hell! I'm not asking your pity, I'm demanding—"

The Dean's lips grew thin and hard, and he sent the shaft home as if to rid himself quickly of a hateful task.

"Well—what did you expect? Juniors must have obstetrical work. Do you think white women patients are going to have a nigger doctor delivering their babies?"

Then Matthew's fury had burst its bounds; he had thrown his certificates, his marks and commendations straight into the drawn white face of the Dean and stumbled out. He came out on Broadway with its wide expanse, and opposite a little park. He turned and glanced up at the gray piles of tan buildings, threatening the sky, which were the University's great medical center. He stared at them. Then with bowed head he plunged down 165th Street. The gray-blue Hudson lay beneath his feet, and above it piled the Palisades upward in gray and green. He walked and walked: down the curving drive between high homes and the Hudson; by graveyard and palace; tomb and restaurant; beauty and smoke. All the afternoon he walked, all night, and into the gray dawn of another morning.

II

In after years when Matthew looked back upon this first sea voyage, he remembered it chiefly as the time of sleep; of days of long, long rest and thought, after work and hurry and rage. He was indeed very tired. A year of the hardest kind of study had been followed by a summer as clerical assistant in a colored industrial insurance office, in the heat of Washington. Thence he had hurried straight to the university with five hundred dollars of tuition money in his pocket; and now he was sailing to Europe.

He had written his mother—that tall, gaunt, brown mother, hard-sinewed and somber-eyed, carrying her years unbroken, who still toiled on the farm in Prince James County, Virginia—

he had written almost curtly: "I'm through. I cannot and will not stand America longer. I'm off. I'll write again soon. Don't worry. I'm well. I love you." Then he had packed his clothes, given away his books and instruments, and sailed in mid-August on the first ship that offered after that long tramp of tears and rage, after days of despair. And so here he was.

Where was he going? He glanced at the pale-faced man who asked him. "I don't know," he answered shortly. The good-natured gentleman stared, nonplused. Matthew turned away. Where *was* he going? The ship was going to Antwerp. But that, to Matthew, was sheer accident. He was going *away* first of all. After that? Well, he had thought of France. There they were at least civilized in their prejudices. But his French was poor. He had studied German because his teachers regarded German medicine as superior to all other. He would then go to Germany. From there? Well, there was Moscow. Perhaps they could use a man in Russia whose heart was hate. Perhaps he would move on to the Near or Far East and find hard work and peace. At any rate he was going somewhere; and suddenly letting his strained nerves go, he dragged his chair to a sheltered nook apart and slept.

To the few who approached him at all, Matthew was boorish and gruff. He knew that he was unfair, but he could not help it. All the little annoyances, which in healthier days he would have laughed away, avoided, or shortly forgotten, now piled themselves on his sore soul. The roommate assigned him discovered that his companion was colored and quickly decamped with his baggage. A Roumanian who spoke little English, and had not learned American customs, replaced him.

Matthew entered the dining-room with nerves a-quiver. Every eye caught him during the meal—some, curiously; some, derisively; some, in half-contemptuous surprise. He felt and measured all, looking steadily into his plate. On one side sat an old and silent man. To the empty seat on the other side he heard acutely a swish of silk approach—a pause and a consultation. The seat remained empty. At the next meal he was placed in a far corner with people too simple or poor or unimportant to protest. He heaved a sigh to have it over and ate thereafter in silence and quickly.

So at last life settled down, soothed by the sea—the rhythm and song of the old, old sea. He slept and read and slept;

stared at the water; lived his life again to its wild climax; put down repeatedly the cold, hard memory; and drifting, slept again.

Yet always, as he rose from the deep seas of sleep and reverie, the silent battle with his fellows went on. Now he yearned fiercely for some one to talk to; to talk and talk and explain and prove and disprove. He glimpsed faces at times, intelligent, masterful. They had brains; if they knew him they would choose him as companion, friend; but they did not know him. They did not want to know him. They glanced at him momentarily and then looked away. They were afraid to be noticed noticing him.

And he? He would have killed himself rather than have them dream he would accept a greeting, much less a confidence. He looked past and through and over them with blank unconcern. So much so that a few—simpler souls, themselves wandering alone hither and thither in this aimless haphazard group of a fugitive week—ventured now and then to understand: "I never saw none of you fellers like you—" began one amiable Italian. "No?" answered Matthew briefly and walked away.

"You're not lonesome?" asked a New England merchant, adding hastily, "I've always been interested in your people."

"Yes," said Matthew with an intonation that stopped further conversation along that line.

"No," he growled at an insulted missionary, "I don't believe in God—never did—do you?"

And yet all the time he was sick at heart and yearning. If but one soul with sense, knowledge, and decency had firmly pierced his awkwardly hung armor, he could have helped make these long hard days human. And one did, a moment, now and then—a little tow-haired girl of five or six with great eyes. She came suddenly on him one day: "Won't you play ball with me?" He started, smiled, and looked down. He loved children. Then he saw the mother approaching—typically Middle-West American, smartly dressed and conscious of her social inferiority. Slowly his smile faded; quickly he walked away. Yet nearly every day thereafter the child smiled shyly at him as though they shared a secret, and he smiled slowly back; but he was careful never to see the elaborate and most exclusive mother.

Thus they came to green Plymouth and passed the fortress walls of Cherbourg and, sailing by merry vessels and white cliffs, rode on to the Scheldt. All day they crept past fields and villages, ships and windmills, up to the slender cathedral tower of Antwerp.

III

Sitting in the Viktoria Café, on the Unter den Linden, Berlin, Matthew looked again at the white leviathan—at that mighty organization of white folk against which he felt himself so bitterly in revolt. It was the same vast, remorseless machine in Berlin as in New York. Of course, there were differences—differences which he felt like a tingling pain. He had on the street here no sense of public insult; he was treated as he was dressed, and today he had dressed carefully, wearing the new suit made for the opening school term; he had on his newest dark crimson tie that burned with the red in his smooth brown face; he carried cane and gloves, and he had walked into this fashionable café with an air. He knew that he would be served, politely and without question.

Yes, in Europe he could at least eat where he wished so long as he paid. Yet the very thought made him angry; conceive a man outcast in his own native land! And even here, how long could he pay, he who sat with but two hundred dollars in the world, with no profession, no work, no friends, no country? Yes, these folks treated him as a man—or rather, they did not, on looking at him, treat him as less than a man. But what of it? They were white. What would they say if he asked for work? Or a chance for his brains? Or a daughter in marriage? There was a blonde and blue-eyed girl at the next table catching his eye. Faugh! She was for public sale and thought him a South American, an Egyptian, or a rajah with money. He turned quickly away.

Oh, he was lonesome; lonesome and homesick with a dreadful homesickness. After all, in leaving white, he had also left black America—all that he loved and knew. God! he never dreamed how much he loved that soft, brown world which he had so carelessly, so unregretfully cast away. What would he not give to clasp a dark hand now, to hear a soft Southern roll of speech, to kiss a brown cheek? To see warm, brown,

crinkly hair and laughing eyes. God—he was lonesome. So
utterly, terribly lonesome. And then—he saw the Princess!

Many, many times in after years he tried to catch and
rebuild that first wildly beautiful phantasy which the girl's
face stirred in him. He knew well that no human being could
be quite as beautiful as she looked to him then. He could never
quite recapture the first ecstasy of the picture, and yet always
even the memory thrilled and revived him. Never after that
first glance was he or the world quite the same.

First and above all came that sense of color: into this
world of pale yellowish and pinkish parchment, that absence
or negation of color, came, suddenly, a glow of golden brown
skin. It was darker than sunlight and gold; it was lighter and
livelier than brown. It was a living, glowing crimson, veiled
beneath brown flesh. It called for no light and suffered no
shadow, but glowed softly of its own inner radiance.

Then came the sense of the woman herself: she was young
and tall even when seated, and she bore herself above all
with a singularly regal air. She was slim and lithe, gracefully
curved. Unseeing, past him and into the struggling, noisy
street, she was looking with eyes that were pools of night—
liquid, translucent, haunting depths—whose brilliance made
her face a glory and a dream.

Matthew pulled himself together and tried to act sensibly.
Here—here in Berlin and but a few tables away, actually sat
a radiantly beautiful woman, and she was colored. He could
see the faultlessness of her dress. There was a hint of something foreign and exotic in her simply draped gown of rich,
creamlike silken stuff and in the graceful coil of her hand-
fashioned turban. Her gloves were hung carelessly over her
arm, and he caught a glimpse of slender-heeled slippers and
sheer clinging hosiery. There was a flash of jewels on her
hands and a murmur of beads in half-hidden necklaces. His
young enthusiasm might overpaint and idealize her, but to the
dullest and the oldest she was beautiful, beautiful. Who was
she? What was she? How came this princess (for in some sense
she must be royal) here in Berlin? Was she American? And
how was he—

Then he became conscious that he had been listening to
words spoken behind him. He caught a slap of American
English from the terrace just back and beyond.

"Look, there's that darky again. See her? Sitting over yonder by the post. Ain't she some pippin? What? Get out! Listen! Bet you a ten-spot I get her number before she leaves this café. You're on! I know niggers, and I don't mean perhaps. Ain't I white. Watch my smoke!"

Matthew gripped the table. All that cold rage which still lay like lead beneath his heart began again to glow and burn. Action, action, it screamed—no running and sulking now—action! There was murder in his mind—murder, riot, and arson. He wanted just once to hit this white American in the jaw—to see him spinning over the tables, and then to walk out with his arm about the princess, through the midst of a gaping, scurrying white throng. He started to rise, and nearly upset his coffee cup.

Then he came to himself. No—no. That would not do. Surely the fellow would not insult the girl. He could count on no public opinion in Berlin as in New York to shield him in such an adventure. He would simply seek to force his company on her in quite a natural way. After all, the café was filling. There were no empty tables, at least in the forward part of the room, and no one person had a right to a whole table; yet to approach any woman thus, when several tables with men offered seats, was to make a subtle advance; and to approach this woman?—puzzled and apprehensive, Matthew sat quietly and watched while he paid his waiter and slowly pulled on his gloves. He saw a young, smooth-faced American circle carelessly from behind him and saunter toward the door. Then he stopped, and turning, slowly came back toward the girl's table. A cold sweat broke out over Matthew. A sickening fear fought with the fury in his heart. Suppose this girl, this beautiful girl, let the fresh American sit down and talk to her? Suppose? After all, who—what was she? To sit alone at a table in a European café—well, Matthew watched. The American approached, paused, looked about the café, and halted beside her table. He looked down and bowed, with his hand on the back of the empty chair.

The lady did not start nor speak. She glanced at him indifferently, unclasped her hands slowly, and then with no haste gathered up her things; she nodded to the waiter, fumbled in her purse, and without another glance at the American, arose and passed slowly out. Matthew could have shouted.

But the American was not easily rebuffed by this show of indifference. Apparently he interpreted the movement quite another way. Waving covertly to his fellows, he arose leisurely, without ordering, tossed a bill to the waiter, and sauntered out after the lady. Matthew rose impetuously, and he felt that whole terrace table of men arise behind him.

The dark lady had left by the Friedrichstrasse door, and paused for the taxi which the gold-laced porter had summoned. She gave an address and already had her foot on the step. In a moment the American was by her side. Deftly he displaced the porter and bent with lifted hat. She turned on him in surprise and raised her little head. Still the American persisted with a smile, but his hand had hardly touched her elbow when Matthew's fist caught him right between the smile and the ear. The American sat down on the sidewalk very suddenly.

Pedestrians paused. There was a commotion at the restaurant door as several men rushed out, but the imposing porter was too quick; he had caught the eye and pocketed the bill of the lady. In a moment, evidently thinking the couple together, he had handed both her and Matthew into the taxi, slammed the door, and they had whirled away. In a trice they fled down the Friedrichstrasse, left across the Französische, again left to Charlotten, and down the Linden. Matthew glanced anxiously back. They had been too quick, and there was apparently no pursuit. He leaned forward and spoke to the chauffeur, and they drove up to the curb near the Brandenburg Gate and stopped.

"Mille remerciements, Monsieur!" said the lady.

Matthew searched his head for the right French answer as he started to step out, but could not remember: "Oh—oh— don't mention it," he stammered.

"Ah—you are English? I thought you were French or Spanish!"

She spoke clear-clipped English with perfect accent, to Matthew's intense relief. Suppose she had spoken only French! He hastened to explain: "I am an American Negro."

"An American Negro?" The lady bent forward in sudden interest and stared at him. "An American Negro!" she repeated. "How singular—how very singular! I have been thinking of American Negroes all day! Please do not leave me yet. Can you spare a moment? Chauffeur, drive on!"

IV

As they sat at tea in the Tiergarten, under the tall black trees, Matthew's story came pouring out:

"I was born in Virginia, Prince James County, where we black folk own most of the land. My mother, now many years a widow, farmed her little forty acres to educate me, her only child. There was a good school there with teachers from Hampton, the great boarding-school not far away. I was young when I finished the course and was sent to Hampton. There I was unhappy. I wanted to study for a profession, and they insisted on making me a farmer. I hated the farm. My mother finally sent me North. I boarded first with a cousin and then with friends in New York and went through high school and through the City College. I specialized on the pre-medical course, and by working nights and summers and playing football (amateur, of course, but paid excellent 'expenses' in fact), I was able to enter the new great medical school of the University of Manhattan, two years ago.

"It was a hard pull, but I plunged the line. I had to have scholarships, and I got them, although one Southern professor gave me the devil of a time."

The lady interrupted. "Southern?" she asked. "What do you mean by 'Southern'?"

"I mean from the former slave States—although the phrase isn't just fair. Some of our most professional Southerners are Northern-born."

The lady still looked puzzled, but Matthew talked on.

"This man didn't mean to be unfair, but he honestly didn't believe 'niggers' had brains, even if he had the evidence before him. He flunked me on principle. I protested and finally had the matter up to the Dean; I showed all my other marks and got a re-examination at an extra cost that deprived me of a new overcoat. I gave him a perfect paper, and he had to acknowledge it was 'good,' although he made careful inquiries to see if I had not in some way cribbed dishonestly.

"At last I got my mark and my scholarship. During my second year there were rumors among the few colored students that we would not be allowed to finish our course, because objection had been made to colored students in the clinical hospital, especially with white women patients. I laughed. It was, I was sure, a put-up rumor to scare us off. I knew black men who had gone through other New York medical schools which had become parts of this great new consolidated school. There had been no real trouble. The patients never objected— only Southern students and the few Southern professors. Some of the trustees had mentioned the matter but had been shamed into silence.

"Then, too, I was firm in my Hampton training; desert and hard work were bound to tell. Prejudice was a miasma that character burned away. I believed this thoroughly. I had literally pounded my triumphant way through school and life. Of course I had met insult and rebuff here and there, but I ignored them, laughed at them, and went my way. Those black people who cringed and cowered, complained of failure and 'no chance,' I despised—weaklings, cowards, fools! Go to work! Make a way! Compel recognition!

"In the medical school there were two other colored men in my class just managing to crawl through. I covertly sneered at them, avoided them. What business had they there with no ability or training? I see differently now. I see there may have been a dozen reasons why Phillips of Mississippi could neither spell nor read correctly and why Jones of Georgia could not count. They had had no hard-working mother, no Hampton, no happy accidents of fortune to help them on.

"While I? I rose to triumph after triumph. Just as in college I had been the leading athlete and had ridden many a time aloft on white students' shoulders, so now, working until two o'clock in the morning and rising at six, I took prize after prize—the Mitchel Honor in physiology, the Welbright medal in pathology, the Shores Prize for biological chemistry. I ranked the second-year class at last commencement, and at our annual dinner at the Hotel Pennsylvania, sat at the head table with the medal men. I remember one classmate. He was from Atlanta, and he hesitated and whispered when he found his seat was beside me. Then he sat down like a man and held out his hand. 'Towns,' he said, 'I never associated with a Negro before who wasn't a servant or laborer; but I've heard

of you, and you've made a damned fine record. I'm proud to
sit by you.'

"I shook his hand and choked. He proved my life-theory.
Character and brains were too much for prejudice. Then the
blow fell. I had slaved all summer. I was worked to a frazzle.
Reckon my hard-headedness had a hand there, too. I wouldn't
take a menial job—Pullman porter, waiter, bell-boy, boat
steward—good money, but I waved them aside. No! Bad for
the soul, and I might meet a white fellow student."

The lady smiled. "Meet a fellow student—did none of them
work, too?"

"O yes, but seldom as menials, while Negroes in America
are always expected to be menials. It's natural, but—no, I
couldn't do it. So at last I got a job in Washington in the
medical statistics department of the National Benefit. This is
one of our big insurance concerns. O yes, we've got a number
of them; prosperous, too. It was hard work, indoors, poor light
and air; but I was interested—worked overtime, learned the
game, and gave my thought and ideas.

"They promoted me and paid me well, and by the middle of
August I had my tuition and book money saved. They wanted
me to stay with them permanently; at least until fall. But I
had other plans. There was a summer school of two terms at
the college, and I figured that if I entered the second term I
could get a big lead in my obstetrical work and stand a better
show for the Junior prizes. I had applied in the spring for
admission to the Stern Maternity Hospital, which occupied
three floors of our center building. My name had been posted
as accepted. I was tired to death, but I rushed back to New
York to register. Perhaps if I had been rested, with cool head
and nerves—well, I wasn't. I made the office of the professor
of obstetrics on a hot afternoon, August 10, I well remember.
He looked at me in surprise.

" 'You can't work in the Stern Hospital—the places are all
taken.'

" 'I have one of the places,' I pointed out. He seemed puzzled
and annoyed.

" 'You'll have to see the Dean,' he said finally.

"I was angry and rushed to the Dean's office. I saw that we
had a new Dean—a Southerner.

"Then the blow fell. Seemingly, during the summer the

trustees had decided gradually to exclude Negroes from the
college. In the case of students already in the course, they were
to be kept from graduation by a refusal to admit them to certain courses, particularly in obstetrics. The Dean was to
break the news by letter as students applied for these courses.
By applying early for the summer course, I had been accepted before the decision; so now he had to tell me. He hated
the task, I could see. But I was too surprised, disgusted, furious. He said that I could not enter, and he told me brutally
why. I threw my papers in his face and left. All my fine
theories of race and prejudice lay in ruins. My life was over-
turned. America was impossible—unthinkable. I ran away,
and here I am."

v

They had sat an hour drinking tea in the Tiergarten, that
mightiest park in Europe with its lofty trees, its cool dark
shade, its sense of withdrawal from the world. He had not
meant to be so voluble, so self-revealing. Perhaps the lady had
deftly encouraged confidences in her high, but gracious way.
Perhaps the mere sight of her smooth brown skin had made
Matthew assume sympathy. There was something at once inviting and aloof in the young woman who sat opposite him.
She had the air and carriage of one used to homage and yet
receiving it indifferently as a right. With all her gentle manner
and thoughtfulness, she had a certain faint air of haughtiness
and was ever slightly remote.

She was "colored" and yet not at all colored in his intimate
sense. Her beauty as he saw it near had seemed even more
striking; those thin, smooth fingers moving about the silver
had known no work; she was carefully groomed from her
purple hair to her slim toe-tips, and yet with few accessories;
he could not tell whether she used paint or powder. Her features were regular and delicate, and there was a tiny diamond
in one nostril. But quite aside from all details of face and
jewels—her pearls, her rings, the old gold bracelet—above and
beyond and much more than the sum of them all was the luminous radiance of her complete beauty, her glow of youth and
strength behind that screen of a grand yet gracious manner.
It was overpowering for Matthew, and yet stimulating. So
his story came pouring out before he knew or cared to whom he

was speaking. All the loneliness of long, lonely days clamored for speech, all the pent-up resentment choked for words.

The lady listened at first with polite but conscious sympathy; then she bent forward more and more eagerly, but always with restraint, with that mastery of body and soul that never for a moment slipped away, and yet with so evident a sympathy and comprehension that it set Matthew's head swimming. She swept him almost imperiously with her eyes— those great wide orbs of darkening light. His own eyes lifted and fell before them; lifted and fell, until at last he looked past them and talked to the tall green and black oaks.

And yet there was never anything personal in her all-sweeping glance or anything self-conscious in the form that bent toward him. She never seemed in the slightest way conscious of herself. She arranged nothing, glanced at no detail of her dress, smoothed no wisp of hair. She seemed at once unconscious of her beauty and charm, and at the same time assuming it as a fact, but of no especial importance. She had no little feminine ways; she used her eyes apparently only for seeing, yet seemed to see all.

Matthew had the feeling that her steady, full, radiant gaze that enveloped and almost burned him, saw not him but the picture he was painting and the thing that the picture meant. He warmed with such an audience and painted with clean, sure lines. Only once or twice did she interrupt, and when he had ended, she still sat full-faced, flooding him with the startling beauty of her eyes. Her hands clasped and unclasped slowly, her lips were slightly parted, the curve of her young bosom rose and fell.

"And you ran away!" she said musingly. Matthew winced and started to explain, but she continued. "Singular," she said. "How singular that I should meet you; and today." There was no coquetry in her tone. It was evidently not of him, the hero, of whom she was thinking, but of him, the group, the fact, the whole drama.

"And you are two—three millions?" she asked.

"Ten or twelve," he answered.

"You ran away," she repeated, half in meditation.

"What else could I do?" he demanded impulsively. "Cringe and crawl?"

"Of course the Negroes have no hospitals?"

"Of course, they have—many, but not attached to the great schools. What can Howard (rated as our best colored school) do with thousands, when whites have millions? And if we come out poorly taught and half equipped, they sneer at 'nigger' doctors."

"And no Negroes are admitted to the hospitals of New York?"

"O yes—hundreds. But if we colored students are confined to colored patients, we surrender a principle."

"What principle?"

"Equality."

"Oh—equality."

She sat for a full moment, frowning and looking at him. Then she fumbled away at her beads and brought out a tiny jeweled box. Absently she took out a cigarette, lighted it, and offered him one. Matthew took it, but he was a little troubled. White women in his experience smoked of course—but colored women? Well—but it was delicious to see her great, somber eyes veiled in hazy blue.

She sighed at last and said: "I do not quite understand. But at any rate I see that you American Negroes are not a mere amorphous handful. You are a nation! I never dreamed— But I must explain. I want you to dine with me and some friends tomorrow night at my apartment. We represent—indeed I may say frankly, we are—a part of a great committee of the darker peoples; of those who suffer under the arrogance and tyranny of the white world."

Matthew leaned forward with an eager thrill. "And you have plans? Some vast emancipation of the world?"

She did not answer directly, but continued: "We have among us spokesmen of nearly all these groups—of them or for them—except American Negroes. Some of us think these former slaves unready for coöperation, but I just returned from Moscow last week. At our last dinner I was telling of a report I read there from America that astounded me and gave me great pleasure—for I almost alone have insisted that your group was worthy of coöperation. In Russia I heard something, and it happened so curiously that—after sharp discussion about your people but last night (for I will not conceal from you that there is still doubt and opposition in our ranks) —that I should meet you today.

"I had gone up to the Palace to see the exhibition of new paintings—you have not seen it? You must. All the time I was thinking absently of Black America, and one picture there intensified and stirred my thoughts—a weird massing of black shepherds and a star. I dropped into the Viktoria, almost unconsciously, because the tea there is good and the muffins quite unequaled. I know that I should not go there unaccompanied, even in the day; white women may, but brown women seem strangely attractive to white men, especially Americans; and this is the open season for them.

"Twice before I have had to put Americans in their place. I went quite unconsciously and noted nothing in particular until that impossible young man sat down at my table. I did not know he had followed me out. Then you knocked him into the gutter quite beautifully. It had never happened before that a stranger of my own color should offer me protection in Europe. I had a curious sense of some great inner meaning to your act—some world movement. It seemed almost that the Powers of Heaven had bent to give me the knowledge which I was groping for; and so I invited you, that I might hear and know more."

She rose, insisted on paying the bill herself. "You are my guest, you see. It is late, and I must go. Then, tomorrow night at eight. My card and address— Oh, I quite forgot. May I have your name?"

Matthew had no card. But he wrote in her tiny memorandum book with its golden filigree, "Matthew Towns, Exile, Hotel Roter Adler."

She held out her hand, half turning to go. Her slenderness made her look taller than she was. The curved line of her flowed sinuously from neck to ankle. She held her right hand high, palm down, the long fingers drooping, and a ruby flamed dark crimson on her forefinger. Matthew reached up and shook the hand heartily. He had, as he did it, a vague feeling that he took her by surprise. Perhaps he shook hands too hard, for her hand was very little and frail. Perhaps she did not mean to shake hands—but then, what did she mean?

She was gone. He took out her card and read it. There was a little coronet and under it, engraved in flowing script, "H.R.H. the Princess Kautilya of Bwodpur, India." Below was written, "Lützower Ufer, No. 12."

VI

Matthew sat in the dining-room of the Princess on Lützower Ufer. Looking about, his heart swelled. For the first time since he had left New York, he felt himself a man, one of those who could help build a world and guide it. He had no regrets. Medicine seemed a far-off, dry-as-dust thing.

The oak paneling of the room went to the ceiling and there broke softly with carven light against white flowers and into long lucent curves. The table below was sheer with lace and linen, sparkling with silver and crystal. The servants moved deftly, and all of them were white save one who stood behind the Princess' high and crimson chair. At her right sat Matthew himself, hardly realizing until long afterward the honor thus done an almost nameless guest.

Fortunately he had the dinner jacket of year before last with him. It was not new, but it fitted his form perfectly, and his was a form worth fitting. He was a bit shocked to note that all the other men but two were in full evening dress. But he did not let this worry him much.

Ten of them sat at the table. On the Princess' left was a Japanese, faultless in dress and manner, evidently a man of importance, as the deference shown him and the orders on his breast indicated. He was quite yellow, short and stocky, with a face which was a delicately handled but perfect mask. There were two Indians, one a man grave, haughty, and old, dressed richly in turban and embroidered tunic, the other, in conventional dress and turban, a young man, handsome and alert, whose eyes were ever on the Princess. There were two Chinese, a young man and a young woman, he in a plain but becoming Chinese costume of heavy blue silk, she in a pretty dress, half Chinese, half European in effect. An Egyptian and his wife came next, he suave, talkative, and polite—just a shade too talkative and a bit too polite, Matthew thought; his wife a big, handsome, silent woman, elegantly jeweled and gowned, with much bare flesh. Beyond them was a cold and rather stiff Arab who spoke seldom, and then abruptly.

Of the food and wine of such dinners, Matthew had read often but never partaken; and the conversation, now floating, now half submerged, gave baffling glimpses of unknown lands, spiritual and physical. It was all something quite new in his experience, the room, the table, the service, the company.

He could not keep his eyes from continually straying side-wise to his hostess. Never had he seen color in human flesh so regally set: the rich and flowing grace of the dress out of which rose so darkly splendid the jeweled flesh. The black and purple hair was heaped up on her little head, and in its depths gleamed a tiny coronet of gold. Her voice and her poise, her self-possession and air of quiet command, kept Matthew staring almost unmannerly, despite the fact that he somehow sensed a shade of resentment in the young and handsome Indian opposite.

They had eaten some delicious tidbits of meat and vegetables and then were served with a delicate soup when the Princess, turning slightly to her right, said:

"You will note, Mr. Towns, that we represent here much of the Darker World. Indeed, when all our circle is present, we represent all of it, save your world of Black Folk."

"All the darker world except the darkest," said the Egyptian.

"A pretty large omission," said Matthew with a smile.

"I agree," said the Chinaman; but the Arab said something abruptly in French. Matthew had said that he knew "some" French. But his French was of the American variety which one excavates from dictionaries and cements with grammar, like bricks. He was astounded at the ease and the fluency with which most of this company used languages, so easily, without groping or hesitation and with light, sure shading. They talked art in French, literature in Italian, politics in German, and everything in clear English.

"M. Ben Ali suggests," said the Princess, "that even you are not black, Mr. Towns."

"My grandfather was, and my soul is. Black blood with us in America is a matter of spirit and not simply of flesh."

"Ah! mixed blood," said the Egyptian.

"Like all of us, especially me," laughed the Princess.

"But, your Royal Highness—not Negro," said the elder Indian in a tone that hinted a protest.

"Essentially," said the Princess lightly, "as our black and curly-haired Lord Buddha testifies in a hundred places. But"—

a bit imperiously—"enough of that. Our point is that Pan-Africa belongs logically with Pan-Asia; and for that reason Mr. Towns is welcomed tonight by you, I am sure, and by me especially. He did me a service as I was returning from the New Palace."

They all looked interested, but the Egyptian broke out:

"Ah, Your Highness, the New Palace, and what is the fad today? What has followed expressionism, cubism, futurism, vorticism? I confess myself at sea. Picasso alarms me. Matisse sets me aflame. But I do not understand them. I prefer the classics."

"The Congo," said the Princess, "is flooding the Acropolis. There is a beautiful Kandinsky on exhibit, and some lovely and startling things by unknown newcomers."

"*Mais,*" replied the Egyptian, dropping into French—and they were all off to the discussion, save the silent Egyptian woman and the taciturn Arab.

Here again Matthew was puzzled. These persons easily penetrated worlds where he was a stranger. Frankly, but for the context he would not have known whether Picasso was a man, a city, or a vegetable. He had never heard of Matisse. Lightly, almost carelessly, as he thought, his companions leapt to unknown subjects. Yet they knew. They knew art, books, and literature, politics of all nations, and not newspaper politics merely, but inner currents and whisperings, unpublished facts.

"Ah, pardon," said the Egyptian, returning to English, "I forgot Monsieur Towns speaks only English and does not interest himself in art."

Perhaps Matthew was sensitive and imagined that the Egyptian and the Indian rather often, if not purposely, strayed to French and subjects beyond him.

"Mr. Towns is a scientist?" asked the Japanese.

"He studies medicine," answered the Princess.

"Ah—a high service," said the Japanese. "I was reading only today of the work on cancer by your Peyton Rous in Carrel's laboratory."

Towns was surprised. "What, has he discovered the etiological factor? I had not heard."

"No, not yet, but he's a step nearer."

For a few moments Matthew was talking eagerly, until a babble of unknown tongues interrupted him across the table.

"Proust is dead, that 'snob of humor'—yes, but his *Recherche du Temps Perdu* is finished and will be published in full. I have only glanced at parts of it. Do you know Gasquet's *Hymnes?*"

"Beraud gets the Prix Goncourt this year. Last year it was the Negro, Maran—"

"I have been reading Croce's *Aesthetic* lately—"

"Yes, I saw the Meyerhold theater in Moscow—gaunt realism—*Howl China* was tremendous."

Then easily, after the crisp brown fowl, the Princess tactfully steered them back to the subject which some seemed willing to avoid.

"And so," she said, "the darker peoples who are dissatisfied—"

She looked at the Japanese and paused as though inviting comment. He bowed courteously.

"If I may presume, your Royal Highness, to suggest," he said slowly, "the two categories are not synonymous. We ourselves know no line of color. Some of us are white, some yellow, some black. Rather, is it not, your Highness, that we have from time to time taken council with the oppressed peoples of the world, many of whom by chance are colored?"

"True, true," said the Princess.

"And yet," said the Chinese lady, "it is dominating Europe which has flung this challenge of the color line, and we cannot avoid it."

"And on either count," said Matthew, "whether we be bound by oppression or by color, surely we Negroes belong in the foremost ranks."

There was a slight pause, a sort of hesitation, and it seemed to Matthew as though all expected the Japanese to speak. He did, slowly and gravely:

"It would be unfair to our guest not to explain with some clarity and precision that the whole question of the Negro race both in Africa and in America is for us not simply a question of suffering and compassion. Need we say that for these peoples we have every human sympathy? But for us here and for the larger company we represent, there is a deeper question—that of the ability, qualifications, and real possibilities of the black race in Africa or elsewhere."

Matthew left the piquant salad and laid down his fork

slowly. Up to this moment he had been quite happy. Despite the feeling of being out of it now and then, he had assumed that this was his world, his people, from the high and beautiful lady whom he worshiped more and more, even to the Egyptians, Indians, and Arab who seemed slightly, but very slightly, aloof or misunderstanding.

Suddenly now there loomed plain and clear the shadow of a color line within a color line, a prejudice within prejudice, and he and his again the sacrifice. His eyes became somber and did not lighten even when the Princess spoke.

"I cannot see that it makes great difference what ability Negroes have. Oppression is oppression. It is our privilege to relieve it."

"Yes," answered the Japanese, "but who will do it? Who can do it but those superior races whose necks now bear the yoke of the inferior rabble of Europe?"

"This," said the Princess, "I have always believed; but as I have told your Excellency, I have received impressions in Moscow which have given me very serious thought—first as to our judgment of the ability of the Negro race, and second"— she paused in thought—"as to the relative ability of all classes and peoples."

Matthew stared at her, as she continued:

"You see, Moscow has reports—careful reports of the world's masses. And the report on the Negroes of America was astonishing. At the time, I doubted its truth: their education, their work, their property, their organizations; and the odds, the terrible, crushing odds against which, inch by inch and heartbreak by heartbreak, they have forged their unfaltering way upward. If the report is true, they are a nation today, a modern nation worthy to stand beside any nation here."

"But can we put any faith in Moscow?" asked the Egyptian. "Are we not keeping dangerous company and leaning on broken reeds?"

"Well," said Matthew, "if they are as sound in everything as in this report from America, they'll bear listening to."

The young Indian spoke gently and evenly, but with bright eyes.

"Naturally," he said, "one can see Mr. Towns needs must agree with the Bolshevik estimate of the lower classes."

Matthew felt the slight slur and winced. He thought he saw

the lips of the Princess tighten ever so little. He started to answer quickly, with aplomb if not actual swagger.

"I reckon," he began—then something changed within him. It was as if he had faced and made a decision, as though some great voice, crying and reverberating within his soul, spoke for him and yet was him. He had started to say, "I reckon there's as much high-born blood among American Negroes as among any people. We've had our kings, presidents, and judges—" He started to say this, but he did not finish. He found himself saying quite calmly and with slightly lifted chin:

"I reckon you're right. We American blacks are very common people. My grandfather was a whipped and driven slave; my father was never really free and died in jail. My mother plows and washes for a living. We come out of the depths— the blood and mud of battle. And from just such depths, I take it, came most of the worth-while things in this old world. If they didn't—God help us."

The table was very still, save for the very faint clink of china as the servants brought in the creamed and iced fruit.

The Princess turned, and he could feel her dark eyes full upon him.

"I wonder—I wonder," she murmured, almost catching her breath.

The Indian frowned. The Japanese smiled, and the Egyptian whispered to the Arab.

"I believe that is true," said the Chinese lady thoughtfully, "and if it is, this world is glorious."

"And if it is not?" asked the Egyptian icily.

"It is perhaps both true and untrue," the Japanese suggested. "Certainly Mr. Towns has expressed a fine and human hope, although I fear that always blood must tell."

"No, it mustn't," cried Matthew, "unless it is allowed to talk. Its speech is accidental today. There is some weak, thin stuff called blood, which not even a crown can make speak intelligently; and at the same time some of the noblest blood God ever made is dumb with chains and poverty."

The elder Indian straightened, with blazing eyes.

"Surely," he said, slowly and calmly, "surely the gentleman does not mean to reflect on royal blood?"

Matthew started, flushed darkly, and glanced quickly at the

Princess. She smiled and said lightly, "Certainly not," and
then with a pause and a look straight across the table to the
turban and tunic, "nor will royal blood offer insult to him."
The Indian bowed to the tablecloth and was silent.

As they rose and sauntered out to coffee in the silk and
golden drawing-room, there was a discussion, started of course
by the Egyptian, first of the style of the elaborate piano case
and then of Schönberg's new and unobtrusive transcription of
Bach's triumphant choral Prelude, "Komm, Gott, Schöpfer."

The Princess sat down. Matthew could not take his eyes
from her. Her fingers idly caressed the keys as her tiny feet
sought the pedals. From white, pearl-embroidered slippers, her
young limbs, smooth in pale, dull silk, swept up in long, low
lines. Even the delicate curving of her knees he saw as she
drew aside her drapery and struck the first warm tones. She
played the phrase in dispute—great chords of aspiration and
vision that melted to soft melody. The Egyptian acknowledged
his fault. "Yes—yes, that was the theme I had forgotten."

Again Matthew felt his lack of culture audible, and not
simply of his own culture, but of all the culture in white
America which he had unconsciously and foolishly, as he now
realized, made his norm. Yet withal Matthew was not unhappy.
If he was a bit out of it, if he sensed divided counsels and
opposition, yet he still felt almost fiercely that that was his
world. Here were culture, wealth, and beauty. Here was
power, and here he had some recognized part. God! If he could
just do his part, any part! And he waited impatiently for the
real talk to begin again.

It began and lasted until after midnight. It started on lines
so familiar to Matthew that he had to shut his eyes and stare
again at their swarthy faces: Superior races—the right to rule
—born to command—inferior breeds—the lower classes—the
rabble. How the Egyptian rolled off his tongue his contempt
for the "r-r-rabble"! How contemptuous was the young Indian
of inferior races! But how humorous it was to Matthew to
see all tables turned; the rabble now was the white workers
of Europe; the inferior races were the ruling whites of Europe
and America. The superior races were yellow and brown.

"You see," said the Japanese, "Mr. Towns, we here are
all agreed and not agreed. We are agreed that the present
white hegemony of the world is nonsense; that the darker peo-

ples are the best—the natural aristocracy, the makers of art,
religion, philosophy, life, everything except brazen machines."

"But why?"

"Because of the longer rule of natural aristocracy among
us. We count our millenniums of history where Europe counts
her centuries. We have our own carefully thought-out philosophy and civilization, while Europe has sought to adopt an
ill-fitting mélange of the cultures of the world."

"But does this not all come out the same gate, with the
majority of mankind serving the minority? And if this is the
only ideal of civilization, does the tint of a skin matter in
the question of who leads?" Thus Matthew summed it up.

"Not a whit—it is the natural inborn superiority that matters," said the Japanese, "and it is that which the present color
bar of Europe is holding back."

"And what guarantees, in the future more than in the past
and with colored more than with white, the wise rule of the
gifted and powerful?"

"Self-interest and the inclusion in their ranks of all really
superior men of all colors—the best of Asia together with the
best of the British aristocracy, of the German Adel, of the
French writers and financiers—of the rulers, artists, and poets
of all peoples."

"And suppose we found that ability and talent and art is
not entirely or even mainly among the reigning aristocrats of
Asia and Europe, but buried among millions of men down in
the great sodden masses of all men and even in Black Africa?"

"It would come forth."

"Would it?"

"Yes," said the Princess, "it would come forth, but when
and how? In slow and tenderly nourished efflorescence, or in
wild and bloody upheaval with all that bitter loss?"

"Pah!" blurted the Egyptian—"pardon, Royal Highness—
but what art ever came from the canaille!"

The blood rushed to Matthew's face. He threw back his
head and closed his eyes, and with the movement he heard
again the Great Song. He saw his father in the old log church
by the river, leading the moaning singers in the Great Song
of Emancipation. Clearly, plainly he heard that mighty voice
and saw the rhythmic swing and beat of the thick brown arm.
Matthew swung his arm and beat the table; the silver tinkled.

Silence dropped on all, and suddenly Matthew found himself singing. His voice full, untrained but mellow, quivered down the first plaintive bar:

"When Israel was in Egypt land—"

Then it gathered depth:

"Let my people go!"

He forgot his audience and saw only the shining river and the bowed and shouting throng:

"Oppressed so hard, they could not stand,
Let my people go."

Then Matthew let go restraint and sang as his people sang in Virginia, twenty years ago. His great voice, gathered in one long deep breath, rolled the Call of God:

"Go down, Moses!
Way down into the Egypt land,
Tell Old Pharaoh
To let my people go!"

He stopped as quickly as he had begun, ashamed, and beads of sweat gathered on his forehead. Still there was silence—silence almost breathless. The voice of the Chinese woman broke it.

"It was an American slave song! I know it. How—how wonderful."

A chorus of approval poured out, led by the Egyptian.

"That," said Matthew, "came out of the black rabble of America." And he trilled his "r." They all smiled as the tension broke.

"You assume then," said the Princess at last, "that the mass of the workers of the world can rule as well as be ruled?"

"Yes—or rather can work as well as be worked, can live as well as be kept alive. America is teaching the world one thing and only one thing of real value, and that is, that ability and capacity for culture is not the hereditary monopoly of a few, but the widespread possibility for the majority of mankind if they only have a decent chance in life."

The Chinaman spoke: "If Mr. Towns' assumption is true, and I believe it is, and recognized, as some time it must be, it will revolutionize the world."

"It will revolutionize the world," smiled the Japanese, "but not—today."

"Nor this *siècle*," growled the Arab.

"Nor the next—and so *in saecula saeculorum*," laughed the Egyptian.

"Well," said the little Chinese lady, "the unexpected happens."

And Matthew added ruefully, "It's about all that does happen!"

He lapsed into blank silence, wondering how he had come to express the astonishing philosophy which had leapt unpremeditated from his lips. Did he himself believe it? As they arose from the table the Princess called him aside.

VII

"I trust you will pardon the interruption at this late hour," said the Japanese. Matthew glanced up in surprise as the Japanese, the two Indians, and the Arab entered his room. "Sure," said he cheerily, "have any seats you can find. Sorry there's so little space."

It was three o'clock in the morning. He was in his shirt sleeves without collar, and he was packing hastily, wondering how on earth all these things had ever come out of his two valises. The little room on the fifth floor of the Roter Adler Hotel did look rather a mess. But his guests smiled and so politely deprecated any excuses or discomfort that he laughed too, and leaned against the window, while they stood about door and bed.

"You had, I believe," continued the Japanese, "an interview with her Royal Highness, the Princess, before you left her home tonight."

"Yes."

"I—er—presume you realize, Mr. Towns, that the Princess of Bwodpur is a lady of very high station—of great wealth and influence."

"I cannot imagine anybody higher."

The elder Indian interrupted. "There are," he said, "of

course, some persons of higher hereditary rank than her Royal Highness, but not many. She is of royal blood by many scores of generations of direct descent. She is a ruling potentate in her own right over six millions of loyal subjects and yields to no human being in the ancient splendor of her heritage. Her income, her wealth in treasure and jewels, is uncounted. Sir, there are few people in the world who have the right to touch the hem of her garment." The Indian drew himself to full height and looked at Matthew.

"I'm strongly inclined to agree with you there," said Matthew, smiling genially.

"I had feared," continued the younger Indian, also looking Matthew squarely in the eye, "that perhaps, being American, and unused to the ceremony of countries of rank, you might misunderstand the democratic graciousness of her Royal Highness toward you. We appreciate, sir, more than we can say," and both Indians bowed low, "your inestimable service to the Princess yesterday, in protecting her royal person from insufferable insult. But the very incident illustrates and explains our errand.

"The Princess is young and headstrong. She delights, in her new European independence, to elude her escort, and has given us moments of greatest solicitude and even fright. Meeting her as you did yesterday, it was natural for you to take her graciousness toward you as the camaraderie of an equal, and —quite unconsciously, I am sure—your attitude toward her has caused us grave misgiving."

"You mean that I have not treated the Princess with courtesy?" asked Matthew in consternation. "In what way? Tell me."

"It is nothing—nothing, now that it is past, and since the Princess was gracious enough to allow it. But you may recall that you never addressed her by her rightful title of 'Royal Highness'; you several times interrupted her conversation and addressed her before being addressed; you occupied the seat of honor without even an attempted refusal and actually shook her Highness' hand, which we are taught to regard as unpardonable familiarity."

Matthew grinned cheerfully. "I reckon if the Princess hadn't liked all that she'd have said so—"

The Japanese quickly intervened. "This is, pardon me, beside our main errand," he said. "We realize that you admire and revere the Princess not only as a supremely beautiful woman of high rank, but as one of rare intelligence and high ideals."

"I certainly do."

"And we assume that anything you could do—any way you could coöperate with us for her safety and future, we could count upon your doing?"

"To my very life."

"Good—excellent—you see, my friends," turning to the still disturbed Indians and the silent, sullen Arab, "it is as I rightly divined."

They did not appear wholly convinced, but the Japanese continued:

"In her interview with you she told you a story she had heard in Moscow, of a widespread and carefully planned uprising of the American blacks. She has intrusted you with a letter to the alleged leader of this organization and asked you to report to her your impressions and recommendations; and even to deliver the letter, if you deem it wise.

"Now, my dear Mr. Towns, consider the situation: First of all, our beloved Princess introduces you, a total stranger, into our counsels and tells you some of our general plans. Fortunately, you prove to be a gentleman who can be trusted; and yet you yourself must admit this procedure was not exactly wise. Further than that, through this letter, our reputations, our very lives, are put in danger by this well-meaning but young and undisciplined lady. Her unfortunate visit to Russia has inoculated her with Bolshevism of a mild but dangerous type. The letter contains money to encourage treason. You know perfectly well that the American Negroes will neither rebel nor fight unless put up to it or led like dumb cattle by whites. You have never even heard of the alleged leader, as you acknowledged to the Princess."

"She is evidently well spied upon."

"She is, and will always be, well guarded," answered the elder Indian tensely.

"Except yesterday," said Matthew.

But the Japanese quickly proceeded. "Why then go on this wild goose chase? Why deliver dynamite to children?"

"Thank you."

"I beg your pardon. I may speak harshly, but I speak frankly. You are an exception among your people."

"I've heard that before. Once I believed it. Now I do not."

"You are generous, but you are an exception, and you know you are."

"Most people are exceptions."

"You know that your people are cowards."

"That's a lie; they are the bravest people fighting for justice today."

"I wish it were so, but I do not believe it, and neither do you. Every report from America—and believe me, we have many—contradicts this statement for you. I am not blaming them, poor things, they were slaves and children of slaves. They can not even begin to rise in a century. We Samurai have been lords a thousand years and more; the ancestors of her Royal Highness have ruled for twenty centuries—how can you think to place yourselves beside us as equals? No—no—restrain your natural anger and distaste for such truth. Our situation is too delicate for niceties. We have been almost betrayed by an impulsive woman, high and royal personage though she be. We have come to get that letter and to ask you to write a report now, to be delivered later, thoroughly disenchanting our dear Princess of this black American chimera."

"And if I refuse?"

The Japanese looked pained but patient. The others moved impatiently, and perceptibly narrowed the circle about Matthew. He was thinking rapidly; the letter was in his coat pocket on the bed beyond the Japanese and within easy reach of the Indians if they had known it. If he jumped out the window, he would be dead, and they would eventually secure the letter. If he fought, they were undoubtedly armed, and four to one. The Japanese was elderly and negligible as an opponent, but the young Indian and the Arab were formidable, and the older Indian dangerous. He might perhaps kill one and disable another and raise enough hullabaloo to arouse the hotel, but how would such a course affect the Princess?

The Japanese watched him sadly.

"Why speak of unpleasant things," he said gently, "or contemplate futilities? We are not barbarians. We are men of thought and culture. Be assured our plans have been laid with care. We know the host of this hotel well. Resistance on your part would be absolutely futile. The back stairs opposite your entrance are quite clear and will be kept clear until we go. And when we go, the letter will go with us."

Matthew set his back firmly against the window. His thoughts raced. They were armed, but they would use their arms only as a last resort; pistols were noisy and knives were messy. Oh, they would use them—one look at their hard, set eyes showed that; but not first. Good! Then first, instead of lurching forward to attack as they might expect, he would do a first-base slide and spike the Japanese in the ankles. It was a mean trick, but anything was fair now. He remembered once when they were playing the DeWitt Clinton High— But he jerked his thoughts back. The Japanese was nearest him; the fiery younger Indian just behind him and a bit to the right, bringing him nearer the bed and blocking the aisle. By the door was the elder Indian, and at the foot of the bed, the Arab.

Good! He would, at the very first movement of the young Indian, who, he instinctively knew, would begin the mêlée, slide feet forward into the ankles of the Japanese, catching him a little to the right so that he would fall or lurch between him and the Indian. Then he would with the same movement slide under the low iron bed and rise with the bed as weapon and shield. But he would not keep it. No; he would hurl it sideways and to the left, pinning the young Indian to the wall and the Japanese to the floor. With the same movement he would attempt a football tackle on the Arab. The Arab was a tough customer—tall, sinewy, and hard. If he turned left, got his knife and struck down, quick and sure, Matthew would be done for. But most probably he would, at Matthew's first movement, turn right toward his fellows. If he did, he was done for. He would go down in the heap, knocking the old Indian against the door.

Beside that door was the electric switch. Matthew would turn it and make a last fight for the door. He might get out, and if he did, the stairs were clear. The coat and letter? Leave them, so long as he got his story to the Princess. It was all a last desperate throw. He calculated that he had good chances against the Japanese's shins, about even chances to get under the bed unscathed, and one in two to tackle the Arab. He had not more than one chance in three of making the door unscathed, but this was the only way. If he surrendered without a fight— That was unthinkable. And after all, what had he to lose? Life? Well, his prospects were not brilliant anyhow. And to die for the Princess—silly, of course, but it made his blood race. For the first time he glimpsed the glory of death. Meantime—he said—be sensible! It would not hurt to spar for time.

He pretended to be weighing the matter.

"Suppose you do steal the letter by force, do you think you can make me write a report?"

"No, a voluntary report would be desirable but not necessary. You left with the Princess, you will remember, a page of directions and information about America to guide her in the trip she is preparing to make and from which we hope to dissuade her. You appended your signature and address. From this it will be easy to draft a report in handwriting so similar to yours as to be indistinguishable by ordinary eyes."

"You add forgery to your many accomplishments."

"In the pursuit of our duty, we do not hesitate at theft or forgery."

Still Matthew parried: "Suppose," he said, "I pretended to acquiesce, gave you the letter and reported to the Princess. Suppose even I told the German newspapers of what I have seen and heard tonight."

There was a faraway look in the eyes of the Japanese as he answered slowly: "We must follow Fate, my dear Mr. Towns, even if Fate leads—to murder. We will not let you communicate with the Princess, and you are leaving Berlin tonight."

The Indians gave a low sigh almost like relief.

Matthew straightened and spoke slowly and firmly.

"Very well. I won't surrender that letter to anybody but the Princess—not while I'm alive. And if I go out of here dead I won't be the only corpse."

Every eye was on the Japanese, and Matthew knew his life was in the balance. The pause was tense; then came the patient voice of the Japanese again.

"You—admire the Princess, do you not?"

"With all my heart."

The Indians winced.

"You would do her a service?"

"To the limit of my strength."

"Very well. Let us assume that I am wrong. Assume that the Negroes are worth freedom and ready to fight for it. Can you not see that the name of this young, beautiful, and high-born lady must under no circumstances be mixed up with them, whether they gain or lose? What would not Great Britain give thus to compromise an Indian ruler?"

"That is for the Princess to decide."

"No! She is a mere woman—an inexperienced girl. You are a man of the world. For the last time, will you rescue her Royal Highness from herself?"

"No. The Princess herself must decide."

"Then—"

"Then," said the Princess' full voice, "the Princess will decide." She stood in the open doorway, the obsequious and scared landlord beside her with his pass-keys. She had thrown an opera cloak over her evening gown, and stood unhatted, white-slippered, and ungloved. She threw one glance at the Indians, and they bowed low with outstretched hands. She stamped her foot angrily, and they went to their knees. She wheeled to the Arab. Without a word, he stalked out. The Japanese alone remained, calm and imperturbable.

"We have failed," he said, with a low bow.

The Princess looked at him.

"You have failed," she said. "I am glad there is no blood on your hands."

"A drop of blood more or less matters little in the great cause for which you and I fight, and if I have incurred your Royal Highness' displeasure tonight, remember that, for the same great cause, I stand ready tomorrow night to repeat the deed and seal it with my life."

The Princess looked at him with troubled eyes. Then she seated herself in the only and rather rickety chair and motioned for her two subjects to arise. Matthew never forgot that scene: he, collarless and in shirt sleeves, with sweat pouring off his face; the room in disorder, mean, narrow, small, and dingy; the Japanese standing in the same place as when he entered, in unruffled evening dress; the Indians on their knees with hidden faces; the Princess, disturbed, yet radiant. She spoke in low tones.

"I may be wrong," she said, "and I know how right, but

infinitely and calmly right, you usually are. But some voice within calls me. I have started to fight for the dark and oppressed peoples of the world; now suddenly I have seen a light. A light which illumines the mass of men and not simply its rulers, white and yellow and black. I want to see if this thing is true, if it can possibly be true that wallowing masses often conceal submerged kings. I have decided not simply to send a messenger to America but shortly to sail myself—perhaps this week on the *Gigantic*. I want to see for myself if slaves can become men in a generation. If they can—well, it makes the world new for you and me."

The Japanese started to speak, but she would not pause:

"There is no need for protest or advice. I am going. Mr. Towns will perform his mission as we agreed, if he is still so minded, and as long as he is in Europe, these two gentlemen," she glanced at the Indians, "will bear his safety on their heads, at my command. Go!"

The Indians bowed and walked out slowly, backward. She turned to the Japanese.

"Your Highness, I bid you good night and good-by. I shall write you."

Gravely the Japanese kissed her hand, bowed, and withdrew. The Princess looked at Matthew. He became acutely conscious of his appearance as she looked at him almost a full minute with her great, haunting eyes.

"I thank you, again," she said slowly. "You are a brave man—and loyal." She held out her hand, low, to shake his.

But the tension of the night broke him; he quivered, and taking her hand in both his, kissed it.

She rose quickly, drew herself up, and looked at first almost affronted; then when she saw his swimming eyes, a kind of startled wonder flashed in hers. Slowly she held out her hand again, regally, palm down and the long fingers drooping.

"You are very young," she said.

He was. He was only twenty-five. The Princess was all of twenty-three.

PART II
THE PULLMAN PORTER

PART II

THE PULLMAN PORTER

September to December, 1923

Fall. Fall of leaf and sigh of wind. Gasp of the world-soul before, in crimson, gold, and gray, it dips beneath the snows. The flame of passing summer slowly dies in the looming shadow of death. Fall on the vast gray-green Atlantic, where waves of all waters heave and groan toward bitter storms to come. Fall in the crowded streets of New York. Fall in the heart of the world.

I

MATTHEW was paring potatoes; paring, paring potatoes. There was a machine in the corner, paring, too. But Matthew was cheaper than the machine and better. It was not hard work. It was just dull—idiotic, dull. He pared mechanically, with humped shoulders and half-closed eyes. Garbage lay about him, and nauseating smells combined of sour and sweet, decay and ferment, offal and delicacy, made his head dizzy and his stomach acrid. The great ship rose, shivered and screamed, dropped in the gray grave of waters, and groaned as the hot hell of its vast belly drove it relentlessly, furiously forward. The terrible, endless rhythm of the thing—paring, rising, falling, groaning, paring, swaying, with the slosh of the greasy dishwater, in the close hot air, set Matthew to dreaming.

He could see again that mother of his—that poor but mighty, purposeful mother—tall, big, and brown. What hands she had —gnarled and knotted; what great, broad feet. How she worked! Yet he seemed never to have realized what work was until now. On the farm—that little forty acres of whitish yellow land, with its tiny grove and river; its sweep of green, white cotton; its geese, its chickens, the cow and the old mule; the low log cabin with its two rooms and wide hall leading to the boarded kitchen behind—how he remembered the build-

ing of that third room just before his father went away—work?
Work on that curious little hell in paradise had not been work
to him; it had been play. He had stopped when he was tired.
But mother and the bent old father, had it been work for them
—hard, hateful, heavy, endless, uninteresting, dull, stupid?
Yes, it must have been like this, save in air and sun; toil must
have dulled and hardened them. God! What did this world—

"That-a nigger did it—I know!—that-a-there damn nigger!"

The Italian bent over him. Matthew looked up at him without interest. His soul was still dreaming far away—rising,
falling, paring, glowing. Somebody was always swearing and
quarreling in the scullery. Funny for him to be here. It had
seemed a matter of honor, life, death, to sail on this particular
ship. He had, with endless courtesy and with less than a hundred dollars in his pocket, assured the Princess that he needed
nothing. And then—fourth class to Hamburg, standing! and
the docks. The *Gigantic* was there. Would she sail on it? He
did not know. He approached the head steward for a job.

"No—no more work." He stood hesitating. A stevedore who
staggered past, raining sweat, dropped his barrow and hobbled
away. Matthew left his bags, seized the heavy barrow, and
trundled it on the ship. It was not difficult to hide until the
boat had swung far down the channel. Then he went to the
head steward again.

"What the hell are you doing here? Just like a damn
nigger!"

Here again it came from the lips of the fat, lumbering
Italian: "—damn-a nigger." And Matthew felt a flat-handed
cuff beside his head that nearly knocked him from the stool.
He arose slowly, folded his arms, and looked at the angry
man. The Italian was a great baby whom the men picked on
and teased and fooled—cruel, senseless sport for people who
took curious delight in tricking others of their kind.

It was to Matthew an amazing situation—one he could not
for the life of him comprehend. These men were at the bottom
of life—scullions. They had no pride of work. Who could
have pride in such work! But they despised themselves. God
was in the first cabin, overeating, guzzling, gambling, sleeping.
They despised what He despised. He despised Negroes. He
despised Italians, unless they were rich and noble. He despised
scullion's work. All these things the scullions despised.

Matthew and the Italian were butts—the Italian openly, Matthew covertly; for they were a little dashed at his silence and carriage. But they sneered and growled at the "nigger" and egged the "dago" on. And the Italian—big, ignorant moron, sweet and childish by nature, wild and bewildered by his strange environment, despised the black man because the others did.

This time their companions had slyly slipped potatoes into the Italian's pan until he had already done twice the work of the others. But the last mess had been too large—he grew suspicious and angry, and he picked on Matthew because he had seen the others sneer at the dark stranger, and he was ready to believe the worst of him.

Matthew stood still and looked at the Italian; with a yell the irate man hurled his bulk forward and aimed a blow which struck Matthew's shoulder. Matthew fell back a step and still stood looking at him. The scullery jeered:

"Fight—the nigger and the dago."

Again the fist leaped out and hit Matthew in the nose, but still Matthew stood and did not lift his hand. Why? He could not have said himself. More or less consciously he sensed what a silly mess it all was. He could not soil his hands on this great idiot. He would not stoop to such a brawl. There was a strange hush in the scullery. Somebody yelled, "Scared stiff!" But they yelled weakly, for Matthew did not look scared. He was taller than the Italian, not so big, but his brown muscles rippled delicately on his lithe form. Even with his swelling nose, he did not look scared or greatly perturbed. Then there was a scramble. The kitchen steward suddenly entered, one of the caste of stewards—the visible revelation of God in the cabin; a splendid man, smooth-coated, who made money and yelled at scullions.

"Chief!"

The Italian ducked, ran and hid, and Matthew was standing alone.

The steward blustered: "Fighting, hey?"

"No."

"I saw you!"

"You lie!"

The scullery held its breath.

The steward, with purple face, started forward with raised

fist and then paused. He was puzzled at that still figure. It wouldn't do to be mauled or killed before scullions. . . .

"All right, nigger—I'll attend to you later. Get to work, all of you," he growled.

Matthew sat down and began paring, paring, again. But now the dreams had gone. His head ached. His soul felt stripped bare. He kept pondering dully over this room, glancing at the shifty eyes, the hunches and grins; smelling the smells, the steam, the grease, the dishwater. There was so little kindness or sympathy for each other here among these men. They loved cruelty. They hated and despised most of their fellows, and they fell like a pack of wolves on the weakest. Yet they all had the common bond of toil; their sweat and the sweat of toilers like them made one vast ocean around the world. Waves of world-sweat droned in Matthew's head dizzily, and naked men were driven drowning through it, yet snapping, snarling, fighting back each other as they wallowed. Well, he wouldn't fight them. That was idiotic. It was human sacrilege. If fight he must, he would fight stewards and cabin gentry—lackeys and gods.

He walked stiffly to his berth and sat half-dressed in a corner of the common bunk room, hating to seek his hot, dark, ill-ventilated bunk. The men were growling, sprawling, drinking, and telling smutty stories. They had, it seemed to Matthew, a marvelous poverty of capacity to enjoy—to be happy and to play.

The door opened. The kitchen steward came in, followed by a dozen men and women, evidently from the first cabin—fat, sleek persons in evening dress, the women gorgeous and bare, the men pasty-faced and swaggering. All were smoking and flushed with wine. Towns started and stared— My God! If one face appeared there—if the Princess came down and saw this, saw him here! He groaned and stood up quickly, with the half-formed design of walking out.

"A ring, men!" called the steward. The scullery glowered, smirked, and shuffled; backed to one side, torn by conflicting motives, hesitating.

"These ladies and gentlemen have given a purse of two hundred dollars to have this fight out between the darky and the dago. Strip, you—but keep on your pants. This gentleman is referee. Come, Towns. Now's the chance for revenge."

The Italian rose, lounged forward, and looked at Towns truculently, furtively. His anger was gone now, and he was not sure Towns had wronged him. Towns looked at him, smiled, and held out his hand. The Italian stared, hesitated, then almost ran and grabbed it. Towns turned to the steward, still watching the door:

"We won't fight," he said.

"We ain't gonna fight," echoed the Italian.

"Throw them into the ring."

"Try it," cried Towns.

"Try it," echoed the Italian.

The steward turned red and green. He saw a fat fee fading.

"So we can't make you rats fight," he sneered.

"Oh, yes, we'll fight," said Matthew, "but we won't fight each other. If rats must fight they fight cats—and dogs—and hogs."

"Wow!" yelled the scullery, and surged.

"Home, James," squeaked a shrill voice, "they ain't gonna be no fight tonight." She had the face of an angel, the clothes of a queen, and the manners of a prostitute. The guests followed her out, giggling, swaying, and swearing.

"S'no plash f'r min'ster's son, nohow," hiccoughed the youth in the rear.

The steward lingered and glanced at Matthew, teetering on heel and toe.

"So that's your game. Trying to stir up something, hey? Planning Bolshevik stuff! D'ye know where I've half a mind to land you in New York? I'll tell you! In jail! D'ye hear? In jail!"

The room was restless. The grumbling stopped gradually. The men looked as though they wanted to talk to him, but Matthew crept to his bunk and pretended to sleep. What was going to happen? What would they do next? Were they going to make him fight his way over? Must he kill somebody? Of all the muddles that a clean, straight life can suddenly fall to, his seemed the worst. He tossed in his narrow, hot bed in an agony of fear and excitement. He slept and dreamed; he was fighting the world. Blood was spurting, heads falling, ghastly eyes were bulging, but he slew and slew until his neighbor yelled:

"Who the hell you hittin'? Are youse crazy?" And the man fled in fear.

Matthew rose early and went to his task—paring, peeling, cutting, paring. Nothing happened. The steward said no further word. The scullery growled, but let Matthew alone. The Italian crept near him like a lost dog, trying in an inarticulate way to say some unspoken word.

So Matthew dropped back to his dreams.

He was groping toward a career. He wanted to get his hand into the tangles of this world. He wanted to understand. His revolt against medicine became suddenly more than resentment at an unforgivable insult—it became ingrained distaste for the whole narrow career, the slavery of mind and body, the ethical chicanery. His sudden love for a woman far above his station was more than romance—it was a longing for action, breadth, helpfulness, great constructive deeds.

And so, rising and falling, working and writhing, dreaming and suffering, he passed his week of days of weeks. He hardly knew when it ended. Only one day, washing dishes, he looked out of the porthole; there was the Statue of Liberty shining. . . .

And Matthew laughed.

II

There is a corner in High Harlem where Seventh Avenue cuts the dark world in two. West rises the noble façade of City College—gray and green. East creeps the sullen Harlem, green and gray. There Matthew stood and looked right and left. Left was the world he had left—there were some pretty parlors there, conventional in furniture and often ghastly in ornament, but warm and homelike in soul. There was his own bedroom; Craigg's restaurant with its glorious biscuit; churches whose music often brushed his ears sweetly, afar; crowded but neat apartments, swaggering but well-dressed lodgers, workers, visitors. He turned from it with a sigh. He had left this world for a season—perhaps forever. It would hardly recognize him, he was sure, for he was unshaven and poorly clothed.

He turned east, and the world turned too—to a more careless and freer movement, louder voices, and easier camaraderie.

By the time Lenox Avenue was reached, the world was gay and vociferous, and shirt sleeves and overalls mingled with tailored trousers and silk hosiery. But Matthew walked on in the gathering gloom: Fifth Avenue—but Fifth Avenue at 135th Street; he knew it vaguely—a loud and unkempt quarter with flashes of poverty and crime. He went on into an ill-kept hall and up dirty and creaking stairs, half-lighted, and knocked on a door. There were loud voices within—loud, continuous quarreling voices. He knocked again.

"Come in, man, don't stand there pounding the door down."

Matthew opened the door. The room was hot with a mélange of smoke, bad air, voices, and gesticulations. Groups were standing and sitting about, lounging, arguing, and talking. Sometimes they shouted and seemed on the point of blows, but blows never came. They appeared tremendously in earnest without a trace of smile or humor. This puzzled Matthew at first until he caught their broad *a*'s and curious singing lilt of phrase. He realized that all or nearly all were West Indians. He knew little of the group. They were to him singular, foreign and funny. He had never been in a group of them before. He looked about.

"Is Mr. Perigua in?"

Some one waved carelessly toward the end of the room without pause in argument or gesture. Matthew discovered there a low platform, a rickety railing, and within, a table and several men. They were talking, if anything, louder and faster than the rest. With difficulty he traced his way toward them.

"Mr. Perigua?"

No response—but another argument of which Matthew understood not a word.

"May I speak with Mr. Perigua?"

A man whirled toward him.

"Don't you see I'm busy, man? Where's the sergeant-at-arms? Why can't you protect the privacy of my office when I'm in conference?"

A short, fat, black man reluctantly broke off an intense declamation and hastened up.

"What can I do for you? Are you a member?"

He seemed a bit suspicious.

"I don't know—I have a message for Mr. Perigua."

"Give it to me."

"It is not written—it is verbal."

"All right, tell me; I'm Mr. Perigua's representative—official sergeant-at-arms of this—" he hesitated and looked suspiciously at Matthew again—"club."

Perigua had heard his name repeated and turned again. He was a thin, yellow man of middle size, with flaming black hair and luminous eyes. He was perpetual motion, talking, gesticulating, smoking.

"What?" he said.

"A message."

"From where?"

"From—abroad."

Perigua leaped to his feet.

"Get out," he cried to his fellows— "State business. Committee will meet again tomorrow night— What? Then Tuesday— No? Well, then tomorrow at noon— You can't— Well, we'll meet without you. Do you think the world must stand still while you guzzle? Come in. Sergeant, I'm engaged. Keep the gate. Well?"

Matthew sat down within the rail on a chair with half a back. The black eyes blazed into his, and the long thin fingers worked. The purple hair writhed out of place.

"I've been in—Berlin."

"Yes?"

"And certain persons—"

"Yes, yes, man. My God! Get on with it."

"—gave me a message—a word of greeting for Mr. Perigua."

"Well, what is it?"

"Are you Mr. Perigua?"

"My God, man—don't you know me? Is there anybody in New York that doesn't know Perigua? Is there anybody in the world? Gentlemen"—he leaped to the rail—"am I Perigua?"

A shout went up.

"Perigua—Perigua forever!" And a song with some indistinguishable rhyme on "Perigua forever" began to roll until he stopped it with an impatient shout and gesture.

"Shut up. I'm busy."

Matthew whispered to him. Perigua listened and rose to his feet with transfigured face.

"Man— My God! Come!" He tore toward the door.

"Le jour de gloire est arrivé! Come, man," he shouted, and dragging Matthew, he reached the door and turned dramatically:

"Men, I have news—great news—the greatest! Salute this Ambassador from the World—who brings salvation. There will be a plenary council tomorrow night. Midnight. Pass the word. Adieu." And as they passed out, Matthew heard the song swell again—"Perigua, Perigua, Per—"

They passed upstairs to another room. It was a bedroom, dirty, disheveled, stuffed with furniture and with stale smells of food and tobacco. A scrawny woman, half-dressed, rose from the bed, and at an impatient sign from Perigua went into the next room and closed the door.

Perigua grasped Matthew by both hands and hugged him.

"Man," he gasped, "man, God knows you're welcome. I am on my last legs. I don't know where to turn. The landlord has dispossessed us, bills are pouring in, and over the country, the world, the brethren are clamoring. Now all is well. We are recognized—recognized by the great leaders of Asia and Africa. Pan-Africa stands at last beside Pan-Asia, and Europe trembles."

Matthew felt his spirits droop. This man was no leader, he was too theatrical. Matthew felt that he must get at the facts before he took any steps.

"But tell me—all about your plans," he said.

"Who are you?" countered Perigua.

Matthew answered frankly.

"I am a Pullman porter. I was a student in medicine, but I quit. I went to Europe and there by accident met people who had heard of you and your plans. They were not agreed, I must say plainly, as to their feasibility, but they commissioned me to investigate."

"Did they send any money?"

"None at present. Later, if my reports are satisfactory, they may."

"And you are a porter? How long have you been in service?"

Matthew answered: "Since this morning. You see, I came back as a scullion. Had some trouble on the boat because I was a stowaway, but despite all, they gave me fifty dollars for my work and offered to hire me permanently. I took the money, bought some clothes, and applied for a Pullman job. It seemed to me that it offered the best opportunity to see and know the Negroes of this land."

"You're right, man, you're right. Have any trouble getting on?"

"Not much."

Perigua pondered. "See here," he said, "I'll make you Inspector of my organization and give you letters to my centers. Travel around as porter. Sound out the country—test out the organization. Make your report soon and get some money. Something must happen, and happen soon."

"But what—" began Matthew.

Matthew never forgot that story. Out of the sordid setting of that room rose the wild head of Perigua, haloed dimly in the low-burning gas. Far out in street and alley groaned, yelled, and sang Harlem. The snore of the woman came fitfully from the next room, and Perigua talked.

Matthew had at first thought him an egotistic fool. But Perigua was no fool. He next put him down as an ignorant fanatic—but he was not ignorant. He was well read, spoke French and Spanish, read German, and knew the politics of the civilized world and current events surprisingly well. Was he insane? In no ordinary sense of the word; wild, irresponsible, impulsive, but with brain and nerves that worked clearly and promptly.

He had a big torn map of the United States on the wall with little black flags clustered over it, chiefly in the South.

"Lynchings," he said briefly. "Lynchings and riots in the last ten years." His eyes burned. "Know how to stop lynching?" he whispered.

"Why—no, except—"

"We know. Dynamite. Dynamite for every lynching mob."

Matthew started and grew uneasy. "But," he objected, "they occur sporadically—seldom or never twice in the same place."

"Always a half dozen in Mississippi and Georgia. Three or four in South Carolina and Florida. There's a lynching belt. We'll blow it to hell with dynamite from airplanes. And then when the Ku Klux Klan meets some time, we'll blow them up. Terrorism, revenge, is our program."

"But—" began Matthew as sweat began to ooze.

Perigua waved. He was a man difficult to interrupt. "We've

got to have messengers continuously traveling to join our groups together and spread news and concert action. The Pullman porters have a new union on old-fashioned lines. I'm trying to infiltrate with the brethren. See? Now you're going to take a job as Inspector and run on a key route. Where are you running now?"

"New York to Atlanta."

"Good! Boys don't like running south. You can do good work there."

"But just a moment—are the Negroes back of you ready for this—this—"

"To a man! That is, the real Negroes—the masses, when they know and understand—most of them are too ignorant and lazy—but when they know! Of course, the nabobs and aristocrats, the college fools and exploiters—they are like the whites."

Matthew thought rapidly. He did not believe a word Perigua said, but the point was to pretend to believe it. He must see. He must investigate. It was wild, unthinkable, terrible. He must see this thing through.

III

"George!"

Inherently there was nothing wrong with the name. It was a good name. The "father" of his country and stepfather of Matthew had rejoiced in it. Thus Matthew argued often with himself.

"George!"

It was the name that had driven Matthew as a student away from the Pullman service. It was not really the name—it was the implications, the tone, the sort of bounder who rejoiced to use it. A scullion, ennobled by transient gold and achieving a sleeping-car berth, proclaimed his kingship to the world by one word:

"George!"

So it seemed at least to oversensitive Matthew. It carried all the meaner implications of menial service and largess of dimes and quarters. All this was involved and implied in the right not only to call a man by his first name, but to choose that name for him and compel him to answer to it.

So Matthew, the porter on the Atlanta car of the Pennsylvania Railroad, No. 183, and Southern Railroad, No. 33, rose in his smart and well-fitting uniform and went forward to the impatiently calling voice.

The work was not hard, but the hours were long, and the personal element of tact and finesse, of estimate of human character and peculiarities, must always be to the fore. Matthew had small choice in taking the job. He had arrived with little money and almost ragged. He had undertaken a mission, and after Perigua's amazing revelation, he felt a compelling duty.

"Do you belong to the porters' union?" asked the official who hired him.

"No, sir."

"Going to join?"

"I had not given it a thought. Don't know much about it."

"Well, let me tell you, if you want your job and good run, keep out of that union. We've got our own company union that serves all purposes, and we're going to get rid gradually of those radicals and Bolsheviks who are stirring up trouble."

Matthew strolled over to the room where the porters were resting and talking. It was in an unfinished dark corner of the station under the stairs, with few facilities and no attempt to make it a club room even of the simplest sort.

"Say," asked Matthew, "what about the union?"

No one answered. Some glanced at him suspiciously. Some went out. Only one finally sidled over and asked what Matthew himself thought of it, but before he could answer, another, passing, whispered in his ear, "Stool pigeon—keep your mouth." Matthew looked after the trim young fellow who warned him. It was Matthew's first sight of Jimmie.

The day had been trying. A fussy old lady had kept him trotting. A woman with two children had made him nurse; four Southern gentlemen gambling in the drawing-room had called him "nigger." He stood by his car at Washington at 9.30 at night, his berths all made. To his delight Jimmie was on the next car, and they soon were chums. Jimmie was Joy. He was not much over twenty-five and so full of jokes and laughter that none, conductors, passengers, or porters, escaped the contagion of his good cheer. His tips were fabulous, and yet he was never merely servile or clownish. He just had bright, straight-eyed good humor, a quick and ready tongue; and he knew his job down to *z*. He was invaluable to the greenness of Matthew.

"Here comes a brownskin," he whispered. "Hustle her to bed if she's got a good berth in the middle of the car, else they'll find a 'mistake' and put her in Lower One," and he sauntered whistling away as the conductor stepped out. The conductor was going in for the diagram.

"Wait till I get back," he called, nodding toward the coming passenger.

The young colored woman approached. She was well dressed but a bit prim. She had Lower Six. Matthew sensed trouble, but remembering Jimmie's admonition, he showed her to her berth. She did not look at him, but he carefully arranged her things.

The conductor came back. "What did you put her there for?" he asked.

"She had a ticket for Six," Matthew answered. Both he and the conductor knew that she had not bought that ticket in person. In Washington, they would never have sold a colored person going south Number Six—she'd have got One or Twelve or nothing. The conductor was mad. It meant trouble for him all next day from every Southerner who boarded the train.

"Tell her there's some mistake—I'll move her later." But Matthew did not tell her. On the contrary, he suggested to her that he make her berth. She knew why he suggested it, and she resented it, but consented without glancing at him. He sympathized even with her resentment. The conductor swore when he came through with the train conductor and found her retired, but he could do nothing, and Matthew merely professed to have misunderstood.

In the morning after an almost sleepless night and without breakfast, Matthew took special care of the dark lady, and when she was ready, carried her bag to the empty drawing-room and let her dress there in comfort. There again he felt and understood the resentment in her attitude. She could not be treated quite like other passengers. Yet she must know it was not his fault, and perhaps she did not know that the extra work of straightening up the drawing-room at the close of a

twenty-four-hour trip was no joke. Still, he smiled in a friendly
way at her as he brought her back to the seat which he had
arranged first, so as to put her to the least unpleasantness from
sitting in some other berth. A woman flounced up and away as
the girl sat down.

She thanked Matthew primly. She was afraid to be familiar
with a porter. He might presume. She was not pretty, but
round-faced, light brown, with black, crinkly hair. She was
dressed with taste, and Matthew judged that she was probably
a teacher or clerk. She had a cold half-defiant air which Matthew understood. This class of his people were being bred that
way by the eternal conflict. Yet, he reflected, they might say
something pleasant and have some genial glow for the en-
couragement of others caught in the same toils.

Then, as ever, his mind flew back to Berlin and to the
woman of his dreams and quest. He wondered where she was
and what she was doing. He had searched the newspapers and
unearthed but one small note in the *New York Sunday Times,*
which proved that the Princess was actually on the *Gigantic:*
"Her Royal Highness, the Princess of Bwodpur, has been
visiting quietly with friends while en route from England to
her home in India, by way of Seattle." He smiled a bit dubi-
ously; what had porters and princesses in common?

He came back to earth and began the daily struggle with
the brushing and the bags through narrow aisles out to the
door; to collect the coats and belongings and carefully brush
the clothes of twenty people; to wait for, take, and appear
thankful for the tip which was wage and yet might be thrown
like alms; to find lost passengers in the smoking-car, toilet, or
dining-room and lost hats, umbrellas, packages, and canes—
Matthew came to dread the end of his journeys more than all
else.

His colored passenger did "not care" to be brushed. As they
rolled slowly through the yards, he glanced at her again.

"Anything I can do for you?" he asked.

"Aren't you a college man?" she asked, rather abruptly.

"I was," he answered, wiping the sweat from his face.

She regarded him severely. "I should think then you'd be
ashamed to be a porter," she said.

He bit his lips and gathered up her bags.

"It's a damned good thing for you that I am," he wanted to

say; but he was silent. He only hoped desperately that she would not offer to tip him. But she did; she gave him fifteen cents. He thanked her.

IV

With a day off in Atlanta, Matthew and Jimmie looked up Perigua's friends. Jimmie laughed at the venture, although Matthew did not tell him much of his plans and reasons.

"Don't worry," grinned Jimmie; "let the white folks worry; it'll all come out right."

They had a difficult time finding any of the persons to whom Perigua had referred Matthew. First, they went down to Decatur Street. It was the first time Matthew had been so far south or so near the black belt. The September heat was intense, and the flood of black folk overwhelmed him. After all, what did he know of these people, of their thoughts, ambitions, hurts, plans? Suppose Perigua really knew and that he who thought he knew was densely ignorant? They walked over to Auburn Avenue. Could any one tell them where the office of the *Arrow* was? It was up "yonder." Matthew and Jimmie climbed to an attic. It was empty, but a notice sent them to a basement three blocks away—empty, too, and without notice. Then they ran across the editor in a barber shop where they were inquiring—a little, silent, black man with sharp eyes. No, the *Arrow* was temporarily suspended and had been for a year. Perigua? Oh, yes.

"Well, there could be a conference tonight at eight in the Odd Fellows Hall!—one of the small rooms."

"At what hour?"

"Well, you know colored people."

If he came at nine he'd be early. Yes, he knew Perigua. No, he couldn't say that Perigua had much of a following in Atlanta, but Perigua had ideas. Perigua had—yes, ideas; well, then, at nine. Jimmie said he'd leave him at that. He had a date, and he didn't like speeches anyhow. They parted, laughing.

Nobody came until nine-thirty; by ten there was the editor, an ironmolder, a college student, a politician, a street cleaner, a young physician, an insurance agent, and two men who might have been idlers, agitators, or plain crooks. It was an ugly

room, incongruously furnished and with no natural center like
a fireplace, a table, or a rostrum. Some of the men smoked,
some did not; there was a certain air of mutual suspicion.
Matthew gathered quickly that this was no regular group, but
a fortuitous meeting of particles arranged by the editor. Instead of listening to a conference, he found himself introduced
as a representative of Mr. Perigua of New York, and they prepared to hear a speech. Matthew was puzzled, nonplused,
almost dumb. He hated speech-making. His folk talked too
easily and glibly in his opinion. They did not mean what they
said—not half—but they said it well. But he must do something; he must test Perigua and his followers. He must know
the truth. So Matthew talked—at first a little vaguely and
haltingly; and then finally he found himself telling them
almost word for word that conversation about American
Negroes in Berlin. He did not say who talked or where it
took place; he just told what was said by certain strangers.
They all listened with deep absorption. The student was the
first to break out with:

"It is the truth; we're punk—useless sheep; and all because
of the cowardice of the old men who are in the saddle. Youth
has no recognition. It is fear that rules. Old slipper is afraid
of missing his tea and toast."

The editor agreed. "No recognition for genius," he said.
"I've published the *Arrow* off and on for three years."

"Usually off," growled the politician.

"And a damn poor paper it is," added the ironmolder.

"I know it, but what can you expect from two hundred and
fifty-eight paid subscribers? If I had five thousand I'd show
you a radical paper."

"Aw, it's no good—niggers won't stay put," returned the
politician.

"You mean they won't stay sold," said some one.

"We're satisfied—that's the trouble," said the editor. "We're
too damn satisfied. We've done so much more with ourselves
than we ever dreamed of doing that we're sitting back licking
our chops and patting each other on the back."

"Well," said the young physician, "we *have* done well,
haven't we?"

"*You* has," growled the ironmolder. "But how 'bout us?
You-all is piling up money, but it don't help us none. If we

had our own foundries, we'd get something like wages 'stead of scabbing to starve white folks."

"Well, you know we are investing," said the insurance agent. "Our company—"

"Hell! That ain't investment, it's gambling."

"That's the trouble," said the scavenger. "We'se strivers; we'se climbing on one another's backs; we'se gittin' up—some of us—by trompin' others down."

"Well, at any rate, some do get up."

"Yes, sure—but the most of us, where is we going? Down, with not only white folks but niggers on top of us."

"Well, what are we going to do about it?"

"What can we do? Merit and thrift will rise," said the physician.

"Nonsense. Selfishness and fraud rise until somebody begins to fight," answered the editor.

"Perigua is fighting."

"Perigua is a fool—Negroes won't fight."

"*You* won't."

"Will *you?*"

"If I get a chance."

"Chance? Hell! Can't any fool fight?" asked the editor.

"Sure, but I ain't no fool—and besides, if I was, how'd I begin?"

"How!" yelled the student. "Clubs, guns, dynamite!"

But the politician sneered. "You couldn't get one nigger in a million to fight at all, and then they'd sell each other out."

"You ought to know."

"I sure do!"

And so it went on. When the meeting broke up, Matthew felt bruised and bewildered.

v

Matthew walked into the church about noon. Jimmie positively refused to go. "Had all the church I need," he said. "Besides, got a date!" The services were just beginning. It was a large auditorium, furnished at considerable expense and with some taste. It gave a sense of space and well-being. The voices of the surpliced choir welled up gloriously, and the tones of the minister rolled in full accents.

Matthew particularly noticed the minister. He remembered the preacher at his own home—an old, bent man, outlandish, with blazing èyes and a fire of inspiration and denunciation that moved every auditor. But this man was young—not much older than Matthew—good-looking, intelligent and educated. This service of mingled music and ceremony was attractive, and the sermon—well, the minister did not say much, but he said it well; and if conventionally and with some tricks of the orator, yet he was pleasing and soothing. His Death was an interpretation of Fall—the approach of looming Winter and the test of good resolutions after the bursting Spring and fruitful Summer.

The audience listened contentedly but with no outbursts of enthusiasm. There were a few "amens" from the faithful near the pulpit, but they followed the cadence of the beautiful voice rather than the impact of his ideas. The audience looked comfortable, well fed and well clothed. What were they really thinking? What did the emancipation of the darker races mean to them?

Matthew lingered after the service, and his tall, well-clad figure attracted attention. A deacon welcomed him. He must meet the pastor; and at the door in his silk robe he did meet him. They liked each other at a glance. The minister insisted on his waiting until most of the crowd had passed. Matthew ventured on his queries.

"I've just returned from Germany—" he began.

The minister beamed: "Well, well! That's fine. Hope to take a trip over myself in a year or two. My people here insist. May get a Walker popularity prize. Now what do people over there think of us here? I mean, of us colored folk?"

It was the opening. Matthew explained at length some of the opinions he had heard expressed. The minister was keen with intelligent interest, but just as he was launching out in comment, they were interrupted.

"Brother Johnson, we're ready now and dinner will be on the table. Mustn't keep the old lady waiting."

The speaker was a big, dark man, healthy-looking and pleasant, carefully tailored with every evidence of prosperity. His car and chauffeur were at the curb—a new Cadillac sedan.

The minister hesitated. "My friend, Mr. Towns—just from Germany—" he began.

"Delighted, I'm sure."

"Yes, this is Brother Jones, president of the Universal Mutual—you've heard, I know, of our greatest insurance society. Mr. Towns is just from Germany. I'll—"

"Bring him along—bring him right along and finish your talk at the table. Always room in the pot for one more. Germany? Well! Well! Are they still licked over there? Been promising the old lady a trip for the last ten years—Germany, France, Italy and all. Like to take in Africa too. But you know how it is—business—" And they were packed into the big car and gliding away.

There was no chance to finish the talk with the young minister. The host started off talking about himself, and nothing could stop him. His home was big and costly—too overdone to be beautiful, but with a good deal of comfort and abundant hospitality. He served a little whiskey to Matthew upstairs with winks and asides about the minister; and then, downstairs and everywhere he talked of himself. He was so naïve and so thoroughly interested in the subject that none had the heart to interrupt him, although his wife, as she fidgeted in and out helping the one rather unskillful maid, would say now and then:

"Now, John—stop boasting!"

John would roar good-naturedly—hand around another helping of chicken or ham, pass the vegetables and hot bread, and begin just where he left off: "And there I was without a cent, and four hundred dollars due. I went to the bank—the First National—old man Jones was my people, his grandfather owned mine. 'Mr. Jones,' says I, 'I want five hundred dollars cash today!'

" 'Well, John,' says he, 'what's the security?'

" 'I'm the security,' says I, and, sir, he handed me the cash! Well, he wasn't out nothing. My check in five figures goes at the bank today—don't it, Reverend?" And so on, and so forth. It was frank, honest self-praise, and his audience hung on his words, although most of them had heard the story a hundred times.

"So you've been to Germany? Well, well! Have they got them radicals in jail yet? Italy's got the dope. Old Moso—what's his name? Mr. Jones was saying the last time I was in the bank, making my weekly deposit—what was it? About six thousand dollars, as I remember—says he, 'John, we need a Mosleny right here in America!' "

"You're not against reform, are you, Mr. Jones?"

"No; no, sir, I'm a great reformer. But no radical. No anarchist or Bolshevik. We've got to protect property."

There was an interruption from some late arrivals.

"What boat did you return on?" asked somebody.

Matthew smiled and hesitated.

"The *Gigantic,*" he said, and he wanted to add, "In the scullery." Could they stand the joke? He looked up and decided they couldn't, for he was looking into the eyes of the latest arrival, and she was the prim young person who had tipped him fifteen cents yesterday morning!

"Mr. Towns, who has just returned from a trip to Germany, Miss Gillespie. Miss Gillespie is our new principal of the recently equipped Jones school—named after the President of the First National."

Matthew smiled, but Miss Gillespie did not. She frankly stared, bowed coldly, and then, after a small mouthful or two, whispered to her neighbor. The neighbor whispered, and then slowly the atmosphere of the table changed. Matthew was embarrassed and amused, and yet how natural it all was—that unfortunate smile of his—that unexplained trip to Germany, and the revelation evidently now running around that he was a Pullman porter. They thought him a liar through and through. It was not simply that he was a railway porter—no, no! Mr. Jones was democratic and all that; but after all one did not make chance porters guests of honor; and Mr. Jones, when the whispering reached him, grew portentously and emphatically silent.

Matthew, now thoroughly upset, rose with the others and made his way straight to the minister.

"Say, I seem to have cut a hog," he said. The minister smiled wanly and said, "I'm afraid I'm to blame—I—"

"No, no," said Matthew, and then tersely he told of his rebuff and flight to Europe and his return to "begin again." "I did not mean really to sail under false colors, and I did come home on the *Gigantic.* I pared potatoes all the way over."

The minister burst into a laugh. They shook hands, and with a hurried farewell to a rather gruff host, Matthew slipped

away. But he left fifteen cents for Miss Gillespie. Jimmie
roared when Matthew told him.

VI

In October, Matthew wrote his first report for the Princess.
He wrote it on his knee and in his one chair, sitting high up in
the narrow furnished bedroom which he had hired in New York
on West Fifty-ninth Street. It was a noisy and dirty region,
but cheap and near his work. There was a bed, a chair, and a
washstand, and he had bought a new trunk, in which he locked
up his clothes and few belongings.

"YOUR ROYAL HIGHNESS:
    "I have at your request made a hasty but careful survey of
the attitude of my people in this country, with regard to the
possibility of their aid to a movement looking toward righting
the present racial inequalities in the world, especially along the
color line.
    "My people are increasing in material prosperity. A few are
even accumulating wealth; large numbers own their homes and
live in cleanliness and a fair degree of comfort. Extreme poverty and crime are decreasing, while intelligence is increasing.
There is still oppression and insult, some lynching, and much
caste and discrimination; but on the whole the Negro has advanced so rapidly and is still advancing at such a rate that
he is more satisfied than complaining. He is astonished and
gratified at his own success, and while he knows that he is
not treated quite as a man and lacks the full freedom of a
white man, he believes that he is daily approaching this goal.
    "This main movement and general feeling is by no means
shared by all. There is bitter revolt in the hearts of a small
intelligentsia who resent color-caste, and among those laborers
who feel in many ways the pinch of economic maladjustment
and see the rich Negroes climbing on their bowed backs. These
classes have no common program, save complaint, protest, and
inner bickering; and only in the case of a small class of im-
migrant West Indians has this complaint reached the stage that
even contemplates violence.
    "Perigua is a man of intelligence and fire—of a certain

honesty and force; but he is not to be trusted as a leader. His organization is a loose mob of incoherent elements united only by anger and poverty. They have their reasons. I first thought Perigua a vain and egotistic fool. But the other night sitting in his dirty flat—a furious, ragged figure of bitter resentment—he told me part of his story; spat it out in bits: of the beautiful wife whom an Englishman seduced; the daughter who became a prostitute; of the promotion refused him in the railroad shops because of color, and his fight with the color line in the army; of his prosecution for 'inciting to riot'; his conviction and toil in jails and chain gangs, with vermin and disease; and of his long, desperate endeavor to stir up revolt in America. Perigua and those whom he represents have a grievance and a remedy, but he will never accomplish anything systematic. Do not think of contributing to his organization. I still have your envelope unopened. He has no real organization. He has only personal followers.

"On the other hand, the Negroes are thick with organizations—they are threaded through with every sort of group movement; but their organizations so far chime and accord with the white world; their business organizations, growing fast, have the same aims and methods as white business; their religion is a replica of white religion, only less snobbish because less wealthy. Even their labor movement is the white trade union, hampered by the fact that white unions discriminate and that colored labor is the wage-hammering adjunct of white capital.

"Your Highness and the friends whom I had the honor of meeting may then well ask, 'Are these folk of any possible use to a movement to abolish the present dictatorship of white Europe?' I answer, yes, a hundred times. American Negroes are a tremendous social force, an economic entity of high importance. Their power is at present partly but not wholly dissipated and dispersed into the forces of the overwhelming nation about them. But only in part. A tremendous striving group force is binding this group together, partly through the outer pounding of prejudice, partly by the growth of inner ideals. What they can and will do in the rebuilding of a better, bigger world is on God's knees and not now clear; but clarity dawns, and so far as we gain self-consciousness today we can be a force tomorrow.

"The burning question is: What help is wanted? What can we do? What are your aims and program? I know well from your own character and thought that you could not encourage mere terrorism and mass murder. The very thought of ten million grandchildren of slaves trying to wrest liberty from ten times their number of rich, shrewder fellows by brute force is, of course, nonsense. On the other hand, with intelligence and forethought, concentrated group action, we can so align ourselves with national and world forces as to gain our own emancipation and help all of the colored races gain theirs.

"Frankly then, what is the Great Plan? How and when can we best coöperate? What part can I take? I am eager to hear from you.

"I am, your Royal Highness"—

There came a knock on the door, and Matthew opened it. A young Japanese stood there who politely asked for Mr. Keswick. No, Mr. Keswick was not here and did not live here. In fact, Matthew had never heard of Mr. Keswick. The Japanese was sorry—very, very sorry for the intrusion. He went softly down the stairs.

VII

There was no answer to Matthew's report. He had given the Princess a temporary address at Perigua's place, and in this report he enclosed this room as a permanent one. He had sent the Princess' letter to her bankers, as agreed on. Still there was no word or sign. Matthew was at first patient. After the second week, he tried to be philosophical. At the end of a month, he was disappointed and puzzled. By the first of December the whole thing began to assume a shape grotesque and unreal. They over there had perhaps succeeded in changing her mind. Perhaps she herself, coming and seeing with her own eyes, had been disillusioned. It would be hard for a stranger to see beneath the unlovely surface of this racial tangle. But somehow he had counted on this woman—on her subtlety and vision; on her own knowledge of the color line.

He did not know what to do. Should he write again? His pride said no, but his loyalty and determination kept him following up Perigua and remaining in touch with him. At least once a week they had conferences and Matthew reported. This week there was, as usual, little to report. He had seen a dozen men—three crazy, three weak, three dishonest, three willing but bewildered, dazed, lost. Broken reeds all. Perigua listened dully, hunched in his chair, chewing an unlit cigar— unkempt, unshaven, ill. His eyes alone lived and flamed as with unquenchable fire.

"Any money yet from abroad?" he asked.

"No."

"Have you asked for any?"

"No."

"So," said Perigua. "You think it useless?" Hitherto Matthew had tried to play his part—to listen and study and say little as to his own thought. Suddenly, now, a pity for the man seized him. He leaned forward and spoke frankly:

"Perigua, you're on the wrong tack. First of all, these people are not ready for revolt. And next, if they were ready, it's a question if revolt is a program of reform today. I know that time has been when only murder, arson, ruin, could uplift; when only destruction could open the path to building. The time must come when, great and pressing as change and betterment may be, they do not involve killing and hurting people."

Perigua glared. "And that time's here, I suppose?"

"I don't know," said Matthew. "But I hope, I almost believe it is. It must be after that hell of ten years ago. At any rate, none of us Negroes are ready for such a program against overwhelming odds—"

"No," yelled Perigua. "We're tame tabbies; we're fawning dogs; we lick and growl and wag our tails; we're so glad to have a white man fling us swill that we wriggle on our bellies and crawl. We slave that they may loll; we hand over our daughters to be their prostitutes; we wallow in dirt and disease that they may be clean and pure and good. We bend and dig and starve and sweat that they may sit in sweet quiet and reflect and contrive and build a world beautiful for themselves to enjoy.

"And we're not ready even to protest, let alone fight. We want to be free, but we don't dare strike for it. We think that the blows of white men—of white laborers, of white women— are blows for us and our freedom! Hell! you damned fool, they have always been fighting for themselves. Now, they're half free, with us niggers to wait on them; we give white carpenters and shop girls their coffee, sugar, tea, spices, cotton, silk, rubber, gold, and diamonds; we give them our knees for scrubbing and our hands for service—we do it and we always shall until we stand and strike."

And Perigua leaped up, struck the table until his clenched hand bled.

Matthew quailed. "I know—I know," he said. "I'm not minimizing it a bit. In a way, I'm as bitter about it all as you—but the practical question is, what to do about it? What will be effective? Would it help, for instance, to kill a couple of dozen people who, if not innocent of intentional harm, are at least unconscious of it?"

"And why unconscious? Because we don't make 'em know. Because you've got to yell in this world when you're hurt; yell and swear and kick and fight. We're dumb. We dare not talk, shout, holler. And why don't we? We're afraid, we're scared; we're congenital idiots and cowards. Don't tell me, you fool—I know you and your kind. Your caution is cowardice inbred for ten generations; you want to talk, talk, talk and argue until somebody in pity and contempt gives you what you dare not take. Go to hell—go to hell—you yellow carrion! From now on I'll go it alone."

"Perigua—Perigua!"

But Perigua was gone.

Matthew was nonplused. All his plans were going awry. Still no word or sign from the Princess, and now he had alienated and perhaps lost touch with Perigua. What next? He paused in the smoky, dirty club rooms and idly thumbed yesterday morning's paper. Again he inquired for mail. Nothing. He stood staring at the paper, and the first thing that leaped at him from a little inconspicuous paragraph on the social page was the departure "for India, yesterday, of her Royal Highness, the Princess of Bwodpur."

He walked out. So this was the end of his great dream—his world romance. This was the end. Whimsically and for the last time, he dreamed his dream again: The Viktoria Café and his clenched fist. The gleaming tea table, the splendid dinner. Again he saw her face—its brave, high beauty, its rapt interest, its lofty resolve. Then came the grave face of the Japanese, the disapproval of the dark Indians, the contempt

of the Arab. They had never believed, and now he himself
doubted. It was not that she or he had failed—it was only
that, from the beginning, it had all been so impossible—so
utterly unthinkable! What had he, a Negro, in common with
what the high world called royal, even if he had been a successful physician—a great surgeon? And how much less had
Matthew Towns, Pullman porter!

A dry sob caught in his throat. It was hard to surrender
his dream, even if it was a dream he had never dared in
reality to face. Well, it was over! She was silent—gone. He
was well out of it, and he walked outdoors. He walked quickly
through 135th Street, past avenue and park. He climbed the
hill and finally came down to the broad Hudson. He walked
along the viaduct looking at the gray water, and then turned
back at 133rd Street. There were garages and old, decaying
buildings in a hollow. He hurried on, past "Old Broadway"
and up a sordid hill to a still terrace, and there he walked
straight into the young Atlanta minister.

"Hello! I *am* glad to see you."

For a moment Matthew couldn't remember—then he saw
the picture of the church—the dinner and the Joneses.

He greeted the minister cordially. "How is Miss Gillespie?"
he asked with a wry grin.

"Married—married to that young physician you met at the
radical conference. Oh, you see we followed you up. They have
gone to Chicago. Well, here I am in New York on a holiday.
Couldn't get off last summer and thought I'd run away just
before the holidays. Been here a week and going back tomorrow. Hoped I might run across you. I feel like a man out
of a strait-jacket. I tell you this being a minister today is—
is—well, it's a hard job."

"My experience is," said Matthew, "that life at best is no
cinch."

The minister smiled sympathetically. "I tell you," he said,
"let's have a good time. I want to go to the theater and see
movies and hear music. I want to sit in a decent part of a
good theater and eat a good dinner in a gilded restaurant, and
then"—he glanced at Matthew—"yes, then I want to see a
cabaret. I've preached about ballrooms and 'haunts of hell,'"
he said with a whimsical smile, "but I've never seen any."

Matthew laughed. "Come on," he said, "and we'll do the

best we can. The first balcony is probably the best we can do
at a theater, and not the best seats there; but in the movies
where 'all God's chillun' are dark, we can have the best. That
gilded restaurant business will be the worst problem. We'd
better compromise with the dining-room at the Pennsylvania
station. There are colored waiters there. At the Grand Central
we'd be fed, but in the side aisles. But what of it? I'm in for
a lark, and I too have a day off.—In fact, it looks as though
I had a life off."

They visited the Metropolitan Art Museum at the minister's
special request; they dined about three at the Grand Central
station, sitting rather cosily back but on one side, at a table
without flowers. Matthew calculated that at this hour they
would be better received than at the more crowded hours.
Then they went at six to the Capitol and sat in the great,
comfortable loge chairs.

The minister was in ecstasy. "White people have everything, don't they?" he mused, as they walked up the Great
White Way slowly, looking at the crowds and shop windows.
"These girls, all dressed up and painted. They look—but—are
many of them for sale?"

"Yes, most of them are for sale—although not quite in the
way you mean. And the men, too," said Matthew.

The minister was a bit puzzled, and as they went into the
Guild Theater, said so. It was an exquisite place and they
had fairly good seats, well forward in the first balcony.

"What do you mean—'for sale'?" he asked.

"I mean that in a great modern city like New York men and
women sell their bodies, souls, and thoughts for luxury and
beauty and the joy of life. They sell their silences and dumb
submissions. They are content to do things and let things be
done; they promise not to ask just what they are doing, or for
whom, or what it costs, or who pays. That explains our
slavery."

"This is not such bad slavery."

"No—not for us; but look around. How many Negroes are
here enjoying this? How many can afford to be here at the
wages with which they must be satisfied if these white folks
are to be rich?"

"You mean that all luxury is built on a foundation of
poverty?"

"I mean that much of the costliest luxury is not only ugly and wasteful in itself but deprives the mass of white men of decent homes, education, and reasonable enjoyment of life; and today this squeezed middle white class is getting its luxuries and necessities by inflicting ignorance, slavery, poverty, and disease on the dark colonies of European and American imperialism. This is the New Poverty and the basis of armies, navies, and war in Nicaragua, the Balkans, Asia, and Africa. Without this starvation and toil of our dark fellows, you and I could not enjoy this."

The minister was silent, for the play began. He only murmured, "We are consenting too," and then he choked—and half an hour later, as the play paused, added, "And what are we going to do about it? That's what gets me. We're in the mess. It's wrong—wrong. What can we do? I can't see the way at all."

Then the play swung on: beautiful rooms; sleek, quiet servants; wealth; a lovely wife loving another man. The husband kills him; the curtain leaves her staring at a corpse with horror in her eyes.

The minister frowned. "Do they always do this sort of thing?" he asked.

"Always," Matthew answered; and the minister added: "Why can't they try other themes—ours for instance; our search for dinner and our reasons for the first balcony. Good dinner and good seats—but with subtle touches, hesitancies, gropings, and refusals that would be interesting; and that woman wasn't interesting."

They rode to Harlem for a midnight lunch and planned afterward to visit a cabaret. The minister was excited. "Don't flutter," said Matthew genially; "it'll either be tame or nasty."

"You see," said the minister, "sex is curiously thrust on us parsons. Men dislike us—either through distrust or fear. Women swarm about us. The Church is Woman. And there I am always, comforting, advising, hearing tales, meeting evil—ducking, dodging, trying not to understand—not understanding—that's the trouble. Towns, what the devil should I know of the temptations—the dirt—the—"

"Look here!" interrupted Towns. They were in a restaurant on Seventh Avenue. It was past midnight. The little half-basement was tasteful and neat, but only a half dozen people were there. The waffles were crisp and delicious. Matthew had bought a morning paper. Glancing at it carelessly, as the minister talked, he shouted, "Look here!" He handed the paper to the minister and pointed to the headlines. The Ku Klux Klan was going to hold a great Christmas celebration in Chicago.

"In Chicago?"

"Yes."

"But Chicago is a stronghold of Catholics."

"I know. But watch. The Klan is planning a comeback. It has suffered severe reverses in the South and in the East; I'll bet a dollar they are going to soft-pedal Rome and Jewry and concentrate on the new hatred and fear of the darker races in the North and in Europe. That's what this meeting means."

The minister frowned and read on. . . . Klansmen from the whole country will meet there. The grand officers and Southern members will go from headquarters at Atlanta on a luxurious special train and meet other Klansmen and foreign guests in Chicago; there they will discuss the repeal of the Fifteenth Amendment, and prepare for a great meeting on the Rising Tide of Color, to be called later in Europe.

"And we sit silent and motionless," said the minister.

"That's it; not only injustice, oppression, insult, a lynching now and then—but they rub it in, they openly flout us. Is there any group on earth, but us, who would lie down to it?"

The minister was silent.

Then he said, "They may be rallying against Rome and liquor rather than against us."

"Nonsense," said Matthew, and added, "What do you think of violence?"

"What do you mean?"

"Suppose Negroes should blow up that convention or that fine de luxe Special and say by this bloody gesture that they didn't propose to stand for this sort of thing any longer?"

The minister quailed. "But what good? What good? Murder, and murder mainly of the innocent; revenge, hatred, and a million 'I told you so's.' 'The Negro is a menace to this land!' "

"Yes, yes, all that; but not simply that. Fear; the hushing of loose slander and insult; the curbing of easy proposals to deprive us of things deeper than life. They look out for the Indian's war whoop, the Italian's knife, the Irishman's club; what else appeals to barbarians but force, blood, war?"

The minister answered slowly: "These things get on our nerves, of course. But you mustn't get morbid and too impatient. We've come a long way in a short time, as time moves. We're rising—we're getting on."

But Matthew brooded: "Are we getting on so far? Aren't the gates slowly, silently closing in our faces? Isn't there widespread, deep, powerful determination to make this a white world?"

The minister shook his head; then he added: "We can only trust in Christ—"

"Christ!" blurted Matthew.

VIII

The cabaret was close, hot, and crowded. There was loud music and louder laughter and the clinking of glasses. More than half the patrons were white, and they were clustered mostly on one side. They had the furtive air of fugitives in a foreign land, out from under the eyes of their acquaintances. Some were drunk and noisy. Others seemed looking expectantly for things that did not happen, but which surely ought to happen in this bizarre outland! The colored patrons seemed more at home and natural. They were just laughing and dancing, although some looked bored.

The minister stared. "Are they having a good time, or just trying to?"

"Some of them are really gay. This girl here—"

The minister recoiled a little as the girl reached their table. She was pale cream, with black eyes and hair; and her body, which she was continuously raising her clothes to reveal, had a sinuous, writhing movement. She danced with body and soul and sang her vulgar "blues" with a harsh, shrill voice that hardly seemed hers at all. She was an astonishing blend of beauty, rhythm, and ugliness. She had collected all the cash in sight on the white side and now came over to the Negroes.

"Come on, baby," she yelled to the minister, as she began singing at their table, and her writhing body curled like a wisp of golden smoke. The minister recoiled, but Matthew looked up and smiled. Some yearning seized him. It seemed so long since a woman's hand had touched him that he scarce saw the dross of this woman. He tossed her a dollar, and as she stooped to gather it, she looked at him impishly and laughed in a softer voice.

"Thanks, Big Boy," she said.

The proprietor with his half-shut eyes and low voice strolled by.

"Would you boys like a drop of something—or perhaps a little game?"

The minister did not understand.

"Whiskey and gambling," grinned Matthew. The minister stirred uneasily and looked at his watch. They stayed on, ordering twenty-five-cent ginger ale at a dollar a bottle and gay sandwiches at seventy-five cents apiece, and a small piece at that.

"Honest," said the minister, "I'm not going to preach against cabarets and dance halls any more. They preach against themselves. There's more real fun in a church festival by the Ladies' Aid!" Then he glanced again at his watch. "Good Lord, I must go—it's three o'clock, and I must leave for Philadelphia at six."

Matthew laughed and they arose. As they passed out, the dancing girl glided by Matthew again and slipped her hand in his.

"Come and dance, Big Boy," she said. Her face was hard and older than her limbs, but her eyes were kind. Matthew hesitated.

"Good-by," he said to the minister, "hope to see you again some time soon."

He went back with the girl.

IX

That trip in his Pullman seemed Matthew's worst. Sometimes as he swung to Atlanta and back he almost forgot himself in the routine, and Jimmie's inexhaustible humor always helped. He became the wooden automaton that his job required. He neither thought nor saw. He had no feelings, no wishes, and yet he was ears and voice, swift in eye and step, accurate and deferential. But at other times all things seemed to happen and he was a quivering bundle of protests, nerves— a great oath of revolt. It seemed particularly so this trip, perhaps because he was so upset about the Princess' departure.

Besides that, Jimmie left him at Atlanta. He had taken a few days off. "Got a date," he grinned.

Matthew was lonesome and tired, and his return trip began with the usual lost article. People always lose something in a Pullman car, and always by direct accusation, glance, or innuendo the black porter is the thief. This time a fat, flashily dressed woman missed her diamond ring.

"—a solitaire worth five hundred dollars. I left it on the window-sill—it has been stolen."

She talked loudly. The whole car turned and listened. The whole car stared at Matthew. It is no pleasant thing to be tacitly charged with theft and to search for vindication under the accusing eyes of two dozen people. Matthew took out the seats, raised the carpet, swept and poked. Then he went and dragged out all the dirty bed-linen from the close-packed closet and went over it inch by inch. He searched the women's toilet room. Then coming back with growling conductor and whispering passengers, he found the ring finally in the spittoon. He got little thanks—indeed he knew quite well that some would think he had concealed it there. The woman gave him fifty cents. Also he missed his breakfast, and his head ached.

The inevitable woman with the baby was furious, for in his search he had forgotten to get the hot milk from the diner, and the cook had used it. A man passed his station because the train conductor had not been notified of the extra stop. The Pullman conductor placed the blame on the porter.

"Damn niggers are good for nothing," said the angry man.

Of course Matthew was supposed to be a walking encyclopedia of the country they were traversing:

"What town is this?"

"Greensboro, madam."

"What mountains are those?"

"The Blue Ridge, sir."

"What creek are we crossing?"

"I don't know, madam."

"Well, don't you know anything?"

Matthew silently continued his dusting.

"Is that the James River?"

"It's a portion of it, madam."

"Is that darky trying to be smart?"

The bell rang furiously. To Matthew's splitting head it

seemed always angry. He brought cup after cup of ice water to people too lazy to take a dozen steps.

"Why the hell don't you answer the bells when they ring?" growled the poker gambler who had the drawing-room. "Bring us some C. & C. ginger ale and be quick about it."

"Sorry, but the—"

"Don't answer me back, nigger."

Matthew went and brought Clicquot Club, the only kind they carried. Apparently the passenger did not know the difference.

It was dinner time and he got a moment to sit down in the end section and dozed off.

"Do you hear?" an elderly man was yelling at him. "Which way is the diner?"

"Straight ahead, sir, second car."

The man looked at him. "Asleep at your post is not the way to get on in this world," he said.

Matthew looked at him. His patience was about at an end, and the man saw something in his eye; he added as he turned away: "Young man, my father fought and died to set you free."

"Well, he did a damned poor job," said Matthew, and he went into the smoking-room and into the toilet and shut and locked the door.

It was nearly ten at night when dinner for the porters was ready, for the passengers had stuffed themselves at lunch and were not hungry until late; the food left was cold and scarce, and the cooks too tired to bother. He was greeted by a chorus when he returned to the car. It began as he passed the drawing-room:

"Where's that porter—George!"—"Can you get me some liquor—any fly girls on the train—how about that one in Lower 5?" Then outside: "Porter, will *you please* make this berth—you've passed it repeatedly. These colored men are too presuming."—"Water!"—"When do we get to ——?"—"What station was that?"—"Please hand me my bag."—"How can I get into that upper? Haven't you a lower?"—"Where's the conductor?"—"What connections can I make?"—"How late are we?"—"When do we change time?"—"When is breakfast?"—"That milk for ba-aby, and right off!"—"Ice water."—"Shoes!"

Matthew left the train with a gasp and took the subway to Harlem. It was after midnight and clear and cold. He wanted warmth and company, and he went straight to the cabaret. He knew he was going, and all day long the yearning for some touch of sympathy and understanding had been overpowering. He wanted to forget everything. He was going to get drunk. He walked by Perigua's place from habit. It was closed and vacant. No one whom he saw could tell him where Perigua was. Matthew turned and walked straight to the cabaret.

"Hello, Big Boy."

He gripped the girl's hand. It was the only handclasp that seemed even friendly that he had had for a long time. She curled her arm about his neck. "What do you say to a drink?" she asked. He drank the stuff that burned and rankled. He danced with the girl, and all the time his head ached and whirled. What could he do? What should he do?

He went out with her at four o'clock in the morning; he scarcely knew when or why. He wanted to forget the world. They whirled away in a taxi, and stumbled up long stairs, and then with a sigh he slipped his clothes off, and clasping his arms around her curving form, fell into dreamless sleep.

X

At the head of the stairs next morning Matthew met Perigua. The girl had looked at his haggard face with something like forgotten shame.

"Good-by, Big Boy," she said, "you ain't built for the sporting game. I wish"—she looked at him uncertainly, her face drawn and coarse in the morning light, her body drooping—"I wish I could help some way. Well, if you ever want a friend, come to me."

"Thank you," he said simply, and kissing her forehead, went. For a long time she stood with that kiss upon her brow.

Then he met Perigua coming out of the door opposite. Was he in Perigua's building? He had been too drunk the night before to notice. No, this was too narrow for 135th Street. He met Perigua, and Perigua blazed at him:

"You're having a hell of a time, ain't you! Prostitutes instead of patriotism." Then he snarled, "Wake up! The time is come! Have you seen this?"

It was an elaborate account of the coming meeting of the
Klan in Chicago. Perigua was trembling with excitement.
Matthew looked at him sharply. Something else was wrong; he
looked hungry and wrought up with drink or excess. Matthew
glanced at the paper. The great Klan Special was leaving
Atlanta for Chicago three days later at 3.40 in the afternoon.
Special cars with certain high guests would join them at various
points and from various cities.

"I'm going to Chicago," said Perigua.

Matthew seized him by the shoulders.

"All right," he said, "but first come and have breakfast."

Perigua hesitated and then morosely yielded. They ate
silently and then smoked.

"Perigua," said Matthew, suddenly, "have you got money to
go to Chicago?"

"Is that any of your damned business?"

"Yes, it is. If you are going to Chicago to look over the
situation, consult with your lieutenants, and lay plans for
future action, you need money. You ought to buy some clothes
and stop at a good hotel."

Matthew knew perfectly well that Perigua was going on
some hare-brained mission and that he might in desperation do
actual harm. He knew, too, that Perigua would like to go, or
to imagine he was going, on some such mission as Matthew
had sketched. Suddenly, Matthew was thinking of that unopened envelope given him by the Princess. Perhaps there lay
the answer to her silence and departure as well as money. The
envelope was to go to Perigua only in case he was found trust-
worthy. But in case he was not and the envelope could not be
returned, what then?

He took a quick resolve. "Come by my room—it's on the
way to the train."

Silently Perigua followed. They went down by Elevated and
soon were sitting in that upper room. Matthew went to his
trunk. It was unlocked. He was startled. He did not remember
leaving it unlocked; he searched hurriedly. Everything seemed
intact, even his bank book and especially the sealed letter
at the bottom, hidden among books. Matthew did not touch
the envelope, but took out his savings bank book, and said:

"I'm going to give you one hundred and fifty dollars to get

some clothes and go to a good hotel in Chicago. Try the Vin-
cennes—I'll write you there."

They went out together to the bank.

Matthew returned feeling that he had done a wise thing.
He had a string on Perigua and could keep in touch with him.
Now for that envelope. The more he thought of it, the more he
was sure that it would throw light on the situation. It was
careless of him to have left his trunk unlocked. The landlady
was all right, but the other lodgers! He drew out the letter
and paused. What did he mean to do? He tore the letter open.
A piece of paper fluttered out. He searched the envelope.
Nothing more. He looked at the paper.

"SIR:

"In unwavering determination to protect the name of a
certain high personage, we have taken the liberty to abstract
her letter and draft. All her letters to you and yours to her
will come to us. Will you not believe this is all for the best
and that we remain, with every assurance of regard,

"YOUR OBEDIENT SERVANTS."

Matthew stared. When and where had it been possible? He
could not conceive. Then he remembered that polite little
Japanese's visit. The Princess had never heard a word from
him. She never would. Then his heart leapt. The Princess had
not deliberately neglected or deserted him! She simply had
not heard from him and could not find him! He had blamed
the Princess for her apparent neglect, when in reality she knew
nothing. He was ashamed of himself. He had yielded to debauchery and drunkenness. Well, he would atone and get back
to his job. Should he write the Princess again? No. The Japanese and Indians were intercepting his letters. He started.
Perhaps they had given her forged reports and sent her home
disillusioned. Never mind. Even then, it would be on his
report, or supposed report, that she was acting. He must get
to work. He must think and plan.

XI

Matthew arose next day saner and clearer-headed and much
less sanguine. It was December fifteenth. The Princess, had

she been in earnest and remembered their meeting, would surely have insisted on seeing him in person and at least greeting him. It must have been curiously easy to make her lose faith. She had in all probability quite forgotten him and his errand. White America had flattered her wealth and beauty. Well, what then? Why, then it was for him to show her and her colleagues that black America counted in the world. But how? How? Then came illumination. He might himself go to Chicago! Without the slightest doubt other observers of the darker world would be on hand. He might go and curb Perigua and watch this meeting.

All the way down to Atlanta he pondered and fidgeted, and decision did not come until, to his great joy, he met Jimmie with his cheerful smile.

"Where've you been, you old cheat?" he cried.

Jimmie laughed. "Running to Chicago now."

"What? Changed your run? Why?"

"Two reasons. First: it's a good run; second—well, that I'll show you later. Come on now and sign up for the Klan Special."

"For the what?"

"The Klan Special. It's on my run, and they want porters. Come and try it."

Matthew stood still. It was just the thing. He'd go to Chicago as a porter and watch. Yes—this was precisely what he would do. With Jimmie he went to the harassed Pullman manager, who was only too glad to get so good a porter on such a train as the Klan Special.

"Had a hell of a time. Boys don't want to wait on the Klan. Damned nonsense. The Klan don't amount to anything. Chiefly a social stunt and gassing for effect. I will put you Car X466 near the end of the train, between Jimmie's compartment car and the observation coach. You are bound to make a pile in tips. So long."

Matthew and Jimmie went out together. Both were overjoyed to see each other. Matthew forgot all about Jimmie's second reason until he noticed that Jimmie was bubbling over with some secret of his own.

"Have dinner with me," Jimmie said. "Got something to show you."

They took the Hunter Street car and rode across town past

the quiet old campus of Atlanta University and through it and then away out by the new Booker Washington High School. Jimmie stopped at a pretty little cream and green cottage. It was tiny, but neat, and there was a yard in front with roses still blooming. Before Matthew could ask what it all meant, out of the house came a girl and the tiniest of babies. Jimmie set up a shout of explanations.

"Been married a year," he said. "Married before I knew you, but the wife was working in Chicago and wouldn't come until I could set up a regular home. But the baby brought her, and I got the home."

She was a little black, sweet-faced girl with lovely skin, crisp hair, and great black eyes—very practical and very loving, and her earth was quite evidently bounded by Jimmie and the baby. Matthew had never seen so small a baby. It was amorphous and dark red-brown and singularly cunning. They had a hilarious dinner, and Jimmie was at the best of his high humor.

He whispered all his romance to Matthew, while his wife washed the dishes.

"Never thought of marrying a black girl," he explained. "I was spending all I could make on a 'high yaller' in Harlem; when she heard I wasn't a banker, merchant, or doctor, she cut me so clean, I fell in two pieces and one landed in Chicago. I met Dolly, and gosh! I couldn't leave her; innocent, sweet, and with sense. O boy, but I got some wife! And that kid!"

Matthew was troubled. Suppose something happened in Chicago or to this train; to this boy with his soul full of joy, and to this sweet-faced little black wife?

The next few days the Klan delegates gathered in Atlanta on special trains from New Orleans and other cities. Jimmie, looking the crowd over with practised eye, prophesied a "hot time," plenty of gambling and liquor and good tips. Matthew was still disturbed, but Jimmie pooh-poohed.

"They're all right. Just don't let yourself get mad. Remember that, for the trip, you are just a machine, a plow or a mule, and I—I'm a savings bank for the kid."

"Jimmie," said Matthew suddenly, "suppose somebody tried to get back at these Klansmen somehow in Chicago."

"Nonsense," said Jimmie carelessly, "niggers dassn't, Catholics and Jews are too long-headed, and the Klan is too well

guarded. Just heard them talking about extra police protection." He was off before Matthew could say more.

Then the rush began. The train was to leave on the twentieth
at three-forty, over the Louisville and Nashville, and for the
last half hour before, Matthew had hardly time to think. His
and Jimmie's cars were at the end of the train; other Pullmans
followed. In the middle of the train was the diner, and the club
car and smoker was far forward.

It was nearly ten o'clock at night before Matthew got his
berths made down and came into Jimmie's car. They started
for the diner. Just as they were passing out of the car, a bell
rang, but Jimmie paid no attention.

"Come on," he said, "there's a flash dame in D who wants
too much attention; I don't trust her. Her husband, or the man
she's with, is up ahead, drunk and gambling. Let her wait."

In the diner with the other porters, they had a gay time.
Jimmie winked at the steward and soon produced a mysterious
flask; immediately they were all drinking to "The Baby" and
listening to some of the choicest of Jimmie's stories.

"Let's go up and see the bunch in the smoker," said Jimmie
when dinner was over. "I hear there's a big game on."

Matthew and Jimmie went forward. They were surely having a wild time in the smoker. The drinking and gambling
were open, and one could see the character of the crowd—
business men, Rotarians, traveling salesmen, clerks—a cross
section of American middle-class life.

"I am going back," said Matthew at last, for he was tired
and not particularly interested.

"Be with you in just a minute," said Jimmie. "I must see
this poker hand through. My God, do you see this flush?
Glance at my car as you go through and see if it is all right;
I'll be back in a jiffy. That fly dame will be yelling for something. Her daddy's in here linin' his grave with greenbacks."

Matthew walked back thinking of Jimmie. That baby! That
mother's face! There were, after all, some strangely beautiful
things in life. He walked through Jimmie's compartment car
and saw that all was quiet. Just as he was leaving, however,
he heard the bell and saw that, sure enough, Compartment D
had rung again. He walked back and knocked lightly.

"Porter!"

Matthew entered.

"It is stifling in here," came a voice from the berth. "Please open the window."

It was warm in Georgia, but the train would soon be in the cooler mountains; nevertheless, Matthew without argument started to open the window at her feet.

"No, this one at my head," insisted the woman, "and for mercy's sake, close the door behind you."

He closed the door softly and then bent over her to raise the window. There came over him at the moment a subtle flash of fear. She was a large woman—opulent and highly colored, and she lay there on her back looking straight up into his eyes. Her breasts were half-covered—one scarcely at all. He could not raise the sash with his hands unaided. He braced his knee on the berth and, using the metal handle for unlocking the upper berth, he bent down hard. The window flew up, but his hand came down lightly on the woman's bosom. Again came that gust of fear. He glanced down. She did not stir, but looked up at him with slightly closed eyes. For a moment, he caught his breath and his heart hammered. Then suddenly the door behind was flung violently open. The woman's face changed in a flash. She screamed shrilly as Matthew started back and drew the sheet close about her:

"Get out of here, you black nigger! How dare you touch me! I asked you to raise the window!"

Matthew, terrified, turned, and with one sweep of his arm fiercely pushed aside the man who was entering. The man went down in a heap, and quickly Matthew passed out into the corridor. He started forward to tell Jimmie, but he heard oncoming footsteps and an opening door. Turning, he ran into his own car, got his pistol from the clothes closet, and stepped into the toilet.

There was a long silence, then a cry, a rush of feet, and hurried voices. Then came a tense quiet. Matthew waited and waited until he could bear it no longer. He stepped out into the washroom and listened. Somewhere he could hear a thump—thump—thump. He raised the window and looked out. Something was dragging and bumping beside the car ahead. He heard a noise behind and turned quickly. A porter staggered in. Matthew recoiled, on guard.

"Anything wrong?" he said, thickly.

"They've lynched Jimmie," said the porter.

Matthew sank suddenly to the lounge. My God! It was
Jimmie he had heard coming. He sat down and vomited. He
stood up again, staggered to the door, and fainted away.

XII

It was morning. Matthew opened his eyes slowly and stared
at the high white walls. There were two blurs before him, one
on either side. Gradually, as he shut his eyes and opened them
again, they resolved themselves into two faces. Then he knew
them. One was Perigua; the other was Jimmie's little black
wife. Where was he? He strove to sit up. He was in a hospital.
He wanted to rage. He wanted to tell Perigua and everybody
that he was a murderer. Poor Jimmie, poor little wife and
baby! Perigua—revenge! All these things he strove to say, but
the nurse glided by and stopped him. She gave him something
to drink, and he fell asleep.

Three days later he left the General Hospital, and he and
Perigua and Jimmie's wife met together in a big brown house
on Fourth Street. He poured out his story, and they listened.
Perigua said nothing. But the little wife put her hand timidly
in his and said: "You are not to blame. It was not your fault."
And then she added: "We had the funeral here in Cincinnati.
I wish you could have been there. There were beautiful flowers.
But they would not open the coffin. They would not let me see
his face." And she repeated, looking up at Matthew: "They
did not let me see his face."

Then Perigua said:

"He didn't have no face."

There rose a shriek in Matthew's throat. It struggled and
surged, and broke to horrid silence within him. The hot tears
burned in his eyes. Something died in Matthew that day. He
put all his savings into the little mother's hand and pushed her
gently out the door.

"Good-by," he said, and "God forgive me!"

Perigua sat down and smoked, and silently showed him
newspaper clippings.

Christmas had passed. The Klan was holding its great meeting in Chicago, and the papers were full of news about it and
of pictures of the members. They seemed to be making a new
campaign against the Catholic Church; they had apparently

dropped the fight on Jews; but they were concentrating on a campaign against colored peoples throughout the world, and the world was listening to them. Moreover, they were adroitly seeking to pit the dark peoples against each other—Japanese against Chinese; Indians against Negroes; Negroes against Arabs; Mulattoes against Blacks. They even had certain Japanese and other Asiatic guests!

"That special train will return in triumph next Monday," said Perigua finally, looking at Matthew, gloomily.

Matthew brooded. "We must do something, Perigua," he said; "we *must* do something—something startling."

Perigua bent forward and glowed. "Something to make the world sit up!"

"Yes," said Matthew, "and my plan is this: I'm going to write and demand a meeting of the national officers of the Porters' Union in Chicago. I'll attend and tell my story of Jimmie's lynching and demand a nation-wide strike of porters until somebody is arrested for this crime."

Perigua's face fell. "Hell!" he said.

XIII

Worn and nervous, Matthew went to the Chicago meeting of the porters. He talked as he had never talked before, in that room with barred doors. With streaming eyes he told the story of Jimmie, of the little black wife, of the baby. He went over the events of that terrible night. He offered to testify in court, if called upon. The porters listened, tense and sympathetic; but they were silent and uneasy over the strike. It was "too risky"; they would "lose their jobs"; "Filipinos would be imported"; white men "at a living wage and no tips" would replace them: the nation would not stand being "held up" by Negroes, and white labor would not back them. "Do you think the white railway unions would raise a finger? I guess not!" said one.

No—a general Pullman strike would never do. Public opinion among Negroes, however, forced them to some action. While the white newspapers had said little about the gruesome lynching, and that little dismissed and excused it because of "an atrocious attack upon a woman," the colored world knew of it to its farthest regions. Once the matter had come up in the Klan Convention and a brazen-throated orator had declared that this was the punishment which would always be meted out to the "black wretches who dared attack Southern womanhood"!

The plan finally agreed on was the utmost Matthew could extract from the union. It confined itself to a porters' strike on the Klan Special. The train was to arrive in Cincinnati at eight at night on the thirtieth of December and leave at eight forty-five. Before the train came in and while it was in the station, the porters were to make up all the berths they could; at eight-forty all the porters were to leave their cars and march out of the train shed to the main waiting-room; there they were to declare a strike, refusing to accompany farther a train on which one of their number, an innocent man, had been lynched, under atrocious circumstances.

Matthew hurried back to Cincinnati to perfect the plans there. Perigua had been in Chicago, but he kept out of the way. No one seemed to know him there, but in his two or three fugitive visits to Matthew he assured him that he was working underground and making sure that none of the porters should see him. He promised to meet Matthew in Cincinnati.

With great fanfare of trumpets and waving of flags, the Klan Special started south. The porters were grim and silent. One of the organizers of the union had a hurried meeting with them just as they left, and on the way down, there were frequent conferences. The train was to leave Cincinnati without a single porter. There was little porter's work to be done at night except making the remaining berths, and this would have to be done by the conductors and the passengers themselves.

It was not, after all, a very bold scheme, or one calling for great courage. Matthew felt how small a gesture it was, and yet just now any protest was something; he knew that even this might not have been feasible, had it not been helped by the fact that none of the porters wanted to go south on this train. Fear, therefore, pushed them to strike for principle when under other circumstances many might have refused. It was extremely unlikely, too, that any porters who were laying over in Cincinnati, or who lived there, would volunteer to take the strikers' places. As the diner was detached at Cincinnati, the waiters would not have to take a stand. They were to disappear quietly, so as not to be asked to serve as porters.

Perigua arrived in Cincinnati three hours before the Klan
Special was due. He and Matthew sat again in the big gloomy
room on Fourth Street.

Matthew looked strained and thin, but he was sanguine.
He detailed his activities.

"Everything's all right here," he said. "I think it's going to
make a big sensation. Newspapers will eat it up, and the whole
of colored Cincinnati is whispering."

Perigua listened in silence and then laughed aloud.

"Well, what's the matter?" asked Matthew, testily.

"They've double-crossed you, you boob," said Perigua at
last.

"Nonsense—they can't as long as the men stick."

"Sure—'as long as.' Know what I've been doing in Chicago?"

"No—what?"

"Working for the Klan. Private messenger and stool pigeon
for Green, the Grand Dragon. Know all the big ikes—Therwald, Bates, Evans. Say, they knew of this strike from a
dozen pigeons before it was planned. They passed the word
to Uncle George. It'll never come off."

"But, say—"

"Shut up—come with me."

Matthew was disturbed but walked silently with Perigua
along Fourth and then over and west on Carlisle Avenue a
couple of blocks, past old brick buildings, smoke-grimed over
the tawdry decorations of a rich, dead generation.

Perigua pointed out a certain large house.

"Go in," he said. "You'll find forty porters lodging there.
'Strike' is the password. They're new men gathered quietly
from all over the South, expenses paid, ready to scab at a
moment's notice. Tell 'em you're inspecting the bunch and
flash this badge on them."

It was as Perigua said. Matthew almost staggered out of
the house, with tears in his eyes.

"I don't care," he cried to Perigua. "We'll strike anyhow.
The men will stick, I know. Let the scabs come—they'll get one
beating!"

"Piffle! They'll never strike. Not a man will budge when
they hear of that bunch waiting for their jobs, and they'll hear
of it before they are well out of Chicago. Uncle George will
see to that."

"But what can we do, Perigua?" We must do something—
God! We *must!*"

"Sure. Listen. Two can play at double-crossing. I brought
Green news of the strike—"

"You?"

"Yes—he heard it from a dozen others. And then, for full
measure, I lied about how Chicago Negroes planned a riot as
the Klan left. He swallowed both tales and gave me a thousand
dollars to push both schemes along; then he tipped off the
Pullman Company and the police."

"But it wasn't true about the riot?"

"Of course it wasn't."

"What did you do?"

"Hung around, filled him with tips and fairy tales, and
finally beat it here!"

"What for?"

Perigua quickly straightened up. "Good-by," he said, holding
out his hand.

"Where are you going?" Matthew asked.

Perigua glared. "I will tell you. I'm taking the next train
south," he said with blazing eyes. Matthew stared.

"But—" he expostulated. "The Klan train will not arrive
for two hours yet!"

"I shall need those hours," said Perigua.

"And you will not see the strike?"

"No—because there won't be no strike."

Matthew gripped Perigua's arm with his own nervous,
shaking fingers.

"What's your plan, Perigua?"

Perigua faced him, speaking slowly and distinctly: "I used
to run on this route from Chicago to Florida through Cum-
berland Gap. Did you see the Gap when you came up?"

Matthew shook his head.

"Well, you come down the valley from Winchester and
Richmond and rush into the hills; suddenly you meet the
mountains, and diving through one great crag, the tunnel
emerges as from a rock wall on to a high trestle which spans
the Powell River! Hm! Great sight! All right. Now for the
great Pullman strike!"

"But Perigua—what have we to do with—with scenery?
And suppose the cowards don't strike?"

Matthew knew the answer before he asked. He saw the heavy black bag which Perigua carried so carefully. He knew the answer. Perigua's mind was made up. He was mad—a desperate fanatic. What—

"Scenery!" laughed Perigua. "Listen, fool: we're mocked, betrayed and double-crossed, your race are born idiots and cowards! Well, I'm going this alone. Get me? Alone! When the Klan Special sees that scenery—when it reaches that trestle, the trestle ain't going to be there!"

"What is going to become of it?" Matthew asked slowly, talking against time and trying to think.

"I am going to blow it up," said Perigua.

"But how can you do it? Where can you stand? How can you fire any charge without elaborate wiring to get yourself far enough away?"

"I am not going to get away," said Perigua. "I am going to sit right on that trestle, and I am going to hell with it."

They looked each other straight in the eye.

"What are you going to do about it?" whispered Perigua.

Matthew hesitated. "Nothing—" he answered slowly.

Perigua approached Matthew, and there was danger in his eyes.

"You'll peach?" he whispered.

"I'll never betray you, Perigua."

"Well, what will you do?"

Matthew was silent.

"Well, speak, man," growled Perigua.

"I'll keep still," said Matthew.

"All right, keep still. But listen, man. It's going to be done, and if you can't be a man, don't be a damned tale-bearing dog!"

He started away. Matthew's thoughts raced. Here was the answer to that sneer of the Japanese. The world would awaken tomorrow to the revolt of black America. His head swam.

He ran after Perigua and gripped his arm. He was all a-tremble. He whispered in Perigua's ear.

"I don't believe what you have said. I don't believe the porters will back down before the scabs, but if they do—"

"Well, if they do, what?" asked Perigua.

"Wait," said Matthew. "How will the world know that this wasn't an accident rather than—revenge?"

"I've got posters that I printed myself."

"Give them to me."

"What for?"

"If the porters strike, I'll destroy them. But if they don't strike, I'll scab with them on the Klan Special—and I'll go to hell with you."

"By the living Christ," said Perigua, "you've got guts!"

"No," said Matthew, "I am a coward. I dare not live."

Perigua gripped his hand.

"I've searched through ten millions," he said, "and found only one who dared. Now I am going. Here! I'll give you half the handbills."

He thrust a bundle into Matthew's hand.

"Placard the cars with these after midnight. And, say—oh, here it is—here's a letter."

XIV

The porters' strike was over before it began. The officials had early wind of the plan, and by the time the Special reached Indianapolis, rumors of the host of strike breakers, ready and willing to work, reached the porters' ears and were industriously circulated by the conductors and stool-pigeons. There was a moment of strained expectancy as the train drew into the depot. Reporters came rushing out, and numbers of colored people who had learned of something unusual stood about. In the waiting-room stood a crowd of porters in new uniforms, together with several Pullman officials, and an unusual number of policemen who bustled about and scattered the crowds.

"Come—clear the way—move on!"

"Where are you going?" one of them asked Matthew, suspiciously. Leaning by the grill and straining his eyes, Matthew had waited in vain for the porters to leave their cars and march out according to the plan agreed on. Not a porter stirred. He saw them standing in their places, some laughing and talking, but most of them silent and grim. Matthew went ashen with pain and anger. He beckoned to some of the men he knew and had talked to. They ignored him.

He leaned dizzily against the cold iron, then started for the gate. A policeman accosted him, roughly seizing his arm.

"I'm joining this train as porter," he explained. "I've been on sick leave." A Pullman official stepped forward.

"I don't know anything about this," he began.

But Matthew spied his conductor.

"Reporting for duty, Cap," he said.

The conductor grinned. "Thought you were leading a strike," he sneered, and then turning to the official he said: "Good porter—came up with me. I was just coming to get an extra man for the smoker."

"All right."

And Matthew passed the gate. He spoke to not a single porter, and none spoke to him. All of them avoided each other. They had failed—they had been defeated without a fight.

"We're damn cowards," muttered Matthew as he climbed aboard.

"Any man's a coward in midwinter when he's got a wife, a mortgage, two children in school, and only one job in sight," answered the old porter who followed him.

"Good," growled Matthew. "Let's all go to hell."

An hour late the Klan Special crawled out of Cincinnati and headed South. The railroad and Pullman officials sighed in relief and laughed. The colored crowd faded away and laughed too, but with different tone.

Matthew donned his uniform slowly, as in a trance. He could not yet realize that his strike had utterly failed. He was numb with the day's experience and still weak from illness. He shrank from work in the smoker with that uproarious, drunken crowd of gamblers. The conductor consented to put him on the last car instead, bringing the willing man from that car to the larger tips of the smoker.

"We've dropped the observation," said the conductor, "and we've got a private car on the end with four compartments and a suite. They're mostly foreign guests of the Klan, and they keep pretty quiet. They are going down to see the South. Afraid you won't make much in tips—but then again you may." And he went forward.

Matthew went back and walked again through the horror of Jimmie's murder. He entered the private car. There was a reception room and a long corridor, but the passengers had apparently all retired. Matthew sat down in the lounge and took from his pocket the package which Perigua had given him; with it was the letter. He looked at it in surprise. He knew immediately whose it was; he saw the coronet; he saw the long slope of the beautiful handwriting; but he did not open it. Slowly he laid it aside with a bitter smile. It could have for him now neither good news nor bad, neither praise nor inquiry, neither disapproval nor cold criticism. No matter what it said, it had come too late. He was at the end of his career. He had started high and sunk to the depths, and now he would close the chapter.

In the first miles of the journey toward Winchester, Matthew was grim; cold and clear ran his thoughts.

"Selig der, den Er in Siegesglänze findet."

He was going out in triumph. He was dying for Death. The world would know that black men dared to die. There came the flash of passing towns with stops here and there to discharge passengers; he helped the porter on the next car, which was overloaded; he was hurrying, helping, and lifting as was his wont. And hurrying, helping, and lifting, he flew by towns and lights. Then coming suddenly back from beneath this dream of loads—from the everyday things—he tried to remember the Exaltation—the Great Thing. What was the Great Thing? And suddenly he remembered. He was going to kill these people. Just a little while and they would be twisted corpses—dead—and some worse than dead—crippled, torn and maimed.

The dark horror of the deed fell hot upon him. He had not seen it before—he had not wholly realized it. Yet he must go on. He could not stop. What had other men thought when they murdered in a great cause? Suddenly he seemed to know. It was not the dead who paid—it was the living; not the killed, but the Killer, who knew and suffered. This was Hell, and he was in it. He must stay in it. He must go through with it. But, Christ! the horror, the infamy, the flaming pain of the thing!

And the world flew by—always, always the world flew by; now in a great blurred rush of sound; now in a white, soft sweep of space and flash of time. Darkness ascended to the stars, and distance that was sight became sound.

It was War. In all ages men had gone forth to kill. But never—never, from Armageddon to the Argonne, had they carried so bitter reasons, so bloody a guerdon. All the enslaved, all the raped, all the lynched, all the "jim-crowed" marched in ranks behind him, bloody with rope and club and iron, crimson with stars and nights. He was going to fight and die for vengeance and freedom. There would be no march of music and stream of banners and whine of vast-voiced trumpets, but it was war, *war*, WAR, and he the grim lone fighter.

But the pity of it—the crippled and hurt—the pain, the great pricks and flashes of pain, the wild screams in the night; the grinding and crushing of body and bone and flesh and limb —and his sweat oozed and dripped in the cold night. He cowered in that dim and swaying room and shook with ague. He was afraid. He was deathly afraid. If he could turn back! If he had but never fallen in with this crazy plan! If he could only die now, quickly and first! Yet he knew he would not flinch. He would go through with it all to the last horror. The cold, white thing within him gripped him—held him hard and fast with all his writhing. He would go through.

The outlines of mountains with snow lay sprinkled here and there. The lights on hill and hollow—on long shining rails and piling shadows paused, came back and forward, curved, and disappeared. He stood stiffly and heard the gay laughter of the smoker, and one shrill voice floated back with war of answering banter.

"Laugh no more!" he whispered, and then his thoughts went racing down to cool places, to summer suns and gay, gleaming eyes. The cars reeled forward, gathered themselves, became one great speeding catapult, and headed toward the last hills. Beside them a little river, silver, whistled softly to the night.

He collected his few pairs of shoes and set them carefully down before him, arranging them mechanically; he smiled— the shoes of the dead—and he strangled as he smiled; strong, big, expensive brogans; soft, sleek, slim calf; patent leather pumps with gaitered sides; slippers of gray suède.

Slowly he got out his shoe brushes, and then paused. His heart throbbed unmercifully and then was cold and still. It was ten o'clock. He put out his hand and felt the letter. Tomorrow she would hear from him. Tomorrow they would know that black America had its men who dared—whose faces were toward the light and who could pay the price.

He laid the letter on the table unopened and took up the rest of the package, the bundle of manifestoes which Perigua had prepared and printed himself. Slowly Matthew read the little six-by-eight poster. It was rhodomontade. It was melodrama, but it told its awful story. Matthew read it and signed his name beneath Perigua's.

VENGEANCE IS MINE

THE WRECK TONIGHT IS TO AVENGE THE LYNCHING OF AN INNOCENT BLACK MAN, JIMMIE GILES, ON THIS TRAIN, DECEMBER 16, 1926, BY MEN WHO SEEK OUR DISFRANCHISEMENT AND SLAVERY.

MURDER FOR MURDERERS

MIGUEL PERIGUA
MATTHEW TOWNS

Matthew folded the posters slowly and held them in his hands.

Murder and death. That was his plan. It did not seem so awful as he faced it. Except by the shedding of blood there was no remission of Sin. Despite deceptive advance, the machinery was being laid to strangle black folk in America and in the world. They must fight or die. There was no use in talk or argument. Here was the challenge. An atrocious lynching; an open, publicly advertised movement to take the first step back to Negro slavery. Kill the men who led it. Kill them openly, publicly, and spectacularly, and advertise the killing and tell why!

Only one thing else, and that was: he must die as they died. It must be no coward's act which brought death to others and escape to himself. He shifted his pistol and pulled it out. It was a big forty-five and loaded with five great bullets. If the wreck did not kill him, this would. He was ready to die. This was all he could do for the cause. He was not worth any other effort —he had tried and failed. He had once a great dream of world alliance in the service of a woman he had almost dared to love.

He laughed aloud. She would not have looked twice even on Dr. Matthew Towns, world-renowned surgeon, save as she saw in him a specimen and a promise. And on a servant and a porter—a porter. He thought of the porters, riding to death. Let the cowards ride. Then he thought of their wives and babies, of Jimmie's wife and child. What difference? No—no—

no! He would not think. That way lay madness. He rushed into the next car.

"Got—nothing to do," he stammered. "Will you lend me some shoes to black?"

"Will I?" answered the astonished and sweating porter. "I sure will! My God! Looks like these birds of mine was centipedes. Never did see so many shoes in mah life. Help yo'self, brother, but careful of the numbers, careful of the numbers."

Matthew carried a dozen pairs to his car. He shuddered as he slowly and meditatively and meticulously sorted them for cleaning and blacking. They would not need these shoes, but he must keep busy; he must keep busy—until midnight. Then he would silently distribute his manifestoes throughout the train. At one o'clock the train would shoot from its hole to the high and narrow trestle. There was only one great deed that he could do for her, for the majority of men, and for the world, and that was to die tonight in a great red protest against wrong. And Matthew hummed a tune, "Oh, brother, you must bow so low!"

Then again he saw the letter lying there. Then again came sudden boundless exaltation. He was riding the wind of a golden morning, the sense of live, rising, leaping horseflesh between his knees, the rush of tempests through his hair, and the pounding of blood—the pounding and pounding of iron and blood as the train roared through the night. He felt his great soul burst its bonds and his body rise in the stirrups as the Hounds of God screamed to the black and silver hills. In both scarred hands he seized his sword and lifted it to the circle of its swing.

Vengeance was his. With one great blow he was striking at the Heart of Hell. His trembling hands flew across the shining shoes, and tears welled in his eyes. On, on, up and on! to kill and maim and hate! to throw his life against the smug liars and lepers, hypocrites and thieves, who leered at him and mocked him! Lay on—the last great whirling crash of Hell . . . and then his heart stopped. Then it was that he noticed the white slippers.

He had seen them before, dimly, unconsciously, out at the edge of the circle of shoes, two little white slippers—two slippers that moved. He did not raise his eyes, but with half-lowered lids and staring pupils, he looked at the slippers—two slippers, far in the rear. They were two white slippers, and he could not remember bringing them in. They stood on the outermost edge of the forest of shoes—he had not seen them move, but he knew they had moved. He was acutely, fearfully conscious of their movement, and his heart stopped.

He saw but the toes, but he knew those slippers—the smooth and shining, high-heeled white kid, embroidered with pearls. Above were silken ankles, and then as he leaped suddenly to his feet and his brushes clattered down, he heard the thin light swish of silk on silk and knew she was standing there before him—the Princess of Bwodpur. His soul clamored and fought within him, raged to know how and where and when, and here of all wild places! He saw her eyes widen with curiosity.

"You—here—Mr. Towns," she said and raised half-involuntarily her jeweled, hanging hand. He did not speak—he could not. She dropped her hand, hesitated a moment, and then, stepping forward: "Have I—offended you in some way?" she said, with that old half-haughty gesture of command, and yet with a certain surprise and pain in her voice.

Matthew stiffened and stood at attention. He touched his cap and said slowly: "I am—the porter on this car," and then again he stood still, silent and yet conscious of every inch of her, from her jeweled feet to the soft clinging of her dress, to the gentle rise of her little breasts, the gold bronze of her bare neck and glowing cheeks, and the purple of her hair. She could not be as beautiful as she always seemed to him—she could not be as beautiful to other eyes. But he caught himself and bit hard on his teeth. He would not forget for a moment that he was a servant and that she knew that he knew he was. But she only said, "Yes?" and waited.

He spoke rapidly. "Your Royal Highness must excuse any apparent negligence. I have received no word from you except one letter, and that only tonight. Indeed—I have not yet read that. I hope I have been of some service. I hope that you and his Excellency have learned something of my people, of their power and desert. I wish I could serve you further and—better, but I can not—"

The Princess sat down on the couch and stared at him with faint surprise in her face. She had listened to what he said, never moving her eyes from his face.

"Why?" she said again, gently.

"Because," he said, "I am—going away."

"Have I offended you in some way?" she asked again.

"I am the offender," he said. "I am all offense. See," he said in sudden excitement, "this is my mission." And he handed her one of Perigua's manifestoes. The Princess read it. He looked on her as she read.

As she read, wrinkling her brows in perplexity, he himself seemed to awake from a nightmare. My God! He was carrying the Princess to death! How in heaven's name had he landed in this predicament? Where was the impulse, the reasoning, the high illumination that seemed to point to a train wreck as the solution of the color problems of the world? Was he mad—had he gone insane?

Whatever he was, his life was done, and done far differently from his last wild dream. There was no escape. He must stop the train. Of course. He must stop it instantly. But how was he to explain to the world his knowledge? He could not pretend a note of warning without producing it, and even then they might ignore it. He could not give details to the conductor lest he betray Perigua.

He did not consciously ask himself the one question: why not let the wreck come after all? He knew why. For a moment he thought of suicide and a dying note. No—they might ignore the warning and think him merely crazy. Already they were flying to make up lost time. No, he must live and spare no effort even to confession until he had stopped that train. First, warning—as a last resort, the bell-rope—and then—jail.

At any cost he must save the Princess and her great cause —God! They might even think her the criminal if anything happened on this train of death. And then he sensed by the silken rustle of garments that the Princess had finished reading and had arisen.

"Read my letter," she said.

His hands shook as he read. She had received and read his reports. They were admirable and enlightening. Her own limited experiences confirmed them in all essentials. The Japanese had joined her and was quite converted. They realized the tremendous possibilities of the American Negro, but they both agreed with Mr. Towns that there was no question of revolt or violence. It was rather the slow, sure, gathering growth of power and vision, expanding and uniting with the thought of the wider, better world.

But she could not understand why he did not answer her specific questions and refused her repeated invitations to call. She wanted to thank him personally, and she had so many questions—so many, many questions to ask. She had twice postponed her return home in order to see him. Now she must go, and curiously enough, she was going to the Ku Klux Klan meeting in Chicago at the invitation of the Japanese, and for reasons she would explain. Would Mr. Towns meet her there? She would be at the Drake and always at home to him. She sensed, as did the Japanese, subtle propaganda, to discount in advance any possible colored world unity, in this invitation to attend this meeting and ride on this special train. They were all the more glad to accept, as he would readily understand. Would he be so good as to wire, if he received this, to the New Willard, Washington?

Matthew was dumb and bewildered. He could not fathom the intricacies of the tactics of the Japanese. His reports had been passed to the Princess, and yet all her letters to him stopped save this. Or had it been Perigua who had rifled his mail? Or the Indians?

But what mattered all this now? It was too late. Everything was too late. Around him like a silent wall of earth and time ranged the symbolic shoes—big and little, slippers and boots, old, new, severe, elegant. He spoke hurriedly. There was no alternative. She had to know all. Time pressed. It was nearly one o'clock, and a cold tremor gripped slowly about his heart. He listened—glanced back at the door. God! If the conductor should come! Then he hurried on.

"I shall stop the wreck; then I am going—away!"

The Princess gave a little gasp and came toward him. He started nervously and listened.

"I must not stay," he said hurriedly, and in a lower voice: "This train will surely be wrecked unless I stop it. I did not dream you were aboard."

She made a little motion with her hands. "Wrecked? *This* train?" she said, and then more slowly, "Oh! Perigua's plan?" Then she stared at him. "And you—on it!"

He smiled. "Wrecked, and I—on it." Then he added slowly: "It was to be a proof—to his Excellency and you. And it was to be more than that: it was revenge." And he told her hurriedly of Jimmie's death.

"But you must stop it. It is a mad thing to do. There are so many sane, fine paths. I was so mistaken. I had thought of you as a nation of outcasts to be hurled forward as shock troops, but you are a nation of modern people. You surely will not follow Perigua?"

"No," he said quietly, "I will not. But let me tell you—"

Then she rose quietly and moved toward him. "And—Perigua must be—betrayed?"

"Never."

"And if—" She stared at him. "And if—"

"Jail," he said quietly, "for long years."

She made a little noise like a sob controlled, but his quick ear caught another sound. "The conductor," he whispered. "Destroy these handbills for me." Quickly he stepped out into the corridor.

"Captain," he said hurriedly, "captain—this train must be stopped—there is danger."

"What do you mean? Is it them damned porters again?"

"No—not they—but, I say—there is danger. Where's the train conductor?"

The Pullman conductor stared at him hard. "He's up in the third car," he said nervously, for it had been a hard trip. "Come with me." Matthew followed.

They stepped in on the conductor in an empty compartment, where he was burrowing in a pile of tickets and stubs.

"Mr. Gray, the porter has a story for you."

"Spit it out—and hurry up," growled the conductor. The train flew on, and faster flew the time.

"You must stop the train," said Matthew.

The conductor glanced up. "What's the matter with you? Are you drunk?"

"I was never so sober."

"What the hell then is the matter?"

"For God's sake stop the train! There's danger ahead."

"Stop the train, already two hours late? You blithering idiot! Have all you black porters gone crazy?"

Matthew stepped out of the compartment and threw his weight on the bell-rope. The conductor swore and struck him aside, but there was a jolt, a low, long, grinding roar, and quickly the train slowed down. The conductor seized Matthew just as some one pounded on the window. A red light flashed ahead. Soon a sweating man rushed aboard.

"Thank God!" he gasped. "That was a narrow squeak. I was afraid I was too late to flag you. You must have got warning before my signal was lighted. There's been an explosion on the trestle. Rails are torn up for a dozen yards."

xv

Matthew Towns blackened shoes. All night long he blackened shoes, cleaning them, polishing them very carefully, and arranging the laces. He was working in a standard Pullman at the forward end of the train, having been hurriedly transferred from the private car after the incident of the night. He gathered more shoes and blackened them, placing them carefully, in the graying dawn, under the appropriate berths. He arranged clean towels in the washrooms and tested the soap cocks. He saw that the toilets were clean and in order, and he carefully dusted the corridor and wiped the windows.

All the time there were two unobtrusive strangers who kept him always in sight. He paid no apparent attention to them but waited, watch in hand, as the train approached Knoxville. Some one asked the time.

"Six-thirty," he whispered.

"We're pretty late."

"Yes, on account of that delay on the road."

"When do we get in Knoxville?"

"About eight-thirty, I imagine. Breakfast will be served as soon as we arrive."

At last he went to some of the berths and pulled the lower sheet gently and then insistently.

"One hour to Knoxville," he said; and again and again. "One hour to Knoxville."

The car aisles began to fill with half-dressed travelers. He brought new bundles of towels and began to make up vacant berths. He worked rapidly and deftly. There was much confusion, and always the two unobtrusive men were near. Some of the returning passengers found their seats in order. Others did not and made sharp remarks, but Matthew pacified them, guided them to resting-places, and began to collect the luggage and to brush the clothes.

The sweat poured off him, but he worked swiftly. When they stopped in the depot, he was at the step in coat and cap, wooden, deferential: "Thank you, sir. All right here, Cap."

They moved out for the swift three-hour run to Atlanta. He finished the other berths, brushed more passengers, stowed dirty linen, swept, dusted, and guided passengers to the dining-car attached at Knoxville.

The train glided into the Atlanta station.

And then it came.

"Towns, step this way—gentleman wants to see you."

He walked back through the train into the lounge of the private car again. On the table lay something under a sheet. About the door, several of the passengers were crowded.

As Matthew entered the car he saw in the vestibule, and for the first time since one awful night, a well-remembered figure— a woman, high-colored, big and boldly handsome, with her lowered eyelids and jeweled hands. Beside her was a weak-looking man, faultlessly tailored, with an old and dissipated face. They were in the waiting throng. The woman looked up. Her eyes widened suddenly, and then quietly she fainted away.

Matthew faltered but an instant and then walked steadily on. He entered the room. The conductor was there, the two quiet men, and a grave-faced stranger. And then came the Princess, the Japanese, and several other guests. They all sat, but Matthew stood silent, his uniform spotless, his head up.

One of the strangers spoke.

"Your name is—"

"Matthew Towns."

"You are a porter?"

"Yes."

"The porters had planned a strike in Cincinnati?"

"Yes."

"Why didn't you strike?"

"I was going to, but I changed my mind."

"Why?"

"Because the others decided not to—and because I heard that this train was going to be wrecked."

"By the porters?"

"Certainly not!"

"By whom?"

"I cannot tell."

"Who told you?"

"I will not say."

"Did the other porters hear this?"

"No, I was the only one."

"How do you know?"

"I am sure."

"When did you hear this?"

"Just before the train started."

"From Chicago?"

"No, from Cincinnati."

"But you were in Chicago?"

"Yes."

"And planned the strike there?"

"Yes. I helped to."

"What did you do when you heard this rumor?"

"I offered to go as porter."

"You offered to go on a train that you knew was going to be wrecked?"

"Yes."

"Why?"

"Well—a porter—my friend—was lynched on this train a week ago. I urged the strike as a protest. When it failed—nothing mattered."

"Did you intend to stop the wreck?"

"At first, no."

"And you—you changed your mind?"

"Yes."

"Why?"

"I cannot tell."

"How did you think you could prevent it?"

"Well—I did prevent it."

"Who told you about this plot?"

"I will not tell."

"Did this man tell you?"

They drew the sheet from Perigua's dead face. Beneath the sheet his body looked queer, humped and broken. But his face was peaceful and smiling. Matthew's face was stone.

"No."

"Do you know him?"

"No."

"Did you ever see him before?"

"Yes."

"Where?"

"In the office of the Grand Dragon of the Ku Klux Klan in the Sherman Hotel, Chicago."

There was a stir among the crowd. A big man with a flat, broad face and little eyes pressed forward and viewed the corpse.

"It may be Sam," he said. "Were any papers or marks found on him?"

"Nothing—absolutely nothing. Not even laundry marks."

"I'm almost sure that's Sam Johnson, who acted as messenger in our Chicago office. If it is," he spoke deliberately, "I'll vouch for him. Excellent character—wouldn't hurt a flea." He glanced at Matthew.

The inquisitor turned back to Matthew.

"Who told you of this wreck?"

"I will not tell."

"Why not?"

"I take all the blame."

"Do you realize your position? You stand between high reward and criminal punishment."

"I know it."

"Who told you of the wreck?"

And then like sudden thunder came the low, clear voice of the Princess:

"I told him!"

XVI

Circuit Judge Windom, presiding over the criminal court of Cook County, Illinois, sat in his chambers with a frown on his face. Beside him sat his son, the gifted young medical student, home from the holidays.

"Certainly I remember Towns," said the younger man. "He was a fine fellow—first-rate brains, fine athlete, and a gentleman. If it had not been for his color, he'd have been sure to make a big reputation, but they drove him out of school. Somebody had kicked about Negroes in the women's clinics. Towns wouldn't beg—he slapped the Dean's face, I heard, and left."

"H'm—violent, even then."

"But, my God, father, Towns was a man—not just a colored man. Why, you remember how he beat me for the Mitchel Prize?"

"Yes, yes—but all that does not clear up this mystery. I can get neither head nor tail of it. Here is an atrocious railroad wreck planned on a leading railway. Half an hour more, and there would have been perhaps five hundred corpses strewn in the river. Awful! Dastardly! The explosion was bungled and premature. The trestle was left intact, but enough damage was done to have made the derailing of the train inevitable had it rushed through the tunnel unwarned. Section hands discovered at the last moment the broken rails and a dead man lying across them. They start to signal too late, but before they start Towns warns the conductor; when the conductor hesitates, Towns himself stops the train. Possibly the signal man might have stopped it eventually, but Towns actually stopped it.

"Now, how did Towns know? Was this the striking porters' plot? Towns and their leaders declare that, far from dreaming of this, they would not quit work for an ordinary strike, and certainly they would hardly have ridden on a train which they expected to be wrecked. An Indian Princess declares that she told Towns of the plot, and taking refuge in her diplomatic immunity, refuses to answer further questions. The English Embassy, which represents her country abroad, backs her reputation, vouches for her integrity, and promises her immediate withdrawal from the country. The dead Negro found on the trestle remains unidentified. Indeed there is no evidence of his connection with the wreck. The chief of the Klan thinks he recognizes the man as a former messenger, vouches for his character, and doubts his connection with any plot; he considers him a victim rather than a conspirator. Says, as of course we know, that Negroes never conspire. And now comes this extraordinary story of Towns himself."

"Well, at any rate, Towns wouldn't lie!" said the son.

"But the point is, he won't tell the truth; and why? It looks dangerous, suspicious. Some red-handed rascals are going free."

"What does Towns say?"

"He says that he did not plan this outrage; that when he

knew that it was planned he assented to it and determined to run on this train and to die in the wreck. Then, for some reason changing his mind and being unable to contemplate the death of all these passengers, he gave warning of the plot. He says that he had met the Princess abroad, had told her of his trouble in the medical school and elicited her sympathy and interest; that he had no knowledge that she was on the train, and no idea she was in the country. He then told her the danger of the train and his dilemma, and in generous sympathy, she had finally sought to direct the blame of guilty foreknowledge to herself.

"In truth, and this he swears before God, the Princess of Bwodpur had not the faintest knowledge of the plot of wrecking the train until he himself told her five minutes before he stopped the train. He begs that she be entirely exonerated, despite her Quixotic attempt to save him, and that he alone bear the full blame and suffer the full penalty."

"Extraordinary—but if Towns says it's true, it's true. It may not be the whole truth, but it contains no falsehood."

"Well, it is full of discrepancies and suspicious omissions. Good heavens! A woman says she knew a train was to be wrecked, and yet rides on it, and tells the porter and not the conductor. The porter declares that she did not know or tell him, but that somebody else did; and yet *he* rides. A man found dead may be the wrecker, but the head of the Ku Klux Klan and one of the threatened party refuses to believe him guilty. The porter refuses to tell where he got his warning and prefers jail rather than reward."

"How did the case get to your court, father?"

"More complications. When Towns was arraigned at Atlanta, the passengers of the Klan Special came forward with a big purse to reward him for his services, and a sharp lawyer. Towns refused the money, but the court listened to the lawyer and held that no crime had been committed by Towns in its jurisdiction.

"Thereupon the District Attorney of this district sought indictment against Towns, charging that the Pullman porter strike was concocted in Chicago and that the wreck was part of the conspiracy. Towns denied this, but offered to come here without extradition papers. The District Attorney expected the help and support of the Pullman Company and the railroads. None of them lifted a finger.

"There is another curious and unexplained angle. You know there was a lynching on that Klan train. There was some dispute as to whether it took place in Georgia or Tennessee. Nothing was ever done about it, not even a coroner's inquest. Well, I have it on the best authority that when the woman who alleged the attack saw Towns face to face at the informal questioning on the train, she fainted away.

"I have tried to get in touch with her and her husband, who is Therwald, a high Klansman. They deny all knowledge and refuse to appear as witnesses. As residents of another state, I can't compel them. Moreover, I find that they have only recently been married, although the newspaper reports of the lynching refer to them as man and wife occupying the same compartment. Now what's behind all this?

"Well, he was indicted for conspiracy and pleaded guilty. He still declared that the porters' union and the Indian Princess knew nothing of the proposed wreck. He admitted that he did and further admitted that he consented to it and started on the journey determined not to betray the arch-conspirators, and then changed his mind and stopped the wreck. Now what can you make of such a Hell's broth?"

"I'm puzzled, father."

"The Princess of Bwodpur herself has come to me, stopping en route, as she explained, to Seattle and India. Evidently a great lady and extraordinarily beautiful, despite her color, which I was born to dislike. I pooh-poohed her story and showed her Towns' sworn confession. There is no doubt of her interest in him. She put up a strong plea, stronger than yours, son, but I was adamant. I had to be. I am not at all sure but that she is the guilty party and that Towns is shielding her. I don't know, my boy—I don't know where the truth lies. But there's more here than meets the eye. I scent a powerful, dangerous movement; and despite all you say, if Towns thinks that a plea of guilty and waiver of jury trial is going to get him mercy in my court, he is mistaken. I am sorry—I hate to do it; but he'll get the limit of the law unless he tells the whole truth."

And the judge sighed wearily and gathered up his books.

XVII

Matthew sat in a solemn hall. It was "across the river"—north of the Loop and west of the Michigan Avenue bridge, in a region of vacant dilapidated buildings, of windows without panes and walls peeling and crumbling. A mighty, gray stone structure covered half the block. The front was wrinkled and uneven, with a shrunken door under an iron balcony. Three elevators with musty, clanking chains faced the door and rolled solemnly up five floors. The lobby was bordered with dark stone; the floor was white and gray and cold, and across one side was a huge sign—"Robert E. Crowe, State's Attorney, Office." Across the other side one read, "Criminal Court."

Within these doors, beyond a narrow, oak-paneled hall, sat Matthew Towns, in a high-ceilinged room. The long narrow windows, with flapping dirty green shades, admitted a faint light. The walls were painted orange-yellow. The lights were hanging from the ceiling in chandeliers of metal once brass-colored, with each light socket in an ornamental oak-leaf holder. The globes were of a bluish-yellow glass, pear-shaped. The Bench, of polished oak, was at the rear of a circular oaken-railed enclosure. The enclosure had tables and chairs for lawyers, clients, and witnesses. Well to the front of this green-carpeted space was the desk of the clerk. To the rear, on a raised platform, were seats for the jury. Raised yet higher was the platform upon which rested the judge's bench; on either side of the bench were doors with signs, "Judge's Chambers," "Jury Rooms." Facing this circular enclosure were long seats in rows for the spectators. The floor here was the same dirty gray, much-worn tile, and the ceiling over the whole, while very high, was noticeable only because it was so soiled and stained.

Soft sunshine filtered in and lighted up the rich polish of the oak. Behind the high desk sat the judge—heavily silked, his grave, gray face looking sternly out upon the world. The strained faces of that world, white, black, and brown, were crowded in the benches below, and some stood in hushed silence. Policemen, bareheaded, moved silently about the throng, and two officials with silver and gilt stood just below the judge. There should have been music, Matthew thought, some slow beat like the Saul death march or the pulse of the Holy Grail. Then the judge spoke:

"Matthew Towns, stand up."

And Matthew rose and stood, center of a thousand eyes, and a sigh and a hiss went through the hall. For he was tall and impressive. The crisp hair curled on his high forehead. The soft brown of his eyes glowed dark on the lighter brown of his smooth skin. His gray suit lay smooth above the muscles and long bones of his close-knit body. He looked the judge full in the face. The eyes of the judge grew somber—but for a tint of skin, but for a curl of hair, but for a fuller curve of lip and cheek, this might have been his own son, this man whom his son had known and honored.

"Matthew Towns," he said in low, slow tones, "you stand accused of an awful crime. With your knowledge and at least tacit consent, some person whom you know and we do not, planned to put a hundred, perhaps five hundred souls to torture, pain, and sudden death. At the last minute, when literally moments counted, you rescued these people from the grave. It may have been a brave—a heroic deed. It may have been a kind of deathbed repentance or even the panic of cowardice. In any case the guilt—the grave and terrible guilt hangs over you for your refusal to reveal the name or names of these blood-guilty plotters of midnight dread—of these enemies of God and man. With the stoicism worthy of a better cause and a cynical hardness, you let these men walk free and take upon yourself all the punishment and shame. It has a certain fineness of sacrifice, I admit; but it is wrong, cruel, hateful to civilization and criminal in effect and intent. There is for you no shadow of real excuse. You are a man of education and culture. You have traveled and read. I know that you have suffered injustice and perhaps insult and that your soul is bitter. But you are to blame if you have let this drown the heart of your manhood. You have no real excuse for this criminal and dangerous silence, and I have but one clear duty before me, and that is to punish you severely. I could pronounce the sentence of death upon you for deliberate conspiracy to maim and murder your fellow men; but I will temper justice with mercy so as still to give you chance for repentance. Matthew Towns, I sentence you to ten years at hard labor in the State Prison at Joliet."

The sun burst clear through the dim windows and lighted the young face of the prisoner.

Some one in the audience sobbed; another started to applaud. Matthew Towns followed the guard into the anteroom, and thither the Princess came, moving quietly to where he stood with shackled hands. The windows all about were barred, and at the farther end of the room stood the stolid officer with a pistol and keys. Down below hummed the traffic.

She took both his manacled hands in hers, and he steeled himself to look the last time at that face and into the deep glory of her eyes. She was simply dressed in black, with one great white pearl in the parting of her breasts.

"You are a brave man, Matthew Towns, brave and great. You have sacrificed your life for me."

Matthew smiled whimsically.

"I am a small man, small and selfish and singularly short of sight. I served myself as well as you, and served us both ill, because I was dreaming selfish little dreams. Now I am content; for life, which was twisting itself beyond my sight and reason, has become suddenly straight and simple. Your Royal Highness"—he saw the pain in her eyes, and he changed: "My Princess," he said, "your path of life is straight before you and clear. You were born to power. Use it. Guide your groping people. You will go back now to the world and begin your great task as the ruler of millions and the councilor of the world's great leaders.

"Your dream of the emancipation of the darker races will come true in time, and you will find allies and helpers everywhere, and nowhere more than in black America. Join the hands of the dark people of the earth. Discover in the masses of groveling, filthy, ignorant black and brown and yellow slaves of modern Europe, the spark of manhood which, fanned with knowledge and health, will light anew a great world-culture. Yours is the great chance—the solemn duty. I had thought once that I might help and in some way stand by the armposts of your throne. That dream is gone. I made a mistake, and now I can only help by bowing beneath the yoke of shame; and by that very deed I am hindered—forever—to help you—or any one—much. I—am proud—infinitely proud to have had at least your friendship."

The Princess spoke, and as she talked slowly, pausing now

and then to search for a word, she seemed to Matthew somehow to change. She was no longer an icon, crimson and splendid, the beautiful perfect thing apart to be worshiped; she
became with every struggling word a striving human soul groping for light, needing help and love and the quiet deep sympathy of great, fine souls. And the more she doffed her royalty
and donned her sweet and fine womanhood, the further, the
more inaccessible, she became to him.

He knew that what she craved and needed for life, he could
not give; that they were eternally parted, not by nature or
wealth or even by birth, but by the great call of her duty and
opportunity, and by the narrow and ever-narrowing limit of
his strength and chance. She did not even look at him now
with that impersonal glance that seemed to look through him
to great spaces beyond and ignore him in the very intensity and remoteness of her gaze. She stood with downcast
eyes and nervous hands, and talked, of herself, of her visit to
America, of her hopes, of him.

"I am afraid," she said, "I seem to you inhuman, but I
have come up out of great waters into the knowledge of life."
She looked up at him sadly: "Were you too proud to accept
from me a little sacrifice that cost me nothing and meant
everything to you?"

"It might have cost you a kingdom and the whole future
of the darker world. It was just some such catastrophe that the
Japanese and Indians rightly feared."

"And so, innocent of crime, you are going to accept the
brand and punishment of a criminal?"

"My innocence is only technical. I was a deliberate co-con-
spirator with Perigua. I—murdered Jimmie!"

"No—no—how can you say this! You did not dream of peril
to your friend, and your pact with Perigua was a counsel of
despair!"

"My moral guilt is real. I should have remembered Jimmie.
I should have guided Perigua."

"But," and she moved nearer, "if the dead man was—
Perigua, what harm now to tell the truth?"

"I will not lay my guilt upon the dead. And, too—if I confessed that much, men might probe—further."

"And so in the end I am the one at fault!"

"No—no."

"Yes, I know it. But, oh, Matthew, are you not conscience-mad? You would have died for your friend had you known, just as now you go to jail for me and my wild errand. But even granted, dear friend, some of the guilt of which you so fantastically accuse yourself—can you not balance against this the good you can do your people and mine if free?"

"I have thought of this, and I much doubt my fitness. I know and feel too much. Dear Jimmie saw no problem that he could not laugh off—he was valuable; indispensable in this stage of our development. He should be living now, but I who am a mass of quivering nerves and all too delicate sensibility— I am liable to be a Perigua or a hesitating complaining fool— untrained or half-trained, fitted for nothing but—jail."

"But—but afterward—after ten little years or perhaps less —you will still be young and strong."

"No, I shall be old and weak. My spirit will be broken and my hope and aspirations gone. I know what jail does to men, especially to black men—my father—"

"You are then deliberately sacrificing your life to me and my cause!"

"I am making the only effective and final atonement that I can to the Great Cause which is ours. I might live and work and do infinitely less."

"You have ten minutes more," said the guard.

"Is there nothing—is there not something I can do for you?"

"Yes—one thing: that is, if you are able—if you are permitted and can do it without involving yourself too much with me and my plight."

"Tell me quickly."

"I would not put this request if I had any other way, if I had any other friend. But I am—alone." She gripped his hands and was silent, looking always straight into his eyes with eyes that never dropped or wavered. "I have a mother in Virginia whom I have forgotten and neglected. She is a great and good woman, and she must know this. Here is a package. It is addressed to her and contains some personal mementoes—my father's watch, my high-school certificate—old gifts. I want her to have them. I want her to see—you. I want you to see her—it will explain; she is a noble woman; old, gnarled, ignorant, but very wise. She lives in a log cabin and smokes a clay pipe. I want you to go to her if you can, and I want you to tell her my story. Tell her gently, but clearly, and as you think best; tell her I am dead or in a far country—or, if you will, the plain truth. She is seventy years old. She will be dead before I leave those walls, if I ever leave them. If she did not realize where I was or why I was silent, she would die of grief. If she knows the truth or thinks she knows it, she will stand up strong and serene before her God. Tell her I failed with a great vision—great, even if wrong. Make her life's end happy for her. Leave her her dreams."

"You have one minute more," said the guard.

The Princess took the package. The policeman turned, watch in hand. They looked at each other. He let his eyes feast on her for the last time—that never, never again should they forget her grace and beauty and even the gray line of suffering that leapt from nose to chin; suddenly she sank to her knees and kissed both his hands, and was gone.

Next day a great steel gate swung to in Joliet, and Matthew Towns was No. 1,277.

PART III

THE CHICAGO POLITICIAN

PART III

THE CHICAGO POLITICIAN

1924, January, to April, 1926

Winter. Winter, jail and death. Winter, three winters long, with only the green of two little springs and the crimson of two short autumns; but ever with hard, cold winter in triumph over all. Cold streets and hard faces; white death in a white world; but underneath the ice, fire from heaven, burning back to life the poor and black and guilty, the hopeless and unbelieving, the suave and terrible. Dirt and frost, slush and diamonds, amid the roar of winter in Chicago.

I

SARA ANDREWS listened to the short trial and sentence of Matthew Towns in Chicago in early January, 1924, with narrowed eyelids, clicking her stenographer's pencil against her teeth. She was not satisfied. She had followed the Klan meeting with professional interest, then the porters' strike and Matthew's peculiar case. There was, she was certain, more here than lay on the surface, and she walked back to Sammy Scott's office in a brown study.

Sara Andrews was thin, small, well tailored. Only at second glance would you notice that she was "colored." She was not beautiful, but she gave an impression of cleanliness, order, cold, clean hardness, and unusual efficiency. She wore a black crêpe dress, with crisp white organdie collar and cuffs, chiffon hose, and short-trimmed hair. Altogether she was pleasing but a trifle disconcerting to look at. Men always turned to gaze at her, but they did not attempt to flirt—at least not more than once.

Miss Andrews was self-made and independent. She had been born in Indiana of the union of a colored chambermaid in the local hotel and a white German cook. The two had been duly married and duly divorced after the cook went on a visit to Germany and never returned. Then her mother died, and this

109

girl fought her way through school; she forced herself into the local business college, and she fought off men with a fierceness and determination that scared them. It became thoroughly understood in Richmond that you couldn't "fool" with Sara Andrews. Local Lotharios gave up trying. Only fresh strangers essayed, and they received direct and final information. She slapped one drummer publicly in the Post Office and nearly upset evening prayer at St. Luke's, to the discomfiture of a pious deacon who sat beside her and was praying with his hands.

For a long time she was the only "colored" person in town, except a few laborers; and although almost without social life or intimate friends, she became stenographer at the dry goods "Emporium" at a salary which was regarded as fabulous for a young woman. Then Southern Negroes began to filter in as laborers, and the color line appeared, broad and clear, in the town. Sara Andrews could have ignored it and walked across so far as soda fountains and movie theaters were concerned, but she wouldn't. A local druggist wanted to marry her and "go away." She refused and suddenly gave up her job and went to Chicago. There, in 1922, she became secretary to the Honorable Sammy Scott.

The Honorable Sammy was a leading colored politician of Chicago. He was a big, handsome, brown man, with smooth black hair, broad shoulders, and a curved belly. He had the most infectious smile and the most cordial handshake in the city and the reputation of never forgetting a face. Behind all this was a keen intelligence, infinite patience, and a beautiful sense of humor. Sammy was a coming man, and he knew it.

He was, in popular parlance, a "politician." In reality he was a super-business man. In the Second Ward with its overflowing Negro population, Sammy began business in 1910 by selling the right to gamble, keep houses of prostitution, and commit petty theft, to certain men, white and black, who paid him in cash. With this cash he bribed the city officials and police to let these people alone and he paid a little army of henchmen to organize the Negro voters and see that they voted for officials who could be bribed.

Sammy did not invent this system—he found it in full blast and he improved it. He replaced white ward heelers with blacks who were more acceptable to the colored voters and were themselves raised from the shadow of crime to well-paid jobs; some even became policemen and treated Negro prisoners with a certain consideration. Some became clerks and civil servants of various sorts.

Then came migration, war, more migration, prohibition, and the Riot. Black Chicago was in continual turmoil, and the black vote more than doubled. Sammy's business expanded enormously; bootlegging became a prime source of graft and there was more gambling, more women for sale, and more crime. Men pushed and jostled each other in their eagerness to pay for the privilege of catering to these appetites. Sammy became Alderman from the Second Ward and committeeman, representing the regular Second Ward Republican organization on the County Central Committee. He made careful alliance with the colored Alderman in the Third Ward and the white Aldermen from the other colored wards. He envisaged a political machine to run all black Chicago.

But there were difficulties—enormous difficulties. Other Negro politicians in his own and other wards, not to mention the swarm of white bosses, had the same vision and ambition as Sammy—they must all be reconciled and brought into one organization. As it was now, Negroes competed with each other and fought each other, and the white party bosses, setting one against the other, got the advantage. It was at this stage of the game that Sara Andrews joined Sammy's staff.

When Sara Andrews applied to the Honorable Sammy for work, he hired her on the spot because she looked unusually ornamental in her immaculate crêpe dress, white silk hose, and short-trimmed hair. She had intelligent, straight gray eyes, too, and Sammy liked both intelligence and gray eyes. Moreover, she could "pass" for white—a decided advantage on errands and interviews.

Sammy's office was on State Street at the corner of Thirty-second. Most of the buildings around there were old frame structures with living-quarters above and stores below. On each corner were brick buildings planned like the others, but now used wholly for stores and offices. The entrance to Sammy's building was on the Thirty-second Street side; a dingy gray wooden door opened into a narrow hall of about three by four feet. Thence rose a flight of stairs which startled by its amazing steepness as well as its darkness. At the top of the

stairs, the hall was dim and narrow, with high ceilings. At the
end was a waiting-room facing State Street. It was finished
with a linoleum rug that did not completely cover the soft
wood floor; its splinters insisted on pulling away as if to avoid
the covering of dark red paint. There were two desks in the
waiting-room, some chairs, and a board upon which were
listed "Apartments for rent." Sash curtains of dingy white,
held up with rods, were at the windows, and above them in
gold letters were painted the names of various persons and
of "Samuel Scott, Attorney at Law."

A railing about three feet high made an inner sanctum, and
beyond was a closed door marked "Private." Back here in
Sammy's private office lay the real center of things, and in
front of this and within the rail, Sammy installed Sara. The
second day she was there, Sammy kissed her. That was four
years ago, and Sammy had not kissed her since. He had not
even tried. Just what happened Sammy never said; he only
grinned, and all his friends ever really knew was that Sammy
and Sara were closeted together for a full half-hour after the
kiss and that Sara did most of the talking. But Sara stayed
at her job, and she stayed because Sammy discovered that she
was a new asset in his business; first of all, that she was a real
stenographer. He did not have to dictate letters, which had
always been a difficult task. He just talked with Sara and
signed what she brought him a few minutes later.

"And believe me," said Sammy, "she writes some letter!"

Indeed Sara brought new impetus and methods into Sammy's
business. When that kiss failed, Sammy was afraid he had got
hold of a mere prude and was resolved to shift her as soon
as possible. Then came her letter-writing and finally her advice. She listened beautifully, and Sammy loved to talk.
She drew out his soul, and gradually he gave her full confidence. He discovered to his delight that Sara Andrews had
no particular scruples or conscience. Lying, stealing, bribery,
gambling, prostitution, were facts that she accepted casually.
Personally honest and physically "pure" almost to prudery,
she could put a lie through the typewriter in so adroit a way
that it sounded better than the truth and was legally fireproof.
She recognized politics as a means of private income, and her
shrewd advice not only increased the office revenue, but

slowly changed it to safer and surer forms. "Colored cabarets are all right," said Sara, "but white railroads pay better."

She pointed out that not only would the World-at-Play pay for privilege and protection, but the World-at-Work would pay even more. Retail merchants, public service corporations, financial exploiters, all wanted either to break the law or to secure more pliable laws; and with post-war inflation, they would set no limit of largesse for the persons who could deliver the goods. Sammy must therefore get in touch with these Agencies in the White World. Sammy was skeptical. He still placed his chief reliance on drunkards, gamblers, and prostitutes. "Moreover," he said, "all that calls not only for more aldermen but more members of the legislature and Negroes on the bench."

"Sure," answered Sara, "and we got to push for Negro aldermen in the Sixth and Seventh Wards, a couple of more members of the legislature, a judge, and a congressman."

"And each one of them will set up as an independent boss, and what can I do with them?"

"Defeat 'em at next election," said Sara, "and that means that you've got to get a better hold on the Negro vote than you've got. Oh, I know you're mighty popular in the policy shops, but you're not so much in the churches. You're corraling the political jobs and ward organizations, but you must get to be popular—get the imagination of the rank and file."

Sammy hooted the suggestion, and Sara said nothing more for a while. But she had set Sammy thinking. She always did that.

In fine, Sara Andrews became indispensable to the Honorable Sammy Scott, and he knew that she was. He would have liked to kiss and cuddle her now and then when they sat closeted together in the den which she had transformed into an impressive, comfortable, and singularly official office. She was always so cool and clean with her slim white hands and perfect clothes. But all she ever allowed was a little pat on the shoulder and an increase in salary. Now and then she accepted jewelry and indicated clearly just what she wanted.

Then for a while Sammy half made up his mind to marry her, and he was about sure she would accept. But he was a little afraid. She was too cold and hard. He had no mind to embrace a cake of ice even if it was well groomed and sleek.

"No," said Sammy to himself and to his friends and even

to Sara in his expansive moments, after a good cocktail, "no, I'm not a marrying man."

Sara was neither a prude nor a flirt. She simply had a good intellect without moral scruples and a clear idea of the communal and social value of virginity, respectability, and good clothes. She saved her money carefully and soon had a respectable bank account and some excellent bonds.

Sammy was born in Mississippi the year that Hayes was elected. He had little education but could talk good English and made a rattling public speech. With Sara's coaching he even attempted something more than ordinary political hokum and on one or two public occasions lately had been commended; even the *Tribune* called him a man of "real information in current events." Sara accordingly bought magazines and read papers carefully. She wrote out his more elaborate speeches; he committed them to his remarkable memory in an hour or so.

Why then should Sammy marry Sara? He had her brains and skill, and nobody could outbid him in salary. Of that he was sure. Why spoil the loyalty of a first-class secretary for the doubtful love of a wife? Then, too, he rather liked the hovering game. He came to his office and his letters with a zest. He discovered the use of letters even in politics. Before Sara's day there was a typewriting machine in Sammy's office, but it was seldom used. Previous clerks had been poor stenographers, and Sammy could not dictate. Besides, why write? Sara showed him why. He touched her finger tips; he brought her flowers and told her all his political secrets. She had no lovers and no prospective lovers. Time enough to marry her if he found he must. Meantime love was cheap in Chicago and secretaries scarce, and, in fine, "I'm not a marrying man," repeated the Honorable Sammy.

Sara smiled coolly and continued:

"I think I see something for us in the Towns case."

Sammy frowned. "Better not touch it," he said. "Bolsheviks are unpopular, especially with railroads. And when it comes to niggers blowing up white folks—well, my advice is, drop it!"

So the matter dropped for a week. Then Sara quietly returned to it: "Listen, Sammy"—Sara was quite informal when they were alone in the sanctum—"I think I see a scoop." Sammy listened. "This Matthew Towns—"

"What Matthew Towns?"

"The man they sent to Joliet."

"Oh! I thought you'd dropped that."

"No, I've just really begun to take it up. This Towns is unusual, intelligent, educated, plucky."

"How do you know?"

"I saw him during the trial, and since then I've been down to Joliet."

"Humph!" said Sammy, lighting his third cigar.

"He is a man that would never forget a service. With such a man added to your machine you might land in Congress."

Sammy laid his cigar down and sat up.

"I keep telling you, Sammy, you've got to be something more than the ordinary colored Chicago politician before you can take the next step. You've got to be popular among respectable people."

"Respectable, hell!" remarked Sammy.

"Precisely," said Sara; "the hell of machine politics has got to be made to look respectable for ordinary consumption. Now you need something to jack you up in popular opinion. Something that will at once appeal to Negro race pride and not scare off the white folks who want to do political business with you. Our weakness as Negro politicians is that we have never been able to get the church people and the young educated men of ability into our game."

"Hypocrites and asses!"

"Quite so, but you'll notice these hypocrites, asses, good lawyers, fine engineers, and pious ministers are all grist to the white man's political machine. He puts forward and sticks into office educated and honest men of ability who can do things, and he only asks that they won't be too damned good and honest to support his main interests in a crisis. Moreover, either we'll get the pious crowd and the educated youngsters in the machine, or some fine day they'll smash it.

"Sammy, have some imagination! Your methods appeal to the same crowd in the same old way. Meantime new crowds are pushing in and old crowds are changing and they want new ways—they are caught by new gags; makes no difference whether they are better or worse than the old—facts are facts, and the fact is that your political methods are not appealing to or holding the younger crowd. Now here's bait for them,

and big bait too. If I am not much mistaken, Towns is a find.
For instance: 'The Honorable Sammy Scott secures the release of Towns. Towns, a self-sacrificing hero, now looms as a
race martyr. Towns says that he owes all to the Honorable
Sammy!' "

"Fine," mocked Sammy, "and niggers wild! But how about
the white folks? 'Sam Scott, the black politician, makes a jail
delivery of the criminal who tried to wreck the Louisville &
Nashville Railway Special. A political shame,' etc., etc."

"Hold up," insisted Sara. "Now see here: the Negroes have
been thoroughly aroused and are bitterly resentful at the Klan
meeting, the lynching of the porter, and Matthew Towns'
incarceration. His release would be a big political asset to the
man who pulled it off. And if you are the man and the white
political and business world know that your new popularity
strengthens your machine and delivers them votes when
wanted, and that instead of dealing with a dozen would-be
bosses, they can just see you—why, Sammy, you'd own black
Chicago!"

"Sounds pretty—but—"

"On the other hand, who would object? I have been talking
to the porters and railroad men and to others. They say the
judge was reluctant to sentence Towns, but saw no legal escape. The railroad and the Pullman Company owe him millions and were willing to reward him handsomely if he had
escaped the law. The Klan owes several hundred lives to him.
None of these will actively oppose a pardon. It remains only
to get one of them actually to ask for it."

"Well—one, which one?" grinned Sammy, touching Sara's
fingers as he reached for another cigar.

"The Klan."

"Are you crazy!"

"I think not. Consider; the Klan is at once criminal and
victim. Its recent activities have been too open and bombastic.
It has suffered political reverses both north and south. It is
accused of mere 'nigger-baiting.' Would it not be a grand wide
gesture of tolerance for the Klan to ask freedom for Towns?
Something like donations to Negro churches, only bigger and
with more advertising value."

"Well, sure; if they had that kind of sense."

"They've got all kinds of sense. Now again, there is something funny about that lynching. I've heard a lot of talk. Towns has let out bits of a strange story, and the porters say he was wild and bitter about the lynching. Suppose, now—I'm only guessing—Towns knows more than he has told about this woman and her carrying on. If so, she might be glad to help him. A favor for keeping his mouth shut. I mention this, because she has married since the Klan convention and her husband is a high official of the Klan."

Sammy still didn't see much in the scheme, but he had a great respect for Sara's shrewdness.

"Well—what do you propose?" he asked.

"I propose to go to Joliet again and have a long talk with Towns. Then I'm going to drop down to Washington. I've always wanted to go there. I'll need a letter of introduction from somebody of importance in Chicago to this woman, Mrs. Therwald."

II

It was a lovely February day as Sara walked down Sixteenth Street, Washington—clear, cool, with silvery sunshine. Sara was appropriately garbed in a squirrel coat and hat, pearl-gray hose, and gray suède slippers. Her gloves matched her eyes, and her manner was sedate. She walked down to Pennsylvania Avenue, looked at the White House casually, and then sauntered on to the New Willard. Her color was so imperceptible that she walked in unhindered and strolled through the lobby. Mrs. Therwald was not in, she was informed by the room clerk. She talked with a bell-boy, and when Mrs. Therwald entered, observed her from afar, carefully and at her leisure. She was a big florid woman, boldly handsome, but beginning to show age. About a quarter of an hour after she had taken the elevator, Sara sent up her card and letter of introduction from the wife of a prominent white Chicago politician.

Mrs. Therwald received her. She was a woman thoroughly bored with life, and Sara looked like a pleasant interlude. They were soon chatting easily. Sara intimated that she wrote for magazines and newspapers and that she had come to see the wife of a celebrity.

"Oh, no—we're nothing."

"Oh, yes—the Klan is a power and bound to grow—if it acts wisely."

"I really don't know much about it. My husband is the one interested."

"I know—and that brings me to the second object of my visit—Matthew Towns."

Mrs. Therwald was silent several seconds—and then: "Matthew Towns? Who—"

"Of course you would not remember," said Sara hastily, for she had noticed that pause, and the tone of the question did not carry conviction. "I mean the porter who was sent to the penitentiary for the attempted wreck of the Klan Special."

"Oh, that—scoundrel."

"Yes. There is, as perhaps you know, a great deal of talk about his silence. He must know—lots of things. I think it rather fine in him to shield—others. I hope he won't break down in jail and talk."

Mrs. Therwald started perceptibly.

"Talk about what?" she asked almost sharply.

Sara was quite satisfied and continued easily.

"Well, about the black conspirators against the Ku Klux Klan—or the white ones, because they are more likely to be white. Or he might gossip and just stir up trouble. But I think he's too big for all that. You know, I saw him and talked to him—really handsome, for a colored man. Oh, by the by—but of course not. I was going to ask if by any possibility you had seen him on the train."

"I—I really don't know."

"Of course you wouldn't remember definitely. But to come to the point of my visit: certain highly placed persons are convinced from new evidence, which cannot be published, that Towns is a victim and not a criminal. They are therefore seeking to have Towns pardoned, and I thought how fine it would be if you could induce your husband and some other high officials of the Klan to sign the petition. How grateful he would be! I think it would be the biggest and fairest gesture the Klan ever made, and frankly, many people are saying so. In that case, if he is a conspirator, he could be watched and traced and his helpers found. And then, too, think of his gratitude to *you!*"

Sara left the petition with Mrs. Therwald, and they talked

on pleasantly and casually for another half-hour. Miss Andrews "would stay to tea"? "But no—so sorry." Sara said that she had stayed already much longer than she had planned, and hoped she had not bored Mrs. Therwald with her gossip. In truth she did not want to let the lady eat with one who, she might later discover, was a "nigger." They parted most cordially.

Mrs. Therwald happened a week later to say casually to her husband:

"That Towns nigger that they sent to jail—don't you think he'd be safer outside than in? He seemed a decent sort of chap on the trip. I was thinking it might be a shrewd gesture for the Klan to help free him."

Her husband looked at her hard and said nothing. But he did some thinking. That very day the white Democrats of Chicago had complained to the Klan that their small but formerly growing Negro vote was disappearing because of the Klan meeting and the Towns incident. Illinois with its growing Negro vote would be no longer a doubtful state politically unless something was done. How would it do to free Towns?

III

Miss Sara Andrews sat in the anteroom of the office of the Grand Dragon of the Ku Klux Klan in Washington. Several persons looked at her curiously.

"I believe she's a nigger," said a stenographer.

"Italian or Spanish, I would say," replied the chief clerk and frowned, for Sara had decided to wait. She said that she must really see Mr. Green personally and privately. After an hour's wait, she saw him. Mr. Green turned toward her a little impatiently, for she was interrupting a full day.

"What can I do for you?" and he glanced at her card and started to say, "Miss Andrews." Then he looked at her slightly olive skin and the suggestion of a curve in her hair and compromised on "Madam."

Miss Andrews began calmly with lowered eyes. She had on a new midnight-blue tailor-made frock with close-fitting felt hat to match, gay-cuffed black kid gloves, gun-metal stockings, and smart black patent leather pumps. On the whole she was pleased with her appearance.

"I am trying to get a pardon for Matthew Towns, and I want your help."

"Who is Matthew Towns?"

The question again did not carry the conviction that Mr. Green did not really remember. But Sara was discreet and carefully rehearsed the case.

"Oh, yes, I remember—well, he got what he deserved, didn't he?"

"No, he saved the train and got what somebody else deserved."

"Why didn't he reveal the real culprits?"

"That is the point. He may be shielding some persons who we might all agree should be shielded. He may be shielding the dead. He may be shielding criminals now free to work and conspire. But in all probability, he does not know who planned the deed. He was a blind tool. In any case he should go free. For surely, Mr. Green, no one is foolish enough to believe this was the plan of a mere porter."

"Have you any new evidence?"

"Not exactly court evidence," said Sara, "and yet I betray no confidence when I say that we have information and it is much in favor of Towns."

"And what do you want of me?"

"I have come to ask you to sign a request for Matthew Towns' pardon. You see, if you do, it will clear up the whole matter." And she looked Mr. Green full in the face. Her eyes were a bit hard, but her voice was almost caressing.

"I am sure," she said, "that the colored people of America are needlessly alarmed over the Klan, and that you are really their friends in the long run. Nothing would prove this more clearly than a fine, generous action on your part like this."

"But do you think it possible that Towns knows—nothing more of the real perpetrators of the plot?"

"If he did, why didn't he talk? Why doesn't he talk now? Reporters would rush to print his story. Indeed, the longer he stays in jail, the more he may *try* to remember. No, Mr. Green, I am sure that Towns either knows nothing more or will never tell it in jail or out."

Mr. Green signed the petition.

A month later, in Chicago, Sammy was close closeted with his congressman.

"This Towns matter: Pullman people are willing; railroads don't object. Even the Klan is asking for it, and the Republicans better move before the Democrats get credit."

Two weeks later the congressman saw the chairman of the National Republican Committee. The matter got to the Governor a week after that. In April it was very quietly announced that because of certain new evidence and other considerations, and at the request of the Ku Klux Klan, Matthew Towns had been pardoned. The Honorable Sammy Scott and his secretary went to Joliet and took the pardon to the prisoner.

IV

The great Jewish synagogue in Chicago, which the African Methodist Church had bought for half a million dollars in mortgages, was packed to its doors, May first, and an almost riotous crowd outside was demanding admittance. The Honorable Sammy Scott promised them an overflow meeting. Within, all the dignitaries of black Chicago were present. And, in addition, the mayor, the congressman from the black belt, and an unusual outpouring of reporters, represented the great white city. On the platform in the center in a high-backed, heavily-upholstered church chair sat the presiding officer, the Right Reverend John Carnes, Presiding Bishop of the District—an inspiring figure, too fat, but black and dignified. At his left sat the mayor, two colored members of the legislature, and several clergymen. At his right sat two aldermen and a congressman, and a tall, thin young man with drawn face and haunted eyes.

Matthew Towns made a figure almost pitiful. He sat drooping forward, half filling the wide chair, and staring blankly at the great audience. At his left was the chairman, and at his right sat the fat old congressman in careless dress, with his shifty eyes; down below the great audience milled and stirred, whispered and quivered.

It was an impressive sight. Every conceivable color of skin glowed and reflected beneath the glare of electricity. There was the strong bronze that burned almost black beneath the light, and the light brown that was a glowing gold. There was every shade of brown, from red oak to copper gold. There were all the shades of gold and cream. And there were yellows that were red and brown; and chalk-like white.

There was every curl and dress of hair. There was every style of clothing, from jewels and evening dress to the rough, clean Sunday coat of the laborer and the blue mohair of his wife. All expressions played on the upturned faces: inquiry, curiosity, eager anticipation, cynical doubt.

The Honorable Sammy was nervous. He did not go on the platform. He hovered back in the rear of the audience, with a hearty handshake here and a slap on the back there.

"*Hel*-lo, old man. Well, well, *well!* And Johnson, as I am alive! My God! but you're looking fit, my boy, *fit*. Well, what's the good word? What do you know? Mother James, as I'm a sinner! Here, Jack! Seat for Mrs. James? *Must* find one. Why, I'd—" etc., etc.

But Sammy was nervous. He didn't "like the look of that bird on the platform." Somehow, he didn't look the part. Why, my God, with that audience he had the cream and pick of black Chicago and the ears of the world. There was one of the *Tribune's* best men, and the *Examiner* and the *News* and the *Post* had reporters. Good Lord, what a scoop, if they could put it over! He had Chicago in the palm of his hand. But "that bird don't look the part!" and Sammy groaned aloud.

Sara had pushed him into this. She was getting too bossy, too domineering; he'd have to put the reins on her, perhaps get rid of her altogether. Well, not that, of course; she was valuable, but she must stop making him do things against his better judgment. He never had quite cottoned to this jailbird, nohow. Who ever heard of a sane man going to jail to save somebody else? It wasn't natural. Something *must* be v/rong with him. Look at those eyes.

Where was Sara? Perhaps she could manage to pump some gumption into him, even at the last moment. If this thing failed, if Towns said the wrong thing or didn't say the right one, he would be knocked into a cocked hat. He had had a hard time bringing the pardon off anyhow. The congressman was skittish; feared the Governor: "Don't like to touch it, Sam. My advice is to drop it."

But Sammy, egged on by Sara, had insisted.

"All right, I'll try it. But look here, nothing else. If I pull this pardon through, that five thousand dollars I promised for the campaign is off. I can't milk the railroads for both."

Sammy had hesitated and consulted Sara.

"Five thousand dollars is five thousand dollars."

And then he would need the cash this fall. But Sara was adamant.

"Five thousand dollars isn't a drop compared with this if we put it over."

And now Sammy groaned again. If he failed—"God damn it to hell!—*Where* is Sara?"

The exercises had opened. A rousing chorus began that raised the roof and hurled its rhythm against the vibrating audience; an impressive and dignified introduction by the Bishop, and a witty, even if somewhat evasive, speech by the mayor. Sammy began to sweat, and his smile wore off.

The congressman started to introduce the "gentleman whom we all are waiting to hear—the hero, the martyr—Matthew Towns!" There was a shout that rose, gathered, and broke. Then a hush fell over the audience. Matthew seemed to hesitate. He started to rise—stopped, looked helplessly about, and then got slowly to his feet and leaned against the pulpit awkwardly.

"O Lord!" groaned Sammy, "O *Lord!*"

"I am not a speaker," said Matthew slowly.

("It's the God's truth," said Sammy.)

"I have really nothing to say." ("And you're sure sayin' it, bo," snarled Sammy.) "And if I had I would not know how to." And then he straightened up and added reflectively, "I am —my speech." The audience rustled and Sammy was faint.

"I was born in Virginia—" And then swiftly and conversationally there came the story of his boyhood and youth; of his father and mother; of the cabin and the farm. He had not meant to talk of this. The speech which Sara had at his request prepared for him had nothing of this, but he was thinking of his home. Then followed naturally the story of his student days, of his work and struggles, of the medical school, of the prizes, of his dreams. The audience sat in strained and almost deathly silence, craned forward, scarcely breathing, at the twice-told human tale that touched every one of them, that they knew by heart, that they had lived through each in its

thousand variations, and which was working unconsciously
to the perfect climax.

("My God"—whispered Sammy—"he's putting it over—
he's putting it over. He's a genius or God's anointed fool!")

Finally Matthew came to that day of return to his junior
medical year. He saw the scene again—he felt the surge of
hot anger; his voice, his great, full, beautiful voice, rose as
again he threw his certificates into the face of the dean. The
house roared and rang with applause—the men shouted, the
women cried, and up from the Amen corner rose the roll and
cadence of the slave song: "Before I'd be a slave I'd be buried
in my grave and go home to my God and be free!"

Sammy leaned against the back wall glowing. It was a diamond stickpin for Sara!

Matthew awoke from the hypnotism of his own words, and
the fierce enthusiasm died suddenly away. Yet he was no
longer afraid of his audience or wanting words. With unconscious artistry he let his climax rest where it was, and he stood
a moment with brooding eyes—a lean, handsome, cadaverous
figure—and told the rest of his story in even, matter-of-fact
tones.

"I ran away from my people and my work. I tried to hide,
but I was sent back. I worked as a porter, and I tried to be a
good porter. And all the time I wanted to help—to do a great
thing for freedom and strike a great blow. I met a man. He
was a fanatic; he was sinking into sin, and worse, he was
planning a terrible deed. I sensed it and tried to dissuade him
from it. I pointed out its impossibility and futility. But he
cursed me for a coward and went on. I could have run away;
I could have betrayed him. I did instead an awful thing—an
awful deed which the death of an innocent man spurred me to.
I do not know whether I was right or wrong, but I resolved
to die on the train that my brave friend was resolved to
wreck—and then—" Matthew paused, and the audience almost
sobbed in suspense. "And then on the rushing train, Something
would not let me do what I had planned. The credit is not
mine. Something hindered; I stopped the train, but I did not
betray my terrible friend. I went to jail. My friend—died—"
He paused and groped; what was it he must say? What was
it to which he must not forget to allude? He stood in silence

and then remembered: "And, tonight, through the efforts of Sammy Scott I am free."

That minute the Honorable Sammy Scott reached the apex of his career. The next day Matthew got a job, and Sara Andrews a diamond stickpin.

v

In jail Matthew Towns had let his spirit die. He had become one with the great gray walls, the dim iron gratings, the thud, thud, thud which was the round of life, which *was* life. Bells and marching, work and meals, meals and work, marching and whistles. Even, unchanging level of life, without interest, memory, or hope.

This at first; then, disturbing little things. As the greater life receded, the lesser took on exaggerated importance. The food, the chapel speaker, this whispered quarrel over less than a trifle; the oath and blows of a keeper.

"When I get out!"

Ten years! Ten years was never. If such a space as ten years ever passed, he would come back again to jail.

"They all do," said the keepers; "if not here, elsewhere."

The seal of crime was on him. It would never lift. It could not; it was ground down deep into his soul. He was nothing, wanted nothing, remembered nothing, and even if he did remember the trailing glory of a cloudlike garment, the music of a voice, the kissing of a drooping, jeweled hand—he murdered the memory and buried it in its own blood.

Then came the miracle. First that neat and self-reliant young woman who tried to make him talk. He was inclined to be surly at first, but suddenly the walls fell away, and he saw great shadowed trees and rich grass. He was bending over a dainty tea-table, and he talked as he had talked once before. But he stopped suddenly, angry at the vision, angry at himself. He became mute, morose. He took leave of Sara Andrews abruptly and went back to his bench. He was working on wood.

Then came the pardon. In a daze and well-nigh wordless, he had traveled to Chicago. He sat in the church like a drowning swimmer who, hurled miraculously to life again, breathed, and sank. He had no illusions left.

He knew Sara and Sammy. They wanted to use him. Well, why not? They had bought him and paid for him. All his enthusiasm, all his hope, all his sense of reality was gone. He saw life as a great, immovable, terrible thing. It had beaten him, ground him to the earth and beneath; this sudden resurrection did not make him dizzy or give him any real hope. He gave up all thought of a career, of leadership, of greatly or essentially changing this world. He would protect himself from hurt. He would be of enough use to others to insure this. He must have money—not wealth—but enough to support himself in simple comfort. He saw a chance for this in politics under the command of Sara and Sammy.

He had no illusions as to American democracy. He had learned as a porter and in jail how America was ruled. He knew the power of organized crime, of self-indulgence, of industry, business, corporations, finance, commerce. They all paid for what they wanted the government to do for them—for their immunity, their appetites; for their incomes, for justice and the police. This trading of permission, license, monopoly, and immunity in return for money was engineered by politicians; and through their hands the pay went to the voters for their votes. Sometimes the pay was in cash, sometimes in jobs, sometimes in "influence," sometimes in better streets, houses, or schools. He deliberately and with his eyes closed made himself a part of this system. Some of this money, paid to master politicians like Sammy Scott, would come to him, some, but not much; he would save it and use it.

He settled in the colored workingmen's quarter of the Second Ward—a thickly populated nest of laborers, lodgers, idlers, and semi-criminals. In an old apartment house he took the topmost flat of four dilapidated rooms and moved in with an old iron bed, a chair, and a bureau.

Then he set out to know his district, to know every man, woman and child in it. He was curiously successful. In a few months scarcely a person passed him on the street who did not greet him. The November elections came, and his district rolled up a phenomenal majority for Scott's men; it was almost unanimous.

He deliberately narrowed his life to his village, as he called it. One side of it lay along State Street in its more dreary and dilapidated quarter. It ran along three blocks and then back three blocks west. Here were nine blocks—old, dirty, crowded —with staggering buildings of brick and wood lining them. The streets were obstructed with bad paving, ashes, and garbage. On one corner was a church. Then followed several places where one could buy food and liquor. On State Street were a dance hall, a movie house, and several billiard parlors, interspersed with more or less regular gambling dens. There were a half-dozen halls where lodges met and where fairs and celebrations were carried on. And all over were the homes—good, bad, indifferent.

He was strangely interested in this little universe of his. It had within a few blocks everything life offered. He could find religion—intense, fanatical, grafting, self-sacrificing. He could find prostitutes and thieves, stevedores, masons, laborers, and porters. Thus his blocks were a pulsing world, and in them there was always plenty to do—a donation to the church when the mortgage interest was about due; charity for the old women whose sons and daughters had wandered off; help and a physician for the sick and those who had fallen and broken hip or leg or had been run over by automobile trucks; shoes and old clothes for school children, bail for criminals; drinks for tramps; rent for the dance hall; food for the wild-eyed wastrels; and always, jobs, jobs, jobs for the workers.

When the new colored grocery was started, Matthew had to corral its customers, many of whom he had bailed out for crime. The police were his especial care. He gave them information, and they tipped him off. He restrained them, or egged them on. He warned the gamblers or got them new quarters. He got jobs for men and women and girls and boys. He helped professional men to get off jury duty. He sent young girls home and found older girls in places worse than home. He did not judge; he did not praise or condemn. He accepted what he saw.

Always, in the midst of this he was organizing and coraling his voters. He knew the voting strength of his district to a man. Nine-tenths of them would do exactly as he said. He did not need to talk to them—a few words and a sign. Orators came to his corners and vociferated and yelled, but his followers watched him. He saw this group of thousands of people as a real and thrilling thing, which he watched, unthrilled, unmoved. Life was always tense and rushing there—a murder,

a happy mother, thieves, strikers, scabs, school children, and
hard workers; a strange face, a man going into business, a girl
going to hell, a woman saved. The whole organism was neither
good nor bad. It was good and bad. Rickety buildings, noise,
smells, noise, work—hard, hard work—

"How's Sammy?" he would hear them say.

"How many votes do you want? Name your man."

Thus he built his political machine. His machine was life,
and he stood close to it—lolling on his favorite corner with
half-closed eyes; yet he saw all of it.

Above it all, on the furthermost corner, on the top floor,
were his bare, cold, and dirty rooms. He could not for the
life of him remember how people kept things clean. It was
extraordinary how dirt accumulated. He never had much
money. Sammy handed him over a roll of bills every now and
then, but he spent it in his charities, in his gifts, in his bribings,
in his bonds. There was never much left. Sometimes there was
hardly enough for his food.

Long past midnight he usually climbed to his bare rooms—
one of them absolutely bare—one with a bed, a chair, and a
bureau—one with an oil stove, a chair, and a table.

Then in time the aspect of his rooms began to change. A
day came when he went in for his usual talk with the secondhand man. Old Gray was black and bent, and part of his business was receiving stolen goods, the other part was quite
legitimate—buying and selling secondhand stuff. Towns
strolled in there and saw a rug. He had forgotten ever having
seen a rug before then. Of course he had—there in Berlin on
the Lützower Ufer there was a rug in the parlor—but he
shook the memory away with a toss of his head.

This rug was marvelous. It burned him with its brilliance.
It sang to his eyes and hands. It was yellow and green—it was
thick and soft; but all this didn't tell the subtle charm of its
weaving and shadows of coloring. He tried to buy it, but Gray
insisted on giving it to him. He declared that it was not stolen,
but Towns was sure that it was. Perhaps Gray was afraid to
keep it, but Towns took it at midday and laid it on the floor
of the barest of his empty rooms. Connors, who was a first-
class carpenter when he was not drunk. was out of work again.
Towns brought him up and had him put a parquet floor in
the bare room. He was afterward half ashamed to take that

money from his constituents, but he paid them back by more careful attention to their demands. Then in succeeding months of little things, the beauty of that room grew.

VI

The Honorable Sammy was by turns surprised, dumb-founded, and elated. He could not decide at various times whether Towns was a new kind of fool or the subtlest of subtle geniuses; but at any rate he was more than satisfied, and the efficiency of his machine was daily growing.

The black population of Chicago was still increasing. Properly organized and led, there were no ordinary limits to its power, except excited race rancor as at the time of the riot, or internal jealousy and bickering. Careful, thoughtful manipulation was the program, and this was the Honorable Sammy's long suit. First of all he had to appease and cajole and wheedle his own race, allay the jealousies of other leaders—professional, religious, and political—and get them to vote as they were told.

This was no easy job. Sammy accomplished it by following Sara's advice; first he refused all the more spectacular political offices; he refused to run again as alderman, declined election to the legislature and the like; he secured instead a state commissionership (whence his "honorable") where he still had power but little display; and of course, he was on the State Central Republican committee; then he "played" the clergy, helping with speeches and contributions of large size to lift their mortgages; he stood behind the colored teachers who were edging into the schools; he belonged to every known fraternal order, and at the same time he continued to protect the cabarets, the bootleggers, the gambling dens, and the "lodging" houses. Slowly in these ways his influence and word became well-nigh supreme in the colored world. Everybody "liked" the Honorable Sammy.

And Sammy found Matthew an invaluable lieutenant.

"By gum, Sara, we have turned a trick. To tell the truth, for a long time I distrusted that bird, even after his great speech. I was afraid he'd be a highbrow and start out reforming. Damned if he ain't the best worker I ever had."

"Yes, he'll do for the legislature," said Sara. Sammy scowled.

That was like Sara. Whenever he yielded an inch, off she skipped with an ell.

"Slow, slow," he said frowning; "we can't push a new man and a jailbird too fast."

"Sammy, you're still a fool. Don't you see that this is the only man we can push, because he's tied to us body and soul?"

"I ain't so sure—"

"Sh!"

Matthew came in. He greeted them diffidently, almost shyly. He always felt naked before these two.

They talked over routine matters, and then without pre-liminaries Matthew said abruptly, "I'd like to take a short vacation. I ought to see my old mother in Virginia."

"Sure," said Sammy cheerfully, and drew out a roll of bills. Matthew hesitated, counted out a few bills, and handed the rest back.

"Thanks!" he said, and with no further word turned and went out.

Sammy's jaw dropped. He stared at the bills in his hand and at the door. "I don't like that handing dough back," he said. "It ain't natural."

"He may be honest," said Sara.

"And in politics? Humph! Wonder just what his game is? I wish he'd grin a little more and do the glad hand act!"

"Do you want the earth?" asked Sara.

It was Christmas time, 1924, when Matthew came back to Virginia after five years of absence. Winter had hardly begun, and the soft glow of Autumn still lingered on the fields. He stopped at the county seat three miles from home and went to the recorder's office. It was as he had thought; his mother's little farm of twenty acres was mortgaged, and only by the good-natured indulgence of the mortgagee was she living there and paying neither interest nor rent.

"Don't want to disturb Sally, you know. She's our folks. Used to belong to my grandfather. So you're her boy, hey? Heard you was dead—then heard you was in jail. Well, well; and what's your business—er—and what's your name? Matthew Towns? Sure, sure, the old family name. Well, Matthew, it'll take near on a thousand dollars to clear that place."

Matthew paid five hundred cash and arranged to pay the

rest and to buy the other twenty acres next year—the twenty acres of tangled forest, hill, and brook that he always had wanted as a boy; but his father strove for the twenty smoother acres—strove and failed.

Then slowly Matthew walked out into the country and into the night. He slept in an empty hut beside the road and listened to creeping things. He heard the wind, the hooting of the owls, and saw the sun rise, pale gold and crimson, over the eastern trees. He washed his face by the roadside and then sat waiting—waiting for the world.

He sat there in the dim, sweet morning and swung his long limbs. He was a boy again, with the world before him. Beyond the forest, it lay magnificent—wonderful—beautiful—beautiful as one unforgettable face. He leaped to the ground and clenched his hand. A wave of red shame smothered his heart. He had not known such a rush of feeling for a year. He thought he had forgotten how to feel. He knew now why he had come here. It was not simply to see that poor old mother. It was to walk in *her* footsteps, to know if she had carried his last message.

A bowed old black man crept down the road.

"Good morning, sir."

"Good mo'nin'—good mo'nin'. Fine mo'nin'. And who might you be, sah? 'Pears like I know you."

"I am Matthew Towns."

The old man slowly came nearer. He stretched out his hand and touched Matthew. And then he said:

"She said you wuzn't dead. She said God couldn't let you die till she put her old hands on your head. And she sits waitin' for you always, waitin' in the cabin do'."

Matthew turned and went down to the brook and crossed it and walked up through the black wood and came to the fence. She was sitting in the door, straight, tall, big and brown. She was singing something low and strong. And her eyes were scanning the highway. Matthew leaped the fence and walked slowly toward her down the lane.

VII

Sara Andrews sat in Matthew's flat in the spring of 1925 and looked around with a calculating glance. It was in her eyes a silly room; a man's room, of course. It was terribly dirty and yet with odd bits—a beautiful but uneven parquet floor, quite new; a glorious and costly rug that had never been swept; old books and pamphlets lay piled about, and in the center was a big dilapidated armchair, sadly needing new upholstery. The room was proof that Matthew needed a home. She would invite him to hers. It might lead to something, and Sara looked him over carefully as he bent over the report which she had brought. Outside his haunted eyes and a certain perpetual lack of enthusiasm, he was very good to look at. Very good. He needed a good barber and a better tailor. Sara's eyes narrowed. She didn't quite like the fact that he never noticed *her* tailor nor hairdresser.

They were expecting the Honorable Sammy to breeze in any moment. They formed a curious troika, these three: Sammy, the horse of guidance in the shafts, was the expert on the underworld—the "boys," liquor, prostitution, and the corresponding parts of the white world. He was the practical politician; he saw that votes were properly counted, jobs distributed where they would do his organization most good; and he handled the funds. Sara was intellectually a step higher; she knew the business interests of the city and what they could and would pay for privilege. She was in touch with public service organizations and chambers of commerce and knew all about the leading banks and corporations. Her letters and advice did tricks and brought a growing stream of gold of which Sammy had never before dreamed. Alas, too, it brought interference with some of his practical plans and promises which annoyed him, although he usually yielded under pressure.

Matthew was quite a different element. On the one hand he knew the life of his section of the Chicago black world as no one else. He had not artificially extracted either the good or the evil for study and use—he took it all in with one comprehensive glance and thus could tell what church and school and labor thought and did, as well as the mind of the underworld. At the other end of the scale was his knowledge of national and international movements; his ability to read and digest reports and recent literature was an invaluable guide for Sara and corrective for Sammy.

Much of all this report and book business was Greek to

Sammy. Sammy never read anything beyond the headlines of newspapers, and they had to be over an inch high to get his undivided attention. Gossip from high and low sources brought him his main information. Sara read the newspapers, and Matthew the magazines and books. Thus Sammy's political bark skimmed before the golden winds with rare speed and accuracy.

Sammy came in, and they got immediately to business.

"I'm stumped by this legislature business," growled Sammy. "Smith picked a hell of a time to die. Still, p'raps it was best. There was a lot of stink over him anyhow. Now here comes a special election, and if we ain't careful it'll tear the machine to pieces. Every big nigger in Chicago wants the job. We need a careful man or hell'll be to pay. I promised the next opening to Corruthers. He expects it and he's earned it; Corruthers will raise hell and spill the beans if I fail him."

"Smith was a fool," said Sara, "and Corruthers is a bag of wind when he's sober and an idiot when he's drunk, which is his usual condition. We've got to can that type. We've got to have a man of brains and knowledge in the legislature this fall, or we'll lose out. We're in fair way to make 'Negro' and 'grafter' synonymous in Illinois office-holding. It won't do. There's some big legislation coming up—street-car consolidation and super-power. Here's a chance, Sammy, to put in our own man, and a man of high type, instead of boosting a rival boss and courting exposure for bribery."

"Well, can't we tell Corruthers how to act and vote? He ought to stay put."

"No. There are some things that can't be told. Corruthers is a born petty grafter. When he sees a dollar, he goes blind to everything else. He has no imagination nor restraint. We can't be at his shoulder at every turn; he'd be sure to sell out for the flash of a hundred-dollar bill any time and lose a thousand and get in jail. Then, too, if he should make good, next year Boss Corruthers would be fighting Boss Scott."

Sammy swore. "If I ditch him, I'll lose this district."

"With a strong nomination there's a chance," said Matthew.

Sara glanced at him and added: "Especially if I organize the women."

Sammy tore at his hair: "Don't touch 'em," he cried. "Let 'em alone! My God! What'd I do with a bunch of skirts dippin' in? Ain't we got 'em gagged like they ought to be? What's the matter with the State Colored Women's Republican Clubs? And the Cook County organization, with their chairman sitting in on the County Central and women on each ward committee?"

But Sara was obdurate. "Don't be a fool, Sammy. You know these women are nothing but 'me-too's,' or worse, for the men. I'm going to have a new organization, independent of the ward bosses and loyal to us. I'm going to call it the Chicago Colored Woman's Council—no, it isn't going to be called Republican, Democratic, or Socialist; just colored. I'm going to make it a real political force independent of the men. The women are in politics already, although they don't know it, and somebody is going to tell them soon. Why not us? And see that they vote right?"

"The white women's clubs are trying to bring the colored clubs in line for a stand on the street-car situation and new working-women laws," added Matthew.

Sammy brooded. "I don't like it. It's dangerous. Once give 'em real power, and who can hold them?"

"I can."

"Yes, and who'll hold you?"

Sara did not answer, and Sammy switched back to the main matter.

"I s'pose we've got to hunt another man for Smith's place. I see a fight ahead."

Matthew's guests left, and he discovered that he had forgotten to get his laundry for now the second week. He stepped down to the Chinaman's for his shirts and a chat.

Then came a shock, as when an uneasy sleeper, drugged with weariness, hears the alarm of dawn. The Chinaman liked him and was grateful for protection against the police and rowdies. He liked the Chinaman for his industry, his cleanliness, his quiet philosophy of life. Once he tried a pipe of opium there, but it frightened him. He saw a Vision.

Tonight the Chinaman was "velly glad" to see him. Had been watching for him several days—had "a flend" who knew him. Matthew looked about curiously, and there in the door stood his young Chinese friend of Berlin. Several times in his life—oh, many many times, that dinner scene had returned vividly to his imagination, but never so vividly as now. It leapt to reality. The sheen of the silver and linen was there before him, the twinkling of cut glass; he heard the low and courteous conversation—the soft tones of the Japanese, the fuller tones of the Egyptian, and then across it all the sweet roll of that clear contralto—dear God!—he gripped himself and hurled the vision back to hell.

"How do you do!" he said calmly, shaking the Chinaman's eager hand.

"I am so glad—so glad to see you," the Chinaman said. "I am hurrying home to China, but I heard you were here, and I had to wait to see you. How—"

But Matthew interrupted hastily, "And how is China?"

The yellow face glowed. "The great Day dawns," he said. "Freedom begins. Russia is helping. We are marching forward. The Revolution is on. To the sea with Europe and European slavery! Oh, I am happy."

"But will it be easy sailing?"

"No, no—hard—hard as hell. We are in for suffering, starvation, revolt and reverse, treason and lying. But we have begun. The beginning is everything. We shall never end until freedom comes, if it takes a thousand years."

"You have been living in America?"

"Six months. I am collecting funds. It heartens one to see how these hard-working patriots give. I have collected two millions of dollars."

"God!" groaned Matthew. "Our N. A. A. C. P. collected seventy-five thousand dollars in two years, and twelve million damn near fainted with the effort."

The Chinaman looked sympathetic.

"Ah," he said hesitatingly. "Doesn't it go so well here?"

"Go? What?"

"Why—Freedom, Emancipation, Uplift—union with all the dark and oppressed."

Matthew smiled thinly. The strange and unfamiliar words seemed to drift back from a thousand forgotten years. He hardly recognized their meaning.

"There's no such movement here," he said.

The Chinaman looked incredulous.

"But," he said—"but you surely have not forgotten the great word you yourself brought us out of the West that night—that word of faith in opportunity for the lowest?"

"Bosh!" growled Matthew harshly. "That was pure poppy-cock. Dog eat dog is all I see; I'm through with all that. Well, I'm glad to have seen you again. So long, and good-by."

The Chinaman looked troubled and almost clung to Matthew's hand.

"The most hopeless of deaths," he said, as Matthew drew away, "is the death of Faith. But pardon me, I go too far. Only one other thing before we part. John here wants me to tell you about some conditions in this district which he thinks you ought to know. Organized crime and debauchery are pressing pretty hard on labor. You have such an opportunity here— I hoped to help by putting you in touch with some of the white laboring folk and their leaders."

"I know them all," said Matthew, "and I'm not running this district as a Sunday School."

He bowed abruptly and hastened away.

VIII

Matthew was uncomfortable. The demon of unrest was stirring drowsily away down in the half-conscious depths of his soul. For the long months since his incarceration he had been content just to be free, to breathe and look at the sunshine. He did not think. He tried not to think. He just lived and narrowed himself to the round of his duties. As those duties expanded, he read and studied, but always in the groove of his work. Sternly he held his mind down and in. No more flights; no more dreams; no more foolishness.

Now, as he felt restless and dissatisfied, he laid it to nerves, lack of physical exercise, some hidden illness. But gradually he began to tell himself the truth. The dream, the woman, was back in his soul. The vision of world work was surging and he must kill it, stifle it now, and sternly, lest it wreck his life again. Still he was restless. He was awakening. He could feel the prickling of life in his thought, his conscience, his body. He was struggling against the return of that old ache—the sense of that void. He was angry and irritated with his apparent lack of control. If he could once fill that void, he could glimpse another life—beauty, music, books, leisure; a home that was refuge and comfort. Something must be done. Then he remembered an almost forgotten engagement.

Soon he was having tea in Sara's flat. He began to feel more comfortable. He looked about. It was machine-made, to be sure, but it was wax-neat and in perfect order. The tea was good, and the cream—he liked cream—thick and sweet. Sara, too, in her immaculate ease was restful. He leaned back in his chair, and the brooding lifted a little from his eyes. He told Sara of a concert he had attended.

"Have you ever happened to hear Ivanoff's 'Caucasian Sketches'?"

Sara had not; but she said suddenly, "How would you like to go to the legislature?"

Matthew laughed carelessly. "I wouldn't like it," he said and sauntered over to look at a new set of books. He asked Sara if she liked Balzac. Sara had just bought the set and had not read a word. She had bought them to fill the space above the writing-desk. It was just twenty-eight inches. She let him talk on and then she gave him some seed-cakes which a neighbor had made for her. He came back and sat down. He tested the cakes, liked them, and ate several. Then Sara took up the legislature again.

"You can talk—you have read, and you have the current political questions at your fingers' ends. Your district will stand with you to a man. Old-timers like Corruthers will knife you, but I can get you every colored woman's vote in the ward, and they can get a number of the white women by trading."

"I don't want the notoriety."

"But you want money—power—ease."

"Yes—I want money, but this will take money, and I have none."

"I have," said Sara. And she added, "We might work together with what I've saved and what we both know."

Matthew got up abruptly, walked over and stared out the window. He had had a similar idea, and he thought it originated in his own head. He had not noticed Sara much hitherto. He had not noticed any woman, since—since— But he knew Sara was intelligent and a hard worker. She looked simple, clean, and capable. She seemed to him noticeably lonely and needing some one to lean on. She could make a home. He never had had just the sort of home he wanted. He wanted a home—something like his own den, but transfigured by capable hands—and devotion. Perhaps a wife would stop this restless longing—this inarticulate Thing in his soul.

Was this not the whole solution? He was living a maimed, unnatural life—no love, no close friendships; always loneliness and brooding. Why not emerge and be complete? Why not marry Sara? Marriage was normal. Marriage stopped secret longings and wild open revolt. It solved the woman problem once and for all. Once married, he would be safe, settled, quiet; with all the furies at rest, calm, satisfied; a reader of old books, a listener to sad and quiet music, a sleeper.

Sara watched him and after a pause said in an even voice:

"You have had a hard shock and you haven't recovered yet. But you're young. With your brains and looks the world is open to you. You can go to the legislature, and if you play your cards right you can go to Congress and be the first colored congressman from the North. Think it over, Mr. Towns."

Towns turned abruptly. "Miss Andrews," he said, "will you marry me?"

"Why—Mr. Towns!" she answered.

He hurried on: "I haven't said anything about love on your side or mine—"

"Don't!" she said, a bit tartly. "I've been fighting the thing men call love all my life, and I don't see much in it. I don't think you are the loving kind—and that suits me. But I do think enlightened self-interest calls us to be partners. And if you really mean this, I am willing."

Matthew went slowly over and took her hand. They looked at each other and she smiled. He had meant to kiss her, but he did not.

IX

It was a grand wedding. Matthew was taken back by Sara's plans. He had thoughts of the little church of his district—and perhaps a quiet flitting away to the Michigan woods, somewhere up about Idlewild. There they might sit in sunshine and long twilights and get acquainted. He would take this lonely little fighting soul in his arms and tell her honestly of that great lost love of his soul, which was now long dead; and then slowly a new, calm communion of souls, a silent understanding, would come, and they would go hand in hand back to the world.

But nothing seemed further from Sara's thought. First she was going to elect Matthew to the legislature, and then in the glory of his triumph there was going to be a wedding that would make black Chicago sit up and even white Chicago take due notice. Thirdly, she was going to reveal to a gaping world that she already owned that nearly new, modern, and beautifully equipped apartment on South Parkway which had just been sold at auction. There was a vague rumor that a Negro had bought it, but none but Sara and her agent knew.

"How on earth did you—" began Matthew.

"I'm not in politics for my health," said Sara, "and you're not going to be, after this. It's got three apartments of seven rooms with sleeping porches, verandas, central heating, and refrigeration. We'll live in the top apartment and rent the other two. We can get easily three thousand a year from them, which will support us and a maid. I've been paying for a car by installments—a Studebaker—and learning to drive, for we can't afford a chauffeur yet."

Matthew sat down slowly.

"Don't you think we might rent the whole and live somewhere—a little more quietly, so we could study and walk and —go to concerts?"

But Sara took no particular notice of this.

"I've been up to Tobey's to select the furniture, and Marshall Field is doing the decorating. We'll keep our engagement dark until after the nomination in the spring. Then we'll have a big wedding, run over to Atlantic City for the honeymoon, and come back fit for the fall campaign."

"Atlantic City? My God!" said Matthew, and then stopped as the door opened to admit the Honorable Sammy Scott.

Sammy was uneasy these days. He was in hot water over this legislature business, and he vaguely scented danger to his power and machine beyond this. First of all he could only square things with Corruthers and his followers by a good lump of money, if Matthew were nominated; and even then, they would try to knife him. Now Sammy's visible source for more money was more laxity in the semi-criminal districts and bribes from interests who wanted bills to pass the legislature. Sammy had given freer rein to the red-light district and doubted if he could do more there or collect much more money without inviting in the reformers. Big business seemed his only

resort, but here he was not sure of Matthew. There might be
a few nominees who were willing to pay a bit for the honor,
but Matthew was not among these. Sara was managing his
campaign, and she was too close and shrewd to cough up much.
Then, too, Sammy was uneasy about Sara and Matthew. They
were mighty thick and chummy and always having conferences. If he himself had been a marrying man—

"Say, Towns," he said genially, "I think I got that nomination cinched, but it's gonna take a pot of dough. Oh, well, what
of it? You've got the inside track."

"Unless Corruthers double-crosses us," said Matthew dryly.

"That's where the dough comes in. Now see here, I've got
a proposition from the traction crowd. They want to ward
off municipal ownership and get a new franchise city-wide with
consolidation. They're going to offer a five-cent car fare and
reversion to the city in forty-nine years, and they're paying
high for support. They're going to control the nomination in
most districts."

"I'll vote against municipal ownership any time," said Matthew.

Sammy was at once relieved and yet troubled anew. He
had an idea that Matthew would get squeamish over this and
would thus lose the nomination. That would force Corruthers
in. Sammy still leaned toward Corruthers. But, on the other
hand, Corruthers would be sure to do some fool trick even if
he were elected, and that or his defeat might ruin Sammy's
own plans for Congress next year. He was glad Matthew was
tractable, and at the same time he suddenly grew suspicious.
Suppose Matthew went to the legislature and made a ripping
record? He might himself dare think of Congress. But no—
Sara was pledged to Sammy's plan for Congress.

"All right!" said Sammy noisily.

"But look here, Sammy," said Matthew. "Things are getting pretty loose and free down in my district. Casey has
opened a new gambling den, and there's a lake of liquor; three
policy wheels are running. The soliciting on the streets is open;
it isn't safe for a working girl after dark."

"Well, ain't they payin' up prompt?"

"Yes—but—"

"Gettin' squeamish?" sneered Sammy. My God! Was the

fool going to cut off the main graft and try to depend on white corporations?

"No—I'm not, but the reformers are. We're just bidding for interference at this rate."

"Hell," said Sammy. "It'll be whore-houses and not Sunday Schools that'll send you to Springfield, if you go." Matthew frowned.

Sara intervened. "I'll see that things are toned down a bit. Sammy will never learn that big business pays better than crime. I'm glad you're going to vote straight on the traction bill."

Matthew still frowned. They both had misunderstood him —curiously. They suspected him of mawkish sentimentality— a conscience against gambling, liquor, and prostitution. Nothing of the sort! He had buried all sentiment, down, down, deep down. He was angry at being even suspected. Why was he angry? Was it because he felt the surge of that old bounding, silly self that once believed and hoped and dreamed— that dead soul, turning slowly and twisting in its grave? No, no, not that—never. He simply meant to warn Sammy that a district too wide open defeated itself and invited outside interference; it cut off political graft; gamblers were cheating gamblers; the liquor on sale was poison; prostitutes were approaching the wrong people—and, well—surely a girl ought to have the right to choose between work and prostitution, and she ought not to be shanghaied.

And then Sara. She assumed too much. If he had the beginning of the unrest of a new conscience—and he had not—it was over these big corporations. He began to see them from behind and underneath. A five-cent fare was a tremendous issue to thousands. The driblets of perpetual tax on light and air and movement meant both poverty and millions. Surely the interests could pay better than gamblers and prostitutes, but was the graft as honest? Was he going on as unquestioningly? He had promised to vote against municipal ownership quickly and easily. Voters were too stupid or too careless to run big business. Municipal ownership, therefore, would only mean corporation control one degree removed and concealed from public view by election bribery. And after all, traction was not the real question. Super-power was that, and he talked his thought aloud to Sara, half-consciously:

"Oh—traction? Sure—that's only camouflage anyway. Back of it is the furnishing of electric power, cornering the waterfalls of America; paying nothing for the right of endless and limitless taxation, and then at last 'financing' the whole thing for a thousand millions and unloading it on the public! That's the real graft. I am going to think a long time over those bills!"

What did he mean by "thinking a long time"? He did not know what he meant. Neither did Sara. But she knew very clearly what she meant. She was silent and pursed her lips. She was already in close understanding with certain quiet and well-dressed gentlemen who represented Public Service and were reaching out toward Super-Power. They had long been distributing money in the Negro districts, but their policy was to encourage rivalry and jealousy between the black bosses and thus make them ineffective. This kept payments down. Sara had arranged for Sammy to make these payments, while the corporations dealt only with him. Also she had raised the price and promised to deliver four votes in the legislature and three in the Board of Aldermen. Finally, she had just arranged to have Sammy's personal representative occupy an office in the elegant suite of the big corporation attorney who advised Public Service and on his payroll as a personal link between Sammy and the big Public Service czar. It was the biggest single deal she had pulled off, and she hadn't yet told Sammy. The selection of that link called for much thought.

x

The house was finished complete with new and shining furniture, each piece standing exactly where it should. Matthew had particularly wanted a fireplace with real logs. He was a little ashamed to confess how much he wanted it. It was a sort of obsession. As long as he could remember, burning wood had meant home to him. Sara said a fireplace was both dirty and dangerous. She had an electric log put in. Matthew hated that log with perfect hatred.

The pictures and ornaments, too, he did not like, and at last, one day, he went downtown and bought a painting which he had long coveted. It was a copy of a master—cleverly and daringly done with a flame of color and a woman's long and naked body. It talked to Matthew of endless strife, of fire and beauty and never-dying flesh. He bought, too, a deliciously ugly Chinese god. Sara looked at both in horror but said nothing. Months afterward when they had been married and had moved home, he searched in vain for the painting and finally inquired.

"That thing? But, Matthew, dear, folks don't have naked women in the parlor! I exchanged it for the big landscape there—it fits the space better and has a much finer frame." Sara let the ugly Chinese god crouch in a dark corner of the library.

The nomination went through smoothly. The "election of Mr. Matthew Towns, the rising young colored politician whose romantic history we all know" (thus *The Conservator*) followed in due and unhindered course, despite the efforts of Corruthers to knife him.

So in June came the wedding. It was a splendid affair. Sara's choice of a tailor was as unerringly correct as her selection of a dressmaker. They made an ideal couple as they marched down the aisle of the Michigan Avenue Baptist Tabernacle. Matthew looked almost distinguished, with that slight impression of remote melancholy; Sara seemed so capable and immaculate.

Sammy, the best man, swore under his breath. "If I'd only been a marrying man!" he confided to the pastor.

The remark was made to Matthew's young ministerial friend, the Reverend Mr. Jameson, formerly of Memphis. He had come with his young shoulders to help lift the huge mortgages of this vast edifice, recently purchased at a fabulous price from a thrifty white congregation; the black invasion of South Side had sent them to worship Jesus Christ on the North Shore.

"Whom God hath joined together let no man put asunder," rolled the rich tones of the minister. Matthew saw two wells of liquid light, a great roll of silken hair that fell across a skin of golden bronze, and below, a single pearl shining at the parting of two little breasts.

"Straighten your tie," whispered Sara's metallic voice, and his soul came plunging back across long spaces and over heavy roads. He looked up and met the politely smiling eyes of the young Memphis school teacher who once gave him fifteen cents. She was among the chief guests with her fat husband, a successful physician. They both beamed. They quite approved of Matthew now.

" 'Tis thy marriage morning, shining in the sun," yelled the choir, with invincible determination. The bridal pair stepped into the new Studebaker with a hired chauffeur and glided away. Matthew looked down at his slim white bride. A tenderness and pity swept over him. He slipped his arm about her shoulders.

"Be careful of the veil," said Sara.

<h2 style="text-align:center">XI</h2>

In Springfield, Matthew was again thrust into the world. He shrank at first and fretted over it. Most of the white legislators put up at the new Abraham Lincoln—a thoroughly modern hostelry, convenient and even beautiful in parts. Matthew did not apply. He knew he would be refused. He did try the Leland, conveniently located and the former rendezvous of the members. He had dinner and luncheon there, and after he discovered the limited boarding-house accommodation of colored Springfield, he asked for rooms—a bedroom and parlor. The management was very sorry—but—

He then went down to the colored hotel on South Eleventh Street. The hotel might do—but the neighborhood!

Finally, he found a colored private home not very far from the capitol. The surroundings were noisy and not pleasant. But the landlady was kindly, the food was excellent, and the bed comfortable. He hired two rooms here. The chief difficulty was a distinct lack of privacy. The landlady wanted to exhibit her guest as part of the family, and the public felt free to drop in early and stay late.

Gradually Matthew got used to this new publicity and began to look about. He met a world that amused and attracted him. First he sorted out two kinds of politicians. Both had one object—money. But to some Money was Power. On it they were climbing warily to dazzling heights—Senatorships, Congress, Empire! Their faces were strained, back of their carven smiles. They were walking a perpetual tight rope. Matthew hated them. Others wanted money, but they used their money with a certain wisdom. They enjoyed life. Some got gloriously and happily drunk. Others gambled, riding upon the great wings of chance to high and fascinating realms of desire. Nearly all of them ogled and played with pretty women.

On the whole, Matthew did not care particularly for their joys. Liquor gave him pleasant sensations, but not more pleasant and not as permanent as green fields or babies. He never played poker without visioning the joys of playing European politics or that high game of world races which his heart had glimpsed for one strange year—one mighty and disastrous year.

And women! If he had not met one woman—one woman who drew and filled all his imagination, all his high romance, all the wild joys and beauty of being—if she had never lived for him, he could have been a rollicking and easily satisfied Lothario and walked sweet nights out of State Street cabarets. Now he was not attracted. He had tried it once in New York. It was ashes. Moreover, he was married now, and all philandering was over. And yet—how curious that marriage should seem— well, to stop love, or arrest its growth instead of stimulating it.

He had not seen much of Sara since marriage. They had been so busy. And there had been no honeymoon, no mysterious romantic nesting; for Matthew had finally balked at Atlantic City. He tried to be gentle about it, but he showed a firmness before which Sara paused. No, he would not go to Atlantic City. He had gone there once—one summer, an age ago. He had been refused food at two restaurants, ordered out of a movie, not allowed to sit in a boardwalk pavilion, and not even permitted to bathe in the ocean.

"I will not go to Atlantic City. If I must go to hell, I'll wait until I'm dead," he burst out bitterly.

Sara let it go. "Oh, I don't really mind," she said, "only I've never been there, and I sort of wanted to see what it's like. Never mind, we'll go somewhere else." But they didn't; they stayed in Chicago.

So now he was a member of the legislature and in Springfield. The politicians came and went. The climbers avoided Matthew. Colored acquaintances were a debit to rising men. The other politicians knew him—jollied him and liked him— even drew him out for a rollicking evening now and then; but voted that he did not quite "belong." He was always a trifle remote—apart. He never could quite let himself go and be wholly one of them. But he liked them. They lived.

There were several members of the House who were not politicians. They did not count. They fluttered about, uttering shrill noises, and beat their wings vainly on unyielding iron bars.

Then there were men in politics who were not members of the legislature. Grave, well-dressed men of business and affairs. They came for confidential conferences with introductions from and connections with high places, governors, brokers, railway presidents, ruling monarchs of steel, oil, and international finance. And from Sammy; especially from Sammy and Sara. Money was nothing to them, and money was all. A thousand dollars—ten thousand—it was astounding, the sums at their command and the ease with which they distributed it. There was no crude bribery as on State Street— but Matthew soon learned that it was curiously easy to wake up a morning a thousand dollars richer than when one went to bed; and no laws broken, no questions asked, no moral code essentially disarranged. Matthew disliked these men esthetically, but he saw much of them and conferred with one or another of them nearly every week. It was his business. They did not live broadly or deeply, but they ruled. There was no sense blinking that fact. Matthew often forwarded registered express packages to Sara.

And he came to realize that legislating was not passing laws; it was mainly keeping laws from being passed.

Then there were the reformers. He held them—most of them —in respectful pity; palliators, surface scratchers. He listened to them endlessly and gravely. He read their tracts conscientiously, but only now and then could he vote as they asked. They were so ignorant—so futile. If only he, as a practical politician, might tell them a little. Birth control? Mothers' pensions? Restricted hours of labor for women and children? He agreed in theory with them all, but why ask his judgment? Why not ask the Rulers who put him in the legislature? And without the consent of these quiet, calm gentlemen who represented Empires, Kingdoms, and Bishoprics, what could he do, who was a mere member of the legislature?

Yet he could not say this, and if he had said it, they would not have understood. They pleaded with him—he that needed no pleas. One was here now—the least attractive—one stocking awry on her big legs, a terrible hat and an ill-fitting gown.

She was president of the Chicago local of the Box-Makers'
Union. Her breasts were flat, her hips impossible, her hair
dead straight, and her face white and red in the wrong places.

"How would you like your daughter down there?" she
bleated.

"I haven't one."

"But if you did have?"

"I'd hate it. But I wouldn't be fool enough to think any law
would take her away."

"Well, what would?"

"Power that lies in the hands of the millionaire owners of
factory stocks and bonds; and the bankers that guide and
advise them. Transfer that power to me or you."

"That's it. Now help us to get this power!"

"How?"

"By voting—"

"Pish!"

"But how else? Are you going to sit down and let these girls
go to death and hell?"

"I'm not responsible for this world, madam."

"Listen—I know a woman—a *woman*—like you. She's just
been elected International President of the Box-Makers. She
can talk. She knows. She's been everywhere. She's a lady and
educated. I'm just a poor, dumb thing. I know what I want—
but I can't say it. But she'll be in Chicago soon—I'm going
to bring her to plead for this bill."

"Spare me," laughed Matthew.

But he kept thinking of that poor reformer. And slowly and
half-consciously—stirred by a thousand silly, incomplete arguments for impossible reform measures—revolt stirred within
him against this political game he was playing. It was not
moral revolt. It was esthetic disquiet. No, the revolt slowly
gathering in Matthew's soul against the political game was
not moral; it was not that he discerned anything practical
for him in uplift or reform, or felt any new revulsion against
political methods in themselves as long as power was power,
and facts, facts. His revolt was against things unsuitable, ill
adjusted, and in bad taste; the illogical lack of fundamental
harmony; the unnecessary dirt and waste—the ugliness of it
all—that revolted him.

He saw no adequate end or aim. Money had been his object,

but money as security for quiet, for protection from hurt and insult, for opening the gates of Beauty. Now money that did none of these was dear, absurdly dear, overpriced. It was barely possible—and that thought kept recurring—it was barely possible that he was being cheated, was paying too high for money. Perhaps there were other things in life that would bring more completely that which he vaguely craved.

It seemed somehow that he was always passive—always waiting—always receptive. He could never get to doing. There was no performance or activity that promised a shining goal. There was no goal. There was no will to create one. Within him, years ago, something—something essential—had died.

Yet he liked to play with words, cynically, on the morals of his situation as a politician. In his office today, he was talking with a rich woman who wanted his vote for limiting campaign funds. He looked at her with narrowed eyes:

"We have got to stop this lying and stealing or the country will die," she said impatiently.

He watched his unlighted cigarette.

"Lying? Stealing? I do not see that they are so objectionable in themselves. Lying is a version of fact, sometimes—often poetic, always creative. Stealing is a transfer of ownership, or an attempted transfer, sometimes from the overfed to the hungry—sometimes from the starving to the apoplectic. It is all relative and conditional—not absolute—not infinite."

"It is laying impious hands on God's truth—it is taking His property."

"I am not sure that God has any truth—that is, any arrangement of facts of which He is finally fond and of which He could not and does not easily conceive better or more fitting arrangement. And as to property, I'm sure He has none. Every time He has come to us, He has been disgustingly poor."

The woman rose and fled. Matthew sighed and went back to his round of thought. Municipal ownership of transportation in Chicago: he had begun to look into it. He was prejudiced against it by his college textbooks and his political experience. But here somehow he scented something else. Back of the demand made to kill the present municipal ownership was another proposal to renew the franchise of the street-car lines with an "Indeterminate Permit," which meant in fact a perpetual charter. There was a powerful lobby of trained lawyers back of this bill, and what struck Matthew was that the same lobby was back of the movement to kill municipal ownership. Were they interested in super-power projects also? Matthew viewed this whole scramble as one who watches a great curdling of waters and begins to sense the current.

He was not evolving a conscience in politics. He was not revolting against graft and deception, but he was beginning to ask just what he was getting for his effort. Money? Some— not so very much. But the thing was—not wrong—no—but unpleasant—ugly. That was the word. He was paying too much for money—money might cost too much. It might cost ugliness, writhing, dirty discomfort of soul and thought. That's it. He was paying too much for even the little money he got. He must pay less—or get more. Matthew sighed and looked at the next card. It was that of the Japanese statesman whom he had met in Berlin. He arose slowly and faced the door.

XII

"I trust I am not intruding," said the Japanese.

Matthew bowed coldly. He gave no sign of recognizing the Japanese, nor did he pretend not to.

"Certainly not—these are my office hours."

The Japanese was equally reticent and yet was just a shade too confidential to be an entire stranger. And again in Matthew's mind flamed and sang that Berlin dinner party. Even the music floated in his ears. But he put it all rudely and brusquely aside.

"What can I do for you, sir? Be seated. Will you smoke?"

The Japanese took a cigarette, tasted it with relish, and leaned back easily in his chair. He glanced at the office. Matthew was ashamed. If he had been white, he would have had a room in the new Abraham Lincoln Hotel; something fine and modern, clean and smart, with service and light. If he had been black, free, and rich, he would perhaps have received his guest in a house of his own—delicately vaulted and soft with color; something beautiful in brick or marble, with high sweep of a curtain and pillar, a possibility of faint music, and silent deferential service. But being black, half slave, and poor, he had the front room of Mrs. Smith's boarding-house, a show room, to be sure, but conglomerate of jarring styles and tastes, overloaded and thick with furnishings; with considerable dust and transient smells and near the noisy street. Matthew was furious with himself for thinking thus apologetically. Whose business was it how he lived or what he had?

Then the Japanese looked at him.

"I have been much interested in noting the increased political power of your people," he said.

"Indeed," said Matthew, noncommittally.

"When I was in the United States twenty years ago—" (So he had been here twenty years ago and interested in Negroes!)—"you were politically negligible. Today in cities and states you have a voice."

Matthew was silent.

"I have been wondering," said the Japanese with the slow voice of one delicately feeling his way—"I have been wondering how far you have unified and set plans—"

"We have none."

"—either for yourselves in this land, or even further, with an eye toward international politics and the future of the darker races?"

"We have little interest in foreign affairs," said Matthew.

The Japanese shifted his position, asked permission, and lighted a second cigarette. He glanced appraisingly at Matthew.

"Some time ago," he continued, "at a conference in Berlin, it was suggested that intelligent coöperation between American Negroes and other oppressed nations of the world might sensibly forward the uplift and emancipation of the darker peoples. I doubted this at the time."

"You may continue to doubt," said Matthew. "The dream at Berlin was false and misleading. We have nothing in common with other peoples. We are fighting out our own battle here in America with more or less success. We are not looking for help beyond our borders, and we need all our strength at home."

It would have been difficult for Matthew to say what prompted him to talk like this. Mainly, of course, it was deepseated and smoldering resentment against this man whose interference, he believed, had wrecked his world. Perhaps, of course, this was not true. Perhaps shipwreck was certain, but— he was determined not to sail for those harbors again, not for a moment even to reconsider the matter; and he repeated as his own the current philosophy of the colored group about him. It sounded false as he spoke, but he talked on. The Japanese watched him as he talked.

"Ah!" he said. "Ah! I am sorry. There were some of us who hoped—"

Matthew's heart leaped. Questions rushed to his lips, and one word clamored for utterance. He beat them back and glanced at his watch.

The Japanese arose. "I am keeping you?" he said.

"No—no—I have a few minutes yet."

The Japanese glanced around, and bending forward, spoke rapidly.

"The Great Council," he said, "of the Darker Peoples will meet in London three months hence. We have given the American Negro full representation; that is, three members on the Board. You are chairman. The other two are—"

Matthew arose abruptly.

"I cannot accept," he said harshly. "I am no longer interested."

"I am sorry," said the Japanese slowly. He paused and pondered, started to speak as Matthew's heart hammered in his throat. But the Japanese remained silent.

He extended his hand. Matthew took it, frowning. They murmured polite words, and the visitor was gone.

Matthew threw himself on the couch with an oath, and through his unwilling head tramped all the old pageant of empire with black and brown and yellow leaders marching ahead.

XIII

Matthew was gray with wrath. Sara was quiet and unmoved.

"Yes," she said. "I promised them your vote, and they paid for it—a good round sum."

Matthew had been a member of the legislature of Illinois about six months. He had made a good record. Everybody conceded that. Nothing spectacular, but his few speeches were to the point and carried weight; his work on committees had been valuable because of his accurate information and willingness to drudge. His votes, curiously enough, while not uniformly pleasing to all, had gained the praise even of the

women's clubs and of some of the reformers, whom he had
chided, while at the same time the politicians regarded
Matthew as a "safe" man. Matthew Towns evidently had a
political future.

Yet Matthew was far from happy or satisfied. Outside his
wider brooding over his career, he had not gained a home by
his marriage. The flat on South Parkway was an immaculate
place which must not be disturbed for mere living purposes
and which blossomed with dignified magnificence. At repeated
intervals crowds burst in for a reception. There was whist and
conversation, dancing as far as space would allow; smoking,
cocktails, and smutty stories back in the den with the men;
whispers and spiteful gossip on the veranda with the ladies;
and endless piles of rich food in the dining-room, served by
expensive caterers.

"Mrs. Matthew Towns' exclusive receptions for the smarter
set" (thus the society reporter of *The Lash*) were "the most
notable in colored Chicago."

And Sara was shrewd enough, while gaining this reputation
for social exclusiveness, to see that no real person of power or
influence in colored Chicago was altogether slighted, so that, at
least once or twice a year, one met everybody.

The result was an astonishing mélange that drove Matthew
nearly crazy. He could have picked a dozen delightful companions—some   educated—some   derelicts—students—politicians—but all human, delightful, fine, with whom a quiet evening would have been a pleasure. But he was never allowed.
Sara always had good reasons of state for including this ward
heeler or that grass widow, or some shrill-voiced young woman
who found herself in company of this sort for the first time
in her life and proclaimed it loudly; and at the same time
Sara found excuse for excluding the "nobodies" who intrigued
his soul.

Matthew's personal relations with his wife filled him with
continual astonishment. He had never dreamed that two
human beings could share the closest of intimacies and remain
unacquainted strangers. He thought that the yielding of a
woman to a man was a matter of body, mind, and soul—a
complete blending. He had never forgot—shamefaced as it
made him—the way that girl in Harlem had twisted her

young, live body about his and soothed his tired, harassed soul and whispered, "There, Big Boy!"

Always he had dreamed of marriage as like that, hallowed by law and love. Having bowed to the law, he tried desperately to give and evoke the love. But behind Sara's calm, cold hardness, he found nothing to evoke. She did not repress passion—she had no passion to repress. She disliked being "mauled" and disarranged, and she did not want any one to be "mushy" about her. Her private life was entirely in public; her clothes, her limbs, her hair and complexion, her well-appointed home, her handsome, well-tailored husband and his career; her reputation for wealth.

Periodically Matthew chided himself that their relations were his fault. He was painfully conscious of his lack of deep affection for her, but he strove to evolve something in its place. He proposed a little home hidden in the country, where, on a small income from their rents, they could raise a garden and live. And then, perhaps—he spoke diffidently—"a baby." Sara had stared at him in uncomprehending astonishment.

"Certainly not!" she had answered. And she went back to the subject of the super-power bills. The legislature had really done little work during the whole session, and now as the last days drew on the real fight loomed. The great hidden powers of finance had three measures: first, to kill municipal ownership of street-car lines; secondly, to unite all the street transportation interests of Chicago into one company with a perpetual franchise or "indeterminate permit"; thirdly, to reorganize, reincorporate, and refinance a vast holding company to conduct their united interests and take final legislative steps enabling them to monopolize electric and water power in the state and in neighboring states.

To Matthew the whole scheme was clear as day. He had promised to vote against municipal ownership, but he had never promised to support all this wider scheme. It meant power and street-car monopoly; millions in new stocks and bonds unloaded on the public; and the soothing of public criticism by lower rates for travel, light, and power, and yet rates high enough to create several generations of millionaires to rule America. He had determined to oppose these bills, not because they were wrong, but because they were unfair. For similar reasons he had driven Casey's gambling den out of business in his district; the roulette wheel and most of the dice were loaded.

But Sara was keen on the matter. Lines were closely drawn; there was strong opposition from reformers, Progressives, and the labor group. Money was plentiful, and Sara had pledged Matthew's votes and been roundly paid for it.

She and Sammy were having a conference on the matter and awaiting Matthew. Sara sensed his opposition; it must be overcome. Sammy was talking.

"Don't understand their game," said Sammy, "but they're lousy with money."

"I understand it," said Sara quietly, "and I've promised Matthew's vote for their bills."

Sammy's eyes narrowed.

Just then, Matthew came in.

"What have you promised?" he asked, looking from one to the other.

Sara quietly gathered up her papers.

"Come home to lunch," she said, "and I'll tell you."

She knew that she had to have this thing out with Matthew, and she had planned for it carefully. Sammy whistled softly to himself and did a little jig after his guests had left. He thought he saw light.

"I didn't think that combination could last long," he said to his new cigar. "Too perfect."

Sara steered her Studebaker deftly through the traffic, bowing to deferential policemen at the traffic signals and recognizing well-dressed acquaintances here and shabby idlers there, who raised their hats elaborately. Matthew sat silent, mechanically lifting his hat, but glancing neither right nor left. They glided up to the curb at home, at exactly the right distance from it, and stopped before the stepping-stone. Sara flooded the carburetor, turned off the switch, and carefully locked it. Matthew handed her down, and with a smile at the staring children, they entered the lofty porch of their house. They opened the dark oaken door with a latchkey and slowly mounted the carpeted stairs. Sara remarked that the carpet was a little worn. She feared it was not as good as Carson-Pirie had represented. She would have to see about it soon.

A brown maid in a white apron smilingly let them into the

apartment and said that lunch was "just ready—yes'm, I found some fine sweet potatoes after you 'phoned, and fried them." Matthew loved fried sweet potatoes. They had a very excellent but rather silent lunch, although Sara talked steadily about various rather inconsequential things. Then they went to the "library," which Matthew never used because its well-bound and carefully arranged books had scarcely a volume in which he had the slightest interest. Sara closed the door and turned on the electric log.

"I promised the super-power crowd," she said, "that you would vote for their bills."

It was then that Matthew went pale with wrath.

"How dared you?"

"Dared? I thought you expected me to conduct your campaign? I promised them your vote, and they paid a lot for it. Of course, it was cloaked in a real-estate transaction, but I gave them a receipt in your name and mine and deposited the money."

Matthew felt for the flashing of a moment that he could kill this pale, hard woman before him. She felt this and inwardly quailed, but outwardly kept her grip.

"I don't see," she said, "any great difference between voting for these bills and against municipal ownership. It is all part of one scheme. I hope," she added, "you're not going to develop a conscience suddenly. As a politician with a future, you can't afford to."

The trouble was that Matthew himself suddenly knew that there was no real difference. It was three steps in the same direction instead of one. But the first was negative and tentative, while the three together were tremendous. They gave a monopoly of transportation and public service in Chicago to a great corporation which aimed at unlimited permission to exploit the water power of a nation forever at any price "the traffic would bear." Of course it was no question of right and wrong. It was possible to buy privilege, as one bought votes; he himself bought votes, but—well, this was different. This privilege could be bought, of course—but not of him. It was cheating mental babies whom he did not represent—whom he did not want to represent.

He was a grafting politician. He knew it and felt no qualms about it. But he had always secretly prided himself that his exchanges were fair. The gamblers who paid him got protection; prostitutes who were straight and open need not fear the police; workers in his district could not be "shaken down" by thieves. Even in the bigger legislative deals, it was square, upstanding give and take between men with their eyes open. But this—there was no use explaining to Sara. She knew the difference as well as he. Or did she? That rankling shaft about "conscience." He was a politician who was directly and indirectly for sale. He had no business with a conscience. He had no conscience. But he had limitations. By God! everybody had some limitations. He must have them. He would sell himself if he wished, but he wouldn't be sold. He was not a bag of inert produce. He refused to be compelled to sell. He was no slave. He must and would be free. He wanted money for freedom. Well, he'd been sold. Where was the money? He wanted money. He must have it. There and there alone lay freedom, and his chains were becoming more than he could bear.

"Where is the money you got?" he said abruptly.

"I've invested it."

"I want it."

"You can't get it—it's tied up in a deal, and to disturb it would be to risk most of our fortune."

"I've put some money in our joint account."

"That's invested too. What's the use of money idle in a savings bank at four per cent when we can make forty?"

"How much are we worth?"

"Oh, not so much," said Sara cautiously. "Put the house minus the first mortgage at, say, fifty thousand—we may have another ten or fifteen thousand more." Thus she figured up.

"Matthew," she added quickly, "be sensible. In a couple of years you'll be in Congress—the greatest market in the land, and we'll be worth at least a hundred thousand. Oppose these bills, and you go to the political ashpile. Sammy won't dare to use you. My mortgagees will squeeze me. The city will come down on us for violations and assessments, and first thing we know we'll be penniless and saddled with piles of brick and mortar. As a congressman you can ignore petty graft and get in 'honestly,' as people say, on big things; in less than ten years, you'll be rich and famous. Now for God's sake, don't be a fool!"

Matthew Towns voted for the traction group of bills, but

they were defeated by an aroused public opinion which neither Republicans nor Democrats dared oppose. Matthew at the same time saved from defeat at the last moment four bills which the Progressives and Labor group were advocating. They were not radical but were entering wedges to reduce the burden on working mothers, lessen the hours of work for women, and establish the eight-hour day. One bill to restrict the power of injunctions in labor disputes failed despite Matthew's efforts.

The result was curious. Matthew was commended by all parties. The machine regarded him as safe but shrewd. The Farmer-Labor group regarded him as beginning to see the light. The Democrats regarded him as approachable. Sara was elated. She determined to begin immediately her campaign to send Matthew to Congress.

XIV

The Honorable Sammy Scott was having the fight of his life and he knew it. It almost wiped the genial smile from his lips, but he screwed it on and metaphorically stripped for the fray. He knew it was the end or a glorious new beginning for Sammy Scott.

Sammy's first real blow had been Sara's wedding. He had settled down to the comfortable fact that if Sara ever married anybody it would be Sammy Scott. At whom else had she ever looked—of whom had she ever thought? He was her hero in shrewdness and accomplishment, and he-preened himself before her. There hung the fruit—the ripe, sleek, dainty fruit at his hand. He had only to reach out and pluck it. He was not a marrying man. But—who could tell? He might want a change. He might make his pile and retire. Or go traveling abroad. Then? Well, he might marry Sara and take her along. Time would tell.

And then—then without warning—without a flash of suspicion, the blow fell. Of course, others had talked and hinted and winked. Sammy laughed and pooh-poohed. He knew Sara. Nobody could take his capable secretary off the Honorable Sammy Scott. No, sir!

After the announcement and through the marriage, Sammy bore up bravely. He never turned a hair, at least to the public. He was best man and general manager at the wedding, and his present of a grand piano, with Ampico attachment, made dark Chicago gasp.

Gradually, Sammy got an idea into his head. Sara was a cool and deep one. Perhaps, perhaps, mused Sammy, as she left him after a long and confidential talk, perhaps this husband business was all a blind. Perhaps after the marriage with a rather dull husband for exhibition purposes, Sara was going to be more approachable. In her despair at not inveigling Sammy himself into marriage—so Sammy argued, waving his patent-leather shoes on their high perch—after her wiles failed, then perhaps she'd decided to have her cake and eat it too. All right—all the same to Sammy. Of course, he might have preferred—but women are curious.

He hinted something of this to Sara and got a cryptic response—a sort of prim silence that made him guffaw and slap his thigh. Of course, he had upbraided her first with disloyalty and quitting; but all this she disclaimed with pained surprise. She gave Sammy distinctly to understand—she did not say it—that she was loyally and eternally his steward forever and ever.

So Sammy was shaken but hopeful, and matters went on as usual until the second blow fell from a clear sky. Sara proposed to resign as his secretary! This brought him to his feet with deep suspicion. Was she double-crossing him? Was she playing him for a sucker? She had been in fact no more approachable to his familiarities since than before marriage—if anything, less. She actually seemed to be putting on airs and assuming a place of importance. If Sammy had dared, he would have dropped her entirely the moment she resigned. But he did not dare, and he knew that Sara knew it. He caught the glint in her gray eyes and almost felt the steel grip of her dainty hand.

Moreover, Sara explained it all very clearly. As the wife of a member of the legislature, it did not look quite the correct thing for her to be just a secretary. She proposed, therefore, to have an office of her own next Sammy's where the work of her women's organization could be done. At the same time, with an assistant, she could still take charge of Sammy's business. Sammy had hopes of that assistant, but before he had any one to propose, Sara had one chosen. She was nothing to look at, but she certainly could make a typewriter talk.

Business went as smoothly as ever, and Sammy couldn't complain.

No, evidently Sara could not be dropped. She knew too much of facts and methods. So, ostensibly, Sammy and Sara were in close alliance and almost daily consultation, and they were at the same time watching each other narrowly.

The trouble culminated over the nomination for Congress. For thirty years, Negroes, deprived of representation in Congress, after White of North Carolina had been counted out, had planned and hoped politically for one end—to put a black man in Congress from the North. The necessary black population had migrated to New York, Philadelphia, and St. Louis; but in Chicago alone did they have not only the numbers but the political machine capable of engineering the deal. It had long been the plan of Sammy's machine to have the white congressman, Doolittle, retire at the end of his present term and Sammy nominated in his stead. This was the ambition of Sammy's life, the crowning of his career. He and Sara had discussed it for years in every detail. Every step was surveyed, every contingency thought out. It was only necessary to wait for enough political power in Sammy's machine to dictate the nomination of one colored candidate among the myriad of aspirants. That time had now come.

Sammy was the recognized colored state boss; three aldermen and three colored members of the legislature took his orders; the colored judge owed his place to Sammy, and, while independent, was friendly. The public service corporations were back of Sammy with money and influence. Four "assistant" corporation counsels named by Sammy were receiving five thousand dollars a year each for duties that, to say the least, were not arduous; while the Civil Service, the Post Office, and the schools had hundreds of colored employees who owed or thought they owed their chance to make a decent living to the Honorable Sammy Scott. Finally, there was Sara's Colored Women's Council, through which for the first time the Negro women loomed as an independent political force.

Thus Sammy was dictator and candidate, and the party machine had definitely and categorically promised. The Negro majority in the First Congressional District was undoubted.

Now, however, and suddenly, matters changed. Since Matthew's success, Sara had definitely determined to kill off

Sammy and send Matthew to Congress. Sammy sensed this, and these politicians began to stalk each other. Sara's task was hardest, and she knew it. Sammy was Heir Apparent by all the rules of the game. But there were pitfalls, and Sara knew them. She was going to make no mistake, but she was watching.

Gradually Sammy became less communicative. He had a number of secret conferences in the early spring of 1926, to which Sara, contrary to custom, was not invited; and his accounts of these meetings were vague.

"Oh, just a get-together—talkee, talkee; nothing important."

But Sara wasn't fooled. She knew that Sammy was in trouble and struggling desperately. The fact was that Sammy was sorely puzzled. First and weightiest, the white party bosses wanted Doolittle for "just one more term." Doolittle held exceedingly important committee places in Congress, and especially as chairman of the Ways and Means Committee of the House, he was a power for tariff legislation. Millions depended on the revision which exporters, farmers, and laborers were demanding more and more loudly. Then there was legislation for the farmers and on the railroads and above all certain nation-wide super-power plans at Niagara, at Muscle Shoals and Boulder Dam. It was no question of "color," the white leaders carefully explained. It was a grave question of party interests. Two years hence, the nomination was Sammy's with bands playing. This year, Doolittle simply *must* go back, and money was no object.

That was reason Number One, and as money always was an object with Sammy, it loomed large in his thought. But that wasn't all. Sammy did not trust Sara, and Sara, by efficiently organizing the colored women, had quietly become the biggest single political force in his colored constituency. Indeed, her new Colored Women's Council was the most perfect piece of smoothly running political machinery that Sammy knew. He couldn't touch it, and he had tried. Now Sara had an uncomfortably popular husband. Matthew was a successful member of the legislature, young and intelligent, with some personal popularity. His very faults—aloofness, absent-mindedness, indifference to money or fame—increased his vogue. If Doolittle were forced to resign, could Sammy land the nomination without Sara's help? And with the knifing of men like Corruthers, who was still sore with Sammy; and particularly without the party slush fund?

Sammy hesitated and all but lost. He pocketed twenty-five thousand dollars for campaign expenses within a few days and consented to Doolittle's renomination. But he did not dare announce it. Sara scented a crisis. She looked over his papers— always kept carelessly—and ran across his bank book. She noticed that twenty-five thousand dollar cash deposit. Then she got busy on the Doolittle end. She knew a maid long connected with the congressman's family. Soon she had inside news. It was going to be announced that Doolittle was not to resign. His health (which was to have been the excuse) had been "greatly improved by a trip to Europe," and the honor of another and strictly final term was to be given this "friend and champion of our race"!

Sara immediately took the high hand. She walked into Sammy's office without knocking and closed the door. She was brief, inaccessible, and coldly indignant. She reminded Sammy of his solemn promise to refuse Doolittle another term; she accused him of being bribed and announced distinctly her withdrawal from all political alliance with the Scott machine!

Sammy was aghast. It was the coldest hold-up he had ever experienced. He promised her office, influence, money, and anything in reason for Matthew. She was adamant. She expressed great sorrow at this breaking of old ties.

"Oh, go to hell!" growled Sammy and slammed the door after her. He knew her game, of course. She was going to run Matthew for Congress, and, by George, she had a chance to win, unless he could kill Matthew off.

Sara immediately gave her story to the newspapers, colored and white, and called meetings of all her clubs. Bedlam broke loose about Sammy's devoted head. He was accused of "Betraying and Selling out his Race to White Politicians!" The Negro papers, by secret information or astonishingly lucky guess, named the exact sum he received—twenty-five thousand dollars. The white papers sneered at Negro grafting politicians and praised the upright and experienced Doolittle. Sammy's appointees and heads of his political machine sat securely on the fence and said and did nothing. They were glad that Sammy had missed the nomination. They were waiting to know just what their share of the slush fund was to be. They were afraid of the popular uproar against Sammy. Above all, they feared Sara. It looked perilously like Sammy's finish.

Sammy was no quitter. When he was "down, he was never out." And now he really began to fight. Sammy turned to the gang he could best trust for underground dirty work. The very respectability which Sara had forced on him in his chief appointments greatly cramped his style. He had to go back to his old cronies and his old methods. He made peace with the Gang. Soon he had around him Corruthers and a dozen like him. Sammy promised the utmost liberality with funds and began by distributing scores of new hundred-dollar bills. They all decided that the case was by no means desperate. Towns could, at worst, defeat Doolittle at the election only by dividing the Republican vote. He himself had small chance for the Republican nomination. And even if he got it, Sammy could also split the vote and defeat him. As long then as the bosses stood pat for Doolittle, Towns' only hope was to run on an independent ticket. Could he win? Probably not. Negroes did not like to scratch a straight Republican ticket. Meantime, however, in order to insure Doolittle's election and keep their machine intact, Towns must be put out of the running altogether. As Sammy said: "We've gotta frame Towns."

"Publish him as a jailbird."

"What, after I got him pardoned as an innocent hero and worked that gag all over the country?"

"Knock him on his fool head," sneered an alderman.

"There's only one thing to do with a bozo like him, and that is to trip him up with a skirt."

"Can't he steal something?"

They went over his career with a fine-tooth comb until at last they came back to that lynching and train wreck and his jail record.

"I remember now," said Sammy thoughtfully, "that Sara unearthed a lot of unpublished stuff."

"We've got to discover new evidence and admit that we were fooled."

Corruthers had been lolling back in his chair, smoking furiously and saying nothing. His red hair blazed, and his brown freckles grew darker. Suddenly now he let the two front legs of his chair down with a bang.

"Oh, to hell with you all!" he snarled. "You don't have to get no new evidence. I've had the dope to kill Towns for six months."

Sammy did not appear to be impressed. He had little faith in Corruthers.

"What is it?" he growled, with half a sneer in his tone.

"It is this. Towns made that attack on the woman for which another porter was lynched on the Klan Special last year."

Sammy sat up quickly. "Like hell!" he snapped.

"Yes, like hell! Towns confessed it to the executive committee of the porters. Said he was in the woman's compartment when the husband discovered them. He knocked the husband down and escaped. The husband thought it was the regular car porter, and he got his friends and lynched him. Towns offered to tell this story to the general meeting of the porters and in court, but the committee wouldn't let him. They let him say only that he knew the lynched porter was innocent, because he wasn't in the car. They figured it would be bad policy to admit that the woman had been attacked by any one. I got this story from the secretary of the committee. After you ditched me for the nomination to the legislature, I tried to get him to come out with it and swear to it, but he wouldn't. He was backing Towns. Then I tried to find the widow of the guy who was lynched. I knew she would tell the truth fast enough. Well, I couldn't get her until the election was over, but I've got her now fast enough. She's in New York, and I've been writing to her.

"And that ain't all. Remember, there was another colored woman mixed up in this. Called herself an Indian princess and got away with it. Princess nothing! I figure she was in the blackmailing game with Towns, double-crossed him, and left him holding the bag. Slip me five hundred for expenses, and I'll go to New York tonight and round up both of these dames. We'll bury Towns so deep he'll never see the outside of jail again."

Sammy hesitated. He didn't like this angle of attack. It was —well, it was hitting below the belt. But, pshaw! politics was politics, and one couldn't be too squeamish. He peeled off five one-hundred-dollar bills.

That night Corruthers went east.

XV

Sara was delighted at Sammy's move in the Doolittle nomination. If he had stuck to his original plan, it would have been difficult for her to refuse him her support. As it was, the chorus of denunciation at Sammy's apostasy was easily turned to a chorus demanding the nomination of Matthew Towns to Congress, before the rival politicians in Sammy's machine could prevent it. It was suggested that if the Republicans refused to nominate him and insisted on Doolittle, he might run independently and get support from the mass of the Negro vote, all the reformers, and, possibly, even the Democrats, in a district where they otherwise had no chance. Sara followed up the suggestion quickly. Club after club in her Colored Women's Council nominated Matthew by acclamation, until almost the solid Negro women's vote apparently stood back of him.

Matthew was astounded. He had never dreamed that Sara could effect his nomination to Congress. He resented her means and methods. He half resolved to refuse utterly, but, after all, it was a great chance, a door to freedom, power. But he would have to pay. He would have to strip his soul of all self-respect and lie and steal his way in. He knew it. What should he do? What could he do?

Sara had immediately taken the matter of Matthew's nomination to the white women's clubs and to the reformers. Here she struck a snag; Matthew had gained applause from the Farmer-Labor group for his support of some of their bills in the legislature; but after all, he was well known as a machine man and had voted at the dictation of big interests in the traction deal. How then could they nominate Towns, unless, of course, he was prepared to cut away from the machine and take a new progressive stand?

It was Mr. Cadwalader, leader of the Progressive group, speaking to Sara. She agreed that Matthew must take a stand. In her own mind it was a first step before she could coerce the Republicans. But how could she induce Matthew to play her game? It would be fairly easy for a trained politician. He would simply say that he was not opposed to municipal ownership but simply to this particular bill, and point out its defects. Defects were always easy to find. Then he would say that he knew that the "indeterminate permit" bill was doomed to defeat and that he could only get support for the other measures by promising to vote for it. This he could say and then make promises for the future, but not too many. But would Matthew do this? Of course not. He had no such subtlety. On the other hand, if he got up and tried to tell the straightforward truth, Sara had a plan that might work. Yes, it was worth trying. She did not see how she could avoid a trial.

"Matthew," she said that night, "I want you to come with me Tuesday and explain frankly to a committee of the Women's City Club your attitude on the super-power projects."

Matthew stared: "And how shall I explain my vote?"

"By telling the truth. Then I'll say a word."

Matthew made no comment. Gradually in his own soul he had made a declaration of independence. He would not in the future, more than in the past, be hemmed in by petty moral scruples. He still honestly believed that burglary was ethically no worse than Big Business. But thereafter in each particular instance he was going to be the judge. He would buy and sell if he so wished, but he would not be bought and sold. He was glad to go before that club and talk openly and cleanly of traction and Super-Power.

The scene inspired him. They sat high up above the roaring city, in a softly beautiful and quiet room. There rose before him intelligent faces—well-groomed and well-carried bodies, mostly of women. He saw clearly, behind their ease and poise, the toiling slavery of colored millions. He was not deceived into assuming that their show of interest would easily survive any real attack on their incomes or comforts. And yet they were willing to listen. Within limits, they wanted reform and the uplift of men.

Matthew knew his subject. He knew it even better than many experts who had spoken there, because he brought in and made real and striking the point of view and the personal interest not simply of the skilled worker, but of the laborer, the ditch-digger, the casual semi-criminal. They listened to him in growing astonishment. Here was a machine politician who had voted deliberately against his own knowledge and convictions, and yet who explained their own belief and aims much better than they could, and who nevertheless—

"Why then did I vote as I did?"

He was about to say frankly that he voted at the dictation of the machine, but that he did not propose to do this again. He would hereafter use his own judgment. His judgment might not always agree with theirs. It might sometimes agree with the machine politicians'. But it would always be his judgment. Before, however, he could say anything, Sara arose. He saw her and hesitated in astonishment.

Sara arose. She looked almost pretty—simply but well gowned, self-possessed and nervously expectant. Matthew never was sure afterward whether she actually was nervous or whether this was not one of her poses.

She arose and said, "May I interrupt right here?"

What could Matthew say? He could hardly tell his own wife in public to shut up, although that was what he wanted to say. He had to bow grimly, even if not politely. The chairman smiled, looked a little astonished, and then explained: "This is perhaps not exactly the place where we would expect an interruption, but as most of you know, this is Mr. Towns' wife, and she wants to say a word right here if he and you are willing."

Many had thought Sara white. Now they all "could see that she was colored"! At least they pretended never really to have been in doubt—that slight curl in her hair—the delicate tint of her skin—the singular gray eyes, etc. But she was unusually well dressed—"yes, quite intelligent, too, they say—yes." But what a singular point at which to interrupt! It would be especially interesting to hear the speaker proceed just here. But Matthew bowed abruptly and sat down. He was curious to see what Sara was up to. Her nimble mind always outran his in unguessed directions.

"He voted as he did because I had promised the politicians that he would, and he was too chivalrous to make me break my word, as he should have."

Matthew gasped and glanced to the door. It was too far off and blocked with silk and fur.

"I know now I was quite wrong, but I did not realize it then. I received my political education, as many of you know, as a member of a political machine, where the first commandment is, Obey. I was and am ambitious for my husband. I was a little scared at his liberal views before I understood his reasons and until we had talked them over. The machine asked his vote against municipal ownership. He gave it. He explained to me as he has to you the case for and against municipal ownership in the present state of Chicago politics. He believed this bill meant indirect corporation control. Then the Interests—the same Interests—came to me about the other two bills. You see," said Sara prettily, "we're partners, and I act as a sort of secretary to the combination and write the letters and see the visitors."

Matthew groaned in spirit, and one lady whispered to another that here was, at least, one ideal family.

"I promised them our support," continued Sara, "without further thought. I probably assumed I knew more than I did, and perhaps I was too eager to curry favor for my husband in high places—"

"And perhaps," whispered Mr. Cadwalader in the rear, "you got damned well paid for it."

Sara proceeded: "I was wrong and my husband was angry, but I pleaded with him. Since then I have come to a clearer realization of the meaning and function of political machines. But I argued then that without the machine, colored people would get no recognition even from respectable and intelligent people; that the machine had elected my husband, and that he owed it support. Finally, he promised to support the bills in loyalty to me, but only on condition that afterward we resign from Sammy Scott's organization. This we have done."

There was prolonged applause. They did not all believe Sara's explanation, but they were willing to forget the past in the face of this seemingly definite commitment for the future. But Matthew gasped. It was the smoothest, coolest lie he had ever heard, and yet it was so near the truth that he had to rub his own inner eyes. He was literally dumb when members of the committees congratulated this ideal couple and promised to turn the support of reformers toward Matthew's independent nomination. Some saw also the wisdom of Sara's delicate suggestion that this—almost domestic misfortune—be not broadcast yet to the public press, and that it only be intimated in a general way that Mr. Towns' attitude was on the whole satisfactory.

XVI

There was war in Chicago—silent, bitter war. It was part of the war throughout the whole nation; it was part of the World War. Money was bursting the coffers of the banks—poor people's savings, rich people's dividends. It must be invested in order to insure principal and interest for the poor and profits for the rich. It had been invested in the past in European restoration and American industry. But difficulties were appearing—far-off signs of danger which bankers knew. European industry could only pay large dividends if it could sell goods largely in the United States. High tariff walls kept those goods out. American industry could pay large dividends only if it could sell goods abroad or secure monopoly prices at home. To sell goods abroad it must receive Europe's goods in payment. This meant lower tariff rates. To keep monopoly at home, prices must be kept up by present or higher tariff rates. It was a dilemma, a cruel dilemma, and bankers, investors, captains of industry, scanned the industrial horizon, while poor people shivered from cold and unknown winds.

There was but one hope in the offing which would at once ward off labor troubles by continued high wages and yet maintain the fabulous rate of profit; and that was new monopoly of rich natural resources. Imperial aggressiveness in the West Indies, Mexico, and Africa held possibilities, when public opinion was properly manipulated. But right here in the United States was White Coal! Black coal, oil, and iron were monopolized and threatened with diminishing returns and world competition. But white coal—the harnessing of the vast unused rivers of the nation; monopolizing free water power to produce dear electricity! Quick! Quick! Act silently and swiftly before the public awakes and sees that it is selling something for nothing. Keep Doolittle in Congress. Keep all the Doolittles in Congress. Let the silent war against agitators, radicals, fools, keep up. Hold the tariff citadel a little longer— then let it crash with the old savings gone but the new investments safe and ready to take new advantage of lower wages and less impudent workers. So there was war in Chicago— World War, and the Republican machine of Cook County was fighting in the van. And in the machine Sammy and Sara and Matthew were little cogs.

A Michigan Avenue 'bus was starting south from Adams Street in early March when two persons, rushing to get on at the same time, collided. Mrs. Beech, president of the Women's City Club, was a little flustered. She ought to have come in her own car, but she did not want to appear too elegant on this visit. She turned and found herself face to face with Mr. Graham, the chairman of the Republican County Central Committee. They lived in the same North Shore suburb, Hubbard Woods, and had met before.

"I beg your pardon," said Mr. Graham hastily. "One has to rush so for these 'buses that it is apt to be dangerous."

Mrs. Beech smiled graciously. She was rather glad to meet Mr. Graham, because she wanted to talk some things out with him. They sat on top and began with the weather and local matters in their suburb. Then Mrs. Beech observed:

"The colored folks are certainly taking the South Side."

"It is astonishing," answered Mr. Graham. "What would the ghosts of the old Chicago aristocracy say?"

"Well, it shows progress, I suppose," said Mrs. Beech.

"I am not so sure about that," said Mr. Graham. "It shows activity and a certain ruthless pushing forward, but I am a little afraid of results. We have a most difficult political problem here."

"So I understand; in fact, I am going to a meeting of one of their women's clubs now."

"Indeed! Well, I hope we may count on your good offices," and Mr. Graham smiled. "I don't mind telling you that we are in trouble in this district. We have got a big Negro vote, well organized under Sammy Scott, of whom perhaps you have heard. Scott and his gang are not easily satisfied. They have been continually raising their demands. First, they wanted money, and indeed they have never got over that; but they demanded money first for what I suspect amounted to direct bribery. This, of course, was coupled with protection for gambling and crime, a deplorable situation, but beyond control. This went on for a while, although the sums handed them from the party coffers were larger and larger. Then they began to want offices, filling appointments as janitors and cleaners at first; then higher and higher until at last the Negroes of Chicago have two aldermen, three members of the legislature, a state senator, a city judge and several commissionerships."

"They are proving apt politicians," smiled Mrs. Beech.

"And they are not through," returned Mr. Graham. "Today they are insisting upon a congressman."

"Well, they deserve some representation, don't they, in Congress?"

"Yes, that's true; but neither they nor we are ready for it just yet. Membership in Congress not only involves a certain social status and duties, but just now in the precarious economic position of the country, we need trained and experienced men in Congress and not mere ward politicians."

"Is Doolittle a man of such high order and ability?"

"No, he is not. Doolittle is an average politician, but he is a white man; he has had long experience; he holds exceedingly important places on the House committees because of his long service; and above all he is willing to carry out the plans of his superiors."

"Or in other words," said Mrs. Beech tartly, "he takes orders from the machine."

"Yes, he does," said Graham, turning toward her and speaking earnestly. "And how are we going to run this country unless thoughtful men furnish the plans and find legislators and workers who are willing to carry them out? We are in difficulties, Mrs. Beech. If the tariff is tinkered with by amateur radicals, your income and mine may easily go to smash. If securities which are now good and the basis of investment are attacked by Bolsheviks, we may have an industrial smash such as the world has seldom seen. We haven't paid our share for the World War yet, and we may have to foot a staggering bill.

"Now, we have farsighted plans for guiding the industrial machine and keeping it steady; Doolittle is a cog, nothing more than a cog, but a dependable cog, in the machine. Now here come the Negroes of this district and demand the fulfillment of a promise, carelessly, and to my mind foolishly given several years ago, that after this term Doolittle was to be replaced in Congress by the head of the black political machine, Sammy Scott. Well, it's impossible. I think you see that, Mrs. Beech. We don't want Scott in Congress representing Chicago. He has neither the brains nor the education—"

Mrs. Beech interrupted. "But I understand," she said, "that there is a young college-bred man who is candidate and who is intellectually rather above the average of our Congressmen."

"There certainly is," said Mr. Graham, bitterly, "and he's got a wife who is one of the most astute politicians in this city."

"Yes," said Mrs. Beech, "I am on my way to one of her meetings. It is at her home on Grand—I mean, South Parkway. I wonder where I should transfer?"

"I will show you," said Mr. Graham. "We have still a little way to go. It would be just like Mrs. Towns to pull all strings in order to get you to her house. She has social aspirations and is the real force behind Towns."

"But Towns himself?" asked Mrs. Beech.

"Towns himself is a radical and has a shady record. He was once in the penitentiary. His wife is trying to keep him in hand, but his appeal is to the very elements among white people and colored people which mean trouble for conservative industry in the United States. He cannot for a moment be considered. I have talked frankly to you, Mrs. Beech. We are coming to your corner now, but I wish we could come to some sort of understanding with the liberal elements that you represent. I do not think that you, Cadwalader, and myself are so far apart. I hope you will help us."

Mrs. Beech descended and Graham rode on.

It was some hours later that Mr. Cadwalader and Mrs. Beech had dinner together. They represented various elements interested in reform. Mr. Cadwalader was the official head of the Farmer-Labor Party in Chicago, while Mrs. Beech represented one element of the old Progressive Party and looked toward alliance with Mr. Cadwalader's group.

"But," Mr. Cadwalader complained over his fish, "we've got an impossible combination. We cannot get any real agreement on anything. You and I, for instance, cannot stand for free trade as a present policy. It would ruin us and our friends. On the other hand, we cannot advocate a high tariff. We and our manufacturing friends want gradual reduction rather than increase of duties. Then, too, our friends among the farmers and the laborers want high and low tariff at the same time, only on different things. The farmers want cheap foreign manufactured goods and high rates on food; the laborers want free food and high manufacturing wages. Finally, we have all got to remember the Socialists and Communists who want to scrap the whole system and begin anew."

"I was talking with Mr. Graham yesterday," said Mrs. Beech, "and he believes that the Republicans and the Farmer-Labor Party could find some common aims."

"I am sure we could, if the Republicans would add to their defense of sound business and investment some thought of the legitimate demands of the farmer and laborer, and then would restrain legislation which directly encourages monopoly."

"True," said Mrs. Beech, "but wouldn't any rapprochement with the Republicans drive out of your ranks the radicals who swell the potential reform vote? And in this case would we not leave them to the guidance of demagogues and emphasize the dangerous directions of their growth?"

"Precisely, precisely. And that is what puzzles me. You know, only last night I was visiting a meeting of one of the newer trade unions, the Box-Makers. It was organized locally in New York in 1919 and now has a national union headquarters there. The union here is only a year old, but it is the center of dangerous radicalism, with lots of Jews, Russians, and other foreigners. They want paternalism of all sorts, with guaranteed wages, restricted ages and hours of work, pensions, long vacations and the like; not to mention wild vaporings about absolute free trade; 'One Big Union'; government ownership of industry, and limitation of wealth. And the trouble also," continued Mr. Cadwalader, "is that this union has some startlingly capable leaders; two representatives from New York were there last night, and a letter from the National President was read which was dangerous in its sheer ability, appeal, and implications.

"I was aghast. I wanted to repudiate the whole thing forthwith, but I was afraid, as you say, that I would drive them bodily over to the Socialists and Communists. In general, I'm beginning to wonder if we could try to marshal this extreme movement back of Matthew Towns. I don't exactly trust him, and I certainly do not trust his wife. But Towns has got sense. He is a practical politician. And it may be that with his leadership we can restrain these radicals and keep them inside a normal liberal movement."

Mrs. Beech pushed her dessert aside and sat for a while in a brown study.

"I am wondering too just how much can be done in this one Chicago congressional district, to use Towns and his wife in order to unite Republicans and Progressives, so as to begin a movement which should liberalize the Republican Party and stabilize the radicals. Unless we do this, or at least begin somewhere to do it, I see little hope for reform in politics. A third party in the United States is impossible on account of the Solid South. They are a dead weight and handicap to all political reform. They have but one shibboleth, and that is the Negro."

"Yes," said Mr. Cadwalader, "and the Democrats play their usual rôle in this campaign. Their positive policies are exactly the same as the Republicans'. They, of course, have no chance of winning in this political district, unless the Republicans split. Now with Towns' revolt, the Republican machine is split and the Democrats are just waiting. If Towns should be nominated they would raise the question of the color line and yell 'nigger.' They might in this way elect one of their own number or some independent. If Doolittle is nominated, it is going to be hard to elect him if Towns runs as an independent; and in that case it might be good politics for the Democrats to back Towns and beat the Republican machine. So there they are on the fence, waiting."

"On the other hand," said Mrs. Beech, "down in the black trenches the war is bitter, as I gathered from my attendance at the meeting this afternoon. Sammy Scott, the boss, and Sara Towns were formerly close associates and know each other's personalities, political methods, and secrets. They are watching each other narrowly and are utterly unhindered by scruples. What sort of personality has this man Matthew Towns? Do you know anything about him?"

"I've been looking up his record. He intrigues me. He had, I find, an excellent record in medical school. Then in silly pique he became a Pullman porter and, I judge, sank pretty low. He does not seem to have committed any crime, but went to jail on a technicality because he wouldn't betray some of his friends. Scott rescued him and used him. He's got brains and education, but he's queer and not easily approachable."

"Well, if I were you," said Mrs. Beech as she arose, "I'd get in touch with Towns and cultivate him. He may be worth while. His wife is a shrewd climber, but even that might be an asset."

And so they parted.

XVII

The Honorable Sammy Scott was carefully planning his lines
of battle and marshaling his forces. First he threw out certain
skirmish lines—feints to veil his main action. Of course they
might unearth or start something, but he was not putting his
main dependence upon them. Of such a nature was Corruthers'
trip to the East. It flattered Corruthers and gave him something to do, and it left Sammy unhindered to arrange his main
campaign. And, of course, it was also possible that something
would come out of this visit.

Among his other skirmishing efforts, Sammy kept looking
for the weak spots in Matthew's armor, but was unable to find
many. Matthew's obvious faults rather increased his popularity. He drank liquor, but not much, and Sammy saw no
chance to make him or keep him drunk. He tried bribery on
Towns from every point of view, but personal graft had not
attracted Towns even when he was in the machine, and
Sammy had little hope that it would now. Nevertheless, he
saw to it that Towns was offered a goodly lump sum to withdraw his candidacy. Of course, Towns refused. The trouble
was, as Sammy argued, that his machine could not afford to
offer enough. He did not dare make the offer to Sara. She was
capable of pocketing cash and candidate too.

Sammy's main dependence was to regain or split the Negro
vote by careful propaganda; to reorganize his own machine
and drive his leaders into line by the free use of money; and
to alienate the church and the women from Towns by digging
up scandals. Of course that second establishment that Matthew
kept down where he used to live was a hopeful bit of scandal,
only Sammy couldn't discover the woman. Naturally there
must be a woman, or why should he keep the rooms?

Finally Sammy would try, wherever possible, strong-arm
methods to intimidate the independents. Beyond this he believed that in some way he might split the Progressive support
back of Matthew by making Matthew say or do something that
was too radical for men like Cadwalader, or by making him
bind himself to a program which was too reactionary for the
radical Laborites.

Sammy envisaged the situation thus: If all efforts failed
and Matthew received the Farmer-Labor nomination and the

support of the majority of the colored vote, nevertheless, a huge sum of money spent at the polls might even then defeat him. If bribery failed and the Negro vote stood solid, the Progressives gave full support and the Democrats secret aid, Matthew would be elected, but Sammy would emerge from the campaign able to tell the party bosses that the fault was theirs; they should have kept their promise and nominated him. After that, with a large campaign fund, he could reorganize his machine and keep watch on Matthew in office. If Matthew failed to do as the machine told him, and this would probably happen, Sammy would succeed to his job. If Matthew succeeded, he could and must be brought back into the Republican fold, which meant into Sammy's machine.

Of course, all these possibilities which hinged on Matthew's successful election were wormwood to Sammy, and he concentrated fiercely on forestalling such success. He was sitting alone in his office this night and thinking things out when there came a 'phone call. It was Corruthers.

Corruthers came in a half hour later. Corruthers was a cadaverous blond, red-headed and freckled, a drunkard, a dope fiend, and a spellbinder, with brains and no self-control, thoroughly dishonest and extremely likable. He claimed to be a nephew of Frederick Douglass. He looked distinctly glum.

Corruthers was one of those who are dangerous only when they are successful. Once he began to lose he gave up. Sammy was quite different. He was most dangerous losing. It was then that he fought furiously and to the last ditch. Evidently Corruthers had been disappointed and was ready to surrender.

"Didn't do a thing," he said, "not a thing. Sure, I found the widow of that porter who was lynched. She's working in New York. But, my God! she thinks Towns is Jesus Christ and won't hear a word against him. Swears that the story that he attacked anybody is a lie and threatens to go to court if anybody says that Towns was responsible for her husband's death. I can't figure her out at all. I know the story that I got was the truth, but I can't prove it.

"Then I tried to get a line on that dead man in the wreck. He may have been a fellow named Perigua. Some say he was, but others declare that Perigua went back to Jamaica and has been seen there since. I don't know how it was. Then I was all wrong about that brown woman. She was an Indian princess, sure enough, a high muck-a-muck and fabulously rich. Probably got interested in Towns because he was a good porter. At any rate, she's gone back to India, so everybody says, although there again some declare that she has returned."

"Well, what did you do about that?" asked Sammy impatiently.

"Well, I found out the address of her bankers and went down to inquire about her. Seems that her little country, Bwodpur, has some sort of commercial agency which I was referred to. It didn't look like much. Mean little office, with two or three Indians sitting around. All I could do was to leave my address and say that I had some news for the lady about Matthew Towns; then I came home."

"Hell!" said Sammy, lighting another cigar.

"But today," continued Corruthers, "an East Indian called at my place and asked about Towns. Said he'd been sent."

"What did you do with him?" asked Sammy.

"He's outside," said Corruthers.

"Well, for God's sake! and gassing about nothing all this time! Bring him in."

"Wait a minute," said Corruthers. "What are we going to say to him? I couldn't think of anything. Of course we might pinch him for a little cash."

"Nix," said Sammy, "leave him to me."

The Indian entered. He looked thin and was poorly dressed. Sammy was disappointed, but he handed him one of his cheap cigars, which the Indian refused.

"I understand," said the Indian in very good English, "that you have some message from or about Matthew Towns."

"Ah," said Sammy. "Not from him, but about him. Er'r, I believe the lady with whom you are connected is friendly with Mr. Towns."

"Her Royal Highness in the past has deigned to express something like that."

"And—still interested?"

"I do not know."

"I mean, she wouldn't let him suffer or get into trouble, and she wouldn't want to get in trouble herself on account of him?"

"Perhaps not, but what is the case?"

Sammy paused thoughtfully and then started.

"You see, it is this way. This fellow Towns got in jail for not peaching on a pal. I got him out. He made a good record in the legislature and his friends persuaded him to run for Congress. I'd be glad to see him go, but his enemies and the enemies of his race are threatening to bring up this old jail case, and they say that your Princess is involved. It would be a nasty mess to have her name dragged in publicly now. The only way out, as I see it, is for somebody to persuade Towns to withdraw. Can you or your lady manage this?"

"Very well," said the Indian, arising.

"But can you do anything?"

"I do not know, I will report."

And then Sammy said: "What! 'way to India?"

"Her Royal Highness is represented in this country. Good day."

Sammy glowered after him.

"Royal Highness! Hell!" said Corruthers. "And see all the time wasted on that guy. We ought to have asked him straight for money."

"Looks like a mare's nest," said Sammy, "but my rule is to try everything. Well, enough of that. Now my plan is to see what we can do to split the Progressives and this Farmer-Labor bunch, which is promising to support Towns. I know Sara's game; she's playing both ends against the middle. But I'm going to break it up."

XVIII

Sara had, of all concerned, the most difficult road and the most brilliant prospects. She saw wealth, power, social triumph ahead if she could elect Matthew to Congress. But she knew just how difficult it would be to beat the Republican machine with its money and organization. Her first task was to hold the Negro vote back of Matthew. That was easy so long as he was a regular Republican. When he bolted Sammy's machine, Sara had to capitalize race pride and resentment against Doolittle and Sammy. She continued to insist that Matthew was a good Republican but not a Sammy Scott henchman. For a while her success here was overwhelming, but could she hold it three months with hungry editors and grafting henchmen?

She concentrated on *The Lash*, whose editors had sharp

tongues and wide pockets and kept them flaying Sammy and the Republicans. She went after her women's clubs and cajoled and encouraged them by every device to stand strong. She made every possible use of the women's organizations connected with the fraternal societies. She already belonged to everything that she could join, and was Grand Worthy Something-or-other in most of them. She pushed the idea of uniforms and rituals, for these things appeal naturally to folk whose lives are gray and uneventful. She had a uniformed Women's Marching Club and a Flying Squadron with secret ritual which she used for political spy work.

All these things carried new dimensions to the lives of a class of colored women who had been hitherto bound chiefly to their kitchens and their churches. Woman's "new sphere," of which they had read something in the papers, had hitherto meant little to them. They were still under the spell of the old housework, except as they raised money in the churches. Here was work newer and more interesting than church work. The colored ministers protested, but were afraid to protest too much, because many of Sara's political followers were still their best church workers, and they dared not say or do too much to alienate them.

Sara worked feverishly during March because she knew perfectly well that the real difficulty lay ahead. The election of Matthew might involve voting not only against Sammy's machine but against the Republican ticket and with the Farmer-Labor group, and possibly even voting with the Democrats. Casual white outsiders cannot understand what this problem is. These colored women were born Republicans, even more than their fathers and brothers, because they knew less of the practical action of politics. Republicanism was as much a part of their heritage as Methodism or the rites of baptism. They were enthusiastic to have a colored man nominated by the Republican Party. But could she so organize and concentrate that enthusiasm that it would carry these women over into the camp of hereditary political and economic enemies?

They looked upon the white labor unions as open enemies because the stronger and better-organized white unions deliberately excluded Negroes. The whole economic history of the Negro in Chicago was a fight for bread against white labor unions. Only in the newer unions just organized chiefly among the foreign-born—and fighting for breath among the unskilled or semi-skilled laborers—only here were colored people welcomed, because they had to be. Of course, the very name of the Democratic Party was anathema to black folk. It stood for slavery and disfranchisement and "Jim Crow" cars. Well, Sara knew that she had a desperate task, and she was fighting hard.

She was in touch with the labor unions and soon sensed their right and left wings. The right wing was easy to understand. They were playing her game and compromising here and there to obtain certain selfish advantages. Sara was sure she could take care of them. The extreme left group was more difficult to understand. She did not know what it was they really wanted, but she quickly sensed that they had astute leadership. The international president of the Box-Makers, who lived in New York, was evidently well educated and keen. Sara had written her in the hope of avoiding contact with the local union. Her answers showed her a desperately earnest woman. Sara did everything to induce her by letter to wield her Chicago influence for Matthew, but so far had seen no signs of success. This left group was meantime clamoring, pushing their claims and asking promises and making inconvenient suggestions. So far Sara had avoided meeting them.

One thing, one very little thing, Sara kept in her mind's eye, and that was Doolittle and his health. If anything happened to Doolittle before the primary election—well, if it happened, Sara wanted to know it and to know it first.

And it was precisely here that Sammy made his second mistake. He calculated that the news of any change in Doolittle's health would reach him first, because Doolittle's valet was a staunch member of his machine. Indeed, he got him the job. Now Sara knew this as well as Sammy, and she worked accordingly. Doolittle lived officially on the South Side but actually in Winnetka, away up on the North Shore, in a lovely great house overlooking the blue lake. Sara had careful and minute knowledge of his household. Of course, his servants were all colored. That was good politics. Sara again had recourse to that maid who had told her first of the plan to renominate Doolittle. She had the maid at tea on one of her Thursday "at homes," and was careful to have in some of her most expensive friends—the doctor's wife, the banker's daughter, the niece of the vice-president of Liberty Life.

Sara did not say that the quiet and well-behaved stranger was simply a maid, and by this very reticence tied the maid to her forever. Also, Sara pumped her assiduously about Doolittle's health without directly asking after it. She easily learned that it was much more precarious than the public believed. Immediately, through the maid, Sara got in touch with the valet. She picked him up downtown in her car and brought him to luncheon one day, when Matthew was away from home.

"I do not want you to think, Mr. Amos, that I have anything against the excellent Mr. Doolittle."

"No, ma'am, no, ma'am, I'm sure you ain't. I am sorry he's running again. He oughtn't to done it. He ain't in no fit condition to make a campaign. He wouldn't of done it if he had been left alone; but there's his wife full of ambition and the big bosses full of plans."

"I do wish Sammy had stood pat and insisted on the nomination," said Sara thoughtfully.

"I'll never forgive him," said Mr. Amos. "It was sheer lack of backbone and an itching palm."

"You are a great friend of his, I know."

"Well," said Mr. Amos, "I don't like him as well as I use to, although I know he got me my job. Tell you what, ma'am, I wish your husband could get the nomination." They talked on. When finally he stood at the front door, Sara was saying:

"I hope, of course, that all will go well, for Doolittle is a deserving old man, but if anything *should* change in his physical condition I'd like to know it *before* anybody else, Mr. Amos; and I'm depending on you." And her dependence was expressed in the shape of a yellow bill which she slipped in Mr. Amos' hand. He took occasion to examine it under the electric light as he was waiting for the bus. It was a bank note for five hundred dollars. Mr. Amos missed two buses looking at it.

Less than a week later, while Sara was at her desk one morning, about to send out notes for one of her innumerable committee meetings, the telephone rang. The low voice of Mr. Amos came over it:

"Mr. Doolittle has had an attack. He is quite ill."

She thanked him softly and hung up.

The next morning Sara went down to Republican head-quarters, where she used to be well known. She was regarded with considerable interest this morning, but remained unperturbed. She asked for a certain gentleman who was always busy, but Sara wrote a note and sent it in to him with a card. He found time to see her.

"Mr. Graham," she said, "what do you think of Congressman Doolittle's health?"

Mr. Graham looked at her sharply, took off his glasses, and polished them carefully, as he continued to look.

"I have every reason to suppose," he said slowly, "that Mr. Doolittle's health is excellent."

"Well, it isn't," said Sara.

"I suppose your source of information—" But Sara interrupted him.

"Frankly, Mr. Graham—suppose that Congressman Doolittle should die before the primary election."

"We'd be in a hell of a muddle," blurted out Mr. Graham.

"You would," said Sara. "You could hardly nominate Sammy, because Sammy is very unpopular just now among colored voters."

"Thanks to you," said Mr. Graham.

"No, Mr. Graham, thanks to you. Now my husband, Mr. Matthew Towns, is both popular and—intelligent."

"Especially," added Mr. Graham, "with the Farmer-Labor reformers and the Bolsheviks."

"Not a bad bunch of votes to bring to the Republican Party just now."

"Well, any colored candidate would have to bring in something to offset the hullabaloo which the Klan would raise in this town if we nominated a Negro and a—one with your husband's record, to Congress."

"Precisely, and I am calculating that the support of the reform groups and the solidarity of the colored vote would much more than offset this and make the election certain."

"In any case, Mrs. Towns, I take it that your husband has been promised the support of the Farmer-Labor group only on condition that he stand on their platform."

"He has given them to understand," said Sara carefully, and with a smile, "that he sympathizes with their ideals."

"Well," said Mr. Graham crisply, "that puts him out of the running for the Republican nomination, even in the extremely unlikely event that Mr. Doolittle for any reason should not or could not receive it."

"I wonder," said Sara. "You know quite well that the intellectuals in the Farmer-Labor group are bound to support Republican policies up to a certain point. Their financial interests compel them; now it would be good politics for the Republicans to go a step beyond that point in order to attract, by some show of liberality, as large a group as possible of the liberals. Then, having split off their leaders and their thinkers, we might let the rest of the radicals go hang. What I am proposing in fine, Mr. Graham, is this: that the nomination of my husband (in the unlikely event that Mr. Doolittle should not be well enough to accept) might be a piece of farsighted politics on your part and bring you the bulk of the liberal vote, while at the same time paralyzing and splitting up the power of the radicals."

Mr. Graham fingered his mustache.

"I will not forget this visit, Mrs. Towns," he said.

Sara walked out; taking a taxi, she quietly slipped over to the Democratic headquarters. She asked to see Mr. Green of Washington.

"Mr. Green?" asked the porter, doubtfully.

"Yes, he is in town temporarily and making his headquarters here. I will not keep him long. Here is my card. I have met him."

After a while another gentleman came out.

"Mr. Green is only calling at this office. Just what is your business with him?"

"Please tell him that once in Washington he signed a petition for me that helped release Matthew Towns from Joliet. Mr. Towns is my husband and is now running for Congress."

A few minutes later Sara was closeted with Mr. Green, a high official of the Klan. He looked at her with interest.

"And what can I do for you this time, madam?"

"You remember me?"

"Perfectly."

"I trust you have not regretted helping me."

"No."

"Have you followed Mr. Towns' career?"

"I know something of it."

"Well, he may be nominated for Congress by the Republicans, and he may not. If he does not get the nomination, he will run independently on the Farmer-Labor ticket. Any help that the Democrats could give us in such a campaign would greatly impede the Republicans."

Mr. Green smiled, but Sara proceeded:

"In the unlikely event that he should be nominated by the Republicans I have come to ask you if it would not be possible for you to restrain any anti-Negro campaign against him or any undue reference to his jail sentence. You see, with the Republican and Farmer-Labor support he would probably be elected, and if that election came with your silent help, he would be even more disposed to look with favor upon you and your help than he is now. And he feels now that he owes you a great deal."

Mr. Green looked at her curiously. Finally, as he arose, he shook hands with her and said:

"I am glad you came to me."

Sara was a little exhausted when she reached home, but she still had some letters to write. The maid said that the telephone had rung and that some Mr. Amos would call her later. Sara sat down by her well-ordered desk and inserted a new penpoint. Soon the telephone rang. Mr. Amos' voice came over the wire:

"Mr. Doolittle is some better, but still in bed."

Sara looked at the clock. It was four. She ordered dinner and went back to her writing. The hours passed slowly. At half-past five Matthew came in, and they ate silently at six. While they were eating the telephone rang again.

"Mr. Doolittle has gone out for a short drive. He is better, but far from well."

They finished dinner. Matthew stood about restlessly a while, smoking. Then with a muttered word he went out. Sara sat down beside the telephone and waited. The messages came at intervals, each shorter than the other.

"Mr. Doolittle has returned."

"He has taken a chill."

"The physicians are working over him."

"He is sinking."

Eight, nine, and ten o'clock chimed on Sara's gilt desk clock, and then:

"Congressman Doolittle is dying."

Sara waited no longer. It was March 20. The primary election was to take place April 8. She took a taxi for Republican headquarters.

XIX

Sammy's campaign was progressing. Its progress was not altogether satisfactory, but Sammy was encouraged. Most of the best colored newspapers had been "seen" and were acting satisfactorily. *The Conservator* had one week a strong defense of the "Grand Old Man and Friend of Our Race, the Honorable Calvin Doolittle!" The next week, it featured a lynching, scored the Democrats, and pointed to Doolittle's vote on the anti-lynching bill. *The Lash,* when Sara refused its last exorbitant demand for cash, started a series of scathing attacks on the white trade unions and accused them of being filled with "nigger-haters" and Catholics. Other smaller sheets followed suit, with regrets that Mr. Towns was being misled into opposition to the Republican leaders who had always been friends, etc. Only one paper, *The Standard,* stood strong for Matthew at a price which Sara could afford; but even that paper avoided all attacks on the Republican Party.

The local clubs and political centers of Sammy's machine gave every evidence of prosperity, while police interference with gambling and prostitution ceased. The prohibition officials apparently stopped all efforts in the main black belt, and there were wild and ceaseless rumors that the Klan was back of a widespread effort to beat the Republicans.

Only the women stood strong. And so strong did they stand under Sara's astute leadership and marshaling care that Sammy was still worried. They were difficult to reach. Sluggers could not break up their meetings. They could easily out-gossip Sammy's sensation-mongers, and against their hold on the churches, the colored newspapers availed nothing. It remained true, therefore, after two months' campaign, that the great majority of Negro voters were still apparently opposed to Sammy and strongly in favor of Matthew's nomination. Nevertheless, with time and money, Sammy was sure he could win. The trouble was, time was pressing. Only two weeks was left before the primary elections.

Reflecting on all this, Sammy Scott after dinner one day took a stroll, smoking and greeting his friends. He dropped in at some of the clubs and had a word of advice or of information. He took drinks in a couple of cabarets; watched a little gambling. As he sat in one of the resorts, he listened to the talk of a young black radical. The fellow was explaining at length what Negroes ought to demand in wages and conditions of labor, how they ought to get into the trade unions, and how they were welcomed by unions like that of the Box-Makers. Sammy sidled over to him. He struck Sammy as the sort of man who might carry on a useful propaganda among some of the colored voters and strengthen the demands made on Matthew to take so radical a position that the Republicans could not accept him.

Sammy talked with him and finally invited him to supper. He was undoubtedly hungry. Then he invited Sammy to come with him to a meeting of the Box-Makers. They went west to that great district where the black belt fades into the white workingmen's belt. In a dingy crowded hall, a number of people were congregated. They were discussing the demands of the Box-Makers, and Sammy listened at the door.

"How many of us," yelled one man, "make as much as fifteen dollars a week, and how can we live on that?"

"Yes," added a woman, "how can we live, even if we women work too? We can make only five or six dollars, and out of work a third of the time."

"Oh, you got it easy even at that. You ought to see where we work, down in damp and unventilated cellars. No porters to keep the shops and the washrooms clean; the stink and gloom and dirt all about us."

"In my shop we never get sunlight a day in the year."

Another one broke in. "And we're working twelve or thirteen hours a day with clean-up on Sunday. It ain't human, and we won't stand it no longer."

Sammy edged in and sat down. Pretty soon the speakers gathered on the stage—the young colored man whom he had met, another colored man whom he did not at first recognize, and several white organizers and delegates. There were long speeches and demands and fiery threats, but Sammy waited because he wanted to talk to that young fellow again. When the meeting was over, the young man came down accompanied by the other colored man, and Sammy noted with a start that it was the Indian with whom he had had conference concerning Matthew. Sammy was puzzled.

What was that Indian doing there on the stage? Especially when he represented aristocracy, at least if what he said about the Princess was to be believed. "Or is it that they are on to me?" thought Sammy. "Is the Princess interfering or not?" Then suddenly he saw a possibility. The Princess or her friends might want Matthew nominated for Congress, but nominated on this radical platform. Good, so did he. Oh, boy! So did he. He got hold of the young colored man and walked away. They had a long conversation about the platform of the radicals and about putting this platform up to Matthew Towns and insisting that he stand on it. Also, Sammy lent the young man twenty-five dollars and told him to come to see him again.

XX

It was late when Sammy got back to his office, after midnight, in fact. As he rushed in hurriedly he saw to his astonishment that Sara Towns was sitting in the outer room. A number of his cronies and henchmen were grouped about, staring, laughing, and smoking. Sara was elaborately ignoring them. She had arranged herself quite becomingly in the best chair with her trim legs in evidence, the light falling right for her costume and not too strongly on her face. The fact was that her face showed some recent signs of wear, despite the beauty parlors. Sammy stopped, swore softly under his breath, and glared. What did it mean? thought Sammy rapidly. Surrender or attack? But he quickly recovered his poise and soon was his smiling, debonair self.

"May I see you a few moments alone?" asked Sara.

"Sure! Excuse me, boys, ladies first."

They went into the inner sanctum and drove out some more of Sammy's lieutenants. Sara closed the door and looked around the inner office with disgust.

"My, but you're dirty here!"

Sammy apologized. "It ain't exactly as clean as it was in your day," he grinned. She dusted a chair, arranged her skirt

and tilted her hat properly, looking into the mirror opposite.
Sammy waited and lighted another cigar.

"Sammy, I came to suggest that we join forces again."

Sammy looked innocent, but did some quick calculations.
Aha! he knew that combination wouldn't last. Wonder what
broke first?

"Well, I don't know," he drawled finally. "You broke it up
yourself, you remember."

"Yes, I did. You see, I thought at the time you were going
to nominate Doolittle for congressman."

"Yes," said Sammy. "And I still am."

"No, you're not," answered Sara. "He just died."

Sammy dropped his cigar. He fumbled for it and got to his
feet. Then he sat down again limply.

"Well, I'll be God damned," he remarked and grabbed the
telephone.

As a matter of fact, Sara had left the house and rushed to
Republican headquarters before Doolittle was actually dead.
Mr. Graham had, of course, been warned of Doolittle's sudden
illness, but he had not heard of his death for the simple reason
that it had not yet taken place. When, therefore, this self-
possessed, gray-eyed little woman came in and announced Doolittle's death, Graham did not believe it. Five minutes later it
was confirmed on the 'phone. But still the thing looked un-
canny, because Sara had only been there five minutes and must
have announced the death at exactly the minute it actually
took place. But she had been quite matter-of-fact and had gone
right to business.

"Can't we get together?" she had said. "Under the circumstances you cannot nominate a white man now. You have no
excuse for doing it after your past promises. Then, too, you
can't nominate Sammy Scott. He is too unpopular, thanks to
you. Even if you try to nominate him, Matthew Towns can
beat him in the primary. If you buy up the primary vote with
a big slush fund, as Sammy plans, Towns, with the support of
the Liberals and perhaps the Democrats, together with the
bolting Negroes, can be elected."

The chairman had sneered in his confusion: "Negroes don't
bolt."

"Not usually," Sara replied, "but they may this time. In
fact," she said, "I think they will."

In his own mind the chairman was afraid she was right.

"Why not nominate Towns?" she asked.

"Well," said the chairman, sparring for time, "first there is Sammy; and secondly, there is the question as to what Towns will do in Congress."

"He will promise to do anything you say," said Sara. "And I am going to see Sammy now." Thus she came and told Sammy the news.

Sammy struggled at the 'phone. The operator was evidently asleep, but he got through to Graham at last. Sure enough, Doolittle was dead! Sammy stared into the instrument. It certainly looked bad for him. Here he had got the most important news of the campaign from headquarters through Sara. Very well. Evidently he must tie up with Sara again. In such an alliance he had everything to gain and nothing to lose. As his political partner, at least she could not continue to attack him. The matter of the nomination would not be settled until the primary was held in April. He had twelve days to work in. He had seen a president made in less time.

Sammy put down the telephone and turned to Sara with a smile, but underneath that smile was grim determination, and Sara, of course, knew it. He was going to fight to the last ditch, but he extended his hand with the most disarming of smiles.

"All right, partner," he said, "we'll start again. Now what's your plan?"

"My plan is," said Sara coolly, "to have you work with me for the nomination of Matthew to Congress."

"Where do I come in?" said Sammy.

"You come in at the head of a united machine with a large campaign fund."

"That wasn't the old plan," said Sammy.

"No, it wasn't," answered Sara, "but who broke up the old plan?"

"Graham tried to," said Sammy, "but God didn't let him."

"True," answered Sara, "and naturally somebody has got to pay for not stopping Graham, and that somebody is you. Still," she said, "the price need not be prohibitive. After Matthew has had a term in Congress, why not Sammy Scott?"

Sammy smiled wryly. "All right," said Sammy. "I'm set. Now what are we going to do?"

"We are going to try and get the Republican and the Farmer-Labor people to unite on the nomination of Matthew."

"Good!" said Sammy. "Here goes."

"Of course," added Sara, "we must be careful not to make our new alliance too open and scare off the Liberals. We must drift together apparently as fast and no faster than these two wings come to an understanding. That understanding I'm going to engineer, and I want your help. First you go to Graham and tell him you'll support Matthew. I've told him you're coming. As soon as I've heard from him that you've seen him, I'll get hold of Cadwalader and tell him the news. We'll work on this toward a final conference just before the primaries."

XXI

Neither to Sammy nor to Sara did their new alliance make any real difference. It healed the open and public split, but Sammy continued to bore into Matthew's support, and Sara continued to strengthen his popularity and defenses. Beyond that, Sammy and Sara had always admired each other. Each was a little at a loss without the other. Neither had many intimate associates or confidants whom they wholly trusted. Both had the highest respect for each other's abilities. They knew that their new alliance was a truce and not a union. Each suspected the other, and each knew the other's suspicions. At the same time, they needed each other's skill and they wanted desperately to confide in each other, as far as they dared.

Sara had suggested that just before the primaries, a conference of Republicans and Liberals might be held in order to come to a final understanding and unite on Matthew's nomination. Sammy had to assent. He had plans of his own for this conference, which he hoped to make a last desperate effort at Matthew's undoing. He knew just what kind of conference would best serve his ends, but he did not dare let Sara know what he wanted.

On one point Sara had of course made up her mind: no agreement between Matthew, Graham, and Cadwalader was going to depend on the chances of a single conference or even of several conferences. She was going to conduct secret negotiations with all parties, until the final conference should find them in such substantial agreement that definitive action would be easy; that is, all except the left-wing labor unions. The surer she became of the main groups, the less did Sara think of these common laborers and foreigners. They could come in at last, when agreement or protest would make little real difference.

Sara hoped that she might come to this agreement by mere verbal fencing. She hoped so, but she knew better. Sooner or later there must be a definite understanding with Graham. Very well, when the crisis came she would meet it.

With her mind then on this closing conference as merely the ratification of agreements practically made, Sara at first settled on something big and impressive: a church or hall mass meeting of all parties and interests, making an overwhelming demand for the election of Matthew Towns as congressman. Sammy listened, his head on one side, his cigar at an impressive angle, his feet elevated, perhaps a bit higher than usual; his coat laid aside.

"Um-um!" he nodded. "Fine; fine big thing. If it could be put over. Smashing publicity." Then he took a long pull at his cigar and looked intently at the glowing end.

"Of course," he said reflectively, "there is one thing: would Matthew make the right kind of speech?" Sammy was really afraid he would; Sara not only did not know whether or not Matthew would make the right kind of speech; she did not even know if he would try. In fact, he might deliberately make the wrong kind of speech, even after agreement had already been reached. Sara's doubt rested on the fact that she and Matthew had had a tilt this very morning, and she at least had had it out. She put the situation before him, frank and stark, with no bandying of words.

"Now see here. You have got this nomination in your hands and on a silver salver, if you want it. But in order to get it you've got to make the kind of statement that will satisfy the Republicans backed by big business, the Democrats backed by big business, and the Farmer-Labor party led by reformers and union labor. You've even got to cater to the radical wing of the trade unions. It will mean straddling and twisting and some careful lying. It will mean promises which it is up to you to fulfill after election, if you want to, and to break if you want to—after election. It will mean half promises and double words and silences to make people think what you are going to do, what you are not going to do, or what you do not know whether you are going to do or not. Unless you do something like this you will lose the nomination.

"Or, what's just as bad, you will lose the Republican nomination. Perhaps you have kidded yourself into thinking that you can make a winning fight with the Farmer-Labor nomination and the independent Negro vote. Well, listen to me. You can't. There isn't such a thing as an independent Negro vote. Or at any rate it is so small as to be negligible. The Negroes are going to fight and yell before election. At the election they are going to trot to the polls and vote the Republican ticket like good darkies. If you want to go to Congress, you have got to get the Republican nomination.

"On the other hand, nothing will clinch this nomination, the election, and the whole-hearted future support of the Republican machine like your ability to poll not simply the Republican vote, but the Farmer-Labor vote and the vote of the independent Democrats and at least a part of the radical vote. You can do this if you don't act like a fool."

Matthew had pushed his breakfast aside and looked out of the window. He saw a few trees and the gray apartment houses beyond. Above lay the leaden sky.

"Suppose," he said, "that instead of making this campaign, I should ask for the part of the money we have made which is mine and give up this game?" Sara's little mouth settled into straight, thin lines. "You wouldn't get it," she said, "because it doesn't belong to you. You didn't earn it; I did. You haven't saved any. You have squandered money, even recently; I don't know what for, and I don't care. But I have drawn out all the money in our joint account and put it in my own account. Everything we have got stands in my name, and it is going to stand there until you get into Congress. And that's that."

Matthew had looked at Sara solemnly with brooding eyes. She was always uncomfortable when he looked at her like that. He seemed to be quite impersonal, as though he were entering lone realms where she could not follow. Soon, some of her assurance had fallen away and her language became less precise:

"Well, what's the idea? What ya glaring at? D'ye think I am going to fail or let you fail after climbing all this distance?"

Apparently he had not heard her. He seemed to be judging her in a far-off sort of way. He was thinking. In a sense Sara was an artist. But she failed in greatness because she lacked the human element, the human sympathy. Now if she had had the abandon, that inner comprehension, of the prostitute who once lived opposite Perigua—but no, no, Sara was respectable. That meant she was a little below average. She was desperately aware of the prevailing judgment of the people about her. She would never be great. She would always be, to him—unendurable. He got up suddenly and silently and walked three miles in the rain. He ended up at his own lodging with its dust and gloom and stood there in the cold and damp thinking of his marriage, six months—six centuries ago.

Again and for a second, for a third, time in his life, he was caught in the iron of circumstance. And he wasn't going to do anything. He couldn't do anything. He was going to be the victim, the sacrifice. Although this time it seemed different from the others. In the first case, of the wreck, he had saved his pride. In the second, the nomination to the legislature, he had sold his body but ransomed his soul, as he hoped. But this time, pride, soul, and body were going.

He looked about at these little trappings of the spirit within him that had grown so thin: gold of the Chinese rug, beneath its dim Chicago dirt; the flame of a genuine Matisse. He had never given up the old rooms of his in the slums, chiefly because Sara would not have the things he had accumulated there in her new and shining house; and he hated to throw them all away. He had always meant to go down and sort them out and store the few things he wanted to keep. But he had been too busy. The rent was nominal, and he had locked the door and left things there.

Only now and then in desperation he went there and sat in the dust and gloom. Today, he went down and waded in. He sat down in the old, shabby easy-chair and thought things out. He was, despite all, more normal and clearer-minded than when he came here out of jail. He was not so cynical. He had found good friends—humble everyday workers, even idlers and loafers whom he trusted and who trusted him. Life was not all evil. He did not need to sell his soul entirely to the devil for bread and butter. Life could be even interesting. There were big jobs, not to be done, but to be attempted, to be interested in. He was not yet prepared to let Sara spoil everything. He began to look upon her with a certain aversion and horror. He planned to live his life by himself as much as possible. She had her virtues, but she was too hard, too selfish, too utterly unscrupulous.

He searched his pockets for money. He went downtown and paid two hundred dollars for a Turkish rug for the bedroom—a silken thing of dark, soft, warm coloring. He lugged it home on the street car and threw it before his old bed and let it vie with the dusky gold of its Chinese mate. He had searched for another Matisse and could not find one, but he had found a copy of a Picasso—a wild, unintelligible, intriguing thing of gray and yellow and black. He paid a hundred dollars for it and hung it on another empty wall. He was half-consciously trying to counteract the ugliness of the congressional campaign.

Long hours he sat in his room. There was no place in Sara's house—it was always Sara's house in his thought—for anything of this, for anything of his: for this big, shabby armchair that put its old worn arms so sympathetically about him. For his pipe. For the books that his fingers had made dirty and torn and dog-eared by reading. For the pamphlets that would not stand straight or regular or in rows. He sat there cold and dark until three o'clock in the morning. Then he stood up suddenly and went to a low bootlegger's dive, a place warm with the stench of human bodies. He sought there feverishly until he found what he wanted—a soul to talk with. There was a mason and builder who came there usually at that hour, especially when he was half drunk and out of work. He was a rare and delicate soul with a whimsical cynicism, with easily remembered tales of lost and undiscovered bits of humanity, with exquisite humor. He played the violin like an angel. Matthew found him. He sat there until dawn. He ordered him to build a fireplace and bathroom in his apartment—something beautiful.

As he sat silently listening to the luscious thrill of the "Spanish Fandango" he determined to do one thing: he would resign from the legislature. Then if he failed in the nomination to Congress, he would be left on the road to freedom. If he gained the nomination, he would gain it with that much less

deception and double-crossing. Of course Sara would be furious.
Well, what of that?

At daybreak he went back to his rooms and started cleaning
up. He swept and dusted, cleaned windows, polished furniture.
He sweated and toiled, then stopped and marveled about Dirt.
Its accumulation, its persistence was astonishing. How could
one attack it? Was it a world symptom? Could machines
abolish it, or only human weariness and nausea?

Late in the afternoon he went out and bought a new big bed
with springs and a soft mattress, a bath robe, pajamas, and
sheets and some crimson hangings. He hid in the wall some of
his money which remained. He knew what he was doing; he
was surrendering to Sara and the Devil and soothing his
bruised soul by physical work and the preparation of a re-
treat where he would spend more and more of his time. He
would save and hide and hoard and some day walk away and
leave everything. But he wrote and mailed his resignation as
member of the legislature. That at least was a symbolic step.

From her interview with Matthew, Sara emerged shaken
but grim. She had no idea what Matthew was going to do.
She had put the screws upon him more ruthlessly than she had
ever dared before. She had cut off his money, his guiding dream
of a comfortable little fortune. She had told him definitely
what he had to think and promise, and he had silently got up
and gone his way. Suppose he never came back, or suppose he
came back and eventually went to this final conference and
"spilled the beans"; threw everything up and over and left
her shamed and prostrate before black and white Chicago?
No, she couldn't risk a mass meeting.

"No," she said in answer to Sammy's query, and looked at
him with a frankness that Sammy half suspected was too frank.
"I don't know what Matthew is going to say or do. And I am
afraid we can't risk a mass meeting."

Sammy was silent. Then he said:

"That resignation was a damn shrewd move."

Sara glanced up.

"What—" She started to ask "What resignation?" but she
paused. "What,—do you think will be its effect?" She would
not let Sammy dream she did not know what he was talking
about.

"Well—it'll mollify the boys. Give me a chance to run Corruthers in at a special election—convince the bosses that Towns is playing square."

Sara was angry but silent. So that fool had resigned from the legislature! Surrendered a sure thing for a chance. Did the idiot think he was already elected to Congress, or was he going to quit entirely?

She took up the morning *Tribune* to hide her agitation and saw the editorial—"a wise move on the part of Towns and shows his independence of the machine."

Sara laid down the paper carefully and thought—tapping her teeth with her pencil. Was it possible that after all— Then she came back to the matter in hand. Sammy would have liked to suggest a real political conference: a secret room with guarded door; cigars and liquor; a dozen men with power and decision, and then, give and take, keen-eyed sparring, measuring of men, and—careful compromise. Out of a conference like that anything might emerge, and Sammy couldn't lose entirely.

But he saw that Sara had the social bee in her head. She wanted a reception, a luncheon, or a dinner. Something that would celebrate a conclusion rather than come to it. He was not averse to this, because he was convinced it would be disastrous to Sara. No social affair of whites and Negroes could come to any real conclusion. It could only celebrate deals already made. Sammy meant to block such deals. But he didn't suggest anything; he let Sara do that, and Sara did. After profound thought, and still clicking her pencil on her teeth, she said:

"A meeting at my home would be the best. A small and intimate thing. A luncheon. No, a dinner, and a good dinner. Let's see, we'll have—"

And then Sara and Sammy selected the personnel. On this they quite agreed. If all went well, Sara suggested that the mass meeting might follow. Sammy cheerfully agreed—if all went well.

Immediately Sara began to prepare for this conference. First she made a number of personal visits, just frank little informal talks with Mr. Graham, with Mr. Cadwalader, with Mrs. Beech and others. Mr. Cadwalader and Mrs. Beech both began by congratulating Sara on Matthew's resignation from the legislature.

"Statesmanlike!" said Cadwalader. "It proves to our people that the reported understanding between him and Scott is untrue."

"Very shrewd," said Mrs. Beech, "to make this open declaration of independence."

"He often takes my advice," said Sara with a cryptic smile, and she explained that when Sammy had approached her, offering coöperation after Doolittle's death, they had, of course, to accept—"to a degree and within limits."

"Of course, of course!" it was agreed.

By her visits she got acquainted with these leaders, measured their wishes, and succeeded fairly well in making them interested in her. She let them do as much talking as possible but also talked herself, clearly and with as much frankness as she dared. She was trying to find out just what the Republicans wanted and just what the reformers demanded.

From time to time she wrote these things down and put the formulas and statements before Matthew, writing them out carefully and precisely in her perfect typewriting. He received them silently and took them away, making no comment. Only once was the resignation from the legislature referred to:

"I'm glad you took my hint about the legislature," said Sara sweetly, one night at dinner.

Matthew stared. When had she hinted, and what?

Sara proceeded further with her plans. She put before Mr. Graham a suggested platform which contained a good many of the Republican demands but even more of the Progressive demands. Mr. Graham immediately rejected it as she expected. He pointed out just how much more he must have and what things he could under no circumstances admit.

Sara tried the same method with Mr. Cadwalader; only in his case she submitted a platform with less of the Progressive demands and more of the Republican. She had more success with him. She could easily see that Mr. Cadwalader after all really leaned considerably toward Republican policies and was Progressive in theory and by the practical necessity of yielding something to the Labor group. But the question Sara quickly saw was, Which Labor group? There were, for instance, the aristocrats in the Labor world; the skilled trade unions connected with the American Federation of Labor; and on the other hand, there was the left wing, the Communist radicals, and there was a string of uncommitted workers between. Mr. Cadwalader consulted the conservative labor unionists and evolved a platform which was not so far from Sara's, and indeed as she compared them, Mr. Graham and Mr. Cadwalader seemed easily reconcilable, at least in words. Sara tried again and brought another modified platform to Mr. Graham. Mr. Graham read it and smiled. So far as words went, there was really little to object to, but he laid it aside and looked Sara squarely in the eye, and Sara looked just as squarely at him. It had come to a showdown, and both knew it. Sara attempted no further fencing. She simply said:

"What is it specifically that you want Matthew to do in Congress? Write it out, and I'll see that he signs it."

He took a piece of paper and wrote a short statement. It had reference to specific bills to be introduced in the next Congress, on the tariff, on farm relief, on railroad consolidation, and on super-power. He even named the persons who were going to introduce the bills. Then he handed the slip to Sara. She read it over carefully, folded it up, and put it in her bag.

"You'll receive this, signed, at or before the final conference."

"Before will be better," suggested Mr. Graham.

"Perhaps," answered Sara, "but on the night of the conference it will be time enough."

Mr. Graham looked almost genial. Sara was the kind of politician that he liked, especially as he saw at present no way to escape a colored candidate, and on the whole he preferred Matthew Towns to Sammy Scott.

"But how about the Radical wing?" he asked. "Are they going to accept this platform?"

"That is the point," said Sara. "I am trying to make the platform broad enough to attract the bulk of the Labor group, but I have not consulted the radicals yet. If they accept what I offer, all right; but even if they do not, we have made sure of the majority of the third party's support."

In this way and by several consultations with Mr. Cadwalader, Mrs. Beech, and their friends, Sara evolved a statement which seemed fair, especially when most of the persons involved began to realize that Matthew Towns on this platform was pretty sure of election.

Sara then turned to the Labor group. Mr. Cadwalader had

smoothed the way for her to meet the labor-union heads, and
it took Sara but a short time to learn how the land lay there.
Eight-hour laws, and anti-injunction legislation, of course; but
above all "down with Negro scabs"! Negroes should be taught
never to take white strikers' jobs.

"Even if white unions bar them before and after the strike,"
thought Sara. But she did not say so. She agreed that scabbing
was reprehensible, and in turn the union leaders unctuously
asserted the "principle" of no color line in the Federation of
Labor. It was quite a love feast, and both Sara and Mr. Cadwalader were elated.

Then Sara finally plucked up courage and visited the headquarters of the left-wing trade unionists. She had anticipated
some unpleasantness, and she was not disappointed. Her earlier
contact with the group had been by letter, and she had been
impressed by the shrewd leadership and evidence of wide
vision. She was prepared for careful mental gymnastics and
careful play of word and phrase. Instead she found a rough
group of painfully frank folk. The surroundings were dirty, and
the people were rude. It was much less attractive than her
visits to the well-furnished headquarters of the Republicans
or to the rooms of the Woman's City Club. But if Sara was
disgusted with the people and surroundings, she was even more
put out with their demands. They came out flat-footed and
assumed facts that were puzzling. She did not altogether understand them, chiefly because she had not taken time to study
them; it was words and personalities that she had come to
probe. The flat demands therefore seemed to her outrageous,
revolutionary.

"Overthrow capital? What do you mean?" she said. "Do you
want to stop industry entirely and go back to barbarism?"

Then all talked at once in that little crowded room, and she
did not pretend to understand:

"What's Towns going to do for municipal ownership of
public services? For raising the income taxes on millionaires?
For regulating and seizing the railroads? For curbing labor
injunctions? For confiscating the unearned increment? For
abolishing private ownership of capital?"

Sara stared; then she gathered up her papers.

"I shall have to ask Mr. Towns," she said, crisply. "We will
have another consultation next week." And she swept out,

vowing to have nothing to do with this gang again. She told Sammy about it and suggested that he hold all further consultations with them.

"It is no place for a lady," she said.

"Lots of them down there," said Sammy.

"You mean those working-women?" said Sara with disgust.

It suited Sammy very well to take charge of further conferences with the Laborites. He had already been engaged in stiffening the demands of the Republicans on the one hand and arousing the suspicion of the colored voters against the trade unionists on the other: and now he was more than willing to push the left wing toward extreme demands. He worked through his young radical friend and now and then saw and talked with the Indian.

Sara was quite sure that he would do something like this, but she did not care. The more radical the left wing was, the fewer votes it would poll and the stronger would be Matthew's hold upon the main bloc of the Progressive group. She was sure of Graham unless Matthew got crazy and went radical. And Matthew seemed to be obeying the whip and bit.

It seemed to Sara the proper time to put Graham's ultimatum before Matthew. She did not argue or expatiate; she simply handed him the statement with the remark:

"Mr. Graham expects to receive this, signed by you, at the conference or before. Your nomination depends upon it."

Then she powdered her nose, put on her things, patted her hat in shape, and walked out. Matthew walked up and down the room. Up and down, up and down, until the walls were too narrow. Then he went out and walked in the streets. It was the last demand, and it was the demand that left him no shred of self-respect. What crazed him was the fact that he knew that he was going to sign it, and that in addition to this, he was going to promise to the Progressives, and perhaps even to the left-wing Laborites, almost exactly opposite and contradictory things. He had reached his nadir. Then he held up his head fiercely. From nadir he would climb! But even as he muttered this half aloud, he did not believe it. From such depths men did not climb. They wallowed there.

Finally, about April first, a week before the primary election, Sara decided that it was time for her final conference. She gave up entirely the idea of a mass meeting. That could come after the primary, when Matthew's nomination was accomplished.

What she really wanted was a dinner conference. There again she hesitated. She was afraid that some of the people whom she was determined to have present, some of the high-placed white folk, might hesitate to accept an invitation to dinner in a colored home. Gradually she evolved something else; a small number of prominent persons were invited to confer personally with Mr. Towns at his home. After the conference, "supper would be served." Sara put this last. If any one felt that they must, for inner or outer compulsions, leave after the conference, they could then withdraw; but Sara proposed to keep them so long and to make the dinner-supper so attractive that it would be, in fact, quite an unusual social occasion. "Quite informal" it was to be, so her written invitations on heavy paper said. But that was not the voice of her dining-room.

XXII

Sara looked across that dining-room and was content. The lace over-cover was very beautiful. The new china had really an exquisite design, and her taste in cut glass was quite vindicated. The flowers were gorgeous. She would have preferred Toles, the expensive white caterer, but, of course, political considerations put that beyond thought. The colored man, Jones, was, after all, not bad and had quite a select white clientele in Chicago. It was a rainy night, but so far not one person invited had declined, and she viewed the scene complacently. She doubted very much if there was another dining-room in Chicago that looked as expensive. Bigger, yes, but not more expensive, in looks at least.

Sara was in no sense evil. Her character had been hardened and sharpened by all that she had met and fought. She craved wealth and position. She got pleasure in having people look with envious eyes upon what she had and did. It was her answer to the world's taunts, jibes, and discriminations. She was always unconsciously showing off, and her nerves quivered if what she did was not noticed. Really, down in her heart, she was sorry for Matthew. He seemed curiously weak and sensitive in the places where he should not have been; she herself was furious if sympathy or sorrow seeped through her armor.

She was ashamed of it. All sympathy, all yielding, all softness, filled her with shame. She hardened herself against it. To-night she looked upon as a step in her great triumph.

There were twenty people in all besides Matthew and Sara. Of these, six were white. There was Mr. Graham, the Republican city boss, and with him a prominent banker and a high state official; Mrs. Beech, the president of the Woman's City Club, was there, and a settlement worker from the stockyard district; and, of course, Mr. Cadwalader. Sara regarded the banker and the president of the City Club as distinct social triumphs for herself. It was something unique in colored Chicago. And especially on a cold and rainy night like this!

Besides these there were fourteen colored persons. First, Sammy and Corruthers. Sara had violently objected to the thin, red-headed and freckled Corruthers, but Sammy solemnly engaged to see that he arrived and departed sober and that he was kept in the background. He made up for this insistence by bringing two of his most intelligent ward leaders with their wives, who were young and pretty, although not particularly talkative, having, in truth, nothing to say. Sara had insisted upon the physician and his wife from Memphis and the minister and his wife. All of these were college-trained and used to social functions. Two colored editors had to be included, and two colored women representing Sara's clubs.

The president of the Trade Unions' City Central was at first included among the guests, but when he heard that the meeting was to be at a colored home and include a supper, he reneged. Mr. Murphy habitually ate with his knife and in his shirt sleeves and he didn't propose "to have no niggers puttin' on airs over him." At the same time the unions must be represented; so the settlement worker was chosen at Mrs. Beech's suggestion.

Sammy had pointed out rather perfunctorily that it might be a mistake not to include some radicals and that in any event they might send a delegation if they heard of the conference. Sara merely shrugged her shoulders, but Sammy saw to it that the left-wing unions *did* hear of the conference and of their exclusion.

The stage was set deftly in the large reception room opening in front on the glass-enclosed veranda. There was a little orchestra concealed here behind the ferns, and it was to play now and then while the company was gathering and afterward while they were eating. There were cigarettes and punch, and as Mr. Corruthers soon discovered, there were two kinds of punch. In the main reception room were soft chairs and a big couch, while thick portières closed off the dining-room and the entrance hall. To the right was the door to the little library, and here Matthew held his interviews, the door standing ajar.

Matthew sat beside a little table in a straight chair. There were pens, blotters, and writing materials, and all over, soft reflected lights. Sara and Sammy had general charge, and both were in their element. The company gathered rather promptly. Sara stood in the main parlor before the portières that veiled the dining-room, where she could receive the guests, entertain them, and send them to consultation with Matthew. Sammy stood between the hall and the reception room where he could welcome the guests, overlook the assembly, and keep his eye on Corruthers.

Everybody was overanxious to please, but the difficulties were enormous. There was no common center of small talk to unite black and white, educated and self-made. The current tittle-tattle of the physician's and minister's wives was not only Greek to the banker and the president of the City Club, but not at all clear to the wives of the colored politicians. The conversation between Mr. Cadwalader and the Republican bosses was a bit forced. Perhaps only in the case of the intelligent white settlement worker and the colored representatives of the Women's Clubs was a new, purely delightful field of common interest discovered.

In Chicago as elsewhere, between white and colored, the obvious common ground was the Negro problem, and this both parties tried desperately to avoid and yet could not. They were always veering toward it. The editor and the banker sought to compare their respective conceptions of finance. But they never really got within understanding distance. Even Sara was at times out of her depth, in a serious definite conversation. With a particular person whom she knew or had measured she could shine. But the light and easy guidance of varied conversation in an assembly of such elements as these was rather beyond her. She hurried here and there, making a very complete and pleasing figure in her flesh-colored chiffon evening frock. But she was not quite at ease.

Sammy's finesse helped to save the day, or rather the night. He had real humor of a kindly sort, and shrewd knowledge of practically everybody present. He supplied the light, frank touch. He subtly separated, grouped, entertained, and re-separated the individuals with rare psychology. He really did his best, and with as little selfishness as he was capable of showing.

The Republican boss, the banker, and the state official were among the earliest arrivals. They sat down with Matthew and entered into earnest conversation. Evidently, they were reading over the latest draft of the proposed platform. Sara was taut and nervous. She tried not to listen, but she could not help watching. She saw Matthew shift the papers until he exposed one that lay at the bottom. The two gentlemen read it and smiled. Quite carelessly and after continued conversation, Mr. Graham absently put the paper in his pocket. By and by they arose and mingled with the other guests. They were all smiling. The boss whispered to Sara that he was satisfied, perfectly satisfied. She knew Matthew had signed the paper.

Sara was radiant. She personally escorted the banker to a seat beside the president of the City Club. She did not know that these two were particularly uncongenial, but they were both well-bred and kept up polite conversation until Mrs. Beech excused herself to talk with Matthew. Matthew was a figure distraught and absent-minded. His dress was much too negligent and careless to suit Sara, although he had put on his dinner jacket. Still, as Sara looked him over now and then, he did not make an altogether bad appearance. There was a certain inherent polish, an evidence of breeding which Sara always recognized with keen delight. It seemed easily to rise to the surface on occasions of this sort. Mr. Cadwalader and Mrs. Beech were now talking with Matthew. They seemed at first a little disturbed, but Sara was pleased to note that Matthew had aroused himself and was talking rather quickly and nervously but impressively. Evidently the two representatives of the liberal groups liked what he said. They called in the settlement worker. When at last they arose, all of them seemed pleased.

"I think," said the president of the City Club, "we have come to a good understanding."

"Really," said Mr. Cadwalader, "much better than I had hoped for. You can count on us."

Sara sighed. The thing was done. Of course, there was the difficulty of those radical Labor people, but these she regarded as on the whole the least difficult of the three groups. She would perhaps approach them again tomorrow. Even if she failed they could not do much harm now.

Sammy had about given up. It looked as though Matthew was going to be triumphantly nominated. In fact, he had just learned that Matthew had made one unexpected move, and whether it was stupid or astute, Sammy was undecided. Corruthers had told him that during that very afternoon the left-wing Labor people had got at Matthew and told him that they had not been included in the negotiations after that first visit of Sara, and that none of their representatives were invited to the conference tonight. Matthew had been closeted with them a couple of hours, but just what was said or done Sammy was unable to learn. Apparently his henchman, the young colored radical, was not present, and he could not find the Indian. His hope then that the radicals would burst in on this conference and make trouble at the last moment seemed groundless. Perhaps Matthew by some hocus-pocus had secured their silent assent. The Labor delegation would probably not arrive at all.

Meantime, this conference must get on. If success was sure, he must be in the band wagon. He gradually gathered his colored politicians out into the dining-room, where there was good liquor. He got the white women and the colored women on the porch in earnest conversation on settlement work for the South Side. The younger women and men, including the Republican boss and his friends, he brought together in the main reception room and started some sprightly conversation. All this was done while Sara had been arranging carefully and not too obviously the personal conferences with Matthew. Well, it was all over.

Then he noticed Corruthers beckoning to him furtively from the half-raised portières that led to the hall. He looked about. Various members of the colored group were talking with the whites, and Matthew had emerged from the little library and seemed to be having a pleasant chat with the minister. Sammy slipped out.

"Say," said Corruthers, "that Labor delegation is here and they want to come in."

Sammy pricked up his ears.

Aha! It looked as though something might happen after all. He walked over to Sara and imparted his news.

"Well, they are not coming in here," said Sara.

"But," expostulated Sammy, "they have evidently been invited."

"Not by me," snapped Sara.

"But I suspect by Matthew. He was with them this afternoon."

Sara started and tapped her foot impatiently. But Sammy went on:

"Don't you think it would be good politics to let them have their say? We don't need to yield to them in any way."

Sara was unwilling, but she saw the point. It was a shame to have this love feast broken into. Then a plan occurred to her. They need not come in here; they could meet Matthew in the little library. The door to the reception room could be closed, and they could enter from the hall. Meantime, Sammy saw Corruthers again beckoning excitedly from the door. He walked over quickly, and Corruthers whispered to him.

"My God!" said Sammy. "Hush, Corruthers, and don't say another word. Here, come and have a drink!"

Then he hurried back to Sara. Sara interrupted him before he could speak.

"Take them into the library. I will have Matthew receive them." She sauntered over to Matthew. "Matthew, dear, some of the Bolsheviks are here and want to talk to you. I have had them taken from the hall directly to the library. You can close the door. They will probably feel more at ease then."

Matthew rose and said a little impatiently: "Why not have them in here?"

"They preferred the smaller room," said Sara. "They are not exactly—dressed for an evening function."

And then, turning, she ordered the portières which concealed the dining-room to be thrown open, and as Matthew stepped into the small library, the blaze of Sara's supper fell upon the company in the reception room.

The table was a goodly sight. The waiters were deft and

silent. The music rose sweetly. The company was hungry, for it was nearly nine. Even Mrs. Beech, who had meant to dine in Hubbard Woods, changed her mind. Little tables with lace, linen, china, and silver were set about, and soon a regular dinner of excellent quality was being served. Tongues loosened, laughter rose, and a feeling of good fellowship began to radiate. Mr. Cadwalader and Mrs. Beech agreed *sotto voce* that really this was quite average in breeding and as a spectacle; they glowed at the rainbow of skins—it was positively exciting.

Sammy was almost hilarious. He could not restrain a wink at Corruthers, and both of them simultaneously bolted for the hall in order to laugh freely and get some more of that other punch. Meantime, Sara's unease increased. Her place and Matthew's had been arranged at the edge of the dining-room at a table with Mrs. Beech and Mr. Graham. The banker, the state official, and the two pretty young politicians' wives were at a table next, and the other tables were arranged as far as possible with at least one bit of color.

But where was Matthew? thought Sara impatiently. It was time for the toast and the great announcement—the culmination of the feast and conference. Mrs. Beech asked for Mr. Towns.

"He's having a last word with the Communists," laughed Sara.

"Oh, are they here?" asked Mr. Cadwalader uneasily—"at the last moment?"

"They wouldn't come in—they are asking about some minor matters of adjustment, I presume."

But Sara knew she must interfere. She distrusted Matthew's mushy indecision. To reopen the argument now might spoil all. She could stand it no longer. She arose easily, a delicate coffee cup in hand, and said a laughing word. She moved to the library door. Sammy watched her. The others sensed in different ways some slight uneasiness in the air.

"Well, Mr. Towns," said Sara, pushing the door wide, "we—"

The light of the greater room poured into the lesser—searching out its shadows. The ugly Chinese god grinned in the corner, and a blue rug glowed on the floor. In the center two figures, twined as one, in close and quivering embrace, leapt, etched in startling outline, on the light.

XXIII

Matthew had turned and started for the library. He had glanced at the reception room. He would not have been human not to be impressed. He was going to be a member of the Congress of the United States. He was going to be the first Negro congressman since the war. No—really the first; all those earlier ones had been exceptions. He was real power. Power and money. Sara should not fool him this time. He understood her. He would have his own funds. He would, of course, follow the machine. He must. He must keep power and get money. But he would have some independence—more and more as time flew. Until— He squared his shoulders, opened the door, and closed it behind him. The room was dimly lighted save the circle under the reading-light on the table. He looked about. No one was there. But there were voices in the hall. He waited.

Then slowly shame overwhelmed him. He was paying a price for power and money. A great, a terrible price. He was lying, cheating, stealing. He was fooling these poor, driven slaves of industry. He had listened to their arguments all this afternoon. He had meant now to meet the delegation brusquely and tell them railingly that they were idiots, that he could do little— something he'd try, but first he must get into Congress.

But he couldn't find the words. He walked slowly over to the table and stood facing the door. It was all done. It was all over. He had sold his soul to the Devil, but this time he had sold it for something. Power? Money? Nonsense! He had sold it for beauty; for ideal beauty, fitness and curve and line; harmony and the words of the wise spoken long ago. He stood in his dinner jacket, sleek but careless, his shirt front rumpled, the satin of his lapel flicked with ash, his eyes tired and red, his hair untidy. He stood and looked at the door. The door opened; he dropped his eyes. He could not look up. He heard not the clumping tramp of a delegation, but the light step of a single person. He almost knew that it was the national president of the Box-Makers, come to make their last appeal. Somehow he had a desperate desire to defend himself before the merciless logic and wide knowledge of this official whom he had never met. She had never even written or answered his letters directly, but only through that dumpy stupid state president. She was to have been present this afternoon. She was not; only her pitiless written arraignment of his platform had been read. He had expected her tonight when he heard the delegation had arrived. But he could not look up. He simply took the paper which was handed to him, sensing the dark veil-like garments and the small hand in its cheap cotton glove. He took the paper which the woman handed him. On it was written:

"Our labor union, in return for its support, asks if you will publicly promise them that on every occasion you will cast your vote in Congress for the interests of the poor man, the employee, and the worker, whenever and wherever these interests are opposed to the interests of the rich, the employer, and the capitalist. For instance—"

Thus the paper began, and Matthew began slowly to read it. It was an absurd request. Matthew almost laughed aloud. He had thought to carry it off with a high hand, to laugh at these oafs and jolly them, insisting that first he must *get* to Congress, and then, of course, he would do what he could. Naturally, he was with them. Was he not a son of generations of workers? Well, then, trust him. But they had not come to argue. They were asking him to sign another paper, and to sign on the line. They could never be trusted to keep such a pledge silent. No, they would publish it to the world. Ha, ha, ha! What ghastly nonsense all this lying was! He stopped and went back to the paper and began reading it again. Something was gripping at him. Some tremendous reminder, and then suddenly the letters started out from the page and burned his gaze, they flamed and spread before him. He saw the strong beauty of the great curves, the breadth and yet delicate uplifting of the capitals, the long, sure sweep of the slurred links. Great God! That writing! He knew it as he knew his own face. His hand had started to his inner pocket—then he tried to whisper, hoarsely—

"Where—who wrote this? Who—" He looked up.

A dark figure stood by the table. An old dun-colored cloak flowed down upon her, and a veil lay across her head. Her thin dark hands, now bare and almost clawlike, gripped each other. They were colored hands. Quickly he stepped forward. And she came like a soft mist, unveiled and uncloaked before him. Always she seemed to come thus suddenly into his life. And yet perhaps it was he himself that supplied the surprise and sudden wonder. Perhaps in reality she had always come quite naturally to him, as she came now.

She was different, yet every difference emphasized something eternally marvelous. Her hair was cut short. All that long, cloudlike hair, the length and breadth of it, was gone; but still it nestled about her head like some halo. Her gown was loose, ill-fitting, straight; her hands, hard, wore no jewels, but were calloused, with broken nails. The small soft beauty of her face had become stronger and set in still lines. Only in the steadfast glory of her eyes showed unchanged the Princess. She watched him gravely as he searched her with his eyes; and then suddenly Matthew awoke.

Then suddenly the intolerable truth gripped him. He lifted his hands to heaven, stretched them to touch the width of the world, and swept her into his tight embrace. He caught her to him so fiercely that her little feet almost left the ground, and her arms curled around his neck as their lips met.

"Kautilya," he sobbed. "Princess of India."

"Matthew," she answered, in a small frightened whisper.

There was a silence as of a thousand years, a silence while again he found her lips and kept them, and his arms crept along the frail, long length of her body, and he cried as he whispered in her ears. Perhaps some murmur from the further rooms came to them, for suddenly they started apart. She would have said the things she had planned to say, but she did not. All the greater things were forgotten. She only said as he stared upon her with wild light in his eyes:

"I am changed."

And he answered:

"The Princess that I worshiped is become the working-woman whom I love. Life has beaten out the gold to this fine stuff." And then with hanging head he said: "But I, ah, I am unchanged. I am the same flying dust."

She walked toward him and put both hands upon his shoulders and said, "Flying dust, that is it. Flying dust that fills the heaven and turns the sunlight into jewels." And then suddenly she stood straight before him. "Matthew, Matthew!" she cried. "See, I came to save you! I came to save your soul from hell."

"Too late," he murmured. "I have sold it to the Devil."

"Then at any price," she cried in passion, "at any price, I will buy it back."

"What shall we do—what can we do?" he whispered, troubled, in her hair.

"We must give up. We must tell all men the truth; we must go out of this Place of Death and this city of the Face of Fear, untrammeled and unbound, walking together hand in hand."

And he cried, "Kautilya, darling!"

And she said, "Matthew, my Man!"

"Your body is Beauty, and Beauty is your Soul, and Soul and Body spell Freedom to my tortured groping life!" he whispered.

"Benediction—I have sought you, man of God, in the depths of hell, to bring your dead faith back to the stars; and now you are mine."

And suddenly there was light.

And suddenly from Matthew dropped all the little hesitancies and cynicisms. The years of disbelief were not. The world was one woman and one cause. And with one arm almost lifting her as she strained toward him, they walked shoulder to shoulder out into that blinding light.

And as they walked there seemed to rise above the startled, puzzled guests some high and monstrous litany, staccato, with moaning monotones, bearing down upon their whisperings, exclamations, movements, words and cries, across the silver and crystal of the service:

"I will not have your nomination."

(What does he mean—who is this woman?)

"I'd rather go to hell than to Congress."

(Is the man mad?)

"I'm through with liars, thieves, and hypocrites."

(This is insulting, shameless, scandalous!)

"The cause that was dead is alive again; the love that I lost is found!"

(A married man and a slut from the streets!)

"Have mercy, have mercy upon us!" whispered the woman.

The company surged to its feet with hiss and oath.

Sara, white to the lips, her hard-clenched hand crushing the fragile china to bits, walked slowly backward before them with blazing eyes.

"I am free!" said Matthew.

The low voice of the Princess floated back again from the crimson curtains of the hall:

"Kyrie Eleison."

The high voice of Sara, like the final fierce upthrusting of the Host, shrilled to a scream:

"You fool—you God-damned fool!"

XXIV

The hall door crashed. The stunned company stared, moved, and rushed hurriedly to get away, with scant formality of leave-taking. It was raining without, a cold wet sleet, but the beautiful apartment vomited its guests upon the sidewalk while taxis rushed to aid.

The president of the Woman's City Club rushed out the door with flushed face.

"These Negroes!" she said to the settlement worker. "They are simply impossible! I have known it all along, but I had begun to hope; such persistent, ineradicable immorality! and flaunted purposely in our very faces! It is intolerable!"

The settlement worker murmured somewhat indistinctly about the world being "well lost" for something, as they climbed into a cab and flew north.

The Republican boss, the state official, and the banker loomed in the doorway, pulling on their gloves, adjusting their coats and cravats, and hailing hurrying taxis.

"Well, of all the damned fiascos," said the banker.

"Niggers in Congress! Well!" said the official.

"It is just as well," said the boss. "In fact it is almost providential. It looked as though we had to send a Negro to Congress. That unpleasant possibility is now indefinitely postponed. Of course, now we'll have to send you."

"Oh!" said the banker softly and deprecatingly.

"It is going to cost something," said the boss shrewdly. "You will have to buy up all these darky newspapers and grease Sammy's paw extraordinarily well. The point is, buying is possible now. They have no comeback. Sammy may have aspirations, but I think we can make even him see that it will be unwise to put up another colored candidate now. No, the thing has turned out extraordinarily well; but I wonder what the devil got hold of Towns, acting as though he was crazy?"

The physician's wife and the lawyer's lingered a little, clustering to one side so as to avoid meeting the white folks; they stared and whispered.

"It is the most indecent thing I have heard of," said the physician's wife. The lawyer's wife moaned in her distress:

"To think of a Negro acting that way, and before these people! And after all this work. Won't we ever amount to anything? Won't we ever get any leaders? I am simply disgusted and discouraged. I'll never work for another Negro leader as long as I live."

And they followed their husbands to the two large sedans that stood darkly groaning, waiting.

The physician snarled to the minister, "And with the streets full of women cheaper and prettier."

The Labor delegation had pushed into the library as Matthew and Kautilya left, and entered the reception room. They stood now staring at the disheveled room and the guests rushing away.

"What's happened?"

"Has he told them what's what?"

"Are they deserting us? Are they running away?"

But the colored club women walked away in silence in the rain. They parted at the corner and one said:

"I'm proud of him, at last."

But the other spit:

"The beast!"

XXV

Sammy's world was tottering, and looking upon its astonishing ruins he could only gasp blankly:

"What t' hell!"

Never before in his long career and wide acquaintanceship with human nature had it behaved in so fantastic and unpredictable a manner. Never had it acted with such incalculable and utter disregard of all rules and wise saws. That a man should cheat, lie, steal, and seduce women, was to Sammy's mind almost normal; that he should tell the truth, give away his money, and stick by his wife was also at times probable.

These things happened. He'd seen them done. But that a man with everything should choose nothing: that a man with high office in his grasp, money ready to pour into his pocket, a home like this, and both a wife *and* a sweetheart, should toss them all away and walk out into the rain without his hat, just for an extra excursion with a skirt—

"What t' hell!" gasped Sammy, groping back into the empty house. Then suddenly he heard the voice of Sara.

He found her standing stark alone, a pitiful, tragic figure amid the empty glitter of her triumph, with her flesh-colored chiffon and her jewels, her smooth stockings and silver slippers. She had stripped the beads from her throat, and they were dripping through her clenched fingers. She had half torn the lace from her breast, and she stood there flushed, trembling, furious with anger, and almost screaming to ears that did not hear and to guests already gone.

"Haven't I been decent? Haven't I fought off you beasts and made me a living and a home with my own hands? Wasn't I married like a respectable woman, and didn't I drag this fool out of jail and make him a man? And what do I get? *What do I get?* Here I am, disgraced and ruined, mocked and robbed, a laughing-stock to all Chicago. What did he want? What did the jackass want, my God? A cabaret instead of a home? A whore instead of a wife? Wasn't I true to him? Did I ever let a man touch me? I made money—sure, I made money. I *had* to make money. *He* couldn't. I made money out of politics. What in hell is politics for, if it isn't for somebody to make money? Must we hand all the graft over to the holy white folks? And now he disgraces me! Just when I win, he throws me over for a common bawd from the streets, and a mess of dirty white laborers; a common slut stealing decent women's husbands. Oh—"

Sammy touched her hesitatingly on the shoulder and pleaded:

"Don't crack, kid. Stand the gaff. I'll see you through."

But she shrank away from him and screamed:

"Get out, don't touch me. Oh, damn him, damn him! I wish I could horsewhip them; I wish I could kill them both."

And suddenly Sara crumpled to the floor, crushing and tearing her silks and scattering her jewels, drawing her knees up tight and gripping them with twitching hands, burying her hair, her head and streaming eyes, in the crimson carpet, and rolling and shaking and struggling with strangling sobs.

While without gray mists lay thin upon a pale and purple city. Through them, like cold, wet tears dripped the slow brown rain. The muffled roar of moving millions thundered low upon the wind, and the blue wind sighed and sank into the black night; and through the chill dripping of the waters, hatless and coatless, moved two shapes, hand in hand, with uplifted heads, singing to the storm.

PART IV
THE MAHARAJAH OF BWODPUR

PART IV

THE MAHARAJAH OF BWODPUR

1926, April—April, 1927

The miracle is Spring. Spring in the heart and throat of the world. Spring in Virginia, Spring in India, Spring in Chicago. Shining rain and crimson song, roll and thunder of symphony in color, shade and tint of flower and vine and budding leaf. Spring—two Springs, with a little Winter between. But what if Spring dip down to Winter and die, shall not a lovelier Spring live again? Love is eternal Spring. Life lifts itself out of the Winter of death. Children sing in mud and rain and wind. Earth climbs aloft and sits astride the weeping skies.

I

THE rain was falling steadily. One could hear its roar and drip and splash upon the roof. All the world was still. Kautilya listened dreamily. There was a sense of warmth and luxury about her. Silk touched and smoothed her skin. Her tired body rested on soft rugs that yielded beneath her and lay gently in every curve and crevice of her body. She heard the low music of the rain above, and the crimson, yellow, and gold of a blazing fire threw its shadows all along the walls and ceiling. The shadows turned happily and secretly, revealing and hiding the wild hues of a great picture, the reflections of a mirror, the flowers and figures of the wall. In silence she lay in strange peace and happiness—not trying to think, but trying to sense the flood of the meaning of that happiness that spread above her. Her head lifted; slowly, noiselessly, with infinite tenderness, she stretched her arms toward Matthew, till his head slipped down upon her shoulder. Then, on great, slow, crimson islands of dream, the world floated away, the rain sang; and she slept again.

217

Long hours afterward in the silence that comes before the dim blue breaking of the dawn, Matthew awoke with a start. The rain had ceased; the fire was dim and low; a vague sense of terror gripped him. His breath struggled dizzily in his throat, and then a little shaft of sunshine, pale, clear, with a certain sweetness from the white dampness of sky and earth, wandered down from the high window and leapt and lay on the face of happiness. She lay very still, so still that at first he scarce could see the slow rise and fall of the soft silk that clung to her breast. And then the surge of joy shook him until he had to bite back the sob and wild laughter.

Hard had been their path to freedom. In his first high courage, Matthew had pictured themselves walking through that door and into the light; a powerful step, a word of defiance against the indignant, astonished, angry wave of the world. Yet in truth they had walked out with hands clasped and faces down, and he never knew what words he said or tried to say. Phrases struck upon their ears.

"—knifed his race—a common bawd—a five-minute infatuation—primitive passion—"

Across the endless length of the parlor they had toiled, and down by the blazing dinner table; out far, far out into the narrowing hall; they had brushed by people who shrank away from them. Coatless, hatless, they had walked into the cold and shivering storm. And then somehow they were warm again. Then happiness had fallen softly upon them. Hand in hand they walked singing through the rain.

"Where are we going, Matthew?" she had whispered long hours later, as her tired feet faltered.

And he lifted her in his arms and raised his face to the water and answered:

"We are going down the King's highway to Beauty and Freedom and Love. I can hear life growing down there in the earth and pulling beneath the hard sidewalks and white bones of the dead. Listen, God's darling, to the singing of the rain; hear the dawn coming afar and see the white wings of the mist, how they beat about us."

And so they had come home and slept in his attic nest.

Slowly Matthew lifted himself, arranged the golden glory of the Chinese rug again around her, tucking in her little feet and drawing it close at the side. Noiselessly he slipped to the fireplace and made the golden flames hiss and sputter and swirl up to the sun-drenched sky, and then he came back and stretched himself beside her, slipping, as she slept on, her head upon the curve of his elbow and looking down upon her face.

It was a magnificent face. Something had come and something had gone since the day when he saw it first. Something had gone of that incomprehensible beauty of color, infinite fineness of texture, richness of curve, loveliness of feature, which made her then to his eyes the loveliest thing in the world. But in its place there lay upon her peaceful, sleeping countenance a certain strength and nobility; a certain decision and calm, that was like beauty swept with life, like sunshine softened with mist. The heavy coils of purple hair had been cut away, and yet the hair still lay thick and strong upon her forehead.

Suddenly he wanted to see her eyes, the eyes that he had never forgotten since first he looked into them, eyes that were pools at once of mystery and revelation, misty with half-sensed desire, and calm with power. He wanted to see her eyes again and see them at once with the high consciousness of birth that belonged to the Princess of the Lützower Ufer and with that look of surrender and selfless love that he had caught in the little room behind the parlor. He wanted again desperately to see those eyes which said all these things; yet he lay very still lest for a single moment he should disturb her. And then he looked down, and her eyes were looking up to him. Slowly and happily she smiled.

"Krishna," she murmured. His mind went racing back through the shadows and he whispered back, "Radha." And again they slept.

When Kautilya awoke again, there was a slow music stealing in from the inner room. It was the andante from Beethoven's Fifth Symphony, infinite in tenderness, triumph, and beauty; and it came from afar so that no scratch of the phonograph or creak of mechanism spoiled its sweet melody. She sat up suddenly with a little cry of joy, throwing aside the great Chinese rug and swathing herself in the silk of the white mandarin's robe that lay ready for her. With this music in her ears she found the bathroom with its tub of steaming water and with its completeness, half plaintive with neglect.

A half hour later she found new silken things in the dressing-room. The rug lay upon the floor, and the old worn easy-chair was drawn before the fire. Beside the flaming dance of the fire was a low, white Turkish taboret; toward it Matthew came, clad in an old green bathrobe which hung carelessly along his tall body. There was a tarboosh with faded tassel upon his head. The hot coffee steamed on the salver, with toast and butter and cream. There was an orange in halves and a little yellow rosebud. She laughed in the sheer delight of it all, and held him long and close before they turned to their eating. The morning sun poured in.

The music was changed to that largo of Dvorák built on the echoing pain of the Negro folksong, which is printed on the other side of the Victor record. Matthew would not let her stir, and after a while from the kitchen came a brave splashing of dishes and a song. He came back soon, bringing an old volume of poems. Without a word, only a long look, they nestled in the chair near the fire and read. And so the day passed half wordless with beauty and sound, full of color and content, until the sunlight went crimson and blue upon the walls and the fire shadows danced again.

"There are so many things I would ask you," said Matthew. And then Kautilya took his head between her hands and laid the breadth of his shoulders upon her knees and said:

"And there are so many things I want to tell you of myself. I want to tell you all the story of my life; of my falling and rising; of my love for you; and of that mother of yours who lives far down in Virginia in the cabin by the wood. Oh, Matthew, you have a wonderful mother. Have you seen her hands? Have you seen the gnarled and knotted glory of her hands?" And then slowly with wide eyes: "Your mother is Kali, the Black One; wife of Siva, Mother of the World!"

II

Matthew was talking in the darkness as they lay together closely entwined in each other's arms.

"You will tell me, dear one, all about yourself? How came you here, masquerading as a trade union leader?"

"It must be a long, long story, Matthew—a Thousand Nights and a Night?"

"So short a tale? Talk on, Scheherazade!"

So Kautilya whispered, nestling in his arms:

"But first, Matthew, sing me that Song of Emancipation—sing that Call of God, 'Go down, Moses!' I believe, though I did not then know it—I believe I began to love you that night."

Matthew sang. Kautilya whispered:

"When you left me and went to jail, I seemed first to awake to real life. From clouds I came suddenly down to earth. I knew the fault was mine and the sacrifice yours. I left the country according to my promise to the government. But it was easy to engineer my quick return from London in a Cunard second cabin, without my title or real name.

"And then came what I shall always know to have been the greatest thing in my life. I saw your mother. No faith nor religion, Matthew, ever dies. I am of the clan and land that gave Gotama, the Buddha, to the world. I know that out of the soul of Brahma come little separations of his perfect and ineffable self and they appear again and again in higher and higher manifestations, as eternal life flows on. And when I saw that old mother of yours standing in the blue shadows of twilight with flowers, cotton, and corn about her, I knew that I was looking upon one of the ancient prophets of India and that she was to lead me out of the depths in which I found myself and up to the atonement for which I yearned. So I started with her upon that path of seven years which I calculated would be, in all likelihood, the measure of your possible imprisonment. We talked it all out together. We prayed to God, hers and mine, and out of her ancient lore she did the sacrifice of flame and blood which was the ceremony of my own great fathers and which came down to her from Shango of Western Africa.

"You had stepped down into menial service at my request—you who knew how hard and dreadful it was. It was now my turn to step down to the bottom of the world and see it for myself. So I put aside my silken garments and cut my hair, and, selling my jewels, I started out on the long path which should lead to you. I did not write you. Why did I need to? You were myself, I knew. But I sent others, who kept watch over you and sent me news.

"First, I went as a servant girl into the family of a Richmond engineer."

Matthew started abruptly, but Kautilya nestled closer to him and watched him with soft eyes.

"It was difficult," she said, "but necessary. I had known, all my life, service, but not servants. I had not been able to imagine what it meant to be a servant. Most of my life I had not dreamed that it meant anything; that servants meant anything to themselves. But now I served. I made beds, I swept, I brought food to the table, sometimes I helped cook it. I hated to clean kitchens amid dirt and heat, and I worked long hours; but at night I slept happily, dear, by the very ache of the new muscles and nerves which my body revealed.

"Then came the thing of which your mother had warned me, but which somehow I did not sense or see coming, until in the blackness of the night suddenly I knew that some one was moving in my room; that some one had entered my unlocked door."

Matthew arose suddenly and paced the floor. Then he came and sat down by the couch and held both her hands.

"Go on," he said.

"I sat up tense and alert and held in my hand the long, light dagger with its curved handle and curious chasing; a dagger which the grandfathers of my grandfather had handed down to me. That night, Matthew, I was near murder, but the white man, my employer, slipped as I lunged, and the dagger caught only the end of his left eye and came down clean across cheek, mouth, and chin, one inch from the great jugular vein, just as the mistress with her electric torch came in.

"Instead of arrest which I thought I surely faced, the man was hurried out and off and the woman came to me in the still morning, worn and pale, and said, 'I thank you. This is perhaps the lesson he needed.'

"She paid me my little wage and I walked away."

"But, Kautilya, why, why did you go through all this? What possible good could it do?"

"Matthew, it was written. I went to Petersburg and worked in a tobacco factory, sitting cramped at a long bench, stripping the soft fragrant tobacco leaf from those rough stems. It was not in itself hard work, but the close air, the cramped position, the endless monotony, made me at times want to scream.

And there were the people about me: some good and broken; some harsh, hard, and wild. Leering men, loud-mouthed women. I stayed there three endless months until it seemed to me that every delicate thought and tender feeling and sense of beauty had been bent and crushed beyond recognition. So I took the train and came to Philadelphia.

"I worked in two restaurants; one on Walnut Street, splendid and beautiful. The patrons usually were kind and thoughtful with only now and then an overdressed woman who had to express her superiority by the loudness of her tones, or a man who was slyly insulting or openly silly. Only the kitchen and the corridors bruised me by their contrast and ugliness. Singularly enough in this place of food and plenty, the only proper food we waitresses could get to eat was stolen food. I hated the stealing, but I was hungry and tired. From there I went down to South Street to a colored restaurant and worked a long time. It was an easy-going place with poor food and poor people, but kind. They crowded in at all hours. They were well-meaning, inquisitive; and if a busy workingman or a well-dressed idler sought to take my hand or touch my body he did it half jokingly and usually not twice."

"Servant, tobacco-hand, waitress; mud, dirt, and servility for the education of a queen," groaned Matthew.

"And is there any field where a queen's education is more neglected? Think what I learned of the mass of men! I got to know the patrons: their habits, hardships, histories. I was the friend of the proprietors, woman and husband; but the enterprise didn't pay. It failed. I cried. But just as it was closing I learned of your release, and after but a year, suddenly I was in heaven. I thought I had already atoned.

"But I knew that yet I must wait. That you must find your way and begin to adjust your life before I dared come into it again. And so I went to New York, that my dream of life and of the meaning of life to the mass of men might be more complete.

"I discovered a paper box-making factory on the lower East Side. It was a non-union shop and I worked in a basement that stank of glue and waste, ten and twelve hours a day for six dollars a week. It was sweated labor of the lowest type, and I was aghast. Then the workers tried to organize— there was a strike. I was beaten and jailed for picketing, but

I did not care. That which was begun as a game and source of experience to me became suddenly real life. I became an agent, organizer, and officer of the union. I knew my fellow laborers, in home and on street, in factory and restaurant. I studied the industry and the law, I traveled, made speeches, and organized. Oh, Matthew—it was life, life, real life, even with the squalor and hard toil."

"Yes, it was life. And the Veil of Color lifted from your eyes as it is lifting even from my blindness. Those people there, these here—they are all alike, all one. They are all foolish, ignorant, and exploited. Their highest ambition is to escape from themselves—from being black, from being poor, from being ugly—into some high heaven from which they can gaze down and despise themselves."

"True, my Matthew, and while I was learning all this which you long knew, you seemed to me striving to unlearn. Oh, how I watched over you! You came down to Virginia. Hidden in the forest, I watched with wet eyes. Hidden in the cabin, I heard your voice. I caught the sob in your throat when your mammy told of my coming. I knew you loved me still, and I wanted to rush into your arms. But, 'Not yet—not yet!' said your wise old mother.

"I was working busily and happily when the second blow fell, the blow that came to deny everything, that seemed to say that you were not self of my own self and life of the life which I was sharing in every pulse with you. You married. I gave up."

"You did not understand, Kautilya. You seemed lost to me forever. I was blindly groping for some counterfeit of peace. If I had only known you were here and caring!"

"I went down again to Virginia and knelt beside your mother, and she only smiled. 'He ain't married,' she said. 'He only thinks he is. He was wild like, and didn't know where to turn or what to do. Wait, wait.'

"I waited. You would not listen to my messengers whom continually I sent to you—the statesmen of Japan, the Chinese, the groping president of the Box-Makers. Like Galahad you would not ask the meaning of the sign. You would not name my name. How could I know, dearest, what I meant to you? And yet my thought and care hovered and watched over you. I knew Sammy and Sara and I saw your slow and sure descent

to hell. I tried to save you by sending human beings to you.
You helped them, but you did not know them. I tried again
when you were sitting in the legislature down at Springfield.
You knew, but you would not understand. You sneered at the
truth. You would not come at my call."

"I did not know it was your voice, Kautilya."

"You knew the voice of our cause, Matthew—was that not
my voice?"

Matthew was silent. Kautilya stroked his hand.

"We met in London, the leaders of a thousand million of
the darker peoples, with, for the first time, black Africa and
black America sitting beside the rest. I was proud of the
Negroes we had chosen after long search. There were to be
forty of us, and, Matthew, only you were absent. I looked for
you to the last. It seemed that you must come. We organized,
we planned, and one great new thing emerged—your word,
Matthew, your prophecy: we recognized democracy as a
method of discovering real aristocracy. We looked frankly
forward to raising not all the dead, sluggish, brutalized masses
of men, but to discovering among them genius, gift, and ability
in far larger number than among the privileged and ruling
classes. Search, weed out, encourage; educate, train, and open
all doors! Democracy is not an end; it is a method of aristocracy. Some day I will show you all we said and planned.

"All the time, until I left for this great meeting I had expected that somehow, some way, all would be well. Some time
suddenly you would come away. You would understand and
burst your bonds and come to us—to me. But as I left
America fear entered my heart—fear for your soul. I began
to feel that I must act—I must take the step, I must rescue
you from the net in which you were floundering.

"I remember the day. Gloom of fog held back the March
spring in London. The crowded, winding streets echoed with
traffic. I heard Big Ben knelling the hour of noon, and a ray
of sunlight struggled dizzily on the mauve Thames. A wireless
came. You were selling your soul for Congress.

"Before, you had stolen for others. You had upheld their
lies—but your own hands were clean, your heart disclaimed
the dirty game. Now you were going to lie and steal for
yourself. I saw the end of our world. I must rescue you at any
cost—at any sacrifice. I rushed back across the sea. Five days

we shivered, rolled, and darted through the storm. Almost we cut a ship in two on the Newfoundland banks, but wrenched away with a mighty groan. I landed Friday morning, and left at two-fifty-five—at nine next morning I was in Chicago. That night I led your soul up from Purgatory—free!

"And here we are, Matthew, my love; and it is long past the hour of sleep; and you are trembling with apprehension at things which did not happen, at pits into which I did not fall, at failures over which we both have triumphed."

The Princess paused, and Matthew started up. There was a loud insistent knocking at the door.

"Go," said the Princess. "Have we not both expected this?"

Matthew hesitated a moment and then walked to the door and opened it. A colored police officer and two white men in citizens' clothes stepped in quickly and started as if to search, until they saw the Princess sitting on the disheveled bed.

"Well?"

"We were hunting for you two," said one of the plain-clothes men.

"And you have found us?" asked Matthew.

"Yes, evidently. We wondered where you were spending the night."

"We were spending the night here, together," said Matthew.

"Together," repeated the Princess.

The other man began to write furiously.

"You admit that," said the first man.

"We admit it," said Matthew, and the Princess bowed her head.

"Perhaps we had better look around a little," said the other man tentatively. But the policeman protested.

"You got what you wanted, ain't ya? Mr. Towns is a friend of mine, and I don't propose to have no monkey business. If you're through, get out." And slowly they all passed through the door.

III

May, and five o'clock in the morning. The sun was whispering to the night, and the mist of its words rose above the park. Matthew and Kautilya swung rapidly along through the dim freshness of the day. They both had knapsacks and knicker-bockers, and shoulder to shoulder, hand in hand, and singing low snatches of song, they hurried through Jackson Park. It was such a morning as when the world began: soft with breezes, warm yet cold, brilliant with the sun, and still dripping with the memory ˙ sweet, clean rain. There was no dust— no noise, no movement. Almost were the great brown earth and heavy, terrible city, still. Singing, quivering, tense with awful happiness, they went through the world. Far out by the lake and in the drowsy afternoon, when they had eaten sausage and bread and herbs and drunk cool water, after Kautilya had read the sacred words of the Rig-veda, she laid aside the books and talked again, straining his back against her knees as they sat beneath a black oak tree, her cheek beside his ear, while together they stared out upon the waving waters of Lake Michigan.

Matthew said:

"Now tell me beautiful things, Scheherazade. Who you are and what? And from what fairyland you came?"

"I cannot tell you, Matthew, for you do not know India. Oh, my dear one, you must know India.

"India! India! Out of black India the world was born. Into the black womb of India the world shall creep to die. All that the world has done, India did, and that more marvelously, more magnificently. The loftiest of mountains, the mightiest of rivers, the widest of plains, the broadest of oceans—these are India.

"Man is there of every shape and kind and hue, and the animal friends of man, of every sort conceivable. The drama of life knows India as it knows no other land, from the tragedy of Almighty God to the laugh of the Bandar-log; from divine Gotama to the sons of Mahmoud and the stepsons of the Christ.

"For leaf and sun, for whiff and whirlwind; for laughter, and for tears; for sacrifice and vision; for stark poverty and jeweled wealth; for toil and song and silence—for all this, know India. Loveliest and weirdest of lands; terrible with flame and ice, beautiful with palm and pine, home of pain and happiness and misery—oh, Matthew, can you not understand? This is India—can you not understand?"

"No, I cannot understand; but I feel your meaning."

"True, true! India must be felt. No man can know India, and yet the shame of it, that men may today be counted learned and yet be ignorant, carelessly ignorant, of India. The shame, that this vast center of human life should be but the daubed footstool of a stodgy island of shopkeepers born with seas and hearts of ice."

"But you know India, darling. Tell India to the world."

"I am India. Forgive me, dearest, if I play with words beyond meaning—beyond the possibility of meaning. Now let me talk of myself—of my little self—"

"That is more to me than all India and all the world besides."

"I was born with the new century. My childhood was a dream, a dream of power, beauty, and delight. Before my face rose every morning the white glory of the high Himalayas, with the crowning mass of Gaurisankar, kissing heaven. Behind me lay the great and golden flood of Holy Ganga. On my left hand stood the Bo of Buddha and on my right the Sacred City of the Maghmela.

"All about me was royal splendor, wealth and jewels and beautiful halls, old and priceless carpets, the music of tinkling fountains, the song and flash of birds; and when I clapped my childish hands, servants crawled to me on their faces. Of course, much that I know now was missing—little comforts of the West; and there were poverty, pain, sickness, and death; but with all this, around me everywhere was marble, gold and jewels, silk and fur, and myriads who danced and sang and served me. For I was the little Princess of Bwodpur, the last of a line that had lived and ruled a thousand years.

"We came out of the black South in ancient days and ruled in Rajputana; and then, scorning the yoke of the Aryan invaders, moved to Bwodpur, and there we gave birth to Buddha, black Buddha of the curly hair. Six million people worshiped us as divine, and my father's revenue was three hundred lakhs of rupees. I had strange and mighty playthings: elephants and lions and tigers, great white oxen and flashing automobiles. Parks there were and palaces, baths and sweet waters, and amid it all I walked a tiny and willful thing, curbed only by my old father, the Maharajah, and my white English governess, whom I passionately loved.

"I had, of course, my furious revolts: wild rebellion at little crossings of my will; wild delight at some of the efforts to amuse me; and then came the culmination when first the flood of life stopped long enough for me to look it full in the face.

"I was twelve and according to the ancient custom of our house I was to be married at a great Durbar. He was a phantom prince, a pale and sickly boy who reached scarcely to my shoulder. But his dominion joined with mine, making a mighty land of twelve million souls; of wealth in gold and jewels, high mineral walls, and valleys fat with cream. All that I liked, and I wanted to be a crowned and reigning Maharanee. But I did not like this thin, scared stick of a boy whose pearls and diamonds seemed to drag him down and make his dark eyes shine terrorstricken beneath his splendid turban.

"I enjoyed the magnificent betrothal ceremony and poked impish fun at the boy who seemed such a child. A tall and crimson Englishman attended him and ran his errands and I felt very grand, riding high on my silken elephant amid applauding thousands. The 'Fringies,' as we called the English, were here in large numbers and always whispering in the background, nodding politely, playing with me gravely, and yielding to my whims. I confess I thought them very wonderful. I set them, unconsciously, above my own people.

"I remember hearing and but half understanding the talk of my guardian and counselors. They were apparently vastly surprised that the English had allowed this marriage. It would seem the English had long resisted the wishes of the people of Sindrabad. They had, you see, more power in his land than in ours. Our land was independent—or at least we thought so. To be sure we sent no ministers to foreign lands—but what did we know or care of foreign lands? To be sure our trade was monopolized by the English, but it was good and profitable trade. Internally we were free and unmolested, save that an unobtrusive Englishman was always at court. He was the Resident. He 'advised' us and spied upon us, as I now know.

"Now it was different in Sindrabad, where my little prince ruled under English advisers. Sindrabad was in the iron grip of the English. They long frowned upon the power of Bwodpur, a native, half-independent Indian state. They refused to countenance a marriage alliance with Sindrabad and continued to refuse; then suddenly something happened. A new English Resident appeared, a commissioner magnificent with medals, well trained, allied to a powerful English family of the nobility, and backed by new regiments of well-armed men. He had lived long in the country of my phantom prince; now he came to us smiling and bringing the little Maharajah by the hand and giving consent and benediction to our marriage.

"I heard my father ask, aside, hesitating and frowning:

" 'What is back of all this?'

"But I only half listened to this talk and intrigue. I wanted the Durbar and the glory of the pageant of this marriage. So in pomp and magnificence beyond anything of which even I, a princess, had dreamed, we were married in the high hills facing the wide glory of the Himalayas; the drums boomed and the soldiers marched; the elephants paraded and the rajahs bowed before me and I was crowned and married, her Royal Highness, the Maharanee of Bwodpur and Sindrabad. There was, I believe, some dispute about this 'royal,' but father was obdurate."

"You mean that you were really a wife, while yet a child?" asked Matthew.

"Oh, no, I was in reality only a betrothed bride and must return to my home for Gauna, that is, to wait for years until I was grown and my bridegroom should come and take me to his home. But he never came. For somehow, I do not remember why, there came a time of darkness and sorrow, when I could not go abroad, when I was hurried with my nurse from palace to palace and got but fleeting glances of my phantom prince even on his rare calls of ceremony. Once I came upon him in a long, cold and marble corridor as, running, I escaped from Nurse. He was standing, thin, pale, and in tears. His brown skin was gray and drawn; he looked upon me with great and frightened eyes and whispered: 'Flee, flee! The English will kill you too.' That was all.

"I do not know how it happened. I know that the English commissioner was transfixed with horror. This bronze boy, just as he had started home, was found in the forest, his face all blood, dead. My father was wan with anger, and, it seemed, all against the English. He did not accuse them directly of this awful deed, but he knew that the death of both these married children, the last of their line, would throw both countries into the control of England. There were wild rumors in the air of the court. In strict compliance with ancient custom, I as a widow should have died with my little bridegroom, but even the priests saw too much power for England in this, and suddenly my father summoned my English companion and sent me with her to England, while he reigned in my name in Sindrabad and in his own right in Bwodpur.

"My governess was a quiet, clear-eyed woman, with a heart full of courage and loyalty. Sometimes I thought that she and my father had loved each other and that because of the hopelessness of this affection she was suddenly sent home and I with her.

"Then came beautiful days. I loved England. I loved the work of my tutors and the intercourse with the new world that spread before me. I stayed two full years, until I was fourteen, and then again came clouds. There was a tall English boy of whom I saw much. We had ridden, run, and played together. He told me he loved me. I was glad. I did not love him, but I wanted him to love me because the other girls had sweethearts. But he was curiously fierce and gruff about it all. He wanted to seize and embrace me and I hated the touch of his hands, for after all he was not of royal blood, which then meant so much to me.

"One day he suddenly asked me to run away and marry him. I laughed.

" 'Yes, I mean to marry you,' he said. 'I am going to have you. I don't care if you are colored.' I gasped in amazement. He didn't *care*. He, a low-born shopman's brat, and I, a princess born. I, *'colored'!* I wanted to strike him with my croquet mallet. I rushed away home.

"It seemed that the scales had fallen from my eyes. I understood a hundred incidents, a dozen veiled allusions and little singular happenings. I suddenly realized that these dull, loud, ugly people actually thought me inferior because my skin was browner than their bleached and roughened hides. They were condescending to me—me, whose fathers were kings a thousand years before theirs were ragpickers.

"I rushed in upon my governess. I opened my lips to rage. She stopped me gently: my father was dead in Bwodpur. I was summoned to India to marry and reign. But I did not go. The news of my father's death came on August first, 1914. When I reached London and the India Office, August fourth, the world was at war.

"There ensued a series of quick moves followed by protracted negotiations; the English explained that it would never do to start their royal charge for India in time of war. Bwodpur retorted that it would never do to have their Maharanee far away in England in time of war. The India Office delicately suggested that the presence in England of an Indian princess of high birth and influence would do much to cement the empire and win the war for civilization, and secretly they whispered that it would be unwise to send to India, when English power was weak, a person who might become a rallying center for independence.

"Bwodpur pointed out that my presence in India was precisely the thing needed to arouse a feeling favorable to England and oppose the disruptive forces of *Swaraj*, which were undermining native dynasties as well as imperial power.

"But after all, England had the advantage in that argument, because I was in England; and while I probably would not have been allowed to return home had I wished, official England put forth every effort to make me want to stay. At first, I was imperious and discontented, remembering that I was 'colored.' But official England took no notice, and with deep-laid plans and imperturbable self-possession proceeded to capture my imagination and gain my affection. England became gracious and kind. London opened its heart and arms to this dark and difficult charge. Even royalty held out a languid hand, and I was presented at court in 1916 and formally received in society.

"I did not yield easily. I sat back upon my rank. I used my wealth. When I was invited out I took the pas from Duchesses as the child of a reigning monarch. I made the county aristocracy cringe and the city snobs almost literally hold my train. All this until my poor foster mother was filled with apprehension. Slowly but surely, however, my defenses were beaten down and I capitulated.

"In the midst of war hysteria, I became the social rage, and I loved it. I forgot suspicion and intrigue. I liked the tall and calm English men, the gracious and well-mannered English women. I loved the stately servants, so efficient, without the eastern servility to which I had been born. I knew for the first time what comfort and modern luxury meant.

"I danced and knitted and nursed and studied. I spent week-ends in storied castles, long days in museums and nights at theaters and concerts, until the War grew harder. Money like water flowed through my careless hands. I gave away gold and jewels. I was a darling of the white gods, and I adored them. I even went to the front in France for temporary duty as a nurse—carefully guarded and pampered.

"Can I make you realize how I was dazed and blinded by the Great White World?"

"Yes," said Matthew. "I quite understand. Singularly enough, we black folk of America are the only ones of the darker world who see white folk and their civilization with level eyes and unquickened pulse. We know them. We were born among them, and while we are often dazzled with their deeds, we are seldom drugged into idealizing them beyond their very human deserts. But you of the forest, swamp, and desert, of the wide and struggling lands beyond the Law—when you first behold the glory that is London, Paris, and Rome, I can see how easily you imagine that you have seen heaven; until disillusion comes—and it comes quickly."

"Yes," sighed Kautilya, with a shudder, "it came quickly. It approached while I was in France in 1917. Suddenly, a bit of the truth leapt through. There, at Arras, an Indian stevedore, one of my own tribe and clan, crazed with pain, bloody, wild, tore at me in the hospital.

" 'Damn you! Black traitor. Selling your soul to these dirty English dogs, while your people die—your people die.'

"I hurried away, pale and shaken, yet heard the echo: 'Your people die!'

"Then I descended into hell; I slipped away unchaperoned, unguarded, and in a Red Cross unit served a month in the fiery rain before I was discovered and courteously returned to England.

"Oh, Vishnu, Incarnate, thou knowest that I saw hell. Dirt and pain, blood and guts, murder and blasphemy, lechery and curses; from these, my eyes and ears were almost never free. For I was not serving officers now in soft retreats, I was toiling for 'niggers' at the front.

"Sick, pale, and shaken to my inmost soul, I was sent back to the English countryside. I was torn in sunder. Was this Europe? Was this civilization? Was this Christianity? I was stupefied—I—"

Shuddering, she drew Matthew's arms close about her and put her cheek beside his and shut her eyes.

Matthew began to talk, low-voiced and quickly, caressing her hair and kissing her closed eyes. The sun fell on the fiery land behind, and the waters darkened.

"We must go now, dearest," she said at length; "we have a long walk." And so they ate bread and milk and swung, singing low, toward the burning city. At Hyde Park she guided him west out toward the stockyard district. In a dilapidated street they stopped where lights showed dimly through dirty windows.

"This is the headquarters of the Box-Makers' Union," he said suddenly and stared at it as at a ghost.

"Will you come in with me?" she asked.

It was a poor, bare room, with benches, a table, and a low platform. Several dozen women and a few men, young and old, white, with a few black, stood about, talking excitedly. A quick blow of silence greeted their entrance; then a whisper, buzz, and clatter of sound.

They surged away and toward and around them. One woman—Matthew recognized the poor shapeless president—ran and threw her arms about Kautilya; but a group in the corner hissed low and swore. The Princess put her hand lovingly on the woman who stood with streaming eyes, and then walked quietly to the platform.

"I am no longer an official or even a member of the international union. I have resigned," she said simply in her low, beautiful voice. A snarl and a sigh answered her.

"I am sorry I had to do what I did. I have in a sense betrayed you and your cause, but I did not act selfishly, but for a greater cause. I hope you will forgive me. Sometime I know you will. I have worked hard for you. Now I go to work harder for you and all men." She paused, and her eyes sought Matthew where he stood, tall and dark, in the background, and she said again in a voice almost a whisper:

"I am going home. I am going to Kali. I am going to the Maharajah of Bwodpur!"

She walked slowly out, but paused to whisper to the president: "That bag—that little leather bag I asked you to keep —will you get it?"

"But you took it with you that—that night."

"Oh, did I? I forgot. I wonder where it is?" and Kautilya joined Matthew and they walked out.

Behind them the Box-Makers' Union sneered and sobbed.

IV

"I do not quite understand," said Matthew. "You have mentioned—twice—the Maharajah of Bwodpur. Did he not die?"

"The King is dead, long live the King! But do not interrupt —listen!"

They were sitting in his den on one side of a little table, facing the fire that glowed in the soft warmth of evening. They had had their benediction of music—the overture to *Wilhelm Tell,* which seemed to picture their lives. Together they hummed the sweet lilt of the music after the storm.

Before them was rice with a curry that Kautilya had made, and a shortcake of biscuit and early strawberries which Matthew had triumphantly concocted. With it, they drank black tea with thin slices of lemon.

"I think," said Kautilya, "that there was nothing in this century so beautiful as the exaltation of mankind in November, 1918. We all stood hand in hand on the mountain top, upon some vaster Everest. We were all brothers. We forgot the horror of that blood-choked interlude. I forgot even the front at Arras. I remember tearing like a maenad, cypress-crowned, through Piccadilly Circus, hand in hand with white strangers.

"I had just had an extraordinary conversation with an Englishman of highest rank. He had bowed over my hand.

" 'Your Highness,' he said, 'when the Emperor saw fit to urge your stay in England, he had hopes that your influence and high birth would do much to win this war for civilization.' I was thrilled. England! Actually to be necessary to this land of enjoyment and power! Perhaps to go back in triumph from this abode of Supermen! To help them win the war, and bring back, as reward, freedom for India!

"Long this member of the cabinet talked while my hostess and chaperon guarded us from interruption. We surveyed the policies and hopes and fears of India. One hour later as he kissed my hand, he whispered: 'Who knows! Your Highness

may take back an English Maharajah to share your throne!'
I looked at him in dumb astonishment; then slowly I saw
light. Long months I pondered over that hint.

"And when the Armistice actually came I had had a glorious vision. I was ready to forgive England and Europe. They
were but masses of shortsighted fallible men, like all of us.
We had all slept. Now the world was awake.

"There was no real line of birth or race or color. I loved
them all. The nightmare was ended. The world was free. The
world was sane. The world was good. The world was Peace.
For the first time in my life and the last, I was English; a
loyal subject of the Emperor then in Buckingham Palace—I
with a thousand years of royalty behind me. I saw New India,
a proud and free nation in the great free sisterhood of the
British Raj."

"Yes, yes, I understand," said Matthew. "There was a
moment then when I loved America. I cannot conceive it now."

"It was so natural that that which happened should happen
just then as I was exalted, blind with ideal fervor, and set
to see God and love everywhere. I saw the man first in Piccadilly on that night of nights. He was my knight in shining
armor: tall, spare, and fair, with cool gray eyes, his arm in
a sling, his khaki smooth and immaculate, his long limbs
golden-booted and silver-spurred. I turned and lighted the
cigarette for him when I saw him fumbling. Then I looked up
at him in startled wonder, unconsciously held out my hand
to him, and he kissed it gravely. I did not then dream that he
knew me and my station.

"We met weeks later and were presented at that country
estate down in Surrey where I had convalesced from my excursion to the front. It was so typically an English traditional
setting—so quiet, sweet, and green; so gracious, restful, and
comforting, cushioned for every curve and edge of body and
soul. Evil, poverty, cruelty seemed so far removed as to be
impossible—some far-off half-mythical giant and ogre about
which one could argue and smile and explain, while deft
servants and endless land and wealth made life a beautiful
and a perfect thing.

"He came down for a week-end. His arm had been ampu-
tated above the elbow and I was desperately sorry for this
maimed fellow, scarcely thirty-five, broken in his very morn-

ing of life. He was neither handsome, witty, nor really educated in any broad way; but his silence and self-repression, his stiff formality, his adherence to his social code, became him. One could imagine depth of thought, fire of emotion, power of command, all sealed and hidden in that fine body. He wooed me in the only way that I was then accessible—not by impetuous word or attempted array of learning, but by silent deference. He was always waiting; always bowing gravely; always rising to his feet and standing at attention with his poor maimed arm, and always insistently arranging my cushions and chairs with his lone hand.

"Then too, to complete the setting and push me by my own pride into what I might otherwise have paused before, there was the young Marchioness of Thorn. She was penniless, plain but stately; and, as every one knew and saw, hopelessly in love with my cavalier, Captain the Honorable Malcolm Fortescue-Dodd. As an earl's youngest son, Malcolm also had naught but his commission. Once I thought he loved her as she loved him. Then I decided not. Perhaps my decision was easier because of her evident dislike for me.

"At first I literally did not notice the Marchioness of Thorn. Then when I sought to atone and be gracious, I realized with astonishment that she was actually trying to be distant and patronizing with me! Patronizing, mind you, to a Maharanee of Bwodpur and Sindrabad! I was at first amused and then half angry, and finally, as guest of honor, I completely ignored this haughty lady and in sheer revenge annexed as my knight Captain the Honorable Malcolm.

"Even then I was startled when with scant delay he formally asked my hand in marriage. It was in a way a singular sort of innovation. Native Indian Princesses were recognized as reigning monarchs by England, but there had never been formal marital alliances, because it would have involved difficulties of rank and religion on both sides; and then, too, our princesses were usually married long before they saw England or knew Englishmen. In this case, however, a scion of ancient English nobility, albeit but a penniless and untitled younger son, was asking a reigning Indian princess in marriage. Should I—could I—accept? Was I lowering my rank? Was I helping or hindering India?

"A discreet emissary of the India Office came down and

discussed matters with me. It seemed that in the new world that was dawning, much of the old order was changing. Indian affairs must soon assume a new status. Should India emerge with new freedom and self-determination as a country entirely separate in race, religion, and politics from Mother England? Or as one allied by interest and even intermarriage? It was an astonishing argument, and—was it not natural?—I was flattered. I saw myself as the first princess of a new order, and while theoretically I held myself the equal of British royalty itself and certainly would have preferred a duke or marquis or even an earl in his own right, yet—and even this was discreetly hinted—earldoms and marquisates were often created for loyal and ambitious servants of the state.

"This very intention again made my head go up in pride. Why should a Maharanee of Bwodpur stoop to English strawberry leaves? I would lift him to my own royal throne, if I so wished. Did I wish it? I felt strangely alone, far from my people and their advice. What would my counselors think? Would they be gratified or alarmed, uplifted or estranged? And then again, was this a high affair of state or a triumph of romantic love? I did not know.

"Yet I was curiously drawn to this tall, silent soldier, with his maimed arm and cold, gray eyes. If only I could draw a light of yearning and passion into those eyes it might bring the answering lightning from my heart and let me, the princess, know such love as peasants only can afford! And so I hesitated and then finally when, through the India Office, the formal assent of my family was handed me, I consented. Formal announcement of the engagement was gazetted and became a nine days' wonder; at Haslemere, some of the great names of England, including British royalty itself, gathered at my betrothal ball.

"I was quite happy. Happy at the gracious reception of my royal blood into the noble blood of England; happy at my consciousness of power. I stood, with my English maidens in attendance, and looked across the ballroom floor—beautiful women, flashing uniforms, stately personages, soft-footed servants; the low hum of word and laughter, the lilt of music.

"Suddenly tears rose in my throat. I was happy, of course, but I wanted love. I had been repressed and cool and haughty toward this wounded man of my choice. I was suddenly yearn-

ing to let my naked heart look unveiled into his eyes and see
if I would flame and his tense cold face kindle in reply. Where
was he? I searched the hall with my glance. He had been
beside me but a quarter of an hour since. A mischievous-eyed
young maiden of my train blushed, smiled, and nodded. I
smiled an answer and turned. There was a draped passage to
the supper room behind us, and looming at the end was that
easily recognized form. I waved my maidens back, and turning, entered noiselessly. I wanted to be alone with love for one
moment, if perchance love were there.

"He was talking to some one I could not see. I stepped
forward and his voice held me motionless.

"It was the Marchioness of Thorn. I froze. I could not move.
His voice came low and tense, with much more feeling in it
than I had ever heard before:

"'What else is there for me, a poor and crippled younger
son? Can you not see, dearest, that this is a command on the
field of battle? Think what it means to have this powerful
buffer state, which we nearly lost, in the hands of a white
English ruler; a wall against Bolshevik Russia, a club for
chaotic China; a pledge for future and wider empire.'

"'But you'll only be her consort.'

"'I shall be Maharajah in my own right. The India Office
has seen to that. I can even divorce her if I will, and I can
name my own successor. Depend upon it, he'll be white.'

"Then came the answering voice, almost shrill:

"'Malcolm, I can't bear the thought of your mating with a
nigger.'

"'Hell! I'm mating with a throne and a fortune. The
darky's a mere makeweight.'

"In those words I died and lived again. The world crashed
about me, but I walked through it; turning, I beckoned my
maidens, who came streaming behind.

"'Malcolm, this is our waltz,' I said as I came into the
light. He stood at attention, and the Marchioness, bowing
slightly, began talking to the women, as we two glided away.
I went through ball and supper, speeded my guests, and let
the Captain kiss my hand in farewell. He paused and lingered
a bit over it and came as near looking perturbed as I ever saw
him; he was not sure how much I had overheard; but I bit
blood from my lips and looked at him serenely. The next day

I left for London and India to prepare for the intra-Imperial and inter-racial wedding."

Matthew and Kautilya had long been walking through the night lights of the crowded streets downtown, hand in hand as she talked. Now she paused and at Michigan and Van Buren they stood awhile shoulder to shoulder, letting the length of their bodies touch lightly. As they waited a chance to cross Michigan, a car snorted and sought to slip by, then came to a wheezy halt.

"Well, well, well!" said the Honorable Sammy, holding out a fat hand and eyeing them quizzically. They greeted him with a smile.

"Say, can't we have a talk?" he asked finally.

"Sure," said Matthew. "Come to my den."

Sammy could not keep his eyes off Kautilya, although there was frank puzzlement in them rather than his usual bold banter. They rode north rapidly in his car, seated together in the rear with close clasped hands. Once at home, Kautilya made Sammy silently welcome and said little. She arranged the small table as Matthew lighted the fire, warmed up a bit of the curry, and brought out a decanter of dark, old crimson wine.

The Honorable Sammy gurgled and expanded.

"What ya gonna do?" he asked. "Gee, this stuff's great—what is it?"

"Indian curry.—We don't know yet."

"Want a job?"

"No," said Matthew slowly, and Kautilya walked over to him softly and slipped an arm about his shoulder.

"Can't coo on air," said Sammy with some difficulty, his mouth being pretty full. "See here! 'Course you and Sara couldn't make it. I never expected you to. She's—well, you're different. Now suppose you just get a divorce. My friend, the judge, will fix it up in a month, and then I can hand you a little job that will help with the bread and butter."

"She can have the divorce," said Matthew.

"But," said Sammy, "you get it, and get it first." Matthew did not answer.

"You see," explained Sammy elaborately, "Sara's funny. Just now she's filled full with hating your lady. She thinks it will hurt you worse to keep you married to her. She thinks you'll tire of this dame and perhaps then come crawling back, so she can kick you good and plenty. See? Now if you begin action for divorce first, for—ah—cruelty—incompatibility—that goes in Illinois—why, she'd fight back like a tiger and divorce you for adultery. See?"

There was an awkward silence. Then Matthew ventured: "And you, Sammy. I hope you are going to Congress?"

Sammy scowled and shoved his plate back.

"No—not this year. You sure mussed that up all right. But wait till we put Bill Thompson back as mayor. Then we'll shuffle again and see."

"I'm sorry," said Matthew.

"Oh, it's all right. 'Course Sara is sore—damned sore and skittish. But it's all right. You just push that divorce and we'll stand together, see?"

Sammy arose, pulled down his cuffs, straightened his tie, and lit a new, long, black cigar.

"Well—so long!" he said, teetering a moment on heel and toe. Then he leered archly at Kautilya, winked at Matthew, and was gone.

For a minute the two stood silently gripping each other close and saying no word. It was as though some evil wind from out the depths of nowhere had chilled their bones.

v

"I want to sit in a deep forest," said Kautilya, "and feel the rain on my face." So they went furtively and separately down the long lanes of men, stepping softly as those who would escape wild beasts in a wilderness. They met in the Art Gallery beside the lake and walked here and there like strangers, and yet happily and deliciously conscious of each other. At last by elaborate accident they sat down together before a great red dream of sun and sky and air and rolling, tossing waters.

Then they went out and climbed on a bus and happened in the same seat and rode wordlessly north. At Evanston they took the electric train and fared further north. Kautilya slipped off her skirt and was in knickerbockers. Matthew slung his knapsack and blankets on his shoulders. The gray clouds rolled in dark arrows on the lake, and at last they sat alone in the dim forest, huddled beneath a mighty elm, and the rain drifted into their faces. They spent the night under the scowling sky with music of soft waters in their ears. At midnight Kautilya turned and nestled and spoke:

"I stopped in London on my way to India, ostensibly for last-minute shopping, but in reality to explore a new world. In that week in the trenches I had met a new India—fierce, young, insurgent souls irreverent toward royalty and white Europe, preaching independence and self-rule for India. They affronted and scared and yet attracted me. They were different from the Indians I knew and more in some respects like the young Europeans I had learned to know. Yet they were never European. I sensed in them revolution—the change long due in Asia. I had one or two addresses, and in London I sought out some of the men whom I had nursed and helped for a month. They knew nothing of my rank and history. They received me gladly as a comrade and assumed my sympathy and knowledge of their revolutionary propaganda. Ten days I went to school to them and emerged transformed. I was not converted, but my eyes and ears were open.

"I was nineteen when I returned to India and found the arrangements for my English wedding far advanced. My people were troubled and silent. The land was brooding. Only the English were busy and blithe. New native regiments appeared with native line officers. New fortifications, new cities, new taxes were planned. New cheap English goods were pouring in, and the looms and hands of the native workers were idle. The trail of death, leading from the far World War, marched through the land and into China, and thence came the noise of upheaval, while from Russia came secret messages and emissaries.

"The four years of my absence had been years of change and turmoil; years when this native buffer state, breasted against Russia and China and in the path of the projected new English empire in Thibet and secured to English power by the marriage of two children, maimed dolls in the thin white hand of the commissioner, was seething with intrigue.

"My own people were split into factions and divided counsels. After all I was a woman, and in strict law a widow. As such I had no rights of succession. On the other hand, I was the last of a long and royal line. I was the only obstacle between native rule and absorption by England in Sindrabad, and the only hope of independence in Bwodpur. I was the foremost living symbol of home rule in all India. The struggle shook the foundations of our politics and religion, but finally, contrary to all precedent, I had been secretly confirmed as reigning Maharanee after the death of my father. Everything now depended on my marriage, which the most reactionary of my subjects saw was inevitable if my twelve million subjects were to maintain their independence against England.

"Immediately I was the center of fierce struggle: England determined to marry me to an English nobleman; young India determined to rally around me, to strip me of wealth, power, and prerogatives, and to set up here in India the first independent state.

"My phantom prince, poor puppet in the hands of England, I soon saw had probably been murdered by the Indian fanatics of *Swaraj*, whom then I hated, although I realized that perhaps Englishmen with ulterior motives had egged them on. Two suitors for my hand and power came forward—a fierce and ugly old rajah from the hills who represented the Indians' determination for self-rule under the form of monarchy, and a handsome devil from the lowlands, tool and ape of England: I hated them both. I could see why in desperation my family had consented to my marriage with Fortescue-Dodd.

"I looked about me and realized my wealth and power from my twelve million subjects and from the pathway of my kingdom between India and China. Widowed even before I was a wife, bearing all the Indian contempt for widowhood; child with the heavy burden of womanhood and royal power, I was like to be torn in two not only by the rising determination of young India to be free of Europe and all hereditary power, but also by the equal determination of England to keep and guard her Indian empire.

"I looked on India with new and frightened eyes. I saw degradation in the cringing of the people, starvation and poverty in my own jewels and wealth, tyranny and ignorance in the absolute rule of my fathers, harsh dogmatism in the transformed word of the great and gentle Buddha and the eternal revelation of Brahma."

"But," cried Matthew, "was there no one to guide and advise this poor child of nineteen?"

"Not at first. My natural advisers were fighting against those who threatened my throne, and young India alone was fighting England. I called my family in counsel. Boldly I took the side of young India against England and called the young educated Indians together, many of them cousins and kinsmen, and offered the weight of my wealth and power to forward their aims. The result was miraculous. Some of my old and reactionary kinsmen stood apart, but they did not actively oppose us. Some very few of the most radical of the advocates of *Swaraj* refused to coöperate with royalty on any terms. But I gathered a great bloc of young trained men and women. Long we planned and contrived and finally with united strength turned on England.

"My own mind was clear. I was to be the visible symbol of the power of New India. With my new council I would rule until such time as I married a prince of royal blood and set my son on the throne as Maharajah of Bwodpur. But I postponed marriage. I wanted light. I wanted to hear what other dark peoples were doing and thinking beneath the dead, white light of European tyranny.

"I called a secret council of the Durbar and laid my plans before them. The splendid wedding ceremony of the proposed English alliance approached. The bridegroom and a host of officials arrived, and from the hills arrived too that ancient and ugly Rajah who was old when he sought my hand in vain seven years before and now had grandchildren older than I.

"The hosts assembled, the ceremony gorgeous in gold and ivory and jewels began: the elephants, painted and caparisoned, marching with slow, sedate, and mighty tread; the old high chariots of the rajahs, with huge wheels and marvelous gilding, drawn by great oxen; the curtained palanquins of the women; the clash of horns and drums and high treble of flutes.

"Then at the height and culmination of the ceremony and before the world of all India and in the face of its conquerors, I took my revenge on the man and nation that had dared to insult a Maharanee of Bwodpur. As Captain the Honorable Malcolm Fortescue-Dodd kneeled in silver and white to kiss my hand, the ancient Rajah from the hills stepped forward and interposed. As the eldest representative of my far-flung family, he announced that this marriage could not be. A plenary council of the chief royal families of India had been held, and it had been decided that it was beneath the dignity of India to accept as consort for a princess of the blood a man without rank or title—unless, he added, 'this alliance was by the will and command of the Maharanee herself.'

"All the world turned toward me and listened as I answered that this marriage was neither of my will nor wish but at the command of my family. Since that command was withdrawn—

"'I do not wish to marry Captain the Honorable Malcolm Fortescue-Dodd.'

"England and English India roared at the insult. There were a hundred conjectures, reasons, explanations, and then sudden silence. After all it was no time for England to take the high hand in India. So it was merely whispered in select circles that the family of Fortescue-Dodd had decided that the women of India were not fit consorts for Englishmen and that they had therefore allowed me gracefully to withdraw. But we of India knew that England was doubly determined to crush Bwodpur.

"Four years went by. Although ruling in my own right, I made that ancient Rajah my guardian and regent and thus put behind my throne all the tradition of old India. Meantime with a growing council of young, enthusiastic followers I began to transform my kingdoms. We mitigated the power of the castes and brought Bwodpur and Sindrabad nearer together. We contrived to spend the major part of the income of the state for the public welfare instead of on ourselves, as was our ancient usage. We began to establish public schools and to send scholars to foreign lands.

"Only in religion and industry were my hands tied—in religion by my own people; in industry by England. We had Hindus and Mohammedans, Buddhists of every shade, and a few more or less sincere Christians. I wanted to clean the slate and go back to the ancient simplicity of Brahma. But, ah! Who can attack the strongholds of superstition and faith!"

"Who indeed!" sighed Matthew. "Our only refuge in America is to stop going to church."

"The church comes to us in India and seizes us. I could only invoke a truce of God to make Allah and Brahma and Buddha sit together in peace, to respect each other as equals.

"In industry my hands were tied by the English power to sell machine goods and drive our artisans from the markets. In vain I joined Mahatma Gandhi and tried to force the boycott over my land. My people were too poor and ignorant. Yet slowly we advanced and there came to us visitors from Egypt, Japan, China, and at last from Russia down across that old and secret highway of the Himalayas, hidden from the world.

"Sitting there in the white shadow of Gaurisankar we conferred with young advanced thinkers of all nations and old upholders of Indian faith and tradition. We conceived a new Empire of India, a new vast union of the darker peoples of the world.

"To further this I started on the Grand Tour of the Darker Worlds. I went secretly by way of Thibet and New China; saw Sun Yat-sen in Pekin. I was three months in Japan, where the firm foundation of our organization of the darker peoples was laid. Then I spent three months in Russia, watching that astonishing experiment in a land which had suffered from tyranny beyond conception. I tried to learn its plans, and I received every assurance of its sympathy. Down by Kiev I came to Odessa and sailed the Black Sea.

"I saw the towers of Constantinople shining in the sun and stood in that great center where once Asia poured the light of her culture into the barbarism of Europe and made it a living soul. I walked around those mighty walls, where Theodosius held back the Nordic and the Hun. I went by old Skutari and its vast city of the dead; down by a slow and winding railway, three hundred and fifty miles westward to Angora. There I sat at the feet of Kemal and heard his plans. Thence overland by slow and devious ways I came through Asia Minor and Syria. Down by the Kizilirmak and the great blue waters of the Tuz Tcholli Gol; over to Kaisapieh and through the dark passes of the Anti-Taurus; then skirting the shining Mediterranean, I saw French Syria at Aleppo, Hamah, and Damascus; I saw Zion and the new Jerusalem and came into the ancient valley of the Nile and into the narrow winding streets of Cairo."

"You have seen the world, Kautilya, the real and darker world. The world that was and is to be."

"It was a mighty revelation, and it culminated fittingly in Egypt, where in a great hall of the old university hung with rugs to keep out both the eavesdropper and the light, the first great congress of the darker nations met under the presidency of Zahglul Pasha. We had all gathered slowly and unobtrusively as tourists, business men, religious leaders, students, and beggars, and we met unnoticed in a city where color of skin is nothing to comment on and where strangers are all too common. We were a thousand strong, and never were Asia, Africa, and the islands represented by stronger, more experienced, and more intelligent men.

"Your people were there, Matthew, but they did not come as Negroes. There were black men who were Egyptians; there were black men who were Turks; there were black men who were Indians, but there were no black men who represented purely and simply the black race and Africa.

"Of all the things we did and planned and said in a series of meetings, I will tell you in other days. Let it suffice now to say that I came back to Europe by Naples and Paris and then went to Berlin. There I sat and planned with a small special committee, and there it was that I brought up the question of American Negroes, of whom I had heard much in Russia. The committee was almost unanimously opposed. They thought of Negroes only as slaves and half-men, and were afraid to risk their coöperation, lest they lose their own dignity and place; but they were not unwilling to let American Negroes, if they would, start some agitation or overt act. Even if it amounted to nothing, as they expected, it would at least focus attention. It would intensify feeling. It would help the coming crisis.

"But who could do this?

"The curious and beautiful accident of our meeting, after my committee had discussed and rejected the Negroes of America as little more than slaves, deeply impressed me. And in the face of strong advice, as you know, I helped you to return to America and report to me on the rumored uprising which had been revealed to me by curious and roundabout ways.

"I was not thinking of you then, Matthew, at least not consciously. I was thinking of the great Cause and I wanted information. I looked at America and tried to understand it. There was here a mystery of the art of living that the world must have in order to have time for life. I saw America and lost you. Almost, in the new intensity of my thinking, I forgot you as a physical fact. You remained only as a spirit which I recognized as part of me and part of the universe. And then suddenly the blow came, falling through open skies, and I saw you facing disgrace and death and locked for ten years in jail.

"Before I saw you, I, with most of the others except the Chinese, had thought of our goal as a substitution of the rule of dark men in the world for the rule of white, because the colored peoples were the noblest and best bred. But you said one word that night at dinner."

"I did not say it—it was said. I opened my mouth and it was filled."

"You remember it! It was a great word that swung back the doors of a world to me. You said that the masses of men of all races might be the best of men simply imprisoned by poverty and ignorance.

"It came to me like a great flash of new light, and you, the son of slaves, were its wonderful revelation. I determined to go to America, to study and see. I began to feel that my dream of the world based on the domination of an ancient royal race and blood might not be all right, but that as Lord Buddha said, and as we do not yet understand, humanity itself was royal.

"Then things happened so rapidly that I lost my grip and balance and sense of right and wrong. I sent you on a wild chase to almost certain death. I planned to go with you to watch and see. The secret, powerful hand of the junta sought to threaten us both and save the great cause. How singularly we fought at cross purposes! They wanted you to go and stir up any kind of wild revolt, but they wanted to keep me and themselves from any possible connection with it in thought and deed. They almost threatened you with death. They pushed you out alone. They tried to keep me from sailing. And finally you went down into the depths, dear heart, almost to the far end." Her voice fell away, and they lay and watched the birth of the new and sun-kissed day.

All that day they wandered and talked and finally late at

night came home. Kautilya was almost ready for bed when she said drowsily:

"Oh, Matthew—the little leather bag I brought—where did you put it, dear?"

"Leather bag? I saw none."

"But it was not at the union headquarters. They said I took it with me to—to Sara's."

"Then you must have left it there. We carried nothing away. Nothing."

"Oh, dear—I must have left it—what shall I do?"

"Was it valuable or just clothes?"

"It was—valuable. Very valuable, intrinsically and—in meaning."

"I am so sorry—may I ask—?"

"Yes—it has many of the crown jewels of Bwodpur."

"The crown jewels!"

"Yes. Some of them always travel with the heir to the throne. I have carried these since father's death. Some of the jewels are beautiful and priceless. Others, like the great ruby, are full of legends and superstitious memory. The great ruby is by legend a drop of Buddha's blood. It anoints the new-born Maharajah. It is worn on his turban. It closes his eyes in death."

"Oh, Kautilya, Kautilya! We must find these things—I will go to Sara's myself. What do you think them worth—I mean would they be worth stealing?"

"Oh, yes, they must be worth at least a hundred lakhs of rupees."

Matthew paused, then started up.

"What—you mean—you don't mean—a million dollars?"

"At least that—but don't be alarmed. They are mostly too large and unusual for sale. They are insured and I have a description. Probably the bag is sitting somewhere unnoticed. Oh, I am so careless; but don't worry. Let me write a note and call a messenger. I have faithful helpers. The bag will soon be found."

The note was dispatched, and Kautilya was soon making a mysterious Indian dish for supper and singing softly.

Matthew was still thinking with astonishment, "The crown jewels of Bwodpur—a million dollars!"

VI

Sammy was uneasy. He had a telegram from Sara announcing her sudden return from New York. She was arriving in the morning. But there was no letter in answer to several urgent ones from Sammy, a bit misspelled and messy, but to the point. He had suggested among other things that Sara remain east until September.

Sara, after the tragic failure of her long-laid plans, had taken a trip to New York. She put on her best clothes and took plenty of funds. She wired to the Plaza for a suite of rooms— a sitting-room, bedroom, and bath. She arrived in the morning of April 10 on the Twentieth Century, had a good lunch, and went to a dressmaker whose name and ability she had learned. She ordered a half-dozen new gowns. She secured, at the hotel, orchestra seats for two good shows—Ziegfeld's Follies and a revue at the Winter Garden, and she also got a seat for *The Jewels of the Madonna* at the Metropolitan. She hired a car with a liveried chauffeur and drove through the park and down the avenue to Washington Square and back to the Plaza and had tea there; she took a walk, went to the Capitol, and dined at the Ritz.

For four or five days Sara tried the joys of free spending and costly amusement. She was desperately lonesome. Then she struck up acquaintance with a lady and her husband whom she met at the Plaza by the accident of sitting at the same table. They were from Texas. Sara was a bit dismayed, but did not flinch. They were as lonesome and distraught as she and grabbed like her at the novelty of a new voice. They played together at theater and dinner, rides to Westchester and Long Island, and at night they went to Texas Guinan's club, accompanied by an extra man whom the husband had picked up somewhere. Sara was sleepy and bored, and the drinks which she tasted made her sick. Her escort when sober danced indifferently and was quite impossible as he got gradually drunk.

Next morning Sara arose late with a headache, reserved a berth to Chicago, and wired Sammy:

"Arrive tomorrow morning at nine."

Sammy had not been expecting this. In fact he had made up his mind that she would be away at least three months and was laying his plans. This sudden turn upset him. He looked about the office helplessly. When the Fall campaign began, he would want Sara back in harness; but he was not ready for her now. First of all, that damned Towns had made no move toward a divorce. There were his belongings which Sara had bundled up hastily and sent to him when she left. They were in the corner of his office now, and Sammy rose and aimlessly looked them over. There was a bundle of clothes, two boxes of books, and two bags. What had Sara written about these bags? Yes, here was the note.

"This smaller bag is not his and doesn't belong in the house. It was sitting in the library. It may belong to some of the guests or to that woman. If it is inquired for, return it. If not, throw it away."

Sammy lifted it. It seemed rather solid. He picked it up and examined it. It was of solid thick leather and tarnished metal, which looked like silver. It was securely locked. There was a small crest stamped on the silver. Yes, it undoubtedly belonged to the Princess. It would be an excuse for another visit to her.

Then Sammy sat down, eyeing the bag idly, and returned to his thoughts. Neither Sara nor Matthew had made the slightest movement toward a divorce. Now it was Sammy's pet idea that Sara should not begin proceedings. He wanted her to pose for some time as the injured victim. He wanted Towns to kill himself beyond redemption by not only deserting Sara but brazenly seeking legal separation. Now that neither made a move Sammy got uneasy. What was the big idea? Was Sara going to hold on to him because she wanted him back or just to thwart the other woman? Did Matthew want his freedom, or was he playing around and ready to return to Sara later? Sammy was stumped. He had spoken to Matthew before the lady, and yet Matthew had neither answered nor taken any steps. Didn't the woman want Matthew divorced?

Then Sammy looked at the bag again. Queer woman—queer bag. Didn't look or feel like a toilet case. No—contents weren't soft enough for clothing. Well—he must get rid of this junk and clean up his office and Sara's and get ready for her tomorrow. Then Sammy looked at that bag again. What was this "Princess," anyhow? What was her game? Here was a chance to find out. He tried to open the bag. It was securely locked. The lock was very curious and was probably a combination and not a key lock, in spite of certain holes. Sammy again felt carefully of the contents—shook the bag, turned it around and upside down. Then suddenly he shut and locked the door and drew the curtain and took out his knife. He attempted to slit the leather. It was very heavy, and once cut, after considerable difficulty, it revealed a fine steel mesh below. Sammy was aroused and beset with curiosity. He got a wire ripper and soon had a hole about two inches long. Through this he drew a small Russian leather box fastened with a gold or gilded clasp. He opened this and found a dozen or more large transparent unset stones that looked like diamonds.

Sammy began to perspire. Then he wiped the sweat from his brow and sat down to think. He examined his own diamond ring. These stones certainly looked genuine. They scratched the window glass. But—it couldn't be! If these were diamonds they'd be worth— Hell! Sammy took out one, closed the box, and inserted it in the bag. He closed the aperture carefully and started with it to the safe. No, suppose Sara asked for it! No, he turned it around and set it carelessly and in full sight in the corner. Then he unlocked the office door and 'phoned Corruthers.

"Say," he said when Corruthers appeared, "take this to Ben and see if it's worth anything."

Corruthers ran his fingers through his red hair.

"Phony," he declared. "Who stuck you with it?"

"Shut up," said Sammy, "and ask Ben and don't try no monkey business neither."

Corruthers was back in a half hour.

"Say," he began excitedly. "Where'd you get this—"

Sammy interrupted. "Send them clothes and books to Towns."

"Sure—but—"

"What's the stone worth?"

"Five thousand dollars."

Sammy bit his cigar in two but managed to keep from swallowing the stub and dropping the end—

"Oh—er—that all?"

"Well—you might get more if you could prove ownership. He says it's an unusual stone. How—"

" 'Tain't mine," said Sammy. "Probably stolen. A bird

wanted to sell it, but I don't know—" and he shooed Corruthers out.

Five thousand! And one of a dozen! And that bag. Again Sammy locked up carefully, drew the shades, and turned on the electric lights. Then he brought the bag to the desk and with a knife and improvised tools, tore it entirely open. There were a half-dozen boxes, several paper bundles, and two or three chamois bags. He spread the contents out on the desk and literally gasped. Such jewels he had never seen. Not only smaller uncut diamonds in profusion, but several large stones in intricate settings, beautiful emeralds, two or three bags of lovely matched pearls, and above all, a great crimson ruby that looked like a huge drop of blood.

Sammy gasped, sat down, stood up, whistled, and whirled about; and whirling, faced, sitting quietly in his own chair, a person who seemed at first an utter stranger. Then Sammy recognized him as the Indian with whom he had had several conferences during the campaign and whom he had met together with the young radical Negro down at the radical Box-Makers' Union.

Sammy suddenly grew furious.

"How the hell—" he began; but the Indian interrupted suavely.

"Through the window there," he said. "You pulled the shade down, but you didn't lock the window. I have been watching there several days."

"Well, by God," and Sammy half turned toward the desk; but the Indian still spoke very quietly.

"I wouldn't if I were you," he said.

Sammy didn't. On the other hand he sat down in another chair and faced the Indian.

"Well, what about it?" he said.

"These jewels," said the Indian, "are, as I presume you suspect, the property of her Royal Highness, the Princess of Bwodpur. In fact they are part of the crown jewels which always accompany the heir to the throne wherever he or she goes. Her Royal Highness is unfortunately very careless. She had the jewels with her when she started to interview Mr. Towns that night, and in the turmoil of the evening, evidently forgot them. Yesterday she sent me a note asking that I find them. I went to the residence of Mrs. Towns and found it locked on account of her absence, but I secured entrance."

"That kind of thing sometimes lands people in jail," said Sammy dryly.

"Yes," said the Indian, "and the theft of jewels like these might land one further in jail and for a longer time."

Sammy didn't answer, and the Indian continued: "I searched the house and was satisfied that the bag was not there, and then I learned that certain things had been delivered at your office. I came down here and saw the bag sitting here. That was early yesterday morning, while the janitor was sweeping."

"Damn him!" said Sammy.

"It wasn't his fault," said the Indian. "I forget what excuse I gave him, but you may be sure it was a legitimate one. Yesterday and today I have spent watching you to be sure of your attitude."

"Well?" said Sammy.

"Well," returned the Indian, "I had hoped that the proof which I have would secure the bag, untampered with and without question or delay."

"What proof?" asked Sammy.

"A careful description of the jewels made by the well-known firm which has insured them and which would at the slightest notice put detectives on their track. Also, a letter from her Royal Highness directing that these jewels be delivered to me."

"And you expect to get these on such trumped-up evidence?"

"Yes," said the Indian.

"And suppose I refuse?"

"I shall persuade you not to."

Sammy thought the matter over. "Say," said he, "can't you and I come to some agreement? Why, here is a fortune. Is there any use wasting it on Matthew and that Princess?"

"We can come to an agreement," said the Indian.

"What?" asked Sammy.

"You have," said the Indian, "an unset diamond in your pocket which, with a certificate of ownership that I could give you, would easily be worth ten thousand dollars. You may keep it."

Sammy rose in a rage. "I can not only keep that," he said, "but I can keep the whole damn shooting-match and—" But he didn't get any further. The Indian had arisen and showed in the folds of his half-Oriental dress a long, wicked-looking dagger.

"I should regret," he said, "the use of violence, but her Royal Highness' orders are peremptory. She would rather avoid, if possible, the police. I am therefore going to take these jewels to her. If afterward you should wish to prosecute her, you can easily do it."

Sammy quickly came to his senses: "Go ahead," he said.

The Indian deftly packed the jewels, always managing to face Sammy in the process. Finally, with a very polite good night, he started to the door.

"Say," said Sammy, "where are you going to take those jewels?"

"I have orders," said the Indian slowly, "to place them in the hands of Matthew Towns."

The door closed softly after him. Sammy seated himself and thought the matter over. He had a very beautiful diamond in his pocket which he examined with interest. His own feeling was that it would make a very splendid engagement ring for Sara. Then he started. . . . Suppose these jewels were given to Matthew, or part of them, and suppose Sara got wind of it? Would she ever give Matthew up? That was a serious matter—a very serious matter. In fact, she must *not* get wind of it. Then Sammy frowned. Good Lord! He had actually had his hands on something that looked like at least one million dollars. Ah, well! It was dangerous business. Only fools stole jewels of that sort.

A messenger boy entered with a telegram.

"Have decided to go to Atlantic City. Do not expect me until I write.

"SARA."

VII

"I've got a job," said Matthew, early in June.

Kautilya turned quickly and looked at him with something of apprehension in her gaze. It was a beautiful day. Kautilya had been arranging and cleaning, singing and smiling to herself, and then stopping suddenly and standing with upturned face as though listening to inner or far-off voices. Matthew had been gone all the morning and now returned laden with bundles and with a sheaf of long-stemmed roses, red and white, which Kautilya seized with a low cry and began to drape like cloud and sun upon the table.

Then she hurried to the phonograph and put a record on, singing with its full voice—a flare of strange music, haunting, alluring, loving. It poured out of the room, and Matthew joined in, and their blended voices dropped on the weary, dirty street. The tired stopped and listened. The children danced. Then at last:

"I've got a job," said Matthew; and answering her look and silence with a caress, he added: "I got it myself—it's just the work of a common laborer. I'm going to dig in the new subway. I shall get four dollars a day."

"I am glad," said Kautilya. "Tell me all about it."

"There is not much. I've noticed the ads and today I went out and applied. There was one of Sammy's gang there. He said I wouldn't like this—that he could get me on as foreman or timekeeper. I told him I wanted to dig."

"To dig, that's it," said Kautilya. "To get down to reality, Matthew. For us now, life begins. Come, my man, we have played and, oh! such sweet and beautiful play. Now the time of work dawns. We must go about our Father's business. Let's talk about it. Let's stand upon the peaks again where once we stood and survey the kingdoms of this world and plot our way and plan our conquests. Oh, Matthew, Matthew, we are rulers and masters! We start to dig, remaking the world. Too long, too long we have stood motionless in darkness and dross. Up! To the work, in air and sun and heaven. How is our world, and when and where?"

They sat down to a simple lunch and Matthew talked.

"We must dig it out with my shovel and your quick wit. Here in America black folk must help overthrow the rule of the rich by distributing wealth more evenly first among themselves and then in alliance with white labor, to establish democratic control of industry. During the process they must keep step and hold tight hands with the other struggling darker peoples."

"Difficult—difficult," mused Kautilya, "for the others have so different a path. In my India, for instance, we must first emancipate ourselves from the subtle and paralyzing misleading of England—which divides our forces, bribes our brains, emphasizes our jealousies, encourages our weaknesses. Then we must learn to rule ourselves politically and to organize our old industry on new modern lines for two objects: our own social uplift and our own defense against Europe and America. Otherwise, Europe and America will continue to enslave us. Can we accomplish this double end in one movement?"

"It is paradoxical, but it must be done," said Matthew. "Our hope lies in the growing multiplicity and world-wide push of movements like ours; the new dark will to self-assertion. China must achieve united and independent nationhood; Japan must stop aping the West and North and throw her lot definitely with the East and South. Egypt must stop looking north for prestige and tourists' tips and look south toward the black Sudan, Uganda, Kenya, and South Africa for a new economic synthesis of the tropics."

"And meantime, Matthew, our very hope of breaking the sinister and fatal power of Europe lies in Europe itself: in its own drear disaster; in negative jealousies, hatreds, and memories; in the positive power of revolutionary Russia, in German Socialism, in French radicalism and English labor. The Power and Will is in the world today. Unending pressure, steadying pull, blow on blow, and the great axis of the world quest will turn from Wealth to Men."

"The mission of the darker peoples, my Kautilya, of black and brown and yellow, is to raise out of their pain, slavery, and humiliation, a beacon to guide manhood to health and happiness and life and away from the morass of hate, poverty, crime, sickness, monopoly, and the mass-murder called war."

Kautilya sat with glowing eyes. She looked at Matthew and whispered:

"Day dawns. We must—start."

Matthew hesitated and faltered. He talked like one exploring the dark:

"I had thought I might dig here in Chicago and that you might write and study, and that we together might live far out somewhere alone with trees and stars and carry on—correspondence with the world; and perhaps—"

"If we only could," she said softly; and in the instant both knew it was an impossible idyl.

A week went by. There grew a certain stillness and apprehension in the air. The hard heat of July was settling on

Chicago. Each morning Matthew put on his overalls and took his dinner-pail and went down into the earth to dig. Each night he came home, bathed, and put on the gorgeous dressing-gown Kautilya had bought him and sat down to the dinner Kautilya had cooked—it was always good, but simple, and he ate enormously. Then there was music, a late stroll beneath the stars, and bed. But always in Kautilya's eyes, the rapt look burned.

And little things were beginning to happen. At first Matthew's old popularity in his district had protected him. He always met nods and greetings as he and Kautilya fared forth and back. Then came reaction—the social tribute of the half-submerged to standards of respectability. Here and there a woman sneered, a child yelled, and a policeman was gruff. As weeks went by, Sammy interfered, and active hostility was evident. Jibes multiplied from chance passers-by who recognized them; the sneers of policemen were open. Then came the question of money, which never occurred to Kautilya, but drove Matthew mad when he tried to stretch his meager wage beyond the simple food to American Beauty roses and new books and bits of silk and gold.

Tonight as they returned from a silent but sweet stroll a bit earlier than their wont, they met a crowd of children, those children who seem never to have a bedtime. The children stared, laughed, jeered, and then stoned them. Matthew would have rushed upon them to tear their flesh, but Kautilya soothed him, and they came breathless home.

They stood awhile clinging in the dark. Then slowly Matthew took her shoulders in his hand and said:

"We have had the day God owed us, Kautilya, and now at last we must face facts frankly. Here and in this way I cannot protect you, I cannot support you, and neither of us can do the great work which is our dream."

"It was brave and good of you, Matthew, to speak first when you knew how hard the duty was to me and how weak I am in presence of our love. Yes, we have had the day God owed us—and now, Matthew, the day of our parting dawns. I am going away."

He knew that she was going to say this, and yet until it was said he kept trying to believe it would not come. Now when it did come it struck upon his ears like doom. The brownness of his face went gray, and his cheeks sagged with sudden age. He looked at her with stricken eyes, and she, sobbing, smiled at him through tears.

"You are a brave man," she went on steadily. "In this last great deed, I will not fear. I go to greet the ghosts of all my fathers, the Maharajahs of Bwodpur. India calls. The black world summons. I must be about my fathers' business. To-morrow, I go. This night, this beautiful night, is ours. Behold the good, sweet moon and the white dripping of the stars. There shall be no fire tonight, save in our twined bodies and in our flaming hearts."

But he only whispered, "Parted!"

"Courage, my darling," she said. "Nothing, not even the high Majesty of Death, shall part us for a moment. There is a sense—a beautiful meaning—in which we two can never part. To all time, we are one wedded soul. The day may dawn when in cries and tears of joy we shall feel each other's arms again. But we must not deceive ourselves. It is possible that now and from this incarnate spirit we part forever. Great currents and waves of forces are rolling down between. It may not lie in human power to breast them. But, Matthew, oh, Matthew, always, always wherever I am and no matter how dark and drear the silence—always I am with you and by your side."

Matthew was calm again and spoke slowly: "No, we will not deceive each other," he said; "I know as well as you that we must part. I am ready for the sacrifice, Kautilya. There was a time when I did not know the meaning of sacrifice, when I interpreted it as surrendering an ounce to get a pound, ex-changing a sunrise for a summer. But now, I know that if I am asked to give up you forever, and nothing, nothing can come in return, I shall do it quietly and with no outcry. I shall work on, doing the very best I know how. I shall keep strong in body and clear in mind and clean in soul. I shall play the great game of life as we have conceived and dreamed it to-gether, and try to dream it further into fact. But in all that living, working, doing, dear Kautilya, I shall be dead. For without you—there is no life for me.

"I suppose that all this feeling is based on the physical urge of sex between us. I suppose that other contacts, other ex-periences, might have altered the world for us two. But the magnificent fact of our love remains, whatever its basis or accident. It rises from the ecstasy of our bodies to the communion of saints, the resurrection of the spirit, and the exquisite crucifixion of God. It is the greatest thing in our world. I sacrifice it, when I must, for nobler worlds that others may enjoy."

"I did not mistake you, Matthew; I knew you would understand. The time is come for infinite wisdom itself to think life out for us. This honeymoon of our high marriage with God alone as priest must end. It was our due. We earned it. But now we must earn a higher, finer thing."

Then hesitatingly she continued and spoke the yet unspoken word:

"First comes your duty to Sara. Even if you had not miraculously returned to me, you would have been forced to let Sara know by some unanswerable cataclysm that you would no longer follow her leading—either this, or your spiritual death; for you knew it, and you were planning revolt and flight. You were frightened at the thought of poverty and unlovely work; but now that you are free and have known love, you must return to Sara and say:

" 'See, I am a laborer: I will not lie and cheat and steal, but I will work in any honest way.' If it still happens that she wants you, wants the *real* you, whom she knows now but partially and must in the end know fully—a man honest to his own hurt, not greedy for wealth; loving all mankind and rejoicing in the simple things of life—if she wants this man, I—I must let you go. For she is a woman; she has her rights."

Matthew answered slowly:

"But she will not want me; I grieve to say it in pity, for I suffer with all women. Sara loves no one but herself. She can never love. To her this world-tangle of the races is a lustful scramble for place and power and show. She is mad because she is handicapped in the scramble. She would gladly trample anything beneath her feet, black, white, yellow, if only she could ride in gleaming triumph at the procession's head. Jealousy, envy, pride, fill the little crevices of her soul. No, she will not want me. But—if you will—as you have said, hers shall be the choice. She must ask divorce, not I. And even beyond that I will offer her fully and freely my whole self."

"And in the meantime," said Kautilya, "there are greater things—greater issues to be tested. We will wait on the high gods to see if maybe they will point the way for us to work together for the emancipation of the world. But if they decide otherwise, then, Matthew—"

"Then," continued Matthew gently, "we are parted, and forever."

"Yes," she whispered. "You and I, apart but eternally one, must walk the long straight path of renunciation in order that the work of the world shall go forward at our hands. We must work. We must work with our hands. We must work with our brains. We must stand before Vishnu; together we will serve.

"For, Matthew, hear my confession. I too face the horror of sacrifice. All is not well in Bwodpur, and each day I hearken for the call of doom across the waters. The old Rajah, my faithful guardian and ruler in my stead, is dead. Tradition, jealousy, intrigue, loom. For of me my people have a right to demand one thing: a Maharajah in Bwodpur, and one—of the blood royal!"

Matthew dropped his hands suddenly. Suddenly he knew that his own proposal of sacrifice was but an empty gesture, for Sara did not want him—would never want him. But Bwodpur wanted—a King!

Kautilya spoke slowly, standing with hanging hands and with face upraised toward the moon.

"We widows of India, even widows who, like me, were never wives, must ever face the flame of Sati. And in living death I go to meet the Maharajah of Bwodpur."

"Go with God, for after all it is not merely me you love, but rather the world through me."

"You are right and wrong, Matthew; I would not love you, did you not signify and typify to me this world and all the burning worlds beyond, the souls of all the living and the dead and of them that are to be. Because of this I love you, you alone. Yet I would love you if there were no world. I shall love you when the world is not."

She continued, after a space:

"I did not tell you, but yesterday my great and good friend, the Japanese baron whom you have met and dislike because you do not know him, came to see me. He knows always where I am and what I do, for it is written that I must tell him. You do not realize him yet, Matthew. He is civilization—he is the high goal toward which the world blindly gropes; high in birth and perfect in courtesy, filled with wide, deep, and intimate knowledge of the world's past—the world, white, black, brown, and yellow: knowing by personal contact and acquaintanceship the present from kings to coolies. He is a man of lofty ideal without the superstition of religion, a man of decision and action. He is our leader, Matthew, the guide and counselor, the great Prime Minister of the Darker World.

"He brought me information—floods of facts: the great conspiracy of England to re-grip the British mastery of the world at any cost; the titanic struggle behind the scenes in Russia between toil and ignorance defending the walls against organized stupidity and greed in Western Europe and America. He tells me of the armies and navies, of new millionaires in Germany and France, of new Caesarism in Italy, of the failing hells of Poland and dismembered Slavdom. The world is a great ripe cherry, gory, rotten—it must be plucked lest it fall and smash.

"My friend talked long of Asia—of my India, of poor Bwodpur. The Dewan who now rules for me, for all his loyalty and ability and his surrounding of young and able men, is distraught with trouble. It is unheard of that a Maharanee without a Maharajah should rule in Bwodpur. Some will not believe that the old Rajah is dead, but say that, shut up within his castle in High Himalaya, that ancient and unselfish man, who was my King in name, still lives as the reincarnate Buddha—lives and rules, and they would worship him. Around this and other superstitions, the continued and inexplicable absence of the Maharanee, the innovations of schools, health training, roads, and mysterious machinery, the neglect of the old religion, looms the intrigue of the English on every side—money, cheap goods, titles, decorations, hospitality, and magnificent Durbars—oh, all is not well in Bwodpur; even the throes of revolution threaten: Moslem and Hindu are at odds, Buddhist and Christian quarrel. Bwodpur needs me, Matthew, but she needs more than me: she needs a Maharajah.

"Facing all this, Matthew, my man, with level eye and clear brain we must drain the cup before us: if return to India severs me from the western world and you—if the dropping of ocean-wide dreams into the little lake of Bwodpur

is my destiny—the will of Vishnu prevail. If your reunion with
Sara is the only step toward the real redemption and emancipation of black America, then, Matthew, drain the cup. But
after all, the day of decision is not yet. And whatever comes,
Love—our love is already eternal."

Matthew pondered and said:

"The paradox is amazing: the only thing that was able to
lift me from cynical selfishness, organized theft and deception,
was that finest thing within me—this love and idealization of
you. If I had not followed it at every cost, I should have sunk
beneath hell. And yet now I am anathema to my people. I
am the Sunday School example of one who sold his soul to
the devil. I am painted as punished with common labor for
following lust and desecrating the home. People who recognize
me all but spit upon me in the street. Oh, Kautilya, what shall
we do against these forces that are pushing, prying, rending us
apart? Is it possible that the great love of a man for a
woman—the perfect friendship and communion of two human
beings—can ever be mere evil?"

They turned toward the room and looked at it. "I cannot
keep these things," he said. "They mean you. They meant you
unconsciously before I knew that I should ever see you again.
The Chinese rug was the splendid coloring of your skin; the
Matisse was the flame of your high spirit; the music was your
voice. I am going to move to one simple, bare room where again
and unhindered by things, I can see this little place of beauty
with you set high in its midst. And I shall picture you still
in its midst. I could not bear to see any one of these things
without you."

She hesitated. "I understand, I think, and the rug and picture shall go with me," she said. "And yet I hate to think of
your living barely and crudely without the bits of beauty you
have placed about you. Yet perhaps it is well. In my land, you
know, men often, in their strong struggle with life, go out and
leave life and strip themselves of everything material that
could impede or weight the soul, and sit naked and alone before
their God. Perhaps, Matthew, it would be well for you to do
this. A little space—a little space."

"How long before—we know?" he said, turning toward her
suddenly and taking both her hands.

"I cannot say," she answered. "Perhaps a few weeks, perhaps a little year. Perhaps until the spirit Vishnu comes down again to earth."

He shivered and said, "Not so long as that, oh, Radha, not so long! And yet if it must be—let it be."

And so they dismantled the room and packed and baled most of the things therein. At last in full day they went down to the Union Station and walked slowly along toward the gates with clasped hands. A beautiful couple, unusual in their height, in the brownness of their skins, in their joy and absorption in each other.

A porter passed by, stopped, and glanced back. He whispered to another: "That's him; and that's the woman." Then others whispered, porters and passengers. A knot of the curious gathered and stared. But the two did not hurry; they did not notice. Some one even hissed, "Shameless!" and some one else said, "Fool!" and still they walked on and through the gates and to the train. He kissed her lips and kissed her hands, and without tears or words she stepped on the train and looked backward as it moved off. Suddenly he lifted both hands on high, and tears rolled down his cheeks.

VIII

Matthew wrote to Kautilya at the New Willard in Washington, and in one of his letters he said: "I am digging a Hole in the Earth. It is singular to think how much of life is and has been just digging holes. All the farmers; all the miners; all of the builders, and how many many others have just dug holes! The bowels of the great crude earth must be pierced and plumbed and explored if we would wrest its secrets from it. I have a sense of reality in this work such as I have never had before—neither in medicine nor travel, neither as porter, prisoner, nor law-maker. What I am actually doing may be little, but it is indispensable. So much can not be said of healing nor writing novels. I am digging not to plant, not to explore, but to make a path for walking, running, and riding; a little round tunnel through which man may send swiftly small sealed boxes full of human souls, from Dan even unto Beersheba.

"Yes, just about that. Just about fifty miles of tunnel we

will have before this new Chicago subway is finished. And you
have no idea of the problems—the sweat, the worry, the toil
of digging this little hole. For it is little, compared to the vast
and brawny body of this mighty earth. It is like the path of
some thin needle in a great football of twine. The earth
resists, frantically, fiercely, tenaciously. We have to fight it;
to outguess it; to know the unknown and measure the
unmeasured.

"There lies the innocent dust and sand of a city street held
down from flying by bits of stone and pressed asphalt. So pliant
and yielding, so vulnerable. But it is watching—watching and
waiting. I can feel it; I can hear it. It will make a bitter
struggle to hide its heart from prying eyes. Its very surrender
is danger; its resistance may be death. I go down girded for a
fight with a hundred others in jerseys and overalls and thick
heavy shoes. We are like hard-limbed Grecian athletes, but less
daintily clad. One can see the same ripple and swelling of
muscles, and I felt at first ashamed of my flabbiness. But this
thing is real, not mere sport: we are not playing. There is no
laughing gallery with waving colors and triumphant cries.
For us this thing is life and death, food and drink, commerce,
education, and art. I am in deadly earnest. I am bare, sweating, untrammeled. My muscles already begin to flow smooth
and unconfined. I have no stomach, either in flesh or spirit.
My body is all life and eagerness, without weight.

"Rain, sun, dust, heat and cold; the well, the sick, the
wounded, and the dead. I saw a man make a little misstep and
jump forward; his head struck the end of a projecting beam
and cracked sickeningly. In fifteen minutes he was forgotten
and the army closed ranks and went forward; I just heard the
echo of the cry of his woman as it sobbed down to the mud
underneath the ground. Yes, it is War, eternal War from the
beginning to the end. We plumb the entrails of the earth.

"The earth below the city is full of secret things. Voices
are there calling day and night from everywhere to everybody.
I did not know before the paths they chose, but now I see them
whispering over long gray bones beneath the streets. Lakes and
rivers flow there, pouring from the hills down to the kitchen
sinks with steady pulse beneath the iron street. Thin blue gas
burns there in leaden pockets to cook and heat, and light is
carried in steel to blaze in parlors above the dark earth. There

is a strange world of secret things—of wire pipes, great demi-johns and caverns, secret closets, and long, silent tunnels here beneath the streets.

"The houses sag, stagger, and reel above us, but they do not fall: we hold them, force them back and prop them up. A slimy sewer breaks and drenches us; we mend it and send its dirty waters on to the canal, the river, and the sea. Gas pipes leak and stifle us. Electricity flashes; but we are curiously armed with such power to command and such faith like mountains that all nature obeys us. Lamps of Aladdin are everywhere and do their miracles for the rubbing: great steel and harnessed Genii, a hundred feet high, lumber blindly along at our beck and call to dig, lift, talk, push, weep, and swear. Yesterday, one of the giants died; fell forward and crumpled into sticks and bits of broken steel; but it shed no blood; it only hissed in horror. We strain in vast contortions underneath the ground. We perform vast surgical operations with insertions of lumber and steel and muscle; we tear down stone with thunder and lightning; we build stone up again with water and cement. We defy every law of nature, swinging a thousand tons above us on nothing; taking away the foundations of the city and leaving it delicately swaying on air, afraid to fall. We dive and soar, defying gravitation. We have built a little world down here below the earth, where we live and dream. Who planned it? Who owns it? We do not know.

"And right here I seem to see the answer to the first question of our world-work: What are you and I trying to do in this world? Not merely to transpose colors; not to demand an eye for an eye. But to straighten out the tangle and put the feet of our people, and all people who will, on the Path. The first step is to reunite thought and physical work. Their divorce has been a primal cause of disaster. They that do the world's work must do it thinking. The thinkers, dreamers, poets of the world must be its workers. Work is God."

Matthew laid down the pen and wrote no further that day. He had a singular sense of physical power and spiritual freedom. There was no doubt in his heart concerning the worth of the work he was doing—of its good, of its need. Never before in his life had he worked without such doubt. He felt here no compulsion to pretend; to believe what he did not believe; or to be that which he did not want to be.

IX

To the woman riding alone into an almost unknown world, all life went suddenly black and tasteless. In a few short years and without dream of such an end, she had violated nearly every tradition of her race, nearly every prejudice of her family, nearly every ideal of her own life. She had sacrificed position, wealth, honor, and virginity on the altar of one far-flaming star. Was it worth it? Was there a chance to win through, and to win to what? What was this horrible, imponderable, unyielding mask of a world which she faced and fought?

The dark despair of loneliness overwhelmed her spirit. The pain of the world lay close upon her like a fitted coffin, airless, dark, silent. Why, why should she struggle on? Was it yet too late? A few words on this bit of yellow paper, and lo! could she not again be a ruling monarch? one whose jewels and motor-cars, gowns and servants, palaces and Durbars would make a whole world babble?

What if she did have to pay for this deep thrill of Life with submission to white Europe, with marriage without love, with power without substance? Could she not still live and dance and sing? Was she not yet young, scarce twenty-six, and big with the lust for life and joy? She could wander in wide and beautiful lands; she could loll, gamble, and flirt at Lido, Deauville, and Scheveningen; she could surround herself with embodied beauty: look on beautiful pictures; walk on priceless carpets; build fairy-tales in wood and stone!

On all this she was trying to turn her back, for what? For the shade of a shadow. For a wan, far-off ideal of a world of justice to people yellow, black, and brown; and even beyond that, for the uplift of maimed and writhing millions. Dirty people and stupid, men who bent and crawled and toiled, cringed and worshiped snakes and gold and gaudy show. What, where, and whither lay the way to all this? It was the perfect love and devotion of one human soul, one whose ideals she tried to think were hers, and hers, his.

Granting the full-blown glory of the dream, was it humanly possible? Was there this possibility of uplift in the masses of men? Was there even in Matthew himself, with all his fineness of soul, the essential strength, the free spirit, the high heart,

and the understanding mind? Had he that great resolve back
of the unswerving deftness of a keen brain which could carry
through Revolution in the world? He was love. Yes, incarnate
love and tenderness, and delicate unselfish devotion of soul.
But was there, under this, the iron for suffering, the thunder
for offense, and the lightning for piercing through the thick-
threaded gloom of the world, and for flashing the seething
crimson of justice to it and beyond? And if in him there lay
such seed of greatness, would it grow? Would it sprout and
grow? Or had servility shriveled it and disappointment chilled
it and surrender to the evil and lying and stealing of life
deadened it at the very core?

Oh, Matthew, Matthew! Did he know just what she had
done and how much she had given and suffered? Did he still
hold the jewel of her love and surrender high in heaven, or was
she after all at this very moment common and degraded in
his sight? Gracious Karma, where was she in truth now? She
of the sacred triple cord, a royal princess of India and incarnate daughter of gods and kings! She who had crossed half
the world to him, fighting like a lioness for her own body.
Where was she now in the eyes and mind of the man whom
she had raised in her soul and set above the world? Only time
would tell. Time and waiting—bitter, empty waiting. Waiting
with hanging hands. And then one other thing, one thing above
all Things, one mighty secret which she had but partially
breathed even to Matthew. For there was a King in India who
sued for her hand. He willed to be Maharajah of Bwodpur. He
would lead Swaraj in India. He would unite India and China
and Japan. He pressed for an answer. Bwodpur pressed. Sindrabad pressed. All the world pressed down on one lone woman.

Then as she sat there crumpled and wan, with tear-swept
eyes and stricken heart, slowly a picture dissolved and swam
and grew faint and plain and clear before her: a little dark
cabin, swathed in clinging vines, nestling beneath great trees
and beside a singing brook; flowers struggled up beside the
door with crimson, blue, and yellow faces; hot sunshine filtered
down between the waving leaves, and winds came gently out
of sunset lands. In the door stood a woman; tall, big, and
brown. Her face seemed hard and seamed at first; but upon
it her great cavernous eyes held in their depths that softness
and understanding which calls to lost souls and strengthens

and comforts them. And Kautilya rose with wet eyes and
stumbled over time and space and went half-blind and groping
to that broad, flat bosom and into those long, enfolding arms.
She strained up into the love of those old, old eyes.

"Mother," she sobbed, "I've come home to wait."

X

"I am tired," wrote Matthew in August, "but I am singularly strong. I think I never knew before what weariness was.
At the day's end I am often dead on my feet, drugged and
staggering. I can scarcely keep awake to eat, and I fall to bed
and die until sudden dawn comes like crashing resurrection.
Yet I have a certain new clearness of head and keenness of
vision. Dreams and fancies, pictures and thoughts, dance
within my head as I work. But I half fear them. They seem
to want to drag me away from this physical emancipation. I
try to drown and forget them. They are in the way. They may
betray and subdue me.

"Where is the fulcrum to uplift our world and roll it forward? More and more I am convinced that it lies in intelligent
digging; the building of subways by architects; the planning
of subways and skyscrapers and states by workingmen.

"Curiously enough, as it is now, we do not need brains
here. Yes, here in a work which at bottom is Thought and
Method and Logic, most of us are required not to think or
reason. Only the machines may think. I wrote of the machines
as our Slaves of the Lamp, but I was wrong. We are the slaves.
We must obey the machines or suffer. Our life is simply lifting. We are lifting the world and moving it. But only the
machines know what we are doing. We are blindfolded. If only
they did not blindfold us! If we could see the Plan and understand; if we could know and thrust and trace in our mind's
eye this little hole in the ground that writhes under Chicago;
how the thrill of this Odyssey would nerve and hearten us!
But no, of the end of what we are doing we can only guess
vaguely. The only thing we really know is this shovelful of
dirt. Or if we dream of the millions of men this hole will shoot
in and out, up and down, back and forth—why will it shoot
them and to whom and from what into what Great End,
whither, whither?

"I could not finish this, and three days have gone. Yesterday I arose with the dawn before work and began reading. It was a revelation of joy. I was fresh and rested and the morning was bright and young. I read Shakespeare's *Hamlet*. I am sure now that I never have read it before. I told people quite confidently that I had. I looked particularly intelligent when *Hamlet* was discussed or alluded to. But if this was the truth, I must have read *Hamlet* with tired mind and weary brain; mechanically, half-comprehendingly. This morning I read as angels read, swooping with the thought, keen and happy with the inner spirit of the thing. Hamlet lived, and he and I suffered together with an all too easily comprehended hesitation at life. I shall do much reading like this. I know now what reading is. I am going to master a hundred books. Nothing common or cheap or trashy, but a hundred master-thoughts. I do not believe the world holds more. These are the days of my purification that I may rise out of selfishness and hesitation and unbelief and depths of mental debauchery to the high and spiritual purity of love.

"Now I must go. I shall walk down to the morning sunlight which is soft and sweet before its midday dust and heat; others gradually will join me as I walk on. On some few faces I shall catch an answering gleam of morning, some anticipation of a great day's work; but on most faces there will be but sodden grayness, a sort of ingrained weariness which no sleep will ever last long enough to drive away—save one sleep, and that the last.

"These faces frighten me. What is it that carves them? What makes my fellow men who work with their hands so sick of life? What ails the world of work? In itself, it is surely good; it is real; it is better than polo, baseball, or golf. It is the Thing Itself. There is beauty in its movement and in the sunshine, storm, and rain that walk beside it. Here is art. An art singularly deep and satisfying. Who does not glow at the touch of this imprisoned lightning that lies inert above the hole? We touch a bit of metal: the sullen rock gives up its soul and flies to a thousand fragments. And yet this glorious thunder of the world strikes on deaf ears and eyes that see nothing. At morning most of us are simply grim; at noon, we are dull; at night, we are automata. Even I cannot entirely escape it. I was free and joyous this morning. Tonight I shall be too tired to think or feel or plan. And after five, ten, fifteen years of this—what?

"I am trying to think through some solution. I see the Plan—our Plan, the great Emancipation—as clearly and truly as ever. I even know what we must aim at, but now the question is where to begin. It's like trying to climb a great mountain. It takes so long to get to the foothills.

"This problem of lifting physical work to its natural level puzzles me. If only I could work and work wildly, unstintingly, hilariously for six full, long hours; after that, while I lie in a warm bath, I should like to hear Tschaikowsky's Fourth Symphony. You know the lilt and cry of it. There must be much other music like it. Then I would like to have clean, soft clothes and fair, fresh food daintily prepared on a shining table. Afterward, a ride in green pastures and beside still waters; a film, a play, a novel, and always you. You, and long, deep arguments of the intricate, beautiful, winding ways of the world; and at last sleep, deep sleep within your arms. Then morning and the fray.

"I would welcome with loud Hosannas the dirt, the strain, the heat, the cold. But as it is, from the high sun of morning I rush, lurch, and crumple to a leaden night. The food in my little, dirty restaurant is rotten and is flung to me by a slatternly waitress who is as tired as I am. My bed is dirty. I'm sorry, dear, but it seems impossible to keep it clean and smooth. And then over all my neighborhood there hangs a great, thick sheet of noise; harsh, continuous, raucous noise like a breath of hell. It seems never to stop. It is there when I go to sleep; it rumbles in my deepest unconsciousness, and thunders in my dreams; it begins with dawn, rising to a shrill crescendo as I awake. There is no beauty in this world about me—no beauty. Or if these people see beauty, they cannot know it. They are not to blame, poor beeves; we are, we are!

"I grow half dead with physical weakness and sleep like death, but my body waxes hard and strong. I refused a clerk's job today, but I have been made a sort of gang foreman. I know the men. There are Finns and Italians, Poles, Slovaks, and Negroes. We do not understand each other's tongues; we have our hates and fancies. But we are one in interest: we are all robbed by the contractors. We know it and we are trying to organize and fight back. I do not know just what I should do in this matter. I never before realized that a labor union

means bread, sleep, and shelter. Can we build one of this helpless, ignorant stuff? I do not know. But this at least I do
know: Work is God."

Matthew wrote no more. He was alone, but he was trying
to think things out. What could really be done? If the task
of the workers were cut in half, would they all work correspondingly harder? Of course they would not. Some would;
he would. Most of them would sit around, dull-eyed, and loaf.
Profits would dwindle and disappear. There might even be
huge deficits. And could one get the men who knew and
thought and planned all this to guide and to lead without the
price of profit?

Oh, yes, some could be got for the sheer joy of fine effort.
They would work gladly for board, clothes, and creation. Some
men would do it because they love the game. But the kind
of men who were spending profits today on the North Shore,
on Fifth Avenue, Regent Street, and the Rue de la Paix—no!
It would call for a kind of man different from them, with a
different scheme of values. Yes, to work without money profit
would demand a different scheme of values in a different kind
of man; and to do full work on half-time in the ditch would
need a different kind of man, with a new dream of living:
perhaps there lay the world's solution: in men who were—
different.

He sat alone and tried to think it all out; but he could think
no further because he was too lonely. He needed the rubbing
of a kindred soul—the answering flash of another pole. His
loneliness was not merely physical; his soul was alone.
Kautilya was not answering his letters, and she had been gone
two great months. Far down within him he was sick at heart.
He could not quite understand why it had seemed to Kautilya
so inevitable that they must part. He kept coming back to
the question as to whether the excuse she gave was real and
complete.

Could it be possible that she must sacrifice herself to a
strange and unloved husband for reasons of state? What after
all was little lost Bwodpur in the great emancipation of races?
What difference whether she ruled as Princess or worked as
worker? What was "royal blood," after all?

And then, too, why this illogical solicitude for Sara's right

to him, after that supreme and utter betrayal and denial of
all right? Was it not possible, more than possible, that he had
disappointed Kautilya, just as he had disappointed himself and
his mother and his people and perhaps some far-off immutable
God?

Kautilya had built a high ideal of manhood and crowned it
with his likeness, and yet when she had seen it face to face,
perhaps it had seemed to crash before her eyes.

Perhaps—and his mind writhed, hesitated; and yet he
pushed it forward to full view—perhaps, after all, there was
unconsciously in Kautilya some borrowed, strained, and seep-
ing prejudice from the dead white world, that made her in her
inner soul and at the touch, shrink from intimate contact with
a man of his race; and perhaps without quite realizing it, they
had faced the end and she had seen life and love and dreams
die; then softly but firmly she had put him by and gone away.

Thus Matthew's dull and tired brain dropped down to
clouds of weariness. But he did not surrender. The old deso-
lation and despair seemed underlaid now by harsher iron built
on sheer physical strength. He did not even rise and undress
lest the ghosts of doubt grip him as he walked and moved. He
slept all night dressed and sitting at his empty deal table, his
head upon his hands. And he dreamed that God was Work.

XI

Sara had at last arrived. Sammy had met her. It was early
in September, and he had not seen her for five months. They
had a good breakfast at the Union Station, and Sammy had
retailed so much news and gossip that Sara was happier and
more alive than she had felt for a long time. She was very
calm and sedate about it, but after all she knew that the Black
Belt of Chicago with its strife, intrigue, defeats, and triumphs
was, for her, Life.

"And where have you been?" asked Sammy.

"New York, Atlantic City, Boston, Newport, and a few
places like that."

"Have a good time?"

"Fair."

Sammy whirled her home in a new Lincoln that was a

dream, and a black chauffeur in brown livery who knew his spaces to the tenth of an inch and glided up Michigan Avenue like smooth and unreverberating lightning.

"New car?" asked Sara.

"Yep! Celebrating."

"Celebrating—what?"

"Saw the old man last Wednesday down at Springfield." And then Sammy adroitly switched into a long and most interesting account of the latest and biggest Jewish tabernacle which her pastor had bought with liberal political donations.

Sara said nothing further about the car and that Springfield interview. Sammy knew she was curious, but just how deeply and personally curious she was, he was not certain. So he waited. In Sara's apartment he wandered about, a bit distrait, while she took her usual good time to dress. The apartment was immaculate and in perfect order. Sammy saw no trace of that scene five months ago. And as for Sara, when she emerged, her simple, close-fitting tailor-made costume was all Sammy could ask or imagine.

"I say, kid, don't you think we might talk this thing out and come to some understanding?"

Sara opened her eyes. "Talk what out?"

"Well, about you and me. You see, you had me going, and I had to do something. I couldn't just stand by and lose everything. So I got busy. I hated to do it, but I had to."

"You didn't do it—God did."

"God nothing! I remembered the woman on that train wreck and I found her."

"It is a lie. She found you," said Sara.

"Well, it was a little like that, but the minute I laid eyes on her I knew she was the woman I had heard about. And I told her all about Matthew; how queer he was, and how he was hesitating, and how no man like him could ever make a politician. Then she laid low, but she came that night. I didn't think she would."

"You're a pretty friend."

"Say, kid, don't be hard. It was a bit tough, I own, for you. But I had you in the plan. Now listen to reason. Matthew was no good. He was going flabby. He's no real politician. He didn't know the game, and he had fool Reform deep in his system. He was just waking up, and he'd 'a' raised Hell in Congress. We never could have controlled him. Now when I get in Congress—"

"Congress!" sneered Sara. "Do you think any nigger has a chance now?"

But Sammy talked on.

"—and when you are my wife—"

"Wife!"

"Sure, I've never been a marrying man, as I have often explained—"

"Not retail," said Sara.

"And wholesale don't count in law; but I need you, I see that now, and I'm damned if I am going to lose you to another half-baked guy."

"I am not divorced yet, and I am not sure—"

"You mean you ain't sure you ain't half in love with him still?"

"I hate the fool. I'd like to horsewhip her in City Hall."

"Too late, kid, she's left him."

"Left him?"

"Sure—gone bag and baggage, and what do you know! He's digging in the subway—a common laborer. Oh, he's up against it, I'll tell the world. Reckon he wouldn't mind visiting the old roost just now." And Sammy glanced about with approval.

Sara looked him over. Sammy was no Adonis. He was approaching middle age and was showing signs of wear and tear. But Sara was lonesome, and between her and Sammy there was a common philosophy, a common humor, and a common understanding. Neither quite trusted the other, and yet they needed each other. Sara had missed Sammy more than she dared acknowledge, while without Sara, Sammy felt one-armed.

Sammy continued:

"No, kid. Your lay is still the quiet, injured wife, shut up at home and in tears, until after the next election. See? A knockout! Matthew is politically dead this minute. Right here I come in. The bosses know that they got to take a dose of black man for Congress sooner or later. They came near getting a crank. But even your fine Italian hand couldn't make him stay put. He never would have been elected."

"Oh, I don't know."

"You got it right there, kid, you *don't* know, and you know you don't. And now here I am. The bosses will have to take me sooner or later. All I need is you."

"But I hear that the governor and Thompson are at outs."

"Sure."

"And you're hitched up with the governor."

"Sure again. I'm opposing Thompson. But after he's elected by a smashing majority, the governor and the Washington crowd will need him, he'll need the governor, and they'll both need me!"

"H'm—I see. Well, you'll have to do some tight-rope walking, my friend."

"Precisely, and that's where you come in."

"Indeed! Now listen to me. Don't think, Sammy, you're going to get both money and office out of these white politicians at the same time. When they pay big they take the big jobs. If you want to go to Congress you'll pay. The only exception to that rule was the game I played and won and then that fool threw away."

Sammy smiled complacently. "Did you see my new Lincoln?"

"Yes, and wondered."

"Four thousand bucks, and the shuffer's gettin' thirty-five per. And say! Remember that big white stone house at Fiftieth and Drexel Boulevard?"

"You mean at Drexel Square, with the big oaks and a fountain?"

"Yep!"

"Yes, I remember it—circular steps, great door with beveled glass, and marble lobby!"

"No different! Driveway and garage; sun parlor, twenty rooms, yard, and big iron fence. Well, that's where we're gonna live. I've bought it."

"Sammy! Why, you must be suddenly rich or crazy!"

"Kid, I've made a killing! While you were leading me a dance for Congress, I got hold of all the dough I could grab and salted it away. Oh, I spent a lot on the boys, but I had a lot to spend. Graham and the Public Service was wild to return Doolittle. I spent a pile, but I didn't spend all by a long shot. I put a hunk into two or three good deals—real estate, bootlegging, and—well—other things. Then when Mat-

thew flew the coop I rushed at the gang and put up such a yell
that they let me in on something big: and listen, sister! little
Sammy is on Easy Street and sittin' pretty! Believe me, I ain't
beggin'—I'm going to buy my way into Congress if it takes
a hundred thousand simoleons."

Sara looked Sammy over.

"And you're counting on me, are you?"

"Sure thing! As soon as election is over, we can have proceedings for a divorce on foot quietly, and it will be over in
a week or so. Meantime, you're my secretary again, and you're
going to name your own salary."

Sara arose and smoothed her frock. She looked so unmoved
and unapproachable that Sammy half lost his nerve.

"Don't let him get you, Sara. Don't let black Chicago think
you're down and out because of one man. What do you say,
kid? You know, I—I always liked you. I was crazy about you
the minute you stepped through that door five years ago. I
figured that nobody but me was ever going to marry you. But
you were so damned stand-offish—I sort-a wanted you to melt
a little first and be human. But now, Lord, kid, I'm crawlin'
and beggin' you on any terms. What do you say? See here!
I'll bet you a diamond as big as a hen's egg against a marriage
license that you'll be happier as my wife than you've been in
ten years. What d'ya say, kid?"

Sara still stood looking at Sammy thoughtfully as she
reached for her vanity case. She turned to the mantel mirror
and was some time powdering her nose. Then she obeyed an
impulse, a thing she had not done for ten years. She turned
deliberately, walked over to Sammy, and kissed him.

"You're on, Sammy," she said.

XII

"Dearest Matthew, my man," wrote Kautilya in September,
"forgive my silence. I am in Virginia with your mother. I could
not stay in Washington. I wanted to sit a space apart and in
quiet to think and hearken and decide. The wind is in the
trees, the strong winds of purpose, the soft winds of infinite
desire; the wide black earth around me is breathing deep with
fancies. There is rain and mud and a certain emptiness. But
somehow I love this land, perhaps because mother loves it so. I

seem to see salvation here, a gate to the world. Here is a
tiny kingdom of tree and wood and hut. Oh, yes, and the
brook, the symphony of the brook. And then there are the
broad old fields as far as we can look toward the impounding
woods.

"Beloved, I am beginning to feel that this place of yours
may be no mere temporary refuge. That it may again be Home
for you. I see this as yet but dimly, but life here seems symbolic. Here is the earth yearning for seed. Here men make food
and clothes. We are at the bottom and beginning of things.
The very first chapter of that great story of industry, wage
and wealth, government, life.

"On such deep founding-stones you may perhaps build. I
can see work transformed. This cabin with little change in its
aspect can be made a place of worship, of beauty and books.
I have even planned a home for you: this old and black and
vine-clad cabin undisturbed but with an L built behind and
above. The twin cabin must run far back and rise a half story
for a broad and peaceful chamber—for life with music and
color floating in it. Perhaps a little lake to woo the brook; and
then, in years, of course, a tower and a secret garden! Yes, I
should like to see a tower, where Muezzins call to God and
His world.

"And this world is really much nearer to our world than I
had thought. This brook dances on to a river fifty miles
away—next door only for a little Ford truck. And the river
winds in stately curve down Jamestown-of-the-Slaves. We went
down the other day, walking part of the way through woods
and dells, toward the great highway of the Atlantic. Think,
Matthew, take your geography and trace it: from Hampton
Roads to Guiana is a world of colored folk, and a world, men
tell me, physically beautiful beyond conception; socially enslaved, industrially ruined, spiritually dead; but ready for the
breath of Life and Resurrection. South is Latin America, east
is Africa, and east of east lies my own Asia. Oh, Matthew,
think this thing through. Your mother prophesies. We sense a
new age.

"This is the age of commerce and industry—of making,
shaping, carrying, buying and selling. We have made manufacturers, railroad men, and merchants rich because we ranked
them highest, and we have helped them in cities for conveni-

ence, and they are white and in white cities. Just suppose we change our ranking. Suppose in our hearts we rate the colored farmers and all discoverers, poets, and dreamers high and even higher and give them space outside of white cities? We would widen the world. It is simply a matter of wanting to. We have bribed white factories with tariffs and monopoly. We are going to bribe black agriculture and poetry. And, Matthew, Work is not God—Love is God and Work is His Prophet."

Hurriedly Matthew wrote back: "No, no, Kautilya of the World, no, no! Think not of home in that breeder of slaves and hate, Virginia. I shudder to find you there even for a season. There is horror there which your dear eyes are not yet focused to see and which the old blindness of my mother forgets. There is evil all about you. Oh, sister, you do not know—you do not dream. Down yonder lurk mob and rape and rope and faggot. Ignorance is King and Hate is High Prime Minister. Men are tyrants or slaves. Women are dolls or sluts. Industry is lying, and government stealing.

"The land is literally accursed with the blood and pain of three hundred years of slavery. Ask mother. Ask her to tell you how many years she has fought and clawed for the honor of her own body. There is a little weal on her breast, a jagged scar upon her knee, a broken finger on one hand. Ask her whence these came. Ask her who imprisoned and killed my father and why and where her other children are buried. Come away, come away, my crumpled bird, as soon as may be, lest they despoil you. You may hide there until our wounds are healed, but then come away to the midst of life. Only in the center of the world can our work be done. We must stand, you and I, even if apart, where beats down the fiercest blaze of Western civilization, and pushing back this hell, raise a black world upon it.

"I ought not to write tonight, for I am in the depths; the sudden change of Chicago's Fall has dropped upon me. I caught cold and was ill a day, and then I arose and did not go to work. Instead, I went down to the art gallery. There was a new exhibit of borrowed paintings from all ends of the world. After mud and filth and grayness, my soul was starving for color and curve and form. I went. And then went back again, day after day. I literally forgot my work for a week and bathed myself in a new world of beauty.

"I saw in Claude Monet what sunrise and sunset on the old cathedral at Rouen might say to a human soul, in pale gold, white, and purple, and in purple, yellow, and gold. I felt the mists of London hiding Big Ben. Rich somber peace and silence fell on me and on the picnic party beneath spreading branches. I walked with that lady about the red flowers of her garden. I reveled in blue seas, faint color-swept fields, riot of sweet flowers, poppies and grain, brooks and villages.

"I saw Pisano's Paris; the colors of Matisse raged in my soul, deluging all form, unbeautiful with rhythm. I delighted in the luscious dark folk of Paul Gaugin, in sun and shade, fruit and sea, palm and totem, and in the color that melts and flows and cries. Then there were the mad brown-gray-green lines of Picasso which swerved and melted into strange faces, forms, and figures, haunting things like their African prototypes; there was a dark little girl by Derain floating in a field of blue with a yellow castle, square and old.

"As from a far flight into the unknown I came back to the lovely coloring of Brangwyn and Cottet. I discovered the lucent blue water of Cézanne, his plunging landscapes and the hard truth of his faces. I saw how lovely Mrs. Samari looked to Renoir and vineyards to Van Gogh.

"At the end of the week I emerged half-ashamed, uncertain in judgment, and yet with added width to my world. I dimly remembered how you all talked of painting there in Berlin; then I knew nothing, nothing. Or rather now I know nothing, and then I did not know how ignorant I was. And, withal, mentally breathless, I returned with a certain peace and slept to dreams of clouds of light. I rose the next morning lightheaded, rested and strong, and went down blithely to that hole in the ground, to the grim, gigantic task. I was a more complete man—a unit of a real democracy.

"Even as I reached for my shovel the boss yelled at me. 'Away a week. You're fired!' Well, somehow I got back again after a few days. After all, I reckon, I am a good worker. But there was still trouble, and the boss had taken a dislike to me. It was like a groom incurring the displeasure of some high lackey in the court of Louis XIV. As I have said, we subway laborers were not yet organized, and emissaries from the trade unions were working among us. I never knew what unions meant before. I think I was a bit prejudiced against them.

They were organizations, to my mind, which took food from the mouths of black men.

"Now suddenly I saw the thing from the other side. Unless we banded ourselves together and as one body against this Leviathan which 'hires and fires,' we were helpless, crushed piecemeal, having no voice as to ourselves and our work. And so I went in for the union. We struck: our hours of work were too long; the overtime was too poorly paid; we were being maimed by accidents and cheated of our insurance; we had no decent luncheon time or place. Well, we struck and we were roundly beaten. There were five hungry men eager to take every place we left vacant; mostly black men—they were hungriest. There were the police and politics against us.

"Again I was 'fired,' and this time for good. It was strange in this great, busy, and rich city, actually to be among the unemployed, and so much work to be done! I never believed in the unemployed before. To me the unemployed were the lazy, the shiftless, the debauched. But I am not lazy. I am eager for work; I am strong and willing and for a week I actually did not know how I was going to earn my bread.

"Several times and in several places I applied for work, and then at last I found a new reason staring in my face. It was Sammy. He strutted over to me as I came out of one employment office, and stood with his legs apart, scowling at me. He had neither smile nor handshake. 'Say, bo,' he blustered, 'why didn't you come around and fix up them divorce papers? See here, I'm tired of your damned stallin'! Think you're going to crawl back to Sara one of these days because your fancy woman has jilted you? Well, by God, you ain't. You're going ahead with this divorce, or I'll damn well drive you out of Chicago.'

"I made no answer. I think I smiled at him a little because he did appear to me pathetically funny; and then I went and got a job. I knew one place where workers were scarce, but I had always shrunk from going there. But that morning I went out to the stockyards and got a job. The world stinks about me. I am lifting rotten food. I am helping to murder things that live. The continual bleating of death beats on my ears and heart. I am drugged with weariness and ugliness. I seem to know as never before what pain and poverty mean. In the world there is only you—only you and that halo about your head which is the world-wide Cause.

"But Sammy's command set me thinking. I dimly see what must be done to restore the balance and coöperation of the white and black worlds: Brain and Brawn must unite in one body. But where shall the work begin? I begin to believe right here in Chicago, crossroads of the world—midway between Atlantic and Pacific, North and South Poles. This is the place. How shall I begin?"

XIII

There had been a long silence. Matthew had set his teeth and written regularly and methodically, words that did not say or reveal much. And then at last out of the South there came one morning a long, clear cry from Kautilya.

"Oh, Matthew, Matthew. The earth here in this October is full with fruit and harvest. The cotton lies dark green, dim crimson, with silver stars above the gray earth; my own hand has carried the cotton basket, and now I sit and know that everywhere seed that is hidden dark, inert, dead, will one day be alive, and here, here! And Matthew, my soul doth magnify the Lord. Within the new twin cabin above the old (for I have built it already, dear—I *had* to) sitting aloft, apart, a bit remote, is a low, dark, and beautiful room of Life, with music and with wide windows toward the rising and the setting of the sun. Outside the sun today is beaten shimmering copper-gold. The corn shocks and the fields are dull yellow; the bare cotton stalks are burning brown. But the earth is rich and full, and Love sits wild and glorious on the world. I have been reading to dear mother. She sits beside me, silent, like some ancient priestess. I read out of the Hebrew scriptures words of cruelty and war, and then in the full happiness of my heart I found that passage in Luke: 'My soul doth magnify the Lord!' And now my spirit is rejoicing, and the ineffable Buddha, blood of the blood of my fathers, seems bowing down to his low and doubting handmaiden. Well-beloved, shall not all generations call us Blessed and do great things for us, and we for them? All children, all mothers, all fathers; all women and all men? Thus do I bend and kiss all the lowly in the name of Him who 'hath put down the Mighty from their seats and exalted them of low degree'!

"We will build a world, Matthew, you and I, where the Hungry shall be fed, and only the Lazy shall be empty. Oh, I am mad, mad, Matthew, this day when the golden earth bows and falls into the death of Winter toward the resurrection of Spring! Life seems suddenly clear to me. I see the Way. Matthew, I am not afraid of Virginia or the white South. I know more of it than you think and mother has told me. It is crude and cruel, but, too, it is warm and beautiful. It is strangely, appealingly human. Nothing so beautiful as Virginia can be wholly hellish. I have my troubles here, mother and I, but we have faced it all and beaten them back with high and steady glance. I see the glory that may come yet to this Mother of Slavery.

"I will not, I can not, be sorry for you, Matthew, for your poor bruised soul and for the awful pit in which your tired feet stand. Courage, my man. Drain the cup. Drain it to the dregs, and, out of this crucifixion, ascend with me to heaven."

XIV

"I am glad, dear Kautilya," wrote Matthew at Christmas time, "that you are happy and content. But I am curious to understand that Way which you see so clearly. As for me, I am sorely puzzled. I believe in democracy. Hitherto I have seen democracy as the corner stone of my new world. But today and with the world, I see myself drifting logically and inevitably toward oligarchy. Baseball, movies, Spain, and Italy are ruled by Tyrants. Russia, England, France, and the Trusts are ruled by oligarchies. And how else? We Common People are so stupid, so forgetful, so selfish. How can we make life good but by compulsion? We cannot choose between monarchy and oligarchy or democracy—no—we can only choose the objects for which we will enthrone tyrannical dictators; it may be dictators for the sake of aristocrats as in Czarist Russia, or dictators for the sake of millionaires as in America, or dictatorship for the factory workers and peasants as in Soviet Russia; but always, everywhere, massed and concentrated power is necessary to accomplish anything worth while doing in this muddled world, hoping for divine Anarchy in some faraway heaven.

"Whether we will or not, some must rule and do for the

people what they are too weak and silly to do for themselves. They must be made to know and feel. It is knowledge and caring that are missing. Some know not and care not. Some know and care not. Some care not and know. We know and care, but, oh! how and where? I am afraid that only great strokes of force—clubs, guns, dynamite in the hands of fanatics—that only such Revolution can bring the Day.

"I wish I could see the solution of world misery in a little Virginia cottage with vines and flowers. I wish I could share the surging happiness which you find there; but I cannot, I am too far from there. I am far in miles, and somehow I seem insensibly to grow farther in spirit. I agree that America is the place for my work, and if America, then Chicago; for Chicago is the epitome of America. New York is a province of England. Virginia, Charleston, and New Orleans are memories, farming and industrial hinterlands. California is just beyond the world. Chicago is the American world and the modern world, and the worst of it. We Americans are caught here in our own machinery; our machines make things and compel us to sell them. We are rich in food and clothes and starved in culture. That fine old accumulation of the courtesies of life with its gracious delicacy which has flowered now and again in other lands is gone—gone and forgotten. We push and shoulder each other on the streets, yell, instead of bowing; we have forgotten 'Please,' 'Excuse me,' 'I beg your pardon,' and 'By your leave' in one vast comprehensive 'Hello!' and 'Sa-ay!'

"Courtesy is dead—and Justice? We strike, steal, curse, mob, and murder, all in the day's work. All delicate feeling sinks beneath floods of mediocrity. The finer culture is lost, lost; maybe lost forever. Is there beauty? Is there God? Is there salvation? Where are the workers so rich and powerful as here in America, and where so arid, artificial, vapid, so charmed and distracted by the low, crude, gawdy, and vulgar? I can only hope that after America has raped this land of its abundant wealth, after Africa breaks its chains and Asia awakes from its long sleep, in the day when Europe is too weak to fight and scheme and make others work for her and not for themselves—that then the world may disintegrate and fall apart and thus from its manure, something new and fair may sprout and slowly begin to grow. If then in Chicago we can kill the thing that America stands for, we emancipate the world.

"Yes, Kautilya, I believe that with fire and sword, blood and whips, we must fight this thing out physically, and literally beat the world into submission and a real civilization. The center of this fight must be America, because in America is the center of the world's sin. There must be developed here that world-tyranny which will impose by brute force a new heaven on this old and rotten earth."

It was almost mid-January when Kautilya's reply to this letter came. It was as ever full of sympathy and love, and yet Matthew thought he saw some beginnings of change.

"If the world is aflame," said Kautilya, "and I feel it flaming—the place of those who would ride the conflagration is truly within and not behind or in front of the Holocaust. Where then is this center, and what shall we who stand there do? Here are my two disagreements with you, dear Matthew. America is not the center of the world's evil. That center today is Asia and Africa. In America is Power. Yonder is Culture, but Culture gone to seed, disintegrated, debased. Yet its re-birth is imminent. America and Europe must not prevent it. Only Asia and Africa, in Asia and Africa, can break the power of America and Europe to throttle the world.

"And, oh, my Matthew, your oligarchy as you conceive it is not the antithesis of democracy—it is democracy, if only the selection of the oligarchs is just and true. Birth is the method of blind fools. Wealth is the gambler's method. Only Talent served from the great Reservoir of All Men of All Races, of All Classes, of All Ages, of Both Sexes—this is real Aristocracy, real Democracy—the only path to that great and final Freedom which you so well call Divine Anarchy.

"And yet this, dear Matthew, you yourself taught me—you and your struggling people here. In Africa and Asia we must work, and yet in Africa and Asia we are outside the world. That is the thing I always felt at home. Outside, and kept outside, the centers of power. Even to us in Europe, the closed circle of power is narrow and straitly entrenched; the stranger can scarce get foothold, and when he gets in, Power is no longer there. It is flown. In America your feet are further within the secret circle of that power that half-consciously rules the world. That is the advantage of America. That is the advantage that your people have had. You are working within. They are standing here in this technical triumph of human power and can use is as a fulcrum to lift earth and seas and stars.

"But to be in the center of power is not enough. You must be free and able to act. You are not free in Chicago nor New York. But here in Virginia you are at the edge of a black world. The black belt of the Congo, the Nile, and the Ganges reaches by way of Guiana, Haiti, and Jamaica, like a red arrow, up into the heart of white America. Thus I see a mighty synthesis: you can work in Africa and Asia right here in America if you work in the Black Belt. For a long time I was puzzled, as I have written you, and hesitated; but now I know. I am exalted, and with my high heart comes illumination. I have been sore bewildered by this mighty America, this ruthless, terrible, intriguing Thing. My home and heart is India. Your heart of hearts is Africa. And now I see through the cloud. You may stand here, Matthew—here, halfway between Maine and Florida, between the Atlantic and the Pacific, with Europe in your face and China at your back; with industry in your right hand and commerce in your left and the Farm beneath your steady feet; and yet be in the Land of the Blacks.

"Dearest, in spite of all you say, I believe, I believe in men; I believe in the unlovely masses of men; I believe in that prophetic word which you spoke in Berlin and which perhaps you only half believed yourself. And why should I not believe? I have seen slaves ruling in Chicago and they did not do nearly as badly as princes in Russia. Gentle culture and the beauty and courtesies of life—they are the real end of all living. But they will not come by the dreaming of the few. Civilization cannot stand on its apex. It must stand on a broad base, supporting its inevitable and eternal apex of fools. The tyranny of which you dream is the true method which I too envisage. But choose well the Tyrants—there is Eternal Life! How truly you have put it! Workers unite, men cry, while in truth always thinkers who do not work have tried to unite workers who do not think. Only working thinkers can unite thinking workers.

"For all that we need, and need alone, Time; the alembic, Time. The slow majestic march of events, unhurried, sure. Do not be in a hurry, dear Matthew, do not be nervous, do not

fret. There is no hurry, Matthew, your mother's Bible puts it
right: 'A day unto the Lord is as a thousand years, and a
thousand years as one day.' "

XV

Endless time! Matthew laughed and wept. Endless time! He
was almost thirty. In a few years he would be forty, and
creative life, real life, would be gone; gone forever. But he
knew; he saw it all; he faced grimly and without flinching
the terrible truth that for seven months he had sought to hide
and veil away from himself. Kautilya did not plan for him in
her life. Almost she did not want him, although perhaps this
last fact she had not quite realized. She had tried him and
his people and found them wanting. It was a sordid mess,
sordid and mean, and she was unconsciously drawing the skirts
of her high-bred soul back from it. She missed—she must
miss—the beauty and wealth, the high courtesy and breath of
life, which was hers by birth and heritage. And she must have
searched in vain and deep disappointment in this muck of
slavery, servility, and make-believe, for life. She had bidden
him drain the cup. He would.

More and more was he convinced that the parting of himself
and Kautilya was forever; that he must look this eventuality
squarely in the face. And looking, he was sure that he had
found himself. With his new physical strength had come a
certain other strength of soul and purpose. Once he had
sought knowledge and fame; once he had sought wealth; once
he had sought comfort. Now he would seek nothing but work,
and work for work's own sake. That work must be in large
degree physical, because it was the physical work of the world
that had to be done as prelude to its thought and beauty. And
then beyond and above all this was the ultimate emancipation
of the world by the uplift of the darker races. He knew what
that uplift involved. He knew where he proposed to work,
despite the ingenuity of Kautilya's argument. He did not yet
see how physical toil would bring the spiritual end he sought,
save only in his own soul. Perhaps—perhaps that would be
enough. No, no! he still rejected such metaphysics.

Meantime one step loomed closer and clearer. He would
follow the word of Kautilya, because there was a certain beauty

and completeness in her desire that he offer himself back to Sara. He saw that it would not be a real offer if it were not really meant. First, of course, Sara must see him as he was and realize him; a man who worked with his hands; a man who did his own thinking, clear and straight, even to his own hurt and poverty. A man working to emancipate the lowest millions. And, because of this and for his own salvation, certain cravings for beauty must be satisfied: simple, clear beauty, without tawdriness, without noise and meaningless imitation. Seeing him thus, perhaps, after all, in her way, in her singular, narrow way, Sara might realize that she had need of him. It was barely possible that, with such love as still oozed thinly in the hard crevices of her efficient soul, she loved him. Very well. If she wanted him as he was, realizing that he had loved some one else as he never could love her, well and good. He would go back to her; he would be a good husband; he would be, in the patois of the respectable, "true," but in a higher and better sense, good.

Matthew saw, too, with increasing clearness, something that Kautilya, he thought, must begin to realize, and that was that her freedom from him and his people—her freedom from this entanglement from which the thoughtful Japanese and Indians had tried to save her—would mean an increased and broader chance for her own work in her own world. And she had a work if she could return to it untrammeled by the trademark of slavery and degradation. She had tried to see a way in America for herself and Matthew to tread together. But all this was self-deception.

Matthew saw clearly, however, that he must give Kautilya no inkling of his own understanding and interpretation of herself. He knew that in her high soul there was that spirit of martyrs which might never let her surrender him voluntarily, that she would seek to stand by him just as long as it seemed the honorable thing to do. And so he would not "wince nor cry aloud," but he would "drain the cup."

That night he telephoned the maid at Sara's house, and learning that Sara was in, went down to see her. It was a hard journey. It was like walking back in time. He went through all the writhings of that period of groping revolt and yearning. He walked up the steps with the same feeling of revulsion and entered that prim and cold atmosphere, that hard, sharp grinding of life. He rang the bell. The maid stared, grinned, and fidgeted.

"Yes—she's in—but I don't think—she said never to—"

She wanted Matthew to push past and go in unannounced, and he meant to, but he couldn't. He stood hesitating.

Sara's clear voice came from within:

"Who is it, Eliza?"

"It is Matthew Towns," said Matthew. "I would like—"

There were quick steps. The maid withdrew. The door banged in his face.

Matthew wrote to Kautilya nothing of this, but only to continue that argument about work and wealth and race. He said:

"Art is long, but industry is longer. Revolution must come, but it must start from within. We must strip to the ground and fight up. Not the colored Farm but the white Factory is the beginning; and the white Office and the Street stand next. The white artisan must teach technique to the colored farmer. White business men must teach him organization; the scholar must teach him how to think, and the banker how to rule. Then, and not till then, will the farmer, colored or white, be the salt of the earth and the beginning of life."

Then in a postscript he added:

"I have had notice of Sara's action for divorce. I shall go in person to the hearing and answer, and I shall assent to whatever she may wish. I hope sincerely you are well. I have feared you might be sick and keep it from me. But even in sickness there is one consolation. Life at its strongest and longest is short. Bad as it is and beautiful as it has been for us, it is soon over. I kiss the little fingers of your hand."

Kautilya replied with a little note that came in early March, scribbled on wrapping-paper, with uncertain curves:

"Matthew, I am afraid. Suddenly I am desperately afraid. Just what I fear I do not know—I cannot say. Perhaps I am ill. I know I am ill. Oh, Matthew, I am afraid. Life is a terrible thing. It looms in dark silence and threatens. It has no bowels of compassion. Its hidden soul neither laughs nor cries—it just is, is, *is!* I am afraid, Matthew—I am in deadly fear. The terror of eternal life is upon me—the Curse of Siva! Come to me, Matthew, come! No, no—do not come until I send. I shall be all right."

Matthew's heart paused in sudden hurt. He knew what must have happened: the Great Decision must be made. She had been summoned to India and must go. He started to pack his suitcase. He telephoned about trains. Then he hesitated. "No, no—do not come until I send." That was her decision. Against her will he must not go. But perhaps already she had changed her mind. Perhaps she was physically ill. Perhaps already Death, cloaked in black, stood in the shadows behind her writhing bed! Or, worse, perhaps she was going away and could not pause to say good-by.

He telegraphed—"May I—" No, he tore that up: "Shall I come?"

The answer came in a few hours.

"No, all is well. I have been very ill, but I am better and I shall be out soon in the sweet springtime. I am going to walk and sew; I am going to be happy: infinitely happy. I want to see the heavy earth curling up before the shining of my plow-share. I want to feel the gray mule dragging off my arms, with the sky for heaven and the earth for love. I want to see seed sink in the dead earth. How can you say that life is short? Life is not short, my darling Matthew, it is endless. You and I will live for a thousand years and then a thousand years more; and then ten thousand years shall be added to that. Oh, man of little faith! Do you not see, heart of me, that without infinite life, life is a joke and a contradiction? Wish and Will are prisoned and manacled in Fact, whatever that fact may be; but with life built on life here on earth, now and not in your silly Christian parlor heaven, the tiny spark that is God thrills through, thrills through to triumph in a billion years; so vast, and vaster, is the Plan."

Matthew humored her mood. She saw the end of their earthly happiness here in time, and she was straining toward eternity. He could not deceive himself or her, and he wrote with a certain sad smile in his heart:

"Infinite and Eternal? Yes, dear Princess of the Winds; the Moonlight Sonata, snow on a high hill, the twitter of birds on boughs in sunlight after rain, health after sickness—God! are not these real, true, good, beautiful, infinite, eternal? Whether Immortal Life, dearest and best, is literal truth or not, I do not care. No one knows whether anything in life or larger than life bursts through to some inconceivable triumph over death. None knows, none can know. But, ah, dear heart, what difference? There is, after all, sunrise and rain, starlight, color, and the surge and beat of sound. And on that night when my body kissed yours, a billion years lived in one heartbeat. What more can I ask? What more have I asked or dreamed, Queen of the World, than that? Already I am Eternal. In thy flesh I have seen God."

And Kautilya answered:

"I know, I know, heart's-ease, but that is not enough: back of it all, back of the flesh, the mold, the dust, there must be Reality; it must be there; and what can reality be but Life, Life Everlasting? If we, we our very selves, do not live forever, Life is a cruel joke."

Yes, Life *was* a cruel joke, and Matthew turned to write of everyday things:

"As I sat last night huddled over my supper—a very greasy pork chop, sodden potatoes, oleomargarine, soggy cornbread, partly cooked cabbage, and weak, cold coffee—as I sat in my grimy overalls and guzzled this mess, some one came and sat at my table with its dirty oil-cloth cover. I did not look up, but a voice, a rather flat, unusual voice, ordered rice. 'Just rice.' Then I looked up at a Chinese woman, and she smiled wanly back.

" 'I prefer,' she said, 'don't you, the cuisine of the Lützower Ufer?'

"It was one of our Chinese friends. I was glad and ashamed to see her. She seemed to notice nothing—made no comment, asked no awkward questions. Principally she talked of China.

" 'Oh, China, China, where shall we find leaders! They rise, they fall, they die, they desert. The men who can do, the men of thought and knowledge, the men who know technique, the unselfish and farseeing—how shall we harness these to the greatest chariot in the world and not have them seduced and stolen by Power, Pleasure, Display, Gluttony? Oh, I know it is the old story of human weakness, but if only we had a little more strength and unity now and then at critical moments, we could climb a step and lift the sodden, smitten mass.

" 'There was Chiang Kai-Shek, so fine and young a warrior! I knew him well. I saw once his golden face alight with the highest ideals, his eyes a Heaven-in-Earth. Today, what is he?

I do not know. Perhaps he does not. Oh, why was it that Sun Yat Sen must die.so soon? But'—she rose from the half-eaten, mushy rice—'we must push on always—on!' And then pausing she said, timidly, 'And you, my friend. Are you pushing—on?'

"I hesitated and then arose and stood before her: 'I am pushing—on!' I said. She looked at me with glad eyes, and touching her forehead, was gone. And I was right, Kautilya, I am pushing on."

And turning from Kautilya's sealed letter, he took another sheet and laboriously wrote a long letter to Sara, saying all there was to be said; explaining, confessing, offering to return to her if she wanted him, but on the conditions which she must already know. He received no answer. Yet once again he wrote and almost pleaded. Again he had no word.

XVI

There was a little court scene on State Street in April, 1927. It looked more like an intimate family party, and everybody seemed in high good humor. The white judge was smiling affably and joking with the Honorable Sammy Scott. Two or three attorneys were grouped about. Hats, canes, and brief-cases were handy, as though no one expected to tarry long. Mrs. Sara Towns came in. Mrs. Towns was a quiet and thoroughly adequate symphony in gray. She had on a gray tailor-made suit, with plain sheath skirt dropping below, but just below, her round knees. There was soft gray silk within and beneath the coat. There were gray stockings and gray suède shoes and gray chamoisette gloves. The tiny hat was gray, and pulled down just a trifle sideways so as to show sometimes one and sometimes two of her cool gray eyes. She looked very competent and very desirable. The Honorable Sammy's eyes sparkled. He liked the way Sara looked. He did not remember ever seeing her look better.

He felt happy, rich, and competent. He just had to tell Corruthers, aside:

"Yes, sir! She just got up of her own accord and gave me a kiss square on the beezer. You could 'a' bowled me over with a feather."

"Oh, she always liked you. She just married Towns for spite."

Sammy expanded. Things were coming very nicely to a head. The new Mayor had just been elected by a landslide; at the same time the Mayor's enemy, the Governor, knew that while Sammy had fought the Mayor in the primaries according to orders, he had nevertheless come out of the election with a machine which was not to be ignored. The pending presidential election was bound to set things going Sammy's way. The Mayor's popularity was probably local and temporary. The Governor had his long fingers on the powerful persons who pull the automata which rule the nation. These automata had been, in Sammy's opinion, quite convinced that no one would do their will in Congress better than the Honorable Sammy Scott. Moreover, the Governor and Mayor were not going to be enemies long. They could not afford it.

In other words, to put it plainly, the slate was being arranged so that after the presidential election of 1928, the succeeding congressional election would put the first colored man from the North in Congress, and it was on the boards that this man was to be Sammy. Meantime and in the three years ensuing, the prospective Mrs. Sara Scott and her husband were going to have a chance to play one of the slickest political games ever played in Chicago. Above all, Sammy was more than well-to-do, and Sara was no pauper. He wasn't merely asking political favors. He was demanding, and he had the cash to pay. Sammy rubbed his hands and gloated over Sara.

A clerk hurried in with a document. The judge, poising his pen, smiled benevolently at Sara. Sara had seated herself in a comfortable-looking chair, holding her knees very close together and yet exhibiting quite a sufficient length of silk stocking of excellent quality.

"Does the defendant make any reply?" asked the judge. And then, without pausing for an answer, he started to write his name. He had finished the first capital when some one walked out of the gloom at the back of the room and came into the circle of the electric light which had to shine in the office even at noontime.

Matthew came forward. He was in overalls and wore a sweater. Yet he was clean, well shaven, and stood upright. He was perhaps not as handsome as he used to be. His face seemed a bit weather-beaten, and his hair was certainly thinner. But he had an extraordinarily strong face, interesting, intriguing. He spoke with some hesitancy, looking first at the judge, then at Sara, to whom he bowed gravely in spite of the fact that after a startled glance, she ignored him. He only glanced at Sammy. Sammy was literally snarling with his long upper lip drawn back from his tobacco-stained teeth. He almost bit through his cigar, threw it away, and brought out another, long, black, and fresh. His hand trembled as he lit it, and he blew a furious cloud of smoke.

"I did not come to answer," said Matthew, "but simply to state my position."

("My God!" thought Sammy. "I'll bet he's got that bag of diamonds!")

"Have you got a lawyer?" asked the judge, gruffly.

"No, and I do not need one. I merely want to say—"

"What's all this about, anyway?" snapped Sammy.

Sara sat stiff and white and looked straight past Matthew to the wall. On the wall was a smirking picture of the late President Harding. An attorney came forward.

"We are willing that the defendant make any statement he wants to. Is there a stenographer here?"

The judge hesitated and then rang impatiently. A white girl walked in languidly and sat down. She stared at the group, took note of Sara's costume, and then turned her shoulder.

"I merely wanted to say," said Matthew, "that the allegations in the petition are true."

"Well, then, what are we waitin' for?" growled Sammy.

"I ran away from my wife and lived with another woman. I did this because I loved that woman and because I hated the life I was living. I shall never go back to that life; but if by any chance Mrs. Towns—my wife needs me or anything I can give or do, I am ready to be her husband again and to—"

He got no further. Sara had risen from her chair.

"This is intolerable," she said. "It is an insult, a low insult. I never want to see this, this—scoundrel, again."

"Very well, very well," said the judge, as he proceeded to sign the decree for absolute divorce. Sara and Sammy disappeared rapidly out the door.

Matthew walked slowly home, and as he walked he read now and then bits of his last letter from Kautilya. He read almost absent-mindedly, for he was meditating on that singularly contradictory feeling of disappointment which he had.

One has a terrible plunge to make into some lurking pool of life. The pool disappears and leaves one dizzy upon a bank which is no longer a bank, but just arid sand.

In the midst of this inchoate feeling of disappointment, he read:

"Are we so far apart, man of God? Are we not veiling the same truth with words? All you say, I say, heaven's darling. Say and feel, want, and want with a want fiercer than death; but, oh, Love, our bodies will fade and grow old and older, and our eyes dim, and our ears deaf, and we shall grope and totter, the shades of shadows, if we cannot survive and surmount and leave decay and death. No, no! Matthew, we live, we shall always live. Our children's children living after us will live with us as living parts of us, as we are parts of God. God lives forever—Brahma, Buddha, Mohammed, Christ—all His infinite incarnations. From God we came, to God we shall return. We are eternal because we are God."

Matthew sat down on the curb, while he waited for the car, and put his back against the hydrant, still reading:

"My beloved, 'Love is God, Love is God and Work is His Prophet'; thus the Lord Buddha spoke."

The street car came by. He climbed aboard and rode wearily home. He could not answer the letter. The revulsion of feeling and long thought-out decision was too great. He had drained the cup. It was not even bitter. It was nauseating. Instead of rising to a great unselfish deed of sacrifice, he had been cast out like a dog on this side and on that. He stared at Kautilya's letter. What had she really wanted? Had she wanted Sara to take him back? Would it not have eased her own hard path and compensated for that wild deed by which she had rescued his soul? Did not her deed rightly end there with that week in heaven? Was not his day of utter renunciation at hand? And if one path had failed, were there not a thousand others? What would be more simple than walking away alone into the world of men, and working silently for the things of which he and Kautilya had dreamed?

XVII

As Matthew reached the landing of his room, four long flights up, he saw a stranger standing in the gloom. Then he

noted that it was an East Indian, richly garbed and bowing
low before him. Matthew stared. Why, yes! It was the younger
of the two Indians of Berlin. Matthew bowed silently and bade
him enter. The room looked musty and dirty, but Matthew
made no excuses, merely throwing up the window and motion-
ing his guest to a seat. But the Indian bowed again courteously
and stood.

"Sir," he said, "I bear a rescript from the Dewan and High
Council of State of the Kingdom of Bwodpur, containing a
command of her Royal Highness, the Maharanee, and addressed, sir, to you. Permit me to read:

"To Matthew Towns, Esquire, of Chicago,
"*Honored Sir:*

"By virtue of the Power entrusted to us and by command of
our sovereign lady, H.R.H. Kautilya, the reigning Maharanee,
we hereby urge and command you to present yourself in person
before the Maharanee, at her court to be holden in Prince
James County, Virginia, U.S.A., at sunrise, May 1, 1927, there
to learn her further pleasure.

"Given at our capital of Khumandat
this 31st day of March, 1927,
at the Maharanee's command.

"BRABAT SINGH,
"Dewan."

"March 31?" asked Matthew.

"Yes," returned the Indian. "It was placed in my hands
this week with the command that it should be presented to you
as soon as the order of divorce was entered. In accordance with
these orders I now present the rescript." Again he bowed and
handed the document to Matthew. Then he straightened again
and said: "I bear also a personal letter from her Royal Highness which I am charged to deliver."

Matthew excused himself, and opening, read it:

"Matthew, Day has dawned. Of course a little Virginia farm
cannot bound your world. Our feet are set in the path of
moving millions.

"I did not—I could not tell you all, Matthew, until now.
The Great Central Committee of Yellow, Brown, and Black
is finally to meet. You are a member. The High Command is
to be chosen. Ten years of preparation are set. Ten more years

of final planning, and then five years of intensive struggle. In 1952, the Dark World goes free—whether in Peace and fostering Friendship with all men, or in Blood and Storm—it is for Them—the Pale Masters of today—to say.

"We are, of course, in factions—that ought to be the most heartening thing in human conference—but with enemies ready to spring and spring again, it scares one.

"One group of us, of whom I am one, believes in the path of Peace and Reason, of coöperation among the best and poorest, of gradual emancipation, self-rule, and world-wide abolition of the color line, and of poverty and war.

"The strongest group among us believes only in Force. Nothing but bloody defeat in a world-wide war of dark against white will, in their opinion, ever beat sense and decency into Europe and America and Australia. They have no faith in mere reason, in alliance with oppressed labor, white and colored; in liberal thought, religion, nothing! Pound their arrogance into submission, they cry; kill them; conquer them; humiliate them.

"They may be right—that's the horror, the nightmare of it: they may be right. But surely, surely we may seek other and less costly ways. Force is not the first word. It is the last—perhaps not even that.

"But, nevertheless, we have started forward. Our chart is laid. Our teeth are set, our star is risen in the East. The 'one far-off divine event' has come to pass, and now, oh, Matthew, Matthew, as soon as both in soul and body you stand free, hurry to us and take counsel with us and see Salvation.

"Last night twenty-five messengers had a preliminary conference in this room, with ancient ceremony of wine and blood and fire. I and my Buddhist priest, a Mohammedan Mullah, and a Hindu leader of Swaraj, were India; Japan was represented by an artisan and the blood of the Shoguns; young China was there and a Lama of Thibet; Persia, Arabia, and Afghanistan; black men from the Sudan, East, West, and South Africa; Indians from Central and South America, brown men from the West Indies, and—yes, Matthew, Black America was there too. Oh, you should have heard the high song of consecration and triumph that shook these rolling hills!

"We came in every guise, at my command when around the world I sent the symbol of the rice dish; we came as laborers, as cotton pickers, as peddlers, as fortune tellers, as travelers and tourists, as merchants, as servants. A month we have been gathering. Three days we have been awaiting you—in a single night we shall all fade away and go, on foot, by boat, by rail and airplane. The Day has dawned, Matthew—the Great Plan is on its way."

Matthew folded the letter slowly. She had summoned him—but to what? To love and marriage? No, to work for the Great Cause. There was no word of personal reunion. He understood and slowly looked up at the Indian. The Indian spoke again:

"Sir, with your permission, I have a final word."

"Proceed."

"I have delivered my messages. You have been summoned to the presence of the Princess. I now ask you—beg of you, not to go. Let me explain. I am, as you know, in the service of her Royal Highness, the Maharanee of Bwodpur. Indeed my fathers have served hers many centuries."

"Yes," said Matthew, without much warmth.

"You will naturally ask why I linger now. I will be frank. It is to make a last appeal to you—to your honor and chivalry. To me, sir, the will of the Maharanee of Bwodpur is law. But above and beyond that law lies her happiness and welfare and the destiny of India. When her Royal Highness first evinced interest in you and your people, we of her entourage foresaw trouble. Our first efforts to forestall it were crude, I admit, and did not take into account your character and ideals. We seriously underrated you. Yet yourself must admit the subsequent events proved us right.

"Once you were in trouble, and, as the Princess rather quixotically assumed, by her fault, it was her nature to dare anything in order to atone. She gave up everything and went down into the depths. It was only with the greatest difficulty that she was prevailed upon not to surrender the Crown itself. As it was, she gave up wealth and caste and accepted only barest rights of protection and guardianship of her person, upon which we had to insist.

"Finally in a last wild excess of frenzy, sir, she sacrificed to you her royal person. Sir, that night I was near murder, and you stood in the presence of death. But duty is duty, and the Princess can do no wrong. To us she is always spotless and forever right. But, sir, I come tonight to make a last plea.

Has she not paid to the uttermost farthing all debts to you,
however vast and fantastic they may appear to her? Can you—
ought you to demand further sacrifice?"

"Sacrifice?"

"Do you realize, sir, the meaning of this summons?"

"I thought I did. It is to attend a meeting which she has
called."

"What I say is from no personal knowledge—I have not seen
her Royal Highness since she left here; but the reason is
indubitable. The day of the coronation of a Maharajah in
Bwodpur is at hand."

Matthew started. "Her Royal Highness is—married?"

"She is to be married."

"And she is summoned to India?"

"She is. Three Indians of highest rank have arrived in this
country, and I believe they have come to fetch her and the
royal ruby."

"And why, then, has she summoned me?"

"Perhaps—she still hesitates between—"

"Love and duty?" said Matthew, dreamily.

"Between self-indulgent phantasy and the salvation of
Bwodpur," cried the Indian passionately.

"And I," said Matthew slowly, "can seal her choice."

"To few it is given to make a higher, finer sacrifice. You are
free. You have but to hint and you can be rich—pardon me—
I know. Well, what more? Will you not, in turn, free the
Princess? Do the fine and generous act; let her go back to
her people."

"Does the Princess wish this freedom?"

"She is one who would not admit it if she did. And yet her
very solicitude concerning Mrs. Towns—did it not suggest to
you that she saw in your reunion with Sara, on a higher and
more congenial plane, a chance for her to renew her own life
and work? Is it possible that she cannot yearn for something
beyond anything you can offer?"

"Yes, that occurred to me, and I made the offer to my
former wife—perhaps too crassly and ungraciously, but with
full sincerity."

"True—and now why not follow further and write the
Princess, definitely and formally withdrawing from her life,
and doing it with such decision that there shall be no doubt

in her mind?" The Indian bent forward with strained and eager face.

"You seem—anxious," said Matthew.

"I am," said the Indian. "You do not realize how our hopes for Bwodpur center on the Princess: an independent sovereignty about which a new Empire of India might gradually gather. Then, her eager and inexperienced mind, reaching out, leapt beyond to All India and All Asia; gradually there came a vision of all the Darker Races in the World—everybody who was not white, no matter what their ability or history or genius, as though color itself were merit.

"And now, now finally, God preserve us, the Princess is stooping to raise the dregs of mankind; laborers, scrubwomen, scavengers, and beggars, into some fancied democracy of the world! It is madness born of pity for you and your unfortunate people.

"With every dilution of our great original idea, the mighty mission of Bwodpur fades. The Princess is mad—mad; and you are the center of her madness. Withdraw—for God's sake and your own—go! Leave us to our destiny. What have you to do with royalty and divinity?"

The Indian was trembling with fervor and excitement, and his black eyes burned into Matthew's heart. "You will forgive me, sir. I have but done my duty as I saw it," he said.

Matthew looked at the Indian thoughtfully.

"I believe you are right," he said. "Quite right. I believe that you and your friends were right from the beginning and that I was—headstrong and blind. Now the problem is to find a way out."

"For the brave," said the Indian, slowly and distinctly, "there is always a way—out."

XVIII

Matthew stood awhile looking at the door where the Indian with low salaam had disappeared. Then, turning hastily, he put a few things into his handbag, and going out, closed the door. He left a note and key under the doormat and started downstairs, almost colliding with a boy who was racing up, two steps at a time.

"Looking for a man named Towns—know where he stays?"

"I'm Matthew Towns."

"Long distance wants you—quick—drugstore—corner." And he flew down, three steps at a time. Matthew stood still a long minute. He could not go away leaving her standing, waiting, listening. No. This thing must be faced, not dodged. He must talk to her. If she asked, he must even go to her. She, too, was no coward. Eye to eye and face to face, she would say the last word: she was summoned home to India. And then the final parting? He could say it—he would. They must work for the world—but she in her high sphere, and he in his, more lowly: forever parted, forever united in soul.

And more: this meeting which she had announced was of the highest importance. He must attend it and make it successful. He must show Kautilya that her return to India need not hinder nor in the slightest degree retard the Great Plan.

He descended slowly and went into the drugstore and into the little booth. How curious that he had never thought of evoking this miracle before in his heavy loneliness! Yet it was well. There was, there could be, but this ending; out of time and space he was calling a memory.

"Hello—hello! New York—hello, Richmond—go ahead."

At first the voices came strained, far-off, unnatural, interrupted with hissings and brazen echoes. Then at last, real, clear, and close, a voice came pouring over the telephone in a tumult of tone:

"Matthew, Matthew! I have heard the great good news. I am happy, very, very happy. And, Matthew, the friends are waiting. They want you here at sunrise."

"But, Kautilya—is it necessary that I come? Is it wise? I have been thinking long, Kautilya—"

"Matthew, Matthew, what is wrong? Why would you wait? Are you ill? Has something happened?"

"No, no, Kautilya, I am well—and if you wish me, I am coming—if the friends insist. But I have been wondering if I could not meet them elsewhere, a little later?"

"Later! Matthew—what do you mean?"

"I mean, Kautilya, that I have a duty to perform toward you and the world."

"Matthew, do you mean that you have changed toward me?"

"Changed? No, never. But I see more clearly—as clearly as you yourself saw when you bade me drain the cup."

"What have you feared, Matthew?"

"Nothing but myself. And now that fear is gone—I have drained the cup."

"Yes, dear one. And yet you knew that never and to no one could I give you up?"

"Rather I knew that each must surrender the other."

"To whom, Matthew?"

"To God and the Maharajah of Bwodpur."

A sound that was a sigh and a sob came over the 'phone. "Oh, God!" it whispered—"the Maharajah of Bwodpur!"

"Listen, Kautilya—I know—all."

"All?" she gasped.

"All! A Maharajah is to be crowned in Bwodpur."

A little cry came over the wire.

"And you have been summoned to the coronation—is it not true?"

"Yes."

"And you must go. Bwodpur—the darker peoples of the world call you. Would it not be easier if—if with this far farewell you left me alone to meet the committee and draft the Plan?"

"No—no—no, Matthew—you do not—you can not understand. You must come—unless—"

"Kautilya, darling, then I will come—of course I will come. I will do anything to make the broad straight path of your duty easier to enter. Only one thing I will not do, neither for Wealth nor Power nor Love; and that is to turn your feet from this broad and terrible way. And so to bid you Godspeed—to greet you with farewell and to hold you on my heart once more ere I give you up to God—I come, Kautilya."

Her voice sang over the wires:

"Oh, Matthew—my beautiful One—my Man—come—come!—and at sunrise."

"I am coming."

"And at sunrise?"

"But—impossible."

"Have you read the rescript? By sunrise, the first of May."

"But, dear, it is April 30. It takes a train—"

"Nonsense. There is an airplane fueled, oiled, and waiting for you at the Maywood flying field. Stop for nothing—go now; quickly, quickly, oh, my beloved."

Click. Silence. Slowly he let the receiver fall and turned away. He would not falter, and yet almost—almost he wished the truth otherwise. It would have been hard enough to surrender a loved one who wanted to be free, but to send away one who clung to him to her own hurt called for bitter, bitter courage; and dark and bitter courage stood staunch within him as he took out his watch. Or, perhaps, she too was full of courage and blithe and ready to part? He shivered. It was ten o'clock at night. The field was far away. He glanced up at his room, then paused no longer.

"Taxi—Maywood flying field. And quick!"

"Good Lord, boss, that's forty miles—it'll cost you near—"

"It's worth twenty-five dollars for me to get there in two hours."

The taxi leapt and roared. . . .

The pilot glanced scowling at the brown face of his lone passenger and climbed aloft. Matthew crawled into the tiny cabin. It was entirely closed in with glass save where up a few steps at the back perched the hard-faced pilot. There were seats for three other passengers, but they were empty.

There arose a roar—a roar that for seven hours never ceased, never hesitated, but crooned and sang and thundered. They moved. The lights of Chicago hurried backward. It was midnight. The lights swayed and swam, and suddenly, with a sick feeling and a shiver of instinctive fright, Matthew realized that they were in the air, off the earth, in the sky—flying, flying in the night.

Slowly and in a great circle they wheeled up and south. The earth lay dark beneath in dim and scattered brilliance. They left the great smudge of the crowded city and swept out over flat fields and sluggish rivers. Fires flew in the world beneath and dizzily marked Chicago. Fires flew in the world above and marked high heaven. Between, the gloom lay thick and heavy. It crushed in upon the plane. The plane roared and rose. Matthew could hear the beating and singing of wings rushing by in the night as though a thousand angels of evil were battling against the dawn. He shrank in his strait cabin and stared. His soul was afraid of this daring, heaven-challenging thing. He was but a tossing, disembodied spirit. There was nothing beneath him—nothing. There was nothing above him, nothing; and beside and everywhere to the earth's ends lay nothing. He was alone in the center of the universe with one hard-faced and silent man.

Then the strange horror drew away. The stars, the "ancient and the everlasting stars," like old and trusted friends, came and stood still above him and looked silently down: the Great Bear, the Virgin, and the Centaur. East curled the Little Bear, Hercules, and Boötes; west swung the Lion, the Twins, and the Little Dog. Vega, Arcturus, and Capella gleamed in faint brilliance.

The plane rocked gently like a cradle. Above the clamor of the engine rose a soft calm. Below, the formless void of earth began to speak with the shades of shadows and flickering, changing lights. That cluster of little jewels that flushed and glowed and dimmed would be a town; that comet below was an express train tearing east; that blackness was a world of farms asleep. In an hour Indianapolis was a golden scintillating glory with shadowy threads of smoke. In another hour Cincinnati—he groped at the map—yes, Cincinnati—lay in pools of light and shade, and the Ohio flowed like ink.

Suddenly the whole thing became symbolic. He was riding Life above the world. He was triumphant over Pain and Death. He remembered death down there where once the head of Jimmie thumped, thumped, on the rails. He heard the wail of that black and beautiful widowed wife. "They didn't show me his face!" He saw Perigua lying still in death with that smile on his lips, and he heard him say, "He didn't have no face!" Then came the slippers, her white and jeweled feet that came down from heaven and opened the gates of hell. Some one touched his shoulder. He knew that touch. It was arrest; arrest and jail. But what did he care? He was flying above the world. He was flying to her.

A soft pale light grew upon the world—a halo, a radiance as of some miraculous virgin birth. Lo! in the east and beneath the glory of the morning star, pale, faintly blushing streamers pierced the dim night. Then over the whole east came a flush. The dawn paused. Mountains loomed, great crags, gashed and broken and crowned with mighty trees. The wind from the mountains shrieked and tore; the plane quivered. A moment it stood still; then it dipped and swerved, swayed and curved, dropped, and shot heavenward like a bird. It pierced the wind-wound mists and rose triumphant above the clouds. The sun sprayed all the heavens with crimson and gold, and the morning stars sang in the vast silence above the roar, the unending roar of the airplane.

Matthew's spirit lifted itself to heaven. He rode triumphant over the universe. He was the God-man, the Everlasting Power, the eternal and undying Soul. He was above everything—Life, Death, Hate, Love. He spurned the pettiness of earth beneath his feet. He tried to sing again the Song of Emancipation—the Call of God—"Go down, Moses!"—but the roar of the pistons made his strong voice a pulsing silence.

The clouds parted, melted, and ran before the gleaming glory of the coming sun. The earth lay spread like a sailing picture—all pale blue, green, and brown; mauve, white, yellow, and gold. He faintly saw cities and their tentacles of roads, rivers like silver ribbons, railroads that shrieked and puffed in black and silent lines. Hill and valley, hut and home, tower and tree, flung them swift obeisance, and down, down, away down on the flat breast of the world, crawled men—tiny, weak, and helpless men: some men, eyes down, crept stealthily along; others, eyes aloft, waved and ran and disappeared.

Out of the golden dust of morning a city gathered itself. Its outstretched arms of roads moved swiftly, violently apart, embracing the countryside. The smudge of its foul breath darkened the bright morning. The living plane circled and spurned it, roared to its greeting thousands, swooped, whirled to a mighty curve, rose, and swooped again. Matthew's heart fell. He grew sick and suddenly tired with the swift careening of the plane. The sorrows of earth seemed to rise and greet him. He was no longer bird or superman; he was only a helpless falling atom—a deaf and weary man. They circled a bare field and fell sickeningly toward it. They dropped. His heart, his courage, his hopes, dropped too. They swooped again and circled, rose, and swooped, until dizzy and deaf they landed on an almost empty field and taxied lightly and unsteadily to a standstill. The engine ceased, and the roar of utter silence arose.

Matthew was on earth again, and on the earth where all its pettiest annoyances rose up to plague him. A half-dozen white men ran out, eager, curious. They greeted the pilot vociferously. Then they stared. Matthew climbed wearily down and stood dizzy, dirty, and deaf. They whispered, laughed, and swore, and turning, took the pilot to his steaming bath and breakfast and left Matthew alone.

Matthew stood irresolute, hatless, coatless in the crisp air, clad only in his jersey and overalls. Then he took a deep breath and walked away. In a wayside brook he bathed. He walked three miles to Richmond and boarded a train at six for his home. He found the Jim Crow car, up by the engine, small, crowded, and dirty. The white baggage men were washing up in it, clad in dirty undershirts. The newsboy was dispossessing two couples of a double seat and piling in his wares, swearing nobly. Matthew found a seat backward by a window. Leaning out, he spied a boy with lunches hurrying up to the white folks' car, and he induced him to pause and bought a piece of fried chicken and some cornbread that tasted delicious. Then he looked out.

The Spring sang in his ears; flowers and leaves, sunshine and shade, young cotton and corn. He could not think. He could not reason. He just sat and saw and felt in a tangled jumble of thoughts and words, feelings and desires, dreams and fears. And above it all lay the high heart of determination.

They rolled and bumped along. He sat seeing nothing and yet acutely conscious of every sound, every movement, every quiver of light, the clamor of hail and farewell, the loud, soft, sweet, and raucous voices. The movement and stopping, the voices and silence, grew to a point so acute that he wanted to cry and sing, walk and rage, scream and dance. He sat tense with half-closed eyes and saw the little old depot dance up from the far horizon, slip near and nearer, and slowly pause with a sighing groan. No one was there. Yes—one old black man who smiled and said:

"Mornin', Matthew, mornin'. How you comin' on?"

But Matthew with a hurried word had stridden on, his satchel in hand, his eyes on the wooded hill beyond. He passed through the village. Few people were astir:

"Hello, Matthew!"

"By God, it's Mat!"

The sounds fell away and died, and his feet were on the path—his Feet were on the Path! and the surge of his soul stifled his breath. He saw the wood, the broc k, the gate. Beyond was the blur of the dim old cabin looking wider and larger.

XIX

He saw her afar; standing at the gate there at the end of the long path home, and by the old black tree—her tall and slender form like a swaying willow. She was dressed in eastern style, royal in coloring, with no concession to Europe. As he neared, he sensed the flash of great jewels nestling on her neck and arms; a king's ransom lay between the naked beauty of her breasts; blood rubies weighed down her ears, and about the slim brown gold of her waist ran a girdle such as emperors fight for. Slowly all the wealth of silk, gold, and jewels revealed itself as he came near and hesitated for words; then suddenly he sensed a little bundle on her outstretched arms. He dragged his startled eyes down from her face and saw a child—a naked baby that lay upon her hands like a palpitating bubble of gold, asleep.

He swayed against the tall black tree and stood still.

"Thy son and mine!" she whispered. "Oh, my beloved!"

With strangled throat and streaming eyes, he went down upon his knees before her and kissed the sandals of her feet and sobbed:

"Princess—oh, Princess of the wide, wide world!"

Then he arose and took her gently to his breast and folded his arms about her and looked at her long. Through the soft and high-bred comeliness of her lovely face had pierced the sharpness of suffering, and Life had carved deeper strong, set lines of character. An inner spirit, immutable, eternal, glorious, was shadowed behind the pools of her great eyes. The high haughtiness of her mien was still there, but it lay loose like some unlaced garment, and through it shone the flesh of a new humility, of some half-frightened appeal leaping forth to know and prove and beg a self-forgetting love equal to that which she was offering.

He kissed the tendrils of her hair and saw silver threads lurking there; he kissed her forehead and her eyes and lingered on her lips. He hid his head in the hollow of her neck and then lifted his face to the treetops and strained her bosom to his, until she thrilled and gasped and held the child away from harm.

And the child awoke; naked, it cooed and crowed with joy on her soft arm and threw its golden limbs up to the golden sun. Matthew shrank a little and trembled to touch it and only whispered:

"Sweetheart! More than wife! Mother of God and my son!"

At last fearfully he took it in his hands, as slowly, with twined arms, they began to walk toward the cabin, their long bodies and limbs touching in rhythm. At first she said no word, but always in grave and silent happiness looked up into his face. Then as they walked they began to speak in whispers.

"Kautilya, why were you silent? This changes the world!"

"Matthew, the Seal was on my lips. We were parted for all time except your son was born of me. That was my fateful secret."

"Yet when first the babe leapt beneath your heart, still you wrote no word!"

"Still was the Silence sealed, for had it been a girl child, I must have left both babe and you. Bwodpur needs not a princess, but a King."

"And yet even with this our Love Incarnate, you waited an endless month!"

"Oh, silly darling, I waited for all—all; for his birth, for news to India, for your freedom. Do you not see? There had to be a Maharajah in Bwodpur of the blood royal; else brown reaction and white intrigue had made it a footstool of England. If I had not borne your son, I must have gone to prostitute my body to a stranger or lose Bwodpur and Sindrabad; India; and all the Darker World. Oh, Matthew—Matthew, I know the tortures of the damned!"

"And without me and alone you went down into the Valley of the Shadow."

"I arose from the dead. I ascended into heaven with the angel of your child at my breast."

"And now Eternal Life makes us One forever."

"Immortal Mission of the Son of Man."

"And its name?" he asked.

" 'Madhu,' of course; which is 'Matthew' in our softer tongue."

Crimson climbing roses, bursting with radiant bloom, almost covered the black logs of wide twin cabins, one rising higher than the other; the darkness of the low and vine-draped hall between caught and reflected the leaping flames of the kitchen within and beyond. Above and behind the roofs, rose a new round tower and a high hedge; the fields were green and white with cotton and corn; the tall trees were softly singing.

Old stone steps worn to ancient hollows led up to the hall and on them loomed slowly Matthew's mother, straight, immense, white-haired, and darkly brown. She took the baby in one great arm, infinite with tenderness, while the child shivered with delight. She kissed Matthew once and then said slowly with a voice that sternly held back its tears:

"And now, son, we'se gwine to make dis little man an hones' chile.—Preacher!"

A short black man appeared in the door and paused. He looked like incarnate Age; a dish of shining water lay in one hand and a worn book in the other. He was clad in rusty black with snowy linen, and his face was rough and hewn in angry lineaments around the deep and sunken islands of his eyes.

The preacher read in the worn book from the seventh chapter of Revelation:

"After these things I saw four Angels standing on the four corners of the earth"—stumbling over the mighty words with strange accent and pronunciation—"and God shall wipe away all tears from their eyes!"

Then in curious short staccato phrases, with pauses in between, he lined out a hymn. His voice was harsh and strong, and his breath whistled; but the voice of the old mother rose clear and singularly sweet an octave above, while at last the baritone of Matthew and the deep contralto of Kautilya joined to make music under the trees:

> "Shall I—be car—
> ried to-oo—the skies
> on flow-ry be-eds
> of ease!"

Thus in the morning they were married, looking at neither mother nor son, preacher nor shining morning, but deep into each other's hungry eyes. The voice of the child rose in shrill sweet obbligato and drowned here and again the rolling periods:

"—you, Kautilya, take this man . . . love, honor and obey—"

"Yes."

"—Towns, take this woman . . . until death do you part?"

"Yes—yes."

"—God hath joined together; let no man put asunder!"

Then the ancient woman stiffened, closed her eyes, and chanted to her God:

"Jesus, take dis child. Make him a man! Make him a man, Lord Jesus—a leader of his people and a lover of his God!

"Gin him a high heart, God, a strong arm and an understandin' mind. Breathe the holy sperrit on his lips and fill his soul with lovin' kindness. Set his feet on the beautiful mountings of Good Tidings and let my heart sing Hallelujah to the Lamb when he brings my lost and stolen people home to heaven; home to you, my little Jesus and my God!"

She paused abruptly, stiffened, and with rapt face whispered the first words of the old slave song of world revolution:

"I am seekin' for a City—for a City into de Kingdom!"

Then with closed and streaming eyes, she danced with slow and stately step before the Lord. Her voice lifted higher and higher, outstriving her upstretched arms, shrilled the strophe, while the antistrophe rolled in the thick throat of the preacher:

The Woman: "Lord, I don't feel no-ways *tired*—"

The Man: "*Children!* Fight Christ's fury, Hallelu*iah!*"

The Woman: "*I'm*—a gonta shout glory when this world's on *fire!*"

The Man: "*Children!* Shout God's glory, Hallelu!"

There fell a silence, and then out of the gloom of the wood moved a pageant. A score of men clothed in white with shining swords walked slowly forward a space, and from their midst came three old men: one black and shaven and magnificent in raiment; one yellow and turbaned, with a white beard that swept his burning flesh; and the last naked save for a scarf about his loins. They carried dishes of rice and sweetmeats, and they chanted as they came.

One voice said, solemn and low:

"Oh, thou that playest on the flute, standing by the water-ghats on the road to Brindaban."

A second voice, still lower, sang:

"Oh, flower of eastern silence, walking in the path of stars, divine, beautiful, whom nothing human makes unclean: bring sunrise, noon and golden night and wordless intercession before the wordless God."

And a third voice rose shrill and clear:

"Oh, Allah, the compassionate, the merciful! who sends his blessing on the Prophet, Our Lord, and on his family and companions and on all to whom he grants salvation."

They gave rice to Matthew and Kautilya, and sweetmeats, and all blessed them as they knelt. Then the Brahmin took the baby from his grandmother and wound a silken turban on its little protesting head—a turban with that mighty ruby that looked like frozen blood. Swaying the babe up and down and east and west, he placed it gently upon Kautilya's outstretched arms. It lay there, a thrill of delight; its little feet, curled petals; its mouth a kiss; its hands like waving prayers. Slowly Kautilya stepped forward and turned her face eastward. She raised her son toward heaven and cried:

"Brahma, Vishnu, and Siva! Lords of Sky and Light and Love! Receive from me, daughter of my fathers back to the hundredth name, his Majesty, Madhu Chandragupta Singh, by the will of God, Maharajah of Bwodpur and Maharajah-dhirajah of Sindrabad."

Then from the forest, with faint and silver applause of trumpets:

"King of the Snows of Gaurisankar!"

"Protector of Ganga the Holy!"

"Incarnate Son of the Buddha!"

"Grand Mughal of Utter India!"

"Messenger and Messiah to all the Darker Worlds!"

Envoy

The tale is done and night is come. Now may all the sprites who, with curled wing and starry eyes, have clustered around my hands and helped me weave this story, lift with deft delicacy from out the crevice where it lines my heavy flesh of fact, that rich and colored gossamer of dream which the Queen of Faërie lent to me for a season. Pleat it to a shining bundle and return it, sweet elves, beneath the moon, to her Mauve Majesty with my low and fond obeisance. Beg her, sometime, somewhere, of her abundant leisure, to tell to us hard humans: Which is really Truth—Fact or Fancy? the Dream of the Spirit or the Pain of the Bone?

Congress Cataloging in Publication Data

 under title:

n to computer architecture.

graphy: p.
es index.
ctronic digital computers. I. Stone,
1938-
           621.3819'5   75-14016
4-18405-8

ments

 acknowledge the following for their permission to reprint the
ed:

Monroe Division, Litton Industries: figure 3-1 (parts *a* and *b*)
Smith-Corona Marchant: figure 3-2
Texas Instruments, Inc.: figures 3-3 and 3-29 (parts *a* and *b*)
Hewlett-Packard: figure 3-39

Introduction to Compute

Harold S. Stone, Editor
University of Massachusetts

Tien Chi Chen
IBM San Jose Research Laboratory

Michael J. Flynn
Stanford University

Samuel H. Fuller
Carnegie-Mellon University

William G. Lane
Department of Computer Science
California State University, Chico

Herschel H. Loomis, Jr.
University of California, Davis

William M. McKeeman
University of California, Santa Cruz

Kay B. Magleby
Cushman Electronics, Inc.

Richard E. Matick
IBM Research

Thomas M. Whitney
Hewlett-Packard Corp.

S R A SCIENCE RESEARCH ASSOCIATES, INC.
Chicago, Palo Alto, Toronto, Henley-on-Th

A Subsidiary of IBM

The SR

Mark E
Mark E
Peter Fr
C. W. G
Stephen
Harold S
Harold S

Library of

Main entry

Introducti

Biblio
Inclu
1. Ele
Harold S.,
QA76.5.I6
ISBN 0-57

Acknowledg

We wish t
materials lis
Courtesy of
Courtesy of
Courtesy of
Courtesy of

Preface

This text is intended as a first book in computer architecture, carrying the student from an elementary overview through a detailed discussion of the various fundamental aspects into several specialized and advanced topics. Part I is suitable for a one-semester course for juniors and seniors. Part II is a natural extension of part I suitable for a second semester for juniors and seniors or as a stand-alone section for first-year graduates. By treating the material in each chapter lightly and selectively, an instructor can cover the bulk of the topics presented here in one semester.

The book is designed to satisfy two types of needs. One need is for the student who wishes to understand the subject of computer architecture because of other interests in computer science but does not wish to learn how to build a computer. This student has generally taken at least one course in programming computers, perhaps in a high-level language, and preferably has learned some of the more intimate details about computers, perhaps through a course on programming in assembly language. He has not had switching theory and logic design and does not have any experience reading logic diagrams or designing with digital logic, but he will be comfortable with most of the text. Some mention of logic design concepts is made in Chapter 3 and again in Chapter 10, but these are not extensive; if the student has used the logic operations AND or OR instructions in programs, he should have no difficulty with this material. He should also have little difficulty with the notation used to describe machine functions, since the notation is purposely made to be similar to ALGOL 60.

The second type of student may have had experience in switching theory and logic design but very little in programming a machine, and very little or no exposure to assembly language. For this student, the book is sprinkled with selected fragments of programs to illustrate the execution of programs and to indicate how this execution is influenced by computer architecture. The examples will teach the student about software considerations while informing him about computer architecture.

This text carries numerous examples of architectural features that have been designed to support specific software functions. Design of this type is possible only if the computer architect has a strong background in both logical design and programming. Material is provided to strengthen a student in both these areas without presuming extensive knowledge in either. University curricula should ideally reflect the intent of this text by providing programs that bridge the two areas.

One of the important features of the book is the inclusion of extensive problems at the end of each chapter. They do not require access to a computer, but often involve the programming of fragments of a computer program. Students should be encouraged to program and debug these program fragments on a real computer, preferably on several different computers.

By starting with calculators, the text provides a natural entry point to the study of computer architecture. Since many different types of calculators are readily available for classroom work, the student can learn first-hand about such considerations as Polish notation versus algebraic entry, programming (on programmable hand calculators and desk calculators), keyboard and display input/output, and floating-point arithmetic. The primary advantage of using calculators for this purpose is that the student can focus attention directly on the topics of importance, since he does not have to learn a great deal of subsidiary knowledge such as system software, job control, program syntax, and logging in to a system. After his introduction to basic subjects by means of hand calculators, he can easily adapt his knowledge to full-fledged computers.

Recommended support for later chapters should include software simulation packages for a number of simple machines. The machines simulated should use a variety of instruction formats to give the student a chance to explore the power of different types of instructions, effective address computations, and encodings. With some additional effort, the simulator may be able to simulate advanced topics treated in later sections, such as parallel and pipelined computer systems and microprogrammed processors. The student should be able to write very simple program fragments for such machines and debug them with the simulator.

Another activity that might be used to support the educational process is a design laboratory in which the student wires together highly integrated functional devices to construct portions of a computer. Another interesting project is a computer design term project. The term project is more ambitious than the design laboratory, because the student need not specify every possible detail and need not face the very difficult task of debugging a design.

Before closing this preface, I wish to thank Steve Mitchell, whose inspiration led to the development of this text. He conceived of the idea, outlined its form, and brought together the co-authors. My role as editor was greatly simplified by the co-authors themselves, who produced the material during pressing times and occasionally met my unrealistic deadlines. Thanks are in order also to George Miller, Amar Mukhopadhyay, Myron Calhoun, Bruce Barnes, Martha Sloan, and Samuel Fuller, whose critical comments on the book helped tie the chapters together and generally improved the exposition itself. Moira Lieberman performed an invaluable service as typist for several chapters, but, more importantly, she was the center of communications that held the writing team together in the crucial final period of manuscript preparation. Paul Kelly and Betty Drury of SRA and Mary Curras, technical editor, were largely responsible for turning a massive pile of unconsecutively numbered pages and hand-scrawled figures into the finished text you see before you. Only through the intense efforts of the people named here and many others unnamed could a project of this complexity reach fruition.

Harold S. Stone
Amherst, Massachusetts

Contents

PART I

Basic Computer Architecture

CHAPTER 1

Introduction

Harold S. Stone

The study of computer architecture is the study of the organization and interconnection of components of computer systems. Computer architects construct computers from basic building blocks such as memories, arithmetic units, and buses. From these building blocks the computer architect can construct any one of a number of different types of computers, ranging from the smallest hand-held pocket calculator to the largest ultra-fast super computer. The functional behavior of the components of one computer are similar to that of any other computer, whether it be ultra-small or ultra-fast. By this we mean that a memory performs the storage function, an adder does addition, and an input/output interface passes data from a processor to the outside world, regardless of the nature of the computer in which they are embedded. The major differences between computers lie in the way the modules are connected together, the performance characteristics of the modules, and the way the computer system is controlled by programs. In short, computer architecture is the discipline devoted to the design of highly specific and individual computers from a collection of common building blocks.

The plan of this text is to take the reader through the central topics of computer architecture, starting with the most general topics to lay the foundation for further reading and leading to more specialized topics involving such areas as the design of the memory and input/output systems and the design of high-speed computers. In this chapter, we present the most elementary aspects of computer design as a brief preparation for what lies ahead. Here we cover a simplified machine architecture as a case study, and, in the course of the design exercise, we expose the several areas that are treated in each of the chapters to follow.

This book is organized into two parts. Part I consists of Chapters 1 through 6 and contains the core introductory material for a good foundation in computer architecture. Part II starts with Chapter 7 and focuses on several specialized topics in depth, carrying them through recent research developments.

In Chapter 2, we investigate computer arithmetic and data representations. The algorithms are given in sufficient detail to make hardware implementation straightforward. Also treated are representations of structures of data as well as the representation of individual numbers.

Chapter 3 treats the design of the smallest and least powerful of all computer systems, hand-held calculators. For these computers, the major design constraints are cost, size, and power consumption. Raw computation speed is often not an important factor simply because the speed at which humans can manipulate calculator buttons is orders of magnitude slower than the basic clock rate of a calculator. The hand calculator is the first major application of the computer-on-a-chip, often known as a *microprocessor*. The chapter serves as an introduction to microprocessor architecture and design.

Chapter 4 begins a series of chapters on conventional computer architecture. In this chapter, we discuss minicomputers, and then follow with chapters on memories and input/output systems. Together, these three chapters contain the material necessary to master the structure of conventional computers.

Minicomputers are used as the example of conventional computers, primarily because minicomputers of today are full-fledged computers in every sense of the word, but they are physically smaller and cost less than their largest counterparts. Minicomputers need not be less powerful than a full-scale computer, and indeed, many minicomputers available today are every bit as powerful as the largest scientific computers of many years ago. The severe cost and size constraints on minicomputers often dictate that the minicomputer be designed with just the bare essentials of functional hardware. Consequently, the architecture of these machines lacks frills and luxuries, resulting in designs of great simplicity and comprehensibility. As such, they make excellent vehicles for the initial study of computers.

Chapters 5 and 6 deal with the subjects of computer memories and input/output, respectively. Both of these chapters are relevant to a variety of advanced types of computers because the differences in memories and input/output systems across the range of computers is small when compared to the differences in processors and in the organization of the system modules. The chapters treat the variations that exist from system to system.

After Chapter 6, we begin Part II and investigate machine architectures emerging in the 1970s. Chapter 7 discusses computers based upon a push-down stack architecture. Such computers appeared in the early 1960s, but relatively few of these computers were installed as compared to the more conventional register-structured computer. Early examples of such machines include the KDF-9 and the Burroughs B-5000. The stack machines have interesting properties that lead to efficiencies of operation, and these are studied in the chapter. The particular advantages of stack machines in some environments, specifically for implementing subroutine calls and parameter passing, have led to a large increase in their popularity today.

High-speed computation has been an area in which advances have generally been attributed to both computer architecture and inherently faster components, and future improvements are likely to be due to architecture rather than raw speed. High-speed computers have deviated from conventional computers to attain greater computational ability, and the speed increase is due mainly to replication of computational hardware to form vector processors, multiprocessors, and pipeline computers. The descriptions of these various types of computers appear in Chapters 8 and 9.

Microprogramming is a hardware implementation technique that is the central theme of Chapter 10. Microprogramming refers to the technique of using a programmed "inner" computer to perform the various control functions required for each instruction in the repertoire of an "outer" computer. Microprogramming uses control programs stored in a separate high-speed control memory to replace wires, flip/flops, and logic gates of a "hard-wired" implementation of the control function. Microprogramming was used as an implementation technique for several of the computers designed in the infancy of computer design in the early 1950s, but it then fell into disuse because of low performance relative to other implementation methods. It became increasingly popular in the middle of the 1960s with the introduction of the IBM System/360. The smaller machines in this series made heavy use of microprogramming because this was the most economical way to implement the large instruction repertoire required for the computer. The faster and more expensive machines in the series also made use of microprogramming, but specialized high-speed logic replaced microprogrammed functions of the smaller machines. In recent years, microprogramming has bloomed into an area of wide interest because of the ease of emulation of different computers and the possibility of tailoring a computer to specific applications to attain greater efficiency.

Chapter 11 explores methods for evaluating the performance of computer systems to enable the computer architect to determine how well his computer system design meets his design goals. The analysis methods include mathematical methods that model the computer system according to principles of queueing theory, event-driven simulation models that predict behavior by crude imitation of the computer system in a realistic environment, and experimental techniques that measure the behavior of the actual system in real working conditions. Each of these techniques is useful, but each has its own weaknesses. Use of a combination of all three techniques is required to guarantee accurate prediction of behavior.

Returning to this introductory chapter, we discuss the simplified conventional computer in the next section, and, in a later section, we explore several issues that are the subject of the chapters sketched above.

1-1. ARCHITECTURE OF AN EARLY COMPUTER

In this section, we examine a highly idealized and simplified computer to serve as a point of departure for material in the remainder of the book. To place this material in a historical context, the machine we describe follows the general description of one of the earliest computers. A paper by Burks, Goldstine, and von Neumann [46] detailed the organization of this computer, and was remarkable in the thoroughness and insight the authors displayed. The influence of the proposal was considerable. Almost all of the computers designed and constructed in the next decade embodied many of the ideas in the proposal. This type of machine is often called a *von Neumann machine* to credit the contribution of John von Neumann to its development. The proposed machine was eventually constructed and run successfully in the early 1950s, but it was not the first computer constructed nor was it the first von Neumann computer to run successfully. Wilkes in England and later Eckert and Mauchly, with the aid of von Neumann's ideas, completed similar machines before von Neumann's reached completion. They generally share the credit for the construction of the first modern electronic computers [Rosen, 69].

Figure 1-1 shows the general structure of the computer proposed by Burks et al. The memory consists of $4096 = 2^{12}$ words, with each word containing 40 bits. Each word is identified with a unique address in the range $0 \leq x \leq 4095$. The proposal by von Neumann and his colleagues called for the storage medium to be a cathode ray storage tube. A particular word is read from this memory by presenting a 12-bit address to the memory address register (identified in the figure as the M register) together with a READ control signal. The selected word is copied from memory and appears somewhat later in the memory data register, denoted S in the figure. Von Neumann anticipated a memory access time from 5 to 50 microseconds in his design, which was quite ambitious for his time. It was roughly a decade before memories with this access time were made widely available through the development of magnetic core technology. Today, of course, with improved core technology and integrated circuit memory technology that is gradually supplanting magnetic cores, memories with access times below .5 microsecond are common.

To write data into the memory, a datum is placed into the S register, the address of the datum is placed in the M register, and a WRITE signal is generated. The memory responds to this signal by copying the datum in the S register into the word selected by the M register. The read and write modes of memory operation described by von Neumann are common to almost all present day computers.

Returning to Figure 1-1, we see that data manipulation is done in the arithmetic unit portion of the computer. This portion consists of an adder/ shifter, an accumulator register (the A register) and an accumulator extension (the B register). Addition and subtraction make use of the adder and the A register. Multiplication requires both shifting and addition, and produces a double length result from two 40-bit operands. The product of a multiplication is captured in the A and B register pair with the most significant bits in A and

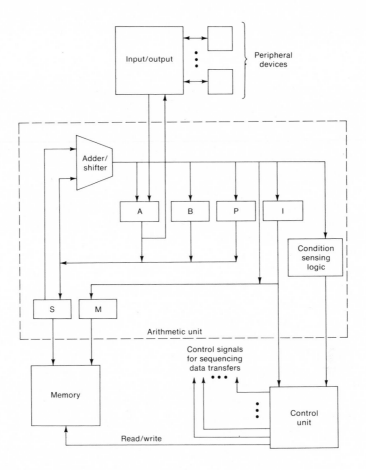

Figure 1-1   Block diagram of a simplified computer

the least significant in B. Division can be done with the hardware given in the figure using an algorithm similar to the familiar algorithm for decimal division.

The control unit shown in Figure 1-1 is the functional unit that forces the memory and arithmetic unit to perform the sequential steps of an algorithm. Von Neumann and his colleagues were well aware of ways of stating an algorithm as a sequence of elementary actions, each elementary action executable by the computer. A computation is done by presenting each step of the algorithm in encoded form to the control unit, which then generates signals that cause the arithmetic unit and memory to perform the necessary actions. The von Neumann machine was among the first to embody the notion that instructions can be stored in the same memory as the data. In this way, instructions can be removed and rewritten into memory much like data, thereby making it possible to change from program to program quite easily. Computers with this property are called *stored-program* computers. As an added property of this instruction storage method, instructions are indistinguishable from data, so they may be manipulated just as data are manipulated. The interpretation of a 40-bit pattern as an instruction or a datum depends on the state of the machine when the item is fetched from memory. If the state of the machine dictates that the item should be passed to the control unit, then the item is interpreted as an instruction. If, on the other hand, the item is transferred to the A register, it is treated as a datum.

The 1946 proposal contains a great insight concerning the use of instructions as data. Special instructions were incorporated in the computer repertoire that modify just the address portion of an instruction. In this way, programs could modify their own instructions during program execution so that a single instruction could be made to operate on many different data. This tended to reduce the memory requirements for programs. This concept was embodied in most computers through the early 1960s when it finally fell into disfavor. By then, much more elaborate and powerful mechanisms to attain the same functional capability had come into use, and these do not suffer from disadvantages of self-modification of instructions. Nevertheless, von Neumann's original idea has been extremely influential, even if not implemented as he originally intended.

Since instructions were stored in the same 40-bit words in which numbers were stored, Burks et al. proposed an instruction encoding compatible with the data format. The encoding is shown in Figure 1-2. Each instruction occupies 20 bits, with two instructions stored in one word. Each 20-bit instruction contains two fields—an 8-bit operation code that determines the function performed, and a 12-bit address of an operand. At this point in our discussion, some of the architectural decisions in the design become apparent. The instruction encoding, the size of the instruction repertoire, the precision of the data representation, and the size of memory are all interrelated. The 40-bit word length gives sufficient precision and numeric range to be useful over a reasonably large class of problems. Longer word lengths increase the cost of the computer, without necessarily giving useful additional numerical precision for typical calculations. A shorter word length might have been possible from the point of view of numerical precision but the instruction encoding would be severely hampered if the word length were shorter. Hence, 40 bits is reasonable for each datum.

Instructions with fewer than 20 bits have less room either for the operation code or for the address, and if either is shortened the computer is severely restricted in power. If, on the other hand, a single instruction is stored per word because a word is too short to contain two instructions, there are too

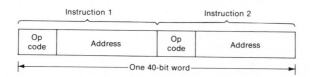

Figure 1-2 Instruction encoding for the von Neumann computer

many bits to specify all of the instructions that could be implemented in von Neumann's machine within the economic constraints of the times. Consequently, some memory is wasted in the instruction encoding.

Within a decade, memories with 2^{15} words became prevalent, so that address fields quickly expanded to 15 bits. Likewise address modification functions such as indexing and indirection were incorporated into the instruction encoding, with the result that one instruction was encoded in as many as 36 bits, and numeric data representations were made equal in length to the length of a single instruction.

Several representative instructions appear in Table 1-1. We also introduce a notation in this table that we follow throughout this text. Machine registers are denoted by their names such as A or B, and memory locations are denoted by the name MEMORY followed by a bracketed expression whose value gives the address of the item in memory. The ALGOL symbol ": =" denotes data flow, and indicates that the value of the expression on the right of the symbol is moved to the register or memory location named on the left. For example, the instruction LOAD X has the effect of copying the contents of a designated memory location into the A register. In the example shown in the table, the contents of MEMORY[X] are moved into the A register, where X is an address in the range $0 \leq X \leq 4095$. Reversing the right side and the left side of the ALGOL-like notation yields MEMORY[X] := A which is a "store" instruction that copies the contents of A into the location X of memory. Note how MEMORY[X] denotes a numeric operand

TABLE 1-1. TYPICAL INSTRUCTIONS FOR A VON NEUMANN COMPUTER

Instruction	Example	Action
Load	LOAD X	A := MEMORY[X];
Store	STORE Y	MEMORY[Y] := A;
Load negative	LNEG Z	A := − MEMORY[Z];
Load absolute value	LABS W	A := ABS(MEMORY[W]);
Add	ADD X	A := A + MEMORY[X];
Subtract	SUB Y	A := A − MEMORY[Y];
Multiply	MUL Z	A := A × MEMORY[Z];
Divide	DIV W	A := A / MEMORY[W];
Branch (left)	BRAL X	P := X, LEFT INSTRUCTION;
Branch (right)	BRAR Y	P := Y, RIGHT INSTRUCTION;
Branch positive (left)	BPOSL X	if A ≥ 0 then P := X, LEFT INSTRUCTION;
Branch positive (right)	BPOSR Y	if A ≥ 0 then P := Y, RIGHT INSTRUCTION;
Store address (left)	STADL X	MEMORY [X], LEFT ADDRESS := A;
Store address (right)	STADR Y	MEMORY [Y], RIGHT ADDRESS := A;
Shift (left)	SHL	A := 2 × A;
Shift (right)	SHR	A := A ÷ 2;

value when it appears on the right of ": =", while it names the destination of a data move when it appears on the left of ": =". This is in keeping with the conventions of most programming languages.

We see from the table that the von Neumann machine could load the A register with the true or negative value of an operand, or with its absolute value. Also, it could add, subtract, multiply, or divide two numbers, one in the A register and one in a designated memory cell.

Two branch instructions and two conditional branch instructions have been included. In each case the two forms of an instruction are used to select between the left and right instructions that lie in a single word. The use of the P register in these instructions is explained below. Address modification in this computer is done by replacing the address part of an instruction with a new address. The two instructions that modify addresses are again identical except that they modify the left and right instruction address fields, respectively. When an address is modified by one of these instructions all other bits in the word are unmodified.

The shift instructions give the programmer access to the hardware for the shift operation, which has to be there to do the multiplication and division. In this example the left shift doubles the numerical value stored in the A register, and the right shift halves that value.

This completes the discussion of the basic instructions of the von Neumann computer. At this point we can touch upon the design of the control unit and describe how the execution of an instruction takes place. In this machine the sequence of steps during execution of a single instruction depends only on the 8-bit operation code of the instruction. The memory address determines the operand of the instruction but does not affect the individual steps in the instruction execution sequence.

The control unit makes use of two registers, P and I, that respectively contain the address of the next instruction to be executed and a copy of the instruction currently in execution. The P register is actually 13 bits in length rather than 12 for it must contain 12 bits to identify a memory address and an additional bit to identify the left or right instruction in the word at that address.

Instruction execution involves the following steps:

1. Read the next instruction into the I register. The address of the instruction is contained in the P register.
2. Update the P register to specify the instruction immediately following the instruction just read.
3. Decode the 8-bit operation code. The decoder output has one output line for each instruction in the computer repertoire. An output line is energized by the decoder if and only if the instruction contains the corresponding operation code.
4. Generate a sequence of data transfers and numerical operations that

constitute the execution of the instruction. The sequence of actions may include reading or writing of memory, transfers of data to and from registers, transfers of data through the adder/shifter for numerical processing, or a change of the value of the P register if the instruction is a branch instruction.

At the close of the last step, the instruction is fully executed. The control unit then returns to Step 1 to execute the next instruction, and continues to repeat the execution of instructions until it stops because of the execution of a HALT instruction or because of a console switch action that forces a halt.

There are various ways of constructing the control unit, and we leave a full discussion of its internal structure to Chapters 3, 4, and 10. Its functional behavior, however, can be determined almost completely from the instruction encoding and the registers and data transfer paths in the remainder of the computer.

The last portion of the computer shown in Figure 1-1 is the input/output unit. The von Neumann proposal was probably less influential on future computer designs in this area than in other areas, but the proposal authors still had sufficient vision to foresee some powerful notions of input/output.

Perhaps the most tantalizing notion is the graphical output they proposed. The Selectron tube, being a cathode ray tube, glowed in positions in which 1 bits were stored and did not glow where 0 bits were stored. By forcing patterns of ones (1s) and zeros (0s) in memory, the programmer could display graphical output on the Selectron tube. Of course, cathode ray output is quite common today, and it bears some resemblance to the Selectron input/output available to von Neumann, but rarely is such a tube used as a memory and as an input/output device simultaneously.

Other input/output media proposed included teletypewriter and an auxiliary magnetic wire memory, which is suggestive of tapes, drums, and disks available today. Data flow for input/output came through the A register to the input/output unit and from there to the peripheral units. Input/output was controlled by the program on a word-by-word basis. For example, to output a data array, each array element had to be loaded into the A register and transmitted to the input/output unit, with each step of this sequence controlled by instructions in the program. There was no capability for overlapping input/output with computation in the proposed machine, primarily because of the high cost of hardware at that time. Nevertheless, early computer designers including the von Neumann team recognized that overlapped input/output and computation was essential for efficient use of a computer. Buffering of input/output to permit this overlap was added to computers while they were still in their infancy.

Now that we have treated the overall design of the von Neumann machine, in the next section we shall reconsider the important aspects of its design and indicate how they have evolved in the past three decades.

1-2. TRENDS IN COMPUTER ARCHITECTURE SINCE VON NEUMANN

The previous section suggests four areas in which we can focus our discussion. They are the:

1. Arithmetic unit
2. Control unit
3. Memory
4. Input/output unit

In this section we give a brief view of how the von Neumann computer has evolved to today's computers in each of these areas.

Arithmetic Unit

The von Neumann computer had an arithmetic unit consisting of an adder and a shifter, each fairly rudimentary. By the late 1950s it was common to include multiplier hardware, fast adders, fast shifters, and in some instances, hardware for automatic floating-point arithmetic, in order to attain higher computation speed. Division then, as now, is seldom done with hardware specialized to division. It is more common to use an iterative algorithm which in turn uses multipliers, adders, and shifters.

By the middle of the 1960s and reaching into the present, the main deviation from the basic design of an arithmetic unit as presented by Burks et al. has been in ultra high-speed computers. Such computers typically use many different arithmetic units, or a single unit timeshared to do parts of many different computations simultaneously. In extreme cases as many as 64 full arithmetic units are combined in one computer as, for example, in the ILLIAC IV computer. Computers consisting of many independent arithmetic units are typically known as *parallel* or *array* computers. Chapter 8 covers various ways of constructing computers of this general class. Computers in which some fast portion of the arithmetic processor can service several computations in different phases at a single time are generally known as *pipeline* computers. These are described in some detail in Chapter 9.

With the rapid development of integrated circuit technology there came the possibility of doing low-speed computation at minimum cost. Hand-held calculators are typical examples of what can be done in this area. For such devices arithmetic operations on whole words simultaneously are far too expensive, and the speed is unnecessary. Consequently, arithmetic is done serially, usually in 1-bit or 4-bit groups. The architecture of these devices appears in Chapter 3.

The algorithms for performing some of the complex arithmetic operations, such as multiplication and division, appear in Chapter 2 together with a discussion of techniques for dealing with negative numbers.

Control Unit

The control unit presents the most interesting challenge for computer architects today because it includes the design of the instruction repertoire, the instruction encoding, and the hardware implementation of the instructions. The control unit is a critical architectural feature of a computer, since it strongly influences the ultimate cost and performance of the machine.

We briefly review the evolution of the design of control units since von Neumann by examining the following areas:

1. Effective address calculations
2. Instruction repertoire
3. Instruction encoding
4. Control unit implementation

Von Neumann, his colleagues, and other early computer designers recognized the need to have one instruction operate on many different data during a calculation. Some computations require a thousand, a million, or even a billion repetitions of a sequence of instructions on different data. It is essential to use a single copy of the instructions for that iteration rather than a different copy for each iteration to eliminate the need to construct and store all of those instructions. The capability to reuse an instruction on different data was built into the von Neumann machine by including instructions that modify the address portion of an item resident in memory. The item modified could be an instruction, and once modified the original form of the modified instruction was lost. Typical iterative loops contained instructions to modify one or more instructions in the loop in preparation for the next repetition of the loop.

The process of instruction modification was both costly in time and error-prone. Moreover, because the instructions were physically changed, a program could not easily be restarted from its beginning at any time during a computation. Restart procedures required a fresh copy of the program to be loaded into memory to begin a computation. This greatly hindered program checkout and debugging.

Automatic address modification techniques were introduced soon after the construction of the first computers. These entailed the use of special registers known then as *b-boxes*, and now as *index registers*, to hold the incremental changes of the addresses in instructions. Let REG[1], REG[2],..., REG[N] denote a set of registers that are part of the arithmetic unit, and let ADDR denote the integer encoded in the address portion of an instruction. Then the operand address of that instruction is MEMORY[ADDR] when no indexing occurs, and is MEMORY[ADDR + REG[INDEX]] when REG[INDEX] is selected as an index register. In many computers the index register is specified by a small field in each instruction. A zero value in this field denotes no indexing, and a nonzero value identifies the index register to be used for indexing. The operand address produced by indexing or by any other address modification is called the *effective address* of the instruction.

Note that the operand address of an instruction may be quite different from the value of ADDR, and that the instruction itself is not altered in memory by the indexing operation. Indexing alters the effect of the instruction without changing the instruction. While indexing is perhaps among the most frequently implemented address modification methods, it is only one of several in use in present computers. For this brief summary we mention another method known as *indirection,* for which the operand address is not ADDR but is found in MEMORY[ADDR], and the operand of the instruction is in MEMORY [MEMORY[ADDR]]. Thus with indirection what would normally be an operand for an instruction becomes the address of the operand.

Computers with both indexing and indirection can combine the methods in either of two ways. *Pre-indexing* does the indexing and then the indirection to give an effective address of MEMORY[ADDR + REG[INDEX]]. *Post-indexing* does the indirection and then the indexing to yield an effective address of the form MEMORY[ADDR] + REG[INDEX]. The more operations done in a single address modification operation, the more specialized is the use of that modification. For example, post-indexing is very useful in the context of a subroutine with an array parameter. The array is to be accessed through an indirect address, and the subroutine computes indices for accesses to array elements. Post-indexing is ideal for this because the address of an array element is obtained by adding the index to the address obtained from an indirect reference. Similar examples exist to show the usefulness of pre-indexing. Perhaps post-indexing is the more useful of the two methods if one had to decide between them, but the designer should make such a decision only after a thorough study of the intended applications of the computer.

This brings us to the next subject of this section, the instruction repertoire. The instruction repertoire of the early computers typically contained fixed-point arithmetic instructions, boolean and shift instructions, and a few primitive instructions for controlling the instruction sequence.

Second generation computers had much larger instruction sets for more flexibility and power. These computers fell into two broad categories: business computers and scientific computers. Business data processing computers were typically character oriented and had extensive capability for manipulating variable length data. Data operations were generally character replacements, or character comparison for sorting and merging, or fixed-point arithmetic. Arithmetic was slow in this type of computer because it was generally done serially by character, and often done decimally rather than in radix 2. The other class of computer at this time was the scientific computer. It had extensive floating-point arithmetic capability, but operated on data of fixed precision. It had very little capability for dealing with variable length strings of characters, and did not have decimal arithmetic, although such operations obviously could be performed through programming at a relatively high cost in computing time. The most prevalent computer series of this type was the IBM 709, 7090, and 7094 series. Toward the end of the second generation the distinction between business data processing and scientific computing tended to fade. Large scientific users often performed computations involving sorting, merging, and character manipulation on variable length data, which were previously classed as business functions. Likewise business applications made use of sophisticated forecasting and inventory control programs which in turn relied heavily on operations typical of scientific computing. Beginning in the middle 1960s manufacturers combined both kinds of functions into one computer.

The major obstacle in the way of having a single general-purpose computer for both business and science was cost. As the number of instructions in a computer repertoire grew then so did the cost of its control unit, which in turn materially affected the cost of the machine to the user. New device technologies and new design techniques brought hardware costs down dramatically in the middle 1960s, thereby making it possible to construct a general-purpose computer to satisfy all users. As computers moved into the third generation, instruction repertoires soon encompassed many different types of instructions designed for a large variety of functions. The number of elementary instructions has presently increased from a few dozen in the early computers to over a hundred, and in a few instances to over two hundred. Arithmetic instructions account for a small portion of these repertoires. Newer functions include instructions for subroutine entry and exit, environment changing, status recording, and memory protection. Even the least expensive minicomputer of the early 1970s can perform upwards of one hundred different elementary instructions.

Current trends in the design of instruction repertoires seem to be aimed in two divergent directions. On the one hand there is the trend to use instructions that are powerful, that do macro or functional high-level operations. Contrary to this is the trend to make use of primitive instructions to attain a high level of flexibility. The flexibility is achieved at the cost of lower computation speed and some wasted memory space for instruction storage. A recent trend that incorporates both power and flexibility is the trend to develop microprogrammed computers whose instructions can easily be altered by the user while executing a program. High-level functions can then be inserted into the repertoire for a short time and removed sometime later. Thus speed and flexibility are both high while the cost is held low. A few microprogrammed computers have essentially no instruction repertoire, and are said to be *soft machines*. These machines have to be provided with instruction sets for the microprogram to interpret when each program is run. Chapter 10 describes these soft machines and other topics related to microprogramming.

Let us turn to the third of the topics mentioned above for the control unit, the evolution of instruction encoding. Von Neumann's computer exemplifies the essence of simplicity in instruction encoding, since each instruction is composed of only two fields, one for the operation and one for an address. Since that time the number of addresses per instruction has evolved considerably. Von Neumann's computer is a *one-address* computer because each instruction contains one address field. Most instructions for von Neumann's computer have either two or three operands, so that some operands in the one-address instruction format are implicit. To be specific consider the instructions

LOAD X

ADD Y

The LOAD instruction has two operands, a source, in this case MEMORY[X], and a destination, which in this case is a machine accumulator A. The destination address A is not explicitly encoded; it is implicitly encoded because A is the unique destination of the LOAD instruction. The ADD instruction has three operands, two source operands which in this example are A and MEMORY[Y], and one destination, which in this case is A. Again the A is not explicitly encoded as the source and destination operand.

The one-address computer was popular through the second generation of computers, but it is becoming less and less common as computers become less expensive and more powerful. By the beginning of the 1960s it became feasible to use multiple accumulators in a processor because hardware costs were dropping rapidly. A computer at that time might have eight or sixteen accumulators instead of a single one. Thus the destination register of the LOAD instruction cannot be encoded implicitly since it might be any of the accumulators. Consequently, the instruction encoding is forced to include an explicit encoding of the destination of the LOAD. The instruction encoding in assembly language might be LOAD 1,X which means copy MEMORY[X] into accumulator REG[1]. In this example at least two addresses per instruction must be designated. But since the register in question is only one of a small number of registers, its field may be encoded in a very few bits.

To encode a three-operand instruction such as ADD Y, we can choose to make one of the source operands the same as the destination operand, so that again only two address fields are required to specify three operands. Thus the instruction may take the form ADD 2,Y which means compute the sum of REG[2] and MEMORY[Y] and store the result into REG[2].

Many computers now use the two-address format because it represents a reasonable compromise in the use of extra memory for instruction encoding to obtain extra flexibility. Of course, since typical arithmetic instructions have three operands, one might be tempted to increase the number of address fields per instruction to three to achieve greater flexibility, but except for some special cases, three-address instructions generally result in more bits expended for instruction encoding. The relatively good encoding efficiency obtained with two-address formats has biased computer designers to favor two-address instructions over three-address instructions. The major exception to this rule is the class of high-speed computers that execute instructions in parallel or out of sequence when conditions permit. For example, to evaluate the expression $A \times B + C \times D$, one can use the sequence of instructions

LOAD 1,A
LOAD 2,B
MUL 1,2,3
LOAD 4,C
LOAD 5,D
MUL 4,5,6
ADD 3,6,7

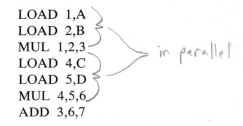

in parallel

The first two instructions are two-address instructions that load REG[1] and REG[2] with operands A and B, respectively. The MUL 1,2,3 instruction forms their product and stores the result in REG[3], because our convention is to use the third address to denote the destination operand. Note that while these operations are going on, the next three operations could proceed concurrently, because they involve a totally separate set of registers. Some computers can actually execute these sequences concurrently, the most notable example of this computer being the CDC 6600.

Having proceeded from one-address to two-address and then to three-address instructions, one might expect the next advance in machine encoding to be four-address instructions. Actually, zero-address instructions have become more popular for arithmetic instructions and may eventually find wide acceptance. For these instructions the three operands are specified implicitly, and need not be specified explicitly in address fields. A complete discussion appears in Chapter 7, with a brief preview of the notion appearing in Chapter 3.

We mention in passing that during the first generation of computers, when primary memories were rotating drums, instructions had the address of the successor instruction encoded within them. This permitted the programmer to arrange instructions on the drum to achieve minimal or near minimal drum latency. This address field is not counted when we speak of instruction formats as being one-address, two-address, and so on. Present day microprocessors use a similar scheme to find the address of the next instruction.

The trend today is to have many different instruction formats in a single computer to provide maximum flexibility with a minimum waste of memory arising from the encoding scheme. Thus an instruction with three operands may well have all three addresses specified if this makes sense for the contexts in which the instruction is used. Instructions with few operands may have fewer address fields, and when operands are specified implicitly, address fields are omitted from the instruction format. One consequence of this encoding practice is that instructions need not all be the same length. Good encoding schemes encode the most often used instructions with the shortest length codes. The great variety of instruction formats within a single computer makes the classification of computers by address format somewhat meaningless today, although such a classification was once possible and meaningful.

The last topic of interest with respect to the control unit concerns the various methods of implementation that have been used during the evolution of computers. Early computer designers recognized the need for modularity in the physical construction of computers. Therefore the natural basic building blocks they used were NAND/NOR logic gates and flip/flops, and these in turn were fabricated from discrete resistors, capacitors, and vacuum tubes. Minimum cost implementations entailed minimization of logic gates. Second generation computers used transistors in place of vacuum tubes, with some simplification in the structure of the logic gates, but the ground rules for designs were similar to those used previously. Registers were relatively expensive during this period so that multiple registers were found mainly in high-speed computers whose users were willing to pay a premium for speed.

During the early 1960s several developments in device technology led to revolutionary changes in the logic devices available and to dramatic decreases in the cost per logic function. The logic devices at this time were fabricated on flat films deposited on a silicon substrate. The resulting silicon chip contained several active transistors, not just a single transistor, yet the cost of chip fabrication was approximately equal to the cost of fabrication of a single transistor using previous fabrication techniques. The new technology was essentially a batch fabrication technology.

The new technology had a major impact on computer design, and then on computer architecture. The costs of batch fabricated devices can be held low if many copies of a chip can be made. For portions of a computer that are naturally iterative such as memories, arithmetic units, and registers, the new technology fit well and adaptation was relatively simple. But for the unstructured logic of the control unit there were significant problems.

The new technology invalidated the former design criteria of minimizing the number of logic gates in a design and replaced it with the need to maximize the repetitive use of a functional chip, or to minimize the number of interconnections between functional partitions of the control unit. This provided in part the impetus for implementing the control portion of a computer with memory rather than with discrete logic to achieve regularity of structure. This in turn contributed to the rise of microprogramming in the middle and late 1960s.

Another alternative implementation method is to use a relatively highly structured array of logic devices that can be easily tailored to various functions. These arrays are called *programmable logic arrays*. A comparison of the three methods for implementing control units, that is discrete logic, microprogramming, and programmable logic arrays, appears in Chapter 3. As we mentioned earlier, Chapter 10 is devoted to a study in depth of microprogramming.

This brings us to the end of the brief history of control unit design. Let us now turn our attention to other portions of the computer.

Memory

Since von Neumann's proposal, three types of memories have come into popular use as the primary memory of a computer. In the 1950s primary memories were principally rotating drums. The late 1950s saw the introduction of second generation computers with magnetic core primary memories. These machines were substantially faster than their predecessors because such memories have random-access capability. That is, any item can be fetched from memory in a fixed time span that is independent of the previous memory reference. Magnetic drums, as we hinted previously, have an inherent rotational latency. If the item accessed is not under the read head of a drum at the time the access request is issued then the request is not granted until the item eventually reaches the read head, and the computer remains idle until the access occurs. The third type of memory in use is the most recently introduced. This is integrated circuit memory. Like magnetic core memories, integrated memories are random-access memories, but they are potentially faster, and in some configurations less expensive, than similar magnetic core memories. Most computers today use either magnetic cores or integrated circuits or both in their primary memory. Magnetic drums have virtually disappeared from use in primary memory, but are still prevalent as auxiliary storage devices. Eventually, with continued rapid technological advances, integrated memories may even supplant magnetic drums as auxiliary memories.

The organization of primary memory today is essentially as von Neumann envisioned. Items have unique physical addresses, and accesses are made by address. The advances in memory architecture since von Neumann's time have been primarily in two related areas:

1. Memory hierarchies consisting of high-speed first-level storage together with much larger and slower second-level and third-level storage appear to the program as a single large memory whose speed is nearly equal to first-level storage. Frequently used items tend to reside in the fastest memory of the hierarchy and automatically drift to the slower parts of the hierarchy when frequency of use diminishes.
2. Several independent programs can run in the same computer system simultaneously, each using a completely separate set of memory locations. Memory address facilities simplify the problem of allocating memory to the individual programs, and guarantee that one program cannot interfere with another program. In some instances, specified data or program segments are permitted to be shared by two or more programs.

Rather than deal with these topics in great detail here, we mention that the first of the two is called *virtual memory,* and is covered in Chapter 5. The second describes a situation known as *multiprogramming* and the address translation facilities are essential for its support. One method for achieving the functional behavior mentioned above is to use a technique called *paging.* It is described in both Chapters 5 and 8. The use of paging for sharing of information appears in Chapter 8.

The major emphasis on memory architecture in this text is at the device and memory organization level rather than at the memory hierarchy level. The emphasis arises because this has received relatively little attention in the computer architecture literature and merits thorough coverage in an intro-ductory text. The bulk of the material on memories appears in Chapter 5.

Input/Output

The first computers were extremely slow by today's standards, but even at their speeds they were much faster than the input/output devices available for them. Almost from the beginning of the development of computers, designers worked out methods for buffering input/output so that input/output could be done concurrently with computation. By the second generation of computers, input/output for high-speed computers was done by functionally independent modules known as *data channels*. These devices do arithmetic computations and access memory, much like the arithmetic processor of a computer. For high-speed computers, the need for faster input/output led to full-fledged processors being dedicated to input/output in support of other processors that are dedicated to numerical computation. The CDC 6600, which appeared in the middle of the 1960s, is an extreme example of this notion because it has ten stored-program input/output processors to service a single ultra high-speed central processor.

Recent advances in input/output architecture have generally been in the regularizing of the input/output system, so that all devices are treated more or less uniformly with respect to the hardware interface and with respect to the software for operating the devices. The use of a single bus for both memory and input/output devices permits input/output devices to be accessed as if they were special locations in memory. This is one method for forcing all devices to fit a uniform pattern, and it has been used successfully in several computers, most notably the DEC PDP-11 minicomputer series. Micro-processors, that is, computers small enough to fit on a few integrated circuit chips, have also been used in the input/output interface of various peripheral devices. In this way all interfaces can be identical in hardware structure, yet they can be specialized to particular input/output devices by changing the program loaded in the microprocessor interface. The most common designs for input/output systems are treated in Chapter 6, and a related discussion of programming considerations for input/output appears in Chapter 4. Among the notable advances in input/output systems since von Neumann is the use of an interrupt system to control input/output. This is described in both Chapters 4 and 6.

1-3. WHAT LIES AHEAD

The trends for the future are clear from the trends of the last several years. Manufacturers will strive to make faster computers, smaller com-puters, and less expensive computers as long as new developments in device technology can support these advances. To give a coherent view of what lies ahead we find it convenient to investigate three different classes of computers, namely, minicomputers and microcomputers at the small end of the range, medium-scale computers in the middle, and high-speed large-scale machines at the other end of the spectrum.

In the area of minicomputers and microcomputers, the range of applications increases considerably as the size and price decreases. Moreover, the number of copies of any given computer that can be sold increases, and this in turn has the effect of decreasing the cost per computer. The challenge here for computer architects is to reduce the cost of a processor to an absolute minimum without sacrificing power and flexibility. The hand-held calculator was the first real venture of the small computer into the consumer market, and it was a successful one. Here the costs fell roughly a factor of 10 in a period of two to three years as the market expanded, and this was coupled with an increase in capability of the instrument.

One of the key points in the design of a successful computer for this market is to make one design fit many applications. Here there is an advantage in the use of microprogramming or programmable logic of some type so that one basic architecture can easily be tailored to several different applications. This type of implementation may be slower than others, but since high replication is essential to reduce cost, the sacrifice in speed may be more than balanced by reductions in cost.

In the medium-scale computer area, cost constraints are greatly relaxed, and the design goal is one of achieving greater computer power and ease of use. Here the trend is to attain these goals through extensive hardware/software facilities. Typical services are rendered through high-level language compilers, extensive application program libraries, data management systems, and data communications. Minicomputers can be used for these applications, but the services provided by a medium-scale computer are different in extent if not in type from those provided by a minicomputer. If there reaches a point when the services available from a minicomputer are indistinguishable from those of a larger and more expensive medium-scale computer, then the medium-scale computer will become difficult to justify as a separate class of computer.

The ultra high-speed computer has found a niche because of a class of problems that require vast amounts of computation. Weather analysis and prediction is one problem in this class, and it clearly has to be done on a computer of sufficient speed to make weather predictions early enough for them to be useful. The air traffic control problem, another problem requiring massive computing facilities, is clearly also a real-time problem. In this range of problems one is willing to pay a premium for high speed, and the premium is often quite high. The challenge for a computer architect in this area is to find reasonable and clever ways of speeding computation beyond the inherent speed of the component devices. We mentioned earlier that parallel and pipeline techniques have been used to this end, and both techniques are likely to continue in the design of high-speed systems. The high-speed computer today is relatively rare, and it may become even rarer as minicomputers and medium-scale computers take over some of the applications that were previously unique to the high-speed computer and find new applications that become feasible as component cost diminishes.

In closing this introduction we wish to call the reader's attention to the literature on computer architecture. Bell and Newell [71] have collected several important papers of the 1950s and 1960s, including descriptions of several computers that appeared during these years. Many of the design techniques, trends, and innovations in computer architecture hinted at in this introduction appear in the original in the Bell and Newell collection. A fairly extensive set of citations to other original source material appear in the bibliography of this book.

CHAPTER 2

Data Representation

Herschel H. Loomis, Jr.

2-1. IMPORTANCE OF DATA REPRESENTATION

One of the powers of the digital computer is its versatility of application to problems in a variety of fields. In making such application, the way in which data to be operated upon is represented in the computer is important. Early numerical applications of modern computers used a binary representation of numbers which had a fixed radix point and a limited, fixed number of significant digits. Since that early beginning, computer words have been interpreted in a variety of ways: as alphanumeric text in business applications; as floating point (scientific notation) numbers for greater programming ease and problem solving flexibility in numerical analysis; and as arrays of light intensities representing an image of the planet Jupiter in an unmanned fly-by, to name just a few.

In all these examples, the fundamental unit of information is the binary digit, called the *bit*. Higher order structures made up of these bits vary widely from example to example. In fact, the difference between a "good" computer program and a "poor" one is often the way in which these higher order structures are put together and interpreted.

In this chapter, we will describe a number of data representations in use and explore several of their characteristics to indicate how to select the best representation for a particular problem.

We will consider information from several different points of view. The most basic aspect examined concerns the representation in nondecimal radices such as octal, decimal, hexadecimal, and the representation of alphanumeric characters. Next we consider the composites of the basic units, such as the *computer word*, which corresponds to a physical unit of information in the computer. Then we examine how words are put together in more complex structures such as arrays or lists. Finally, we take a brief look at the relation between these interpretations and the physical characteristics of the computer implied by these interpretations.

Throughout all of this discussion, one very important point must be kept in mind: the meaning of a set of data depends on the interpretation of the bits that represent the data in memory.

2-2. NUMBERS

Fixed Radix-point Representation

One of the major advances of civilization was the development of the positional number system, which is quite different from the nonpositional system used by the early Romans. In the positional system, each digit position has a value called a *weight* associated with it. To obtain the value of such a number, each digit value is multiplied by the weight of its position, and then all such products are summed. We can formally represent such a number and its numerical value as shown in Equation 2-1:

$$X = X_{n-1} X_{n-2} \cdots X_0 . X_{-1} \cdots X_{-m}$$
$$V(X) = \sum_{i=0}^{n-1} X_i R^i + \sum_{i=-1}^{-m} X_i R^i \qquad (2\text{-}1)$$

$$\underset{\text{portion)}}{\text{(Integer}} \qquad \underset{\text{portion)}}{\text{(Fractional}}$$

where $V(X)$ denotes the numerical value of X.

The point between X_0 and X_{-1} is the so called *radix point* and separates the integer (whole number) portion of the number from the fractional portion.

Normally, some constraints are placed upon the values that X_i may assume, that is:

$$0 \leq X_i < R$$

(Some number systems, such as the signed digit system proposed by Avizienis [64] allow other values, but those systems are rarely used.)

The value R associated with such a system is called the *radix* or *base* of the number system. Various common values of R are in use. When R is 10 we have the standard *decimal* system, the one in which most of us carry out day-to-day computations. Base 2 systems are called *binary* and form the foundation for all electronic computer systems in current use. Other commonly used systems are ones which have a power of 2 as the base such as base 8 or *octal*, base 16 or *hexadecimal*. These systems are used because they are simply related to the binary system.

The binary system is the fundamental system used by all electronic computer systems known to the author, because electronic devices which operate in more than two states are difficult to achieve. Theoretical studies have been made of systems having a fundamental base larger than base 2, in particular of the *ternary* or base 3 system, although no practical application of this theory has yet been made.

Working with numbers in different bases requires the conversion of numbers from one base to another and involves performing some arithmetic. Normally in such a conversion, the arithmetic is performed in the familiar base. Because of this, conversion from a familiar base to a foreign base involves different algorithms than the one used to convert from a foreign base to the familiar one.

Conversion of integers from a foreign base to the familiar base is relatively simple; it involves substitution of the appropriate values into the integer portion of Equation 2-1.

$$V(X) = \sum_{i=-0}^{n-1} X_i R^i$$

For example

$$V(3507_8) = 7 + 0(8) + 5(8^2) + 3(8^3) \text{ (arithmetic base 10)}$$
$$= 1863_{10}$$

The subscript signifies the base.

$$V(110110_2) = 0 + 1(2) + 1(2^2) + 0(2^3) + 1(2^4) + 1(2^5)$$
$$= 54_{10}$$
$$V(110.111_2) = 0 + 1(2) + 1(2^2) + 1(2^{-1}) + 1(2^{-2}) + 1(2^{-3})$$
$$= 6.875_{10}$$
$$V(.325_6) = 3(6^{-1}) + 2(6^{-2}) + 5(6^{-3})$$
$$= (.578703703\ldots\ldots\ldots)_{10}$$

In the last example note that some numbers that are exact in one radix representation cannot be represented exactly in a finite number of digits in some other base. This occurs when the reciprocal of the foreign base itself cannot be represented in a finite number of digits in the familiar base.

Equation 2-1 can also be used when the familiar base is other than 10.

Example 2-1:

Convert 31_{10} to base 2.   (arithmetic base 2)
$$V(31_{10}) = [1 + 3\,(10)]_{10}$$
$$= [1 + (011)\,(1010)]_2$$
$$= [1 + (11110)]_2$$
$$= 11111_2$$

Conversion to a foreign base can be accomplished using the following methods, one for integer, the other for fractional numbers.

Consider the integer number in some familiar base R' and consider how to compute its base R representation using arithmetic in base R'.

$$V(X) = X_0 + X_1(R) + X_2(R^2) + \cdots + X_{n-1}(R^{n-1})$$
$$\text{(arithmetic base } R')$$

Divide $V(X)$ by R yielding

$$V(X)/R = \frac{X_0}{R} + [X_1 + X_2(R) \cdots + X_{n-1}(R^{n-2})]$$

$$= \frac{X_0}{R} + Q_1$$

The portion in brackets, Q_1, is the quotient of the division operation, and since X_0 is less than R, X_0 is the remainder. Note that the remainder gives X_0, and the quotient Q_1 can be used to compute the remaining X_is. Thus, in general, we have

$$Q_i/R = \frac{X_i}{R} + [X_{i+1} + \cdots + X_{n-1}(R^{n-2-i})] = \frac{X_i}{R} + Q_{i+1}$$

This suggests a repeated process of division by R where each remainder yields a digit of the representation, and each successive quotient is the dividend of the next step. The process terminates when a quotient of 0 is produced.

Example 2-2:

| Convert 63_{10} to base 5 | (arithmetic base 10) |

$$63/5 = \frac{3}{5} + 12 \qquad\qquad X_0 = 3$$

$$12/5 = \frac{2}{5} + 2 \qquad\qquad X_1 = 2$$

$$2/5 = \frac{2}{5} + 0 \qquad\qquad X_2 = 2$$

$$63_{10} = 223_5$$

Example 2-3:

| Convert 1863_{10} to base 8 | (arithmetic base 10) |

$$1863/8 = \frac{7}{8} + 232 \qquad\qquad X_0 = 7$$

$$232/8 = \frac{0}{8} + 29 \qquad\qquad X_1 = 0$$

$$29/8 = \frac{5}{8} + 3 \qquad\qquad X_2 = 5$$

$$3/8 = \frac{3}{8} + 0 \qquad\qquad X_3 = 3$$

$$1863_{10} = 3507_8$$

The conversion of fractional numbers likewise proceeds from an interpretation of the fractional portion of Equation 2-1. Let X be such a fractional number.

$$V(X) = \frac{X_{-1}}{R} + \frac{X_{-2}}{R^2} + \cdots + \frac{X_{-m}}{R^m}$$

Multiplying $V(X)$ by R yields

$$R \times V(X) = X_{-1} + \frac{X_{-2}}{R} + \cdots + \frac{X_{-m}}{R^{m-1}} = X_{-1} + F_1$$

Now, $R \times V(X)$ has an integer part, namely X_{-1} and a fractional part. Multiplying F_1 by R yields

$$R \times F_1 = X_{-2} + \frac{X_{-3}}{R} + \cdots + \frac{X_{-m}}{R^{m-2}} = X_{-2} + F_2$$

In general,

$$R \times F_j = X_{-(j+1)} + \frac{X_{-(j+2)}}{R} + \cdots + \frac{X_{-m}}{R^{m-(j+1)}} = X_{-(j+1)} + F_{j+1}$$

The process terminates when (and if) a fractional part of 0 is achieved.

Example 2-4:   Convert $.63671875_{10}$ to hexadecimal: *

$$16 \times (.63671875) = 10.1875 \qquad X_{-1} = 10 \text{ or } A$$
$$F_1 = .1875$$
$$16 \times (.1875) = 3.0 \qquad\qquad X_{-2} = 3$$
$$F_2 = 0$$

so the process terminates and

$$.63671875_{10} = .A3_{16}$$

---

*Note that symbols for hexadecimal digit values of 10 through 15 are represented by the symbols A through F, respectively.

Example 2-5: Convert $.5125_{10}$ to base 8.

$$8 \times (.5125) = 4.1 \qquad X_{-1} = 4, F_1 = .1$$
$$8 \times (.1) \quad = 0.8 \qquad X_{-2} = 0, F_2 = .8$$
$$8 \times (.8) \quad = 6.4 \qquad X_{-3} = 6, F_3 = .4$$
$$8 \times (.4) \quad = 3.2 \qquad X_{-4} = 3, F_4 = .2$$
$$8 \times (.2) \quad = 1.6 \qquad X_{-5} = 1, F_5 = .6$$
$$8 \times (.6) \quad = 4.8 \qquad X_{-6} = 4, F_6 = .8$$

A repetition of a previously arrived at fractional part has occurred $(F_2 = F_6)$, so the conversion process does not terminate. Therefore the result is:

$$.5125_{10} = (.4063146314\ldots\ldots)_8$$

Finally, in connection with the conversion to a foreign base, note that mixed numbers are converted by separating the number into its integer and fractional parts and converting each separately and then combining the result.

The binary system currently is firmly established as a fundamental system for representing data in electronic digital computers. It is however inconvenient to represent large binary numbers in positional binary because of the large number of digits required. Consider the 24-bit binary integer number.

$$(010011001111000001011010.)_2 = (5,042,266)_{10}$$

We can group the bits in groups of b bits, and each b-bit group can be considered as a digit of a radix 2^b representation of the number. For example, if $b = 3$, our number becomes

$$(010,011,001,111,000,001,011,010.)_2$$
$$\text{or } (23170132)_8$$

This is the octal system, and is seen to be three times as economical in positions. Also, octal numbers are easier to remember for short periods of time than are binary numbers. If $b = 4$, we have the hexadecimal representation, and in our example,

$$(0100,1100,1111,0000,0101,1010.)_2$$

becomes

$$(4CF05A)_{16}$$

In principle, other values of b could be used, but the octal and hexadecimal systems are the only ones used in practice. First and second generation computers tend to use the octal system, whereas more recent systems have used hexadecimal.

Earlier in this section, the question of foreign radix arithmetic was raised and the method was discussed as being unwieldy for human use in radix conversion. Nevertheless, computers generally operate in a radix system that is either binary or some other power of 2, and the rules for doing arithmetic in these systems become important when one designs logic circuits for the performance of arithmetic. We consider the rules for simple arithmetic in the next section.

Rules for Simple Arithmetic   In order to define the rules for arithmetic in a foreign radix, we make use of the basic relation that the result of an operation on two numbers represented in radix R must have the same value as the result of the same operation on the radix 10 representation of the same numbers. Thus we can define digit addition tables for the various systems as shown in Figure 2-1. Since the operation of addition is *commutative* ($x+y = y+x$, for all choices of x,y), the portion of the base 8 and 10 tables below the diagonal is not shown.

One can easily verify that these tables follow the rule stated above for constructing foreign radix arithmetic rules. For example, we note: $(2 + 3)_4 = 11_4$ corresponds to $(2 + 3)_{10} = 5_{10}$ and $11_4 = 5_{10}$.

Base 2

+	0	1
0	0	1
1	1	10

Base 3

+	0	1	2
0	0	1	2
1	1	2	10
2	2	10	11

Base 8

+	0	1	2	3	4	5	6	7
0	0	1	2	3	4	5	6	7
1		2	3	4	5	6	7	10
2			4	5	6	7	10	11
3				6	7	10	11	12
4					10	11	12	13
5						12	13	14
6							14	15
7								16

Base 10

+	0	1	2	3	4	5	6	7	8	9
0	0	1	2	3	4	5	6	7	8	9
1		2	3	4	5	6	7	8	9	10
2			4	5	6	7	8	9	10	11
3				6	7	8	9	10	11	12
4					8	9	10	11	12	13
5						10	11	12	13	14
6							12	13	14	15
7								14	15	16
8									16	17
9										18

Figure 2-1   Addition tables for various bases

Likewise:

$(8 + C)_{16} = (14)_{16}$ corresponds to $(8 + 12)_{10} = 20_{10}$ and 14_{16}
$= 16 + 4 = 20_{10}.$

Next, let us apply our knowledge of addition of n-digit numbers in the decimal system to infer some general rules for adding n-digit numbers in any radix. We note that whenever the sum in a given digit position exceeds 9 (radix-1), we keep the least significant digit of the sum in the current digit position and generate a *carry* into the next digit position, as we work in the number from least significant to most significant portions of the number. We show the carries as ones (1s) and zeros (0s) above the column to which they are added. As examples consider

Radix 10:

```
    1 1 1 0   Carries
    5 1 7 3
  +2 8 4 7
    8 0 2 0
```

Radix 2:

```
1 1 1 1 1 0 1 0   Carries
  1 0 1 1 1 0 1
  0 1 0 0 1 0 1
1 0 0 0 0 0 1 0
```

Radix 16:

```
    1 0 1 0   Carries
    A 1 3 F
    2 F 9 3
    D 0 D 2
```

Subtraction, or the minus ($-$) operation can be similarly determined using individual tables for each digit and a *borrow* which propagates from least to most significant digit. Figure 2-2 shows subtraction tables for radix 2, 8, and 10. The form of these tables is such that the less significant digit represents the difference and the more significant, if a 1, represents a borrow. When subtracting multi-digit numbers, one way to deal with the borrow is to use it to decrement the minuend in the next digit position before applying the table to find the difference and borrow. Note the following examples :

	Radix 10	Radix 2	Radix 8
Borrow	0 0 1 0 0	0 1 0 0 0 1 1 0	1 1 1 0 0
Minuend	2 5 1 9	1 0 1 1 1 0 0	2 7 1 5
Subtrahend	1 4 2 8	0 1 0 1 0 0 1	2 7 3 4
Difference	1 0 9 1	0 1 1 0 0 1 1	····7 7 7 6 1

Radix 2   Subtrahend
Minuend   –   0   1

↓	0	11
0	0	11
1	1	0

Radix 8   Radix 10
Subtrahend →   Subtrahend →

Minuend

–	0	1	2	3	4	5	6	7
0	0	17	16	15	14	13	12	11
1	1	0	17	16	15	14	13	12
2	2	1	0	17	16	15	14	13
3	3	2	1	0	17	16	15	14
4	4	3	2	1	0	17	16	15
5	5	4	3	2	1	0	17	16
6	6	5	4	3	2	1	0	17
7	7	6	5	4	3	2	1	0

–	0	1	2	3	4	5	6	7	8	9
0	0	19	18	17	16	15	14	13	12	11
1	1	0	19	18	17	16	15	14	13	12
2	2	1	0	19	18	17	16	15	14	13
3	3	2	1	0	19	18	17	16	15	14
4	4	3	2	1	0	19	18	17	16	15
5	5	4	3	2	1	0	19	18	17	16
6	6	5	4	3	2	1	0	19	18	17
7	7	6	5	4	3	2	1	0	19	18
8	8	7	6	5	4	3	2	1	0	19
9	9	8	7	6	5	4	3	2	1	0

Note: Entry in table is either borrow and difference or just difference.

Figure 2-2   Subtraction tables for various bases

In the example for radix 8, we note that a borrow is generated by the most significant digit subtraction, a direct consequence of the fact that the subtrahend is greater than the minuend and we therefore expect a negative result. In fact, we humans normally rearrange the problem to subtract the smaller from the larger and then affix a minus sign to the result. Some computers may in fact handle the situation in this way, however, there are other ways to represent and operate on negative numbers; some of these ways are covered in the next section.

Negative Number Representation   The most familiar way for dealing with negative numbers is by means of the *signed-magnitude* representation. In this representation, $-b$, where b is a string of digits forming a fixed point number, is that number which when added to b, forms the sum 0. In other words, $-b$ is the *additive inverse* of b. Now we have a mechanism for dealing with the case where a larger positive number is subtracted from a smaller positive number: invert the order of subtraction and affix a minus sign to the difference.

The presence of so called negative numbers now increases the number of cases we must consider in developing rules for handling addition. To examine the various cases, let B and C be the numbers to be added and b and c be magnitudes of these numbers, respectively, as shown in Table 2-1. Note that

TABLE 2-1. ADDITION OF NEGATIVE NUMBERS

B	C	b/c	B+C	Operation Required
b	c	don't care	$b+c$	addition
$-b$	$-c$	don't care	$-(b+c)$	addition
b	$-c$	≥ 1	$b-c$	subtraction
b	$-c$	< 1	$-(c-b)$	subtraction

the operation is commutative, so the two cases for $B = -b$ and $C = c$ are not illustrated.

Treating negative numbers in this way means that we need both an adder and a subtractor together with some input switching and testing to permit the four basically different cases of Table 2-1 to be implemented.

Two other ways of representing negative numbers that do not require the special case handling of the signed-magnitude method are the *radix-complement* (RC) and the *diminished radix-complement* (DRC) methods.

To develop the *radix-complement*, assume that we have an n-digit, radix R number. An n-digit, radix R number can have any one of R^n possible values and hence has a representation in R^n. If a positive number b is to be represented, so long as $b < R^n/2$, its n-digit, radix R representation is used. To represent $-b$, we use instead $R^n - b$ as the representation.

Using this system, let us examine the result of forming the sum of b and $-c$, as shown in Table 2-2

TABLE 2-2. FORMING THE SUM OF b AND $-c$

B	C	b/c	B+C	Sign of Result
b	R^n-c	$b/c>1$	$R^n+(b-c)$	Positive
b	R^n-c	$b/c \leq 1$	$R^n-(c-b)$	Negative

In Table 2-2 we see two cases explored, if $b \leq c$, then $R^n - (c-b)$ is less than R^n, and the result is still negative, as shown in Example 2-6.

Example 2-6: If $R = 10$, $n = 3$, find $053 - 101$ by the radix-complement method.

$$053 + (10^3-101) = 053 + (899)*$$

$$= 952$$

$$= (10^3-48)$$

$$(53-101 = -48)$$

If $b > c$, then $R^n + (b-c)$ is greater than or equal to R^n, and this fact is signalled by the generation of a carry out of the nth digit position of the

---

*$X_n \geq R/2$ is frequently interpreted to mean the number is negative, and hence in radix-complement form.

adder. An n-digit adder normally ignores this carry, since it adds modulo R^n. Thus all multiples of R^n are ignored.

Example 2-7:   If $R = 10$, $n = 3$, find $101 - 053$ by the radix-complement method.

$$101 + (10^3 - 053) = 101 + (947)$$
$$= 1048$$

but we ignore the 1 in the 10^3 column since our addition is modulo 10^3, so our result is 048.

Because the most significant digit of the number carries the information about sign, we will slightly change the interpretation of the most significant digit to permit more convenient representation and evaluation of negative numbers. We add the nth digit or sign which can contain only a 0 or a 1, and we do our addition there modulo 2. Now we are able to generalize our positional number to the radix-complement method. Let us allow a sign, n integer digits, and m fractional digits as follows:

$$X_n X_{n-1} \ldots \ldots X_0 X_{-1} \ldots \ldots X_{-m}$$
$$0 \le X_n \le 1, 0 \le X_i < R \quad \text{for } -m \le i < n.$$

How do we evaluate such a number? If $X_n = 0$, that is if the number is positive, it is no different from the previous value as found in Equation 2-1.

Let x be the magnitude of a number; let X be its representation and $\hat{X}$ its $n - 1$ digit representation. Consider the representation of $-x$.

$$V_{RC}(X) = -x$$
$$\text{but } V(\hat{X}) = R^n - x \quad \text{and } X_n = 1.$$

Therefore we must assign $V(X_n)$ to be $-R^n$, and we have

$$V_{RC}(X) = (-X_n R^n) + \sum_{-m}^{n-1} X_i R^i \tag{2-2}$$

Applying this to our previous examples, adding a sign digit, we find:

$$V_{RC}(1899) = -1000 + 800 + 90 + 9 = -101$$
$$V_{RC}(0101) = -0 + 100 + 1 = 101$$

At this point we see that there are at least two ways for determining the sign of the number being represented. We can allow the most significant digit its full range of R values (0 to $R - 1$), interpreting values greater than or equal to $R/2$ as meaning the entire number is a radix-complement representation of a negative number and interpreting values less than $R/2$ as meaning the entire number is positive. The other way is to treat the sign digit as a binary digit regardless of R, interpreting a 1 as meaning that the rest of the number is a radix-complement representation of a negative number and a 0 as meaning that the rest of the number is positive. Both methods are in use, although the second method has the advantage that a simple evaluation-expression exists for numbers irrespective of sign. Also, the first method is awkward when R is odd, although there are no odd radices in common use. Finally, note that both methods are the same if $R = 2$, the most important case.

In the decimal system, the radix complement is called *ten's complement* and in the binary system, *two's complement*.

Now let us examine a variation of the radix-complement method, the *diminished radix-complement* method. This variation is designed to make the conversion from positive to negative numbers easier than with the radix-complement method. The ith digit of the diminished radix-complement of a number, x, is $R - 1 - X_i$. The sign of the number is positive if the number is less than $1/2R^n$. As before, the sign can be indicated by a separate binary sign digit, in which case the sign digit of the diminished radix-complement of number is $1 - X_n$. For example, let $R = 10$, $n = 3$, $m = 0$ and assume a binary sign digit:

$$DRC(0053) = 1946$$
$$DRC(0931) = 1068$$
$$DRC(0000) = 1999$$

One criterion that the negative number system must fulfill is that the sum of X and the representation of minus X must be a representation of 0.

Thus, $X + DRC(X) = Z$ such that $V_{DRC}(Z) = 0$. Now let us take a number X and add to it its DRC. $X_{-m} + (R - 1 - X_{-m}) = R - 1$, with no carry. Then, for each $i > -m$, we find the ith digit sum is $X_i + (R - 1 - X_i) = R - 1$, generating no carry. Finally with no carry into the sign, $X_n + (1 - X_n) = 1$. Thus the representation of the sum of a number and its diminished radix-complement is $1(R - 1)(R - 1)...(R - 1)$, which is one representation of the number 0, that is, the diminished radix-complement of 0. Therefore, if we compare the radix complement of a number and its diminished radix-complement, we note that the DRC is one less than the RC in the least significant digit position, or $RC(x) - R^{-m} = DRC(X)$. Thus, we can state that the value of the general signed diminished radix-complement number is given by the following expression:

$$V_{DRC}(X) = X_n(-R^n + R^{-m}) + \sum_{i=-m}^{n-1} X_i R^i \tag{2-3}$$

Applying this to the previous examples for $n = 3$, $m = 0$, $R = 10$ yields

$$V_{DRC}(1946) = (-10^3 + 10^0) + 900 + 40 + 6$$
$$= -53$$

$$V_{\text{DRC}}(1999) = (-10^3 + 10^0) + 900 + 90 + 9$$
$$= 0$$
$$V_{\text{DRC}}(0053) = 0 + 50 + 3$$
$$= 53$$

Finally, as a result of Equation 2-3 we now have a simpler method for taking the radix-complement of a number than performing R^n minus the number. We form the diminished radix-complement (digit-by-digit DRC), and add R^{-m} (a 1 in the least significant digit position).

Now let us leave the general case and look specifically at the binary case. Here, the diminished radix-complement is called the *ones' complement*.

Let us first treat the two's complement or radix-complement representation. We assume the leading digit, X_n, is the sign digit. Addition falls into four basic categories, each of which should be examined: 1) both summands positive, 2) both negative, 3) one positive and one negative with the negative one greater in magnitude, and finally, 4) one positive and one negative with the positive greater than or equal in magnitude to the negative.

Case 1. Both positive, that is, $B = b, C = c$.

   a) $b + c < 2^n$ implies that the result is representable.
   b) $b + c \geq 2^n$ generates a carry into sign bit, which is an *overflow* condition.

Case 2. Both negative, that is, $B = 1, (2^n - b); C = 1, (2^n - c)$.

   a) $b + c \leq 2^n$ implies that

$$(2^n - b) + (2^n - c) = 2^{n+1} - (b + c) \geq 2^n$$

and hence the addition generates a carry into the sign bit position, preserving a sign of 1.

   b) $b + c > 2^n$ implies that

$$(2^n - b) + (2^n - c) = 2^{n+1} - (b + c) < 2^n$$

and hence the addition generates no carry into the sign bit position, causing an apparent sign change, which is an *overflow*.

Case 3. One positive, one negative, that is, $B = 1, (2^n - b), C = c, b > c$

$$c + 2^n - b = 2^n - (b - c) < 2^n$$

and hence no carry occurs into the sign bit and the result is negative.

Case 4. Same as case 3 except $b \leq c$

$$c + 2^n - b = 2^n + (c - b) \geq 2^n$$

and a carry occurs into the sign bit, yielding a positive result.

Example 2-8: $R = 2, n = 5, m = 0$

Case 1. *a)*   001100   Carries (Shown above the column to which they are
                 ‾‾‾‾‾‾
                 010010. $(+18)$ added)
                 000111. $(+ 7)$
                 ‾‾‾‾‾‾
                 011001. $(+25)$ No overflow

        b)   111100   Carries
                 ‾‾‾‾‾‾
                 010010. $(+18)$
                 001111. $(+15)$
                 ‾‾‾‾‾‾
                 100001. $(+33)$ Overflow, $33 > 32$

Case 2. *a)* 1100000   Carries
                 ‾‾‾‾‾‾
                 110010. (-14)
                 111001. $(- 7)$
                 ‾‾‾‾‾‾
                 101011. (-21) No overflow

        b) 1000100   Carries
                 ‾‾‾‾‾‾
                 110010. (-14)
                 100011. (-29)
                 ‾‾‾‾‾‾
                 010101. (-43) Overflow, $43 > 32$

Case 3.   0000000   Carries
                 ‾‾‾‾‾‾
                 110010. (-14)
                 001101. $(+13)$
                 ‾‾‾‾‾‾
                 111111. $(- 1)$ No carry out of sign, no sign change.

Case 4.   1110000   Carries
                 ‾‾‾‾‾‾
                 111001. $(- 7)$
                 001010. $(+10)$
                 ‾‾‾‾‾‾
                 000011. $(+ 3)$ Carry out of sign, sign change.

The ones' complement number system has a similar set of categories and, in addition, the question of a correction in certain cases enters the picture. The need for this correction can be seen from the fact that the negative ones' complement must have an additional 2^{-m} added to obtain its value (see Equation 2-3), so if two negative numbers are added, the sum will be off by 2×2^{-m}. Thus some correction is needed in the addition process.

Case 1. The same as for two's complement

Case 2.

    a) $B = 1, (2^n - b - 2^{-m}); C = 1, (2^n - c - 2^{-m})$

       $b + c < 2^n$ implies

$$2^n - b - 2^{-m} + 2^n - c - 2^{-m} = 2^{n+1} - (b + c) - 2 \times 2^{-m}$$

This result will not be the correct ones' complement unless a correction is added to the least significant bit position:

    Then $2^{n+1} - (b + c) - 2^{-m+1} + 2^{-m} \geq 2^n$

and a carry is generated into the sign digit position. Also a carry is generated out of the sign digit position. There is no overflow.

    b) $b + c \geq 2^n$ implies

$$2^n - b - 2^{-m} + 2^n - c - 2^{-m} + 2^{-m} =$$
$$2^{n+1} - (c + b) - 2^{-m} < 2^n$$

$$\uparrow$$
$$\text{correction}$$

Thus no carry is generated into the sign digit but a carry is generated out of the sign digit.

Case 3. $B = 1, (2^n - b - 2^{-m}); C = c$

    $b \geq c$
    $c + 2^n - b - 2^{-m} = 2^n - (b - c) - 2^{-m} < 2^n$

No carry is generated into the sign bit and no carry is generated out of the sign. The result is negative and no correction need be made.

Case 4. $B = 1, (2^n - b - 2^{-m}); C = c$

    $b < c$
    $c + 2^n - b - 2^{-m} = 2^n + (c - b) - 2^{-m} \geq 2^n$

Thus a carry is generated into the sign bit and causes a sign change to a positive number. Also a carry then propagates out of the sign bit. Since the sign is now positive, 2^{-m} must be added as a correction to make the result look like $2^n + (c - b)$ which is congruent modulo 2^n to $c - b$, the desired result.

    Now the question arises, how do we effect the correction needed in cases 2 and 4? Note that in both of these cases, the carry out of the sign bit is 1 and in cases where no correction is needed, that same carry is 0. Thus the correction needed is simply to use the carry out of the sign as the carry into the least significant bit. This correction is called the *end-around carry*.

Example 2-9: $R = 2, n = 0, m = 3$

Case 2. *a*) $(1) 1 1^{11}$ Carries

$$
\begin{array}{ll}
\overline{1.011} & (-0.500) \\
1.110 & (-0.125) \\
\overline{1.010} & (-0.625)
\end{array}
$$

b) $(1) 0^{\ 11}$ Carries

$$
\begin{array}{lll}
\overline{1.001} & (-0.750) \\
1.011 & (-0.500) \\
\overline{0.101} & (-1.250) & \text{Overflow}
\end{array}
$$

Case 3. $(0) 0 1 1^{\ 0}$ Carries

$$
\begin{array}{lll}
\overline{1.011} & (-0.500) \\
0.011 & (+0.375) \\
\overline{1.110} & (-0.125) & \text{No end-around carry}
\end{array}
$$

Case 4. $(1) 1 1^{\ 01}$ Carries

$$
\begin{array}{lll}
\overline{1.010} & (-0.625) \\
0.110 & (+0.750) \\
\overline{0.001} & (+0.125) & \text{End-around carry}
\end{array}
$$

To conclude our discussion of the specifics of addition of the various forms of binary numbers, we finally consider the simple hardware realizations of these addition algorithms.

The basis for the simple parallel adder is the *full adder*, which is the implementation of the addition rules for one bit position. This logic circuit accepts two inputs from the appropriate bits of the addend and augend and one input from the carry generated by the next less significant bit addition. Its outputs are the sum bit for this bit position and the carry into the next more significant bit position.

A two's complement adder can be obtained by connecting the adder as just described, with the carry into the least significant bit set at 0 and the carry out of the most significant bit ignored (except perhaps for overflow detection). The ones' complement adder is formed similarly, except the carry input to the least significant bit is connected to the carry output of the most significant bit (end-around carry). Part *a* of Figure 2-3 shows a symbol for a full adder along with its truth table. The connection of six of these adders to form a six-bit, two's complement adder is shown in Part *b* of Figure 2-3 along with labelling of each signal line to correspond to the two's complement adder example, case 4 [$(-7) + (+10)$]. Part *c* of Figure 2-3 shows four full adders connected to form a ones' complement adder. The signal lines in this figure are labelled to correspond to the example of case 4 [$(-0.625) + (0.750)$].

We have examined three different ways of treating negative numbers, the

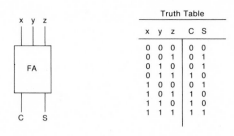

Truth Table				
x	y	z	C	S
0	0	0	0	0
0	0	1	0	1
0	1	0	0	1
0	1	1	1	0
1	0	0	0	1
1	0	1	1	0
1	1	0	1	0
1	1	1	1	1

a) Binary full adder

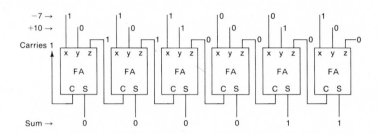

b) Six bit, two's complement adder

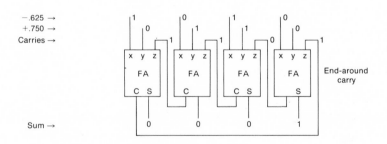

c) Four bit, ones' complement adder

Figure 2-3   Binary adders

signed-magnitude method, radix-complement (RC) method, and the diminished radix-complement (DRC) method. We have seen how each relates to the operations of addition, sign conversion and evaluation, and we have seen simple implementations of two's and ones' complement adders.

With respect to addition, the signed-magnitude method is most difficult to implement, with DRC being considerably simpler, and RC the simplest since it lacks the end-around carry.

For the process of negation, sign and magnitude seems most suited, DRC is next, with RC being most expensive. In practice, the difficulties with signed-magnitude addition outweigh its advantages and it is not used as commonly as the RC or DRC method.

More Arithmetic   Three arithmetic operations are examined in this section, those of shifting, multiplication, and division of binary numbers (base 2). These topics could be treated in more generality, but the complexity of the treatment does not appear to be justified in view of the preponderance of the binary system.

Our intuition tells us that a left-shift corresponds to multiplication by a power of 2. Let us therefore multiply Equation 2-1 by 2^p and interpret the result.

$$2^p V(X) = \sum_{i=-m}^{n-1} X_i 2^{i+p} = \sum_{j=-m+p}^{n-1+p} X_{j-p} 2^j = \sum_{j=-m+p}^{n-1} X_{j-p} 2^j + \sum_{j=n}^{n-1+p} X_{j-p} 2^j \quad (2\text{-}4)$$

Thus we find the original ith bit of the word associated with the 2^{i+p} value bit position, or the original X_i has to be moved to the $i + p$ position. The following table shows this for $n = 5, m = 5$.

Bit No.	4	3	2	1	0	·	−1	−2	−3	−4	−5
Before	X_4	X_3	X_2	X_1	X_0	·	X_{-1}	X_{-2}	X_{-3}	X_{-4}	X_{-5}
After 4 left	X_0	X_{-1}	X_{-2}	X_{-3}	X_{-4}	·	X_{-5}	—	—	—	—

Thus a shift left of p places is effectively multiplication by 2^p provided the bits lost off the most significant end are all 0 and provided that 0s are shifted into the p least significant position of the word, otherwise an *overflow* condition results. For the rest of this discussion, we let $m = 0$, and hence are dealing with integer words. Figure 2-4 shows the basic 1-bit shift for signed-magnitude numbers.

Division by 2^p yields the following modification to Equation 2-1:

$$2^{-p} V(X) = \sum_{i=0}^{n-1} X_i 2^{i-p} = \sum_{j=-p}^{n-1-p} X_{j+p} 2^j$$

or

$$2^{-p} V(X) = \sum_{j=0}^{n-1-p} X_{j+p} 2^j + \sum_{j=-p}^{-1} X_{j+p} 2^j \quad (2\text{-}5)$$

The second part of the expression contains those bits which are "shifted off" the right end of the number, and correspond to the remainder when $V(X)$ is divided by 2^p. The p most significant bits of the resulting word must be 0.

Thus with signed-magnitude representation, a left shift of p bits corresponds to multiplication by 2^p, with overflow a result if any of the p most significant bits of X are 1. A right shift corresponds to taking the quotient of the original word and 2^p, if dealing with integer representations.

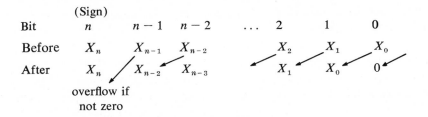

a) Left shift, one bit, signed-magnitude

Bit	n	$n-1$	$n-2$	$\cdots$	2	1	0	
Before	X_n	X_{n-1}	X_{n-2}		X_2	X_1	X_0	
After	X_n	0	X_{n-1}		X_3	X_2	X_1	lost

b) Right shift, one bit, signed-magnitude

Figure 2-4 Signed-magnitude shifts

When dealing with signed ones' complement or two's complement number systems, the sign bit enters into the operation in a direct way. Figure 2-5 shows how two's complement numbers are shifted.

Let us first consider a left shift of the negative number X by p bits by simple application of Part a of Figure 2-5 p *times*.

As long as bits X_n through X_{n-p} are all one, then the original number has a value of at least $-2^n + 2^{n-1} + \cdots + 2^{n-p} = -2^{n-p}$. Therefore we would expect no information to be lost by a left shift of p bits; that is no overflow occurs. If this condition exists, then a p-bit left shift of a negative number in two's complement form corresponds to multiplication by 2^p.

Bit	n	$n-1$	$n-2$	$\cdots$	2	1	0	
Before	X_n	X_{n-1}	X_{n-2}		X_2	X_1	X_0	
After	X_{n-1}	X_{n-2}	X_{n-3}		X_1	X_0	0	

a) Left shift, one bit, two's complement

Bit	n	$n-1$	$n-2$	$\cdots$	2	1	0	
Before	X_n	X_{n-1}	X_{n-2}		X_2	X_1	X_0	
After	X_n	X_n	X_{n-1}		X_3	X_2	X_1	lost

b) Right shift, one bit, two's complement

Figure 2-5 Two's complement shifts

In a right shift of p bits (see Part b of Figure 2-5), the sign bit must be maintained, to preserve the sign of the word. Information shifted off the right end of the word is lost, causing the value of the shifted word to be less than or equal to the value of the original word divided by 2^p, or equal to $\lfloor V_{2C}(X)/2^p \rfloor$.* When X is positive, this is the normal quotient, but when X is negative, it has a slightly different interpretation.

Example 2-10: $R = 2, n = 15, m = 0$: $\quad X_{15}X_{14}X_{13} \cdots X_1X_0$

Signed Magnitude: Same as plus numbers in two's complement

Two's Complement:

	15 14 13 12	11 10 9 8	7 6 5 4	3 2 1 0
Positive numbers:				
$+1123_{10}$	0 0 0 0	0 1 0 0	0 1 1 0	0 0 1 1
Shift right 3 or $140 = \lfloor 1123/8 \rfloor$	0 0 0 0	0 0 0 0	1 0 0 0	1 1 0 0
Shift left 4 or $17968 = (16 \times 1123)$	0 1 0 0	0 1 1 0	0 0 1 1	0 0 0 0
Shift left 5 with overflow (looks like negative number)	1 0 0 0	1 1 0 0	0 1 1 0	0 0 0 0
Negative numbers:				
-1123_{10}	1 1 1 1	1 0 1 1	1 0 0 1	1 1 0 1
Shift right 3 or $-141 = \lfloor -1123/8 \rfloor$	1 1 1 1	1 1 1 1	0 1 1 1	0 0 1 1
Shift left 4 or $-17968 = (-1123 \times 16)$	1 0 1 1	1 0 0 1	1 1 0 1	0 0 0 0
Shift left 5 with overflow (looks like positive number)	0 1 1 1	0 0 1 1	1 0 1 0	0 0 0 0

The examples just shown illustrate one result of the lack of symmetry in positive and negative numbers in the two's complement system. Note that the value of 0000 0100 0110 0011 after shifting right three places is $\lfloor 1123/8 \rfloor$ or 140 whereas the value of 1111 1011 1001 1101 after a 3-bit right shift is $\lfloor -1123/8 \rfloor$ or -141. This latter result, since humans divide in signed-magni-

---

*$\lfloor Z \rfloor$ means the largest integer less than or equal to Z.

tude form, is not the quotient of -1123 and 8. As a consequence, shifting right by p bits and converting to the two's complement value does not yield the same result as converting to the two's complement, then shifting right by p bits. This lack of consistency in results when interchanging the order of some operations can create serious problems because it may not be expected by the user and may create errors that are difficult to find.

The effect of shifting in the ones' complement system is easier to understand because the negative of a number is simply its bit-by-bit complement. Thus if a positive number is shifted right by p bits, this corresponds to taking the integer part of the original number divided by 2^p, provided 0s are shifted from the sign bit X_n into X_{n-1}. Then to obtain a similar result for negative numbers, we shift right by p bits, shifting 1s from X_n into X_{n-1}.

When shifting positive numbers left, if we do not shift a 1 into the sign bit and if 0s are shifted into the least significant bit position, then a p-bit shift corresponds to multiplication by 2^p. For negative numbers, no overflow occurs if we do not shift a 0 into X_n, and we must shift 1s into the least significant bit position. Figure 2-6 (on the following page) shows the basic 1-bit shift operation for ones' complement numbers.

Example 2-11: $R = 2, n = 15, m = 0: X_{15}X_{14} \cdots X_1X_0$.

Positive numbers: (same as two's complement)

	15 14 13 12	11 10 9 8	7 6 5 4	3 2 1 0
$+2880$	0 0 0 0	1 0 1 1	0 1 0 0	0 0 0 0
Shift right 3 $360 = \lfloor 2880/8 \rfloor$	0 0 0 0	0 0 0 1	0 1 1 0	1 0 0 0
Shift right 7 $22 = \lfloor 2880/128 \rfloor$	0 0 0 0	0 0 0 0	0 0 0 1	0 1 1 0
Shift left 3 $23040 = 2880 \times 8$	0 1 0 1	1 0 1 0	0 0 0 0	0 0 0 0

Negative numbers:

	15 14 13 12	11 10 9 8	7 6 5 4	3 2 1 0
-2880	1 1 1 1	0 1 0 0	1 0 1 1	1 1 1 1
Shift right 7 $-22 = -\lfloor 2880/128 \rfloor$	1 1 1 1	1 1 1 1	1 1 1 0	1 0 0 1
Shift left 3 $-23040 = 2880 \times 8$	1 0 1 0	0 1 0 1	1 1 1 1	1 1 1 1

In summary, then, positive numbers are treated the same in all three number systems. A right arithmetic shift of p bits shifts 0 into X_{n-1}, and produces $\lfloor V(X)/2^p \rfloor$. A left arithmetic shift of p bits shifts 0 into X_0 and produces $2^p \times$

Bit	n	$n-1$	$n-2$		2	1	0	
Before	X_n	X_{n-1}	X_{n-2}		X_2	X_1	X_0	from X_n
After	X_{n-1}	X_{n-2}	X_{n-3}		X_1	X_0	X_n	
	to bit 0							

a) Left shift, one bit, ones' complement

Bit	n	$n-1$	$n-2$		2	1	0	
Before	X_n	X_{n-1}	X_{n-2}	$\ldots$	X_2	X_1	X_0	
After	X_n	X_n	X_{n-1}	$\ldots$	X_3	X_2	X_1	lost

b) Right shift, one bit, ones' complement

Figure 2-6   Ones' complement shifts

$V(X)$ so long as X_n is not changed; that is, so long as no overflow occurs. Negative signed-magnitude numbers are treated as if the number were positive.

An arithmetic right shift of negative two's complement numbers involves shifting the sign (X_n) into X_{n-1}, preserving the sign and a p-bit shift produces $\lfloor V_{2C}(X)/2^p \rfloor$. The right shift of negative ones' complement numbers is performed the same way but the result is $-\lfloor |V_{1C}(X)|/2^p \rfloor$. A left shift of a two's complement number involves shifting 0s into X_0, whereas for ones' complement, X_n, the sign is shifted into X_0. The result is the same in both cases, however, so long as no overflow occurs, $V(X) \times 2^p$.

Next in our consideration of arithmetic, we move to the process of multiplication. We consider the process as applied to signed magnitude, two's complement and then ones' complement numbers.

Signed-magnitude multiplication is easiest to perform and to understand because the signs and the magnitudes can be treated separately. Thus the sign of the product is minus if either (but not both) the multiplier or the multiplicand is negative. Otherwise, the sign of the product is positive. Let us next consider the magnitude of the product. Let $\hat{B}$ be the magnitude portion of B, the multiplier and $\hat{C}$ the magnitude for C, the multiplicand. Using Equation 2-1 and the requirement that the value of the product of $B \times C$ must be the product of the values, we obtain Equation 2-6.

$$V(\hat{B} \times \hat{C}) = V(\hat{B}) \times V(\hat{C}) = \left(\sum_{i=-m}^{n-1} B_i 2^i \right) \times \left(\sum_{i=-m}^{n-1} C_i 2^i \right)$$

$$= \sum_{i=-m}^{n-1} \left(B_i 2^i \times \left(\sum_{j=-m}^{n-1} C_j 2^j \right) \right) \qquad (2\text{-}6)$$

Equation 2-6 concerns the value of the product and the value of $\hat{B}$, $\hat{C}$. Let us for the moment move into the realm of binary arithmetic and see what operation *in binary* on $\hat{B}$ and $\hat{C}$ in Equation 2-6 might be related to:

$$\hat{B} \times \hat{C} = \sum_{i=-m}^{n-1} B_i \times (2^i \times (C)) \tag{2-7}$$

Equation 2-7 suggests that if these operations are carried out properly, then Equation 2-6 will hold. But $(2^i \times \hat{C})$ simply represents an arithmetic left shift of i bit positions, assuming no overflow or truncation. Since B_i is either 1 or 0, if B_i is 1, $B_i \times (2^i \times (\hat{C}))$ corresponds to the selection of the appropriate shifted value of $\hat{C}$ and if B_i is 0, that shifted value of $\hat{C}$ is ignored. To produce the summation, the selected, shifted values of $\hat{C}$ are added together.

Because overflow and truncation can occur in shifting, destroying the simple arithmetic significance of the shift, attention must be paid to allowing the proper number of bits to appear in the partial products and developing sum as it develops. Consider numbers with n integer bits and m fractional bits, that is, $X_{n-1} X_{n-2} \cdots X_0 . X_{-1} \cdots X_{-m}$. The smallest representable number is 2^{-m}, and the product of two such numbers is 2^{-2m}; hence the product must have provision for $2m$ fractional digits. The largest representable number is $2^n - 2^{-m}$ in this system and the product of two such numbers is

$$2^{2n-1} - 2^{-2m} < 2^{2n} - 2 \cdot 2^{n-m} + 2^{-2m} < 2^{2n} - 2^{-2m}$$

Thus $2n - 1$ integer bits are insufficient, and $2n$ bits are sufficient to represent the product. This is a satisfying result, since we probably recall a rule-of-thumb that suggests that the product of two p-digit numbers is a $2p$-digit number (the sign bit doesn't count). Our product has the form

$$P_{2n-1} P_{2n-2} \cdots P_1 P_0 . P_{-1} \cdots P_{2-m}$$

Thus, the process of multiplying two $(m + n)$-bit numbers is as follows:

1. Set $i = -m$, set $\hat{P}(-m) =$ all zeros ($2n + 2m$ bits).
2. Extend the word of $\hat{C}$ by adding $n + m$ bit positions of zeros at the more significant end of the word, producing $\hat{C}(-m)$.
3. If $B_i = 1, \hat{P}(i + 1) = \hat{P}(i) + \hat{C}(i)$, else $\hat{P}(i + 1) = \hat{P}(i)$
4. $i = i + 1$, shift $\hat{C}(i)$ one bit left to produce $\hat{C}(i + 1)$
5. If $i = n$, go to 6, else go to 3.
6. $\hat{P}(n)$ contains the $2n + 2m$ bit product.

In this algorithm, the $\hat{P}(i)$ corresponds to the accumulating sum of the partial products, the selected $\hat{C}(i)$. In normal hand multiplication, we would write down the appropriate partial products in correct registration and then add them all up to obtain the product. Usually a computer does this in an accumulating register, adding partial products one at a time to the developing product.

Example 2-12: $n = 5, m = 0$

	9 8 7 6 5 4 3 2 1 0	
B:	0 - - - - 0 1 1 0 1	$V(B) = 13$
C:	0 - - - - 1 0 1 1 1	$V(C) = 23$
$\hat{C}(0)$	0 0 0 0 0 1 0 1 1 1	
$\hat{P}(0)$	0 0 0 0 0 0 0 0 0 0	
$+\hat{C}(0)$	0 0 0 0 0 1 0 1 1 1	
$\hat{P}(1)$	0 0 0 0 0 1 0 1 1 1	
$+0$		
$\hat{P}(2)$	0 0 0 0 0 1 0 1 1 1	
$+\hat{C}(2)$	0 1 0 1 1 1 0 0	
$\hat{P}(3)$	0 0 0 1 1 1 0 0 1 1	
$+\hat{C}(3)$	0 1 0 1 1 1 0 0 0	
$\hat{P}(4)$	0 1 0 0 1 0 1 0 1 1	
$+0$		
$\hat{P}(5)$	0 1 0 0 1 0 1 0 1 1	$V(\hat{P}) = 299 = 13 \times 23$

Example 2-13: $n = 0, m = 4$

B:	.0 1 0 1	$V(B) = .3125$
A:	.1 1 0 1	$V(C) = .8125$
$\hat{C}(-4)$	. 0 0 0 0 1 1 0 1	
$\hat{P}(-4)$	. 0 0 0 0 0 0 0 0	
$+\hat{C}(-4)$	. 0 0 0 0 1 1 0 1	
$\hat{P}(-3)$	. 0 0 0 0 1 1 0 1	
$+0$		
$\hat{P}(-2)$	. 0 0 0 0 1 1 0 1	
$+\hat{C}(-2)$	. 0 0 1 1 0 1 0 0	
$\hat{P}(-1)$	. 0 1 0 0 0 0 0 1	
$+0$		
$\hat{P}(0)$	. 0 1 0 0 0 0 0 1	$V(\hat{P}) = .25390625$
		$= (.3125) \times (.8125)$

Multiplication of signed two's complement numbers could be treated in a similar fashion; that is, by converting to magnitudes, multiplying, and then recomplementing the result if negative. There are also direct ways to handle the operation without explicitly recognizing that either positive or negative numbers are involved. One such method for two's complement multiplication is discussed here.

We will proceed from the expression for the two's complement value of the number $B_n B_{n-1} \cdots B_0 . B_{-1} \cdots B_{-m}$, where B_n is the sign bit, to represent the product of two numbers, B and C:

$$V_{2C}(B \times C) = V_{2C}(B) \times V_{2C}(C) = \left[(-B_n 2^n) + \sum_{i=-m}^{n-1} B_i 2^i \right]$$

$$\times \left[(-C_n 2^n) + \sum_{i=-m}^{n-1} C_j 2^j \right]$$

$$= (-B_n 2^n) \times \left[(-C_n 2^n) + \sum_{j=-m}^{n-1} C_j 2^j \right]$$

$$+ \sum_{i=-m}^{n-1} B_i 2^i \times \left[(-C_n 2^n) + \sum_{j=-m}^{n-1} C_j 2^j \right]$$

$$(2\text{-}8)$$

Interpreting Equation 2-8 as we did Equation 2-6 in light of our knowledge of two's complement arithmetic operations, we can deduce that

$$\widehat{B} \times \widehat{C} = \sum_{i=-m}^{n-1} B_i \times (2^i \times (\widehat{C})) - B_n \times (2^n \times \widehat{C})$$

where $\widehat{B}$ is in full two's complement representation. Now if we use arithmetic shifts to obtain the product of 2^i and $\widehat{C}$, properly allowing the requisite digits to be added so that overflow and truncations do not occur, and add or not to the partial product depending or whether or not B_i is equal to 1 for i from $-m$ to $n - 1$, we will account for all but the sign of B. Finally, *subtracting* $2^n C$ (add $-2^n C$) from the partial product takes proper account of the effect of the sign bit.

The number of bits required in the final product is $2n + 2m + 1$, and is more than needed to handle positive numbers, since a bit is needed for the sign. This is sufficient to deal with all cases but that of multiplying two values of -2^n together, whose product, 2^{2n} is greater than $2^{2n} - 2^{-m}$, the largest representable positive number. This special case must be tested for and an overflow signaled if it occurs. It should be noted however, that this is related to the basic difficulty with the two's complement system that one representable number (-2^n) has no representable negative $(+2^n)$.

The algorithm for multiplication based on the foregoing is as follows:

1. Set $i = -m$, set $\widehat{P}(-m) = $ all 0s $(2n + 2m + 1$ bits).
2. Extend C into $\widehat{C}(-m)$ by copying $n + m$ sign bits (C_n) at the more significant end of the word. (Note: This yields $2^{-m}C$ in a word where $n' = 2n$ and $m' = 2m$).

3. Use the ith multiplier bit, B_i, to form the next partial product, $C(i)$, then accumulate:
   If $B_i = 1$, form $\widehat{P}(i + 1) \leftarrow \widehat{P}(i) + \widehat{C}(i)$, else $\widehat{P}(i + 1) \leftarrow \widehat{P}(i)$.
4. $i \leftarrow i + 1$ and form $\widehat{C}(i + 1)$ by shifting $\widehat{C}(i)$ one bit left.
5. If $i = n$ go to 6, else go to 3.
6. If the sign of the multiplier is 1, subtract the multiplicand from the accumulated product:
   If $B_n = 1$, form $\widehat{P}(n + 1) \leftarrow \widehat{P}(n) - \widehat{C}(n)$, else $\widehat{P}(n + 1) \leftarrow \widehat{P}(n)$.
7. $\widehat{P}_{n+1}$ contains the $2n + 2m + 1$ bit signed, two's complement product.

Example 2-14: $n = 4, m = 0$

		B negative, C positive
Step		8 7 6 5 4 3 2 1 0
	B:	- - - - 1 0 1 0 0 . $\quad V_{2C}(B) = -12$
	C:	- - - - 0 1 1 0 1 . $\quad V_{2C}(C) = +13$
2	$\widehat{C}(0)$	0 0 0 0 0 1 1 0 1 . $\quad V_{2C}(\widehat{C}(0)) = +13$
1	$\widehat{P}(0)$	0 0 0 0 0 0 0 0 0 .
3	$+0$	_____
	$\widehat{P}(1)$	0 0 0 0 0 0 0 0 0 .
3	$+0$	_____
	$\widehat{P}(2)$	0 0 0 0 0 0 0 0 0 .
4	$+\widehat{C}(2)$	0 0 0 1 1 0 1 0 0 .
3	$\widehat{P}(3)$	0 0 0 1 1 0 1 0 0 .
	$+0$	_____
3	$\widehat{P}(4)$	0 0 0 1 1 0 1 0 0 .
6	$-\widehat{C}(4)$	
	$(+$ two's comp.$)$	1 0 0 1 1 0 0 0 0
7	$\widehat{P}(5)$	1 0 1 1 0 0 1 0 0 $\quad V_{2C}(\widehat{P}(5)) = -156$
		$= (-12) \times (13)$

Direct implementation of the foregoing algorithm would require a $2n + 2m + 1$ bit shift register to hold the shifting multiplicand $\widehat{C}(i)$, a $2n + 2m + 1$ bit adder and a $2n + 2m + 1$ bit product register. A slight rearrangement of Equation 2-7 yields:

$$\widehat{B} \times \widehat{C} = \sum_{i=-m}^{n-1} 2^{-(n-i)} (B_i \times (2^n \widehat{C})) - B_n \times (2^n \widehat{C}) \qquad (2\text{-}8)$$

This form suggests that for each bit of the multiplier that is a 1, the multiplicand is added to the most significant end of the developing product and the product is shifted right as more significant bits of the multiplier are considered. This algorithm is as follows:

1. Set $i = -m$ and set $\widehat{P}(-m) = 0$
   ($\widehat{P}$ has sign, $2n$ integer bits, and $2m$ fractional bits)
2. Use the ith multiplier bit to form the next partial product and accumulate:
   If B_i form $\widehat{P}(i + 1) \leftarrow 2^{-1}\widehat{P}(i) + \widehat{C}(i)$, else $\widehat{P}(i + 1) \leftarrow 2^{-1}\widehat{P}(i)$
3. $i \leftarrow i + 1$
4. If $i = n$ go to 5, else go to 2.
5. If the sign of the multiplier is 1, subtract the multiplicand from the shifted, accumulated product:
   If $B_n = 1$, $\widehat{P}(n + 1) \leftarrow 2^{-1}\widehat{P}(n) - \widehat{C}(n)$, else $\widehat{P}(n + 1) \leftarrow 2^{-1}\widehat{P}(n)$.
6. $P(n + 1)$ contains the $2n + 2m + 1$ bit signed, two's complement product.

Figure 2-7 shows a block diagram of a realization of this algorithm, and the following example illustrates the operation of this realization. Initially, the $n + m$ most significant bit of $\widehat{P}$ are cleared and the multiplier is inserted in the $n + m + 1$ least significant bits. The least significant bit of the P register contains the bit of the multiplier being examined and controls whether C or 0 is added to the feed-back, shifted accumulated product. The final step, when bit n of the multiplier, the sign, is in the least significant position of the product register, is to *subtract* the partial product from the shifted accumulated product by selecting the subtract function of the adder/subtractor.

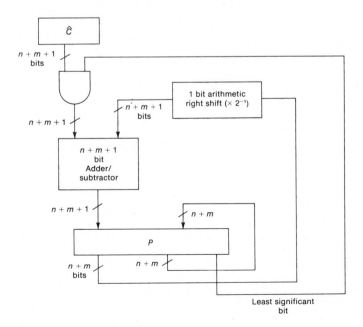

Figure 2-7   Block diagram of a two's complement sequential multiplier

Example 2-15: $n = 4, m = 0$ $\qquad$ *B* negative, *C* positive

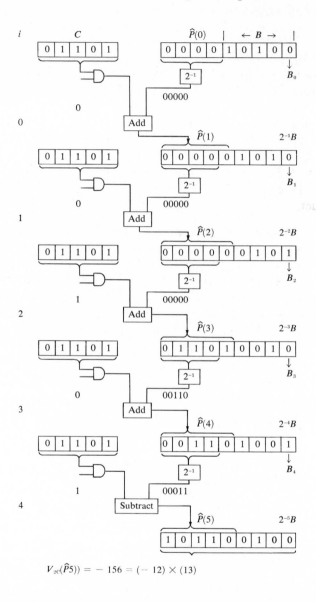

$$V_{2c}(\hat{P}5)) = -156 = (-12) \times (13)$$

A similar analysis will yield a similar algorithm for the multiplication of two ones' complement numbers.

In this portion of this chapter, we have presented multiplication algorithms for signed-magnitude and two's complement number systems. Other algorithms are presented in the book by Chu [62] and in the paper on arithmetic by MacSorley [61].

The final arithmetic process which we will treat in this chapter is division. Rather than follow the previous pattern of developing algorithm for sign and magnitude, two's complement and ones' complement schemes, we shall treat only the two's complement method. Normally, division is formulated on the basis of integer divisor and dividend yielding an integer quotient and remainder. This interpretation is consistent with the Euclidian division algorithm, presented in Equation 2-9. Other interpretations of the numbers involved are possible, but are not common and will not be treated here. An integer Y divided by an integer D has an integer quotient Q and remainder R defined by:

$$Y/D = Q + R/D \tag{2-9}$$

where normally $0 \leq R < D$

We develop our algorithm in a rather ad-hoc way by means of additions, subtractions, and shifts in order to produce the desired remainder. The series of operations is then interpreted to deduce the quotient.

We will produce in this process a method which avoids the process of restoring the partial remainder after subtracting too far, that is, causing a sign change in the partial remainder. Thus, this method is called *nonrestoring*. The basis for this is the fact that if 2^jD is subtracted from a number, and then $2^{j-1}D, 2^{j-2}D \cdots, 2^kD$ are all added, this is equivalent to subtracting 2^kD from the number, since

$$-2^j + \sum_{i=k}^{j-1} 2^i = -2^k.$$

The method is informally presented and illustrated by example, then analyzed, formally stated, and illustrated by other examples. We consider the case where $m = 0$, Y is the dividend, D the divisor, Q the quotient and $R(j)$ the jth partial remainder.

To avoid needless complication while attempting to explain the basic principles behind nonrestoring division, we will first treat the division of two positive numbers. The process is as follows:

1. Increase the number of digits in the dividend by $n - 1$ zeros, producing the initial partial remainder, and place the divisor below the extended dividend, aligning the leftmost digits. This establishes the first trial division as $2^{n-1}D$.
2. Subtract the trial divisor from the partial remainder.
3. If the result is positive, enter a 1 in the appropriate quotient bit. If the result is negative, we really should not have subtracted, so enter a 0 in the appropriate quotient bit.
4. If the partial remainder is positive, shift and subtract the trial divisor; if the partial remainder is negative, shift and add the trial divisor. Repeat step 3.

The final step in the operation, either adding or subtracting 2^0D, will not be corrected by subsequent operations to effect the equivalent of a restore operation if the final remainder is negative, so some special correction will be needed.

Let us analyze the meaning of the quotient bits and relate them to Equation 2-9.

$$R(n - 1) = Y - 2^{n-1}D \text{ always}$$

$$R(j - 1) = R(j) - 2^{j-1}D \text{ if } Q_j \text{ is } 1$$

$$R(j - 1) = R(j) + 2^{j-1}D \text{ if } Q_j \text{ is } 0$$

$$\text{or } R(j - 1) = R(j) + (1 - 2Q_j)2^{j-1}D$$

Combining these equations successively yields

$$R(0) = Y - \left[2^{n-1} - \sum_{j=1}^{n-1} (1 - 2Q_j)2^{i-1} \right] D$$

$$= Y - \left[2^{n-1} - \sum_{j=1}^{n-1} (2^{i-1}) + \sum_{j=1}^{n-1} Q_i 2^j \right] D$$

$$= Y - \left[\sum_{j=1}^{n-1} 2^j Q_j + 2^0 \right] D$$

Thus $R(0) = Y - Q \times D$ provided a 1 is inserted into bit Q_0, to account for the 2^0 appearing in the equation for $R(0)$. Therefore, the final step of the algorithm is as follows:

5. $Q_0 \leftarrow 1$, end.

Example 2-16:   Divide $D = 00110$ (6_{10}) into $Y = 10010$ (18_{10}).

	43210 ← Bit position		
	00011	Step	Operation
00110 ⟌	000010010	1	Trial Divisor $= 2^4 D$
	001100000	2	Subtract $2^4 D$
$R(4)$ −	1001110	3	Negative so $Q_4 \leftarrow 0$
	00110000	4	Shift and add $2^3 D$
$R(3)$ −	11110	3	Negative so $Q_3 \leftarrow 0$
	0011000	4	Shift and add $2^2 D$
$R(2)$ −	00110	3	Negative so $Q_2 \leftarrow 0$
	001100	4	Shift and add $2^1 D$
$R(1)$ +	00110	3	Positive so $Q_1 \leftarrow 1$
	00110	4	Shift and subtract $2^0 D$
$R(0)$	000	5	$Q_0 \leftarrow 1$

The method just presented suffers from several shortcomings: it does not properly handle the case of negative representations nor does it always leave a remainder which is nonnegative, the conventional situation in the implementation of the Euclidian division algorithm as shown in Equation 2-9. The modifications necessary to remove these shortcomings are incorporated in the following algorithm for nonrestoring division of signed, two's complement integers:

1. Extend the sign of the dividend by $n - 1$ digits, producing the initial partial remainder, and place the divisor below the extended dividend, aligning the leftmost digits. This establishes the first trial divisor as $2^{n-1}D$.
2. If the signs of the divisor and dividend are the same, the result will be positive, so enter 0 into Q_n; otherwise; the sign of the quotient will be negative and a 1 should be entered in Q_n.
3. If the signs of the partial remainder and the trial divisor are the same, subtract the trial divisor (add its two's complement); if the signs of the partial remainder and the trial divisor differ, add the trial divisor.
4. If the new partial remainder and the trial divisor have the same sign, enter a 1 in the appropriate quotient bit and shift the trial divisor; if these signs differ, enter a 0 in the appropriate bit position of the quotient and shift the trial divisor.
5. If the signs of the partial remainder and the trial divisor are the same, subtract the trial divisor (add its two's complement); if the signs differ, add the trial divisor.
6. Repeat steps 4 and 5 until 2^0D has been added or subtracted, then go to step 7.

At this point we have $R(0)$ and bits n through 1 of the quotient. Analyzing $R(0)$ as before we find:

$$R(n - 1) = Y - (1 - 2Q_n)\, 2^{n-1}D$$
$$R(j - 1) = R(j)\, (-1 + 2Q)2^{j-1}D$$

thus

$$R(0) = Y - \left[(1 - 2Q_n)2^{n-1} + \sum_{j=1}^{n-1}(-1 + 2Q_j)2^{j-1}\right]D$$
$$= Y - \left[2^0 - 2^nQ_n + \sum_{j=1}^{n-1}2^jQ_j\right]D$$
$$= Y - Q \times D$$

and the correct quotient, as in the unsigned case, is obtained by making $Q_0 = 1$.

Now note also that the sign of the partial remainder may be of opposite sign to the original dividend as seen in the truth table shown in Table 2-3.

TABLE 2-3. TRUTH TABLE

Sign of Divisor	Sign of Dividend	Sign of $R(0)$	Correction Needed
+	+	+	None
+	+	−	$R \leftarrow R(0)+D$; $Q_0 \leftarrow 0$
−	−	−	None
−	−	+	$R \leftarrow R(0)+D$; $Q_0 \leftarrow 0$
+	−	−	None
+	−	+	$R \leftarrow R(0)-D$; $Q \leftarrow Q+1$
−	+	+	None
−	+	−	$R(0) \leftarrow R(0)-D$; $Q \leftarrow Q+1$

The addition of 1 to the quotient in two of the cases is required since an additional $2_0 D$ is subtracted from the partial remainder to obtain the correct final remainder. Thus, the final steps of the algorithm are:

7. Set $Q_0 = 1$
8. If the sign of the dividend and of $R(0)$ agree, the algorithm is complete, otherwise go to step 9.
9. If the sign of the divisor and $R(0)$ agree, subtract the divisor from $R(0)$ to form the final remainder R and make the final quotient be $Q + 1$. If the signs of the divisor and $R(0)$ disagree, add the divisor to $R(0)$ to form the final remainder and make $Q_0 = 0$. End.

Example 2-17:   Divide $D = 001110$ $(+14_{10})$ into $Y = 011110$ $(+30_{10})$ for $n = 5$.

Quotient bit position → 543210

	Step	Operation
000011		
001110 | 0000011110	1	Trial divisor $= 2^4 D$
−$2^4 D$ 1100100000	2	Signs same ∴ $Q_5 \leftarrow 0$
$R(4)$ 1100111110	3	Signs same ∴ add −$2^4 D$
+$2^3 D$ 0001110000	4	Signs differ ∴ $Q_4 \leftarrow 0$
	5	add $2^3 D$
$R(3)$ 1110101110	4	Signs differ ∴ $Q_3 \leftarrow 0$
+$2^2 D$ 0000111000	5	add $2^2 D$
$R(2)$ 1111100110	4	Signs differ ∴ $Q_2 \leftarrow 0$
+$2^1 D$ 0000011100	5	add $2^1 D$
$R(1)$ 0000000010	4	Signs same ∴ $Q_1 \leftarrow 1$
−$2^0 D$ 1111110010	5	add −$2^0 D$
$R(0)$ 1111110100	7	$Q_0 \leftarrow 1$
	8	Signs of $R(0)$ and Y disagree
Q: 000011	9	Signs of D and $R(0)$ disagree
0000001110		∴ add $2^0 D$
R 0000000010	$= 2_{10}$	$Q_0 \leftarrow 0$

Q: $000010 = 2_{10}$

$2 = 30 - (2)(14)$

Other Number Representations

In addition to the simple binary representation studied in the previous section, other codes are used to represent numbers in certain circumstances. Sometimes decimal numbers are represented by using a 4-bit binary code for each digit. Nonweighted codes are also used. These are codes in which the value of a number cannot be expressed as a sum of weighted digit values such as in Equations 2-1, 2-2, or 2-3. Finally, extra digits can be added to provide error detecting or correcting capability.

The most straight forward of the decimal codes is the *Binary-Coded Decimal* representation. In this representation, each digit is represented by a conventional 4-bit binary number. Thus, each 4-bit digit can take on the value 0000 to 1001. Values larger than 1001 are not allowed since they correspond to values greater than or equal to $R = 10$. As an example, the 4-digit BCD version of the number 5830 is represented as

$$X_{33}\ X_{32}\ X_{31}\ X_{30} \qquad X_{23}\ X_{22}\ X_{21}\ X_{20} \qquad X_{13}\ X_{12}\ X_{11}\ X_{10} \qquad X_{03}\ X_{02}\ X_{01}\ X_{00}$$
$$0\quad 1\quad 0\quad 1 \qquad\quad 1\quad 0\quad 0\quad 0 \qquad\quad 0\quad 0\quad 1\quad 1 \qquad\quad 0\quad 0\quad 0\quad 0$$

Arithmetic involving BCD representation of numbers is based primarily on the rules of decimal arithmetic and secondarily on performing binary operations on the digits themselves. For example, the addition of two n-digit BCD numbers involves for each digit, the addition of the binary versions of the digits and the one bit decimal carrry into the digit. If the sum is 1001 or less, the decimal carry out of that digit is 0 and the binary sum is the BCD version of the sum digit. If the sum is 1010 or greater, a decimal carry out of 1 is generated and the decimal sum is generated by subtracting 1010 (10) or by adding 0110 modulo 16. .

Another commonly used code is the *excess-three code*. In this case a normal decimal digit is represented by the binary code for the digit plus 3. Thus, 0_{10} has code 0011, 6_{10} is 1001, and 9_{10} is 1100. This code has the advantage that the negative of a number in this representation is easily related to its ones' complement. A more practical advantage appears when addition of two excess-three numbers is considered. If the codes for two digits and a decimal carry into the digit are all added together, the binary value of the sum is 6 greater than the actual sum; 0110 corresponds to 0_{10} and 1111 to 9_{10}. Thus, if the sum is 16 or greater, a decimal carry is needed and the sum is corrected by adding 3. Otherwise the sum is corrected by subtracting 3 or by adding 13 modulo 16.

Another basically binary code which finds some application is a so-called Gray code. This is a code whose characteristic is that if you arrange the code representations in numerical order of their value, adjacent code words differ in one and only one bit position. Table 2-4 shows one such code for four bits. The next code word in sequence is obtained by changing one bit in the least significant position which yields a new code word.

TABLE 2-4. SIXTEEN-VALUE GRAY CODE

Value	Code	Value	Code
0	0000	8	1100
1	0001	9	1101
2	0011	10	1111
3	0010	11	1110
4	0110	12	1010
5	0111	13	1011
6	0101	14	1001
7	0100	15	1000

One very important application for such codes is in mechanical to binary encoders such as shaft encoders, which give a binary representation of some parameter such as angle. For example, in the case of the shaft encoder, if a normal binary code were used, at some point, a transition between 0001111 and 0010000 must be made. Since the angle may vary continuously, the point of transition must occur exactly at the same place for each bit or else some code such as 0000100 representing a drastically different angle may be spuriously generated. Table 2-5 shows a code suitable for a shaft encoder with 10 degree resolution.

Other codes for the representation of numbers have been devised with specific purposes in mind. One important purpose, especially for data held in auxiliary storage media, is for detection and correction of errors. These codes involve the use of parity checks or other redundant data with which the validity of the information can be determined. A complete description of this topic is beyond the scope of this chapter, so we mention in passing codes for correction of single errors [Hamming, 50] and codes for correction of errors in arithmetic units [Massey and Garcia, 72]. In each type of code some patterns of 0s and 1s correspond to valid data and other patterns to data contaminated by errors. The idea is that any highly probable error will change a valid datum into a pattern that is invalid, and validity can be easily checked. The Hamming codes can detect combination of two random changes, and can correct any single random change. The arithmetic codes are similar except the changes produced by an error in a carry circuit are additive and effect the entire carry chain rather than a single bit. The interested reader should consult the references for more information.

TABLE 2-5. CODE FOR A 36-ELEMENT GRAY CODE SHAFT ENCODER

Angle (degrees)	Code	Angle (degrees)	Code	Angle (degrees)	Code	Angle (degrees)	Code
0-10	000000	90-100	001101	180-190	110000	270-280	111101
10-20	000001	100-110	001111	190-200	110001	280-290	111111
20-30	000011	110-120	001110	200-210	110011	290-300	111110
30-40	000010	120-130	001010	210-220	110010	300-310	111010
40-50	000110	130-140	001011	220-230	110110	310-320	111011
50-60	000111	140-150	001001	230-240	110111	320-330	111001
60-70	000101	150-160	011001	240-250	110101	330-340	111000
70-80	000100	160-170	011000	250-260	110100	340-350	101000
80-90	001100	170-180	010000	260-270	111100	350-360	100000

Word Size

The size of a word, the unit of information containing the representation of a number, is usually fixed by physical constraints for a particular computer system. Thus, selection of a word size of b for a computer causes the fixed point number to have b significant bits. Sometimes, however, it is desired to have operations and data be defined for a higher degree of significance. One way to accomplish this is to define a class of double-word arithmetic instructions, where each number is represented by two standard words. One physical word contains the less significant half and the other word contains the more significant half. One application of this can be seen in multiplication, where the product of two n-bit words produces a $2n$-bit product. This multiplication of two single words produces a double word. The programmer must decide or must equip the program to decide how to round the result back to n bits or whether to carry on the calculations in double words. The decision to discard significant bits normally has to be made at some level of significance, for if we choose to carry double words as our standard, the product of two double words yields a quadruple word, and so on.

Also, there are operations that reduce the level of significance of a result. For example, the difference of two nearly equal large numbers has many fewer significant digits than either number. To cope with this problem of increasing and decreasing significance during the course of a computation, a variable word length can be used. This word length normally is some multiple of the actual physical word length. Thus we might address the least significant end of a number, encountering first a number which specifies the number of words which make up the number. The arithmetic then proceeds reading successive physical words until the specified number of words are processed. This scheme on the face of it would appear to be quite reasonable, however, problems involved with detection of the sign of the number, and the position of the radix point may make the whole process more cumbersome than the result justifies [Avizienis, 64]. Instead, what is commonly done is to allow for two degrees of significance for numbers and to then use whichever one is justified to meet the particular needs of the program.

Another problem with the fixed radix-point and fixed word size representations is overflow in addition and multiplication. Overflow prevention requires proper scaling of the values throughout the program and the detection of overflow to signal an incorrect result. The most common solution to the problem is to use the computer equivalent of scientific notation, *floating-point numbers*, which are treated in detail in the next section of this chapter.

Floating-point Numbers

A floating-point number is similar to the familiar scientific notation that allows human representation of very large or very small numbers. The number consists of two parts—a signed, fixed-point (usually fractional) part often called the *mantissa* and a part called the *exponent*. In scientific notation the exponent is the power of 10 by which the mantissa is multiplied to produce the true value of the number. In floating-point representation the same relation holds except the exponent is the power of the radix of the representation.

Types of Floating-point Representation   There exists a large number of possible variations in the way that floating-point numbers are represented, such as variation in radix, treatment of exponent and matissa signs, and word size. Rather than treat all possible variations in general, we consider only those few in common use and try to call attention to some general principles as they emerge in this treatment.

The simplest type is the one used by the Burroughs B-5500 where both the mantissa and the exponent are in signed-magnitude form. An example of this type of number follows, where the radix is 8 and numbers are shown in their octal representation.

Sign	Sign of exponent	Exponent	Mantissa
(1 bit)	(1 bit)	(6 bits)	(39 bits)
0	0	03	0000000000053

which converts to $53_8 \times 8^3 = 53,000_8$ or $5 \times 4096 + 3 \times 512 = 22,016_{10}$.

Many other types of representation exist; Table 2-6 shows the characteristics of some of the commonly used representations, based on a word whose bits are numbered starting with bit 0 at the left hand end.

TABLE 2-6. FLOATING NUMBER SYSTEMS

	Xerox Data Systems, Sigma Series	IBM SYSTEM 360, 370	Burroughs B-5500	Control Data 6000, 7000 Series
Bits used:	0-31	0-31 ·	1-47	0-59
Radix:	16	16	8	8
Radix point:	Before first digit	Before first digit	After last digit	After last digit
Mantissa				
Sign position	0	0	1	0
Value position	8-31	8-31	9-47	12-59
Representation	Two's complement of *entire word*	Signed magnitude	Signed magnitude	Ones' complement of *entire word*
Exponent				
Sign position	1	1	2	1
Value position	1-7	1-7	3-8	1-11
Representation	Value plus 64	Value plus 64	Signed magnitude	*Value + 1024 if ≥ 0; Value + 1023 if < 0.
Range of value	−64 to +63	−64 to +63	−63 to +63	−1023 to +1023

* An exponent of 1777_8 is possible; it corresponds to −0 exponent and denotes an "indefinite" operand.

There are several characteristics of floating-point words to consider, a primary one being the lack of uniqueness of representation; for example 5.3×10^1 is the same number as 0.53×10^2, but yet it has these two different representations (and others). This characteristic makes comparison of numbers difficult and consequently floating-point numbers are usually represented in normalized form, where the mantissa is always represented with a nonzero most significant digit. Thus our first example showing a Burroughs floating-point number is not normalized. Another advantage of normalization is that the maximum number of significant digits is retained in the representation.

Example 2-18: For the decimal number -53851_{10} the signed magnitude representation in various systems is:

Hexadecimal	$-$D25B
Binary	$-1101, 0010, 0101, 1011.$
Binary	$-001, 101, 001, 001, 011, 011.$
Octal	$-\ 1\quad 5\quad 1\quad 1\quad 3\quad 3\ .$
Hexadecimal scientific	$-.D25B \times 10^{+4}$
Octal scientific	$-.151133 \times 10^{+6}$
Bit positions	0 1234567 8901,2345,6789,0123,4567,8901
IBM representation	1 1000100 1101,0010,0101,1011,0000,0000
XDS representation (Two's complement of bits 1-31 of IBM)	1 0111011 0010 1101 1010 0101 0000 0000

Example 2-19: For the decimal number $+.011718750$, the signed magnitude representation in various systems is:

Hexadecimal	$+.03$
Binary	$+.0000,0011$
Binary	$+.000,000,110$
Octal	$+.006$
Hexadecimal scientific	$+.3 \times 10^{-1}$
Octal scientific	$+.6 \times 10^{-2}$
Bit positions	0 1234567 8901,2345,6789,0123,4567,8901
IBM representation	0 0111111 0011,0000,0000,0000,0000,0000
XDS representation	0 0111111 0011 0000 0000 0000 0000 0000

Floating-point numbers represent a much wider range of values than do fixed-point numbers. For example, the largest IBM floating-point number is approximately 1×16^{63} whereas the smallest normalized nonzero positive number is approximately 1×16^{-64}.

The number 0 deserves some special attention in this system. For any value of x, $0 = 0 \times 10^x$, and furthermore, one cannot normalize the number in the usual way. Consequently, a convention for the representation of 0 is usually adopted. Zero is represented as a zero fraction with the most negative exponent. In many computer systems, this is the all zero word $00 \ldots 0_2$, a convenient number to detect.

Arithmetic Floating-point arithmetic is relatively simply performed in the cases of multiplication and division, since the mantissa and exponents can be treated separately. However, in addition and subtraction, the exponents of the two numbers must be made equal by "unnormalizing" the number with the smaller exponent. In any of the cases overflow conditions and underflow conditions must be watched and treated properly to obtain correct results.

Let us consider multiplication and addition in some detail. First, though, we adopt some simple conventions:

> B,C refer to the binary words
> $m(B)$ refers to the mantissa of B
> $e(B)$ refers to the value of the exponent of B

The multiplication procedure is as follows:

1. Form the double length product $m(B) \times m(C)$
2. Form the sum $e(B) + e(C) = e(P)$
3. Special cases
    a) If $e(P)$ produced an overflow when formed and $V(e(P)) > 0$: force $m(P)$ to be largest magnitude fraction with proper sign, and force $e(P)$ to be largest positive exponent.
    b) If $e(P)$ produced an overflow bit and $V(e(P)) < 0$: force $m(P)$ to be 0 and $e(P)$ to be most negative value.
    c) If $m(B)$ or $m(C)$ is 0: force $m(P)$ to be 0, and $e(P)$ to be most negative value. (Force P to be true zero.)
4. Normalize the double-length product.
5. Round the product to the appropriate word size, renormalizing if necessary.

The process for division is similar, but will not be covered in detail.

Example 2-20: (Using IBM 360 representation)

1. Form the product of B and C:

$$m(B) = -.A58300_{16} \qquad e(B) = -35_8$$
$$m(C) = .100000_{16} \qquad e(C) = -50_8$$
$$m(B) \times m(C) = -.0A583_{16} \qquad e(B) + e(C) = -105_8 < -100_8$$

Case 3*b* occurs.

$$m(P) = 0 \qquad\qquad e(P) = -100_8 \ (P=0)$$

2. Form the product of B and C:

$$m(B) = -.A58300_{16} \qquad\qquad e(B) = -35_8$$
$$m(C) = -.100001_{16} \qquad\qquad e(C) = 50_8$$
$$m(P) = .0A58300 + .000000A583$$
$$\qquad = .0A5830A583 \qquad\qquad e(P) = 13_8$$

Normalize:

$$m(P') = .A5830A583 \qquad e(P') = 12_8$$

Round:

$$m(P'') = .A5830A \qquad e(P'') = 12_8$$

The addition procedure is as follows (assume $e(B) \le e(C)$):

1. Form $e(C) - e(B)$ and shift $m(B)$ to the right by that number of digits. (Alignment)
2. Form $m(S) = m(B) + m(C)$
3. One of three special cases may occur:
   a) If $m(S) = 0$, make $e(S) = $ smallest negative value.
   b) If $m(S)$ has overflow: form $m(S')$ by a right shift of $m(S)$, properly shifting in overflow, and $e(S') = e(S) + 1$.
   c) After b, $e(S')$ may have overflowed, then set S to largest value.
4. Normalize S, setting $S = 0$ if exponent becomes less than smallest negative value.

Example 2-21: (Using IBM representation)

1. Add B and C:

$$m(B) = -.A53800_{16} \qquad e(B) = 0_8$$
$$m(C) = +.A57AFF_{16} \qquad e(C) = 0_8$$
$$m(S) = .0042FF_{16} \qquad e(S) = 0_8$$

Normalize:

$$m(S') = .42FF00_{16} \qquad e(S') = -2_8$$

2. Add B and C:

$$m(B) = +.250000_{16} \qquad e(B) = -2_8$$
$$m(C) = +.FFE301_{16} \qquad e(C) = 0_8$$
$$m(s) = .FFE301_{16} + 002500_{16} = 1.001801 \qquad \text{(overflow)}$$

Case $3b$ has occurred.

$$m(S') = .100180_{16} \qquad e(S') = 1_8$$

The first of the previous examples shows how a loss of significance can occur when two opposite signed, nearly equal magnitude numbers are added. The second illustrates how the case of an overflow in addition is handled.

Floating-point numbers when implemented in a computer system are a tremendous help to the programmer, for now he does not need to concern himself with a large number of special error cases which can regularly occur in fixed-point arithmetic. Instead, he need only worry about the case of overflow and underflow in the very unlikely cases where the extremes in exponents are reached and must concern himself about whether or not sufficient precision has been used to avoid the effects of loss of significance in addition.

2-3. CODES FOR CHARACTER REPRESENTATION

As was mentioned earlier, computers are used for processing information in many kinds of situations not restricted to simple arithmetic involving numbers. Sorting and classifying text information, parsing statements in a computer language (what a compiler does), information retrieval—these are all examples of applications where the computer processes information consisting of letters, numbers, punctuation and special symbols. Two considerations associated with this type of data emerge: 1) the coding or representation in bits of each desired symbol, and, 2) the method of packing character codes into physical computer words to make efficient use of storage.

The first task in selecting a code for character information is to establish a set of desired symbols to be represented. A set of 64 possible code words, for example, is sufficient to represent 26 letters, 10 digits, and 28 special symbols (such as * / + − @ % #), punctuation (. , : ; ! ?) and functions (tab, space, change color, and so on). For these 64 codes we require 6 binary bits, and in the early days of computer usage this field length was universally accepted for character representations. Computers such as the IBM 7090 have a word size of 36 bits, dividing nicely into six 6-bit characters. This 6-bit character code is sometimes called the *BCD code* since it is a logical extension of the 4-bit BCD code discussed earlier. Table 2-7 shows this code. Six-bit codes are also in use in the CDC 6000 series and 7600 machines.

In spite of its wide-spread use, the 6-bit code has become insufficient for many computer applications. For example, it does not permit the expression of upper and lower case letters, nor can Greek symbols be included, to mention a few of the limitations. To avoid some of these limitations, a standard character size of eight bits was proposed and two different standard 8-bit character sets are currently in use: ASCII (American Standards Committee on Information Interchange) was developed as the standard code for the data communications and computer industries and EBCDIC (Extended Binary Coded Decimal Interchange Code) was developed by IBM. Table 2-7 also shows the portion of these codes corresponding to those characters represented in BCD. To distinguish the 8-bit character from its 6-bit predecessor, the 8-bit character is usually called a *byte*. In an excess of cuteness, it has even been suggested that 4-bit digits be called "nibbles", since two of them make up a byte. As a consequence of the change to an 8-bit standard code, most current computers (the CDC 6000 series and 7600 computers are exceptions)

TABLE 2-7. SIX-BIT BCD CODE AND A PORTION OF THE EIGHT-BIT ASCII AND EBCDIC CODES

Character	BCD code	EBCDIC code	ASCII code
blank	110 000	0100 0000	0100 0000
.	011 011	0100 1011	0100 1110
(	111 100	0100 1101	0100 1000
+	010 000	0100 1110	0100 1011
$	101 011	0101 1011	0100 0100
*	101 100	0101 1100	0100 1010
)	011 100	0101 1101	0100 1001
—	100 000	0110 0000	0100 1101
/	110 001	0110 0001	0100 1111
'	111 011	0110 1011	0100 1100
,	001 100	0111 1101	0100 0111
=	001 011	0111 1110	0101 1101
A	010 001	1100 0001	1010 0001
B	010 010	1100 0010	1010 0010
C	010 011	1100 0011	1010 0011
D	010 100	1100 0100	1010 0100
E	010 101	1100 0101	1010 0101
F	010 110	1100 0110	1010 0110
G	010 111	1100 0111	1010 0111
H	011 000	1100 1000	1010 1000
I	011 001	1100 1001	1010 1001
J	100 001	1101 0001	1010 1010
K	100 010	1101 0010	1010 1011
L	100 011	1101 0011	1010 1100
M	100 100	1101 0100	1010 1101
N	100 101	1101 0101	1010 1110
O	100 110	1101 0110	1010 1111
P	100 111	1101 0111	1011 0000
Q	101 000	1101 1000	1011 0001
R	101 001	1101 1001	1011 0010
S	110 010	1110 0010	1011 0011
T	110 011	1110 0011	1011 0100
U	110 100	1110 0100	1011 0101
V	110 101	1110 0101	1011 0110
W	110 110	1110 0110	1011 0111
X	110 111	1110 0111	1011 1000
Y	111 000	1110 1000	1011 1001
Z	111 001	1110 1001	1011 1010
0	000 000	1111 0000	0101 0000
1	000 001	1111 0001	0101 0001
2	000 010	1111 0010	0101 0010
3	000 011	1111 0011	0101 0011
4	000 100	1111 0100	0101 0100
5	000 101	1111 0101	0101 0101
6	000 110	1111 0110	0101 0110
7	000 111	1111 0111	0101 0111
8	001 000	1111 1000	0101 1000
9	001 001	1111 1001	0101 1001

now have a word size which is a multiple of eight bits. In fact most minicomputers are 16-bit machines, and the first two microprocessors have word sizes of four and eight bits.

At this point, it is appropriate to emphasize an important point made earlier: most computers store information as binary bits packed into words of fixed size. What in fact these bits mean depends on how they are used. They may be a collection of 8-bit characters; they certainly are treated as such if transmitted to a line printer by the input/output system. They may be integer numbers, and are so processed when sent to the integer adder. They may be computer instructions, and are interpreted this way by the computer control section. Sometimes, it is difficult to make a decision as to what the information represents and in fact, sometimes the interpretation of words may change during a computation.

2-4. DATA STRUCTURES

The physical considerations which govern a computer structure also impose constraints on the way in which data is organized for use by the computer. We have already seen how the basic elements of data representation (bits, digits, and characters) are related to the interpretation of information in a computer word. Additionally, we have seen that nonnumerical interpretations are possible. To compound the problem, the physical resource of main memory is expensive, too expensive to be large enough to service the storage needs of large programs. Recourse is then made to cheaper storage such as tape, disk, or drum. The different character of these storage media force a hierarchical approach to data management. At one extreme, the programmer is forced to deal explicitly with each type of memory, performing the detailed exchanges of information between working memory and bulk storage. On the other hand, so-called *virtual* schemes have been implemented which automatically manage the transfer of needed blocks of data between bulk storage and working storage, making the programmer think that he has unlimited, virtual, main memory. Here he has traded programming simplicity for operating system complexity and degradation of system performance. All of these issues are important to the design of computers and to their intelligent use; they are, however, the proper subject of later chapters. In the balance of this section, we deal with the various types of higher order structures from the point of view of ease of manipulation for various classes of problems.

Basic Consideration

We consider here a number of different types of data structures. Each type has a particular advantage for applying the computer to certain classes of problems. For example, data may be presented to a computer in one order, but it may be desired to process it in a different order, depending on other interrelationships of the data. One may then have the choice of sorting the data first before processing to establish new interrelationships, or one may set up a more complex cross-reference scheme. The ultimate decision depends on the details of the problem and the performance requirements. We make no conclusions regarding the use of these schemes, but merely present these schemes and show some of their applications. Also, some of these schemes are automatically embodied in programs produced by certain compilers, and certain specialized languages are used to facilitate use of particular data structures. In some cases, the use of special classes of data structures may be actually built into the hardware. In the remainder of this section, we describe arrays, sparse arrays, stacks, and lists.

Arrays

Arrays (or subscripted variables) are a basic storage mode for all high-level languages. They are advantageous when implementing algorithms in which the position arrangement of data is important. An example of an array used in a FORTRAN program is shown in Figure 2-8. This program segment shows the type of index computation for which arrays are well-suited.

Array storage has several drawbacks including the need for a fixed pre-declared size (in some languages at least), and the need for a fixed word size for the elements of the array. The advantages are that many problems, particularly numerical analysis problems, have solutions that are easily understood and specified through the manipulation of arrays. This fact also has given rise to the development of hardware to process arrays automatically for improved processing performance. Chapters 8 and 9 treat such developments in detail.

Sparse Arrays and Scatter Tables

In some cases, the range of indices for an array is large, and only a small percentage of the elements of the large array are actually used. In such cases, storage of the entire array can be wasteful of memory space. One common way to solve this problem is to store only those elements actually needed, finding

```
      DIMENSION A(10,10),B(10,10),C(10,10)
         .
         .
   C     CALCULATE MATRIX PRODUCT C OF A*B
         DO 10 I=1,10
            DO 10 J=1,10
               Z=0.
               DO 9 K=1,10
   9                Z=Z + A(I,K)*B(K,J)
   10          C(I,J)=Z
         .
         .
         END
```

Figure 2-8 Example of a FORTRAN program using arrays

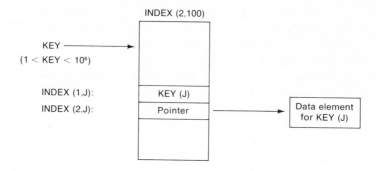

Figure 2-9 Scatter storage scheme

them by means of an *index*, sometimes called a *scatter table*. Figure 2-9 shows the scheme of using an index to find 1 of 100 actual elements stored in a 1,000,000 element vector. The subscript of the desired element is the key used to find the proper index entry, which in turn either contains the desired data or points to the desired data. The index is searched until KEY(J) matches the desired key. Several methods for finding the desired index can be used: simple search, binary search, or hash code computation.

Simple search is the easiest to program but the most time consuming for search. Whenever an entry is desired, the index is searched from the beginning until either the desired entry is found or the index is exhausted.

Binary search is more complex to program and requires that the index be sorted each time an entry is added, but is much less time consuming than the previous type. The average length of search for simple search is proportional to one-half the index length whereas the maximum length of search for the binary search is proportional to the logarithm base 2 of the index size.

In the binary search, the index is in numerical order of keys. The first look at the table is at the middle word; if the key is less than the mid-point index, the desired value is in the upper half. If greater, the desired value is in the lower half. If equal, the desired index is found and the search terminates. The next step is to search the portion known to contain the value (if it is included in the index) by dividing it in half and examining the middle index entry. Thus the region of interest is halved at each step, giving rise to the $\log_2 N$ dependency of maximum search length.

The *hash code* method of indexing is slightly more complex than the binary search algorithm but appears to be the most economical in time required for entry, deletion, and recovery of items from a sparse array. Morris presents some different methods of hash coding and contrasts their efficiencies [Morris, 68].

In the hash coding scheme, a function is used to compute an index, J, based on the key word which is the actual name of the entry. Various functions can be used, but they usually have the one desirable feature of producing differing hash codes for closely similar key words. Thus, for example, in Figure 2-9, if the hash code for 953,252 were 53, then 953,252 would be placed in INDEX (1,53) and a pointer to the 953,252 element of the array would be placed in INDEX (2,53). Other keys are often used, such as the coded representation of names.

One problem can arise in the use of hash coding techniques. By its very nature, a hash code must map some number of keys into one hash code, and although efforts are taken to reduce the likelihood of two keys having the same code, as the index fills up, the likelihood of a conflict increases. Several schemes for handling a conflict are possible, the simplest being: if a conflict occurs (same hash code, different key), look in adjacent index locations for the desired key or for an empty position to put the desired key.

In general, hash-coding schemes require more programming effort than simple search techniques, but the savings in search time usually justify the use of the method.

Stacks and Queues

Stacks and queues are extremely useful data structures when data is to be accessed in an order directly related to the order in which entries were made. The *stack* or *push-down-stack* is a first-in, last-out structure. One important application of stack structures is for the storing of arguments and return points in recursive procedures. Some compilers and interpreters construct stacks automatically when a recursive procedure is encountered. Some Burroughs computers even go to the extent of directly implementing the stack structure in the hardware [Hauck and Dent, 68].

The basic operations on a stack are to add an item to the top of the stack (push) and to remove an entry from the stack (pop). Also the top element of the stack can be read while leaving the stack unchanged.

Implementation of a stack structure can take one of many forms. In addition to the variety of stack format, there is also variability in the amount of special hardware capabilities included in the machine to facilitate stack operations, although we will not discuss those hardware tradeoffs here.

The simplest form of stack structure is the one in which a fixed area of memory is allocated to the stack. If the current top of the stack index is equal to the dimension of the array in which the stack is stored, the stack is full and no further items may be "pushed" onto the stack.

Figure 2-10 shows the stack data structure in which a fixed-size block of storage is allocated to each stack. Two additional registers are needed to implement this scheme: one contains the value of the array subscript for the current top of the stack and the other contains a number related to the maximum number of stack entries such as the value of the array subscript cor-

S(3) | YELLOW   STKSZ: N
S(2) | ORANGE   STKIND: 3
S(1) | RED

a) Stack S contains
     three entries.

S(4) | GREEN
S(3) | YELLOW   STKSZ: N
S(2) | ORANGE   STKIND: 4
S(1) | RED

b) Stack S after GREEN has been
     "pushed" onto stack.

S(2) | ORANGE   STKSZ: N
S(1) | RED   STKIND: 2

c) Stack S after the stack has
     been "popped". The data
     YELLOW is the result of the
     operation.

Figure 2-10   Simple stack structure

responding to a full stack. An empty stack is signified by a subscript value of 1 less than the minimum subscript for an array.

The FORTRAN subprograms in Figure 2-11 illustrate the mechanics of "pushing" and "popping" stacks. This illustration shows a somewhat primitive system in that the main program must allocate appropriate arrays for the stacks used by the functions and must properly initialize their indices prior to first call of either one.

Automatic expansion of the stacks and reallocation of space when stacks fill up can be accomplished in fixed stack systems, but some overhead is required. In fact, the overhead increases dramatically as the number of stacks involved increases and as the amount of unused stack space decreases.

One way to avoid these difficulties is to use linked lists as the means to implement the stack. This scheme will be discussed in the section on linked lists.

Queues are lists in which the first element entered in the list is the first element withdrawn; they are sometimes called first-in, first-out (FIFO) lists. Queues derive their name from the queue of people waiting to board a bus or receive service of some kind, and the analogy between queues as data structures and these examples from day-to-day life is useful. Many activities within the operation of computer programs correspond to queues for service and a whole body of theory surrounding the behavior of queues has developed and goes by the name of *queueing theory*.

Items of data arriving for processing by the computer, where processing is to be accomplished in the order received, are often stored in queues. For example, information read from an input device is read into a queue, and a processing procedure extracts information from the queue as needed.

Implementation of the queue in a linear list arrangement in a program is not quite so simple as it is for stacks. Our first inclination might be to assign

```
      LOGICAL FUNCTION  PUSH(DATA, STACK, STKIND, STKSZ)
C     RETURNS  .TRUE.   IF STACK FULL
      DIMENSION  STACK (STKSZ)
      INTEGER STKIND, STKSZ
      IF (STKIND .GE. STKSZ) GO TO 100
C     PUT DATA IN NEXT WORD OF STACK
      STKIND = STKIND + 1
      STACK (STKIND) = DATA
      PUSH = .FALSE.
      RETURN
C     RETURN ERROR- STACK FULL
  100 PUSH = .TRUE.
      RETURN
      END



      LOGICAL FUNCTION  POP  (DATA, STACK, STKIND, STKSZ)
C     RETURNS TRUE IF STACK EMPTY
      DIMENSION  STACK (STKSZ)
      INTEGER  STKIND, STKSZ
      IF (STKIND .LE. 0) GO TO 100
C     REMOVE ONE ITEM FROM STACK
      DATA = STACK (STKIND)
      STKIND = STKIND-1
      POP = .FALSE.
      RETURN
C     STACK EMPTY
  100 POP = .TRUE.
      RETURN
      END
```

Figure 2-11   FORTRAN subprograms for stack operations

an array of some size N to the queue. Let I be the index of the location in the array to receive the next item entered in the queue and let J be the index of the location in the array next to be removed. Note that $I = J$ means that the queue is empty and that $J = N + 1$ means that the end of array allocated for the queue has been reached. With this kind of system, the following operations are defined:

1. Enter A into the queue
   IF (I .EQ. N + 1) GO TO 20   (queue space used up)
   Q(I) = A
   $I = I + 1$
2. Remove an element from the queue and put it in A
   IF (I .EQ. J) GO TO 10   (queue empty)
   A = Q(J)
   $J = J + 1$

As items are entered into the queue and serviced or removed from the queue, the actual areas containing data entered but unserviced moves through the array allocated to the queue, and eventually J will become N + 1 and the simple process must halt, even though the unserviced items may number considerably less than N, the size of the array.

Several solutions to this rather inefficient use of space are possible. One can move the active queue back to the beginning of the array every time the end of the queue is encountered when an item is entered. This solution, however, requires many overhead operations to carry out the move. Another way to solve the problem is to effectively make the array into a circular array, where the top, Q(N) is logically adjacent to the bottom, Q(1). I, J, and N are as before, and I = J will continue to mean that the queue is empty. J = I + 1 signifies that the tail of the queue has wrapped around and is in danger of bumping into the head of the queue so no more items can be added; otherwise the queue status will be mis-read as being empty. The special case when J = N and I = 1 is also an impending collision of the tail with the head. Figure 2-12 shows the status of the queue as various items are added and with-

	Empty Queue	After the addition of 11 12, 13, 14 in that order	After the removal of 11, 12
Q(1):	-- *	11	11*
Q(2):	-- *	12	12*
Q(3):	-- *	13	13
Q(4):	-- *	14	14
Q(5):	-- *	-- *	-- *
Q(6):	-- *	-- *	-- *
I:	1	1	3
J:	1	5	5
N:	6	6	6

	After the addition of 15, 16, 17 (list full)	After the removal of 13, 14, 15, 16	After the addition of 18, 19, 20, 21 (list full)
Q(1):	17	17	17
Q(2):	12*	12*	18
Q(3):	13	13*	19
Q(4):	14	14*	20
Q(5):	15	15*	21
Q(6):	16	16*	16*
I:	3	1	1
J:	2	2	6
N:	6	6	6

* Items left from previous entries.

Figure 2-12   Queue status

drawn. Notice that the full queue conditions actually leave one location vacant, but to effectively utilize all locations of the array requires a more complex way of keeping track of the head and tail pointers.

Linked Lists

Linked lists are the most complex of the data structures to be considered in this chapter. In this type of structure, related data elements are interconnected by means of pointers. Starting with the reference pointer for the list, successive pointers are followed (following the links) until either the desired data is found, or the point is found to insert a new data element. In this latter case, the pointers are adjusted so that the new element is pointed to from an old member of the list and the pointer in the added element points back to the next member of the old list. Part *a* of Figure 2-13 shows this basic structure. Part *b* shows the process of inserting new data in between items 1 and 2 in the original list. One important property of list structures is that only a limited amount of moving of data is necessary when data is inserted or deleted from a list. If data were instead actually stored in order in a fixed array, all data from the point of alteration to the end of the list would have to be moved. With the linked list structure, actual position in storage is not important, so insertion or deletion can be accomplished by changing a few pointers.

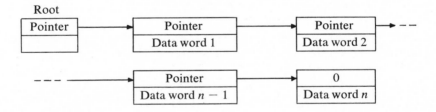

a) Basic linked list

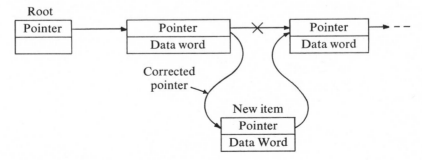

b) Insertion of new data into a linked list

Figure 2-13 Linked list operations

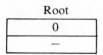

a) Empty stack

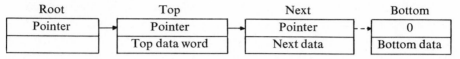

b) Nonempty stack

Figure 2-14   Use of linked lists for stacks

Linked lists can be used to implement stacks. Each stack has an associated root location which is a link to the top data element in the stack. Successive elements in the stack point to the next element down in the stack until a null pointer is reached, signifying the bottom of the stack. Figure 2-14 shows this structure.

More sophisticated list structures are used in some cases. Having two-way pointers eases the problem of searching backwards. Data may actually belong to several intertwined lists, each list having its own common property and meaning of order. This latter complication leads to the terminology "knotted list." Sutherland was one of the first to extensively employ the use of the digital computer to process spatial information contained in an engineering drawing and his program "sketchpad" made extensive use of knotted lists [Sutherland, 63]. A good general discussion of list structures is presented in a paper by Wilkes [64].

Several special compilers are available to carry on automatically the functions required to create and use lists. LISP [Berkeley and Bobrow, 64] is such a language, which in particular has very high power in dealing with list structures. On the other hand, most common high level languages have no built-in list processing facility at all, so some basic list processing routines must be added to the library of available routines if list processing is to be employed.

This has been a somewhat brief look at some of the important types of data structures. For a more thorough treatment of the subject, the reader is referred to the specific references given, or to a good general text on data structures such as Knuth [68], or Stone [72].

2-5. SUMMARY

This chapter has presented a study of the techniques of data representation. The properties, representation, and arithmetic involving different number systems were considered in substantial detail. Nonnumerical representations were next considered and then, finally, the higher order data structures of arrays, stacks and lists were treated. Specific implications of data representation considerations on hardware will appear in all of the later chapters.

2-6. PROBLEMS

2-1. Convert the following numbers from the indicated base to the decimal system:
  a) 15.21_6
  b) 15.21_8
  c) 1101.101_2
  d) $AF.3_{16}$

2-2. Convert the following numbers from the decimal system to the designated base:
  a) 78_{10} to hexadecimal
  b) 105_{10} to ternary (base 3)
  c) $.5625_{10}$ to binary
  d) 33.51_{10} to base 5
  e) $.05_{10}$ to binary

2-3. Convert 23170132_8 and $4CF05A_{16}$ to decimal to demonstrate their equivalence to the binary number 010011001111000001011010.

2-4. Convert 7352.7_8 to hexadecimal.

2-5. Using A and B for the digit values 10 and 11 respectively, construct an addition table for the duo-decimal system (radix 12). Perform the addition in radix 12 of the following pair of numbers:

$$10A_{12} + BB1_{12}$$

2-6. Give in your own words two different methods to find the radix complement of a number.

2-7. Express the following magnitudes as negative numbers in 1) signed-magnitude; 2) signed diminished-radix complement; and 3) signed radix complement form using as the radix that which the number is given in. The most significant digit is to be a binary sign digit.
  a) 153_{10}
  b) 101101_2
  c) 28.35_{10}
  d) $.7725_8$
  e) $.AA10_{16}$

2-8. Find the indicated value (the leading digit is a binary sign digit):
  a) $V_{SM}(19053_{10}) =$
  b) $V_{DRC}(19053_{10}) =$

c) $V_{RC}(19053_{10}) =$

d) $V_{RC}(0A.5F_{16}) =$

e) $V_{DRC}(17\ 52_8) =$

2-9. What happens when two negative RC numbers are added together where the sum of the magnitude equals R^n?

2-10. Convert $V(A) = 1953_{10}$ and $V(B) = 785_{10}$ to 4-digit hexadecimal positive numbers and perform the following operations in the system indicated. Check your results by operating in the decimal system and comparing the results.

a) $A + B$ SM

b) $A - B$ RC

c) $B - A$ RC

d) $A - B$ DRC

2-11. We are working with binary numbers containing fifteen information bits plus one sign bit of the form: $X_{15}X_{14} \cdot \ldots \cdot X_0$.

a) Two's complement negative numbers:

    1) Represent $V_{2C}(B) = 2583_{10}$ and $V_{2C}(C) = -2583_{10}$.

    2) Shift B and C five bits right arithmetically and evaluate results.

    3) Shift B and C three bits left and evaluate results.

    4) Comment on the results of 2 and 3.

b) Repeat part a for the ones' complement system.

2-12. Multiply the following numbers in the indicated systems. Check your results. The most significant bit in all cases is the sign.

a) $(1.320_8) \times (0.510_8)$ in two's complement $(n=0, m=9)$

b) $(01101110_2) \times (10011001_2)$ in ones' complement $(n=7, m=0)$

c) $(FF0A_{16} \times (003F_{16})$ in two's complement $(n = 15, m = 0)$

Hint: Convert all numbers to binary before proceeding.

2-13. Carry out the division algorithm for the following sets of numbers, $n=5$, check the results:

a) $D = 010000$ $Y = 001110$

b) $D = 111001$ $Y = 100111$

c) $D = 0001000$ $Y = 101011$

2-14. Present a formal algorithm for the addition of two n-digit, excess-three numbers, using as the basic operation the binary addition of two 4-bit numbers.

2-15. Devise a 1-digit multiplication table in BCD. Show the partial product as a 2-digit BCD number.

2-16. By trial and error, construct a Gray code to encode a shaft into 16 sectors.

2-17. *a)* State the algorithm for floating-point division, in a manner similar to that given in the chapter

b) Illustrate B/C where $M(B) = +7531_8$, $e(B) = +25_8$ and $m(C) = -653_8$, $e(C) = +37_8$.

2-18. Assume a computer program has read in the five decimal digits of an integer starting with the most significant in $J(5)$ and ending with the least significant in $J(1)$. Each digit is in EBCDIC Code. Write a FORTRAN program to produce the binary integer representation of the number in the variable N. Assume only digits 0 through 9 are allowed.

2-19. Another scheme for dealing with conflicts in a hash coding algorithm makes use of a linked list for all entries with the same hash code. Describe in detail such a data structure.

CHAPTER 3

Introduction to Calculators

Thomas M. Whitney

3-1. INTRODUCTION

Throughout history man has devised tools to help him in the fundamental task of counting. He has progressed from readily available aids such as fingers and toes to gigantic computers such as ILLIAC IV. Before computers all mechanical calculating devices were called "calculators." There were no formal courses of study on how to design such machines but rather the skills of design were passed on in the internal company secrets of early leaders in the trade. With World War II and the birth of stored program computers, the new disciplines of Computer Science and Computer Engineering began to appear in college curricula but calculator design remained mechanical and specialized until the electronic era of the 1970s.

Until recently the difference between calculators and computers was obvious. Calculators sat on desks and had keys, levers, switches, and sometimes cranks; computers came in nineteen-inch racks or seven-foot cabinets, had blinking lights and were connected to whirling, clacking tape units, hammering, clanking line printers, and spinning, glittering disks.

Today the extremes of each product type are also clearly differentiated. Some calculators are hand-held devices that fit in shirt pockets; some computers fill medium-sized rooms and are connected to hundreds of peripheral devices. However, the dividing line is blurred when products called "calculators" interface to disk memories, card readers, line printers, and CRT terminals and communicate in a traditional computer language like BASIC. What are the fundamental characteristics that differentiate calculators from computers?

Perhaps the most distinguishable feature of a calculator is its keyboard inputs and its display or printer output. Numbers are entered directly into the processing portion of the calculator by simply pressing keys; answers are returned immediately in decimal readable form on the display or the printer. Computers on the other hand, are inherently too fast and too costly to be idle while awaiting human response. Thus calculators are designed for continual human interaction, while computers, with their powerful and rapid input/output capability, communicate mainly with other machines.

Because the input to calculators is digit serial and the output is decimal, calculators represent numbers internally in binary coded decimal (BCD) form. Arithmetic is done in BCD, while nearly all computers use a binary representation. Although the binary format is inherently more efficient in terms of storage requirements, the need for BCD-to-binary and binary-to-BCD conversion and the loss of accuracy that can result from conversions outweighs the apparent savings for calculators.

Calculator processors are slow by computer standards, operating at kilohertz clock rates rather than megahertz. But why go faster? For non-programmable calculators the critical time factor is the time it takes a person to press two keys in succession, about 100 milliseconds. This important difference in processing speed is one of the fundamental and essential factors differentiating calculator design from computer design. The calculator trades speed for cost; the result to the consumer is a product for tens or hundreds of dollars rather than thousands or millions.

Another characteristic differentiating calculators from computers is the amount of memory available. Simple calculators only have enough registers to perform addition, subtraction, multiplication, and division; more sophisticated models have one location for a constant or a temporary answer. Even the most costly programmable calculators generally do not have capacity to add as much memory as the least expensive computers. This memory limitation is overcome by calculator users in a variety of ways, ranging from the crude to the elegant. With the availability of human interactions it is always possible to force the user to write down intermediate answers for later reentry when memory is unavailable. Once again the trade-off between time and cost is evident.

Keyboard and display input/output, serial decimal arithmetic, slower processor speed, and limited memory are four characteristics that make calculator design different from computer design. One does not take computer architecture, scale it down and have a good calculator architecture.

In this chapter we begin with a look at calculator history, with emphasis placed on the time since 1964 when electronic calculators first appeared. In section 3-3 we discuss the factors to be considered before design begins, particularly the importance of complete external operating specifications. In section 3-4 the essentials of calculator design are explained. The primary design task is the organization of the control function; several currently used methods are explained in detail with example problems. We also treat the arithmetic unit, registers, and input/output portions. Section 3-5 examines two types of calculator systems. The first is a basic four-function, one-chip calculator circuit, as typified by the Texas Instrument TMS 1802. The second is an architecture for more complex calculators that require more storage and transcendental function calculations. An example of this type of system is the Hewlett-Packard HP-35.

Finally, problems are included to stimulate discussion and to demonstrate the material covered in the chapter.

This chapter stops short of several calculator-related topics, primarily in the design of programmable machines. In this area many of the procedures from the later chapters of this book apply, particularly the chapters on minicomputers, input/output processing, and microprogramming.

3-2. HISTORY OF CALCULATORS

Counting is such an inherent part of human existence that man has always looked for ways to speed the process and ease the drudgery. Although fingers and toes have the advantages of portability and low power, two features currently advertised for hand-held calculators, the range is limited and the procedure impractical in cold weather. We can view calculating device history as occurring in four stages: the manual, the mechanical, the electrical, and the electronic.

The abacus appeared in China during the 12th century. It consists of columns of beads on rods, five below a divider and two above. Two important numerical concepts are represented. First, an object, a bead, is used as a counter, just as fingers were used earlier. This object is an abstract representation for *quantity*. The second principal is the use of *position* to show different kinds of quantities. The beads above the divider and to the left represent greater quantities. The abacus is still in use in oriental countries today.

John Napier was a Scottish mathematician who lived from 1550-1617. He invented and named the logarithm, which allowed multiplication to be done by the simpler operation of addition. He transcribed the results of multiplication onto rods of ivory which became known as "Napier's bones." By arranging the rods side by side and matching up the numbers he wanted to multiply, Napier could read off the answer very quickly without having to do any calculations at all. In 1617 Robert Bissaker transferred Napier's results to sliding wooden strips, thus creating the first slide rule.

The mechanical era was begun in 1642 by Blaise Pascal. His machine had numbers printed on rotating gears. When a gear was rotated one time, it advanced the adjacent gear one tenth of a revolution. Such mechanical counters are still in use today by supermarket shoppers. In 1694 Von Leibniz improved Pascal's machine to allow multiplication by repeated addition.

Neither Pascal nor Von Leibniz were engineers and their machines were often inaccurate and unreliable, due to poor construction techniques. By the early 19th century, mechanical technology had advanced to where several types of mechanical adders were in use. However, technology could still not implement the advanced ideas of Charles Babbage, an English mathematician who is rightly called the grandfather of modern computing (von Neumann is credited as the father). By 1823 Babbage had evolved the basic ideas of memory, automatic input and output of data, and branching to solve different problems in different ways. He even knew about parallel processing since he foresaw his "Difference Engine" as being composed of several smaller engines each working with the other to perform its own special task. Unfortunately his ideas were never implemented successfully as the engineering difficulties of such a complex machine could not then be overcome.

The late 19th century was a rich time for the mechanical inventor applying himself to calculating machines. The increase of business and commerce brought a substantial need for faster and more accurate methods of calculating and recording the results. In 1872, E. D. Barbour patented the first printing calculator. However it required the operator to daub ink on a hinged platen, place paper under it, and depress the platen, a very messy process. In 1875 Frank Baldwin invented a practical printing calculator. He later (1912) joined with Jay Monroe to form the Monroe Calculating Machine Company. Their calculator, one of the first mass produced, is shown in part *a* of Figure 3-1. It could subtract and divide as readily as add or multiply without the necessity of shifting levers or resetting the mecha-

Figure 3-1   (Part *a*) Early Monroe calculator (1912)

Figure 3-1 (Part *b*) Modern Monroe printing and display calculator (1972)

nism. Compare it to Monroe's 1972 integrated circuit model in part *b* of Figure 3-1. Meanwhile, in 1892 William S. Burroughs perfected a printing calculator. By 1926 his company had produced one million calculators.

Another early leader was Marchant. Their 1918 "Pony" model is shown in Figure 3-2. It came in either 13- or 18-digit versions, was almost noiseless, and featured a short crank handle for highest operating speed with minimum effort.

The era of mechanical calculator development is explained in almost painful detail in a book by Turck [72]. His final closing quote is interesting, "... It is a safe conjecture, however, that in the present high state of the Art (of calculator design) it will tax the wits of high class engineers to offer any substantial and broadly new feature which will be heralded as a noticeable step in the Art. And that, as in the past, thousands of mistakes and impractical as well as inoperative machines will be made and patented,

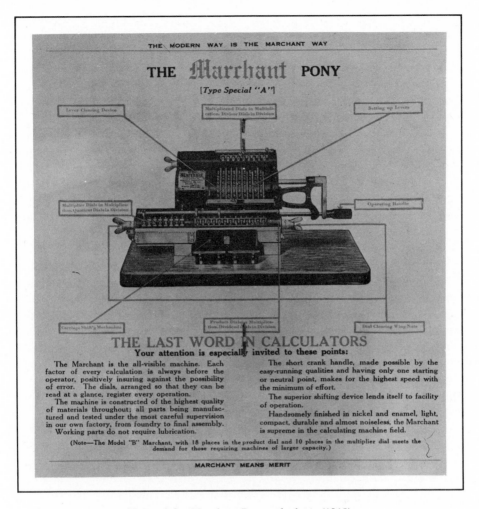

Figure 3-2   Marchant Pony calculator (1918)

to one that will hold real value." In some ways his statement was true for 50 years, until the advent of integrated circuits.

By 1930 the electric or motor driven calculator era was in full swing. Machines from Monroe, Burroughs, Marchant, and Friden were mass produced. Special features such as square root, subtotal, and nonadd keys became popular. New models were introduced for the banking, railroad and insurance industries. In 1935 Burroughs had 450 standard models in their product line. Mechanical designers were strained for greater reliability and lower cost models.

It was during this period that computers were born. There was a need to replace huge rooms of accountants and mathematicians preparing tables of insurance rates or ballistic data with their desk calculators with more efficient and more accurate methods. While computer technology progressed through relay, vacuum tube, and transistor versions, calculators were still made from levers, springs, gears, and motors. Even in the late 1950s and early 1960s, calculators were mechanical and electrical and not transistorized. An array of $10 and $1 transistors could not compete with a mass-produced gear when speed of operation was not the overwhelming design criteria.

It was 1963 before transistors decreased to a price that made electronic calculators practical. Introduced that year were the Anita, a simple machine built by Lanson Industries of England, and the Mathatronics programmable desk top computer with core memory, printer, and peripheral capability. This revolutionary $3940 product started to fill the large gap during that era between the most powerful $1500 rotary calculators and the $20,000 smallest computers. The next year saw the Friden 130 with a Cathode Ray Tube display and a 4-register stack, the Monroe Epic 2000, with "Learn-mode" programming, and the electronic awakening in Japan with a machine by Sony. These were also sophisticated calculators costing several thousand dollars, never intended to compete with the workhorse mechanical desk calculator. In October 1965, the electronics industry was excited by the announcement of the Victor 3900, the first metal-oxide semiconductor large-scale integrated circuit (MOS/LSI) calculator. It cost $1825 and contained 29 integrated circuit chips with 21,000 integrated transistors. Unfortunately, like Babbage's Difference Engine, the technology was not quite ready. The low yield of the MOS circuits prevented the machine from ever being marketed.

The idea was too "right" to be idle long however, and in 1967 Japan's Hayakawa Electric (now Sharp Company) introduced a manufacturable integrated circuit calculator. It contained 50 integrated circuits, 43 discrete transistors, 200 germanium diodes, an 8-digit display, consumed 6 watts and weighed 6 pounds. The era of integrated circuit calculators had arrived, and progressed much faster than most American mechanical calculator manufacturers thought possible.

In 1969-1970, Sharp's 8-digit Microcompet calculator was in great demand. With four United States-made integrated circuits (made by Rockwell International, then Autonetics) the Japanese rapidly captured the world-wide calculator market. Traditional calculator users no longer were dissuaded by "MOS unreliability," especially when prices on electronic calculators were dropping by a factor of two each year, and the reliability proved to be superior to mechanical calculators which require periodic maintenance.

In 1970 the Electronics Arrays Company introduced a set of six "standard" calculator integrated circuits. Prior to this, calculator circuits were "custom" designed by a semiconductor company for one calculator manufacturer. The custom design process was expensive, time consuming, and required extensive knowledge about calculator design. With the standard circuits almost anyone could be in the calculator business. Just connect a keyboard and a display to the six circuits, mold a plastic case, add a simple power supply, and develop a marketing plan. Especially develop a marketing plan. Many were using the same circuits to produce functionally identical calculators; marketing made the difference in sales.

The next major advancement came in January, 1971 when MOSTEK, a newly formed company in Dallas, Texas, announced the first single integrated circuit calculator chip. Under contract to Busicom, a Japanese calculator company, the single circuit replaced a previous electronic calculator containing two printed circuit boards and 700 gate functions. MOSTEK's press release described the MOS/LSI circuit size as approximately "three grains of rice." Its power consumption was so low (50 milliwatts) that a pocket-sized calculator operated from penlight batteries could be made. Later in 1971 and 1972, several other semiconductor companies introduced one-chip calculator circuits and the industry underwent immense changes. Figure 3-3 shows the effect of the rapid progress in

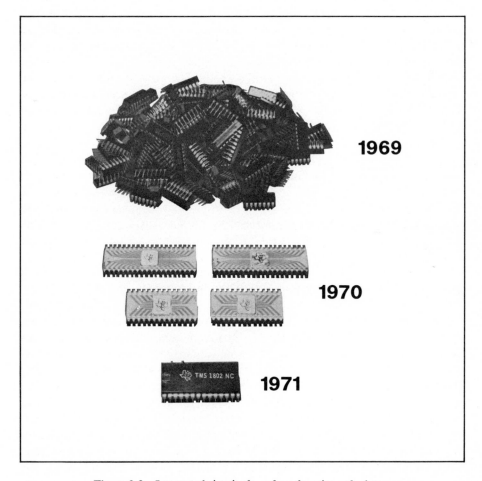

Figure 3-3   Integrated circuits for a four-function calculator

integrated circuits; from 1969 to 1971 the IC package count for a simple calculator went from about forty, to four, to one. Perhaps no other product has ever had the rapid price decreases of single chip portable calculators. The first models available in the fall of 1971 were priced at $395. In the late fall of 1973 the same calculating capability was available for $39.50.

The one-chip calculator was important because it meant the production cost of a calculator was more material intensive than labor intensive. The advantage of a low labor rate became less significant and American manufacturers began to regain a larger share of the calculator market.

As the cost of electronics decreased, the dividing line between calculators and computers began to get blurred. Just as minicomputers reached consumers in 1965, the Olivetti Programma 101 calculator was announced. It was a discrete transistor machine which featured programming with branching capability and a magnetic card input/output mechanism to enter and record programs. Competing machines appeared from Wang (1965), Smith Corona Marchant (1965), Wyle Laboratories (1966), and Hewlett-Packard (1968). Some of these were truly "computers with keyboards."

In 1968 an experimental electronic digital slide rule was described by General Electric [Schmid and Busch, 68]. Based on a digital rate-multiplier computing technique, it never became a commercial product, probably because its cost was too great to entice engineers to throw away their trusty slipsticks. In January 1972, Hewlett-Packard introduced an MOS Electronic Slide Rule, the HP-35. Its success at $395 proved that engineers would pay for the accuracy and convenience of a more powerful tool.

The electronic era of calculators will eventually give way to some other era as man's need to calculate provides a rich ground for the practical implementation of new technologies. Turck's statement of 1921, that the "high state of the Art" will "tax the wits of high class engineers" for improvements will not hold for the next 50 years. Just as he could not have forseen microelectronics, we cannot foresee the new technologies soon to be discovered.

3-3. PRELIMINARY DESIGN CONSIDERATIONS

Before beginning the details of calculator architecture it may be well to obtain an overview of the task facing the designer. Calculators may be characterized in a variety of ways. From a marketing point of view there are the different classes of buyers—scientists, businessmen, and the general consumer—and there are the different ways calculators are sold—through dedicated calculator sales forces, office machine dealers, department stores, and by direct mail.

From a design point of view, calculators may be classified by performance as standard four functions, expanded function (square root, percent, reciprocal, square, memory, and constant), special purpose, (slide rules for engineers and scientists, or interest and statistics functions for businessmen), simple learn-mode programmable, and on up to higher-level language programmable with expanded input/output capability. Classification also might be made by size, (pocket, portable, or desk top), or by power source, (AC only, rechargeable DC, primary cell DC, or a combination of the above).

Any design must begin with a careful, complete external specification of what type of calculator is required for what type of market. Because the calculator is an interactive product, in which man and machine must constantly communicate, the designer must be aware of human factor considerations that provide convenient and efficient operation. The human factor design criteria are more difficult to specify and more often overlooked than the more obvious electrical specifications.

Functions, Languages, and Algorithms

The first factor specifying the calculator organization is whether or not it is programmable. A simple form of programming, called "learn-mode" or straight line programming can be implemented by storing key codes in a shift register as they are entered from the keyboard with a mode switch in a LEARN position. This type of operation differs from true programming since no branching is allowed and there are no means of editing. A calculator with branching must have several extra keys, such as GO TO and IF $x \leq y$, and must provide for more sophisticated control procedures.

The programming decision is coupled with the type of input/output capability to be provided. Any input/output more complex than keyboard and display forces an expanded processor organization. Fully programmable calculators may have separate input/output processors similar to those in computers.

The next critical decision concerns the type of arithmetic functions to be executed. Add, subtract, multiply, and divide can be done in three registers with a simplified control scheme. More complex functions, such as the transcendental functions found on electronic slide rules require a fourth register and provision for the storage of constants. The type of algorithm used to implement complex functions also affects both the calculator hardware and software. In addition, the calculator's speed of operation is likely to become a factor. Depending upon the algorithm implemented, computation of transcendental functions may take hundreds of multiply operations. Even if each multiply requires only tens of milliseconds the accumulated time may stretch to several seconds. This is an annoying interval when we have come to expect "instant" response from electronic devices. The time factor is even more critical on programmable calculators and some special function calculators that solve implicit equations by iterative methods.

Another consideration is the type of keyboard language to be implemented. Three basic types of keyboards are found in currently available calculators: algebraic, reverse Polish, and plus-equals. Each is described in more detail.

The algebraic keyboard has five operator keys: $+$, $-$, $\times$, $\div$, and $=$.

Problems are solved in the way they are written algebraically as shown in part *a* of Figure 3-4. Note the inconvenience encountered when solving the sum of products problem $(a \times b) + (c \times d)$. If the sequence of $a \times b + c \times d =$ is used, the computed answer is $((a \times b) + c)d$. To circumvent this, one of four methods is used: 1) the user manually writes down the intermediate answer $a \times b$, then reenters it and adds after the product $c \times d$ is computed, 2) if the calculator has a separate memory, the user stores $a \times b$, then recalls it and adds to $c \times d$, 3) the calculator may have an "operator precedence" built in such that additions and subtractions are not executed until all multiplications and divisions have been completed; this places the burden on the calculator by requiring additional internal storage and bookkeeping, and makes the product of sums problem come out wrong, 4) the calculator may have two additional operator keys, a right and left parenthesis; the user must then use additional keystrokes and the calculator bookkeeping is more complex, but there is no confusion as to what equation is being executed. These four methods amply demonstrate how calculator architecture is dependent upon keyboard language.

The second language is called reverse Polish, sometimes referred to as the "adding machine language." In 1951 a book on formal logic by Jan Łukasiewicz [51] demonstrated that arbitrary expressions could be specified unambiguously without parentheses by placing operators immediately before or after their operands. For example, the expression $(a + b) \times (c + d)$ is

a) Algebraic		b) Reverse Polish		c) Plus-equals	
Key	Display	Key	Display	Key	Display
Clear	0	Clear	0	Clear	0
a	a	a	a	a	a
$+$	a	$\uparrow$	a	$\underset{=}{+}$	a
b	b	b	b	b	b
$-$	$a+b$	$+$	$a+b$	$\underset{=}{+}$	$a+b$
c	c	c	c	c	c
$\times$	$a+b-c$	$-$	$a+b-c$	$=$	$a+b-c$
d	d	d	d	$\times$	$a+b-c$
$\div$	$(a+b-c)d$	$\times$	$(a+b-c)d$	d	d
e	e	e	e	$\div$	$(a+b-c)d$
$-$	$(a+b-c)d/e$	$\div$	$(a+b-c)d/e$	e	e
f	f	f	f	$\underset{=}{+}$	$(a+b-c)d/e$
$=$	$(a+b-c)d/e-f$	$-$	$(a+b-c)d/e-f$	f	f
				$\underset{=}{-}$	$(a+b-c)d/e-f$

Figure 3-4   Evaluation of the expression $\dfrac{(a+b-c) \times d}{e} - f$ by three keyboard languages

specified in operator prefix notation as: $\times + ab + cd$. This may be read as: find the product ($\times$) of the sum ($+$) of a and b and the sum ($+$) of c and d. Similarly the expression can be specified in postfix with the same meaning as: $ab + cd + \times$. In honor of Łukasiewicz, prefix and postfix notation became known as Polish and reverse Polish, respectively.

There are several advantages to reverse Polish notation. As the expression is scanned left to right, every operator encountered may be executed immediately. Second, if a push-down stack (that is, a last-in, first-out memory) is used to store operands as a reverse Polish expression is evaluated, the operands that an operator requires are always at the top of the stack. Most compilers for FORTRAN or ALGOL convert expressions to reverse Polish for internal execution.

Calculators with reverse Polish keyboards generally have a key called ENTER or ↑ to load operands onto the stack. The $+$ key can also be used if each expression is preceded by a CLEAR operation. Part b of Figure 3-4 shows how a complex expression would be evaluated in reverse Polish. Note that in all cases operations are executed upon the last two operands immediately when the operation key is depressed. The product of sums problem would be executed by the sequence $a \uparrow b + c \uparrow d + \times$.

The last language is a combination of algebraic and reverse Polish and is called the adding machine, accounting, or "plus-equals" method. It is popular on lower cost calculators since only four keys are required:

$$\div, \times, \underline{\underline{-}}, \text{ and } \underline{\underline{+}}$$

Basically the addition and subtraction operations are postfix, allowing operations like an adding machine, while multiplication and division are algebraic. Part c of Figure 3-4 shows a problem executed in this language. Evaluation of a product of sums would require one of the four solution methods mentioned for algebraic keyboards.

Thus calculator architecture is closely linked to the algorithms and functions to be evaluated and the type of keyboard language implemented. It is imperative that the design of the calculator hardware and software proceed together in close cooperation for efficient design.

Input/Output

All calculators have a keyboard input and a display or printer output as a minimum complement of input/output devices; some have much more and in fact the peripheral assortment for some programmable calculators is as extensive as that of computer systems. The capability to treat extended input/output effectively and efficiently cannot be added late in the design process. It is extremely important to have a complete specification for input/output requirements before design commences.

The keyboard is where man and machine come together. This is not intended to be a design chapter about keyboards, but calculator electronics are involved in keyboard features. Besides the important mechanical qualities such as depression force and tactile feedback, provision must be made either mechanically or electronically so that key bounce does not execute the same key twice.

In calculators to be used by accountants and others trained to operate "by touch" the electronics must provide for two features called *N-key rollover* and *buffering.*

Fast operators sometimes press a second digit key before the first is released. It is desirable for the calculator to allow both keys to be entered; this is called *2-key rollover.* In some instances *3-key rollover* may be required.

When adding a column of numbers on a printing calculator some operators can "beat the machine," that is, they can enter numbers and hit the add key faster than the calculator can process inputs and print the results. In this case, provision for keyboard buffering allows the keyboard entries to be placed in a queue (registers inside the calculator). Entries are processed sequentially by the processor as it needs them. No keystroke is lost; the calculator may rattle on for several seconds after the operator has stopped.

In addition to the standard keys, most calculators have sense switches or rotary switches. These switches are used to select operating modes such as degrees or radians operation, fixed point or scientific display format, or the decimal point setting. In some four-function calculators, a lock-in type switch is used to designate that a constant multiplier or divisor is being entered.

Another design decision is the type of output. For office work a printer tape is often required; for less professional use, it is satisfactory that answers be copied from a display. The most convenient design, and most expensive, is to have both a printer and display such as in part *b* of Figure 3-1. This allows the user to see numbers as they are entered and prevents many errors.

Printers consume power, are bulky and, depending on technology, may be noisy. These factors may prevent their inclusion in many products.

Most calculator displays are scanned rather than driven continuously. The eye will not detect flicker at refresh rates greater than 100 per second. This allows a matrix addressing arrangement which greatly reduces the required number of leads. Decisions must be made concerning a serial or parallel output from the processor, the scanning rate, and the synchronization of the calculator with the display interface.

Packaging

It may seem unusual to mention packaging in a text on architecture but the subjects are definitely related. In some projects the case size is the first parameter specified, for instance for a pocket-sized calculator. It then becomes the task of the designer to partition the system into the number of printed circuit boards and integrated circuits that fit in the required package (or press for a larger package). A multichip or hybrid packaging arrangement in which several integrated circuits and possibly some passive components are placed on the same ceramic substrate, may be required to decrease size.

Power

A last important consideration too often passed over lightly is the question of power consumption. It is wise for any designer to understand early in his career that power consumption is expensive in any product. Prior to MOS/LSI devices, many calculators had to be cooled with internal fans. In the design of today's battery powered portables each milliwatt is scrutinized in order to reduce the size and number of batteries required, and to lengthen the operating time.

The Calculator as a System

In summary, the student of calculator architecture should remember that a calculator processor is not an end in itself but rather a part of a system. Decisions that may be wise from the digital electronics standpoint may make operation of the calculator clumsy; product features requested after marketing surveys may be too difficult and expensive to implement. Cooperation of all involved becomes a keynote of practical products. The central concern must be that it is the design of the system to be optimized and not merely one of the components.

3-4. CALCULATOR ARCHITECTURE

In the previous section we considered the various aspects of calculator design from a general viewpoint; in this section we will examine specific techniques for implementing the digital components that comprise a calculator. Our primary emphasis is on methods of generating a control sequence. It is in this area that the impact of integrated circuit technology has been especially pronounced. Also, conventional courses in switching theory are sometimes remote from design practices commonly used in industry.

System Design

We begin by reviewing the familiar block diagram for a general purpose computer as in part *a* of Figure 3-5. There are five main blocks connected by data paths (double lines) and control and status lines (single lines). In higher performance systems a direct data path to memory independent of the arithmetic logic unit (ALU) is usually provided.

The block diagram for a nonprogrammable calculator is shown in part *b* of Figure 3-5. It differs from the general purpose computer in the lack of a separate memory section. Working data registers are part of the ALU. Also the input and output blocks have been specialized to keyboard and display although as discussed earlier, other variations are common, particularly in more expensive systems.

As in any engineering task, a complete and careful specification of the

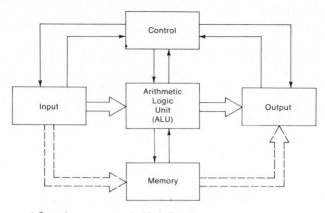

a) General-purpose computer block diagram

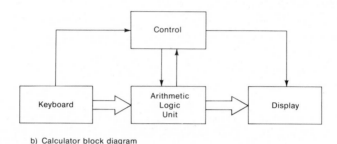

b) Calculator block diagram

Figure 3-5 Block diagrams

design requirements is an important prerequisite to beginning work. This usually starts with a block diagram of the system showing the data paths and instruction and status lines.

Control Techniques

We prefer to consider the design of a control sequence as similar to writing a computer program. The designer has in mind data stored in registers, the connecting paths between them, and the list of instructions or commands that transfer the data and perform logical operations. Simply stated, it is only necessary to devise a correct sequence of instructions to meet the specific requirements. Hence we approach control from the point of view of a flow chart, and then examine ways to obtain hardware implementations of the flow chart.

Classical Sequential Machine One way of representing the classical sequential machine is shown in Figure 3-6. It is composed of storage elements or memory, sometimes called the state, and combinational logic elements, with feedback

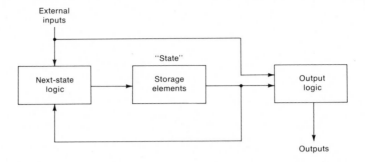

Figure 3-6   Classical sequential machine

from the storage outputs back to the input. We will be concerned only with synchronous or clocked systems, in which the memory transitions all occur in response to a clock. In asynchronous systems, the moment of transition of a memory element may be dependent upon a completion signal from another part of the system.

In discussing control we refer back to the classical sequential machine to place the different design methods in context.

Flowchart   A flowchart is a graphic representation of the definition or solution of a problem or sequence of events in which symbols are used to represent operations, data, or decisions. It is a valuable tool that aids the designer in organizing his requirements; whether it be a computer program or a digital system control sequencer. For our purposes in control design, we can think of the flowchart as representing the next state function and output function for a sequential machine. We examine how a design requirement for a sequential machine, once specified by a flowchart, can be implemented directly in different hardware configurations.

Many different types of flowcharts exist. We use three different symbols as shown in Figure 3-7.

The *instruction block*, sometimes called a *state block*, is represented by a rectangle. Inside are placed instructions, often represented by mnemonics. A

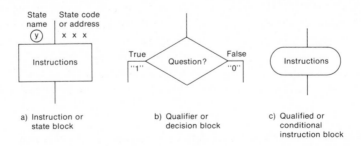

Figure 3-7   Flowchart symbols

state block has a state name, often a letter, and a state code or address. The instructions correspond to the outputs in the sequential machine of Figure 3-6.

The *qualifier* or *decision block* is shown as a diamond and is used to represent branching in a sequence. Only one qualifier may be used in each block and the next path followed is dependent upon whether the answer to the "question" asked by the qualifier is true or false. Qualifiers are also called status signals and correspond to the external inputs in Figure 3-6.

The *qualified* or *conditional instruction block* is represented by an oval; it always follows a decision block in a flowchart. Inside are written instructions that are *dependent* upon the qualifier in the preceding qualifier block. Some examples will help to clarify how these symbols are used.

The first example (part *a* of Figure 3-8) shows a simple state transition in which no qualifiers are tested. The machine makes the transition from State A where instructions I1 and I2 are executed, to State B where instructions I1 and I3 are executed.

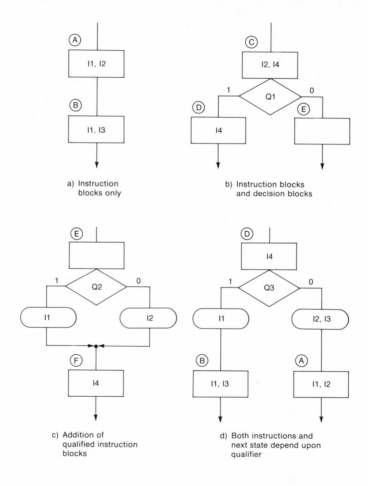

a) Instruction blocks only

b) Instruction blocks and decision blocks

c) Addition of qualified instruction blocks

d) Both instructions and next state depend upon qualifier

Figure 3-8 Uses of flowchart symbols

In part *b*, the qualifier Q1 is tested; if true, the "1" path is followed and the next state is D where I4 is executed. If Q1 is false, or in the "0" logic state, the next machine state is E where no instructions are executed. Note that Q1 is tested while the machine is in state C and at the same time instructions I2 and I4 are active. The decision block is *always* associated with a state block.

The term Algorithmic State Machine (ASM) chart is used by Clare [73] to distinguish the fact that a state block and a decision block are executed simultaneously as opposed to a conventional program flowchart in which each block is executed sequentially. In flowcharts we will always draw a state block and a decision touching to represent this important difference.

The third example (part *c*) uses the conditional instruction block. The two conditional instructions, I1 and I2, are both associated with State E. If Q2 is true, I1 is active in State E, if false, I2 is active. In either case, the next state is F.

The final example in Figure 3-8 (part *d*) combines examples of parts *b* and *c*. Both the next state and some of the instructions are dependent upon Q3.

Figure 3-9 combines the segments of Figure 3-8 into a complete six state flowchart for a sequential machine with three external inputs, that is, qualifiers Q1, Q2, and Q3, and four outputs, instructions I1 through I4.

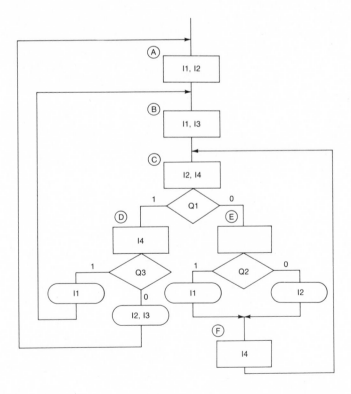

Figure 3-9 Flowchart for six-state sequential machine

It is necessary to tie the concept of "state" to the principle of synchronous operation. In a synchronous sequential machine the memory elements are changed only in response to a clock signal and all change at once. This change corresponds on the flowchart to movement to the next state block. In fact, each state is represented in hardware by a state code or address realized by memory elements. Thus if the machine is "in" State C, the next clock signal moves the machine to either state D or E depending upon input Q1. Q1 is associated with State C because it is tested during the State C period of time.

Example 3-1: Represent a 3-bit Gray code counter by a flowchart. If an input x is true, the counter is to go through only the first four states. An output Z1 is required at binary counts 2 and 6 and at count 3 if x is active, and an output Z2 at counts 1 ,3, 5, 7.

> *Solution to Example 3-1:* We let each state code correspond to the binary number for that state. Remember in a Gray code, only one variable changes at each state transition.

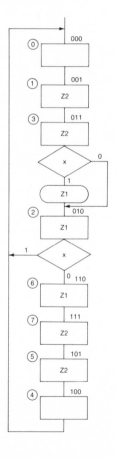

Solution: Example 3-1

Counter-Decoder Control Methods A commonly used control technique is the use of a multiphase clock, implemented with a counter and decoded outputs as instructions. Synthesis procedures for counters are covered in basic courses on sequential circuit design and are not explained here. Clare [73] explains the procedures in a manner tied closely to the other material of this chapter.

Figure 3-10 shows a four-phase clock. By logically ANDing the clock signals with qualifier inputs, instructions can be activated sequentially. Figure 3-11 shows a generalized control unit for a digital system. Note the correspondence to the classical sequential machine of Figure 3-6. The block called external memory represents the registers and single flip/flops present in the system. Some of the outputs from these memory elements are returned to the counter input logic to determine the next state. These are examples of qualifiers appearing on flowcharts.

The counter can be designed to count in any manner required; the modulo 4 counter of Figure 3-10 is a simple example. In most systems it is inefficient always to proceed through four phases. By proper design of the input logic the counter can vary its sequence. As in most digital system designs, the

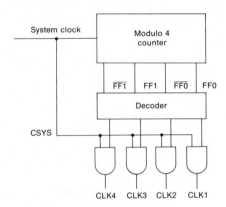

a) Hardware for a four-phase clock

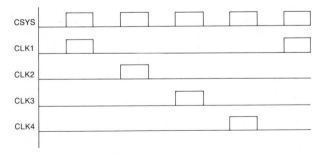

b) Timing signals for a four-phase clock

Figure 3-10   Four-phase clock

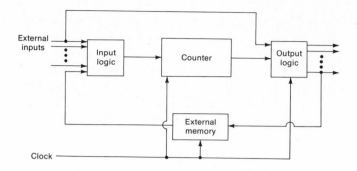

Figure 3-11   Controller implemented with a counter and logic

tradeoff is between speed and cost; the variable counter provides higher system speed, but costs more than the simple fixed multiphase clock.

Example 3-2:   Implement the flowchart of Example 3-1 using clocked D-type (Delay) flip/flops.

Solution to Example 3-2:   Since there are eight states, we need three state flip/flops. The first step is the construction of a next-state and output table:

Present State				Input	Next State			Output	
	F3	F2	F1	x	F3	F2	F1	Z1	Z2
0	0	0	0	—	0	0	1	0	0
1	0	0	1	—	0	1	1	0	1
3	0	1	1	0	0	1	0	0	1
3	0	1	1	1	0	1	0	1	1
2	0	1	0	1	0	0	0	1	0
2	0	1	0	0	1	1	0	1	0
6	1	1	0	—	1	1	1	1	0
7	1	1	1	—	1	0	1	0	1
5	1	0	1	—	1	0	0	0	1
4	1	0	0	—	0	0	0	0	0

The excitation function for delay flip/flops is particularly simple since the output follows the input after each clock signal. The inputs D1, D2, D3 are therefore equivalent to the next state. The solution is shown in Karnaugh maps for the three inputs and the two outputs.

This example may seem trivial. Actually all the principles of counter-gate control method design are embodied in it.

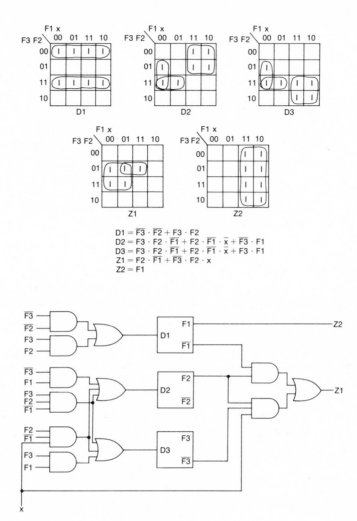

$D1 = \overline{F3} \cdot \overline{F2} + F3 \cdot F2$
$D2 = F3 \cdot F2 \cdot \overline{F1} + F2 \cdot \overline{F1} \cdot \overline{x} + \overline{F3} \cdot F1$
$D3 = F3 \cdot F2 \cdot \overline{F1} + F2 \cdot \overline{F1} \cdot \overline{x} + F3 \cdot F1$
$Z1 = F2 \cdot \overline{F1} + \overline{F3} \cdot F2 \cdot x$
$Z2 = F1$

Solution: Example 3-2

Read-Only Memory Based Methods The greatest change in recent years in control unit design has been the increased use of Read-Only Memory, or ROM. The advantage of read-only memory has been known since the earliest computers but the costs were prohibitive until the advent of integrated circuits. Actually the third generation IBM 360 computers used read-only memory extensively, but at an increased cost over conventional counter-gate or random logic design. Some of these increased costs were regained in other benefits such as flexibility and ease of design change.

The simplest way to think of a read-only memory is as an array of OR gates. Each output from the memory can be selected by several of the input

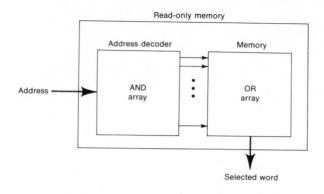

Figure 3-12   Read-only-memory as an AND-OR array

or address lines. Associated with the memory is a decoder which is an array of AND gates. Figure 3-12 shows the two parts of a read-only memory. An address consists of an n-bit code, which can select one of 2^n words in memory when fully decoded. Words consist of fixed-length bit strings that are set or programmed at the time of manufacture and are not intended to be altered during machine operation, hence the term "read-only." However, depending upon the technology that implements the memory, there may be ways to change the stored patterns, either electrically or manually, but always at a speed slower than the speed of access.

To illustrate the use of read-only memory as an AND-OR array, we can implement a pair of simple three-variable Boolean equations.

Example 3-3:   Implement the following two equations in a three-input, two output ROM.

$$X = \bar{B}C + ABC$$
$$Y = AC + AB\bar{C}$$

Solution to Example 3-3:   We can expand the two equations to become

$$X = (A+\bar{A})\bar{B}C + ABC = A\bar{B}C + \bar{A}\bar{B}C + ABC$$
$$Y = A(B+\bar{B})C + AB\bar{C} = ABC + A\bar{B}C + AB\bar{C}$$

Now consider C to be the least significant bit of a binary address. Then X is true for addresses 101, 001, 111 or 5, 1, 7 and Y is true for 7 or 5 or 6. The diagram shows connections for these lines.

In this example the three inputs are fully decoded to eight lines. However, no outputs are required for addresses 0, 2, 3, 4 so part of the decoder can be deleted. This concept forms the basis of the programmable logic array, or PLA, described in the following text.

The programmable logic array is simply a partially decoded read-only

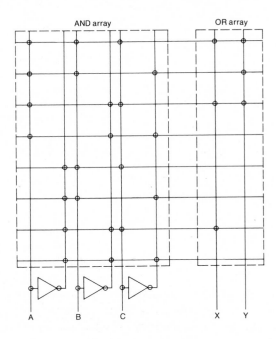

Solution: Example 3-3

memory. The term *programmable logic array* is perhaps inappropriate since all read-only memories are programmable in some manner and are also logic arrays. By convention a programmable logic array has come to mean a read-only memory with n inputs and less than 2^n words, in which both AND and OR arrays are programmed [Reyling, 74].

In a programmable array, the address is made up of two types of inputs: 1) the state inputs as fed back from the memory elements of the sequential machine, and 2) qualifier or status inputs from throughout the digital system. Logical combinations of the next state and status inputs are ANDed in the decoder portion of the PLA to form "product terms." Groups of product terms are ORed together to form the outputs or instructions from the PLA, just as example 3-3 is implemented.

The programmable logic array may be implemented as a part of an integrated circuit, such as the control section for a one-chip calculator, (see Section 3-4) or it may be fabricated as a separate component. A programmable logic array made by Hewlett Packard for in-house use is an integrated circuit with eight state flip/flops within the circuit, 16 status inputs, 72 product terms, and 30 outputs. This circuit is shown in Figure 3-13. One of the disadvantages of this type of separate integrated circuit is the large number of leads: 46 are required plus power and clock inputs. When realized as an internal part of an integrated circuit the inputs and outputs do not all require connection to the outside world, and this disadvantage is eliminated.

The National Semiconductor PLA (part No. DM 7575) has 14 inputs, 8

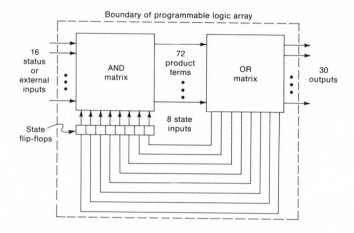

Figure 3-13   Block diagram of an integrated programming logic array

outputs, 96 allowable product terms, and no internal state flip/flops. The circuit is provided in a 24-pin dual in-line package.

When implemented with integrated circuits, the determination of a connection in the AND and OR matrices is specified by a single masking operation. This allows one basic circuit design to be customized or programmed at little incremental expense.

Example 3-4:   Implement the counter of Example 3-1 with a PLA.

Solution to Example 3-4:   The equations from the solution to Example 3-1 are repeated here:

$$D1 = \overline{F3} \cdot \overline{F2} + F3 \cdot F2$$

$$
\begin{aligned}
D2 &= \\
D3 &=
\end{aligned}
\left\{ F3 \cdot F2 \cdot \overline{F1} \right\} + \left\{ F2 \cdot \overline{F1} \cdot \overline{X} \right\}
\begin{aligned}
&+ \overline{F3} \cdot F1 \\
&+ F3 \cdot F1
\end{aligned}
$$

$$Z1 = F2 \cdot \overline{F1} + \overline{F3} \cdot F2 \cdot X$$

$$Z2 = F1$$

The solution has four inputs, (X, D3, D2, D1), nine product terms, and five outputs. Note that product terms appearing in more than one output can be shared, just as AND gates are shared in Example 3-1. In this simple example the Z2 output can be eliminated by taking Z2 directly from F1. This also eliminates one product term. Another trick is to form Z2 from the product term $\overline{F3} \cdot F1$ and $F3 \cdot F1$, that is, $Z2 = F1 = \overline{F3} \cdot F1 + F3 \cdot F1 = (\overline{F3} + F3) = F1$.

One other important point should be mentioned. In Example 3-1 we use Karnaugh maps to simplify the logic equations; this reduces the number of

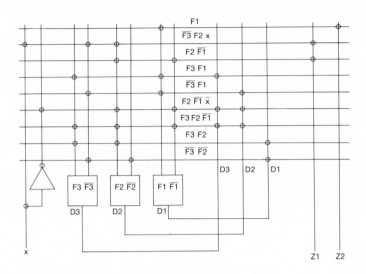

Solution: Example 3-4

gates and the number of inputs per gate to lower the cost of implementations. In the AND matrix of Example 3-4, all connections points are available and there is no need to minimize the logical equations, except to reduce the number of products to fall within the capacity of the PLA.

Microprogramming Microprogramming is the term applied to the use of read-only memory for the implementation of the control function in a digital system. Programming is required as part of this technique since techniques analogous to computer programming are used, such as stored instructions, subroutines, conditional branches, and program description by flowcharts. It is "micro" in the sense that a more detailed control of the hardware of the machine is implied than with higher level programming. Chapter 10 contains a detailed description of microprogramming. Most small and medium-sized computers and many large ones use microprogramming for the control. Since calculators and small processors are also increasing their use of this technique we introduce it here, but only in its simplest form and in relation to other calculator control techniques.

In the classical sequential machine in Figure 3-6, there are two logic sections: one to decode the next state and the other to determine the appropriate outputs. There are two types of inputs to the logic: external and present state. In a programmable logic array, the next state logic and the output logic are combined into a single AND-OR (that is, decoder-memory) array. The user customizes or programs both the decoder and the memory. This has the advantage that all the logic of the system can be realized in one low-cost part.

However, a system typically has many external inputs, corresponding to various status conditions and special flip/flops, but fewer than all combinations of external inputs are required as product term lines. The programmable logic array of Figure 3-13 has a total of 24 inputs so that $2^{24} = 16,777,216$ product terms are possible, yet only 72 are built into the array. A standard read-only memory cannot be used efficiently as a programmable logic array for this reason. A memory with ten address inputs would have 1024 words which correspond to product terms, but typically less than ten percent of these would be used with the programmable logic array approach to logic design.

A fully decoded read-only memory is more widely manufactured because of the optimum usage of pins and because of its application to microprogramming. Microprogramming is a method for efficiently using low-cost read-only memory for most of the logic of the system. In Figure 3-14 the total logic of a microprogrammed system has been separated into two parts. All dependence upon external inputs is concentrated in gate logic shown as "next-state logic." The system outputs are implemented by read-only memory and thus depend only upon the present state of the system as represented by the address input to the memory. External inputs help select the next address, but are not part of the address itself, as in a programmable logic array.

In any digital system there are some instruction lines that are never active simultaneously, that is, they are mutually exclusive. Two examples of such signals would be the commands to the arithmetic logic unit and the selection of data for a bus. In the first example, an arithmetic logic unit may have capability to add, subtract, invert, and so on, but it can only do one of these at a time. In the second example, several registers may have access to a bus, but only one can be connected at a time. A group of 2^n mutually exclusive outputs can be encoded into n lines to narrow the width of a microprogrammed word. An external decoder is then required, but the additional decoder cost is less than the additional ROM cost. The decoder, however, adds a time delay which may be objectionable in some high speed systems.

Determining a microprogramming word format is similar to selecting a word format for the assembly language of a computer. Wider words with two or more possible next addresses provide flexibility but at the cost of additional ROM. Some typical formats are considered in the following text.

In many microprograms, most of the time the next address is one more than the present address. In this instance a simple implementation is for the ROM address register to be a counter. When a branch is to be made, an

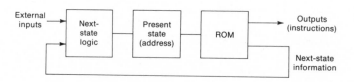

Figure 3-14   Microprogramming model of a sequential machine

address is forced into the counter from the output of the memory. Figure 3-15 shows this implementation. The word format for this example has only two fields, one for the jump address and one for instructions. If the jump line is true the jump address is loaded into the counter (address register), otherwise the counter is incremented.

A modification of Figure 3-15 is to perform the increment operation by an adder and gate the incremented address and the jump address into a switch, the output to be selected by a branch signal.

Figure 3-15 makes no provision for the generation of the jump signal. The most common technique is to add a third field to the word format called the *test field*. At each state of the machine one qualifier is selected as the jump signal, or if no branch is required, the jump signal is forced false. Figure 3-16 shows an implementation with these added features. The switch, selector, and adder are all part of the next state logic implemented with gates. Compare Figures 3-14 and 3-16 to see the correspondence.

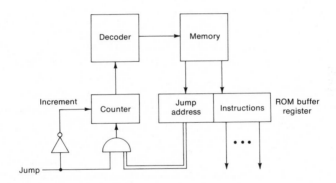

Figure 3-15   Microprogrammed control implementation—two-field word format

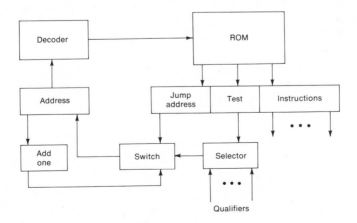

Figure 3-16   Microprogrammed control implementation—three-field word format

Another common word format is the two-address format. The two possible next addresses are included in the read-only memory output. This provides greater flexibility since any of two addresses can be selected at each state, but at added cost since the ROM word becomes wider. Part *a* of Figure 3-17 shows this format.

One last format is worthy of mention. The size of read-only memory is determined by the number of words and the width of each word. A technique to keep the width small is to use a variable word format. With two different formats, one bit of each word is used as a format indicator. As shown in part *b* of Figure 3-17, Format 1 has all but the format bit used as instruction bits as indicated by the left-most bit equal to 1. Format 2 has two fields, a test and a jump address, and the format indicator is 0. With Format 2, the control lines are all disabled. With Format 1 the next address is assumed one more than the present address.

Variations and combinations of these formats are possible. More discussion of their relative merits is contained in Chapter 10. In general, wider words result in a faster machine, but at an added cost.

The type of flowchart drawn for an algorithm or sequential machine is necessarily related to the type of implementation. With a programmable array it is possible to test more than one qualifier per state and to implement conditional outputs. Conditional outputs are not allowed in a microprogrammed approach since the ROM output lines are not conditioned by the external or qualifier inputs. All requirements can still be implemented by microprogramming but a corresponding flowchart must be used. To test two qualifiers in one microstep with microprogramming, two test fields are necessary in the word format and in general, there must be provision for branching four ways. This is so costly that a restriction to a single qualifier per state time is generally made.

The address of a ROM word corresponds to the state code on a flowchart. A state is described by the contents of a ROM word. Generally it includes instructions to be executed, the qualifier to be tested, if any, and the next state choices.

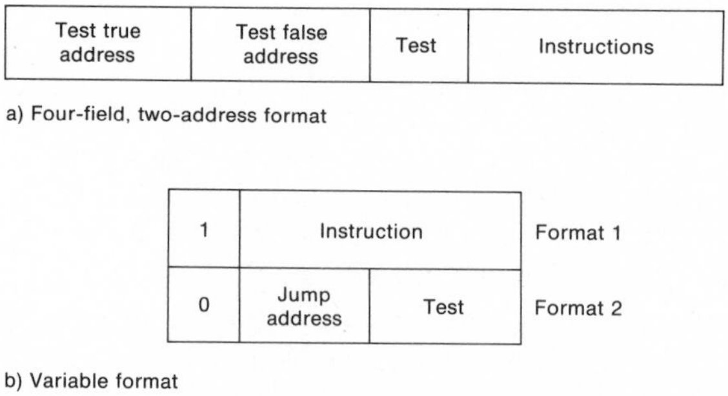

Test true address	Test false address	Test	Instructions

a) Four-field, two-address format

1	Instruction		Format 1
0	Jump address	Test	Format 2

b) Variable format

Figure 3-17 Two types of word formats

Example 3-5: Redraw the flowchart of Figure 3-9 and implement it with microprogramming using the two address format of Figure 3-17, part *a.*

Solution to Example 3-5: (illustrated on the following page)

False addr.	True addr.	Test	Inst.
xxxx	xxxx	xx	I4 I3 I2 I1

There are ten states so a 4-bit address is required. The four qualifier conditions Q1, Q2, Q3 and "no test" can be encoded into two bits. The complete memory map is shown below.

Qualifier Encoding		Address Assignment		Memory Map				
				false	true	qual	I4 I3 I2 I1
no test	00	A	0000	0001	0001	00	0 0 1 1
Q1	01	B	0001	0010	0010	00	0 1 0 1
Q2	10	C	0010	0100	0011	01	1 0 1 0
Q3	11	D	0011	0111	0110	11	1 0 0 0
		E	0100	1001	1000	10	0 0 0 0
		F	0101	0010	0010	00	1 0 0 0
		G	0110	0001	0001	00	0 0 0 1
		H	0111	0000	0000	00	0 1 1 0
		I	1000	0101	0101	00	0 0 0 1
		J	1001	0101	0101	00	0 0 1 0

The removal of the conditional outputs is accomplished by the addition of four states. The specific physical situation would have to be examined to see if the desired system behavior is still represented. For instance, in the original flowchart in state D, the instruction I1 is active *at the same time as* I4 if Q3 is true. In Example 3-5, I1 becomes active in state G and I4 is no longer active. The ultimate acceptance of this revision depends upon what I4 and I1 actually do. It may be necessary, for instance, to carry I4 over into state G to obtain equivalence.

Arithmetic Unit Design

Before examining the details of the design of an arithmetic unit, we review the arithmetic requirements of a calculator. What is it intended to do? In its basic configuration, a calculator requires interaction with an operator; numbers and operations are entered from the keyboard; answers appear on a display or a printer. A calculator is never far from human interaction and humans prefer decimal input and decimal output. Execution can be relatively slow, by electronic standards. These last two characteristics imply a design based on decimal arithmetic internally. For although BCD-to-binary con-

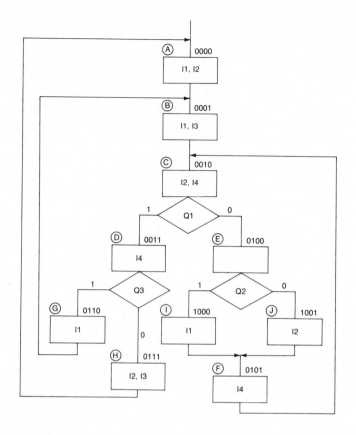

Solution: Example 3-5

verters can be easily designed, their cost overshadows the inherent savings possibly realized from greater efficiency using binary representation of numbers. Decimal arithmetic is conveniently handled in a serial-by-digit manner. Serial techniques are slow and less expensive, characteristics well suited to calculator design.

Once decimal arithmetic has been decided upon we must select a code to represent a digit. There are eighty-eight 4-bit weighted decimal codes, seventeen with positive weights, and four with the convenient self-complementing feature. Many of these have been used in specialized digital systems. Although certain advantages can be realized from other codes, the 8421 code has become the standard. Each digit is uniquely represented, a binary adder can be used for the first stage of a decimal adder, and many standard integrated circuits, such as BCD-to-7-segment display decoders, are designed around its use. It is *the* BCD code.

Arithmetic Unit Operations The first requirement for a calculator is to add. Along with the requirement for addition comes subtraction. A decimal subtraction can be implemented by standard logic design techniques but the more common procedure is to subtract by adding the complement of the number as discussed in the previous chapter. This produces a requirement for a complementing operation; we then need only complementation and addition to perform subtraction.

In practice it is generally economical to include some hardware to speed up subtraction when implemented as a complement and add operation. For instance, the complement of an addend can be gated directly into the adder with one additional set of gates.

A calculator differs from an adding machine in its ability to multiply and divide. Multiplication can be implemented several ways, such as looking up products in a stored table. A lower cost and more common method is through repeated addition. The simplest method is for the multiplicand to be added to itself the multiplier number of times, that is 53×127 is accomplished by adding 127 to itself 53 times. After each addition the multiplier is decremented by one. Although this is a subtraction of one, decrement is used so often we prefer to think of it as a fundamental operation and to provide a special method of implementation. With the realization that multiplication by 10 involves only a left shift by one digit position, the number of additions is reduced to the sum of the multiplier digits, when we include another fundamental operation, the left shift. That is, 53×127 equals $50 \times 127 + 3 \times 127 = 5 \times 10 \times 127 + 3 \times 127 = 5 \times 1270 + 3 \times 127$.

The common method of division is to successively subtract the divisor from the dividend until an overdraft occurs. The divisor must then be shifted right and subtracted from the partial remainder. A right shift can be performed by a circular left shift but this takes longer. We thus need both right and left shift operations to perform arithmetic effectively. However, not all registers need this capability.

In performing the subtraction during division we are actually comparing the divisor and the dividend to determine which is larger. In many instances we wish to compare two numbers, but not destroy either. Since most additions are arranged to replace either the addend or augend with the sum, we need another operation called *compare*, which subtracts two numbers but simply sets or clears a status bit to designate the result. Often we wish to compare to zero so we also need a capability of clearing.

In all of these operations we must fetch data into registers prior to an operation and move data from registers after an operation. The operations of transfer and exchange are thus necessary.

The following list shows the fundamental arithmetic operations we have found useful:

Add	Circular right shift	Transfer
Subtract	Left shift	Exchange
Complement	Decrement	Clear
Right shift	Increment	Compare

This list is not intended to be minimal or optimal, but simply typical.

Implementation of a Decimal Adder   The following text describes the implementation of a decimal adder. This can be done by counters, as a parallel adder, or a serial adder.

Consider two clocked four-bit BCD *counters*, as shown in Figure 3-18. One method to add two BCD digits is to place the addend in one counter, the augend in the other, and to decrement one counter while incrementing the other. When the decremented counter reaches 0, the incremented counter contains the sum. A carry flip/flop is provided to treat sums larger than 9 and to input a carry from a lower decade.

Two possible flowcharts for Figure 3-18 are shown in parts *a* and *b* of Figure 3-19. Care must be taken in these flowcharts that the counter does not get decremented one time too many, as would happen in the incorrect flowchart in part *c*.

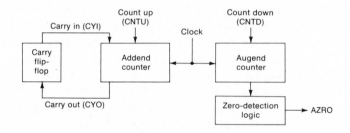

Figure 3-18   BCD adder implemented with counters

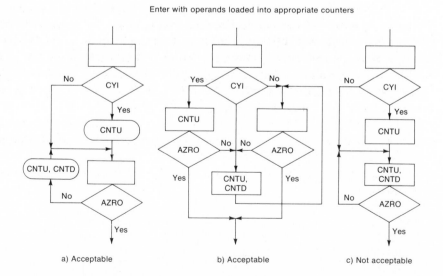

Figure 3-19   Flowcharts for a BCD adder implemented with counters

In Figure 3-19, part *a* makes use of conditional instructions and only two states are required. Part *b* uses no conditional instructions but needs four states. In Part *c*, an attempt was made to combine the testing and counting operations into one state. One count too many will always occur, as can be verified by letting the augend start at 0.

The counter method can also be used for subtraction by decrementing both counters until one reaches 0.

This form of addition has been used in at least one calculator (HP 9100). It is slower than a combinational logic adder and the execution time is dependent on the augend (an augend of all nines is the worst case). It is an interesting technique, however, as it is one of the least expensive implementations, particularly if the counters can be used for other purposes.

Figure 3-20 shows the general configuration for a single decade *parallel decimal adder*. There are nine inputs, that is, four bits from both addend and augend, and the carry-in from the previous decade, and five outputs, the four sum bits and the carry to the next higher decade. It is possible to consider this a standard 9-input problem in combinational logic design, make a truth table with 512 rows and five outputs, and proceed to write and minimize Boolean equations. This might be termed the brute force method and is not practical. It is better to be a little clever.

We are operating in a modulo 10 (decimal) mode with a code (8421) which is basically modulo 16. If we perform a straight binary addition, the sum is correct if less than 10. Sums 10 through 19 can be corrected by adding 6, that is, by forcing modulo 10 instead of modulo 16. We see we must add 6 whenever a decimal carry occurs. The logical equation for this is:

$$\text{Add } 6 = C_2 = K_3 + U_3 \cdot U_2 + U_3 \cdot U_1$$

where the variables are defined in Figure 3-21. The figure shows a decimal parallel adder implemented with one-bit binary adders. The K_3 term generates an add 6 signal for sums 16-19, the $U_3 \cdot U_1$ term for sums 10 and 11 and the $U_3 \cdot U_2$ term for sums 12-15.

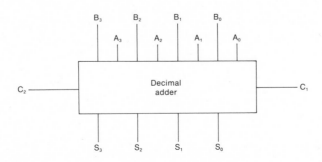

Figure 3-20   Diagram of a decimal adder

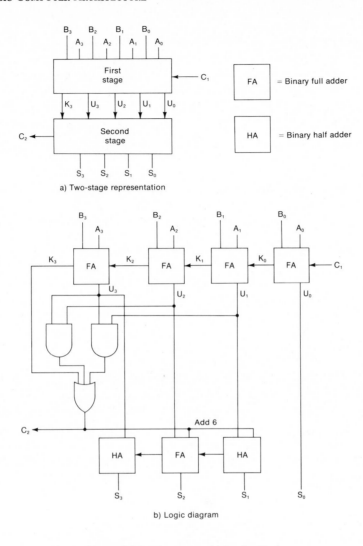

a) Two-stage representation

b) Logic diagram

Figure 3-21   Two-stage decimal adder

A serial binary adder consists of simply a full adder circuit and a carry flip-flop. A *serial decimal adder* presents an interesting problem since the correction by adding 6 cannot be determined until the most significant bits of both addend and augend are present. Still, the advantages of a serial structure for calculators warrants consideration of the serial decimal adder.

One solution is presented in Figure 3-22. The N digits of the augend and addend are held in shift registers A and B. Register A is recirculated while the sum is stored in B. During the first three bit times for each digit, the operand bits are input to the adder and the sum bits U_0, U_1, U_2 stored serially in B. Figure 3-22 shows the data during bit time 4. The correction is made by

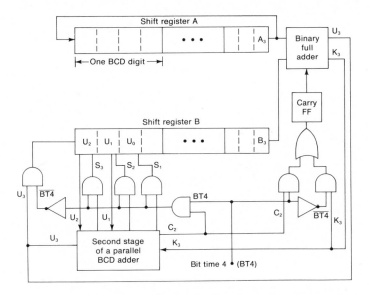

Figure 3-22 Serial decimal adder (data shown during bit time 4)

loading the last three bits of B in parallel from the correction logic, which happens to be equivalent to the second stage of the parallel decimal adder. The least significant bit of the sum is always correct. At bit time 4, the carry FF is loaded with the true decade carry, C_2 instead of the binary carry K_3.

It may appear from the amount of logic in the serial adder, that a better solution is to bring the data to the adder serially, convert to parallel into the adder, then convert back to serial. In some situations this may be true.

Register Configurations Most computers are designed to function in environments where performance is largely measured by processor speed. The parallel binary adder, with various carry look-ahead techniques to further increase speed, can be justified in such applications, but the multi-digit parallel decimal adder is never used in calculators. Its use would contradict one of the most important calculator design criteria—low cost—without providing added performance except in the most sophisticated programmable machines. Moreover, arithmetic units are more efficient if the inputs to the adder can come from more than two registers. The gating required in a totally parallel system is prohibitive.

The remaining components of the arithmetic logic unit are the registers. A minimum configuration would contain three registers. Figure 3-23 shows a typical arrangement for a calculator. The A input to the adder can come from registers X or Z, the B input from Y or Z. The sum can be stored in any of the registers.

Three registers are required to perform multiplication: a multiplier register,

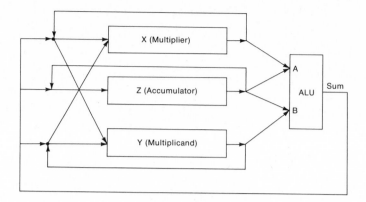

Figure 3-23 Registers and arithmetic logic unit (ALU) for a serial processor

a multiplicand register, and an accumulator for the partial product. A double length product can be accumulated by shifting the least significant digits of the product into the multiplier register as the multiplier digits are used.

The arithmetic logic unit (ALU) must also contain logic to implement the selected instruction set such as complementing logic for one of the inputs. Transfer instructions can be implemented by addition with one input disabled and storing the sum in a different register. Additional gating is required for exchanges. In Figure 3-23, the X and Y registers can be exchanged.

The adder within the ALU of Figure 3-23 can be either the parallel adder of Figure 3-21 or the serial adder of Figure 3-22. With a serial structure each of the inputs to the adder requires only one gate, and associated complementing and disabling logic is also reduced. This is why the serial structure may be preferred even though the speed of the serial adder is less than one fourth that of the parallel adder. If words are processed serial-by-digit, as shown in Figures 3-22 and 3-23, there is no speed advantage to a parallel decimal adder.

A technique borrowed from computer architecture used in the HP-35 calculator is the push-down (or push-up) stack, or last-in, first-out (LIFO) memory [Whitney, 72]. In the HP-35, the arithmetic unit operates on the bottom two members of the stack, as shown in Figure 3-24. As intermediate answers are computed they are pushed up the stack. In an operation involving two operands, R1 and R2, the result is placed in R1, R2 is lost, and the stack drops, that is, $R_i \leftarrow R_{i+1}$, $2 \leq i \leq N - 1$. This brings a previously computed result into R2, ready to be combined with R1 in another operation. New data enters R1 and pushes the stack up, $R1 \rightarrow R2$, and so on, so that a new entry can be easily combined with a previous result.

The stack is particularly efficient when the calculator keyboard "language" is reverse Polish, as explained in Section 3-2. Many calculators and an increasing number of computers use a stack internally, even though it may never be apparent to the user.

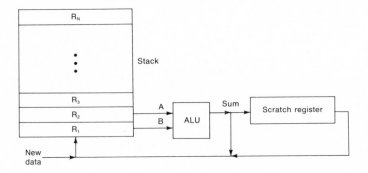

Figure 3-24   Stack and ALU for a serial processor

Example 3-6:   Using the three register configurations of Figure 3-23 and an algebraic keyboard language, determine the arithmetic and transfer instructions that permit chain calculations, that is, the result of one calculation to be used as an entry in a new calculation. Consider multiply and data entry as subroutines. Show the sequence of register contents as the calculation $(12 + 6) \times 4 - 3 =$ is performed.

Solution to Example 3-6:

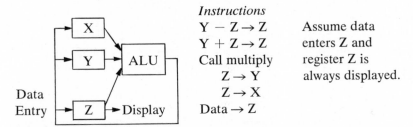

Instructions

$Y - Z \rightarrow Z$ Assume data
$Y + Z \rightarrow Z$ enters Z and
Call multiply   register Z is
$Z \rightarrow Y$ always displayed.
$Z \rightarrow X$
Data $\rightarrow Z$

Z	Y	X	Operation	Instruction or Subroutine
12			Enter 12	Data $\rightarrow$ Z
12	12		+	Z $\rightarrow$ Y
6	12		Enter 6	Data $\rightarrow$ Z
18	18		$\times$	$Z + Y \rightarrow Z, Z \rightarrow Y$
4	18	4	4	Data $\rightarrow$ Z, Z $\rightarrow$ X
72	72		—	Call multiply, Z $\rightarrow$ Y
3	72		3	Data $\rightarrow$ Z
69	72		=	$Y - Z \rightarrow Z$

Example 3-7:   Repeat Example 3-6 using the stack architecture of Figure 3-24 and Reverse Polish notation. Add a scratch register to hold the multiplier.

Solution to Example 3-7:

R1	R2	Scratch	Operation	Instructions
12			12	Data entry
12	12		↑	Push stack
6	12		6	Data entry
18			+	Add
4	18		4	Data entry
72		4	×	R1 → Scratch, multiply
3	72		3	Data entry
69			−	Subtract

Subroutines	*Instructions*
Data entry (automatic push stack)	Push stack
Multiply	Add (drops stack)
	R1 → Scratch
	Subtract
	Data entry
	Multiply

Each language requires eight steps to complete. In problems requiring the temporary storage of an intermediate answer, the reverse Polish notation benefits because the automated "stack drop" feature saves a recall from storage. However, the algebraic notation may be more "natural" since we are accustomed to seeing equations containing an equals sign.

A complete algebraic keyboard requires the use of parentheses. These calculators are frequently designed with a stack architecture internally.

Calculator Input

The only data input mechanisms available on most calculators are the keyboard and the mode switches. Programmable calculators generally have other inputs such as a magnetic card reader, punched card reader or tape cassette, and often have general input/output capability that allows interfacing typewriters, teletype, digitizers, magnetic disks, or external tape transport much in the manner of a digital computer. We concern ourselves only with keyboard interface. Computer input/output techniques (see Chapter 6) are frequently applicable for the more complex calculators.

The keyboard circuitry has several functions as listed below.

1. Scan keys — the input keys must be periodically scanned to determine a key down signal. A desirable feature is to allow an additional key (or keys) to be depressed (rollover).
2. Buffer keys — since the user may get ahead of the machine, particularly with printing calculators, keystrokes must be stored internally.

3. Lock out — when the maximum number of keys have been detected, all other keys must be locked out until the processor has a chance to catch up.

4. Code conversion — each key must be uniquely identified by a key code.

5. Key debounce — switches, being imperfect, have a tendency to bounce, both on depression and release. This bounce may last tens of milliseconds, depending upon the keyboard technology. The calculator system must be designed to separate successive key depressions, such as entering the number 99, from a single depression and release.

6. Input mode switches — almost every calculator has at least one mode switch—also called *static switches* or *status switches*. The most common are decimal point settings on business or general purpose calculators, and a degree-radian switch on scientific calculators.

One simple keyboard technique is to use a single input line for each key. This has the advantage of simple decoding and control circuitry but the disadvantage of requiring the maximum number of input lines. More commonly some form of encoding is used.

A common method used with single-chip MOS integrated circuit calculators is to use a diode keyboard encoder external to the chip. Figure 3-25 shows a portion of the keyboard input circuitry used with the MOSTEK MK5010P, 10-digit calculator circuit. A negative voltage represents a logic 1. Note the code 0000 represents "no digit" and 1010 represents the 0 key. The circuit

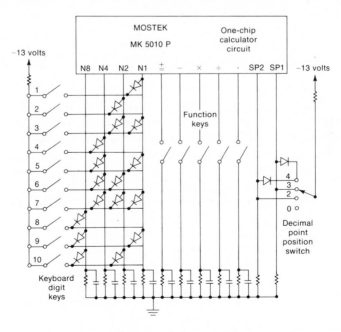

Figure 3-25 Keyboard input with diode encoding

shows four options for the decimal point setting encoded by the two inputs SP1 and SP2.

External diodes are too expensive and bulky for some calculator applications. Another technique is to arrange a keyboard as an $m \times n$ matrix of switches. This type of keyboard is connected as if it had m inputs and n outputs. The circuitry external to the keyboard must then have m output or drive lines, and n input or receiver lines. The column and row lines are activated sequentially or "scanned," that is one column is selected, then all row lines scanned searching for a key closure, then the next columns driven and another row scan and so on. Figure 3-26 shows the 11×4 matrix used in the MOSTEK 5020A calculator circuit. The 44 key locations require 15 calculator connections. A more square array, such as 7×7, allows 49 keys with only 14 connections but the functional separation of the keys allows simpler control circuitry in the calculator. Also, the column drive lines can be used in the display circuitry as discussed in the next section.

The calculator keyboard circuitry must convert a matrix, row-column location into a code to be used by the control circuitry to specify the calculator operation. In microprogrammed calculators the encoded key location is often used as a starting address in read-only memory.

A convenient scanning-encoding scheme used in the HP-35 design is shown in Figure 3-27. A 6-bit counter is divided into the upper three bits, Q3-Q5, which are decoded to drive the rows, and the three lower bits, Q2-Q0, which select, via a multiplexor, the one column to examine at each count. When a switch closure is detected, the counter state is transferred to a register. This counter state becomes the address of a subroutine in the microprogram to process the operation specified by the key.

Prevention of errors due to key debounce can be accomplished by hardware or software or a combination of the two. The keyboard inputs in Figure 3-25 have capacitors to reduce electrical bounce noise. Many microprogrammed calculators use programmed lockouts and status bits to ensure a key is really down and to indicate a key has been released. This is a good example of the many types of tradeoffs between hardware and software that can be realized. Extremely close cooperation between hardware designers

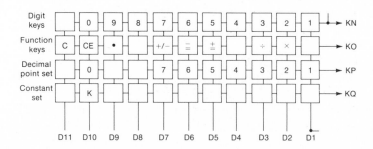

Figure 3-26 Matrix keyboard input

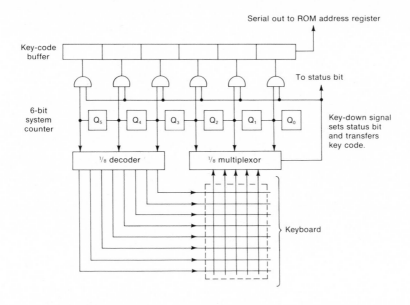

Figure 3-27   Counter-scanner method of keyboard interrogation

and programmers is required to capitalize on these tradeoffs and at least one person in a design team must keep the overall picture constantly in mind.

Calculator Output

The primary calculator output device is a display. A printer is also frequently used but printer interface requirements are specialized and will not be discussed here. Although mechanical printers have dominated the market in the past, other techniques such as thermal printers are becoming more popular. Output techniques for programmable calculators do not differ significantly from techniques used with computers.

Calculators almost always use some form of seven or eight-segment display (Figure 3-28). Easily readable numerics can be made from the seven segments, particularly if the segments are stylized to different lengths and shapes. The eighth segment, the decimal point, is generally placed to the lower right of the digit although left-hand decimal points are also commercially available. Many display technologies are competing for the huge calculator display market. Light-emitting diode (LED), liquid crystal, plasma, and cold cathode gas tubes are currently the leading technologies with LED's holding the edge.

Scanning is used to drive the display in most calculators. The main reason for this is to reduce the number of connections to the display. Consider an eight-digit calculator with eight segments per digit. With no scanning, 64 connections are required. With the display treated as an 8 × 8 matrix, only 16 connections are required plus segment and digit switches as shown in

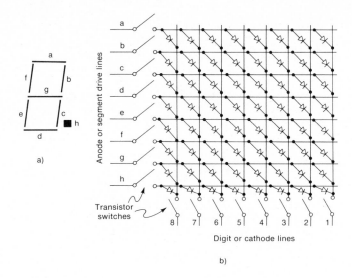

Figure 3-28 Eight 8-segment numerics shown as a light emitting diode array

Figure 3-28b. A scanned display requires additional electronics, but with integrated circuits, electronics are often less expensive than connections.

There is another clever design technique that can be implemented in calculators having scanned keyboards and displays. The column scanning lines for the keyboard can also function as the digit scanning lines for the display. Commercially available integrated circuitry exists to interface the calculator to the type of display used, such as the light emitting diodes of Figure 3-28b. For example, for the Texas Instrument TMS 0100 single-chip calculator, there is available the TI SN25391 segment-driver circuit and the SN25392 digit-driver circuit.

Other calculator outputs, such as alarm signals, may use special single indicator lamps or the same seven-segment display. An error signal can be indicated by a capital E. On some calculators, the display is flashed to indicate an error, and all decimal points are lit to indicate low-battery, thus no additional display devices are required.

The flashing display is accomplished through microprogramming with hardware already required for the basic function of the calculator. This is another instance in which software can take over a task usually assigned to hardware. Because read-only memory is so relatively inexpensive, this is often an efficient tradeoff.

3-5. EXAMPLE SYSTEMS

Two types of calculator systems are discussed in this section. The first, the Texas Instruments (TI) single-chip TMS 1802 circuit, allows implementation of a standard eight- or ten-digit, four- to seven-function calculator with only the one integrated circuit. Programmable variations, made by changing one of the integrated circuit masks, are possible so that the circuit in reality represents a family of products. Functions such as $1/x$, x^2, and square root of x can be programmed, but the primary purpose of the circuit is for four-function, lowest cost calculators.

The second system, the Hewlett Packard HP-35, is a scientific multifunction calculator with five user registers, ten mantissa and two exponent digits and a cost four to five times that of a four-function calculator. Because of the added architectural and programming complexity, the system requires three MOS circuits.

The technology which made both of these products possible is Metal-Oxide-Semiconductor/Large-Scale-Integration (MOS/LSI). It was the commercial application to calculators that gave this technology the incentive it needed in 1968–69. There are many variations of the basic technology; TI uses a silicon nitride process to permit lower supply voltage; HP calculator circuits use an ion implantation technique for the same purpose.

The systems described here were originally designed in 1970–71 and represent the state of the art in circuit and packaging density of that time. Rapid advances in semiconductor technology usually require manufacturers to redesign calculator integrated circuits every three years to remain competitive. The newest technology must be used to provide the highest performance for a given cost, or the lowest cost for a given performance. Architecturally, however, the techniques do not change significantly, so these systems are still instructive and applicable for study here.

The Texas Instruments TMS 1802

The TMS 1802 is a large chip by 1972 integrated circuit standards, measuring 0.23 inches on a side. It contains the equivalent of 6000 discrete transistors. It is packaged in a 28 pin dual-in-line plastic package. Part *a* of Figure 3-29 shows the chip, the equivalent transistors, and a calculator, the TI SR-11, which uses the chip; part *b* shows the chip being placed on a 28-pin lead frame prior to encapsulation.

Control Function   Control within the chip is partitioned into two levels; the highest level, called *program control*, is contained in a 320-word, 11 bits per word, read-only memory. The next level of control is implemented with programmable logic arrays.

Table 3-1 shows the instruction set for the TMS-1802. Normally the program memory address is incremented each word time; a branch to any of the 320 locations can be made by testing the ALU output for positive or negative. There is no subroutine capability. The arithmetic instructions make use of a timing mask that allows the instruction to operate on the data during one of eight time windows. The remaining codes are used for the testing and setting of flags, loading constants, and various other bookkeeping tasks.

Figure 3-29 (Part *a*) Texas Instrument TMS 1802 chip, the transistors it replaces, and the TI SR-11 calculator that uses it

The next level of control is implemented by three separate programmable arrays as shown in Figure 3-30. The control array decodes arithmetic operation codes from the program memory and controls the ALU and the timing

Figure 3-29 (Part *b*) Texas Instrument TMS 1802 chip being placed on 28-pin lead frame

TABLE 3-1. TMS 1802 INSTRUCTION SET

Instruction	Format				
Branch on Positive	1 1	9-Bit Address			
Branch on Negative	1 0	9-Bit Address			
		3	2	2	2
Arithmetic	0 1	Timing Mask	Op Code	ALU In	ALU Out
Other; Load Constant, Test Flag, Data Transfer, Clear, etc.	0 0	9-Bit Operation			

array. The timing array provides enable signals to the ALU and scans the keyboard and the digit lines of the display. The output array decodes BCD data into the desired display format, such as 7- or 8-segment display.

Each of these four control sections is gate-mask-programmable to allow maximum flexibility in making changes to the operating characteristics of the calculator without a redesign of the logic circuitry. Some of the variations

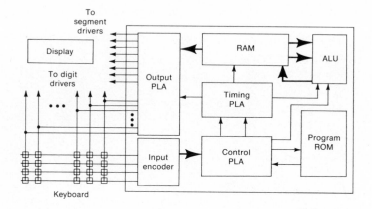

Figure 3-30   Block diagram of the TMS 1802 calculator chip

that are possible include keyboard language (algebraic or accounting), input data format, key location, keyboard debounce method, display round-up or round-down, leading zero suppression, output data format, and fixed and floating display format.

Arithmetic Logic Unit   The data storage in the system consists of a 182-bit dynamic random-access memory (RAM) organized as three registers of 13 BCD digits each and two 13-bit status or flag registers. Addressing for the memory is by a 13-bit shift register that contains a single 1 and twelve 0s. The digits must be addressed sequentially so that the RAM is architecturally similar to a shift register. However, because of the integrated circuit techniques employed, the dynamic memory actually occupies less area than a shift register, and area minimization is one objective of the design. All data paths are four bits wide. The ALU contains four bits of storage that allows both right and left-shift operations to be performed. The flag registers can be compared with masks from the program memory to allow branching operations.

Input/Output   The keyboard consists of an 11×4 matrix of switches as shown in Figure 3-26. One row is for digits, one for functions, one for decimal point setting, and one for the "constant" mode switch. The keyboard does not have to be fully populated and key locations can be changed by mask programming. The eleven column-scanning signals are also used to scan the display as shown in Figure 3-30. The input sensing program provides protection against transient noise, double entry, leading-edge bounce, and trailing edge bounce.

The output PLA has the functions of driving the display segments. In addition to segment code, both interdigit blanking and segment polarity can be selected.

The Hewlett Packard HP-35

The architecture of the HP-35 processor differs in many aspects from the TI TMS 1802. This is understandable since the type of calculators that use the processors are very different. The HP-35 and HP-45 are scientific calculators that calculate trigonometric and logarithmic functions to ten-digit accuracy in less than a second. With the addition of a program storage circuit and interface circuitry to a magnetic card reader, the same basic design is used in the HP-65 programmable calculator. This requires a structure of computer-like capability.

System Design   A block diagram of the processor portion of the calculator is shown in Figure 3-31. The system is partitioned into three MOS arrays called Control and Timing (C & T), read-only-memory (ROM) and Arithmetic and Registers (A & R). The three arrays are interconnected by three basic bus lines called SYNC (synchronization), I_s (Instruction), and WS (Word select). All operations occur on a 56-bit word cycle that corresponds to fourteen 4-bit BCD digits.

The partitioning into three arrays requires further explanation. As shown in Figure 3-31, the read-only memory is arranged on a bus structure so that from one to eight may be present in any calculator. This allows design flexibility in terms of the amount of programming to be included (the HP-35 has three ROM circuits; the HP-80 has seven). There are other reasons for

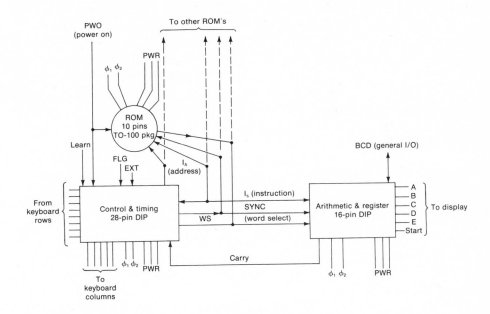

Figure 3-31   HP-35 processor diagram block

separating the ROM function. During the calculator design phase, several reiterations of ROM patterns are typically required. With the ROM separate, the control and arithmetic functions need not be affected while a smaller circuit, the ROM, is changed.

Since the control and timing and arithmetic functions are the same for each calculator built with this processor, it is logical to combine these functions on one chip. However, they are complex arrays, and the technology available when they were designed did not permit this much circuitry to be packed into one reasonably sized chip.

The HP-35 is a totally serial processor. The adder is a BCD serial type similar to Figure 3-22. The serial structure means less integrated circuit area must be allocated to interconnection lines and gating functions and an interesting trade off occurs. A bit-serial, digit-serial architecture is inherently one fourth the speed of a bit-parallel, digit-serial structure such as that used in the TMS 1802 and many other calculators. But the basic clock rate for a bit-serial structure can sometimes be increased since additional area can be allocated for larger integrated devices that are necessary for greater speed. In the HP-35, the execution time of the most complex functions is under one second, while the serial architecture permits an increased circuit complexity.

The system and logic designer working with MOS/LSI technology must understand enough about integrated circuits to make intelligent design trade offs. The classical academic minimization criteria of reducing gates and gate inputs are no longer valid.

Instructions in the HP-35 are transferred serially from the active read-only memory to the arithmetic and control circuits and to other ROMs if present. The SYNC line is used to synchronize the calculator system. It consists of one 10-bit wide window each word time that also functions as an enable signal for the I_s line. Figure 3-32 is a timing diagram that shows these relationships.

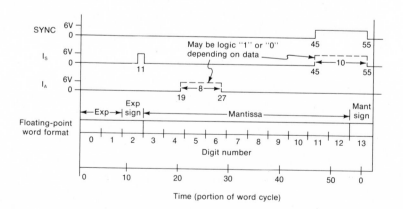

Figure 3-32   HP-35 timing signals

The 10-bit instruction is received by each circuit in the system and decoded to determine if the instruction is applicable to that circuit. Although this duplicates decoding, the number of instruction leads is reduced to one.

The remaining bus line, WS, is generated either on a ROM or on the control circuit, and is used as an enable signal on the instruction being executed by the arithmetic circuit. The signal selects a portion of the 14-digit word during which the arithmetic instruction is active. For example, it is possible to enable an add instruction for just the mantissa portion of a floating-point word passing through the adder.

The instruction address is transferred from the control circuit to each ROM over the I_a line. The primary qualifier or status line is the carry signal from the arithmetic unit. It is one of the bases for instruction branching.

Control and Timing Circuit   The main components of the control and timing circuit are shown in Figure 3-33. Shown are the master system counter, the keyboard scanner, a 28-bit shift register divided into three fields, a 1-bit adder, a 4-bit "pointer" register, an address buffer register, a 6-bit counter, and a microprogrammed controller.

The controller is a 58 word by 25 bit/word ROM that receives qualifier

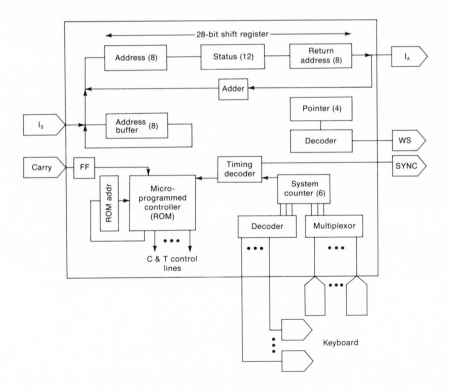

Figure 3-33   Control and timing circuit block diagram

or status conditions from throughout the circuit and from the CARRY line. It sequentially outputs signals to control the flow of data within the control circuit.

In a serial system based on a 56-bit word, something has to count to 56. The 6-bit system counter has this responsibility and also functions as the keyboard scanner as described in Figure 3-27. Timing signals are decoded and sent to the controller as qualifiers and the SYNC signal is sent to the other circuits.

The 28-bit shift register circulates twice each word time. It is used to store the current address (8 bits), the return address from a subroutine call (8 bits) and 12 status bits. The instruction address is incremented each word time and transferred to the ROMs unless a branch instruction is executed (when the branch address from the address buffer is substituted), or when a "Return from Subroutine" is executed (when the return address is substituted). The twelve status bits may be individually set, cleared, and tested for branching purposes.

The pointer is a four-bit register. It is used in conjunction with the WS line to "point" to *one* of the fourteen digits in the arithmetic unit. This is useful for comparing or adding single digits. The pointer can be incremented, decremented, and tested.

Other inputs to the control and timing circuit allow external setting of status bits, external serial entry of key codes, power-on in a cleared condition, and three power and two clock inputs.

Read-Only Memory Circuit   The ROM circuits (Figure 3-34) store the programs to execute the functions required of the calculator such as add, subtract, multiply, divide, transcendental functions, and so on. They correspond to programs stored in the main memory of a digital computer. It is important to understand the different functions of the read-only memory within the control circuit and the main external ROM. The control circuit ROM is a microprogrammed controller as discussed in section 3-4. The main ROM stores a program in a language similar to the assembly language of a computer system

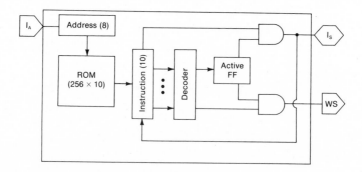

Figure 3-34   Read-only-memory block diagram

and is not a microprogram in the conventional sense of the term even though the programs are in read-only memory.

The read-only memories are organized as 256 words of 10 bits each.[1] The eight-bit address is received by each memory in the system but only one memory is "active" at a time and allowed to output an instruction on the I_s line. A separate "ROM Select" instruction is used to change control from one ROM to another by setting the "ACTIVE" flip/flop on the selected ROM and resetting it on the currently active ROM. Thus I_s is a bidirectional line: an output from the active ROM and an input to all other ROMs.

The ROM also decodes some of the word-select instructions (those not dependent upon the pointer.)

Because of the serial nature of the processor, the system does not need to operate with separate instruction-execute phases typically found in computers, but rather instruction fetch and execution take place simultaneously. The actual implementation or action of an instruction is delayed one word time from the receipt of that instruction by the circuit. This is shown on the timing diagram in Figure 3-35.

Arithmetic and Register Circuit (A&R)   The third MOS circuit provides the arithmetic and data storage function for the calculator system. It receives instructions from the ROMs over the I_s line, gating signals from ROMs or the control circuit over the WS line, and sends information back to the control circuit via the CARRY line. The arithmetic circuit also partially decodes

---

[1] In later designs, that is, the HP-45, 55, 65, and 70, a 1024-word by 10-bit ROM is used. The A&R and C&T circuits remain the same.

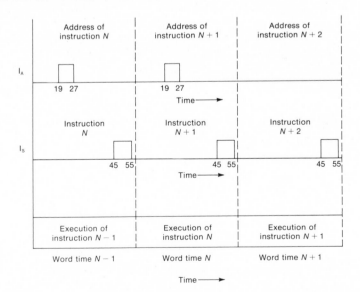

Figure 3-35   Timing for addresses, instruction, and instruction execution

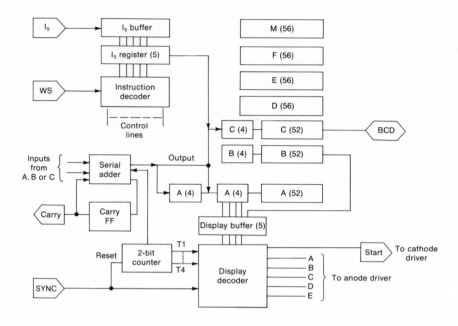

Figure 3-36 Arithmetic and register circuit block diagram

the BCD data into seven-segment display data and transmits it to the anode display driver via the A, B, C, D, and E lines. START is a pulse to synchronize the display with the calculator. Figures 3-36 and 3-37 show the arithmetic and register circuit block diagram and register transfer paths.

The power and flexibility of an instruction set is partially determined by the variety of data paths available. One of the advantages of a serial structure is that additional paths are not very costly in terms of area, as measured by gates or metal lines in the integrated circuit. The arithmetic circuit structure is optimized for the type of transcendental function algorithms required of a scientific calculator.

The seven 56-bit shift registers are organized as three working registers, A, B, and C, a four-high stack C, D, E, and F, and a separate memory register, M. Register A has an additional four bits to allow a left shift operation. The stack is useful since the keyboard language is reverse Polish as described in Section 3-2. Register C communicates to the outside world via the BCD line, allowing the attachment of external memory (which exists in the HP-45 calculator).

The thirty-two arithmetic instructions are listed in Table 3-2. Each of these can be combined with one of eight word-select signals to provide an actual instruction set of 256 instructions. The last column of Table 3-2 indicates compare instructions. They allow registers to be tested against zero, each other, or unity without destroying one of the registers, as a subtract operation

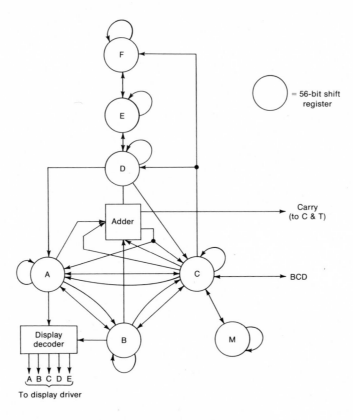

Figure 3-37   Register data paths

does. Results of the test are sent to the control circuit via the carry signal, and instruction branches as made based on these tests.

Instruction Set   The instructions in the HP-35 processor are shown in Table 3-3. One interesting feature is that no unconditional jump, that is, GOTO, instruction is included. This important operation is achieved by using a *con-*

TABLE 3-2. HP-35 ARITHMETIC INSTRUCTIONS

$0 \rightarrow A$	$A+C \rightarrow C$	$A+1 \rightarrow A$	$0 - B$
$0 \rightarrow B$	$A-C \rightarrow C$	$C+1 \rightarrow C$	$0 - C$
$0 \rightarrow C$	$A+B \rightarrow A$	$A-1 \rightarrow A$	$A - C$
$A \rightarrow B$	$A-B \rightarrow A$	$A-1 \rightarrow C$	$A - B$
$B \rightarrow C$	$A+C \rightarrow A$	Shift A Right	$A - 1$
$C \rightarrow A$	$A-C \rightarrow A$	Shift B Right	$C - 1$
$A \leftrightarrow B$	$C+C \rightarrow A$	Shift C Right	
$B \leftrightarrow C$	$0-C \rightarrow A$	Shift A Left	
$C \leftrightarrow A$	$0-C-1 \rightarrow C$		

TABLE 3-3. HP-35 Instruction Set

Type	Fields					

Jump Subroutine

	8			
Subroutine	Address		0	1

Conditional Branch

	8			
Branch	Address		1	1

Arithmetic

5	3		
Operation	Word Select	1	0

Status Register

4	2				
Bit #	Operation	0	1	0	0

Pointer Register

4	2				
Digit #	Operation	1	1	0	0

Load constant, Stack operations

6					
X X X	X X X	1	0	0	0

Miscellaneous ROM Select, subroutine return, external entry No operation, etc.

6					
X X X	X X X	0	0	0	0

ditional branch, but with the branch condition known. The programmer must be careful, but this feature helped to allow the word length to be cut from eleven to ten bits.

The serial nature of the processor encourages an encoding format rarely used in more powerful parallel structures. The instruction bits are decoded serially from right to left. If the first bit is a 1, the last eight bits are an address and hence "don't care" bits to the arithmetic unit. Additional instruction types follow a similar format with leading zeros. This type of encoding permits simpler serial decoding.

Auxiliary Circuitry The display is driven by a bipolar anode driver and a bipolar cathode driver as shown in the HP-35 system diagram, Figure 3-38. The cathode driver is a shift register that sequentially selects a digit under command of stepping signals from the anode driver. The anode driver receives partially decoded eight-segment data on five lines, completes the decoding, and scans the light-emitting diode segments. The anode driver also generates a two-phase clock.

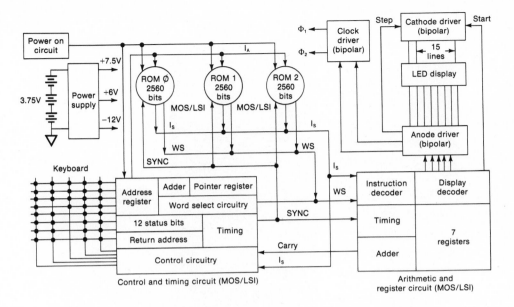

Figure 3-38 HP-35 system block diagram

The bipolar clock driver receives the two-phase clock signal and translates the voltages to MOS levels, $+6$ and -12 volts.

The power supply generates three voltages; $+6$ and -12 for the MOS, and $+7.5$ for the display drivers. The power-on circuit provides a delay signal to the control circuit to prevent ROM turn on until a valid clock is available, thus ensuring the calculator comes on in a cleared state. Packaging is on two circuit boards, as shown in Figure 3-39.

3-6. TRENDS IN CALCULATOR DESIGN

Certain types of products are more receptive to changes in technology than others; those with a lower price can be replaced more easily by the owner to obtain the newest, most up to date product. Calculators are a particularly fertile area for the application of new technology. The market size is huge; every person beyond first grade does some computation daily. Just as the transistor radio has become a throw-away product, that is, it is cheaper to buy a new one than to repair a malfunction, the calculator at $19.95 (or less) cannot be economically repaired.

Thus we can expect that any advancement in the semiconductor industry will rapidly find application in new calculators. Here we discuss what types of improvements might be made in terms of four types of calculators: 1) four function, 2) expanded function, 3) professional or specialized, and 4) programmable.

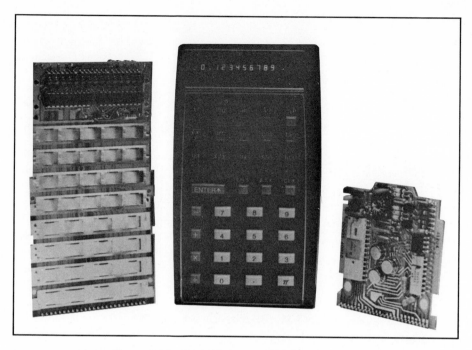

Figure 3-39   Packaging for the HP-35 calculator

For the majority of buyers, a calculator that can add, subtract, multiply and divide is all that is really needed; a main incentive to buy is low cost. Four-function calculator material costs are dominated by the semiconductor electronics and the display. The price decreases these parts have experienced in recent years are leveling off, since basic material costs and labor rates offset advances in yield and productivity. As prices stabilize, customers are lured by additional features that cost little to incorporate into a design. Thus the pure four-function calculator will probably die and be replaced by the expanded, seven- or eight-function calculator.

Some features can be added with no design changes, for examples, the TI TMS 1802 calculator can be reprogrammed to execute reciprocal, square, and square root with no additional ROM. Other features, such as registers for storage may cost little but require design changes. Expanded function calculators will benefit from lower power integrated circuits which will allow, like electronic watches, operation for a year or more on one set of replaceable batteries (giant steps in battery technology will also be made).

As more and more features are added, the designer's dilemma of inventing a machine that is widely useful, and at the same time easy to use emerges. It soon becomes advantageous to provide a family of products, each specialized to a particular user or profession.

Specialized calculators were designed during the mechanical era for the banking and accounting fields. Early specialized electronic calculators were designed for the engineering and scientific market, bridging the broad gap between mechanical desk calculators and computers. The first specialized hand-held calculator was the electronic slide rule, with models for investment, surveying, statistics, and navigation soon to follow.

These machines can benefit from advances in display technology that will allow alphanumeric displays at a reasonable cost. Keystroke sequences for complex problems may become lengthy, and a display that operates in English (or Japanese or whatever) can lead the user through a problem, identify the answers by name, and point out his errors in a language he will remember! Such sophistication in display requires additional read-only memory, which is fortunate since decreases in cost of read-only memory are a certainty.

It would be convenient to be able to specialize small calculators by replacing one part, such as some desk top calculators can be specialized by the substitution of a ROM. As calculator prices decrease this becomes less feasible, since the cost of the replaceable part is a larger fraction of the total cost, and the system cost of "replaceability" can make buying two specialized machines cheaper than one main calculator with replaceable programs.

Other changes will come in packaging .The calculator in a wrist watch is a certainty as is the calculator in a telephone; and the calculator in the automobile and perhaps the calculating pencil.

Programmable calculators will see rapid price decreases, with IC technology providing more performance in terms of faster processing and more memory. The HP-65 programmable hand-held calculator has made powerful computer-like products a personal possession. Computer languages such as FORTRAN and APL will appear in calculators; or perhaps we should say, calculator-like ease of use will appear in computers. In either case future customers for computing devices will be even more confused as to whether to turn to the telephone book yellow pages under **Calculators** or **Computers**.

3-7. PROBLEMS

3-1.   Will calculating devices such as the abacus, slide rule, and mechanical adding machine ever pass completely out of use? Why or why not?

3-2.   Obtain a mechanical adding machine or calculator and examine the mechanical principles of its design. What reliability and maintenance problems might be expected?

3-3.   Imagine yourself the president of a calculator company in 1962. What would you want to know before beginning a development project for an electronic calculator?

3-4.   List four ways calculators generally differ from computers. Are these differences always valid? Do advances in technology tend to close or separate the boundary between calculators and computers? List some major technological changes and how they have affected calculators.

3-5. It is sometimes stated that people who solve problems on programmable calculators must be more ingenious and clever than those who use computers because of the limitations on program and data memory and speed. What do you think?

3-6. Specify a calculator keyboard for a special user group such as electrical engineers, computer programmers, surveyors, bankers, gamblers, and so on. What keys should be included? How many data registers are required? Demonstrate the keystrokes to solve some sample problems. What price could you most likely receive for your calculator?

3-7. How many multiply operations would be required to calculate a sine function to five-digit accuracy using a Taylor's series expansion? How many registers and constants are required? Repeat using a continued fraction expansion. If you have access to a programmable calculator or a computer, try your solutions.

3-8. Consider three operating environments: an office, a traveling salesman, and the home. What calculator features or characteristics are important in each situation?

3-9. Derive an algorithm for evaluating equations in reverse Polish language. The key decision is when to load an operand onto the stack and when to execute an operation.

3-10. Draw a flowchart similar to Figure 3-19 for a counter type of BCD adder in which the augend is *always* incremented before the addition process begins. What problems might you foresee for this method?

3-11. Implement a BCD subtractor with counters. Draw a flowchart for your design.

3-12. Write a flowchart for a key debounce routine. Consider transient noise, leading edge bounce, and trailing edge bounce. Also consider that the user may want to push the same key twice in succession.

3-13. Explain why connections to integrated circuit packages and to printed circuit boards may be more expensive than additional electronics.

3-14. Is there any purpose to minimizing logical equations implemented in a PLA? Does the answer depend upon the technology used?

3-15. Design a traffic light controller for a four-way intersection using a PLA. Include left turn lanes on one of the streets. Draw a flowchart to specify your design.

3-16. Design a digital machine to play the card game Blackjack (or Twenty-one) using a microprogrammed controller. Define the necessary hardware qualifiers and control signals; show a flowchart. The external appearance of the machine should be a slot to accept cards and three lights labeled STAND, HIT, and BROKE.

3-17. The question sometimes arises as to the minimum number of arithmetic instructions required to implement a four-function BCD calculator. Design a three-register architecture and determine what instructions are necessary to execute the following instructions in floating point:

a) Add
b) Subtract
c) Multiply
d) Divide

You will need to construct flowcharts to define the algorithms. Try your algorithms on a programmable calculator or computer.

3-18. How can the HP-35 system operate with more than 256 ROM words with only an 8-bit address? How are subroutine calls done and what restrictions are imposed?

3-19. Explain why the TI TMS 1802 RAM is analogous to a shift register memory.

3-20. How might an alphanumeric display be used in pocket calculators? How much more do you believe such a calculator would be worth?

CHAPTER 4

Introduction to Minicomputers

Kay B. Magleby

4-1. EARLY MINICOMPUTERS

Minicomputers as we know them today first became an important part of the computer community about 1965 when economic and technical factors combined to make small, dedicated processors attractive for a large number of applications. By almost any definition that excludes cost, most early computers would be classified as minicomputers. Classifications such as computing power, memory size, system configuration, and so on, fail to distinguish the salient features of the minicomputer from their larger, more expensive counterparts.

The minicomputer industry has grown from essentially no sales in 1965 to over $300 million annual sales in 1973, and is continuing to grow at a 20 percent to 30 percent per year rate. What have been the economic and technical factors that have produced this remarkable industry? What distinguishes a minicomputer from other classes of computers? What applications have contributed to the rapid growth of minicomputer sales?

In the early days of computing, computer designers found that the computing power per dollar of hardware cost increased as the size of computer increased. This resulted from the fact that there was a large cost for input/output equipment, memory interfaces, and basic control logic that represented fixed overhead independent of computing power. To take advantage of these economic considerations, hardware and software systems were organized as powerful computing systems that executed a large number of fairly simple computing tasks. As the cost of computing hardware was reduced by advanced semiconductor technology in the early 1960s for many viable applications of computers, hardware costs were no longer the major cost elements in providing a solution. Data acquisition and display, programming costs, and real time requirements assumed a major role. The computer could be dedicated to a specific task even if it were not utilized 100 percent of the time.

One of the first successful minicomputers, the PDP-8, was introduced by Digital Equipment Corporation (DEC). This computer had the same command structure as an earlier computer offered by DEC, the PDP-5, but was about a third the size, half the cost, and three times as fast. While only about 100 PDP-5 computers were sold, over 10,000 PDP-8 computers have been sold. The PDP-8 provided the right computing power for a large number of applications. Shortly after the introduction of the PDP-8, a number of other manufacturers introduced minicomputers that were also quite widely used. These include the Hewlett-Packard 2116, the Varian 620, the Data General NOVA, and others. Computers such as the Control Data 1200 and the IBM 1130 have about the same computational power, but due to size and cost they are not generally considered to be in the minicomputer class.

The features that are usually used to classify a computer as a minicomputer are as follows:

1. Size
2. Computational power
3. Cost
4. Application

Most minicomputers are part of systems dedicated to a specific task. The physical size of the computer is normally less than half a conventional 19 inch by 6 foot equipment rack. Often, the system must be portable, and both physical size and weight are important. While some of the early computers were of the same computational power as minicomputers in the late 1960s, they occupied hundreds of square feet and were certainly not portable.

Computational power is not as easily defined as the horsepower of an engine, but certain figures of merit have been established which are indications of computational power. The primary figures of merit that determine computational power are word length and memory speed. With fixed instruction set effectiveness, these two features determine the rate at which a computer solves problems. The available memory size often determines the size of a problem that a computer can tackle, but not the rate at which the computer solves the problem, and hence is not a useful factor in determining computational power. The generally accepted computational power of a minicomputer implies a word length of 16 bits or less and a memory cycle time of 2 microseconds or less. While some 18-bit computers have been classified as minicomputers, they are the exception rather than the rule.

Perhaps the most significant factor in the rapid growth of the minicomputer industry is system cost. When computers are expensive, the efficient use of the machine is the most important economic consideration. The users must stand in line, and cannot be allowed to sit at the console and think while the machine stands idle. As the machine costs drop, the salary of the user becomes more significant, and efficient use of his time becomes important enough to allow the machine to stand idle for long periods of time. This cost threshold was reached in the mid 1960s when the cost of a computing system with minimum input/output capability and enough memory to perform a variety of useful work (about 4K 16-bit words) dropped below $25,000. Today this threshold is higher due to inflation and other economic factors. One can forecast when minicomputers will become economically viable in developing nations as the time when the ratio of per capita GNP and machine costs is the same as in the United States in the mid 1960s.

The proper size, computational power, and cost determine when it is practical to use a minicomputer for a given application. The key factor that makes a minicomputer so attractive for a large number of applications is the ability to dedicate the machine to a given job or group of jobs. Usually the data is collected and processed as it occurs "in real time" as opposed to being gathered, encoded, and run at the convenience of the computer in "batch mode." Often, since the data is available in real time, results of analysis of current data can be used to control the external system and minimize the amount of data that is collected and analyzed. As an example, a minicomputer system was used to analyze the structure of a crystal in real time in less than half the time required by a computer with 50 times the computational power operating in batch mode. The minicomputer controlled the position of the crystal and analyzed only useful data and did not need to make 100 readings to determine which single reading was significant.

Table 4-1 summarizes the definition of a minicomputer.

TABLE 4-1. DEFINITION OF A MINICOMPUTER

Feature	Requirement
Size	Less than ½ of a 19″ x 6′ equipment rack
Power	16-bit word 2-μsec cycle time
Cost	Less than $25,000 for a 4K system with minimum input/output
Application	Dedicated to one job or a group of related jobs

4-2. FACTORS CREATING THE MINICOMPUTER REVOLUTION

As the cost/performance ratio of minicomputers improved, it became practical for minicomputers to be useful in so many applications that an actual revolution occurred. This revolution took place in much the same way that the industrial revolution took place when machine power became less expensive than manpower. Several developments happened almost simultaneously to cause the rapid improvement in the cost/performance ratio, which in turn created the explosive growth in the minicomputer usage. These developments are:

1. Economic factors
2. Technological advances
3. New system structures
4. Software development
5. System development

While these developments occurred simultaneously, they are considered here one at a time. The critical economic factor that led to the development of the minicomputer occurred when the cost of a machine with significant computing dropped below the salaries of the people using it. This created the situation where it was more important to maximize the utilization of people rather than the machine, and enough machines could be made available for convenient usage even if some machines were idle some of the time. During the period from 1950 to 1965, small computers improved in performance by a factor of 1,000 while their cost dropped by a factor of 100. This produced an improvement in the cost/performance ratio of 10^5. A similar improvement in the automobile industry would have produced a personal vehicle that could travel at the speed of sound and whose cost would be less than one dollar. The availability of such a vehicle would change our concept of travel and the way in which travel affects our lives. In a similar way, the availability of small economical computers has changed our concepts of how computers can be used in many applications.

This improved performance at low cost was made possible by a number of significant technical advances. Early computers used vacuum tubes for logic elements and had limited memory capacities. Later, rotating drum memories served as the main working memory and discrete transistors and diodes served as logic elements. A significant improvement in performance occurred when the drum memory was replaced by magnetic core memories. (Chapter 5 describes these storage devices.) The first minicomputers became possible when the logic elements were produced using integrated circuits and low cost core memory systems were developed. This occurred in about 1965. In about 1970, large scale integrated circuits made solid state memories more economical than core memories with a factor of from 2 to 5 improvement in performance.

For each technology, computer designers found a different system structure was optimum. Early minicomputer designers minimized the number and size of solid state storage registers and simplified the logic structure with restricted commands which placed a greater burden on software development. With large scale integration, the cost of logic decreased and arrays of storage registers were available at low prices. In the early 1970s minicomputers were developed with extensive command structures and several multi-purpose registers. Large scale integration also made read-only memories economically attractive in place of the fixed logic that earlier minicomputers used to implement command structures. This led to microprogrammed processors in which microcode stored in a read-only memory implements the command structure.

In Figure 4-1, a typical integrated circuit contains a J-K flip/flop or four 2-input AND gates. The central processing unit excluding main memory and input/output control requires about 400 integrated circuits. In Figure 4-2, a typical LSI circuit is a 16-bit flip/flop array or an 8-bit arithmetic and logic unit with more functions than the simple list in the ALU shown in Figure 4-1. The control and timing is still done with fixed logic, but up to four times as

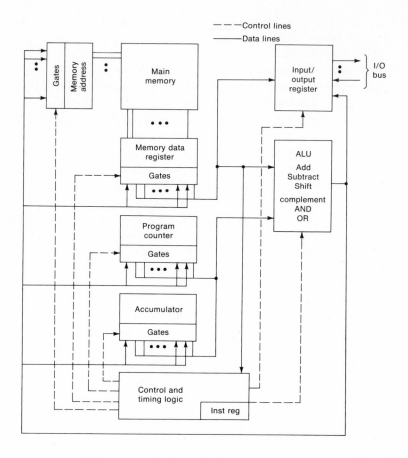

Figure 4-1 Early minicomputer structure

many gates are included in a single circuit such as a 4-bit decoder. The central processing unit is reduced to about 100 LSI circuits. Figure 4-3 gives the block diagram of a microprogrammed computer that uses a read-only memory to replace the fixed logic to implement the control function.

As the technology used by computer designers developed, corresponding advances were being made in the development of software for minicomputers. Early minicomputer manufacturers supplied only a symbolic assembler and a few utility routines such as a symbolic editor and a simple math library. This placed a heavy burden on the user to develop his own operating system as well as solve his basic programming tasks in a primitive language. In about 1967 minicomputer suppliers developed compilers that could operate in the restricted core memories used in most minicomputers (4K or 8K 16-bit words). By 1970 most suppliers offered several compilers, such as ALGOL and

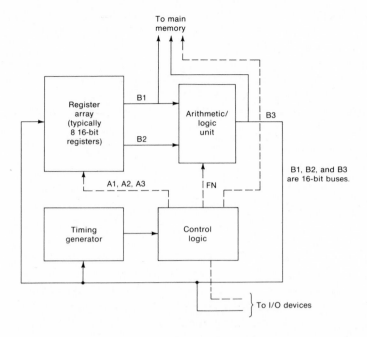

Figure 4-2 Minicomputer structure using integrated circuits

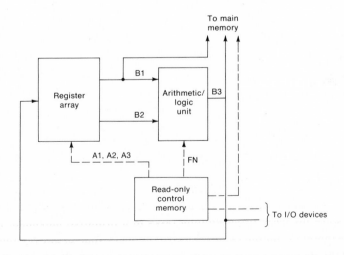

Figure 4-3 Microprogrammed minicomputer structure

FORTRAN, and interpreters such as BASIC. During this same time, simple operating systems were developed which greatly simplified the task for the user. By 1970 the minicomputer user had most of the conveniences offered to the large computer user.

The development of powerful programming aids greatly accelerated the use of minicomputers in a diverse range of applications. The following list gives some applications where minicomputers are commonly used.

- Real-time instrument systems
- Online control systems
- Data communication controllers
- Engineering problem solving
- Production control and scheduling
- Data acquisition systems

The broad spectrum of usage caused the market for minicomputers to develop rapidly. Figure 4-4 shows the growth of minicomputer sales and a forecast of the market through 1980. During the period up to 1972 the market was lightly penetrated with sales limited by the availability of equipment and application programmers. As a result, sales growth was steady and relatively insensitive to economic conditions. Note that during the 1970-71 recession minicomputers sales continued to grow.

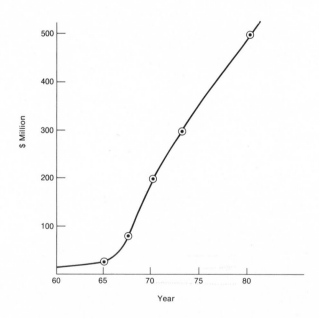

Figure 4-4　Growth of the minicomputer market

4-3. SYSTEM STRUCTURES

All minicomputer organizations can be classified by the basic bus structure that interconnects the major elements of the system. The three types of bus structures are:

1. Single bus structure
2. Compatible input/output bus
3. Multiple bus structure

Each type is described in more detail in the following text.

Single Bus Structure

The simplest system organization is to have a single bus that interconnects system elements as shown in Figure 4-5. A typical assignment of signal lines

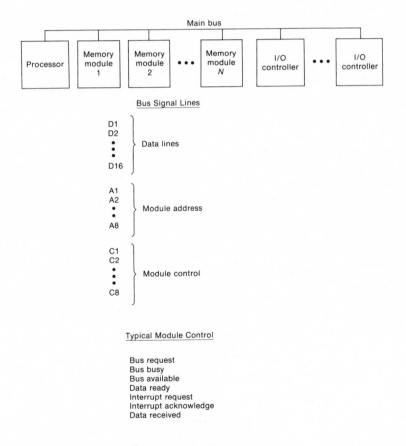

Figure 4-5 Single bus structure

in the bus is also shown. These lines may be directional, that is, duplicated for signals in each direction. To evaluate the performance of a bus structure, the following parameters are defined:

1. *Programmed input/output data rate* — the rate of input or output that can be accomplished via the processor under program control.
2. *Interrupt-processed input/output data rates* — the speed that processing of input/output data can be accomplished via the processor under interrupt control.
3. *Direct memory access speed* — the rate that data can pass from an input/output device directly to memory.
4. *Memory bus data rate* — the speed of the bus that interconnects the main memory and processor.
5. *Input/output bus data rate* — the speed of the bus that connects the input/output devices to the processor.
6. *System overloading* — a condition that occurs when a task cannot be performed by a major element because the bus is busy.
7. *System overhead* — any activity required to accomplish data transmission between two major elements of the system, but that is not part of the data transmission.

In a single bus system, the data transmission protocol between any two elements of the system is identical. A typical protocol is as follows:

1. The element that is to initiate action signals the bus controller via the bus request line.
2. When the bus is free, the bus controller signals the waiting element via the bus available line.
3. The element takes control of the bus and signals all other elements via the bus busy line that the line is unavailable.
4. The element requests or transmits data via the data lines to the element identified on the module address bus.
5. The destination element indicates that it has received the data via the data received line.
6. Steps 4 and 5 are repeated until the data transfer is complete.
7. The initiator signals the bus controller that it has completed the transfer by clearing the bus busy line.

Since the protocol is identical for all elements, data can be transferred between any pair of elements. This makes it possible for two input/output devices to communicate with each other without processor control or for input/output devices to send or receive data directly to or from the memory. Programmed input/output is accomplished by a data transfer from an input/output device to the processor and then a transfer from the processor to memory.

The bus speed must be fast enough to allow the processor-memory communication to take place at speeds limited by the memory speeds. Since the bus must be long enough to interconnect all elements of the system and each interconnection adds capacity on the lines, each element must have relatively expensive low impedance drivers to drive the bus. Input/output controllers must be fast enough to operate at the same data transfer rate as the processor-memory rate, which may add to the cost of the controllers.

An analysis for each parameter defined for a single bus structure follows.

Programmed Input/Output Data Rate   The best way to determine the fastest input/output rate possible under program control is to write a benchmark program using a typical command structure for a minicomputer and to estimate the running time. An assumed assembly language is used with a description of the function of each instruction. (For a more detailed description of assembly language, refer to Stone [72].) The following is an example of such a benchmark with the estimated running time:

			Number of Cycles	
READ	SDS		1	(Skip if data ready flag set) Test to see if data ready flag is set. If set skip next instruction.
	JMP	*-1	1	Jump back to previous instruction and loop until flag is set.
	LIA	N	1	(Load input into accumulator) Load data from device N into accumulator.
	STA	M, I	3	(Store, indirect) Store data into the memory location given by the contents of memory location M.
	ISZ	M	2	(Increment, skip if zero) Increment memory location M.
	ISZ	CTR	2	Increment and test CTR for zero *Note:* CTR must contain $(-L)$ where L is length of record to be read.
	JMP	READ	1	Repeat reading until L reads occur.
	Total cycles		11	

(handwritten annotation: cycles / read ; time / cycle)

At the maximum input/output rate the JMP *-1 instruction would not be executed so the maximum data rate would be 1/10 CT where CT is the memory cycle time. Thus, for a 1 microsecond memory the maximum data rate is 100,000 words per second.

Interrupt-Processed Input/Output   Again, a simple benchmark program is the best way to estimate the maximum data rate for an interrupt processor. Assume the main program is running when an interrupt occurs and hardware diverts the computer to begin processing the interrupt in the interrupt service routine. The service routine is as follows:

		Number of Cycles	
SVS	NOP		Location to save return address to main program.
	STA T1	2	Save contents of accumulator in memory location T1.
	LIA N	2	Load data from device N into accumulator.
	STA M, I	3	Store data into memory location given by contents of memory location M.
	ISZ M	2	Increment memory location M.
	LDA T1	2	Restore accumulator.
	JMP SVS, I	2	Return to Main Program.
		13	

At least two cycle times are typically used to divert the execution to the service routine by the interrupt hardware. Thus, at least 15 cycle times are required. Again, for a 1 microsecond memory the maximum data rate is $10^6/15 = 66,666$ words per second.

Direct Memory Access (DMA) Input/Output Rate *Direct memory access* refers to a hardware controller for routing data directly between memory and an input/output device without requiring intervention by the processor to control each datum transferred. This is discussed in more detail in Chapter 6.

In a single bus structure, the direct memory access data rate is limited by the memory cycle time and the bus speed. For a 1 microsecond memory and a system with a bus cycle time of less than 1 microsecond, the maximum input/output rate is one million words per second. Few peripheral devices require input/output rates approaching this speed. Also, when the rate exceeds 500,000 words per second, the DMA channel uses every bus cycle, and the processor must suspend operation.

Memory Bus Speed To gain a performance advantage over other bus structures, the memory bus should be at least twice as fast as the main memory. This allows one DMA transfer to take place in one memory module while the processor accesses another memory module without degrading system performance. Thus, for a 1 microsecond memory, a 500 nanosecond bus cycle time is required. Systems exist that do not conform to this guideline, and system performance is then limited by the bus speed.

Input/Output Bus Speed The preceding paragraph on memory bus speed also applies here.

System Overloading In a system that does ~~not~~ meet the memory bus speed requirement described previously, system overloading can occur. For example, a system with a bus speed of 1 microsecond and a 1 microsecond memory cycle time suffers a 50 percent rate reduction in program execution if data is transferred from one input/output device to another at a rate of 500,000 words per second.

System Overhead The input/output method that usually has high overhead is interrupt-processed data. In the benchmark interrupt service program, the two cycles required to divert execution to the interrupt service routine plus the four cycles to save and restore the register plus the two cycles to return to the main program all represent system overhead. In this the overhead is eight out of fifteen cycles or 53 percent. Other types of interrupt systems may require many cycles to determine which device is requesting service or the device priority.

For low data rate devices, programmed input/output can have high overhead since the processor must wait for the external device.

Relative Merits of Single Bus Structure The single bus structure offers a single type of interface for all system modules. All module controllers can be almost identical. Since the processor can address peripheral devices in the same way that memory locations are addressed, no special input/output instructions are needed. This provides a powerful input/output instruction set. Due to the universal nature of addressing modules on the bus, device-to-device data transfers are straightforward.

Since the bus must be physically long enough to interconnect all the system elements, it is generally expensive to make it extremely fast. Its speed can be a limiting factor in system performance.

Example of a Single Bus Minicomputer The PDP-11 manufactured by the Digital Equipment Corporation is an outstanding example of a single bus minicomputer. The PDP-11 uses a "Unibus" consisting of 56 signal lines to interconnect all system modules. Each module interface to the Unibus has a module address and intermodule communication can take place between any two modules (see Chapter 6 for the signal assignments and communications protocol for the Unibus).

Compatible Input/Output Structure

The compatible input/output bus structure uses two buses to interconnect system modules. The input/output bus is similar to the single bus and is used to connect the processor and direct memory access modules to the various peripheral devices. A second bus is added to interconnect the processor and direct memory access units to the memory modules. Figure 4-6 gives a block diagram of a compatible input/output bus structure.

The peripheral devices can communicate with each other via the input/output bus, and can communicate with either the processor or DMA unit. A single device interface can operate either under programmed input/output or via the DMA channel since a compatible communication protocol is used. The device interface does not need to know whether the processor or the DMA unit is responding to its input/output requests.

The centralized direct memory access unit simplifies the device interface requirements and lowers interface hardware costs. This unit performs packing and unpacking of characters into words and contains starting address of the memory locations holding data to be transferred, the block length, and the address of the next word in memory to be accessed.

Since there are two buses, the memory bus can be physically short, permitting a low-cost high-speed bus while the longer input/output bus can be slow. Typically, the input/output bus is 2 to 4 times slower than the memory bus.

The evaluation of the basic parameters is as follows.

1. *Programmed input/output data rate* — same as for the single bus structure.
2. *Interrupt-processed input/output data rate* — same as for the single bus structure.
3. *Direct memory access data rate* — typically one half to one quarter of the memory speed since the input/output bus is typically slower; for a 1 microsecond memory the DMA rate is 250,000 to 500,000 words per second.

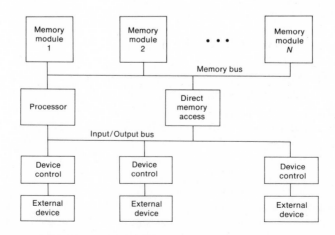

Figure 4-6   Compatible input/output bus structure

4. *Memory bus cycle time* — no more than half the main memory cycle time to allow processor access to one memory module while the DMA unit is accessing another without degrading system performance; for a 1 microsecond memory the memory bus cycle time should be less than 500 nanoseconds.

5. *Input/output bus speed* — determined by the data rate required for the fastest peripheral devices. Typically from one half to one quarter the speed of the main memory is fast enough to provide for the transfer rate required by a disk memory which usually is the fastest peripheral device used in a minicomputer system. Disk memory transfer rates are usually below 4,000,000 bits per second which requires a 250,000 word per second transfer rate for a 16-bit minicomputer.

6. *System overloading* — Data can be transferred between input/output devices without affecting processor execution. The memory bus need only be tied up during transfers between input/output devices and memory, and then for less than one memory cycle since the memory bus can be fast. System overloading usually does not occur with this bus structure.

7. *System overhead* — as with the single bus structure, the system overhead for interrupt processed data is about 53 percent or higher depending on the type of interrupt system used.

Multiple Bus Structure

This is the most common minicomputer structure. A typical multiple bus minicomputer is shown in Figure 4-7. A separate input/output bus is used for each input/output method. A programmed input/output bus is used for devices that use interrupt processing or programmed input/output. A DMA bus is used for devices that transfer blocks of data to or from memory. The speed of these buses corresponds to the required data transfer rate for each method of input/output.

Each device interface is preassigned to a particular input/output method, and it is not possible for a single device to operate using both programmed input/output and DMA. This restriction allows the system designer to optimize the signaling protocol for each bus.

Most of the basic parameters are the same as those for the compatible input/output bus. DMA data transfer rate is typically faster, approaching the speed of main memory since the added bus speed does not affect the speed or cost of the programmed input/output bus. System overloading is also less likely to occur than with either of the other two bus structures.

4-4. HARDWARE STRUCTURE

The organization of the processor is more important in determining system cost/performance than any other minicomputer system consideration. The

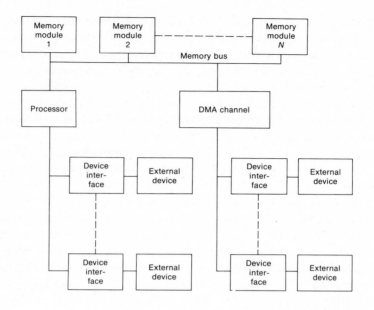

Figure 4-7 Multiple bus structure

parameters that a system designer uses to determine processor organization are as follows:

1. Internal bus structure
2. Register structure and function
3. Arithmetic and logic unit function
4. Instruction set
5. System timing

Figure 4-8 gives a block diagram of a typical processor showing the basic elements of the processor. The register block contains the general purpose registers accessible to the programmer as well as the system registers such as

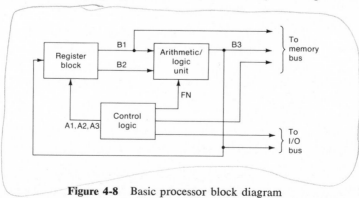

Figure 4-8 Basic processor block diagram

the program counter, auxiliary registers used in computation, and stack pointers. The instruction is usually held in a register contained in the control logic while being decoded.

The processor buses, B1, B2, and B3 are single-direction buses that contain one line for each bit in the word length of the processor. The register block can gate the contents of any register onto B1 or B2, and load any register from B3. The register to be gated onto or loaded from a bus is given by addresses A1, A2, and A3 from the control logic. A typical control cycle begins by placing the contents of two different registers on B1 and B2 terminals with the results of an operation loaded back into one of the registers. Interconnection to the outside buses is done via B1, B3, and lines from the control logic. A1, A2, and A3 select either an external bus or a register in the register block.

The arithmetic and logic unit accepts the two input buses, B1 and B2, and performs the function given by the FN code from the control logic. The results of the function are available for transfer to an external bus or a processor register via B3. Typical functions performed by the arithmetic and logic unit are:

Minimal Functions	Extended Functions
ADD	SUBTRACT
LOGICAL AND	SHIFT RIGHT 2
LOGICAL OR	SHIFT LEFT 2
EXCLUSIVE OR	SHIFT RIGHT 4
SHIFT RIGHT	SHIFT LEFT 4
SHIFT LEFT	COMPARE
ROTATE RIGHT	INCREMENT
ROTATE LEFT	DECREMENT

Note: Shift operations drop the bit shifted out of the register and insert a 0 into the opposite end of the register. Rotate operations insert the bit shifted out of the register into the opposite end.

The control logic and timing contains the sequence control logic that determines the basic phase that the processor is executing. The processor typically has the following phases:

1. Instruction fetch — read the next instruction to be executed from the main memory location given by the contents of the program counter.
2. Instruction execute — perform the functions indicated by the instruction that has been fetched in Step 1.
3. Indirect — if the instruction uses indirect addressing, the indirect phase is needed to get the address of the operand.
4. Interrupt — this phase is entered if an external device has requested an interrupt.

Figure 4-9 gives the sequence of the various phases that can occur.

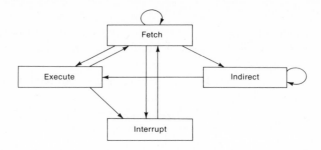

Figure 4-9   Sequence control

The control logic decodes the instruction and provides the proper signals to the other units to execute it. An ADD instruction, which adds two registers and places the results in a third, is executed by sending the appropriate operand addresses to the register blocks and sending the ADD function code to the arithmetic and logic unit.

There are five different classes of instructions that may be included in any minicomputer's instruction list. These classes are:

1. *Memory reference instructions* — instructions for which one of the operands is contained in the main memory.
2. *Register reference instructions* — instructions that perform operations on the contents of a single register.
3. *Register-register instructions* — instructions that perform operations between two registers.
4. *Input/output instructions* — those instructions that transfer data to or from an input/output device or control some function of the input/output device.
5. *Macroinstructions* — complex instructions that typically take several main memory cycles; these instructions may involve several main memory locations.

A more detailed description of typical instruction formats and examples of each type of instructions follows.

Memory Reference Instructions

A typical word format for the memory reference class of instructions is as follows:

Operation Code	Register	Address Mode	Displacement
4	2	2	8

The number under each field indicates the number of bits in the field. This instruction format is the most difficult to establish since a critical trade-off between three basic parameters must be performed. These parameters are: 1) instruction power, that is, the number of bits in the operation code field; 2) addressing power, that is, the number of bits in the address mode field; and 3) address range, that is, the number of words addressable by the displacement field.

The following text includes a more detailed description of each field. This material is followed by a brief discussion on stack addressing, since it uses memory reference instructions.

Operation Code Field   Since the instructions included in this class reference memory, typical operations involve one memory location and one register. Two exceptions to this rule are typically included in the memory reference class. They are the increment memory location and jump instructions.

Typical operation codes are:

ADD	COMPARE
SUBTRACT	MULTIPLY
LOAD REGISTER	DIVIDE
STORE REGISTER	INCREMENT MEMORY
AND	DECREMENT MEMORY
OR	JUMP
EXCLUSIVE OR	JUMP TO SUBROUTINE

Register Field   One of a number of registers in the register block shown in Figure 4-8 may be selected as one of the operands that is involved in the instruction. This field determines which register is selected. In some mini-computers, only one register is involved in memory reference instructions, and this field is omitted. For those instructions that do not reference a register, that is, increment and jump instructions, this field has no meaning and can be decoded to indicate which instruction is specified.

Address Mode Field   The displacement field can be interpreted in several ways. The address mode field is used to specify how the displacement field is to be used in computing the effective address of the second operand of the instruction. There are a number of different address modes that have been used in minicomputers, and while most of them are described here, no single minicomputer uses all of these modes. Controversy exists among computer designers over the relative merit of each mode, and selection of the set of address modes to be included in a given machine is often one of the most difficult tasks of the designer.

To indicate the address calculation that the address mode specifies, the following definitions are helpful:

- Effective address (EA)—the address that results from the address mode calculation; the EA is the absolute address in memory of the operand.
- Contents of a given memory location (MEMORY [X])—the contents of memory location X, where X represents an integer address.
- Displacement (D)—the contents of the displacement field of the instruction.
- Register contents (REG [1])—the contents of register 1; the value of the program counter is denoted as PC, and the value of the stack pointer is denoted as SP.

Typical address modes used in minicomputers are:

1. *Direct addressing*—the displacement field specifies the absolute address of the operand. No other register is used and no computation performed.

$$EA: = D$$

2. *Indirect addressing*—the displacement field or the result of some address computation gives the memory location that contains the address of the operand.

$$EA: = MEMORY [D]$$

3. *Indexed addressing*—the displacement field is added to or subtracted from an index register, IR, that may be automatically incremented after each operation. Indexed addressing is useful for manipulating arrays or blocks of data.

$$EA: = D + REG [IR]$$

4. *Relative addressing*—the effective address is obtained by adding the displacement field to one of the other registers in the computer. Typical registers that are used are:

- *Program counter*—used in relative jump instructions or to set up parameters to be passed to subroutine.

$$EA: = D + PC$$

- *Base register*—used in programs where the base register BASE contains the starting address of the segment assigned to the data or instruction being referenced.

$$EA: = D + REG [BASE]$$

- *Stack pointer*—used to reference a push-down stack location other than the top of the stack.

$$EA: = D + SP$$

Displacement Field This field gives the displacement from a memory location. The location itself is determined by the address mode used to find the address of the operand. The displacement is either a positive integer or a signed integer. Often the address-mode field must be interpreted to determine if the displacement field is used as a positive integer or a signed integer. For example, in direct addressing the displacement is usually a positive integer while in relative addressing from the program counter the displacement is usually a signed integer.

Address Range The address range of a particular address mode is defined as the number of words in main memory that can be addressed. Since the address range is usually a power of 2, often the range is given as a multiple of 1024 (2^{10}) where K is used to mean 1024. Thus an address range of 4K indicates that the particular address mode has a 12-bit address and can address 4096 words.

The displacement field in the instruction format given previously has 8 bits. This has an absolute range of 256 words. Some method of address range extension is needed to provide a useful number of main memory locations that can be addressed. The register or memory location used to contain the other elements of address calculations are usually the same length as a computer word. Thus, in a 16-bit computer, most other addressing modes have a range of 2^{16} or 64K.

The program counter and other registers contain 16 bits, so the relative address modes can also address up to 64K words.

Stack Addressing The use of last-in, first-out stacks (push-down stacks) can add significant power to the address capability of minicomputers. Instructions that operate with stacks reference either a register or a memory location as one operand and the top of the stack for the other. As data are loaded onto the stack, a stack pointer that indicates the top of the stack is automatically incremented to point to the new top of the stack. Operations that use the top of the stack as an operand automatically increment the stack pointer. Some operations use the top two elements of the stack as operands leaving the result as the top of the stack. The use of stacks allows reentrant subroutines and can reduce the running time of arithmetic operations.

Reentrant subroutines are used to avoid the need for a separate copy of common programs used by several different main programs that are running under interrupt control. As a simple example, assume two interrupt service routines reference the same subroutine (see Figure 4-10). The lower priority interrupt service routine is executing the common subroutine (see Figure 4-11) when the higher priority interrupt occurs. The sample code in Figures 4-10 and 4-11 indicates how the use of a stack simplifies reentrant programming.

The higher priority interrupt occurs when the lower priority service routine is executing SUM. The arrow in Figure 4-11 indicates the instruction being

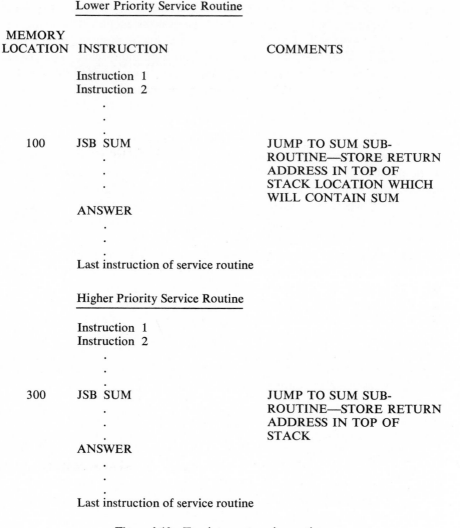

Figure 4-10   Two interrupt service routines

executed when the higher priority interrupt occurs. Figure 4-12 traces the contents of the stack.

The remaining part of the service routine may use the stack, but it returns the stack to the last condition shown before it returns to the lower priority service routine. It is assumed that interrupt return addresses are stored in a way which does not involve the stack.

Another use of stacks is in reducing the number of instructions needed in arithmetic operations. Consider, for example, the computation performed by SUM with addressing not involving a stack as shown in Figure 4-13.

The code in Figure 4-13 uses 14 instructions, while the code for SUM using stack addressing uses 11 instructions; stack addressing results in a 27 percent

THE SUM SUBROUTINE EVALUATES: SUM = AB + CD + EF + GH

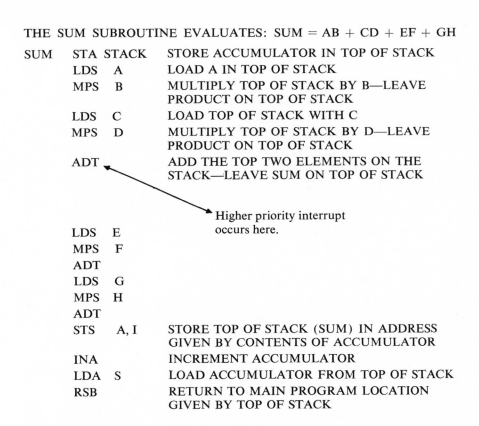

SUM	STA STACK	STORE ACCUMULATOR IN TOP OF STACK
	LDS A	LOAD A IN TOP OF STACK
	MPS B	MULTIPLY TOP OF STACK BY B—LEAVE PRODUCT ON TOP OF STACK
	LDS C	LOAD TOP OF STACK WITH C
	MPS D	MULTIPLY TOP OF STACK BY D—LEAVE PRODUCT ON TOP OF STACK
	ADT	ADD THE TOP TWO ELEMENTS ON THE STACK—LEAVE SUM ON TOP OF STACK

Higher priority interrupt occurs here.

	LDS E	
	MPS F	
	ADT	
	LDS G	
	MPS H	
	ADT	
	STS A, I	STORE TOP OF STACK (SUM) IN ADDRESS GIVEN BY CONTENTS OF ACCUMULATOR
	INA	INCREMENT ACCUMULATOR
	LDA S	LOAD ACCUMULATOR FROM TOP OF STACK
	RSB	RETURN TO MAIN PROGRAM LOCATION GIVEN BY TOP OF STACK

Figure 4-11   Common subroutine

TOS→
| AB + CD |
| ACCUM |
| RET. ADDRESS$_1$ |

CONTENTS OF STACK WHEN HIGHER PRIORITY INTERRUPT OCCURS; THE INTEGER 1 INDICATES FIRST ENTRY INTO SUBROUTINE

TOS→
| RET. ADDRESS$_2$ |
| AB + CD |
| ACCUM |
| RET. ADDRESS$_1$ |

CONTENTS OF STACK WHEN HIGHER PRIORITY SERVICE ROUTINE ENTERS SUM

TOS→
| AB + CD + EF + GH |
| RETURN ADDRESS$_2$ |
| AB + CD |
| ACCUM |
| RET. ADDRESS$_1$ |

CONTENTS OF STACK AFTER SUM FOR HIGHER PRIORITY ROUTINE IS CALCULATED

TOS→
| AB + CD |
| ACCUM |
| RET. ADDRESS$_1$ |

CONTENTS OF STACK AFTER HIGHER PRIORITY SERVICE ROUTINE EXITS SUM

(TOS = TOP OF STACK)

Figure 4-12. Stack Contents

SUM = AB + CD + EF + GH

INSTRUCTION		COMMENT
LDA	A	LOAD ACCUMULATOR WITH A
MPA	B	MULTIPLY ACCUMULATOR BY B
STA	TEM 1	STORE PRODUCT IN TEMP. LOCATION
LDA	C	LOAD ACCUMULATOR WITH C
MPA	D	MULTIPLY ACCUMULATOR BY D
ADA	TEM 1	ADD PRODUCTS AB AND CD
STA	TEM 2	STORE AB + CD IN TEMP. LOCATION
LDA	E	
MPA	F	
ADA	TEM 1	
STA	TEM 1	
LDA	G	
MPA	H	
ADA	TEM 1	

Figure 4-13 Subroutine without a stack

reduction in instruction count. A more thorough discussion of stack machines appears in Chapter 7.

In the interrupt example, the return address for interrupts must be saved in a way that does not involve the stack. One way to accomplish this is to store the return address for interrupts in the first location of the interrupt service routine. The last instruction in the interrupt service routine is an indirect jump instruction using the first location of the service subroutine. Figure 4-14 illustrates this method.

MAIN PROGRAM
1000 —
1001 — ←——— Interrupt occurs while this instruction is being executed
1002 —

INTERRUPT SERVICE ROUTINE
SVS 1002   Contains return address after the interrupt was granted
—

—
JMP SVS, I   Returns to main program at location 1002

Note: This routine assumes that the interrupt is serviced before the same device interrupts again. If this assumption cannot be guaranteed, the service routine must store the return address on the stack and restore it before the exit jump.

Figure 4-14 Saving return addresses

Register Reference Instructions

Register reference instructions operate on a single register. Typical instruction format is as follows:

Class Code	Register	Operation	Parameter
4	3	5	4

The class code specifies that this instruction is a register reference instruction. Other class codes indicate input/output instructions, register-register instructions, and macro instructions.

The register field specifies which register in the register block is being referenced. In the example shown here, up to eight of the registers in the register block may be referenced by this type of instruction.

The operation field is decoded to indicate what function is to be performed on the register. In this example, up to 32 different operations may be specified. Typical operations for register reference instructions are given in Table 4-2.

The parameter field is used as a constant that can be loaded into the register, added to the register, or subtracted from the register; it can also specify a number of times to shift or rotate the register.

Many variations exist in this class of instructions, but the same typical operations are performed.

TABLE 4-2. REGISTER REFERENCE INSTRUCTIONS

Clear register
Complement register
Increment register
Decrement register
Skip if register is 0
Shift register N bits to right
Shift register N bits to left
Rotate register N bits right
Rotate register N bits left
Clear bit N
Set bit N
Complement bit N
Skip if bit N is 0
Skip if bit N is 1
Rotate register and carry N bits right
Rotate register and carry N bits left
Arithmetic shift right N bits
Arithmetic shift left N bits
Long shift right N bits (two adjacent registers lighted)
Long shift left N bits
Load register with N
Add N to register
Subtract N from register

Note: N is given by the parameter field.

Register-Register Instructions

Register-register instructions involve two of the registers in the register block. One method of achieving this class of instructions is to assign the first n main memory locations to refer to the n registers in the register block. Then the operations specified in the memory reference class can also be performed between registers. This method, however, makes the first n main memory locations unaccessible. A more common method of implementing this instruction is shown in the following instruction format:

Class Code	Register 1	Register 2	Operation
4	3	3	6

As with the register reference commands, the class code specifies the instruction as a register-register type. The two register fields give the registers to be operated upon. Note that if both register fields specify the same register, a register reference command results, that is, an instruction that references only one register. The operation field defines the function to be performed on the two registers. With this format, there is a simple implementation using the structure shown in Figure 4-8. Usually the result of the operation replaces one of the operands, say register 2. Then the three addresses in A1, A2, A3 become register 1, register 2, register 2. The FN in Figure 4-8 is derived directly from the operation field for many of the instructions. Typical operations for this class of instructions are given in Table 4-3.

This class of instructions in some minicomputers has been extended to function as memory reference instructions or memory-memory instructions by interpreting the contents of the named register as the address of the operand rather than the operand. In this case the format of the instruction becomes:

Class Code	Mode	Register 1	Mode	Register 2	Operation
2	2	3	2	3	4

The mode field specifies how to interpret the register field. Typical mode specifications are:

1. Direct—use the contents of the register as an operand.
2. Indirect—use the contents of the register as the address of the operand.
3. Indirect, increment—use the contents of the register as the address of the operand and increment after use.
4. Decrement, indirect—decrement the contents of the register and use the contents as the address of the operand after decrementing.

The increment and decrement modes make processing arrays much more

TABLE 4-3. REGISTER-REGISTER INSTRUCTIONS

Operation	Definition
ADD	Add the two registers—result in register 2
AND	Logical AND the registers—result in register 2
OR	Logical OR—result in register 2
XOR	Exclusive OR—result in register 2
COMP	Compare—skip next instruction if equal
MPY	Multiply—result in register 2
DIV	Divide—result in register 2
SHIFT	Linked shift of the two registers
ROTATE	Linked rotate of the two registers
MOVE	Transfer the contents from register 1 to register 2
XCHANG	Exchange the contents of the two registers

efficient. Incrementing the register after its use and decrementing the register before its use simplifies the implementation of stack operation.

Input/Output Instructions

Input/output instructions transfer data between the computer and external devices and control the operations of these devices. A typical instruction format is as follows:

Class Code	Device Select Code	Register	Operation
2	8	3	3

The class code specifies the input/output class of instructions. The device select code identifies which external device is being addressed. The device select code usually is sent directly to an external bus that allows external device controllers to detect that they are being accessed. The operation field specifies the action being taken. Typical external device interfaces contain two flip/flops to indicate the status of the external device. (See Chapter 6 for more detailed descriptions of programmed input/output.) Figure 4-15 gives a simplified device interface.

A typical transfer of data is accomplished via the following sequence:

1. Data is transferred from a processor register to the device register by gating the data onto the data bus and the desired external device selector code onto the select code bus.
2. The busy flip/flop is set to signal to the external device that data is ready.
3. The device accepts and processes the data, and sets the done flip/flop.
4. The device control logic clears the busy flip/flop and the processor can send more data.

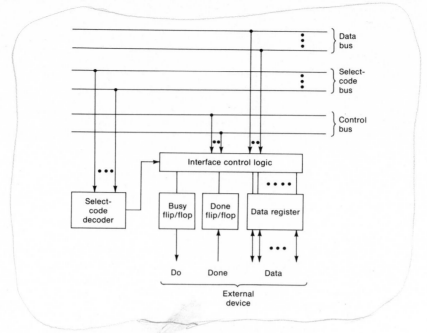

Figure 4-15   Input/output device interface

To perform these functions, the typical input/output instructions are as given in Table 4-4.

TABLE 4-4. INPUT/OUTPUT INSTRUCTIONS

Operation	Definition
LIR	Load data into register
MIR	Inclusive OR data into register
OTR	Output data from register
STB	Set busy flip/flop
SBS	Skip next instruction if busy flip/flop is set
STD	Set done flip/flop
SDS	Skip next instruction if done flip/flop is set
CLB	Clear busy flip/flop
CLD	Clear done flip/flop

Macroinstructions

Macroinstructions are the more complex instructions that are typically implemented using a read-only control memory. They usually take several main memory cycles and may involve several registers and memory locations. Typically they have a one-word format for those operations relating to the registers and a two-word format for those operations that reference memory. The typical one-word format is:

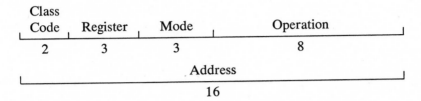

Code Class	Register 1	Register 2	Operation
2	3	3	8

The typical two-word format is:

Class Code	Register	Mode	Operation
2	3	3	8
Address			
16			

The various fields are interpreted in essentially the same way as their corresponding fields are interpreted in other instructions. Some typical macroinstructions are given in Table 4-5.

TABLE 4-5. MACROINSTRUCTIONS

Operation	Definition
One-Word Format	
FLT	Convert the number contained in the two registers to a floating-point number
FIX	Convert the number contained in the two registers to a fixed-point number
BCD	Convert the number contained in the two registers to binary
Two-Word Format	
FMPY	Floating-point multiply
FADD	Floating-point add
FSUB	Floating-point subtract
FDIV	Floating-point divide
FLOAD	Load two registers with contents of successive memory locations
FSTORE	Transfer contents of two registers to successive memory locations

[handwritten annotation: reference memory]

Overview of Basic Instruction Structure

During the preceding discussion we have indicated that all the instructions except the memory reference instructions are specified by a class code. How does the minicomputer determine that a memory reference instruction is specified? There are two methods used to separate instruction types:

1. Reserved operation code method
2. Class code method

The reserved operation code method reserves some of the operation codes of the memory reference class to specify the other classes. This is the method used in the typical instruction formats given previously to separate instruction types. Table 4-6 summarizes the instruction formats using the reserved operation code method.

TABLE 4-6. RESERVED OPERATION CODE METHOD

Format				Instruction Class
Operation Code	Register	Mode	Displacement	Memory reference (Operation code 0000 not used)
Class Code 0000	00	Device Select Code	Register | Operation	Input/output
Class Code 0000	01	Register 1 | Register 2	Operation	Register-register
Class Code 0000	10	Register | Operation | Parameter		Register reference
Class Code 0000	11	Register 1 | Register 2	Operation	Macroinstruction

The class code method uses a separate fixed-length class code field to specify the different instruction types. This uses some of the bits in the memory reference instruction to specify its class, but uses fewer bits in the other instruction types. This method usually results in a poor balance between the memory reference class and the other instruction types. An example of the class code method is shown in Table 4-7.

TABLE 4-7. CLASS CODE METHOD

Format					Instruction Class
Class Code 00	Operation Code	Register | Mode	Displacement		Memory reference
Class Code 01	Operation Code	Mode | Register 1	Mode | Register 2		Register-register
Class Code 10	Operation Code	Register	Operation Code | Parameter		Register reference
Class Code 11	Operation Code	Device Select Code			Input/output

4-5. INPUT/OUTPUT STRUCTURE

Chapter 6 gives a detailed discussion of input/output systems, both for minicomputers and larger computing systems. In this section the input/output problem and solution for the minicomputer is summarized and the reader is referred to Chapter 6 for more details.

In minicomputer systems, performance is usually sacrificed to obtain lower cost. The system designer has a trade-off for performance versus cost in the input/output system. In this system a minimal hardware interface with external devices can be used with either many of the functions of the device controller performed via software or a more expensive high performance controller. In most cases, minicomputer designers place a greater burden on software implementation of controller functions. Perhaps the only exception to this rule is in the interrupt system provided by minicomputer designers. Since many of the applications of minicomputers are in real-time systems, interrupt response time is an important parameter in evaluating minicomputers.

There are three methods presently used to transfer data from peripheral devices and the main memory. These are:

1. Programmed input/output
2. Direct memory access
3. Interrupt-processed input/output

These methods are defined earlier in this chapter and are described in detail in Chapter 6. A short classification of interrupt systems and the relative merits of each type of interrupt system is presented here. There are basically four types of interrupt systems:

1. Single-priority, polled
2. Single-priority, selective
3. Multi-priority, polled
4. Multi-priority, selective

Single-Priority, Polled Interrupt Systems

In a single-priority, polled minicomputer system, the interrupt request input is the INCLUSIVE OR of all of the interrupt requests of the external devices. No hardware assistance is provided in determining which device is requesting service, or, in the case of simultaneous interrupts, which device is most important. These latter tasks must be determined via software by polling the devices with instructions that sample each device individually. While this saves in hardware cost, system performance is greatly degraded since a high system overhead is incurred in performing the device recognition and priority determination functions.

Single-Priority, Selective Interrupt Systems

In selective minicomputer interrupt systems, the interrupt system also has an INCLUSIVE OR for all device interrupt requests; in addition, a device select code bus is included to allow the device requesting service to identify itself. If another device requests service before the first request is granted, the processor receives the interrupt request, but the second device is inhibited from placing its select code on the bus until acknowledgement is sent to the first device indicating that the processor has its request. After all requests have been received, the processor determines priority and starts to service the device with highest priority. As in the first case, if another device of lower priority requests service, the higher priority service routine is interrupted until the processor determines the priority of the new request. System overhead is greatly reduced, but high priority devices may still be interrupted for short periods by low priority devices and thus degrade overall system performance.

Multi-Priority, Polled Interrupt Systems

In a multi-priority, polled interrupt system, the device interrupts are combined via an INCLUSIVE OR gate. The devices are arranged in an a priori priority arrangement with the highest priority devices "closest" to the processor. The lower priority devices must be inhibited from interrupting while higher priority devices are being serviced. One method is to use an interrupt request chain that passes through each device interface. If a device is granted an interrupt it breaks the chain and this inhibits lower priority devices from requesting an interrupt. This is shown in Figure 4-16. When an interrupt is posted, the processor must poll the devices to determine which device caused the interrupt.

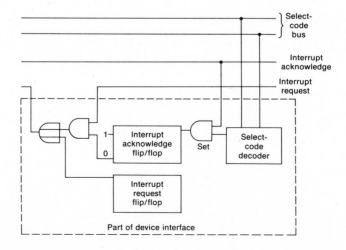

Figure 4-16 Device interface for multi-priority interrupt system

Multi-Priority, Selective Interrupt System

This is the most powerful interrupt system used in minicomputer systems. Its implementation is essentially a combination of the two previously described systems. The external devices are arranged in order of their priority along the input/output bus. The priority chain is as shown in Figure 4-16. When the interrupt acknowledge flip/flop is set, the acknowledged device puts its select code on a select code input bus to indicate to the processor which device caused the interrupt.

Summary

Table 4-8 summarizes the relative merits of each interrupt system.

TABLE 4-8. INTERRUPT SYSTEM SUMMARY

	Response Time	System Overhead	Maximum Data Rate	Cost
Single-priority, Polled	Slow	High	Low	Low
Single-priority, Selective	Fast	High	Moderate	Moderate
Multi-priority, Polled	Slow	Moderate	Low	Low
Multi-priority, Selective	Fastest	Low	High	High

4-6. SOME APPLICATIONS OF MINICOMPUTERS

The rapid growth of the minicomputer market is due to the many different applications that can benefit from dedicated real-time computation. A typical example is an instrumentation system which is designed to characterize a device or subsystem. Such a system is shown in Figure 4-17.

Typical applications for a system of this type range from production testing of printed circuit boards to checkout of the complete electronic system of a fighter aircraft on the flight line. In the printed circuit board testing application, programmable signal sources simulate the inputs normally supplied to the device from the other parts of the system. Under program control, the parameters of the signal sources such as level, frequency, and rise time are varied to determine sensitivity to parameter variations. For complete printed circuit boards, manual parameter sensitivity testing is costly and is normally not performed. When the device does not give the correct response, the computer enters into a diagnostic mode to provide the test technician with enough data to locate the faulty components. Upon completion of the testing, a report is printed giving the major test results that can be used as production records. These results are also stored and can be analyzed by the computer to give statistical trends to help production management control production quality.

To assist in writing programs to control the system and analyze the data, an extension of the BASIC programming language is used. Commands are added

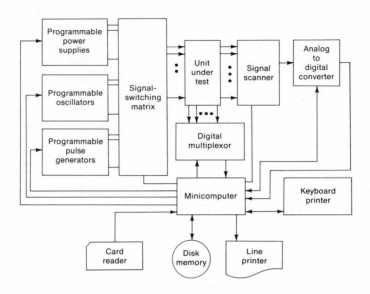

Figure 4-17   Real-time instrumentation system

that specify a given signal type with certain parameters to be applied to a given input pin on the printed circuit board. The BASIC compiler generates the necessary instructions to set up the programmable signal sources and the signal switching matrix to accomplish the application of the desired signal to the specified input pin. A sample command to set up a measurement is shown in Figure 4-18.

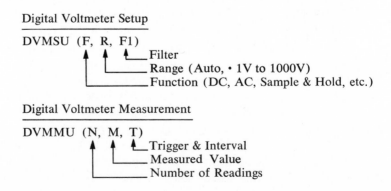

Figure 4-18   Measurement commands

4-7. PROBLEMS

4-1. For a 16-bit general purpose computer of your design:

a) List the types of instructions and give basic instruction formats.

b) For the input/output type of instructions, give a complete list and coding of those instructions.

c) Give a block diagram of a typical input/output interface.

4-2. Draw a block diagram of a microprogrammed machine and give the microcode to implement the input/output instructions given in the answer to problem 4-1.

4-3. A minicomputer uses a push-down stack to provide for reentrant programs. The computer's compiler uses a left-to-right scan for arithmetic functions. The computer's command list includes the following operations between the top of the stack and memory:

LDS A —Load the top of the stack with the contents of memory location A and push the previous contents of the stack down.

STS B —Store the top of the stack into memory location B and pull the remaining contents of the stack up.

ADS C —Add the contents of memory location C to the top of the stack.

SBS D —Subtract the contents of memory location D from the top of the stack.

MPS E —Multiply the top of the stack by the contents of memory location E.

DVS F —Divide the top of the stack by the contents of memory location F.

The computer also has the following commands which operate on the top two elements in the stack and leave the result in the top of the stack; the remaining elements are pulled up after the arithmetic operation is complete.

ADT, SBT, MPT, DVT

For the equation

$$Y = A[(BC + EF) + 6]$$

a) Give the coding to evaluate the equation and indicate the contents of the stack after each operation.

b) If the coding given in a) is reentered three times, what is the minimum number of words on the stack that must be reserved?

4-4. A computer has the following memory reference instruction format:

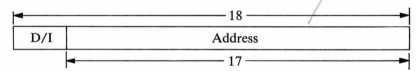

The address is a two's complement signed number.

The address modes are as follows:

    D/I—Direct or indirect
    MODE
        0—Absolute
        1—Relative to data base register
        2—Relative to program counter
        3—Immediate
    X
        0—No indexing
        1—Relative to IR-1
        2—Relative to IR-2
        3—Relative to IR-3

The registers have the following lengths:

    Index register—8 bits
    Program counter—16 bits
    Base register—14 bits

The indirect address is interpreted as follows:

D/I	Address

For each of the following address modes give the equation to compute the effective address and sketch a memory map to give the number of words and relative location that can be addressed:

  a) Direct, absolute, no indexing
  b) Direct, absolute, indexed
  c) Direct, relative to program counter, no indexing
  d) Direct, relative to program counter, indexed
  e) Direct, relative to data base register, no indexing
  f) Direct, relative to data base register, indexed
  g) Direct, immediate, no indexing

h) Direct, immediate, indexing
i) Indirect, absolute, no indexing
j) Indirect, absolute, indexed
k) Indirect, relative to program counter, no indexing
l) Indirect, relative to program counter, indexed
m) Indirect, relative to data base register, no indexing
n) Indirect, relative to data base register, indexed
o) Indirect, immediate, no indexing
p) Indirect, immediate, indexed

For the immediate modes, give the range of the operand that can be addressed. Use the following symbols

EA	—Effective address which results after address computation
A	—Address given in instruction
DB	—Data base register contents
PC	—Program counter contents
IR	—Index register contents
MEMORY [N]	—Contents of memory location N

4-5. Two computers A and B have the following interrupt systems:

Computer A—When an interrupt occurs, computer A completes the next instruction in the current program; it then stores the program counter contents in memory location 0 and executes a JUMP INDIRECT with memory location 1 containing the indirect address.

Computer B—When an interrupt occurs, computer B completes the current memory cycle; it then executes the contents of the memory location given by the device select code. For multicycle instructions, the program counter contains the location of instructions whose execution was interrupted. For single cycle instructions, the program counter contains the location of the next instruction to be executed.

For each computer, determine the minimum interrupt response time (the time from the occurrence of an interrupt until the interrupt service routine is started) and maximum data transfer rate under interrupt processing. Assume that direct and indirect addressing capabilities and the following instructions are included in the computer's instruction set:

JSB N   —Jump subroutine, stores contents of the program counter in memory location N and starts execution at location N + 1.

JMP   —Unconditional jump to location N.

LIA N$^{H/C}$ —Load into accumulator contents of device N data and either hold or clear the device flag.

OTA N$^{H/C}$—Output to device N the contents of the accumulator and either hold or clear the device flag.

STC N   —Set control bit and start device action.

CLC N   —Clear control bit and idle device.

SFS N   —Skip if flag on device N is set.

SFC N   —Skip if flag on device N is clear.

LDA N   —Load accumulator from memory location N.

STA N   —Store accumulator into memory location N.

4-6. Draw a block diagram for three different bus structures. For each list its advantages and disadvantages.

4-7. Draw a block diagram of a microprogrammed processor and give a format for the control memory instructions.

4-8. A computer has the following instruction formats:

Memory Reference

D/I	Operation Code	Mode	Address
1	4	3	8

The address modes are:

000—Direct/indirect
001—Relative to base register
010—Relative to program counter
011—Immediate, data is in address field
100—Index register 1, automatically increment
101—Index register 2, automatically increment
110—Index register 1, no change
111—Index register 2, no change

The operation codes are:

ADA—Add to register A
ADB—Add to register B
ISZ —Increment, test memory, skip if 0
JMP —Jump to effective address (EA)
JSB —Save program counter in EA, jump to EA + 1
LDA —Load register A
LDB —Load register B
STA —Store register A
STB —Store register B

CPA —Compare register A to EA, skip if not equal
LXA —Load index-register 1
LXB —Load index-register 2
SXA —Store index-register 1
SXB —Store index-register 2

Register Reference

Class Code	Register	Operation
5	2	9

The operations are:

CLR —Clear register
CMR—Complement register
SRR —Shift right
SLR —Shift left
RRR —Rotate right
RRL —Rotate left
SRZ —Skip if register 0
SLZ —Skip if LSB is 0
SMZ —Skip if MSB is 0
SRN —Skip if register is not 0
SLN —Skip if LSB is not 0
SMN —Skip if MSB is not 0

Input/Output

Class Code	Register	Operation Code
5	2	9

The operation codes are:

LIA —Load input/output data into A
LIB —Load input/output data into B
OTA—Output register A to input/output register
OTB—Output register A to input/output register
STF —Set flag on input/output device
CLF —Clear flag on input/output device
SFZ —Skip if input/output flag is 0
SFN —Skip if input/output flag is not 0
STC —Set control bit
CLC—Clear control bit
SCZ —Skip if control bit is 0
SCN —Skip if control bit is not 0

Notes: The computer sets control bit to start device or signal that data is ready for device.

The device sets flag to interrupt computer or signal that data is ready for computer.

a) Write an input/output service routine which:
   1) Saves the contents of the registers used.
   2) Read an 80-character record from device (8-bit characters).
   3) Tests for an alarm condition and branches to subroutine if alarm character 11110000 is detected.
   4) Compares each character for a match in a table of 20 characters. Store in Table 1 if a match is found, in Table 2 if no match.
   5) Restores registers.

b) Write a version of the program and determine the execution time in cycles for the following addressing modes:
   1) Direct/indirect only.
   2) Direct/indirect, relative to a base register, and relative to the program counter.
   3) Direct/indexed only.
   4) Any combination which gives the shortest execution time.

c) For each program determine the number of memory locations used and the execution time in cycles.

CHAPTER 5

Memory and Storage

Richard E. Matick

The architecture of memory and storage is so vast and poorly defined that a general approach would bypass the underlying reasons for the seemingly complex systems encountered in practice. For that reason, this chapter is aimed primarily at providing the student with a fundamental understanding of the various storage devices, their technology, and organizations which are the basic principles required for dealing with storage architecture in the real world. This chapter, for the most part, attempts to show how the organization of various memory systems evolves from the fundamental requirement of storage and retrieval. In some cases the details of device hardware are included in order to provide a more complete understanding. The terms *storage* and *memory* are often used interchangeably to mean the online storage system.

5-1. PROCESSOR VERSUS MEMORY SPEED

All mathematical computations, either mental, mechanical, or electronic, require a storage system of some kind, whether it be numbers written (stored) on paper, in our brain, on the mechanical cogs of a gear, as holes in paper, as electronic circuits, or any other. In fact, the minimum storage requirements of any computing or calculating system are:

1. An internal storage capability for temporarily holding the numbers to be processed, the intermediate results, and the final answer
2. External storage for holding the input numbers to be processed
3. External storage for permanent (or semi-permanent) recording of answers for further use

In some cases, 1 and 3 or 2 and 3 are the same media but not always. The power of a computer or calculator is dependent to a large extent on the size and speed of its associated storage capabilities, both internal and external. Thus, it is not surprising that the development of storage has played a significant role in the development of calculators and computers.

The major problems of computer design and architecture have shifted away from the central processor per se toward other areas in which memory and storage are a significant part. In the early years of computers, the processor and storage were the main preoccupation of hardware designers. They were faced with improving both the raw speed of the central processing unit as well as the speed of memory since there was little difference between these two. In addition, there was little memory needed or available even on large systems. For instance, the first large scientific computer, the IBM 704, had a basic machine cycle time of $12\mu s$ and a main memory cycle time of $12\mu s$. The cycle time was dictated by the main memory cycle time and the basic logic and memory device speeds were comparable. As technology progressed, logic speed increased greatly while memory speed at reasonable costs increased much less relative to logic until there existed a large gap measured in orders of magnitude between logic delay and memory cycle time. For example, the **IBM System 360 Model 195** has a logic delay of 5ns per stage, a basic machine cycle time of 54ns and a main memory (magnetic cores) of $0.756\mu s$ cycle time;* the CDC 7600 computer has a basic cycle time of 27.5ns and a main memory of $0.275\mu s$. Faster main memories in the range of $0.2\mu s$ have been built from other technologies such as plated wires and magnetic thin films but still are rather expensive.

Thus, processor cycle times and logic delays have improved dramatically while main memory has improved at a much slower rate, leaving a significant gap. This discrepancy in speed can be seen in Table 5-1 where some of the important parameters are plotted for several of the larger commercial computers. A comparison of the two columns Main Memory Cycle Time and Processor Cycle Time shows the latter improving much more so than the former. In fact, the difficulty with main memory can be seen even more dramatically by plotting main memory cycle time versus year as in Figure 5-1. This shows that main memory cycle time has reached a lower limit and represents the current status with respect to ferrite core technology. It is, of course, possible to further increase the speed of main memory but the difficulties arise in two ways—first, it is necessary to go to new technologies which are more expensive compared to the 20-year-old, well-developed core technology; second, the size of main memory required on a given system has increased dramatically (see Table 5-1, third column) which in itself tends to increase the cycle time just by increasing the array propagation delay.

---

*The gap between logic and main memory is bridged by a small cache memory paged out of main memory; see Section 5-11.

TABLE 5-1. PROCESSOR AND MEMORY PARAMETERS FOR VARIOUS COMMERCIAL COMPUTERS

Computer	Year Delivered	Main Memory Capacity	Word Size (bits)	Main Memory Cycle Time (μs)	Processor Cycle Time (μs)	Address Size (bits)
IBM 650 (drum main)	1954	1-2 K words	60 5 bits/digit	4.8 max. rotation	7.8 μs/pulse but serial by digit	20 (4 decimal digits)
IBM 704	1955	4-32 K' words	36	12	12	15
IBM 7090	1960	32 K' words	36	2.2	2.2	15
IBM 7030 Stretch	1961	16-256 K' words	64	2.1	0.6	18
CDC 6600	1964	32-128 K' words	60	1.0	0.1	18
Univac 1108	1965	64-256 K' words	36 (6 Char.)	0.75	0.125	18
IBM 360/75	1965	256-1024 K' Bytes	64 (8 Bytes)	0.75	0.195	24
RCA Spectra 70/55	1966	64-512 K' Bytes	32	0.84	—	16
IBM 360/85	1969	512-4096 K' Bytes + 16-32 K' cache	128	0.96 (0.08 cache)	0.08	24
CDC 7600	1969	64-512 K' words	60	0.275	0.0275	30
IBM 360/195	1971	1-4 M' Bytes + 32 K' cache	128	0.756 (cores) 0.054 (cache) 0.162 effective storage	0.054	24
Burroughs B7700	1972	128-1024 K' words	48	1.5	0.0625	20
Univac 1110	1972	131-1024 K' words 32-256 K' plated wire	36 (6 Char.)	1.5 (core) .52 write, .38 read, plated wire	0.075	24

where K = 1000, K' = 1024, M' = K'K'

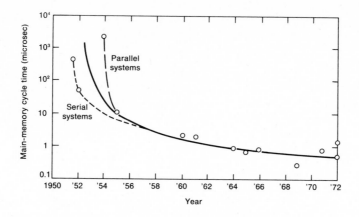

Figure 5-1 Main memory cycle time versus year

Technology is faced with the need to decrease the cost per bit while increasing speed or decreasing cycle time, a rather formidable task. This chapter attempts to put the various storage technologies into perspective to show the basic principles, the fundamental trade-offs between speed, size, and cost (in terms of transducer or circuit count, not absolute) and some of the fundamental retrieval difficulties that arise in the process of storing and retrieving information in nonrandom storage systems [see also Matick, 76].

5-2. TYPES OF MEMORY SYSTEMS AND COST/PERFORMANCE GAPS

While there has often been much debate and speculation about what kind of storage is needed for practical computers, it should be understood that with few exceptions, the fundamental requirements for computer applications of all types can be reduced simply to large, random-access, writable, high-speed (both read and write) memory systems at very low cost. (Small associative memories are useful in "paged" memory hierarchies as discussed in section 5-11). While the size of main memory attached to computing systems has greatly increased in capacity along with a substantial decrease in cost per bit, nevertheless, the storage requirements for most systems still are so large so as to make the use of only main memory technology much too expensive. The appearance of a variety of hardware as well as software systems, storage allocation, data management, buffering, paging, and so on is simply a result of the fact that in most cases, the trade-offs between cost, speed, and size can be made more attractive by combining various hardware systems coupled with these special features. Thus, technology and systems requirements have together produced a variety of memory types. While the definitions of these various types of memories are somewhat arbitrary, there are five classes that are readily identifiable and frequently encountered in practice, namely:

1. Random access
2. Direct access
3. Sequential
4. Associative
5. Read only (postable and nonpostable)

Classes of Memory Systems

We shall describe each class and then proceed to give some rules of thumb for the access time and the cost of various systems relative to one another.

1. *Random-access memory* — a memory for which any location (word, bit, byte, record) of relatively small size has a unique, physically wired-in addressing mechanism and is retrieved in one memory cycle time interval. The time to retrieve any given location is made to be the same for all locations.

2. *Direct-access storage* — a storage system for which any location (word, record, and so on) is not physically wired-in and addressing is accomplished by a combination of direct access to reach a general vicinity plus sequential searching, counting, or waiting to find the final location. The access time depends on the physical location of the record at any given time; thus access time can vary considerably both from record to record, as well as to a given record when accessed at a different time. Since addressing is not wired-in, the storage media must contain a certain amount of information to assist in the location of the desired data. This will be referred to as *stored addressing information* throughout this chapter.

3. *Sequential access storage* — a memory for which the stored words or records do not have a unique address and are stored and retrieved entirely sequentially. Stored addressing information in the form of simple interrecord gaps is used to separate records and assist in retrieval. Access time varies with the record being accessed as with direct access, however, sequential accessing may require searching of every record in the memory before the correct one is located.

4. *Associative (content-addressable) memory* — a random access type of memory which in addition to having a conventional wired-in addressing mechanism also has wired-in logic that enables one to make a comparison of desired bit locations for a specified match and do this for all words simultaneously during one memory cycle time. Thus, the specific address of a desired word need not be known since only a portion of its contents can be used to access the word. All words that

match the specified bit locations are flagged and can then be addressed on subsequent memory cycles.

5. *Read-only memory (ROM)* — a memory that has permanently stored information programmed during the manufacturing process and can only be read and never destroyed. There are several variations of ROM. Postable or programmable ROM is one for which the stored information need not be written in during the manufacturing process but can be written at any time, even while the system is in use, that is, can be posted at any time. However, once written, the media cannot be erased and rewritten, that is, read-only after written. Another variation is a fast-read, slow-write memory for which writing is an order of magnitude slower than reading. In one such case, the writing is done much as in random-access memory but very slowly to permit use of low cost devices. Another version of slow-write memory is one with a changeable or replaceable storage medium, for example, magnets on a card, wires or metal plates (capacitors) punched with holes. These are read-only memories which are programmable at any time but require considerable time (minutes to hours) to change.

Cost and Access Time Comparisons

The primary reason for the large variety of memories is cost and cost is related to the memory access time. A short access time can only be obtained at a high cost and conversely, inexpensive memories have slower access times. Approximate rules of thumb for cost and access time comparisons of specific memory and storage are as follows (where B = bytes):

Cost or Price

$$\text{Cache } \text{¢/B} = 10 \times \text{Main ¢/B} = 10^4 \times \text{Disk ¢/B} = 10^7 \times \text{Tape ¢/B}$$

$$\text{(Off line)} \tag{5-1}$$

$$\text{Gap} \approx 10^3 \qquad \approx 10^5 \times \text{Tape ¢/B}$$

$$\text{(Online)}$$

Access Time

$$\text{Cache } T_c = 10^{-1} \text{ Main } T_m = 10^{-6} \text{ Disk } T_d = 10^{-9} \text{ Tape } T_t$$

$$\tag{5-2}$$

$$\text{Gap} \approx 10^5 \qquad \text{Gap} \approx 10^3$$

We see some large gaps between main memory and disks as well as disk and tapes. These large gaps in cost can only be brought about by large gaps in access time as indicated in Equation 5-2. The access time is sacrificed in order to achieve economy. This is an inherent characteristic that can only be understood by considering the physical system requirements for storing and retrieving information. This is covered in Section 5-3 where it is shown that addressing small pieces of information at high speed requires large numbers of transducers (decoders, drivers, sense amplifiers) and therefore is expensive. The cost can be lowered by sharing transducers which necessitates a slower system. Thus, we shall be faced continually with the trade-offs between cost and speed which, along with size, provides a spectrum of memory systems to meet increasing storage requirements.

Storage System Parameters

In any storage system, the most important parameters are the capacity of a given module, the access time to find any piece of stored information, the data rate at which the stored information can be read out (once found), the cycle time (how frequently the system can be accessed for new information), and the cost to implement all these functions.

The *capacity* is simply the maximum number of bits, bytes, or words one can assemble in one basic operating module which is totally self contained, for example, 4K bytes of core memory, 29M bytes on a disk pack. Access time can vary depending on definition; the definition used here is different for random and nonrandom access storage.

For random-access memory, the *access time* is the time span from the instant a request appears in an address register until the desired information appears in the output buffer register or proper location in memory where it can now be further processed. For nonrandom-access memory, the access time is the time span from the instant an instruction is decoded asking for nonrandom-access memory until the desired information is found but not read. Thus, access time is a different quantity for random- and nonrandom-access memory. In fact, it is the access time that distinguishes the two as is evident by the definitions above. Access time is made constant on random-access memory whereas on nonrandom storage, access time can and does vary substantially depending on the location of information being sought and the current position of the storage system relative to that information.

Data rate is the rate, usually bits per second, bytes per second, or words per second, at which data can be read out of a storage device. Data transfer time for reading or writing equals the product of the data rate and the size of the information being transferred. Data rate is usually associated with nonrandom-access memory where large pieces of information are stored and read serially. Since an entire word is read out of random-access memory in parallel, data rate has no significance for such memories. Data rate is a constant for a given system but the data transfer time depends on the length of the data.

Cycle time is the rate at which a memory can be accessed, that is, the number of accesses per unit time and is applicable primarily to random-access storage. It does not necessarily equal the access time for various reasons. If a random-access memory works in the destructive read out mode, the information must be regenerated before another access can be made.

This will cause a wide disparity between access and cycle time. Even if non-destructive read out is used, there are quite often transients which must be allowed to die out. For example, drivers or sense amplifiers must recover from large transients that drive them into saturation; ringing, which is caused by multiple pulse reflections on the array line, must be allowed to die out, and so on. Thus, cycle time is often substantially larger than access time. Cycle time has little meaning on nonrandom serial storage; cycle time is essentially the access time plus data transfer time, both of which can vary widely on a given storage system as a function of time and of data being accessed.

5-3. FUNDAMENTAL SYSTEM REQUIREMENTS FOR STORING AND RETRIEVING INFORMATION

In order to be able to store information and subsequently find and retrieve it, a memory system must have the following four basic requirements:

1. Media for storing energy
2. Energy source for writing the information, that is, write transducers on word and bit lines
3. Energy sources and sensors to read, that is, read and sense transducers
4. Information addressing capability, that is, address selection mechanism for reading and writing

The fourth requirement implicitly includes some coincidence mechanisms somewhere within the memory to bring the necessary energy to the proper position on the media for writing, and likewise, a coincidence mechanism for associating the sensed information with the proper location during reading. In random-access memory, this coincidence is provided by the coincidence of electrical pulses within the storage cell whereas in nonrandom-access storage, it is commonly provided by the coincidence of an electrical signal with mechanical position. In many cases, the write energy source serves as the read energy source as well, thus leaving only sense transducers for the third requirement. Nevertheless a read energy source is still a basic requirement.

The differences between all memory systems lies only in the manner by which the preceding four requirements are implemented and more specifically, in the number of transducers that are required to achieve these necessary functions. Here a *transducer* denotes any type of device (for example, magnetic head, laser, transistor circuits) that generates the necessary energies for reading and writing, senses stored energy and generates a sense signal, or provides the decoding for address selection. We will now consider some of the similarities and differences between various memory systems based on the four requirements.

In all storage systems in use today, with the possible exception of holographic systems which are mainly experimental, all storage is accomplished by means of storing some discrete quantity such as magnetic moments, current, conduction paths, and others, so the physical or mechanistic storage attributes of the media are not a distinguishing feature separating main memory from other storage. In fact, the same physical phenomena such as magnetism can be and has been used for both. The magnetic phenomenon of magnetic recording systems is identical in principle, although different in detail from that used in ferrite cores, flat film, and plated wire main memories. The fundamental difference between main memory and other storage lies in the second, third, and fourth requirement, namely in the writing, sensing, and addressing mechanisms as shown in Figure 5-2.

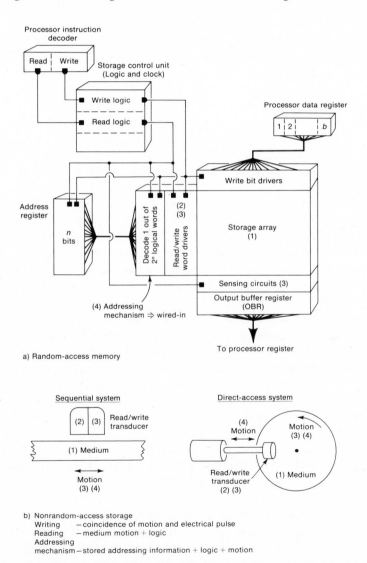

Figure 5-2   Differences in writing, reading, and addressing mechanisms for random- and nonrandom-access storage systems

In order to achieve a high-speed main memory system, it is necessary for each bit location to be electrically "wired" in order to receive (writing) and send (sensing) energy locally. The storage medium is stationary with all writing and reading transducers hard-wired to the memory. Thus, the read/write transducers cannot be shared but rather are on constant alert to serve the bits designated in the hard-wired design. This is essential in order to provide high speed reading and writing. Since these transducers are expensive, they constitute a substantial part of the memory cost. This is considered in detail in Section 5-5.

In contrast to this, the read/write transducers in nonrandom-access systems are generally shared over a large number of bits, thus greatly reducing the cost per bit. This is precisely what is done in disk, tape, and drums, where usually one read/write head assembly, sometimes having heads arranged in physical clusters, is used to reduce cost, and the storage media is moved to provide access to large areas of storage. This results in substantial cost savings but also greatly increased access time. The degradation in access time for such storage is primarily a result of the mechanical nature involved in moving the media (and the slider head assembly in disk technology). If it were possible to design mechanical systems with time constants comparable to electronic systems, this access time limitation could be removed. This does not appear to be possible.

Just from these simple considerations of sharing transducers, it becomes apparent that mass storage will tend to be cheaper than main memory, indicating storage systems are likely to be around for some time. Even if main memory were reduced to the cost of mass storage, there would still be problems in data organization which results from the way people use data. Suppose we have a large file such as an inventory part number file—how do we organize this for best access? Since such files can and do change, we need a way to delete records, increase record size, and cross correlate between various pieces of data in a file which requires additional information stored along with the data. Since there is no way to determine such cross correlations and changes ahead of time, the physical, logical, and addressing means must allow for this and in a very general way. Thus, it is becoming more evident that a spectrum of storage systems is a natural consequence of storage requirements and will become increasingly more evident.

5-4. FUNDAMENTAL REQUIREMENTS FOR A REVERSIBLE BINARY STORAGE MEDIA

In order to construct a binary memory system of any type, it is first necessary to have a storage media (first requirement described previously) to store energy in terms of some discrete physical quantity, such as magnetic moment or circulating current, that is a symbolic representation of the two binary states of a bit of information. In order to accomplish this, a potential medium for binary storage must have at least:

1. Two stable (or semi-stable) energy states separated by a high energy barrier (Figure 5-3)
2. Capability of switching between these two stable states an infinite number of times by the application of external energy
3. Capability for sensing the two energy states with an external energy source
4. Energy losses during the writing process for reliable storage [Swanson, 60; Landauer, 61; Landauer, 62; Freiser and Marcus, 69]

The two energy states mentioned in item 1 do not necessarily have to be stable for an infinite time; in fact it is only necessary that they remain stable with time constants much larger than the time needed to "refresh" or rewrite the information. Many such storage systems have been built this way, see Section 5-10.

In order to implement this storage medium in a system, the additional requirements 2, 3, and 4 in Section 5-3 are necessary. These additional three requirements of the system can be fashioned around the media in numerous ways and the different ways chosen represent another fundamental distinction between various memory systems.

In random-access memory, a considerable amount of the four system requirements is contained within the medium or basic storage cell itself, whereas in nonrandom-access memory, very little is contained within the medium or cell and hence must be provided by other parts of the system. Thus in random-access storage, the cell is rather complex, more difficult to fabricate and more costly. This cost, added to the substantial "stand-by" transducers needed, yields an array that is high in cost but also fast and easily addressed. This complexity is duplicated many times within each cell and spread over a large

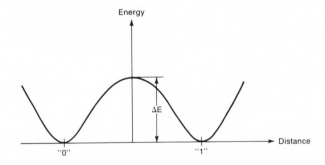

Figure 5-3 Bistable potential well for storing binary information

array. While the array is more complex internally, from an external point of view it is quite simple, requiring very little external components for actual operation; for example, see a semiconductor memory module (as shown later in Figure 5-11). On the other hand, nonrandom-access memory cells are relatively simple and hence cheap but the remainder of the system must be more complex. However, this complexity is not duplicated and the cost per bit can be reduced by having the additional complexity serve many inexpensive cells.

The device requirement for energy losses mentioned previously does not require losses in the quiescent or steady state since magnetic cores and tapes have no losses until the magnetization is reversed. However, for reliable writing of information, the storage medium must exhibit a certain fundamental amount of losses during the writing process and this represents a conflict between fundamental and practical constraints. Energy losses produce heat which often entails serious cooling problems in memory design. Hence one strives for cells or media with as little heat loss as possible while realizing the heat loss can never be totally eliminated. The exact value for this fundamental limit has not been amenable to calculation from first principles but it is known that losses in practical devices are still orders of magnitude larger than any fundamental limit [Keyes, 69; Keyes, 70].

5-5. RANDOM-ACCESS MEMORY ORGANIZATION AND TRANSDUCER COUNT

While technically speaking, randomly accessed memory can refer to a number of specific memory types such as cache, read-only, and associative memory, we shall concentrate here primarily on main memory and more specifically on ferrite cores and transistor memories. This section discusses primarily the 2-dimensional or 2D, 2½-dimensional or 2½D, and 3-dimensional or 3D organizations and implementations of random-access memory.

The 2D and 3D organizations are derived from the number of functional terminals or degrees of freedom provided by the basic storage cell (Section 5-6). The former is the most simple and the latter the most complex organization with 2½D falling in-between. Random-access memory is organized into words with a given number of bits per word, b. During a read cycle, one entire word is fetched. Hence the reading process must select a word and sense each bit of that word. During writing, individual bits of a particular word must be selected to store the desired data. Hence the writing process requires simultaneous selection at the word and bit level. The various organizations simply reflect the physical interconnections between the storage cells in the array as well as between the array and the external circuitry required to accomplish reading and writing. It can be deduced that the minimum functional array lines which must be provided to each storage cell are a word line, a bit line, and a sense line. However, only two of these are ever required at any one time, namely the word and bit line for writing, and the word and sense line for reading. Hence the bit and sense line may be a common conductor in the array but must be capable of performing two separate functions on the storage cell. It follows that this storage cell must then have a minimum of two independent functional terminals. This represents the minimum cell and array requirements as implemented in the 2D organization. In the more complex 3D organization, the minimum functional array lines which must be provided to each storage cell are two word lines for coincident word selection, one bit line and one sense line. Only three of these are ever required for any operation, namely two word lines and one bit line for writing, and two word lines and one sense line for reading. As before, the bit and sense line may be a common conductor as is often the case. It follows that this storage cell must have a minimum of three independent functional terminals.

In the operation of any random-access memory system, the processor generates at least two basic pieces of information which are used by the memory subsystem to produce the required result. As shown in Figure 5-2, these two pieces of processor information are

1. An address of n bits specifying the referenced word
2. An indication whether the operation is read or write

Once this information is provided, it is the job of the storage system control unit to perform all the subsequently required gating and timing functions to decode the address and read or write the specified word. The total number of separately distinguishable memory entities, E, possible is equal to the total possible number of combinations of the n address bits which is simply 2^n or

$$E = 2^n \qquad (5\text{-}3)$$

Since E is usually given, the required number of address bits can be explicitly obtained by taking the $\log_2$ of both sides of the above equation to get

$$n = \log_2 E \qquad (5\text{-}4)$$

In such a system, n must always be an integer (even or odd) since we can either have an address bit or not; there are no fractional address bits in a binary system. Such being the case, 2^n or E must be an even number. In some cases, $E = W$, the number of total words in a memory and in other cases, $E = \sqrt{W}$ or other fractional part. In any case, it is to be noted that these parts which are really physical conductors of some sort, must be grouped in even numbers and in fact, in groups of 2^n to make efficient use of the address bits. In randomly accessed memory systems, n is independent of the organization of the memory and E is thus equal to the total number of logical words that can appear in the output buffer register. This is a fundamental relationship: the various memory organizations make use of these n bits in different ways as we shall see. In some cases, n is divided into two equal groups of $n/2$ for 3D organization, or into (usually, not always) unequal groups for 2½ D or not divided at all for 2D organization.

The sequence of operations performed by the control unit for writing are as follows:

1. Gate the n bits from the address register into the decoder to select one out of 2^n words
2. Gate the data out of the processor data register to the bit drivers
3. Gate the word and bit drivers to write the data; the word pulse in coincidence with a 1-bit pulse will write a 1 in that cell whereas the word pulse in coincidence with a 0-bit pulse will write a 0. In some storage arrays, all cells along a word must first be set to 0 by a clear cycle and then only 1s are written. This type of operation is used in core arrays.

In some cases, 1 and 2 can be done simultaneously. In all cases, sufficient time must be allowed between sequential gating functions for the worst case delays and electrical transient decays.

The sequence of operations performed by the control unit for reading are as follows:

1. Gate the n bits from the address register into the decoder to select one out of 2^n words
2. Gate word sense drivers to energize the selected word
3. Strobe sense circuits to latch the data into the output buffer register. (In some cases the output buffer may set on the sense signal without need for logic control by the control unit.)
4. Signal that data is available for use by the processor. Sufficient time must be allowed between steps 2 and 4 for worst case driver, array and sensing delays. In addition, before a subsequent read or write cycle can be initiated, sufficient time must be allowed for all transient electrical signals both in the array and circuits, to decay to acceptable levels. This often is a relatively long time, particularly after writing where sense amplifiers are overdriven by noise and require a long recovery time. This is the fundamental problem which makes access time different from the cycle time. Data can usually be completely read from the array and available in the output buffer register but a new cycle cannot be initiated for a certain recovery-time period. Some typical values for a core memory array are 300 ns read access time, 500 ns read/write cycle time.

The gating functions performed by the control unit are accomplished by means of clock pulses which trigger the various circuits. The clock pulses are usually obtained from the basic processor clock. However, since memory is slower than processor logic, there are fewer memory clock pulses per second or a longer period between pulses. The memory clock period is typically two to ten times that of the processor. These two clocks are synchronized to one another to minimize logic complexity. The worst case reading or writing delay is known to be a fixed number of clock periods so a simple counter can be used by the processor to determine, after the initiation of a read cycle for instance, exactly when the data is available for processing. In some complex systems, for example, pipelined processors, the clocks are still synchronous but as soon as the control unit signals that data from a read cycle is available, the processor uses it on the next processor clock period.

It should be recognized that during the writing process, many storage cells receive a small excitation called a half-select or *disturb* signal which is insufficient to switch the cell. At cell locations where a half-select bit and half-select word signal *coincide,* the two signals add to give a full-select excitation which switches the storage cell. This coincidence of signals occurs only at the desired cells. Ideally, all other cells which receive a half-select signal should remain in the previous state. However, in actual operation since one cell may receive millions of such half-select disturb signals, the effects must not be accumulative, that is, the summation of all the small disturbs must not cause the device to switch. If it does, the device is disturb sensitive and therefore not workable in a random-access memory.

The above discussions apply generally to all random access memory organizations. Before discussing the various organizations, it is important to define the meaning of a memory word and in fact, it is necessary to define two words, namely *physical word* and *logical word.* In 3D and 2D organization, they are identical but in 2½D which is becoming more common, physical and logical words are quite different. A logical word is the total number of bits that are retrieved and delivered to the output buffer register in one memory cycle or fetch. A physical word (in 2½D) is several logical words but only one is delivered to the output register; in other words, a physical word is the maximum number of bits energized during reading but only a portion of these, the logical word, is fully switched or gated into the sense amplifiers and output register. Thus,

> *Physical word* = total number of bits along word line
> *Logical word* = total number of bits sensed and gated into output register

If one knows the size of the output register, the maximum size of the logical word is known. In some systems one can fetch a portion of a (full) logical word rather than the entire word. (For example, IBM System 360 and System 370 can fetch a byte (8 bits) which is part of a full logical word.) Thus, the physical word is the largest entity which must be addressed by the word addressing mechanism.

The major reason for the appearance of various random access memory organizations is the attempt to design an array with the highest speed, largest capacity and lowest peripheral circuit and array cost. We saw in Section 5-3 that this was the same motivating factor separating random- from nonrandom-access memory and leads to the sharing of transducers in the latter, but a sacrifice in access time. The various organizations of random-access memory can also reduce cost by sharing circuits, but also at a sacrifice in speed as we shall now see.

2D Organization

The simplest method organizationally, but most expensive for achieving word and bit addressing is the so-called 2-Dimensional (2D) or "word" organized geometry shown in Figure 5-4. This scheme has a number of advantages and disadvantages. The word selection mechanism uses linear decoding. The physical and logical word are identical so if there are W logical words, the decoding must select 1 out of W physical word lines. For large arrays, this becomes complex and expensive as shown later. The reason for the term "2 Dimensional Organization" is that the array has only length and breadth, W and b. The number of circuits that must be connected to a 2D array is easily deduced. Each word line must have one driver and each driver must be connected to one logic gate on the last stage of the decoder of Figure 5-4. Thus, a minimum of W drivers and W gates in the last stage of the decoder are required. Of course, more than W gates are required in the entire decoder. One driver is required for each bit line and one sense amp for each sense line, thus b bit drivers and b sense amps are required. These circuit requirements are summarized in Table 5-2.

In Figure 5-4, it is apparent that if a memory consists of a very large num-

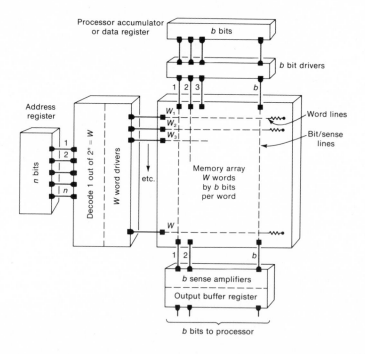

Figure 5-4   Schematic of 2D organized memory

TABLE 5-2. TRANSDUCERS REQUIRED FOR VARIOUS RANDOM-ACCESS
MEMORY ORGANIZATIONS

Function	2D	2½D 2-terminal cell	2½D 3-terminal cell	3D
Decoding Gates				
Word lines (last level)	W	$W_l/s = W_p$	$W_l/s = W_p$	$2\sqrt{W}$
Bit lines	0	design dependent	design dependent	0
Read/Write Drivers				
Word lines	W	$W_l/s = W_p$	$W_l/s = W_p$	$2\sqrt{W}$
Bit lines	b	b $+ sb$ switches	b $+ sb$ switches	b
Sense amps	b	b $+ sb$ switches	b	b

where
W_l = total number of logical words W_p = total number of physical words
b = bits per logical word s = segmentation (2½D) = W_l/W_p

ber of words, for example 65K, and relatively few bits per word, for example 64, then the physical structure is very long but narrow, that is, very long in the direction of bit/sense lines but short in the direction of the word lines. This arrangement, while being most simple conceptually, is not only cumbersome in structure, but is extremely costly in terms of the number of circuits required for its operation. From Table 5-2, W drivers and W decode gates (last stage) are required which can be the dominating factor in a large memory. For example, a 500K word memory would require more than 500K drivers and 500K decode gates. This circuit count is larger than that of most large size computers. However, it is also to be noted that, at least for small memory sizes where the array is more nearly square, 2D organization gives the minimum wiring delay. Thus, this type of organization is most suitable for small memories where speed is important.

In 2D, no decoding is required on the bit lines since the processor data register in Figure 5-4, contains one bit for each bit of the memory word so there exists a one-to-one correspondence between the data bits and bit drivers. The data bits from the register either turn on or hold off the corresponding bit drives. Likewise, no decoding is required on the sense lines since there is one sense amplifier for each bit in a word. The number of bits in the address register is then simply given by Equation 5-4 where E now is W.

2½D Organization

In nearly all main memory systems, the number of logical words required greatly exceeds the number of bits per logical word. In 2D organization, this produces a long, thin array. For minimization of wiring delay, it is desirable to keep the array as nearly square as possible. One way to do this is to simply fabricate more physical bits per physical word. However, the number of logical bits per word is fixed by the processor architecture design and seldom exceeds 100 bits per logical word whereas the number of words is seldom less than 1000; thus a factor of 10 is a bare minimum for W/b. One way to overcome this is to make a physical word contain more than one logical word; that is, segment the 2D memory and piece it together to make a more nearly square structure. This produces what is commonly referred to as a $2\frac{1}{2}$ D memory organization [Russell, Whalen, and Leilich, 68].

Some authors (for example, Gilligan, 66) use a slightly different definition of $2\frac{1}{2}$ D memory that depends on accessing characteristics of the memory organization and not on the distinction between physical and logical words. To include this view in our definition, we note that a 2D memory is a two-dimensional memory in which one dimension is used exclusively for word selection and the other dimension used exclusively for bit selection. A $2\frac{1}{2}$ D memory is a two-dimensional memory in which one dimension is used exclusively for word selection, and the second dimension is used both for additional selectivity of words and for bit selection. In our terminology the first dimension selects one physical word from memory, and the second dimension selects a logical word from a physical word and selects the individual bits of the logical word.

Since we started initially with a 2D organization before reconfiguring it, 2-terminal cells are implicitly assumed. Thus the bit and sense line are functionally one terminal or physical line. There are two basic ways these bit/sense lines can be grouped together to form the logical words as shown in Figure 5-5. In part (a) the first bits of all logical words are grouped together, likewise the second bit and so on, to form b bit groups of s bits per group. One bit/sense line must be selected from each bit group both during reading and writing giving b bits written or read at any one time as desired. In part (b), the bit/sense lines are arranged into segments, each segment representing one logical word. One segment of b bit/sense lines must be selected from the total of s segments during reading and writing, again giving b bits per logical word. A more complete system implementation of the word organized segmentation of part (b) of Figure 5-5 is shown in Figure 5-6. In either case, the reading and writing of the cells is identical to that of the 2D organization but external selection circuits at the data end of the bit line and sense end of the sense line ensure that only the desired cells receive data pulses or are gated into the sense amplifiers. There are fewer physical words than in 2D and hence fewer address bits required for decoding the word lines. The extra bits, n_2, in Figure 5-6 are used to decode the proper segment required for the desired logical word. The circuit count becomes more complex since, while there are still b bit drivers and b sense amplifiers, a substantial number of switches is required to gate the proper segment into the b drivers or sense amplifiers. In Table 5-2, an additional number of switches (equal to sb on each end) is specified and each of the switches may contain several transistors.

During reading in the above $2\frac{1}{2}$ D organization using 2-terminal cells, the read pulse must be supplied entirely by the word line with signals propagating on the bit/sense line. Because of this 2-terminal characteristic, all cells along a given physical word line are sensed and signals propagate on all sb sense lines. The selection circuitry decodes the proper sense lines for the correct

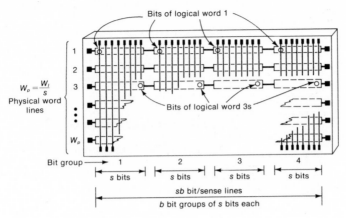

a) Bit-organized groups of bit/sense lines

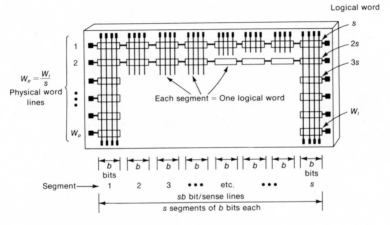

b) Word-organized groups of bit/sense lines

Figure 5-5 Schematic of 2½D organization using 2-terminal cells showing case for $b = 4, s = 8, W_l = 48$

logical word. Note, however, that if the cells are destructively read so that the stored information is lost, all sb cells along the physical word must be regenerated which requires much redundant circuitry, particularly when s is large. Thus it is almost a necessity that 2-terminal cells for 2½D organization possess nondestructive read-out characteristics. The transistor flip/flop cell of Figure 5-15 shown later in this chapter possesses this capability and is often used in this type of organization.

It is possible to remove the requirement of a nondestructive read-out cell in 2½D by using a 3-terminal cell. In this case, the writing process is identical to that described previously but reading is slightly modified. The bit and sense lines are functionally separate so that a coincidence of a bit pulse with word pulse can be used during reading as well as during writing. This coincident reading is applied to only b cells during a read cycle, hence only b cells

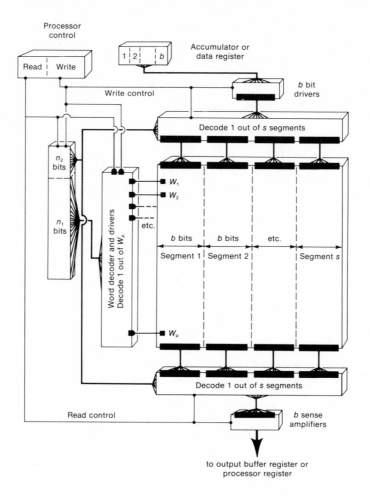

Figure 5-6 Schematic of 2½D organization using 2-terminal cells and word-organized segments of bit/sense lines

are switched destructively and only these must be regenerated, a substantial saving. There is another advantage provided by the 3-terminal cell. Since the sense lines are functionally separate and since only b cells switch at any one time, the sense lines within a given bit group can in principle be connected in series or parallel. The series connection for the two bit groupings of Figure 5-5 (both parts a and b) are shown for corresponding cases in Figure 5-7 (parts a and b). There are now only b sense lines, each with only one sense signal during a read cycle. Thus no selection circuitry is required on the sense lines as indicated by the absence of sb switches on the sense lines in Table 5-2. This represents another advantage of this 3-terminal configuration over the 2-terminal case. Selection circuitry is still needed on the bit lines, however. The 2½D—3-terminal cell organization with series sense lines is that

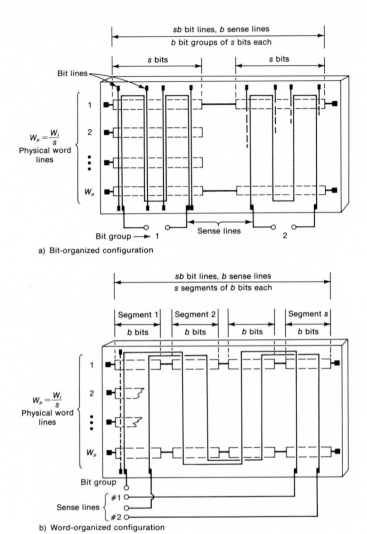

Figure 5-7 Schematic of series sense line configuration for 2½D organization using 3-terminal cells showing case for $b = 2$, $s = 4$, $W_l = 16$

most commonly used for 2½D ferrite core memories and has been an important factor in the design of the higher speed core memories.

The sense lines in the above case could, in principle, be connected in parallel giving even smaller propagation delay. This is done whenever possible or practical which generally is not so for ferrite core memories but is possible with semiconductor integrated circuit chips.

The overall organization of such a 2½D memory using 3-terminal cells grouped in word-organized segments is very similar to the 2-terminal cell case of Figure 5-6 except for two changes. First, the sense line decoding circuitry can be removed, and second, the bit drivers and bit decoding must be activated on the read cycle instead of just on the write cycle. Since the bit-organized grouping is identical to the word-organized segmentation except for the method of connecting the sense lines, the same changes are necessary for both configurations of the 2½D organization using the 3-terminal cells. The bit-organized grouping of part *a* of Figure 5-7 is most commonly used in core and transistor memories resulting in the so-called "one bit per chip" organization for the latter. It should be noted that this bit-grouping is similar to the bit-plane concept in 3D and in fact, these two organizations are closely related.

3D Organization

The most economical, in terms of transducer count, and slowest type of random access memory is the 3D organization which has long been used with ferrite core arrays. This organization is one of the oldest and also one of the important factors that put magnetic cores in the forefront of main memory technology. The basic selection principle of the 3D organization consists of a set of X select, Y select and inhibit lines as shown in Figure 5-8. The X and Y select lines are connected in series between the horizontal bit planes while the inhibit lines are each a single series wire threading all cores within a horizontal bit plane. The planes of each X and Y select line intersect in one vertical column as shown in part *a* of Figure 5-8 to select an entire physical word. This stores all 1s in that word. To store a 0, an inhibit current that cancels or inhibits the field of the Y lines is applied in the inhibit wire that threads through only the XY plane, one such wire for each plane or bit. To store all 0s requires eight inhibit lines to be energized in this case. For reading, the energizing of an X and Y line selects one word, each bit being sensed by one sense line passing through each bit as shown. Thus there is one inhibit and one sense line in each XY plane, or one for each bit per word.

For 3D organization, the physical words are themselves coincidently selected, not just a bit, but the entire word. In order to minimize the peripheral circuit count, the array is broken into a square of $\sqrt{W} \times \sqrt{W}$ words by b bits/word. The number of address bits required for decoding is still $n = \log_2 W$. However, while there is no saving in address bits, there is considerable saving in total circuit count in the decoding circuitry. For instance, on the last stage of the decoder for a selection of one out of W logical words, W gates are needed for a straight forward 2D array. However, for 3D, the total number of gates in the last stage of the decoder is $2\sqrt{W}$ which is quite a substantial saving since now a considerable amount of the decoding is done within the memory device. This puts additional stringent requirements on the device which is costly and may offset the saving in decode gates in some cases.

It should be clear from the preceding discussion why a 3D organization requires a memory cell structure with at least three independent, isolated inputs for X, Y, and inhibit or X, Y, and sense functions (inhibit and sense could conceptually be the same line). As a result, magnetic cores lend them-

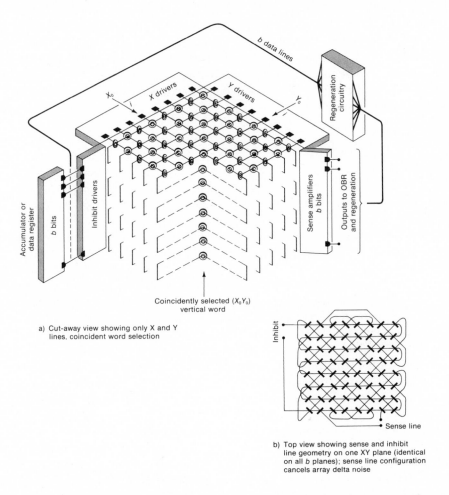

a) Cut-away view showing only X and Y
lines, coincident word selection

b) Top view showing sense and inhibit
line geometry on one XY plane (identical
on all *b* planes); sense line configuration
cancels array delta noise

Figure 5-8 Schematic of 3D 4-wire core memory of 8 × 8 words by 8 bits/word showing case for $W = 64$, $b = 8$

selves readily to 3D organization. While flat films can be used in 3D, such devices are more attractive in the 2-terminal cell structure that requires 2 or 2½D organization. This results from the fact that flat films are desirable mainly for high speed and high density. Under such circumstances, the limitations of planar fabrication coupled with the mode of device operation produces severe noise problems that force the designer toward a 2-terminal cell. Similarly the more common, less expensive transistor memory cells are in essence 2-terminal structures having word and bit-sense inputs only. An additional input can be added for 3D organization but only at substantial cost (Section 5-6) although transistor cells have been used in 3D. Since cores are operated in a destructive read-out mode, a word fetched from the array must be rewritten each time it is read. This requires a regeneration cycle and hardware which increases the cycle time. In theory, we could parallel all *X* lines along a given vertical plane and likewise all Y lines to achieve a much faster array. However cores require large currents for switching and the practical problems of a high speed, high current driver forces series operation and long delays. Evolution of $2\frac{1}{2}$D organized core memories has been motivated primarily by the need to reduce the long delay inherent in 3D organizations with series line connections.

Transistor Memory Organization

Transistor memories have a distinct advantage over nearly all other technologies in that the peripheral circuits can be made from the same fabrication processes and in many cases, the decoding, driving, and sensing can be done with the identical device with simple changes in the fabrication procedure. There are two main differences between transistors and core arrays that give rise to slightly different array configurations:

1. The number of storage cells (bits) per chip that can be successfully fabricated is considerably less than the total memory size, and also considerably less than even the number of total words desired in most memories.
2. The basic 2-terminal MOS cell of Figure 5-15 allows coincident selection (word and bit line) for writing, but no coincident selection for reading as discussed in Section 5-6; a 3-terminal cell can be made but at the expense of a more complex cell, lower density and higher cost.

These two inherent problems affect the organization of arrays in different, but interrelated ways. The basic question is, given a specific number of cells per chip, how do we organize this chip into words of memory? In 3D core arrays, we have to put the X lines and Y lines all in series from one plane to another because of the large current requirement since cores are current sensitive elements. In transistor memories, large currents are not required so that a 3D parallel conductor selection scheme is possible and, in fact, used in actual memory systems. The same paralleling of selection conductors can be used in $2\frac{1}{2}$D organization and, since $2\frac{1}{2}$D requires a simpler and hence cheaper cell, this is often the organization, with slight modification, used in commercial systems.

When Number of Logical Words Equals Bits Per Chips In order to understand transistor memory organization, let us take a 2-terminal cell and consider the possible ways one could organize a memory system. First consider building a 256 word 4 bits per word memory from chips containing 256 bits each. The chip is an array of 16 word lines by 16 bit/sense line pairs. Thus 4 chips are needed. The question now is, how do we connect these chips together to form the 256 word by 4 bits per word memory? If we use the $2\frac{1}{2}$D organization of Figure 5-6, then the organizations of Figure 5-9 result. In part a, 4 segments are used, requiring $W_p = W_t/s = 256/4 = 64$ physical words.

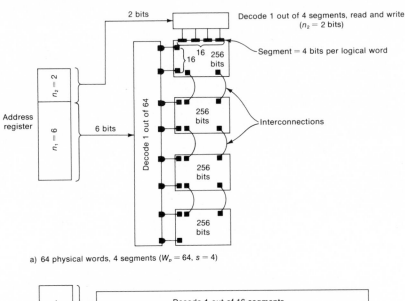

a) 64 physical words, 4 segments ($W_p = 64$, $s = 4$)

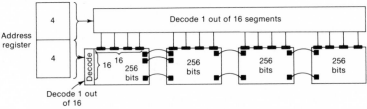

b) 16 physical words, 16 segments ($W_p = 16$, $s = 16$)

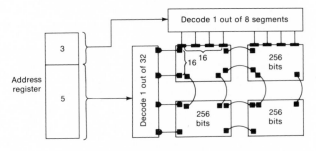

c) 32 physical words, 8 segments ($W_p = 32$, $s = 8$)

Figure 5-9   Possible 2½D organization for 256 word by 4 bits/word transistor memory using chips of 16×16 bits

These 64 words are decoded by an off-chip decoder, as are the 4 segments (only one decoder is shown for the segments for simplicity whereas in general, one for the bit lines and one for sense segment decoding is needed). In part *b*, the 16 word lines are tied in series from chip to chip, and 16 segments are used. Since the word decoder must decode only 1 out of 16 words, this decoder could be put on chips to simplify fabrication. The 16 bit/sense line pairs on each chip are now segmented into 4 groups of 4 bits per group or 16 total groups. In part *c,* a square organization is used with 32 physical words, 8 segments, and a 1 out of 32 word decoder as shown. These organizations are feasible, in principle, with certain drawbacks and difficulties. First, the word line is rather long in part *b* as is the sense line in part *a* not only giving a long delay (if delay is important) but the pulse deterioration could be significant—this cannot be decided without a specific embodiment and known parameters. Next, two types of chips are required in part *b*: one with decoders and one without decoders. This is very expensive and even more important, the chips without decoders in all three parts require many interconnections and pins, in this case 16 pins for the word lines on two sides in part *b,* 16 for sense lines in part *a* and on all four sides in part *c.* This is difficult to do and becomes even more difficult and unreliable as the density increases. It is desirable to remove pin connections as much as possible. The decoded chip only requires 4 pin connections for the input to word lines, suggesting that each chip be decoded with on-chip decoders. Another important consideration, especially if one wishes to buy or supply "off the shelf" components, is the word length variation. Suppose we wish to increase the word length to 6 bits instead of 4 bits. This would be expensive and cumbersome to do with the preceding three organizations.

All these objections can be removed with the organization of Figure 5-10 (only 2 bits are shown). This example is organized as 1 bit per chip which

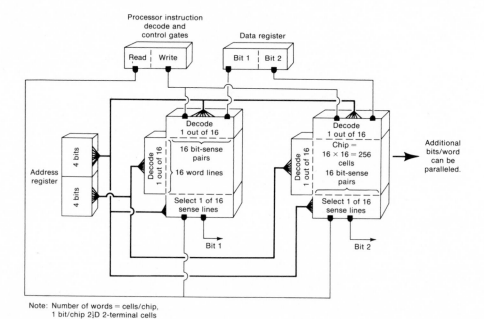

Note: Number of words = cells/chip,
1 bit/chip 2½D 2-terminal cells

Figure 5-10   Typical bit-organized, integrated circuit memory of 256 words by 2 bits/word using fully decoded semi-conductor chips

is quite commonly used. Additional bits can be added to the system for any specified number of bits per word with very little change—word lines and various data paths are paralleled as shown.

The input lines into the bit decoder and sense decoder which specify read or write are single digit (one bit on or off) lines which activate or do not activate the ENABLE input of the decoder. This ENABLE is a very important function since it provides another level of decoding and will be used in the next example as a "chip selector" which allows easy expansion of a standard chip of small size into a large memory array.

When Number of Logical Words Exceeds Bits Per Chip   When the total number of bits exceeds that of a single chip, it is necessary to interconnect a large number of chips in two directions to achieve the required words and number of bits per word. In the previous case, we essentially had to expand only in one direction to achieve the bits per word. Suppose we now wish to build a 1024 word by 8 bits per word memory using the above chips of 256 bits each. Since the cells are 2-terminal structures, the organization must still be 2½ D.

As with the previous case of 256 words, many organizations similar to those in Figure 5-9 can be conceived which implement a larger array. However, they suffer from the same shortcomings as before, such as long lines, excessive pin connections, not easily expanded, and so on. The previous organization of 1 bit per chip can circumvent many of these difficulties except that now there will be 10 address register bits, requiring additional decoding. The fundamental question is, where and how will this additional decoding be done? If we wish to maintain a uniformity of the chip layout to allow easy expansion and versatility, and also minimize pin connections to each chip, it is desirable to use chips as described previously: namely a decoder to select 1 out of 16 word lines, and a decoder to select 1 out of 16 (pair) bit/sense lines. This requires only eight address bits as before so two bits must be decoded elsewhere. We can use the ENABLE function, previously described, to select a given chip. One possible way to do this is shown in Figure 5-11. The 4-bit word address attempts to decode all four groups of chips simultaneously but the ENABLE input to each chip, which is itself decoded off-chip by two bits, allows selection of only one of the four groups or one out of $4 \times 16 = 64$. This enable function is easily provided and is present on most decoders.

Interleaving Architecture   For a specified cycle time, any given technology can provide a self-contained memory module of only a limited size. Such modules are generally referred to as basic systems modules. For instance, a typical core module such as in Figure 5-8 might consist of 4096 words by 64 bits per word with all circuitry self contained. A larger memory capacity is obtained by piecing together many such modules with additional control logic. The access and cycle time of these modules are identical but the operation of each is completely independent of all others. As a result, it is possible in some

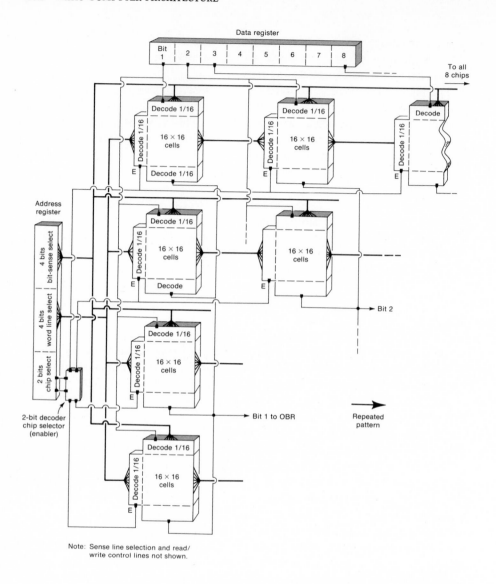

Note: Sense line selection and read/
write control lines not shown.

Figure 5-11 Integrated circuit memory organization with number of words larger than cells/chip: 1024 words by 8 bits/word (modified 2½D with 1 bit/chip)

cases to achieve an apparent increase in the total system cycle time by staggering the initiation of fetches to each module. This is possible because the processor cycle time is faster than the memory module cycle time. Suppose, for instance, that for each cycle time of a module, four processor cycles are completed. If four memory references are required, and if they are resident in four separate modules, then one module can be referenced on the first processor clock period, a second module can be referenced on the second processor clock period and so on for the other two. Each module itself can only be referenced every four processor clock periods but in the above case, the total system is now working at the processor clock rate. If the memory references were all read cycles, a new word would appear every processor cycle time as desired in an ideal memory. This technique is known as *interleaving* and is used particularly in high speed computers. In general computation, it is not possible to organize data such that subsequent references are always to a different module. Perfect interleaving is not possible in general. However, in paged virtual memory systems (Section 5-11) interleaving can be quite useful, particularly in the cache-main memory hierarchy. The structure of the pages is known so successive words of a page are stored in separate modules of the main memory. When a page transfer to the faster cache is required, the words can be read out of main memory at a rate equal to the cycle time divided by the amount of interleaving. Thus for the four-way interleaving described previously, the page can be read out of main memory four times faster than it could be out of a single module or non-interleaved system.

5-6. RANDOM-ACCESS MEMORY DEVICES AND CELLS

Numerous technologies have been pursued as well as implemented in random access memory but only ferrite cores and transistor integrated circuits have had a large impact on computing systems. Thus we shall concentrate on devices and cells for these two with only casual treatment of other technologies [see also Matick, 74; Scott, 70].

It was shown in Section 5-5 that the storage cell must have two or three independent functional terminals to be useful in a random access array. Two-terminal cells can be used in the 2D or one form of 2½D organizations. Three-terminal cells can be used in the 3D or a second form of 2½D organizations. The number of terminals on the cell is a very important consideration in cell design since more terminals require a larger, more complex cell structure. This is particularly true in the newer integrated circuit type of memory cells. It is also true for ferrite cores but to a lesser extent; more wires only require a large core, the cell complexity remaining nearly constant.

Magnetic ferrite cores remained the dominant memory technology for about twenty years from the mid 1950s to the mid 1970s. The medium itself consists of a small torroid made from pressed and subsequently fired ceramic-like magnetic ferrite material. The magnetic properties are ideally characterized as having a "square" BH loop. "Square" means nearly vertical sides and horizontal top meeting at a 90° (square) angle. No material has a truly "square" loop but some come reasonably close. In this BH loop, B is the magnetic flux density which is the stored quantity and H is the magnetizing force which is proportional to the current. H is proportional to i through Ampere's circuital law $\oint H \cdot d\ell = Ni$ while the flux density B is proportional to $\int V(dt)$ since $V = N (d\phi/dt) \propto (dB/dt)$ where $\phi = BA$, $\phi = $ flux, $A = $ cross-sectional area.

This loop describes the change in flux density as the drive current is changed. We can see this loop if we put two windings on the torroid as in Figure 5-12, apply a slowly varying current in one winding, integrate the sense voltage, and apply these two signals to the x and y deflection inputs of an oscilloscope. Let us check to see if this has the requirements of Section 5-3 for binary devices. Two stable states are provided by the points plus and minus B_r. Switching between these two is easily accomplished by controlling the polarity and amplitude of the applied current. A current of $i \geq i_c$ or $i \leq -i_c$ and subsequent removal of current will switch the core into the $+B_r$ or $-B_r$ state respectively. Coincident selection is accomplished by using two separate writing windings as for example in Figure 5-13, applying $i_c/2$ in each winding and having only one core or one group of cores in the array residing at the intersection (coincidence) of the two windings (see Section 5-5).

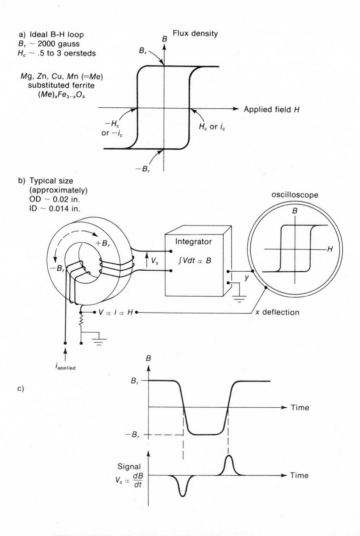

Figure 5-12 Basic principles of magnetic cores

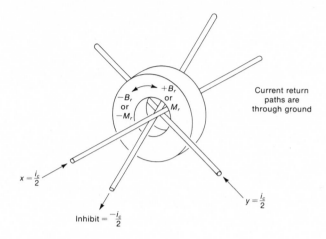

Figure 5-13 Magnetic core storage cell (one bit)

Sensing is provided by the fact that upon switching from $+B_r$ (say a stored 1) to $-B_r$ (stored 0) the sense signal V_s is a negative pulse, while switching from $-B_r$ to $+B_r$ gives a positive pulse as shown in part c of Figure 5-12. Thus sensing 0 and 1 is possible. The energy loss per unit volume is proportional to the area of the BH loop. Thus, it satisfies all the medium requirements. This core has a well defined threshold H_c and this threshold makes it insensitive to disturb pulses on unselected cells. The two or three functional terminals are easily provided by separate windings along the torroid. In fact, this ease of having any number of separate terminals or windings was a major advantage of this technology in its early history since a single straight wire through the core opening serves as a single turn winding as shown in Figure 5-13.

Plated magnetic wires have been pursued as an alternative to the hand-stringing problem involved in magnetic cores. The basic operating principles are very similar in principle with some small variations [Mathias and Fedde, 69]. Thin magnetic films were pursued for many years as a planar, integrated technology to replace cores. The basic idea is again nearly the same as in cores but with slight differences in detail. [Raffel et al., 61; Pohm and Zingg, 68; Higashi, 66; Jones and Bittmann, 67.]

Transistor integrated circuit cells have begun to have major impact in main memory. While many circuit configurations exist for the storage cell, the basic bistable flip-flop was initially the most widely used. It is interesting to note that the internal storage used on some of the original computers (ENIAC at the University of Pennsylvania by Eckert and Mauchly) consisted of the same basic flip/flop circuit but implemented in vacuum-tube technology. The high cost, large space, and large power dissipation of such devices precluded their being used for large memory arrays and hence other technologies were developed (cores, drums, and so on). Now that large-scale integrated electronics has overcome the original problems, memory has completed a full cycle back to the original concept. We shall start with the basic flip/flop of Figure 5-14 which is essentially a one terminal device, then add necessary components to make it a two terminal device as required for random access memory. Additional complexity can then be added to provide a 3-terminal device.

The operation of the basic flip/flop of Figure 5-14 is as follows: the two transistors T_0 and T_1 form a normal flip/flop circuit as shown in part *a*. When T_0 is turned on, current flows through T_0 to ground, putting node point A at ground. This in turn puts the base of T_1 at ground potential and thus holds T_1 off. If T_1 is off, node B must be at a voltage V_a, which is also at the base of T_0, holding it on. Thus, this would be a stable state with T_0 on and T_1 off, storing an arbitrarily labeled 0 (could just as well be called 1). To store the opposite state, that is, a 1, it is necessary to bring node B to 0 voltage and node A to V_a. This can easily be done with additional transistors tied to nodes A and B. This is exactly what is done in the common cell in use and the manner in which this is done leads to many different cell designs. To understand this, note that the flip/flop has only two access points, A and B, both of which are functionally the same, that is to turn off one and turn on the other transistor. While there are two physical terminals A and B, only one can be controlled externally at a given time, hence it is in essence a 1-terminal device as far as memory organization is concerned. For useful memory implementation, the minimal requirement is that points A and B must be coincidentally selected, each by two additional terminals. This requires at least one additional transistor for each node, and perhaps additional components. This represents the major differences in flip/flop type cell design, namely how points A and B are selected. If A and B are each selected for writing by coincidence of pulses on two terminals as suggested above, then

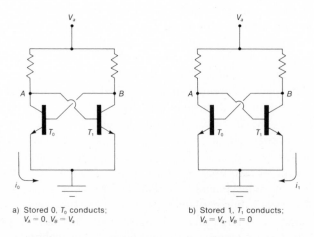

a) Stored 0, T_0 conducts;
   $V_A = 0$, $V_B = V_a$

b) Stored 1, T_1 conducts;
   $V_A = V_a$, $V_B = 0$

Figure 5-14 Basic transistor storage flip/flop

a device for 2D or 2½D organization results. For 3D organization, three terminals for selecting A or B must be provided, adding to cell complexity.

A basic MOS storage cell with two terminals to select either node A or B (2½D organization) is shown in Figure 5-15. A coincidence of pulses on a word line and either of the two bit/sense lines results in a coincident writing into the cell. The operation of this cell is as follows. Recall that to write a 0 or a 1, one node A or B must be brought to ground while the other node, B or A must be floating or high. This is exactly what is done in Figure 5-15. To write a 0 regardless of the initial state of the cell, a positive voltage is applied to the word line in coincidence with a pulse to bring bit/sense line write 0 to ground potential. The word pulse on the gate of T_2 turns it on and its source is at ground, so point A is brought to ground potential. If the cell was initially in the 0 state, point A would already be at ground so nothing would happen. If the cell was initially in the 1 state, point A would initially have been at $+V_a$ and T_1 conducting. In this case, A is brought to ground, causing T_1 to turn off, point B to increase to V_a which causes T_0 to conduct and hence the cell switches to the 0 state. To write a 1, a word pulse is

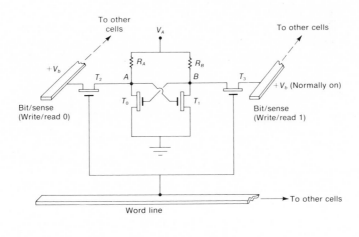

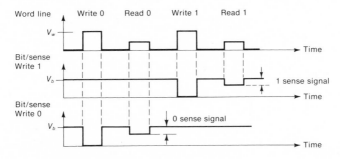

Figure 5-15   Two-terminal MOS transistor flip/flop storage cell (*n*-channel enhancement MOSFET)

applied in coincidence with bringing bit/sense line write 1 to ground and the same sort of behavior results. For reading, a small word pulse is applied *without* a bit pulse which causes both T_2 and T_3 to conduct slightly. Whichever node A or B is at 0 volts will cause a small current to flow through its respective transistor giving a small decrease in the normally "high" bit/sense line voltage. The sensing is nondestructive since the word read pulse only strobes to see which state the cell is in. Note that for reading, the bit/sense lines are the "sense lines." Thus, while there is an inherent capability for coincident selection for writing into such a cell, there is no coincident selection capability for reading, that is, all cells along a pulsed word line are read and hence this cell can only be used in 2D or 2½D organization. This is a result of the fact that in essence, this cell is a two-input structure, that is, word and bit/sense. While there are actually three inputs, word, write 0, and write 1, the latter two serve only one function at a time, namely bit writing or bit sensing.

In order to make the cell of Figure 5-15 a 3-terminal cell, another "functional" selection terminal must be added that allows coincident selection of cells for reading in addition to writing. This can be easily accomplished by inserting another pair of transistors between the node points A and B and their respective bit/sense lines such as those shown in Figure 5-16. We have, in essence, added another word line to the previous cell. These two lines, now labeled X and Y lines, require a coincidence of pulses before either writing or sensing can take place through the bit/sense lines. The operation of the

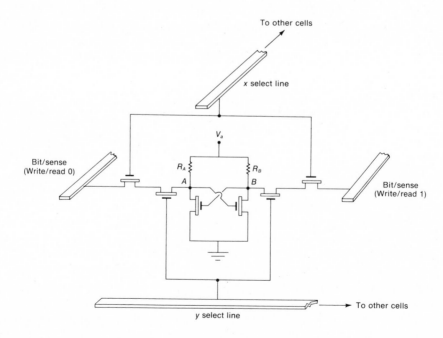

Figure 5-16 Three-terminal MOS transistor flip/flop storage cell

cell is basically the same as previously. The three terminals of the cell are
1) X select, 2) Y select, and 3) bit/sense line pair. The digit (bit/sense) lines
are isolated from nodes A and B by two transistors and a direct connection
is only made on a selected cell; the presence of a single, or half-select pulse
on adjacent cells still leaves the digit lines isolated by at least one transistor
and hence this cell is undisturbed.

The same basic ideas above concerning random-access memory cells using
MOS transistor flip/flops can be applied to bipolar transistor cells as well. In
fact, the first commercial, mass-produced semiconductor memory cell used
bipolar transistors in a 3-terminal cell configuration [Farber and Schlig, 72;
Ayling and Moore, 71]. The primary difference between bipolar and MOS
devices for memory is that the former are fast but the fabrication process is
complex, requiring double diffusion and many steps which are expensive. MOS
devices are much simpler to fabricate but are slower than bipolar devices.

The 6-device, 2-terminal cell of Figure 5-15 can be simplified by using
MOS transistors in a shift-register arrangement. This type of operation is
referred to as dynamic MOS memory and is quite common for low cost, low
speed operation. MOS as well as other shift registers are described in Section 5-10.

5-7. DIGITAL MAGNETIC RECORDING PRINCIPLES

Digital magnetic recording is extremely attractive for data processing
since the media (tapes and disks) can be removed, are unaffected by normal
environmental changes, can be transported as a means of data file transfer,
and can be reused many times with no processing or development steps
necessary. Since magnetic recording is nonrandom access, it does not require
wired-in array hardware so there is no cell or array configuration and only
the fundamental system and medium requirements of Sections 5-2 and 5-3
are necessary. The essential parts of a simplified but nevertheless complete
magnetic recording system are shown in Figure 5-17 and consists of a controller (sometimes a large computer) to perform all the logic functions as well
as write current generation and signal detection; serial/parallel conversion
registers; a read/write head with an air gap to provide the magnetic field for
writing and to sense the stored flux during reading; and finally the medium.
The wired-in cells, array, and transducers of random-access memory have
been replaced by one read/write transducer which is shared among all stored
bits, and a controller, also shared.

The recording medium is very similar in principle to ferrite material used
in cores with the exception that the coercive force is very large, on the order
of 300 oersteds, which is necessary to resist the self-demagnetizing field inherent in small magnetic bits. The common material for all digital magnetic
recording is Fe_2O_3 and has remained essentially unchanged for many years
except for reductions in the particle size, smoother surfaces and thinner, more
uniform coating, all necessary for higher density. The operation of the

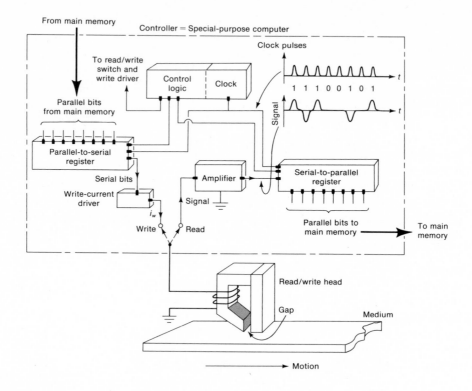

Figure 5-17 Essential features of simplified, complete magnetic recording system

medium is nearly identical in principle to that of the ferrite core, that is, there are two stable states $+B_r$ and $-B_r$ for storing binary bits* and the medium can be switched an infinite number of times by a field produced by the write head which exceeds the coercive force. The stored information is sensed not by switching as in random-access memory but by moving the magnetic bits at constant velocity under the read gap to induce a time changing flux and hence induce a sense voltage. The energy losses during writing are once again represented by the area of the *BH* loop for a given bit. Hence the fundamental media requirements are all satisfied.

With respect to the system requirements, the first one, a medium for storing energy has already been discussed. The energy source for writing is the "write" driver circuits within the controller and the write transducer, both of which are shared. The energy source for reading is the motor which drives the medium and converts the static field of a stored bit into a time changing field, $d\phi/dt$.

The essence of magnetic recording consists of being able to write very

---

*True in principle but in practice the codes used do not allow a direct correspondence between the magnetization direction and stored bit.

small binary bits, to place these bits as close together as possible, to obtain an unambiguous read-back voltage from these bits, and to convert this continuously varying voltage into discrete binary signals. In the following discussions, we shall see that to obtain small bits, and to be able to convert back to digital form, special codes are expedient. We shall also see that the write head is not a major factor in determining density since the writing is done by the trailing edge of the write field.

The minimum size of one stored bit is determined by the minimum transition length required within the medium to change from $+M_r$ to $-M_r$ without self-demagnetizing; the smaller the transition length, the larger will be the self-demagnetizing field. The minimum spacing at which adjacent bits can now be placed with respect to a given bit is not governed by an increase in self-demagnetization but rather, the bit spacing is determined mainly by the distortion of the sense signal when adjacent bits are too close, referred to as *bit crowding*. This results from the overlapping of the fringe field from adjacent bits when they are too close and this total, overlapped magnetic field is picked up in the read-head giving a different induced signal compared to that produced by a single transition. Conversion of the analog read-back signal to digital form requires accurate clocking which in turn requires clocking information to be built into the coded information, particularly at higher densities.

Neglecting clocking and analog-to-digital conversion problems for the moment, the signals obtained during a read cycle are just a continuous series of 1s and 0s. A precise means for identifying the exact beginning and end of the desired string of data is necessary, and furthermore, some means for identifying various parts within the data string is often desirable. Since the only available way to recognize particular pieces of stored information is through the sequence of pulse patterns, special sequences of patterns such as gaps, address markers, and numerous other coded patterns are inserted into the data. These can be recognized by the logic hardware built into the controller which, in all cases is a special purpose computer attached to the storage unit. These special recorded patterns, along with other types of aids, are referred to as the *stored addressing information* and constitute at least a part of the addressing mechanism.

Coding schemes are chosen primarily to increase the linear bit density. The particular coding scheme used determines the frequency content of the write currents and read-back signals. Different codes place different requirements on the mode of operation and frequency response of various parts of the system such as clocking techniques, timing accuracy, head time-constant, medium response, and others. Each of these can influence the density in different ways but in the overall design, the trade-offs are made in the direction of higher density. Special coding schemes are thus not fundamentally necessary, but only practical. For instance, it is possible to store bits in much the same way as in random-access memory where say $+M_r$ is a stored 1 and $-M_r$ a stored 0 as in part *a* of Figure 5-18. The transition region in-between is assumed to have $M = 0$ except for the small regions of north and south

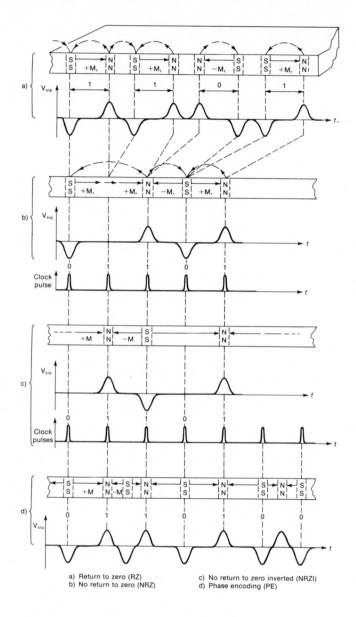

Figure 5-18 Cross-sectional view of magnetic recording medium showing stored bits, signal patterns, and clocking for various codes

poles on the edges as shown. As the medium is moved past the read head, a signal proportional to dM/dt or dM/dx is induced so the north poles induce, say a positive signal and south poles a negative signal as shown (polarity arbitrary). This code is known as return to zero or RZ since the magnetization returns to 0 after each bit. Each bit has one north and one south pole region so two pulses per bit result. Not only are two pulses per bit redundant, but considerable space is wasted on the medium for regions separating stored bits. It is possible to just push these bits closer together as in part *b* of Figure 5-18 such that the magnetization does not return to 0 when two successive bits are identical as shown for the first two 1s, hence the name *nonreturn to zero* or NRZ. When a region previously separating bits contains alternate north and south poles, the poles cancel when pushed closer together whereas when the region previously contained both north or both south, the poles remain upon being pushed together. The result is then only one transition region and one signal pulse per bit. By adjusting the clocking pulses to coincide with the signal peaks, we have a coding scheme in which 0s are always negative signals and 1s are always positive (part *b* of Figure 5-18). The difficulty is that only a change from 1 to 0 or 0 to 1 produces a pulse, a string of only 1s or only 0s produces no signals. This requires considerable logic and accurate clocking in the controller to avoid accumulated errors as well as to separate 1s from 0s. One popular coding scheme is a slightly revised version of the above, namely *nonreturn to zero* inverted or NRZI in which all 1s are recorded as a transition (signal pulse) and all 0s are no transition as shown in part *c* of Figure 5-18. There is no ambiguity between 1s and 0s but again a string of 0s produces no pulses. A double clock scheme, where two clocks, each triggered by the peaks of alternate signals is used to set the clock timing period to the following strobe point. NRZI is a very common coding scheme for magnetic tapes used at medium density. For high density such as 1600 bits/inch the clocking and sensing become critical so *phase encoding*, or PE, shown in part *d* of Figure 5-18 is often used since 1s give a positive signal, 0s give a negative signal and hence a signal is available for every bit. Phase encoding requires additional transitions within the medium, for example, between successive 1s or successive 0s as shown but density is most often limited by sensing, clocking, and other problems rather than the medium capability.

For magnetic disk recording, a double frequency NRZI code is often used. This is obtained from NRZI by adding an additional transition just prior to each stored bit; the additional pulse generated serves as a clocking pulse and hence a well specified window between bits is provided to avoid clocking problems when a string of 0s is encountered.

Writing of the transitions (north or south) in the medium is done by the trailing edge of the fringe field produced by the write head. To see this, suppose we have a medium initially saturated throughout in the $+M_r$ direction as in the bottom view (part *c*) of Figure 5-19 (shown displaced for clarity) and we wish to write one transition. The medium is moving to the left and at some time t_0, we suddenly apply the write current i_{write} with polarity shown in the head (part *a*). The H_x versus x curve in part *d* shows the x or horizontal component of magnetic field created by the head on the medium. (Magnetic media are made to respond primarily to the x component of the applied field.)

Stopping the motion momentarily, we see that the field acting on the medium varies with distance and any portion with the field equal to or greater

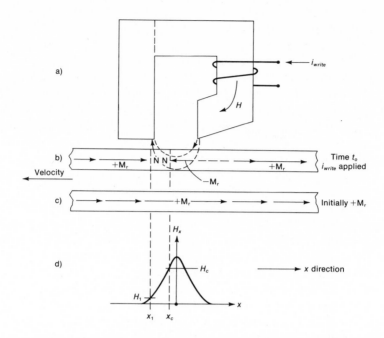

Figure 5-19 Writing process showing writing by trailing edge of head

than H_c (coercive force) will be nearly saturated at $-M_r$. However, for small values of H_x, less than H_c, the medium will respond with small changes in M; for instance, a field of H_1 in part d can start reversing toward $-M_r$ but the medium cannot be fully reversed until a field of H_c is reached. Thus the transition (stored bit) extends approximately from x_1 to x_c. A similar transition exists on the right side of the gap but if we now let the medium move to the left, the right hand transition never appears as a stored bit since it is essentially held at a fixed position with respect to the head, or moves to the right within the medium. Thus, we are left with one permanent transition of distance $x_1 - x_c$ written by the trailing edge of the write head.

Read back signals can best be understood in terms of the reciprocity theorem of mutual inductance. This theorem states that for any two coils in a linear medium, the mutual inductance from coil 1 to 2 is the same as that from coil 2 to 1. When applied to magnetic recording, the net result is that the signal as a function of time observed across the read head winding induced by a step function magnetization transition or $x_1 - x_c$ very small, has a shape which is identical to the curve H_x versus x in part d for the same position of the medium below the head gap. It is only necessary to replace x by vt where v = velocity for the translation of the x scale on H_x to the time scale on V_{sig} versus t. The H_x versus x curve with a multiplication factor is often referred to as the *sensitivity function*.

We saw above that the writing of a transition was done by only one small portion of the H_x versus x curve whereas the sense signal is determined by the entire shape of H_x versus x; that is, the signal is spread out. It is this fact that gives rise to *bit crowding* which makes the read-back process more detrimental in limiting density than the writing process. To understand bit crowding, suppose we have two step function transitions of north and south poles separated by some distance L as in Figure 5-20. When these transitions are far apart, their individual sense signals shown by the dotted lines appear at the read winding. However, as L becomes small, the signals begin to overlap and, in fact, subtract from each other* giving both a reduction in peak amplitude as well as a time shift in the peak position as shown. This represents, to a large extent, the actual situation in practice—the transitions can be written closer together than they can be read.

Clocking or strobing of the serial data as it comes from the head to convert it to digital characters is another fundamental problem. If perfect clock circuits with no drift and hence no accumulated error could be made, the clocking problem would disappear. But all circuits have tolerances and as the bit density increases, the time between bits becomes comparable to the drift in clock cycle times. Since the drift can be different during reading compared to what it was during writing, serious detection errors could result. Hence it is necessary to have some clocking information contained within the stored patterns. This is what is accomplished by the PE and double frequency NRZI codes discussed above. Clocking and its influence on the code, logic circuits, head design, and so on represents an area of important system design in practice.

---

*Linear superposition is possible since the air gap makes the read head linear.

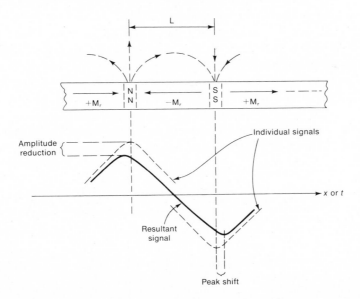

Figure 5-20   Bit-crowding on read-back showing amplitude reduction and peak shift

Density improvements in the past and future require the scaling of three major parameters to smaller dimensions and closer tolerances, namely the head air-gap, head-to-medium spacing, and medium thickness.

Locating previously stored data, as well as determining where to write new data is much more complicated in magnetic recording systems than in random-access memory since there is no addressing capability within each bit cell.

5-8. SEQUENTIAL-ACCESS STORAGE SYSTEMS

Flexible tape represents the primary sequential system, the most common one being ½ inch wide mylar tape about 1 mil thick coated with about 0.5 mil thick magnetic oxide. The tape, usually 2400 ft. long, is contained on a reel about 10.5 inches in diameter. There are either 7 or 9 tracks written across the width of the tape and hence either 7 or 9 read/write heads to store one complete character or byte at a time. Tape tracks and hence bit cell widths are in the range of 0.04 inches wide but bit spacing along a track is much smaller. The latter is approximately the reciprocal of the linear density or 1.25×10^{-3} inches for 800 bits per inch system. The actual transition lengths (N or S regions in Figure 5-18) are generally about half this bit cell spacing. In many systems there are separate read and write gaps as shown in Figure 5-21 in order to check the reliability of the recording by reading immediately after writing. The tape is mechanically moved back and forth in contact with the heads, all under the direction of a controller. The tape and heads both wear from abrasive contact and must eventually be replaced.

The most commonly used code for tapes below 1600 bits per inch is NRZI. Phase encoding which is self-clocking on all channels is necessary for 1600 bits per inch and higher. Even though a string of zeros in the NRZI code gives no signal pulse for clocking, the use of odd parity across each recorded character ensures that at least one pulse is obtained for every character. This provides a type of self-clocking although not on all channels simultaneously and thus is not adequate at high density.

The important operational parameters are: tape speed, data rate, linear density and rewind time. Some typical values are given in Table 5-3.

TABLE 5-3. TYPICAL PARAMETERS OF COMMON TAPE SYSTEMS

Tape Speed (inches/sec)	Data Rate (K bytes/sec)	Linear Density (bits/inch)	Rewind Time (sec)	Recording Code
18.75	15	800	minutes	NRZI
37.5	30	800	minutes	NRZI
75	120/60/41.5/15	1600/800/556/200	45-100	PE/NRZI
100	160/80	1600/800	72	PE/NRZI
112.5	180/90	1600/800	55-97	PE/NRZI
125	200/100	1600/800	55	PE/NRZI
200	320/160/111.2	1600/800/556	45-60	PE/NRZI/NRZI
250	800	3200	45	PE

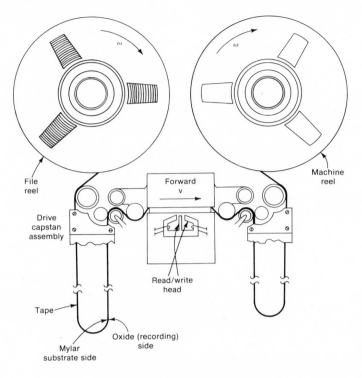

File reel

Machine reel

Forward
v

Drive capstan assembly

Read/write head

Tape

Oxide (recording) side

Mylar substrate side

Reference track 1 nearest front (observer) end

Figure 5-21   Schematic of sequential-access (tape) system

The stored addressing information in tapes is relatively simple, consisting of specially coded bits and tape marks in addition to inter-record gaps (IRG). The latter are blank spaces on tape that provide space for the tape to accelerate and decelerate between records since reading and writing can only be done at constant velocity. The common gap sizes are 0.6 and .75 inches for specifications shown in Table 5-4. Typically, a tape recorded at 800 bits per inch, with 8 tracks plus parity storing records of 1K bytes each, and a gap

TABLE 5-4. TAPE INTER RECORD GAPS (IRG)

Density (bits/inch)	IRG* (inches)
7 Tracks	
200	
556	0.75
800	
9 Tracks	
800	
1600	0.6

*Standard accepted by most of the industry

of 0.6 inches between each record can hold over 10^8 bits of data. Even though this represents a large capacity, the gap spaces consume nearly 50 percent of the tape surface, a rather extravagant amount. In order to increase efficiency, records are often combined into groups known as blocks as shown in Figure 5-22. Since the system can only stop and start at an inter-record gap, the entire block is read into main memory for further processing during one read operation.

5-9. DIRECT-ACCESS STORAGE SYSTEMS

There are two major types of direct access systems, namely 1) moveable head (arm) disks and 2) fixed, one head per track systems, the latter including both disks and drums as shown in Figure 5-23. The former is more common since it is less expensive as a result of a greater sharing of the read/write heads and we shall concentrate mainly on this type. The recording head usually consists of one gap which is used for both reading and writing. This head is "flown" on an air cushion above the disk surface at separations in the neighborhood of 50 to 100 microinches for different systems. A well controlled separation is vital to ensure reliable recording.

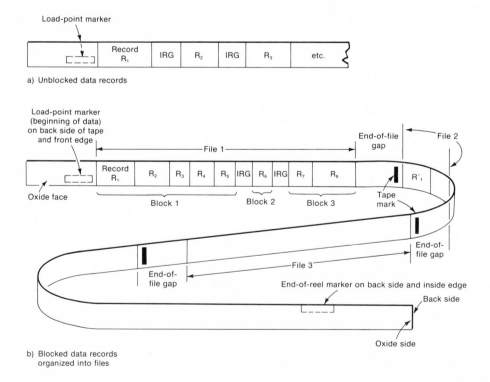

Figure 5-22   Data record formatting on tape

The medium consists of very thin coatings (about 100 microinches) of the same magnetic material used on tapes but applied to polished aluminum disks. Several disks are usually mounted on one shaft, all rotated in unison and each surface is serviced by one head as in part *a* of Figure 5-23. The arms and heads are moved mechanically along a radial line and each fixed position sweeps out a track on each surface, the entire group of heads sweeping out a cylinder. A typical bit cell is 5×10^{-3} inches wide by $(2000)^{-1} = 0.5 \times 10^{-3}$ inches long for a 2000 bits per inch linear density. The transition length is about half this size. Some typical disk parameters are:

Linear density:	1100 to 4040 bits per inch
Track density:	100 to 200 tracks per inch
Rotation speed:	1200 to 3600 revolutions per minute
Arm movement:	50 ms between adjacent tracks
	100 ms across all tracks
Data rates:	0.5 to 10 million bits per second

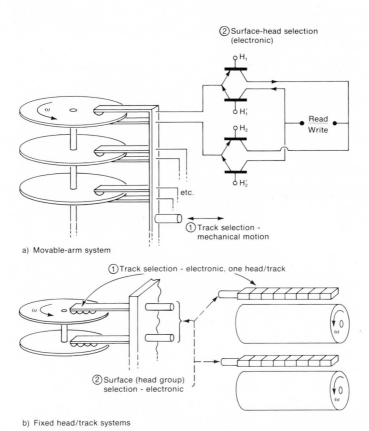

Figure 5-23   Schematic of direct-access systems

The fundamental difference between various disk systems centers on the addressing mechanisms provided. In order to store and subsequently retrieve information there are six fundamental requirements for all systems. We must be able to:

1. Select a given track on a specific surface either by moving the heads in a movable head system or electronically selecting one head on a specific track in a fixed head system.
2. Verify the position (track and surface); in a movable head system the mechanical positioning can be in error and require verification. In a fixed head system, mechanical tolerance and other problems can result in misalignment or excessive head-medium spacing between a head and its associated tracks. (The user in principle does not have to verify position each time but can take his chances; should an error occur, it is a fundamental requirement that position verification be possible to find the error. In addition, position verification can be used, in some cases, as an address for stored records.)
3. Have a well defined reference point for counting, clocking, and position reference for searching.
4. Record (write) data at various positions along a track.
5. Locate a previously stored record.
6. Determine when to start reading the located record and when to stop the data transfer to main memory.

All of these should be done efficiently, with as little wasted storage as possible. This is not a fundamental requirement but a practical one. In addition, there are numerous other requirements such as error detection and correction, identification of unusable tracks and alternates, and so on, that are important but arise because of practical problems. In principle, they can be eliminated with higher cost and special precautions. The preceding six requirements are more basic and cannot be eliminated even in principle, since they are an intimate part of the storage and retrieval problem. The manner in which these requirements are implemented not only varies substantially from one manufacturer to another, but also between various systems of a given manufacturer.

With respect to requirement 1, track selection, it has already been pointed out that both movable and fixed heads are used and the selection is either electronic or mechanical. With the exception of requirement 1, all the other requirements necessitate some form of stored addressing information; this can be essentially of two forms, one is very close or adjacent to the data it is associated with, or the other is remote such as on another track or surface. This latter might consist of index markers, clock and sector marks or other information recorded on separate surfaces. It should be clear that they all serve the same function, namely to help write and locate data.

Position verification (requirement 2) is handled in different ways but in all cases, it is adjacent to its associated data since we may wish to read or write immediately after verification and hence should verify as close to the data as possible.

With respect to requirement 3, in order to write records with a known point of origin as well as have a point for future and continuing reference, all systems have some form of index marker to signify the start of the track. This is an arbitrary reference point, placed anywhere initially, and can be a physical mark (but electronically sensed) or stored bits on a separate surface. This is not as critical as position verification and hence can be done with remotely stored addressing information.

With respect to requirements 4 and 5, writing and locating records at various positions along a track can be accomplished with either adjacent or remote addressing information and both methods are used in practice.

For requirement 6, knowing when to start and stop reading requires some adjacent addressing information. The exact method for implementing these requirements dictates, in a very general way, the format of the records stored on direct-access storage. The format then chosen determines the storage utilization efficiency of the system which in turn depends on the particular application, being good for some and less efficient for others.

All formats for stored records require the use of some adjacent addressing information as shown in part *a* of Figure 5-24. Some form of information must precede the data to separate the various pieces of data from one another and as an aid in addressing specific records. In the most simple formats, part *b*, the stored addressing information is a gap, usually but not always of fixed length with some coded information that is interpreted by the controller. This gap is usually not accessible to the programmer. The advantages of this sys-

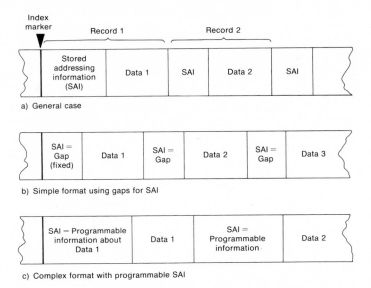

a) General case

b) Simple format using gaps for SAI

c) Complex format with programmable SAI

Figure 5-24 Fundamental track formats for direct-access storage systems

tem are that few bits are required for the addressing information and the overall electronics can be relatively simple. The disadvantage is mainly that this system is more suitable to well-organized data such as that found in scientific or engineering calculations. This is essentially the format in use, with slight modifications, on CDC 66XX and 76XX disk systems. It is not intended for use in a general environment where highly variable data lengths are continually encountered.

For an environment that must accommodate wide variations in data organization, a more complex format such as that shown in part c of Figure 5-24 is desirable. Here, the adjacent addressing information contains gaps such as in part b but more importantly, contains areas which are programmed information about the data and are valuable aids in locating records. In essence this is the format used in IBM System 360/370 disk systems but with substantially more detail in the actual systems. The disadvantage of this format is the large amount of storage space consumed by the stored addressing information in many cases, and the additional complexity of the electronics and programming when the full capabilities are used.

The amount and complexity of the remote addressing information also determines the flexibility of the system and the cost. The addition of a separate disk for indexing, clocking, and sector addressing can greatly aid in implementing requirements 3, 4, and 5 but requires considerably more hardware and logic capability in the system.

5-10. ELECTRONIC SHIFT REGISTER MEMORIES

Magnetic Bubble and Charge-Coupled Semiconductor Arrays

Shift registers represent one of the oldest concepts for electronically storing and retrieving information in digital computers. The first experimental stored program computer with address modification, the EDSAC, and the first commercial computer, the UNIVAC, both used ultrasonic mercury delay line type of shift registers for high speed storage. The major appeal of shift register storage is the relative simplicity and high density with low cost. The low cost is achieved by sharing the input and output transducers as in magnetic recording on disks or tapes. In the latter, the stored information is shifted mechanically by moving the storage medium while in electronic shift registers, the medium remains stationary and the bits are moved within. The major disadvantage is the nonrandom accessing requirement which is quite time consuming. This shortcoming can often be improved by using the concept of dynamic reordering [Beausoleil, Brown, and Phelps, 72] which makes use of the idea of clustering of memory references to the most recently used pages as described in section 5-11. Dynamic reordering keeps the most recently used pages nearest the input/output transducers and when reference is made to one of these pages, little shifting is required which can greatly improve access time. This technique is applicable only to shift registers which can remain static for long periods as well as shift backwards and forwards on command.

The basic idea of a shift register is identical in principle to that used for performing logic in the central processor except for capacity and some minor details. Many schemes have been proposed over the years which attempt to circumvent the mechanical difficulties associated with magnetic recording, while at the same time making use of the concept of sharing transducers and thereby avoiding the high cost of random-access storage. Several recent magnetic schemes are magnetic bubbles, domain tip storage [Spain, 66], and Dynabit [Computer Design, 70]. All three replace the mechanical motion of a moving medium with the intrinsically faster movement of magnetic domains in a stationary media, making use of a shift register configuration to share transducers. The latter two are more recent versions of long tried schemes to use small magnetic domains with magnetization (easy axis) in the plane of the film and magnetic field applied within the same plane [Broadbend and McClung, 60; Rubinstein, McCormack, and Fuller, 61]. Bubbles are different in that the magnetization is perpendicular to the plane of the film while the external moving fields are applied parallel to the plane. The resulting devices and structures have a number of advantages that will become clear later. Another recent type of electronic shift register is the charge-coupled device which is similar in its fabrication to integrated circuits. Magnetic bubbles and charge-coupled semiconductor devices have potential for high density at reasonable speeds and thus will be described with the exclusion of other devices.*

Magnetic Bubble Domain Storage

In order to understand the nature of a magnetic bubble, let us trace the series of events necessary for forming bubbles in an actual experimental case. Bubbles can only exist in thin platelets of suitable magnetic materials with large magnetic crystalline anisotropy and an easy axis of magnetization normal to the plane of the platelet. Such a material in its isolated condition consists of a serpentine pattern of domains as in Figure 5-25 with equal areas of oppositely magnetized domains in the totally demagnetized case so there is no net average remanence. If a bias field, H_b is applied normal to the platelet as shown in Figure 5-26, the domains with magnetization M in the direction of H_b will grow in size while those with M opposite to H_b will decrease in length, eventually reaching a nearly circular "bubble" shape. Thus a bubble is just a small region with one polarity of magnetization which is maintained within another larger region of opposite polarity of magnetization. A bias field of roughly 20 to 50 oerstads opposing the magnetization in the bubble domain is required for typical operation. The bubble diameter depends critically on the amplitude of the bias field, collapsing to zero if the field is too large, or reverting back to the serpentine pattern if too small.

The diameter and hence density of bubbles within the medium is depend-

---

*See Matick, 72 for a review of other devices and additional information.

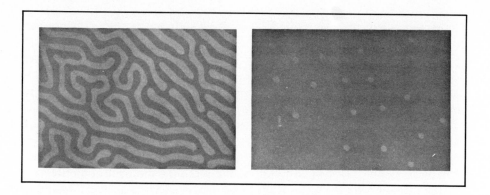

Figure 5-25 Microphotograph of magnetic bubbles as seen using faraday rotation

ent entirely upon the magnetic material parameters. This is very unlike other open flux magnetic film devices in which the bit size and hence packing density is determined by the geometry in combination with the magnetic parameters. Thus bubble density can only be increased with better materials, assuming they can be found. A good rule of thumb for speed determination is that irrespective of actual diameter, bubbles require roughly 100ns per shift from one given location to an adjacent one as described below.

For useful devices, we must be able to generate and annihilate bubbles when and where desired, shift them in some sort of storage register, and sense them, also when desired. Propagation or shifting of bubbles can be done in many ways but one attractive way for shift register storage makes use of magnetic permalloy overlays in the form of T and I bars. The generators and annihilators are also permalloy overlays but of a slightly different shape. These overlays are placed directly on the surface of the thin platelet as shown in

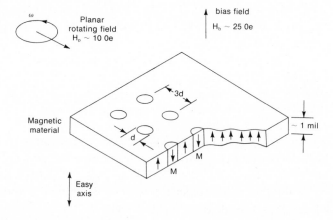

Figure 5-26 Schematic of bubble formation and magnetization pattern in suitable magnetic material

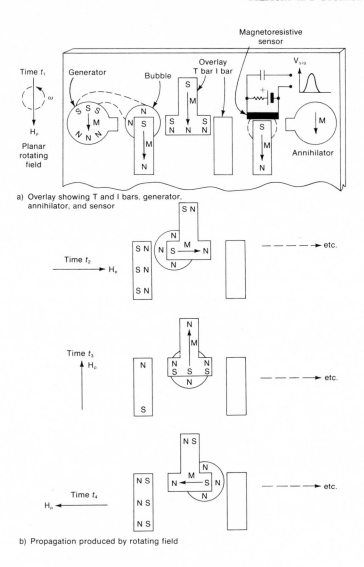

a) Overlay showing T and I bars, generator, annihilator, and sensor

b) Propagation produced by rotating field

Figure 5-27   Magnetic bubble shift register

Figure 5-27 and form a multi-stage shift register, one position for each T and I bar. The operation of the register is based on the fact that the bubbles have a given polarity of magnetic pole on the top surface, let us assume north poles, N, and the opposite pole polarity, south, on the bottom surface. If a magnetic field, H_p, is applied within the plane as shown, the permalloy overlays become magnetized with N and S poles. The dimensions of the T and I bars are chosen such that the long dimension is about three bubble diameters in length whereas the width of the permalloy is smaller than a bubble diameter. The S poles induced along a long dimension on any permalloy bar attract the N poles of any nearby bubble whereas the N poles induced in the permalloy repel a bubble.

The poles induced across the short dimension of any permalloy bar are too closely spaced to have any effect on the bubbles. Thus a bubble will reside, for instance, at the induced south poles on the left-most I bar of part *a* Figure 5-27. Motion of this bubble from left to right is obtained by rotating the applied field H_p as shown. The rotating field induces successive S poles in successive T and I bars with the induced S poles moving from left to right. Each 90 degree rotation of the applied planar field causes the induced south poles to move to the various position shown at times t_1, t_2, and so on. The north poles of the bubble follow these south poles and hence a shift register results.

Generation of the bubbles is illustrated in Figure 5-28. The circular permalloy pattern traps a bubble and causes it to smear out to a larger, noncircular shape. The rotating field induces rotating N and S poles on the circular

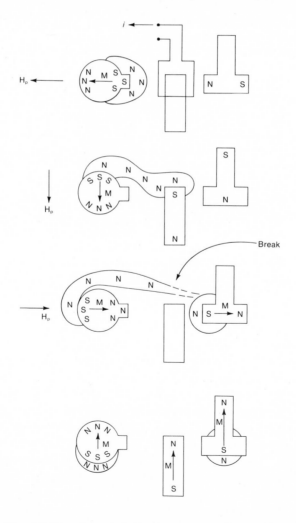

Figure 5-28   Magnetic bubble generation and annihilation

overlay and the trapped bubble rotates continuously. When the generator has a protrusion which is near an I bar as shown, this trapped bubble on the circular overlay can be elongated and eventually breaks, giving a new bubble within the T I shift register. A new bubble is formed for each 360 degree rotation of the planar field. If a bubble represents a stored 1, then a stored 0 is obtained by preventing the expansion of the bubble. This is done by means of a current loop on the first permalloy overlay. A current is applied in this loop which cancels the induced S pole on the first I bar. Hence the first jump of the bubble from the generator to the I bar is prevented, leaving a vacant spot in the shift register.

Annihilation of bubbles is the direct analog of generation; if the direction of rotation of the applied field H_p is reversed in Figure 5-28 the bubbles travel from right to left and the generator now accepts and annihilates the bubbles.

The last basic function needed for a storage system is a sensing transducer. The most convenient means to do this is to use a magneto-restrictive sensing element [Almasi et al., 70] as shown in Figure 5-27. This element is just a small strip of permalloy with electrical connections on the ends. When a bubble appears under the element, the magnetic field of the bubble causes about a 1 to 3 percent change in the resistance of the sensor. A small, DC current passed through this element experiences an increase in resistance causing a decrease in current. When the circuit is capacitively coupled to the sense amplifier, only the change in current, or a pulse, is sensed for each bubble which passes under the sensing element.

These basic functions can be fabricated into many intricate shift register memories. Other functions such as shifting between channels, address decoding, and even logic can be performed with the same basic idea [Chang et al., 71]. Thus all the memory functions of decoding, generation, switching, and sensing can be done with the same technology and all these can be placed on the same chip to minimize interconnections. This is another distinct advantage of bubbles over other magnetic technologies.

Shift register operation has been demonstrated in laboratory models at a shift rate of over 0.3 megabits per second and a density of greater than 1.55×10^5 bits/cm^2) (10^6 bits/in^2) [Bobeck, 70a]. Much faster operation at 3 megabits per second has been achieved using current loops [Bobeck, 70b] to drive the bubble rather than the slower T and I bar configuration. The major problem at present concerns obtaining crystals with adequate parameters. Several classes of materials such as orthoferrites, magnetoplumbites and rare earth garnets can exhibit bubble behavior but only a few presently known materials within these classes have attractive parameters for reasonable density and speed.

MOS Shift Register Storage

Random-access transistor flip/flop cells such as those in Figures 5-15 and 5-16 generally consist of 4 to 6 devices per bit, giving low density compared to other technologies, in addition to increased cost and decreased reliability. Low density has a direct impact on cost since in semiconductor device fabrication, the cost is determined by the number of processing steps for a given silicon wafer, independent of the number of devices fabricated. The individual device size is determined by the limits of photolithography so the bit density decreases and cost increases significantly as the number of devices per bit increases. It is possible to reduce the required storage cell area on a chip by using field effect transistors alone or in combination with diodes, in a shift register configuration as shown in Figure 5-29, for instance. Note that for each bit stored in the register, a minimum of three transistors and three diodes is required, in addition to a three-phase pulsing system. A two-phase and four-phase system are also possible with the latter being widely used. These configurations store data in the form of charge on the gate capacitance of the transistor. This charge cannot remain on the gate for a very long time since it leaks off through the circuit connection to the previous device. Hence the information must be continually regenerated. This is most easily accomplished by continually circulating the data. This type of operation is referred to as dynamic MOS memory and is quite common for low cost, low speed operation.

Charge-Coupled Semiconductor Storage

Another technology which has potential for further cost reductions makes use of the charge-coupled (or charge-transfer) device [Boyle and Smith, 70; Amelio, Tompsett, and Smith, 70]. The essential idea is to store information in the form of electric charge in potential wells which are created at the surface of a semiconductor. These wells are created by means of a dielectric (oxide) interface between the semiconductor and metal electrodes to form a string of capacitors. As the voltages applied to these capacitors are properly sequenced, the underlying potential wells and associated charge can be moved from one well to adjacent ones. The device works as a shift register: minority

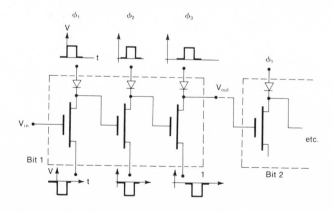

Figure 5-29 MOSFET-diode shift register in a three-phase system (enhancements FETs)

carriers are introduced at some specific point and are then moved around from one electrode to another. In this sense, they are completely analogous to magnetic bubbles where magnetic charges are generated at one point (or not generated) and moved around in a shift register fashion. One difference is that magnetic bubbles are nonvolatile as long as the bias field is present which can easily be done via permanent magnets whereas this device is volatile upon removal of power. In addition, the "stored charge" has a finite lifetime, on the order of a fraction of a second to many seconds as a result of recombination processes.

The basic device is shown in Figure 5-30. A low resistivity silicon wafer is oxidized to form a thin (about 1-kÅ) dielectric layer and metal electrodes are placed on top of this. All electrodes can be operated at some bias V_b. If charges, previously introduced through, say, a p-n junction are now located under electrodes A, then when a potential of say $2V_b$ is applied to electrode B, thereby creating a deeper depletion region, the charges from A will spill over into region B. Electrode B is then returned to voltage V_b and so on. Three electrodes forming a three-phase system are required for each bit location in storage since the charges can move in either direction thereby necessitating at least two electrodes between each separate charge bundle for isolation. One major advantage over other technologies is that the fabrication process is very simple. One disadvantage is that the system must be three-phase with each third conductor connected together forming three separate groups, one for each phase. One serious problem with this device is the charge transfer efficiency from one bit location to the next. Because of traps that can occur near the semiconductor surface and in the interface surface states, it is quite likely that not all the charge from one potential well will be transferred to the next. Charge-transfer efficiencies of 99.9 and 99.98 percent per electrode have been reported [Tompsett, Amelio, and Smith, 70; Gelberger and Salma, 72]. The higher efficiency requires special clocking techniques, so

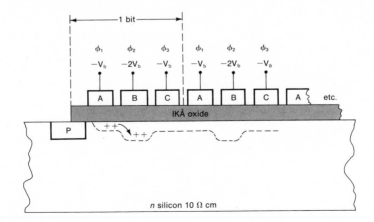

Figure 5-30   Schematic cross section of charge-coupled shift register

the lower number is more realistic. For a memory with 1000 electrodes (1000 shifts) the last bit position can receive only 37.2 percent of the initial charge $(0.999^{1000} = .3676)$. For a three-phase system, 1000 electrodes store 333 bits, and for a two-phase system, store 500 bits. If we specify a maximum attenuation of say 50 percent then this would limit the register length to 700 electrodes (assuming 99.9 percent efficiency) or 233 bits for a three-phase system. For mass memory, a 512-bit shift register seems a bare minimum so additional amplifiers and circuits would have to be inserted at specific intervals which increases the complexity and cost, although not very significantly.

Both magnetic bubbles and charge-coupled devices require additional development to compete with existing technologies but both offer sufficient potential to warrant continued investigations.

5-11. MEMORY HIERARCHIES AND VIRTUAL MEMORY SYSTEMS

The main memory available to a processor has never had sufficient capacity to serve as the sole storage medium. Hence, computers have used memory hierarchies from the earliest times. The first electronic calculator, ENIAC, had a hierarchy consisting of 20 storage registers and 312 words of externally alterable read-only memory all backed up with punched cards for storing intermediate results. The main storage was inadequate for complex calculations so the first computer with address modification, EDSAC, was designed around a 1024-word mercury delay line backed by 4608 words of magnetic drum storage.

Since then, numerous technologies for storage have appeared as well as numerous memory hierarchies. The early hierarchies were generally of the most simple kind; main memory was loaded from, say, a tape and the processor worked on this until all processing was completed. Upon completion, a new segment of information was loaded into main memory and the computation continued. It was the job of the programmer to divide a large problem into independent segments such that no two segments were required at any one time. Thus, for large, complex programs, the burden was placed on the user to ensure efficient operation of the system. As systems have grown both in size and complexity, this burden has become too great and the memory hierarchy has taken on a new character, known as *virtual memory*. Thus, the term "memory hierarchy" is ambiguous and can mean any of many possible configurations. The two most significant types of hierarchies in use today are the main memory-disk virtual store and the cache-main memory hierarchy. While the former is referred to as "virtual store," the latter makes use of the identical principles and concepts, and hence is a virtual store in a fundamental sense. The essential difference is that the cache-main memory hierarchy is aimed only at bridging the speed gap between main memory and the processor whereas the main memory-disk hierarchy is aimed at bridging the capacity gap between main memory and secondary storage. The use of the term "virtual storage" for the latter is thus more descriptive but tends to mask the similarity to the cache concept. In the following discussion of virtual storage, we shall outline the general fundamental requirements applicable both to the cache-main memory and the main memory-disk hierarchy. The differences lie mainly in the implementation of the required functions with cache systems having more hardware for high speed and virtual systems implemented more in software by way of complex operating systems, mainly for low cost and greater flexibility.

Fundamentals of Virtual Memory Systems

The primary goal of a virtual memory system is to combine two storage systems that have widely different capacities, speeds, and costs such that the combined system has very nearly the speed of the smaller, faster memory but the cost and storage capacity of the more inexpensive and slower memory. We will often refer to the smaller, faster memory as primary storage and the slower, larger, less expensive backing store as secondary storage. However, for sake of expediency, the basic ideas will be described using main memory as the primary store and disk as the secondary store except where noted. The storage capacity of the primary store is thus virtually extended by proper interconnection with the larger store. This extension of capacity implies a very important characteristic of a virtual system, namely each user must separately be able to address the virtual memory in terms of the total size of the processor logical address, irrespective of the actual capacity of the smaller memory.

The goals of a virtual system can be achieved only as a result of the fortuitous fact that for most processing, the references to memory tend to be highly localized or cluster in small groups at any given time and the regions tend to change relatively slowly during the course of instruction execution. Virtual memory systems can make use of the locality property of program references by providing a way for the active regions to be available in primary memory, and by exchanging regions between primary and secondary memory when activity shifts from region to region during program execution. It is most reasonable to break a program into many program segments, where a segment is a portion of a program that becomes active or inactive as a whole unit. However for ease of implementation programs are usually partitioned into equal-size portions known as *pages*.

Both primary and secondary memory are partitioned into equal-size pages, and the partition of a program into pages includes both the program instructions and data. If a page reference is made to a page in secondary memory, then it is first transferred to primary memory before the reference is honored. To make room for the new page, an inactive page is transferred from primary memory to secondary memory. This operation requires some replacement algorithm to determine which page in main memory is to be removed, and a mapping function to determine where it is to be relocated on secondary storage. Also an address translation function is required to locate the desired page in secondary storage and keep track of the actual location of both pages that have been moved. Page replacement is a time consuming operation, so in the multiuser systems, the processor transfers operation to a different user whose pages are in main memory and continues processing.

The degree of localization or clustering of memory references within the entire system is specified by the *hit ratio* which is the ratio of the total number of memory references that are found to be already present, divided by the total number of memory references, for a given period of time. The *miss ratio* is 1 minus the hit ratio. The hit ratio cannot be determined from first principles but rather is obtained by simulation of the memory references of actual programs [Mattson et al., 70; Murphy and Wade, 70] or by direct measurements on running programs. Figure 5-31 shows one example of the former type. The memory references required by a large number of typical FORTRAN programs, called the *job stream*, are analyzed with the aid of a computer program. In this case, referenced pages not found in primary storage when needed are assumed to be relocated and one is removed based on least recent use as described later. The analysis program counts the total number of misses and total number of memory references to give the miss ratio.

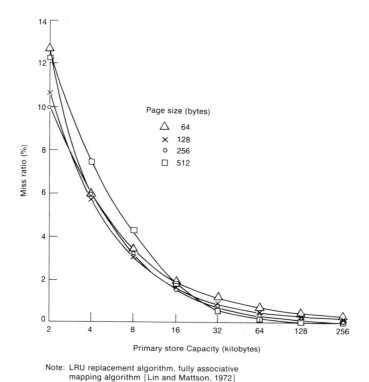

Note: LRU replacement algorithm, fully associative
mapping algorithm [Lin and Mattson, 1972]

Figure 5-31 Miss ratio versus primary storage capacity for various page sizes in a typical FORTRAN job stream

Figure 5-31 shows that the miss ratio decreases or improves as the size of primary storage increases as would be expected. It also shows that for small primary stores below about 16K, better miss ratios are obtained for a smaller page size whereas above this, larger page sizes give better results. For a cache-main memory system, an average miss ratio of 1 to a few percent is usually quite adequate indicating a cache of 16K to 32K is desirable with a page size of 64 to 128 bytes. For a main memory-disk virtual store, a much better miss ratio of about 100 times smaller is needed. While not easily seen in Figure 5-31, this requires a primary (main memory) capacity of 128K or larger and pages of 512 bytes or larger. It should be noted that the results of Figure 5-31 are merely statistical distributions of memory references and are independent of all memory access, cycle, and transfer times. The results clearly show that memory references do indeed cluster and that for certain assumed characteristics, there is a preferred page size.

With this knowledge of preferred page size and achievable miss ratios, the implementation of a paged virtual memory system requires the following minimum fundamental functions:

1. Word addressing within a page
2. Page mapping function
3. Page address translation
4. Page replacement algorithm

A mapping function is required to compress the larger virtual store into a smaller primary memory. This virtual store is "real" in that the storage capacity must exist and ideally must all eventually reside in primary storage. Since it cannot all reside in primary storage at one time, a method for sharing page slots is required. This is done by the mapping function which maps virtual pages into main memory during a page request and also relocates the "real" page removed from main memory and converts it into a virtual page within the virtual store. There are a number of different mapping functions possible. *Direct mapping* results if a given page has a fixed location in virtual storage and can only be located in one specific page slot in main memory as in Figure 5-32. Since there are many more pages in the virtual store, many virtual pages will have to share the same slot in main memory at different times. This can result in contention problems and a low hit ratio if two pages which share the same slot are required on successive memory references. This contention problem can be eliminated by allowing a virtual page to occupy more than one page slot in primary storage. The number of such slots which a virtual page can occupy is called a set and the mapping is known as *set associative*. The case for two primary storage slots per set is shown in Figure 5-33. If both page 1 and 2 are needed simultaneously for instance, they can both be present in the first set. Contention problems still can arise but are now less likely. For instance, if page 1, 2 and Q + 1 are simultaneously needed for a calculation, considerable page swapping may be required because of the high miss ratio in such a case.

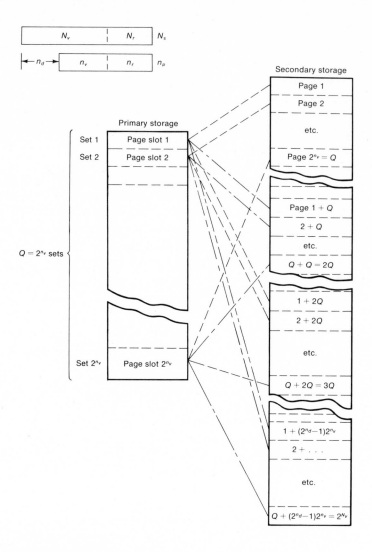

Figure 5-32 Direct mapping schematic with one page slot per set and $Q = 2^{n_v}$ sets in primary storage

Contention problems can be further reduced by increasing the number of page slots per set. The contention problem is minimized when the size of the set becomes equal to the number of page slots in primary storage. This represents the case of *fully associative* mapping as shown in Figure 5-34. In the general associative case, a given page can have any location in either secondary or primary memory and the exact location in either can be different at different times, that is, after page relocations. This places a greater burden on the mapping function and address translation to locate available spaces as well as keep track of the actual location of all pages. The advantages are greater flexibility and higher hit ratios. This fully associative mapping is

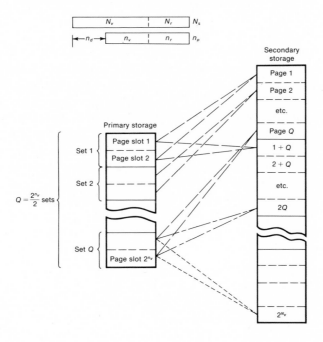

Figure 5-33   Schematic of set associative mapping showing two page slots per set

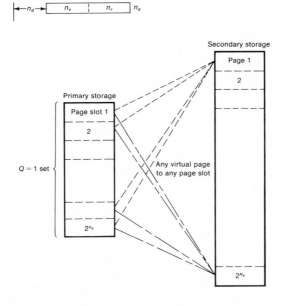

Figure 5-34   Schematic of fully associative mapping with 2^{n_v} page slots per set

used in the IBM System 360/370 virtual systems, Honeywell MULTICS, and DEC PDP-10 Tenex Systems.

The direct and fully associative mapping schemes represent the two extreme cases of set associative mapping, with limited and unlimited flexibility respectively in page replacement and relocation. There is another class of mapping functions known as *sectoring* in which sectors or groups of pages are associatively mapped into main memory. This is similar to Figure 5-34 except the page slots become groups of page slots and only those virtual pages needed at any one time reside in primary storage. The set associative is the more common mapping function used for cache-main memory hierarchies. The IBM computer models 195, 155, 165 use set associative mapping while the model 85 uses sector mapping for the cache .

In order to understand the page address translation function and word addressing, consider the general case of a multiuser virtual system in which there are U users, each with a total capacity of $2^{N'_v}$ virtual pages as in Figure 5-35. The effective address bit strings necessary to address the virtual

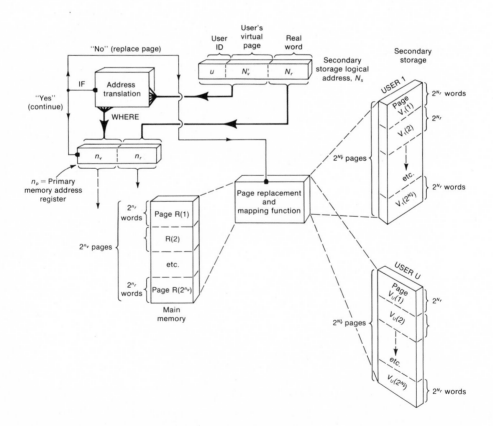

Figure 5-35 Schematic of basic functional requirements in multi-user virtual memory system showing information flow during translation of virtual address to main memory address

system are shown in Figure 5-36. The total secondary address length is N_s as in part a, the total primary address length is n_p as in part b and when secondary storage is a disk, the address must ultimately specify the drive, head track, and record number as in part c or other appropriate means for finding the required page on disk. The N_s bits in part a are broken into a virtual part N_v which represents the total number of pages in the virtual store, and a real part N_r which represents the number of words per page. Each user has $2^{N'_v}$ pages: U users require u address bits where $U = 2^u$. The total virtual store thus contains $2^u \, 2^{N'_v} = U \, 2^{N'_v}$ pages. In a single user system, $u = 0$ in which case $N_v = N'_v$ and only $2^{N'_v}$ pages are present in the virtual store. In the general case with $u \neq 0$ the secondary address bit string of N_s bits as shown in part a of Figure 5-36 must be converted to either n_p bits as in part b with $n_p < N_s$ or into an equal number as in part c but with a different string of bits for the correct disk address. In all cases, the word addressing within a page is simple because the lower order N_r address bits are real and are used directly as the lower order n_r bits in part b or as N_r in part c. Thus the address translation must only be concerned with translating the N_v virtual page bits into n_v in part b or into a different string of N_v bits in part c. The latter conversion from part a to part c is relatively easy. Pages are usually allowed to reside wherever convenient in disk so a table, to be called *Table I* is used for address translation. In such a case, the address bits in part a point to the word in the table which contains the correct address string such as in part c.

The address translation function converting N_s bits into n_p is the most difficult and the one that has the greatest effect on the overall efficiency of the system. For the general case, given N_s bits, then all these bits must be used to address the word in primary storage. However, the primary storage register

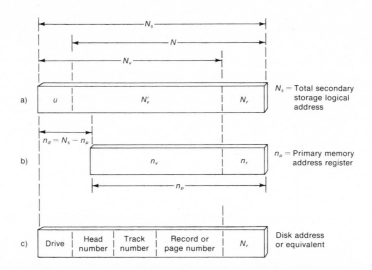

Figure 5-36 Definition of address bits at various levels of a virtual storage system

commonly contains n_p bits, hence the deficient bits $n_d = N_s - n_p$ must be decoded separately. In fact, the essence of the translation is that a portion of the N_v bits are converted to n_v bits which specifies WHERE the desired virtual page should be and the same N_v bits are separately decoded to indicate IF the required page does, in fact, reside at that location in primary storage. This IF function need be only a 1-bit control on the primary memory address register and also on the page relocation function as in Figures 5-35 and 5-37. A "yes" on the IF part of the address translation would allow the primary memory to proceed. A "no" on the IF control would transfer control to initiate relocation of the desired page. The IF and WHERE parts of the address translation function are accomplished in two different ways, namely by the use of a page table, called *Table II,* or by the use of tags which are stored in a small directory which is separate but adjacent to the primary memory. The table translation is very slow but is relatively inexpensive and flexible. The tag scheme is potentially very fast, but expensive and less flexible. Table translation is usually used for main memory-disk virtual systems because of the need for low cost and flexibility. In some cases, the slow speed of table translation can be greatly improved by supplementing it with a small amount of tag decoding as will be seen in Example 5-1 to follow. The tag directory is used in cache-main memory hierarchies such as that in Example 5-2.

The fourth requirement above, namely the page replacement function, has the responsibility of keeping track of the page usage in main memory and must constantly provide the identity of the page which can next be removed. When a requested page does not reside in main memory and main memory is full, the removal of a page to provide a slot for the new page is most often based on the criterion of removing the page least recently used (LRU) or an approximation to LRU.

All these necessary functions are shown in separate boxes for clarity in

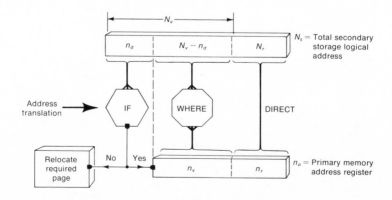

Figure 5-37 Schematic of translation of virtual logical address, N_s, into primary address showing IF, WHERE, and DIRECT components

Figure 5-35 but it should be remembered that they can be implemented in software, hardware, or a combination. Also, the functions are often so closely interrelated that one cannot always put them in separate boxes as we shall see.

The fundamental distinction between the main memory-disk virtual store and the cache-main memory virtual hierarchy is the hardware versus software implementation of the four basic functions. In the cache system most of the basic functions are implemented in hardware for speed while in the main memory-disk system, some additional hardware, which is directed and supplemented by considerable software, is used for lower cost. Since even in the latter, certain functions such as mapping tables must exist physically, the software makes use of the main memory for storing needed information. Thus, while no special hardware is needed for certain functions, additional hardware is required in the form of main memory which cannot be addressed by the users.

Segmentation Versus Paging

A decision of central importance in virtual memory design is whether the "chunk" of information transferred from secondary to primary storage should be of fixed size, called a *page,* or of variable size, generally called a *segment.* In early, simple hierarchies where independent problem segments were simply overlayed in main memory as needed, these segments were always of variable size. Hence it seemed natural that the size of the referenced chunk of data in a virtual memory system should also be allowed to vary for optimum results. The first commercial, virtually addressed system, the Burroughs B5000, provided for such variable segment size. However, segmentation tends to be very wasteful of physical memory space for the following reason. A segment, as originally conceived, requires contiguous words in primary storage and the segments can all be of varying lengths. A request for transfer of a new segment into primary storage requires locating an empty region of the proper minimum size. The empty regions currently available may not singly be large enough even though their sum may be more than sufficient. However, since they are not necessarily contiguous, they cannot be used and on a statistical basis, many regions of primary storage can remain unused. An attempt to use a replacement algorithm for removing one segment to allow for the new one only further complicates the memory allocation process. In a paged system using pages of a fixed size, contiguous words are still required in storage. However, transferring a new page only requires finding or creating an empty page slot in primary storage. Since the slots are all of the same size, this is considerably easier than finding or creating contiguous segments of varying size. Thus memory management and the resulting efficiency is considerably better and represents one of the major reasons for the use of fixed pages in all virtual memory systems.

In a fixed page virtual system, a single table (that is, Table II previously defined) can be used for the address translation function as shown in Figure 5-38. In such a case, the user ID register provides the origin address in primary storage for that user's page table. The N'_v bits of the virtual address increment from this origin address to the entry in the table which contains the correct address for this desired page. The correct higher order n_v bits are then read out of the table and placed in the primary storage address register. A subsequent reference is made to the word specified by the n_v bits catenated to the real n_r bits as shown. It should be understood in this figure that all references to primary storage, including the user ID bits and N'_v virtual address must first go either to the primary address register or be paralleled with this register into the word decoder. A considerable amount of additional logical control, not shown, must be included to perform these operations.

The single table scheme of Figure 5-38 for address translation is fundamentally workable. However, there are a number of practical problems which lead to the use of a two-level table translation scheme. Two serious practical problems are:

a) The total size of all user's Tables II which must remain in primary storage for efficient operation can consume large amounts of primary storage.

b) Sharing of common data, subroutines, and so on among several users becomes complex.

With respect to *a*), a multiuser system requires tables of a number of users to be present simultaneously in primary storage for efficient task switching during a page miss. As shown in Example 5-1 below, these tables could become quite large unless organized differently.

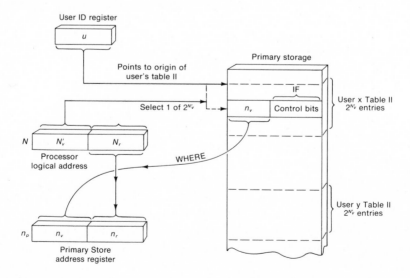

Figure 5-38   Virtual memory address translation using single page table for each user

With respect to *b*), sharing common pages with only a single Table II per user presents formidable updating problems. Several users sharing a common page which is present in primary storage requires each user's Table II to contain the real address of the common page. If this page is moved out of primary storage and subsequently brought back to a new page slot, all common user's Table II must be updated. Some of these users may not be current in the system, that is, their Table II may be in the secondary storage. An elaborate tracking and updating procedure would be required. Hence Table II is usually broken into two parts, a small so-called segment table and a larger page table. The segment tables are small enough so that they can be left in primary storage for all users. Sharing can be done at the segment level by allowing common users to have identical page table references in some of the segment table entries. Updating these segments is considerably simplified since they are always current in primary storage.

Example 5-1: In order to better understand the above general principles, let us consider the IBM System/360 Model 67 Virtual machine and see how the various fundamentals components are implemented. This system consists of essentially the components of Figure 5-35 but with the user address bits, *u,* contained in control register 0 which is a full 32-bit register. In principle, this allows a very large number of users but the permissible number is limited by practical constraints such as the size of main memory and speed of the processor. Part of the address translation and part of the mapping function uses tables stored in main memory. Hence the actual size of main memory available for storing pages is less than the total main memory capacity. (There are other, even more significant overhead factors in the operating system which significantly reduce the usable size of the main memory.) These tables are required for keeping track of actual page locations since a fully associative mapping of pages in both main memory and disk is used. These two tables, Table I and Table II, were previously described, with the former storing actual locations of all pages on disk and the latter storing the actual locations of pages that reside in main memory at any given time.

Parameters for the Model 67 are:

$N = 24$ bits or $2^{24} = 16$ M′ bytes of virtual storage per user on disk

$N_r = 12$ bits or 4K′ bytes per page

$N_v' = N - N_r = 12$ bits or 4K′ pages of virtual storage per user
where $N = N_s - u$ is the processor logical address length exculsive of the user ID bits and represents the normal address length on nonvirtual IBM 360/370 systems.

The size of main memory varies with the particular installation but 256K′ usable bytes is a reasonable value giving $n_p = 18$ bits and $n_p - N_r = 6$ bits or $2^6 = 64$ pages total in main memory. Let us trace through the sequence of events as they occur in the addressing of a page. Assume the memory is completely full and a page request enters the logical address along with a user ID in the register. Referring to Figure 5-35, the lower order 12 bits of N_s are the lower order address bits of the real word and are directly executable. Determination of whether the requested page is in main memory is done in two different ways. Strictly for purpose of speed, a set of eight associative registers which contain the eight most recently used pages is searched in one cycle to find if the requested page is in main memory. If so, the correct binary address for the n_v or 12 higher order address bits of that page is immediately provided. If the page is not one of the eight most recently used, it could still be in main memory so a Table II similar to that of Figure 5-38 is used to find IF so and WHERE it is located. In the actual case, the associative search and Table II look-up are initiated simultaneously for increased speed since the latter is slower. Since Table II is implemented in main memory hardware, it cannot be associatively searched. Furthermore, any searching routines such as binary search would be time consuming. Instead, the exact address of the table word which will contain the information is obtained as follows. Each user has his own Table II: the user ID located in the ID register specifies the origin or beginning address of this table. Table II actually consists of two tables called segment and page tables to form a type of table hierarchy as shown in Figure 5-39. This is done for reasons previously indicated, namely to allow sharing and to save memory capacity. The latter can be quite significant. Suppose, for instance, that there were 32 users on the system and each user was occupying 4K' pages which is the limit. Since each page requires a 2-byte word entry in Table II, this would mean 32 tables of 8K' bytes each or 256K' bytes just in Table II which consumes the entire memory. This number can be greatly reduced in the following way. While the total size of this table must remain constant, the statistics of the problem execution requires that only a portion need reside in main memory at any one time. This results from the fact that no user requires all 4K' pages simultaneously. A small portion of Table II can be stored in main memory with the remainder stored on disk and transferred in as needed. This is accomplished by breaking N'_v into two parts of 4 and 8 bits respectively. The 4-bit part refers to one of 16 segments while the 8-bit part refers to one of 256 pages. The origin location obtained from the user's ID register points to the segment table of that user as in Figure 5-39. The 4-bit segment portion of N'_v specifies how deep into the segment table the required entry is located. The segment table entry so-located contains the page table origin for that segment. To speed up the translation process, an 8 bit "length" field is stored in the segment table; this field specifies how long the page table really is: the page table can contain between 1 and 256 entries, depending on past usage. The binary value of the 8-bit page portion of N'_v is compared with the binary value of the length field. If the former is larger than the latter, the required page cannot be present and a page relocation cycle is initiated. If the former is smaller or equal to the latter, then the page table origin obtained from the segment table is concatenated to the 8-bit page address of N'_v to access the page table and obtain the correct value of the higher order page address bits, n_v as shown. A sub-

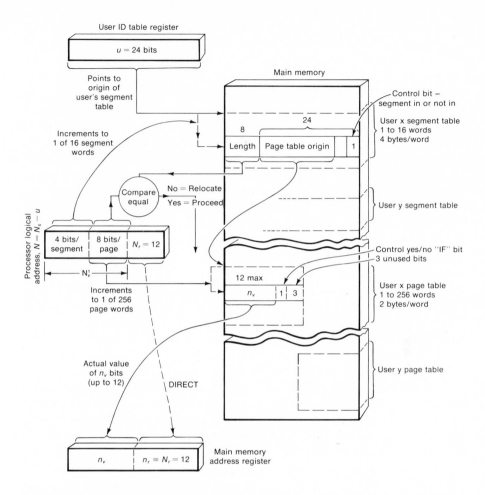

Figure 5-39 IBM System 360/370 virtual memory system showing address translation using segment and page tables

sequent fetch using these bits and the n_r real bits finally produces the required word. Both the segment table and page table entries contain a control bit which indicates whether or not the segment or page is actually present. If either control bit is off, a relocation cycle is initiated.

Note that the address translation via these tables requires two main memory cycles to obtain the correct address and a third cycle to reference the desired word. All memory references require address translation and since this table look-up is slow, the small, fast associative registers mentioned previously are very important in obtaining a high average speed.

The two tables work in principle the same as the single table of Figure 5-38 except the word capacity required to store one segment is reduced to $16 + 256 = 272$ words per user and only 17,408 words for 32 users instead of the entire memory. Since the segment table is small, consisting of 16 words or segments per user, it is stored permanently in main memory for all users and allows easy sharing of pages. The total page table for each user consists of 256 words for each of the 16 segments. However, only those segments required by the immediate task solution are maintained in main memory.

It should be noted that these segments can have a variable number of pages and thus in this sense are similar to the original concept of a variable segment size. However, this similarity is artificial; these segments must contain an integer number of pages and since pages are typically large, the segments are not continuously variable as in the original concept. In addition, the use of separate segment and page tables is not motivated by a need for variable segment size. Since any user can have as many pages as needed, up to $2^{N'}_{v}$, the single table scheme of Figure 5-38 provides this flexibility just as well as a two-level segment and page table scheme.

The same overall characteristics as described previously are also used on the IBM System/370 Models 158 and 168 but with some changes in detail. These systems also use a cache-main memory hierarchy in addition to the main memory-disk virtual store.

The Honeywell MULTICS system also has overall characteristics very similar to those already described although there are considerable differences in detail. One important distinction between MULTICS and the IBM 360/67 paging and segmentation scheme is that segments are independent entities in MULTICS but not in the IBM system. Since the IBM addressing scheme uses conventional addresses and extracts page and segment information from these addresses, the last item in a given segment is contiguous to the first item of the next segment. Indexing operations in a given segment that run beyond the end of that segment automatically overflow into the next segment. In the MULTICS system segments are logically distinct entities and are not contiguous. An indexing operation that runs beyond the end of a segment generates an invalid address rather than an address in the next segment. Segments are designed to be shareable quantities in the MULTICS system, and more easily shared among programs than segments of the IBM-type because of the differences outlined here.

Example 5-2: A cache-main memory hierarchy is designed mainly for speed and hence the address translation is often different than that described for Example 5-1. As an example, we consider the IBM System/360 Model 195 and concentrate on the fundamental differences from the previous examples. The cache memory consists of a random access array using 3-terminal storage cells. The cycle time of the array itself is 54ns which equals the basic processor clock cycle time. Pages are 64 bytes in length since this is more optimum as previously described. The cache capacity is 32K′ bytes or 512 pages maximum. Main memory is 1M′ to 4M′ bytes of core storage with 756ns cycle time. A physical word consists of eight bytes so one access to a physical word every 756ns produces eight logical words or eight bytes. The mapping function is set associative with four page slots per set.

In order for the cache system to be efficient, two design factors are important. First, it is desirable to perform a memory reference with only one access time to the cache plus a minimum of other delays for address translation and data transfer. Second, when a miss occurs, the processor does not transfer to a new user but rather continues with the current task. So it is desirable to have a fast page transfer from main memory to cache in order to minimize any possible subsequent delays.

In order to allow fast address translation but without consuming large amounts of additional storage for the translation, the address translation is quite different from that in the previous example of Figure 5-39. Irrespective of implementation, the address translation must determine WHERE a desired page might reside in the cache and IF in fact it does. The IF part is performed by the n_d bits which are stored as tags in the directory. The WHERE function is partly predetermined since a given page can only reside in one set. Since there are 512 total pages in the cache and four possible page slots per set, there are only 128 sets available. Let us decode these sets with address bits called q' in Figure 5-40 where $q' = 7$. The decoding of q' is a built-in hardware algorithm. The selection of one of the four slots within a set, rep-

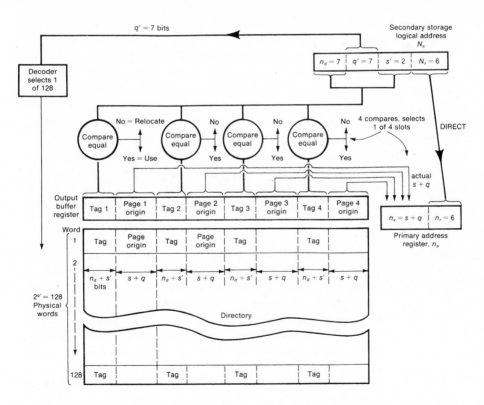

Figure 5-40 Tag directory system showing various components for address translation of N_s to n_p with set associative mapping and four page slots per set

resented by the address bits s', where $s' = 2$, must be done separately as must be the decoding of the n_d bits.* This is done as follows. The directory contains 128 physical words with each word storing four tags of length $n_d + s'$ and the correct $s + q$ bits associated with each tag. The virtual q' bits are used to select one of the 128 words which is read into the output buffer register, yielding four possible locations. Four logical compares are simultaneously made of the $n_d + s'$ bits of the virtual address with each of the four tags. Only one of these can give an equal or "yes" which then gates the correct page location bits, $s + q$, into the primary address register. The n_r bits are real so the translation is complete. If none of the four compares gives a "yes," then a page transfer is initiated.

While the cycle time of the cache array itself is 54ns and equals the processor clock cycle time, a full cycle of the system requires three such cycles for a total access time of 162ns. The first cycle gates the information from the processor to the directory, accesses the directory and obtains the correct page address. The second cycle accesses the cache and produces the required word in the cache output buffer register, which is contained in the buffer storage control unit. The third cycle gates this data to the proper section of the processor as required for processing. Hence the data is then available for use by the processor only after three machine cycles. Note that since the correct beginning location of a page is contained only in the directory, pages can reside anywhere in the cache and need not be contiguous.

The replacement algorithm is the least recently used or LRU procedure previously discussed. However, since a page can only reside in one of four possible page slots, the least recently used page of each set must be tracked. In any design using LRU replacement, the order of all elements relative to one another in terms of usage must be stored. For four elements or four pages per set, there are 4! or 24 possible combinations of relative usage. This requires five bits minimum plus an occupancy control bit, giving a 6-bit usage control function for each of the 128 sets. These are stored in special hardware with separate control and updating algorithms. Two other important features which increase the overall speed are store-through and load-through. Store-through merely ensures that when a change is made in the cache, it is simultaneously made in main memory. Thus when a miss occurs and the cache is full, the least recently used page within the proper set can just be erased without having to write possible changes back into main memory. Load-through arranges the order of page data transfer when a miss occurs. It ensures that the referenced 8-byte word of the desired page is retrieved first and simultaneously provided to the processor and cache. The remaining bytes are then transferred. This provides the necessary information to the processor with minimum delay.

Performance evaluation of the IBM 360/195 cache [Murphy and Wade,

---

*Note, $q' = q$ and $s' = s$; these symbols represent only the number of address bits in each part and are not the actual bit sequences themselves. The prime is used to indicate that the secondary address bit sequence is different from that in the primary address although the number of bits is identical for equivalent parts.

70] using 17 job segments indicates that the effective cycle time of the hierarchy is about 162ns with occasional increases to 175ns. The 17-segment job stream contained a mixture of commercial processing with moderate amounts of decimal arithmetic, scientific, engineering, and systems type processing (sorting, assembling, compile, link, edit). The average hit ratio for the buffer during simulated processing of these 17 segments was 99.6 percent. Smaller hit ratios still give quite adequate performance.

5-12. PROBLEMS

5-1. Show that the number of bits needed in the address register for addressing main memory of a given size is identical whether it is organized as 2D, 2½D, or 3D.

5-2. Organize in block diagram form all the essential features of a read-only memory, indicating the fundamental device requirements needed for operation and the potential circuit saving as in Table 5-2.

5-3. Show the general structure of a 4096-word, 64-bit per word core memory when organized as 2D, 2½D, and 3D. Assuming one circuit per line, indicate the number and type of circuit (driver, sense amp) for each organization.

5-4. Organize, as in Figure 5-9 or 5-10, a 2D random-access memory of 32 words by 32 bits per word using the chips of Figure 5-9.

5-5. Organize a 3D random access memory of 4K words by 64 bits/word using the undecoded chips of Figure 5-9. What are some of the disadvantages of such an organization?

5-6. Show how 2-terminal cells with no inherent coincidence selection capabilities can still be used in a random-access storage array.

5-7. Given a magnetic recording system using NRZI coding, show the type of distortion that bit crowding introduces into the read signal when a long string of 1 bits is read.

5-8. For a single magnetic recording channel, determine the linear bit density and transport speed necessary to achieve a data rate of 10^7 bits per second.

5-9. For a magnetic tape reel of 2400 feet in length, recorded at 1600 bits per inch and containing a single file, determine the block size necessary to give an 80 percent recorded surface utilization.

5-10. Indicate for a virtual memory system where an associative memory could be used and the approximate size for the system of Figure 5-39 with $N = 24$ bits $N_r = N_v = 12$.

5-11. Show how the T and I bar patterns of Figure 5-27 can be used to produce a continuously circulating bubble shift register.
Hint: Use a similar second channel adjacent to the first and connect the ends properly.

CHAPTER 6

Input/Output Processing

William G. Lane

In their "Preliminary Discussion of the Logical Design of an Electronic Computing Instrument," Burks, Goldstine and von Neumann specified: ". . . there must exist devices, the input and output organ, whereby the human operator and the machine can communicate with each other." [Burks et al., 46]. They envisioned this communication in support of a one-man, one-machine environment, allowing for the possibility of simultaneous input/output and computation, for storage of data in secondary memory (part of the input/output function), and for interactive operation between the user and his program.

But, since that time, the function of input/output processing has been greatly expanded and has become increasingly complex. Now, in contrast to the early days when each user was directly responsible, the input/output function has been largely turned over to the computer. Multiprogramming and timesharing systems allow multiple users to access both private and shared programs and data files simultaneously. Users compete for computer resources, sometimes by order of request and sometimes on a priority basis. Jobs are broken up, interleaved, and completed as time becomes available. And through it all, it is the input/output processor function that must ultimately "put it back together" and ensure that each user program receives all inputs and transmits all outputs to the proper places, in the proper order.

6-1. THE INPUT/OUTPUT PROBLEM

Before attempting a discussion of the architecture pertaining to input/output processing, it is appropriate to examine the factors that contribute to the problem of interfacing the central processor to the outside world.

Codes

Codes exist in computer systems at three levels: internal to the central processor, in the interconnect link between the processor and external devices, and within the external devices themselves. Except for the American Standard Code for Information Interchange (ASCII), which has been established for character-set-sensitive input/output communication, there is little code standardization. Internally, the processor may be binary or decimal, word or character oriented; Bell and Newell [71] list twenty-five different internal word lengths (from 4 to 128 bits) for machines built through 1970. Externally, the peripheral devices usually reflect the code of the storage or input medium as shown in Table 6-1. In between, the communication links conform to the physical restrictions of the devices and the central processor. Code translation may and can be required between each level. A recent survey of 313 teleprints and CRT terminals manufactured by 75 companies showed the communication code distribution shown in Table 6-2 [Bowers, 73].

Operating Rates

The difference in response speeds of physical processes, man (the user), mechanical devices, and electronic circuits constrains computers to operate in an asynchronous mode such that each part of the system works at its own

TABLE 6-1. COMMON CODES OF EXTERNAL DEVICES

Device	Codes*
Magnetic tape	Word binary, character, byte
Magnetic disk/drum	Word binary, character, byte
Punched paper tape	Word binary, character (5, 6, 7, 8)
Punched cards	Column binary (12), Hollerith, character
Printers	Character, byte
Plotters	Character, byte
Interactive terminals (keyboard, light pen, cursor, tablet, and so on)	Character, byte
Instrumentation, Analog/digital converter	Signed binary word (10 to 15 bits), character, byte

*Unless otherwise noted, word binary is assumed to be the internal code length of the processor; character, 6 bits; byte, 8 bits. All others show the bit count variations in parentheses.

TABLE 6-2. COMMUNICATIONS CODE DISTRIBUTION

Code	Number of Terminals*
ASCII	278
EBCDIC	29
IBM Selectric Correspondence	12
BAUDOT	9
Unspecified Special	10

*Several terminals allow selection between more than one code.

TABLE 6-3. RANGE OF TYPICAL OPERATING SPEEDS

Device	Device Rate	Character Equivalents per Second (Maximum Rate)
Physical systems measurement		
Individual	.01 to 50 samples per second	100
Multiplexed ADC	up to 50,000 samples per second	100,000
Man	up to 5 actions per second	5
Communication modems	300 to 9,600 bits per second	960
Interactive terminals		
Teleprinters	10 to 60 characters per second*	60
CRT	10 to 240 characters per second	240
Graphic Plotters		
Incremental	up to 700 characters per second	700
Vector	up to 200 characters per second	200
Paper tape (5, 6, 7, 8 level)		
Readers	10 to 1,000 characters per second	1,000
Punches	10 to 150 characters per second	150
Cards (80 and 96 columns)		
Readers	100 to 2,000 cards per minute	3,200
Punches	100 to 250 cards per min	400
Line printers (80 to 132 columns)		
Impact	100 to 3,000 lines per minute	6,600
Electrostatic	100 to 40,000 lines per minute	88,000
Magnetic tape (7, 9 track)	15K to 320K characters per second	320,000
Magnetic disk/drum	30K to 1.5M characters per second	1,500,000

*Characters are assumed to be interchangeable with bytes in indication of operating rates for device classes.

pace. As can be seen from Table 6-3, this range is quite wide, necessitating speed compensation at the boundaries between each system level.

Timing and Control

In normal operating environments, the central processor may be simultaneously communicating with one or more of its external devices (of the same or different types), but seldom with all and usually not in any predictable pattern. This, coupled with the speed change requirement noted previously, requires the establishment of timing and control procedures to effect a proper connection and to provide "momentary interface synchronization" for the individual message pulses.

For communication initiated by the central processor, a simplified but typical sequence of events is:

1. The processor selects the desired external device and determines its operational status.

2. The processor signals the device to connect itself to the processor.
3. The processor requests the device to initiate an input operation and to transmit the message elements to the processor.
4. As each element is ready for transmission, the device signals the processor that a new datum is ready and must be read (this step is repeated until the entire message has been transmitted).
5. The processor detects the end of message and logically disconnects the device.

The reverse case, with transmissions initiated by devices, is similar except that the device must signal the processor to interrupt the program flow. A program interrupt in the input/output function is a signal to the processor that an external device needs attention, much like the ringing of a doorbell or a telephone signals that someone outside wishes to communicate with those within. During the program interruption, the processor reestablishes the connection before transmission begins. In both cases, action must be taken in a timely fashion to ensure that no signal is unintentionally disregarded and no data are lost.

Communication Link Structures

Internal data communication within a processor is, in general, word-parallel where each bit requires a separate wire (or equivalent blocks of 2, 4, or 8 characters). The structure of the link between a processor and an external device, on the other hand, may be configured to allow transmission serial-by-bit, quasi-parallel (serial-by-character), or fully parallel (serial-by-word). The choice depends on the required speed of operation of the device, its proximity to the processor, and the projected cost of the link. Telephone charges, for example, with multiple circuit (parallel) transmission from remote devices can outweigh any speed advantage gained over single circuit (serial) transmission. Device control, selection, status, and timing synchronization must be transmitted over additional parallel circuits or can be imbedded within the message pulse train itself. In parallel transmission, care must be taken to ensure that all lines in the parallel path exhibit the same electrical characteristics so that individual data and control bits are not delayed (skewed) with respect to each other. In serial transmission, a constant clock rate must be maintained over the duration of the message to avoid loss of synchronization.

Errors

Errors, particularly in input/output and other data transfer operations, present major problems in computer systems [Gray, 72; Mills, 72]. Their detection, location, and correction depends, in part, on successful prediction and analysis of the probable manner and frequency of failure, not only in the thousands of components and circuits that make up the hardware, but also in the almost limitless combinations of instructions and codes that make up the program and data structures. To further complicate matters, errors may be intermittent or continuous in duration, single or multiple in occurrence, repeated or random in pattern and internal or external in origin. The following list gives some of the sources of error and some insight into the magnitude of the detection problem:

1. Environment
   a) Dirt, moisture and pollutants in optical, magnetic, and mechanical storage media and devices
   b) Temperature and humidity
   c) Electromagnetic radiation, lightning, and so on
   d) Electrical power surges and transients
2. Component aging and misadjustment
   a) Circuit parameter drift
   b) Mechanical wear
   c) Skew
3. Previously undetected system "bugs"
   a) Unanticipated instruction sequences and code combinations
   b) Incorrect memory allocation and input/output buffer size
   c) Incompletely planned and tested combinations of system modules
4. User and operator mistakes
   a) Missequenced programs
   b) Incorrect operating and user procedures
   c) Unmounted or mismounted data storage media

The goal is to be able to locate and identify the malfunctioning component or procedure so that it may be replaced or corrected. In the meantime, the system should be able to recover from any errors that have occurred and, if possible, function on diminished power while the correction is in progress. All of this requires some form of redundancy, not only in hardware but also in code, with increasing amounts being required as the complexity level of detection/automatic correction rises. Redundancy can be in the form of additional checking bits in each code (parity), reverse transmission of the received message (echo), retransmission of defective messages, and so on. (Chapter 2 treats error protection techniques based on parity check codes.) Further, to be complete, the error checking provisions must also include circuitry and routines for checking the checking provisions.

Record Structures

Input and output in a computer system is generally not continuous. Rather, it is usually broken up into blocks of characters, called records, much like the text of a page is broken up into lines. The length of records is determined by several factors:

1. The physical capacity of the peripheral device (line length, card type, segment size, and so on)
2. The structural restrictions imposed by the programming language being used
3. The amount of memory that can be made available for assignment to input/output buffering
4. The probability of error occurrence during message transmission

The first three of these are self explanatory; the fourth, probability of error occurrence, requires some explanation.

Present transmission error handling procedures depend, in large part, on the almost absolute ability of the system to detect (through analysis of parity checks) the existence of errors in a message and then to recover from errors through retransmission of the defective block. Burton and Sullivan [72] estimate that the probability of having an undetected error in an 800-bit code with 16 parity bits is less than 10^{-8}. The simplest and presently most used method of error detection requires a "handshaking" between the transmitting and receiving devices where the transmitter waits after each message for a positive or negative acknowledgement before sending another message. If the acknowledgement is positive, the next message is sent; if negative, the same message is repeated. This procedure is common in both remote (for example, telephone) and local (for example, magnetic tape read/retry) communications. It is known as the stop-and-wait ARQ (automatic-repeat-request) and has an approximate throughput rate given by Burton and Sullivan [72]:

$$T = \frac{n(1 - P(n))}{n + c \cdot v} \cdot v \qquad (6\text{-}1)$$

where T is the throughput rate in bits per second

n is the block length in bits

$P(n)$ is the probability that a block of n bits will contain an error

c is the time delay between the transmission of one block and the beginning of transmission of the next; it includes the line turn-around and signaling times of the communication links, the start/stop and access times of the devices, and so on

v is the transmission rate of the link or the device

$c \cdot v$ represents the number of bits that could have been sent during the wait period

It can be seen from Equation 6-1 that the throughput rate is influenced by the block error probability ($P(n)$) and by the relationship of the wait-time-bit-length ($c \cdot v$) to the block length (n) of the messages. But since the mean error rate may vary from 10^{-4} to 10^{-12} depending on the type of link or device, the probability of having a single block contain an error becomes a direct function of the block length assigned.

Efficiency

In the noncomputer world, machines are rated by their work output capacity and by their relative energy conversion efficiency. Measurements of power, torque, and speed are independent of the devices attached and are, in most cases, relatively easy to make since both input and output are homogeneous. Efficiency is measured by the ratio

$$\eta = \frac{\text{work output}}{\text{work input}} = \frac{\text{work output}}{\text{work output} + \text{losses}} \tag{6-2}$$

The second form of Equation 6-2 is the most important to the machine designer since, if he can identify the losses, he may be able to identify the areas of greatest opportunity and make changes that will reduce the factors toward zero.

A somewhat similar situation exists with input/output processors. But here, work input and output are not nearly so easy to measure. In the absolute sense, work input can be considered to be the power (time, in memory cycles) drained from the central processor resources. This includes time to:

1. Set up and initiate the transfer
2. Read, assemble, and convert each element into the required code format
3. Detect and correct any transmission errors
4. Terminate the transfer

In the total sense, concern must also be given to the time that the peripheral device is necessarily idle (waiting for control response, data access) and not available to the processor for other uses.

Work output, on the other hand, consists solely of the useful work (in operations) required by the user in reading or writing data elements needed by his mathematical or logical equations. (Each element may be several digits or characters in length.) Operations required by the computer to access and process the input/output driver program do not contribute to the direct solution of specified equation and although necessary, are not useful in that sense. Hence, while it may take many memory cycles to actually read or write a single drum, the actual number of useful operations (the work output) is only one. Therefore it is possible to define a measure of input/output efficiency as

$$\eta_{I/O} = \frac{\text{number of data elements transferred}}{\text{total number of processor memory cycles required}} \tag{6-3}$$

As with noncomputer machines, this equation provides the criteria for suggesting design trade-offs at all three levels: the processor, the device, and the interconnecting link. Accordingly, those factors that make up the denominator must be examined in detail for each combination class. An expanded derivation of computer efficiency equations is given by Lane [72].

6-2. COMPUTER RESIDENT INPUT/OUTPUT PROCESSORS

Most present-day architectures provide, within the central processor, all the hardware and software features required for communication with the outside world. And, while they may vary from machine to machine in specific detail, all are functionally similar.

Conventional Architectures

Conventional third generation single processor computers have the organization shown in Figure 6-1. Except for the operator's control panel, all com-

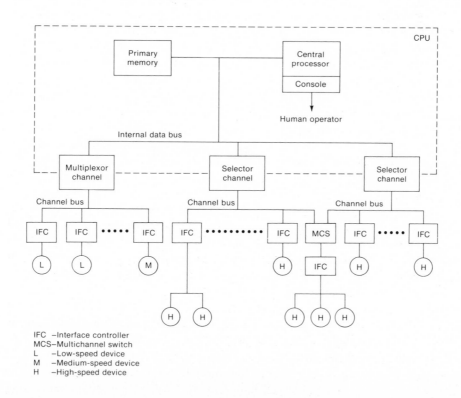

IFC –Interface controller
MCS–Multichannel switch
L   –Low-speed device
M   –Medium-speed device
H   –High-speed device

Figure 6-1 Conventional computer architecture

munication with external devices is through one or more of the following types of special input/output controllers known as channels:

1. Selector: The selector channel operates in a single mode, providing a momentarily exclusive input/output path for a single high-speed, program selected external device. The channel is totally dedicated to the selected device, and until released, may not be used for another input/output function, regardless of whether the device is actually transferring data or merely waiting to access the requested record.

2. Character multiplexor: This channel has two modes. One provides a momentarily exclusive input/output path for a single medium-speed program selected device (burst mode); the other provides a time shared, character interleaved path for several low-speed devices (multiplex mode). In burst mode, the multiplexor acts like a medium speed selector channel and, as such, must be totally dedicated to one device. In the multiplex mode, the channel serves several devices at the same time. Here, for example, the resultant character string from three devices with different rates and individual streams $A_1A_2A_3A_4 \ldots$, $B_1B_2B_3B_4 \ldots$, and $C_1C_2C_3C_4 \ldots$ might be $A_1B_1C_1A_2C_2A_3B_2C_3A_4$ and so on.

3. Block multiplexor: This channel also has two modes. One provides a momentarily exclusive path to a single high-speed program selected device (selector mode); the other provides a time shared, block interleaved path for several high-speed or buffered devices (multiplex mode). The multiplex mode allows the channel to be logically disconnected from the selected external devices during access wait times. Reconnection occurs after an interrupt signifying record ready for transfer. A recent study by IBM showed that the block multiplexor was able to deliver approximately three times the throughput at a third the response time of a conventional selector channel [Brown and Eibsen, 72].

All channels are bidirectional, operate either totally or semi-independently as small special purpose processors, each with separate access to primary memory. Since transmission is time dependent and often from mechanical devices, the channels are given priority and allowed to "steal" the necessary memory cycles to accomplish the transfer while the internal program waits. Channel programs may be resident within the channel hardware or may be stored in main memory. Interface controllers are required to electrically connect each device to its respective channel.

In systems with more peripheral devices than selector channels, alternate input/output paths can be provided, through use of program selectable multichannel switches. Each switch connects one or more devices to several channels and permits the connection to be made to any available channel. This feature is not required with block multiplexors. Each of the three channels is described in detail and diagrammed in the following text.

Selector Channel   The selector channel is a small limited-purpose processor that controls the transfer of single blocks of data between memory and selected external devices. Initialization requires location of the first word in primary memory, the length of the block to be transferred and the device identification. After initialization, the selector channel functionally provides:

1. The next memory location to be accessed in the input/output buffer (updated automatically with each datum transferred)
2. The number of words remaining to be transferred in the block (updated automatically with each datum transferred)
3. Assembly (or disassembly) of data elements to match the requirements of internal and external data bus structures
4. Parity checking of or parity insertion into transferred data characters
5. Channel and device status reporting and interrupt generation
6. Data pulse synchronization

Figure 6-2 shows the functional organization of a typical selector channel. The word assembly register (WAR) stores the contents of the word currently

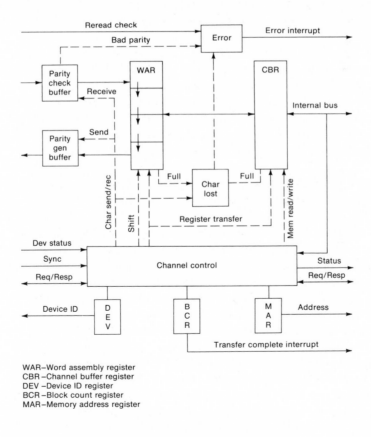

WAR—Word assembly register
CBR —Channel buffer register
DEV —Device ID register
BCR —Block count register
MAR—Memory address register

Figure 6-2   Selector channel architecture

being transferred to (or received from) the external device. It operates in a character shift mode. If the link between the channel and the device interface controller is halfword or fullword in width rather than one character, the shift pattern of the WAR is modified to correspond. The WAR is synchronized with the connected device's interface controller. As shown, it transmits or receives data from the devices in a single character mode while communicating with primary memory in a word mode (four characters at a time). The channel buffer register (CBR) contains the word currently being received from (or transferred to) main memory and operates in synchronization with the processor's internal clock. Transfer between the WAR and the CBR is parallel and asynchronous, thus providing the required speed change between the operating rates of the device and the processor. Channels could be constructed with only one data register for both word assembly and channel buffering, but the next character might arrive before the present word has been transferred to memory and hence, the character would be lost.

The current memory address is stored in the memory address register (MAR); the remaining block length, in the block count register (BCR). After word transfer between the channel and main memory, the MAR and CBR are respectively incremented and decremented by 1 to reflect the proper address and count.

A "transfer-complete" interrupt is generated when the block count becomes zero. Error interrupts are generated on detection of bad parity, on receipt of an error signal from the device interface controller, or if a character is lost on input. But, not all channel configurations are equipped with character-lost detector logic and care must be taken to ensure that the aggregate channel capacity does not exceed the effective capability rate of main memory to avoid character loss on input.

Character Multiplexor Channel   The character multiplexor channel, shown in Figure 6-3, can be viewed as a set of low-speed selector channels (subchannels). The connection to any one subchannel can be totally dedicated for a burst transfer or it can be passed cyclically to each subchannel in succession, only long enough to send or receive one character. In either mode, the channel controls the input/output operation in essentially the same manner as with a selector channel except that incrementing of memory addresses and decrementing of block counts are on a character rather than word basis. Further, addresses and block counts for each subchannel are stored in fixed location lists in primary memory rather than in hardware registers in the channel.

In multiplex mode, the scan control polls the request flag flip/flop of each subchannel in succession. If the flag is set, a service request is issued and the subchannel mode flip/flop is interrogated to determine whether the requested operation is input or output. For output, the control sequence is:

1. Read, increment, and restore the character location address from the memory address list.

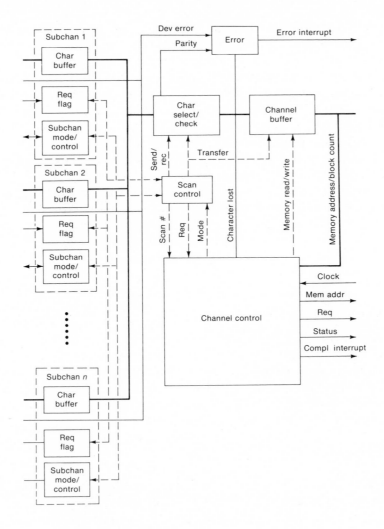

Figure 6-3 Character multiplexor channel architecture

2. Transfer the character from memory to the channel buffer.
3. Transfer the character to the subchannel character buffer; read and decrement the block count and test for zero.
4. Poll the next subchannel.

Input is similar except that the character must be transferred to the channel buffer at the same time the character address is read (Step 1) to enable its transfer to memory in Step 2. And, as with the selector channel, interrupts are generated on detected error and on block count equal to zero (transfer-complete).

In burst mode, the scan control is stopped at the selected channel and transfer proceeds in a manner similar to that in a selector channel.

Block Multiplexor Channel The block multiplexor channel combines the desirable features of both the selector and character multiplexor channels. Like the selector channel, the memory address and block count registers are implemented in channel-resident hardware, allowing incrementing and decrementing to take place simultaneously with data transfer. And, like the character multiplexor, all subchannels can time share the input/output path, except the interleaving is by block rather than by character.

But, unlike the selector channel, the block multiplexor is able to disconnect and service other requests during the time that a selector channel must wait for a disk or other similar device to access the desired record. Reconnection is then made at the beginning of the addressed sector on an interrupt signal to the channel from the device interface controller. Provision must be made in the device controller logic to re-interrupt the channel on the next revolution if the channel is busy when reconnect is requested.

Interrupt generation in the central processor is on detection of an error or on completion of block transfer.

Minicomputer Architectures

Minicomputers do not have independent channels in the conventional machine sense. Rather, each device interface controller connects directly into a universal input/output bus that can, under program selection, look to each device either like a character (or word) multiplexor or like a conventional high speed selector channel. Figure 6-4 shows this functional organization which allows input/output operations in two modes:

1. Program controlled input/output: data transfer through the arithmetic accumulator, one character (or word) at a time
2. Direct memory access (DMA): data transfer directly to memory on a cycle stealing basis without program intervention from the processor

Since DMA and program controlled input/output are features that apply equally well to any input/output bus position, any device may be controlled under either mode. Further, since DMA logic is duplicated in most minicomputers, all peripheral devices also have the equivalent of multichannel switches.

Program Controlled Input/Output The program controlled input/output of the minicomputer is the most flexible, yet the most inefficient, of all the input/output procedures presently available. It depends on the availability of basic input/output instructions in the processor repertoire that will move data elements between the arithmetic accumulator and an individual program selected device on the input/output bus. This is done in much the same manner that the load/store instructions move data elements between the accumulator and a program selected location in memory. But, since all input/output control is also the responsibility of the program, additional processor instructions must be provided to set and clear the various flag and control registers required by the handshaking sequence between the device and the processor.

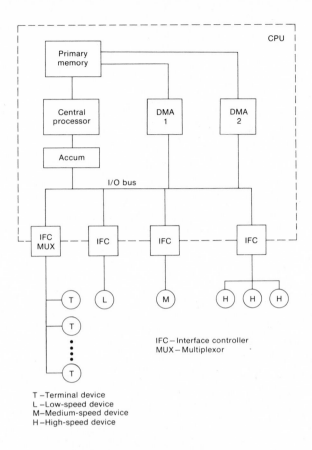

Figure 6-4   Minicomputer architecture

Figure 6-5 shows the functional diagram of a typical program controlled interface controller. (Most minicomputer manufacturers supply detailed peripheral and interfacing manuals that specify the design procedures to be followed in building custom interface controllers.) The buffer matches the processor internal structure format on one side and the device/interface link on the other. Unused bits are dropped on output and filled with zeros on input. Code checking, word assembly, memory address incrementing, block count decrementing, and data element transfer must all be accomplished under a special (to each type of device) input/output driver subroutine using the arithmetic accumulator. Preemptive use of the arithmetic registers for input/output operations also requires that the contents of the accumulator be saved before and restored after each character is transferred.

For output with a program controlled interface, a typical sequence is:

1. Transfer the output word from memory to the accumulator (the memory address of the word is contained, either directly or indirectly, within the load accumulator instruction).

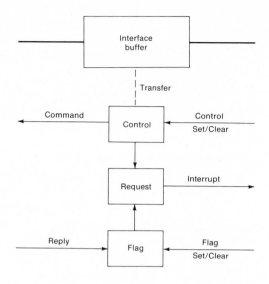

Figure 6-5   Program controlled input/output interface controller

2. Transfer the word from the accumulator to the interface buffer (the device identification is contained within the move instruction as a bus address).
3. Issue a start device command to the same bus address, specifying the operation as output.
4. Increment the memory address, decrement the block count and enter a "wait-loop" or return to the main program while the interface issues a start command to the device, transmits the character (or word), waits for a character received reply and generates a program interrupt.
5. Return to Step 1 or, if the block transfer is complete, issue a clear command to the interface to stop further device activity.

Input is similar except that the transfer is in reverse.

Since several instructions must be processed for each character transferred, the total bandwidth of this path is severely limited. Further, since character-lost detection is usually not included within the interface logic, the probability of invalid input occurring rises sharply as the input/output activity increases.

Direct Memory Access   The DMA input/output control is an internal mini-computer hardware feature that connects the processor internal memory bus to the processor input/output bus in the same manner that a selector channel connects its interfaced devices to the internal data bus of a conventional machine. It includes separate memory address, block control, and device identification registers plus the necessary logic to steal memory cycles from the central processor and to control the transfer of data elements from the device interface through the input/output and internal data busses to memory. However, several differences exist between the DMA and a selector channel:

1. Since the DMA channels connect the internal data bus to its own input/ output bus rather than to just a specific subset of external devices, any one device can be served by any DMA channel (equivalent to a multi-channel switch per device).
2. Any DMA channel is capable of preempting consecutive memory cycles and of controlling transfer at full memory rate. The actual rate is dictated by the connected device.
3. Interrupt is generated only on block count zero (transfer complete) and not on error conditions since no error checking is provided within the DMA channel itself. Hardware error detection, if any, must be included in the device interface controller.

Interrupt Structures

To avoid having to "lock up" the computer while it is waiting for a character or word to be actually transferred and received, and to have the capability of accessing several low speed devices at the same time, it is necessary that the processor be provided with a multilevel interrupt structure that allows ready identification of any connected device needing service. Figure 6-6 shows a simplified schematic representation of a two-dimensional interrupt array. In practice, the line switches are latching circuits that are set (latched) by a signal from the device (unless previously inhibited by a device interrupt mask command) and cleared under program control at the end of the interrupt service routine. Level switches, on the other hand, are latched on receipt of any interrupt on the level and are automatically reset when all those pending on that line are cleared.

Operation of the system, for a single interrupt, includes the following sequence:

1. An interrupt signal from the peripheral device latches the device interrupt switch, which in turn latches the level switch, thereby creating a direct and unique signal path between the device and the processor.
2. The processor returns an interrupt acknowledge signal to the device over a parallel path controlled by the same switch array.
3. The device transmits its interrupt identifier to the processor over the input/output bus. The interrupt identifier may be the actual address of the interrupt service routine or it may be used as a pointer to a location in a table of service routine addresses.
4. The interrupt service routine stores the present contents of the program registers, services the device, restores the registers, clears the device latch switch, and returns control to the interrupted program in the manner detailed in section 6-4 of this chapter.

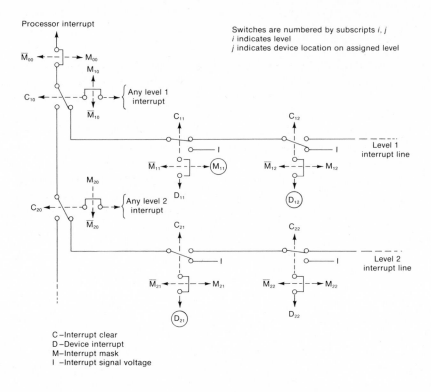

C –Interrupt clear
D –Device interrupt
M –Interrupt mask
I –Interrupt signal voltage

Figure 6-6 Multi-level interrupt array
Mask M_{11} and device interrupts D_{12} and D_{21} shown actuated

Multiple interrupts are handled in the same way, in order of their priority (by level and line distance from the processor), regardless of the sequence in which they are received. Masking is provided for the total system, for a particular level, or for an individual device by setting the interrupt mask register bit positions associated with the specific switches whose latching is to be inhibited. Interrupt clear is under program control on an individual device, particular level, or total system basis.

Communication Multiplexor

Communication between a keyboard terminal and a minicomputer can be through a serial interface controller (one per terminal) or, if a number of terminals are to be served, through a character multiplexor interface controller connected to the processor input/output bus. Functionally, this interface is the same as the character multiplexor shown in Figure 6-3 except that front-end circuitry must be added for each subchannel as shown in Figure 6-7. The character assembly register (CAR) converts the serial bit stream to a parallel character format. Synchronization is provided through use of start/ stop bits inserted into each character transmitted and bit rates are selected

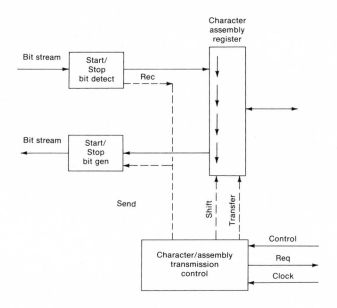

Figure 6-7   Bit serial interface controller

either by hardware switch or by program control depending on the specific connected device.

Communication with the processor may be under program control or through the DMA channel depending on the specific design. If program control is used, the multiplexor generates an interrupt for each character received or sent and transfer to memory is software directed. In this case, the transferred word is packed to contain both the received character and the subchannel number to enable later reconstruction of the input information. With DMA, the multiplexor interrupts only on block count or on the receipt of a specific control character such as carriage return signifying end of line. And, as with the character multiplexor, memory addresses and block counts for each subchannel are contained in fixed location lists in memory.

Example: The Unibus

Three things must be present at instruction execution time to accomplish a data transfer:

1. The data element itself
2. The transfer destination
3. The link control, priority and timing procedures

Unfortunately, these are usually different in format and specification for different types of destinations depending on whether they are processor registers, primary memory, or peripheral devices. One specific structure exists, however, that allows the same form of communication between all bus-connected devices, regardless of whether they are internal or external. As such, it deserves examination.

Digital Equipment Corporation has implemented the Unibus (universal bus) in the PDP-11 series minicomputers. The Unibus includes a wider bus structure than is needed for the data element itself (16 data bits plus 2 parity bits). Additional parallel lines indicate the element destination (device or memory address) as well as transmit the control and timing signals necessary to effect the transfer. As implemented, the Unibus is a ribbon cable, 120 conductors wide, with 56 signal lines bounded and alternated between 64 ground wires. Signal functions include those listed in Table 6-4.

TABLE 6-4. PDP-11 UNIBUS SIGNALS

Name	Number of Lines	Function
Address	18	Identifies data element destination (memory location or device bus address)
Data	16 plus 2 parity	Data element value
Type	2	Type of transfer (data in, data in pause, data out, and data out byte)
Control	8	Bus control, timing and status
Bus request	5	Priority interrupt request
Bus grant	5*	Bus grant (assign)

*All lines are bidirectional except the five bus grant lines which are unidirectional.

Uniform addressing is accomplished by reserving, for peripheral devices and internal registers, the upper 4K words of the 32K that are instruction addressable in 16 bits. (Normally, 16 bits would allow 64K addresses except in the PDP-11 which allows addressing at the byte (halfword) level.) The address bus allows 18 bits for total effective address, a feature that is used with the optional memory management hardware. Uniform interrupt servicing is accomplished by equipping each peripheral device with a hardware pointer to a specific pair of words in primary memory. These words contain the new processor status information and the address of the device service routine. Thus, priority allowing, control is passed to the device service routine almost immediately upon receipt of interrupt. Five primary interrupt priorities are provided with secondary priority (for the same primary priority) assigned according to the electrical line distance from the central processor (closest devices have highest secondary priority).

Communication is in a master/slave mode, with the master controlling the transfer timing and synchronization. In a disk/memory transfer, for ex-

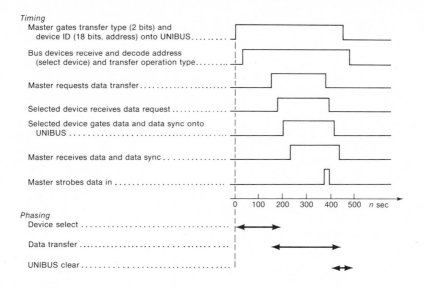

Figure 6-8 PDP-11 Unibus master/slave data input timing and phase diagram

ample, the disk is the master since it is most time sensitive and the memory is the slave. The data element, its destination address, and the device control are all transmitted simultaneously and asynchronously (at the next available time slot) as shown in the timing diagrams of Figure 6-8. Further, because all the attributes necessary for the transfer are included with each data element, the transfer does not require intervention of the central processor. DMA transfers to memory, for example, can be interleaved between instruction executions that involve register-to-register operations. This feature increases the actual bus bandwidth (maximum transfer rate) to a value greater than that for any one device on the bus. For the PDP-11, the bus bandwidth is approximately 2.5 times the maximum memory rate.

Finally, the parallel structure also lends itself readily to extension and interconnection by use of the following types of bus modules:

1. Bus extender: adds 19 input/output slots per added section
2. Bus switch: allows a bus and the peripherals connected to it to be switched under program control and shared between two processors
3. Bus window: a high speed interbus channel that allows communication between any two devices connected to either bus; control circuitry is provided to allow automatic address displacement (that is, a 39-word block addressed from 4010 through 4056 on one bus might translate to the addresses from 47,521 through 47,567 on the other)
4. Bus link: a high speed half-duplex DMA controlled channel that allows communications between two processors

6-3. SYSTEM RESIDENT INPUT/OUTPUT PROCESSORS

The input/output problem, as outlined in section 6-1, includes several functional requirements that are not treated by the conventional processor resident input/output channel:

1. Code conversion where the external code does not match the internal (for example, ASCII to binary, BCD to binary, unpacked to packed decimal)
2. Error detection or correction where the error is only detectible over a span of characters or words
3. Format control where the input/output record must be assembled or disassembled before transmission or processing

These, coupled with the channel initialization and some control operations, require additional program intervention and a dedication of processor resources that might be better spent on internal program processing.

Because of this drain, computers are considered to be *input/output bound* when input/output operations require a disproportionate share of the memory cycles available. One solution, often advanced by computer salesmen, is to change to higher priced and faster peripherals. Another is to add a small front-end processor between the central processor and the peripheral devices. This latter method allows the separation and dedication of tasks—input/output operations to the front-end processor and internal program processing to the central computer.

Disk Coupled Systems

Figure 6-9 shows the organization of a typical disk coupled system. The front-end processor controls all the low-to-medium speed input and output, buffers and assembles records from each device, checks for errors (requesting retransmission, if necessary), performs code conversions and line editing, and stores or retrieves programs and data structures on the shared disk. Control of the job flow through the system generally rests with the front-end processor. Its operating system manages the program queue and allocates the disk space for each program processed.

This organization is probably the easiest to implement, particularly where the two machines are built by different manufacturers. Neither must conform to the physical bus timing and control restrictions of the other. Programs operate independently without cycle-stealing and the only information that is required to be passed directly between the two machines consits of the program attributes (type, storage location, number of blocks). On input, this information is transmitted from the front-end processor when the program is ready for processing by the central processor. Conversely, when the central processor is finished, it notifies the front-end processor of the attributes of the output files that it has stored on the shared disk. Requests are honored

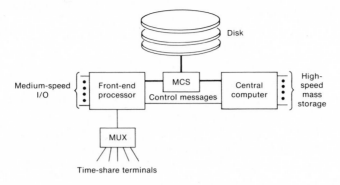

MCS - Multichannel switch

Figure 6-9   Disk coupled front-end processor

by the shared disk controller from either machine on a first-received basis if the disk is not busy, at the end of the block being transferred if the disk is busy, or from the front-end processor only if a preemptive wait request has been issued to clear a rapidly filling input queue in front-end memory.

Direct Coupled Shared Memory Systems

The direct-coupled organization requires that both the front-end and central processors have essentially the same architecture and internal timing/control characteristics. In this configuration, primary memory replaces the disk as the shared coupling but with information exchange interleaved on a word rather than block basis.

Depending on the specific machine, connection may be any of the following:

1. DMA to DMA between separate computers where each channel is initialized by its respective processor and where each machine can access selected memory locations in the other
2. Through separate ports of primary memory where the front-end processor is given access priority in the case of simultaneous requests
3. Through a shared wide-band bus where the front-end processor is given bus control priority in an asynchronous control environment or alternate cycles in the synchronous case

The general advantage of the direct coupled system over the shared disk system is the ability to pass data and programs without the need for intermediate disk store. (Intermediate disk store may be required if queue lengths exceed available memory.) In the latter two connections, even the memory-to-memory move is eliminated.

Integrated Systems

The shared disk and direct coupled connections provide a way, after the architecture has been established, to configure a system that is essentially a dual processor. Larger architectures, on the other hand, are initially designed as multiprocessors with integrated input/output computers. Typical of these is the CDC 6600, shown in Figure 6-10. It has ten peripheral processors which are given access to central memory on a polled, time-slice basis. System control is resident in one of these, which, in turn, controls all activities of the central machine as well as the nine other peripheral units. All are architecturally the same, all have access to any of the twelve input/output channels, and any can be assigned as master control.

6-4. PROGRAMMING CONSIDERATIONS

The architectural features described in the previous sections have been designed to facilitate the transfer of data elements between the central processor and its external devices. They accomplish the serial/parallel transformation, parity checking of individual characters, interleaved transfer of words to and from memory, modification of next memory address and remaining

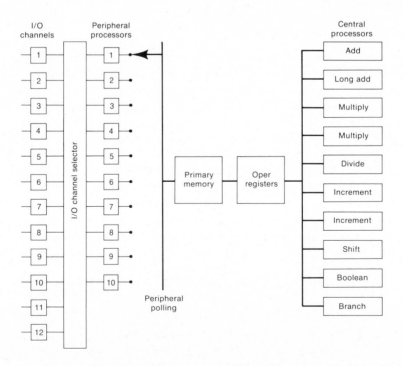

Figure 6-10 CDC 6600 structure

block counts, and the generation of program interrupts on detected errors and at the end of block transfer. These functions have been converted to hardware because their characteristics are stable and readily defined and because their implementation does not impair program flexibility.

The linkage to user programs, however, must be made by input/output driver subroutines in the operating system. These are normally reentrant, so that one routine can serve several devices, with the possibility that several could be active simultaneously, and interruptable, so that a higher priority device can be serviced without wait. The user specifies his required record length, its buffer location and the device to be accessed. Additionally, he specifies other information such as record formating requirements, file name, sector location, and character code. Everything else required to establish the program/hardware interface is the system architect's responsibility.

Input/Output Driver Linkage

Each input/output driver must provide for the return linkage to the main-line program by storing the return address and the contents of any status registers that would otherwise be lost by the routine. These are stored in dedicated memory locations unique to the routine and to the particular device. They are restored prior to return.

Interrupt Servicing

On recognition of a program interrupt, an immediate indirect jump is made to the subroutine address specified for the device. The return linkage is generated and stored (other interrupts are inhibited during this phase to prevent loss of return address) and the interrupt is serviced. If a higher priority interrupt is received before the routine is finished, it is acknowledged and serviced, providing return linkage to the first interrupt routine. If the second interrupt is of the same or lower priority, it is locked out until the first routine is finished and control is returned to the main-line program. Other interrupts are also inhibited during the return jump to allow proper restoration prior to servicing the next. The second interrupt is then serviced in the normal manner.

Channel Initialization

Channel initialization requires that the device identification code, first memory buffer address, and the block count be transferred, either to the channel itself or to specific list locations in memory, before a channel input/output operation can begin. Since it is possible that either the device or the channel might be busy, program provision must be made to allow for some form of a device/channel available signal or a connect retry after a time lapse. Provision must also be made to disconnect and reset the interface prior to routine exit.

Communication Subchannel Initialization

Initialization of a communication subchannel is similar to the setup of a conventional channel except that the transmission rate of the device must also be selected. (One minicomputer multiplexor allows 256 different program selectable frequencies, but a review of available terminals shows that 10 frequencies should be sufficient for most configurations: 110, 150, 220, 300, 440, 600, 880, 1200, 1760, and 2400 Hz.) If the connection is by telephone dial-up from an unknown device, it may also be necessary to test several frequencies to determine the correct one. This can be done through a simultaneous multiple frequency sampling or through the repeated transmission of a known character until a proper match is made.

Format and Code Conversion

Format and code conversions are specified by the programmer and dictated by the characteristics of the device and the processor. This is accomplished in several steps, each of which requires a separate subroutine section. For input, these are:

1. Conversion of the external code to an intermediate internal code; if the external device has more than one code, such as in a card reader, then the input/output driver has to be structured to deal with each one
2. Separation of format specified characters into separate strings representing single data elements
3. Transformation of the adjacent characters of the string into a complete data word
4. Transfer of the data word to its appropriate location in memory

Output is in reverse order except that the record buffer must be initialized and the necessary device control characters added.

Buffer Allocation

Each active device must be assigned one or more blocks of memory (buffers) to serve as intermediate storage for record assembly during input/output operations. The specific type and length depends on the characteristics of the external device, the application, the restrictions of the programming language, and the space available in primary memory. Figure 6-11 shows several possible structures:

1. Fixed-length linear buffers — generally used if the record lengths are easily predictable as in card devices, line printers, and so on
2. Multiple (ping-pong) linear buffers — assigned to smooth the input/

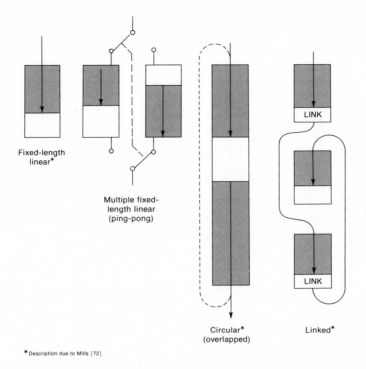

Figure 6-11 Input/output buffer structures

output flow and allow the device to be reading the next record while the program is processing the present

3. Circular buffers — used where several records (of fixed or variable length) may be in the input queue before they are processed; this structure is common with keyboard terminals where records may contain editing information affecting portions of the same or previous record (for example, backspace, delete previous line)

4. Linked buffers — used when record lengths within a sequence of transmissions are random and vary over a wide range

Error Detection and Recovery

Error circuitry is commonly provided to check and interrupt on the following conditions:

1. Detection of bad character parity at the channel on input and at the peripheral on output

2. Loss of input characters

3. Reread mismatch in buffered peripherals equipped with reread stations, such as:

Card reader	Read, reread
Card punch	Write, reread
Magnetic tape	$\begin{cases} \text{Read, reread} \\ \text{Write, reread} \end{cases}$

Unfortunately, single bit character parity will not detect errors involving an even number of bits, where the individual character codes still appear to be valid (invalid codes can be detected during code conversion). Detection and correction of errors of this type require an additional longitudinal check code spanning all characters in the record. Unfortunately, space does not permit a full discussion of error detection, analysis and control. The reader is therefore referred to Chapter 2 for a brief description of error protection techniques and to recent papers by Burton and Sullivan [72] for errors in communication networks and by Chien [73] for errors in mass storage devices.

Once an error is detected, the action taken depends on the characteristics of the device, the type of error detected, the importance of the data, and the time available for analysis. In real-time control systems, for example, measurements are repeated each time period and, as such, control programs are not generally susceptible to errors in individual readings so long as they are identified as incorrect. The value or trend of the previous period is accepted until the next regular sampling occurs.

The error handling procedures may attempt to correct the error based on the information at hand (if a sufficient number of check bits are available) or they may request a limited number of retries. The method used to retry the operation depends on the device. For magnetic tape this requires a backspace and reread (or write); for magnetic disk, a reread on the next revolution; for cards, operator intervention and card reinsertion; and for remote communications, a retransmission not only of the record containing the error but also all of the records transmitted subsequent to it.

6-5. EFFICIENCY LOSS FACTORS

The input/output system provides the interface between the central processor and the outside world as well as between the processor and its secondary storage devices. The input/output system serves as an interface between levels to reconcile differences in speed, code, and word size; it also checks for read, write, and transmission errors as well as for special control characters, and controls the operations of all channels and peripheral devices. The efficiency level depends on the number of operations that are required per data element transferred.

Program Operations

Because of the nature of the internal architecture in today's machines, memory cycles must be expended for all instruction accesses and for most instruction executions. Most of these, however, do not contribute to the actual transfer of complete data elements, the only useful work that can be assigned to the input/output system. Program operations and memory cycles that contribute to an efficiency loss include those used for the following processes:

1. Initialization and termination of a channel block transfer
2. Program control of packing and unpacking characters in records
3. Code conversion and format editing
4. Error checking and recovery
5. Interrupt servicing, including operations to store and restore any preempted registers

Channel Operations

In some architectures, memory cycles are required by the channel for operations other than the actual transfer of data elements; such operations include:

1. Transfer of wide format data elements (for example, some floating-point words)
2. Modification of next memory address and remaining block count if these are stored in fixed locations in memory
3. Waits for device or channel busy, for record access (latency and seek) and for intersector and interrecord gaps. (These factors contribute to a loss in efficiency only if the processor is not able to perform other tasks during the wait periods.)

6-6. SUMMARY AND ADDITIONAL OPPORTUNITIES

The input/output system is usually designed to be "transparent to the user" and, as such, is often overlooked as an opportunity for architectural improvement. In actual fact, however, it can be one of the most time consuming links in the chain between program input and result output. Several opportunities for architectural advances exist.

Microprogrammed Input/Output Control

Many of the current program operations are fully amenable to microprogramming techniques which are discussed in Chapter 10. If converted, microroutines should be able to eliminate most of the instruction accesses currently used for these purposes.

Transmission by Exception

Most input/output records contain filler characters (blanks or zeros) to complete a fixed length format equal to the length of a disk sector, printed line, or punched card. Most languages and systems require the transmission of these characters with each record, even though they are unchanged for the duration of the particular format control. An additional bit identifying which characters are fixed and which are variable would allow all records, after the first, to be transferred in compact form with the record containing only the variable information (each record would have to be preceded by a control character signifying whether it was a new or continued format).

Synchronous Input/Output

The interfaces between levels are complicated by the need for speed and word length conversion. Multiple track recording formats (parallel by word or by character) and clocking from the disk during data transfer would allow the disk to be matched to the internal memory rate and structure. Further, if the record is to be processed as a vector, it may be possible to eliminate the presently required intermediate unproductive transfer to memory by routing the data stream directly to the processor. (This variation changes the normally passive secondary memory and allows it to act the same as active primary memory for selected applications).

The Future

With the advent of reliable low cost integrated circuitry, the next generation of input/output channels should be capable of many of the functions now distributed between the processor (and its programs), the channel, and the device controller. Microprogramming will make the channel/controller more adaptable and universal, capable of accepting different word structures and character codes, and of detecting and correcting errors including retransmission on either the same or alternate data paths.

6-7. PROBLEMS

6-1. Analyze the peripheral devices attached to the computer available to you.
  a) How is each device logically connected to the computer?
  b) How is it physically connected?
  c) What codes are utilized?
  d) What transmission structure is required?
  e) What hardware buffering is supplied?
  f) What data rates are available?

  g) What error checking provisions are included at the peripheral?
  h) What device control is allowed to the user?
  i) Is it modifiable and if so, how?

6-2. Analyze the word structure of the computer available.
  a) What codes are implemented in hardware?
  b) How many more are in software?
  c) What conversions are necessary to match the peripherals attached?

6-3. Obtain channel architecture diagrams for any computer available to you.
  a) How do they compare to those described in this chapter?
  b) What features do they lack?
  c) What additional features do they have?

6-4. Analyze the interrupt structure of the computer available to you.
  a) How is the device priority determined?
  b) What masking is available?
  c) How is it invoked?
  d) How are multiple interrupts handled?

6-5. Obtain the structural diagram of a communications multiplexor. Determine the procedural steps necessary to input or output one record from each peripheral device attached, assuming all devices are active simultaneously.

6-6. Review recent publications of the ACM and IEEE for papers on front-end processors. Analyze the functions assigned and the methods of communicating with the central processor.
  a) What changes are indicated?
  b) What additional architectural features are needed?
  c) What additional software features?

6-7. Analyze the input/output driver routines of any conventional computer available to you.
  a) What portion of the instruction set is used for this purpose?
  b) What is the efficiency of each routine?

6-8. Analyze the input/output driver routines of any minicomputer available to you.
  a) What portion of the instruction set is used for this purpose?
  b) What is the efficiency of each routine?

6-9. Determine the additional costs for the input/output hardware features required for the routines of problems 6-7 and 6-8. What are the relative efficiencies and cost effectiveness of each structure?

6-10. *a)* How does the instruction subset determined in problems 7 and 8 differ from the instruction set in a general purpose computer?

b) Would this subset be appropriate as the total instruction set of a front-end processor?

c) What additional instructions would be necessary? Why?

6-11. Analyze the input/output buffer structures used by the operating system of the computer available to you.

a) Are they appropriate or should they be changed?

b) What would be the effect of doubling the memory allocated to each?

c) How would the added resource be used?

d) Would the structure change?

6-12. Analyze a circular buffer in which several records can be stored at one time. Assume transmission from a keyboard terminal. What provisions must be made to implement line editing such as tab, backspace, delete previous word, and delete previous line.

PART II

Advanced Topics

CHAPTER 7

Stack Computers

William M. McKeeman

7-1. INTRODUCTION AND NOTATION

A stack is a last-in, first-out data structure. In its simplest form it can be accessed by only two operations, PUSH and POP. A PUSH operation takes new data and pushes it onto the top of the stack where it is saved. Subsequent PUSH operations bury the item deeper and deeper under each new data entry. A POP operation removes the most recently pushed data item from the stack, uncovering the item underneath.

The appellation "stack computer" designates a class of computers using one or more stacks. The stacks can be understood in isolation, and can be used independently of, and even in the absence of, the other stacks. An actual implementation, however, almost always mixes a number of conceptually different stacks into one rather tightly bound, interleaved structure. The easiest road to understanding is to describe each stack individually and then to show how they may be combined into more complex mixtures. Each stack is described as a data structure, in terms of the data contained within it, and as a reflection of the programming language and machine language that drive it.

The use of stacks in computers is not a particularly recent event. Randell and Russell [64], in a major contribution to the literature, give detailed diagrams and flowcharts for a very elaborate stack structure. Much of that structure is to be found in the Atlas Computer and had a heavy influence on the Burroughs B6700. Wirth and Weber [66] also present an elegant stack architecture in an implementation of the programming language Euler. The Burroughs B5000 was designed at the same time as the programming language ALGOL 60 appeared, and was marketed in 1963. It proved to be a reliable and versatile computer, partly due to its reliance on stack hardware.

The Burroughs B5500, B5700, and B6700 carried this development further, perhaps even near the practical limit of complexity for such mechanisms. The Hewlett-Packard HP3000 was marketed starting in 1972 with a somewhat less elaborate stack structure tailored more specifically to a demand for real-time response. The Burroughs B1700 appeared about the same time with a stack mechanism controlled by a writeable microcode. The Digital Equipment Corporation PDP-11, also of this period, has some rudimentary stack construction and uses facilities cleverly merged into a more conventional set of operations. Many of the ideas found in these computers were first expressed by Barton [61] and Iliffe [68]. Wortman [72] has carried the ideas further in his doctoral thesis and also provided a basis for evaluating the relative merits of alternative mechanisms. The concept of tagged data, an important adjunct to stack organization, is discussed by Wirth and Weber [66]. Organick [73] gives a detailed treatment of the Burroughs B5700/B6700 series of computers.

To avoid misunderstandings arising out of the various colloquial interpretations of the word "stack," we now offer some definitions which we will then use throughout this chapter. A datum of width n is visualized as a contiguous vector of n independent bits of information, thus any one of 2^n values. A datum is the basic unit for transactions between the various parts of the computer. The width of data is bounded from below by the number of values in the range of the data being represented, and from above by economic considerations of bus width, memory size, adder precision, and the like. The data pushed onto the stack comes from somewhere in the computer; the data popped goes somewhere. Thus each action involving PUSH and POP takes the form of an assignment. An assignment *to* a stack implies a PUSH onto the stack; and assignment *from* a stack implies a POP from the stack. Thus if X is some other place in the computer (for example, a register), and S is a stack, the assignments in Table 7-1 are the prototypes for stack access.

TABLE 7-1. ASSIGNMENTS FOR PUSH AND POP

Action	Assignment
PUSH from register X to stack S	$S := X$
POP from stack S to register X	$X := S$

Suppose X, Y, and Z are registers containing the integers 2, 3, and 4, respectively, and that S is an initially empty stack. Then the contents of the stack and the registers will be changed by assignments between them as depicted in the example in Figure 7-1.

When the stack is empty, a POP operation is pathological since there is no data to be removed. Also, stacks do not have infinite capacity. When a stack is full, a PUSH operation is pathological since there is no room for the new data item. These two conditions are called *stack underflow* and *stack overflow* and the computer must be able to detect them and respond reasonably. Since there is no response that will preserve stack properties, computers usu-

Action		Registers			Stack	
		X	Y	Z	S	
Initial		2	3	4	(empty)	
S := X	(PUSH)	2	3	4	2	
S := Y	(PUSH)	2	3	4	3 / 2	
X := S	(POP)	3	3	4	2	
S := Z	(PUSH)	3	3	4	4 / 2	
S := X	(PUSH)	3	3	4	3 / 4 / 2	
Z := S	(POP)	3	3	3	4 / 2	
Y := S	(POP)	3	4	3	2	
Z := S	(POP)	3	4	2	(empty)	

Figure 7-1 Sequence of stack assignments

ally react by interrupting the sequence of instruction execution much as when arithmetic overflow occurs in the adding circuitry.

There are some uses of stacks that require more elaborate access methods than just PUSH and POP. They will be described in following sections.

Since there are likely to be several stacks in one computer, we will normally use subscripted variables of the form S_i to denote stacks. To be more precise about the foregoing notation, whenever an S_i is used in a context where its value is needed, we imply the value is "popped" off of the stack, leaving the next value in the stack exposed. Whenever an S_i is used in a context where its location is needed, we imply that a value is "pushed" onto the stack. For example,

$$S_i := 3$$

implies the value 3 is pushed onto the stack S_i;

$$S_i := S_i - S_i$$

implies the top two items are popped off S_i, subtracted, and then the result is pushed back onto S_i. As in Iverson's language APL [Iverson, 62], the assignments are read from *right to left,* so that the rightmost S_i in the example is the data item at the top of the stack before the assignment is executed and thus represents the subtrahend of the expression.

Certain other data structures are used in conjunction with stacks and therefore must be representable in the notation being described. In particular, main memory is a large array accessed via subscripts (absolute addresses). We would write

$$\text{MEMORY}[3] := S_i$$

to signify popping a value off S_i and placing it into location 3 of main memory. We will use other similar notations freely.

The reader must be constantly aware of the different meaning given to the expressions involving stacks, just described, and the expressions appearing in programming languages which will also be used as examples throughout this chapter. In programming language expressions the sequence of evaluation of operators is normally left-to-right except as modified by operator hierarchy and parenthisization. Furthermore, the order of evaluation of operands is usually immaterial since their value is constant during the evaluation of the expression. The stacks, S_i, are, on the other hand, operands with side effects. The use of the value of an operand signals not only the use of the value but also the removal of that value from the stack. Here the order of evaluation of operands matters very much, and must be right-to-left for the reasons given previously.

7-2. STACKS

Simple stacks can be implemented in a variety of ways; more complex versions are generally restricted to being placed in main memory and being accessed through index registers. For example, if A, B, and C are index registers, and MEMORY is an array representing the main memory of the computer, we can define a stack as lying between A and C, with push-pop point B, as depicted in Figure 7-2.

The PUSH and POP of Table 7-1 can be defined as shown in Table 7-2. The advantage in expanding the detail of the stack mechanism in this way is that later some more elaborate access methods must be defined in terms of the given registers, and other registers must be added in a similar fashion for some even more complex stacks.

The disadvantage is in giving the reader an incorrect intuition about the sequentiality of the details of stack access. Note that all three actions in the expanded PUSH and POP of Table 7-2 can be done simultaneously. Designers have achieved substantial gains in efficiency by taking such opportunities for parallel processing into account.

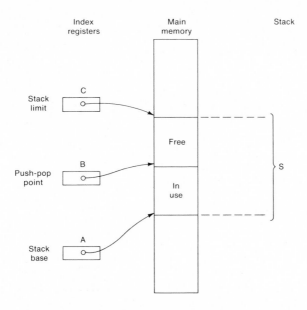

Figure 7-2   Stack in main memory

Suppose, as another example, we need a stack as depicted in Figure 7-3. The signals POP and PUSH are pulse controls that cause the corresponding actions to take place. Prior to pulsing PUSH, the datum must be on the bundle IN. The bundle TOP gives a continuous readout of the top of the stack except for a brief period after a PUSH or POP pulse. The outputs OFLO and UFLO signal the corresponding failures in the stack mechanism.

The main components in the stack will be some 32-bit shift registers as illustrated in Figure 7-4. The input lines SHL and SHR are pulse controls that cause the corresponding actions to take place. The bit to be shifted in must be ready on the lines LI or RI respectively when the control pulse arrives. The lines LO and RO give continuous readout of the bits at the ends of the shift register. (LI signifies left-end input, LO signifies left-end output, etc.)

TABLE 7-2. THE ACTIONS PUSH AND POP DEFINED IN TERMS OF A, B, C, AND MEMORY

Action	Notation
PUSH S := X	IF B > = C THEN STACK__OVERFLOW; MEMORY[B] := X; B := B + 1;
POP X := S	IF B < = A THEN STACK__UNDERFLOW; X := MEMORY[B − 1]; B := B − 1;

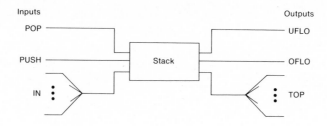

Figure 7-3 Simple stack mechanism

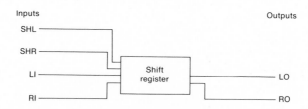

Figure 7-4 Shift register

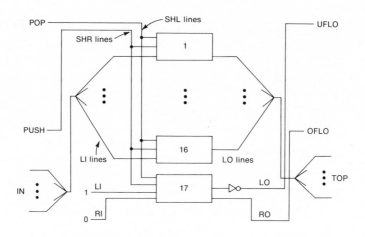

Figure 7-5 Simple stack built out of shift registers

For a stack of width 16, seventeen shift registers are combined as shown in Figure 7-5. The seventeenth register carries the underflow and overflow information. A logical 1 is gated into the left end with each PUSH and a logical 0 is gated into the right end with each POP. Whenever a 0 propagates all the way from right to left, we have an overflow situation and similarly for a 1 from the left. The seventeenth register must be initialized to all 0s for which no mechanism is displayed.

7-3. ARITHMETIC EVALUATION STACKS S, S_i, S_r, S_a

The evaluation of arithmetic expressions is the most easily understood use of stacks. It is also perhaps the least consequential use of stacks, but does form a convenient starting point. Suppose we have an expression in programming notation:

$$2 + (3 - 1) - 4$$

It can be evaluated using a stack by the following sequence of actions:

1. Push 2 onto the stack
2. Push 3 onto the stack
3. Push 1 onto the stack
4. *a*) Pop the two top data items off the stack
   b) Subtract the first to appear from the second
   c) Push the result back onto the stack
5. *a*) Pop the two top items off the stack
   b) Add them
   c) Push the result back onto the stack
6. Push 4 onto the stack
7. *a*) Pop the two top items off the stack
   b) Subtract the first to appear from the second
   c) Push the result back onto the stack

This leaves the result, 0, as the only value on the stack.

The actions fall into two classes: placing operands on the stack and operating on the operands at the top of the stack. We can unambiguously indicate a sequence of such actions with a sequence of corresponding symbols. It is sufficient to denote PUSH actions with the value to be pushed and operations with the corresponding arithmetic operator. The sequence is called Reverse Polish (properly, operator-late, parenthesis-free notation of Łukasiewicz). The sequence of actions for the expression given previously is, for example, denoted by:

$$2\ 3\ 1\ -\ +\ 4\ -$$

Any expression can be translated into Reverse Polish. One can find a variety of translation algorithms in texts such as Randell and Russell [64].

Combining the Reverse Polish and the stack manipulation notation for the example above, we get Table 7-3.

Conventional computers accomplish arithmetic evaluation with registers instead of stacks. The instructions name registers and memory locations which must contain the operands. As in the case of Polish, there are many published algorithms which translate expressions to conventional instruction

TABLE 7-3. Reverse Polish and Stack Actions for the
Expression $2 + (3 - 1) - 4$

Reverse Polish Symbol	Stack Action
2	$S := 2$
3	$S := 3$
1	$S := 1$
−	$S := S - S$
+	$S := S + S$
4	$S := 4$
−	$S := S - S$

sequences. They are, however, considerably more complicated than those for Polish code. In particular the value in a register may not be needed as soon as it is calculated while the register may be needed for further calculations. As a result the translator must create a temporary variable for the value in the register which must then be recovered later when it is needed. The use of a stack eliminates the need for explicit temporary stores.

The situation is more complicated when different types of data are mixed in the programming notation, for example, the integer 2 and the real number 2.0. The data may have different widths and the bit patterns have different interpretations. There are many solutions of which we shall examine three.

The first is to let the larger width determine the width of the stack and to inject explicit type conversion instructions into the Reverse Polish. Such extension of the notation needs a new name; we shall call it Generalized Reverse Polish or simply *Polish*. We must also have different operators for the different types of data. We let "r" stand for conversion to type real, "$+_r$" stand for addition on type real, and so on. Then the expression

$$2 + (3.0 - 1.6) + 1$$

yields the Polish

$$2 \text{ r } 3.0 \text{ } 1.6 \text{ } -_r \text{ } +_r \text{ } 1 \text{ r } +_r$$

where everything is forced to type real before computation. The corresponding sequence of events is shown in Figure 7-6.

A second approach is to provide a separate evaluation stack for each

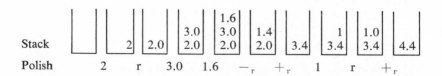

Figure 7-6 Sequence of stack configurations during evaluation of the expression $2+(3.0-1.6)-1$

data type. In this case the type conversion operator "r" causes data to be popped off the integer stack converted to real format, then pushed onto the real stack. The two different stacks may have different width and format, thus avoiding the necessity of coming up with a compromise for the single stack. In terms of the stack notation, the solutions are shown in Table 7-4.

TABLE 7-4. MIXED TYPE ARITHMETIC EVALUATION

Polish	One-Stack Solution	Two-Stack Solution
2	$S := 2$	$S_i := 2$
r	$S := r(S)$	$S_r := r(S_i)$
3.0	$S := 3.0$	$S_r := 3.0$
1.6	$S := 1.6$	$S_r := 1.6$
$-_r$	$S := S -_r S$	$S_r := S_r -_r S_r$
$+_r$	$S := S +_r S$	$S_r := S_r +_r S_r$
1	$S := 1$	$S_i := 1$
r	$S := r(S)$	$S_r := r(S_i)$
$+_r$	$S := S +_r S$	$S_r := S_r +_r S_r$

The third solution is to use tagged data. Suppose that the data type is intrinsically recognizable (usually by the addition of some extra bits called *tag bits* as in Euler [Wirth and Weber, 66]). Then Polish can ignore the type differences by depending on the arithmetic algorithms of the hardware recognizing what to do just before the actual operations take place. We write (r, 2.0) to denote a value 2.0 tagged as being in real format and (i, 2) to denote a value 2 tagged as being in integer format. Then the Polish

$$2 \ 3.0 \ 1.6 \ - \ + \ 1 \ +$$

results in the sequence of stack actions in Table 7-5.

TABLE 7-5. TAGGED DATA SOLUTION
FOR MIXED TYPE ARITHMETIC

Polish	Tagged Data Solution
2	$S := (i, 2)$
3.0	$S := (r, 3.0)$
1.6	$S := (r, 1.6)$
$-$	$S := S - S$
$+$	$S := S + S$
1	$S := (i, 1)$
$+$	$S := S + S$

The values in the stack are illustrated in Figure 7-7. Note that while the first subtraction finds two operands of type real and thus simply performs a real addition, the following addition must first convert the integer 2 into real format before doing the addition.

Stack	$\begin{vmatrix} \\ \\ (i, 2) \end{vmatrix}$	$\begin{vmatrix} (r, 3.0) \\ (i, 2) \end{vmatrix}$	$\begin{vmatrix} (r, 1.6) \\ (r, 3.0) \\ (i,2) \end{vmatrix}$	$\begin{vmatrix} (r, 1.4) \\ (i, 2) \end{vmatrix}$	$\begin{vmatrix} \\ \\ (r, 3.4) \end{vmatrix}$	$\begin{vmatrix} (i, 1) \\ (r, 3.4) \end{vmatrix}$	$\begin{vmatrix} \\ \\ (r,4.4) \end{vmatrix}$
Polish	2	3.0	1.6	$-$	$+$	1	$+$

Figure 7-7 Stack configurations during evaluation of $2+(3.0-1.6)+1$ using tagged data

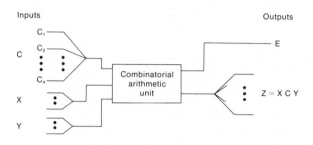

Figure 7-8 Arithmetic unit

Suppose, for example, that we have a purely combinatorial arithmetic unit as shown in Figure 7-8. The two bundles X and Y are the input data and bundle C is a set of control lines C_1, C_2, ... C_n, one for each arithmetic operation the unit can perform. When one control line is up, and the data are on X and Y, the unit will, after a certain delay, provide on the bundle Z the result, X C Y = Z. A failure in the arithmetic algorithms (for example, integer overflow) is signaled on line E.

We want to combine the arithmetic unit in Figure 7-8 with the stack in Figure 7-3 to make a stack arithmetic processor, as shown in Figure 7-9. It is a sequential circuit and we assume the availability of appropriately timed clock pulses on $CLOCK_1$, $CLOCK_2$, ... The control line C_0 signifies a PUSH action from a datum ready on A. The other control lines C signal the operations corresponding to the arithmetic operations the arithmetic unit can carry out. B contains the contents of the top of the stack except during the period where the stack is being changed. The output lines E, UFLO, and OFLO signal the failures of the corresponding internal units.

We will require some registers of the form shown in Figure 7-10. The control line R is pulsed to cause the datum on the U to be read into the register. The V is the current value of the register except for a short period after the line R is pulsed.

Figure 7-11 depicts a solution to the problem. The PUSH action is initiated on pulse $CLOCK_3$ in both the case where C_0 signals a PUSH and the case where two operands have been operated on and the result needs to be placed back into the stack. Pulses $CLOCK_1$ and $CLOCK_2$ trigger POP actions,

Inputs   Outputs

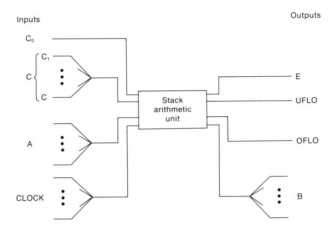

Figure 7-9   Stack arithmetic unit

Inputs   Outputs

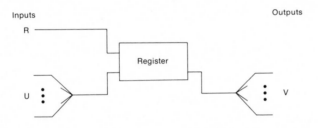

Figure 7-10   A register

the first to feed the data to the arithmetic unit and the second to discard the unneeded datum from the top of the stack to make room for the computed result. R_1 and R_2 are registers of a size consistent with the rest of the unit.

Variables can also appear in expressions. From the viewpoint of the arithmetic unit, a variable is identified with an area of main memory. How that identification is made, and how it is encoded, are deferred to the section on storage allocation. For the moment, we represent a variable by its name with the understanding that it ultimately is mapped into an absolute memory address. As a datum, an address is simply another type (denoted "a"). Each of the three previous solutions can be extended to handle assignment statements. For example, suppose we have the assignment

$$X := X + 3$$

where X is of type real. We have to fetch a value from memory, convert 3 to real format, add them and replace the value in memory.

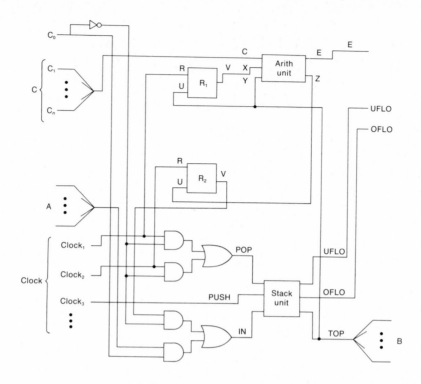

Figure 7-11 Stack arithmetic unit in detail

The transactions with memory depend upon how the values are encoded in memory itself. For purposes of discussion here, we assume that in memory the values have the same format they have in the evaluation stack. This implies that if we have a tagged evaluation stack, we have tagged values in memory. The Polish and interpretation of the preceding assignment are given in Table 7-6.

The data in the stacks S_a and S_i may be identical in format. If so, they may be combined to simplify the three-stack solution in terms of hardware components. It is left to the reader to reformulate the stack notation description for this possibility.

Suppose we have a memory device as depicted in Figure 7-12. The bundles MA and MI must contain the memory address and memory datum respectively. When the control line W is pulsed, the contents of MI are placed in location MA. MO has the value of the contents of location MA except for a brief period after W is pulsed, or MA is changed. Then the stack arithmetic unit in Figure 7-9 and the memory unit in Figure 7-12 can be combined to provide for execution of the Polish in Table 7-6.

The arithmetic operations of the stack processor are relatively complex, requiring the relatively complex sequential implementation in Figure 7-11. They could be expanded into more primitive operations (for example, two

TABLE 7-6. STACK ACTIONS TO EVALUATE
THE ASSIGNMENT $X : = X + 3$

Polish	One-Stack Solution
X	$S : = X$
X	$S : = X$
M	$S : = MEMORY[S]$
3	$S : = 3$
r	$S : = r(S)$
$+_r$	$S : = S +_r S$
$: =$	$MEMORY[S] : = S$

Polish	Three-Stack Solution
X	$S_a : = X$
X	$S_a : = X$
M	$S_r : = MEMORY[S_a]$
3	$S_i : = 3$
r	$S_r : = r(S_i)$
$+_r$	$S_r : = S_r +_r S_r$
$: =$	$MEMORY[S_a] : = S_r$

Polish	Tagged Data Solution
S	$S : = (a, X)$
X	$S : = (a, X)$
M	$S : = MEMORY[S]$
3	$S : = (i, 3)$
$+$	$S : = S + S$
$: =$	$MEMORY[S] : = S$

POPs into different registers, a register-register arithmetic operation, and
finally a PUSH of the result) which would expand the Polish form of the
program but simplify the processor. If the Polish is expanded, it is difficult

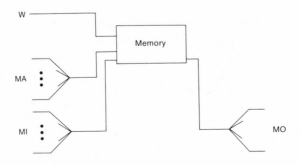

Figure 7-12 A memory

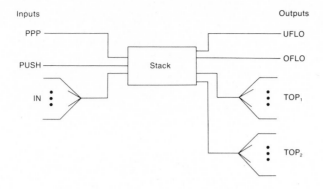

Figure 7-13   Stack for arithmetic evaluation

or impossible for the designer to detect and take advantage of the inherent parallelism of the algorithms.

An alternative form for a stack to be used in arithmetic evaluation is shown in Figure 7-13. It has two output bundles, TOP_1 and TOP_2 (compare with Figure 7-3) which are, respectively, the top and the next-to-top items on the stack. In addition, the control line POP is replaced with a line PPP which has the effect of POP POP PUSH, discarding the two top items on the stack and replacing them with the datum on line IN.

7-4. CONTROL STACK S_c

During execution of a program, the machine code resides in main memory and the control point is defined by the contents of a processor register called the program counter (denoted PC). During normal sequencing, PC is incremented by the current instruction width to establish a new control point for the next instruction. The main use of PC is to permit the processor to access memory for its instructions. PC also serves in special ways for the implementation of subroutines, loops, the consequences of conditional tests, and so on.

There are two control points important for subroutine entry and exit. PC must first be set to the entry point and execution of the subroutine allowed to proceed. At some point in time the subroutine is finished and PC must be reset to the value it had prior to the entry, allowing the calling routine to proceed. Suppose that Q is the entry point for a subroutine. The action

CALL Q

is used to get to the entry point, and the action

RETURN FROM Q

is used inside the subroutine to terminate its execution and return control to the point of call. All of this can be accomplished as shown in Table 7-7, using a stack called the control stack and denoted S_c. The value of PC is saved on S_c when a CALL is executed and the top of S_c is popped off and placed back in PC when a RETURN is executed. A stack can be used because CALL-RETURN pairs are nested in time.

TABLE 7-7. ALGORITHMS USING S_c FOR CALL AND RETURN

CALL Q	RETURN FROM Q
$S_c := PC$ $PC := Q$	$PC := S_c$

The mechanism implied in Table 7-7 has a number of advantages. First, of course, is that the subroutines may call each other to any depth and in any order, even recursively. Also, no more storage is used than actually needed. In a nonrecursive program, the maximum size of S_c can be computed prior to execution by, for example, a compiler. The control stack S_c is separate from, and independent of, other stacks. The importance of this comment is illustrated by the function subroutine, which uses the arithmetic stacks for its computations and the control stack for CALL and RETURN. The delicate sequencing of preparing the value to be returned, returning, and using the returned value is simplified by implementing the mechanisms in separate structures.

Table 7-7 was constructed on the assumption that the machine code for a program was fixed in memory during execution, and the absolute address of the entry point could be recorded in the program code itself. It is, however, sometimes desirable to permit the code for a subroutine to be moved about in memory during execution (called dynamic relocation). The absolute addresses are kept in an execution-time table (called the program reference table or PRT) and updated by the operating system that is doing the relocation. The machine code for the program then has only a record of the index of the entry point in that table. The solution is to save in S_c two values: the index of the calling routine in the PRT and the control point in the calling routine *relative* to its entry. If P is the PRT index of the calling routine, and Q is the PRT index of the called routine, and T and U are registers in the processor, then the extended algorithm in Table 7-8 is sufficient.

TABLE 7-8. ALGORITHMS USING S_c FOR CALL AND RETURN WITH
DYNAMICALLY RELOCATABLE CODE

CALL Q FROM P	RETURN FROM Q TO P
$S_c := (P, PC - PRT[P])$ $PC := PRT[Q]$	$(T, U) := S_c$ $PC := PRT[T] + U$

There are other actions besides sequential instruction processing and sub-routines affecting PC: loops and decisions. On conventional computers these actions are implemented through branches, conditional branches, and indexed branches. Sometimes the conventional solutions are carried over into stack computers. One can, instead, continue to elaborate on the constructs affecting S_c to achieve the same effects. The latter solutions are more in the spirit of the stack computer and are discussed below.

For example, there is a more powerful interpretation of the RETURN FROM construct than that implied by Table 7-8. Suppose that Q is a sub-routine for a complicated algorithm and that Q has called various other subroutines to carry out the details. Deep into a sequence of such calls a condition is encountered that requires that the whole algorithm Q be abandoned. Then, in some other subroutine R, the construct

RETURN FROM Q

would be interpreted as a whole sequence of RETURN actions, eventually returning control to the point from which Q was called. Such a situation is sketched in Figure 7-14.

The appropriate multi-level return is combined with the dynamically relocatable solution in Figure 7-15. Note that it would be difficult to extend the solution in Table 7-7 in this way. As before, T and U are processor registers and Q is the index of the address of the entry of Q in the PRT.

A similar algorithm can be devised for the construct RETURN TO P. The interpretation for this construct depends on P having been called and not yet exited prior to the encounter of the RETURN TO P action. P has called some other subroutine which may have called others to an arbitrary depth. RETURN TO P has the effect of terminating all subroutines including the last one P called and returning control to P at the point where the last one would have normally returned.

```
P:    PROCEDURE OPTIONS (MAIN);
      S:    PROCEDURE(N);
                /*arbitrary algorithm;*/
            END S;
      R:    PROCEDURE(N);
                IF N < 0 THEN RETURN FROM Q;
                CALL S(5);
            END R;
      Q:    PROCEDURE(N);
                IF N < 3 THEN CALL R(N);
                CALL S(10);
            END Q;
            CALL Q(-1);
      END P;
```

Figure 7-14 PL/I program requiring a multi-level RETURN

$$T := nil$$
$$while\ T \neq Q\ do\ (T, U) :\ = S_c$$
$$(T, U) :\ = S_c$$
$$PC :\ = PRT[T] + U$$

Figure 7-15   Algorithm for multi-level RETURN from Q initiated outside of Q

In the discussion on arithmetic stacks, it was clear that single operators in Polish corresponded to single lines of the stack manipulation notation. In the algorithms effecting S_c however, conceptually monolithic operations CALL and RETURN have a more complicated appearance. It is proper to consider the alternatives for the Polish forms of CALL and RETURN since complicated monolithic operations imply complicated monolithic hardware.

In the case of an arithmetic operation, it was possible to factor the whole action into a sequence of simpler actions. The situation for CALL and RETURN is complicated by the fact that PC is being changed. In each algorithm in this section, the change of PC is the last thing done which is a condition on being able to factor the whole operation into simpler operations.

The most straightforward Polish form for CALL is a CALL operator followed in the Polish by the absolute address of the entry (Table 7-7) or the index into the PRT (Table 7-8). Alternatively, the address could be pushed onto an arithmetic stack (say S_a) and the CALL operator would find it there.

One advantage of separating the access of the entry point address from the actual transfer of control is in allowing other ways of getting the address. The access to PRT (Table 7-8 and Figure 7-15) could in fact be done with whatever mechanism is used for subscripts in the arithmetic processor (see section 7-6). There are two other programming language constructs that can also be implemented by subscribing into a table of entry points and then calling the selected subroutine.

The CASE statement of PASCAL [Wirth, 71] and ALGOL-W [Wirth and Hoare, 66] takes the form shown in Figure 7-16. It is interpreted to mean that the CASE expression is computed, and then the corresponding statement

CASE *n* OF

BEGIN

    statement__1;

    statement__2;

      . . .

    statement__*m*;

END

Figure 7-16   ALGOL-W CASE statement selecting the *n*th out of *m* statements

in the following block is selected and executed. It is equivalent to the SWITCH construct in most programming languages except that after execution of the selected statment, control is automatically returned to the point beyond the end of the block of statements.

If each statement _k is treated as a separate subroutine, then the CASE statement can be implemented by tabulating the entry points for the m statements, using n to index into that table and fetch the corresponding entry point to stack S_c whereupon the solution to problem 7-13 can be used. A RETURN operator must, of course, be appended to each of the m subroutines to cause the RETURN action to take place.

The IF-THEN-ELSE construct is in fact a subcase of the CASE statement where the only value of the selection expression are TRUE and FALSE (that is, 1 and 0). It can be implemented exactly as the CASE statement is implemented. The effect is to avoid altogether the familiar branching logic normally associated with the IF-THEN-ELSE and CASE constructs.

Loops can also be expressed in terms of manipulations of S_c. Suppose the body of a loop is also a subroutine. It is entered via a CALL. At some point in the loop (usually the bottom), it is discovered that the loop must be repeated. Then the algorithm in Figure 7-17 can be used. It corresponds to the algorithms for CALL and RETURN in Table 7-3. The body of the loop is identified by index Q into the PRT. All that is needed is to reset PC to the address of the entry point.

The multi-level RETURN and REPEAT operations are of indeterminate duration. Thus during their operation, they must inhibit the operation of the rest of the computer.

For the same reasons that a multi-level RETURN is needed, a multi-level REPEAT is needed. It is a combination of the RETURN TO action with the REPEAT action. The action

$$REPEAT \ Q$$

is interpreted as

$$RETURN \ TO \ Q \ AND \ REPEAT \ IT$$

Figure 7-18 gives an algorithm combining the result of problem 7-13 and Figure 7-17. The interpretation of the symbols is as before.

PC := PRT[Q]

T := nil
while T ≠ Q do (T, U) := S_c
PC := PRT[Q]

Figure 7-17 REPEAT operation   **Figure 7-18** Multi-level REPEAT operation

7-5. STORAGE FOR SIMPLE VARIABLES S_v, S_m

The concept of local variable arises from the combination of the programming language concepts of variable and subroutine. The local variables of a subroutine have the property that they can be accessed only from statements within the subroutine. They have undefined values when the subroutine is entered and whatever values they have acquired during the execution of the subroutine are lost when control is returned to the point of call.

All variables can be considered local variables if the program itself is a subroutine (called by some more global authority such as an executive program).

When the definition of one subroutine is nested within the definition of another, the inner subroutine has access to the variables local to the containing subroutine but not vice versa. The scope of a variable (the set of places from which it may be accessed) is the body of the subroutine to which it is local and all subroutines defined within that subroutine. The PL/I program in Figure 7-19 illustrates the possibilities. The scope of the variable A is the whole program; the scope of variable B is the body of SUBROUTINE only.

As a consequence of local variables being undefined whenever the corresponding subroutine is not being executed, storage need not even be assigned to the local variables until the subroutine is called, and may be freed for other uses as soon as control has left the subroutine via a RETURN. This can be accomplished by using a stack, S_v, for the local variables. A PUSH on S_v corresponds to the allocation of storage for a local variable and a POP corresponds to deallocation. Generally speaking, one would expect several PUSH operations with each CALL action corresponding to the several local variables

```
PROGRAM:
  PROCEDURE OPTIONS(MAIN);
    DECLARE A FIXED;
  SUBROUTINE:
    PROCEDURE;
      DECLARE B FIXED;
      B = 2;
      A = B;
      RETURN;
    END SUBROUTINE;
    A = 1;
    /* AN ASSIGNMENT TO B HERE
       WOULD VIOLATE SCOPE RULES */
    CALL SUBROUTINE;
    RETURN;
  END PROGRAM;
```

Figure 7-19   Illustration of scope in PL/I

in the called subroutine. As it happens, it is more convenient to associate the PUSH actions with the local variables of the calling routine.

The local variables are not, however, accessed by PUSH and POP as in the case of the previous stacks. They are accessed at random at any time during the execution of their scope. Thus other access methods must be added to PUSH and POP. It is the need for other access methods, together with the fact that S_v may be rather large and thus best kept in main memory, that indicates a solution along the lines of Figure 7-2. The base registers can be used for more random access as well as for control of PUSH and POP.

Suppose that we are willing to settle for the ability to access only the most global variables (those in the main program subroutine which contains all the other subroutine definitions) and the local variables of the subroutine currently being executed. The global variables are the first to be allocated in S_v and are thus at the bottom of it; the most local variables are the last to be allocated in S_v and are thus at the top of it. Such a solution is illustrated in Figure 7-20. The configuration of S_v corresponds to the situation when execution has just begun of the body of procedure R in Figure 7-14. The register G points to the base of the area containing the global variables and the register L points to the base of the area containing the most local variables (those of procedure R).

There are six actions that must be defined in terms of MEMORY and the registers G and L: scope entry, scope exit, and four kinds of variable access. They are given in Table 7-9. V(P) stands for the number of local variables in subroutine P. The data accesses are to the Kth variable in the corresponding scope. No mention is made of checking L for stack overflow or underflow but it should be done.

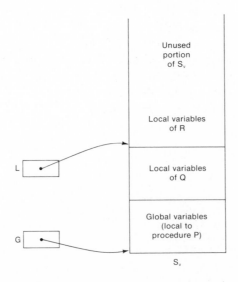

Figure 7-20 Implementation of S_v

TABLE 7-9. SIX ACCESSES TO S_v

Action	Algorithm
SCOPE—ENTRY	$L := L + V(P)$
SCOPE—EXIT	$L := L - V(P)$
PUSH the Kth global variable onto stack S	$S := MEMORY[G+K]$
POP a value from stack S into the Kth global variable	$MEMORY[G+K] := S$
PUSH the Kth local variable onto stack S	$S := MEMORY[L+K]$
POP a value from stack S into the Kth local variable	$MEMORY[L+K] := S$

The actions for SCOPE—ENTRY and SCOPE—EXIT are usually associated with the actions for CALL and RETURN. Note that the change to L can be thought of as protecting the local variables of the calling procedure by making them inaccessible. Dynamically, the sequence of events must be:

> SCOPE—ENTRY
> CALL
> . . .
> RETURN
> SCOPE—EXIT

Since the CALL operator must be in the code of the calling subroutine, so must SCOPE—ENTRY. RETURN is the last thing done in the body of the called subroutine, thus SCOPE—EXIT must also be in the code of the calling subroutine. That is convenient since the value $V(P)$ (Table 7-9) is more readily available to P, the calling routine. The combined operators (Tables 7-7 and 7-9), ENTER_SCOPE, CALL, EXIT_SCOPE in the calling subroutine, and RETURN in the called subroutine, are shown in Table 7-10.

A subroutine with no local variables needs a CALL without a SCOPE—ENTRY; a block in ALGOL or PL/I needs a SCOPE—ENTRY without a CALL. The implication is that the action in Table 7-10 may, or may not, be a candidate for a single operator in Polish.

TABLE 7-10. CALLING AND RETURNING ALGORITHMS

Action	Algorithm
Actions in the calling subroutine P to enter the scope of Q, call it, and exit the scope after return	$L := L + V(P)$ $S_c := PC$ $PC := Q$ $L := L - V(P)$
Action to return from Q to P	$PC := S_c$

One further stack, S_m, the marker stack, can be of use. If the value of L is saved in S_m prior to changing L, then SCOPE_EXIT need only POP the needed value off S_m and back into L. The implication is that the number of locals $(V(P))$ is needed only at the time of call. This solution is analogous to the use of S_c for saving PC instead of tabulating all the potential return addresses ahead of time to be selected by the RETURN operator. As it happens, the contents of S_m are needed for the general case of nested addressing (as opposed to the global/local solution discussed previously).

The actions for accessing variables (Table 7-9) must also be reflected in Polish. The Polish must carry the information of either global or local, and the offset K from the appropriate base register. Recall that the memory access in the Polish of section 7-3 required that the address be on the stack. The same solution applies to S_v if the computations $G+K$ and $L+K$ are carried out by the arithmetic unit. That would imply a sequence of actions something like

$$S := G$$
$$S := K$$
$$S := S + S$$
$$S := MEMORY[S]$$

to access a simple variable. In an application where keeping the processor simple is paramount, it may be the proper solution. On the other hand, one can devise special hardware and a richer Polish so that the shorter sequence

$$S := G + K$$
$$S := MEMORY[S]$$

or even

$$S := MEMORY[G + K]$$

is used.

There are arbitrarily many ways of encoding the Polish. Some are better than others, depending upon which measure of efficiency one decides to apply. The following example is intended to tie together the preceding material in a reasonable way, making definite choices from the many possibilities. It is also intended to be a good choice under some circumstances, thus providing in addition a certain amount of intuitive guidance in choosing encodings.

Let us suppose that 8 bits is a convenient unit for encoding the Polish (due to memory structure or some other arbitrary external constraint). There is a stack S of width 16 combining the functions of S and S_c as well as a stack S_v. One half of the 256 patterns will be reserved for the action $S := C$, where C is a constant in the range $0 \leq C < 128$. The remaining 128 patterns represent operations such as ADD and CALL. If we need more (unlikely), then one operation code can be set aside to signify that the next 8 bits are to be used.

TABLE 7-11. PARTIAL SET OF POLISH OPERATORS

Operator	Algorithm
LIT0, LIT1, ... LIT127	Constant in the range 0 to 127 is placed on the stack
SHL7	$S := S * 128$
ADD	$S := S + S$
LOADG	$S := MEMORY[G + S]$
LOADL	$S := MEMORY[L + S]$
STOREL	$MEMORY[L + S] := S$
STOREG	$MEMORY[G + S] := S$
CALL	$PC := PRT[S]; S := PC$
RETURN	$PC := S$
SCOPE_ENTRY	$L := L + S$
SCOPE_EXIT	$L := L - S$

Except for the choice made for the constants, the assignment of actual bit patterns is irrelevant, thus we will use mnemonic names instead of the patterns in the discussion. A partial table of operators is given in Table 7-11. It is assumed that PRT, S, and S_v are properly set up prior to initiating execution of a program. Figure 7-21 gives the machine code using Table 7-11 for the program in Figure 7-19. Figure 7-22 shows an elaborate stack computer which uses separate stacks for S_c and S_m.

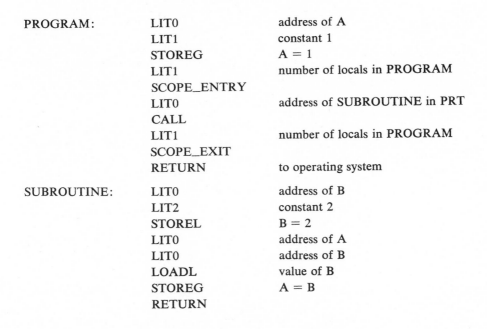

PROGRAM:

	LIT0	address of A
	LIT1	constant 1
	STOREG	A = 1
	LIT1	number of locals in PROGRAM
	SCOPE_ENTRY	
	LIT0	address of SUBROUTINE in PRT
	CALL	
	LIT1	number of locals in PROGRAM
	SCOPE_EXIT	
	RETURN	to operating system

SUBROUTINE:

	LIT0	address of B
	LIT2	constant 2
	STOREL	B = 2
	LIT0	address of A
	LIT0	address of B
	LOADL	value of B
	STOREG	A = B
	RETURN	

Figure 7-21   Polish form of a PL/I program

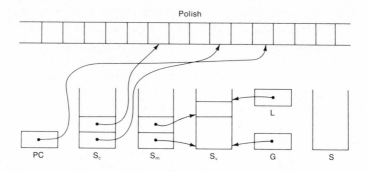

Figure 7-22 Elaborate stack computer

The global/local solution just described is less general than needed for ALGOL 60 and later languages that have adopted nested scope structures. The nesting of subroutine definitions mentioned at the start of this section may continue to arbitrary depth, permitting the innermost subroutine to access its own variables, those of the subroutine within which it is defined, and so on, out to the most global subroutine. The global/local solution can be used for such languages but it fails to implement all the accessing implied by the nesting. There are, for a subroutine nested k levels into the program, k separate areas in S_v that should be accessible, requiring k base registers in the place of the two bases G and L.

This new set of base registers is called the display [Wirth and Hoare, 66] and is designated by the identifier D. The Kth register is denoted D[K].

When a subroutine defined at the Kth level of nesting is being executed, each of the containing subroutines has been called and has an area allocated on S_v. The most global is pointed to by D[0] and the local variables (nested in level K) are pointed to by D[K]. The intervening registers point to the locals defined on the corresponding level of nesting. D[0] thus takes the place of G and D[K] takes the place of L (see Figure 7-23).

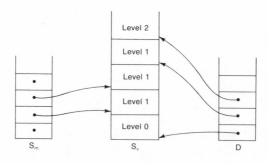

Figure 7-23 Local variable stack addressed through the display

The values of D change only on scope entry and exit so we may expand the SCOPE_ENTRY and SCOPE_EXIT operators to set them. Furthermore, assuming that we have used the stack S_m, all the values that must be in D are already somewhere in S_m.

There are two cases: either the programming language allows subroutines to be passed as parameters to other subroutines or it does not. If it does not, a subroutine can be called only if it is defined in one of the scopes that is accessible to the calling routine. The implication is that D is already correctly set except for the newly allocated local variables of the called routine. If the language does allow parametric subroutines, then the name of the subroutine can be passed to the body of a subroutine that is not accessible to the calling subroutine, thus all of D (except D[0]) might have to be changed.

The algorithms for scope entry and exit are shown in Table 7-12. The calling subroutine is P. The number of locals in P is $V(P)$, P is nested K levels deep, the called subroutine is nested J levels deep (Note: $J \leq K + 1$). S_m now holds only the values of D that are saved instead of all values of L. Referring again to Figure 7-23, note that some entries in S_m do not point to S_v (denoted by the missing arrows). The first entry into a level of scope causes a save of D[K] where it did not have a valid previous setting.

If there is a parametric subroutine defined on level J, we must have enough information to set all the bases D[1] to D[J − 1]. The information is in S_m and D but it is difficult to access at the time the parametric subroutine is called unless some preparations have been made beforehand. There are many solutions to the problem [see, for example, Randell and Russell, 64].

One solution is to save the needed information as the parametric subroutine is passed, in effect making the registers D[1] to D[J − 1] parameters to be passed along with the subroutine itself. Further discussion is deferred to section 7-9, after the treatment of parameter passing in section 7-8.

TABLE 7-12. SCOPE ENTRY AND EXIT USING D
(NO PARAMETRIC SUBROUTINES)

Action	Algorithm
SCOPE_ENTRY	$S_m := D[J]$
	$D[J] := D[K] + V(P)$
SCOPE_EXIT	$D[J] := S_m$

7-6. STORAGE FOR STRUCTURED VARIABLES

A structured variable is a collection of simple variables together with some predetermined method of data access. Arrays, strings, lists, records, tables, structures, queues, stacks, trees, sets, and ordered sets are types of structured variables that have appeared in programming languages. Some languages have, in addition, mechanisms for programmer definition of other types of structured variables.

The access of structured variables involves computations on addresses. While arbitrary computations could be envisioned, in practice they are almost entirely limited to indirection, addition, subtraction, and multiplication. Some types, such as arrays, are so commonly used that special hardware has been devised for their allocation and access.

The time that the size of a variable is known is an important consideration. In languages such as FORTRAN the size of all data structures is known and fixed at the time the program is compiled. In ALGOL 60 the size is fixed after SCOPE_ENTRY but before the first statement is executed in this scope. In PL/I the size of a string may vary dynamically within predetermined limits. In LISP the size of a list may vary dynamically with no predetermined limits.

The topic of data structures could fill a book [Stone, 72]. The best we can do here is to take some examples and indicate what mechanisms might be useful for their implementation. We will consider arrays first.

Suppose that the sizes of all arrays in a scope are known prior to the possible entry of another scope (not true of PL/I or ALGOL 60 since the array bound computation may cause a SCOPE_ENTRY), and that the arrays are allocated in S_v. Then the needed increment $V(P)$ to the local variable base register (L or D[K]) is known and the allocation algorithms in Tables 7-9 and 7-12 can be used. Notice that by considering the SCOPE_ENTRY action as protecting the local variables of the calling subroutine we can wait longer to know how much storage needs allocation.

Even if not all the arrays are allocated when a new scope is entered, the algorithms mentioned can be used if no new allocations can take place prior to the SCOPE_EXIT leading back to the scope in question. It simply means that $V(P)$ must be determined anew prior to each SCOPE_ENTRY. This solution is adequate for ALGOL 60 and PL/I arrays with attribute AUTO-MATIC. It is not adequate for the PL/I ALLOCATE and FREE statements.

Even though arrays could be accessed directly by doing arithmetic on register L, there is advantage in allocating a single word for each array along with the other local variables in a scope. The word is called the *array descriptor* and contains the information needed to access the array. Such a situation is shown in Figure 7-24. There are four local variables, one of which has the value 3, and the others are arrays. The pointers in the array descriptors must be filled in when the array is allocated. In the normal case, size information is discovered immediately after scope entry by the initial code of the called subroutine.

Suppose that we have an indexing operation INX (which may just be an ADD in the simplest case) which combines the subscript value and the array descriptor to give a new descriptor pointing to the selected cell in an array. Then the Polish in Table 7-13 is an example of accessing arrays.

The array descriptor may contain more information than just the address in S_v. If the array data is tagged, the descriptor may contain the tag so that

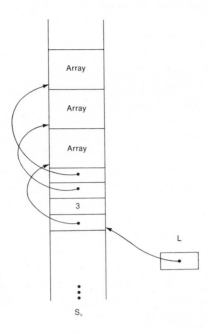

Figure 7-24   Arrays allocated in S_v accessed through pointers

the data type can be discovered without accessing the data itself. This is of critical importance if the address arithmetic for access depends on the type of the data, or if the data memory has no provision for tags.

The descriptor may also contain information allowing the access to be checked for array bounds violations. If special hardware is dedicated to the checking task it can be done with no penalty in execution speed or code density. Since array overrun is a source of very troublesome programming errors, and since programmers will not tolerate the inefficiency of bounds checking in conventional instruction sets, the designer of bounds checking hardware will find the cost/benefit ratio very much in his favor.

TABLE 7-13. POLISH FOR THE ASSIGNMENT A[I] := 2

Polish	Action	Comment
I	S := I	Offset from L of the cell for I in S_v
LOADL	S := MEMORY[L+S]	Value of I
A	S := A	Offset from L of the cell containing the descriptor for A
LOADL	S := MEMORY[L+S]	Descriptor for A
INX	S := S INX S	Access operation given a new descriptor for the subscripted variable A[I]
2	S := 2	Value 2
STORE	MEMORY[S] := S	Store 2 in A[I] (Note that a new absolute store operation has been introduced.)

TABLE 7-14. ALGORITHM FOR SUBSCRIPTING: S : = S INX S

Action	Comment
(T, U) : = S; W : = S; if W < 0 or W ≥ T then ERROR; S : = W + U;	T is the size; U is the address W is the value of the subscript Address of the subscripted variable

If the array descriptor contains the size of the array then the indexing operation INX in Table 7-13 can be expanded into the algorithm in Table 7-14. The descriptor is encoded as a pair (size, address) and the array is presumed to start with position 0. If the array does not start with position 0, the lower bound of the array must be subtracted from the subscript prior to the INX operation. Note that the checking and the address computation can proceed in parallel.

The stack S_v is not the only place where arrays can be allocated. They could have their own stack. In that case S_v might be tagged and the array stack untagged since the array data all have the same tag which can then be added once to the descriptor for the whole array.

Arrays might also be allocated in a free storage area (an ALGOL 68 heap). In this case the descriptors point out into the free area and the storage allocation actions imply the presence of an answering mechanism managing the free area. Again, there is no need for tags on the data in the free area since the tag can reside in the descriptor. The data is dynamically relocatable since only the descriptor contains the actual address. The free area solution is also capable of implementing the ALLOCATE and FREE statements of PL/I which are asynchronous with the scope entry and exit actions of the Polish.

Allocating arrays in a free area has an additional property: storage need not be allocated until the data is actually accessed. Thus if information is added to the descriptor indicating whether the data has been accessed, the INX operation can include the call to the storage allocation actions.

Although S_v is stack-like in nature, it is possible to allocate the storage for each local variable area in a free area much as though they were arrays. The scope entry action, instead of incrementing the pointer into S_v would use the address supplied by the allocation routine. The only information connecting the Polish and the variables is then the base registers D (or G and L).

In some programming languages, it is possible to multiprogram several subroutines in a common environment. That is, several subroutines may execute asynchronously while accessing common data in S_v. The mechanisms in the previous sections can be used if the subroutines have their own private stacks S_c and S_m. In effect, the display D is used to bind the active subroutine to S_v. When execution is suspended in one subroutine so that another can resume, D is saved for the suspended subroutine and restored to the state it had when the resumed subroutine was last suspended. In effect the stack S_v becomes a tree with a common area near the root and then branching out where the multiprogramming begins.

7-7. THE PARAMETER PREPARATION STACK S_p

The preparation and passing of parameters to a subroutine is yet another function for which a stack is an appropriate mechanism. Parameters are, in effect, initialized local variables of the called subroutine. The values may be data types of the language or they may be pointers to variables that are local to other subroutines, most usually the calling subroutine.

While control is still in the calling subroutine, the parameter values must be computed in a stack S_p. The values in S_p must ultimately end up in the local variable stack, S_v, so the format of S_p should be the same as that of S_v. In particular, if S_v contains tagged data, then S_p must also. After the parameters are prepared, the new scope is entered, and stacked parameters are moved into the new area in S_v.

It is possible to have prepared some of the parameters to a subroutine and suddenly be forced to prepare a second set before the first set is used. The expression

$$MAX (A, MIN (B, C, D), E)$$

is an example. After the value of A is stacked, the parameters B, C, and D must be prepared for subroutine MIN. Finally, after its value is available and stacked on S_p, the value of E must be stacked and then MAX invoked.

The value of a function can also be considered a parameter; the direction of information flow is just different. The function subroutine may place its value on S_p so that the calling routine may recover it there. When the returned value is arithmetic, the effect is to move the value from the arithmetic evaluation stack to S_p and back to the arithmetic evaluation stack.

The returned value may, however, be more complex. If the value returned is structured (for example, an array) more than one value must be moved while it is sufficient to place in S_p only a descriptor of the information. The effect is to leave the values in place in S_v until after the RETURN, but prior to any ill effects of the SCOPE_EXIT. The values then are moved from the area of S_v about to be abandoned to the local (structured) variable which is its destination.

If the subroutine is multi-valued, the values may be held in S_p to be used upon completion of RETURN and SCOPE_EXIT. For example, the function invocation

$$A, B, C = F (X + 1, Y - 1, X + Y)$$

where F is the (extended) PL/I procedure

```
F:  PROCEDURE (P1, P2, P3) RETURNS (FLOAT, FLOAT, FLOAT);
        RETURN (P1 + P2, P2 + P3, P3 + P1);
    END F;
```

requires a place to hold the value P1 + P2 while the remaining values are computed.

The necessary Polish operators are SAVE_PARAMETER, which places a computed value on S_p; USE_PARAMETER, which reverses the process; and INITIALIZE, which move values from S_p into S_v. Algorithms for them are given in Table 7-15. S is an arithmetic evaluation stack, T is an intermediate register, L is the register pointing to the local variable area in the newly called subroutine (perhaps one of the D registers). INITIALIZE needs the number of parameters; it has been left on S.

TABLE 7-15. Operators Using the Parameter Preparation Stack

Operator	Algorithm
SAVE_PARAMETER (prior to CALL)	$S_p := S$
USE_PARAMETER (after RETURN)	$S := S_p$

INITIALIZE	$T := S;$
	while $T > 0$ do
	begin
	$\quad T := T - 1;$
	$\quad$ MEMORY[L+T] $:= S_p;$
	end

7-8. ADDRESSING VARIABLES IN LANGUAGES WITH NESTED, RECURSIVE PARAMETRIC SUBROUTINES

If subroutines are not recursive, storage can be allocated statically at the time the program is loaded into memory. If subroutines are not nested then they need only address global and local variables. If there are no parametric subroutines, the display described in section 7-5 provides for addressing of local variables. But there are languages with all three properties for subroutines. We will present a solution which is identical to the display of section 7-5 except for special provisions for setting the display for parametric subroutines. The intuitive reason behind this solution is that parametric subroutines are an infrequent construct; thus we can afford to make their use more costly so long as we do not interfere with the normal mechanisms for CALL and RETURN.

When the parametric subroutine is called, the problem is to set the display to the state it had when the subroutine was first passed as a parameter. The solution is to pass as parameters the relevant part of the display along with the subroutine and keep it in S_v to be used each time the parametric subroutine is called.

There are three actions: PTPAPS, which prepares to pass a parametric subroutine; PTCAPS, which prepares to call a parametric subroutine; and RECAPS, which recovers from calling a parametric subroutine. The latter two are similar in function to SCOPE_ENTRY and SCOPE_EXIT.

The actions are shown in Table 7-16. P denotes the index into PRT for the parametric subroutine. We suppose that all static information about the subroutine (its address, level of nesting, number of local variables, number of parameters) is also recorded in the PRT and is available to the algorithms. The mechanisms for transferring the information would, of course, have to be specified, but that detail would only obscure the algorithms as presented here.

TABLE 7-16. SPECIAL POLISH OPERATORS FOR NESTED
RECURSIVE, PARAMETRIC SUBROUTINES

Operator	Algorithm

PTPAPS: Pass P (defined on level J) to Q (defined on level K)	$S_p := P; T := 1;$
	while $T < J$ do
	begin
	    $S_p := D[T];$
	    $T := T + 1;$
	end;

PTCAPS: Call P (defined on level J) from Q (defined on level K)	$T := 1; U := D[K];$
	while $T < J$ do
	begin
	    $S_m := D[T];$
	    $D[T] := MEMORY[U + T];$
	    $T := T + 1;$
	end;

RECAPS: Recover from calling P (defined on level J) from Q (defined on level K)	$T := J - 1;$
	while $T > 0$ do
	begin
	    $D[T] := S_m$
	    $T := T - 1;$
	end;

7-9. EVALUATION CRITERIA

Compilers, operating systems, and programs in general are both easier to write and run more reliably on stack computers than on conventional computers. As a result the efficiency of such a design cannot be properly measured without taking into account programmer costs, the costs of unreliability, and so on. The difficulty in answering such a broadly stated evaluation question has often precluded the use of stack hardware.

Stack computers are more complex and therefore more expensive to manufacture. And they usually do not run as fast as conventional computers for typical short pieces of code. The question is whether the improvements in the programming systems are worth the costs. The answer to this question changes with time. Hardware costs are decreasing and speed is increasing. Programmers, however, seem to cost more. One can conclude then that the case for stack computers, whatever it was, is getting increasingly better. In some applications, such as intelligent computer terminals, it is already the case that the cost of the hardware is negligible and the speed of the hardware is such that the computer is idle most of the time.

One question that must be answered for any computer design, most especially those that depart from the established norm, is whether or not the needed programming systems can be implemented. The simplest approach is to collect a set of very simple programs, each of which uses either one programming primitive or, where there is interaction between primitives, two or more. Collectively they should span the programming language. Each program should then be rendered, by hand, into machine language. Assuming that the translation is successful, then one should estimate how difficult it is going to be for the compiler to do it and how efficient the result is going to be.

Any further evaluation begins to be a substantial effort. For a stack computer the cost of an evaluation would probably exceed the cost of designing. We sum up this chapter by referring the reader to problems 7-51 through 7-54; these problems are intended as much to illustrate the scope of the evaluation problem as to give specific answers to evaluation questions.

7-10. PROBLEMS

7-1. Specify, at some level of detail, another hardware implementation of a simple stack. How many different answers can you find for this exercise? Give a brief discussion of the effect each implementation has on the external characteristics of the stack, such as speed, overflow, underflow, and so on. (Refer to Figure 7-5.)

7-2. Using components available today, design a stack functionally identical to that in Figure 7-5. Provide for the initialization of the overflow/underflow register either in the hardware or as a sequence of external commands.

7-3. Which of the solutions in Tables 7-4 and 7-5 is implemented by Figure 7-11?

7-4. Design a stack arithmetic processor similar to the one in Figure 7-11 using two stacks as in Table 7-4. Assume the availability of two arithmetic units, one for each type, and specify two stacks to be used with them.

7-5. What problems arise in designing an implementation for the solution in Table 7-5? What additional pieces of hardware would be useful? How much depends upon how the data types are recognized?

7-6. Along the lines of Figure 7-11, combine a memory unit and a stack arithmetic unit to process Polish with variables.

7-7. Design a multi-register arithmetic processor that uses a stack for temporary results in arithmetic expressions. The action PUSH X_n and

POP X_n are the transactions between the stack and register X_n. The action LOAD X_n, A puts the contents of memory cell A into register X_n. The action OP X_n, X_m replaces the contents of register X_n with the value X_n OP X_m. Compare your results with Figure 7-11 in every meaningful way you can. Under what circumstances would you recommend each design?

7-8. Design, along the lines of Figure 7-5, an implementation of the stack in Figure 7-13.

7-9. Design, along the lines of Figure 7-11, a stack arithmetic processor using the stack in Figure 7-13 instead of the one in Figure 7-3. How much gain in performance is there over the stack processor in Figure 7-11? Can you discover a general rule behind the change that can be applied to other design problems?

7-10. Register PC can be eliminated if the top of the stack S_c is itself the program counter. Using the notation

$$: = S_c$$

to denote that the top value of S_c is popped and discarded, redo Table 7-7.

7-11. Assuming that PC is held in a register of the form shown in Figure 7-10, and that S_c has the form shown in Figure 7-3, design to implement Table 7-7. The circuit should have two control lines CALL and RETURN and may use CLOCK lines such as those used in Figure 7-11. The value Q can be assumed to be available on an input bundle.

7-12. Suppose that a whole memory box like the one in Figure 7-12 is dedicated to the PRT. Combine the arithmetic unit (Figure 7-8), the memory, and a stack to redo problem 7-11 according to the algorithms in Table 7-8. The values P and Q can be assumed to be present on input bundles. (Where do they come from?).

7-13. Redo Figure 7-15 for a RETURN TO construct.

7-14. Recall the results of problem 7-10, and also the three-stack arithmetic processor in Table 7-6. If the entry address is pushed onto S_a, then the CALL takes the form of a stack-to-stack transfer identical in form to a type transfer. Redo Table 7-7 using this solution. What would the interpretation of the "type transfer" be if Table 7-8 were redone?

7-15. Extend the result in problem 7-12 to handle multi-level RETURN and REPEAT. You may assume the presence of a control line TICTOC which carries a sequence of suitably spaced pulses. An output, IN-HIBIT, must be up until it is certain that the operation will be finally complete in some further fixed amount of time.

7-16. Consider a looping construct in some programming language (for example, DO in FORTRAN or WHILE in ALGOL). Can it be implemented without any branching instructions beyond those in section 7-4? Is either a multi-level RETURN or REPEAT needed?

7-17. Suppose that integers and memory addresses both have the same format. Design a circuit that combines the algorithms for S_c in Table 7-7 and the arithmetic evaluation stack in Figure 7-9, using only one internal stack.

7-18. The combination in problem 7-17 does not work smoothly for languages with function subroutines but does otherwise. Show why this is the case.

7-19. Suppose that there is a single stack containing tagged data words, and that values of PC have a unique tag, c, (that is, CALL pushes a value of the form (c, P, PC − PRT[P] into the stack). Design a circuit combining the stack in Figure 7-9 and the multi-level CALL, RETURN FROM, RETURN TO, and REPEAT operations (Refer to Figures 7-15 and 7-18).

7-20. Design a circuit with inputs V, SCOPE_ENTRY, and SCOPE_EXIT; and outputs G, L, SV_OFLO, and SV_UFLO. Whenever SCOPE_ENTRY or SCOPE_EXIT are pulsed, the appropriate action is taken to update L with the value on V.

7-21. One may design S_v so that the SCOPE_ENTRY and SCOPE_EXIT operations are executed in the called subroutine instead of in the caller. In this case one has the more direct view that SCOPE_ENTRY is allocating space for the variables of the called subroutine (as opposed to the previous view that SCOPE_ENTRY protects the variables of the calling subroutine). Do so. What are the trade-offs?

7-22. S_c and S_v can be combined. SCOPE_ENTRY must allocate one extra cell for the saved program counter. Rewrite Table 7-10 for this possibility.

7-23. Combine Table 7-9 with Table 7-8 and Figure 7-17 to give an analog for Table 7-10.

7-24. Table 7-9 cannot be easily combined with Figures 7-15 and 7-18. The problem is that the value V(P) for each subroutine that is exited must be available in one place. Propose a change to PRT that would facilitate the combination, and then work out the details.

7-25. Rework problem 7-24 using S_m.

7-26. Using the operators in Figure 7-21 write Polish to compute each of:

$$2 + 2$$
$$2 + 127$$
$$2 + 128$$

2 + A where A is the 7th global variable

B + A where B is the 3rd local variable

and A is the 130th global variable

7-27. Give the Polish for Figure 7-14, extending Table 7-11 as necessary.

7-28. Redo Table 7-11 using separate stacks for S_c and S_m (refer to Figure 7-22).

7-29. Redo Table 7-11 using only one stack for all of S, S_c, S_v, and S_m.

7-30. Redo problem 7-29 to allow multi-level RETURN and REPEAT.

7-31. Using as many different tags as necessary, redo Table 7-11 based on an untagged arithmetic stack S_i and a tagged combination of S_c, S_v and S_m.

7-32. Redo problems 7-28 through 7-31 using the scope entry and exit in Table 7-12. Where do the values J, K, and V(P) come from?

7-33. Present an argument that Table 7-12 correctly implements PL/I except for parametric procedures. Does it matter what the initial value of D is?

7-34. Find an example where calling a parametric subroutine is inconsistent with the algorithms in Table 7-12.

7-35. Discuss the problem of making the value V(P), the size of the local area of subroutine P, available to increment register L in the case of dynamically allocated arrays (ALGOL 60 or PL/I).

7-36. Redo Table 7-14 to provide an operation INXLOAD with combines INX and LOAD. Is this likely to have a high payoff in performance?

7-37. Pick another data structure (list, stack, queue, structure) and design an accessing operator for it. Does it make sense to include bounding information and checking?

7-38. Assume that values in memory are tagged, integers with tag *i*, descriptors with tag *d*. Redo Table 7-14 checking to see that only descriptors are indexed, and only with integer values. What tag should be given to the result of the INX operation?

7-39. Redo Table 7-14 for descriptors of the form (*tag, presence, indexed, size, address*) where the *tag* is *d*, *presence* is a bit indicating whether or not the array is already in memory, *indexed* is a bit indicating whether the descriptor points to an array or an array element, *size* is the upper bound of the array if *indexed* is false and is the index value if *indexed* is true, *address* is the address of the array in memory if *presence* is true and the address of the array on the disk if *presence* is false (an address of 0 indicates the array has never been accessed at all, hence has no address at all).

7-40. Considering the solution to problem 7-39, what additional mechanisms must be supplied for scope entry and exit, LOAD and STORE, and the initialization of the descriptors in S_v?

7-41. Devise a circuit to implement S_p according to Table 7-15.

7-42. Write in Polish (Tables 7-13 and 7-15) code for functions MAX and MIN then write code for the expression MAX (A, MIN (B, C, D), E).

7-43. If S and S_p are combined, the Polish is simplified and the number of internal data transfers is reduced. It cannot be done, however, if we have multiple arithmetic evaluation stacks unless we have multiple S_p also. Discuss the trade-offs.

7-44. Pick a programming language and show whether or not, for a given program, an upper bound on the size of S_p can be computed.

7-45. S_p can be combined with S_v. The parameters are eventually going to reside in the area in S_v directly above the present local variables, and can be placed there directly. The problem is that a function subroutine may be called during the computation of a latter parameter, causing a SCOPE_ENTRY action and therefore destroying the partially prepared parameter list in S_v. Rework Table 7-15 to combine S_p and S_v using whatever additional mechanisms you need.

7-46. S, S_p, S_v, and S_c can all be combined. To avoid unnecessary complications assume either only one kind of arithmetic (that is, integer) or a tagged stack and rework Table 7-15.

7-47. Redo problem 7-46 and add the functions of S_m to the combined stack. What additional mechanisms do you have to use for multi-level RETURN?

7-48. The operator PTPAPS in Table 7-16 copies a part of the display when the subroutine is first passed. It may happen that the parametric subroutine is passed again to another subroutine, and so on. No additional provision need be made in the Polish for this possibility. Why?

7-49. Write a short program that has a nested, recursive, parametric subroutine.

7-50. Show that the combination of the algorithms in Tables 7-12 and 7-15 work for the scope rules of PL/I or ALGOL 60. Hint: Show that paired actions (for example, SCOPE_ENTRY and SCOPE_EXIT) always leave the addressing state invariant and that the display is always correct when CALL is executed.

7-51. Formalize the evaluation problem in terms of a cost/benefit difference. Be specific about the variables that must be known to make a decision. Use monetary units.

7-52. Assuming that one of your terms in problem 7-51 is memory residence for a program during its execution, outline a method for comparing the average memory residence demanded by a stack computer and a conventional computer. Both program and data must be considered.

7-53. Assume that the failure rate for two computers consists of two terms, one for hardware failures and one for software failures. Outline a method for comparing the failure rates of two different computers where you suspect that the additional hardware on one will cause more hardware failures but eliminate some software failures. Does the trade-off chosen depend on the cost of a failure to the user?

7-54. Taking reasonable estimates of cost for programmer salaries and computer rental, what is the trade-off in monthly rental in a typical medium scale computer shop, in eliminating all subscripting errors (an effect of the descriptor logic in section 7-6)?

CHAPTER 8

Parallel Computers

Harold S. Stone

8-1. INTRODUCTION

Computation speed has increased by orders of magnitude over the past three decades, with a major share of the increase in speed attributable to inherently faster electronic parts. The earliest electronic computers used vacuum tubes for logic functions and magnetic drums for central memory. Significant speed increases came when drums were replaced by magnetic cores, when vacuum tubes were replaced by transistors, and when both transistors and cores were replaced by integrated components. Today, we cannot obtain speed increases as we have done in the past by increasing the basic speed of the logic components, but we must necessarily take other approaches.

To understand why this is so, in looking over the past contributions of device technology, we see that new technologies not only improved switching speeds, but improved other factors that affect computation speed as well. Consider, for example, propagation delays, which are a significant factor in determining the speed of computation and can become a dominant factor when delays exceed the switching time of logic devices. Since logic signals travel at the speed of light, or approximately 30 centimeters per nanosecond, a signal requires over 3 nanoseconds to travel between two gates a meter apart. Thus, propagation delays for this distance are the same order of magnitude as the switching time of fast logic. Note that if switching delays are reduced to zero, propagation delays will dominate and determine computation speed.

In the transition from vacuum tubes to integrated circuits, the physical dimensions of logic have decreased substantially so that integrated components are packaged with greater density than vacuum tubes, thereby reducing propagation delays while the technology produced faster switching times. Any further increase in computation speed that comes from new device technologies must likewise be done through both increased switching speed and increased circuit density. Such breakthroughs are less likely to come in the future because circuit densities are fast approaching the limits of optical resolution. Hence, even if switching times become instantaneous, distances between components may not become small enough to decrease propagation delays enough to make a material increase in computation speed. To achieve even faster computers in the future we must take new approaches that do not depend on breakthroughs in device technology, but rather on imaginative application of the skills of computer architecture.

One obvious approach is to increase speed through parallelism. The throughput potential of N identical computers is N times that of a single computer. The issue at hand is whether it is possible and economical to tie N computers together to form a super computer. If this question can be answered affirmatively then speed increases of 10, 100, or 1000 to 1 may be obtainable at reasonable cost using today's logic technology. The use of a large degree of parallelism in a computer was once prohibitively expensive and economically unjustifiable. In terms of the parameter N indicating the degree of parallelism, N = 2, 3, or 4 was conceivable, but N = 100 or 1000 was wildly unrealistic. Today the picture has changed drastically. A $10,000 minicomputer has virtually the same computer power as a large-scale $1 million computer of the late 1950s, so that we have gained roughly a factor of 100 in cost effectiveness for these selected computers. Thus it is potentially realistic to connect 100 of these minicomputers together to form a super computer whose cost is roughly equal to that of the computer of the 1950s. There is some economy achieved by replication, so that 100 copies of our minicomputer should cost somewhat less than 100 times the cost of a single minicomputer. Current trends indicate that the $10,000 minicomputer may cost only $1,000 in a few years. Consequently, by this argument even N = 1000 is potentially realistic while keeping the total cost of the computer system fixed at $1 million. Although these figures are somewhat contrived, the trend to cheaper logic is clearly evident, which tends to justify parallelism as a suitable means to attain higher speed.

Parallelism in various forms has appeared in computers produced during the 1960s, and has proved to be an effective approach. Nevertheless, the parallelism was limited by the cost of logic, so that the computers were substantially serial with only moderate capability to support parallel operation. As the cost of components decreased drastically in the past decade, computers have become more and more complex to achieve higher computation speed. By the end of the 1960s and the beginning of the 1970s several projects were underway for the development of truly parallel computers.

When the development projects for parallel computers were first initiated the notion of parallel computers was still in its infancy. Several competitive methods were proposed for organizing a parallel computer, but there was very little evidence as to which design was superior, nor was there sufficient knowledge on which to make a careful evaluation. Flynn [66] helped initiate an organized study of high-speed computer architecture by showing that computer systems fall naturally into four classes. Within this classification system, it is possible to make some nontrivial observations about the utility of a computer system and its relative cost-effectiveness on specific types of problems. In the next section we describe these classes in greater detail, but at this point it suffices to say that there are two classes that represent the two divergent methods for organizing parallel computers. One type of parallel computer, the *array computer*, operates on vectors as basic units of information. Instruction execution in this type of machine is quite similar to a conventional computer in the sense that the array computer executes a single instruction stream with loads, adds, stores, and branches much like the instruction stream of a serial computer. However, the instructions manipulate vectors of data simultaneously, rather than single items. The alternative architecture is the *multiprocessor*. This type of computer consists of N computers plus interconnections for passing data and control information among the computers. This type of computer has N independent instruction streams in execution simultaneously. In the remainder of this chapter we treat the two types of parallel computer in great detail. We should mention here that there is a technique for increasing parallelism of execution known as *pipelining* that achieves parallelism through a combination of logic replication and sharing of logic. This topic is of sufficient importance to deserve a chapter by itself, and is the subject of Chapter 9.

The present chapter is organized as follows. In the next section we describe the Flynn classes, then show crude models of the array processor and the multiprocessor, together with typical programs for which they are well suited. In later sections the discussion focuses first on the array computer then the multiprocessor computer. We shall discover that the array computer is the less flexible, and more specialized computer because all programs for it must be cast into a sequence of vector instructions. While many numerical problems are easily adapted to this type of computer, there is a large class of problems that do not appear to be well-suited to this computer. Thus the central issues behind the development of an array computer deal with efficient execution of specific problems.

In contrast to the constraints of the array computer, the multiprocessor is relatively flexible. Since this machine consists of N complete computers, N independent computations can be supported simultaneously. The problems lie in making the N computers cooperate so that one problem can be partitioned among the N processors of a multiprocessor and can be solved with greater speed than it could be done on a single processor. To make the multiprocessor effective we need the speed increase to be substantial, hopefully N times the speed of a single processor computation.

We investigate specifically the problem of synchronizing the processors in a multiprocessor, of sharing data among the processors, and of scheduling computations to utilize the computer system resources efficiently.

We find in the course of this chapter that the questions of interest for vector processors are questions concerning the implementation of specific algorithms. For multiprocessors the questions are much less specific to the application but rather deal with the general question of resource allocation and cooperation among processors. The techniques for programming multiprocessors for efficient parallel operation are much less developed than the corresponding techniques for vector processors. There are many research questions that remain to be answered before multiprocessors with a large number of processors can be utilized well.

8-2. CLASSES OF COMPUTER SYSTEMS

In dealing with parallel computer systems it is quite natural to classify computers in terms of parallelism within the instruction stream and parallelism within the data stream. In this context, by *instruction stream*, we mean the sequence of instructions that are executed in a processing unit. By *data stream* we mean the sequence of operands that are manipulated in a processor. A serial computer is a computer with a single instruction stream and a single data stream. At any given time during the execution of the program, there is at most one instruction in execution and that instruction affects at most one datum. This is shown in part *a* of Figure 8-1.

Flynn [66] observed that the methods for achieving parallel operation depended on replicating the instruction stream and the data stream. This gives rise naturally to four classes of computers. The *single-instruction single-data stream* (SISD) computer is the serial computer mentioned above. By replicating the data stream we create a processor of the type shown in part *b* of Figure 8-1. This processor, known as the *single-instruction multiple-data stream* (SIMD) computer, is a vector processor in the sense that each instruction operates on a data vector rather than on a single operand. The multiple data stream is the collection of individual data streams containing the elements of the vector operands. Each processor in part *b* of Figure 8-1 has a complete arithmetic unit and its own memory. However, each processor differs from a conventional serial processor because the processor in an SIMD computer does not have the ability to generate its own instruction stream. Not shown in part *b* of Figure 8-1 are the necessary interconnections among processors to enable data to be shared among the processors. We shall describe these interconnections in section 8-3.

Instead of parallelism in the data stream, it is conceivably possible to have parallelism in the instruction stream. Part *c* of Figure 8-1 depicts a *multiple-instruction single-data stream* (MISD) computer. In this case each operand is operated upon simultaneously by several instructions. This mode of operation is generally unrealistic for parallel computers, but there is at least one example of such a processor. The example we have in mind is a punched card processor which has the capability of operating on several data fields of a punched card simultaneously. The program for such a processor consists of subprograms of instructions for each data field, where each subprogram generates one instruction stream, and the instruction streams are processed

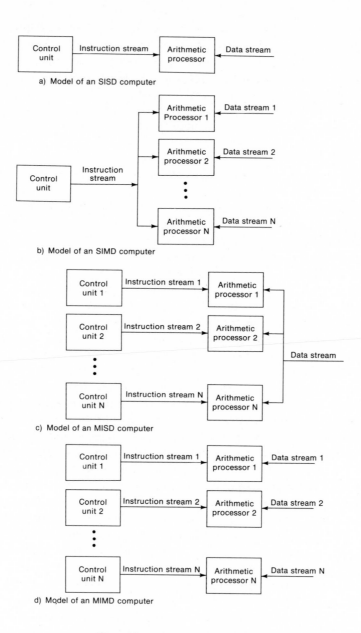

a) Model of an SISD computer

b) Model of an SIMD computer

c) Model of an MISD computer

d) Model of an MIMD computer

Figure 8-1   Computer models

simultaneously. The single operand in this case is a punched card. Admittedly this is stretching the definition of single datum somewhat, because the information on a punched card can equally well be viewed as a collection of several data. Nevertheless, this example is useful because it illustrates the notion of a MISD computer.

By combining parallelism in the instruction and data streams, we have the *multiple-instruction multiple-data stream* (MIMD) computer shown in part *d* of Figure 8-1. This computer is composed of N processors, each of which is a complete computer, with the processors connected together to provide a means for cooperating during a computation. Since there are N independent serial processors in an MIMD computer there are N instruction streams and N data streams, one data stream per instruction stream.

Most serial computers manufactured today have data channels (sometimes called *direct memory access* or *DMA channels*) which are in a sense independent processors. Thus, a computer with one processor and one data channel fits the model of an MIMD computer. The processor and the data channel are not identical, but then there is no reason to insist that all of the processors in an MIMD computer be identical. The completion performed by an independent data channel is quite different from that done by the central processing unit, but nevertheless, the parallelism of operation can result in higher overall computation speed than can be attained without separate data channels. The challenge for the present is to design MIMD computers, not just with 2 to 4 processors, as in the case of a serial computer with data channels, but with 100 or 1000 processors. The MIMD computer with the largest number of independent processors to date is under development at Carnegie-Mellon University. The computer can support sixteen independent processors, but will probably be used with roughly half that number for the near future [Wulf and Bell, 73].

Of the four types of computers, the two of immediate interest are the SIMD and MIMD computer. These types of computers are vastly different in how they attain parallelism of operation. The SIMD computer is well suited to computations that can be broken up into a sequence of vector operations. A problem with no particular vector structure or other natural iterative structure may have a great deal of potential parallelism that can be exploited by an MIMD computer because the MIMD computer is not constrained to a repertoire of vector instructions. Parallelism for this type of computer generally consists of doing independent tasks in parallel on independent computers; then the results of the execution of the independent tasks are combined. The following example illustrates the relative characteristics of SIMD and MIMD computers, and should help clarify some of the points made in this discussion. We have chosen a problem that can be executed on both types of computers to strengthen the distinction between the computers.

The example is a matrix multiplication. We wish to compute $C = A \times B$, where A, B, and C are $N \times N$ matrices. To do the computation we use the well-known formula:

$$c_{i,j} = \sum_{k=0}^{N-1} a_{i,k} \times b_{k,j}$$

where A, B, and C have elements $a_{i,j}$, $b_{k,j}$, and $c_{i,j}$, respectively.

We show a program for the SIMD computation in an ALGOL-like language. In this program we assume that there are N processors in the system, and that we can process a row vector of a matrix in a single vector operation. In the program the notation $(0 \le k \le N - 1)$ indicates that the operations are carried out for all indices k in the given interval simultaneously.

Matrix Multiplication Algorithm (SIMD)

for $i := 0$ **step** 1 **until** N − 1 **do**
  begin comment initialize the sums to 0;
    $C[i, k] := 0, (0 \le k \le N - 1)$;
    for $j := 0$ **step** 1 **until** N − 1 **do**
      begin comment add the next term to the sum;
        $C[i, k] := C[i, k] + A[i, j] \times B[j, k], (0 \le k \le N - 1)$;
      end of j loop;
  end of i loop;

In this algorithm, we compute all of the elements in the ith row simultaneously. Note that each element of the product matrix is a summation, and the summations are done serially rather than in parallel. However, because N summations are computed simultaneously, only N^2 vector multiplications are required for this algorithm as compared to the N^3 scalar multiplications required for the usual matrix multiplication.

The statement in the inner loop of this algorithm indicates that we must have the capability to multiply each element of the jth row of B by the constant $A[i, j]$, that is, we must be able to multiply a vector by a scalar. This suggests that it is highly desirable to be able to broadcast a single element (in this case $A[i, j]$) simultaneously to all processors to be used as an operand. Thus we see some need for communication among processors in an SIMD computer to enhance its capabilities. Interprocessor communication and data access are central problems in both the design and programming of SIMD computers. In the next section we investigate these problems more thoroughly.

To perform the same process on an MIMD computer, we must somehow parcel out the computation to the individual processors in the system. Conway [63] proposed an ingenious method for doing this by means of two primitive machine operations he calls FORK and JOIN. If NEXT is a label in a program, then execution of the instruction FORK NEXT causes an independent computation to be initiated at the label NEXT. In the meantime, the computation containing the FORK instruction continues execution at the instruction immediately following the FORK, so that a FORK splits one instruction stream into two instruction streams that can be executed simultaneously on independent processors.

The JOIN instruction is something like the inverse of a FORK because it brings instruction streams together. The statement JOIN N causes N independent instruction streams to merge into a single stream. In actual operation, the execution of instructions following a JOIN N instruction will not take place until the Nth independent process has executed the JOIN instruction.

With these two primitive operations we examine the execution of a matrix multiplication algorithm on an MIMD computer.

Matrix Multiplication Algorithm (MIMD)

```
comment spawn N − 1 independent processes, each with
    a different value of k;
for k : = 0 step 1 until N − 2 do
    FORK NEXT;
comment in the one process that reaches this point,
    set k to N − 1 in order to process the Nth element
    of a row;
k : = N − 1;
NEXT: comment N different processes reach this point,
        each with a different value of k
for i : = 0 step 1 until N − 1 do
    begin comment initialize each sum to 0;
    C[i, k] : = 0;
    for j : = 0 step 1 until N − 1 do
        begin comment add each successive term to the sum;
        C[i,k] : = C[i,k] + A[i,j] × B[j,k];
        end of j loop;
    end of i loop;
    JOIN N;
```

We have purposely written the MIMD program so that the actions performed on the independent processors are exactly analogous to the actions performed by the individual processors of an SIMD computer. There are important differences, however, that distinguish the two computations. Recall that the processors of an SIMD computer are synchronized, instruction for instruction, while a program is in execution. On the contrary, the processes spawned in the example for the MIMD computer need not be synchronized. In fact, they need not be identical instruction sequences, even though we have shown such an example for comparison purposes.

Another major attribute of MIMD computers is that they are insensitive to the number of processors that are available in a computer system. The reason for this is that Conway has proposed a way of executing the FORK and JOIN instructions so that processors are automatically assigned to computation tasks spawned by FORKS, and they are automatically reassigned when a computation task dies at a JOIN. Conway suggests that the FORK NEXT instruction be executed as follows:

1. At the point of execution, some or all of the memory accessible to the executing process is identified as common to the spawned process.

Pointers required to access this memory are created and allocated to the spawned process.

2. The task that executed the FORK instruction continues execution on the processor originally allocated to it.

3. If a processor is available, it is allocated to the spawned process. If no processor is available, the spawned process is queued and awaits execution.

When there are insufficient processors to do all N tasks simultaneously, one or more of the tasks are queued during the computation. How does a task become assigned to a processor and executed at a later time? This occurs when a JOIN instruction is executed. The interpretation of a JOIN is as follows:

1. For each JOIN instruction there is a unique counter which is initialized to 0. When the instruction JOIN N is executed, the counter is incremented and compared to N.

2. If the value of the counter is N, then this is the Nth task to pass the JOIN instruction. This task continues execution past the JOIN instruction on its assigned processor.

3. If the value of the counter is less than N, then more tasks must reach the JOIN instruction before proceeding beyond. The task now in execution is terminated. All resources private to it are returned to a resource pool. The processor assigned to the task is then reassigned to some other task that has been queued for service.

Since tasks are terminated at JOIN instructions, these instructions are natural points for the reallocation of processors to other tasks. Figure 8-2 shows a schematic diagram of the execution of the MIMD matrix multiplication

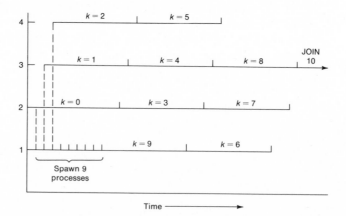

Figure 8-2 Timing diagram for MIMD matrix multiplication

algorithm in execution for a four-processor machine when N = 10. Each task is identified with a value k. Note that tasks with $k = 0$, $k = 1$, and $k = 2$ are executed first because they are spawned first. Tasks for $k = 3$ through 8 are queued. The task for $k = 9$ proceeds before these because it has a processor assigned to it during the spawning process, and this processor is not relinquished until the JOIN instruction is reached. The order of execution of tasks for $k = 3$ to 9 is first-in, first-out in this example, but for more general problems any ordering consistent with precedence constraints is acceptable. We have not indicated how precedence constraints can be incorporated into a program for an MIMD computer, but we discuss this question further in a later section. In the meantime we mention that first-in, first-out scheduling of computational tasks is not necessarily the best, and again we defer detailed discussion on this question until later.

Note that problems of scheduling and of satisfying precedence constraints are absent in SIMD computers. Thus the increased flexibility of MIMD computers comes at an increased cost in their programming and control. The cost-effectiveness of MIMD computers relative to both SIMD and SISD computers depends heavily on the effectiveness of the scheduling algorithm, the overhead for scheduling, and also on how precedence constraints can be met wtihout interfering with the parallelism attainable by the computer system. These problems have not been solved in a practical sense to date, and remain important problems for research in computer architecture.

8-3. SIMD COMPUTERS

The previous section illustrates the general form of SIMD and MIMD computers, but at a rather gross level of detail. In this section we examine SIMD computers in depth. We shall see that the central questions for this class of computers relate to the ability to cast a computation into a vector computation. The SIMD computer system must be designed with data access and communication capabilities to facilitate typical vector computations to make the computer system useful for as large a variety of computations as possible. We investigate several useful capabilities that have been proposed or implemented, but the reader should take note that this area is still under investigation.

To evaluate the effectiveness of SIMD computers, we look into techniques for performing typical computations with vector instructions. A number of surprising conclusions result from this evaluation. Perhaps most surprising is that an efficient algorithm for a serial computer may lead to a relatively inefficient algorithm for a vector computer. In fact, the best vector algorithm for a given problem may be adapted from an algorithm that is known to be inefficient for serial execution. Thus we cannot generally rely on knowledge of serial computation when we construct vector algorithms.

In the first part of this section we give a block diagram for a hypothetical SIMD computer that greatly resembles the ILLIAC IV computer. A description of the use of the various features of the computer system appears in the second section. There we investigate fragments of instruction sequences that implement frequently used program constructs. We also investigate the data communication requirements of vector programs to determine the most valuable interconnections to design into the computer system. We also investigate methods for storing data structures to take best advantage of the parallelism of the computer.

In the last part of this section we look into general techniques for transforming a serial program into a vector program. Even though the techniques are said to be general, they apply only to a class of programs that exhibit a strong iterative structure. We show how to apply these techniques to certain algorithms that appear to be sequential to obtain new vector algorithms with a good degree of parallelism. Unfortunately, these techniques do not work for all highly iterative programs, nor do there exist widely applicable techniques for casting programs into parallel form when they do not have an iterative structure.

Organization of an SIMD Computer

For this description we shall draw heavily upon the ILLIAC IV computer [Barnes et al., 68]. However, this discussion differs in many specific details from the ILLIAC IV, and we generally use a different and more suggestive terminology.

A block diagram of the SIMD computer appears in Figure 8-3. Note the N independent processors, and N independent memories as we have described earlier. The memories are all connected to a high-speed data bus whose bandwidth is compatible with both the processor and input/output bandwidths. The maximum bandwidth is one word from each memory per memory cycle, which is N times the bandwidth of an individual memory. Note that there is a permutation network included here that permits information to be exchanged among the processors. The permutation network is shown as a processor-to-processor interconnection, but it may also be a memory-to-memory interconnection.

The control unit in this scheme is itself a computer, with its own high-speed registers, local memory, and arithmetic unit. The crucial difference between this processor and the others is that this processor can execute conditional branch instructions and can thereby determine the order in which instructions are executed. The arithmetic processors do not have this ability because the processors must always be in synchronization, and therefore cannot take different actions after a conditional branch instruction.

As in conventional machines, the instructions are stored in main memory together with data. The main memory in this system is the collective memory in the N processors. Hence, the instructions are fetched from the processors into an instruction buffer in the control processor. The communication path for this transfer is shown in the figure.

The flow of instructions and data during execution is rather interesting in

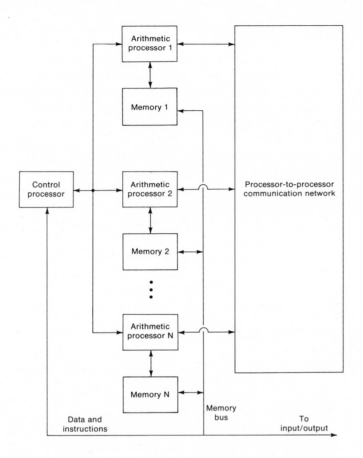

Figure 8-3   Block diagram of an SIMD computer

this type of computer. Each instruction is either a control instruction and is executed entirely within the control unit, or it is a vector instruction and is executed in the processor array. But the instruction stream is much like a conventional serial instruction stream with each instruction executed in sequence. A primary function of the control processor is to examine each instruction as it is to be executed and to determine where the execution should take place. If the instruction is a control instruction, then it is executed in the control unit itself, otherwise it is broadcast to the processor array for execution there. In either case the instruction passes through the control processor, so that the control processor has the exclusive privilege to determine which instruction to execute next.

We shall shortly trace the execution of a fragment of a sequence of instructions to make these points clearer. Before doing so, we shall introduce just one more feature of the computer, and treat this in the example as well. Our intuition tells us that index registers should be at least as important to have in this type of computer system as in a conventional system. We can place index registers either in the control processor or in the processing array, and, in fact, they exist in both places in the ILLIAC IV. The index registers in the control processor have a global action on the processor array because a single indexing operation affects all processors. During the execution of an instruction intended for the processor array, the operand address contained in this instruction can be modified by indexing in the control processor before it is passed to the processing array for final execution. This indexing action is completely analogous to indexing in an SISD processor.

Indexing can also be done in the individual processor. The effect of indexing in the processor array is to force each processor to modify the broadcast operand address by an amount determined individually by each processor. The index offsets may all be the same or all different, or any arbitrary combination. We shall see later that this feature is invaluable for accessing rows and columns of an array with equal ease.

Having gone through a block diagram of the SIMD computer, we are ready to examine the execution of a typical instruction sequence. For this example, let us return to the matrix multiplication example, and show the instruction sequence that implements the statement

> **for** j : $= 0$ **step** 1 **until** $N - 1$ **do**
> **begin**
> $\quad$ $C[i,k]$: $= C[i,k] + A[i,j] \times B[j,k], (0 \leq k \leq N - 1)$;
> **end**

The matrices A, B, and C are stored as shown in Figure 8-4. In this example each operand address in the vector instructions is the address of a row vector of A, B, or C. Addresses of rows increase sequentially by unit amounts as shown in the figure.

Let us assume that each data processor in the array is a single accumulator computer, and we denote the accumulator in the kth processor as $ACC[k]$. The index registers in the control processor are denoted $INDEX[i]$. Then the vector instructions we need for this example are in Table 8-1.

TABLE 8-1. VECTOR INSTRUCTIONS

Instruction	Example	Action
Vector load	LOAD A	$ACC[k]$: $= A[0,k]$, $(0 \leq k \leq N-1)$;
Vector load (indexed)	LOAD A[i]	$ACC[k]$: $= A[INDEX[i],k]$, $(0 \leq k \leq N-1)$;
Vector store	STO A	$A[0,k]$: $= ACC[k]$, $(0 \leq k \leq N-1)$;
Vector add	ADD A	$ACC[k]$: $= ACC[k] + A[0,k], (0 \leq k \leq N-1)$;
Vector multiply	MUL A	$ACC[k]$: $= ACC[k] \times A[0,k], (0 \leq k \leq N-1)$;
Broadcast scalar	BCAST i	$ACC[k]$: $= ACC[INDEX[i]]$, $(0 \leq k \leq N-1)$;

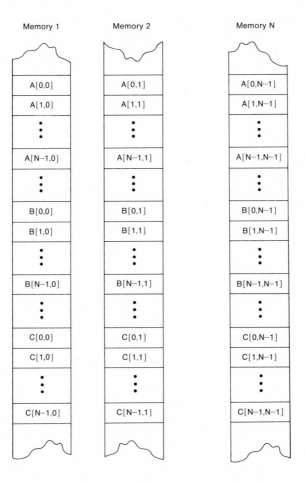

Figure 8-4 Storage format of matrices A, B, and C

The vector instructions are extensions of a scalar instruction set. When the A, B, or C appears by itself as operand address, we obtain the address of row 0 of the respective matrix. When the address appears in the form A[i] or B[j], the address is indexed by index register i or j, respectively, where the index register is understood to be an index register in the control processor. This is consistent with the conventions for assembly language for SISD computers. The vector store, add, and multiply instructions are not shown in their indexed form, but we assume that they have indexed modes similar to the indexed vector load instruction. The broadcast scalar instruction selects one of the N accumulators and broadcasts its contents to all other accumulators. To simplify the explanation, the operand of this instruction is an index register, but in actual implementation the operand might be an integer or an indexed integer.

There remains to exhibit the loop control and counting instructions that are executed in the control processor. These instructions are functionally identical to similar instructions found in any SISD computer repertoire. For this example we shall use the instructions in Table 8-2 but any equivalent repertoire may be substituted instead.

TABLE 8-2. INDEXING INSTRUCTIONS

Instruction	Example	Action
Enter index constant	ENXC i,1	INDEX[i] := 1;
Load index	LDNX i,Y	INDEX[i] := MEMORY[Y];
Increment index by constant	ICNX i,1	INDEX[i] := INDEX[i] + 1;
Compare index, branch if low	CPNX i,j,LABEL	**if** INDEX[i] $<$ INDEX[j] **then** **go to** LABEL;

The instructions in Table 8-2 should require no additional explanation except for the MEMORY[Y] notation in the action of the load index instruction. Since there are N memories in the processor array and one more in the control processor, we should further specify which of the N+1 memories is the intended reference of MEMORY[Y]. There are a number of standard ways to resolve the ambiguity in the encoding of the instruction, such as the use of special fields in the instruction, or to use separate and distinguishable addresses for the different memories, or to use address interpretation dependent upon processor state. For the purpose of this example, we do not need to resolve the ambiguity.

The encoding of the inner loop appears in Figure 8-5. Note that it is quite straightforward. We assume that index registers in the control processor have been allocated to hold the current values of i, as well as the value of N to terminate the looping. The encoding is remarkably like the encoding for an SISD computer, except that N operations occur while each arithmetic instruction is executed instead of just one. The only visible difference here is the inclusion of the broadcast instruction BCAST to communicate the value of A[i,j] to all processors. This example is typical of many SIMD codes in that the inner loop of an SISD computation (in this case, the loop on k) is replaced by vector operations.

The flow of information during the execution of the instructions in Figure 8-5 is again rather similar to instruction execution in SISD computers. Because SIMD computers are built to attain ultra-high speeds, the instruction stream should be highly buffered in the control processor, and the memory bandwidth between the control processor and processor array should be high enough to minimize the effect of instruction fetches on computation speed. As an example of these considerations, the ILLIAC IV fetches sixteen instructions at a time from the processor array into its instruction buffer. Moreover, instruction fetches are overlapped with operations that do not use main

	LABEL	INSTRUCTION	OPERANDS	COMMENTS
1.		ENXC	j, 0	INDEX[j]: = 0
2.		LDNX	LIM, N	Place the value of N from memory into INDEX[LIM], the upper-limit register.
3.	JLOOP	LOAD	A[i]	ACC[k]: = A[i, k], $(0 \leq k \leq N - 1)$;
4.		BCAST	j	ACC[k]: = ACC[INDEX[j]], $(0 \leq k \leq N - 1)$; Each accumulator now contains the value of A[i, j].
5.		MUL	B[j]	ACC[k]: = ACC[k] $\times$ B[j, k], $(0 \leq k \leq N - 1)$; ACC[k] now contains A[i, j] $\times$ B[j, k].
6.		ADD	C[i]	ACC[k] now contains C[i, k] + A[i, j] $\times$ B[j, k].
7.		STO	C[i]	Update the value of the sum for C[i, j] in memory.
8.		ICNX	j, 1	INDEX[j]: = INDEX[j] + 1;
9.		CPNX	j, LIM, JLOOP	Branch to JLOOP and repeat until this loop is done N times.

Figure 8-5 Program for matrix multiplication on an SIMD computer

memory. These two design features, plus sufficient buffering in the control processor, guarantee that the ILLIAC IV rarely has to wait for an instruction fetch.

The facilities described thus far for the SIMD computer are sufficient for a very restricted class of computations. A few additional facilities greatly enlarge the class of problems that can be done efficiently. Perhaps the most important of these is the facility to mask computations to force some of the N processors into an idle state. We take this up presently. Later we discuss data communication and data structures.

Masking for Conditional Branching

Conditional branches are particularly vexing in an SIMD computer. Suppose, for example, a conditional branch instruction provides two alternatives, say, branch if the accumulator contains zero, otherwise no branch. What should happen if some but not all of the N accumulators contain zeros? The ideal solution is for the instruction stream to split into two streams, one for the zero accumulator case and one for the nonzero accumulator case. Unfortunately, the SIMD computer, as we have described it, cannot support two

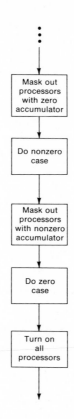

Figure 8-6 Masking operations that simulate a conditional branch

independent instruction streams simultaneously. Within the constraints imposed, the two cases must be done sequentially, as indicated in Figure 8-6. There is still parallelism, but the effective parallelism is roughly N/2 rather than N. We assume in Figure 8-6 that we can deactivate a selected subset of the processors, and execute an execution sequence in parallel on the subset that remains active. This way we can execute instructions for the zero accumulator branch of the program on the processors with the zero accumulators. Then we execute the branch for the nonzero accumulators on the complementary set of processors. Figure 8-6 shows an example in which the two separate instruction streams merge into a single stream after their execution.

Note that in this example an SISD computer does one branch or another for each datum, but never both. The SIMD computer, on the other hand, takes the time to execute both branches for each datum, although each processor is deactivated for one of the two alternatives.

To implement this type of masking facility, we use a 1-bit register in each processor of the array. It is convenient to treat the N 1-bit registers as a single N-bit register, which we call MASK. Then MASK[i] denotes the mask register of the ith processor. When MASK[i] contains a 1, processor i obeys the instruction broadcast by the control processor. When the mask bit is 0, processor i does nothing. In actual practice, it is both more economical and less likely to have an adverse affect on processor speed to disable only part of a processor when its mask bit is off. For example, memory storage operations and accumulator modification may be inhibited, but other portions of the processor, including portions not described here, may be left activated without changing the functional effect of the mask bit.

It is crucial that the mask bit of a processor be determined by data dependent conditions. Moreover, it is often necessary to compute active subsets of processors through a series of set union, intersection, and complementation operations, which therefore require that mask settings be computable by logical operations on mask vectors. Here we follow the philosophy of the ILLIAC IV instruction repertoire. We assume that a vector of conditional tests with binary outcomes can be performed simultaneously in the N processors, and the resulting vector of bits is placed in a designated index register of the control processor. Thus, after a test is performed, the ith bit of the result indicates the outcome of the test in the ith processor.

To compute subsets of active processors, we include instructions for performing logical operations on the control processor's index registers, and we make provision for loading and storing the mask register from these index registers. It is also useful for some calculations to compute masks by using shift operations as well as logical operations. We simply assume that a full set of shift operations is included in the instruction repertoire without giving them explicit.

A typical set of instructions for accomplishing the mask computations appears in Table 8-3, together with conditional branch instructions for testing the outcome of a comparison operation. For subset computations there are three natural conditions to sense. Have none, any, or all of the processors satisfied the test? We show conditional branch instructions for each of these possibilities. We also show instructions for extracting the first 1 bit of a bit vector, and the index of the first bit of the vector. The ability to extract the first 1 bit is useful when a subset of processors responds to a conditional test, and we must then scan the responding processors sequentially, while performing some calculation in each one. We include the facility to compute the index of the first bit so that this index can be used in a broadcast instruction. Thus, a conditional test can be used to identify a subset of processors, after which the data from one of the selected subset can be broadcast to all processors.

A trivial example illustrates the use of the mask registers and condition setting instructions. For this example we look at a normalization operation. Suppose that A is an $N \times N$ matrix, and that we wish to normalize the rows of A by replacing $A[i,j]$ by $A[i,j] / A[0,j]$, provided that $A[0,j] \neq 0$. In an ALGOL-like notation this becomes:

> **for** $i :=$ 1 **step** 1 **until** N $-$ 1 **do**
>   **if** $A[0, j] \neq 0$ **then** $A[i, j] := A[i, j] / A[0, j], (0 \leq j \leq N - 1)$;

TABLE 8-3. CONDITION SETTING, MASKING, AND BRANCHING INSTRUCTIONS

Instruction	Example	Action
Vector compare less	CLSS A,i	if ACC[k] < A[k] then INDEX[i], kth bit, is set to 1, otherwise reset to 0.
Vector compare equal	CEQL A,i	Similar to CLSS above.
Branch all	BRALL i,LOOP	Branch to LOOP if INDEX[i] has all 1 bits.
Branch any	BRANY i,LOOP	Branch to LOOP if INDEX[i] has any 1 bit.
Branch none	BRNON i,LOOP	Branch to LOOP if INDEX[i] has all 0 bits.
Logical AND of index	AND i,j	INDEX[i] := INDEX[i] AND INDEX[j];
Logical OR of index	OR i,j	Similar to AND above
Complement index	CMP i	INDEX[i] := NOT INDEX[i];
Load mask from index	LDMSK i	MASK := INDEX[i];
Store mask in index	STMSK i	INDEX[i] := MASK;
Set first 1 bit	FIRST i	If INDEX[i] has no 1 bits it is unchanged. Otherwise, all but the first 1 bit are reset.
Index of first 1 bit	NXFIR i	Replace the contents of INDEX[i] with bit index of its first 1 bit. Set INDEX[i] to all 1s if it originally contains no 1 bits.

We actually perform the computation by computing the mask for A[0,j] $\neq$ 0 before entering the loop, then we loop through the array without making any further tests. In the ALGOL-like notation, we can change the condition in parentheses from (0 $\leq j \leq$ N $-$ 1), which indicates all processors are activated, to the condition (M[j]), which indicates that the only processors active are those corresponding to the 1 bits of the bit vector M. The statement now reads:

```
    comment M is a vector of bits;
    M[k] : = A[0,k] ≠ 0,(0 ≤ k ≤ N− 1);
    comment M now contains the result of the comparison;
    for i : = 1 step 1 until N − 1 do
        begin comment normalize A[i,j] only in
            the processors activated by M;
            A[i,j] : = A[i,j] / A[i,j], (M[j]);
        end;
```

The implementation of this program fragment in machine instructions appears in Figure 8-7. The loop control instructions follow the now familiar form as shown in Table 8-2. Note the comparison is made and the mask register loaded prior to entry to the loop. We make use of a DIV instruction in the loop to do the division, and we note that it behaves similarly to the MUL and ADD instruction described in Table 8-1.

The examples of the program fragments for an SIMD computer suggest that programming such a computer is only slightly different from programming SISD computers. There tend to be fewer conditional branches in SIMD programs, however, as we see from the present example. The masking operation here replaces one conditional branch and the simultaneous computation on N items replaces an inner loop with its conditional branch at the end of the loop. Hence the SIMD code in this case eliminates two conditional branches that would normally occur in SISD code.

	LABEL	INSTRUCTION	OPERANDS	COMMENTS
1.		ENXC	i, 0	INDEX[i]: = 0;
2.		SUB	ACC	ACC[k]: = ACC[k] − ACC[k], $(0 \le k \le N − 1)$; This clears all ACC registers.
3.		CEQL	A, M	if ACC[k] = A[0,k] then set kth bit of INDEX[M], otherwise reset the kth bit. This creates the complement of the mask we require.
4.		CMP	M	INDEX[M] : = NOT INDEX[M]; Complement the mask.
5.		LDMSK	M	Load the mask register.
6.		LDNX	LIM, N	Set the loop limit in INDEX[LIM]. INDEX [j] has already been initialized to 0.
7.	ILOOP	LOAD	A[i]	ACC[j] : = A[i,j] if the jth mask bit is 1.
8.		DIV	A	ACC[j] : = ACC[j]/A[0,j] if the jth mask bit is 1.
9.		STO	A[i]	A[i,j] : = ACC[j] if the jth mask bit is 1.
10.		ICNX	i, 1	INDEX[i] : = INDEX[i] + 1;
11.		CPNX	i,LIM,ILOOP	Branch to ILOOP and repeat until this loop is done N times.

Figure 8-7 Normalization program

This completes the discussion of conditional control and branching in an SIMD computer. The next section investigates interprocessor communication and data structures.

Interprocessor Communications

We have previously acknowledged that, in general, we shall permit data transfers to occur between processors in an SIMD computer. Recall from Figure 8-3, the network that interconnects the processors to support such exchanges. The question that we investigate in this section concerns the design of this interconnection network.

In order to maximize the parallelism of computer in an SIMD computer, we must utilize as much of the available memory and processor bandwidths as is possible. To use the memory bandwidth, we must store data to avoid memory accessing conflicts. Two items that are accessed simultaneously must be placed in physically distinct memory modules, because a single module can access no more than one item at a time. Efficient utilization of memory bandwidth requires careful structuring of the data in its storage format. We discuss some approaches to the problem in the next section.

With respect to the utilization of the processor bandwidth, we request that data, after being accessed from memory, be permuted if necessary to bring operand pairs together at available processors. As we shall see in the next section, data structuring for efficient memory access frequently causes data to be fetched so that operand pairs are not properly aligned for parallel manipulation. To solve this problem we must install a permutation network in the processor. The nature of the permutation network is the subject of this section.

The permutation network that has the most flexibility is one in which every processor is directly connected to every other processor. This network is called a *complete interconnection* network. It is not difficult to show that in this network there are $N(N-1)/2$ bidirectional links between N processors, where each link presumably has sufficient bandwidth to pass one operand per unit time. Note that the cost of a complete interconnection network increases proportionally to the square of N. This suggests that complete interconnection networks are not realistic for large N because the performance of an SIMD computer increases at most linearly in N. For sufficiently large N the cost of the complete interconnection network dominates the cost of the computer, and its cost could be so high as to nullify the gains from the parallelism. One argument in favor of the complete interconnection scheme is that logic costs can diminish to near zero so that the cost of large networks may not be so high after all. Nevertheless, even with free hardware, it is worthwhile to use simple interconnections if they do not restrict the functional utility of the computer. With fewer interconnection paths, we can attain fewer gate delays in transmission paths with an attendant increase in the basic cycle speed.

We should add that we can retain full interconnection flexibility with logic growth proportional to $N \log_2 N$ if we permit the time to do a permutation to grow as $\log_2 N$ instead of being constant. Such schemes have been studied in the context of telephone switching networks [Clos, 53; Benes, 65]. We choose rather to reduce the number and hence the flexibility of the interconnections, while maintaining as high a speed as possible.

The best way to proceed in designing an interconnection network is to study the interconnections that are most widely used in SIMD algorithms. Unfortunately, this is rather difficult to do. There is relatively little known about SIMD algorithms when compared to the extensive knowledge of SISD algorithms. Secondly, the SIMD algorithms that do exist have been strongly influenced by the interconnections available on the few existing SIMD computers. A particular calculation might be done in several different ways on an SIMD computer, depending on the particular interconnections available.

We shall proceed here to exhibit interconnection patterns that are known to be good for a broad class of problems. In particular, we shall examine patterns that shift data cyclically among processors, and that route the data by a permutation known as the *perfect shuffle*.

The cyclic shift interconnection pattern is shown in Figure 8-8. If we number the processors $0, 1, 2, \ldots, N - 1$, the processor i is directly connected to processors $i - 1$ and $i + 1$, where the addition is done modulo N. Note that for each i and j, processor i is connected to processor j, although not necessarily directly connected to processor j. Between the processors there are i-j-1 mod N processors through which data must flow while moving from processor i to processor j. A cyclic shift of data by an arbitrary amount can be done as a sequence of cyclic shifts by a unit amount.

The motivation for including the cyclic shift comes from the many parallel algorithms that contain assignment statements of the following type:

$$X[i] := X[i - 1] + X[i + 1] - 2 \times X[i];$$

These statements usually occur in a context in which every element of the X

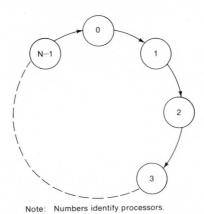

Note: Numbers identify processors.

Figure 8-8   Cyclically connected collection of processors

vector has to be updated simultaneously, except for the first and last elements, which require special processing. Thus the vector operations require that the vector be aligned with itself shifted by a unit shift. Permutation interconnections of a unit cyclic shift in either direction is useful to perform this type of operation. To perform the special processing required for the first and last elements of the vector, we need the mask facilities mentioned previously. These do not detract from the general usefulness of the cyclic shift in this application.

This type of calculation shows up frequently in one-dimensional partial differential equations. Typically we have an iterative calculation in the sense that the vector updating process is done repeatedly until the result converges to a solution.

Computations in two and three dimensions entail generalizations of the basic iteration mentioned above. For example, a typical iteration might take the form

$$X[i, j] := X[i + 1, j] + X[i - 1, j] + X[i, j - 1]$$
$$+ X[i, j + 1] - 4 \times X[i, j];$$

The pattern for updating $X[i, j]$ is shown in Figure 8-9 and is often called the

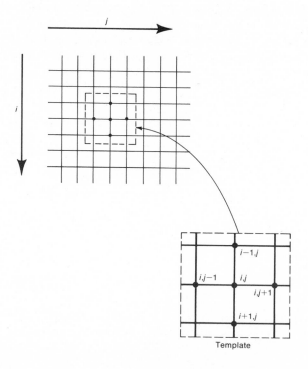

Template

Figure 8-9   Grid with a calculation template for an iterative calculation

four-point iteration. In three dimensions the pattern has six points. Although modern methods for solving partial differential equations have led to the development of iteration formulas with faster convergence properties than the ones given here, the formulas all have the characteristics in common of updating information at a grid point as a function of the values at nearby grid points. This observation led Slotnick to propose an array computer whose interconnection structure has the near neighbor connections built in as shown in Figure 8-10.

The interconnection pattern that was actually built into the ILLIAC IV computer was greatly influenced by Slotnick's early research. The 64 processors are interconnected so that processor i is connected to processors $i - 1$, $i + 1$, $i - 8$ and $i + 8$ with addition taken modulo 64. When the processors are numbered as shown in Figure 8-10, we obtain the interior connections shown there plus others necessary to make cyclic shifts of unit distance or of distance 8.

Note that with these four cyclic shifts, any arbitrary cyclic shift can be constructed, and the number of shifts required is small compared to 64. In fact, every shift can be realized by a combination of no more than seven of the built-in cyclic shifts.

No other interconnection patterns are available in the ILLIAC IV. Thus, algorithms that require noncyclic data permutations are rather inefficient. While it is true that every permutation is realizable with the ILLIAC IV shifting and masking capability, some permutations may require as many as N mask-and-shift operations. Such permutations are effectively performed serially rather than in parallel, and thus are ill-suited to this type of interconnection pattern.

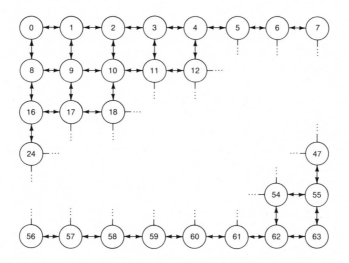

Figure 8-10   An 8 × 8 processor array designed for the four-point iteration

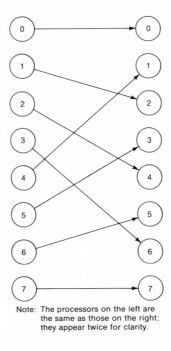

Note: The processors on the left are
the same as those on the right;
they appear twice for clarity.

Figure 8-11   The perfect shuffle interconnection of eight processors

A more promising interconnection pattern is shown in Figure 8-11. It is called the *perfect shuffle*. The name is derived from the fact that the cards in a deck of playing cards undergo a perfect shuffle when the deck is divided into equal halves and the halves are combined into a single deck by interlacing the cards. Since bidirectional interconnections are normally not much more expensive than unidirectional interconnections we shall assume that the *inverse perfect shuffle* is available if the perfect shuffle is implemented. This is the permutation obtained by reversing the arrows in Figure 8-11. In many applications either will work as well; in some applications one of the two is favored over the other. For greatest flexibility, it appears to be advantageous to have both available.

There are a number of parallel algorithms that make effective use of the perfect shuffle or its inverse. The algorithms include algorithms for performing Fourier Transforms, sorting, and matrix transposition. They are too specialized to include in full detail here, but we shall try to summarize the pertinent aspects. For a more complete discussion the reader is referred to Stone [71].

In the following discussion we shall assume that we have interconnections available for cyclic shifts of $+1$ and -1, and for the perfect shuffle and its inverse. The computations that we investigate have the general form shown here:

for $j := 1$ **step** 1 **until** $\log_2 N$ **do**
    begin
        Shuffle (Y);
        $Y[i] := F(Y[i + 1], Y[i])$,   (even i);
        $Y[i + 1] := G(Y[i + 1], Y[i])$,   (even i);
    end;

In this form, the vector Y is shuffled to align its components for parallel arithmetic. Then the functions F and G are applied to the even and odd elements, respectively. In general, the F and G functions depend both on i, the subscript of the arguments, and on j, the iteration number. Here we assume that F and G both operate on the old values of Y to produce the new values of Y, and that they act simultaneously.

A data flow graph for this type of computation appears in Figure 8-12. Note that the data is stored initially in the N registers on the left of the figure.

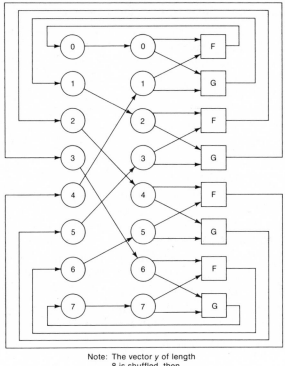

Note: The vector y of length
8 is shuffled, then
passed through the
F and G functions
and returned to place.

Figure 8-12   Data flow in parallel algorithm

The data traverses a perfect shuffle network, whereupon the F and G functions operate on pairs of data that occupy adjacent even-odd locations. The results of the computation are returned to the registers on the left for another iteration. Figure 8-12 is intended to show the data flow, rather than the actual processor structure. We expect that an SIMD computer will not have functional modules that evaluate F and G, but rather will compute these by means of stored programs as we have illustrated previously. The data transfer between odd-even adjacent pairs is implementable with cyclical shifts of $+1$ and -1.

To give a feeling of the complexity of the perfect shuffle algorithms, we just give examples of F and G functions for various algorithms, and omit all derivations and proofs.

For the Fast Fourier Transforms, F and G produce weighted sums of their arguments. In both cases the inputs are complex numbers, and the functions multiply the inputs by complex coefficients prior to the summation operation. F forms the weighted complex sum, and G forms the weighted complex difference of the numbers. The weighting factors used at each step can be computed dynamically rather than stored ahead of time, by using a functional algorithm of the type shown in Figure 8-12 [Cyre and Lipovski, 72]. The vector obtained after $\log_2 N$ steps is the Fourier transform, except that the elements are scrambled in what is known as *reverse binary order*. Apart from the unscrambling required at the end, the algorithm sketched here attains maximum speed for the computation of the Fourier Transform. There are several ways to unscramble the data from reverse binary order to normal order with fairly good efficiency, but the details are beyond the scope of this discussion. This version of the Fourier Transform due to Pease [68] requires much less routing than the best available Fourier Transform for computers limited to cyclic interconnection.

Another important application, sorting, uses the same data flow pattern, but uses three different pairs of F and G functions. Either F returns $Y[i]$ and G returns $Y[i + 1]$, or F returns the maximum of its two values and G the minimum, or F returns the minimum and G the maximum. The selection of the proper pair can be done by using mask bits, where the mask bits at each iteration are computed by passing the mask bit vector of the previous iteration through the perfect shuffle network. Sorting requires $(\log_2 N)^2$ rather than $\log_2 N$ iterations. Again, this algorithm is faster than any known sorting algorithm for cyclic interconnections. The algorithm is described in Stone [71] and is due originally to Batcher [68].

For matrix transposition, we assume that $N = 2^{2m}$, that is, N is an even power of two, and that we wish to transpose a matrix A, of size $2^m \times 2^m$. In this case we store the two-dimensional matrix A as elements of a one-dimensional vector, Y, of length 2^{2m}. If A has elements $A[i, j]$, $0 \le i, j \le 2^m - 1$, then A is stored by rows in Y if $A[i, j]$ is stored in $Y[2^m \times i + j]$. In this storage format, the ordering of the elements of A in Y is $A[0, 0]$, $A[0, 1], \ldots, A[0, 2^m - 1], A[1, 0], A[1, 1], \ldots, A[1, 2^m - 1], \ldots,$ $A[2^m - 1, 2^m - 2], A[2^m - 1, 2^m - 1]$. Thus the elements in each row are stored in adjacent cells in memory. The transpose of A is a format in which the columns of A are adjacent in memory. In this format, $A[i, j]$ is stored in $Y[i + 2^m \times j]$. The F and G functions for transposing the matrix are simply $F(Y[i], Y[i + 1]) = Y[i]$ and $G(Y[i], Y[i + 1]) = Y[i + 1]$. The matrix transpose permutation then requires exactly m iterations through the network in Figure 8-12. Matrices of other sizes can also be transposed by the same type of algorithm, with the greatest efficiency for matrices whose sizes are powers of 2. This algorithm is due to Stone [71].

There is a large class of problems for which the inverse of a perfect shuffle appears to be an attractive interconnection pattern. This class is a special class of recurrence problems. To show how the inverse of the perfect shuffle is used, we first illustrate the form of the recurrence equations, then show typical solutions for an SIMD computer.

First we show a simple example. To find the sum of the elements in a vector A, we typically use an iteration described by the following ALGOL-like statements:

$$SUM : = 0;$$
$$\textbf{for } i : = 0 \textbf{ step } 1 \textbf{ until } N - 1 \textbf{ do}$$
$$SUM : = SUM + A[i];$$

Now let y_i denote the value of sum after the ith iteration, and note that the iteration can be phrased mathematically in the form:

$$i_{-1} = 0$$
$$y_i = y_{i-1} + A[i];$$

This form of the algorithm is quite sequential, since we compute y_i before y_{i-1} for each i. It clearly requires time proportional to N to compute the value of y_N. But intuitively, we know that we can compute y_N in only $\log_2 N$ steps on a parallel computer with N processors, provided that we use a scheme like the one shown in Figure 8-13 for N = 8. Note that at the first step, we sum even-odd adjacent pairs of elements, then we sum the sums of pairs at the second step, and finally sum the sums of four adjacent elements at the last step. Figure 8-14 shows a variation of the previous figure (8-13) in which we compute all values of y_i, $1 \leq i \leq 8$, simultaneously in $\log_2 N$ steps. In this case, at some steps in the scheme slightly different functions have to be applied so that the process is not a vector arithmetic process in the strict sense. The nodes indicated by open circles do not form the sum of two operands, but merely pass through a single operand. Since this happens currently with the summation of operand pairs in other processors, the circled nodes indicate instances in which masking is required to fit the computation into a vector computation.

The computation scheme shown in Figure 8-14 can be implemented in the processor shown in Figure 8-15, which uses the inverse perfect shuffle

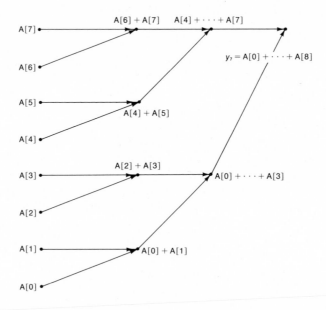

Figure 8-13 Parallel calculation of the sum of all elements of a vector

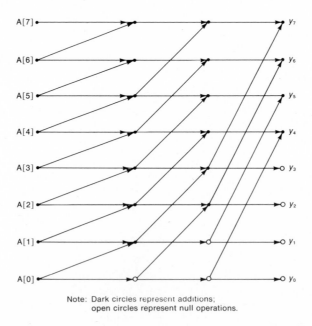

Note: Dark circles represent additions;
open circles represent null operations.

Figure 8-14 Parallel calculation of $y_i = y_{i-1} + A[i]$

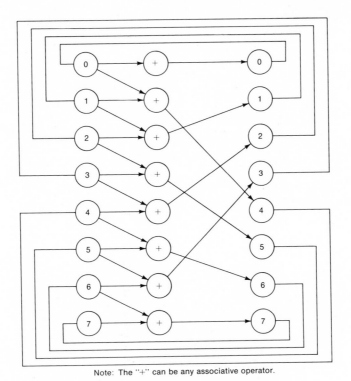

Note: The "+" can be any associative operator.

Figure 8-15 Data flow for recursive doubling algorithms

and a cyclic shift interconnection pattern. The general flow of the algorithm through this form of SIMD computer is:

> initialize Y[i] to A[i], ($0 \leq i \leq N - 1$);
> **for** j : = 1 **step** 1 **until** $\log_2$ N **do**
>   **begin** Y[i] : = Y[i] + Y[$i - 1$], (mask);
>   inverse shuffle (Y);
>   compute new mask;
> **end;**

In the first iteration, the sums computed are sums of adjacent elements of A. That is, Y[i] = A[i] + A[$i - 1$]. After the inverse shuffle, the items that are adjacent in Y are pairs of sums of elements of A, and, in fact, are A[i] + A[$i - 1$] and A[$i - 2$] + A[$i - 3$] in adjacent words, except in a few elements that represent special cases. Figure 8-16 shows the intermediate results in the computation, and the intermediate values of the mask bits. It is rather interesting that the mask bits for one iteration can be computed from the mask bits for the preceding calculation by first replicating the present

Y	Mask	Sum	Inverse Shuffle	Mask	Sum	Inverse Shuffle	Mask	Sum	Inverse Shuffle
$A[0] = Y_{0,0}$	0	$Y_{0,0}$	$Y_{0,0}$	0	$Y_{0,0}$	$Y_{0,0}$	0	$Y_{0,0}$	$Y_{0,0} = y_0$
$A[1] = Y_{1,1}$	1	$Y_{0,1}$	$Y_{1,2}$	1	$Y_{0,2}$	$Y_{1,4}$	1	$Y_{0,4}$	$Y_{0,1} = y_1$
$A[2] = Y_{2,2}$	1	$Y_{1,2}$	$Y_{3,4}$	1	$Y_{1,4}$	$Y_{0,1}$	0	$Y_{0,1}$	$Y_{0,2} = y_2$
$A[3] = Y_{3,3}$	1	$Y_{2,3}$	$Y_{5,6}$	0	$Y_{3,6}$	$Y_{2,5}$	1	$Y_{0,5}$	$Y_{0,3} = y_3$
$A[4] = Y_{4,4}$	1	$Y_{3,4}$	$Y_{0,1}$	1	$Y_{0,1}$	$Y_{0,2}$	0	$Y_{0,2}$	$Y_{0,4} = y_4$
$A[5] = Y_{5,5}$	1	$Y_{4,5}$	$Y_{2,3}$	1	$Y_{0,3}$	$Y_{3,6}$	1	$Y_{0,6}$	$Y_{0,5} = y_5$
$A[6] = Y_{6,6}$	1	$Y_{5,6}$	$Y_{4,5}$	1	$Y_{2,5}$	$Y_{0,3}$	0	$Y_{0,3}$	$Y_{0,6} = y_6$
$A[7] = Y_{7,7}$	1	$Y_{6,7}$	$Y_{6,7}$	1	$Y_{4,7}$	$Y_{4,7}$	1	$Y_{0,7}$	$Y_{0,7} = y_7$

Figure 8-16 Intermediate values for recursive doubling calculation of $y_i = y_{i-1} + A[i]$

$$Y_{i,\,j} \text{ denotes } \sum_{k=i}^{k=j} A[k]$$

mask bits with a cyclic shift, followed by passing the resulting vector through the inverse perfect shuffle network. The exact details of the mask calculation are left to the reader.

The reason this form of recurrence relation can be done in $\log_2 N$ steps is because addition is associative. That is, we can form the sum of N items by parenthesizing the N items arbitrarily into pairs without changing the result of the summation (provided that arithmetic is assumed to be done exactly). If the addition operation is replaced by any other associative operation, the same general form of the computation is still valid. Thus we can compute the product, the maximum, or the minimum of the elements of a vector in $\log_2 N$ steps because these operations involve associative operators. A list of the important cases appears in Table 8-4. Other forms, which do not have an associative operator also appear in the table. Nevertheless, these forms can be placed in the same general framework as well. For example the linear recurrence $x_i = a_i x_{i-1} + b_i x_{i-2}$ can be changed into the form

$$\begin{bmatrix} x_i \\ x_{i-1} \end{bmatrix} = \begin{bmatrix} a_i & b_i \\ 1 & 0 \end{bmatrix} \begin{bmatrix} x_{i-1} \\ x_{i-2} \end{bmatrix}$$

This is a vector-matrix equation which can be expressed as:

$$X_i = A_i \times X_{i-1}$$

where the vector $X_i = (x_i, x_{i-1})^t$, and A_i is the corresponding 2×2 matrix shown above. The latter form has a single operator, a matrix multiplication, which is associative. The latter form can be fitted directly into the form we require. This schema has been called *recursive doubling* by Stone [73] and by Kogge and Stone [73]. The interested reader should refer to these papers for further details.

TABLE 8-4. FUNCTIONS SUITABLE FOR RECURSIVE DOUBLING

Function	Description
$X_i = X_{i-1} + a_i$	Sum the elements of a vector
$X_i = X_{i-1} \times a_i$	Multiply the elements of a vector
$X_i = \min(X_{i-1}, a_i)$	Find the minimum
$X_i = \max(X_{i-1}, a_i)$	Find the maximum
$X_i = a_i X_{i-1} + b_i$	First order linear recurrence, inhomogeneous
$X_i = a_i X_{i-1} + b_i X_{i-2}$	Second order linear recurrence
$X_i = a_i X_{i-1} + b_i X_{i-2} + \cdots$	Any order linear recurrence, homogeneous or inhomogeneous
$X_i = (a_i X_{i-1} + b_i)/(a_i X_{i-1} + d_i)$	First order rational fraction recurrence
$X_i = a_i + b_i/X_{i-1}$	Special case of first order rational fraction
$X_i = \sqrt{(X_{i-1})^2 + (a_i)^2}$	Vector norm

At this point we have shown algorithms that make use of various permutations including cyclic shifts, the perfect shuffle, and the inverse perfect shuffle. These interconnections appear to be the most important ones to include because they are well suited to a wide variety of applications. At this writing there are no other permutations that appear to be as useful for most applications, although research efforts may expose new ones. We expect that the next generation of SIMD computers will have the interconnections mentioned here or variations of them. Processors that are dedicated to highly specialized algorithms should have interconnections designed for these algorithms, and such interconnections may be quite different from the ones mentioned here.

Data Structures

We mentioned earlier that high-efficiency utilization of SIMD computers requires high-utilization of both the processor and memory bandwidths. The interconnection study of the previous section was motivated primarily to show how to use the processor bandwidth. In this section we show techniques for efficient utilization of the memory bandwidth.

The basic constraint that hampers peak utilization of memory is that each memory module can honor at most one request per memory cycle. In the most favorable circumstances, the N operands of a vector instruction lie in distinct memory modules, and thus can be fetched simultaneously. In the least favorable circumstances, the N operands all lie in a single memory module, and must be fetched sequentially. In the latter case, N memory cycles rather than a single cycle are required, and the efficiency of computation is quite low.

For one-dimensional computations, the vectors involved are normally stored so that vectors of length N have one element in each memory module. Longer vectors are stored analogously by distributing the elements cyclically among the memories.

Serious difficulties arise in two-dimensional problems. Typical of these are matrix computations in which rows and columns are both treated as vectors in intermediate calculations. Suppose, for example, that A is an $N \times N$ matrix to be processed on an N-processor SIMD computer. Part a of Figure 8-17 shows A stored in a format known as straight storage. The figure shows a four-processor system, and the four memories are indicated as the four columns. Each memory is assumed to have its locations addressed starting at 0, and increasing as shown. Thus, there are N different words with address i, one in each of the memories, and this collection of N words forms the ith row of the system of N memories.

The figure shows rows of A stored in rows of memory, with rows aligned. In this case any row of A can be accessed in a single cycle, but N cycles are required to access a column because each column lies completely within one memory module.

Part b of Figure 8-17, on the other hand, shows a storage format known

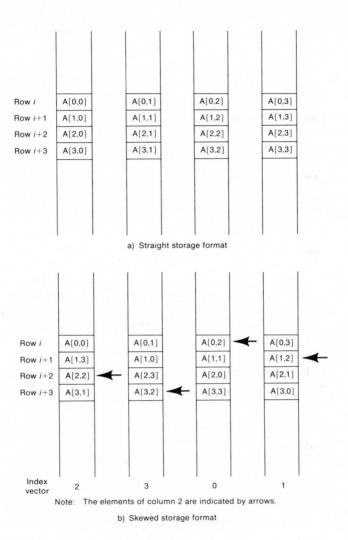

a) Straight storage format

Note: The elements of column 2 are indicated by arrows.

b) Skewed storage format

Figure 8-17   Storage formats

as *skewed storage,* in which successive rows of A are cyclically shifted by unit amounts. In this figure we see that each column of A is distributed across the memory modules. Hence, both the rows of A and the columns of A can be accessed in unit time. In row mode, instructions broadcast a row address, and that address is the actual address of an access. For column instructions, the vector of index registers private to each processor is preloaded with the index vector 0, 1, 2, . . . , N-1. To access column *i*, this index vector is shifted cyclically until the 0 is aligned to memory *i*. Then the address of row 0 of the matrix is broadcast with an indication that indexing is required. The broad-cast address is then modified locally at each processor, by adding the index element to it, and access is made at the indexed address. The figure shows the access of column 2. After both row and column access, the vector obtained is a cyclic shift of what normally is desired. The leading element of row or col-umn i is in processor i, not in processor 0. A cyclic shift is often required to bring the data into position for further manipulation.

Skewed storage is a relatively inexpensive way to enhance the processing capability of an SIMD computer. It requires each processor to have a private index register, and it requires cyclic interconnections. Both of these features are reasonable to include in the computer on other grounds, so that they should not be viewed as part of the extra expense of skewed storage. There is an additional cost in processing time if skewed storage is used in place of straight storage, because shifting the index vector, modifying the broadcast address, and then reshifting the fetched vector are processes that need not be done with straight storage. Thus if an algorithm requires access to rows only, straight storage is slightly preferable to skewed storage. On the other hand, if both rows and columns need to be accessed as vectors, then skewed storage is strongly preferable to straight storage. If access to columns is the only requirement, then the matrix should be stored in transposed form in the straight-storage format.

Historically speaking, the skewed storage concept as we described it here is due to David Kuck in unpublished work during the early development of the ILLIAC IV. More recently, investigators have been looking into other types of access that permit access to other useful portions of a matrix such as diagonals or small square submatrices. Budnik and Kuck [71] examined a scheme in which N processors are connected to N + 1 memories, and found some interesting solutions. However, a much more practical scheme was ex-plored by Lawrie [73] for the SIMD computer of the type discussed here. He discovered that when N is a power of 2, the perfect shuffle and cyclic interconnection structure can be used together with new storage schemes to gain access to all of the most often needed collections of elements of two-dimensional arrays.

At this point we turn attention to the use of the features described above in realistic program fragments. To give some intuition about the use of data structures in algorithms for SIMD computers, we show various ways of solv-ing simple examples. The examples illustrate the importance of skewed stor-age, cyclic shifting, and individual indexing in the construction of good parallel algorithms.

The first example illustrates how to form row sums and column sums for two-dimensional arrays. Let A be an $N \times N$ matrix, and LOCINDEX[k] be the local index register in processor k. In straight storage format, the column sums of A can be computed and stored in the vector SUM by means of the following program.

```
SUM[k] : = 0, (0 ≤ k ≤ N − 1);
for i : = 0 step 1 until N − 1 do
    begin
    LOCINDEX[k] : = i, (0 ≤ k ≤ N − 1);
    comment the index is set to obtain the next element of each
        column by fetching the next row;
    SUM[k] : = SUM[k] + A[LOCINDEX[k], k], (0 ≤ k ≤ N−1);
    comment A[i, k] is added to the sum;
    end;
```

In this algorithm, the N sums are computed simultaneously as a collection of sums, but each individual sum is computed sequentially by marching down a column. Each memory fetch brings up a row of N items, and each item fetched is added to its respective column sums.

To form row sums, we might fetch one row at a time, then the elements in that row. The algorithm takes the form:

```
for i : = 0 step 1 until N − 1 do
    begin
        TEMP[k] : = A[i, k], (0 ≤ k ≤ N − 1);
        for j : = 1 step j until N − 1 do
            begin
            comment compute the sum elements in row i of A by
                recursive doubling;
            TEMP[k] : = TEMP[k] + TEMP[k − j], (j ≤ k ≤ N − 1);
            end j loop;
        SUM[i] : = TEMP[N − 1];
    end i loop;
```

This algorithm requires a time that grows as $N \log_2 N$ rather than as N, which makes it much slower than the column sum algorithms. However, a small change to the first algorithm yields a row sum algorithm of nearly equal speed. The trick is to observe that we can change the indexing to fetch diagonals of A rather than rows of A. We can distribute the elements of a diagonal to N row sums and thus obtain a high degree of parallelism. The fast row sum algorithm is the following:

```
LOCINDEX[k] : = k, (0 ≤ k ≤ N − 1);
comment LOCINDEX has been initialized to enable us to fetch
    diagonals;
SUM[k] : = 0, (0 ≤ k ≤ N − 1);
for i : = 0 step 1 until N − 1 do
    begin
        SUM[k] : = SUM[k] + A[LOCINDEX[k], (k − i) mod N],
            (0 ≤ k ≤ N − 1);
        comment shift LOCINDEX cyclically to obtain next diagonal;
    end;
```

In this example we show the subscript calculations with the modulo N operation given explicitly. Since N is a power of 2, no division is required, and this is easily implemented automatically in the hardware. Note that the summation adds a diagonal to the running sum. The diagonal is cyclically shifted i places by the expression $(k - i)$ mod N to associate the element in row i with the running sum for row i which is maintained in processor i. If the expression for the diagonal were changed to read A[LOCINDEX[k], k], we would obtain column sums rather than row sums.

Through these examples we see that the interconnections, the local index registers, and the data structure all come into play together in the construction of algorithms of high efficiency. It is an interesting exercise to repeat these examples for the same matrix stored in skewed form.

SIMD Computers in Perspective

The discussion thus far illustrates the general characteristics of SIMD computers, plus several techniques for increasing their effectiveness on particular problems. For a problem to be suitable for SIMD computation, it must have at least the three characteristics:

1. The computation must be describable by vector instructions such that a majority of computation time is spent with many identical operations in action simultaneously on different data.
2. High-speed data routing between processors must be possible to do with the available processor interconnections.
3. Operands manipulated simultaneously must be capable of being fetched simultaneously from memory.

If any of these three conditions is not satisfied for a computation, then the computation may run essentially serially in an SIMD computer. The class of computations that satisfy the points above is a special class of computations, and hence the SIMD computer is not a general-purpose computer. Given this fact, we note that SIMD computers appear to be suitable for a greater variety of computations than had once been believed to be the case. The advances in research in this area have found new algorithms, new interconnection patterns, and new methods for structuring data to satisfy the conditions for high-speed processing on SIMD machines. Presently there are a sufficient number of important problems that lend themselves to SIMD computation to justify the construction of a few and possibly many of these machines. The number constructed beyond the first few depends strongly on how broadly the machines can be applied.

The use of SIMD computers for information retrieval is unlikely to be a major application as once projected. It is tempting to believe that an N-processor system can search a file N times faster than a one-processor system, because N pieces of data can be inspected simultaneously. However, serial searching can be so efficient, due to techniques known as binary searching and hash-addressing, that the actual speed-increase for searching is more like $\log_2 N$. Karp and Miranker [68] observed this behavior for a typical search problem with realistic assumptions. Many search problems fit their assumptions, so their results hold for a broad class of search problems. For small N, a speed increase of $\log_2 N$ is acceptable but for large N it is intolerably low. This suggests that information retrieval and other applications involving file searching are better suited to serial computers or to computers with a low degree of parallelism. Of course, some search problems fail to satisfy the Karp and Miranker assumptions, and these may be quite reasonable to implement on parallel computers. A search for the nearest match in an unstructured data base has these qualities, and thus may be suited for parallel computers. Air traffic control is representative of an application involving this type of search since a major problem of control is the identification of aircraft in close proximity to each other.

8-4. MIMD COMPUTERS

In this section we look into the organization and control of multiple-instruction, multiple-data stream (MIMD) computers. The interesting problems to consider for these machines are quite different from those of the SIMD computers. MIMD processors are suitable for a much larger class of computations than SIMD computers because they are inherently more flexible. It is relatively straightforward to fit a computation to an MIMD machine, which is not the case for SIMD computations. However, to attain high-efficiency computation in an MIMD environment, one has to be careful in synchronizing computations on different processors, and in allocating tasks among the processors. In some cases, improper synchronization can lead to gross inefficiency, and, in extreme situations, computation may cease entirely. This is in sharp contrast to SIMD computers. Synchronization of the processors in an SIMD system is done automatically as each instruction is executed, and there is no task allocation problem because all of the processors do the same task. Thus the constraints of SIMD computers guarantee that some problems of parallel computation do not arise. However these problems are eliminated at the cost of flexibility. In the MIMD system, the added flexibility comes with the need to solve the synchronization and allocation problems.

In this section we look at the overall details of an MIMD computer system and investigate methods for performing computation on such a machine. We focus most strongly on the portions of the system, both hardware and software, that are necessary to support parallelism, and we presume that the portions of the system not described here are much like the corresponding portions of conventional serial computers. In the first part of the section we describe the hardware structure of an MIMD computer. This illustrates a realistic way, but not the only way, of structuring such a system. Next we investigate the problems of controlling parallel computation and look further into the FORK and JOIN instructions as well as other synchronization operations. In the final portion we investigate problems of resource sharing that are inherent in this type of system.

Structure of MIMD Computer Systems

The notion of connecting several processors together to create a powerful parallel processor appears to be primarily a problem in hardware design. But this appearance is deceptive, because the implementation of such an idea requires thorough consideration of hardware, operating systems, and ultimate applications to make an effective parallel computer. The architecture of an MIMD computer strongly reflects the need to support parallelism, synchronization, and data communication among programs. In this section we describe a model for an MIMD computer, which is abstracted largely from an MIMD computer in development at Carnegie-Mellon University [Wulf and Bell, 73]. We focus principally on the hardware here, but make some attempt to motivate how various facilities are used. In later sections we concentrate on parallel computations and how they are executed within the structure described here. In the later sections the functional use of most of the hardware features will be made clear. This section closes with a brief discussion of an alternative MIMD computer structure proposed by Flynn et al. [70].

Let us begin the description of an MIMD computer system with the graph of the structure shown in Figure 8-18. There are N processors in this system, M memories, and P input/output channels. The interconnections among these modules as shown in the system are extensive. A switch connects every processor to every memory. The switch is shown as an $N \times M$ crossbar where each of the NM crosspoints is a potential connection. As usual, we assume that each memory can respond to a single request during one memory cycle, so that no memory services two processors simultaneously. However, one processor may transmit data simultaneously to two memories simultaneously, if there is some reason to do so. A similar crossbar of size $N \times P$ connects the processors to the data channels and input/output buses.

Although we show only a single line from each memory and processor into the $N \times M$ crossbar, this line has sufficient bandwidth to transmit whole words of data instead of single bits. Thus one line in the figure may correspond to roughly 32 to 64 wires in a real implementation. Similarly the $N \times P$ crossbar has many wires per interconnection. These crossbars have a complexity that is enormous as N and M become large. For purposes of efficiency, we expect M to be at least as large as N, so that all N processors can gain access to the memories simultaneously. Thus the number of crosspoints in the $N \times M$ switch grows as N^2, for sufficiently large N, the cost of the crossbar dominates the cost of the N processors and M memories. This phenomenon was mentioned earlier in the discussion of SIMD computers, where we proposed that the interconnections between processors be severely limited. For SIMD computers the alternatives to complete interconnectivity are more clear cut because the class of application problems is more limited. For MIMD computers the question of alternatives to complete interconnec-

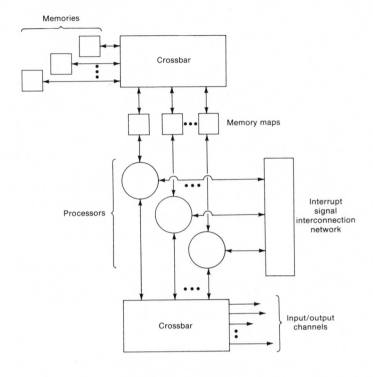

Figure 8-18   MIMD computer

tivity is still open because so little is known about the communication that takes place within a parallel computation on such a computer. We shall assume here that we retain the complete interconnectivity of the crossbar switch, while recognizing that such a decision places a severe limitation on the number of processors that can be justifiably included in the system.

Returning to the figure, we note that there is another complete interconnection network. This is the interrupt signal network. Each processor can transmit an interrupt signal to any other processor. This network need not be as costly as a crossbar because its bandwidth is rather low. A bus consisting of a single wire can serve as an interrupt line for each processor, and every processor can signal on each bus. Further study and experience may show that even less is required. A single bus or a few buses may satisfy the requirements for the system provided the interrupt signals are multiplexed on the buses.

Between the processors and the memory we show a memory address translator known as a *memory map*. Its function is to translate addresses produced by the processor into addresses in physical memory. There are many purposes served by the memory map, not all of them pertinent to parallel processing. In the context of the MIMD machine, it is unreasonable to permit programs to generate physical memory addresses during their execution, because this implies that the physical storage occupied by a program and by its data must be fixed during the execution of the program. In a parallel computation environment, the precise memory requirements for a group of programs in execution cannot be predicted ahead of time, so that in case we reach an impasse in execution, we must be able to remove items from physical memory or to move and compact items in physical memory in order to make room for new items. The memory map provides the necessary facilities for dealing with the memory problem, because the actual physical locations occupied by a program and data can be varied dynamically during program execution. Should data be moved, then the map function must be changed to reflect this change so that the memory map can correctly translate program references to the data into true physical addresses.

The memory map also gives a convenient facility for permitting two or more programs to share data. The technique here is to set the map functions of two or more cooperating programs so that references to the shared data are mapped into the same physical addresses of the data. References to private data are mapped into physical locations that are distinct for each program. Thus, programs can share specific data without sharing all data.

A reasonably straightforward implementation of memory mapping is the paging scheme illustrated in Figure 8-19. Here the addresses presented to the map are divided into two fields, one denoted *page number* and the other denoted *displacement within page*. The page number is used as the index of a page register whose contents replace the page number of the address. The

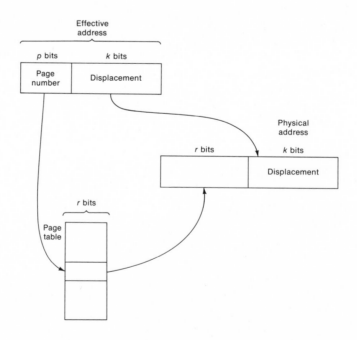

Figure 8-19 Typical paging scheme

resulting address is the physical address of the reference. Note that if the displacement field has k bits, then the displacement within page can vary from 0 to $2^k - 1$. Also, since the map function cannot alter displacement, items can be moved in memory only in contiguous blocks of size 2^k, which are called *pages* and from which this mapping scheme takes its name. This scheme is consistent with the scheme used in the Carnegie-Mellon processor, but other memory mapping schemes are potential candidates.

The physical address reported at the output of the memory map is decoded by the crossbar into the index of a specific memory module, and an address within that module. The memory module index selects a specific crosspoint, and the address within module is routed through the selected crosspoint. The decoding function has remarkable influence on the reliability, expandability, and efficiency of execution of an MIMD system so that it is worthwhile to investigate this more thoroughly. Figure 8-20 shows three different decoding functions. The first function (part a) extracts the leading m bits from a physical address and uses this as an index to select a particular module from a collection of 2^m modules. The remaining bits select an address within the module. This scheme is similar to the paging scheme with contiguous storage addresses of size 2^r lying within one module. The second scheme (part b) reverses the order of the two fields so that the least significant m bits are used to select a module and the remaining high-order r bits select an address within the module. Now addresses are interlaced among the memories so that a sequence of increasing physical addresses selects each of the memory modules in cyclical fashion in the order of increasing module index.

These two schemes represent extremes in the choice of decoding functions.

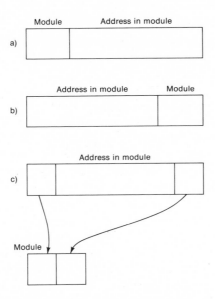

Figure 8-20   Module selection schemes for decoding physical addresses

The first scheme places contiguous addresses together, increasing the probability of memory access conflict when two or more processors are sharing data within the memory. The second scheme tends to reduce storage conflicts, particularly accesses by several processors to shared data, because a data structure tends to be uniformly distributed among the modules. Sharing a single item, of course, produces conflicts if that item is accessed with great frequency, but this type of sharing is less usual than the sharing of moderate-size programs and data structures.

On the basis of the memory conflict argument, the second of the two schemes appears to be preferable. However, this scheme is susceptible to reliability and expandability problems. Note that there must be 2^m modules present for this scheme to work, and a failure of any single module almost certainly causes total failure of the entire system. In the first system, the system can work with any number of modules from 1 to 2^r, provided that the decoding function identifies invalid address references should a nonexistent module index be generated. Also, if a module fails, then the system can still function with less memory and degraded performance. The failed module is simply removed from the system and the system is restarted in such a way that the memory maps never produce physical addresses in the failed module.

Each of these schemes has been used in practice, and no single philosophy dominates. The third scheme (part c of Figure 8-20) suggests a possible compromise. Here the module-index field is the concatenation of two fields, one from the least significant bits and one from the most significant bits. If the least significant field has size k, addresses are interlaced among groups of 2^k memory modules, which tends to reduce memory access conflicts to a block of shared data. System expansion has to be done in blocks of 2^k modules, and a single failure disables an entire block of 2^k modules. If we use $k = 2$ or $k = 3$, then we obtain a reasonable compromise while the total number of memory modules in the system can be much larger than 2^k. The third decoding scheme or a variant of it appears to be the most attractive for systems with a large number of modules.

The processors, memories, and other components of the MIMD system described here are conventional components, with only a few capabilities not ordinarily found in components of purely sequential systems.

Before turning to the investigations of computation on an MIMD computer, we present a brief view of an alternate proposal for an MIMD computer due to Flynn, et al. [70]. Rather than present all of his ideas, we focus on the implementation of the multiple-instruction stream. Flynn's proposed system has many notable features including pipeline implementation, but most of the details are not pertinent to this discussion. The central idea of his proposal is to interconnect several independent processors, each of which executes an independent instruction stream. The unconventional notion in this proposal is to make processors "skeleton" processors, by removing from them all arithmetic functions and computational logic. These functions are performed by highly specialized high-speed processors that are shared among the skeleton processors as shown in Figure 8-21. Consider, for example, the events that

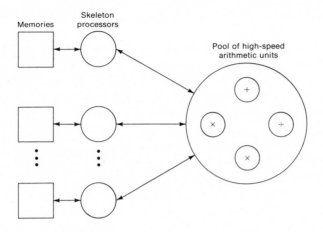

Figure 8-21   MIMD computer with skeleton processors and centralized computation facilities

occur when a skeleton processor generates an ADD instruction. After obtaining the operands for the instruction, the processor requests access to a high-speed adder. If one is available, the operation is performed and the result returned to the skeleton processor. In case of conflict, the request for computation can be queued or the request can be repeated until an adder is free.

This processor exhibits an interesting behavior pattern when the skeleton processors are executing the same program, but with different data. Suppose, for example the program contains the instruction sequence

LOAD
ADD
MUL
SUB
DIV
STORE

where the operands for these instructions have been omitted. If three computations execute this sequence, and are initially synchronized, then conflicting access requests for the arithmetic hardware soon forces them out of synchronization. The interesting thing is that once out of synchronization, the processors can repeat this sequence iteratively with very little conflict. For example, if each type of arithmetic instruction takes roughly the same amount of time, then it is possible for the processors to be one instruction apart as they execute the program. Thus, processor 1 can do a multiply, while processor 2 does a subtraction, and processor 3 does a division. A favorable situation is shown in Figure 8-22. To execute this program sequence at full speed iteratively with three processors, we require two independent adders, a multiplier, and a divider.

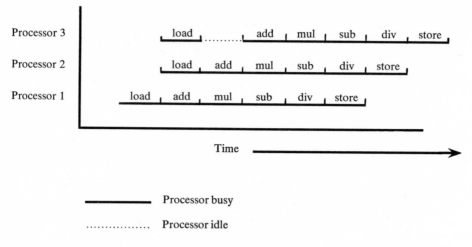

Figure 8-22 Timing of a sample program for Flynn's MIMD computer

The advantage of this system is that it creates a multiple-instruction stream processor, but the expensive components of the processors are not replicated to the same extent that the processors are replicated. We obtain high speed from specialized arithmetic units, and utilization is high because of sharing. A single processor cannot make efficient use of all its functional hardware if the hardware is specialized to high-speed arithmetic operations. For example, a high-speed multiplier lies idle while an addition is in progress. The Flynn approach permits the system to support extremely high-speed multipliers, dividers, and other functional hardware that would ordinarily be too expensive to replicate.

Besides Flynn's approach, there are many variations of MIMD computers possible, and we do not intend this discussion to be exhaustive. Another approach worthy of mention is that of Kuck et al., [72], which presumes an MIMD system much more closely coupled than those described here. We now turn to the problem of controlling computation in an MIMD system.

Interprocessor Control

Two major questions have received the greatest attention in the literature and have influenced present approaches. The first question concerns what happens when two or more processors are in execution concurrently and must cooperate during the computation. Here the concern lies in synchronization of the processors so that the parallel computation can be carried out correctly. The second problem area concerns techniques for creating and terminating parallel execution paths within a program. We explore each of these and indicate how the problem can be solved using similar tools.

To illustrate an example of the first problem, we propose to perform a summation in parallel. The program is to form the sum of the elements of the vector V[i], $1 \leq i \leq N$. To do this in parallel, the program forks into N branches, each of which performs the statement

$$SUM : = SUM + V[i]$$

for i varying from 1 to N. Without synchronization and interlocking, the execution may occur as follows:

1. Processor 1 fetches the value of SUM from memory.
2. Processor 2 fetches the value of SUM from memory.
3. Processor 1 adds $V[i_1]$ to its private value of SUM and restores the new value of SUM in memory.
4. Processor 2 adds $V[i_2]$ to its private value of SUM, different now from the value in memory, and restores it in memory.

This sequence of instructions produces incorrect results because the effect of adding $V[i_1]$ by processor 1 has been lost. The problem arises because processors 1 and 2 copy the value of SUM and operate on their private copies. For the program to be correct, a processor must be able to fetch and update the value of SUM without any intervening memory references to SUM.

The synchronization and interlocking can be done relatively simply, provided we implement certain instructions. Before describing the instructions, and techniques for solving the problem, we find it convenient to define informally some terms that we have occasion to use. We shall call a sequence of instructions a *program,* which coincides with our usage on SISD and SIMD computers. By *process*, we mean an instance of a program that is in execution. In MIMD computers, one program may be in execution simultaneously on each of N processors, so that we call each one of these instances of execution a process. The system is assumed to be able to suspend a process and restart it, not necessarily on the processor on which it was suspended.

We propose to solve the problem at hand by using operations called WAIT and SIGNAL. Both of these operations supply a memory address as a parameter, and operate on the contents of that memory address. The operations must be indivisible in the sense that once initiated, the item at the special address is fetched, modified, and returned to memory, and no other process may access the same memory cell until this operation is complete. The idea is to use WAIT and SIGNAL as shown below.

```
comment several processors may be executing statements before
    the WAIT and after SIGNAL;
WAIT (FLAG);
comment critical section. At most one processor reaches this
    point at any time;
SUM : = SUM + V[i];
SIGNAL (FLAG);
```

We call the statements between WAIT and its corresponding SIGNAL a *critical section*. The variable FLAG controls access to the critical section. The WAIT operation on FLAG permits a process to continue if and only if no other process is in the critical section. When a process leaves the critical section, it signals, which in turn permits exactly one of the waiting processes to enter the critical section. In an ALGOL-like notation, the actions of WAIT and SIGNAL are:

```
procedure  WAIT (FLAG);
   begin   FLAG : = FLAG − 1;
           if FLAG < 0 then suspend this process;
   and;
procedure  SIGNAL (FLAG);
   begin   FLAG : = FLAG + 1;
           if FLAG ≤ 0 then awaken a program suspended by
             this flag;
   end;
```

We initialize FLAG to the value 1 at the beginning of the computation. Except during execution of WAIT and SIGNAL, the value of FLAG is the negative of the number of processes waiting to enter the critical section. Thus if a process performs a WAIT and discovers the value of FLAG to be 0, it can proceed. Likewise, if a process performs a SIGNAL and finds the value of FLAG to be less than −1, at least one other process is waiting.

To obtain high efficiency utilization of processors during a computation, it is usual to remove a suspended process from a processor and permit another process to proceed on the newly available processor. This suggests the need for an operating system to control the change-over of processes, and to queue suspended processes as well as ready-to-run processes when there are more ready-to-run processes than processors available. This matter is pursued in greater depth in Chapter 11, but we cover the pertinent details here.

First we assume that the actions of process suspension and process awakening invoke an operating system program. We also assume that variables used as arguments of WAIT and SIGNAL have special attributes. For convenience, we call a variable of this type a *semaphore*. Each semaphore has a queue associated with it, and the entries in the queue refer to processes suspended because of the value of that variable. The operating system suspends a process by storing its state, and by enqueuing sufficient information to restart the program.

When a process is suspended, its processor is available for other computation. The operating system selects a process from a queue of ready-to-run processes, and restarts its computation.

During a SIGNAL operation, if a process needs to be restarted, the operating system selects one from the queue associated with the queue semaphore and places this on the ready-to-run queue. Figure 8-23 depicts the operating system actions for execution of WAIT and SIGNAL.

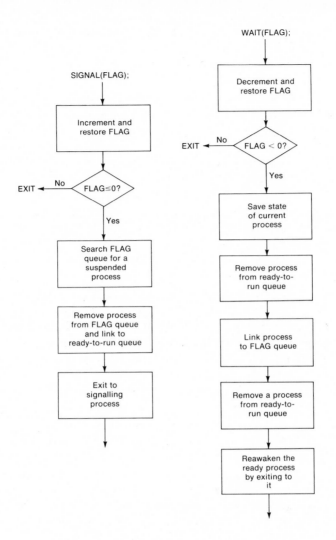

Figure 8-23   Flowcharts for SIGNAL and WAIT

This methodology permits a cooperation among several processes and provides for smooth switching of processors among the ready-to-run processes. The philosophy described here is principally due to Dijkstra [68].

The WAIT and SIGNAL operations can be used as general purpose operations for synchronizing processes not necessarily just to create critical sections. As an example of their use of synchronizing a pair of asynchronous processes, consider a program in which one process produces data to be consumed by another process. The producer and consumer are supposed to work independently, and asynchronously. We constrain the consumer to deal with items in the same order they are produced. Because the execution time of producer and consumer can vary arbitrarily, we assume that the producer places items in a queue, and that the consumer removes items from the queue. The problem is to suspend the producer when the queue is full and to suspend the consumer when the queue is empty.

A solution to this problem is illustrated graphically in Figure 8-24. The procedures for inserting into and deleting from a queue are named PUT and GET, respectively. The variable NOTFULL permits a WAIT to pass it if and only if fewer than *n* items are in the queue. Similarly, the variable NOTEMP-TY permits a WAIT to pass it if and only if the queue has at least one item. This example is drawn from Wirth [69].

In actual implementation, WAIT and SIGNAL are usually implemented to interrupt with a call on the operating system when a process is to be suspended or awakened, because such an operation is more reasonable to implement as an operating system function than as a machine operation. The crucial point in the implementation is that a variable must be fetched from storage, modified, and returned to storage without any other intervening access to that variable by another process. To perform this access we need special machine instructions, and special interlocks on the memories. Using the philosophy described here, an INCREMENT AND TRAP and DECRE-MENT AND TRAP instruction are suitable for WAIT and SIGNAL, where the trap conditions are given in the descriptions of WAIT and SIGNAL.

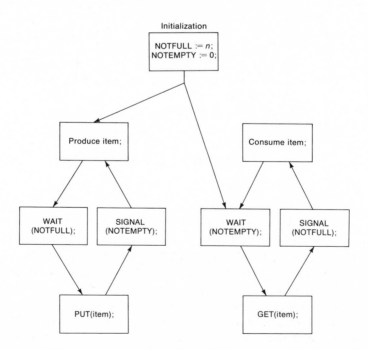

Figure 8-24   WAIT and SIGNAL operations for controlling a queue

In the remainder of this section we investigate the FORK and JOIN operations and their implementations. These are similar to Conway's FORK and JOIN as described in section 8-1, but are slightly modified to take advantage of the Dijkstra synchronization operators.

Recall that a FORK initiates an independent process. The question at hand is how is this to be implemented? To answer the question, consider how processes are initiated and how they are represented. The representation of a process is the collection of all of the data required to initiate or continue execution of the process. It includes among other things the state of all registers controlled by the process, the memory map function of the process, the links to the program instructions for executing the process, and the data that specifies all access rights and resources owned by the process. A FORK action creates a new process by creating a collection of data that represents the new process. Then the new process is linked to the ready-to-run queue, as if it were just awakened.

If process A creates a new process, say process B, through the action of a FORK, then B need not be an exact duplicate of A, and, in fact, at least the program counter of B should differ from A's program counter. Note that the ability to create an identical process with a different program counter is inherent in the notion of a FORK to a label in a program. Also of concern are the other items in the representation of a process. In most implementations, process A can specify how to construct the representation of process B in all of its detail. For access rights and privileges, process A can specify that the new process has all of its rights or fewer, but no more. Similarly, process A can limit the resources available to the new process. Finally process A is recorded as the owner of process B, with the ability to terminate B under program control.

Of specific concern is the problem of sharing information among processes. In our examples we have instances in which one process for each value of i is created where $1 \leq i \leq N$. In actual implementation, the value of i for a newly spawned process is placed in a specific machine register, and this register is part of the stored state of the process prior to its first execution. When the process is awakened for the first time, the variable is loaded into a physical machine register where it is available to the process. This is an example of information sharing in a limited sense, but no information can be communicated by this mechanism after forking creates a new process.

For more general sharing of information, the obvious method is share memory by means of the mapping function. We set the map of a new process so that some pages in its memory are identical to physical memory locations of its creator process. This is shown in Figure 8-25. References to page 2 by both processes A and B are directed to the same page in memory, but the other pages of these processes are physically distinct. Note that the flags used by WAIT and SIGNAL operations must be in shared memory because they are normally accessed by two or more processes.

From this discussion of FORK, and from our earlier discussions of WAIT and SIGNAL, it is relatively straightforward to construct an implementation

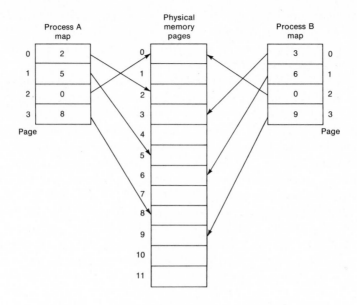

Figure 8-25   Example of paging used for memory sharing

of the JOIN instruction. Following the implementation philosophy of WAIT and SIGNAL, we choose to implement JOIN with a single parameter, a semaphore. A typical example of the JOIN in a program is the following:

> FLAG : = − (N − 1);
> **comment** the semaphore FLAG is initialized to terminate N − 1
> of the N processes that pass the JOIN statement;
> **for** i : = 1 **step** 1 **until** N − 1 **do**
> FORK to NEXT
>
> . . .
> NEXT: **comment** N processes pass this point;
> COMPUTEFUNCTION(i);
> JOIN (FLAG);

Only the last process to pass the JOIN statement continues past it. The first N-1 processes terminate when they reach it. In an ALGOL-like notation, JOIN might be implemented as follows:

> **procedure** JOIN (FLAG);
> **begin**
> FLAG : = FLAG + 1;
> **if** FLAG ≤ 0 **then** terminate process;
> **end;**

Note that this is quite similar to the SIGNAL operation. In fact, both can be implemented with the same instruction, INCREMENT AND TRAP, which interrupts if the memory cell is not greater than zero. Again we stress that the memory operations on FLAG must be noninterruptible.

At this point it should be clear that many other control operations can be implemented within the framework given here. The operations include control functions that operate on two or more semaphores, conditional execution of synchronization, and similar complex functions. The major issue in computer architecture is to support these operations at the hardware level with noninterruptible operations on semaphores, and operations for switching quickly from process to process. It is also necessary to design the system to support operating system functions because an operating system is essential for MIMD systems. The latter problem is not specific to MIMD computers, however, but is a problem of major importance in itself.

This completes the discussion of synchronization and control in MIMD computer systems. There remains the problems of resource allocation and scheduling, which are treated in the next section.

Resource Sharing and Scheduling

In this section we examine some difficult problems of resource allocation that can severely hamper computations on MIMD computers. The problems are generally characteristic of medium- and large-scale serial computer systems. The effect of parallelism within multiprocessor systems, particularly multiprocessors with a large number of processors, is to compound these problems to the point that computational efficiency can drop to intolerably low levels. It is unlikely that a highly parallel MIMD computer can be built until these problems are solved satisfactorily.

Of primary importance is the notion that processes operating in a computer system are sharing a pool of resources which includes processors, memories, input/output channels, and bulk storage. The pool must be allocated among the processes to honor all requests, and to do so in a way compatible with the constraints on the system resources so as to achieve high utilization of the resources. For example, requests for central memory at any given time may exceed the total amount of central memory in the system. Thus, one or more requests must be queued and granted at a later time. The allocation problem is to determine which requests to grant and which to enqueue, and to do so on an intelligent basis. One of the major points of this discussion is that some allocation strategies can lead to a situation known as *deadlock*, a situation in which part of the system shuts down in a permanent wait state that can be broken only by an external control. In many real situations, deadlock can be broken only by halting computation and reinitializing the machine.

In the first part of this section we look into typical scheduling techniques, and explore the problem of deadlock. We show that policies for the requesting and granting of resources can avoid deadlocks. In the latter part of this section, we turn to strategies for scheduling for maximum efficiency. However, the general problem of optimal scheduling has still not been solved. In fact, there is good reason to believe that no satisfactory optimum scheduling algorithm can be found, and that other suboptimal approaches to the scheduling problem will be the most practical ones to implement.

To begin the discussion of deadlock, we present an example of system deadlock involving only processor and memory resources. We shall assume that processes request memory in large chunks, each chunk large enough to hold a sufficient number of pages to contain the bulk of memory references for a computation. A collection of pages with this property is often called a *working set*. After a process owns a working set of pages of central memory, it can be executed on a processor and can run for a relatively long time before needing access to other data. Occasionally, the process suspends itself while synchronizing with another process or with input/output operations. Thus, during the lifetime of a process that owns a working set of memory, it periodically owns a processor while executing and gives up ownership of a processor while suspended. This level of detail is sufficient for our discussion, but we note that in systems with a hierarchy of memories, a process can own memory of several different types, and, contrary, to our assumptions, can be swapped completely out of central memory.

Now to construct an example of a system deadlock, suppose that a process, process A, requires exclusive control of half of central memory for its first phase of execution, then requires exclusive control of a total of three-fourths of central memory for its second and final phase of execution. The process executes by requesting half of memory, and then proceeds through the first phase. When more memory is required, the process requests an additional one-fourth of memory, without returning control of the memory owned by it. The process enters its second phase and completes execution when the request for additional memory is granted. Note that the resources required by process A do not exceed the total resources of the system, and that the two requests made by process A are each permissible requests in themselves, although we shall shortly demonstrate that they may not be permissible in certain contexts.

Now let us assume that process B, an identical copy of process A, executes simultaneously, and asynchronously with process A. The following sequence of events might occur :

1. Process A requests and is granted exclusive control of half of memory.
2. Process B requests and is granted exclusive control of the remaining half of memory.
3. Process A requests an additional one-fourth of memory. Process A is suspended, pending the availability of memory. Memory owned by process A is assumed to be unavailable for use by process B.
4. Process B requests an additional one-fourth of memory, and it too is suspended pending available memory. As above, the memory owned by B is unavailable to other processes.

At this point, all of available memory is owned jointly by A and B. Process A cannot proceed until B terminates, and B cannot proceed until A terminates. An impasse exists, and the system is unable to do any useful computation.

The state of affairs that exists in this example is one in which a nonsatisfiable circular set of constraints exists. That is, A depends on B, but B in turn depends on A. This situation is called a *deadlock* or *deadly embrace* [Dijkstra, 68], and is a situation that has arisen frequently in systems that have not been designed specifically to avoid the problem. In this particular example, there are several possible methods to prevent deadlock. For example, we could force a suspended task to relinquish central memory by moving the task to back-up memory. However, instead of attempting to prevent each particular instance of the deadly embrace, we can formulate some global policy statements that can prevent the problem in general.

A careful analysis of deadlock shows that all of the following conditions must be satisfied:

1. A process must have exclusive control of some system resources.
2. A process continues to hold exclusive control of some resources while a request for more resources is pending. Moreover, the resources owned by the process cannot be removed from its ownership until the process specifically releases its control.
3. There exists a circular chain of ownership such that A_2 holds some resources required by A_1, A_3 holds resources required by A_2, and so on with A_1 holding resources required by A_n.

Note that each of these conditions is present in the example. The resources in question can be central memory, as in the example, or processors, specific input/output devices, data channels, or even crosspoints in the crossbar switches. A simple way to prevent deadlock is to construct a system in which at least one of the conditions is never satisfied. For example, any of the following rules is a valid way to prevent deadlock:

1. A suspended process cannot retain control over a resource. It may be forced to relinquish control and request a renewal of ownership at a later time.
2. A process must place a single request for all resources it needs. It holds no resources until its request is granted, at which point it obtains everything required.
3. The resources are ordered as R_1, R_2, ..., R_n, and processes must request R_i before R_j if i is less than j. No circular constraint exists under this rule.

The point in giving the rules is to show that deadlock can be prevented with a global strategy. The example above indicates that deadlock can occur from a sequence of requests, each of which is reasonable in itself. Consequently, a strategy for deadlock prevention almost certainly cannot be a function of individual requests, but must necessarily be based upon the global context of requests. Thus, the problem of preventing deadlocks is traditionally treated by an operating system, which has global information rather than by special hardware, which processes individual requests. The reader should consult the article by Coffman et al. [71] for more complete information.

At this point we see that the process of granting requests for system resources is a hazardous one, for poor strategies can cause a disastrous deadlock in the midst of computation. But the question of good strategies for resource allocation goes beyond deadlock, for a system must be efficient as well as deadlock free. Efficiency can vary dramatically under different allocation strategies, even when the strategies under comparison are deadlock free.

In general, MIMD computer systems have all of the resource allocation problems of conventional serial computers, but processor scheduling and memory allocation tend to be the dominant problems. In the area of processor scheduling, fast optimum scheduling algorithms are available only for a few highly restricted cases, and the most realistic cases can be scheduled optimally only by algorithms that are basically enumerative. Lawler and Moore [69], for example, give efficient algorithms for several multiprocessor problems. Recent results in the study of complexity of algorithms indicates that processor scheduling and memory allocation problems may be so complex inherently that there is no hope of solving them with fast algorithms. The problem of performing optimum memory allocation and processor scheduling jointly appears to be even more difficult than doing these processes individually. In most cases, actual implementations of operating systems are often priority driven, in the sense that resource allocation and processor scheduling tends to favor processes with high priority. Priority-driven algorithms tend to be very fast, and in many situations produce allocations that are reasonable if not optimal. The algorithms do not solve the resource allocation problem, of course, because they are sensitive to the priorities assigned by the user. Thus the resource allocation problem is placed back in the hands of the user to solve through his assignment of relative priorities. Until good solutions to the resource allocation problem are known, priority-driven allocations appear to be as suitable as any other ad hoc allocation method.

MIMD Computer System in Perspective

In closing this section, we should summarize the principal problem areas for MIMD computers, for these are the problems that must be solved before MIMD computers are likely to be effective computing systems. The major issue is that of partitioning a problem into many processes that can be executed in parallel on an MIMD computer. For a small number of processors, say two to four processors, this problem is not a significant one because the parallelism available is not significant. For several processors, say sixteen or thirty-two, the problem is extremely difficult. Programs without a specific iterative structure are seldom so complex that they have sixteen to thirty-two distinct subprocesses. Programs with an iterative structure are likely to be better suited to SIMD computers and execute with somewhat lower efficiency on MIMD computers because of resource allocation and synchronization overhead. At Carnegie-Mellon University, a speech recognition program runs in the MIMD environment. Here the independent processes involve numerical analysis of speech waveforms, correlation, modeling of the context of the speech, hypothesis generation, and selection of the most likely words or phrases. The speech analysis program appears to be well-suited to MIMD computation because each of the processes can be done in parallel, yet each process is a distinct process so SIMD computation is not possible.

Presently there exists virtually no practical techniques suitable for partitioning computations into parallel processes for MIMD computers. A number of researchers have found algorithms for establishing precedence constraints among portions of a computer program, particularly among standard FORTRAN programs [Kuck et al., 72]. However, these techniques do not consider the high overhead of process synchronization and processor reassignment, nor do they take into consideration the fact that algorithms for MIMD computers may be substantially different from algorithms for the same problem for SIMD computers.

The development of MIMD computers is several years behind the development of SIMD computers. For the next few years SIMD computers will have the greatest degree of parallelism, although MIMD computers with two or four processors may become more prevalent. MIMD computers with many processors can become a reality only if resource allocation, synchronization, and problem partitioning can be solved on a practical basis.

8-5. PROBLEMS

8-1. Consider the perfect shuffle of $N = 2^m$ items where the items are indexed from 0 to $N - 1$. Let the integer i have the binary expansion $i_{m-1} \times 2^{m-1} + i_{m-2} \times 2^{m-2} + \ldots + i_0 \times 2^0$. Show that item i is moved by the shuffle to the position formerly occupied by item $i_{m-2} \times 2^{m-1} + i_{m-3} \times 2^{m-2} + \ldots + i_0 \times 2^1 + i_{m-1} \times 2^0$.

8-2. Show that an 8×8 matrix can be transposed in three perfect shuffles when it is stored as vector of length 64. Prove that m shuffles transpose a matrix of size $2^m \times 2^m$.

8-3. Consider the reverse binary permutation, which is required for Fourier transforms. In this permutation item $i = i_{m-1} \times 2^{m-1} + i_{m-2} \times 2^{m-2} + \ldots + i_0 \times 2^0$ is moved to position $i_{m-1} \times 2^0 + i_{m-2} \times 2^1 + \ldots + i_0 \times 2^{m-1}$, that is, to the integer obtained by reversing the digits in its binary expansion. Show that a vector of length 2^{2r} can be permuted into reverse binary ordering by 2^r cyclic shifts with masking.

8-4. Find a recursive doubling algorithm to solve the recursion
$$x_i = (a_i x_{i-1} + b_i)/(c_i x_{i-1} + d_i)$$
when x_0 is given as a boundary condition. Assume $c_i x_{i-1} + d_i \neq 0$ for all i.

8-5. Find a recursive doubling algorithm to solve the recursion
$$x_i = a_i + 1/x_{i-1}$$
when given x_0 as a boundary condition. Assume $x_i \neq 0$ for all i.

8-6. Formulate algorithms in an ALGOL-like language for multiplying two $N \times N$ matrices on an SIMD computer with N processors when the matrices are stored in the following formats.
   a) Both skewed
   b) First operand skewed and the second stored straight by rows
   c) First operand skewed and second operand stored straight by columns

8-7. Rewrite one matrix multiplication algorithm from the previous problem in assembly language. [Invent new instructions as you need them.]

8-8. Analyze Figure 8-24, and establish the correctness of the logical flow. Formulate an algorithm for which two producers feed one consumer. Assume that the consumer need not consume items in the exact order they are produced, provided that subsets of items associated with each producer are consumed in order.

8-9. Prove that each of the three rules for avoiding deadlock is a correct solution.

CHAPTER 9

Overlap and Pipeline Processing

Tien Chi Chen

A computer system takes a job description, that is, a program, and handles the job by mobilizing its own resources to produce the desired result. For a given program, there are many different ways to get the job done. As long as the machine user gets the proper result at reasonable cost, he is not overtly concerned with the manner of internal processing.

Instead of doing one thing at a time, the machine can run faster by subdividing the work over a number of concurrently operating units. This is possible even if the work subdivisions show a precedence dependence. The designer's challenge is to allow this type of multiprocessing to enhance performance, yet give the same results as serial monoprocessing.

This chapter surveys the principles of overlap and pipelining as general techniques for handling precedence dependent tasks, operative at several levels of machine design. Pipelining has a time synchronism reminiscent of SIMD parallel processing; the latter is also discussed in the broader context of synchro-parallelism.

Since multiprocessing aims to increase performance, it is desirable and important to quantify the increase. A simple geometric theory of tight coupling is developed, revealing the strong dependence on job homogeneity. This dependence can be broken through associative control, as seen in an example with memory interleaving. The principles are illustrated by actual machines based on overlap and pipelining techniques.

9-1. DIVISION OF LABOR AND ASSEMBLY LINES

A significant aspect of our civilization is the division of labor. Major engineering achievements are based on subdividing the total work into individual tasks which can be handled despite their inter-dependencies.

Overlap and pipelining are essentially operation management techniques based on job subdivisions under a precedence constraint. An understanding of the problem and feasible solution can be reached through the industrial manufacturing process.

The Partition of Workload

If a job takes a single worker eight hours to complete, can it be done with eight workers in one hour? With 480 (8×60) workers in a minute? With 28.8 billion ($8 \times 60 \times 60 \times 10^6$) workers in one microsecond?

The last number is absurd not so much because it exceeds the current world population, but because human reaction times tend to be large multiples of one microsecond. The number of worker-hours may be a convenient way to talk about a job requirement, but is almost never a sufficient description. The qualification of the workers and the manner in which the work is distributed are critically important factors in getting the job done. Subtle requirements emerge upon detailed analysis, when we try to distribute the given job over many working units.

Partition Modes

A job may be partitioned symmetrically into equal components, each to be assigned to separate but identically constituted working units. On the other hand, the components making up the job may be quantitatively different and the partition then would have to be unsymmetric.

Example 9-1.   The construction of three identical houses can be assigned symmetrically to three separate teams of equal qualification. If the job is to construct a house, a road, and a swimming pool, three separate specialized teams may be involved.

Often certain aspects of a job can be partitioned, but the rest cannot; or the partition of some parts can be symmetric, but the rest can only be unsymmetric. Also, a job (or aspects thereof) may be partitionable in several mutually exclusive ways; the fact that it can be done one way does not imply that it must be so partitioned. The actual choice depends on the resources available, as well as the size of investment, projected pay-off, and the delay involved.

Example 9-2.   The construction of three identical houses can be entrusted unsymmetrically to four teams: a foundation working team, a frame and wall team, a roofing team, and an interior trim team. The foundation team must start working ahead of the others.

Synchro-Parallelism

For a symmetric partition, the assigned teams can be synchronized in time, making the control of the progress easier; this is a general form of SIMD parallel processing (see Chapter 8), but there may not be an explicit "instruction stream." We shall use the broader term "synchro-parallelism" to denote the phenomenon of identical teams working in unison. SIMD parallelism is then a special case.

Example 9-3. A collection of eight checking units are said to run in synchro-parallelism if each can simultaneously detect the parity of a distinct 8-bit byte in a 64-bit information word.

Example 9-4. A collection of 100 complete computers are said to run in synchro-parallelism if 2 of them are disabled and the remaining 98 happen to be *decoding and executing* floating-point divisions in the same way with different data.

Example 9-5. A collection of memory units are said to be linked in a synchro-parallel manner if all are activated together to deliver words based on supplied addresses.

Precedence Requirements

Ideally, in a synchro-parallel computer the work is distributed via a multiple FORK operation to the teams, which then proceed completely independently until finished. Then, the results are brought together with a JOIN, as described in Chapter 8. Actually, such clean-cut fan-merge situations are rare, and interdependencies do occur. In fact, many tasks have a precedence constraint such that some parts may not begin until some others are finished. For instance, a worker on an automobile assembly line cannot put wheels on until the chassis has been installed to receive them.

Precedence dependence is more characteristic for unsymmetric than symmetric work subdivisions, tending to become more explicitly revealed as the subdivisions become finer.

If the job involves many similar tasks, concurrent handling is actually possible despite the precedence requirements. We need only to learn from a production assembly line; an abbreviated account is given here, the kth task being denoted by T^k.

Assume the typical task involves M processes, then

$$T^k = \{T^k_0, T^k_1, \ldots, T^k_{M-1}\}.$$

Note the superscripts refer to the task, the subscripts refer to process. Some, or all of these processes may obey precedence rules. The completion of one task may involve many processes, possibly in precedence sequence so that T^k_j cannot start until some earlier task T^k_i finishes, $i < j$; also each process is applicable to many tasks.

Example 9-6.   In an automobile factory the kth task is the manufacture of the kth car. The processes might be:

0 — frame and power chain installation
1 — bolting the body on the frame
2 — mounting the engine
3 — putting on the seats and wheels

with explicit precedence requirements.

Let S_j be the processing stage designed to handle the process for any task. The handling of T^k starts by S^o handling T_0^o. When this is finished, the task T^0 enters S_1, where the subtask T_1^o is performed; the vacancy created in S_0 can be filled by the task T^1, and so on. After a length of time, all M stages can be fully engaged.

This is not guaranteed, however. When S_j has completed T_j^k, it is ready to accept T^{k+1}, but the precedence requirements may forbid this until S_{j-1} has done T_{j-1}^{k+1} first. Meanwhile, S_j may have to remain idle.

Example 9-7.   Stage S_2, responsible for the engine mounting process, may have finished T_2^{997}, but it cannot process T_2^{998} until S_1 has completed the body bolting for task T^{998}.

The preceding is the basis of internal overlap and pipelining; more of this is discussed throughout the present chapter.

Assembly Lines

If the processing stages are designed so that each completes a process in a fixed quantum of time τ (a cycle), then with sufficient tasks, we can easily have

$$S_j \quad \text{handling } T_j^m \text{ at time } t$$
$$\text{handling } T_j^{m+1} \text{ at time } t + \tau$$

for all j. There are no delays to cause idleness in any stages, and precedence is honored; this is the principle of pipelining, which is discussed in more detail in secton 9-3.

Example 9-8.   In the automobile assembly at some time t,

Stage 0   may be building the chassis of car 1000
Stage 1   may be bolting on the body of car 999
Stage 2   may be mounting the engine of car 998
Stage 3   may be installing the seats and wheels for car 997;

See part *a* of Figure 9-1.

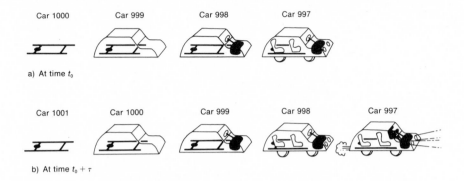

a) At time t_0

b) At time $t_0 + \tau$

Figure 9-1   Automobile frame assembly

At time $t + \tau$,

Stage 0   may be handling car 1001
Stage 1   may be handling car 1000
Stage 2   may be handling car 999
Stage 3   may be handling car 998

The car 999, previously at stage 1, is now being serviced at stage 2, in strict precedence order. See part *b* of Figure 9-1. The precedence is not only honored, but is actually exploited.

9-2. OVERLAP DESIGNS

Overlap is the phenomenon of concurrent processing, often towards some well-defined common goal. In a computer system there can be overlap between the processor and the input/output unit; within the processor, there can be overlap between instruction processing (I), and execution (E). More detailed overlaps can be designed into the machine; a simple example is the simultaneous running of many input/output devices. The I and E processes each can also be further subdivided into smaller pieces, each entrusted to a different sub-unit.

Input/Output Overlap

"Asynchronous input/output," namely the overlapping of input/output operations with processor operations, is natural and relatively straightforward in principle. There is no detailed precedence issue; the division of labor is that of "fork, then join." (See Figure 9-2.) The processor issues a command to start an input/output process; subsequently both the processor and the input/output unit proceed concurrently until the latter finishes. The processor is then alerted by an input/output completion signal. See Chapter 6, for example, for more details.

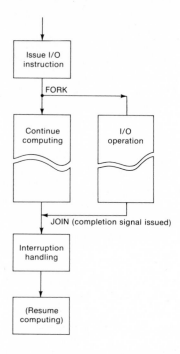

Figure 9-2 Asynchronous input/output

In practice input/output activity requires the sharing of machine resources, notably memory and data path. In the original von Neumann design, the path between memory and input/output units runs through the multiply-quotient register; as a result during input/output processing no multiply, divide, or long shifts can occur. The processor is so crippled that one should consider the machine to have no input/output overlap.

True asynchronous input/output is already found in the UNIVAC I machine, first delivered in 1951. Since then, as machines became more and more complex, the number and variety of concurrently operating input/output devices grew correspondingly. In the 1960s the cost of input/output equipment began to dominate the total hardware system cost, and input/output overlap and interruption handling became a major concern. However, even now in many applications the actual overlap between input/output and processor is small. Often the processor spends a significant portion of time waiting for the completion of some special input/output task, before it can proceed further.

Instruction Preparation/Execution Overlap

Overlap designs can deal with precedence-dependent tasks in a computer processor; an example is the concurrent handling of instruction preparation

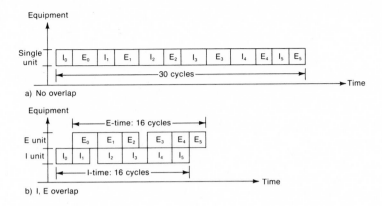

a) No overlap

b) I, E overlap

Figure 9-3 Overlap of instruction preparation (I_j) with execution (E_{j-1})

(instruction fetch, decode, effective address generation, and operand fetch if any) with execution, the latter using the results of instruction preparation (see Figure 9-3).

We shall first discuss the precedence requirements in handling a program. For the jth instruction, let I_j be the instruction preparation and E_j the subsequent execution, and $T(I_j)$, $T(E_j)$ the corresponding handling times. For all meaningful j, the following precedence rule is necessary:

1. I_j precedes E_j.

Further, it is easy to guarantee correctness by demanding

2. I_j precedes I_{j+1}.

Conventional wisdom suggests a third relationship,

3. E_j precedes I_{j+1}, leading to a simple, straightforward design.

There is hardware economy; the same equipment can be shared between I and E.

The overall processing time for n instructions for this conventional processor is:

$$T = \sum_{j=1}^{n} [T(I_j) + T(E_j)] \tag{9-1}$$

In the I/E overlap design, item 3 is altered to read

3'. E_j precedes I_{j+2}

and concurrency of E_j and I_{j+1} is allowed. The overall processing time now becomes

$$T' = \sum_{j=0}^{n} \max \left[T(E_j), T(I_{j+1}) \right] \tag{9-2}$$

with the convention $T(E_0) = T(I_{n+1}) = 0$.

If the I/E times are nearly equal, the overlapped processor tends to be twice as fast as a conventional design. If, however, either of the two times dominates, the gain through overlap may not merit the effort.

Equipment sharing between I and E is difficult to achieve in the overlap design. For example, in the IBM 7094 the index adder is distinct from the arithmetic adder. Also, interlocks are needed to enforce the precedence rules. The simplest mechanisms are signals from one unit to the other units, indicating the completion of a task and the validation of registers at the interfaces. An alternative technique is to put validation flags on the interface registers; the contents of the latter are used only when the flags are on. These flags must be marked invalid between periods of validity.

Multiple processor overlap designs potentially can raise the performance of the machine several fold, but the interlocks can be complex. Pipelining (section 9-3) is an extreme form of multiple overlap, in which the completion signals are replaced by synchronizing time clock pulses.

Logical State of the Machine Processor

From time to time, it is necessary to summarize the status of program execution in a processor. For example, when a program interruption occurs, the execution of the current program is halted, possibly to be resumed later; meanwhile, another program with higher priority (here the monitor program) is invoked. The status of the interrupted program is saved upon interruption, to be reestablished upon resumption of execution. The halting often has to occur within a very short time interval after the interruption signal has been given.

This poses a problem for the designers of overlapped machines, which usually have two or more instructions in various stages of execution. The actual physical state of execution can be extremely tedious to describe; and worse, the exact re-creation of the state may be impossible.

The usual solution is to create on demand logical states which correspond to a snapshot of the physical state of an unoverlapped machine. It usually suffices to provide only logical states referring to the moment between instruction executions, where description is most concise, and subsequent resumption is easiest. When the machine is poised to start the instruction at location L, the instruction counter contents (L) and the contents of a few data registers then can define the logical state of the processor.

Although the overlapped machine can, and usually does, treat several instructions at one time, the logical state is nevertheless a valid (though uncommon) physical state. The automatic transition from the usual overlapped state to the logical state may be needed at any time (say upon an input/output completion signal), and must be carefully designed to avoid errors.

Upon receiving a signal to produce a logical state, the machine selects a point in time in the execution and creates the effect of an unoverlapped machine halting at this point. Every processor action leading up to this point must be completed, and no action after this point (advance processing) should appear to have been started. Very often, the point in time is so chosen that some units have to continue processing forward to reach it; but some other units need to invalidate the work in progress, and revert to an earlier physical state. The creation of the logical state thus amounts to a temporary disabling of the overlap in the processor.

Example 9-8. In a machine with I/E overlap, when a halt signal is received, the I-unit may have finished decoding instruction 1000, and the E-unit may be executing the arithmetic for instruction 999. More often than not, the I-unit is altered to appear as if the decoding for 1000 has not been performed, and the instruction processing is said to halt exactly after instruction 999 and before the start of 1000.

Overlap Machines

Realistic constraints often cause deviations from the simple prescriptions of I/E overlap. An example is the IBM 7094 machine (Figure 9-4). It is of

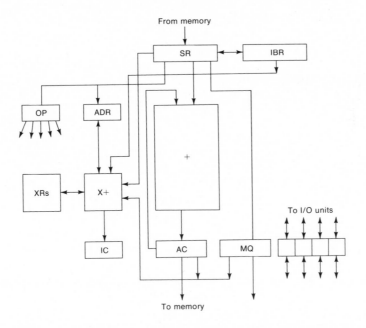

Figure 9-4   IBM 7094 computer schematic

interest particularly because it is a late model in a long series of systematic machine developments, starting with the vacuum tube, fixed-point, unover-lapped 701, first shipped in 1953 (see Table 9-1) [Rosen, 69].

The 7094 is a word-oriented machine using 36-bit words. Each instruction occupies one word. The machine uses a 72-bit wide memory (twice the 36-bit word length); each memory operation actually involves a pair of consecutively addressed words starting with the even-addressed word. In a memory fetch, the word accompanying the target word often can be exploited in the immediate future. In terms of "locality" (see Chapter 11), both the target word and the companion word tend to belong to the same working set.

In the 7094, the instruction fetch is a special I-operation, done independently of the other I-operations. As instructions are usually handled in strict instruction-address sequence, unless the even-addressed instruction currently being decoded turns out to be a (successful) branch, the odd-addressed instruction word, which follows it, is usually immediately needed. In the 7094, the latter is already present, and need not cost an extra memory fetch. When a branch occurs, the target should preferably be an even-addressed location; an odd-addressed target tends to be less rewarding.

A data fetch from memory also accesses two consecutively addressed words. Storing is done by first fetching the even-odd addressed pair, replacing the intended word, then storing back the word-pair.

Double-precision arithmetic, dealing with a word-pair at a time, makes use of the full memory width, and is also featured in this design.

Figure 9-5 shows a typical processing pattern of the 7094, executing a sequence of fixed-point add instructions. Here each instruction requires a duration of at least 2 cycles, but the average cost for one instruction is 1.5

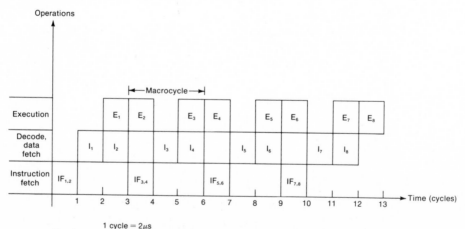

1 cycle = 2μs
Macrocycle = 3 cycles for 2 instructions

Figure 9-5 I/E overlap in the 7094 for repeated adds

TABLE 9-1. 701-704-709-7090-7094 II Machines

Machine	Circuit Technology	Memory Cycle	Processor Cycle	Number of Index Registers	Floating-Point Arithmetic	Double-Precision Floating Point	Input/ Output Overlap	I/E Overlap	Approx. First Deliver
701	Tubes	Williams/cores	$12\mu s$	0	No	No	No	No	1953
704	Tubes	Cores ($12\mu s$)	$12\mu s$	3	Yes	No	No	No	1955
709	Tubes	Cores ($12\mu s$)	$12\mu s$	3	Yes	No	Yes	No	1958
7090	Transistors	Cores ($2.2\mu s$)	$2.2\mu s$	3	Yes	No	Yes	No	1960
7094	Transistors	Cores ($2.0\mu s$)	$2.0\mu s$	7	Yes	Yes	Yes	Yes	1962
7094 II	Transistors	Cores ($1.4\mu s$)	$1.4\mu s$	7	Yes	Yes	Yes	Yes	1964

cycles. A cycle, incidentally, is 2.0 microseconds, equalling a memory cycle time. As the I-time includes a data fetch (DF), the memory unit is busy at every cycle, though gaps appear for I-procesing and E-processing.

Thus, the bottleneck in Figure 9-5 is actually memory access; to better the performance it would be necessary to have two banks of memories, capable of being accessed independently. The 7094 II separates memory words with even-addressed and odd-addressed banks (interleave by 2) and can achieve a rate of one instruction per cycle for the same problem. The cycle is still matched with memory cycle time, both improved to 1.4 microseconds. The two-fold interleaving is adopted also by the IBM System 360 Model 75; a memory word there is actually a 72-bit double word [Brown et al., 64]. See Figure 9-6.

Perhaps the most systematic overlap machine is the UNIVAC LARC, which uses interleaved memory and a four-fold overlap (instruction fetch, indexing, data fetch ,and execution), and can run one floating add per cycle of four microseconds. A conditional branch, however, takes three cycles. At such a level of systematicity the machine is hard to distinguish from a pipelined system, but for the heavy emphasis on crosslinking and exception handling, and completion-signaling mechanisms.

9-3. PIPELINING AS A DESIGN PRINCIPLE

Pipelining is a processing technique aiming for a steady throughput. The processing power is decentralized, distributed more or less uniformly over the processing path. Consider a number of micro-processors, each capable of doing its own work in a fixed cycle time. By stringing many of these together, one can observe precedence dependence and still achieve a total work

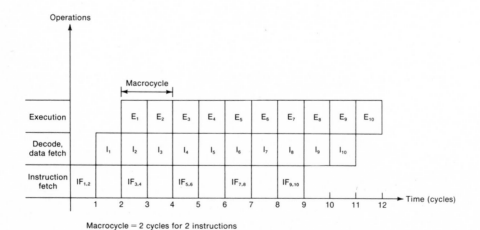

Figure 9-6   I/E overlap in the Model 75 for repeated adds

rate equal to the sum of the work rate of each of the micro-processors. This is the principle of pipelining.

The pipeline is closely related to the industrial assembly line in section 9-1. Like the assembly line, precedence is automatically honored; also, it takes time to fill the pipeline before full efficiency per cycle is reached, and also time to drain a pipeline totally.

Execution of Normalized Floating-Point Addition

Let us consider the execution of a normalized floating-point addition between $A = a \times 2^p$ and $B = b \times 2^q$ with a, $b < 1$. The operations are required to follow the equation

$$a \times 2^p + b \times 2^q = (a \times 2^{p-r} + b \times 2^{q-r}) \times 2^r \qquad (9\text{-}3)$$
$$= c \times 2^r = d \times 2^s$$

where $r = \max(p, q)$, and $1 > d \geq 0.5$.

Because binary arithmetic is used, multiplication and division by powers of 2 are done conveniently by shifting.

The detailed steps are the following (see Figure 9-7):

a. Exponent subtraction: Put (a, p, b, q) in registers (A, P, B, Q) respectively. Compute $x = p - q$.
b. Fraction interchange: If $x < 0$, the contents of A and B are interchanged, and Q replaces P. No action needs to be taken otherwise.
c. Fraction preshift: $B \times 2^{-x}$ replaces B; this is done by a binary shift. B must be a double-length register to house the result.
d. Fraction addition: $A + B$ replaces B.
e. Post-normalization: One of four possible actions is taken:

   1) If $|B| \geq 1$, $B \times 2^{-1}$ replaces B, and $P + 1$ replaces P.
   2) If $1 > |B| \geq 0.5$, no special action need be performed.
   3) If $0.5 > B > 0$, find $z =$ the number of leading zero bits after the binary point, and then $B \times 2^z$ replaces B, also $P - z$ replaces P.
   4) If B is zero, P and possibly B are replaced by special constants to signify a "true zero." This is the commonest scheme to treat order of magnitude zeros; a full discussion is outside the scope of the present text.

f. Exponent exception treatment: Under normal situations, B after proper truncation is identified with the final fraction d, and P is identified with the final exponent s. However, the post-normalization may lead to exponent overflow or underflow, and special treatment is then needed. Usually exponent underflow causes the replacement of the entire result by true zeros, and exponent overflow leads to a program interruption.

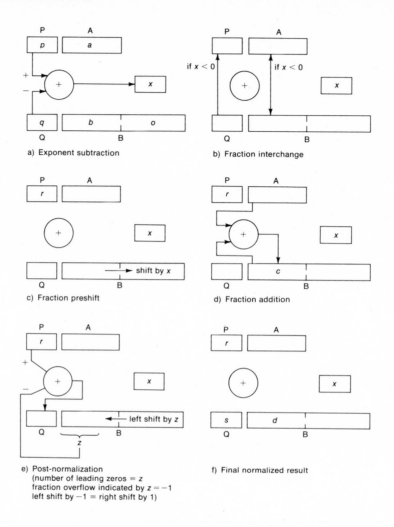

Figure 9-7   Steps in normalized floating-point addition in a conventional (simplex) machine (exception handling not shown)

Designs for the Floating-Point Adder

In the conventional simplex design, only one execution at a time is allowed in the floating add unit. The exponent subtraction and the fraction addition can share the same adder; likewise all shifting can be assigned to the same hardware. This was shown in Figure 9-7. On the other hand, each of the steps outlined in the previous section can be implemented by a separate unit. Then all six units can be overlapped in execution; at any time each unit is handling a different execution.

In doing this, we notice that it is necessary not only to have a separate exponent adder distinct from the fraction adder, a separate pre-shifter distinct from the post-shifter, but each stage has to have its own set of registers to house the operands, and mechanisms are required to move operands from stage to stage. The inflation of register requirements was a deterrent against overlap and pipeline designs, until recently when the cost of registers is no longer prohibitive.

In the overlap design, each stage processes at its own pace and total execution time can be variable. This flexibility is paid for by extra inter-stage signalling protocol and leads to lower usage factors overall.

The ideal nonsimplex design for floating addition is a pipeline design. In a pipeline design the stages are selected to consume equal processing time, and inter-stage signalling is done by the timing pulse. The equal duration per stage implies rigid overall handling, and we can no longer allow operand-dependent processing times.

Figure 9-8 gives a sketch for a pipelined floating point adder. The exponent

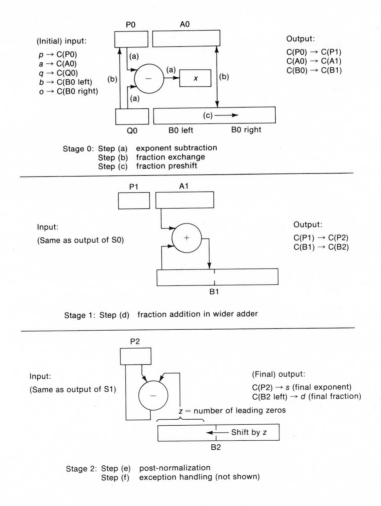

Figure 9-8 Pipelined floating-point adder

subtraction, fraction interchange, and preshift are done in the same stage (Stage 0), fraction add is done in Stage 1, and post-normalization and automatic corrections of formats in Stage 2. While each of the stages may not exactly take equal time originally, the longest duration defines the pipeline cycle. The "early finisher" stages must be designed to wait for the time pulse without losing information.

A mechanism to facilitate this waiting process is called a *latch*. It holds an operand indefinitely, releasing its contents only when triggered, in the pipeline case, by a time signal.

Steady-State Behavior of a Linear Pipeline

We now define a linear pipeline as a time-synchronized assembly line. Let us consider a collection of M special purpose microprocessing stages (S_0, S_1, ... S_{M-1}), the jth member can accept a_j bits of input, and do local work w_j within the time interval τ_j (see Figure 9-9). At the end of the interval it produces b_j bits of output, and is ready to accept new inputs. Then one can string these M units together one after the other for the purpose of doing work Σw_j. It is noted that the term "work" is used loosely; it is not necessarily the same term used in physics.

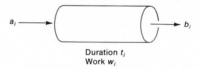

Duration t_j
Work w_j

Figure 9-9   Microprocessing stage (S_j)

With the data width matching conditions,

$$b_j = a_{j+1} \tag{9-4}$$

and the time matching conditions

$$\tau_j = \tau = \text{constant for all } j \tag{9-5}$$

a linear pipeline (Figure 9-10) results, with

Time duration $= T = M\tau$
Cycle time $= \tau$
Number of stages $= M$
Global input $= a_0$ bits per cycle
Global output $= b_{M-1}$ bits per cycle
Work done $= \Sigma w_j$ per cycle (9-6)

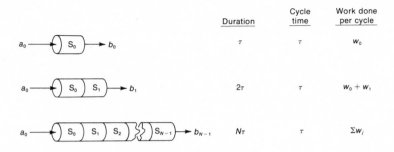

	Duration	Cycle time	Work done per cycle
	τ	τ	w_0
	2τ	τ	$w_0 + w_1$
	$N\tau$	τ	Σw_i

Figure 9-10 Linear pipelines

Although the preceding pipeline uses building blocks with equal traversal times, this is not strictly necessary. For example, two pipe segments of duration T and T′ can be joined into a pipeline of global duration (T + T′) if both are pulsed at the same rate, and if the data widths do match at the juncture.

Starting this pipeline at time t_0, and supplying a_0 bits at every cycle, a steady state is reached at time $(t_0 + M\tau)$; subsequently a new set of b_{N-1} bits is obtained at every cycle. See Figure 9-11.

After the pipeline is filled, and before it is drained, every stage is busy with the jth state doing work w_j. The steady-state work rate is therefore Σw_j per cycle.

In terms of the notation used in section 9-1 the global input at time $(t_0 + k\tau)$ characterizes the task T^k, which moves through the pipeline at a steady rate. During the time interval $[t_0 + (k + j) \tau, t_0 + (k + j + 1) \tau]$, this

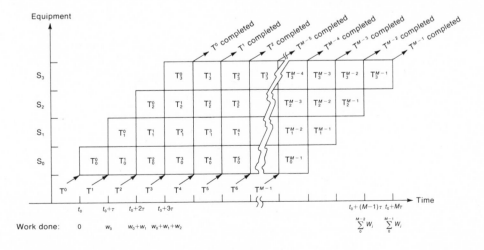

Figure 9-11 Tasks {T^k} through a four-segment pipeline

task is being handled by pipeline stage S_j for the subtask T_j^k; upon its completion the amount of work w_j has been performed. When the task emerges at time $[t_0 + (k + M)\tau]$, the total subtasks done on T^k is the union of all subtasks $\{T_j^k\}\, j = 0, 1, \ldots M - 1$, and the work done due to T^k is Σw_j, equal to the pipeline work done per cycle.

While the duration of processing increases with the number (M) of pipeline stages, the throughput, in terms of work done per cycle, also increases correspondingly.

In terms of the rate of task handling, the pipeline at the steady-state finishes one task per cycle, independent of the number of stages required, and hence is insensitive to distance effects.

Example 9-9.   In the automobile factory assembly line discussed earlier, if every one of the four stages takes exactly 10 minutes, task number 997 entering the assembly line at time $t_0 + 9970$ minutes emerges as the 997th completed automobile at time $t_0 + 10010$ minutes. Within the 10 minute cycle, the total work done is measured by one completed car, which is also equal to the work done on the 997th task. The fact that there are four stages has an influence on the total time duration (40 minutes), but not the throughput (1/10 car per minute).

Limitations to Performance

Signals cannot travel faster than c, the speed of light in a vacuum, which is about 3×10^{10} centimeters per second. A task to be processed in an apparatus of length D, cannot be completed in less than D/c seconds. Let the work entailed in the task be W. If only one task is permitted in the apparatus at any time, then the work accomplished per second (computing power) cannot exceed Wc/D, limited by the finite speed of light.

However, in pipelining, the apparatus is subdivided into M subunits, each of which is permitted to process an amount of work, and if a_0 bits are input at the end of every cycle, b_{M-1} bits emerge at the end of every cycle. The work done is W per cycle, and the cycle is limited by D/Mc, which is M times smaller than the single-task limit. As long as one is free to choose M, the speed of light poses no obstacle to performance. The more finely the task is subdivided, the higher is the overall performance of the pipeline.

One concedes, however, that there is a limit also to the subdividing process. The time-energy uncertainty principle states that

$$\Delta E \times \Delta t \gtrsim \hbar \quad (\hbar = 1.05 \times 10^{-27} \text{ erg-seconds})$$

or, the product of the uncertainty in energy multiplied by the uncertainties in time is roughly bounded by $\hbar$, the Planck-Dirac constant. Then, if the time subdivision τ is arbitrarily small, then the uncertainty in time must be even less than τ; one may have to provide an arbitrarily large energy source to ensure the proper handling, but this is impossible.

Another limiting factor is the atomicity of the physical world. To do a meaningful subtask, one probably needs the physical space (D/N) of at least one hydrogen atom diameter ($2a_H$). Then the speed of light leads to a time interval limitation of $2a_H/c = 0.3 \times 10^{-18}$ seconds.

Before these restrictions become meaningful, however, there already are other limitations due to technology and engineering considerations.

Engineering Considerations

The transition from a conventional device, which is devoted to doing one thing at a time for a duration of many cycles, into a pipeline producing an output at every cycle, usually requires the systematic insertion of latching circuits to preclude mutual interference. These circuits hold the output from a given pipeline stage, freeing the latter to accommodate a new set of inputs, while furnishing input to the next stage on line. These latches mark the stage boundaries of a pipeline. It may sometimes be impossible or inconvenient to insert latches, and the duration of processing may also increase due to latching. Hallin and Flynn [72] have shown, however, that for arithmetic functions using combinational circuits, a latch devised by J. Earle entails no additional delays if the cycle time is equal to 4 or more logical delays.

In general, when pipelining is possible, the cost for the latching is low, and the potential gain high. The question is whether the steady-state occurs frequently enough to merit the special design. This is discussed in the later sections. For additional information on the engineering aspects, see Cotten [65] and Loomis [66].

9-4. PIPELINE NETWORKS

The number of interesting applications using solely linear pipelines is small, and we now consider more general organizations which still follow the pulsed delivery principle. In so doing, a number of important arithmetic designs can be put in perspective, and more powerful interconnected systems can be studied.

Lateral Input/Output

Linear pipelines can be generalized to allow lateral flow. In this design the jth pipeline stage has lateral input c_j, and lateral output d_j (see Figure 9-12). An example is a high-performance pipelined digit-adder mechanism [Hallin and Flynn, 72], where a pipeline stage is a digit-adder stage (see Figure 9-13). The pipeline inputs are the carries from the previous stage, the outputs are carries for the next stage. The lateral inputs are the operand digits, and the lateral outputs are the sum digits. At time $(t_0 + k\tau)$, stage j produces sum digit d_j for the $(k-j-1)$st addition. In the steady state, a complete sum is obtained at every cycle, yet the sum digits appear one at a time for any particular sum; the pipeline produces one digit from each addition.

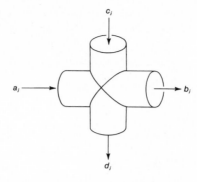

Figure 9-12   Pipeline stage with lateral input/output

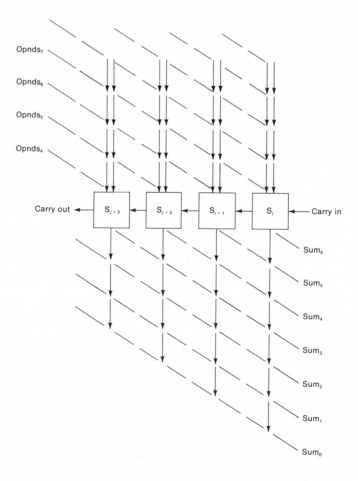

Figure 9-13   Pipelined adder based on digit adder stages

Crosslinking

A two-dimensional pipelined network can result from a system of N pipelines with lateral input and outputs. If the jth stage of the kth pipe is S_{jk}, accomplishing work w_{jk}, with lengthwise input a_{jk}, lengthwise output b_{jk}, lateral input c_{jk}, and lateral output d_{jk}, then by cross-linking we can make d_{jk} also the input $c_{j,k+1}$ of the $(k + 1)$th pipeline. This way pipeline flow is achieved in the c,d direction as well as the a,b direction, with total work done per cycle equal to $\Sigma\Sigma w_{jk}$.

A pipelined multiplier, generating one product per cycle, can be cast into a two-dimensional network (see Figure 9-14). This multiplier handles $\{(abcd)_k \times (efgh)_k\}$ at time $t_0 + (k + 5)\tau$. Each diagonal line represents a different multiply; the kth multiply is being done by inputting $(ag)_k$, $(cf)_k$. Each vertical path is a pipeline as well as each horizontal path. The total duration is 9τ. The pipelined version of the Wallace tree multiplier, which shortens the total duration as well, is best viewed in three dimensions.

Complex Networks

Two directions for the development of pipeline networks are possible. The organization can become more and more systematic, using simpler and simpler stages, so that the layout can be identified with the branch of abstract design

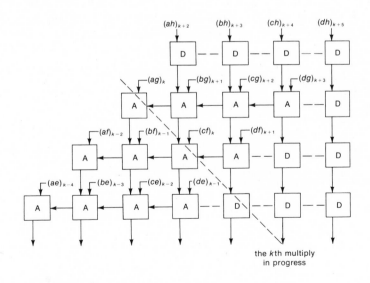

D–Delay unit
A–Full adder with simultaneous production
   of carry (horizontal) and sum (vertical) bits

Figure 9-14   Four-bit pipelined multiplier

known as cellular logic. The other direction, which has proved to be more realistic, calls for lower systematicity, and more ad hoc linkages. This point is illustrated in sections 9-8 and 9-9.

Example 9-10.   The automobile frame assembly discussed earlier, requires the supply of bodies, which can come from another assembly line. The combination is then a Y-shaped flow. When other processes are taken fully into account, the total flow resembles a river system with tributaries and subtributaries. The geometry is that of a tree, rather than a rectangular network.

9-5. MULTIPROCESSING BY TIGHT-COUPLING

Both pipelining and synchro-parallelism typify multiprocessing by tight-coupling, in that all units perform in unison. Work completion in one unit implies work completion in all others. The next three subsections are devoted to a discussion of tight-coupling schemes, their common advantages and disadvantages, their mutual dependence, and techniques for throughput enhancement.

Synchro-Parallelism and Pipelining

Synchro-parallelism is the general phenomenon where enabled units do the same detailed work at any given time t with N different processing resources. The underlying principle is symmetric job partition into identical tasks. Pipelining, on the other hand, exploits unsymmetric job partition with fixed precedence; the tasks flowing through are given identical treatment, with time lag $(k\tau)$ for the kth task.

Mutual Enhancement

Far from being opposing extremes, synchro-parallelism and pipelining actually complement each other. The regular two-dimensional pipeline network (see Figure 9-14), for example, is based on the linking of a number of synchro-parallel pipelines, pulsed in synchronism.

Synchro-parallelism can even augment linear pipelines when some processes, by their physical nature, cannot be subdivided in time as finely as desired. A set of K units, each taking K cycles to complete its work, can be connected in parallel to deliver one result per cycle. One needs only to install fast switches before and after, to select the parallel resources in cyclic order to synchronize with the pipeline pulses. (The matching of fast circuitry with interleaved, slower core memories, creates an interesting problem to be addressed in section 9-7, "Example: Memory Interleaving.")

Also, a good way to increase the utilization of equipment in a synchro-parallel machine is to entrust the collective arithmetic requirements of the latter to a sophisticated pipeline design. Such a scheme was advocated by Senzig and Smith [65], see section 9-9, "The VAMP Study."

The Job in Equipment-Time Space

The proper handling of a computer job requires the use of equipment over intervals of time. We can thus represent a job by enclosed areas in a space-time diagram; the space here refers to various equipment arranged in some prescribed way. With skillful arrangement, the profiles of a single job can be shown as one continuous area; this shall be assumed to simplify discussions. The processing of a job is represented by covering the profile with actual equipment usage in time.

The job may have inherent repetitions; within a time interval many processes may be identical in form, though different in operands used, and thus could take advantage of synchro-parallelism. The number of such identical processes is the inherent *width* of the job within the time interval. As an example, in the addition of two 32-element vectors to produce a third vector, the inherent width is 32.

The Model Job and a Measure of Repetition

A model job profile, amenable to simple analysis yet retaining salient properties of actual jobs, is one involving only two contiguous rectangular areas (panels). These areas have the inherent width unity and W, with corresponding time durations τ_1 and τ_2 (see Figure 9-15).

The job profile is said to be perfectly repetitious if $\tau_1 = 0$, when it becomes a rectangular block of width W. We define a repetition ratio for multiprocessing by

$$\rho \equiv (\text{area of the tall rectangle})/(\text{area of job profile}) \qquad (9\text{-}7)$$

$$= W\tau_2/(\tau_1 + W\tau_2)$$

then
$$\rho = 1 \text{ if } \tau_1 = 0 \quad \text{(perfectly repetitious profile)} \qquad (9\text{-}8)$$

$$= 0 \text{ if } \tau_2 = 0 \quad \text{(non-repeating profile)}$$

We can write $\rho = 1/(r + 1)$ \qquad (9-9)

where $\qquad r \equiv \tau_1/W\tau_2.$ \qquad (9-10)

Synchro-Parallel Processing

To process the job represented in Figure 9-15, a synchro-parallel processor of multiplicity N behaves like a steadily moving plow of width N, sweeping over the job profile to achieve complete coverage. The part of the profile with a narrow width is covered completely (though wastefully) in one sweep,

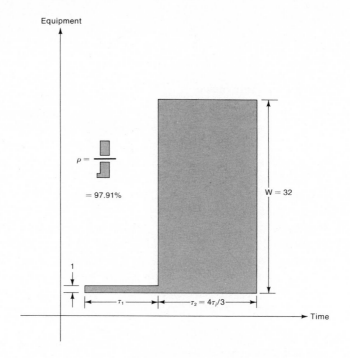

Figure 9-15 Space time diagram of a model job profile

and the extra processors have to be disabled. Those parts wider than N need to be swept more than once, each time at a different height.

In general, the number of sweeps is n, where

$$n = \lceil W/N \rceil \text{ (the ceiling of W/N)} \tag{9-11}$$

namely $n = W/N$ if N divides W, else it is equal to $[1 + $ (the integer part of W/N)]. We note that

$$x + 1 > \lceil x \rceil \geq x \tag{9-12}$$

hence

$$W/n \leq N \tag{9-13}$$

Figure 9-16 shows the case with $W = 32$, $N = 20$, hence $n = 2$.

The most efficient processing occurs when the job width is a multiple of N, when all processors are gainfully employed; otherwise inefficiency results from under-employment, and there is an additional cost to disable the part of processors not usefully engaged. Resweeping also requires bookkeeping actions

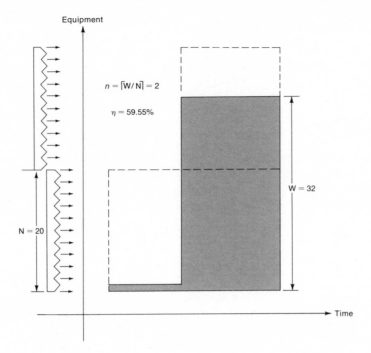

Figure 9-16 Sweeping by parallel processor to cover the job profile

(such as branching back to the top of a loop in SIMD processing). Both the bookkeeping cost and the cost to disable a subset are disregarded here.

9-6. PERFORMANCE OF TIGHTLY COUPLED SYSTEMS

The performance of tightly coupled systems, that is, the product between the number of processors and the overall efficiency, is straightforward to analyze in terms of a performance function $\phi\,(\rho, x)$. A surprise outcome is the very strong dependence on the inhomogeneity of the job, as measured by the repetition ratio. A way to improve performance is described in the next subsection.

General Performance Function

The performance function

$$\phi\,(\rho, x) \equiv 1/[(1 - \rho) + \rho/x] \tag{9-14}$$

has wide applications in subsequent pages, and is briefly studied here.

a. Transformations:

$$\phi\,(\rho,\,x) = \phi\,(1 - 1/x,\, 1/(1 - \rho)) = x\phi\,(1 - \rho,\, 1/x) \qquad (9\text{-}15)$$

$$a/\,(b + c) = \phi\,(1 - b/a,\, (a - b)/c) \qquad (9\text{-}16)$$

$$\phi\,(\rho,\,x) - \phi\,(\rho,\,y) = \phi\,(\rho,\,x)\,\phi\,(\rho,\,\text{y})\,(1/y - 1/x)\rho \qquad (9\text{-}17)$$

b. Inequalities:

   For the important range $0 \le \rho \le 1$, $1 \le x \le \infty$, $\phi\,(\rho,\,x)$ is an increasing function of ρ, also an increasing function of x (see Figure 9-17). Hence:

$$1 = \phi\,(0,\,x) \le \phi\,(\rho,\,x) \le \phi\,(1,\,x) = x \qquad (9\text{-}18)$$

$$1 = \phi\,(\rho,\,1) \le \phi\,(\rho,\,x) \le \phi\,(\rho,\,\infty) = 1/(1 - \rho) \qquad (9\text{-}19)$$

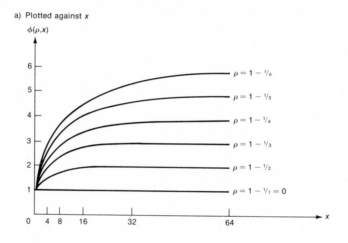

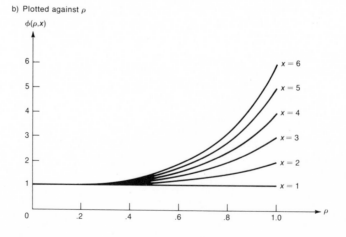

Figure 9-17   Performance function $\phi(\rho,x)$

Efficiency

The total efficiency of the synchro-parallel system is η

$$\eta = \frac{\text{total area of job}}{\text{total area swept by multiprocessor}}$$
$$= (\tau_1 + W\tau_2)/N \, (\tau_1 + n\tau_2)$$
$$= (r + 1)/(Nr + Nn/W)$$
$$= \phi \, (1/ \, (r + 1), W/n)/N \qquad \text{(using Equation 9-16)} \qquad (9\text{-}20)$$

We have immediately, using Equation 9-9

$$\eta = \phi \, (\rho, W/n)/N \qquad (9\text{-}21)$$
$$= 1/N[(1 - \rho) + \rho n/W]$$

Hence
$$\eta = 1/N, \qquad \rho = 0 \qquad (9\text{-}22)$$
$$= W/nN, \qquad \rho = 1 \qquad (9\text{-}23)$$

and
$$\eta \le \phi \, (\rho, N)/N \equiv \eta_N. \qquad (9\text{-}24)$$

The equal sign holds when N divides W exactly, also is asymptotical when W becomes arbitrarily large.

Figure 9-18 shows a plot of η_N versus ρ with N = 32, corresponding to η with W = N = 32, n = 1. The minimum value (1/N) at $\rho = 0$ and the

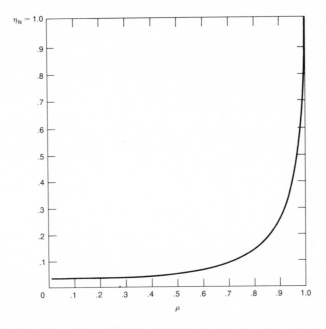

Figure 9-18 Efficiency (η_N) in parallel processing for N = 32 ($\eta = \eta_N$ if W = N)

maximum value (unity) at $\rho = 1$ are expected. However, it would be a mistake to assume a straight-line relationship; the curve instead is well approximated by *two*, almost mutually perpendicular, straight line segments.

Indeed, η_N drops precipitously when ρ deviates but slightly from unity. The situation is little helped by increasing N; η_N simply falls more rapidly. In the present case ($\rho = 0.979$, N = 32) η_N is only 0.61. In general, η_N drops to ½ when ρ is still as high as $1 - 1/(N - 1)$. By Equation 9-24, for given ρ, N if η_N is already small, η itself can only be smaller. The best efficiency (though obviously not the best speed) is achieved using the least positive definite integer N, namely unity.

The preceding discussions show the danger of unlimited synchro-parallelism. Surely economy is realized through mass-production of repetitious modules, and some loss of efficiency can thus be tolerated. What is shown here, however, is the law of diminishing returns for jobs with $\rho < 1$. As will be seen later, "Performance," not only efficiency, but performance also, obeys such an inexorable law. A good synchro-parallel design therefore should consider finite multiplicity, specialized problems (such as partial differential equations), and hardware-software techniques to increase the effective ρ,

Efficiency for General Job Profiles

The jobs encountered in practice usually do not resemble the model job explicitly. However, some are just repetitions of the model job pattern, and the formulas in the last sections would still apply.

More irregular job profiles can be treated after having been fully covered by the synchro-parallel processor. Then parts of the profile and/or parts of the cover can often be rearranged into the model form. These rearrangements do not change the area of either the profile or the cover; and the efficiency, being the ratio between two unchanged areas, remains invariant, but is now easier to compute.

One can also treat an irregular profile directly in terms of the constituent panels. Let the kth panel height be W_k, the duration be τ_k, and (using a synchro-parallel processor of multiplicity N) the sweep count be $n_k = \lceil W_k/N \rceil$. The efficiency for sweeping over the panel is $\eta_k = W_k/Nn_k$, and the overall efficiency is

$$\eta = \Sigma\eta_k n_k \tau_k / \Sigma n_k \tau_k = \Sigma W_k \tau_k / \Sigma n_k \tau_k.$$

This way even panels with zero height can be handled.

Performance

The expected performance P is defined as the number of effective processors. We have

$$P \equiv N\eta = \phi (\rho, W/n) \tag{9-25}$$
$$= 1/[(1 - \rho) + \rho n/W] \tag{9-26}$$

The performance intuitively can exceed neither the invested multiplicity N, nor the maximum job width W. As $\phi(\rho, x)$ increases monotonically with ρ and x, we have

$$P = \phi(\rho, W/n) \leq \phi(1, N) = N \tag{9-27}$$
$$\leq \phi(1, W) = W \tag{9-28}$$

since $W/n \geq N$, $n \geq 1$, and $1 \geq \rho$.

More powerful bounds are readily derived, based on Equations 9-18 and 9-19:

$$P = \phi(\rho, W/n) \leq \phi(1, W/n) = W/n \tag{9-29}$$
$$\leq \phi(\rho, W) \equiv P_W \tag{9-30}$$
$$\leq \phi(\rho, N) \equiv P_N \tag{9-31}$$
$$\leq \phi(\rho, \infty) = \frac{1}{1-\rho} \equiv \hat{P} \tag{9-32}$$

Equation 9-32 is especially significant in that P is completely bounded by the gross property ρ in the job profile, not explicitly involving either N or W. The bound P_W is reached whenever $N \geq W$, for then $n = 1$ exactly. P_N is also reached whenever N divides W.

Table 9-2 shows the values of P and P_N as functions of N, with $W = 32$ for the two cases:

a) $\rho = 0.9791$ and $P_W = 18.71$, $P = 43.67$
b) $\rho = 0.9000$ and $P_W = \;\;7.80$, $P = 10.00$

Case b) is plotted in Figure 9-19 to show the interrelations among $\hat{P}$, P_N, P_W and the actual performance P.

TABLE 9-2. PARALLEL PROCESSING PERFORMANCE FOR TWO PARALLELISM RATIOS
AND FIXED JOB PROFILES WITH $W = 32$

MULTIPLICITY (N)	a) $\rho = 0.9791$		b) $\rho = 0.9000$	
	$P = \phi(\rho, W/n)$	$P_N = \phi(\rho, N)$	$P = \phi(\rho, W/n)$	$P_N = \phi(\rho, N)$
1	1.00	1.00	1.00	1.00
2	1.96	1.96	1.82	1.82
3	2.79	2.87	2.44	2.50
4	3.74	3.74	3.08	3.08
5	4.23	4.58	3.37	3.57
6	4.85	5.38	3.72	4.00
7	5.70	6.15	4.16	4.38
8-10	6.90	6.90-8.29	4.71	4.71-5.26
11-15	8.73	8.95-11.36	5.42	5.50-6.25
16-31	11.91	11.91-18.38	6.40	6.40-7.75
32-63	18.71	18.71-26.03	7.80	7.80-8.75
64-1000	18.71	26.20-41.88	7.80	8.77-9.89
∞	$18.71 = P_W$	$43.67 = \hat{P}$	$7.80 = P_W$	$10.00 = \hat{P}$

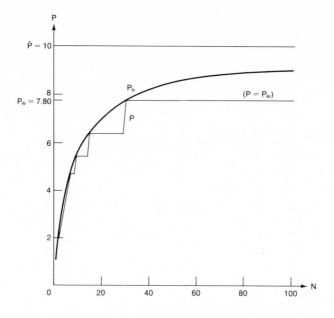

Figure 9-19   Performance versus multiplicity for $\rho = 0.90$ and W $= 32$

Is there a minimum repetition requirement for a prescribed minimum performance?

Since	$P \leq \phi(\rho, \infty) = 1/(1 - \rho)$
we have	$\rho \geq 1 - 1/P$
thus for	$P \geq 2, 4, 8, 16, 32$
we need correspondingly	$\rho \geq \frac{1}{2}, \frac{3}{4}, \frac{7}{8}, \frac{15}{16}, \frac{31}{32}$
	respectively.

Therefore, regardless of the maximum width W and investment N, there is no way to achieve a performance equal to 16 processors, unless ρ is as large as $15/16 = 0.9375$.

The preceding shows that a small admixture of serial computing can have a very sizable negative effect on overall performance of synchro-parallel mechanisms. As the next subsection shows, a similar effect occurs for pipeline processing as well. The difficulty lies in the rigidity in the processing for all tight-coupling systems, and judicial relaxing of the interfacial coupling can improve efficiency by a large factor.

In the ILLIAC IV computer, an SIMD machine, a separate overlapped control unit is used for bookkeeping functions, so as to expose the synchro-parallel multiprocessor only to the wide portions of the jobs. For example in a matrix computation the control unit will do all initializing and loop testing, while concurrently the processors devote themselves to arithmetic, in a generalization of I/E overlap.

Connection with Pipeline Efficiency

We shall now study the interrelation between synchro-parallelism and pipelining. Let us consider the central part in Gaussian eliminations involving the typical program sequence

$$\{L_j, M_j, A_j, S_j\} \qquad j = 0,1 \ldots (W - 1)$$

with L_j, M_j, A_j, S_j representing the jth load, multiply, add, and store, respectively. We have ignored the indexing requirements, and shall assume the four instructions to form a task of $M = 4$ procedures, each consuming equal time to simplify discussions.

When seen by a synchro-parallel SIMD multiprocessor (see Figure 9-20), the collection of N tasks form a perfectly repetitious job profile of width W; ρ is therefore unity. However, there is still residual inefficiency due to the N, W mismatch.

$$\eta = (W/N)/\lceil W/N \rceil = W/Nn \qquad (9\text{-}33)$$

which varies from unity to W/(W + N − 1).

While the SIMD processor acts as a simultaneous multi-task machine and executes the different procedures for all identical tasks in strict sequence, the

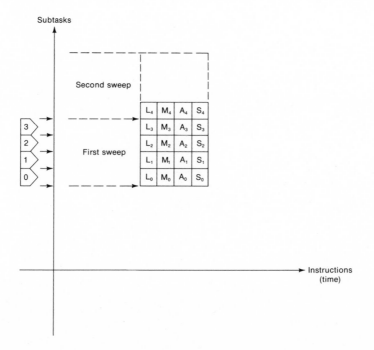

Figure 9-20   Synchro-parallel handling of array job

pipeline processor is a simultaneous multi-procedure machine and each procedure handler steps over the sequence of tasks in one cycle. Without changing the job profile, a pipeline multiprocessor of $M (= 4)$ stages can be represented by a slanted jagged edge which moves from south to north, orthogonal to the synchro-parallel processor movement, and achieves complete coverage by sweeping over a jagged parallelogram. In Figure 9-21 the stages consist of a loader L, a multiplier M, and adder A, and a storer S.

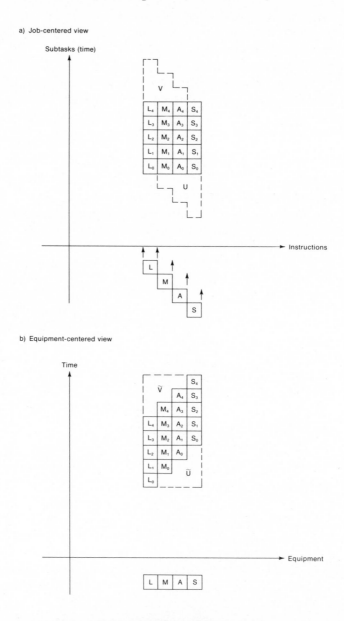

Figure 9-21 Pipeline handling of array job

In the beginning cycle, only L_0 is handled by the leading stage of the slanting edge; other stages, though available, are not being used.

One cycle later, L_1 and M_0 are simultaneously processed, still another cycle later L_2, M_1, and A_0. Then starting with the fourth cycle, all stages of the pipeline are gainfully employed, and the processing enters the steady state. At the kth cycle (time interval $[t_0 + k\tau, t_0 + (k + 1)\tau]$) L_k, M_{k-1}, A_{k-2}, and S_{k-3} are treated together.

At the Wth cycle, the supply of tasks begins to run out, and the pipeline depletes at a constant rate, with an extra unused stage per cycle; total pipeline drain occurs at cycle $(W + M - 1)$, with $M = 4$ in this case.

Effective Repetitiousness for Pipelines

Let us now study the pipeline performance for array processing.

In Figure 9-21 the pipeline efficiency is obscured by the jagged contours. However, using the rearrangement technique, the triangular cover area U can be moved over to fit the contour of V, resulting in a rectangle of size M by $(W + M - 1)$ as shown in Figure 9-22. Since total cover is unchanged in area, there is no change in the processing efficiency, thus

$$\eta \equiv \frac{\text{job area}}{\text{area swept by pipeline}} = \frac{WM}{(W + M - 1) M} = \frac{W}{W + M - 1} \quad (9\text{-}34)$$
$$= \phi (1/W, 1/M) = \phi (1 - 1/W, M)/M,$$

using Equation 9-15. Then the performance becomes

$$\begin{aligned}\widetilde{P} &= \widetilde{M}\eta \\ &= \phi (1 - 1/W, M) \\ &= W/[1 + (W - 1)/M] \end{aligned} \quad (9\text{-}35)$$

This corresponds to an effective parallel processor with

$$M_{\text{eff}} = M \quad (9\text{-}36)$$

which sweeps horizontally over a job characterized by

$$W_{\text{eff}} = M, \rho_{\text{eff}} = 1 - 1/W \quad (9\text{-}37)$$

This is seen explicitly in Figure 9-23 with the part P moved over to Q, resulting in a model job profile.

Figure 9-24 shows a superposition of two efficiency curves. The efficiency of the parallel scheme for fixed N and increasing W is shown in the jagged plot; the parts with positive slope are exactly rendered by

$$\widetilde{\eta} = W/Nn \quad n = \lceil W/N \rceil = 1, 2, \ldots \quad (9\text{-}38)$$

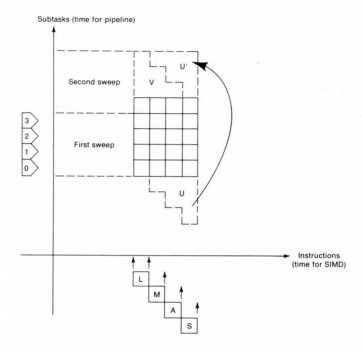

Figure 9-22 Pipelining versus SIMD (efficiencies are equal)

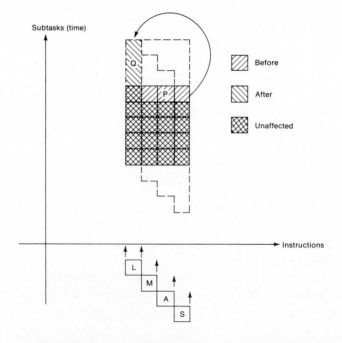

Figure 9-23 Conceptual creation of a model job profile

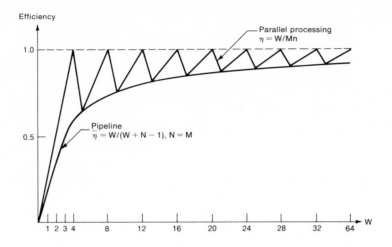

Figure 9-24   Efficiencies in array processing with increasing number of tasks

The smooth curve represents $\widetilde{\eta}$, the efficiency of the pipeline multiprocessor with stage count $M = N$. Intriguingly, the pipeline efficiency curve furnishes a lower envelope to the zigzag-shaped efficiency curve for parallel processing.

$$\widetilde{\eta} \leq \eta \qquad \text{for } M = N$$
$$= \eta \qquad \text{if M divides } (W - 1). \qquad (9\text{-}39)$$

9-7. THROUGHPUT ENHANCEMENT FOR TIGHT-COUPLING

The input queue of jobs describes the users' needs; though the server system is designed to satisfy these needs, it usually follows a different rhythm. Given a sufficiently large sample and assuming a proper design, the gross profile of the jobs may be satisfiable by the server. But to maintain processing efficiency, one must aim for agreement in detail. This is roughly achievable if dispatching of the incoming tasks can be made to satisfy the server's inherent abilities, rather than blindly following the arrival sequence; this is the topic of the present section.

Common Complications

In terms of instructions executed in conventional machines, perfect repetitiousness is not at all common. Jobs may appear superficially suited for synchro-parallelism and pipelining, yet may involve excessive hidden monoprocessing requirements. Further, smooth processing may be disrupted due to the inability to exercise uniform control over the processing phases.

In SIMD parallel processing, conditional operations link the operands to the control mechanism, precluding smooth synchro-parallelism. A test on operand signs, for example, potentially subdivides the SIMD collection of processors into two distinct classes for separate treatment.

Even if most operands react identically, one lone exception out of N predestines the system to do some future monoprocessing with $(N - 1)$ other processing units disabled. Skillful programming, indeed, is needed to encourage effective homogeneous treatment.

There is further an intercommunication problem; if processor j needs to send information to processor k, it does not follow that processor $(j + q)$ needs to send to $(k + q)$, for all q. But arbitrary intercommunication flow patterns may not be permitted; even if permitted, they would be hard to enforce without loss of efficiency.

These difficulties have their counterparts in pipeline processing, increasing control cost while decreasing average efficiency. While the SIMD machine is based on a collection of complete arithmetic units, a linear pipeline is a sequence of special purpose processing stages. Tasks are permitted to enter a pipeline only if all anticipated exceptional events are known to be resolvable automatically within the pipeline. Combination of events which lead to unresolvable problems may have to be handled by temporary blockage of input. For example, the task T^k may create a condition upon being processed by stage S_j, and this condition may affect the processing of T^{k+1}. The pipeline feeder control may keep T^{k+1} outside the pipeline until the condition becomes known; by then the stages S_0 through S_j will be idle.

To avoid too frequent pipeline drains of the type mentioned, the stages can be made more flexible, or more stages can be inserted to handle complex happenings. A way to give different treatment to differently signed operands is to define certain flag bits at some pipeline stage, then subsequent stages may operate on the operand or not, dependent on the flag. This scheme roughly corresponds to the SIMD disabling of processors based on operand sign.

Intercommunications among pipeline stages can in principle exploit lateral signals; this is seldom done in practice.

The problems in designing high-performance systems are discussed in section 9-8.

Micro-Multiprogramming

Though minor irregularities in one job profile may lead to excessive unused capacity, other jobs can be summoned to take up the slack; this is multiprogramming. Very high performance systems can subdivide a given task into locally independent streamlets, to be handled concurrently without conflict, shaping the streamlets to fit the internal processing patterns, to enhance the average performance. Such multiprogramming in the small can be called *micro-multiprogramming*.

The workload characteristic being exploited is local independence rather than task regularity. While the control problem is harder, the technique is essentially generalized table management by associative control. The hardware requirement consists of memories to buffer waiting tasks and coded signals, and capabilities for selection and switching; in other words associative memory with supporting logic.

Micro-multiprogramming is an example of self-optimization of the machine in real time. Like all self-optimization processes, the resultant processing is hard to anticipate in detail, yet the time-averaged performance becomes predictably better.

Example: Memory Interleaving

A good example of micro-multiprogramming is found in the handling of the memory-interleaving problem. A pipelined memory access system, confronted with imperfect time quantization, can employ synchro-parallelism using, say K memory units each taking K processor cycles to deliver a word. To avoid undue extra cost, no actual memory replication is used; only the addresses are interleaved so that all addresses equaling (j modulo K) refer to memory box j. This way the input stream of addresses now behaves like an admixture of K different streams; nevertheless consecutive addresses refer to different boxes, and the one-output-per-cycle maximum rate can be honored for well-chosen address request sequences. The (normalized) performance of the interleaved system is P_{int} = (throughput of the interleaved system)/ (throughput of the single box memory) and $1 \leq P_{int} \leq K$.

In the long run, the total addresses are expected to distribute more or less uniformly over the K available boxes. However, the overall average behavior does not guarantee average behavior in the small. Within several cycles a given box may be selected much more often than others. It can be shown that for random requests, the average throughput in a first-come, first-served system is proportional roughly to the square-root of K, rather than K itself. A 100-fold interleaved system gives only a 12.21-fold average throughput. The performance roughly equals $K^{0.56}$ [Hellerman, 73]; a better approximation is $(\pi K/2)^{1/2} - 0.28$ (see Table 9-3).*

The lower limit of the interleaved system performance is 1; this occurs for a data memory in delivering a column of elements of a $(K \times K)$ matrix, stored by row. The maximum performance K is reachable in the delivery of a row of elements in the same matrix. In actual computations with instructions and data stored in the same K-fold interleaved system, the performance due to interleaving should be somewhat better than $K^{0.56}$, because the instruction fetches tend to be sequential. It suffices to say that the delivery delay is uncertain, and the overall performance is not proportional to the interleave count K.

The difficulty can be overcome by buffering the input queue and honoring

---

*The throughput is actually equal to Knuth's Q-function, the series expansion of which is $\sqrt{(\pi K/2)} - 1/3 + 0(K^{-1/2})$. (Private communications with H. S. Stone, University of Massachusetts, and D. E. Knuth and G. Rau, Stanford University.)

TABLE 9-3. AVERAGE THROUGHPUT OF A K-FOLD INTERLEAVED MEMORY UNDER RANDOM REQUESTS

N	Average Through-put	Approximation A $(\pi N/2)^{.5} - .28$		Approximation B $(\pi N/2)^{.5} - 1/3$		Approximation C $(\pi N/2)^{.5}$		Approximation D $N^{.56}$	
		Value	% Error	Value	% Error	Value	% Error	Value	% Error
1	1.000	0.973	2.67	0.920	8.00	1.250	−25.33	1.000	0
2	1.500	1.492	.50	1.439	4.06	1.772	−18.16	1.474	1.72
4	2.219	2.227	−.34	2.173	2.06	2.507	−12.96	2.174	2.05
8	3.245	3.265	−.61	3.212	1.03	3.545	−9.24	3.204	1.25
16	4.704	4.733	−.62	4.680	0.51	5.013	−6.57	4.724	.42
32	6.774	6.810	−.53	6.756	0.26	7.090	−4.66	6.964	−2.81
64	9.706	9.746	−.42	9.693	0.13	10.027	−3.30	10.267	−5.78
128	13.855	10.390	−.32	13.846	0.06	14.180	−2.34	15.137	−9.25
256	19.726	19.773	−.24	19.720	0.03	20.053	−1.66	22.316	−11.31
512	28.030	28.079	−.18	28.026	0.01	28.357	−1.17	32.900	−17.37
1024	39.776	39.826	−.13	39.773	0.008	40.106	−0.83	48.503	−21.94

Approximation A: Adjusted two-term formula

Approximation B: First two terms in the asymptotic formula

Approximation C: First term in the asymptotic formula

Approximation D: Hellerman's approximation designed for N $\leq$ 64

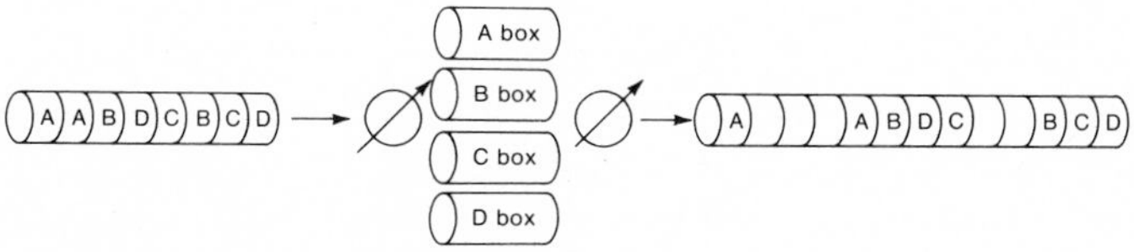

a) Memory interleaving based on first-come, first-served discipline. Note bubbles in output pipeline.

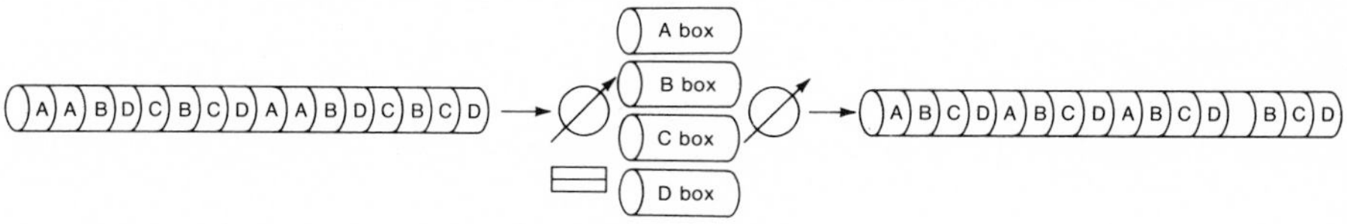

b) Buffer memory associated with the switching. The output pipeline is nearly full but delivery is out of sequence.

Figure 9-25   Memory interleaving

only those requests which refer to boxes currently available. Figure 9-25 shows the dramatic improvement with only two buffer registers. The system, with the queue-selection mechanism, optimizes its own throughput.

Typically, the ordering of the delivery sequence does not mirror the input sequence, and detailed prediction of system behavior becomes difficult. The gross behavior, on the other hand, now tends to exhibit a good local statistical distribution.

The sequence can be resorted using a set of buffer registers; there will, however, additional delays. Whether resorted or not, there must be temporary identifiers associated with the delivered words to compensate for the out-of-sequence delivery.

9-8. CONSIDERATIONS IN HIGH PERFORMANCE MACHINE DESIGNS

We shall now discuss the use of performance-enhancement techniques in a large general purpose computer. To obey the user-prescribed instructions, the computer must have a flexibility absent in simple synchro-parallel or pipeline designs, unless accompanied by very flexible interfaces, with self-optimization capabilities. It will be seen that the major issue is not the supply of hardware for arithmetic, but rather the interlocks to honor the causal precedence specified by the user program.

Example of high performance machines are the CDC 6600 and 7600, the IBM System/360 Models 91 and 195.

Requirements for Instruction Processing

To lend concreteness to the discussions which follow, we shall assume a computer processor with the data-flow indicated in the diagram in Figure 9-26. (The notation is defined in Table 9-4.) Further, we assume that all instructions are of equal length, and that there are only four kinds of CPU instructions:

1. Arithmetic (including load, $+$, $-$, $\times$, $\div$, shifts)—the result goes to AC and sets pertinent bits in the condition register CR
2. Index arithmetic—the result goes to an index register and sets CR
3. Store from the accumulator
4. Conditional branch based on a 1-bit condition: branch if and only if the bit is 1

This machine, though capable of meaningful processing, does not correspond to any existent system. For a high performance machine, many of the assumptions, for example, the limited access to index registers, are too confining; nevertheless, the design is adequate for a discussion of the performance enhancement principles.

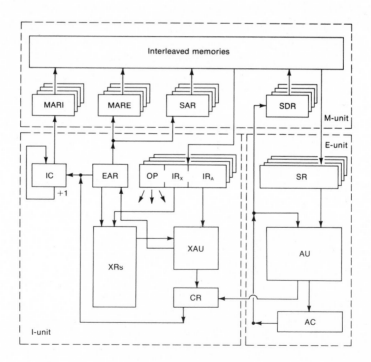

Figure 9-26   Hypothetical processor

TABLE 9-4. NOTATION USED IN FIGURES 9-26 AND 9-27

Notation	Meaning
AC	Accumulator
AC_s	Sign field of AC
AU	Arithmetic unit
CR	Condition register
EAR	Effective address register
EAR_s	Sign field of EAR
IC	Instruction counter
INDEX [IR_x]	Index register selected by IR_x
IR	Instruction register(s)
IR_A	Address field of IR
IR_x	Index field of IR
MARE	Memory address register(s) for E-unit
MARI	Memory address register(s) for I-unit
MEMORY [MARE]	Data word based fetched from memory based on MARE
MEMORY [MARI]	Instruction word fetched from memory based on MARI
OP	Operation code field of IR
SAR	Store address register(s)
SDR	Storage data register(s)
SR	Storage (output) register(s)
XAU	Index arithmetic unit
XR	Index register

A schematic for handling the instructions is given in Figure 9-27. It turns out to be simpler to consider the updating of the instruction counter to follow the decoding of an instruction, rather than to occur at the beginning of the next.

On examining the schematic, we note that all instructions call for essentially the same initial handling. It is desirable to design a powerful decode-indexing mechanism for that. And, in order to ensure precedence sequencing, the decoding should be in the order specified in the program.

After initial decoding, the instructions can be subdivided into semi-independent streams of tasks and assigned to the proper units. The M-unit performs all memory activities, the E-unit handles arithmetic and logic; the index arithmetic and branching are done largely within the I-unit.

Within the E-unit, the instructions tasks are executed in the program-directed order; the same is roughly true in the I-unit. The M-unit need not follow any processing sequence for fetch operations; a special precaution, however, needs to be exercised for stores, as will be shown in "Store-Fetch Interlock" later in this section.

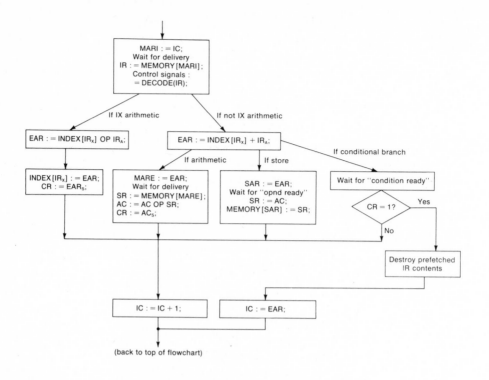

Figure 9-27 Instruction handling schematic

Memory Accessing Mechanisms

If the memory cycle time is K processor cycles, K boxes interleaved together can supply a maximum rate of one word per cycle. To avoid the "square-root catastrophe" one needs the countermeasure of an associative buffer. It is therefore desirable to have several each of MARI, MARE, SAR, and SDR, and powerful associative selection mechanisms.

The buffered interleaved memory delivers the contents in box availability sequence which seldom matches the request sequence. To avoid mis-identification, either the delivery should be accompanied by the original address in some encoded form or it must be to a pre-arranged specific location. The out-of-sequence delivery makes it difficult to predict the detailed output sequence from the memory bus. It is often desirable to buffer the memory output (by having several IRs and SRs) so that the desired sequence can be selected, again by associative search.

Fast integrated circuit memories now can match the processor in cycle time. However, being costlier than conventional memory, they tend to be provided only in small quantities, serving as the cache in a memory hierarchy (see Chapter 5). Deliveries from the cache can be fast, and in sequence, and the hit-ratio tends to exceed 90 percent for most computations. When a request refers to a word absent from the cache, however, a slow memory fetch must be made; meanwhile if there are to be useful cache deliveries, the delivery will not be in the memory request sequence.

The decoding action is relatively fast; independent indexing requires the use of a separate index arithmetic mechanism.

Arithmetic Mechanisms

The high performance machine usually has several units for arithmetic; the CDC 6600, for example, has 10. The most common computer pipelines are in arithmetic units. A 48-bit Wallace tree multiplier, for example, can be handled by a pipeline to generate one product per cycle of 4 logical delays [Hallin and Flynn, 72]. The mechanism uses one cycle to pre-encode the multiplier operand, one cycle to halve the number of addends to 24, then three cycles to compress the number of addends from 24 to 3, and finally, four more cycles for a 3-input addition.

An entire floating-point execution can be pipelined. Many simple instructions, such as fixed-point addition and logical instructions, can be executed in one cycle; they can be considered pipelined.

Some high performance machines use identical processing units to achieve arithmetic performance. The CDC 6600, for example, uses two independent multiplier units. Some machines practice synchro-parallelism; one ILLIAC IV processor can do a pair of floating adds simultaneously.

To avoid unnecessary interlocks among independent arithmetic processes, one accumulator may not suffice. (See however "Internal Forwarding" later in this section.)

Despite the relative ease to achieve arithmetic pipelining, the usage frequency may not really justify full pipelining. The IBM System/360 Model 91 has a pipelined 2-cycle floating adder, but a nonpipelined 3-cycle multiplier; the CDC 7600 has a 5-cycle multiplier pipelined every 2 cycles, mainly to replace the pair of unpipelined multipliers in the preceding 6600 computer. Divisions occur less frequently than multiply, and are seldom pipelined.

A characteristic of a pipeline is that the pulsed delivery is independent of the number of stages. The insertion of a checking stage, for example, will not affect the steady state throughput.

Bridging the Information Gap

In handling causally chained subtasks, overlap and pipelining may increase the overall throughput, but the total processing duration may increase. Therefore, in handling computer instructions, the time lag between the instruction fetch and the final execution may be long relative to the average time interval between instruction completions (often several times). This lag constitutes an information gap, within which the outcome of the instruction is uncertain.

Fortunately, for most instructions it is not necessary to know the complete outcome before starting the subsequent instructions. The arithmetic instructions, for example, usually can be dispatched as fast as the decoder and the memory can allow, with little chance of conflict.

However, with the conditional branches, the next instruction fetch depends on the previous arithmetic outcome. The most important thing to avoid is fetching and using the wrong instruction, based on premature use of the condition register, before the true information arrives.

A safe strategy is to wait for arithmetic completion before proceeding with the branching action. This takes nearly the full information gap time, corresponding possibly to two memory accesses, many processing stages, and many buffer-emptying actions.

The IBM System/360 Model 91 fills this gap partly by prefetching the branch target instruction and some subsequent instructions, holding them in abeyance until the arithmetic outcome is clear. Then, one of the two alternatives is selected, and the other ignored.

Logical State Revisited

By waiting for arithmetic completion in an overlap machine, the conditional branch instruction physically realizes a logical machine state in which the arithmetic execution has completed, but no specific branching action has yet been taken. Because it does not refer to an instruction boundary, this state differs from the one used in program interruption (section 9-2, "Logical State of the Machine"). The branch instruction has been decoded, and no attempt is made to invalidate this fact. All the same, the high overlap in the system is replaced by mono-processing, for a duration, and the hardware capability is under-utilized.

Logical states in a highly overlapped processor are much harder to generate than the simple I/E overlap case. There are not only more unfinished instructions to consider, the probability is much greater for some of the operations to have altered the machine irrevocably, and thus cannot be invalidated. To run all unfinished instructions to completion may appear to be a safe strategy, but this sometimes takes too long a time to be worthwhile.

In general we can conclude that the frequent creation of logical states is detrimental to highly overlap mechanisms.

Store-Fetch Interlock

Consider the instruction sequence below:

$$AC : = AC + MEMORY [1024];$$
$$MEMORY [1000] : = AC;$$
$$AC : = AC - MEMORY [1000];$$

Since the arithmetic execution is among the last subtask in the instruction, the first instruction may not be ready to produce the result for several cycles after the second instruction, a store, has been decoded. The latter must not be performed until the proper data is residing in the accumulator.

That one instruction has to wait does not mean postponement in toto.

Instructions normally should be allowed to proceed, unless forbidden by a precedence requirement.

Such a requirement is found in the third instruction. The implied data fetch operation is normally done in the middle part of the instruction processing, and thus may precede the arithmetic execution of the previous few instructions. In other words, the third instruction could fetch something from location 1000, even before the store action has started. Such a premature fetch would give a wrong answer in the accumulator (the right answer, incidentally, is 0). We therefore have the following constraints:

1. Store should not proceed until the source register has been correctly affected by all previous instructions. Any pending action which may change its value implies a postponement of store action.
2. Fetch should not proceed until the source memory location has completed all pending stores.

These two are special cases of the same principle: do not copy (or remove) until all pending alterations on the original have been made. A feasible scheme to handle the interlock is the following:

1. On encountering a store instruction, put the store address in an SAR, and mark it valid. The store instruction is treated as a pseudo-arithmetic instruction and follows the arithmetic execution sequence to ensure the proper handling of the accumulator operand. The store execution time thus logically follows the execution time of the previous arithmetic instruction.
2. All fetch actions are done by first matching the fetch address against the contents of all valid SARs. The fetch can proceed normally if and only if there exists no match. Those that do match are suspended.
3. When the accumulator has a valid content to be stored, its contents are copied into an SDR corresponding to the valid SAR; this frees the arithmetic unit to proceed freely.
4. The true store action occurs between SAR, SDR, and memory. Afterwards, the pertinent SAR and SDR are marked invalid.

The interested reader should sequence through a few instructions to convince himself that this scheme works correctly. We also note that the third instruction can be replaced by any fetch-type instruction, and there could be a number of instructions between the first and second, also the second and third. The rules remain valid. With a sufficient number of instructions separating a precedence pair, however, the precedence is automatically honored.

Internal Forwarding

The completion of the third instruction using the rules in the previous section requires the sequence:

1. Wait for the proper information to arrive at the SDR

2. Store into memory: MEMORY [1000] : = SDR
3. Fetch from memory: SR : = MEMORY [1000]

Rather than waiting for two slow memory operations to complete, we note that the correct information resides in SDR, and step 3 can be short-circuited by the fast register-to-register transfer known as store-fetch forwarding:

$$SR : = SDR$$

as practiced in the IBM STRETCH computer in 1961.
More generally,

$$B : = A \quad \text{then } C : = B$$

could be replaced by

$$C : = B : = A$$

This is known as internal forwarding. See Figure 9-28. The high-speed cache in a memory hierarchy actually is a systematic generalization of the internal forwarding principle.
A related observation is that,

$$B : = A, \quad \text{then } B : = C$$

can be replaced by $B : = C$ alone, the second act rendering the first one redundant.

An interesting example of internal forwarding occurs in the IBM System/360 Model 91 computer [Chen, 64], [Tomasulo, 67]. Consider the Gaussian elimination sequence in ALGOL

for J : = K **step** 1 **until** N **do**
      A [I, J] : = A [I, J] $\times$ Q $-$ A [K, J];

which is compiled into something like

$$R0 : = A_{IJ};$$
$$R0 : = R0 \times Q;$$
$$R0 : = R0 - A_{KJ};$$
$$A_{IJ} : = R0;$$

where R0 is register 0.

This can be represented by the linear graph in Figure 9-29, where the multiplier and the adder are represented by points A and M.

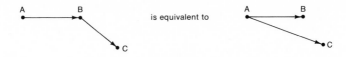

Figure 9-28   Internal forwarding

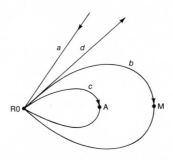

Figure 9-29   Linear graph for Gaussian elimination

Now if the fetch, store, multiply, and add operations, as well as all intermediate paths are pipelined for one output per cycle, the total sequence still cannot achieve the full rate, because of the heavy demands on R0.

Using the principle of internal forwarding (and the fact that there are operand registers in front of each of the arithmetic operator unit), the Model 91 in fact handles a macro-instruction not unlike the original ALGOL statement (see Figure 9-30):

$$
\begin{array}{l}
\cancel{R0} : = A_{IJ}; \\
\cancel{R0} : = \cancel{R0} \times Q; \\
\cancel{R0} : = \cancel{R0} - A_{KJ}; \\
A_{IJ} : = \cancel{R0} \\
\hline
A_{IJ} : = A_{IJ} \times Q - A_{KJ};
\end{array}
$$

The cancellations occur at the interface between two instructions, and the routing is achieved by a common data bus [Tomasulo, 67].

9-9. SPECIAL PIPELINE MACHINES

We now discuss special machines with special instructions to effect pipelining. To some extent, we can ignore the design practices of conventional general purpose machines when designing a pipeline machine, if the main

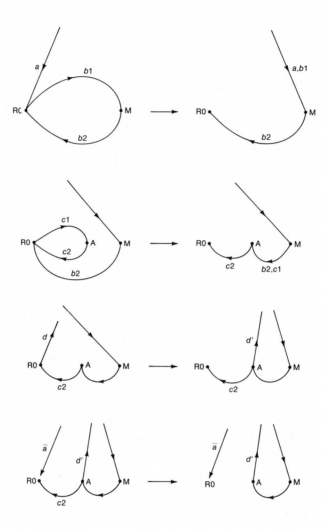

Figure 9-30 Internal forwarding in the Model 91

application is superfast numerical processing. However, a general purpose computing ability, though not emphasized, is nevertheless needed. The square-root catastrophe (sections 9-7 and 9-8) is just as important to consider here as for conventional machines. In the CDC STAR-100 and the Texas Instruments ASC, the interface between the interleaved memory and the processing area is a flexible buffer.

The VAMP design [Senzig and Smith, 65] aims to achieve the effects of SIMD parallelism through pipeline arithmetic. Each processor element is just a few registers; the memory and the arithmetic units are centralized. This idea has a strong influence on subsequent pipeline machines.

The VAMP Study

Senzig and Smith [65] observed that in the design of a vector machine of the SIMD parallel type, the cost of the arithmetic units can be reduced by using a small number of powerful pipelined arithmetic units to be shared by the N processors. In the VAMP (Vector Arithmetic Multi-Processor) study, all memories are also collected into one large interleaved unit. Each processor is represented by a few registers; the processors are only "virtual," with enough resources to specify the tasks, but not the true processing capability to do the work. An N-bit (screening) register specifies the active subset of virtual processors; another N-bit (condition) register summarizes the condition of a previous execution. The condition register contents can be sent to the screening register for masking the subsequent instructions.

Each of the pipeline units delivers one output per cycle. The processors are served in succession, and there is no true simultaneity of action. But simultaneity may never be needed, and the successive use of the processors implies that the memory words can also be delivered with fixed time lags, by time-sharing a small set of memory buses. The separation of the memory from the processors also allows easier access to the same word by a number of processors.

We note that the disabled virtual processors need not affect the arithmetic resources, in contrast to the SIMD case, when an unused PE *withdraws* its share of processing power from the system. The effective arithmetic speed per PE is dependent on the number of active elements in the VAMP. A comparison with SIMD parallelism is summarized in Table 9-5.

Senzig and Smith also were the first to suggest using the primitives of APL as part of the machine language.

The study went a long way to clarify the interrelations between pipelining and SIMD parallelism, and has a lasting impact on subsequent pipeline designs. Flynn et al. [70], for example, incorporate this form of pipeline operation in their MIMD design (see Chapter 8).

TABLE 9-5. CONTRAST BETWEEN SIMD AND VAMP

	SIMD	VAMP
Memory	Distributed, one unit per PE	Centralized but interleaved
Arithmetic unit	Distributed, one unit per PE	Centralized, powerful pipeline
PE	Memory + arithmetic unit + registers	Registers
Arithmetic time duration	Local arithmetic time	Proportional to the number of enabled PEs
Simultaneity	Exact	Systematic time lag
Masking action	Disables PE, performance withdrawn	Fewer entries into pipeline, arithmetic performance retained

The Array Processor

The IBM 2938 array processor [Ruggiero and Coryell, 69] is a stored program computer with a limited fast memory but high processing power. It is normally used as an attachment to an input/output channel of a general purpose computer (initially the IBM System/360 Model 44). The instructions are pseudo-input/output control words, each of which can specify a complete algorithm, such as fast Fourier transform.

Figure 9-31 gives a general data flow of the array processor. A typical arithmetic operation is the element-wise vector multiply-add

$$Y_i := Y_i + U_i \times X_i \text{ for all } i \leq \min (\text{X-limit, Y-limit, U-limit})$$

in 32-bit normalized hexadecimal floating-point arithmetic. The pipeline is pulsed every 200ns, although the multiply and add duration is 800ns. The actual computing bottleneck tends to be the delivery of data from the host machine, which on the Model 44 is 1200ns per word.

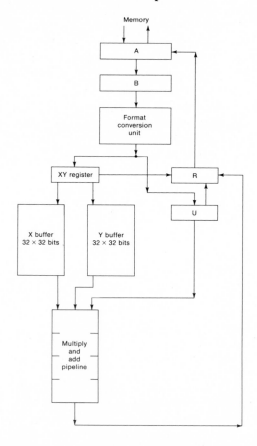

Figure 9-31 Data flow of the IBM 2938 Array Processor

The Control Data STAR-100

The STAR-100 had its origin in 1965, as a counter-proposal to the U.S. Atomic Energy Commission's request for a parallel design. It has a basic pipeline cycle time of 40ns. Floating-point additions and multiplications of short operations (32-bit) can be handled at the rate of 100 million instructions per second, with up to four operations done simultaneously per cycle. For long (64-bit) operands the maximum rates are 50 million adds, and 25 million multiplies per second. Division and square-root each have a rate half that of the multiplies of the corresponding word lengths.

The STAR memory has a cycle time of 1280ns, is interleaved by 32, and can deliver 512 bits every 40ns. The 512 bits are subdivided into four streams, each of 128 bits; two for the operands, one for the result, and the last for input/output requests and control-vector references. Memory buffers are used to even out address conflicts of the interleaved memory.

The 64-bit floating add is implemented in four pipeline stages (exponent compare, fraction shift, fraction add, then renormalize and transmission). Like the ILLIAC IV, the same unit can handle two short adds in parallel instead of one long add, and the two adder units behave like four short adder units.

A 32-bit multiply employs a Wallace tree (section 9-4, "Complex Networks"), terminating in an add operation using half of an adder. A 64-bit multiply is done using standard double-precision arithmetic and uses four short multiplies.

The machine has two sets of arithmetic units which are not identical (see Figure 9-32). One of these units also handles the divide and square-root functions. A multipurpose unit handles sparse vector, bit-string, and character string operations.

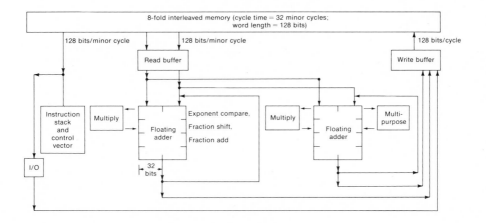

Figure 9-32   The CDC STAR-100

The STAR instruction design is similar to the VAMP, and makes use of APL primitives.

The Texas Instruments Advanced Scientific Computer

Texas Instruments started on the Advanced Scientific Computer in 1966, aiming for the seismic computations needed by the oil industry to determine underground geological structures. The result is a general purpose pipeline machine with particular accommodations for modularity; see Figure 9-33 [Watson, 72; Texas Instruments, Inc., 72].

The memory control unit (MCU) is asynchronous in design so as to be able to attach to any memory or processing units. The overall memory rate is 640 million 32-bit words per second. The central memory has a 160ns cycle time, and delivers eight informational words as a unit, with a Hamming error correction code for each of the eight words. Interleaving is by eight units. A back-up semiconductor storage with 1 microsecond cycle time is available. A paging technique with a protection scheme is employed.

The central processor employs both scalar and vector instructions, allowing 16-, 32-, and 64-bit operands. The processor can have up to four pipeline arithmetic units and each can deliver a result every 60ns. The pipeline has eight stages, with bypasses to generate a number of arithmetic functions (see Figure 9-34), in either fixed- or floating-point format. The vector instructions feature automatic three-dimensional indexing. Each pipeline unit has a memory buffer unit to even out the memory deliveries, and can reuse the outcome of a previous execution.

The peripheral processor consists of eight virtual processors, which are eight skeleton processors that share a common processing, control and data storage resource.

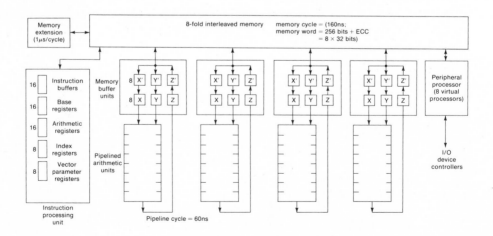

Figure 9-33   The Texas Instrument ASC

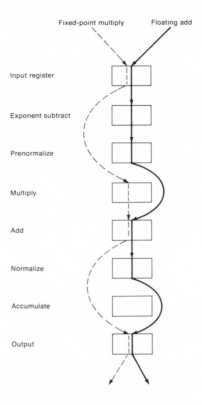

Fixed-point multiply   Floating add

Input register

Exponent subtract

Prenormalize

Multiply

Add

Normalize

Accumulate

Output

Figure 9-34   ASC arithmetic pipeline

Historical Note

The word "pipeline" and a sister term "streaming" probably originated within the **IBM STRETCH** project (1954-62), which developed both the highly overlapped **STRETCH** and the radically different 7950 Stream Processor [Campbell et al., 62]. The latter is probably the first comprehensive pipelined machine. Figure 9-35 shows a simplified data flow schematic. The computation is handled by two automatic input data streams P, Q, and an automatic return stream R. Each stream has its own buffer, is indexed separately, and can perform array indexing. The operations include byte processing, table look-up, inflight monitoring, and responses (adjustments) to unusual stimuli.

9-10. CONCLUSIONS

Overlap and pipelining are important multiprocessing tools for the improvement of system performance, especially when the task subdivisions show distinct precedence linkages.

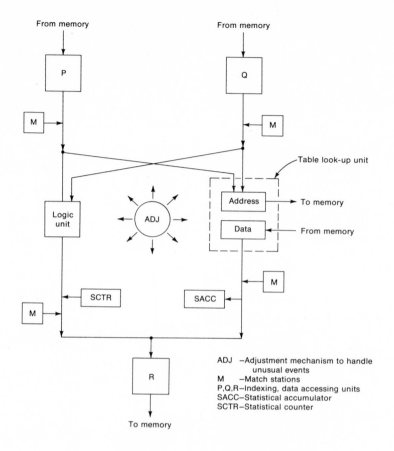

Figure 9-35　IBM 7950 schematic

In overlap processing, interfacial protocol grows and the benefits tend to drop as the multiplicity increases. Pipelining unifies the protocols by time pulses, requiring all stages to produce results at fixed time intervals. This way it resembles synchro-parallelism (typified by SIMD parallelism), since both belong to the same categories of tight-coupled multiprocessing.

Indeed, pipelining and synchro-parallelism complement each other; they merge in pipeline networks. Job inhomogeneity can create an alarming performance loss, calling for the insertion of associative mechanisms to optimize the deployment of internal resources.

Special pipeline machines have been constructed for well-formed problems as early as 1962. For more conventional processing in large computers, high order overlap among autonomous units tends to prevail; synchro-parallelism and pipelining are seldom used exclusively. Some machines are even able to create internal multiprogramming to optimize its own processing.

As large-scale integration technology advances, associative control will become more inexpensive and self-optimization should become common practice. On the other hand, systematic structures akin to cellular logic may also become economical enough to merit serious study.

9-11. PROBLEMS

9-1. Sketch one example each of

a) Synchro-parallel multiprocessing
b) Precedence dependent multiprocessing

9-2. Consider the I/E overlap processing of 100 instructions, $j \equiv 4k + m =$ 1, 2, ... 100. Compute the overlap ratio and comment on the result in each of the following cases:

a) $T(I_j) = T(E_j) = 1$
b) $T(I_j) = 0.5, T(E_j) = 1$
c) $T(I_j) = T(E_j) = 0.5, 0.75, 1.25, 1.5$ for $m = 0, 1, 2, 3$, respectively
d) $T(I_j)$ is the same as above; $T(E_j) = 0.75, 1.25, 1.5, 0.5$ for $m = 0, 1, 2, 3$ respectively
e) $T(I_j)$ is the same as above; $T(E_j) = 1.5, 1.25, 0.75, 0.5$ for $m = 0, 1, 2, 3$ respectively

9-3. The kth width in an efficiency curve can be defined as the deviation of ρ from unity in order for the efficiency to drop to $1/k$. Find

a) The kth width for $\eta_N(\rho)$
b) The kth width for $\eta(\rho)$. Can it be zero for $k > 1$?
c) Obtain the half-width for $\eta_N(\rho)$ and $\eta(\rho)$, assuming
   (1) $N = W = 32, \rho = 97.91$
   (2) $N = 20, W = 32, \rho = 97.91$.

9-4. Prove that $(\partial P_N/\partial N)\rho = \eta_N^2 \rho \leq \rho$.

9-5. For a given fixed job profile, the multiplicity of the synchro-parallel processor may increase from N to N', leading to a change of sweep count from n to n', efficiency from η to η' and performance from P to P'. The average performance gain due to multiplicity increase can be measured by

$$\frac{\Delta P}{\Delta N} \equiv \frac{P' - P}{N' - N}$$

Prove $\quad \Delta P/\Delta N = 0 \quad$ for $n = n'$
$\qquad\qquad = \eta \eta' \rho$ for $n \neq n'$ and $W = nN = n'N'$
$\qquad\qquad \leq \eta \eta' \rho$ in general

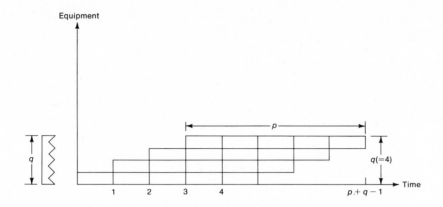

Figure 9-36 Ladder job profile to be handled by a synchro-parallel mechanism with matching height

9-6. In Figure 9-36 the job profile consists of q horizontal strips each with height $= 1$, and length $= p$, extending from $t = 0$ to $t = p + q - 1$. Using a synchro-parallel processor of multiplicity q,

 a) Prove that $\eta = p/(p + q - 1)$.
 b) Can you construct, by displacement, a profile with $\rho = 1$ to give the same efficiency?
 c) Can you construct, by displacement, a profile with $\rho = 1 - 1/q$ to yield the same efficiency?
 d) What is the relation to pipeline processing?

9-7. One of the fastest and most accurate methods to add 2^{k+1} normalized floating-point numbers together is to group the operands in pairs, do the individual adds, then repeat by regrouping, until only one operand remains. [See Linz, 70; Gregory, 72.] Comment on the result if the adds are done using synchro-parallelism with

 a) $N = 2^k$
 b) $N = 2^{k-1}$

9-8. W. Kahan proposed a highly accurate technique for summing N floating-point numbers $\{A_k\}$ together, using adders which normalize results before rounding or chopping. The method in FORTRAN reads as follows:

```
        S = 0.0
        D = 0.0
        DO   4   K = 1, N
        D = D + A (K)
        T = S + D
        D = (S − T) + D
   4    S = T
```

where D is an error estimate and $T = S$ is the accumulated sum. [See Kahan, 65.]

Is it possible to construct a pipeline to do the summation, in which every stage contains exactly one full floating-point adder? Explain.

9-9. Assume a VAMP multiplier with a duration of four cycles, pipelined at one cycle per multiply. The raw processing rate of eight of these is equivalent to 64 fully engaged, non-pipelined multipliers, each consuming eight cycles.

   a) How many actual multipliers would be needed to serve 64 virtual processors, if all of them need to start and finish within the same eight-cycle interval? Ignore the time needed for operand pick-up from and return to the virtual processors.
   b) Comment on possible changes in the design assumption in a) to reduce hardware cost.

NOTATION

$\phi(\rho, x)$	Performance function; $\phi(\rho, x) \equiv 1/[(1 - \rho) + \rho/x)]$
ρ	Repetition ratio
N	Multiplicity; number of unit processors
W	Width of job
$n = \lceil W/M \rceil$	Number of sweeps
η	Efficiency of synchro-parallel processor
$\tilde{\eta}$	Efficiency of pipeline processor
$P(= N\eta)$	Performance of synchro-parallel processor
$\tilde{P}(= M\tilde{\eta})$	Performance of pipeline processor
I_j	Preparation of the jth instruction; $T(I_j)$: preparation time.
E_j	Execution of the jth instruction; $T(E_j)$: execution time.
M	Number of pipeline stages
S_j	Pipeline stage for process designated j
T^k	kth task
T^k_j	jth process in handling the kth task

CHAPTER 10

Interpretation, Microprogramming, and the Control of a Computer

Michael J. Flynn

10-1. INTRODUCTION

Many readers are familiar with the concept of an interpretive program. This is simply a set of routines that directly process or operate on a source program—statement by statement—rather than first translating it into some intermediate form. The instruction set of a compuer is also a "source language," and programs written in this language are interpreted—instruction by instruction—in the process that is commonly known as execution. In section 10-2 of this chapter we review the familiar notion of instruction covered earlier in Chapters 3 and 4.

The routines that interpret instruction execution represent the internal computer control and they can be implemented in a number of different ways. One approach is to write an interpretive subroutine for each of the machine instructions. These subroutines are stored in a special control storage and consist of control descriptors called microinstructions. This chapter is basically concerned with this viewpoint: that control of the computer rests in primitive operations which are sequenced in subroutine fashion in order to interpret an instruction (microprogramming).

In section 10-3 we deal with the nature of these primitive operations. Just as with higher level operations, primitive operations have two aspects: operational and sequential. For its operational part the primitive operation must connect the appropriate source unit to its corresponding destination. It also must ensure that no unwanted information transfers take place. This operation takes place in a primitive time quantum called the computer's *cycle time*. Section 10-3 describes methods of connecting functional units to one another and various types and forms of cycles. Thus, section 10-3 is devoted to *what* has to be controlled. The actual primitive itself, the micro-instruction, can take on a number of different forms. These give rise to different microinstruction forms—the *how* of instruction control. In section 10-4 we discuss microprogramming using subroutines of these microinstruction to interpret conventional machine instructions. Microprogramming is not a new concept, it dates back over twenty years. Just as notions of control in languages have changed radically over this time, so too has the notion of microprogramming. Most important is the understanding that this technique can be generalized so that one particular machine is able to interpret more than one instruction set. This process is called emulation.

Of course, technology itself has influenced the development of control techniques. Most notable is the recent availability of fast read/write control storage. The fact that this storage can be easily altered, demonstrates its more general use as a new level of storage hierarchy. This not only influences computer architecture—the way the computer is organized as discussed in section 10-5—but also substantially changes our notions of fixed instruction sets and points to the development of special interpretive languages (each of which would cater to a specific higher level language) as discussed in section 10-6.

10-2. INTERPRETATION OF CONVENTIONAL INSTRUCTIONS

The machine instruction provides the basic specification for all control action. As mentioned in Chapters 3 and 4, a machine instruction usually takes two source operands and produces a single result. An operand may not correspond exactly to a physical word in storage. It may, in fact, refer to multiple words. For our purposes we can assume that a typical reference operand is a single physical word. The instruction, then, has five logical functions:

1. Specify the address (location in storage) of the first source operand
2. Specify the address of the second source operand
3. Specify the address of the result of the operation
4. Specify the operation to be performed on the two source operands
5. Specify the address of the next instruction in the sequence

Thus a general instruction appears as:

Operation	Source address 1	Source address 2	Result address	Next instruction address

These specifications may or may not be made explicitly. The location of many of these addresses can be implied (by use of an accumulator, assuming instruction locations lie "inline," and so on), saving space in the instruction at the expense of additional instructions in the sequence required to by-pass or reinitialize the implication.

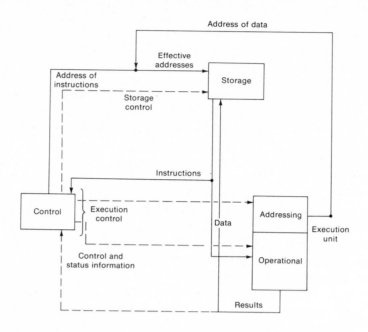

Figure 10-1 Flow of information in a computer

In order to get a better understanding of the execution unit and the control mechanism, consider the functional units that make up the processor: control unit, execution unit, and storage. As shown in Figure 10-1, the execution unit may be further broken down into two basic pieces—addressing and operation. The storage module is a conventional memory with data retrieval by address. All system elements are activated by the control unit acting on registers in the processor. The registers are distinct from storage and the execution unit. By separating the registers in this way we remove the facility to store or hold exit data information from storage and the execution unit. Thus, the operation of any functional system element involves transfer of information from one or two registers through the execution unit and the return of a result to another register (perhaps one of the source registers). We illustrate several typical operations in the example in section 10-4.

After an instruction has been transmitted to the instruction register (IR) from storage, the operation part of the instruction drives the control unit through a sequence of control steps. The first control steps calculate a source address and fetch the source datum into the storage register. Following that, a sequence of steps is performed with two registers as sources and another register as a result. These control signals are, in part, determined by the operation itself. Certain test signals are examined continually by the control unit to

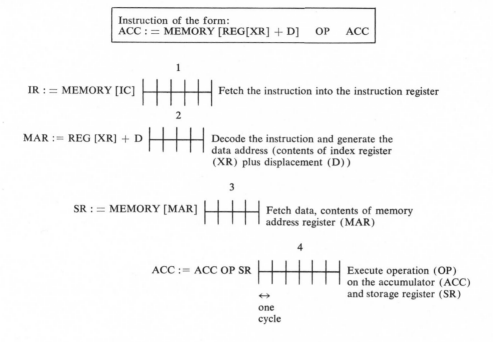

Figure 10-2   Sequencing through a simple instruction

determine the next control command. The instruction is executed in four basic phases as shown in Figure 10-2.

The operation in the execution phase might be an ADD, for example. In order to accomplish this, however, a number of suboperations are necessary as shown in Figure 10-3.

First, the sign of each of the source data has to be inspected. If a complement of the operation is required, it may involve the injection of an additional 1 into the least significant data position (as in two's complement arithmetic). Finally, after the ADD there is a potential recomplementation (again, depending upon the representation) and an inspection for overflow.

10-3. BASIC CONCEPTS IN CONTROL

Before discussing techniques of control, consider what is being controlled. Information is processed within a functional unit by a specific configuration of logic gates (combinatorial logic) in a single time unit or by a sequence of steps through such logical configurations (sequential logic). The communication may be transmitted and transformed in one time unit cycle by the data paths of the system. A sequence of transmissions from register to

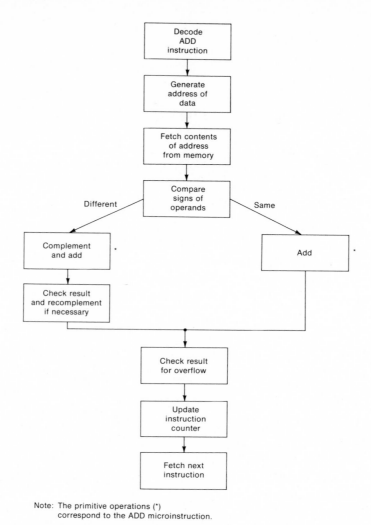

Note: The primitive operations (*)
correspond to the ADD microinstruction.

Figure 10-3 ADD machine instruction with sign and magnitude data representation

register requires multiple cycles. This section investigates the general requirements for controlling both the data paths of the system as well as various
kinds and forms of internal cycles which the designer may use.

Data Paths and Control Points

A machine, exclusive of control, consists largely of registers and combinatorial execution logic (adders, shifters, and so on). Each register position in
the system can be gated to one of a number of other registers during one
cycle. The inter-register connections together with the registers and resources
are referred to as the "data paths" of the system. The output of each register

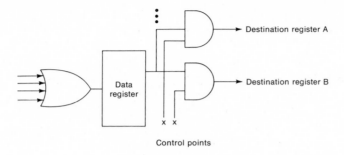

Figure 10-4   Gating logic

activates AND gates which are directed to each of the destinations reachable from the source register in one cycle (see Figure 10-4).

There are two types of data paths as shown in Figure 10-5:

1. Those paths that connect the source register to a destination register (perhaps itself) without any intervening transformational logic
2. Those paths connected from a source register into an execution unit and then directed to a destination register

Figure 10-5 shows the data paths for the ith bit of a storage register, an adder, and an accumulator. In this example, the accumulator register is added to a word from memory which has been placed in the storage register; the sum is returned to the accumulator. This occurs during the execute phase shown in Figure 10-3. A simple ADD instruction may have a three-cycle execute. One cycle is used for inspection of the signs of each of the operands before the addition, the second cycle is used for the addition, and the third cycle is used for sign and overflow inspection. During the second cycle the information in bit i of the storage register is gated to bit i of the adder, activated by an appropriate control signal, labeled *SR-to-adder*. This allows the information from bit i of the accumulator and bit i of the storage register to activate the two inputs to the ith position of the adder. This together with the carry from bit $i - 1$ position determines the sum, which is gated through OR into the accumulator. The accumulator does not actually change its value upon receiving this signal, but at the end of the cycle a sample pulse is used to set this new information into the accumulator. At the same time, new information can be entered into the storage register. If the instruction, instead of being an ADD instruction, were a SHIFT instruction, we would use a path from each bit of the accumulator to its neighbor. Notice that operations involving the adder require a substantial number of logic decisions before the final value can be determined and set into the accumulator, while the SHIFT operation involves only two decisions.

In general, if the execution unit (for example, the adder in the preceding example) has internal storage, it may be treated as a multiple-cycle operation.

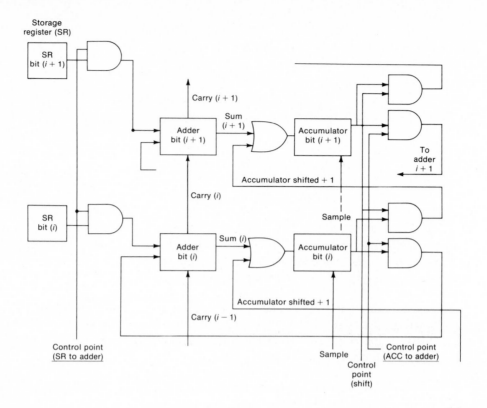

Figure 10-5 Two types of data paths

If it does not, then the time required to direct information from a register through the execution unit and back to a register defines the register cycle time. Combinatorial logic has no memory by itself; all information is lost at the end of one cycle, unless it is stored in a register.

Control points are the hardware locations at which the output of the processor instruction decoder activates specific registers and operation units. Control points basically govern intracycle register-to-register communications. For each register in the processor there are a fixed number of other registers to which data may be transmitted in one cycle. For each such possibility, a separate AND circuit is placed on the output of each bit of the source register with the entry into the destination register being collected from all possible sources by an OR circuit.

For example, consider a 32-bit computer with eight registers. Assume that each register can communicate with three other registers in one cycle. The number of control points required for register communication is therefore $3 \times 8 \times 32$ or 768. In addition, assume the machine has three execution units, each of whose 32-bit outputs can be gated to one of four registers. This accounts for an additional $3 \times 4 \times 32$ or 364 control points. There are additional control points for the selection of a particular function within a designed module. This might account for 100 more control points. Thus, there are a total of somewhat over 1200 control points that must be established each cycle by the output of the instruction decoder. Fortunately, in most computer design situations, many of these control points are not independent. For example, bit 7 of a certain register is not separately gated to another register, but rather the entire contents of the register is gated to its destination register. Since only one line is required to control these multiple control points, the total number of outputs required can be significantly reduced. These outputs are then referred to as independent control points. For the hypothetical system described, there might be anywhere from 50 to 200 independent control points depending upon the variety of instructions.

The operation code specifies the operation to be performed; by itself it is insufficient to specify multiple control steps for the execution of an instruction; some additional counting mechanism is also required. If the control implementation is to be done with hardware implementation—using a combinatorial network—then a counter is used to sequence through the control steps to transmit signals to control points. This counter identifies the particular step of the instruction that is executed at any moment. The combination of the sequence count and the operation is the input to the network which then describes the exact state of each control point at each cycle of every instruction (see Figure 10-6).

Cycle Time

The cycle time of a computer is the time required to change the information in a set of registers. This is also sometimes referred to as a *state transition* time. The internal cycle time may not be of constant value; there are basically three different ways of clocking a processor:

1. *Synchronous fixed* — In this scheme all operations are composed of one or more clock cycles with the fundamental time quantum being fixed by the design. Such systems are also referred to as clocked, since usually a master oscillator (or clock) is used to distribute and define these cycles.

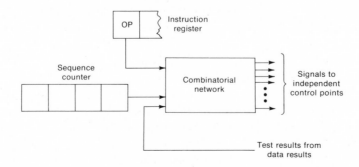

Figure 10-6   Hardware control

2. *Synchronous variable* — This is a slight variation of the former scheme in which certain long operations are allowed to take multiple cycles without causing a register state transition. In such systems there may be several different cycle lengths. For example, a register-to-register transfer of information might represent one basic cycle while a transfer from a register to an adder with return to a register requires perhaps two or three basic cycles. The fundamental difference between these two schemes is that the synchronous fixed scheme stores information into registers at the end of every cycle time, while the synchronous variable scheme sets information into registers after a number of cycles depending upon the type of operation being performed.

3. *Asynchronous operation* — In a completely asynchronous machine there is no clock or external mechanism that determines a state transition. Rather the logic of the system is arranged in stages. When the output value of one stage has been stabilized, the input at the stage can admit a new pair of operands.

Asynchronous operation is advantageous when the variation in cycle time is significant since a synchronous scheme must always wait for the worst possible delay in the definition of the time quantum required. On the other hand, when logic delays are predictable, synchronous techniques have an advantage since several additional stages of logic are required in the asynchronous scheme to signal completion of an operation. In actual practice, most systems are basically synchronous (either fixed or variable) with some asynchronous operations used for particular functions of the machine, such as in accessing main memory.

The cycle itself is composed of two components: 1) the time necessary to decode the control information and set up the control points, and 2) the time necessary to transmit and transform the data (the data state transition). In simple machines the cycle is the sum of the control decoding time and the data state transition time. In second generation computers with hard-wired control logic, control time was approximately 35 percent of the entire cycle and the data state transition was the remaining 65 percent of the cycle (see Figure 10-7). With the use of microprogram store for the implementation of the control function in third generation computers, the control time increased and overlapping of the two became more prevalent.

10-4. MICROPROGRAMMING

Control of a processor involves the specification at each cycle of all of the control points of the system. In this section we review the concept of microprogramming and the evolution of that concept over the past several years. We then look at specific types of microinstructions: methods of specifying the various kinds and types of control mechanisms which may be required. (This is really a generalization of the discussions of control in Chapters 3 and 4.)

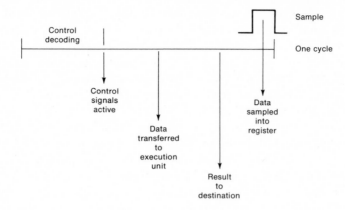

Figure 10-7   Action within a cycle

Finally we give a detailed example of a simple processor to see how all of these control aspects are realized via a microprogram and microinstructions.

Concept of Microprogramming

Microprogramming refers to the use of storage to implement the control unit. In the simplest cases, the operation code in the machine instruction is used as a (partial) address to the first microinstruction. This microinstruction contains the required control point values as well as the address of the following microinstruction in the sequence for operation interpretation. Alternatively the microinstructions can be placed inline and the address field is eliminated. Microprogramming, like many other words in a computer system vocabulary, has come to take on a variety of meanings in different applications situations. In fact, it has evolved in several stages since it was first used over twenty years ago by M. V. Wilkes [51]. Many different interpretations of the word stem from a variety of equivalent implications that are derived from basically the same principles.

Originally microprogramming meant the systematic and orderly control of the functional modules of a computer system. We must distinguish between microinstructions and instructions: functions performed by a microinstruction can usually be implemneted by combinational logic. Equivalently, we may view the machine instruction as causing a main memory state change while the microinstruction causes a register state change. Since the exact nature of main memory and registers is not always well defined (a register referred to by instructions may actually be locations in main memory and vice-versa), reliance on this distinction for a microinstruction can lead to confusion. For example, an operation, from a physical point of view, is the primitive logic that performs an absolute binary addition. It does not include such associated functions as sign detection, complementation, check for overflow, and so on. Inclusion of these features into the ADD operation is a sequence of steps (microinstructions) which together make up a machine language instruction.

Other notions of microprogramming are possible. Many of these notions rest on one of the properties or applications of microprogrammed systems which are possible with a storage implementation of the control function of an instruction and are not possible with direct boolean implementation. One such application is emulation. Informally, emulation is the interpretation of a machine language (machine instruction set). Since a nonmicroprogrammed computer can only emulate the machine code for its own boolean equations, it is limited to the emulation of the single language for which it is designed [Rosin, 69]. In a microprogrammed system, since the control function is implemented via storage, if the storage is changed with a different set of sequences, the control unit of the system can be used to emulate more than one machine language, although not necessarily with equivalent efficiency.

Evolution of Microprogramming

Microprogramming has evolved through three distinct phases. The earliest phase involved the use of microprogramming for engineering convenience. The control storage contains simple descriptions of the gating patterns of each of the control points for each cycle. Ease of engineering change and design were important considerations. For these early microprogram implementations, diode matrix technology was well suited. Microprogrammed implementation of control during this era is best illustrated perhaps by Wilkes's ideas as shown in Figure 10-8. Wilkes viewed the microprogrammed control store as consisting of two diode matrices. The first matrix determines the control information for the data paths while the second matrix determines, at least in part, the next microinstruction to be selected to continue the interpretation of the given instruction. The choice of the next microinstruction is influenced by a selected datum, for example the sign bit of an accumulator. If the sign is negative, one microinstruction might be called; if the sign is positive another might be invoked. This is required so that proper complementation rules can be used for addition and subtraction.

The decoding tree shown in Figure 10-8 has the function of transforming a pattern of n bits into a unique selection of 1 out of 2^n possible outputs. Thus, for example, a 4-bit binary input into a decoder tree has four input variables. These define sixteen possible configurations, from 0000 to 1111. The output of the tree has sixteen lines or possible events. Each output line corresponds to one and only one of the input configurations and is activated when that input configuration is present. When an output line is activated, it also activates all lines out of the matrix that are connected to it via diodes. The diode action essentially allows current to pass from the drive or input line into the output line. Of course, if no diode connects an input line to an output line, no current is transmitted and that line remains in a zero state.

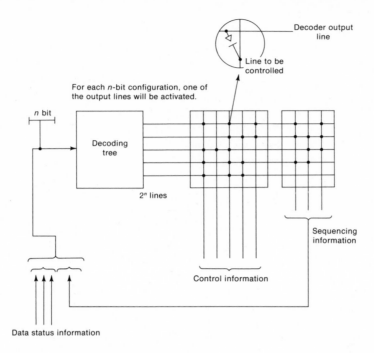

Figure 10-8   Wilkes's microprogrammed control storage

These diode arrays give a simple and regular implementation to the control function. However, speed could be a problem. In early implementations no speed problems developed since main memory was quite slow, about 10 microseconds cycle time, and the diode matrix had an access time of under .5 microsecond. The ratio of control access time to main memory access time is an important one. When there are a large number of internal cycles in each memory cycle, the microprogramming task is relatively simple and straightforward. One register-to-register transformation is performed per internal cycle and performance is essentially limited by the main memory cycle. As the main memory access time decreases, however, microprogramming techniques must become correspondingly more sophisticated. If there are only one or two internal machine cycles for each main memory cycle, it is necessary to have multiple data transfers in each machine cycle. That is, the microinstruction must simultaneously control a number of resources internal to the system. This gives rise to a type of internal parallelism within the processor.

During the decade of the fifties the relatively slow main memory technology made implementation of the microstorage array rather easy. A variety of technologies were used for control storage but diode storage was probably the most widely used. During the latter 1950s, with the introduction of magnetic core memories, the ratio of access time between main memory and the control storage dropped rather sharply making it difficult to execute the required number of microinstructions for a machine instruction in a reasonable amount of time. This became especially important in light of the development of fast transistorized logic circuits for the wired implementation of the control function.

The second generation of microprogrammed systems is distinguished by its small number of internal machine cycles per main memory cycle. By the early 1960s main memory speed had dropped to under 1 microsecond; yet the technology for control store had not noticeably improved—the best access for read-only store varied between 200 and 400 nanoseconds. In addition, read-only storage technologies tended to be exotic. The technology was not common with any other part of the machine and not always reliable. However, by this time the arguments for using microprogramming went well beyond the reasons cited by Wilkes. In the beginning of 1964, with the announcement of the IBM System/360, an important application for microprogramming had been added—*emulation* of multiple machines on a single host system [Tucker, 67 and 65]. It made the customer's transition from an old to a new system much more palatable in that the customer could, with one system, support his old software, as well as develop new applications with new programming languages and facilities.

The third generation of microprogramming dates from about 1970 with the advent of fast read-write control store. With the development of bi-polar monolithic technology, one has a storage medium with the same access time as combinatorial logic gating delays since they are made out of the same material. The writable capability of control store represents an important transition, since now the control store becomes a true member of the memory hierarchy. In section 10-5, we examine the impact of third generation microprogramming on computer architecture.

Microinstructions

The microinstruction is, by definition, the control mechanism that causes each data register change. The flexibility in this interaction allows a variety of possible microinstructions. Terms such as horizontal and vertical formats, nanoinstruction, and packed or unpacked microinstructions have been used, sometimes ambiguously, to describe certain differences.

If, within the control of a single functional unit, the microinstruction contains a separate description of each independent control point in the resource (for example, the true description of the control gating) that activation is said to be an *unpacked* or *exploded form* of the microinstruction (see part *a* of Figure 10-9). This form is most expensive in terms of space but provides the ultimate flexibility in that any combination of control point values may be specified at any time. As an alternate to this unpacked form, a specific number of combinations of control point values may be chosen. These combinations then, may be coded into a few bits and, through use of a decoder, be regenerated when the microinstruction is executed (see part *b* of Figure 10-9). This is called a *packed form*, and it saves space at the expense of flexibility. Occasionally the distinction between packed and unpacked forms of micro-

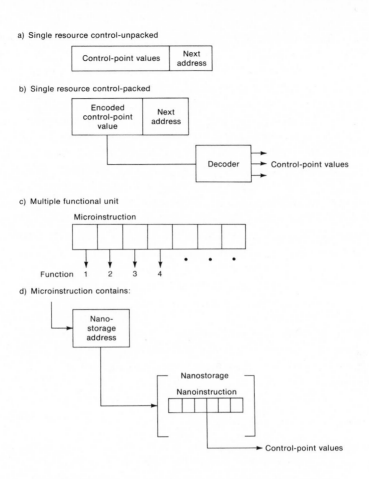

Figure 10-9   Some concepts used in microinstruction formats

instructions are referred to as *vertical* and *horizontal* microinstructions, respectively.

As previously mentioned, the functional units of the system may be partitioned into a number of independent units that can be used simultaneously. If this partitioning is done, then the microinstruction that activates these functions simultaneously contains a separate information field for each function. Thus in part *c* of Figure 10-9, a functional unit might be an adder, a shifter, a unit for loading and/or storing information into a register, or a test and branch unit. Notice that these can be operated simultaneously, as long as they do not make conflicting use of a datum. The distinction between the control of a single function through the use of control points, whether by packed or unpacked microinstructions, and the multiple functional unit control illustrated in part *c* of Figure 10-9 is worth noting. Of course, the control for an adder still requires a set of control points, whether or not a microinstruction specifies the adder action only or the action of multiple units. This simultaneous use of resources gives rise to a type of internal parallelism that is explicit—visible to the microprogrammer—yet within the single instruction stream. This is unlike the type of internal parallelism of certain highly pipe-lined machines such as the CDC 6600 and IBM 360 Model 91 whose paral-lelism is transparent to the programmer, as described in Chapter 9. In any event, this use of the microinstruction for identification of possible simul-taneous use of resources has also been referred to as a *horizontal microinstruc-tion*. The alternative is to use a universal single resource. Its corresponding control mechanism is sometimes referred to as a *vertical microinstruction*.

Some computers use a level of control which interprets a microinstruction, sometimes referred to as a nanolevel. In this mode a packed microinstruction, usually with only one or two fields, is used as the basic control mechanism. However, instead of driving the resources directly, it indirectly refers to the resources through another interpretative level, the nanoinstruction. The nano-instruction is usually a horizontal instruction that contains the exploded form of the control description. This technique is used on machines such as the Nanodata Corporation QM1.

Part *d* of Figure 10-9 illustrates this concept in which the *microinstructions* represent a sequence of addresses and each address points to a nanoinstruc-tion. A nanoinstruction may be horizontal with multiple resource specifica-tions. The purpose of this technique is to reduce the size of the storage needed to represent the microprogram.

Simple Microprogrammed Machine

In this section we illustrate the use of a read-only memory to control a simple digital machine. This fictitious machine was organized by Rosin [69] to exemplify the use of microprogramming for implementing the control function of a computer. Rosin proposed an emulator but left the "control" details unspecified. The proposed machine used a single accumulator with 4096 words of main memory and a maximum of 16 operations. All opera-tions are of the form: ACC : = ACC OP MEMORY [ADDRESS].

The machine has the following memories and registers:

SR:	storage register	MM:	main memory
MAR:	memory address register	CM:	control (read-only) memory
AC:	accumulator	MIC:	microinstruction counter
IC:	instruction counter	MIR:	microinstruction register
IR:	instruction register	REG:	added output register

In addition, the following transformational resources are provided:

Adder (with no storage resources)
Increment unit
Clear unit

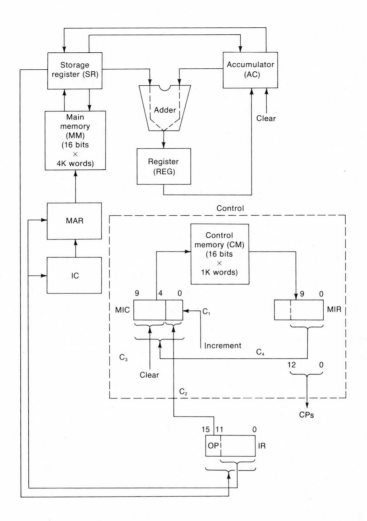

Figure 10-10 Block diagram of a simple digital machine

The block diagram of the machine is given in Figure 10-10. The machine has thirteen independent control points. Each point controls a particular line. The lines that have to be controlled are:

$0 - \text{AC} \rightarrow \text{SR}$ $7 - \text{IR}_{0-11} \rightarrow \text{MAR}$

$1 - \text{SR} \rightarrow \text{ADD} \rightarrow \text{REG}$ $8 - \text{IC} \rightarrow \text{MAR}$

$2 - \text{AC} \rightarrow \text{ADD} \rightarrow \text{REG}$ $9 - \text{INC IC}$

$3 - \text{REG} \rightarrow \text{AC}$ $10 - \text{MM READ}$

$4 - \text{SR} \rightarrow \text{AC}$ $11 - \text{MM WRITE}$

$5 - \text{SR} \rightarrow \text{IR}$ $12 - \text{CLEAR AC}$

$6 - \text{IR}_{0-11} \rightarrow \text{IC}$

When control point 0 is activated, the following transformation takes place: SR : = AC. Following this, the contents of AC are gated onto the AC-SR data path and read into SR at the end of the cycle. When control point 1 is activated, the contents of SR are gated through the left port of the adder at the beginning of the cycle. If control point 2 is simultaneously activated, the following transformation is completed at the end of the cycle: REG : = SR + AC. If control point 2 is not activated then the contents of SR are gated through the adder and transferred to REG at the end of the cycle. Activation of control point 9 causes the following transformation to take place: IC : = IC + 1. Activation of control point 10 causes SR : = MEMORY [MAR] after one main memory cycle, and activation of control point 11 causes the transformation MEMORY [MAR] : = SR. Activation of control point 12 has the following effect: AC : = 0.

The control part has a read-only memory 16 bits $\times$ 1K words. Each machine instruction is interpreted by a routine residing in control memory. When an instruction I is fetched in IR, contents of IR_{12-15} are transferred to MIC_{0-3}. This causes the contents of one of the locations 0 through 15 to be placed in MIR. A jump is then executed to a particular location in control memory where the routine that will interpret I resides. The words in control memory can have two possible formats.

Format 1:

	15	14	13	12			0
	1	X	X			. . .	

In Format 1, bits 0 to 12 indicate the control points to be activated. Bit $i = 1 => $ control point i is to be activated. Whenever a word with Format 1 is brought into the MIR, the specified control points are activated during that microcycle.

Format 2:

	15	14	13	12	11	10	9		0
	0							(address)	

Format 2 indicates how MIC is to be modified and is used for branching.

Control memory is accessed whenever a change in the contents of MIC takes place. Let us assume there are two bits, a and b, which are set depending on the contents of AC: a equals 1 if and only if AC is zero and b equals 1 if and only if AC is negative. Let B_i denote the ith bit of MIR. Then $T = (B_{15}, B_{14}, B_{13}, B_{11}, B_{10}, a, b)$ denotes a particular value of the seven bits. Four control lines C_1, C_2, C_3, and C_4 are used to change MIC. A cycle starts with a decode of the seven bits (see Figure 10-11).

The T = 1XX00XX path indicates the usual case of a microinstruction of format 1 type whose ICP's are to be applied immediately and whose completion will occur in one cycle.

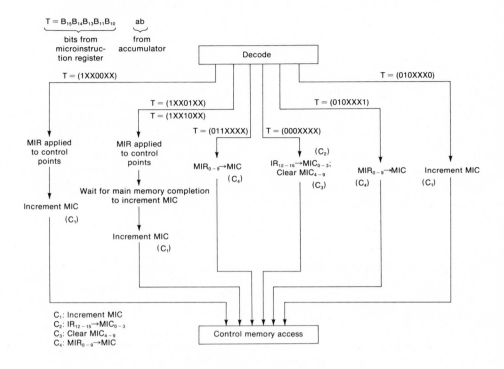

Figure 10-11   Bit decoding

The T = 1XX01XX or 1XX10XX are the main memory and read and write cases—these will not complete in one cycle. Here, the MIC is not incremented until a Main Memory completion signal is received.

Bit 15 selects between format 1 and 2. Bits 14 and 13 are used only in the format 2 case—they distinguish between three types of branching situation: $B_{15} B_{14} B_{13} = 011$ is used for an unconditional transfer (i.e., the low ten bits of the current microinstruction are entered into the microinstruction counter); 000 indicates an unconditional transfer initiated from the OP code field of the instruction register. These four bits are used to enter a table in the first 16 words of control storage. Each entry in the table is a transfer to the corresponding microprogrammed routine. Finally, the 010 case indicates a conditional transfer—if the b bit (from the accumulator) is 0 (i.e., accumulator positive), normal MIC incrementing is performed; if b = 1 (accumulator negative), the transfer takes place. The a bit is tested by $B_{15} B_{14} B_{13} = 001$. This is not used in any of the examples indicated; however, transfer on zero functions and more extensive detailing of the machine would use this feature.

We can now define a microcycle as the time taken for the first path: decode, MIR applied to control points, increment MIC, control memory

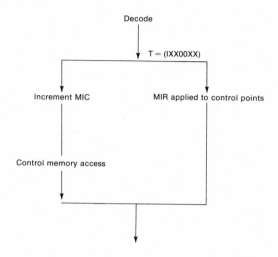

Figure 10-12   Control cycle with parallel execution

access. In fact the cycle time can be reduced considerably by doing two operations in parallel as shown in Figure 10-12.

Care must be taken to ensure that **MIR** is undisturbed by the access until application of signals to the control points is complete. The microcycle then becomes the time required to decode, increment MIC, then access control memory. The decoding is done by the combinatorial network COMB and the required control lines C_1, C_2, C_3, and C_4 are activated. The control memory thus operates in a completely synchronous manner except when a read or write to main memory is in progress. In this case, control memory is not accessed until a signal is given to C_1 by the main memory indicating that it has completed its cycle.

Figure 10-13 gives the contents of control store for IFETCH (instruction fetch), ADD, and STORE ZERO. This figure also gives the routines for interpreting the following machine instructions: IFETCH (0000), ADD (0001), CLA (0010), STORE AC (0011), TRA (0100), TRA IF ACNEG (0101), and STORE ZERO (0110). CLA denotes "clear and add" and TRA denotes "transfer." Figure 10-14 gives timing diagrams for IFETCH, ADD, and STORE ZERO.

For simplicity this example ignores input/output control and certain aspects of sign control for arithmetic routines.

10-5. MICROPROGRAMMING AND COMPUTER ARCHITECTURE

As mentioned earlier, the replacement of read-only storage for implementing control with read/write storage can substantially impact not only the way control is implemented but also the overall architecture—the instruction set—

LOC	CONTROL STORE 15	14	13	12	11	10	9	8	7	6	5	4	3	2	1	0	COMMENTS
0 0 0 0 0 0	0	1	1	X	X	X	0	0	0	0	0	1	0	0	0	0	Jump to microroutine for interpreting IFETCH
0 0 0 0 0 1	0	1	1	X	X	X	0	0	0	0	0	1	0	1	0	0	Jump to microroutine for interpreting ADD
0 0 0 0 1 0	0	1	1	X	X	X	0	0	0	0	0	1	1	0	0	1	Jump to microroutine for interpreting CLA
0 0 0 0 1 1	0	1	1	X	X	X	0	0	0	0	0	0	1	1	0	1	Jump to microroutine for interpreting STORE AC
0 0 0 1 0 0	0	1	1	X	X	X	0	0	0	0	1	0	0	0	1	0	Jump to microroutine for interpreting TRA
0 0 0 1 0 1	0	1	1	X	X	X	0	0	0	0	0	0	0	1	0	0	Jump to microroutine for interpreting TRA IF AC NEG
0 0 0 1 1 0	0	1	1	X	X	X	0	0	0	0	0	1	0	1	0	0	Jump to microroutine for interpreting STORE ZERO
0 0 1 0 0 0	0	1	1	X	X	X	0	0	0	0	0	0	0	0	0	0	
0 0 1 0 0 1	0	1	1	X	X	X	0	0	0	0	0	0	0	0	0	0	
. . .																	
0 0 1 1 0 0	0	1	1	X	X	X	0	0	0	0	0	0	0	0	0	0	
0 0 1 1 0 1	0	1	1	X	X	X	0	0	0	0	0	0	0	0	0	0	IFETCH
0 1 0 0 0 0	1	X	X	0	0	0	1	1	0	0	0	0	0	0	0	0	START IFETCH, (control line 9), IC→ MAR
0 1 0 0 0 1	1	X	X	0	0	1	1	0	0	0	0	0	0	0	0	0	MM READ (MIC not incremental till read complete); INC IC
0 1 0 0 1 0	1	X	X	0	0	0	0	0	0	0	1	0	0	0	0	0	SR→IR
0 1 0 0 1 1	0	0	X	X	X	X	X	X	X	X	X	X	X	X	X	X	IR_{12-15}→MIC_{0-3}
																	ADD
0 1 0 1 0 0	1	X	X	0	0	0	1	0	0	0	0	0	0	0	0	0	START ADD, IR_{0-11}→ MAR
0 1 0 1 0 1	1	X	X	0	0	1	0	0	0	0	0	0	0	0	0	0	MM READ (MIC not incremented till read complete)
0 1 0 1 1 0	1	X	X	0	0	0	0	0	0	0	0	0	0	1	1	0	SR→ADD→ REG; AC→ ADD→REG
0 1 0 1 1 1	1	X	X	0	0	0	0	0	0	0	0	0	1	0	0	0	REG→AC
0 1 1 0 0 0	0	1	X	0	0	0	0	0	0	0	0	0	0	0	0	0	GO TO IFETCH

Figure 10-13 (Part 1 of 2) Map of microstorage for control of Rosin's machine

CONTROL STORE

LOC	15	14	13	12	11	10	9	8	7	6	5	4	3	2	1	0	COMMENTS
																	CLA
0 1 1 0 0 1 0 1	1	X	X	0	0	0	0	0	1	0	0	0	0	0	0	0	START CLA, IR$_{0-11}$→MAR
0 1 1 0 0 0 1 0	1	X	X	0	0	1	0	0	0	0	0	0	0	0	0	0	MM READ
0 0 1 1 0 1 1 0	1	X	X	0	0	0	0	0	0	0	0	1	0	0	0	0	SR→AC
0 0 1 1 1 0 0 0	0	1	1	X	X	X	0	0	0	0	1	1	0	0	0	0	GO TO IFETCH
																	STORE AC
0 1 1 1 0 1 0 1	1	X	X	0	0	0	0	0	1	0	0	0	0	0	0	1	START STORE AC, IR$_{0-11}$→ MAR; AC→SR
0 1 1 1 1 0 1 0	1	X	X	0	1	0	0	0	0	0	0	0	1	0	0	0	MM WRITE (MIC not incremented till write complete)
0 1 1 1 1 1 1 0	0	1	1	X	X	X	0	0	0	1	0	0	0	0	0	0	GO TO IFETCH
																	TRA
1 0 0 0 0 1 0 1	1	X	X	0	0	0	0	0	0	1	0	0	0	0	0	0	START TRA, IR$_{0-11}$→IC
1 0 0 0 0 0 0 1	0	1	1	X	X	X	0	0	0	0	0	1	0	0	0	0	GO TO IFETCH
																	TRA IF ACNEG
1 0 0 0 1 0 1 0	0	1	0	X	X	X	0	0	0	0	1	0	0	0	0	0	START TRA IF ACNEG, IF ACNEG, GO TO TRA
1 0 0 0 1 1 1 0	0	1	1	X	X	X	0	0	0	1	0	0	0	0	0	0	GO TO IFETCH
																	STORE ZERO
1 0 0 1 0 1 0 0	1	X	X	0	0	0	0	0	0	0	0	0	0	1	0	0	START STORE ZERO, AC→ADD→REG
1 0 0 1 0 0 1 0	1	X	X	1	0	0	0	0	0	0	0	0	0	0	0	0	CLEAR AC
1 0 0 1 1 1 1 0	1	X	X	0	0	0	0	0	1	0	0	0	0	0	0	1	IR$_{0-11}$→MAR; AC→SR
1 0 0 1 1 0 1 0	1	X	X	1	1	0	0	0	0	0	0	1	0	0	0	0	MM WRITE; REG→AC
1 0 1 0 0 0 0 0	0	1	1	X	X	X	0	0	0	0	0	0	0	0	0	0	GO TO IFETCH

Figure 10-13 (Part 2 of 2) Map of microstorage for control of Rosin's machine

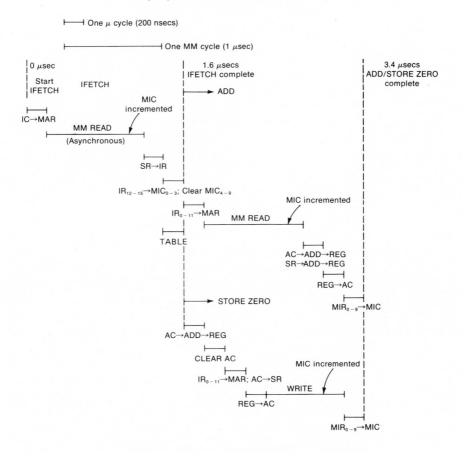

Figure 10-14 Timing diagrams for Rosin's machine

of a particular computer. In this section we investigate the impact of read/write control storage on computer organization. We then look at some aspects of modern microprogrammable computers, as well as an example of how third generation microprogramming impacts computer design.

Read/Write Microstorage and Its Impact

The availability of read/write storage can be expected to play an important role in the evolution of machine organizations. Traditional second and third generation machines have been organized about two artifacts.

1. The importance of the machine language as a ready interface with the human.

2. The access time to main memory for data and instructions is quite long compared to the time required to move data from a register.

The machine language programmer of today does not program in assembly code for convenience, but for efficiency; that is, in order to control the physical resources of the system in an optimal way. Ease of use is not as important a consideration in the development of a machine language as efficiency. This, coupled with improved memory, implies a significant change in computer organization: an increase in parallel or concurrent use of resources. Since hardware is inexpensive, units are made independent in the hope that their concurrent operation will produce more efficient programs.

How can the control of the resources of the system be made visible to the programmer? First let us review the evolution of the machine language as we know it and compare it with microprogrammed systems.

An essential feature of a microprogrammed system is the availability of a fast storage medium. "Fast" in this sense implies that the memory access time is approximately the same as the cycle time of the processor. Another important attribute of modern microprogrammed systems is that this fast storage is also writable, establishing the concept of "dynamic microprogramming." Consider the timing chart for a conventional (nonmicroprogram) machine instruction as shown in Figure 10-2. Due to slow access of instruction and data, a substantial amount of instruction execution time is spent in overhead operations. Contrast this with the situation shown in Figure 10-15, which illustrates the execution of an instruction on a microprogrammed computer.

In Figure 10-15 there is an implicit assumption of homogeneity of memory, that is, that data and operands are all located in the same fast storage as the instructions. The latter assumption is valid when the control storage is writable.

The preceding implies another significant difference in conventional machines and microprogrammed machines, namely, the power of the operation to be performed by the instruction. In the conventional machine, given the substantial accessing overhead, it is important to design each instruction to do as much as possible in computing a result. To do otherwise would necessarily involve a number of additional accessing overheads; thus we have the evolution of the rich and powerful instruction sets of the second and third generation machines. With dynamically microprogrammed systems, the basic

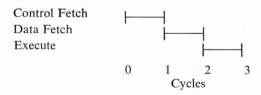

Control Fetch
Data Fetch
Execute

0   1   2   3
Cycles

Figure 10-15   Microinstruction execution

operations will tend to remain (as in simpler microprogrammed processors) primitive.

In the traditional system, since the overhead penalties are so significant,
there is little real advantage in keeping the functional units of the system
simultaneously busy or active. Use of such parallelism may result in 17 or 18
cycles required to perform an instruction rather than 20—hardly worth the
effort. A much different situation arises with the advent of dynamic microprogramming. Consider a system partitioned into functional units such as
adder, storage, address calculation unit, and so on. If the microinstruction
can be designed in such a fashion that during a single execution one can
control the flow of data internal to each one of these units, as well as communicate between them (see Figure 10-16), significant performance advantages can be derived. Of course these advantages come at the expense of a
wider microinstruction. Microinstruction size can be reduced by the residual
control technique: filtering the relatively static control information out of the
microinstruction into a register which is set up by a microinstruction.

In any event computer architecture cannot help but be influenced strongly
by the availability of fast memory and the use of primitive (one cycle) operations and explicit parallelism to simultaneously control the resources of the
system. The third generation microprogrammed systems can achieve a performance speed-up in direct proportion to the reduced access time of microstor-

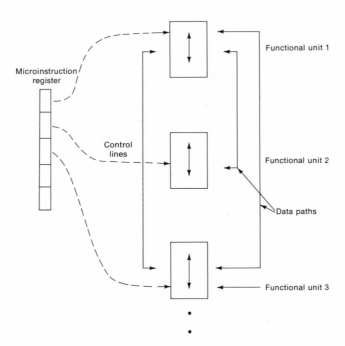

Figure 10-16   Simultaneous resource management

age memory over main memory. Thus, a factor of up to an order of magnitude performance improvement is potentially available, since the main memory accessing time represented such a considerable portion of instruction execution.

Some Microprogrammable Machines

In order to clarify some of the previously presented concepts, we present some of the details of four currently produced microprogrammable machines:

1. Nanodata QM1
2. Burroughs D Machine
3. Digital Scientific Meta 4
4. Burroughs B1700

A number of interesting characteristics serve to distinguish and characterize modern microprogrammable systems.

1. The "soft" computer architecture — as represented by the Nanodata QM1 and the Burroughs B1700 especially and the Burroughs D Machine to a lesser degree. These systems take advantage of the fast read/write control store capability with a number of processor features to allow easy emulation across a variety of image or target machines and hence are called "soft" architectures. These processor features usually include:

   a) Extensive field handling and selections facilities
   b) High speed shifting abilities
   c) Extensive bit testing
   d) Flexible specification of data paths (residual control)

2. Storage of data in the read/write control storage — the direct execution of data operands from the high speed control store allows the possibility of high performance since the access time to the data is shortened by having the data present in this high speed storage media. The control store acts as a type of explicit cache. None of the four machines presented here can directly access operands from their base control storage. However, this technique is used in the sample organization presented at the end of this section.

3. The alterability of the control store in certain systems — the control store may be altered, but only from an external peripheral device thus restricting the introduction of new interpretive routines. The Meta 4 is the most restrictive machine in this regard, while the QM1 allows modification of its control at two levels from main storage and hence is the most flexible of the four machines presented.

4. The characteristics of the micro/nanoinstructions — the QM1 and the D Machine have a two-level control structure with unpacked nanoin-

structions and packed microinstructions while the B1700 and Meta 4 use packed microinstructions and have no second level of control.

It should be pointed out that there are a number of other microprogramming machines. Some are quite interesting especially from a control point of view. These would include the IBM 125, 135, and 145. However, these are not included here since user microcode is not supported or encouraged by the manufacturer.

Nanodata QM1 The Nanodata QM1 is probably the most flexible machine of the four. It is a two-level machine. Among its notable features are a dynamically alterable bus structure and extensive residual control facilities through the F register. The nanoinstructions are sequentially executed (four to a nanoword). Automatic looping within the four occurs without an explicit or specific branch. There are a total of twelve buses that connect local storage to the data and address lines of the other storage units and to the input of the arithmetic logic unit and shifter. The QM1 has no fixed microinstruction set; the set is variable depending upon the interpretation of the nanoprogram.

It is reported that the QM1 emulates System/360 at Model 40 to Model 50 performance.

Figure 10-17 summarizes the characteristics of the QM1.

Burroughs D Machine The Burroughs D Machine has two control levels with one unpacked nanoinstruction per nanoword at the base level. The nanostorage, however, is read only and cannot be changed dynamically. Thus only fixed interpretations of the microcode are allowable. There are two predetermined microinstruction formats.

Among the more interesting features of the D Machine is its modularity. Main memory word sizes vary from 8 to 64 bits. The arithmetic logical unit is also implemented in a modular fashion so that up to eight logic unit sections each 8 bits wide may be concatenated in a parallel fashion.

The D Machine was designed not only to emulate a variety of conventional image machines but also to interpret special languages which correspond to various higher level languages.

Figure 10-18 summarizes the characteristics of the D Machine.

Digital Scientific Meta 4 The Digital Scientific Meta 4 uses a read-only storage for its one-level control memory. This is an interesting transitional machine between second and third generation microprogram systems. The Meta 4 includes a flexible bus structure controlled by packed vertical microinstructions. Since data must be separated from microinstructions, scratch pad and local store are included to hold operand values.

The Meta 4 has been used extensively to emulate the IBM 1130 and for certain graphical applications.

Figure 10-19 summarizes the characteristics of the Meta 4.

| Memory | Control Levels | | Operand Storage | Main Memory |
	Nano	Micro	(Two local register arrays)	
Addressable units	360 bits per word	18 bits	18 bits	8-bit byte
Size	1024	32K (maximum)	32 per array	512K
Cycle time	75 nanoseconds	75 nanoseconds	75 nanoseconds	750 nanoseconds
Use	Each word contains four nanoinstructions plus one residual control constant	Contains pointers to nanostorage and operands	Contains operands	Data and source program
Operational				
Instruction specifies	Control point values	ALU/shift and data transfers		Variable format
Cycle time	75 nanoseconds	150–300 nanoseconds (typical)		—
Use	Defines gating structure; flexible (12 buses)	Implement basic operations		

Figure 10-17 Summary of QM1 characteristics

Memory

	Control Levels		Operand Storage	Main Memory
	Nano	Micro	(Local registers)	
Addressable unit	54 bits	16 bits	8–64 bits	8–64 bits
Size	4096	—	4	64K or 2^{32}
Cycle time	80 nanoseconds (ROM)	80 nanoseconds	80 nanoseconds	640 nanoseconds
Use	Contains one nanoinstruction	Two formats: a) Pointer to nanostorage b) Immediate data and ALU operations	Data	Data and source programs

Operational

	Nano	Micro		Main Memory
Instruction specifies	Control point values	ALU/shifter data transfers		Variable format
Cycle time	90 nanoseconds	160–320 nanoseconds (typical)		—
Use	Defines gating structure and conditional tests	Implement basic operations		Has modular size

Figure 10-18 Summary of D machine characteristics

Memory	Control Levels Micro	Operand Storage Local	Operand Storage Scratch Pad	Main Memory
Addressable unit	16 bits	16 bits	16 bits	18 bits
Size	4096	4–32	16–256	65K
Cycle time	90 nanoseconds (ROM)	90 nanoseconds	90 nanoseconds	900 nanoseconds
Use	Four formats packed One microinstruction per word	Operands	Operands	Data and program
Operational				
Instruction specifies	ALU/skew units and bus			Pointer to micro-instruction
Cycle time	90 nanoseconds			
Use	Flexible bus structure			(Fixed format)

Figure 10-19   Summary of Meta 4 characteristics

Burroughs B1700   The Burroughs B1700 uses a one-level read-write control storage which is homogeneous with main storage; that is, the lower 4000 bytes of main storage can be replaced with the high speed control storage. The microinstructions are of the packed vertical type. Main memory is bit address-able up to 16 million bits. Variable length field selection from 0 to 24 bits for operands is possible coupled with variable length operations in the arith-metic logic unit ($1 - 24$ bits). The B1700 uses a combination of stack, scratch pad, and general purpose registers to hold operands, operations, and field descriptions.

The B1700 supports a variety of specific interpreters for source languages. Thus there is a separate interpreter for RPG, FORTRAN, ALGOL, and COBOL. These are called the S languages. The B1700 has not been widely used for traditional machine language emulation.

Figure 10-20 summarizes the characteristics of the B1700.

Sample Computer Architecture Using Read/Write Control Storage

Consider the organization outlined in Figure 10-21 [Cook and Flynn, 70; Neuhauser, 73]. This is a simple two-level control structure with a packed field microinstruction. There is no nanocontrol level; however, there are multiple functional units to be controlled.

Let the general purpose registers contain the instruction register and in-struction counter. Assume that the instruction width is the same as the data word width (perhaps 32 bits). A typical host instruction is partitioned into three fragments, each of which is essentially a primitive instruction (see Figure 10-21):

1. A register-to-register operation (OP) — the contents of R_1 and the con-tents of R_2 are used as arguments and the result is placed in R_1.

$$\text{REG}[R_1] : = \text{REG}[R_1] \quad \text{OP} \quad \text{REG}[R_2]$$

This can be thought of as the functional part of the instruction (F-part).

2. A load or store from microstorage into the general purpose registers — immediate values can be contained in the address field in this instruc-tion fragment.

$$\text{MEMORY} : = \text{REG}[R_3]$$
$$\text{or}$$
$$\text{REG}[R_2] : = \text{MEMORY}$$
$$\text{or}$$
Other operation
specified by OP

This is the memory management (M-part) of the instruction.

Memory	Control Levels	Operand Storage			Main Memory
	Micro	General Purpose Register	Stack	Scratch Pad	
Addressable unit	16 bit	24 bits	48 bits	24 bits	1 bit to 24
Size	2096	4	16	16	16 million bits
Cycle time	166 nanoseconds	166 nanoseconds	166 nanoseconds	166 nanoseconds	667 nanoseconds
Use	Contains one microinstruction	Operands	Address/operands	Field descriptors, pointers, and so on	Source program and data
Operational					
Instruction specifies	ALU and memory control				Variable format
Cycle time	166 nanoseconds				
Use	Variable field operands 0–24 bits starting at any bit in memory				—

Figure 10-20   Summary of B1700 characteristics

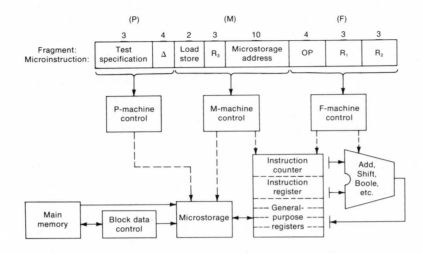

Figure 10-21   Computer architecture diagram

3. The branch instruction — this includes specification of a test mask and an offset value (Δ) relative to the location counter (*) and defines the procedural (P-part) of the instruction.

$$REG\ [0] :\ = REG\ [0] \pm \Delta$$

or

$$REG\ [0] :\ = REG\ [0] + 1$$
<u>on condition</u>

The net effect is to simultaneously control the operation of three finite state machines; actually it will not always be possible to have concurrent operations for two reasons. Inconsistent use of the registers by two of the fragments could cause a conflict. Also it may not be possible to write code which uses all three fields in many instances. In any event, the foregoing instruction resembles a familiar microinstruction. It executes in essentially one machine cycle—perhaps 200 nanoseconds using ordinary circuitry. Depending upon the arrangement of microstorage, conflicts between the load/store fragment and the next instruction fetch mechanism could double the instruction execution time.

For transfer of input data to and from main memory, an alternate instruction format is used. This instruction format is block oriented and asynchronously moves blocks of data between microstorage and main storage. Thus main memory is in many ways treated as an input/output device. Notice that this treatment of control store, except for its explicit nature, is very similar to cache-based memory systems already in use.

The reader is encouraged to complete the details of this organization (Problems 10-15 through 10-20) and compare it with Rosin's machine, described in section 10-4.

10-6. INTERPRETATION

Interpretation, Emulation, and Directly Executable Languages

An *interpreter* is a program that executes each statement in a source program as it is produced. It also selects the statements of the source program in a sequence determined by the execution and presents the results of the executed program as final output [Iverson and Brooks, 65].

In general then, the interpreter has two parts: 1) the set of routines that correspond to the set of operations in the source program language and 2) a control routine that sequences source statements and selects operations for decoding. The arguments for each operational routine should be defined or be readily computable at the time control is passed to the routine. An interpretive program is a one-pass program that does not require any further scanning of the source text before its execution can be begun.

The notion of emulation and interpretation are completely equivalent. However, by common usage, the interpreter usually is thought of as a program residing in main storage while the emulator is thought of as a program residing in microstorage. This, of course, is an arbitrary distinction, therefore we use these terms synonymously.

A *directly executable language* (DEL) is a source language for an interpreter. It is difficult to formalize the DEL notion since it is actually a property of, or a restriction to, a language which provides for the efficient execution of interpretive or emulator programs. Notice that many familiar languages do not have the DEL property. Languages not directly executable include symbolic assembly code, relocatable binary code, and assembly programs with macros since either the arguments or the operations are not completely defined as they are first encountered. On the other hand, one could conceive of a rather high-level language suitable as a DEL. That is, a directly executable language that resembles our familiar higher level languages but which obeys the argument definition and operator decoding restrictions and provides for orderly sequencing of source statements.

Emulation of Conventional Machine Languages

In the past, emulation has mainly involved the interpretation of machine language with the use of microprogramming techniques for instruction interpretation [Rosin, 69]. It is relatively easy for a single physical system to interpret more than one machine language. The physical machine as defined by its microinstructions and their actions is called the *host machine*. Machine languages emulated by sets of microprogrammed routines are called *image machines* or *target machines*. It is possible to write an emulator for one image machine in terms of another image machine language; thus one can conceive of layers of emulators. However, more common usage of the term emulator implies that the interpretive set of programs are written in the microlanguage of the host processor.

Probably the most widely known use of emulation is the IBM System/360. Most of the models of System/360 and System/370 are microprogrammed [Tucker, 67] (with the notable exception of the Models 91 and 195). Each model of the System/360 and System/370 series is quite a distinct machine with widely differing performance characteristics, data paths, size, and price. However, each has the common machine language of the System/360. In all of the microprogrammed models of the System/360, the interpretation of the machine language is done by an emulator that resides in microstorage. This emulator consists of a series of routines and each routine represents a particular System/360 instruction.

The emulation of a machine other than a 360 is not so straightforward [Tucker, 65]. Consider a Model 65 that emulates a 7090. The "emulation" of a 7090 on a Model 65 is more accurately described as a simulation of the 7090 using a combination of techniques including 360 instructions, special instructions, and 7090-type instructions. The hybrid approach to emulation reduces the size of microstorage needed to provide emulation for the 360 and the 7090. In the Model 65, each 7090 instruction is interpreted by an emulation subroutine contained in main memory as shown in Figure 10-22. This subroutine uses special instructions as well as conventional System/360 instructions. One of the special instructions is the DIL (do interpretive loop)

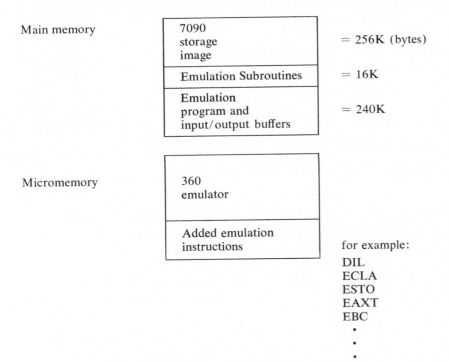

Figure 10-22 Configuration of main memory and micromemory in a System/360 Model 65 emulating an IBM 7090

7090 Instruction	360 Emulation Routine	
AXT	EAXT	Microroutine that does AXT
Address to index true	DIL	Microroutine that does fetch and interpretation of next instruction
AXC	LCR	360 instruction that complements the address
Address to index complemented	EAXT	Microroutine
	DIL	Microroutine
TMI	ESTO	Microroutine that puts the value into a work area of the 360 (the simulated accumulator)
Transfer on minus	TM	360 instruction, test under mask (to obtain the sign bit)
	EBC	Microroutine that does a 7090 branch if the test is satisfied

Figure 10-23 Emulation of IBM 7090 instructions

which is a microprogrammed routine that does a fetch and interpretation of the next 7090 instruction. In addition to the DIL instruction, a number of other subroutines are added to the microstorage to assist in emulating specific 7090 instructions. The configuration of main storage during emulation is shown in Figure 10-22; the emulation of three 7090 instructions is shown in Figure 10-23 [Tucker, 65].

Principles of Directly Executable Languages

The problem of developing efficient directly executable languages is similar to the problem of developing an efficient language of any type, be it higher level or machine language. For example, the design of such languages depends upon assumed user behavior and the physical characteristics of the base machine. A directly executable language should be developed with reference to particular higher level languages which will be translated and particular base machines which will be used to interpret programs of that language. Some general statements concerning the nature of efficient DELs can be made.

An efficient DEL should be a reasonably good output medium for compilers. It should require a relatively small amount of space for DEL program representation. Also, the time required for transformation of a higher level source language fragment into an equivalent DEL fragment should be minimal. Since the compiler has many functions which are dependent upon the source program, rather than the object DEL, we must consider such functions separately: for example, functions such as source program scanning and global optimization are source language dependent while code generation is DEL dependent. DEL aspects include:

1. The DEL should facilitate efficient code generation. The input to the code generator is assumed to be a tree structure of a source program. The code generator must assign appropriate "macros" to each node in the source program tree. Thus the number of potential macros for a typical node in the tree should be small. This minimizes the time needed to decide which macro should be expanded and also possibly minimize the amount of "contextual information" needed to make "optimal" decisions.

2. The DEL should avoid needlessly restricting critical physical resources that must be allocated by the compiler. In a 360 language, for example, there is a fixed identification of the number of general purpose registers (16). This is so, in spite of the fact that on some machines there are physically no such registers (they are, in fact, part of the physical memory) and in other implementations there are many more than 16 registers (buffers and cache memory). This particular DEL then, interferes with the compiler's responsibility to assign variables to storage space.

3. Where possible the DEL should allow for the generation of code which is independent of the structure of the data on which it operates. That is, it should be possible to execute a DEL routine against several different types of input data organizations simply by redefining the data structure and without recompiling the entire program.

So far we have looked at the characteristics of the DEL from the compiler point of view. The DEL must also be a good input medium for an emulator or an interpreter. It should be easy to interpret a particular DEL operation; for example, to decode this operation to locate the proper arguments and invoke the interpretative routine which actually executes the particular operation. From the emulator point of view a number of efficient DEL forms are possible. Without elaborate arguments, we would propose that the following are reasonable characteristics of an efficient DEL:

1. The general form of the DEL should probably include a Polish suffix notation. This form is well known for its ease of interpretation.

2. Individual operations in DEL may have several formats in order to allow operands to be selected either implicity from an evaluation stack or explicitly from a given storage cell or register.

3. Implementation of procedure entry and exit should be efficient. Frequently there is a penalty in register oriented machine languages due to saving and restoring the registers upon entry and exit from a procedure. Similarly with stack oriented machines, the problem is to retain a locally allocated block of storage after procedure exit, or to free (or reallocate)

a global variable within a nested procedure. Ignoring the physical problems for a moment, it seems that the language should not prohibit the allocation of register cells upon entry to a new block and should maintain separate evaluation and control stacks.

4. The formats in each of the DEL operations must be easy to decode. This simply means that the operation code should be easy to associate with the microroutine which interprets its function and that explicit operands are easy to recognize and pass as parameters to the appropriate interpreting routine.

The rapid interpretation of the DEL program representation is of considerable importance. Here a number of conditions must be satisfied:

1. The number of microinstructions required to decode and to implement a given DEL operation should be minimized.
2. The number of microinstructions required to pass parameters into an interpretative routine must be minimized.
3. The average width of a DEL statement is minimized; that is, through variable length instructions and literal specification.

Notice the contrast between the DEL and the microinstruction language. The microinstruction is designed to make physical resources available (or visible) for potential control. Hopefully this may be done with concurrency and independence of operation. In any event, the rapid execution of all fragments of the microinstruction is of prime importance together with efficient coding and representation of each of the resources identified.

The DEL has a much more general function. It must be a good output medium from the compiler, as well as a good interpretive medium. Thus different source languages will have different DELs. The DEL must be extensible and changeable to suit a variety of source language environments. On the other hand, the DEL will be more valuable if it can serve as an input medium for more than one interpretative emulation. Thus, it should allow identification of required logical resources in a flexible, hierarchial manner. That is, it should allow for a variety of interpretations based upon the actual physical resources which are available in a particular host machine. It is the responsibility of the interpreter to provide for this. Existing machine languages are deficient in many respects. They limit compiler specification of the storage space. The interpreters for these languages are static, as are the languages themselves. The full interpreter for the entire language must be available regardless of the characteristics of the source language. No new interpretative routines can be easily introduced into the machine language repertoire.

Recent Developments in Directly Executable Languages

DELs can be generated for implementing specific compilers or as output from specific higher level language compilers for interpretation and execution

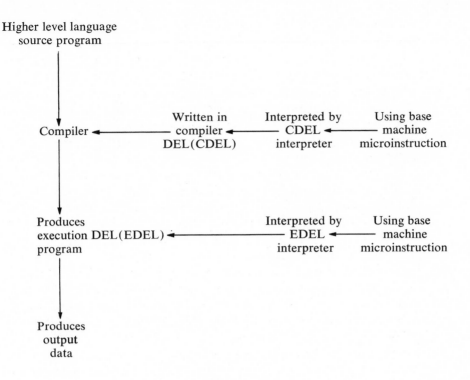

Figure 10-24   Use of DELs

(see Figure 10-24). Thus we can have a compiler DEL (CDEL) as well as an execution DEL (EDEL); either of these DELs might be trivial in a particular implementation. That is, either the compiler may be written completely in microinstructions or the compiler produces microinstructions as its output. For a DEL to have value it must represent an efficiency savings in space or time (or both) during the compilation or execution of the program.

The most notable early work in the area of DELs is that of Weber [67] who experimented with the high level source language EULER, a subset of ALGOL, and used an IBM System/360 Model 30 as the base machine. Weber chose to implement the EULER compiler completely in Model 30 microinstructions. This compiler produces a string-like execution DEL which resembles EULER. The Model 30 itself has a 50-bit microinstruction arranged in packed, multiple fields. Approximately 2000 microwords are used for the System/360 interpretation in the Model 30. Weber reports that in his experimental system, the EULER compiler uses 500 microinstructions (that is, 500 microwords). The string language execution interpreter uses 2500 microwords. He reports a performance improvement in the execution of the compiled programs of up to a factor of 10, depending upon the nature of the source program. He also reports that the compiler performance is limited by input/output.

The Burroughs Corporation at Paoli, Pennsylvania has developed the TRANSLANG, a translator language which is an assembler for interpreter macroprograms, in conjunction with the development of their D Machine. Reigel et al., in their paper [72] on the D Machine (the "interpreter"), cite the need for specific interpretive languages or execution DELs which correspond to the various higher level languages. He indicates that an ALGOL execution DEL might resemble machine code for the Burroughs B5500 while a COBOL execution DEL might resemble the B3500-type machine code.

The Burroughs B1700 reduces Reigel's speculations to practice. The B1700 uses a series of interpretive S languages—one each for FORTRAN, RPG, COBOL, and BASIC. These S languages operate under control of a master control program; in addition, there is a more general system description language, SDL, which has its own interpreter. Since the B1700 makes active use of the read/write control store only the interpreter required by the user program is present in the control store at execution time.

While not concerned with microprogramming per se, it should be pointed out that there have been two other notable attempts at "high level language machines." Both were hard-wired and tailored to specific languages. The earliest is a study by Bashkow [67] in which he proposed a FORTRAN machine and his study projects an order of magnitude performance improvement in execution time over traditional instruction sets. Rice [71], on the other hand, develops a new language called SYMBOL and a hardwired implementation of this machine has been constructed and is under study. The non-dynamic nature of these hardwired implementations seem to be a serious limitation to their general acceptance.

10-7. SUMMARY

Various methods have been used in the past to implement the control of a computer. Control in this sense is synonymous with interpreting an instruction through a sequence of steps that take specified source data and produce a proper result and then transfer control to the next instruction in an orderly manner. One of the most interesting implementations of internal computer control is the use of storage. Special storage media have been developed over the years to allow control of the computer resources on a step-by-step basis. This step or state transition within the computer is controlled by a *microinstruction*. The ensemble of microinstructions which interpret a machine instruction is called a *microprogram* for that particular instruction.

As the storage media which contains the microprogram became increasingly more sophisticated, microprogramming assumed additional functions. In particular, the possibility of the *host machine* being able to interpret a number of different machine languages was introduced. This process is called *emulation*. Emulation in a general sense is synonymous with interpretation.

With the advent of fast read-write microstorage, new developments in both architecture and in machine languages become possible. The traditional role of the microinstruction remains the same: to make visible the physical resources of the computer. However, it is no longer necessary to remain with fixed and restricted machine languages. In the larger sense, the term microprogramming is really inadequate. What we really mean is interpretative programming. The source language for the interpreter is a *directly executable language*. The object language is the microinstruction. The novelty in all this is that the DELs may be flexible, changeable, and extensible.

Machines that support the implementation of variable DELs have been termed *soft* computer architectures. In emulation of specific image machines, these architectures may not give as good performance as comparable machines designed specifically for that instruction set. However, their performance should be reasonable and their other advantages noteworthy. In particular these advantages include:

1. The ability to emulate a wide variety of image machines
2. The ability to support as yet undefined intermediate interpretive language levels—directly executable languages which may be expanded or changed dynamically
3. The ability to get significant performance improvements on specific problems where that application is programmed completely at the microinstruction level

REFERENCE COMMENTS

The introduction to interpretation in Chapter 10 follows the Iverson and Brooks [65] treatment. The sample computer described is a variation of a 16-bit machine discussed in considerable detail in Cook and Flynn [70]. The sample 32-bit machine described at the end of section 10-5 is actually a version of a system under development at The Johns Hopkins University [Neuhauser, 73]. Certain sections, definitions, and discussion contained in this chapter have been abstracted from earlier works of the author, especially articles on microprogramming and emulation [Ralston, 74; Flynn and Rosin, 71]. The treatment of emulation and DELs is strongly influenced by Hoevel [73], Tucker [67 and 65], and Weber [67]. The papers of Tucker and Weber are frequently cited and well known in the field. They merit reading by any serious student of the area. For general references the reader is recommended to Wilkes's original paper on microprogramming [Wilkes, 51] and Rosin's survey and exposition [Rosin, 69]. The simple microprogramming machine used in this chapter comes from Rosin. For a detailed treatment of microprogrammed machine structure, see Husson [72]. Further bibliographic material can be found in Wilkes [69], Jones [72 and 73], and Davies [72]. Jones is quite comprehensive, while Davies is an especially useful guide to key papers in the field.

10-8. PROBLEMS

10-1. Refer to Figure 10-2 of this chapter. Assume a simple processor has a memory with a 5-cycle access time and a 5-cycle regeneration time. Draw a sequencing diagram and a flowchart for:
   a) A STORE instruction (that is, MEMORY [REG [XR] + D]: = ACC)
   b) A conditional branch instruction

10-2. Refer to Figure 10-7. Show how control overlap might operate between sequential cycles.

Problems 10-3 through 10-9 refer to Rosin's Machine (see section 10-4, "Simple Microprogrammed Machine").

10-3. Write the Boolean equations for the decode of microinstruction Format 2.

10-4. Show the intracycle operations required for non-overlapped operations. Repeat for the overlapped case.

10-5. Write the microprogram for multiplying the accumulator by a memory operand. (Note: define additional ICP's.)

10-6. Write the microprogram for a shift instruction whose format specifies a right or left shift of the accumulator by an amount specified in the instruction (up to 16). (Note: define additional ICP's.)

10-7. Show a memory map for microstorage for problem 10-6.

10-8. Show a timing diagram for problem 10-6.

10-9. How would you revise Rosin's machine to allow for a read/write control storage?

10-10. It is desired to emulate the action of the serial decimal adder discussed in Chapter 4 (Figure 10-8 on Rosin's Machine). Discuss how this implementation might be accomplished.

10-11. How does emulation differ from simulation? Discuss this in the context of the IBM Model 65 emulating the IBM 7090.

10-12. Identify five restrictions that would have to be made to FORTRAN to make it more useful as a DEL.

10-13. Discuss the differences between FORTRAN and BASIC as to their suitability as DELs.

10-14. Identify five limitations in System/360 machine code that restrict its value as a DEL.

The following problems refer to the sample architecture at the end of section 10-5.

10-15. Specify a typical set of operations (see Chapter 4 for hints).

10-16. Specify the required load and store operations.

10-17. Specify a typical set of test specifications.

10-18. Write an emulator for the ADD instruction of Rosin's machine.

10-19. Draw a timing diagram for problem 10-18 and compare with the example in Figure 10-11 assuming the same cycle times.

10-20. Discuss the nature of the cycle time in Rosin's machine as compared with the sample machine.

Acknowledgments

I am especially indebted to Mr. Lee Hoevel and Mr. T. Agerwala, colleagues at The Johns Hopkins University, for their helpful suggestions and assistance with certain sections of this chapter.

CHAPTER 11

Performance Evaluation

Samuel H. Fuller

11-1. INTRODUCTION

Computer systems have evolved into a wide variety of structures, some of which are complex systems that often include more than one central processor; many input/output channels or processors; and 30, 60, or more drum and disk storage units. Even as late as the early 1960s, almost 20 years after the stored-program computer was originally developed, technology, programming systems, and user demand had not pushed computer systems beyond simple, one-job-at-a-time processing. Engineers and systems programmers received a rude awakening when multiprogramming systems were introduced with the third generation of computer systems (circa 1964) and they did not perform up to expectations. Many current computer systems are as complex as such other artificial systems as high-performance aircraft or modern skyscrapers. The discouraging fact is not the complexity of computer systems but that computer engineers do not have the range of tools to evaluate performance that aeronautical or civil engineers do.

Since performance evaluation in its current state of development is as much an art or skill as it is a science, this chapter has two purposes: first, to present those aspects of performance evaluation that are well understood and to develop the associated mathematics; and second, to discuss the additional concepts that underlie pragmatic efforts to evaluate complex computer systems. However, the best way to learn the pragmatic aspects of performance evaluation, as with any skill, is to apply the basic concepts to real problems. For this reason, several of the problems at the end of this chapter outline small performance monitors that have relatively simple implementations. The effort spent developing one of these monitors, using it in the context of a measurement experiment to collect data, and finally critically analyzing the resultant data is probably the most effective way to learn the rudiments of performance evaluation.

The need for performance evaluation does not come from a single source. Therefore, a set of measures or analysis techniques adequate for one performance evaluation study may well be inadequate for another study. In the following discussion, we partition the need for performance evaluation into three primary areas: *design, purchase,* and *optimization* studies. While this division is somewhat arbitrary, it does provide a useful vehicle to discuss the various needs for evaluation.

Performance evaluation needs to play a strong, continuing role in the design and development of new computer systems. Engineers and systems programmers should use projections of performance to guide the development of their system; designs based solely on the elegance of the system architecture or the processor's instruction set are at best academic exercises. In this realm of performance evaluation we include not only the evaluation of new central processor designs, but also the evaluation of other hardware components (for example, secondary storage units, input/output processors, etc.) as well as software systems (for example, schedulers, memory management systems, and compilers).

The second area of performance evaluation concerns the decision to purchase or lease a unit of hardware or software, or a complete computer system. This differs from performance evaluation for design in at least two important respects: configurations of the systems in question are usually operational and available for inspection and testing, and the options open to the purchaser are limited to several announced product lines rather than the complex space of alternatives open to the designer of a new product. For these reasons, performance evaluation for a purchasing decision can rely more on measurement and detailed (but costly) models than can performance evaluation for design.

The third area of performance evaluation is concerned with the optimization of a specific computer system. Typical questions in this area include: how many disk drives do we need and how should they be distributed over the available input/output channels; how much of the operating system should reside in primary memory and what modules should reside on the drum? These questions begin to multiply when we consider multiprocessor systems and we must decide which processors should control which input/output devices. The nature of performance evaluation for optimization is fundamentally different from the two other areas of performance evaluation. It most often involves incremental, and often reversible, changes. There is the opportunity for continual monitoring of a system's performance to detect trends or shifts in user demands. As inefficiencies in the system are discovered, new scheduling or memory management policies can be tried as well as reconfigurations of the hardware.

Before proceeding to discuss the elements of performance evaluation, let us first put the problem in proper perspective. When we use a computer system, the central processor is usually the most visible component of the system.

However, when we must consider the performance of a computer system, the central processor is only one of many components, albeit an important one, in the system. Moreover, the hardware is only a portion of the total cost of running a computation facility. A survey of 1974 budgets for data processing centers helps to put these comments in clearer focus [McLaughlin, 74]. The survey included 194 centers. An average of 39 percent of the total data processing budget was allocated for hardware. The remaining money was spent on salaries (47 percent) and miscellaneous costs (14 percent) such as supplies, consulting, training, and so on. Of the money spent for hardware, typically less than half was spent for the central processor and main memory; the majority of the hardware budget was spent on such peripherals as drums, disks, card readers, and line printers. These figures should indicate that maximum utilization of the central processor does not necessarily minimize the cost of running a computing system. A useful cost/performance analysis of any computer system must include some measure of the productivity of the programmer.

Following this introduction, we present the fundamental concepts of performance evaluation in section 11-2. We begin with the classic measures and parameters that are used to characterize computer systems. The shortcomings of these simple measures are discussed, but there is real value in measures that quickly enable anyone to grasp the scope of a computer system's capabilities. Careful application of these simple measures can be useful. Next, a range of stochastic, or queueing models are developed. These analytical models are excellent tools when there is a need to explore a wide range of possible alternatives or when you must understand the functional dependence of the system's performance on the controllable parameters of the system. Simulation models are discussed in section 11-4 and emphasis is placed on the construction and verification of these models. Finally, the basic techniques for evaluating the performance of an operational system are presented.

Section 11-6 covers fundamental measurement techniques. Although measurement techniques per se are of limited interest, analytical and simulation models require measurements to guide in the selection of realistic assumptions. Moreover, studies of operational systems require a set of measurement tools in order to monitor the results of the tests.

11-2. MEASURES AND PARAMETERS OF PERFORMANCE

No single measure of performance can give a truly accurate measure of a computer system's performance, just as no single parameter can characterize the performance of an aircraft or the utility of a skyscraper. Different measures are needed to characterize a computer used in the numerical solution of a set of simultaneous linear equations and a computer used in an airline reservation system. Moreover, even when we are considering a single application on a particular computer system, it is misleading to use a single parameter of performance. The following discussion considers this problem in detail.

Memory Bandwidth and Add Times

In an effort to characterize a computer system's performance, a natural tendency is to enumerate a small number of the parameters of the system. For example, the cycle time of main memory or the time to execute an add instruction have traditionally been used to indicate the power of a computer system. These parameters, however, can be misleading. A PDP 11/05 has a memory cycle time of 0.9 microseconds and an IBM 370/155 has a cycle time of 2.1 microseconds, yet by any rational measure of computational power the IBM 370/155 ranks above the PDP-11/05. Similarly, the Varian 620f has an add time of 1.5 microseconds and the Burroughs 5500 has an add time of 3 microseconds and again the Burroughs machine is a more powerful computer than is the Varian machine. The fact that we neglect such parameters as word length, the number of ways the memory is interleaved, and the structure of the input/output system resulted in the misleading comparisons. As we include more parameters in our comparison we get a better idea of the actual performance of a particular system. A list of current computers, along with a set of their major parameters, is provided on a periodic basis by *Computer Review* [74].

Although memory cycle times and add times are poor measures of performance, it is sometimes necessary to determine a small but reasonable set of performance measures based solely on parameters of the hardware structure. It is interesting to approach this problem, as have Bell and Newell [71], by asking what is the single most accurate measure or parameter of computer performance, the second most meaningful, and so on? This exercise is of limited utility but there are times when a quick gauge of a computer system's capabilities is needed and the only information available is its hardware parameters. The notion of *instructions per second*, called MIPS (millions of instructions per second), has often been proposed as an indicator of a computer system's performance. It captures the gains realized by many of the more advanced techniques of processor design, for example, cache memories, pipelining, and interleaved memory units. However, MIPS is inappropriate for vector machines such as the CDC STAR or the Texas Instrument ASC since a single instruction often results in many operand fetches and stores. The same phenomenon is seen in computers designed to interpret high-level languages: again the ratio of instruction fetches to operand fetches is low. (In fact, how low this ratio is provides a good measure of how effectively a machine embodies the concepts of high-level language emulation.) To correct this deficiency with the MIPS measure, MOPS (millions of operands per second) is sometimes used. As measures of performance, MIPS and MOPS share the same serious deficiency with the cycle time of memory—they neglect word length. As a result, the millions of bits per second (memory bandwidth) that are processed by the central processor is generally more appropriate. If we must choose a single, simple parameter of performance, memory bandwidth is a good choice.

Candidates for the second and third most critical measures of a computer system's performance are the size of main memory and the size of secondary memory in bits, respectively. It is questionable how productive it is to put this enumeration any further. However, it should be pointed out that all three measures suggested do not explicitly take into account the word length. For example, when we are processing relatively small integers 16-bit words may be sufficient to hold all numbers of interest. Hence machines with 32 bits per word, while having twice the memory bandwidth and twice as large a main memory as a machine with 16 bits per word, may perform at essentially the same rate for some applications.

MIPS, MOPS, and memory bandwidth are actually upper bounds on performance rather than first-order estimates when these measures are derived strictly from hardware characteristics of the machines. For example, the IBM 360/91 is capable of initiating a new instruction every 60 nanoseconds and hence has a potential instruction execution rate of 16.7 MIPS. However, measurements of the IBM 360/91 indicate it rarely exceeds 3 or 4 MIPS. The discrepancy results from instruction sequences that branch in such ways that the 360/91 central processor cannot fetch the instructions fast enough to keep the instruction decoder saturated. Similarly, multiprocessor systems have memory bandwidths higher than the "effective" bandwidths measured in operations. Here the difference arises from the processors not cooperating perfectly and as several processors attempt to access the same memory bank, queueing results and performance degrades. The correction of these measures from maximum (and often unattainable) rates to effective, or average rates, takes us into the evaluation of operational computer systems. We reserve the rest of this discussion for section 11-5 where we consider the evaluation of operational computer systems.

Fundamental Measures

In the remainder of this section we consider more complex, and hopefully more accurate, representations of performance than we have just seen in our discussion of hardware parameters. As we go beyond the basic parameters of a computer system in an effort to find more meaningful measures of performance, we face a bewildering array of possible measures. Although it is neither possible nor practical to give a definitive list of all the appropriate measures for computer systems, it is useful to note that measures fall into the two fundamental classes, *response time* and *throughput* measures. Response time measures, sometimes also called *waiting time* or *turnaround time*, are measured in seconds and describe the length of time from a request for service until the request is completed. For example, many time-sharing systems attempt to minimize the response time seen by the user to a request from his terminal. In this case, the response time is defined as the time from the carriage return until the user receives a response. The response time, or access time, of a disk is often an important parameter of a computer system. This is the time from when the central processor requests a record from the disk until it receives an interrupt saying that the record has been transferred into primary memory. In batch processing systems the response time is more often called turnaround time and is the interval of time from submission of the job until the job is completed and the result printed on the line printer, stored on tape, or saved on some other storage medium.

Throughput measures attempt to gauge how well the capacity of the system is being used rather than how responsive the system is to the demands of the user. The classic measure of throughput is the number of jobs per day processed. This is often refined by partitioning the jobs into classes and measuring the throughput for each class. A measure closely related to throughput is *utilization*, that is, the fraction of time a specified component is busy. Utilization is an even more direct indicator of how much of the capacity of the system is being used. Utilization studies have sometimes centered on the central processor, but a meaningful analysis of the effective utilization of the system must consider all the components of the computer system.

11-3. STOCHASTIC MODELS

When our need to describe the performance of a computer system, or a series of computer systems, exceeds the simple parameters of the hardware structure discussed in the previous section we must address the underlying stochastic nature of operating computer systems. Requests for service arriving at processors within the computer system often can only be modeled as a random process. The amount of computation required by a process (task) at a central processor or input/output processor is commonly modeled as a random variable because of the data-dependent branching in the program and the variety of tasks processors are called upon to service. The area of mathematics often called *queueing theory* encompasses the set of analytical models that most adequately describe computer systems. Over a dozen books exist on the theory of queues and over a thousand papers have been written that analyze queueing structures. On the other hand, a discouragingly large fraction of practical queueing systems continue to elude exact analysis. Some people find these two observations a pessimistic commentary on work in queueing theory; however, the situation in queueing theory is consistent with most other areas of applied mathematics: real systems are very complex mechanisms and tractable mathematical models must often be simplified approximations to the real systems.

Because of the limitations of queueing theory, simulation techniques, which are discussed in the next section, must be used in conjunction with queueing models. Simulation models can accurately model more complex structures than can queueing models. Queueing models find their main utility when qualitative and approximate quantitative answers are needed. They describe the performance of a queueing structure as a function of a number of parameters, for example, the number of central processors, speed of secondary storage, or number of on-line terminals. To get the same information through simulation often takes many hours of computation and even then the performance of the computer system over a wide range of configurations and job mixes is still not as clear as when an analytical solution is available. As stated in the introduction of this chapter, both queueing and simulation models play an essential role in the performance evaluation of computer systems.

It would be impractical to cover all of the significant results in queueing theory here. This section analyzes several queueing models that are fundamental, have application to the analysis of computer systems, and form the basis for many of the more advanced models. Every queueing structure includes the following four characteristics:

1. *Arrival process* — A mechanism that generates requests to be serviced by the processor. The delivery of card decks to a computing system is an example of an arrival process. An important statistic of the arrival process is the modeling of the interarrival times. The models discussed in this section assume that the interarrival times are random variables drawn from an *arrival-time* probability distribution function.
2. *Service mechanism* — The primary demand of a task after it arrives is that it be serviced by one or more of the processors in the system. Like interarrival times, service times are modeled as random variables with a *service-time* probability distribution function.
3. *Queueing discipline* — When requests for service arrive at a processor faster than they can be serviced, a line, or queue, forms and a policy is needed to determine the order in which outstanding requests will be processed.
4. *Routing* — When there is more than one processor in the queueing system, the manner in which requests circulate among the queues must be specified.

In the models that follow we make a number of different assumptions concerning these four characteristics in order to develop models that cover a wide range of computer structures.

Interarrival-Time and Service-Time Distributions

Poisson Arrival Process   The simplest arrival process to treat mathematically is the Poisson (completely random) arrival process. X-rays striking a Geiger counter, automobiles entering an intersection, and incoming calls to a telephone exchange, are all examples of processes that have been successfully modeled as Poisson processes. In computer systems, the arrival of individuals at a card reader, the failure of circuits in a central processor, and requests from terminals in a time-sharing system are processes that are essentially Poisson in nature. The property common to all these examples is that the events are generated from a very large population and each member in the population is acting independently. The arrival at a past or future instant does not affect the arrival, or non-arrival, at the present instant. This lack of dependence on the past (and future) is commonly called the *Markovian*, or *memoryless*, property. As we will see in a moment the independence of arrivals is responsible for the simplicity of the analysis of the Poisson arrival process.

To describe a Poisson process we begin by letting λ be the average arrival rate of the Poisson process. Figure 11-1 is a time line with marks at several *epochs*, or points in time, to denote arrivals. The fundamental assumption that an arrival is independent of all other arrivals can be stated with the following two postulates:

1. The probability of an arrival between epochs t and $t + \delta t$ is $\lambda \delta t + o(\delta t)$, where $o(\delta t)$ denotes a quantity of smaller order of magnitude than δt.*
2. The probability of more than one arrival between epochs t and $t + \delta t$ is $o(\delta t)$.

From these postulates it is possible to derive an expression for $P_n(t)$, the probability of n arrivals during an interval of duration t.

First consider $P_0(t + \delta t)$, the probability that there are no arrivals during time $t + \delta t$. The postulates indicate the number of arrivals in disjoint intervals are independent events and hence:

$$P_0(t + \delta t) = P_0(t)\, P_0(\delta t)$$
$$= P_0(t)\, \{1 - [\lambda \delta t + o(\delta t)]\}. \tag{11-1}$$

Since $o(\delta t)$ simply indicates a quantity whose order of magnitude is less than δt, multiplying $o(\delta t)$ by any coefficient independent of δt is still $o(\delta t)$. Hence Equation 11-1 reduces to:

$$P_0(t + \delta t) = P_0(t)\,[1 - \lambda \delta t] + o(\delta t)$$

and dividing by δt we obtain:

$$\frac{P_0(t + \delta t) - P_0(t)}{\delta t} = -\lambda P_0(t) + \frac{o(\delta t)}{\delta t}$$

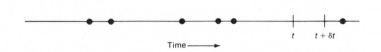

Time ⟶

Figure 11-1   Arrival epochs for a Poisson process

---

*More precisely, $\lim_{\delta t \to 0} o(\delta t)/\delta t = 0$.

Now taking the limit as $\delta t \to 0$ we are left with the simple first-order ordinary differential equation:

$$\frac{d}{dt} P_0(t) = -\lambda_0 P_0(t) \tag{11-2}$$

From the boundary condition that $P_0(0) = 1$ we find:

$$P_0(t) = e^{-\lambda t} \tag{11-3}$$

We follow a similar development to find $P_n(t)$ for $n > 0$. Now, however, there are two nonnegligible ways to get n arrivals in the interval $t + \delta t$: either n arrivals during interval t and none during interval δt or $n - 1$ arrivals during interval t and a single arrival during interval δt. Recall that the probability of more than one arrival during δt is $o(\delta t)$. The general restatement of Equation 11-1 is:

$$\begin{aligned} P_n(t + \delta t) &= P_n(t)P_0(\delta t) + P_{n-1}(t) P_1(\delta t) + o(\delta t) \\ &= P_n(t)[1 - \lambda \delta t] + P_{n-1}(t) \lambda \delta t + o(\delta t) \end{aligned} \tag{11-4}$$

$$\frac{P_n(t + \delta t) - P_n(t)}{\delta t} = -\lambda P_n(t) + \lambda P_{n-1}(t) + \frac{o(\delta t)}{(\delta t)}$$

Again, taking the limit as $\delta t \to 0$:

$$\frac{d}{dt} P_n(t) = -\lambda P_n(t) + \lambda P_{n-1}(t), n > 0 \tag{11-5}$$

For $n = 1$ the preceding differential equation reduces to

$$\frac{d}{dt} P_1(t) = -\lambda P_1(t) + \lambda e^{-\lambda t}$$

Since $P_1(0) = 0$ we find:

$$P_1(t) = \lambda t e^{-\lambda t}$$

Similarly, we can solve Equation 11-5 for $n = 2, 3, \ldots$ and in general we have:

$$P_n(t) = \frac{(\lambda t)^n}{n!} e^{-\lambda t} \tag{11-6}$$

Figure 11-2 shows the Poisson probability function for $\lambda = \frac{1}{2}, 1, 4,$ and 10. Both the mean and standard deviation of the Poisson function are λt.

Exponential Service-Time Distribution   In the previous discussion on the Poisson process we stated that the fact that it enjoys the memoryless, or Markovian, property is very important for our later work. The same statement applies

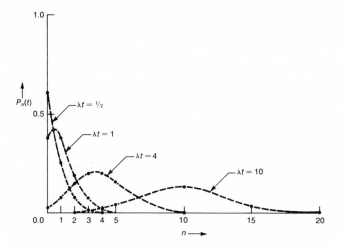

Figure 11-2 The Poisson probability density function

to service-time distributions; the exponential distribution is the only distribution to enjoy the Markovian property. Let μ be the average rate of service completions at a processor. We now make an assumption similar to the central assumption used in the development of the Poisson arrival process: let the probability of the completion of service between epochs t and $t + \delta t$ be $\mu \delta t + o(\delta t)$. The exponential distribution directly follows from this single postulate.

Let $S(t)$ be the probability that service is not completed by time t. Then

$$
\begin{aligned}
S(t + \delta t) &= S(t)\, S(\delta t) \\
&= S(t)\,\{1 - [\mu \delta t + o(\delta t)]\} \\
&= S(t)\,[1 - \mu \delta t] + o(\delta t)
\end{aligned}
$$

$$
\frac{S(t + \delta t) - S(t)}{\delta t} = -\mu S(t) + \frac{o(\delta t)}{\delta t}
$$

Taking the limit as $\delta(t) \to 0$ we obtain:

$$
\frac{d}{dt}\, S(t) = -\mu S(t)
$$

Using the boundary condition that $S(0) = 1$, we find:

$$
S(t) = e^{-\mu t}, \quad t \geq 0 \tag{11-7}
$$

$S(t)$ is commonly known as the *survivor function*; the more commonly used *distribution function* $F(t)$ is just $1 - S(t)$, so:

$$
F(t) = 1 - e^{-\mu t} \tag{11-8}
$$

and the *density function, f (t)*, is

$$f(t) = \frac{d}{dt} F(t)$$
$$= \mu e^{-\mu t}, t > 0 \tag{11-9}$$

In other words, $f(t)$ is the probability that service is completed between t and $t + \delta t$.

The similarity between the Poisson arrival process and the exponential service-time distribution is seen when we recognize that the interarrival times of a Poisson process are exponentially distributed with parameter λ.

$$Pr \{\text{interarrival time} \leq t\} = 1 - P_0(t)$$
$$= 1 - e^{-\lambda t}$$

Simple Queueing Structure: A Single Processor with Exponential Interarrival and Service Times

The first queueing structure we analyze is simple: Poisson arrival process, exponential service time, and a single, simple processor. This queueing structure is often denoted as the $M/M/1$ case.* Figure 11-3 is a schematic representation of this case. The fact that the model is simple does not imply it is of little use. On the contrary, the model is often a good initial approximation to a number of computer structures.

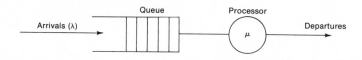

Figure 11-3 $M/M/1$ queueing structure

For example, consider a data concentrator, or preprocessor, for a computation center. The data concentrator supports the remote terminals communicating with the computation center. The function of the data concentrator is to buffer the messages from the terminals in its local memory, perform some initial preprocessing of the message, and then either pass the message on to the central computer for further processing or respond directly to the terminal

---

*This notation follows from a general scheme in the queueing literature that has the basic form $A/B/n$ where A identifies the arrival process: M for a Markovian, or Poisson process, G for a general distribution, and D for constant interarrival times; B identifies the service-time distribution and uses the same abbreviations as the arrival process; and n represents the number of processors serving the queue.

if the request is relatively trivial. A number of important questions must be answered when designing such a data concentrator:

1. How much time can be spent preprocessing each message? The answer to this question will have a direct impact on the power of the processor that is required as well as the extent of preprocessing attempted.
2. How much memory must be included in the concentrator to ensure that the message buffers overflow infrequently?
3. What is the average response time seen by terminals to trivial requests handled by the data concentrator? What is the variance of the response times?
4. What fraction of time will the data concentrator be free to either handle low priority input/output devices or run maintenance routines?

We assume the arrival of messages forms a Poisson process with an average arrival rate of λ messages per second. In addition, we model the processing time per message as an exponentially distributed random variable. For the moment let us assume a first-in, first-out (FIFO) queue discipline. (Later in this section we take a more detailed look at alternative scheduling disciplines.)

To begin the analysis of our simple queueing structure we return to the elementary considerations used in the development of the Poisson process and the exponential service-time distribution. Let $p_n(t)$ denote the probability of the system being in state E_n at epoch t, that is, having n messages in service or waiting for service. First consider $p_0(t + \delta t)$, the probability of the data concentrator being idle (in state E_0) at epoch $t + \delta t$. There are two significant ways for the processor to be idle at epoch $t + \delta t$: either to be in state E_0 at epoch t and have no new messages arrive during the interval δt, or to be in state E_1, have the single message complete its processing, and have no new messages arrive during the interval δt. The chance of more than a single arrival or departure is of order $o(\delta t)$. We can express these observations in the equation:

$$p_0(t + \delta t) = (1 - \lambda \delta t)\, p_0(t) + (1 - \lambda \delta t)\, \mu \delta t p_1(t) + o(\delta t)$$
$$= (1 - \lambda \delta t)\, p_0(t) + \mu \delta t p_1(t) + o(\delta t)$$

We can also write similar equations for $p_n(t + \delta t)$ where $n > 0$. However, it is now also possible to get to state E_n at epoch $t + \delta t$ by being in state E_{n-1} at epoch t and having an arrival during the interval δt. Hence we see:

$$p_n(t + \delta t) = (1 - \mu \delta t)\, \lambda \delta t p_{n-1}(t) + (1 - \mu \delta t)(1 - \lambda \delta t) p_n(t)$$
$$+ (1 - \lambda \delta t)\, \mu \delta t p_{n+1}(t) + o(\delta t)$$
$$= \lambda \delta t p_{n-1}(t) + [1 - \lambda \delta t - \mu \delta t] p_n(t) + \mu \delta(t) p_{n+1}(t) + o(\delta t), n = 1, 2, \dots$$

By collecting terms and taking the limit as $\delta t \to 0$ we are left with the following set of simultaneous, ordinary differential equations:

$$\frac{d}{dt} \, p_0(t) = - \lambda p_0 \, (t) + \mu p_1 \, (t) \tag{11-10a}$$

$$\frac{d}{dt} \, p_n(t) = \lambda p_{n-1} \, (t) - (\lambda + \mu) \, p_n \, (t) + \mu p_{n+1} \, (t), \quad n = 1, 2, \ldots \tag{11-10b}$$

If we supply sufficient boundary conditions, the preceding set of differential equations can be solved to give the complete, time-dependent performance of the queue [Cox and Smith, 61, Chapter 3]. Such time-dependent information can be very important if bursts of arrivals, or other severe congestion, is probable and we wish to find the expected length of time needed to alleviate the congestion.

We are often more interested in the long-term, or steady-state, behavior of the system than we are in any initial transient behavior. We can find the steady state solution by setting $d/dt \, (p_n \, (t)) = 0$ for $n = 0, 1, 2, \ldots$ When we do this we are left with the simple recurrence relations:

$$\lambda p_0 = \mu p_1 \tag{11-11a}$$

and

$$(\lambda + \mu) \, p_n = \lambda p_{n-1} + \mu p_{n+1}, \quad n = 1, 2, \ldots \tag{11-11b}$$

These recurrence relations are called the *balance equation, steady-state equation*, or *equilibrium equations* of the queueing structure. Note we have replaced $p_n \, (t)$ by p_n in Equation 11-11 since p_n is the equilibrium probability of being in state E_n, independent of any particular epoch t in time. From Equation 11-11a we find

$$p_1 = \frac{\lambda}{\mu} p_0$$

or if we define ρ to be the ratio λ/μ:

$$p_1 = \rho p_0 \tag{11-12}$$

and from Equation 11-11b we get:

$$\mu p_2 = (\lambda + \mu) \, p_1 - \lambda p_0$$

$$p_2 = \rho^2 p_0$$

Continuing to use Equation 11-11b we see in general:

$$p_n = \rho^n p_0 \tag{11-13}$$

In order to explicitly solve for p_n in terms of λ and μ, the parameters of the model, we will need to use the normalizing equation:

$$\sum_{0 \leq i < \infty} p_i = 1 \tag{11-14}$$

In other words, the equilibrium probabilities relate to disjoint events that span the space of all possibilities and hence must sum to unity.

$$\sum_{0 \le i < \infty} p_i = \sum_{0 \le i < \infty} \rho^i p_0$$

$$= p_0 \sum_{0 \le i < \infty} \rho^i$$

$$= \frac{p_0}{1 - \rho}, \rho < 1 \tag{11-15}$$

Now from Equations 11-14 and 11-15 we get:

$$p_0 = 1 - \rho$$

and in general

$$p_n = \rho^n (1 - \rho), n = 0, 1, 2, \ldots \tag{11-16}$$

Note Equation 11-15 is undefined for $\rho = 1$. In fact, the preceding analysis of the steady-state behavior of the $M/M/1$ queueing system is only meaningful for $\rho < 1$. Recall that $\rho = \lambda/\mu$ and unless $\lambda < \mu$, requests are arriving at a faster rate than they can be serviced. In other words, for $\rho \ge 1$ no steady-state solution exists because the arrival process has saturated the processor and the queue is growing without bound. ρ plays a prominent role in queueing theory and is commonly referred to as the *traffic intensity* of the queueing system.

Now let us return to the example of the data concentrator and apply the results of our analysis to answer the questions about its performance.

Equation 11-16 directly answers some of our original questions. Since p_0 indicates the processor has no outstanding messages to process we find the data concentrator free $(1 - \rho)$ of the time. The observation that the processor is *not* idle with probability ρ is a general fact that transcends the simple $M/M/1$ case: in *any* single processor queueing structure, the utilization of the processor is equal to the ratio of the arrival rate to the service rate.

Equation 11-16 also provides some guidelines on the amount of memory required to buffer memory. Suppose the data concentrator has sufficient local memory to hold k messages. Then our model indicates that the probability we need more than k buffers is:

$$Pr \{\text{more than } k \text{ buffers required}\} = \sum_{k < n < \infty} p_n$$

$$= \sum_{k+1 \le n < \infty} \rho^n (1 - \rho)$$

$$= \rho^{k+1} (1 - \rho) \sum_{0 \le n' < \infty} \rho^{n'}$$

$$= \rho^{k+1} \tag{11-17}$$

We may also be interested in the average queue length, $\bar{Q}$,

$$\bar{Q} = \sum_{0 \leq n < \infty} n p_n$$

$$= (1 - \rho) \sum_{1 \leq n < \infty} n \rho^n$$

$$= \rho (1 - \rho) \sum_{1 \leq n < \infty} n \rho^{n-1}$$

$$= \rho (1 - \rho) \sum_{1 \leq n < \infty} \frac{d}{d\rho} (\rho^n)$$

$$= \rho (1 - \rho) \frac{d}{d\rho} \left\{ \sum_{1 \leq n \leq \infty} \rho^n \right\}$$

$$= \rho (1 - \rho) \frac{d}{d\rho} \left\{ \frac{\rho}{1 - \rho} \right\}$$

$$= \frac{\rho}{1 - \rho} \qquad (11\text{-}18)$$

In order to find $\bar{W}$, the average waiting time (or response time) of a message at the data concentrator, we make use of one of the most important identities in the theory of queues: $\bar{Q} = \lambda \bar{W}$. See Conway, Maxwell, and Miller [67] or Jewell [67] for proof of this fundamental result. Hence the expected waiting time of a message at the data concentrator is simply:

$$\bar{W} = \lambda^{-1} \frac{\rho}{1 - \rho}$$

$$= \frac{1}{\mu - \lambda} \qquad (11\text{-}19)$$

It should now be clear the effect that the processing time per message has on the performance of the concentrator. Strictly speaking, as long as $\mu < \lambda$, or equivalently $\rho < 1$, the system performs properly. Figure 11-4 shows the waiting time, queue size, processor utilization, and buffer memory overflow as a function of ρ.

The expected waiting time for the $M/M/1$ case, Equation 11-19, is a special instance of the more general formula for the $M/G/1$ case:

$$\bar{W} = \lambda^{-1} \left\{ \rho + \frac{\rho^2 (1 + C^2)}{2 (1 - \rho)} \right\} \qquad (11\text{-}20)$$

where C is the coefficient of variation of service time, that is, the ratio of the standard deviation of service time to the expected service time. ($C = 1$ for an exponential distribution.) Equation 11-20 is called the *Pollaczek-Khinchine formula*.

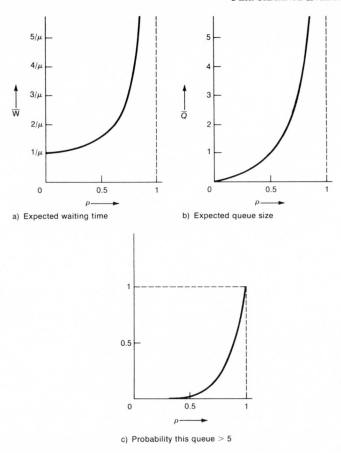

a) Expected waiting time   b) Expected queue size

c) Probability this queue > 5

Figure 11-4   Performance of the $M/M/1$ queueing structure

In the remainder of this section we develop extensions of the simple $M/M/1$ queueing structure that are often of practical interest. First we consider a few generalizations of the single processor and service-time assumptions. Later we consider some simple queueing networks that provide better models of computer systems than does the Poisson arrival assumption. Finally, we study some variations of the third major aspect of a queueing structure: the scheduling discipline.

Multiprocessors and Service-Time Distributions That Are a Function of Queue Length

We now consider the generalization of the $M/M/1$ model to an $M/M/n$ model for the analysis of multiprocessor systems. Although there has been interest in multiprocessor systems for many years, the advent of the very low cost minicomputer has resulted in the relatively recent application of multiprocessor systems to practical problems. An example of this is the development of multiprocessor message processors for the ARPA network and the construction of a multiprocessor at Carnegie-Mellon University that has (up to) 16 miniprocessors. For more information, see the discussion of multi-processors in Chapter 8.

Suppose we decide to investigate the possibility of using several processors to perform message processing. The decision to consider a multiprocessor system can stem from more than performance considerations. Stringent reliability considerations often dictate a multiprocessor design. In some cases the requirement for a highly modular system results in a multiprocessor de-sign so that each installation of the system can tailor the processing power to its local requirements.

There are several ways to organize a multiprocessor version of the data concentrator. For example, each processor could have its own private queue with each message being routed to one of the processor queues upon arrival. The messages could be assigned to the processors simply on a cyclic basis, that is, the ith message goes to processor i modulo n, where n is the number of processors in the system. A queueing structure with a higher performance is shown in Figure 11-5 where there is only a single queue and whenever a processor finishes serving a message it goes to the queue for another message.

In this model we maintain the assumptions of our first example: a Poisson arrival process, an exponential service-time distribution for processing each message, and a FIFO scheduling discipline.

First, consider a two-, or dual-, processor system. The equilibrium equation relating p_0 and p_1 is identical to the single process case, Equation 11-11a, because even though we have two processors, we assume only one processor can be used to service a single message. But now consider the next equi-librium equation:

$$(\lambda + \mu) p_1 = \lambda p_0 + 2\mu p_2 \tag{11-21}$$

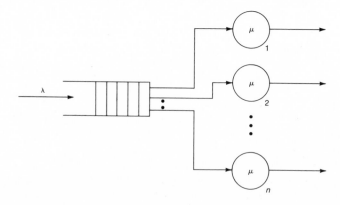

Figure 11-5 The $M/M/n$ queueing structure: a simple model of a multiprocessor

This equation is different than the corresponding equation for the single processor case. The difference is the coefficient of 2 before the last μ and indicates that when there is more than one outstanding request, both processors can service a message and the probability of a service completion from one of the processors is $2\mu\delta t$. The following recurrence relation describes the remaining equilibrium equations:

$$(\lambda + 2\mu)\, p_n = \lambda p_{n-1} + 2\mu p_{n+1}, \ n = 2, 3, \ldots \tag{11-22}$$

Solving this set of equilibrium equations in a manner analogous to the single processor case we find:

$$p_0 = \frac{1 - \rho'}{1 + \rho'} \tag{11-23a}$$

$$p_n = 2(\rho')^n \frac{1 - \rho'}{1 + \rho'}, \ n = 1, 2, \ldots$$

where $\rho' = \dfrac{\lambda}{2\mu}$ \hfill (11-23b)

We now generalize this two-processor case to the k-processor case. Let:

$$\rho' = \frac{\lambda}{k\mu}$$

and

$$S = 1 + k\rho' + \frac{(k\rho')^2}{2!} + \ldots + \frac{(k\rho')^{k-1}}{(k-1)!} + \frac{(k\rho')^k}{k!\,(1 - \rho')}$$

Then we find:

$$p_i = \frac{(k\rho')^i}{i!\,S}, \ i = 0, 1, \ldots, k - 1 \tag{11-24}$$

$$p_i = \frac{k^k(\rho')^i}{k!\,S}, \ i = k, k + 1, \ldots \tag{11-25}$$

Figure 11-6 shows the expected waiting time, expected number of processors free to do other tasks, and the probability of the queue exceeding four messages for systems with one, two, three, five, and eight processors.

An example of a queueing structure with queue dependent service-time is the following model of the shortest-latency-time-first (SLTF) file drum shown in Figure 11-7. The drum rotates at a constant angular velocity, with period τ, and the read/write heads are fixed. Blocks of information, often called records, or files, are read or written onto the surface of the drum as the appropriate portion of the drum passes under the read/write heads. Once a

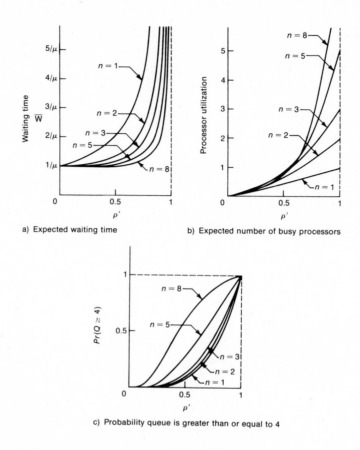

a) Expected waiting time

b) Expected number of busy processors

c) Probability queue is greater than or equal to 4

Figure 11-6 Performance of $M/M/n$ queueing structures where $n=1,2,3,5,8$

decision has been made to process a particular record, the time spent waiting for the record to come under the read/write heads is called *rotational latency* or just *latency*. With a drum storage unit organized as a file drum we do not constrain the records to be of any particular length nor do we impose restrictions on the starting position of records. For convenience, let our unit of length be the circumference of the drum.

As indicated by its name, the SLTF file drum uses the shortest-latency-time-first scheduling discipline. At all times, an SLTF policy schedules the record that will come under the read/write heads first as the next record to be transmitted. For example, in Figure 11-7, assuming the drum is not transmitting record 2, an SLTF policy will schedule record 5 as the next record to be processed.

The measurement of an operational computer system [Fuller and Baskett, 75] has shown that the starting addresses of successive input/output requests can be realistically modeled as independent random variables uniformly distributed around the circumference of the drum, and the length of the records

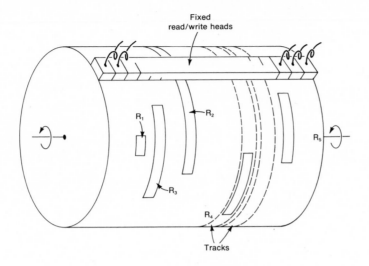

Figure 11-7   File drum

can be approximated by exponentially distributed random variables with a mean of one third of the drum's circumference.

Although more complex models of the SLTF file drum have been analyzed [Abate and Dubner, 69; Fuller and Baskett, 75], a simplification that results in a surprisingly good approximation is to model the service-time and the immediately preceding latency interval with the following queue-size dependent service rate:

$$\frac{1}{\mu_n} = \frac{\tau}{n+1} + \frac{1}{\mu} = \frac{\mu\tau + n + 1}{\mu(n+1)} \qquad (11\text{-}26)$$

where μ^{-1} is the mean record transmission time and n is the number of outstanding requests. It follows in a straightforward manner that the equilibrium equations for the drum model are:

$$\lambda p_0 = \mu_1 p_1, \qquad (11\text{-}27a)$$

$$(\lambda + \mu_n) p_n = \lambda p_{n-1} + \mu_{n+1} p_{n+1}, \quad n = 1, 2, \dots \qquad (11\text{-}27b)$$

The preceding set of recurrence relations can be solved directly with forward substitution and yield:

$$p_1 = \frac{\lambda}{\mu_1} p_0$$

$$p_2 = \frac{\lambda^2}{\mu_1 \mu_2} p_0$$

and in general:

$$p_n = \frac{\rho^n}{\mu\tau + 1}\left(\begin{array}{c}\mu\tau + n + 1\\n + 1\end{array}\right)p_0, \quad n = 0, 1, \ldots \quad (11\text{-}28)$$

where $\rho = \lambda\mu$ and $\left(\begin{array}{c}m\\n\end{array}\right)$ is the binominal coefficient $m!/n!\,(m - n)!$.

As before, the sequence $\{p_n\}$ must sum to unity, that is:

$$\sum_{0 \leq n < \infty} \frac{\rho^n}{\mu\tau + 1}\left(\begin{array}{c}\mu\tau + n + 1\\n + 1\end{array}\right)p_0 = 1$$

and hence

$$p_0^{-1} = \frac{1}{\mu\tau + 1}\sum_{0 \leq n < \infty}\left(\begin{array}{c}\mu\tau + n + 1\\n + 1\end{array}\right)\rho^n \quad (11\text{-}29)$$

With the aid of the binominal theorem the preceding relation reduces to

$$p_0 = \frac{\rho\,(\mu\tau + 1)\,(1 - \rho)^{\mu\tau+1}}{1 - (1 - \rho)^{\mu\tau+1}} \quad (11\text{-}30)$$

Using Equations 11-28 and 11-30 we see that for this model of a SLTF file drum we are able to get explicit expressions for p_n, the probability of being in state E_n. Now to find the average waiting time we use Little's formula ($\bar{Q} = \lambda\bar{W}$) where $\bar{Q}$ is the average queue length given by:

$$\bar{Q} = \sum_{0 \leq n < \infty} np_n$$

$$= \frac{\rho\,(\mu\tau + 1)}{(1 - \rho)\,(1 - (1 - \rho)^{\mu\tau+1})} - 1 \quad (11\text{-}31)$$

Finite Population Queueing Systems

We now turn our attention to modifications of the Poisson arrival assumption that has been common to our earlier models. In the queueing structure of Figure 11-8 we do not have an arrival process at all, but a closed network consisting of a single central processor and a finite number of "sources." In this model assume all the source servers have exponential service-time distributions. Note there is no queueing for the source servers, a source server has an exponential holding time and then requests service from the central server. After possibly queueing for and then receiving service at the central processor, control is returned to the source server and it begins another holding time.

The queueing structure of Figure 11-8 is a classic queueing model and is termed the *machine repairman* or simply the *finite-source* model. It is called the machine repairman model because it is often explained as a model of a set of machines in a job shop that occasionally fail. Upon failure they require

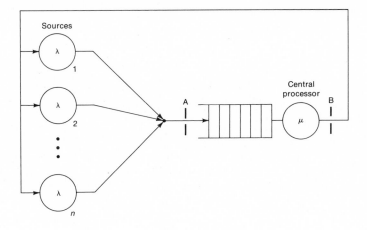

Figure 11-8   Finite-source queueing structure

maintenance from a repairman and the machines may need to queue for repair if another machine breaks down before the repairman can fix the first machine. In this context, λ is the failure rate of an operating machine and μ is the repair rate of the repairman.

The importance of the finite-source model for the analysis of computer systems becomes clear when we consider a time-shared or remote-job-entry system. The sources now correspond to users at terminals and the central server, or repairman, corresponds to the central processor.

To begin with, let us assume FIFO queueing at the central processor and in the next section we investigate more effective scheduling disciplines.

As in our previous examples, we can write down a set of equilibrium equations. Now, however, it is the arrival rate that is a function of the queue size rather than the service-time. Specifically, if k sources are processing requests at epoch t, the probability of an arrival during the interval between t and $t + \delta t$ is $k\lambda\delta t$. As before, the chance of more than one arrival during δt is $o(\delta t)$. The equilibrium equations are:

$$n\lambda p_0 = \mu p_1 \tag{11-32a}$$

$$[(n - i)\lambda + \mu]\, p_i = (n - i + 1)\,\lambda p_{i-1} + \mu p_{i+1} \quad i = 1, 2, \ldots, n - 1 \tag{11-32b}$$

$$\mu p_n = \lambda p_{n-1} \tag{11-32c}$$

Solving the preceding set of equations we get:

$$p_i = \frac{n!}{(n - i)!}\rho^i p_0$$

Using the normalizing equation, that is, $p_0 + p_1 + \ldots + p_n$ must sum to unity, we find:

$$p_0 = \left\{ \sum_{0 \le i \le n} \frac{n!}{(n-i)!} \rho^i \right\}^{-1} \tag{11-33}$$

An important application of the finite source model is the modeling to time-sharing systems. An important characteristic of interactive time-sharing systems is the response time of the central processor as seen by a user. The response time is the waiting time (W) of a request at the central processor.

Let us solve for $\overline{W}$, the average response time seen by the user of a time-sharing system. An observation that significantly simplifies our analyses is that the arrival rate of requests to the central processor (point A in Figure 11-8) must be equal to the departure rate (point B in Figure 11-8). Each user, or source, has an average "think" time of λ^{-1}, sees an average response time of $\overline{W}$, and hence is in the thinking state the following fraction of time:

$$\frac{\lambda^{-1}}{\lambda^{-1} + \overline{W}}$$

There are n users and each generates requests at a rate of λ requests per second when in the thinking state. Therefore, the central processor sees a total arrival rate at point A of:

$$n\lambda \frac{\lambda^{-1}}{\lambda^{-1} + \overline{W}}$$

or more simply:

$$\frac{n\lambda}{1 + \lambda\overline{W}} \tag{11-34}$$

There is a departure rate of μ from the central processor as long as at least one user has an outstanding request. Hence we have an average departure rate of:

$$\mu(1 - p_0) \tag{11-35}$$

Equating Equation 11-34 to Equation 11-35 and solving for $\overline{W}$ we obtain:

$$\overline{W} = \frac{n}{\mu(1 - p_0)} - \frac{1}{\lambda} \tag{11-36}$$

In Figure 11-9 $\overline{W}$ is plotted as a function of n, the number of users. It can

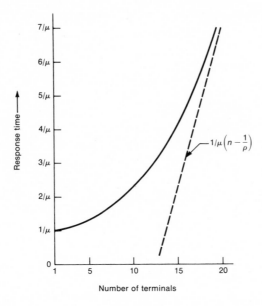

Figure 11-9 Response time for a JOSS-like system using FIFO scheduling in the central processor

be shown (see problem 11-7) that as n grows $\overline{W}$ approaches the asymptotic expression

$$\frac{1}{\mu}\left\{n - \frac{1}{\rho}\right\} \tag{11-37}$$

where $\rho = \lambda/\mu$.

These results were first discussed by Scherr [67] where he found this simple model predicted remarkably well the measured performance of MIT'S CTSS time-sharing systems. Measurements of the JOSS system [Bryan, 67] add credibility to this model because user think times were shown to be exponentially distributed (with a mean of 30 seconds).

Now consider the system in Figure 11-10 where both servers have queues. In particular Figure 11-10 is a model of a central processor/SLTF file drum system where the service time of the processor is assumed exponential and the model of the drum is the same as developed previously. A fixed number of tasks, denoted here as m, circulate in the system, alternately requesting service at the processor and the drum. Typically, the behavior of real multiprogrammed systems is more complex than this, but this model does allow us to study the feedback effects inherent in such systems. Since drums are often used as secondary storage device, this processor/drum model captures

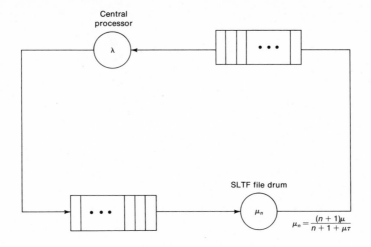

Figure 11-10   Cyclic queue model of processor/drum system

the major points of congestion, and hence the queueing behavior of many actual multiprogramming systems. The balance equations for this model are:

$$\lambda p_0 = \mu_1 \, p_1 \qquad\qquad\qquad (11\text{-}38a)$$

$$(\lambda + \mu_n) \, p_n = \lambda p_{n-1} + \mu_{n+1} \, p_{n+1}, \quad 1 \leq n < m \qquad (11\text{-}38b)$$

$$\mu_m \, p_m = \lambda p_{m-1} \qquad\qquad\qquad (11\text{-}38c)$$

The solution to these equilibrium equations follows our previous work. We see:

$$p_n = \left(\prod_{i=1}^{n} \rho_i \right) p_0, \; 1 \leq n \leq m \qquad (11\text{-}39)$$

$$p_0 = \left\{ \sum_{i=0}^{m} \prod_{j=0}^{i} \rho_j \right\}^{-1} \qquad (11\text{-}40)$$

where $\rho_n = \lambda / \mu_n$.

These expressions for the equilibrium probabilities are used to plot the expected response times of the drum and the utilization of the processor for various degrees of multiprogramming as shown in Figures 11-11 and 11-12.

Scheduling Disciplines

We now consider the third aspect of a queueing system: the scheduling discipline. In our previous discussions we have assumed a first-in, first-out (FIFO) policy. In fact, our previous results apply to more than just the FIFO

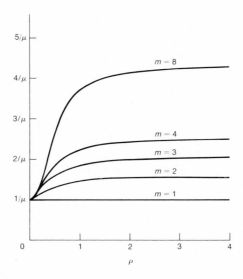

Figure 11-11   Expected waiting time at drum
$$\mu\tau = \frac{1}{3}$$

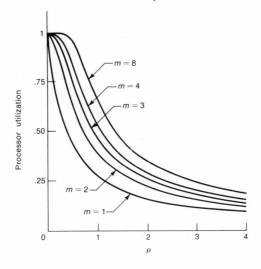

Figure 11-12   Expected processor utilization with SLTF drum
$$\rho = \frac{\lambda}{\mu}; \tau = 1; \mu = \frac{1}{3}\tau$$
$m =$ degree of multiprogramming

case. The *expected* waiting time and *expected* queue length are independent of the particular queueing discipline used, provided the policy is not based on the service-time requirements of the requests. Therefore, expressions for $\overline{W}$

and $\overline{Q}$ that we have already derived apply to such service-time independent scheduling disciplines as last-in, first-out (LIFO) and random, as well as FIFO.

Although FIFO, LIFO, and random policies share the same expression for expected waiting time and queue size, they differ in the other statistical measures of their performance. For example, the second moment, $E[W^2]$, of the waiting time for the three disciplines is:

$$E_{\text{FIFO}}[W^2] = \frac{\rho(1 + \rho)}{(1 - \rho)^2} \tag{11-41a}$$

$$E_{\text{LIFO}}[W^2] = \frac{1}{1 - \rho} E_{\text{FIFO}}[W^2] \tag{11-41b}$$

$$E_{\text{Random}}[W^2] = \frac{1}{(1 - \rho/2)} E_{\text{FIFO}}[W^2] \tag{11-41c}$$

These simple expressions can be used to find the variance, or *central* second moment from the formula:

$$\text{Var}[W] = E[W^2] - (W)^2$$

In systems operating under deadlines or attempting to minimize the frustration level of users, it is important to minimize the variance as well as the mean of the waiting time. An example of a scheduling policy that is not independent of service-time requirements is the shortest-remaining-processing-time (SRPT) policy for drums. It is a straightforward exercise in order statistics (see problem 11-10) to show that $\overline{W}$ for the simple $M/M/1$ queueing structure using SRPT is:

$$\overline{W}(t) = \frac{\rho/\mu}{1 - \rho[1 - e^{-\mu t}(1 + \mu t)]^2} \tag{11-42}$$

Note that the expected waiting time under SRPT is less than the waiting time for FIFO and the other service-time independent policies. It is a general result of scheduling theory that SRPT is the scheduling policy that minimizes the expected waiting time (and from Little's formula, the expected queue size).

Unfortunately, in most computer systems we cannot use SRPT discipline because we do not have a priori knowledge of the service-time requirements of outstanding requests. Many of the priority schemes used in computer systems attempt to approximate an SRPT discipline. For example, in the simplest case we might try to partition requests arriving at the processor into two classes: compute-bound and input/output-bound. We expect the compute-bound tasks to require a substantially longer service time than input/output bound tasks, which ask for only a small amount of processor time before requiring further service at a disk, terminal, printer, or other device. Now if we give input/output bound tasks priority over compute-bound tasks, we might expect, from analogy with the SRPT case, that we reduce the waiting time at the processor. We can generalize this two-level priority example to an *n*-level priority scheme if we have sufficient information about the tasks to classify the jobs into the *n* classes.

To understand the effect of an *n*-level priority scheduling discipline let us return to the simple *M/M/*1 queueing structure: a Poisson arrival process, exponential service-time, and a single server. Now, however, let arrivals have a priority associated with them. Arrivals with priority *i* form a Poisson arrival sequence with arrival rate λ_i and have an exponential service-time distribution with parameter μ_i, $1 \le i \le n$. In the following analysis we derive the primary performance parameters of the system.

First we make the observation that it is sufficient to analyze only a two-priority system. In any priority system with preemption, the highest priority level, call it level 1, behaves as if it were the only class of requests queueing for service. In other words, Equations 11-18 and 11-19 express $\bar{Q}_1$ and $\bar{W}_1$, respectively, if we simply replace λ by λ_1 and μ by μ_1.

For a priority level greater than 1 (denoted here as *i*), we treat an arrival in one of three ways based on the priority level (*k*) of the request. If the arrival has a lower priority than the priority level of interest ($k > i$) then it can be ignored for the same reason given when $i = 1$. The remaining arrivals are partitioned into either *low* priority requests ($k = i$) or *high* priority requests ($k < i$). In the set of equilibrium equations that follow let $\rho_{j,k}$ denote the equilibrium probability of being in a state with *j* high priority requests and *k* low priority requests. The equilibrium equation can be stated as:

$$\{\lambda_H + \lambda_L + \mu_H\zeta(j) + \mu_L\zeta(k)[1 - \zeta(j)]\}p_{j,k}$$
$$= \lambda_H p_{j-1,k} + \lambda_L p_{j,k-1} + \mu_H p_{j+1,k} + \mu_L[1 - \zeta(j)]p_{j,k+1} \quad (11\text{-}43)$$

where $j = 0, 1, 2, \ldots$; $k = 0, 1, 2, \ldots$; $\zeta(m)$ is the unit step function that is unity if $m > 0$ and zero otherwise; and any $p_{j,k}$ with a negative subscript is zero. The solution of this set of simultaneous equations defined by Equation 11-43 can be solved in a direct manner to obtain:

$$\bar{Q}_L = \sum_{0 \le j < \infty} \sum_{0 \le k < \infty} kp_{j,k} = \frac{\rho_L}{1 - \rho_L - \rho_H}\left[1 + \frac{\mu_L\rho_H}{\mu_H(1 - \rho_H)}\right] \quad (11\text{-}44)$$

Little's formula can be used to find $\bar{W}$.

A class of scheduling disciplines generally termed *feedback scheduling disciplines* have been found to be very useful in many time-sharing and multiprogramming systems. The purpose of using feedback disciplines is to give better service to short requests than is possible with simple service-time independent disciplines like FIFO and LIFO. SRPT gives the quickest service to the short requests, but as we have said before, in most computing systems we do not know the service-time requirements of the requests queued for service.

The simplest feedback scheduling discipline is the round-robin (RR) dis-

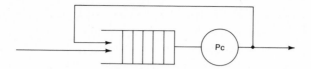

Figure 11-13 Round-robin (RR) model

cipline illustrated in Figure 11-13. In RR scheduling, arriving requests enter a FIFO queue to await service. When the request begins service at the processor it is allowed to continue for a maximum quantum of time, denoted as q. If the request completes service within the quantum it simply leaves the system. However, if service is not complete by the end of the quantum, the request is preempted from the processor and put back on the end of the queue. Service is resumed only after the request again reaches the front of the FIFO queue. This sequence is continued until the request finally completes service. The analysis of the general RR queueing model is beyond the scope of this text, but Figure 11-14 indicates its practical value.

If we let $q \to 0$ in an RR system, we get what is called *processor-sharing*. The main value of the processor-sharing model is that it results in some surprisingly simple results that aid in our understanding of feedback scheduling

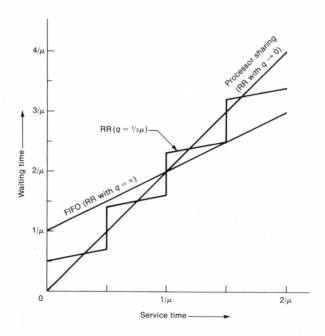

Figure 11-14 Expected waiting time for RR

schemes. Specifically, the expected waiting time for a processor-shared RR system is:

$$\overline{W}(t) = \frac{t}{1 - \rho} \tag{11-45}$$

where t interval of service required a task.

Now note that our original $M/M/1$ FIFO model can be considered the limiting case of the RR scheme when $q \rightarrow \infty$ so that Equations 11-19 and 11-45 bound the range of performance attainable with RR scheduling disciplines.

Discrete-Time Markov Chains

In the preceding queueing models of this section we have started with a set of difference equations and through a limiting process derived a set of differential equations that in turn lead to the equilibrium equations we have found so useful. Many interesting problems can be studied by using the difference equations directly and here we consider two such examples to illustrate this method.

Analysis of a Processor's Stack   A number of central processors are designed around the concept of a stack rather than a simple accumulator or set of general-purpose registers. (Chapter 7 has a detailed discussion of such pro-cessors.) Probably the best known of the stack machines is the Burroughs B5500, and its successors. The stack in the B5500 is essentially a last-in, first-out queue, or push-down stack, and the primary operations of the processor include pushing operands onto the stack; popping operands from the stack; performing unary operations on the operand at the top of the stack (for example, complement, branch of operand zero, and so on); and binary operations between the two top elements of the stack, leaving the result on top of the stack.

In order to increase the efficiency of a stack processor, the top few elements of the stack are kept in high-speed registers in the processor and the re-mainder of the stack is kept in main storage. A stack pointer in the processor identifies the beginning of the stack that resides in main storage. Figure 11-15 shows the effect of a push, pop, complement, and add on a processor with the two top elements of the stack implemented as high-speed registers. Note that when an operand is popped from the stack and when the two operands of the binary operation are replaced by the single result, an operand from main memory is not brought into memory to keep the stack registers in the processor full. In general, the best scheme is to only push or pop operands into the portion of the stack in main memory when absolutely necessary and to use flags associated with the high-speed registers to indicate which registers contain an operand. An important option in the design of a stack

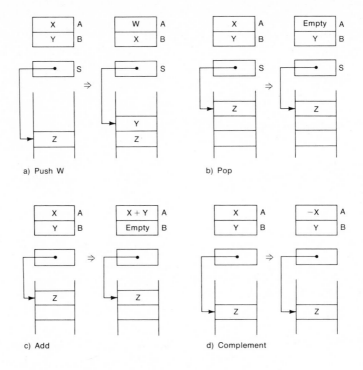

Figure 11-15   Operation of the B5500 stack

processor concerns how much of the stack should be implemented in high-speed registers. The following analysis should help answer this question.

We begin by partitioning the instructions of the stack processor into the classes listed in Table 11-1. The parameter associated with each class represents the frequency of occurrence of instructions in that class.

TABLE 11-1. INSTRUCTION CLASSES FOR A STACK PROCESSOR

Instruction Type	Frequency of Occurrence
Push	α
Pop	β
Unary operator	γ
Binary operator	δ
Other	ϵ

Now let E_n be defined as the state in which n of the high-speed registers contain operands and $P_n(i)$ be the probability that the processor is in E_n immediately after the completion of instruction i. We can now write the set of difference equations that describe the time variations in the contents of the stack.

$$P_0(i + 1) = (\epsilon + \beta)P_0(i) + \beta\, P_1(i)$$
$$P_1(i + 1) = (\alpha + \gamma)P_0(i) + (\gamma + \epsilon)P_1(i) + (\beta + \delta)P_2(i)$$
$$P_k(i + 1) = \alpha\, P_{k-1}(i) + (\gamma + \epsilon)P_k(i) + (\beta + \delta)P_{k+1}(i)$$

$$\cdot$$
$$\cdot$$
$$\cdot$$

$$P_N(i + 1) = \alpha\, P_{N-1}(i) + (\alpha + \gamma + \epsilon)P_N(i) \qquad (11\text{-}46)$$

Now in equilibrium $P_k(i + 1) = P_k(i)$ for $k = 0,1,\ldots,N$. Let p_k denote the equilibrium probability of being in state E_k. Now the preceding time-dependent equations reduce to:

$$p_k = [\alpha + \Delta_{0,k}\,\gamma]p_{k-1} + [\epsilon + a\Delta_{N,k} + \beta\Delta_{0,k} + \gamma(1 - \Delta_{0,k})]p_k$$
$$+ [\beta + (1 - \Delta_{0,k}\,\delta)]p_{k+1} \; k = 0, 1, \ldots, N \qquad (11\text{-}47)$$

where $\Delta_{i,j}$ is the Kronecker delta function that is unity if the two parameters are equal and zero otherwise.

The preceding set of simultaneous equations can be solved when we use the additional normalizing equation:

$$\sum_{0 \le k \le N} p_i = 1$$

More important than the probability of being in each of the states is the expected number of accesses to the part of the stack in main memory that are required per instruction.

$$E[\text{accesses/instruction}] = (\underline{\gamma} + 2\underline{\delta})p_0 + \underline{\delta}p_1 + \underline{\alpha}p_N \qquad (11\text{-}48)$$

Multiprocessor Model   Discrete-time Markov chain models can be used to model queueing structures for which we want to approximate service times as constant rather than exponentially distributed. For example, consider the dual processor system of Figure 11-16. In this model we assume the service times of each of the memory modules is constant and for simplicity, we assume all

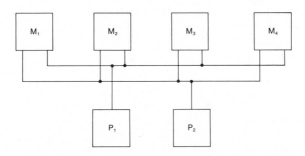

Figure 11-16   Physical structure of multiprocessor

the modules are synchronized. As in the stack processor, we now have epochs that recur at regular instants in time. If we further assume the processors are fast enough to issue new requests immediately after receiving their current request, we can model the multiprocessor by the queueing structure of Figure 11-17. Let the probability that each processor requests a word from any memory module be ¼ and assume successive requests to memory are independent events.

In the dual processor system $p_{i,j,k,l}$ denotes the equilibrium probability that there are i requests outstanding to memory module 1, j to memory module 2, k to memory module 3, and l to memory module 4. Since there are two processors:

$$i + j + k + l = 2$$

This constraint limits the number of states in the model to 10. The four states $(2, 0, 0, 0)$, $(0, 2, 0, 0)$, $(0, 0, 2, 0)$, and $(0, 0, 0, 2)$ represent the situation where both processors attempt to access the same memory module. In these cases only one request is satisfied and the other is queued and served during the following memory cycle. Because the two processors are identical and the four memory modules are also identical we see from symmetry:

$$p_{2000} = p_{0200} = p_{0020} = p_{0002}$$

and

$$p_{1,1,0,0} = p_{1010} = p_{1001} = p_{0110} = p_{0101} = p_{0011}$$

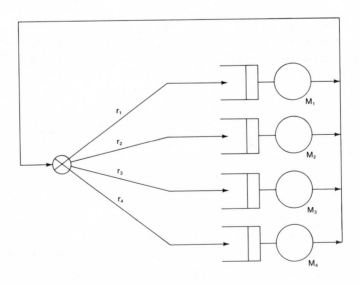

Figure 11-17 Queueing structure of multiprocessor

Therefore the two equilibrium equations for p_{2000} and p_{1100} can be simplified as follows:

$$p_{2000} = \frac{1}{4} p_{2000} + \frac{1}{16} (p_{1100} + p_{1010} + p_{1001} + p_{0110} + p_{0101} + p_{0011})$$

$$= \frac{1}{4} p_{2000} + \frac{3}{8} p_{1100}$$

$$p_{1100} = \frac{1}{4} (p_{2000} + p_{0200}) + \frac{1}{8} (p_{1100} + p_{1010} + p_{1001} + p_{0110} + p_{0101} + p_{0011})$$

$$= \frac{1}{2} p_{2000} + \frac{3}{4} p_{1100}$$

Now using the additional constraint that the equilibrium probabilities must sum to unity we obtain

$$p_{2000} = \frac{1}{16}$$

$$p_{1100} = \frac{1}{8}$$

From the equilibrium probabilities we see that the average memory bandwidth is 7/4 accesses per memory cycle (as opposed to the maximum memory bandwidth of 2).

This completes our introduction to analytical models of computer systems. There are many more queueing models that have application to computer systems and in section 11-7 we give a guide to these more advanced techniques and models. Many of the assumptions we have made here are quite restrictive and there is the question of whether they can be considered approximations to the behavior of real computer systems. In fact, a number of studies cited in this section have addressed this criticism of queueing models and they show that queueing models, when appropriately applied, can be good approximations to the actual behavior of computer systems. On the other hand, there are numerous instances when a more detailed analysis of system behavior is needed than can be provided with queueing models and other instances when the system to be analyzed simply does not lend itself to being approximated by known analytical methods. In these cases simulation techniques, which are discussed in the next section, are the most appropriate tools available to the analyst.

11-4. SIMULATION MODELS

We now turn our attention to another modeling technique: *discrete-event simulation*. While simulation models can be built to analyze arbitrarily complex structures, the price we must pay is time. For example, to reproduce the results of Figures 11-11 and 11-12 with a simulation model required approximately eight minutes of central processor time. (The model was written in ALGOL and run on a PDP-10.) If a model is simple enough to be analyzed with queueing models, it is certainly advantageous to use them.

Let's backtrack for a moment to see exactly what type of simulation is appropriate for the study of computer systems. Simulation techniques are used to study phenomena ranging from the flight dynamics of supersonic aircraft to theories of cognitive processes. The type of simulation discussed here, and the type that is most appropriate for the modeling of computer systems, is discrete event simulation. Figure 11-18 is a flowchart of the basic structure of such a simulation. A central feature is the fact that it is event driven. In other words, time is not a continuous variable nor is it incremented by uniform intervals in the model (as it might be in weather simulation or the simulation of the flight of the supersonic aircraft) but it is always advanced to the time of the next event. The execution of an event is simply the updating of the state of the simulation to reflect the occurrence of the event. After the event is processed the simulation clock is updated to the next event and the process is repeated. In many cases the length of intervals in the simulation, for

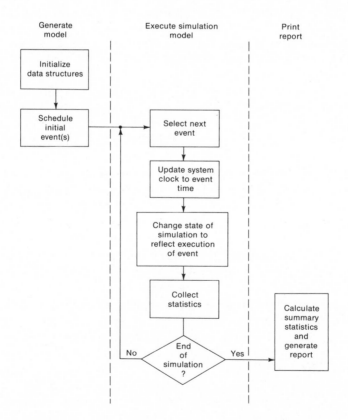

Figure 11-18   Basic structure of a simulation program

example, interarrival time of jobs, length of processing time required by jobs, number of disk accesses, and so on, are modeled as random variables, hence the term *Monte Carlo simulation*.

The other discrete-event simulators discussed here that are appropriate for the modeling of computer systems, are *trace-driven simulators*. In this case the trace-driven simulator is driven by a stream of measured event times. For example, rather than assume the arrival of jobs to a computer system to be a Poisson process with mean arrival rate λ, we record the actual interarrival times of jobs at an operational computer system and use this trace of event times to drive the simulation model. We do not consider trace-driven simulators any further here but for more details see Cheng [69] and Sherman et al. [72].

Random Number Generation

The generation of a stream of random variates is central to any Monte Carlo simulation. This section discusses the major aspects of generating a sequence of random numbers on a digital computer. The pedagogically appealing methods of flipping coins, shuffling cards, or spinning roulette wheels are too slow to be seriously considered, but analog units that use as their underlining random process the "shot" or thermal noise of an electron tube were used in early attempts to generate a sequence of random numbers. However, although relatively fast, the process produced by these analog devices shares a deficiency with coin tossing, it is nonreproducible. If we wish to run a set of simulations comparing different implementations of some policy, it is advantageous to subject them to the same random number stream. Another reason for using reproducible generators is that debugging simulation programs is much easier when random number sequences can be repeated. The RAND Corporation addressed this reproducibility issue when they published their widely used table of one million random digits [RAND, 55]. These tables are available on magnetic tape and have been used to produce random sequences for a number of applications.

It is a bit cumbersome to read streams of random digits from magnetic tape. Early in the development of computers attention turned to searching for an algorithmic method of generating a stream of random numbers. This technique was first suggested by von Neumann in 1946 when he proposed the *middle-squares* method of generating a random sequence. In the middle squares method the next random digit in the sequence is generated by squaring the current random digit and using the middle digits of the product as the next random number. Although von Neumann's middle-squares method does not generate a very random sequence [Knuth, 69], another simple recurrence does; it is the *linear congruential generator*. The general form of a linear congruential generator is:

$$x_{i+1} = (ax_i + b) \text{ modulo } m \qquad (11\text{-}49)$$

It has become common to call a linear congruential generator with $b = 0$ a *multiplicative congruential generator* and with $b \neq 0$ a *mixed congruential generator*. Although the general form of congruential generators is simple, care must be taken in the choice of a, b, and m to avoid the use of a poor generator. A famous example concerned the multiplicative generator:

$$x_{i+1} = (262147 \, x_i) \text{ modulo } 2^{35} \qquad (11\text{-}50)$$

This generator was widely used on early IBM machines [Greenburger, 66] and has the attractive property that 262147 is $2^{18} + 3$ and hence the multiplication can be implemented as one shift and three adds. The deficiency of Equation 11-50 becomes apparent when we express a random digit in the sequence as a function of the two previous digits rather than simply the immediately preceding digit.

$$x_{i+2} = (6x_{i+1} - 9x_i) \text{ modulo } 2^{35}$$

If we plot x_{i+2} as a function of x_i and x_{i+1} we see the points fall onto one of a small number of planes. Any simulation that uses Equation 11-50 to generate "random" points in three-dimensional space, for example, is likely to yield very misleading results.

With this example serving as a cautionary note, let us consider how to construct a good random number generator. There exists a rich literature on the subject of random number generators, and linear congruential generators in particular, but Knuth [69] has done an excellent job of summarizing what is known about these generators. He has suggested that the following six rules be used when constructing a linear congruential random number generator:

1. x_0, the starting random digit (sometimes called the *seed*), may be chosen arbitrarily. The last random digit of a previous simulation or the time of day are common choices for the value of x_0.
2. The number m should be large and is usually chosen to make the modulo operation trivial (for example, 2^{31} or 2^{35}).
3. If m is a power of 2, let a modulo $8 = 5$. If m is a power of 10, let a modulo $200 = 21$.
4. $\sqrt{m} < a < m - \sqrt{m}$ and $m/100 < a$.
5. Let b be an odd number (and not a multiple of 5 if m is a power of 10) such that:

$$b \cong \frac{3 - \sqrt{3}}{6} m$$

6. The least significant digits of x_i are less random than the most significant digits. Therefore, to generate a sequence of random integers from

the range $0, 1, \ldots, k - 1$, multiply x_i by k and then truncate rather than use x_i modulo k.

After selecting a specific random number generator it is important to verify that the generator is producing reasonable random sequences. There are a number of statistical tests, for example, the chi-squared test and the Kolmogorov-Smirnov test, as well as a number of empirical tests to validate the randomness of the generated sequence. The reader is encouraged to apply a number of these tests before blindly accepting a new random number generator. Knuth [69] and Fishman [74] give good introductions to the problem of verifying random number generators. The most common word lengths for computers are 32 and 36 bits, and random number generators that meet the above six rules, as well as a reasonable battery of statistical and empirical tests are:

$$x_{i+1} = (314159269 \, x_i + 453806245) \text{ modulo } 2^{31}$$

$$x_{i+1} = (3141592653 \, x_i + 2718281829) \text{ modulo } 2^{35}$$

The final point that we make here is how to generate a random variate from an arbitrary distribution. Suppose we must draw random variates from some arbitrary cumulative distribution function $F(x)$. We know that $F(x)$ is a monotonically nondecreasing function that rises from 0 to 1. Suppose we use a random number generator to obtain a random variate, call it v, uniformly distributed over the interval $[0, 1)$. Now find x_i

$$x_i = F^{-1}(v_i)$$

where $F^{-1}(v_i)$ is the inverse of $F(x)$ and is a mapping from the unit interval $[0, 1)$ into the domain of x (see Figure 11-19). It is easy to show that the sequence $\{x_i\}$ is a sequence of numbers randomly distributed with distribu-

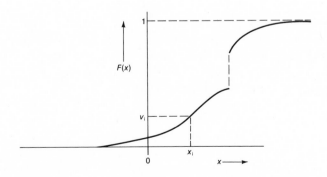

Figure 11-19   Generation of nonuniform random numbers

tion function $F(x)$. An important example is the generation of exponentially distributed random variables. If $\{v_i\}$ is a sequence of uniformly distributed random variables, then we can generate $\{e_i\}$, a sequence of exponentially distributed random variables with distribution function:

$$F(x) = 1 - e^{-\lambda x}$$

with the relation:

$$e_i = -\lambda^{-1} \ln v_i$$

Another important distribution is the normal distribution (with mean 0 and standard deviation 1)

$$F(x) = \frac{1}{\sqrt{2\pi}} \int_{-\infty}^{x} e^{-u^2/2} \, du$$

Since $F(x)$ involves an integral the preceding technique is not easily applied. However, from the Central Limit Theorem we know the sum of N independent and identically distributed random variates with mean μ and variance σ^2 is also a random variate that approaches the normal distribution for large N with mean $N\mu$ and variance $N\sigma^2$. Therefore we can approximate a normally distributed random variate with mean μ and variance σ^2 from N uniformly distributed random variates:

$$y = \sigma \left(\frac{12}{N} \right)^{1/2} \left(\sum_{1 \le i \le N} v_i - \frac{N}{2} \right) + \mu$$

To eliminate the square root computation, 12 has been a popular choice for N.

An Example: A Simple Multiprogramming System

Now that we have seen the general structure of a discrete system simulation, as outlined in Figure 11-18, and we know how to generate a sequence of random numbers to drive the simulation, let us now work through a simple example. The simulation is written in a general purpose programming language, that is, ALGOL; later we consider languages specially designed to facilitate the construction of simulation models.

Figure 11-20 is a diagram of a simple multiprogramming system. The example simulation of this system illustrates the central points of most simulation models. The computer system consists of two units: the central processor and a paging drum. The system allows at most M jobs to enter the processing loop simultaneously and if jobs arrive for service at the computer system and there are already M jobs being processed by the system, the arriving jobs enter a FIFO queue. Once a job is in the multiprogramming

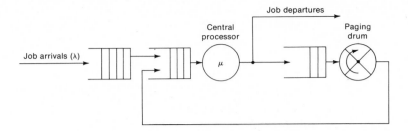

Job departures

Central
processor

Paging
drum

Job arrivals (λ)

μ

Figure 11-20   Simple multiprogramming system

mix, it alternately requests a burst of processor time and then requires a page
be accessed (read or written) from the paging drum. A job continues to alternately receive service from the central processor and drum unitl it has received
a predetermined amount of central processor time. Upon receiving the total
amount of processor time it leaves the computer system, triggering another
job to enter the multiprogramming mix if the job queue is not empty.

The following assumptions completely specify the simulation model and
will allow us to discuss the details of its structure:

1. The arriving jobs form a Poisson process with mean arrival rate λ.
2. The maximum degree of multiprogramming is M.
3. The total computation time required by a job is a random variable with
   mean η, variance σ^2, and having a log normal distribution.
4. The computation time required by a job between input/output requests
   is an exponentially distributed variable with mean μ.
5. The paging drum contains K sectors and rotates with a constant angular
   velocity with period T. When a job accesses a page from the drum it
   directs its request for a page to any one of the K sectors with equal
   probability.
6. All the queueing in the model is FIFO.

An ALGOL program that simulates the multiprogramming system model
just defined is presented here along with some discussion of the more important parts of the program. The primary data structures that define the "state,"
or current status, of the model are given in lines 102–109.

```
100   begin
101      comment Major state variables of simulation model;
102         real clock,cpumark,drummark;
103         integer seed;
104         real array arrival,cputime,eventtime[1:101];
105         integer array type,linkof[1:101];
106         integer array freeQ,jobQ,cpuQ,drumQ[0:1];
107         boolean cpubusy, drumbusy;
```

```
108        integer active,current,M,i,next,nextevent;
109        real cpuburst,latency,transfertime;
110
111     comment Parameters and Constants;
112        integer newjob,requestIO,starttransfer,transfercomplete,
113               jobcomplete,simulationdone,head,tail,null,maxN;
114        real lambda,mu,drumperiod,T,pi;
115        integer sectors,multiprog;
116
117     comment Variables for collecting summary statistics;
118        integer N,Nstart;
119        real waitsum, maxwait, waitsumsq,wait,cputotal,cpusqtotal;
120        integer cpurequests,cpuintervals;
121        real cpubusytime,cpusqbusytime,cpuservicetime,cpusqservicetime;
122        integer transfers,drumintervals;
123        real drumbusytime,drumsqbusytime,drumservicetime,drumsqservicetime;
```

These data structures are illustrated in Figure 11-21. There are several scalar variables: the *clock* that indicates the current epoch in time being simulated; the *seed* that is the last pseudo-random variate generated and is the basis for generating the next random variate; *cpubusy* and *drumbusy* are boolean variables that are true if the processor and drum, respectively, are busy processing a job.

The large array in Figure 11-21 describes the state of the jobs in the simulation. The entries in each row specify:

arrival: The time the job arrived at the computer system for service

cputime: The total amount of processor time required to complete servicing the job

eventtime: The time the event associated with the job is scheduled to occur. Some jobs will not have an event time if they are queued waiting to begin service on a processor.

eventtype: For those jobs that have an event pending, this field specifies the type of the event. The six event types are 1) arrival of a new job, 2) termination of a processor interval to request an input/output operation, 3) termination of a processor interval because the job is done, 4) the current record being read or written from the paging drum is in position to begin its transfer, 5) the paging drum has completed its current transfer, and 6) the simulation is completed.

The *link* field of each job used to link it into one of the five lists of the simulation. The event queue is a list of events, linked in order of pending execution. The head of the list is pointed to by event queue and the first element in the queue is the next event processed.

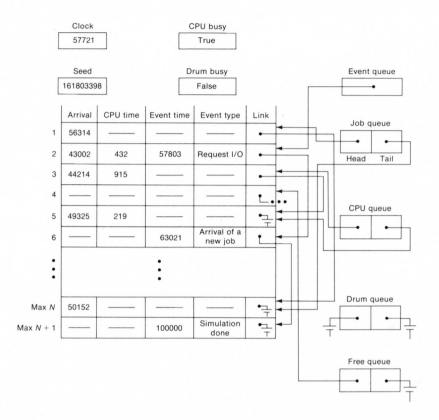

Figure 11-21   Major variables and data structures of simulation model

The job queue, processor (cpu) queue, and drum queue are the three queues shown in Figure 11-21. All are FIFO and have pointers to their head as well as their tail to facilitate the addition and deletion, respectively, from these queues.

The free queue is simply a list of unused job descriptions. Jobs are taken from this queue when they are scheduled to arrive at the computer system and are added to the free queue upon departure from the system. For simplicity, this queue is implemented here as a stack, or last-in, first-out queue.

The remaining variables declared in lines 112 through 123 are needed to collect summary statistics and define several useful constants.

The procedures *Draw, Negexp,* and *Lognormal* generate random variates with the indicated distributions. The kernel of each of these procedures is a uniform random number generator. The random number generator is not defined since it is discussed in the previous section and is machine dependent.

The fact that we use random variates to drive the simulation in this example implies our example is a Monte Carlo simulation. We could modify this example to be an event-driven simulation simply by replacing these procedures by a set of routines that read the duration of compute intervals, location of input/output requests, and so on from a tape of trace information.

```
200    real procedure Random (seed); integer seed;
201        begin
202            comment This procedure returns with equal probability
203                a real value between 0 and 1;
                .
                .
                .
206        end;
207
208    real procedure Negexp (a, seed); real a; value a; integer seed;
209    begin
210        comment This procedure generates a random variate
211            from the negative exponential distribution
212            by the method described in Section 11-4;
213        Negexp : = − a* ln (Random (seed))
214    end;
215
216    integer procedure Draw (k, seed); integer k; value k; integer seed;
217        begin
218            comment This procedure returns with equal probability one
219                of the integers 0, 1, . . . , k − 1;
220            Draw : = k * Random (seed)
221        end;
222
223    real procedure Lognormal (a, b, seed); real a, b; integer seed;
224        begin real x, y;
225            comment This procedure returns a random variate
226                from the lognormal distribution with mean a
227                and standard deviation b;
228            x : = ln (a↑4/(a↑2 + b↑2))/2;
229            y : = ln((b/a)↑2 + 1);
230            Lognormal : = exp (x + sqrt (− 2*y* ln (Random (seed)))*
231                            sin (2*pi* Random (seed)))
232        end;
```

The next part of our example consists of a set of procedures to facilitate the maintenance of the queues in the model. The *Schedule* procedure adds a job to the queue of pending events and *Queue* adds a job to one of the three FIFO queues. The procedure *Cpuidle* is used to put the central processor to sleep when no jobs are awaiting service and *Dispatch* initiates the processing of a job on the processor.

```
300    procedure Schedule (k); integer k; value k;
301        begin comment Insert process [k] into calender of pending events;
302        if linkof [nextevent] = null        or
303            eventtime [k] ≤ eventtime [nextevent] then
304            begin
305                if eventtime [nextevent] < eventtime [k] then
```

```
306            begin linkof [nextevent] : = k; linkof [k] : = null end
307            else begin linkof [k] : = nextevent; nextevent : = k end
308         end
309      else
310         begin
311            integer i, lasti;
312            lasti : = nextevent;
313            i : = linkof [nextevent];
314            while (eventtime [i] ≤ eventtime [k] ∧ linkof [i] ≠ null) do
315               begin
316                  lasti : = i;
317                  i : = linkof [i]
318               end;
319            if linkof [i] = null ∧ eventtime [i] ≤ eventtime [k] then
320               begin last i : = i; i : = linkof [i] end;
321            linkof [lasti] : =  k;
322            linkof [k] : = i
323         end
324      end;
325
326   procedure Queue (Qname, i); integer array Qname; value i; integer i;
327      begin comment Add process [i] to end of queue;
328         if Qname [head] = null then Qname [head] : = Qname [tail] : = i
329         else Qname [tail] : = linkof [Qname [tail]] : = i;
330         linkof [i] : = null
331      end;
332
333   procedure Cpuidle;
334      begin real busytime;
335         comment Change Cpu to idle condition because no jobs
336            are waiting for service;
337         cpbusy : = false;
338         comment collect statistics;
339            busytime : = clock − cpumark;
340            cpumark : = clock;
341            cpubusytime : = cpubusytime + busytime;
342            cpusqbusytime : = cpusqbusytime + busytime↑2;
343            cpuintervals : = cpuintervals + 1
344      end;
345
346   procedure Dispatch;
347      begin integer next; comment Dispatch next process on cpu;
348         next : = cpuQ [head];
349         cpuQ [head] : = linkof [next];
350         cpuburst : = Negexp (mu, seed);
351         if cpuburst < cputime [next] then
352            begin
353               comment cpu execution will be interrupted by a request
354                  to the drum;
355               cputime [next] : = cputime [next] − cpuburst;
356               eventtime [next] : = clock + cpuburst;
```

```
357              type [next] : = request IO
358          end
359       else
360          begin
361             comment job will run to completion;
362             cpuburst : = cputime [next];
363             eventtime [next] : = clock + cputime [next];
364             type [next] : = job complete
365          end;
366       Schedule (next);
367       comment collect statistics;
368          cpurequests : = cpurequests + 1;
369          cpuservicetime : = cpuservicetime + cpuburst;
370          cpusqservicetime : = cpusqservicetime + cpuburst↑2
371    end;
```

Lines 400 through 449 initialize the state of the simulation and begin the simulation by scheduling the arrival of the first job.

```
400    comment Define constants;
401       pi: = 3.141592;
402       null : = 0;
403       head : = 0;
404       tail : = 1;
405
406       newjob        : = 1;   transfercomplete : = 4;
407       request IO    : = 2;   jobcomplete      : = 5;
408       starttransfer : = 3;   simulationdone   : = 6;
409
410    comment Initialize the state of the simulation model;
411       clock : = 0;
412       M : = 0;
413       jobQ [head] : = cpuQ [head] : = drumQ [head] : = null;
414       freeQ [head] : = 1;
415       jobQ [tail] : = cpuQ [tail] : = drumQ [tail] : = null;
416       freeQ [tail] : = maxN;
417       for i : = 1 until maxN − 1 do linkof [i] : = i + 1;
418       linkof [maxN] : = null;
419    comment lambda        = mean time between arrivals.
420            mu            = mean cpu service time between I/O requests.
421            mu*10         = mean cpu service required by each job.
422            drumperiod    = time required for 1 drum revolution.
423            sectors       = number of sectors (pages) that can be
424                            transfered in a revolution of the drum.
425            seed          = random number generator seed.
426            T             = Total length of simulation.
427            multiprog     = allowable degree of multiprogramming;
428       Read (lambda, mu, drumperiod, sectors, seed, T, multiprog);
429       eventtime [maxN + 1] : = T;
430       linkof [maxN + 1] : = null;
431       nextevent : = maxN + 1;
432       type [nextevent] : = simulationdone;
```

```
433
434        comment Initialize accumulators;
435            N : = Nstart : = 0;
436            waitsum : = maxwait : = waitsumsq : = 0;
437            cpuintervals : = cpurequests : = 0;
438            cpubusytime : = cpusqbusytime : = 0;
439            cpuservicetime : = cpusqservicetime : = 0;
440            drumintervals : = transfers : = 0;
441            drumbusytime : = drumsqbusytime : = 0;
442            drumservicetime : = drumsqservicetime : = 0;
443
444        comment Schedule arrival of first job to start simulation;
445            active : = freeQ [head];
446            freeQ [head] : = linkof [active];
447            eventtime [active] : = Negexp (lambda, seed);
448            type [active] : = newjob;
449            Schedule (active);
```

Lines 501 through 629 constitute the main loop of the simulation model. Figure 11-22 is a flowchart that shows major computations of the simulations and is an expanded version of Figure 11-18. Note that for each of the five events, the simulation updates the state of the model, schedules a future event or queues a request when the unit is busy, and collects summary statistics.

```
500        comment Enter main loop of simulation;
501        while type [nextevent] ≠ simulationdone do
502          begin
503            comment Update calender and clock and select next event:
504                active : = nextevent;
505                nextevent : = linkof [active];
506                clock : = eventtime [active];
507
508            if type [active] = newjob then
509              begin comment A new job has arrived at computer system;
510                  arrival [active] : = clock;
511                  cputime [active] : = Lognormal (10*mu,4*mu,seed);
512                  comment collect statistics;
513                      cputotal : = cputotal + cputime [active];
514                      cpusqtotal : = cpusqtotal + cputime [active]↑2;
515                      Nstart : = Nstart + 1;
516                  comment put the new job in cpu system if possible;
517                  if multiprog = M then Queue (jobQ,active)
518                    else begin
519                      M : = M + 1;
520                        comment wakeup cpu if necessary;
521                        if cpubusy then Queue (cpuQ,active)
522                          else
523                            begin
524                                cpubusy : = true;
525                                cpumark : = clock;
```

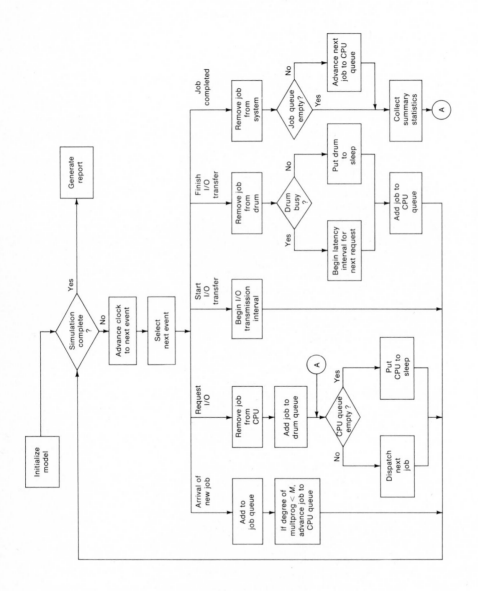

Figure 11-22 Flowchart of multiprogramming model

```
526                            Queue (cpuQ,active);
527                            Dispatch
528                        end
529                    end;
530                comment Setup arrival of next job;
531                if freeQ [head] ≠ null then
532                    begin
533                        current : = freeQ [head];
534                        type [current] : = newjob;
535                        eventtime [current] : = clock + Negexp(lambda,seed);
536                        freeQ [head] : = linkof [current];
537                        Schedule(current)
538                    end
539                else begin
540                        Write ("Insufficient number of jobs declared.");
541                        goto Terminate
542                    end
543            end

545        else if type [active] = request IO then
546            begin comment Process currently executing on cpu has requested
547                        a page of information from the drum;
548                comment Update state of cpu;
549                    if cpuQ [head] = null then Cpuidle else Dispatch;
550                comment Move current process to drum;
551                    if drumbusy then Queue(drumQ,active)
552                    else begin
553                            drumbusy : = true;
554                            drummark : = clock;
555                            latency : = drumperiod*Draw(sectors,seed)/sectors;
556                            eventtime [active] : = clock + latency;
557                            type [active] : = starttransfer;
558                            Schedule (active)
559                        end
560            end

562        else if type [active] = starttransfer then
563            begin comment Drum is in position to begin transfer
564                    a page to main memory;
565                transfertime : =drumperiod/sectors;
566                eventtime [active] : = clock + transfertime;
567                comment record statistics;
568                    drumservicetime : = drumservicetime + (transfertime + latency);
569                    drumsqservicetime : = drumsqservicetime + (transfertime + latency) ↑2;
570                    transfers : = transfers + 1;
571                type [active] : = transfercomplete;
572                Schedule (active)
573            end

575        else if type [active] = transfercomplete then
576            begin comment Move job to cpu and update drum state;
```

```
577          if drumQ [head] = null then
578            begin real busytime;
579              drumbusy : = false;
580              comment record drum statistics;
581                drumintervals : = drumintervals + 1;
582                busytime : = clock − drummark;
583                drumbusytime : = drumbusytime + busytime;
584                drumsqbusytime : = drumsqbusytime + busytime↑2
585            end
586          else begin
587              next : = drumQ [head];
588              drumQ [head] : = linkof [next];
589              latency : = drumperiod*Draw(sectors,seed)/sectors;
590              eventtime [next] : = clock + latency;
591              type [next] : = starttransfer;
592              Schedule(next)
593            end;
594          if cpubusy then Queue(cpuQ,active)
595          else
596            begin
597              cpubusy : =true;
598              cpumark : =clock;
599              Queue(cpuQ,active);
600              Dispatch
601            end
602        end

603
604        else if type [active] = jobcomplete then
605          begin comment Job has completed all its processing time
606                and is leaving the computer system;
607            Queue(freeQ,active);
608            comment Promote a job from the jobQ to cpuQ if possible;
609              if jobQ [head] ≠ null then
610                begin
611                  current : = jobQ [head];
612                  jobQ [head] : = linkof [current];
613                  linkof [current] : = null;
614                  Queue(cpuQ,current);
615                  Dispatch
616                end
617              else begin
618                  M : = M − 1;
619                  if cpuQ [head] ≠ null then Dispatch else Cpuidle
620                end;
621            comment Collect the waiting time for departing job;
622              N : = N + 1;
623            wait : = clock − arrival [active];
624            waitsum : = waitsum + wait;
625            waitsumsq : = waitsumsq + wait↑2;
626            if wait > maxwait then maxwait : = wait
627          end

628
629      end;
630
631
```

```
632   Terminate:
633      comment obtain final statistics;
634        if cpubusy then
635          begin
636            busytime : = T − cpumark;
637            cpubusytime : = cpubusytime + busytime;
638            cpusqbusytime : = cpusqbusytime + busytime↑2;
639            cpuintervals : = cpuintervals + 1
640          end;
641        if drumbusy then
642          begin
643            busytime : = T − drummark;
644            drumbusytime : = drumbusytime + busytime;
645            drumsqbusytime : = drumsqbusytime + busytime↑2;
646            drumintervals : = drumintervals + 1
647          end;
648
649      comment Calculate summary statistics;
650      begin real cpuutil,cpubusyave,cpubusySD,cpuserviceave,cpuserviceSD,
651          drumutil,drumbusyave,drumbusySD,drumserviceave,drumserviceSD,
652          waitave,waitSD,cpujobave,cpujobSD;
653
654        procedure MeanSD(sum,sumsq,N,ave,SD);
655        real sum,sumsq,ave,SD; integer N; value sum,sumsq,N;
656        begin
657          comment compute the mean and standard deviation of the given values;
658          if N ≠ 0 then
659            begin
660              ave : = sum/N;
661              SD : = sqrt(sumsq/N − ave↑2)
662            end
663          else begin ave : = 0; SD : = 0 end
664        end;
665
666        cpuutil : = cpubusyime/T;
667        MeanSD(cpubusytime,cpusqbusytime,cpuintervals,cpubusyave,cpubusySD);
668        MeanSD(cpuservicetime,cpusqservicetime,cpurequests,cpuserviceave,cpuserviceSD):
669
670        drumutil : = drumbusytime/T;
671        MeanSD(drumbusytime,drumsqbusytime,drumintervals,drumbusyave,drumbusySD);
672        MeanSD(drumservicetime,drumsqservicetime,transfers,drumserviceave,drumserviceSD);
673
674        MeanSD(waitsum,waitsumsq,N,waitave,waitSD);
675        MeanSD(cputotal,cpusqtotal,Nstart,cpujobave,cpujobSD);
676
677      comment Print summary of statistics;
            .
            .
            .

699      end
700   end
```

Lines 649 through 700 calculate and print a summary of the statistics that are collected.

After considering the simulation model of this section it can be seen that there are a number of procedures and concepts that are common to any simulation. A number of languages have been developed in recognition of this fact and the simulation languages that have gained the widest acceptance include GPSS, SIMSCRIPT, and SIMULA. While it is beyond the scope of this chapter to go into these languages in any detail, we should recognize that they all are designed to accomplish the following things:

1. Generate random variates
2. Create, modify, and generally describe processes (jobs) that move through the simulation
3. Delimit and sequence the phases of a process
4. Facilitate the queueing of processes
5. Collect, generate, and display summary statistics

Estimation of the Variance of Simulation Results

A difficult problem concerning any simulation is the determination of the variance of the summary statistics. In any well-designed simulation, the precision of the results improves as the simulation is run longer, and this phenomenon is called *stochastic convergence*. The problem with stochastic convergence is that it is slow. In order to double the precision of a simulation, we must quadruple the number of events simulated. This section discusses the number of input/output requests that must be simulated in order to achieve a specified precision in the summary statistics.

The simulation's estimate of the mean of a random variable, call it $\overline{X}$, is:

$$\overline{X} = \frac{1}{n} \sum_{1 \le i \le n} X_i \tag{11-51}$$

where n is the number of times an event is simulated, and X_i is the value of the ith occurrence of the event. A few examples of events of interest in this simulation are: job waiting time, input/output waiting time, central processor busy intervals, and input/output busy intervals.

Let σ^2 be the variance of X and $\sigma_{\overline{X}}^2$ be the variance of the sample mean $\overline{X}$. Suppose the X_is are independent, identically distributed (iid) random variables; successive processor busy intervals, or input/output busy intervals are good examples of iid random variables. The Central Limit Theorem guarantees that the sample mean is normally distributed for large n and from the linearity properties of the expectation operation [Parzen, 60, p. 206] it follows that:

$$\sigma_{\overline{X}}^2 = \frac{\sigma^2}{n} \tag{11-52}$$

Hence the standard deviation of $\overline{X}$, which we will use as a convenient measure of precision, is:

$$\sigma_{\overline{x}} = \frac{\sigma}{\sqrt{n}} \tag{11-53}$$

This expresses a fundamental property of Monte Carlo simulation: the accuracy of the simulation's summary statistics is proportional to $1/\sqrt{n}$. While this simple analysis is sufficient for the estimation of busy intervals and processor utilization, it is not adequate to estimate the precision of the sample mean of events that are correlated; input/output waiting time is an important instance of a random event that is likely to be correlated with a neighboring event. In other words, if request i experiences an unusually long wait time, it is very likely request $i + 1$ will also have a long wait time since it was probably queued behind request i for most of i's waiting time. A simple analysis, however, can still be of value even in this more complex case. The variance of the sum of a set of identically distributed random variables is:

$$\sigma_{\overline{X}}^2 = \frac{\sigma^2}{n} + \frac{2\sigma^2}{n} \sum_{k=1}^{n} \left(1 - \frac{k}{n}\right) \rho\,(k) \qquad (11\text{-}54)$$

where $\rho(k)$ is the autocorrelation coefficient and:

$$\rho\,(k) = \frac{1}{\sigma^2\,(n-k)} \sum_{1 \le i \le n-k} (X_i - \overline{X})\,(X_{i+k} - \overline{X}) \qquad (11\text{-}55)$$

Simulators described in this chapter are capable of executing 10^4 to 10^6 events in a reasonable amount of time, and the autocorrelation coefficient for waiting times is usually insignificant for lags greater than 100. Therefore, Equation 11-55 can be approximated by:

$$\sigma_{\overline{X}}^2 = \frac{\sigma^2}{n} \left[1 + 2 \sum_{1 \le k \le 100} \rho\,(k) \right] \qquad (11\text{-}56)$$

This expression for $\sigma_{\overline{X}}^2$ can be used in place of Equation 11-52 to estimate the precision of the simulation results. Another practical technique for estimating the precision of the sample mean of correlated events is discussed by Fishman [67]. He suggests a spectral analysis of the time series of successive events to determine the equivalent number of independent events, and then uses Equation 11-52 to estimate the precision of the sample mean.

11-5. EVALUATION OF OPERATIONAL COMPUTER SYSTEMS

We now consider the evaluation of actual computer systems, not models of computer systems. The analytical and simulation methods we have just discussed are valuable tools to the analyst: however, it should also be recognized that the limitations of present analytical techniques and the expense of detailed simulations require these modeling techniques be augmented with the evaluation of an actual system processing program. Moreover, it is often useful to compare the results of analytical or simulation models with the performance of the real system in order to test the validity of the assumptions underlying the mathematical models.

We begin this section with the old but still valuable technique of determining the instruction mix of a processor. An instruction mix simply enumerates the relative frequency of the use of each instruction in a processor's instruction set. After our discussion of instruction mixes, we move on to consider kernels, benchmarks, and synthetic jobs. These techniques provide a more complete measure of a computer system's performance and are widely used in the evaluation of machines for purchase as well as aids in configuration and optimization studies.

Instruction Mixes

Instruction mixes address the need to get a more accurate measure of a central processor's performance than is given by the add time. Instruction mixes are also an important indicator of the way a specific architecture is used and can be a valuable guide to engineers concerned with implementing a particular processor architecture. Table 11-2 lists five instruction mixes. Care must be exercised when we attempt to extrapolate general conclusions from these mixes. A particular instruction mix is often the result of a hopefully representative, but nonetheless restricted, set of programs. Probably the most striking observation from Table 11-2 is the small percentage of the time spent doing arithmetic operations and the large percentage of time spent doing loads, stores, and branching. These instruction mixes further highlight the weakness of using the add time (or worse, the multiply or divide time) as a measure of the performance of the processor. The frequency of branch instructions, roughly a third, explains why pipelined instruction execution units are so difficult to design. It is very hard to predict what instructions will be executed very far into the future.

With the instruction mix, and the timing for each instruction, we can estimate the MIPS rate for a processor. As already indicated, however, it is difficult to account for such advanced processor features as caches, pipelining, and interleaved memory in the MIPS calculation. The performance of these processors is dependent on the order in which instructions are executed and not just instruction frequencies.

The main value of the instruction mix seems to be in processor design. The efficient implementation of any architecture can be significantly aided by knowledge of the instruction mix. Particularly when designing a processor via microprogramming, the instruction mix gives a good indication which instructions need to be implemented as efficiently as possible and which instructions can be implemented so as to conserve the size of the microprocessor's control store. Mixes are also useful for high performance machines such as the IBM 360/91, CDC 6600, and their descendants. These processors contain several

TABLE 11-2. INSTRUCTION MIXES

Instruction Type	IBM 704/650, Gibson [70]	Arbuckle [69]	PDP-10, Lunde [74]	Scientific Mix, Knight [66]	Commercial Mix, Knight [66]	CDC 3600, Foster et al. [71]
Data Transmission						
Load, Store, and Move	31.2	28.5	42.4			30.0
Data Manipulation						
Fixed						
Add/Subtract	6.1		12.4	10.0	25.0	1.2
Multiply	0.6		1.1	6.0	1.0	0.1
Divide	0.2		0.5	2.0		0.1
Floating						
Add/Subtract	6.9	9.5	4.9	10.0		0.5
Multiply	3.8	5.6	2.6			0.5
Divide	1.5	2.0	1.1			0.2
Other (for example shift, logical operations)	6.0	22.5	4.9			2.7
Indexing						13.4
Program Control						
Compares	3.8					1.2
Branches	16.6	13.2	28.2			38.3
Other		18.7	1.9	72.0	74.0	11.9

independent arithmetic units and instruction mixes indicate the traffic these units must handle.

Kernels, Benchmarks, and Synthetic Jobs

Since we generally lack models of program behavior more sophisticated than instruction mixes, when we need more information than instruction mixes can provide we turn to actual programs, or fragments of programs, and run these on the computer, or simulator if we are designing a new computer. The techniques discussed here are the set of tools, in order of increasing generality, available for determining the performance of an operational computer system.

Kernels   A kernel is a fragment of a program that includes the most frequently executed portion, or "inner loop"; it is the nucleus of the program. Examples of typical kernels include the inversion of a matrix, calculation of a Social Security tax, or editing and formatting a page of text for a line printer.

We may use kernels to overcome some of the shortcomings of instruction mixes to evaluate the instruction set of a processor. For example, processors such as the Burroughs B6700 have a sophisticated stack mechanism that significantly simplifies the programming of procedure calls, coroutine linkage, and so on. Simply examining the instruction mix of the B6700 gives us no measure of the gains in performance with this mechanism. (Many features of the instruction set, such as the B6700 stack, were implemented to ease the programming of the machine, not primarily to increase performance. However, it is important to know the performance of a processor architecture, as well as its "elegance.") Another good example is the PDP-11. This minicomputer has a substantially richer set of addressing modes than most other minicomputers and the value of these addressing modes cannot be measured by simply comparing the instruction execution rate to several other minicomputers. The PDP-11 may well need to use fewer instructions than other minicomputers to execute an algorithm.

In addition to the power of the instruction, kernels can also provide an effective measure of many of the more advanced implementation concepts found in many processors, for example, cache memories, pipelined instruction execution schemes, and multiple arithmetic units. The effectiveness of all these concepts is dependent on the details of the sequence of instructions executed such as the size of the inner loop calculations, the relative positions of branch instructions, and the memory reference pattern used to access operands. All of these details of program execution are reflected in the kernel's performance.

Although kernels offer a significant improvement over instruction mixes, they have a number of limitations. First, and most significantly, they measure the performance of the processor/main memory and do not evaluate the input/output structure at all. If the performance of the entire computer system is needed, other methods must be used. While it is relatively easy to measure the instruction mix over a wide range of programs and apply the mix to processors with similar architectures, kernels often must be specifically coded for the machine under evaluation and consequently it is usually not feasible to collect more than five to ten kernels for a given study. The question then arises as to how to choose the small set of kernel programs that constitute a representative mix for the machine. To some extent this can be answered by comparing the instruction mix generated by the kernels with the instruction mix of the operational computer (if an operational version of the computer exists).

Benchmarks   To provide a measure of the performance of the entire computer system, rather than just the central processor and main memory, we must extend the concept of a kernel to include input/output operations. Specifically, a benchmark is a complete program considered to be representative of a class of programs and that includes input/output operations. Examples of benchmarks include polytape sort programs, chess playing programs, and program compilation.

There are a few other important differences between kernels and benchmarks than simply the inclusion of input/output operations in benchmarks. Note the performance of a benchmark may be signficantly affected by the configuration of the computer system in addition to the speed of the processor. It is no longer sufficient to say we have run our benchmark on a CDC 6600; we now need to include the amount of main storage available, which may well affect the number of overlays required or page faults sustained, as well as the configuration of the input/output devices. This is certainly useful when you must decide between two or more specific systems, but often is troublesome when you are interested in comparing two computer organizations, independent of a specific installation configuration.

To more accurately estimate the performance of a computer system executing an actual job stream, it is necessary to measure a mix of benchmarks on the system rather than running one benchmark at a time. The overhead incurred by the operating system in managing the multiprogramming of the mix is a good estimate of the overhead that will actually be incurred. The need to run a mix of benchmarks can also apply to the evaluation of the processor. For example, the effectiveness of a cache is in part dependent on how often a running program goes through a transient phase where performance is poor while the cache is loaded.

Benchmarks share a number of disadvantages with kernels. It is often costly to collect a set of benchmarks and after they are collected we face the problem of evaluating how representative they are of the real workload.

It is a common practice to program benchmarks in a higher-level language such as FORTRAN, ALGOL, or PL/I. This not only eases the task of generating the benchmarks, it makes them somewhat machine independent, so one set can suffice for the evaluation of a number of systems with different architectures. It also provides a measure of the effectiveness of the language translators provided on the machines in question.

Synthetic Jobs   A synthetic job, as the name implies, is an artificial program that exercises all the components of the computer system in a manner representative of an actual job mix. Synthetic jobs are highly parameterized to allow them to simulate a wide range of program performance. They do not refine the evaluation of computer system performance provided by a benchmark; they exist only to ease the construction of benchmark-like jobs.

A number of articles [Hamilton and Kernighan, 73; Sreenivasan and Kleinman, 74] describe techniques for measuring an operational system's average number of disk accesses, processor demands, and storage requirements per job. These measurements are then used to construct a representative set of synthetic jobs.

The actual algorithm performed, as well as the information read and written from secondary storage units, is often trivial. For example, the kernel of the synthetic job is commonly nothing more than a sum of integers. The important point is that all phases of the synthetic job be parameterized, and cyclic, so that arbitrary ratios of compute to input/output operations can be generated.

There is little reason to write synthetic jobs in anything but a high-level language. Buchholtz [69] has written a synthetic job that is less than 100 PL/I statements but is able to mimic a wide range of benchmarks.

Before leaving this section let us return to our earlier statement in section 11-2 that said if we have to use a single measure to characterize computer performance, the maximum memory bandwidth is the leading candidate. If we now allow ourselves to consider measures besides direct parameters of the hardware, the performance of a set of benchmarks, or an abstraction to a set of synthetic jobs, is probably a more accurate single measure than is the maximum memory bandwidth.

11-6. MEASUREMENT

We conclude this chapter on performance evaluation with a discussion of measurement techniques. The analytical and simulation methods we have already discussed are of little utility if their initial assumptions do not model reality. For example, Figure 11-23 is the cumulative distribution function of the duration of compute time intervals between input/output operations [Pinkerton, 69]. Highly skewed distributions such as this have been measured on a variety of computer systems; models that assume the compute-time intervals are drawn from a normal distribution, particularly when the mean is greater than the standard deviation, are immediately suspect. On the other hand, blind measurement, without a model or hypothesis to guide the measurement experiment, is also of little use. It is easy to collect many tape reels

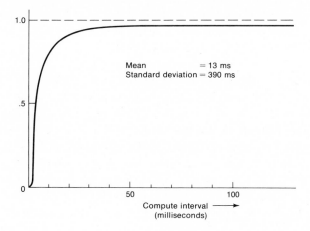

Figure 11-23 Cumulative distribution of compute time between input/output requests

of largely meaningless data if measurements are not guided by some model of system behavior.

The simplest measurement tools are often the processor console and maintenance panels. For example, most processors include a wait light on the console. This light indicates the state of the wait flag in the processor state. When the wait flag is set, the processor is idle. (Older processors simply went into a tight one- or two-instruction loop whenever there was nothing to do, but this degrades performance if other processors or input/output channels are trying to access memory and must wait for the "idle" processor to fetch its instructions.) Simply watching this wait light can give a crude approximation of the utilization of the processor. A common refinement is to attach the simple circuit shown in Figure 11-24 to the wait light. It is an averaging circuit that

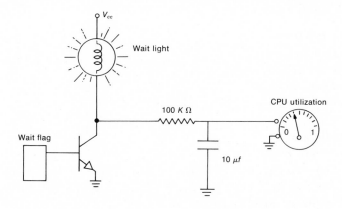

Figure 11-24 Wait meter

measures the voltage across the wait light with a voltmeter. Another simple but sometimes useful technique is to watch the maintenance panel, or read/ write arms of the drum or disk storage units. If one disk appears to be experiencing considerably heavier traffic than another disk, reallocation of files between the two devices may improve system throughput. While these observations of system performance are fun and easy to perform (everyone enjoys a short walk around the machine room occasionally), they are no substitute for more thorough and precise measurements. Display lights on the console and other panels are generally indicators of utilization, and there are many other measures that are of interest. The next section outlines the range of performance monitors that is of practical interest.

Types of Performance Monitors

Performance monitors for computer systems are available in many forms, often designed to measure very different parameters of the operational system. In order to provide some context in which to discuss the array of monitors that have been developed, the following distinctions between performance monitors are enumerated.

1. *Hardware versus software monitors* — This is the most obvious distinction between performance monitors. Hardware monitors are implemented as free-standing hardware units capable of sensing bits and words of the computer system's state. Software monitors are a collection of system routines capable of interrogating software structures such as queues, control locks, and so on. The hardware/software boundary has begun to blur as processors include features, often in microcode, to enhance the measurement process and hardware monitors are connected as peripherals to the host (measured) machine.

2. *Event-driven versus sampling techniques* — This distinction refers to the monitor's basic data collection technique. In event-driven systems the monitor is notified every time a significant event is triggered. Sampling systems use an interval timer to wake up the monitor periodically and, when awake, the monitor looks at important lists and variables to determine the state of the system. Sampling techniques have the advantage for software monitors that the overhead can be made arbitrarily small by selecting a sufficiently large sampling interval.

3. *Tracing versus summary data* — Tracing monitors generate and save a record of information for every significant event or sample. Summary monitors typically only maintain running averages.

4. *Real-time processing versus post-processing* — Some monitors simply display their results at a terminal or via a set of meters for use by the operator or users of the system. Other real-time monitors attempt to use performance information collected to dynamically adjust a set of system parameters, for example, the HASP execution task monitor [Strauss, 73]. Post-processing monitors are more common and they spool their

data onto tape for later analysis by programs that are not under any serious deadline constraints.

5. *User versus system information* — There is often a significant difference in the implementation of monitors designed to evaluate user programs and those designed to evaluate the global performance of the computer system. The monitor for the user program can often call upon many standard system routines for assistance. Measurement hooks can be inserted at compile time, the runtime environment of the program can be modified to collect information, or the user program can be run interpretively. Although these techniques are not usually available to system monitors, the use of virtual machines [Goldberg, 73] may prove to be of considerable assistance.

In this section on measurement techniques, we use the event-driven/sampling dimension to organize our discussion. However, we will point out further differences along the other dimensions as the opportunity arises.

Event-Driven Techniques

There exists a range of event-driven measurement techniques. Let us begin with the typical hardware monitor shown in Figure 11-25. The figure shows the major components of the monitor.

The heart of the monitor is the bank of 8 to 32 counters. In some of the early monitors the most significant bits of the counters were electromechanical but now they are entirely electronic. Each counter consists of an 8-12 digit BCD counter. A BCD counter makes less efficient use of the bits in the counter but this does allow the counter to be more easily read visually.

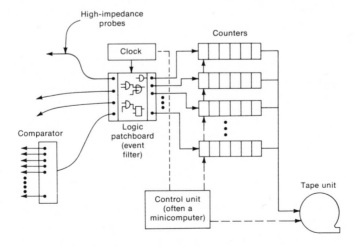

Figure 11-25   Conventional hardware monitor

The counters are driven by lines from a logic patchboard that consists of an assorted collection of gates, latches, and decoders. The inputs to the patchboard are probes that monitor the signals of interest in the computer system. Generally the probes are designed much like an oscilloscope probe to impose a minimal electrical load on the monitored signal so they can be freely moved around the system without fear of disturbing the hardware. Figure 11-26 shows a simplified patchboard and how it might be wired to count the number of disk arm movements on disks A, B, and C. It is useful to consider the logic patchboard as an event "collector" and "filter." In other words, it takes a collection of primitive signals from the host machine and manipulates and interprets them so meaningful signals are sent to the counters.

The purpose of the clock in the hardware monitor is to allow the counters to measure the duration of intervals, as well as simply the number of intervals. Note in part *b* of Figure 11-26 the circuit of part *a* has been modified to measure the utilization of the input/output channel. Actually counter 1 is accumulating the total number of clock ticks generated so the ratio of counter 1 to counter 0 will give the utilization of the input/output channel.

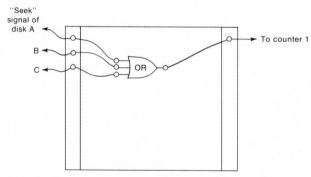

a) Count number of arm movements on disks A, B and C

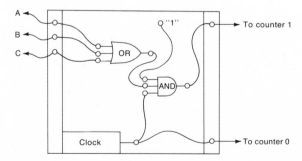

b) Count duration of arm-movement intervals

Figure 11-26   Logic patchboard

Latches, or flip/flops, are included on the patchboard to allow the measurement of sequences of events. For example, you may suspect inefficiencies in the disk arm movement algorithm and want to know the fraction of disk transmissions that are preceded by arm movement. Another item found on most patchboards is a 3-to-8 or 4-to-16 decoder. These are useful when information must be unpacked from fields in instructions or operands.

Another important component shown in the block diagram of the hardware monitor is the comparator. This unit connects to a bus, usually the address bus, and compares the contents of the bus to an internal register. The result of the comparison is then fed to the patchboard for use in defining events for the counters. A pair of comparators is often used to bracket an interval of memory. This allows the monitor to discriminate between the utilization of different software systems. Finally, the monitor must have some storage facility such as a magnetic tape to periodically dump the contents of the counters.

The hardware monitor that has just been described might well be termed a "first generation" monitor. A number of these monitors are offered commercially, but this basic organization was first used in 1962 by IBM's Basic Counting Unit that was designed to measure the performance of IBM 7090 systems.

First generation hardware monitors were summary monitors and could only provide average measures, for example, drum transmissions per second or utilization of the central processor. A number of more recent monitors include the ability to generate histograms. For example, you may be interested in the distribution of transmission times to an input/output device, not just the average transmission time. Histograms can be generated by first timing the transmission interval, as before, and on termination of the transmission using the contents of the counter as an index into a table in a buffer storage. Every time an entry in the table is accessed the index is incremented.

In order to provide this histogram generation capability, and other more advanced features, second generation hardware monitors often include a minicomputer as the primary control element. Now a range of peripherals can be supported: disk, tape, display scope, and others, and a modest amount of preprocessing of the data collected by the hardware monitor can be done before being stored on disk or tape. The minicomputer can also indicate the progress of a measurement experiment via a teletype or display terminal so the analyst can verify he has not made some gross error in configuring the measurement session.

So far we have presented the hardware monitor as a device that passively observes the host machine. While there are cases where this situation is adequate, and even desirable, the more common situation is to allow the hardware monitor to interrupt the host processor or otherwise feed information back to the host machine (possibly via storing information in the main memory of the host computer for use by the processor in memory management or scheduling decisions). Another important advantage of linking the hardware monitor to the host machine is that the host machine can now control the measurement experiment. The ability to use the hardware monitor as an extension of software monitoring routines can substantially increase the utility of the hardware monitor [Fuller, Swan, and Wulf, 73].

Much of the capability that has just been described for the hardware monitor can also be provided in a suitably designed software monitor. Input/output devices such as drums, disks, tape units, and terminals usually operate at a rate several orders of magnitude slower than the instruction rate of the processor. It is not uncommon for a processor to operate at 10^6 instructions per second and a disk to have an arm seek time of from 10 to 100 milliseconds and a data transmission time of from 1 to 50 milliseconds. Hence with very little overhead, instructions can be inserted into the disk control routines to count the number of seeks per second, utilization of the disk input/output channel, and many other summary statistics commonly associated with a hardware monitor. Moreover, a software monitor can measure such things as the queue length to the disk and the time from the instant the program requests a record from the disk and enters the disk queue until its request is granted. In other words, many times we are interested in the status of software structure, such as the disk queue, and a software monitor is the most direct way to measure it. Due to the difference in the basic speeds of the processor and the disk we should be able to effectively monitor it without putting an appreciable load on the processor. If the overhead for measurement does become objectionable, however, we can turn to the sampling techniques that are discussed later. Another very important measure of system performance is memory utilization and in many computer systems the only effective way to monitor the utilization of this valuable resource is through a software monitor.

The question may now arise as to why ever use a hardware monitor. Portability and high bandwidth are the outstanding features of a hardware monitor. Regardless of the operating system, and every installation has a different version of the operating system even if they share the same basic system, a hardware monitor can relatively quickly measure many of the basic parameters of system performance. The other feature of a hardware monitor is its high data rate from the measurement probes and into the counters. (The data rate from the counters to tape is not substantially different from the data rate available to software monitors from main storage to tape drives.) Particularly in studies where we are evaluating the performance of hardware components of the system these high data rates may be necessary. For example, to measure the amount of conflict for memory associated with input/output processors with the central processor attempting to access the same memory module can only be measured with a hardware monitor.

In addition to the collection summary statistics, event-driven monitors are also capable of collecting event traces. Many operating systems include an event tracing facility to aid in debugging. Table 11-3 lists the most commonly traced information. This information can be used in the diagnosis of a system's failure. The operating system usually includes a moderate-size buffer to hold the most recent events. When the table is full the tracing process simply over-

TABLE 11-3. COMMONLY TRACED EVENTS

Event	Record Content
Input/output interrupt	Device signaling interrupt, type of interrupt (for example done, error, and so on), time of day
Request input/output service	Name of process requesting service, type of request, device requested, time of day
Start input/output device	(Essentially the same information as for request input/output service)
Dispatch process on processor	Name of process, time of day
Trap to supervisor from user program	Name of process, service requested of supervisor, time of day

writes the previous information. If such a trace facility exists in the operating system, to get a tracing monitor we need only divert the trace information onto a tape instead of letting it be overwritten in the memory-resident table. A trace facility has a number of important uses in the performance evaluation of computers. It can be used as the input to the trace-driven simulation discussed in section 11-4 or it may be used to measure the service-time distribution of the various processors in the system for use in developing one of the queueing models of section 11-3.

Another type of event-tracing monitor is needed in studies of memory hierarchies and it deals with the sequence of accesses to main memory by the central processor rather than the coarser events of the trace facility just discussed. The alternative implementations in this case are to interpret the instruction execution process or alter the microcode to capture all references to memory [Saal and Shustek, 72]. Needless to say, this is a time-consuming process; often a 50 to 200 degradation in processing power is suffered when instruction execution must be emulated.

Sampling Techniques

There are many instances in the measurement of computer systems where we can get an accurate measure of performance without counting every event. For example, if we are interested in the utilization of an input/output channel, we can test whether or not the channel is busy several times a second for several minutes and we will get as accurate an estimate of channel utilization as if we trapped every start and stop of the channel and integrated the results over the measurement interval.

A common example of a sampling monitor is illustrated in Figure 11-27. The figure is a histogram of a user program activity. To generate this histogram the sampling monitor periodically interrupted the processor and examined the contents of the program counter. If the program counter was within the program's region, the content of the program counter was used to index into an array, the selected element in the array was incremented, and after

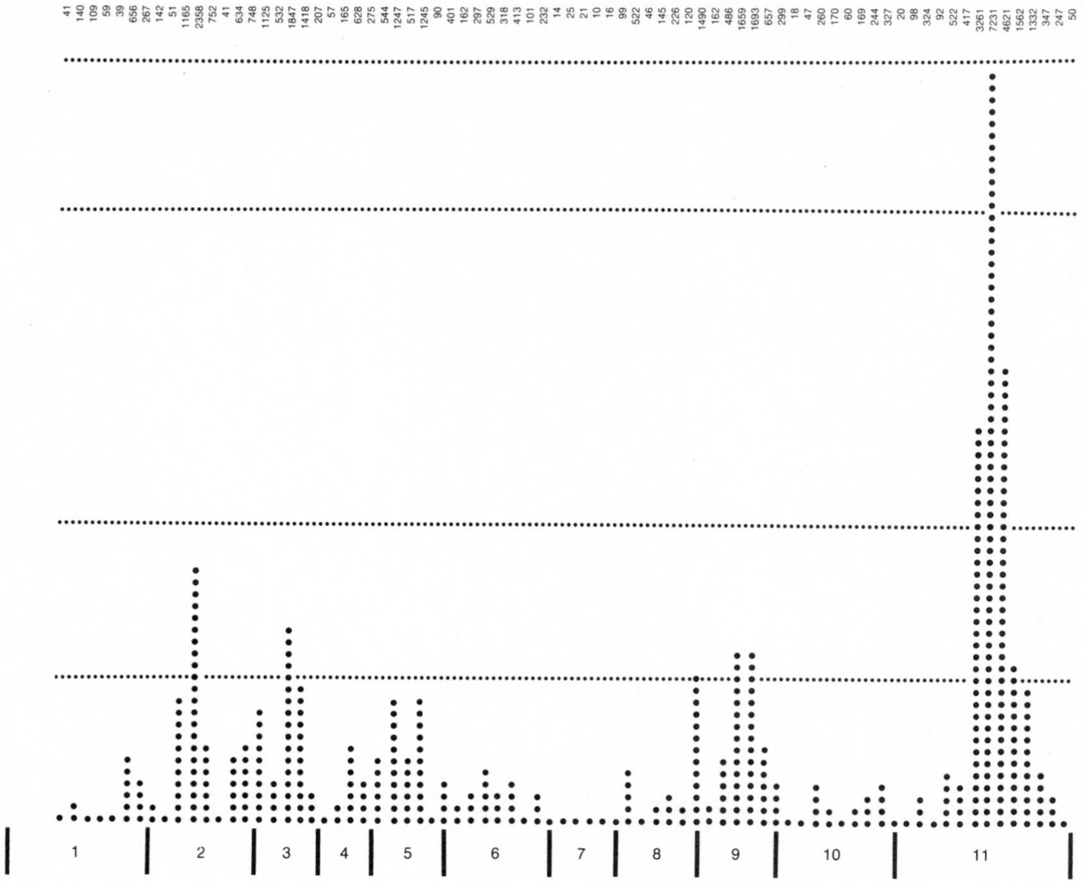

Figure 11-27 Histogram of program activity

the measurement interval was completed the array was used to generate the histogram. From this histogram, if the efficiency of the program must be improved, segment 11 should be closely studied for possible recoding, but segments 1, 7, and 10 are of little consequence.

The histogram in Figure 11-27 has an interesting shortcoming. Examination of the source listing shows segment 11 is a subroutine called from segments 2, 5, and 7 and the histogram offers no information as to which calling site initiated the most invocations of the subroutine. A useful modification to the sampling monitor would be to allow it to examine the run-time environment of the program and credit a particular sample to the calling site as well as the subroutine itself.

The edited output of a sampling monitor designed to collect system information is shown in Figure 11-28. The first part of the output itemizes the sampling rates for the various measurements. The minimum sampling interval for this run is 0.5 seconds. which was used to measure input/output channel

```
MACHINE ACTIVITY AT A GLANCE - MONITORING COMPLETED
                    DATE:   71.306
                   ENDED:   12.29.41
          TIME MONITORED:   3.00 MINUTES


                    PARAMETERS
          CYCLE RANGE
             MAIN MEMORY        5
             SOFTWARE MODULES   5
             QUEUES             2
             I/O DEVICES        1
             CHANNELS           1
          CYCLE TIME        0.50 SECONDS

                    ACTIVITY
          CPU UTILIZATION                       61.10%
          ANY I/O SELECTOR CHANNEL BUSY         87.32%
          I/O ACTIVITY INDEX            81,729
          I/O INTERRUPTS                19,262          6,421 PER MINUTE
          DEVICES USED                     41
          TOTAL SUPERVISOR CALLS (SVC'S)  99,193       33,064 PER MINUTE
             EXECUTE I/O CHANNEL PROGRAMS 39,283       13,094 PER MINUTE
             OPENING OF FILES               81            27 PER MINUTE

                    POSSIBLE BOTTLENECKS
          128K REGION AVAILABLE                79.71%
          AVERAGE CORE WASTED                    132K
          TAPE CONTROL UNIT WAITING            12.97%
          DISK CONTROL UNIT WAITING             .58%

                    PROGRAMS USED
          PROCESSORS
                   FORTRANG          2
                   FORTRANH          2
                   LINKEDIT          2
                   MAIN              1
                   SORT              2
                   UTILITY           6
          STEPS INITIATED                  17
```

Figure 11-28   Summary statistics of a system monitor

and device utilization, and the maximum sampling interval was 2.5 seconds and was in the measurement of main memory and software module utilization. A few of the measurements shown in Figure 11-28 are event-driven measures. For example, the input/output interrupt rate and SVC rate are derived from event-triggered routines and in fact it would be very difficult to estimate these rates with a sampling technique. The statistics shown in Figure 11-28 were collected with very little overhead—on the order of 2 percent. Therefore, while these statistics may not include all you need to know about the system's performance, it costs so little that it can be used on a continuing basis to monitor performance.

11-7. GUIDE TO FURTHER READING

Although there has always been the desire to build higher performance computer systems since the first days of computing, as we mentioned in the introduction to this chapter, the need for sophisticated performance analysis did not become critical until the third generation of computer systems was introduced in the early 1960s. Calingaert [67] wrote one of the earliest papers to outline and discuss the main issues in the area of performance evaluation. More recently, Johnson [70] has given an interesting overview of the area and attempts to focus on some of the inadequacies in our current methods. The articles by Lucas [71] and Lynch [72] also provide an introductory discussion of performance evaluation.

The method of difference-differential equations used to study the stochastic models in this chapter has many applications and can be generalized to many situations beyond those discussed in this chapter. See Morse [58] for further discussions of this technique. Although beyond the scope of this chapter, more powerful techniques exist for the analysis of queueing structures. These other techniques that employ generating functions and Laplace-Stieltjes transforms are introduced in a number of texts such as Kleinrock [75]; Cox and Smith [61]. Several texts now exist that treat queueing theory strictly in the context of modeling computer systems [Coffman and Denning, 73; Kleinrock, 75].

The literature on the analysis of priority disciplines is extensive, but Conway, Maxwell, and Miller [67] provide a good general introduction to the topic and McKinney [69] as well as Coffman and Kleinrock [68] survey the more specialized area of feedback scheduling policies. See Fuller and Baskett [75] for a discussion of queueing models of drums and disks.

An area of queueing theory that has important applications in the modeling of computer systems is the analysis of networks of queues. These more general models are capable of capturing many of the important properties of large computing systems that have many processors and input/output devices. Jackson [63] and Gordon and Newell [67] wrote the important original papers in this area and Conway, Maxwell, and Miller [67] and Kleinrock [75] provide a clear introduction to the subject. Buzen [73] has written an article that applies the main results of queueing networks to several common types of computer systems.

The theory of discrete-time simulation is covered in a number of introductory texts [Fishman, 74; Gordon, 69; Naylor, 66]. The problem of verifying the results of simulation experiments is still a topic of active research and the reader is directed to a number of articles for further discussion [Fishman, 67; Crane and Iglehart, 74]. While it is possible to write simulations in general purpose programming languages, as we did in section 11-4, several languages have been specifically developed for simulation. SIMULA is an ALGOL-based language designed for simulation and an article by Dahl and Nygaard [66] describes SIMULA's central features. Nielson [67] and McDougall [68] describe specific simulations of computer systems that provide good examples of simulations of computer systems.

The evaluation of operational computer systems and measurement techniques are relatively pragmatic topics that have not received much attention in the literature. References on benchmarks and synthetic jobs were given in section 11-5 [Buchholz, 69; Hamilton and Kernighan, 73; Sreenivasan and Kleinman, 71]. A number of articles outline the structure of specific hardware monitors [Estrin et al., 67; Aschenbrenner et al., 71] and software monitors [Cantrell and Ellison, 68]. Drummond [73] gives a detailed discussion of standard practices in the evaluation and measurement of computer systems.

11-8. PROBLEMS

11-1. Suppose there are two independent Poisson processes with parameters λ_1 and λ_2. Consider the "compound" process that is defined to have an arrival when either of the two Poisson processes has an arrival. Find the probability function for this compound process that is analogous to Equations 11-6 and 11-3 for the simple Poisson process.

11-2. A processor with an exponential service-time distribution is modified as shown:

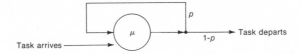

Upon every completion of service at the processor, a task is immediately routed back to the processor for another interval of service with probability p. This choice of being recycled occurs after every service so that a single task may receive 1, 2, 3, . . . service times. Find the service-time distribution of the modified processor with feedback.

11-3. Verify Equation 11-44, the expression for the expected queue size of a model with an n-level priority queueing discipline.

11-4. Consider the following $M/M/1$ model of an IBM 360/50 computer system:

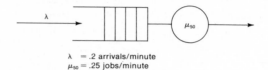

λ = .2 arrivals/minute
μ_{50} = .25 jobs/minute

a) Find the average waiting time of a job (that is, the time from arrival of the job in the queue until service is complete and the job leaves the system), the average queue size, and the utilization of the 360/50.

b) The preceding IBM 360/50 system is easily saturated and in order to reduce the waiting time an IBM 370/155 was purchased and added to the system. In other words:

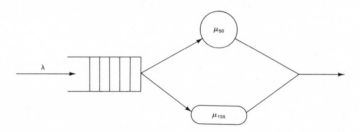

In this configuration assume that once a job begins on either the 360/50 or 370/155 it runs to completion on that machine. The IBM 370/155 is much faster than the IBM 360/50, and in fact assume the 370/155 is α times faster than the 360/50. Derive an expression for the waiting time (from job arrival to completion), queue size, utilization of the 360/50 and utilization of 370/155.

c) Suppose we want to minimize the average waiting time (from job arrival to completion); at what point (that is, for what α and λ) will it be better to just pull the plug on the 360/50 and run all the jobs on the 370/155? (In reality, the IBM 370/155 is from 3.5 to 4.0 times as fast as an IBM 360/50.)

11-5. Show that the departure process of an $M/M/1$ queueing system is a Poisson process.

11-6. Consider the following $M/M/1$ queueing system with a queue limited to N positions:

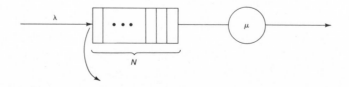

a) Find the equilibrium probability an arriving request is rejected because the queue is full and the probability the processor is idle.

b) Given that a request has been accepted into the queue find its average waiting time.

11-7. [Kleinrock, 69]. Show that for *n* sufficiently large, Equation 11-37 becomes a good approximation to Equation 11-36, the average response time of a time-sharing system with *n* users. (Hint: Use the original definition of p_0 as given in Equation 11-33 and use the fact that $\rho n \gg 1$ for large *n*.)

11-8. A number of terminals now on the market each contain an 8-bit microcomputer. One major advantage of these terminals is that they do not need to interrupt the central processor after every keystroke; they can buffer a line of input and interrupt the central processor only on a carriage return, break character, other control character, or if the buffer in the microcomputer is full. An important question with regard to these terminals is how does the rate of interrupts to the central processor vary as a function of the microcomputer's buffer size? Let us model this problem as a Markov chain where an epoch is defined at each keystroke and E_i is the state in which *i* characters of the buffer are full. Suppose:

$$Pr \text{ \{standard alphanumeric character\}} = .90$$
$$Pr \text{ \{backspace (that is, rubout)\}} = .07$$
$$Pr \text{ \{control character\}} = .03$$

a) Find an expression for the expected number of keystrokes per interrupt as a function of *N*, the number of characters in the microcomputer's buffer.

b) Assume the amount of time the central processor needs to service an interrupt is:

$$300 + 15c \text{ microseconds}$$

where *c* is the number of characters being transmitted from the terminal to the central processor, plot the number of terminals needed to saturate the central processor as a function of the buffer size. Assume five keystrokes per second per terminal.

11-9. Design a mixed, linear congruential random number generator for a 16-bit per word minicomputer. Assume the minicomputer uses a standard two's complement number representation and hence let $m = 2^{15}$. Point out the advantages, deficiencies, and possible remedies for the generator you specified.

11-10. In many cases, a hyper-exponential distribution function with the density function:

$$f(x) = a\mu_1 e^{-\mu_1 t} + (1 - a)\mu_2 e^{-\mu_2 t}$$

is a better approximation to actual service-times in computer systems than is the exponential distribution. Write an ALGOL procedure that generates random variates with a hyper-exponential distribution to replace *Negexp* starting in line 16 of the ALGOL program given in section 11-4.

11-11. Write a simulation of the simple $M/M/1$ queueing structure presented in section 11-3. Instrument your simulation to estimate the mean and variance of the waiting time.

a) Plot the mean and variance of the waiting time as a function of the number of arrivals simulated. Compare these curves with expressions derived from the analytical model.

b) Estimate the autocorrelation of the sequence of successive waiting times as suggested in Equation 11-56. Add bars to the plot for part *a* that extend $\pm \sigma_{\bar{x}}$ from the estimate of the average waiting time. How many requests must be simulated until you are "confident" that the estimate of the waiting time is accurate to two significant (decimal) digits?

11-12. In many computer systems the data collected for accounting purposes is also useful for performance evaluation studies. In this exercise we use accounting data to validate the Markovian assumptions of the queueing models discussed in section 11-3. Write a program that will scan several days (or weeks) of accounting data and construct histograms of interarrival times and service-time requirements for the tasks run on the system.

a) Compare these histograms to exponential density functions that have the same mean as the histograms. How do the histograms compare to the density functions?

b) What is the coefficient of variation (C) of the service time? Use the Pollaczek-Khinchine formula (Equation 11-20) to estimate the turnaround time, that is, W, of tasks. How does this compare with the observed turnaround time? Why?

11-13. The common way to collect an instruction mix for a particular type of processor is to first write an interpreter for the processor's instruction set. Then measurement procedures are added to the interpreter to tabulate the frequency of instructions executed. This technique has a number of disadvantages: the interpreter can be time-consuming to write and debug and since the interpreted programs run anywhere from 20 to 200 times slower than programs run directly on the actual processor, the interaction of input/output transfers and program execution time is severely distorted. An alternate approach is to write a sampling monitor that periodically interrupts the central processor, finds the instruction the processor was going to execute next, increments the appropriate entries in the instruction mix table, and continues this sampling process until a sufficient number of instructions have been measured. Build a sampling monitor to tabulate an instruction mix for the processor of a locally available system. Use the validation techniques given in the last part of section 11-4 to estimate the precision of the instruction mix collected. Compare your results to the instruction mixes in Table 11-2.

11-14. Implement a very low overhead, summary, software monitor to measure the utilization of the central processor(s), input/output channels or processors, and the major input/output devices. Use either a sampling or trace-driven technique; whichever is most convenient to implement on your local system. Plot the utilization of the processors and devices as a function of time. Use both a small grain, say a report every one to five minutes, and a large grain, one every half hour to hour. Look for idle devices, saturated or bottleneck units, and how utilization of the units varies as a function of the time of day. It is a rare computer system that does not display some anomolous behavior that when corrected results in a nontrivial improvements in system performance.

Bibliography

Abate, J., and Dubner, H. 1969. Optimizing the performance of a drum-like storage. *IEEE Trans. on Comp.* C-18:992–97.

Almasi, G. S.; Keefe, G. E.; Lin, Y.; and Thompson, D. A. 1970. A magnetoresistive detector for bubble domains. Presented at 16th Annual Conference on Magnetism and Magnetic Materials, November 1970, at Miami Beach, Fla.

Amelio, G. F.; Tompsett, M. F.; and Smith, G. E. 1970. Experimental verification of the charge coupled device concept. *Bell Sys. Tech. J.* 49:593–600.

Arbuckle, R. A. 1966. Computer analysis and thruput evaluation. *Computers and Automation*, January 1966, pp. 12–15.

Aschenbrenner, R. A.; Amiot, L.; and Natarajan, N. K. 1971. The neurotron monitor system. *AFIPS Conf. Proc., 1971 FJCC* 39:31–37. Montvale, N. J.: AFIPS Press.

Avizienis, A. 1964. Binary-compatible signed-digit arithmetic. *AFIPS Conf. Proc., 1964 FJCC* 26:663–72. Baltimore: Spartan Books.

Ayling, J., and Moore, R. D. 1971. Main monolithic memory. *IEEE J. of Solid-State Circuits* SC-6:276–79.

Barnes, G. H. et al. 1968. The ILLIAC IV computer. *IEEE Trans. on Comp.* C-17:746–57.

Barton, R. 1961. A new approach to the functional design of a digital computer. *AFIPS Conf. Proc., 1961 Western Joint Computer Conference* 19:393.

Bashkow, T. R. et al. 1967. System design of a FORTRAN machine. *IEEE Trans. on Comp.* EC-16:485–99.

Batcher, K. E. 1968. Sorting networks and their applications. *AFIPS Conf. Proc., 1968 SJCC* 32:307–14. Washington, D.C.: Thompson Books.

Beausoleil, W. F.; Brown, D. T.; and Phelps, B. E. 1972. Magnetic bubble memory organization. *IBM J. of Res. and Dev.* 16:587–91.

Belady, L. A. 1966. A study of replacement algorithms for a virtual-storage computer. *IBM Sys. J.* 5:78–101.

Bell, C. G., and Newell, A. 1971. *Computer structures: readings and examples.* New York: McGraw-Hill.

Benes, V. E. 1965. *Mathematical theory of connecting networks and telephone traffic*. New York: Academic Press.

Berkeley, E. C., and Bobrow, D. G., eds. 1964. *The programming language LISP: its operation and applications*. Cambridge, Mass.: Information International, Inc.

Bobeck, A. H. 1970*a*. The magnetic bubble. *Bell Lab. Record*, June/July 1970, p. 163.

————. 1970*b*. A second look at magnetic bubbles. *IEEE Trans. on Magnetics* MAG-6:445–46.

Bobeck, A., and Scovil, H. 1971. Magnetic bubbles. *Scientific American*, June 1971, pp. 78–90.

Bobeck, A. H. et al. 1969. Application of orthoferrites to domain-wall devices. *IEEE Trans. on Magnetics* MAG-5:544–53.

Bobeck, A. H. et al. 1970. Uniaxial magnetic garnets for domain wall "bubble" devices. *Applied Physics Letters* 17:131–34.

Bowers, D. M. 1973. An analysis of computer terminals. *The Office*, October 1973, pp. 89–124.

Boyle, W. S., and Smith, G. E. 1970. Charge coupled semiconductor devices. *Bell Sys. Tech. J.* 49:587–93.

Bremer, J. 1970. A survey of mainframe semiconductor memories. *Computer Design*, May 1970, pp. 63–73.

Broadbent, K. D., and McClung, F. J. 1960. A thin magnetic-film shift register. *1960 International Solid-State Circuits Conference, Digest of Technical Papers*, February 1960, pp. 24–25.

Brown, D. T.; Eibsen, R. L.; and Thorn, C. A. 1972. Channel and direct access device architecture. *IBM Sys. J.* 11:186–99.

Brown, J. L. et al. 1964. IBM System/360 engineering. *AFIPS Conf. Proc., 1964 FJCC* 26: 205–32. Baltimore: Spartan Books.

Bryan, G. E. 1967. JOSS: 20,000 hours at a console—a statistical summary. *AFIPS Conf. Proc., 1967 FJCC* 31:769–77. Montvale, N.J.: AFIPS Press.

Buchholz, W. 1969. A synthetic job for measuring system performance. *IBM Sys. J.* 8:309–18.

Budnick, P., and Kuck, D. J. 1971. The organization and use of parallel memories. *IEEE Trans. on Comp.* C-20:1566–69.

Burks, A. W.; Goldstine, H. H.; and von Neumann, J. 1946. Preliminary discussion of the logical design of an electronic computing instrument. *U.S. Army Ordnance Department Report*, 1946. Reprinted in Bell and Newell (1971), pp. 92–119.

Burton, H. O., and Sullivan, D. D. 1972. Errors and error control. *Proc. of the IEEE* 60:1293–1301.

Buzen, J. P. 1973. Computational algorithms for closed queueing networks with exponential servers. *Communications of the ACM* 16:527–31.

Calingaert, P. 1967. System performance evaluation: survey and appraisal. *Communications of the ACM* 10:12–18.

Campbell, S. G.; Herwitz, P. S.; and Pomerene, J. H. 1962. A nonarithmetical system extension. In *Planning a computer system: Project STRETCH*, edited by W. Buchholz, pp. 254–71. New York: McGraw-Hill.

Cantrell, H. N., and Ellison, A. L. 1968. Multiprogramming system performance measurement and analysis. *AFIPS Conf. Proc., 1968 SJCC* 32:213–21. Washington, D.C.: Thompson Books.

Chang, H.; Fox, J.; Lu, D.; and Rosier, L. L. 1971. A self-contained magnetic bubble domain memory chip. Presented at the Solid-State Circuits Conference, February 1971, at Philadelphia, Pa.

Chen, T. C. 1964. The overlap design of the IBM System/360 Model 92 central processing unit. *AFIPS Conf. Proc., 1964 FJCC* 26 (part II): 73–80. Washington, D.C.: Spartan Books.

———. 1971. Parallelism, pipelining, and computer efficiency. *Computer Design*, January 1971, pp. 69–74.

Cheng, P. S. 1969. Trace-driven system modeling. *IBM Sys. J.* 8:280–89.

Chien, R. T. 1973. Memory error control: beyond parity. *IEEE Spectrum*, July 1973, pp. 18–23.

Christopherson, W. 1961. Matrix switch and drive system for a low-cost magnetic-core memory. *IRE Trans. on Elec. Comp.* EC-10:238–46.

Chu, Y. 1962. *Digital computer design fundamentals.* New York: McGraw-Hill.

Clare, C. R. 1973. *Designing logic systems using state machines.* New York: McGraw-Hill.

Clos, C. 1953. A study of non-blocking switching networks. *Bell Sys. Tech. J.* 32:406–24.

Coffman, E. G., and Denning, P. J. 1973. *Operating systems theory.* pp. 209–18. Englewood Cliffs, N.J.: Prentice-Hall.

Coffman, E. G., Jr.; Elphick, M. J.; and Shoshani, A. 1971. System deadlocks. *Computing Surveys* 3:67–78.

Coffman, E. G., and Kleinrock, L. 1968. Feedback queueing models for time-shared systems. *J. of the ACM* 15:549–76.

Collins, D. R. et al. 1974. Electrical characteristics of 500-bit $Al — Al_2 O_3 — Al$ CCD shift registers. *Proc. of the IEEE* 62:282–84.

Computer Design 1970. Faster, simpler magnetic memory developed. May 1970, p. 26.

Computer Review. 1974. Lexington, Mass.: GML Corporation.

Conti, C. 1969. Concepts for buffer storage. *IEEE Comp. Group News*, March 1969, pp. 9–13.

Conway, M. E. 1963. A multiprocessor system design. *AFIPS Conf. Proc., 1963 FJCC* 24:139–46. Baltimore: Spartan Books.

Conway, R. W.; Maxwell, W. L.; and Miller, L. W. 1967. *Theory of scheduling.* Reading, Mass.: Addison-Wesley.

Cook, R. W., and Flynn, M. J. 1970. System design of a dynamic microprocessor. *IEEE Trans. on Comp.* C-19:213–22.

Cotten, L. W. 1965. Circuit implementation of high-speed pipeline systems. *AFIPS Conf. Proc., 1965 FJCC* 27 (part I): 489–504. Washington, D.C.: Spartan Books.

Cox, D. R., and Smith, W. L. 1961. *Queues.* London: Methuen and Co. Ltd. and New York: Wiley.

Crane, M. A., and Iglehart, D. L. 1974a. Simulating stable stochastic systems, II: Markov chains. *J. of the ACM* 21:114–23.

———. 1974b. Simulating stable stochastic systems, I: general multiserver queues. *J. of the ACM* 21:103–13.

Cyre, W. R., and Lipovski, G. J. 1972. On generating multipliers for a cellular fast Fourier transform processor. *IEEE Trans. on Comp.* C-21:83–87.

Dahl, O., and Nygaard, K. 1966. SIMULA—an ALGOL-based simulation language. *Communications of the ACM* 9:671–78.

Davies, P. M. 1972. Readings in microprogramming. *IBM Sys. J.* 11:16–40.

Denning, P. 1970. Virtual memory. *Computing Surveys* 2:153–89.

Digital Equipment Corporation. 1973. *PDP-11 peripherals handbook*. Maynard, Mass.

Dijkstra, E. W. 1968. Cooperating sequential processes. In *Programming Languages,* edited by F. Genuys, pp. 43–112. New York: Academic Press.

Drummond, M. E. 1973. *Evaluation and measurement techniques for digital computer systems*. Englewood Cliffs, N.J.: Prentice-Hall.

Estrin, G.; Hopkins, D.; Coggan, B.; and Crocker, S. D. 1967. Snuper computer—a computer in instrumentation automation. *AFIPS Conf. Proc., 1967 SJCC* 30:645–56. Washington, D.C.: Thompson Books.

Farber, A. S., and Schlig, E. S. 1972. A novel high-performance bipolar monolithic memory cell. *IEEE J. of Solid-State Circuits* SC-7:297–98.

Farina, M. V. 1968. *Programming in Basic, the time-sharing language*. Englewood Cliffs, N.J.: Prentice-Hall.

Fishman, G. S. 1967. Problems in the statistical analysis of simulation experiments: the comparison of means and the length of sample records. *Communications of the ACM* 10:94–99.

————. 1973. *Concepts and methods in discrete event digital simulation*. New York: Wiley.

Flynn, M. J. 1966. Very high-speed computing systems. *Proc. of the IEEE* 54:1901–9.

Flynn, M. J.; Podvin, A.; and Shimizu, K. 1970. A multiple instruction stream with shared resources. In *Parallel Processor Systems, Technologies, and Applications*, edited by L. C. Hobbs, pp. 251–86. Washington, D.C.: Spartan Books.

Flynn, M. J., and Rosin, R. F. 1971. Microprogramming: an introduction and a viewpoint. *IEEE Trans. on Comp.* C-20:727–31.

Foster, C. C. 1970. *Computer architecture*. New York: Van Nostrand Reinhold.

Foster, C. C.; Gonter, R. H.; and Riseman, E. M. 1971. Measures of op-code utilization. *IEEE Trans. on Comp.* C-20:582–84.

Freiser, M., and Marcus, P. 1969. A survey of some physical limitations on computer elements. *IEEE Trans. on Magnetics* MAG-5:82–90.

Fuller, S. H., and Baskett, F. 1975. An analysis of drum storage units. *J. of the ACM* 22:83–105.

Fuller, S. H.; Swan, R. J.; and Wulf, W. A. 1973. The instrumentation of C. mmp, a multi-(mini)processor. *COMPCON 73 Digest of Papers*, 7th Annual IEEE Computer Society International Conference, March 1973, San Francisco. pp. 173–76.

Gear, C. W. 1969. *Computer organization and programming*. New York: McGraw-Hill.

Gelberger, P. P., and Salama, C. A. T. 1972. A uniphase charge-coupled device. *Proc. of the IEEE* 60:721–22.

Gibson, J. C. 1970. The Gibson mix. TR 00.2043. Systems Development Div., IBM Corp., Poughkeepsie, N.Y., June 18, 1970.

Gilligan, T. J. 1966. 2½D high speed memory systems—past, present, and future. *IEEE Trans. on Elec. Comp.* EC-15:475–85.

Goldberg, R. P. 1973. Architecture of virtual machines. *AFIPS Conf. Proc., 1973 National Computer Conference* 42:309–18. Montvale, N.J.: AFIPS Press.

Gordon, G. 1969. *System simulation*. Englewood Cliffs, N.J.: Prentice-Hall.

Gordon, W. J., and Newell, G. F. 1967. Closed queueing systems with exponential servers. *Operations Research* 15:254–65.

Gray, J. P. 1972. Line control procedures. *Proc. of the IEEE* 60:1301–12.

Greenberger, M. 1965. Method in randomness. *Communications of the ACM* 8:177–79.

Gregory, J. 1972. A comparison of floating point summation methods. *Communications of the ACM* 15:838.

Hallin, T. G., and Flynn, M. J. 1972. Pipelining of arithmetic functions. *IEEE Trans. on Comp.* C-21:880–86.

Hamilton, P. A., and Kernighan, B. W. 1973. Synthetically generated performance test loads for operating systems. *1st Annual SIGME Symposium on Measurement and Evaluation*, February 1973, Palo Alto, Calif., pp. 121–26.

Hamming, R. W. 1950. Error detecting and error correcting codes. *Bell Sys. Tech. J.* 29:147–60.

Hauck, E. A., and Dent, B. A. 1968. Burroughs' B6500/B7500 stack mechanism. *AFIPS Conf. Proc., 1968 SJCC* 32:245–51. Washington, D.C.: Thompson Books.

Hellerman, H. 1973. *Digital computer system principles.* 2nd ed. p. 245. New York: McGraw-Hill.

Higashi, P. 1966. A thin-film rod memory for the NCR 315 RMC computer. *IEEE Trans. on Elec. Comp.* EC-15:459–67.

Hintz, R. G., and Tate, D. P. 1972. Control Data STAR-100 processor design. *COMPCON 72 Digest of Papers*, 6th Annual IEEE Computer Society International Conference, September 1972, San Francisco, pp. 1–4.

Hoagland, A. 1963. *Digital magnetic recording.* New York: Wiley.

———. 1972. Mass storage—past, present, and future. *AFIPS Conf. Proc., 1972 FJCC* 41 (part II):985–91. Montvale, N.J.: AFIPS Press.

Hodges, D. A. 1968. Large-capacity semiconductor memory. *Proc. of the IEEE* 56:1148–62.

Hoevel, L. 1973. Ideal directly executed languages: an analytic argument for emulation. Computer Research Report #29. Electrical Engineering Dept., The Johns Hopkins University, Baltimore, Md., December 1973.

Husson, S. S. 1970. *Microprogramming: principles and practices.* Englewood Cliffs, N.J.: Prentice-Hall.

IEEE Trans. on Comp. 1971 and 1973. Special issues on fault tolerant computing. Vols. C-20 and C-22.

Iliffe, J. 1968. *Basic machine principles.* New York: American Elsevier.

Iverson, K. 1962. *A programming language.* New York: Wiley.

Iverson, K., and Brooks, F. P. 1965. *Automatic data processing.* Ch. 8, Sect. 2. New York: Wiley.

Jackson, J. R. 1963. Jobshop-like queueing systems. *Management Science*, October 1963, pp. 131–42.

Jewell, W. S. 1967. A simple proof of: $L = \lambda W$. *Operations Research* 15:1109–16.

Johnson, R. R. 1970. Needed: a measure for measure. *Datamation*, December 15, 1970, pp. 22–30.

Jones, L. H. et al. 1972. An annotated bibliography on microprogramming: late 1969–early 1972. *SIGMICRO Newsletter (ACM)*, July 1972, pp. 39–55.

Jones, L. H., and Carvin, K. 1973. An annotated bibliography on microprogramming II: early 1972–early 1973. *SIGMICRO Newsletter (ACM)*, July 1973, pp. 7–18.

Jones, R., and Bittmann, E. 1967. The B8500-microsecond thin-film memory. *AFIPS Conf. Proc., 1967 FJCC* 31:347–52. Montvale, N.J.: AFIPS Press.

Joslin, E. O. 1965. Application benchmarks: the key to meaningful computer evaluation. *ACM Proc. of the 20th National Conf.*, August 1965, Cleveland, Ohio, pp. 27–37.

Kahan, W. 1965. Further remarks on reducing truncation errors. *Communications of the ACM* 8:40.

Karp, R. M., and Miranker, W. L. 1968. Parallel minimax search for a maximum. *J. of Combinatorial Theory* 4:19–35.

Katzan, H., Jr. 1973. *Computer data security.* Sect. 2.5, Virtual Storage. New York: Van Nostrand Reinhold.

Keyes, R. 1969. Physical problems and limits in computer logic. *IEEE Spectrum*, May 1969, pp. 36–45.

———. 1972. Physical problems of small structures in electronics. *Proc. of the IEEE* 60:1055–61.

Kleinrock, L. 1969. Certain analytic results for time-shared processors. *Proc. of the 1969 IFIP Congress*, pp. 838–45.

———. 1975. *Queueing systems: theory and applications.* New York: Wiley.

Knight, K. E. 1966. Changes in computer performance. *Datamation*, September 1966, pp. 40–54.

Knuth, D. E. 1968. *The art of computer programming, vol. 1: fundamental algorithms.* Reading, Mass.: Addison-Wesley.

———. 1969. *The art of computer programming, vol. 2: seminumerical algorithms.* Reading, Mass.: Addison-Wesley.

———. 1971. An empirical study of FORTRAN programs. *Software: Practice and Experience* 1:105–33.

Kogge, P. M., and Stone, H. S. 1973. A parallel algorithm for the efficient solution of a general class of recurrence equations. *IEEE Trans. on Comp.* C-22:786–93.

Kosonocky, W. F., and Carnes, J. E. 1971. Charge-coupled digital circuits. *IEEE J. of Solid-State Circuits* SC-6:314–22.

Krambeck, R. H.; Walden, R. H.; and Pickar, K. A. 1971. Implanted barrier two-phase CCD. Presented at the IEEE International Electron Device Meeting, October 1971, Washington, D.C.

Kuck, D. J.; Muraoka, Y.; and Chen, S. C. 1972. On the number of operations simultaneously executable in FORTRAN-like programs and their resulting speedup. *IEEE Trans. on Comp.* C-21:1293–1310.

Landauer, R. 1961. Irreversibility and heat generation in the computing process. *IBM J. of Res. and Dev.* 5:183–91.

———. 1962. Fluctuations in bistable tunnel diode circuits. *J. of Applied Physics* 33:2209–16.

Lane, W. G. 1972. Pipeline array processing, an efficient architectural alternative. Ph.D. dissertation, University of California, Davis.

Lawler, E. L., and Moore, J. M. 1969. A functional equation and its application to resource allocation and sequencing problems. *Management Science*, September 1969, pp. 77–84.

Lawrie, D. H. 1973. Memory-processor connection networks. Report No. UIUCDCS-R-73-557. Dept. of Computer Science, University of Illinois, Urbana, February 1973.

Lin, Y. S., and Mattson, R. L. 1972. Cost-performance evaluation of memory hierarchies. *IEEE Trans. on Magnetics* MAG-8:390–92.

Linz, P. 1970. Accurate floating-point summation. *Communications of the ACM* 13:361–62.

Little, J. D. C. 1961. A proof for the queueing formula: $L = \lambda W$. *Operations Research* 9:383–87.

Loomis, H. H., Jr. 1966. The maximum rate accumulator. *IEEE Trans. on Elec. Comp.* EC-15:628–39.

Lucas, H. C. 1971. Performance evaluation and monitoring. *Computing Surveys* 3:79–91.

Lukasiewicz, J. 1951. *Aristotle's syllogistic: from the standpoint of modern formal logic*. Oxford: Oxford University Press.

Lunde, A. 1974. Evaluation of instruction set processor architecture by program tracing. Dept. of Computer Science Technical Report, Carnegie-Mellon University, Pittsburgh, Pa., July 1974.

Lynch, W. C. 1972. Operating system performance. *Communications of the ACM* 15:579–85.

MacDougall, M. H. 1970. Computer system simulation: an introduction. *Computing Surveys* 2:191–209.

McKinney, J. M. 1969. A survey of analytical time-sharing models. *Computing Surveys* 1:105–16.

McLaughlin, R. A. 1974. A survey of 1974 DP budgets. *Datamation*, February 1974, pp. 52–56.

MacSorley, O. L. 1961. High-speed arithmetic in binary computers. *Proc. of the IRE* 49:67–91.

Massey, J. L., and Garcia, O. N. 1972. Error-correcting codes in computer arithmetic. In *Advances in Information Systems Science, vol. 4*, edited by J. T. Tou, pp. 273–326. New York: Plenum Press.

Mathias, J., and Fedde, G. 1969. Plated-wire technology: a critical review. *IEEE Trans. on Magnetics* MAG-5:728–51.

Matick, R. E. 1972. Review of current proposed technologies for mass storage systems. *Proc. of the IEEE* 60:266–89.

———. 1976. Computer storage systems and technology. New York: Wiley, forthcoming.

Mattson, R. et al. 1970. Evaluation techniques for storage hierarchies. *IBM Sys. J.* 9:78–117.

Meade, R. 1971. Design approaches for cache memory control. *Computer Design*, January 1971, pp. 87–93.

Mills, D. L. 1972. Communication software. *Proc. of the IEEE* 60:1333–41.

Morris, R. 1968. Scatter storage techniques. *Communications of the ACM* 11:38–44.

Morrison, P., and Morrison, E., eds. 1961. *Charles Babbage and his calculating engines: selected writings*. New York: Dover Publications.

Morse, P. M. 1958. *Queues, inventories, and maintenance*. New York: Wiley.

Mrazek, D., and Morris, M. 1973. How to design with programmable logic arrays. Application Note AN-89. National Semiconductor Corp., August 1973.

Murphy, J., and Wade, R. M. 1970. The IBM 360/195. *Datamation*, April 1970, pp. 72–79.

Naylor, T. H. et al. 1966. *Computer simulation techniques*. New York: Wiley.

Neuhauser, C. 1973. An emulation oriented, dynamic microprogrammable processor. Computer Research Report #28. Electrical Engineering Dept., The Johns Hopkins University, Baltimore, Md., November 1973.

Nielsen, N. R. 1967. The simulation of time sharing systems. *Communications of the ACM* 10:397–412.

Organick, E. 1973. *Computer system organization: the B5700/B6700 series*. New York: Academic Press.

Parhami, B. 1973. Associative memories and processors: an overview and selected bibliography. *Proc. of the IEEE* 61:722–30.

Parzen, E. 1960. *Modern probability theory and its applications.* New York: Wiley.

Pear, C. B., ed. 1967. *Magnetic recording in science and industry.* New York: Reinhold Publishing.

Pease, M. C. 1968. An adaptation of the fast Fourier transform for parallel processing. *J. of the ACM* 15:252–64.

Pinkerton, T. B. 1969. Performance monitoring in a time-sharing system. *Communications of the ACM* 12:608–10.

Pohm, A., and Zingg, R. 1968. Magnetic film memory systems. *IEEE Trans. on Magnetics* MAG-4:146–52.

Pugh, E. 1971. Storage hierarchies: gaps, cliffs, and trends. *IEEE Trans. on Magnetics* MAG-7:810.

Raffel, J. et al. 1961. Magnetic film memory design. *Proc. of the IEEE* 49:155.

Ralston, A., ed. 1974. *Auerbach computer encyclopedia.* Philadelphia: Auerbach.

RAND Corporation. 1955. *A million random digits with 100,000 normal deviates.* Glencoe, Ill.: The Free Press.

Randell, B., and Russell, L. 1964. *ALGOL 60 implementation.* New York: Academic Press.

Reigel, E. W.; Faber, U.; and Fisher, D. A. 1972. The interpreter—a microprogrammable building block system. *AFIPS Conf. Proc., 1972 SJCC* 40:705–23. Montvale, N.J.: AFIPS Press.

Reyling, G. 1974. PLAs enhance digital processor speed and cut component count. *Electronics*, August 8, 1974, pp. 109–14.

Rice, R., and Smith, W. R. 1971. SYMBOL—a major departure from classic software dominated von Neumann computing systems. *AFIPS Conf. Proc., 1971 SJCC* 38:575–87. Montvale, N.J.: AFIPS Press.

Rosen, S. 1969. Electronic computers: a historical survey. *Computing Surveys* 1:7–36.

Rosin, R. F. 1969. Contemporary concepts of microprogramming and emulation. *Computing Surveys* 1:197–212.

Rubenstein, H.; McCormack, T. L.; and Fuller, H. W. 1961. The application of domain wall motion to storage devices. *1961 International Solid-State Circuits Conference, Digest of Technical Papers*, February 1961, pp. 64–65.

Ruggiero, J. F., and Coryell, D. A. 1969. An auxiliary processing system for array calculations. *IBM Sys. J.* 8:118–35.

Rusch, R. B. 1969. *Computers: their history and how they work.* New York: Simon & Schuster.

Russell, L.; Whalen, R.; and Leilich, H. 1968. Ferrite memory systems. *IEEE Trans. on Magnetics* MAG-4:134–45.

Saal, H. J., and Shustek, L. J. 1972. Microprogramming implementation of computer measurement techniques. SLAC-PUB-1072. Stanford University, Stanford, Calif., July 1972.

Scherr, A. L. 1967. *An analysis of time-shared computer systems.* Cambridge, Mass.: MIT Press.

Schmid, H., and Busch, D. S. 1968. An electronic digital slide rule. *The Electronic Engineer*, July 1968, pp. 54–64.

Scott, N. 1970. *Electronic computer technology.* Ch. 10. New York: McGraw-Hill.

Senzig, D. N., and Smith, R. V. 1965. Computer organization for array processing.

AFIPS Conf. Proc., 1965 FJCC 27 (part I):117–28. Washington, D.C.: Spartan Books.

Sherman, S.; Baskett, F.; and Browne, J. C. 1972. Trace-driven modeling and analysis of CPU scheduling in a multiprogramming system. *Communications of the ACM* 15:1063–69.

Spain, R. J. 1966. Domain tip propagation logic. *IEEE Trans. on Magnetics* MAG-2:347–51.

Sreenivasan, K., and Kleinman, A. J. 1974. On the construction of a representative synthetic workload. *Communications of the ACM* 17:127–33.

Stone, H. S. 1971. Parallel processing with the perfect shuffle. *IEEE Trans. on Comp.* C-20:153–61.

———. 1972. *Introduction to computer organization and data structures.* New York: McGraw-Hill.

———. 1973. An efficient parallel algorithm for the solution of a tridiagonal linear system of equations. *J. of the ACM* 20:27–38.

Strauss, J. C. 1973. An analytic model of the HASP task monitor. *1st Annual SIGME Symposium on Measurement and Evaluation*, February 1973, Palo Alto, Calif., pp. 22–28.

Sutherland, I. E. 1963. Sketchpad: a man-machine graphical communication system. *AFIPS Conf. Proc., 1963 SJCC* 23:329–46. Baltimore: Spartan Books.

Swanson, J. A. 1960. Physical vs. logical coupling in memory systems. *IBM J. of Res. and Dev.* 4:305.

Texas Instruments, Inc. 1972. ASC, a description of the Advanced Scientific Computer System. Publication M1001P. Texas Instruments, Inc., April 1972.

———. Undated. Texas Instruments TMS 1802 NC one-chip calculator circuit. Bulletin CB-143. Texas Instruments, Inc.

Thiele, A. A. 1969. The theory of cylindrical magnetic domains. *Bell Sys. Tech. J.* 48:3287–335.

———. 1970. Theory of the static stability of cylindrical domains in uniaxial platelets. *J. of Applied Physics* 41:1139–45.

Tomasulo, R. M. 1967. An efficient algorithm for exploiting multiple arithmetic units. *IBM J. of Res. and Dev.* 11:25–33.

Tompsett, M. F.; Amelio, G. F.; and Smith, G. E. 1970. Charge coupled 8-bit shift register. *Applied Physics Letters* 17:111–15.

Tucker, S. G. 1965. Emulation of large systems. *Communications of the ACM* 8:753–61.

———. 1967. Microprogram control for System/360. *IBM Sys. J.* 6:222–41.

Turck, J. A. V. 1972. *Origin of modern calculating machines.* Reprint of 1921 edition. New York: Arno Press.

Vadasz, L.; Chua, H.; and Grove, A. 1971. Semiconductor random-access memories. *IEEE Spectrum*, May 1971, pp. 40–48.

Watson, W. J. 1972. The TI ASC—a highly modular and flexible super computer architecture. *AFIPS Conf. Proc., 1972 FJCC* 41 (part I):221–28. Montvale, N.J.: AFIPS Press.

Weber, H. 1967. A microprogrammed implementation of EULER on IBM System/360 Model 30. *Communications of the ACM* 10:549–58.

Whitney, T. M.; Rode, F.; and Tung, C. 1972. The powerful pocketful: an electronic calculator challenges the slide rule. *Hewlett-Packard Journal*, June 1972, pp. 2–9.

Wilkes, M. V. 1951. The best way to design an automatic calculating machine. Manchester University Computer Inaugural Conference, July 1951, pp. 16–18.

———. 1964. Lists and why they are useful. *ACM Proc. of the 19th National Conf.*, August 1964, Philadelphia, pp. F1-1 through F1-5.

———. 1969. The growth of interest in microprogramming: a literature survey. *Computing Surveys* 1:139–45.

Wirth, N. 1969. On multiprogramming, machine coding, and computer organization. *Communications of the ACM* 12:489–98.

———.1971. The programming language Pascal. *Acta Informatica* 1:35–63.

Wirth, N., and Hoare, C. A. R. 1966. A contribution to the development of ALGOL. *Communications of the ACM* 9:413–32.

Wirth, N., and Weber, H. 1966. EULER: a generalization of ALGOL, and its formal definition: part I. *Communications of the ACM* 9:13–25.

Wortman, D. 1972. A study of language directed computer design. Ph.D. thesis, Stanford University.

Wulf, W. A., and Bell, C. G. 1973. C.mmp—a multi-mini-processor. *AFIPS Conf. Proc., 1972 FJCC* 41 (part II):765–77. Montvale, N.J.: AFIPS Press.

Young, L. H. 1967. Uncalculated risks keep calculator on the shelf. *Electronics*, March 6, 1967, pp. 231–34.

Index

This book was set in Times Roman
by Typographic Service Co. of Los Angeles
illustrated by John Foster
and
printed and bound by Kingsport Press

67/32